Insect Pathology

Insect Pathology

Second Edition

Fernando E. Vega
Sustainable Perennial Crops Laboratory
United States Department of Agriculture
Agricultural Research Service
Beltsville, Maryland

Harry K. Kaya
Department of Nematology
University of California
Davis, California

AMSTERDAM • BOSTON • HEIDELBERG • LONDON • NEW YORK • OXFORD • PARIS
SAN DIEGO • SAN FRANCISCO • SINGAPORE • SYDNEY • TOKYO

Academic Press is an Imprint of Elsevier

Academic Press is an imprint of Elsevier
32 Jamestown Road, London NW1 7BY, UK
225 Wyman Street, Waltham, MA 02451, USA
525 B Street, Suite 1800, San Diego, CA 92101-4495, USA

First edition 1993
Second edition 2012

Notice
No responsibility is assumed by the publisher for any injury and/or damage to persons or property as a matter of
products liability, negligence or otherwise, or from any use or operation of any methods, products, instructions or
ideas contained in the material herein. Because of rapid advances in the medical sciences, in particular, independent
verification of diagnoses and drug dosages should be made

British Library Cataloguing-in-Publication Data
A catalogue record for this book is available from the British Library

Library of Congress Cataloging-in-Publication Data
A catalog record for this book is available from the Library of Congress

ISBN: 978-0-12-384984-7

For information on all Academic Press publications
visit our website at elsevierdirect.com

Typeset by TNQ Books and Journals Pvt Ltd.

www.tnq.co.in

Printed and bound in China

12 13 14 15 16   10 9 8 7 6 5 4 3 2 1

Working together to grow
libraries in developing countries

www.elsevier.com | www.bookaid.org | www.sabre.org

ELSEVIER   BOOK AID International   Sabre Foundation

We dedicate this second edition of "Insect Pathology" to Dr. Lawrence (Lerry) A. Lacey in recognition of his outstanding contributions to insect pathology and microbial control. Lerry received his PhD from the University of California in Riverside in 1978, and has been at the forefront of fundamental and applied research with various entomopathogens, including baculoviruses, bacteria, fungi, and nematodes to control key insect pests of agricultural, medical, and veterinary importance. He has distinguished himself nationally and internationally with his numerous scientific achievements and has mentored a number of students and international scientists. Internationally, his accomplishments include (1) the use of *Bacillus thuringiensis israelensis* for the control of the black fly, *Simulium damnosum*, the principal vector of onchocerciasis in West Africa, (2) basic and applied research to select the most appropriate microbial control agents for the suppression of the Japanese beetle, *Popillia japonica*, in the Azores, Portugal, and (3) foreign exploration and research on fungal entomopathogens of the whitefly, *Bemisia tabaci* type B. Nationally, his comprehensive studies on the granulovirus of the codling moth have highlighted the utility of this pathogen in organic and conventional apple and pear orchards and revealed various factors that influence its efficacy. His research has also shown that nematodes can be used effectively to control the overwintering stage of the codling moth in fruit bins. In addition to more than 140 peer-reviewed publications, he has edited "Manual of Techniques in Insect Pathology" and co-edited "Safety of Microbial Insecticides" and "Field Manual of Techniques in Invertebrate Pathology", which are widely used throughout the world. Lerry was designated a fellow of the Entomological Society of America in 2008 and has served in many different capacities for the Society for Invertebrate Pathology and the Entomological Society of America. He is the quintessential insect pathologist and microbial control researcher. We honor Lerry for his many noteworthy scientific accomplishments, but beyond these achievements, he has contributed in many other ways to insect pathology and microbial control. As a leader and spokesperson, he has brought integrity and wisdom to insect pathology and microbial control and has helped growers to implement the use of microbial control agents for controlling insect pests.

Contents

4. Baculoviruses and Other Occluded Insect Viruses 73

Robert Harrison and Kelli Hoover

12. From Silkworms to Bees: Diseases of Beneficial Insects 425

Rosalind R. James and Zengzhi Li

13. Physiology and Ecology of Host Defense Against Microbial Invaders 461

Jonathan G. Lundgren and Juan Luis Jurat-Fuentes

Insect pathology is an essential component of entomology and provides a non-chemical alternative for insect pest management. There are several groups of organisms that can infect and kill insects, including viruses, fungi, microsporidia, bacteria, protists, and nematodes. The dilemma in insect pathology has been that even though there has been a steady increase in our knowledge of insect pathogens, their share of the market for insect control remains minuscule. Why is this? Is it perhaps due to a lack of understanding on specific aspects of the pathogens, e.g., their ecology in the field? Or is the current state of our knowledge — even after more than 100 years of research — a limiting factor in making these microbes effective, economically acceptable, and widely accepted?

The twentieth and early twenty-first centuries have brought incredible advances to the field of insect pathology. From the landmark books by Edward Steinhaus (Chapters 1 and 2) and his founding in 1959 of a scientific journal fully devoted to insect and invertebrate pathology, which to date has published over 5000 peer-reviewed articles, we have seen the development of molecular biology concepts and tools as the most revolutionary scientific breakthrough of our era. These advances have led to the sequencing of hundreds of genomes, including those of various viruses, bacteria, and fungi of importance to insect pathology (Chapter 2). Molecular biology has also led to the elucidation of phylogenetic relatedness that has clearly challenged many of our basic assumptions, such as in the placement of microsporidia in the kingdom Fungi.

Our discipline has moved very rapidly since the publication of "Insect Diseases", edited by G. E. Cantwell (1974), "Insect Pathology" by Y. Tanada and H. K. Kaya (1993), "Principles of Insect Pathology" by D. G. Boucias and J. C. Pendland (1998), and the second edition of "Controle Microbiano de Insetos", edited by Sérgio Batista Alves (1998). A number of other edited books in insect and invertebrate pathology have also been published with a focus on more specific areas, such as those edited or co-edited by L. A. Lacey, to whom this second edition is dedicated. The vast amount of new concepts and technologies, as well as the enormous complexity of insect pathogens, make it nearly impossible for any specific scientist to master the field and write a book covering the topic. The large database has led to this edited second edition of "Insect Pathology", which consists of 13 chapters. The book should satisfy the needs of insect pathologists, entomologists, mycologists, nematologists, protistologists, ecologists, biological control researchers, and students interested in all aspects of insect pathology in a single up-to-date information source. It should also serve as a resource for others in any of the biological sciences who are interested in insect pathology

We thank Y. Tanada for allowing the use of the title of the book as a second edition of "Insect Pathology" and all the authors for their tireless efforts in contributing to the book. All royalties from this book will be donated to the Chris J. Lomer Memorial Fund, which supports scientists from developing countries to attend the Society for Invertebrate Pathology annual meetings. We express our deepest appreciation to the following chapter reviewers: T. G. Andreadis, J. S. Cory, D. Cox-Foster, M. S. Goettel, G. Hoch, M. S. Hunter, A. Koppenhöfer, J. C. Lord, and R. D. Possee. Special thanks to Pat Gonzalez and Kristi Gomez at Elsevier for their help and support throughout this project. Thanks also to Ann Simpkins for meticulously cross-checking references in each chapter without losing her good sense of humor. Finally, F. E. Vega and H. K. Kaya are most grateful to Wendy S. Higgins, Ian G. Vega, and Joanne Kaya for their loving support.

REFERENCES

Alves, S. B. (Ed.). (1998). *Controle Microbiano de Insetos* (2nd ed.). *Biblioteca de Ciências Agrárias Luiz de Queiroz*, Vol. 4. Piracicaba.

Boucias, D. G., & Pendland, J. C. (1998). *Principles of Insect Pathology*. Boston: Kluwer Academic.

Cantwell, G. E. (Ed.). (1974). *Insect Diseases*, Vols. 1 and 2. New York: Marcel Dekker.

Tanada, Y., & Kaya, H. K. (1993). *Insect Pathology*. San Diego: Academic Press.

Fernando E. Vega
Harry K. Kaya

Insect pathology is a biological discipline that is unique in many ways. Pathology, the study of diseases and disorders, is an ancient pursuit at least with respect to human body, dating back almost 25 centuries to Hippocrates, father of Western medicine. Discovering what has gone awry in a body has been key to figuring out ways of fixing the problems in a wide range of organisms. Thus, there are veterinary pathologists and plant pathologists dedicated to identifying causes of diseases with the ultimate goal of opening a path to restoring their non-human animal and photosynthesizing organisms of interest, respectively, to good health. But insect pathologists are, for want of a better phrase, uniquely bipolar with respect to their subjects; whereas many share the goal with other kinds of pathologists of finding ways to right whatever has gone wrong, the vast majority of insect pathologists are dedicated to making things go wrong, notably by identifying disease-causing agents that will dispatch their subjects quickly, efficiently, and specifically.

Underlying the bipolar nature of the discipline is the tremendous diversity of the taxon comprising the targets of study. With over 900,000 species described to date, it is not at all surprising that insects interact with humans in more different ways than perhaps any other group of organisms. Some insects, such as the domesticated silkworm or western honey bee, provide goods and services upon which entire economies depend; others, however, ravage crops, destroy buildings, spread deadly pathogens, and destroy nations. The tremendous diversity of ways in which humans interact with insects is another dimension of insect pathology that distinguishes it from other biological disciplines: its almost unimaginable complexity.

But the complexity of the human–insect interaction masks even greater complexity that must be addressed by insect pathologists. Each of those 900,000-plus species hosts its own unique microflora and fauna that can influence its health and welfare, so the insect pathologist literally has millions of combinations and permutations of organisms. A diverse range of invertebrate taxa, including many insect species, is infected with obligate rickettsial endosymbionts in the genus *Wolbachia*; entomopathogenic nematodes depend on their symbiotic bacteria to infect and kill their hosts. Thus, insect pathologists not only have to become expert entomologists, they also have to become microbiologists to differentiate between pathogens that destroy their hosts and symbionts that keep them alive. And traditional microbiological training will not do: the various and sundry pathogens that cause disease in insects, including viruses, fungi, bacteria, protists, and nematodes, and virtually every other kind of disease-causing organism, for the most part are specific to insects and have no effect on organisms other than insects. This arrangement works well for designing pest controls with minimal impacts on non-target species but it also requires insect pathologists to develop their own body of knowledge and their own methods and techniques, rather than relying on comparative pathology for enlightenment.

Because an important component of insect pathology, particularly in the context of biological control of insect pests, is the ability to predict outbreaks and epidemics, insect pathologists also to some extent have to become ecologists. Understanding the forces influencing insect population fluctuations — density, weather, other environmental factors — provides valuable insights into disease dynamics and helps to inform forecasting and modeling. Knowledge about insect physiology and toxicology helps the pathologist to understand the origins and progression of diseases, and a basic knowledge of evolution is essential for identifying mechanisms underlying differential resistance and susceptibility. An insect pathologist, then, is a little bit of just about every kind of life scientist.

Despite its status as a specialized field (that requires broad training), insect pathology has contributed to improving the quality of human life in ways that are disproportionate to the number of its practitioners throughout history. It is clear that people benefit when agricultural pests can be stopped by disease, but the benefits of insect pathology to agriculture have been magnified with molecular methods — today, for example, 63% of American corn and 73% of American cotton is genetically engineered to produce an insecticidal toxin from an insect-specific bacterium described by insect pathologists. And the impact of insect pathology extends well beyond the farm. In some respects, insect pathology was a necessary component for refining the germ theory of disease: the legendary French microbiologist Louis Pasteur, charged with stopping pébrine, a mysterious epidemic that was destroying France's silkworms, provided the first practical demonstration that killing germs (a microsporidian parasite, in the case of pébrine) could cure disease. Insect pathologists, in the course of studying viruses that kill insects, gained insights that led them to devise remarkably efficient way of genetically altering these so-called baculoviruses and co-opt them for manufacturing all kinds of recombinant proteins, today used to manufacture a dizzying diversity of human therapeutics, including, most recently, vaccines to protect people against bird flu.

It would seem, given the wide-ranging scope of the field and seemingly limitless applications of the knowledge it generates, that becoming an insect pathologist could be insuperably challenging. Fortunately, the editors of this volume are not easily daunted; every page helps to untangle the complexity and to illuminate the amazing interactions among insects, microbes, and humans. Fernando E. Vega and Harry K. Kaya are to be commended for putting together a book that is not only enormously useful to those specifically involved in the field but also incredibly fascinating even to those with no stake in either lengthening the life expectancy of useful insects or shortening the life expectancy of pestiferous ones.

May R. Berenbaum

Numbers in parentheses indicate the pages on which the authors' contributions begin.

James J. Becnel (133, 221), Center for Medical, Agricultural & Veterinary Entomology, United States Department of Agriculture, Agricultural Research Service, Gainesville, Florida 32608, USA

May R. Berenbaum (xv), Department of Entomology, University of Illinois at Urbana-Champaign, Urbana, Illinois 61801, USA

Meredith Blackwell (171), Department of Biological Sciences, Louisiana State University, Baton Rouge, Louisiana 70803, USA

Denny J. Bruck (29), Horticulture Crops Research Laboratory, United States Department of Agriculture, Agricultural Research Service, Corvallis, OR 97330, USA

Yan Ping (Judy) Chen (133), Bee Research Laboratory, United States Department of Agriculture, Agricultural Research Service, Beltsville, MD 20705, USA

David J. Clarke (395), Department of Microbiology and Alimentary Pharmabiotic Centre, University College Cork, Cork, Ireland

Elizabeth W. Davidson (13), Arizona State University, School of Life Sciences, Tempe, Arizona 85287, USA

Robert Harrison (73), Invasive Insect Biocontrol and Behavior Laboratory, United States Department of Agriculture, Agricultural Research Service, Beltsville, MD 20705, USA

Kelli Hoover (73), Department of Entomology, Pennsylvania State University, University Park, Pennsylvania 16802, USA

Grant L. Hughes (351), College of Agricultural Sciences, Department of Entomology, Pennsylvania State University, University Park, Pennsylvania 16802, USA

Trevor A. Jackson (265), AgResearch, Biocontrol and Biosecurity, Lincoln Research Centre, Canterbury, New Zealand

Rosalind R. James (425), Pollinating Insects − Biology, Management and Systematics Research, United States Department of Agriculture, Agricultural Research Service, Logan, Utah 84322, USA

Juan Luis Jurat-Fuentes (265, 461), Department of Entomology and Plant Pathology, University of Tennessee, Knoxville, Tennessee 37996, USA

Harry K. Kaya (1), Department of Nematology, University of California, Davis, California 95616, USA

Lawrence A. Lacey (29), IP Consulting International, Yakima, Washington 98908, USA

Carlos E. Lange (367), Comisión de Investigaciones Científicas (CIC) de la provincia de Buenos Aires CCT La Plata, CEPAVE-CONICET−UNLP, La Plata 1900, Argentina

Edwin E. Lewis (395), Department of Entomology and Department of Nematology, University of California, Davis, California 95616, USA

Zengzhi Li (425), Department of Forestry, Anhui Agricultural University, Hefei, Anjui 230036, PR China

Jeffrey C. Lord (367), Stored Product Insect Research Unit, Grain Marketing & Production Research Center, United States Department of Agriculture, Agricultural Research Service, Manhattan, Kansas 66502, USA

Janet Jennifer Luangsa-ard (171), Phylogenetics Laboratory, BIOTEC, Thailand Science Park, Khlong Luang, Pathum Thani 12120, Thailand

Jonathan G. Lundgren (461), North Central Agricultural Research Laboratory, United States Department of Agriculture, Agricultural Research Service, Brookings, South Dakota 57006, USA

Nicolai V. Meyling (171), Department of Ecology, University of Copenhagen, Frederiksberg C, DK-1871, Denmark

David H. Oi (221), Center for Medical, Agricultural & Veterinary Entomology, United States Department of Agriculture, Agricultural Research Service, Gainesville, Florida 32608, USA

Jason L. Rasgon (351), College of Agricultural Sciences, Department of Entomology, Pennsylvania State University, University Park, Pennsylvania 16802, USA

David I. Shapiro-Ilan (29), United States Department of Agriculture, Agricultural Research Service, Byron, Georgia 31008, USA

Leellen F. Solter (221), Illinois Natural History Survey, Prairie Research Institute, University of Illinois, Champaign, Illinois 61801, USA

Steven M. Valles (133), Center for Medical, Agricultural & Veterinary Entomology, United States Department of Agriculture, Agricultural Research Service, Gainesville, Florida 32608, USA

Fernando E. Vega (1, 171), Sustainable Perennial Crops Laboratory, United States Department of Agriculture, Agricultural Research Service, Beltsville, Maryland 20705, USA

Chapter 1

Scope and Basic Principles of Insect Pathology

Harry K. Kaya* and Fernando E. Vega[†]

*University of California, Davis, California, USA, [†] United States Department of Agriculture, Agricultural Research Service, Beltsville, Maryland, USA

Chapter Outline

SUMMARY

Insects are the dominant animals in the world, with more than one million described species. The vast majority of insects are innocuous or beneficial to humans, but a small percentage are pests that require a significant amount of our time, effort, and funds to reduce their negative effects on food production and our health and welfare. One environmentally acceptable method to control these insect pests is to use pathogens. The study of pathogens infecting insects is referred to as "insect pathology". Insect pathology is the study of anything that goes wrong with an insect and, therefore, includes non-pathogenic and pathogenic causes. The present focus is on pathogens that can be used as microbial control agents of insects. Here, the basic principles in insect pathology including the microorganisms that cause diseases, their classification and phylogeny, portal of entry, infectivity, pathogenicity and virulence, course of disease, Koch's postulates, and diagnosis are covered.

1.1. INTRODUCTION

Insects represent three-quarters of all animal species in the world, with the vast majority of them being terrestrial and/or occurring in freshwater systems (Daly et al., 1998). More than 99% of the approximately one million described insect species (Grimaldi and Engel, 2005) are either innocuous or beneficial to humans, such as the silkworm (Bombyx mori), cochineal scale (Dactylopius coccus), pollinators, and parasitoids and predators (Gullan and Cranston, 2005; Pedigo and Rice, 2009). To place this in another context, only 1% of the known insect species are our competitors that vie for our crops and stored products, damage our belongings, serve as vectors for plant pathogens, or are of medical or veterinarian importance by feeding on us or our livestock and pets and, in some cases, serving as vectors of disease agents to humans and other vertebrates (Pedigo and Rice, 2009). Even though the number of insect pest species is small compared to the number of described insect species, they require a significant amount of our time, effort, and funds to reduce their negative effects on food production and our health and welfare.

One of the main tactics to control insect pests is the use of chemical pesticides. Unfortunately, the application of chemical pesticides can (1) have a negative effect on human health and the environment; (2) result in resistance of the pest

Insect Pathology. DOI: 10.1016/B978-0-12-384984-7.00001-4

species to pesticides; and (3) kill or negatively affect non-target organisms. An alternative to chemical control is biological control (or biocontrol), which is the study and use of living organisms for the suppression of population densities of pest insects (Eilenberg *et al.*, 2001). The living organisms are predators, parasitoids and entomopathogens (meaning microorganisms capable of causing diseases in insects; from the Greek *entoma* = insect, *pathos* = suffering, *gennaein* = to produce, synonymous with "insect pathogens").

The use of microorganisms for biological control is commonly referred to as microbial control, an approach that includes four strategies: classical, inoculation, inundation, and conservation (Eilenberg *et al.*, 2001). Classical biological control involves the intentional importation, release and establishment of entomopathogens into a new environment. Inoculation biological control deals with the release of an entomopathogen with the expectation that it will multiply and will provide temporary control. This approach is sometimes referred to as augmentation biological control. Inundative biological control is dependent upon the release of significant amounts of inoculum of an entomopathogen to provide immediate control of the pest; control ensues from the release inoculum and not from its progeny. Conservation biological control involves the modification of the environment to protect and enhance an established entomopathogen. An entomopathogen can also be judiciously used with chemical pesticides in an integrated pest management program (Tanada and Kaya, 1993) or combined with a chemical substance(s) that enhances its effectiveness as a microbial control agent (Koppenhöfer *et al.*, 2002; Koppenhöfer and Fuzy, 2008).

Before the development of microbial control agents as management tools for insect pests, maladies had been recorded from beneficial insects for a long time. Thus, the discipline of insect pathology dates back to over 2000 years ago when the Chinese recorded diseases in the silkworm and the Greeks noted diseases in the honey bee, *Apis mellifera* (see Chapter 2). Through the ages, many diseases have been discovered and described from both beneficial and pestiferous insects. This early development of the discipline can be attributed to (1) the curiosity of scientists in describing and ascertaining the cause of pathological conditions in insects; (2) the need to find cures for diseases that afflicted beneficial insects; and (3) the potential use of pathogens to control insect pests.

What is insect pathology? Broadly defined, insect pathology is the study of anything that goes wrong [i.e., disease ("lack of ease")] with an insect. Disease is a process that represents the response of the insect's body to insult or injury (Steinhaus, 1949, 1963c). Often, it is not easy to separate a healthy insect from one that is diseased owing to the absence of symptoms. Steinhaus (1963c) differentiates a healthy insect from a diseased one as follows: "A healthy insect is one so well adjusted in its internal environment and to its external environment that is capable of carrying on the functions necessary for its maintenance, growth, and multiplication with the least expenditure of energy. A diseased insect is simply one that is not healthy; it is an insect that can no longer tolerate an injury or hardship without having an abnormal strain placed upon it."

The scope of insect pathology encompasses many subdisciplines in entomology (Fig. 1.1). In ecology, for example, epizootics of viral, bacterial, fungal,

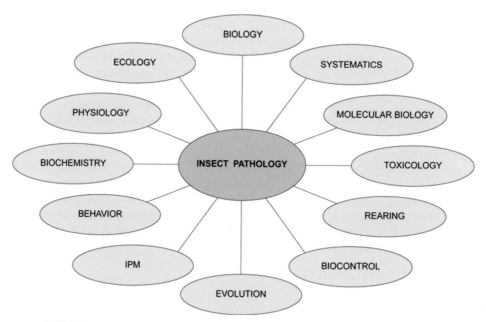

FIGURE 1.1 Relationship between insect pathology and other subdisciplines in entomology.

microsporidian, and nematode diseases can cause significant mortality to devastate insect populations. In physiology, biochemistry, and toxicology, diseased insects will affect the results of the experiments with differences in enzyme, lipid, or protein profiles compared to healthy insects, and diseased insects will be more susceptible than healthy insects to pesticides. Moreover, our knowledge of insect pathology has made significant contributions in (1) the control and eradication of diseases in laboratory insect colonies reared for research, sterile insect technique programs, pet food sales, educational purposes, exhibits (e.g., insect zoos), and for the sale of beneficial insects such as the silkworm and honey bee (see Chapter 12) and parasitoids and predators (Inglis and Sikorowski, 2009a, b); (2) the investigation of the intracellular (Chapter 9) and extracellular symbiosis including the disruption of mutualistic relationships between insects and microbes (Douglas, 2007, 2010); (3) the development of *Bacillus thuringiensis* (Bt) and other bacteria for microbial control (Chapter 8); and (4) the diagnosis or identification of etiological agents that cause insect diseases in the laboratory and field (Hukuhara, 1987; Inglis and Sikorowski, 2009b). In human medicine, baculoviruses in tissue culture systems have been used for the production of papillomavirus, influenza vaccines (Safdar and Cox, 2007; Einstein *et al.*, 2009) and research for malaria vaccines (Blagborough *et al.*, 2010). Steinhaus (1963c) indicated that the basic elements of insect pathology embrace etiology, pathogenesis, symptomatology, morphopathology, physiopathology, and epizootiology. Thus, insect pathology contributes to many other disciplines beyond entomology such as microbiology, veterinary and human medicine, agriculture, and basic biology.

1.2. CATEGORIES OF DISEASE

Insects are exposed to a wide array of non-living (abiotic) or living (biotic) factors (i.e., causal agents) that can result in disease. The factors leading to a disease state can be referred to as non-infectious or infectious (Steinhaus, 1963c), making it clear that both types of maladies are an integral part of insect pathology. Furthermore, Steinhaus (1963c) separated the non-infectious diseases using the following categories: (1) mechanical injuries; (2) injuries caused by physical agents; (3) injuries caused by poisons or chemical agents; (4) diseases caused by nutritional disturbances or deficiencies of proper nutriments; (5) diseases caused by deranged physiology and metabolism; (6) genetic diseases or inherited abnormal conditions; (7) congenital anomalies and malformations, non-generic teratologies; (8) certain tumors and neoplasm (i.e., those not associated with microbes); (9) disturbances in development and in regenerative capacity of tissues; and (10) injuries caused by parasitization or infestation by other insects or arachnids or by predation. For more detailed information on diseases caused by non-infectious agents, the reader is referred to Steinhaus (1963a, b), Cantwell (1974), and Tanada and Kaya (1993).

Tanada and Kaya (1993) summarized the categories of insect diseases following the classification scheme used by Steinhaus (1949) that included both non-infectious and infectious diseases. These were: (1) the presence or absence of an infectious microorganism (i.e., diseases caused by infectious and non-infectious agents); (2) the extent of the disease (i.e., local, focal, or systemic disease); (3) the location or site of the disease (i.e., midgut, fat body, nerve, hemocyte, hypodermis, etc.); (4) the course of the disease (i.e., chronic, subacute, acute); (5) the source of the infectious agent (i.e., exogenous, endogenous, or idiopathic); (6) the etiological or causal agent (virus, bacterium, fungus, protist, or nematode); (7) the distribution or prevalence of the disease in an insect population (i.e., sporadic, enzootic, or epizootic); (8) the method of transmission (i.e., direct contact, vector, *per os*, transovum, or transovarial); and (9) the basis of sequence (i.e., primary, secondary, attenuated, progressive, mixed, or multiple). Although these categories are useful, Tanada and Kaya (1993) used the broader categories of diseases as those caused by amicrobial (non-infectious) agents and those caused by microbial (infectious) agents.

The focus of this book is on microbes that cause diseases in insects, with emphasis on their use as microbial control agents. In line with this focus, the book includes the historical development of insect pathology and microbial control (see Chapter 2), the principles of microbial control and epizootiology (Chapter 3), the various pathogen groups infecting insects (Chapters 4–8, 10, and 11), and resistance to entomopathogens (Chapter 13). It also covers *Wolbachia*, a genus of obligate bacteria, to control arthropods using transinfection into novel hosts (Chapter 9), and pathogens of beneficial insects, especially of silkworms and bees (Chapter 12).

1.3. BASIC PRINCIPLES IN INSECT PATHOLOGY

Basic principles in insect pathology include: (1) the microorganisms that cause diseases (entomopathogens); (2) understanding the classification and phylogeny of entomopathogens; (3) how the microorganisms invade an insect host (portal of entry); (4) whether toxins are involved in the disease process (microbial toxins); (5) infectivity of the microorganisms (infectivity); (6) the disease-producing power of the microorganisms (pathogenicity and virulence); (7) the number of microorganisms needed to cause an infection (dosage); (8) the manifestation of disease (signs, symptoms, and syndromes); (9) the progress of the infection (course of disease); (10) types of infection (acute, chronic, and latent); (11) the proof that a given microorganism is the cause of the disease (Koch's postulates); and (12) how to determine and/or identify the causal agent (diagnosis).

The definition of a pathogen as used in this book is "A microorganism capable of producing disease under normal conditions of host resistance and rarely living in close association with the host without producing disease" (Steinhaus and Martignoni, 1970; also see Martignoni *et al.*, 1984 and Onstad *et al.*, 2006). Even though the term parasite is often used interchangeably or synonymously with pathogen, the term parasite is used in this book as defined by Onstad *et al.* (2006) as "an organism that lives at its host's expense, obtaining nutriment from the living substance of the latter, depriving it of useful substance, or exerting other harmful influence upon the host". The term parasitic is functionally distinct from pathogenic in that the parasite usually does not cause the mortality of the host, whereas the pathogen is routinely lethal to the host (Steinhaus and Martignoni, 1970; Onstad *et al.*, 2006). Thus, there is a distinction between parasites and pathogens. The term "entomopathogen" is used in the context as defined earlier as a microorganism capable of producing a disease in insects, and will be used interchangeably with "insect pathogen" and "pathogen." Therefore, entomopathogenic bacteria would refer to bacteria that produce diseases in their insect hosts and have the capability of being lethal to them. Yet, in the case of many entomopathogenic protists, they will infect their insect hosts but are not lethal to them. Thus, use of the term entomopathogenic can be paradoxical, and the reader should be aware of this situation. Furthermore, for nematodes (see Chapter 11), some (i.e., mermithids) are referred to as parasites as the second stage juveniles enter a host and no reproduction takes place, whereas others (i.e., steinernematids and heterorhabditids) are referred to as entomopathogens because they kill their host and reproduce (see Section 1.3.1).

1.3.1. Entomopathogens

The infectious agents, entomopathogens, are microorganisms that invade and reproduce in an insect and spread to infect other insects. These entomopathogens include noncellular agents (viruses), prokaryotes (bacteria), eukaryotes (fungi and protists), and multicellular animals (nematodes). In the latter group, nematodes differ from the other entomopathogens by having digestive, reproductive, nervous, and excretory systems, and characteristics of parasitoids and predators. However, they have no functional response and often produce pathologies similar to other entomopathogens, and many, especially steinernematid and heterorhabditid nematodes, can invade and

reproduce in an insect, and can spread to infect other insects (Kaya and Gaugler, 1993; Grewal *et al.*, 2005).

Not all microorganisms cause infection even after they enter the insect's hemocoel. The lack of an infection may be due to the resistant characteristics of the host or to the inability of the microbe to survive and reproduce in the host. Many entomopathogens show a high degree of specificity and will infect only one or several insect species, whereas others are generalists and infect a number of insect species in different orders and may infect species in different phyla. These infectious microorganisms can be separated into four broad categories of opportunistic, potential, facultative, and obligate pathogens, keeping in mind that they do not infect all insects. The following definitions are from Onstad *et al.* (2006):

- Opportunistic pathogen: "A microorganism which does not ordinarily cause disease but which, under certain conditions (e.g., impaired host immunity), becomes pathogenic" (e.g., *Aspergillus flavus*).
- Potential pathogen: "1) A microorganism that has no method of invading or infecting a host but can multiply and cause disease if it gains entrance, for example, though a wound; potential pathogens generally grow readily in culture and do not cause specific diseases in specific hosts. 2) a secondary invader" (e.g., *Serratia marcescens*).
- Facultative pathogen: "A pathogen that can infect and multiply in host animals but is also capable of multiplying in the environment; facultative pathogens generally are readily cultured *in vitro*" (e.g., *Bacillus thuringiensis, Beauveria bassiana*).
- Obligate pathogen: "A pathogen that can multiply in nature only within the bodies of specific hosts in which it causes specific diseases. Obligate pathogens usually have a narrow host range and can be cultured *in vitro* only with difficulty, if at all; therefore, some mechanism must exist for their transmission from one host generation to another" (e.g., *Paenibacillus popilliae*, microsporidia, baculoviruses).

1.3.2. Some Major Classification and Taxonomic Changes

Major changes in the classification of some entomopathogens have occurred since the publication of the first edition of "Insect Pathology" by Tanada and Kaya (1993). Some of these changes include the exclusion of Oomycota from the Kingdom Fungi (see Chapter 6), the Microsporidia were in the Phylum Protozoa but are now in the Kingdom Fungi (Chapter 7), and the Protozoa were in the Kingdom Animalia and have been reclassified to the Kingdom Protista (Chapter 10). Several major reclassifications have also occurred at various levels within most of the pathogen groups (see "Classification and Phylogeny" in each of the pathogen chapters).

At the generic and species level, recent taxonomic revisions within various entomopathogen groups can confuse the reader not familiar with the older literature. With viruses, for example, the International Committee for the Taxonomy of Viruses (ICTV) (http://www.ictvonline. org/index.asp?bhcp=1) has adopted the following guidelines for taxonomic nomenclature of viruses. Italicization occurs in formal taxonomic usage, when the writer is explicitly referring to a taxon (e.g., family *Baculoviridae* or *Reoviridae* and genus *Alphabaculovirus* or *Cypovirus*), but no italicization is used when the viruses are referred in the vernacular (e.g., baculovirus infection, alphabaculoviruses, or cypoviruses) (see Chapters 4 and 5). A name referring to any virus species (not just the type species of a genus or family) is fully italicized. For example, references to the species *Autographa californica multiple nuclepolyhedrovirus* are printed in italics. In some cases, the type species uses a common or descriptive name such as *Yellow fever virus* (family: *Flaviviridae*; genus: *Flavivirus*) or *Deformed wing virus* (family: *Iflaviridae*; genus: *Iflavirus*), which is in italics (Chapters 5 and 12).

Examples from fungi include *Paecilomyces fumosoroseus* and *P. farinosus*, which have been reclassified as *Isaria fumosorosea* and *I. farinosa*, respectively; and *Verticillium lecanii* was reclassified into several species in the new genus *Lecanicillium*, which includes *L. lecanii* (see Chapter 6). In addition, *Metarhizium* (Bischoff *et al.*, 2009) and *Beauveria* (Rehner *et al.*, 2011) have undergone major changes based on phylogenetic analyses. An example of a generic reclassification within the microsporidia is the new genus *Endoreticulatus* replacing *Pleistophora*. Although *Nosema* is still a valid genus, a well-known species, *N. locustae*, used as a microbial control agent of locusts, is now *Paranosema locustae*.

With bacteria, some *Bacillus* species that are obligate pathogens of insects have been reclassified into the genus *Paenibacillus*. In the older literature, *Bacillus popilliae* and *B. larvae*, pathogens of scarab larvae and honey bee larvae, respectively, are now called *Paenibacillus popilliae* (see Chapter 8) and *P. larvae* (Chapter 12). In another example, the genus of the facultative mosquito pathogen, *Bacillus sphaericus*, has been reclassified to the genus *Lysinibacillus*. In this case, the older name has been retained in Chapter 8 because it is still commonly used even after the taxonomic change.

1.3.3. Portal of Entry

The portals of entry are the sites through which an entomopathogen invades or gains entry into an insect host (Fig. 1.2). The most likely portals of entry into the insect host are through the mouth (*per os*) or integument. The entomopathogens, especially nematodes, may also invade through the anus (per anal) or spiracles. Other routes of invasion for pathogens include through wounds or injuries to the integument, congenital passage within the ova (transovarial transmission) or on the ova (transovum transmission), and the contaminated ovipositor of parasitoids.

The usual portal of entry for most entomopathogens (viruses, bacteria, protists, microsporidia, some fungi, and some nematodes), especially for insects with chewing, chewing/lapping, and sponging mouthparts, is *per os*, or for most entomopathogenic fungi and many nematodes, it is through the integument. For *per os* entry, once the entomopathogen gets into the gut, it may reproduce in the digestive lumen (e.g., bacteria) or it may penetrate through the peritrophic membrane, infect the midgut cells (e.g., viruses, bacteria, microsporidia, some fungi, and protists) and/or invade directly into the hemocoel (e.g., nematodes). Those entomopathogens that invade into the midgut cells and then into the hemocoel can infect and reproduce in specific tissues (e.g., some granuloviruses only infect the fat body), reproduce in the hemolymph (e.g., *Paenibacillus popilliae*), or cause a systemic infection of many different tissues (e.g., nucleopolyhedroviruses, many microsporidia, and protists). For entry through the integument, fungi use enzymes, specialized structures, and pressure to penetrate through the cuticle into the hemocoel (see Chapter 6), and nematodes use a stylet or tooth to penetrate directly into the hemocoel (Chapter 11).

Once the entomopathogen becomes established and reproduces in the insect, it usually produces a resistant stage that can survive in the environment for varying periods. The resistant stages include occlusion bodies for baculoviruses, cypoviruses, and entomopoxviruses, spores for bacteria and protistans, environmental spores for microsporidia, resting spores or sclerotia for fungi, and infective juveniles for steinernematid and heterorhabditid nematodes. These resistant stages are usually the infective stages that will infect a new host. Some non-resistant stages such as non-occluded virions, vegetative bacterial rods and fungal conidia will infect insects but do not survive long in nature. In other cases, the entomopathogen may survive in an alternate host, require an alternate host for successful completion of its life cycle, or remain in its usual host through the adult stage when it is transmitted to the next generation. Thus, a pathogen may have one or more mechanisms of survival. When a susceptible host is encountered and the pathogen is in its infectious state, it can initiate a new infection in a healthy host (Fig. 1.2).

1.3.4. Microbial Toxins

In some cases, the entomopathogen (e.g., some bacteria) does not need to infect cells or invade into the hemocoel to

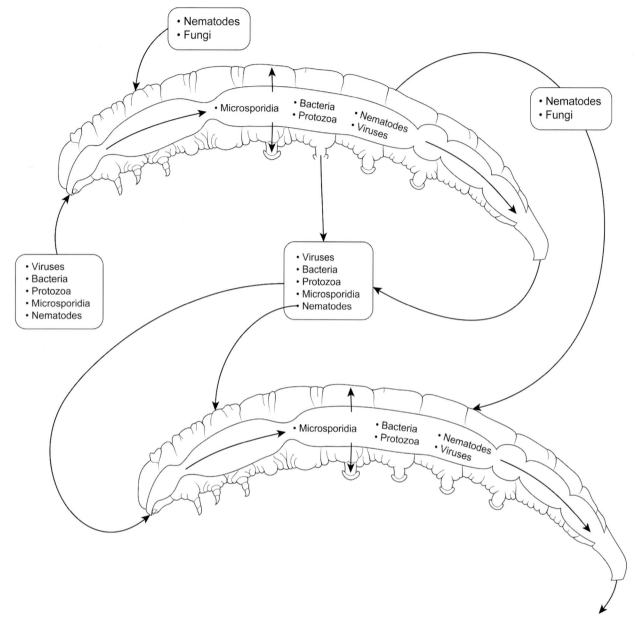

FIGURE 1.2 The primary portal of entry for most entomopathogens (viruses, bacteria, protists, microsporidia, and some nematodes) is primarily through the mouth (*per os*), whereas for fungi and some nematodes, the primary portal of entry is through the cuticle. The entomopathogens that enter through the mouth can infect the midgut cells and/or penetrate into the hemocoel to cause a systemic infection. Similarly, the entomopathogens that penetrate through the integument may initiate the infection on the cuticle (fungi) with subsequent penetration into the hemocoel or penetrate directly into the hemocoel (some protists and nematodes). In some cases, a few fungal species may enter a host through the mouth or nematodes may enter the anus or spiracles (not shown in figure). Once the entomopathogens reproduce in the first host, the infective propagules are released or leave from a dead or living host and can initiate a new infection in another susceptible host. (Modified after Tanada and Kaya, 1993.)

cause disease. It is confined to the digestive tract, produces dysentery, and causes the insect to shrink in size (brachyosis), resulting in death (Bucher, 1961; Jackson *et al.*, 2001). In such cases, the insects appear to be affected by toxins that are produced by the bacterium. Hence, a disease can be brought about in a susceptible insect host by the pathogen through the actions of chemical or toxic substances.

Two types of toxins, catabolic and anabolic, can be produced by entomopathogens (Tanada and Kaya, 1993). Catabolic toxins result from decomposition brought about by the activity of the pathogen, whereas anabolic toxins are substances synthesized by the pathogen. The breakdown of proteins, carbohydrates, and lipids by the pathogen may produce toxic alcohols, acids, mercaptans, alkaloids, etc. In the case of anabolic toxins, the substances synthesized by the pathogen can be classified as exotoxins and endotoxins. Exotoxins are excreted or passed out of the cells of the pathogen during reproduction and have been isolated from entomopathogens, especially bacteria and fungi. Endotoxins produced by the pathogen are not excreted but confined within the cell. These endotoxins are liberated when the pathogen forms a resistant stage, dies, or degenerates. The best known endotoxin in insect pathology is produced by Bt (see Chapter 8). In this case, as the bacterium initiates the sporulation process, excess proteins are produced, resulting in the formation of a proteinaceous crystal (δ-endotoxin) within the sporangial wall adjacent to the spore. The sporangial wall is easily ruptured, releasing the δ-endotoxin and spore into the environment. The δ-endotoxin is a protoxin, and when it is placed in a high pH solution as occurs in the insect gut, the toxic component becomes activated and adversely affects the midgut cells. Thus, "intoxication" is brought about by the activity of the pathogen in the form of toxins.

Not all toxins produced by microorganisms are in the realm of insect pathology. For example, at least two bacteria (actinomycetes), *Streptomyces avermitilis* and *Saccharopolyspora spinosa*, produce exotoxins that have been developed into the insecticides, avermectins and spinosad, respectively. In contrast to Bt, these actinomycetes are not covered in this book because the toxins alone are used with no living pathogenic microorganisms involved. However, Bt spore or vegetative rod does not need to be present to cause the disease because the δ-endotoxin alone can kill a susceptible insect host. Although this statement appears contradictory with what was stated above for the actinomycetes, the application of Bt is with the δ-endotoxin and the spore. Transgenic plants containing the Bt δ-endotoxin, on the other hand, are not considered to be part of insect pathology *per se*, but are aspects of host plant resistance. Yet, with the frequent application of Bt as an organism in agricultural systems and the planting of transgenic plants with the Bt toxin genes in major agricultural crops (e.g., corn and cotton), the resistance to Bt can and has occurred even before transgenic crops were available. Accordingly, insect resistance to pathogens including managing resistance in Bt transgenic crops is covered in Chapter 13.

1.3.5. Infectivity

Infectivity is the ability of a microorganism to enter the body of a susceptible insect and produce an infection. However, many insect species harbor microorganisms that are beneficial rather than harmful. These mutualistic microorganisms infect and reproduce in the insects but are contained within specific cells (mycetocytes or bacteriocytes) or tissues (mycetomes or bacteriome) and are beneficial by producing some useful product for their insect

hosts. The relationship is mutualistic because the microorganisms also benefit by obtaining nutrients and protection from their insect hosts. Thus, an infection may result in a non-diseased condition. The focus here is on the diseased condition, but the reader should be aware that many microorganisms are mutualistically associated with insects (Bourtzis and Miller, 2003, 2006, 2009). This book does include *Wolbachia* (Chapter 9) because they have the potential to be manipulated and used as control agents of some insect pest species.

When infection results in disease, it generally causes detectable pathological effects such as injuries or dysfunctions (i.e., impairments in function, especially of a bodily system or organ). There are two main factors associated with a disease: invasiveness and pathologies resulting in abnormalities or dysfunctions. In some cases where toxins are involved, invasiveness of the pathogen into cells and tissues or into the hemocoel need not occur.

When an infectious agent is transmitted naturally by direct contact to an insect, the resultant disease is called "contagious" or "communicable". Contagious diseases are common in insects and occur with all major pathogen groups (i.e., viruses, bacteria, microsporidia, fungi, protists, and nematodes). Infection and "contamination" are not the same, as a susceptible insect may be contaminated or be harboring a pathogen without being infected. A non-susceptible insect, other organisms or objects may also be contaminated with a pathogen. In both situations, the pathogen is a potential source to infect a susceptible host.

1.3.6. Pathogenicity and Virulence

Pathogenicity and virulence are two terms used regularly in insect pathology. These two terms have spurred some debate among scientists (see Thomas and Elkinton, 2004; Shapiro-Ilan *et al.*, 2005), but the definitions by Onstad *et al.* (2006) and Tanada and Kaya (1993) and recommended by Shapiro-Ilan *et al.* (2005) are used in this book. Pathogenicity is defined as "the quality or state or being pathogenic, the potential or ability to produce disease", whereas virulence is defined as "the disease producing power of an organism, the degree of pathogenicity within a group or species" (Shapiro-Ilan *et al.*, 2005). Thus, pathogenicity is a qualitative term and for a given host and pathogen, it is absolute, whereas virulence quantifies pathogenicity and is variable owing to the strain of the pathogen or to environmental effects. Furthermore, pathogenicity can be considered as an all-or-none response; that is, the microorganism is either pathogenic to a host or not and is applied to groups or species (Shapiro-Ilan *et al.*, 2005). Virulence is a measurable characteristic of the ability of the microorganism to cause disease and is intended for within-group or within-species comparisons.

Perhaps the best way to illustrate the difference between these two terms is to provide an example of each. The nematode—bacterium complex of *Steinernema carpocapsae—Xenorhabdus nematophila* shows pathogenicity to some non-insect arthropods (e.g., ticks) (Samish *et al.*, 2000) but not to vertebrates (Akhurst and Smith, 2002). The fungus *Beauveria bassiana* GA strain is pathogenic to the pecan weevil, *Curculio caryae*, but *B. bassiana* MS1 strain is not as virulent to this host (Shapiro-Ilan *et al.*, 2003). Finally, the *Agrotis ipsilon* nucleopolyhedrovirus (NPV) is more virulent to its original host, the black cutworm (*Agrotis ipsilon*) from which it was isolated, compared to *Autographa californica* NPV, which has a wide lepidopterous host range but is less virulent to *Agrotis ipsilon* (Boughton *et al.*, 1999).

1.3.7. Dosage

One pathogen can infect and kill a host, as has been demonstrated with the infective juvenile of the nematode—bacterium complex (*Steinernema—Xenorhabdus* or *Heterorhabditis—Photorhabdus*) (Kaya and Gaugler, 1993). In general, however, with other entomopathogens, a minimal number of infective propagules is needed to pass through the portal of entry for infection to occur. This number is referred to as a dose that can be defined as the quantity of an active agent (i.e., entomopathogen) to which an insect is exposed at any one time. Dosage can be expressed quantitatively depending upon the host susceptibility in terms of mortality, infection, or time to death as lethal dose (LD), effective dose (ED), or lethal time to death (LT). The 50% or 90% level of response is assigned as LD_{50} or LD_{90}, ED_{50} or ED_{90}, or LT_{50} or LT_{90}. The 50% level of response is referred to as the median lethal dose, median effective dose, or median lethal time. To obtain this type of information, a bioassay is conducted with a minimum of five dosage levels of the infective stage (i.e., occlusion bodies, spores, conidia, infective juveniles) plus a control treatment administered to a given stage of the host, and the response (i.e., mortality, infection, or time to death) is plotted with dosage on the *x*-axis and the response on the *y*-axis. If host mortality occurs in the control, Abbott's formula can be used to correct for the mortality in the treatments (Abbott, 1925). Ideally, the range of the dosages should provide a response level between 10% and 90%. The lowest and highest dosages should not give a 0% or 100% response, respectively. The range of dosages obtained from the bioassay when plotted against the host response (i.e., mortality, infection, or time to death) should give an S-shaped curve which is transformed into a straight line by converting the response to a probit scale, and the dosage to the log scale. The level of response can be obtained by going up the probit scale on the *y*-axis and reading across to where the slope of the line intersects the dosage scale on the *x*-axis, which then provides the dosage

for that response. The bioassay should be conducted at least twice with a different cohort of insects and entomopathogens to demonstrate that the data are reproducible. For further information on conducting bioassays with entomopathogens, the reader should refer to Burges and Thomson (1971) and Navon and Ascher (2000).

Dose is a precise number of the infective stage to which the bioassay insects are subjected to and usually attaining this level of precision is not possible. That is, it is not possible to determine the dosages quantitatively because of the size of the insect, the bioassay system (e.g., placing the infective stage on the diet surface and not knowing the amount of entomopathogen acquired), or because the insects live in an aqueous habitat where the dose cannot be calculated. In such cases, median concentration of the entomopathogen which produces a response in half of the test insects is used because the experimental method is not sufficiently accurate to determine the precise dose to which the test insects were exposed. For example, the median lethal concentration would then be expressed as LC_{50} or the median effective concentration would be EC_{50}.

1.3.8. Signs, Symptoms, and Syndromes

Diseased insects exhibit characteristic aberrations or dysfunctions that are designated as signs and symptoms. When there is a physical or structural abnormality, the term sign is used, whereas when there is a functional or behavioral aberration, the term symptom is used. A sign is indicated by abnormalities in the morphology or structure such as color, malformed appendages or body segments, fragility of the integument, etc., whereas a symptom may be expressed by abnormal movement, abnormal response to stimuli, digestive disturbances (vomiting or diarrhea), inability to mate, etc.

A particular disease has a group of characteristics signs and symptoms which is called a syndrome. The syndrome refers to a system complex or a particular combination or sequence of signs and symptoms. Sometimes, the syndrome is very characteristic and specific for a disease caused by an entomopathogen. More often, the same syndrome occurs with many different diseases. Vomiting and diarrhea may develop from ingestion of chemical pesticides, bacterial toxins, or entomopathogenic bacteria, viruses or protistans. A distinct syndrome occurs when a silkworm (*B. mori*) larva ingests Bt subspecies *sotto* spores and δ-endotoxin. The larva stops feeding in a few minutes, becomes sluggish in about 10 min, the pH of the blood increases and the pH of the midgut decreases, and within an hour, the larva becomes moribund and dies.

1.3.9. Course of Infection

After an entomopathogen invades a healthy, susceptible insect, it starts to reproduce and the course of infection can be partitioned into various phases or stages: the incubation period, the beginning of disease with the appearance of the first signs and/or symptoms, and peak of disease (Fig. 1.3). The incubation period is the time from when the entomopathogen infects a host until the development of signs and/or symptoms. The appearance of the signs and/or symptoms indicates that there is a patent or frank infection including the production of toxins with some entomopathogens marking the beginning of disease. As the entomopathogen reproduces, the disease manifests itself to the fullest extent in the host reaching the peak of disease. The peak of the disease is when the signs and symptoms are most severe and either start to abate or attain a steady state. Also at this time, the entomopathogen has usually reached its highest level of reproduction, and if toxins are produced, they are present in the greatest amount. However, many entomopathogens, especially bacteria, fungi, and steinernematid and heterorhabditid nematodes, continue to develop and reproduce after the death of the host. When the peak of disease is reached, the insect may recover as the result of an immune response or die because of the absence of an immune response or the presence of only a weak immune response. In the case where the entomopathogen has low virulence, the insect may have a chronic infection and survive to adulthood and reproduce.

The time from when the insect enters and infects the host until the insect dies is referred to as the period of lethal infection, or if the insect recovers as the period of infection. A short period of lethal infection indicates that the entomopathogen has high virulence, whereas a long period of lethal infection indicates that it has low virulence. An entomopathogen with high virulence is not necessarily the most infectious. For example, Bt subspecies *kurstaki* has high virulence to cabbage butterfly (*Pieris rapae*) larvae with a period of lethal infection of 48 h. But high virulence does not mean that the entomopathogen is highly contagious, and Bt subspecies *kurstaki* is not easily transmitted from one insect to another. Conversely, an entomopathogen with low virulence (i.e., period of lethal infection of several weeks as occurs with some microsporidian infections) may be highly contagious, with the pathogen being easily transmitted from one insect to another.

1.3.10. Acute, Chronic, and Latent Infections

Infections in insects may vary from latent to chronic to acute. Acute infections are the most apparent because of the distinct and characteristic response of the insect to the pathogen causing the disease. Acute infections are of short duration and usually result in the death of the host

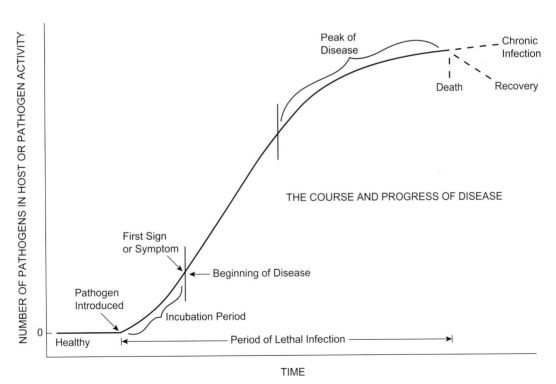

FIGURE 1.3 When a healthy, susceptible insect acquires an infectious concentration of an entomopathogen, it sets into motion a sequence of events referred to as the course and progress of disease. As the pathogen reproduces and/or produces toxins, the intensity of disease increases. The time from when the insect acquires the pathogen until its death is referred to as the "period of lethal infection". (Modified after Tanada and Kaya, 1993.)

(i.e., the period of lethal infection is short). Chronic infections are often overlooked because their manifestations tend to be less striking and apparent to the observer until very late in the infection process. Chronic infections are of long duration and the hosts may or may not die. If the host dies, the period of lethal infection is long, whereas if the host does not die, it may reach adulthood and reproduce. Latent infections in insects have been detected primarily with viruses. In such cases, the term latent or occult viral infection is used, and the virus is referred to as an occult virus and not as a latent virus (Onstad *et al.*, 2006). Recently, an occult nodavirus belonging to the genus *Alphanodavirus* in the family *Nodaviridae* was described from an insect cell line (Li *et al.*, 2007).

1.3.11. Koch's Postulates

One of the basic tenets in pathology for establishing the etiological or causal agent of a disease involving microorganisms is the application of Koch's postulates. Robert Koch (1843−1910), a German physician who is considered one of the founders of microbiology, made brilliant discoveries on the causal agents of anthrax, tuberculosis, and cholera through the application of postulates that bear his name.

Koch's postulates as stated in Tanada and Kaya (1993) are:

1 A specific pathogenic organism must be seen in all cases of a disease.
2 This organism must be attained in pure culture.
3 The organism from the pure culture must reproduce the disease in experimental animals.
4 The same organism must be recovered from the experimental animal.

It is not always possible to fulfill Koch's postulates because one or more steps cannot be performed or can be performed only with great difficulty. For step 1, some organisms are difficult to detect, although the electron microscope has made it possible to view submicroscopic organisms such as viruses, rickettsia, rickettsiella, and mollicutes putatively in infected insects. However, the mere presence of these organisms in a host does not prove that they are the causal agent of a particular disease without further experimentation by following Koch's postulates. For step 2, certain microorganisms cannot be isolated in pure culture with the present available laboratory methods or are very difficult to culture. These microorganisms are mainly obligate pathogens requiring living cells or tissues. For some obligate microorganisms, insect tissue cultures can be used to fulfill step 2. For step 3, different pathogens may produce similar signs, symptoms, and syndromes which may mistakenly result in designating the wrong organism as the etiological agent.

The contributions of molecular biology have opened up the concept of "molecular Koch's postulates" to help characterize whether a particular microbial gene is an essential constituent of the ability of a microorganism to infect and cause disease in a given host (Falkow, 1988). Recently, Falkow (2004) summarized the application of the molecular Koch's postulates to bacterial pathogenicity in animal systems. Table 1.1 shows Koch's original three postulates with the proposed molecular Koch's postulates (Falkow, 2004). Koch's original three postulates are essentially the same four steps as stated by Tanada and Kaya (1993). The molecular Koch's postulates are used to demonstrate that a particular gene or set of genes is the virulence factor and inactivation of this gene(s) results in the loss of pathogenicity. Restoration of pathogenicity should occur with the reintroduction of the gene into the microorganism. To the authors' knowledge, the molecular Koch's postulates as proposed by Falkow (2004) are not intentionally used in insect pathology. However, the insecticidal protein genes of Bt have been cloned into other Bt subspecies or other bacterial species and have proven to be efficacious against the target insects (Park and Federici, 2009).

TABLE 1.1 Comparison of Koch's Original Postulates and the Molecular Koch's Postulates

Koch's Original Three Postulates[a]	Molecular Koch's Postulates
The microorganism occurs in every case of the disease and can account for the pathological changes and course of the disease	The phenotype or property under investigation should be associated with a pathogenic microorganism of a genus or strain of a species
The microorganism occurs in no other disease as a fortuitous and non-pathogenic microbe	Specific inactivation or deletion of the gene(s) from the microorganism should lead to the loss of function in the clone
After being fully isolated from the host's body and grown in pure culture, the microorganism can induce the same disease anew	Restoration of pathogenicity should occur with the reintroduction of the wild-type gene

[a]*Koch's original three postulates state essentially the same four steps as used by Tanada and Kaya (1993).*
Source: Modified after Falkow (2004)

1.3.12. Diagnosis

Diagnosis is a fundamental branch of insect pathology which involves the process by which one disease is distinguished from another. The identification of the etiological or causal agent alone is not diagnosis, but only one of a series of steps in the operation to determine the cause of the disease. To conduct a proper diagnosis, a study has to be made of the etiology, symptomatology, pathogenesis, pathologies, and epizootiology of the disease. Steinhaus (1963d) stressed that "The importance of diagnosis in insect pathology lies in the fact that one must know the nature of the disease and what ails or has killed an insect before the disease can be properly studied, controlled, or suppressed, used as a microbial control measure, its potential for natural spread determined, or its role in the ecological life of an insect species ascertained". Steinhaus (1963d), Tanada and Kaya (1993) and Inglis and Sikorowski (2009b) provide detailed procedures on how to submit diseased insect specimens and how to perform a disease diagnosis properly.

Some helpful references for diagnosing diseased insects and identifying the entomopathogens include "An Atlas of Insect Diseases" by Weiser (1969), "Laboratory Guide to Insect Pathogens and Parasites" by Poinar and Thomas (1984) and "Manual of Techniques in Insect Pathology" edited by Lacey (1997). Inglis and Sikorowski (2009b) provide an excellent discussion on diagnostic techniques, and Stock et al. (2009) edited a book dealing with molecular diagnostic techniques entitled "Insect Pathogens: Molecular Approaches and Techniques." The reader is referred to these excellent sources to identify some of the more important entomopathogens that infect insects.

REFERENCES

Abbott, W. S. (1925). A method of computing the effectiveness of insecticides. *J. Econ. Entomol., 18,* 265–267.

Akhurst, R., & Smith, K. (2002). Regulation and safety. In R. Gaugler (Ed.), *Entomopathogenic Nematology* (pp. 311–332). Wallingford: CABI.

Bischoff, J. F., Rehner, S. A., & Humber, R. A. (2009). A multilocus phylogeny of the *Metarhizium anisopliae* lineage. *Mycologia, 101,* 512–530.

Blagborough, A. M., Yoshida, S., Sattabongkot, J., Tsuboi, T., & Sinden, R. E. (2010). Intranasal and intramuscular immunization with baculovirus dual expression system-based Pvs25 vaccine substantially blocks *Plasmodium vivax* transmission. *Vaccine, 28,* 6014–6020.

Boughton, A. J., Harrison, R. L., Lewis, L. C., & Bonning, B. C. (1999). Characterization of a nucleopolyhedrovirus from the black cutworm, *Agrotis ipsilon* (Lepidoptera: Noctuidae). *J. Invertebr. Pathol., 74,* 289–294.

Bourtzis, K., & Miller, T. A. (Eds.). (2003). *Insect Symbiosis.* Boca Raton: CRC Press.

Bourtzis, K., & Miller, T. A. (Eds.). (2006). *Insect Symbiosis, Vol. 2.* Boca Raton: CRC Press.

Bourtzis, K., & Miller, T. A. (Eds.). (2009). *Insect Symbiosis, Vol. 3.* Boca Raton: CRC Press.

Bucher, G. E. (1961). Artificial culture of *Clostridium brevifaciens* n. sp. and *C. malacosomae* n. sp., the causes of brachytosis of tent caterpillars. *Can. J. Microbiol., 71,* 223–229.

Burges, H. D., & Thomson, E. M. (1971). Standardization and assay of microbial insecticides. In H. D. Burges & N. W. Hussey (Eds.), *Microbial Control of Insects and Mites* (pp. 591–622). New York: Academic Press.

Cantwell, G. E. (Ed.). (1974). *Insect Diseases,* (Vols. 1–2). New York: Marcel Dekker.

Daly, H. V., Doyen, J. T., & Purcell, A. H., III (1998). *Introduction to Insect Biology and Diversity* (2nd ed.). Oxford: Oxford University Press.

Douglas, A. E. (2007). Symbiotic microorganisms: untapped resources for insect pest control. *Trends Biotechnol., 25,* 338–342.

Douglas, A. E. (2010). *The Symbiotic Habit.* Princeton: Princeton University Press.

Eilenberg, J., Hajek, A., & Lomer, C. (2001). Suggestions for unifying the terminology in biological control. *BioControl, 46,* 387–400.

Einstein, M. H., Baron, M., Levin, M. J., Chatterjee, A., Edwards, R. P., Zepp, F., Carletti, I., Dessy, F. J., Trofa, A. F., Schuind, A., & Dubin, G. (2009). Comparison of the immunogenicity and safety of *Cervarix*™ and *Gardasil*® human papillomavirus (HPV) cervical cancer vaccines in healthy women aged 18–45 years. *Hum. Vacc., 5,* 705–719.

Falkow, S. (1988). Molecular Koch's postulates applied to microbial pathogenicity. *Rev. Infect. Dis., 10,* 8274–8276.

Falkow, S. (2004). Molecular Koch's postulates applied to bacterial pathogenicity – a personal recollection 15 years later. *Nat. Rev. Microbiol., 2,* 67–72.

Grewal, P. S., Ehlers, R.-U., & Shapiro-Ilan, D. (Eds.). (2005). *Nematodes as Biocontrol Agents,* Wallingford: CABI.

Grimaldi, D., & Engel, M. S. (2005). *Evolution of the Insects.* New York: Cambridge University Press.

Gullan, P. J., & Cranston, P. S. (2005). *The Insects: An Outline of Entomology* (3rd ed.). Oxford: Blackwell.

Hukuhara, T. (1987). Epizootiology: prevention of insect diseases. In J. R. Fuxa & Y. Tanada (Eds.), *Epizootiology of Insect Diseases* (pp. 497–512). New York: John Wiley & Sons.

Inglis, G. D., & Sikorowski, P. P. (2009a). Microbial contamination and insect rearing. In J. C. Schneider (Ed.), *Principles and Procedures for Rearing High Quality Insects* (pp. 150–222). Mississippi State: Mississippi State University.

Inglis, G. D., & Sikorowski, P. P. (2009b). Entomopathogens and insect rearing. In J. C. Schneider (Ed.), *Principles and Procedures for Rearing High Quality Insects* (pp. 224–288). Mississippi State: Mississippi State University.

Jackson, T. A., Boucias, D. G., & Thaler, J. O. (2001). Pathobiology of amber disease, caused by *Serratia* spp., in the New Zealand grass grub, *Costelytra zealandica. J. Invertebr. Pathol., 78,* 232–243.

Kaya, H. K., & Gaugler, R. (1993). Entomopathogenic nematodes. *Annu. Rev. Entomol., 38,* 181–206.

Koppenhöfer, A. M., & Fuzy, E. M. (2008). Effect of the anthranilic diamide insecticide, chlorantraniliproe, on *Heterorhabditis bacteriophora* (Rhabditida: Heterorhabditidae) efficacy against white grubs (Coleoptera: Scarabaeidae). *Biol. Control, 45*, 93–102.

Koppenhöfer, A. M., Cowles, R. S., Cowles, E. A., Fuzy, E. M., & Baumgartner, L. (2002). Comparison of neonicotinoid insecticides as synergists for entomopathogenic nematodes. *Biol. Control, 24*, 90–97.

Lacey, L. A. (Ed.). (1997). *Manual of Techniques in Insect Pathology.* New York: Academic Press.

Li, T.-C., Scotti, P. D., Miyamura, T., & Takeda, N. (2007). Latent infection of a new Alphanodavirus in an insect cell line. *J. Virol., 61*, 10890–10896.

Martignoni, M.E., Krieg, A., Rossmoore, H.W., & Vago, C. (1984). Terms Used in Invertebrate Pathology in Five Languages: English, French, German, Italian, Spanish. US Forest Service. Pacific Northwest Forest and Range Experiment Station General Technical Report PNW-169, Portland.

Navon, A. & Ascher, K. R. S. (Eds.). (2000). *Bioassays of Entomopathogenic Microbes and Nematodes.* Wallingford: CABI.

Onstad, D. W., Fuxa, J. R., Humber, R. A., Oestergaard, J., Shapiro-Ilan, D. I., Gouli, V. V., Anderson, R. S., Andreadis, T. G., & Lacey, L. A. (2006). *An Abridged Glossary of Terms Used in Invertebrate Pathology,* (3rd ed.). *Society for Invertebrate Pathology.* http://www.sipweb.org/glossary.

Park, H.-W., & Federici, B. A. (2009). Genetic engineering of bacteria to improve efficacy using the insecticidal proteins of *Bacillus* species. In S. P. Stock, J. Vandenberg, I. Glazer & N. Boemare (Eds.), *Insect Pathogens: Molecular Approaches and Techniques* (pp. 275–305). Wallingford: CABI.

Pedigo, L. P., & Rice, M. E. (2009). *Entomology and Pest Management* (6th ed.). Upper Saddle River: Pearson Prentice Hall.

Poinar, G. O., Jr., & Thomas, G. M. (1984). *Laboratory Guide to Insect Pathogens and Parasites.* New York: Plenum.

Rehner, S. A., Minnis, A. M., Sung, G.-M., Luangsa-ard, J. J., Devotto, L., & Humber, R. A. (2011). Phylogeny and systematic of the anamorphic, entomopathogenic genus *Beauveria. Mycologia, 103*, 1055–1073.

Safdar, A., & Cox, M. M. (2007). Baculovirus-expressed influenza vaccine: a novel technology for safe and expeditious vaccine production for human use. *Expert. Opin. Investig. Drugs, 16*, 927–934.

Samish, M., Alekseev, E., & Glazer, I. (2000). Mortality rate of adult ticks due to infection by entomopathogenic nematodes. *J. Parasitol., 86*, 679–684.

Shapiro-Ilan, D. I., Gardner, W. A., Fuxa, J. R., Wood, B. W., Nguyen, K. B., Adams, B. J., Humber, R. A., & Hall, M. J. (2003). Survey of entomopathogenic nematodes and fungi endemic to pecan orchards of the southeastern United States and their virulence to the pecan weevil (Coleoptera: Curculionidae). *Environ. Entomol., 32*, 187–195.

Shapiro-Ilan, D. I., Fuxa, J. R., Lacey, L. A., Onstad, D. W., & Kaya, H. K. (2005). Definitions of pathogenicity and virulence in invertebrate pathology. *J. Invertebr. Pathol., 88*, 1–7.

Steinhaus, E. A. (1949). *Principles of Insect Pathology.* New York: McGraw-Hill.

Steinhaus, E. A. (Ed.). (1963a). *Insect Pathology: An Advanced Treatise, Vol. 1.* New York: Academic Press.

Steinhaus, E. A. (Ed.). (1963b). *Insect Pathology: An Advanced Treatise, Vol. 2.* New York: Academic Press.

Steinhaus, E. A. (1963c). Introduction. In E. A. Steinhaus (Ed.), *Insect Pathology: Vol. 1. An Advanced Treatise* (pp. 1–27). New York: Academic Press.

Steinhaus, E. A. (1963d). Background for the diagnosis of insect diseases. In E. A. Steinhaus (Ed.), *Insect Pathology: Vol. 2. An Advanced Treatise* (pp. 549–589). New York: Academic Press.

Steinhaus, E. A., & Martignoni, M. E. (1970). *An Abridged Glossary of Terms Used in Invertebrate Pathology,* (2nd ed.). *Pacific Northwest Forest and Range Experimental Station* Portland: USDA Forest Service.

Stock, S. P., Vandenberg, J., Glazer, I., & Boemare, N. (Eds.). (2009). *Insect Pathogens: Molecular Approaches and Techniques.* Wallingford: CABI.

Tanada, Y., & Kaya, H. K. (1993). *Insect Pathology.* San Diego: Academic Press.

Thomas, S. R., & Elkinton, J. S. (2004). Pathogenicity and virulence. *J. Invertebr. Pathol., 85*, 146–151.

Weiser, J. (1969). *An Atlas of Insect Diseases.* Shannon: Irish University Press.

Chapter 2

History of Insect Pathology

Elizabeth W. Davidson

Arizona State University, Tempe, Arizona, USA

Chapter Outline

SUMMARY

Diseases of insects were recorded more than 2000 years ago. This chapter explores the history of discoveries of insect pathogens and the scientists who were responsible for the discoveries. In addition, the development of insect pathology as a discipline including the early commercialization of microbial control products, events that brought scientists together and advanced the field, and some novel and unexpected uses of insect pathogens are covered.

2.1. INTRODUCTION

Insects and microorganisms have very old and complex relationships. Studies of insects preserved in amber between 15 and 200 million years ago revealed that they contained insect pathogens (entomopathogens) such as nematodes, cytoplasmic polyhedrosis viruses (= cypoviruses), nucleo-polyhedrovirus (NPV), trypanosomes, and fungi (Poinar, 1984; Poinar and Poinar, 2005). However, only in the past *ca.* 200 years have scientists begun to understand these associations and to make use of them to improve our food supply and for many other uses. Insect pathogens continue

to provide viable, economical, and reliable alternatives to other pest management strategies. This chapter provides a brief look at the important accomplishments of scientists from around the world who became fascinated with pathogens of insects and their potential uses, and the current status of these accomplishments (Table 2.1).

2.2. EARLY HISTORY OF THE DISCOVERY OF INSECT PATHOGENS

Two beneficial insects, honey bees (*Apis mellifera*) and silkworms (*Bombyx mori*), have been important to human lives for thousands of years. Archaeological evidence suggests that honey was harvested from wild bees as early as 6000 BC, while silkworms were domesticated in China for the production of silk as early as 700 BC (Aristotle, transl. 1878; Steinhaus, 1975; Tanada and Kaya, 1993).

2.2.1. Honey Bees

Honey bees provide several important functions to humans, in particular honey as a sweetener for foods and pollination

Insect Pathology. DOI: 10.1016/B978-0-12-384984-7.00002-6

TABLE 2.1 Timeline: Events Influencing Insect Pathology

6000 BC	Humans harvest honey from honey bees
700 BC	Silkworms domesticated in China
330–334 BC	Aristotle describes diseases of honey bees
37–39 BC	Virgil records honey bee diseases
77 AD	Pliny records honey bee diseases
555 AD	Silkworm is introduced to Europe and Middle East
1527	Marco Girolamo Vida publishes the poem "De Bombyce"
1602	Aldrovandi describes worms emerging from grasshoppers
1679	Maria Sibylla Merian describes melted caterpillars
1734–1742	Réamur finds *Cordyceps* in silkworms, nematodes in bumble bees
1747	Gould describes nematodes in ants
1758	Linnaeus describes nematodes in invertebrates in *Systema Naturae*
1771	Schirach describes foulbrood disease of honey bees
1808	Nysten publishes the first book on silkworm diseases
1808–1813	Bassi studies fungus disease of silkworms
1822	Kirby and Spence publish *Diseases of Insects* in *An Introduction to Entomology: or Elements of the Natural History of Nematodes*
1835	Bassi publishes his discovery (*Beauveria bassiana*) in *Del mal del segno, calcinaccio o moscardino*
1855	Cohn describes *Empusa* fungus in house fly
1856	Maestri and Cornalia observe nuclear polyhedrosis virus [= nucleopolyhedrovirus (NPV)] "crystals"
1857	Naegeli describes *Nosema bombycis,* from silkworms
1870	Pasteur publishes *Etudes sur la Maladie des Vers a Soie*
1873	LeConte proposes microbial control to American Association for the Advancement of Science meeting
1878	Metchnikoff finds green muscardine fungus
1883	Sorokin names green muscardine *Metarrhizium* (= *Metrahizium*) *anisopliae*
1885	Cheshire and Cheyne show that foulbrood is caused by *Bacillus alvei*
1887	Institut Pasteur (Pasteur Institute) is founded
1888	Krassilstchik attempts biocontrol with *Metarhizium anisopliae*
1888	Lugger and Snow attempt control of chinch bug with *Beauveria bassiana*
1894	Bolle observes NPV polyhedra soluble in gut juices
1898	Ishiwata discovers *Bacillus sotto* (*thuringiensis*)
1901, 1905	Ishiwata publishes discoveries of *Bacillus sotto*
1906	G. F. White finds two foulbrood diseases in honey bees
1909	Zander describes *Nosema apis* from honey bees
1909	Berliner discovers *B. thuringiensis* in flour moth
1911	d'Herelle observed bacterial epizootics in grasshoppers
1912	White finds European foulbrood to be caused by *Bacillus* (*Melissococcus*) *pluton*
1912	Fantham and Porter describe structure of *Nosema apis*
1913, 1916	Maassen describes chalkbrood in honey bees
1913	Glaser and Chapman demonstrate that wilt disease in gypsy moth is a virus

TABLE 2.1 Timeline: Events Influencing Insect Pathology—cont'd

1915, 1916	Aoki and Chigasaki, Mitani and Watarai find *B. thuringiensis* alkaline extract toxic
1916	Japanese beetle accidently introduced into USA
1917	F. G. White describes the first virus in honey bees
1918	Glaser demonstrates serial transmission of polyhedrosis to gypsy moth
1918	Paillot describes *Thelohania mesnili* from the cabbage butterfly *Pieris brassicae*
1921	Keilin describes first fungus in aquatic insect (*Coelomomyces* in mosquitoes)
1923	Steiner describes first species of entomopathogenic nematode [*Neoaplectana* (*Steinernema*) *kraussei*]
1924	Kudo publishes *A Study of Microsporidia*
1927	Mattes isolates *B. thuringiensis* from flour moth
1928–1932	Husz, Vouk, Metalnikov, Chorine set up field trials of *B. thuringiensis* against corn borer
1929	Glaser finds *Neoaplectana* (*Steinernema*) *glaseri* in Japanese beetle larvae
1929	*Bacillus thuringiensis* and *Metarhizium* discussed at International Corn Borer Conference in Paris
1931	First *in vitro* culture of *Neoaplectana* (*Steinernema*) *glaseri* by Glaser
1933	Paillot publishes *L'infection chez les insectes: immunité et symbiose*, describing granulosis virus (= granulovirus)
1934	Ishimori describes first cytoplasmic polyhedrosis virus (= cypovirus)
1934	Filipjev suggests use of nematodes for biocontrol
1935	Hawley and White isolate bacteria from Japanese beetles
1935	First releases of *Steinernema glaseri* against Japanese beetles
1936	Jirovec publishes *Studien über Microsporidien* (*Studies on Microsporidia*)
1938	First commercial *B. thuringiensis* product (Sporeine) in France
1940	Dutky finds *Bacillus* (= *Paenibacillus*) *popilliae* and *B.* (*P.*) *lentimorbus* causing milky disease in Japanese beetles
1942	Steinhaus publishes *Catalog of Bacteria Associated Extracellularly with Insects and Ticks*
1946	Steinhaus publishes *Insect Microbiology*
1947	Bergold publishes first electron micrographs of NPV
1948	Fawcett distributes *Aschersonia* sp., for control of whiteflies
1948	*Bacillus* (= *Paenibacillus*) *popilliae* registered for use against Japanese beetle in USA
1949	Steinhaus publishes *Principles of Insect Pathology*
1950	Smith and Wyckoff describe cypovirus
1953	Hannay finds parasporal inclusions in *B. thuringiensis* responsible for activity
1953	Canning describes *Nosema locustae*
1954	Angus shows parasporal inclusions contain *B. thuringiensis* toxin
1956	Steinhaus publishes "Living insecticides" in *Scientific American*
1956	First International Conference on Insect Pathology and Biological Control, Montreal
1958	International Conference on Insect Pathology and Biological Control, Prague
1959	*Journal of Insect Pathology* first published
1961	First *B. thuringiensis* products registered in USA (Thuricide, Biocontrol)
1962	Kurstak isolates *B. thuringiensis* subspecies *kurstaki*
1962	de Barjac *et al.* develop serotyping of *B. thuringiensis*
1962	Grace develops insect cell culture

(Continued)

TABLE 2.1 Timeline: Events Influencing Insect Pathology—cont'd

1963	Vago describes entomopoxviruses
1963	Tanada finds and describes granulovirus in codling moth
1963	Steinhaus publishes *Insect Pathology: An Advanced Treatise*
1965	Kellen *et al.* report pathogenic *Bacillus sphaericus* in mosquitoes
1965	*Journal of Insect Pathology* renamed *Journal of Invertebrate Pathology*
1966	Huger discovers virus in rhinoceros beetle
1966	Poinar and Thomas describe the mutualistic bacterium, *Achromobacter nematophilus* (= *Xenorhabdus nematophila*) from neoaplectanid (= steinernematid) nematodes
1967	Burges organizes *B. thuringiensis* standards
1967	Society for Invertebrate Pathology is established
1970	Dulmage isolates *B. thuringiensis* strain HD-1
1970, 1973	Singer *et al.* isolate active strains of *B. sphaericus*
1973	*Helicoverpa zea* NPV registered, first virus product (Viron/H, Elcar)
1973	Hink develops plaque assay for NPV
1976	TM BioControl-1 registered (Douglas fir tussock moth NPV)
1976	*Bacillus thuringiensis israelensis* pathogenic to mosquitoes discovered by Goldberg and Margalit
1978	Gypchek NPV registered against gypsy moth
1980	*Nosema locustae* approved for control of grasshoppers
1981	*Hirsutella thompsonii* registered for use against citrus mite in USA
1981	Schnepf and Whitely report first *B. thuringiensis* toxin gene sequence
1981	*Bacillus thuringiensis israelensis* registered for use against mosquitoes
1983	Mullis develops polymerase chain reaction (PCR)
1983	Smith, Fraser and Summers manipulate NPV genes
1984	Poinar finds nematodes in insects preserved in amber
1984–1985	*Black beetle virus* genome published by Dasgupta *et al.* and Dasmahapatra *et al.*
1987	First transgenic plants with *B. thuringiensis* genes reported
1988	Baculovirus expression system patented
1988	*Bacillus thuringiensis tenebrionis* registered for use against Colorado potato beetle in USA
1990s	*Metarhizium* spp. introduced against locusts and grasshoppers in Africa
1994	First NPV (*Autographa californica* NPV) genome is fully sequenced by Ayres *et al.*
1995	Transgenic plants with *B. thuringiensis* genes marketed
1996	Microsporidia found to be related to fungi
2003	Genome of *Photorhabdus luminescens* (heterorhabditid nematode symbiont) published by Duchaud *et al.*
2004	Genome of *Wolbachia pipientis* (insect symbiont) published by Wu *et al.*
2006	Genome of honey bee pathogens, *Paenibacillus larvae* and *Ascosphaera apis*, published by Qin *et al.*
2007	*Bacillus thuringiensis* genome published by Challacombe *et al.*
2009	*Nosema ceranae* genome published by Cornman *et al.*
2011	Genomes of *Metarhizium anisopliae* (= *M. robertsii*) and *M. acridum* published by Gao *et al.*

of an important portion of our food supply. Diseases of bees were described more than 2000 years ago by Aristotle in *Historia Animalium* (334–330 BC), and were later recorded by Virgil (39–37 BC) and by Pliny, the Elder (AD 77) (Pliny, 1855; Virgil, 1941). The early bee diseases were often attributed to sick plants that the bees had visited (Aristotle, transl. 1878). With the development of microscopes in the late seventeenth and early eighteenth centuries, scientists began to associate bee diseases with microorganisms. In 1771, Adam G. Schirach (1724–1773) described a disease that he called foulbrood because of the "stench" of the decomposing larvae (Schirach, 1771). Cheshire and Cheyne (1885) proved experimentally that foulbrood was a transmissible pathogen caused by a bacterium called *Bacillus alvei* (Cheshire, 1884). By 1906, G. F. White (1906) demonstrated that there were, in fact, two foulbrood diseases. The first, American foulbrood (*Bacillus larvae*, now called *Paenibacillus larvae*), was found in many countries in Europe as well as in North America. The second, European foulbrood, thought by Cheshire and Cheyne (1885) to be caused by *Bacillus alvei*, was found by White (1912) to be caused by *Bacillus pluton* (now called *Melissococcus pluton*) (Shimanuki, 1990). In the early twentieth century, microorganisms other than bacteria were also found to be associated with diseases of bees. Zander (1909) first described spores of *Nosema apis* in ventricular epithelial cells of bees and the structure and homology of *N. apis* were described by Fantham and Porter (1912a, b). Maassen (1913, 1916) described chalkbrood (*Ascosphaera apis*), the first fungus disease in bees, and White (1917) described the first viruses in bees. Honey bees are now known to suffer from over 30 different infectious diseases as well as parasites such as tracheal and varroa mites (Gochnauer, 1990; Furgala and Mussen, 1990; Gilliam and Vandenberg, 1990; Bailey and Ball, 1991).

2.2.2. Silkworms

In ancient Chinese sericulture, silkworms were housed in crowded, warm trays and fed fresh mulberry leaves harvested from the field; as a result they suffered from many diseases. *Cordyceps*, which produces large growths on the host called Chinese plant-worms, was the first fungus identified in ancient Chinese silk culture, and is still in use in traditional Chinese medicine (Wang, 1965; Tanada and Kaya, 1993; Paterson, 2008). Many other early reports of *Cordyceps* are described by Steinhaus (1975).

The silkworm was introduced to Europe and the Middle East around AD 555, and became an important source of revenue in Italy and France beginning in the twelfth and thirteenth centuries. Diseases of the silkworm were investigated by scientists, whose discoveries led to a new understanding of infectious diseases. The first scientific treatise on silkworm diseases, *Recherches sur les Maladies des Vers a Soie*, was published by P. H. Nysten (1808).

Agostino Bassi (1773–1856) is considered to be the "Father of Insect Pathology." Bassi, an Italian lawyer and scientist, studied a disease of silkworms that was plaguing the Italian silk industry. This disease was called "calcinaccio" because the caterpillars became covered with a white calcium-like powder. After a series of experiments from 1808 to 1813, Bassi (1835; Yarrow, 1958) showed that the disease was caused by a "vegetable parasite", a fungus now called *Beauveria bassiana*, which could be transmitted by contact or inoculation or by contaminated leaves fed to the caterpillars. Bassi's research was the first to confirm that a microorganism can cause a disease and was an important contribution in disproving the theory of spontaneous generation.

An outbreak of disease that threatened the French silk industry in the mid-nineteenth century led to the recruitment of a chemist, Louis Pasteur (1822–1895), to study the causes and potential elimination of these diseases. Although Pasteur was reluctant to accept this position as he knew nothing about silkworms, he was persuaded by his friend and former teacher, Senator Jean-Baptist Dumas, to move to the village of Alès in the south of France, where he consulted with the famous entomologist Jean Henri Fabre (1823–1915) and began studies of the disease called pébrine. Pébrine is characterized by tiny black spots on the surface of caterpillars. During more than two years of research, Pasteur found that pébrine appeared to be caused by small microscopic organisms described by Nägeli (1857) as *Nosema bombycis*. Pasteur showed that the disease could be passed by contact with diseased caterpillars, by contaminated food, and even from mother to offspring through the egg. This finding was the first demonstration of vertical transmission of a pathogen (Pasteur, 1870). Pasteur developed a technique for screening the female moth for pébrine before the eggs hatched that is still in use in the sericulture industry. He also studied a second disease, flacherie, which he thought to be caused by a bacterium. Pasteur's seminal work on pébrine and flacherie was published as two volumes, *Etudes sur la Maladie des Vers a Soie*, in 1870. Pasteur's work on silkworms sparked his interest in human and animal diseases, leading to landmark research on infectious diseases at the Institut Pasteur in Paris, founded by Pasteur in 1887 (Jehle, 2009).

2.3. EARLY DISCOVERIES OF OTHER PATHOGENS

2.3.1. Viruses

Signs and symptoms of virus diseases in insects were observed by many early scientists. Wilting and "melting" of silkworm caterpillars was described by the Italian Marco Girolamo Vida (*ca.* 1490–1566) in a poem entitled

"De Bombyce", first published in 1527 (Vida, 1527), and similar signs and symptoms were described by the German illustrator Maria Sibylla Merian (1647—1717) in 1679 (Merian, 1679). Nysten (1808) described the symptoms as "jaundice" of silkworms. Maestri (1856) and Cornalia (1856) both reported refractive crystal-like bodies in the cells of jaundiced silkworms, the first microscopic description of what we now know as NPV. Bolle confirmed the activity of these crystals (Bolle, 1894) and observed that they were soluble in the alkaline gut juices of the caterpillar. Polyhedral (= occlusion) bodies were found by von Tubeuf (1892) in larvae of the nun moth as they were dying of "wilt diseases", although he did not make the connection between these bodies and the cause of the disease. Glaser and Chapman (1913) used filters to demonstrate that the wilt disease was actually caused by a virus, because it was capable of passing through filters that removed larger microorganisms (Benz, 1986), and Glaser (1918) first demonstrated serial transmission of polyhedral disease in gypsy moth using filtrates from infected larvae. With the development of high-quality compound microscopes and the electron microscope in the 1940s and 1950s, we began to understand the complexity of these pathogens. Gernot Bergold (1947) published the first electron micrographs of baculoviruses (NPVs) and developed new techniques to purify the viruses (Benz, 1986; Arif, 2005).

A granulovirus was first discovered in cabbage butterfly larvae by André Paillot, along with several other types of viruses. His book "L'Infection Chez les Insectes" (1933) contains beautiful hand-drawn illustrations of his observations. Other early descriptions of disease signs and symptoms likely caused by viruses are chronicled by Steinhaus (1975). In his book "Principles of Insect Pathology" (1949), Steinhaus proposed the first classification of insect viruses, giving scientific names to the genera and type species. A third type of insect virus, now known as cypovirus, was described by Ishimori (1934) and later by Smith and Wyckoff (1950). Entomopoxviruses in the European cockchafer, *Melolontha melolontha*, were first described by Constantin Vago (1963).

2.3.2. Bacteria

With the development of commercially available microscopes in the late nineteenth and early twentieth centuries, scientists began to observe many other insect pathogens, especially bacteria. The valuable Japanese silk industry observed sudden deaths of caterpillars in the late nineteenth century, leading to the discovery in 1898 of a spore-forming bacterium by the Japanese scientist Sigetane Ishiwata (1868—1941) (Aizawa, 2001). Ishiwata called the bacterium *Bacillus sotto* to describe the sudden death that occurred within hours after caterpillars consumed the bacteria (Ishiwata, 1901). In Japanese the disease was called "sotto-byo-kin" (collapse-disease-microorganism) or "sotto-kin." Ishiwata performed sophisticated experiments that demonstrated that the bacterium could survive for many years and that it was most lethal when the cultures were at least one week old (Ishiwata, 1905). Other Japanese scientists, including Aoki and Chigasaki (1915) and Mitani and Watarai (1916), confirmed Ishiwata's discoveries and found that the filtrate of a culture dissolved in alkaline solution was lethal to silkworms, the first demonstration of a toxin (Beegle and Yamamoto, 1992; Aizawa, 2001). Across the world, in Thuringia, Germany, Ernst Berliner discovered in 1909 a similar bacterium that killed the flour moth, *Ephestia kuhniella*, in flour mills. Berliner named the bacterium *Bacillus thuringiensis* (Berliner, 1915). This name has been retained for the species to the present and includes *B. sotto* and certain strains originally called *Bacillus cereus* (Beegle and Yamamoto, 1992). Berliner was able to show that the bacterium was toxic when fed to insects and suggested that it could be used to control insects. However, his strains of *B. thuringiensis* were lost. Fortunately, another German scientist isolated another strain of *B. thuringiensis* from the flour moth (Mattes, 1927) and found promising results for this isolate against the European corn borer which eventually led to the first commercial product in 1938 (Beegle and Yamamoto, 1992; Milner, 1994).

2.3.3. Fungi

As mentioned above, because of their unusual and very visible appearance, fungal infections by *Cordyceps* were observed in silkworms in the early Chinese silk industry. The first published record of *Cordyceps* was a report of "vegetable growths" by the French scientist René-Antoine Ferchault de Réamur (1683—1757) (Réamur, 1734—1742). Hagen (1879) observed an epizootic of fungi in a dung fly that occurred in 1867. Cohn (1855) described a fungus that he named *Empusa* on the house fly, but there was debate over the name because *Empusa* is also a genus of orchid. The German physician J. B. Georg W. Fresenius (1808—1866) proposed the name *Entomophthora* (Fresenius, 1856). The Russian scientist Eli Metchnikoff (beginning in 1878) found a fungus, which he called green muscardine (*Entomophthora anisopliae*), on the wheat cockchafer, a serious pest in Russia (Metchnikoff, 1879) (later named *Metarrhizium anisopliae* by Sorokin (1883) and now spelled *Metarhizium*). Metchnikoff found that larvae could be infected by being placed in soil contaminated with conidia. He cultured the fungus on artificial medium consisting of sterilized beer mash, and put forth some of the earliest proposals to use a pathogen for the control of insects (Steinhaus, 1975). Several other European scientists suggested the use of fungi against flies, the nun moth (*Lymantria monacha*), grasshoppers, and others (Tanada and Kaya, 1993). The first fungal infection of an aquatic insect, *Coelomomyces* in mosquito larvae, was described by Keilin (1921).

2.3.4. Microsporidia

Microsporidia are currently classified as fungi (Corradi and Keeling, 2009) (see Chapter 7), but they are considered separately here because of the significance they have in insect populations. As mentioned above, Pasteur performed seminal experiments on transmission of pébrine in the silk industry, described in his book in 1870. *Nosema apis* was described from honey bees by Zander (1909). Nearly a decade later, Paillot (1918) described *Thelohania mesnili* from the European cabbage worm, *Pieris brassicae*. In the following years, many different groups within microsporidia were described. Roksabro Kudo (1924) compiled *A Study of Microsporidia*, which became the basis for future work. Otto Jirovec (1936) compiled a further list of newly described microsporidia. Jaroslav Weiser (Fig. 2.1) has described the early research on these organisms in a chapter in "Insect Pathology, An Advanced Treatise, Volume 2" (Weiser, 1963) and more recently in Weiser (2005).

The taxonomy of microsporidia has undergone many changes over the years. Originally thought to be protozoa (Sporozoa, Microspora, Microsporea), in 1993 microsporidia were reclassified as Archezoa (Cavalier-Smith, 1993). Edlind *et al.* (1996) found that microsporidia were phylogenetically related to fungi, which was later supported by further studies based on protein and genetic analyses (Tanabe *et al.*, 2002).

2.3.5. Nematodes

The final major group of insect pathogens, or perhaps more correctly parasites, is nematodes. Aldrovandi described "worms" emerging from grasshoppers in *De Animalibus*

FIGURE 2.1 Jaroslav Weiser (left) and Edward Steinhaus (right) at the First International Colloquium on Invertebrate Pathology, Prague, 1958.

Insectis (1602), the first description of nematodes in insects. Nematodes were later described by Réamur in bumble bees (Réamur, 1734–1742), by Gould in ants (Gould, 1747), and by Linnaeus in both vertebrates and invertebrates in *Systema Naturae* (Linnaei, 1758). Using improved microscopy, several other insect parasitic nematodes were described in the nineteenth century, summarized by Kirby and Spence (1822) in *An Introduction to Entomology: Or Elements of the Natural History of Insects*. The first species of entomopathogenic nematodes, *Aplectana kraussei* (= *Steinernema kraussei*), was described by Steiner (1923) from sawflies. The first successful *in vitro* culture of an entomopathogenic nematode, *Neoaplectana glaseri* (= *Steinernema glaseri*), was accomplished by Glaser (1931).

2.4. DEVELOPMENT OF INSECT PATHOGENS FOR BIOLOGICAL CONTROL

2.4.1. Viruses

When insecticidal baculoviruses (i.e., NPVs and granuloviruses) were found to be safe for vertebrates and very active against target insects, they became attractive potential biological control agents. The first product, the *Helicoverpa zea* (corn earworm) NPV, was registered in 1973 under trade names Viron/H and later Elcar (Huber, 1986; Erlandson, 2008) and was re-registered in 1991. Outbreaks of the Douglas fir tussock moth, *Orgyia pseudotsugata*, were first reported in western Canada in 1916, and quickly spread across North America (Martignoni, 1999). In 1947, R. L. Furniss, working with the United States Department of Agriculture (USDA) Forest Service, found an NPV in tussock moth caterpillars collected in Oregon (Steinhaus, 1951; Martignoni, 1999). Following many years of research in the USDA and other laboratories, a product was registered in 1976 under the name TM BioControl-1 (Martignoni, 1999). Another major forestry target was *Neodriprion sertifer*, the European pine sawfly, an introduced pest of North American forests. F. T. Bird and colleagues in Ontario obtained an NPV from Sweden in 1949 (Bird and Whalen, 1954; Stairs, 1971), which became a product that is still widely used in forestry in Europe and North America (Lord, 2005). Although not a commercial product, the virus (nudivirus) discovered by Alois Huger in the rhinoceros beetle, *Oryctes rhinoceros*, in the Pacific islands (Huger, 1966) became a successful classical biological control agent against this major pest of coconut and other palm crops (Young, 1986). In 1978, an NPV was registered by the US Environmental Protection Agency for use against the gypsy moth, *Lymantria dispar*, under the name Gypchek. The product was applied by aerial spraying, with good results (Rollinson *et al.*, 1965;

Cunningham, 1982), and continues to be commercially available. A granulovirus was found in the codling moth, *Cydia pomonella*, in Mexico and described by Yoshinori Tanada in 1963 (Tanada, 1964) and products are currently used in North America, Europe, Argentina, Australia, New Zealand, and South Africa (Tanada and Kaya, 1993; Cross *et al.*, 1999). Several other virus products are commercially available for use against orchard pests (Lacey *et al.*, 2008; Lacey and Shapiro-Ilan, 2008). Currently, several different insect virus-based insecticides are registered or in the experimental stage (Huber, 1986; Erlandson, 2008).

2.4.2. Bacteria

The first attempt to use a bacterium to control insects is credited to Félix d'Herelle, who observed epizootics in Mexican grasshoppers (d'Herelle, 1911, 1914). He isolated a bacterium that he called *Coccobacillus acridorum* and applied it in several Latin American countries (d'Herelle, 1912). Others who attempted to use this bacterium were often unable to confirm that it was a useful biological control agent, and in 1959 Gordon Bucher obtained a culture of *C. acridorum* which he identified as a common saprophyte in the grasshopper gut (Bucher, 1959). The true identity of d'Herelle's bacterium was never determined (Steinhaus, 1975).

Bacillus (Paenibacillus) popilliae

The introduction of the Japanese beetle, *Popillia japonica*, to the USA in 1916 led to the discovery of two pathogens that became important insect control agents. The USDA set up a Japanese beetle laboratory in 1917 in Riverton, New Jersey (St. Julian and Bulla, 1973; Fleming, 1976), where G. E. Spencer isolated several bacterial cultures from diseased beetle grubs (Hawley and White, 1935). I. M. Hawley and G. F. White later found that infected beetle larvae could be divided into three groups, those that turned either black or white when infected with bacteria, plus larvae that died from fungal infections (Hawley and White, 1935). Research focused on the white disease, called "milky disease", later confirmed by Samuel Dutky (1940) to be caused by two different bacterial species, *Bacillus popilliae* and *Bacillus lentimorbus*. Because *B. popilliae* has resisted efforts to culture it on artificial medium, it must be propagated in living beetle larvae. Products based on *B.* (now called *Paenibacillus*) *popilliae* were distributed by the US government in 1939, became registered for use in the USA in 1948 (Engler and Rogoff, 1980; Burges, 1981), and are still available (Federici, 2005; Lord, 2005).

Bacillus thuringiensis

When *B. thuringiensis* was first discovered, it was considered a potential danger to the silkworm (Glare and

O'Callaghan, 2000). However, field trials against the European corn borer, *Ostrinia nubilalis*, were attempted in the late 1920s by Husz in Hungary (Husz, 1929) and by Metalnikov and Chorine (1928) and Vouk (1932) in Yugoslavia, with mixed results (Beegle and Yamamoto, 1992; Glare and O'Callaghan, 2000; Lord, 2005). The first commercial product, Sporeine, was produced by the French company Laboratoire Libec in 1938; unfortunately World War II brought production to a halt (Luthy *et al.*, 1982). Interest in *B. thuringiensis* as a commercial product was reawakened by Edward A. Steinhaus (Fig. 2.1), then a young professor at the University of California, Berkeley. In 1956, he published an article, "Living insecticides", in *Scientific American* that stimulated interest in the possibility of commercial microbial control products (Steinhaus, 1956). He met Robert Fisher at Pacific Yeast Products (later Biofirm) in 1956, and with his encouragement the company produced the first US product, Thuricide, in 1957, which was registered for use in the USA in 1961 (Engler and Rogoff, 1980; Burges, 1981). Another firm, Nutrilite Products, soon entered the market with Biotrol (Steinhaus, 1975; Milner, 1994; Glare and O'Callaghan, 2000).

The first commercial products suffered from unpredictable activity because spore count did not accurately predict activity. In 1953, Steinhaus sought the advice of Christopher Hannay at the Canada Department of Agriculture, concerning the odd particles that were seen alongside the spores, and Hannay soon determined that these bodies, parasporal inclusions (= δ-endotoxin or crystal), were important in the activity of the product (Hannay, 1953). Hannay's colleague, Thomas Angus, proved in 1954 that the parasporal inclusion was the source of a toxin that was responsible for the rapid activity (Angus, 1954). This discovery opened the door to many years of research on the complex toxins associated with *B. thuringiensis* (Steinhaus, 1975; Beegle and Yamamoto, 1992; Milner, 1994)

All the original commercial *B. thuringiensis* products were from the strain (later to become subspecies) *thuringiensis*. However, scientists soon began to discover *B. thuringiensis* in many different insects. Edouard Kurstak found an isolate from *Ephestia khuniella* in 1962 that had a much broader host range (Kurstak, 1962, 1964), and a few years later, Howard Dulmage obtained a similar one from *Pectinophora gossypiella* that he designated HD-1 (Dulmage, 1970). Huguette de Barjac and colleagues at the Institut Pasteur in Paris developed a serotyping technique based on flagellar antigens (H-antigens) that allowed various strains to be classified (de Barjac and Bonnefoi, 1962, 1968). The HD-1 strain was designated *kurstaki*, and rapidly became the primary subspecies for commercial products against lepidopteran pests. In 1967, H. Denis Burges and collaborators organized a system of bioassay standards that greatly improved the quality of commercial products (Burges, 1967; Dulmage *et al.*, 1971).

In the next decades, many different *B. thuringiensis* strains and subspecies were isolated, including such notable ones as *B. thuringiensis israelensis*, active against mosquitoes and black flies, isolated by Leonard Goldberg and Joel Margalit in the Negev Desert in 1976 (Goldberg and Margalit, 1977; Margalit and Dean, 1985). *Bacillus thuringiensis israelensis* was registered for use in 1981 (Tanada and Kaya, 1993) and was rapidly developed in the 1990s into products used for mosquito control in many countries. *Bacillus thuringiensis israelensis* became important in attempts in the 1980s and 1990s to control black flies, which serve as a vector for the nematode that causes onchocerciasis in West Africa (Federici, 2005). *Bacillus thuringiensis tenebrionis*, active against beetles, was discovered by Alois Krieg and colleagues (Krieg *et al.*, 1983). This subspecies was registered for use in 1988 and entered commercial production in the 1990s (Tanada and Kaya, 1993; Lord, 2005). The first gene sequence for a *B. thuringiensis* toxin, a 130 kDa toxin from HD-1, was reported by H. E. Schnepf and Helen Whitely (1981). Further immunological and genetic research has now permitted classification of *B. thuringiensis* toxins separate from the strain identification and has led to a better understanding of the activity and host range of strains. More than 50 years after the first commercial products, over 200 *B. thuringiensis* products have been commercialized under different names (Glare and O'Callaghan, 2000), and the entire *B. thuringiensis* genome was published by Challacombe *et al.* (2007).

Bacillus sphaericus

In 1975, the World Health Organization (WHO) initiated the Tropical Disease Research Group with the goal of finding biological control agents for mosquitoes and black flies, vectors of malaria and onchocerciasis, respectively. Many different parasites and pathogens were explored, and out of this program came the discovery of *B. thuringiensis israelensis* in 1976, as mentioned above. Another bacterium, *Bacillus sphaericus*, was first found in mosquitoes in 1965 (Kellen *et al.*, 1965). In 1970 and 1973, Samuel Singer and colleagues, working with the WHO group, isolated the highly active strains SSII-1, 1404 and 1593 (Singer, 1973, 1977; Ramoska *et al.*, 1977). A third strain, 2362, isolated by Jaroslav Weiser from Nigerian black fly samples, also proved highly effective (Weiser, 1984). *Bacillus sphaericus* was shown to carry a complex of toxins that retained their activity for significant periods in the mosquito habitat, although its highest activity was restricted to *Anopheles* and *Culex* mosquito species (Davidson, 1982). Commercial products based on *B. sphaericus* continue to be used for mosquito control. The *B. sphaericus* genome was published by Hu *et al.* (2008).

2.4.3. Fungi

Fungi became the first pathogens to be considered for biological control of insects. As early as 1835, Bassi suggested that *Beauveria bassiana* might be used to control pest species, as he had successfully transferred the fungus from silkworms to other caterpillars (Steinhaus, 1975).

The first large-scale production of a microbial pesticide and the first actual field trials are credited to a colleague of Elie Metchnikoff, Isaak Krassilstchik, who produced *Metarhizium anisopliae* on beer mash and distributed it into fields around Kiev in 1888 to act against the cockchafer (Krassilstchik, 1888). *Beauveria bassiana* was disseminated for control of the chinch bug in Minnesota in 1888 (Lugger, 1888), while in the same year Francis Snow began efforts on fungus introduction in Kansas that led to the establishment of a field station that distributed *B. bassiana* to growers in eight states (Snow, 1891). Observations of fungal disease on scale insects and whiteflies in the 1940s led H. S. Fawcett to develop culture methods for fungi, particularly *Aschersonia*, and to distribute the cultures through the Florida State Experimental Station for control of whiteflies (Fawcett, 1948). Unfortunately, the unpredictability of fungi to control the chinch bug and scale insects in the late nineteenth and early twentieth centuries led to reduced interest in the use of microbial control agents in general until the mid-twentieth century.

Hirsutella thompsonii was the first fungus registered for use in the USA against the citrus rust mite in 1981 (Tanada and Kaya, 1993). More recently, there have been several successful products including those based upon *Beauveria bassiana* against Masson's pine caterpillar (Wang *et al.*, 2004), *Metarhizium* spp. against grasshoppers in the 1990s (Lomer *et al.*, 1997; Goettel and Johnson, 1997), *Aschersonia* spp. against citrus pests (McCoy *et al.*, 1988), and several fungi against the silverleaf whitefly (Faria and Wraight, 2001).

2.4.4. Microsporidia

Although a number of microsporidia were initially considered to be promising candidates for microbial control (McLaughlin, 1971), only one, *Nosema locustae*, has become widely used (Streett, 2000). *Nosema locustae* was described in African locusts by Elizabeth Canning (1953) and is also found in North American grasshoppers (Steinhaus, 1951). Research at the USDA by John Henry and colleagues led to its approval of use for control of grasshoppers (Henry, 1981; Anon., 1992). *Nosema locustae* is produced in living grasshoppers and formulated as bait (Henry *et al.*, 1978; Henry, 1981; Johnson, 1997).

Commercial products based on *N. locustae* are currently sold by several companies.

2.4.5. Nematodes

The Russian scientist Filipjev (1934) and others, including Steinhaus (1949), suggested the use of nematodes for control of agricultural insects (Welch, 1963). The impetus for their use really began when Rudolph W. Glaser discovered heavily infected Japanese beetle larvae on a New Jersey golf course in 1929 (Glaser, 1931, 1932). Gotthold Steiner named the nematode *Neoaplectana* (now *Steinernema*) *glaseri* (Steiner, 1929). Glaser succeeded in culturing the nematodes on an artificial medium (Glaser, 1931; Poinar, 1991) and in 1935 carried out the first releases of nematodes for biological control of beetle grubs in golf courses (Glaser and Farrell, 1935). Parasitized grubs were later found in the treated areas, and over the following years many more hectares were treated with the nematodes. As recently as 2001, nematodes were found once again in Japanese beetle larvae in New Jersey golf courses (Stock and Koppenhüfer, 2003). In 1966, George O. Poinar, Jr. and Gerard M. Thomas showed that entomopathogenic nematodes rely on symbiotic bacteria to kill the host by septicemia (Poinar and Thomas, 1966; Gaugler *et al.*, 1992). Commercial products based on *S. glaseri* were available until recent years, and currently several companies produce and market products based on other nematode species such as *S. carpocapsae*, *S. riobrave*, *S. feltiae*, *Heterorhabditis bacteriophora*, and *H. megidis* (Kaya and Gaugler, 1993; Stock, 2005; Lord, 2005).

2.5. EVENTS THAT BROUGHT SCIENTISTS TOGETHER AND ADVANCED THE FIELD

Certain events or individuals can have a great impact on a field of science. Among the most important are scientific societies and conferences that bring scientists together and lead to collaborations, courses that train the next generation of scientists, and the establishment of journals that publish the results of novel research. J. L. LeConte presented a talk to the American Association for the Advancement of Science in 1873 in which he proposed that muscardine fungus should be useful in controlling caterpillars; this talk became an important landmark in the use of these agents in the USA (LeConte, 1874; Steinhaus, 1975). Reports of disease appeared in early proceedings of the American Beekeeping Association and the California Silk Culture Association (Steinhaus, 1975). The potential usefulness of *B. thuringiensis* and *Metarhizium* for corn borer control was discussed at the International Corn Borer Conference held in Paris in 1929 (Steinhaus, 1975). The first

International Conference (Colloquium) on Insect Pathology and Biological Control was held during the International Congress of Entomology in Montreal, Canada, in 1956, where Edward Steinhaus met a Czech colleague, Jaroslav Weiser, who later hosted a similar conference, the First International Colloquium on Invertebrate Pathology, in Prague in 1958 (Fig. 2.1) (Steinhaus, 1975).

Edward Steinhaus (1914–1969) had a great impact on the field of insect pathology. His dissertation at Ohio State University on microbes associated with insects was published as *Catalogue of Bacteria Associated Extracellularly with Insects and Ticks* (1942). He taught the first course on insect pathology at University of California, Berkeley, in 1947 and mentored several of the scientists who later established laboratories that moved the field forward. He wrote textbooks including "Insect Microbiology" (1946), "Principles of Insect Pathology" (1949), "Laboratory Exercises in Insect Microbiology and Insect Pathology" (Martignoni and Steinhaus, 1961), and "Insect Pathology, An Advanced Treatise" (2 volumes, 1963). "Disease in a Minor Chord", partially completed at his death in 1969 and published in 1975, contains a detailed history of the field across the world and through the ages. In 1959, Steinhaus established of the *Journal of Insect* (changed to *Invertebrate* in 1965) *Pathology* and was its first editor. Over 5000 scientific articles were published in this journal between 1959 and 2011. In 1967, Steinhaus brought together a group of colleagues to initiate the Society for Invertebrate Pathology, which first met at Ohio State University in 1968. The Society has members from over 50 countries who come together annually to discuss current research and establish international collaborations (Steinhaus, 1975; Davidson and Burges, 2005). Steinhaus is considered by his peers to be the "Founder of Modern Insect Pathology" (Tanada and Kaya, 1993).

2.6. UNEXPECTED PRODUCTS OF RESEARCH ON INSECT PATHOGENS

The early 1980s saw amazing advances in our understanding of the genetics and biochemistry of insect pathogens, due in large part to the development of new techniques such as the polymerase chain reaction in 1983 (Mullis *et al.*, 1986). The first *B. thuringiensis* crystal protein genes were cloned into *Escherichia coli* by H. Ernest Schnepf and Helen Whitely in 1981 (Schnepf and Whiteley, 1981). Transgenic plants containing *B. thuringiensis* genes were developed in the late 1980s and early 1990s (Fischoff *et al.*, 1987; Perlak *et al.*, 1991; Koziel *et al.*, 1993) and by 1995 field corn containing *B. thuringiensis* genes to control *H. zea* was on the market. Currently, a large proportion of the field corn, cotton, and soybeans grown in the USA contain genes from *B. thuringiensis* (Glare and O'Callaghan, 2000; Federici, 2005).

The development of insect cell culture by T. D. C. Grace (Grace, 1962) and plaque assay for NPV by W. Fred Hink and Pat V. Vail (Hink and Vail, 1973), along with the efforts of many other scientists (Maramorosch, 1991), led to the exploration of genes involved in the activity of baculoviruses (i.e., NPVs). The polyhedron gene of NPV was found to be non-essential, and could be replaced with another gene that would be strongly expressed by the modified virus in cell culture (O'Reilly, 1997). Gail Smith, Matt Fraser, and Max Summers published two papers in 1983 describing the first manipulation of the *Autographa californica* NPV, and production of a human immunological protein, beta-interferon, in insect cells infected with modified NPV (Smith *et al.*, 1983a, b). This research, along with research of investigators in Oxford, UK (Matsuuray *et al.*, 1987), and elsewhere, led to patenting of the baculovirus expression system in 1988 (Smith and Summers, 1988). Thousands of medically and commercially important proteins, including experimental vaccines for human immunodeficiency virus, malaria, and human papilloma virus, have since been produced in the baculovirus expression system (Arif, 2005). Malaria continues to be a major health concern in many countries. In 2011, Fang *et al.* (2011) produced recombinant strains of *Metarhizium anisopliae* that express molecules which block production of the malaria parasite, *Plasmodium falciparum*. When mosquitoes were infected with this strain, sporozoite counts were significantly reduced, potentially leading to a new method to reduce malaria transmission.

2.7. CONCLUSIONS

The long history of insect pathology, from the observations of Aristotle more than 2000 years ago to the many useful products derived from insect pathogens today, gives us an appreciation of the potential utility of these organisms. However, it is important to remember that these developments do not happen on their own but are the result of the hard work and imagination of scientists around the world studying diseases of insects, only a small percentage of which are mentioned here. With the rapid development of molecular methods, many insect pathogens have been fully sequenced, beginning with the black beetle virus in 1985 (Dasgupta *et al.*, 1984; Dasmahapatra *et al.*, 1985) and the first NPV (*Autographa californica*) in 1994 (Ayres *et al.*, 1994). Nematode and insect symbionts, *Photorhabdus* and *Wolbachia*, were fully sequenced in 2003 and 2004, respectively (Duchaud *et al.*, 2003; Wu *et al.*, 2004), and the *B. thuringiensis* genome was published in 2007 (Challacombe *et al.*, 2007). The genomes of two bee pathogens, the fungus *Ascosphaera apis* and the bacterium *Paenibacillus larvae*, were published in 2006 (Qin *et al.*, 2006) and the

genome of the microsporidian bee pathogen, *Nosema ceranae*, was published in 2009 (Cornman *et al.*, 2009). The genomes of *Metarhizium anisopliae* (now *M. robertsii*) and *M. acridum* were published in 2011 (Gao *et al.*, 2011).

With the costs of sequencing becoming more affordable every year, the day will come when many insect pathogens will be fully sequenced. This will allow for a better understanding of the evolution of pathogenicity, the presence of virulence genes, metabolite biosynthesis, genetic relatedness with other organisms and, perhaps, the development of more effective delivery methods in the field.

REFERENCES

Aizawa, K. (2001). Shigetane Ishiwata: his discovery of sotto-kin (*Bacillus thuringiensis*) in 1901 and subsequent investigations in Japan. *Proceedings of a Centennial Symposium Commemorating Ishiwata's Discovery of Bacillus thuringiensis*. Japan: Kurume. November 1−3, 2001.

Aldrovandi, U. (1602). De animalibus insectis libri septem, cum singulorum iconibus ad vivum expressis. Bologna.

Angus, T. (1954). A bacterial toxin paralyzing silkworm larvae. *Nature*, *173*, 545.

Anonymous (1992). RED Facts, *Nosema locustae*. US Environmental Protection Agency EPA738-F-92-011, September, 1992.

Aoki, K., & Chigasaki, Y. (1915). Uber die Pathogenität der sog. Sotto-Bacillen (Ishiwata) bei Seidenraupen. *Mitteil. Der Med. Fakult. Der Kaisser Univ. zu Tokyo, 13*, 419−440.

Arif, B. (2005). A brief journey with insect viruses with emphasis on baculoviruses. *J. Invertebr. Pathol., 89*, 39−45.

Aristotle. (1878). *Historia Animalium. History of Animals, in 10 books, translated by R. Cresswell*. London: G. Bell and Son.

Ayres, M. D., Howard, S. C., Kuzio, J., Lopez-Ferber, M., & Possee, R. D. (1994). The complete DNA sequence of *Autographa californica* nuclear polyhedrosis virus. *Virology, 202*, 586−605.

Bailey, L., & Ball, B. V. (1991). *Honey Bee Pathology* (2nd ed.). New York: Academic Press.

Bassi, A. (1835). *Del Mal del Segno*. Lodi: Tipografia Orcesi.

Beegle, C. C., & Yamamoto, T. (1992). Invitation paper (C.P. Alexander Fund): History of *Bacillus thuringiensis* Berliner research and development. *Can. Entomol., 124*, 587−616.

Benz, G. (1986). Introduction: historical perspectives. In R. Granados & B. A. Federici (Eds.), *The Biology of Baculoviruses, Vol. 2. Practical Application for Insect Control* (pp. 1−35). Boca Raton: CRC Press.

Bergold, G. (1947). Die Isolierung des Polyeder-Virus und der Natur der Polyeder. *Z. Naturforsch., 2b*, 122−143.

Berliner, E. (1915). Über die Schlaffsucht der Mehlmottenraupe (*Ephestia kuhniella* Zell.) und irhen Erreger *Bacillus thuringiensis* n.sp. *Z. Angew. Entomol., 2*, 29−56.

Bird, F. T., & Whalen, M. M. (1954). Stages in the development of two insect viruses. *Can. J. Microbiol., 1*, 170−174.

Bolle, G. (1894). Il giallume od il mal del grasso del baco da seta. Communicazione preliminare. *Atti. E Mem. Dell' I.R. Soc. Agric. Gorizia, 34*, 133−136.

Bucher, G. E. (1959). The bacterium *Coccobacillus acridiorum* d'Herelle: its taxonomic position and status as a pathogen of locusts and grasshoppers. *J. Insect Pathol., 1*, 331−346.

Burges, H. D. (1967). The standardization of products based on *Bacillus thuringiensis*. In P. A. van der Laan (Ed.), *Proceedings of the International Colloquium on Insect Pathology and Microbial Control, Wageningen* (pp. 306–314). Amsterdam: North-Holland.

Burges, H. D. (1981). Safety, safety testing and quality control of microbial pesticides. In H. D. Burges (Ed.), *Microbial Control of Pests and Plant Diseases 1970–1980* (pp. 737–767). London: Academic Press.

Canning, E. U. (1953). A new microsporidian, *Nosema locustae* n.s., from the fat body of the African migratory locust, *Locusta migratoria migratorioides* (R. and F.). *Parasitology, 43*, 287–290.

Cavalier-Smith, T. (1993). Kingdom Protozoa and its 18 phyla. *Microbiol. Rev., 57*, 953–994.

Challacombe, J. F., Altherr, M. R., Xie, G., Bhotika, S. S., Brown, N., Bruce, D., Campbell, C. S., Campbell, M. L., Chen, J., Chertkov, O., Cleland, C., Dimitrijevic, M., Doggett, N. A., Fawcett, J. J., Glavina, T., Goodwin, L. A., Green, L. D., Han, C. S., Hill, K. K., Hitchcock, P., Jackson, P. J., Keim, P., Kewalramani, A. R., Longmire, J., Lucas, S., Malfatti, S., Martinez, D., McMurry, K., Meincke, L. J., Misra, M., Moseman, B. L., Mundt, M., Munk, A. C., Okinaka, R. T., Parson-Quintana, B., Reilly, L. P., Richardson, P., Robinson, D. L., Saunders, E., Tapia, R., Tesmer, J. G., Thayer, N., Thompson, L. S., Tice, H., Ticknor, L. O., Wills, P. L., Gilna, P., & Brettin, T. S. (2007). The complete genome sequence of *Bacillus thuringiensis* Al Hakam. *J. Bacteriol., 189*, 3680–3681.

Cheshire, F. R. (1884). Foul brood (not *Micrococcus* but *Bacillus*), the means of its propagation and the method of its cure. *Br. Bee J., 12*, 256–263.

Cheshire, F. R., & Cheyne, W. W. (1885). The pathogenic history and history under cultivation of a new bacillus (*B. alvei*), the cause of a disease of the hive bee hitherto known as foul brood. Ser. II. *J.R. Microsc. Soc., 5*, 581–601.

Cohn, F. (1855). *Empusa muscae* und die Krankheit den durch parasitische Pilze charakterisiten Epidemieen. *Verh. Kaisrl. Leopald-Carolin. Naturforscher, 25*, 300–360.

Cornalia, E. (1856). Monografia del bombice del gelso. *Mem. R. Istit. Lombardo Sci. Lett. Arte, 6*, 3–387.

Cornman, R. S., Chen., Y. P., Schatz, M. C., Street, C., Zhao, Y., Desany, B., Egholm, M., Hutchison, S., Pettis, J. S., Lipkin, W. I., & Evans, J. D. (2009). Genomic analyses of the microsporidian *Nosema ceranae*, an emergent pathogen of honey bees. *PLoS Pathogens 5*, e1000466.

Corradi, N., & Keeling, P. J. (2009). Microsporidia: a journey through radical taxonomical revisions. *Fungal Biol. Rev., 23*, 1–8.

Cross, J. V., Solomon, M. G., Chandler, D., Jarrett, P., Richardson, P. N., Winstanley, D., Bathon, H., Huber, J., Keller, B., Langenbruch, G. A., & Zimmermann, G. (1999). Biocontrol of pests of apples and pears in northern and central Europe. I. Microbial agents and nematodes. *Biocontrol Sci. Technol., 9*, 125–149.

Cunningham, J. C. (1982). Field trials with baculoviruses: control of forest insect pests. In E. Kurstak (Ed.), *Microbial and Viral Pesticides* (pp. 335–386). New York: Marcel Dekker.

de Barjac, H., & Bonnefoi, A. (1962). Essai de classification biochemique et serologique de 24 souches de *Bacillus* du type *B. thuringiensis*. *Entomophaga, 7*, 5–31.

de Barjac, H., & Bonnefoi, A. (1968). A classification of strains of *Bacillus thuringiensis* Berliner with a key to their differentiation. *J. Invertebr. Pathol., 11*, 335–347.

Dasgupta, R., Ghosh, A., Dasmahapatra, B., Guarino, L. A., & Kaesberg, P. (1984). Primary and secondary structure of black beetle virus RNA2, the genomic messenger for BBV coat protein precursor. *Nucl. Acids Res., 12*, 7215–7223.

Dasmahapatra, B., Dasgupta, R., Ghosh, A., & Kaesberg, P. (1985). Structure of the black beetle virus genome and its functional implications. *J. Molec. Biol., 182*, 183–189.

Davidson, E. W. (1982). Bacteria for the control of arthropod vectors of human and animal disease. In E. Kurstak (Ed.), *Microbial and Viral Pesticides* (pp. 289–316). New York: Marcel Dekker.

Davidson, E. W., & Burges, H. D. (2005). History of the Society for Invertebrate Pathology. *J. Invertebr. Pathol., 89*, 2–11.

Duchaud, E., Rusniok, C., Frangeul, L., Buchrieser, C., Givaudan, A., Taourit, S., Bocs, S., Boursaux-Eude, C., Chandler, M., Charles, J.-F., Dassa, E., Derose, R., Derzelle, S., Freyssinet, G., Gaudriault, S., Médigue, C., Lanois, A., Powell, K., Siguier, P., Vincent, R., Wingate, V., Zouine, M., Glaser, P., Boemare, N., Danchin, A., & Kunst, F. (2003). The genome sequence of the entomopathogenic bacterium *Photorhabdus luminescens*. *Nat. Biotechnol., 21*, 1307–1313.

Dulmage, H. T. (1970). Insecticidal activity of HD-1, a new isolate of *Bacillus thuringiensis* var. *alesti*. *J. Invertebr. Pathol., 15*, 232–239.

Dulmage, H. T., Boening, O. P., Rehnborg, C. S., & Hansen, G. D. (1971). A proposed standardized bioassay for formulations of *Bacillus thuringiensis* based on the international unit. *J. Invertebr. Pathol., 18*, 240–245.

Dutky, S. R. (1940). Two new spore-forming bacteria causing milky diseases of Japanese beetle larvae. *J. Agric. Res., 61*, 57–68.

d'Herelle, F. (1911). Sur une épizootie de nature bacterienne sévissant sur les sauterelles au Mexique. *C.R. Acad. Sci. Paris Ser. D, 152*, 1413–1415.

d'Herelle, F. (1912). Sur la propagation, dans la République Argentine, de l'épizootie des sauterelles du Mexique. *C.R. Acad. Sci. Paris Ser. D, 154*, 623–625.

d'Herelle, F. (1914). Le coccobacille des sauterelles. *Ann. Inst. Pasteur Paris, 28*, 280–328, 387–407.

de Réamur, R.-A. F. (1734–1742). *Mémoires pour server á l'historie des insects*. (Vols. 1–6). Imprimerie Royale, Paris.

Edlind, T. D., Li, J., Visvesbara, G. S., Vodkin, M. H., McLaughlin, G. L., & Katiyar, S. K. (1996). Phylogenetic analysis of β-tubulin sequences from amitochondrial protozoa. *Molec. Phylogenet. Evol., 5*, 359–367.

Engler, R., & Rogoff, M. H. (1980). Registration and regulation of microbial pesticides. *Biotechnol. Bioeng., 22*, 1441–1448.

Erlandson, M. (2008). Insect pest control by viruses. In B. W. J. Mahy & M. H. V. van Regenmortel (Eds.), *Encyclopedia of Virology* (3rd ed.)., (Vol. 3) (pp. 125–133). Amsterdam: Elsevier.

Fang, W., Vega-Rodríguez, J., Ghosh, A. K., Jacobs-Lorena, M., Kang, A., & St. Leger, R. J. (2011). Development of transgenic fungi that kill human malaria parasites in mosquitoes. *Science, 331*, 1074–1077.

Fantham, H. B., & Porter, A. (1912a). Microsporidiosis, a protozoal disease of bees due to *Nosema apis*, and popularly known as Isle of Wight disease. *Ann. Trop. Med. Parasitol., 6*, 145–162.

Fantham, H. B., & Porter, A. (1912b). The structure and homology of the microsporidian spore, as seen in *Nosema apis*. *Proc. Camb. Philos. Soc., 16*, 580–583.

Faria, M., & Wraight, S. P. (2001). Biological control of *Bemisia tabaci* with fungi. *Crop Prot., 20*, 767–778.

Fawcett, H. S. (1948). Biological control of citrus insects by parasitic fungi and bacteria. In L. D. Batchelor & H. J. Webber (Eds.), *The Citrus Industry*, (Vol. 2) (pp. 628–664). Berkeley: University of California Press.

Federici, B. A. (2005). Insecticidal bacteria: an overwhelming success for invertebrate pathology. *J. Invertebr. Pathol., 89*, 30–38.

Filipjev, I. N. (1934). Harmful and Useful Nematodes in Rural Economy. Moscow and Leningrad. (In Russian.)

Fischoff, D. A., Bowdish, K. S., Perlak, F. J., Marrone, P. G., McCormick, S. M., Niedermyer, J. G., Dean, D. A., Kuzankretzmer, K., Meyer, E. J., Rochester, D. E., Rogers, S. G., & Fraley, R. T. (1987). Insect tolerant tomato plants. *Bio/Technology, 5*, 807–813.

Fleming, W. E. (1976). Integrating control of the Japanese beetle – a historical review. *U.S. Dept. Agric. Tech. Bull.*, 1545.

Fresenius, G. (1856). Notiz, Insekten-Pilze betreffend. *Bot. Ztg., 14*, 882–883.

Furgala, B., & Mussen, E. C. (1990). Protozoa. In R. A. Morse & R. Nowogrodzki (Eds.), *Honey Bee Pests, Predators and Diseases* (2nd ed.). (pp. 48–63). Ithaca: Cornell University Press.

Gao, Q., Jin, K., Ying, S.-H., Zhang, Y., Xiao, G., Shang, Y., Duan, Z., Hu, X., Xi, X.-Q., Zhou, G., Peng, G., Luo, Z., Huang, W., Wang, B., Fang, W., Wang, S., Zhong, Y., Ma, L.-J., St. Leger, R. J., Zhao, G.-P., Pei, Y., Feng, M.-G., Xia, Y., & Wang, C. (2011). Genome sequencing and comparative transcriptomics of the model entomopathogenic fungi *Metarhizium anisopliae* and *M. acridum*. *PLoS Genetics, 7*, e1001264.

Gaugler, R., Campbell, J. F., Selvan, S., & Lewis, E. E. (1992). Large-scale inoculative releases of the entomopathogenic nematode *Steinernema glaseri*: assessment 50 years later. *Biol. Control, 2*, 181–187.

Gilliam, M., & Vandenberg, J. (1990). Fungi. In R. A. Morse & H. Nowogrodski (Eds.), *Honey Bee Pests, Predators and Diseases* (2nd ed.). (pp. 64–90). Ithaca: Cornell University Press.

Glare, T. R., & O'Callaghan, M. (2000). *Bacillus thuringiensis: Biology, Ecology and Safety*. Chichester: John Wiley & Sons.

Glaser, R. W. (1918). The polyhedral virus of insects with a theoretical consideration of filterable viruses generally. *Science, 48*, 301–302.

Glaser, R. W. (1931). The cultivation of a nematode parasite of an insect. *Science, 73*, 614–615.

Glaser, R. W. (1932). Studies on *Neoaplectana glaseri*, a nematode parasite of the Japanese beetle (*Popillia japonica*). NJ Department of Agriculture, Circ. No. 211.

Glaser, R. W., & Chapman, J. W. (1913). The wilt disease of gypsy moth caterpillars. *J. Econ. Entomol., 6*, 479–488.

Glaser, R. W., & Farrel, C. C. (1935). Field experiments with the Japanese beetle and its nematode parasite. *J.N.Y. Entomol. Soc., 43*, 345–371.

Gochnauer, T. A. (1990). Viruses. In R. A. Morse & R. Nowogrodzki (Eds.), *Honey Bee Pests, Predators and Diseases* (2nd ed.). (pp. 12–47) Ithaca: Cornell University Press.

Goettel, M. S., and Johnson, D. L. (Eds.) (1997). Microbial Control of Grasshoppers and Locusts. *Mem. Entomol. Soc. Canada, 171*, 1–400.

Goldberg, L. J., & Margalit, J. (1977). A bacterial spore demonstrating rapid larvicidal activity against *Anopheles sergentii, Uranotaenia unguiculata, Culex univitattus, Aedes aegypti* and *Culex pipiens*. *Mosq. News, 37*, 355–358.

Gould, W. (1747). An Account of English Ants. A. Millar, London.

Grace, T. D. C. (1962). Establishment of four strains of cells from insect tissues grown *in vitro*. *Nature, 195*, 788–789.

Hagen, H. A. (1879). Obnoxious pests – suggestions relative to their destruction. *Can. Entomol., 11*, 110–114.

Hannay, C. L. (1953). Crystalline inclusions in aerobic spore-forming bacteria. *Nature, 172*, 1004–1006.

Hawley, I. M., & White, F. G. (1935). Preliminary studies on the diseases of larvae of the Japanese beetle (*Popillia japonica* Newm.). *J.N.Y. Entomol. Soc., 43*, 405–412.

Henry, J. E. (1981). Natural and applied control of insects by protozoa. *Annu. Rev. Entomol., 26*, 49–73.

Henry, J. E., Oma, E. A., & Onsager, J. A. (1978). Relative effectiveness of ULV spray applications of spores of *Nosema locustae* against grasshoppers. *J. Econ. Entomol., 71*, 629–632.

Hink, W. F., & Vail, P. V. (1973). A plaque assay for titration of alfalfa looper nuclear polyhedrosis virus in a cabbage looper (TN-368) cell line. *J. Invertebr. Pathol., 22*, 168–174.

Hu, X., Fan, W., Han, B., Liu, H., Zheng, D., Li, Q., Dong, W., Yan, J., Gao, M., Berry, C., & Yuan, Z. (2008). Complete genome sequence of the mosquitocidal bacterium *Bacillus sphaericus* C3-41 and comparison with those of closely related *Bacillus* species. *J. Bacteriol., 190*, 2892–2902.

Huber, J. (1986). Use of baculoviruses in pest management programs. In R. Granados & B. A. Federici (Eds.), *The Biology of Baculoviruses* (pp. 181–202). Boca Raton: CRC Press.

Huger, A. (1966). A virus disease of the Indian rhinoceros beetle, *Oryctes rhinoceros* (Linnaeus) caused by a new type of insect virus, *Rhabdionvirus oryctes* gen.n., sp.n. *J. Invertebr. Pathol., 8*, 38–51.

Husz, B. (1929). On the use of *Bacillus thuringiensis* in the fight against the corn borer. *Int. Corn Borer Invest. Sci. Rept., 2*, 99–110.

Ishimori, N. (1934). Contribution a l'étude de la grasserie du ver á soie (*Bombyx mori*). *C.R. Seances Soc. Biol. Soc Franco-Japonaise Biol., 116*, 1169–1170.

Ishiwata, S. (1901). On a kind of severe flacherie (sotto disease). *Dainihon Sanshi Kaiho, 114*, 1–5.

Ishiwata, S. (1905). About "Sottokin", a bacillus of a disease of the silkworm. *Rept. Sericult. Assoc. Jpn., 160*, 1–8, No. 161, 1–5.

Jehle, J. A. (2009). André Paillot (1885–1944): his work lives on. *J. Invertebr. Pathol., 101*, 162–168.

Jirovec, O. (1936). Studien über Microsporidien. *Vestn. Cesk. Spol. Zool., 4*, 1–75.

Johnson, D. L. (1997). Nosematidae and other protozoa as agents for control of grasshoppers and locusts: current status and prospects. *Mem. Entomol. Soc. Canada, 171*, 375–389.

Kaya, H. K., & Gaugler, R. (1993). Entomopathogenic nematodes. *Annu. Rev. Entomol., 38*, 181–206.

Keilin, D. (1921). On a new type of fungus: *Coelomomyces stegomyiae*, n.g., n.sp., parasitic in the body-cavity of the larva of *Stegomyia scutellaris* Walker (Diptera, Nematocera, Culicidae). *Parasitology, 13*, 225–234.

Kellen, W. R., Clark, T. B., Lindegren, J. E., & Ho, B. C. (1965). *Bacillus sphaericus* Neide as a pathogen of mosquitoes. *J. Invertebr. Pathol., 7*, 442–448.

Kirby, W., & Spence, W. (1822). *Diseases of insects*. In: *An Introduction to Entomology: or Elements of the Natural History of Nematodes*, (Vol. 4). Hurst, Rees, Orme and Brown, London: Longman. 197–232.

Koziel, M. G., Beland, G. L., Bowman, C., Carozzi, N. B., Crenshaw, R., Crossland, L., Dawson, J., Desai, N., Hill, M., Kadwell, S.,

Launis, K., Lewis, K., Maddox, D., McPherson, K., Meghji, M. R., Merlin, E., Rhodes, R., Warren, G. W., Wright, M., & Evola, S. V. (1993). Field performance of elite transgenic maize plants expressing an insecticidal protein derived from *Bacillus thuringiensis*. *Nat. Biotechnol., 11*, 194–200.

Krassiltstchik, I. M. (1888). La production industrielle des parasites végétaux pour la destruction des insectes nuisibles. *Bull. Sci. France Belg., 19*, 461–472.

Krieg, A., Huger, A. M., Langenbruch, G. A., & Schnetter, W. (1983). *Bacillus thuringiensis* var. *tenebrionis*: a new pathotype effective against larvae of Coleoptera. *Z. Angew. Entomol., 96*, 500–508.

Kudo, R. (1924). A Study of the Microsporidia. *Illinois Biol. Monog.* (Vol. 9). Univ. of Illinois. Nos. 1 and 2

Kurstak, E. (1962). Donnees sur l'epizootie bacterienne naturelle provoquee par un *Bacillus* du type *Bacillus thuringiensis* sur *Ephestia kuhniella* Zeller. *Entomophaga Mem. Hors* Ser., 2, 245–247.

Kurstak, E. (1964). Le processus de l'infection par *Bacillus thuringiensis* Berl. d' Ephestia kühniella Zell. déclenché par le parasitisme de *Nemeritis canescens* Grav (Ichneumonidae). C.R. Hebd. Séanc. *Acad. Sci. Paris, 259*, 211–212.

Lacey, L. A., & Shapiro-Ilan, D. I. (2008). Microbial control of insect pests in temperate orchard systems: potential for incorporation into IPM. *Annu. Rev. Entomol., 53*, 121–144.

Lacey, L. A., Thomson, D., Vincent, C., & Arthurs, S. P. (2008). Codling moth granulovirus: a comprehensive review. *Biocontrol Sci. Technol., 18*, 639–663.

LeConte, J. L. (1874). Hints for the promotion of economic entomology. *Am. Assoc. Adv. Sci. Proc., 22*, 10–22.

Linnaei, C. (1758). Systema naturae per regna tria naturae, secundum classes, ordines, genera species, cum characteribus differentiis, synonymis, locis. Editio Decima, Reformata. L. Salvii, Homiae.

Lomer, C. J., Prior, C., & Kooyman, C. (1997). Development of *Metarhizium* spp. for the control of grasshoppers and locusts. *Mem. Entomol. Soc. Canada, 171*, 265–286.

Lord, J. C. (2005). From Metchnikoff to Monsanto and beyond: the path of microbial control. *J. Invertebr. Pathol., 89*, 19–29.

Lugger, O. (1888). Fungi which kill insects. *Univ. Minnesota Agric. Exp. Sta. Tech. Bull., 4*.

Luthy, P., Cordier, J., & Fischer, H. (1982). *Bacillus thuringiensis* as a bacterial insecticide: basic considerations and applications. In E. Kurstak (Ed.), *Microbial and Viral Pesticides* (pp. 35–74). New York: Marcel Dekker.

Maassen, A. (1913). Weitere Mitteilungen uber die seuchenhaften Brutkrankheiten der Bienen (Further communication on the epidemic brood diseases of bees). *Mitteilungen aus der Kaiserlichen Biologischen Anstalt fur Land- und Forstwirtsch., 14*, 48–58.

Maassen, A. (1916). Uber Bienenkrankheiten (On bee diseases). *Mitteilungen aus der Kaiserlichen Biologischen Anstalt fur Land- und Forstwirtsch., 16*, 51–58.

Maestri, A. (1856). *Frammenti anatomici, fisiologici e patologici sul baco da seta*. Pavia: Fratelli Fusi.

Maramorosch, K. (1991). Thomas D.C. Grace – insect tissue culture pioneer. *J. Invertebr. Pathol., 58*, 151–156.

Margalit, J., & Dean, D. (1985). The story of *Bacillus thuringiensis israelensis* (B.t.i.). *J. Am. Mosq. Control Assoc., 1*, 1–7.

Martignoni, M. E. (1999). History of TM BioControl-1: the first registered virus-based product for control of a forest insect. *Am. Entomol., 45*, 30–37.

Martignoni, M. E., & Steinhaus, E. A. (1961). *Laboratory Exercises in Insect Microbiology and Insect Pathology*. Minneapolis: Burgess.

Matsuuray, Y., Possee, R. D., Overton, H. A., & Bishop, D. H. L. (1987). Baculovirus expression vectors – the requirements for high-level expression of proteins, including glycoproteins. *J. Gen. Virol., 68*, 1233–1250.

Mattes, O. (1927). Parasitäre Krenkheiten der Mehlmottenlarven und Versuch über ihre Verwendbarkeit als biologisches Bekämpfungsmittel. *Ges. Beford. Gesamte Naturwiss. Marburg, 62*, 381–417.

McCoy, C. W., Samson, R. A., & Boucias, D. G. (1988). Entomogenous fungi. In C. Ignoffo & N. B. Mandava (Eds.), *CRC Handbook of Natural Pesticides, Vol. 5, Microbial Insecticides, Part A, Entomogenous Protozoa and Fungi* (pp. 151–234). Boca Raton: CRC Press.

McLaughlin, R. E. (1971). Use of protozoans for microbial control of insects. In H. D. Burges & N. W. Hussey (Eds.), *Microbial Control of Insects and Mites* (pp. 151–172). New York: Academic Press.

Merian, M. S. (1679). Der Raupen wunderbare Verwandelung und sonderbare Blumen-nahrung. J.A. Graff, Nürnberg.

Metalnikov, S., & Chorine, V. (1928). Maladies bacteriennes chez les chenilles de la pyrale du mais (*Pyrausta nubialis* Hbn.). *C. R. Seances Hebd. Acad. Sci., 186*, 546–549.

Metchnikoff, E. (1879). O boleznach litchinok khlebnogo zhuka. Zapiski Imperatorskogo Obschestva sel' skogo khoziaistva luzhnoi Rossii. (pp. 21–50). Odessa.

Milner, R. J. (1994). History of *Bacillus thuringiensis*. *Agric. Ecosyst. Environ., 49*, 9–13.

Mitani, K., & Watarai, J. (1916). A new method to isolate the toxin of *Bacillus sotto* Ishiwata by passing through a bacterial filter and a preliminary report on the toxic action of this toxin to the silkworm larva. *Aichi Gensanshu Seizojo Hokoku, 3*, 33–42.

Mullis, K., Faloona, F., Scharf, S., Saiki, R., & Erlich, H. (1986). Specific enzymatic amplification of DNA *in vitro*: the polymerase chain reaction. *Cold Spring Harb. Symp. Quant. Biol., 51*, 263–273.

Nägeli, K. W. (1857). Über die neue Krankheit der Seidenraupe und verwandte Organismen. *Bot. Zeitung, 15*, 760–761.

Nysten, P. H. (1808). *Recherches sur les maladies des vers a soie et les moyens de les prévenir*. Paris: De L'Imprimerie Impériale.

O'Reilly, D. R. (1997). Auxiliary genes of baculoviruses. In L. K. Miller (Ed.), *The Baculoviruses* (pp. 267–300). New York: Plenum Press.

Paillot, A. (1918). Deux microsporidies nouvelles parasites des chenilles de. *Pieris brassicae. C.R. Soc. Biol., 81*, 66–68.

Paillot, A. (1933). *L'infection chez les insectes: immunité et symbiose*. Trévoux: G. Patissier.

Pasteur, L. (1870). *Études sur la Maladie des Vers à Soie*. Paris: Tome I et II. Gauthier-Villars.

Paterson, R. R. M. (2008). *Cordyceps*, a traditional Chinese medicine and another fungal biofactory? *Phytochemistry, 69*, 1469–1495.

Perlak, F. J., Fuchs, R. L., Dean, D. A., McPherson, S. L., & Fishhoff, D. A. (1991). Modification of coding sequence enhances plant expression of insect control protein genes. *Proc. Natl. Acad. Sci. USA, 88*, 3325–3328.

Pliny, the Elder (Caius Plinius Secundus). (1855). *The Natural History of Pliny*. Translated, with Copious Notes and Illustrations by the Late John Bostock and H.T. Riley. London: H.G. Bohn.

Poinar, G. O., Jr. (1984). First fossil record of parasitism by insect parasitic Tylenchida (Allantonematidae: Nematoda). *J. Parasitol., 70*, 306–308.

Poinar, G. O., Jr. (1991). Rudolph W. Glaser (1888—1947) — a pioneer of steinernematid nematodes. *J. Invertebr. Pathol., 60,* 1—4.

Poinar, G., Jr., & Poinar, R. (2005). Fossil evidence of insect pathogens. *J. Invertebr. Pathol., 89,* 243—250.

Poinar, G. O., Jr., & Thomas, G. M. (1966). The nature of *Achromobacter nematophilus* as an insect pathogen. *J. Invertebr. Pathol., 9,* 510—514.

Qin, X., Evans, J. D., Aronstein, K. A., Murray, K. D., & Weinstock, G. M. (2006). Genome sequences of the honey bee pathogens *Paenibacillus larvae* and. *Ascosphaera apis. Insect Mol. Biol., 15,* 715—718.

Ramoska, W. A., Singer, S., & Levy, R. (1977). Bioassay of three strains of *Bacillus sphaericus* on field-collected mosquito larvae. *J. Invertebr. Pathol., 30,* 151—154.

Rollinson, W. D., Lewis, F. B., & Waters, W. E. (1965). The successful use of a nuclear-polyhedrosis virus against the gypsy moth. *J. Invertebr. Pathol., 7,* 515—517.

Schirach, A. G. (1771). Histoire naturelle de la riene des abeilles, avec l'art de former des essaims. The Hague.

Schnepf, H. E., & Whiteley, H. R. (1981). Cloning and expression of the *Bacillus thuringiensis* crystal protein gene in *Escherichia coli. Proc. Natl. Acad. Sci. USA, 78,* 2893—2897.

Shimanuki, H. (1990). Bacteria. In R. A. Morse & R. Nowogrodzki (Eds.), *Honey Bee Pests, Predators, and Diseases* (2nd ed.). (pp. 27—47) Ithaca: Cornell University Press.

Singer, S. (1973). Insecticidal activity of recent bacterial isolates and their toxins against mosquito larvae. *Nature, 244,* 110—111.

Singer, S. (1977). Isolation and development of bacterial pathogens of vectors. In J.D. Briggs, (Ed.), Biological Regulation of Vectors (pp. 3—18). US Department of Health, Education, and Welfare Publication No. (NIH) 77-1180.

Smith, G.E., and Summers, M. D. (1988). Method for producing a recombinant baculovirus expression vector. US Patent No. 4,745,051.

Smith, G. E., Fraser, M. J., & Summers, M. D. (1983a). Molecular engineering of the *Autographa californica* nuclear polyhedrosis virus genome: deletion mutants within the polyhedron gene. *J. Virol., 46,* 584—593.

Smith, G. E., Summers, M. D., & Fraser, M. J. (1983b). Production of human beta interferon in insect cells infected with a baculovirus expression vector. *Mol. Cell. Biol., 3,* 2156—2165.

Smith, K. M., & Wyckoff, R. W. G. (1950). Structure within polyhedra associated with insect virus diseases. *Nature, 166,* 861—862.

Snow, F. H. (1891). Chinch-bugs. Experiments in 1890 for their destruction in the field by the artificial introduction of contagious disease. *7th Bienn. Rpt. Kansas State Board Agric., 12,* 184—188.

Sorokin, N. V. (1883). Izdanie glavnogo Voenno-Meditsinskago Upraveleneia. St. Petersburg, 544 pp. Pervoe prilozhenie k Voenno-Meditsinskomu Zhurnalu za 1883 (First supplement to the Journal of Military Medicine for the year 1883), (pp. 168—198).

St. Julian, G., & Bulla, L. A., Jr. (1973). Milky disease. In T. C. Cheng (Ed.), *Current Topics in Comparative Pathobiology,* (Vol. 2) (pp. 57—84). New York: Academic Press.

Stairs, G. R. (1971). Use of viruses for microbial control of insects. In H. D. Burges & N. W. Hussey (Eds.), *Microbial Control of Insects and Mites* (pp. 97—124). London: Academic Press.

Steiner, G. (1923). *Aplectana kraussei* n.sp., eine in der Blattwespe *Lyda* sp. parasitierende Nematodenform, nebst Bemerkungen über das Seitenorgan der parasitischen Nematoden. 59. *Zbl. Bakt. Parasitenk. Infetionskrank. Hyg. Abt, 1,* 14—18.

Steiner, G. (1929). *Neoaplectana glaseri,* n.g., n.sp. (Oxyuridae), a new nemic parasite of the Japanese beetle (*Popillia japonica* Newm.). *J. Wash. Acad. Sci., 19,* 436—440.

Steinhaus, E. A. (1942). *Catalog of Bacteria Associated Extracellularly with Insects and Ticks.* Minneapolis: Burgess.

Steinhaus, E. A. (1946). *Insect Microbiology.* Ithaca: Comstock Publishing.

Steinhaus, E. A. (1949). *Principles of Insect Pathology* New York: McGraw-Hill.

Steinhaus, E. A. (1951). Report on diagnosis of diseases of insects, 1944—1950. *Hilgardia, 20,* 629—678.

Steinhaus, E. A. (1956). Living insecticides. *Sci. Am., 195,* 96—103.

Steinhaus, E. A. (1963). *Insect Pathology. An Advanced Treatise,* (Vols. 1 and 2). New York: Academic Press.

Steinhaus, E. A. (1975). *Disease in a Minor Chord.* Columbus: Ohio State University Press.

Stock, S. P. (2005). Insect-parasitic nematodes: from lab curiosities to model organisms. *J. Invertebr. Pathol., 89,* 57—66.

Stock, S. P., & Koppenhöfer, A. M. (2003). *Steinernema scarabaei,* n.sp. (Rhabditida: Steinemematidae), a natural pathogen of scarab beetle larvae (Coleoptera: Scarabeidae) from New Jersey, USA. *Nematology, 5,* 191—204.

Grasshoppers Streett, D. A. (2000). Their biology, identification and management. In Grasshopper Integrated Pest Management User Handbook   www.sidney.ars.usda.gov/grasshopper/Handbook/I/i_2. htm.

Tanabe, Y., Watanabe, M. M., & Sugiyama, J. (2002). Are Microsporidia really related to Fungi? a reappraisal based on additional gene sequences from basal fungi. *Mycol. Res., 106,* 1380—1391.

Tanada, Y. (1964). A granulosis virus of the codling moth, *Carpocapsa pomonella* (Linnaeus). *J. Insect Pathol., 6,* 378—380.

Tanada, Y., & Kaya, H. K. (1993). *Insect Pathology.* New York: Academic Press.

von Tubeuf, C. (1892). Die Krankheiten der Nonne (*Liparis monacha*). *Forstlich- Naturwiss. Zeitschr., 1,* 34—37.

Vago, C. (1963). A new type of insect virus. *J. Insect Pathol., 5,* 275—276.

Vida, M. H. (1527). The Silkworm (Bombycum): a poem in two books. Translated into English by S. Pullein (1750). S. Powell, Dublin.

Virgil (Publius Vergilius Maro). (1941). *The Georgics of Virgil.* translated by C. Day Lewis. London: Jonathan Cape.

Vouk, V. (1932). Rad botanickog institute universiteta u Zagrebu na izucavanju kukuruznog crva. *Acta Botanica, Inst. Bot. Univ. Zagrebensis, 7,* 129—144.

Wang, Z. (1965). Knowledge on the control of silkworm disease in ancient China. *Symp. Sci. Hist. (China), 8,* 15—21.

Wang, C., Fan, M., Li, Z., & Butt, T. M. (2004). Molecular monitoring and evaluation of the application of the insect-pathogenic fungus *Beauveria bassiana* in southeast China. *J. Appl. Microbiol., 96,* 861—870.

Weiser, J. (1963). Sporozoan infections. In E. A. Steinhaus (Ed.), *Insect Pathology. An Advanced Treatise,* (Vol. 2) (pp. 291—334). New York: Academic Press.

Weiser, J. (1984). A mosquito virulent strain of *Bacillus sphaericus* in adult *Simulium damnosum* from northern Nigeria. *Zbl. Mikrobiol., 139,* 57—60.

Weiser, J. (2005). Microsporidia and the Society for Invertebrate Pathology: a personal point of view. *J. Invertebr. Pathol., 89,* 12—18.

Welch, H. E. (1963). Nematode infections. In E. A. Steinhaus (Ed.), *Insect Pathology. An Advanced Treatise*, (Vol. 2) (pp. 363—392). New York: Academic Press.

White, G. F. (1906). The bacteria of the apiary, with special reference to bee diseases. US Department of Agriculture, Bureau of Entomology, Technical Series, No. 14.

White, G. F. (1912). The cause of European foulbrood. US Department of Agriculture, Bureau of Entomology, Circ. No. 157.

White, G. F. (1917). *Sacbrood. U.S. Dept. Agric. Bull., 431*, 1—55.

Wu, M., Sun, L. V., Vamathevan, J., Riegler, M., Deboy, R., Brownlie, J. C., McGraw, E. A., Martin, W., Esser, C., Ahmadinejad, N., Wiegand, C., Madupu, R., Beanan, M. J., Brinkac, L. M., Daugherty, S. C., Durkin, A. S., Kolonay, J. F.,

Nelson, W. C., Mohamoud, Y., Lee, P., Berry, K., Young, M. B., Utterback, T., Weidman, J., Nierman, W. C., Paulsen, I. T., Nelson, K. E., Tettelin, H., O'Neill, S. L., & Eisen, J. A. (2004). Phylogenomics of the reproductive parasite *Wolbachia pipientis* *w*Mel: a streamlined genome overrun by mobile genetic elements. *PLoS Biol., 2*, 327—341.

Yarrow, P. J. (1958). *On the Mark Disease, Calcinnacio or Muscardine, a disease that affects silkworms.* translation of Bassi, Del Mal del Segno. Baltimore: American Phytopathological Society.

Young, E. C. (1986). The Rhinoceros Beetle Project: history and review of the research programme. *Agric. Ecosyst. Environ., 15*, 149—166.

Zander, E. (1909). Tierische Parasiten als Krankheitserreger bei der Biene. *Leipziger Bienenzeitung, 24*, 147—150, 164—166.

Principles of Epizootiology and Microbial Control

David I. Shapiro-Ilan*, Denny J. Bruck[†] and Lawrence A. Lacey**

* *United States Department of Agriculture, Agricultural Research Service, Byron, Georgia, USA,* [†] *United States Department of Agriculture, Agricultural Research Service, Corvallis, Oregon, USA,* ** *IP Consulting International, Yakima, Washington, USA*

SUMMARY

An epizootic is defined as an outbreak of disease in which there is an unusually large number of cases. A central question in insect pathology is: what are the factors that cause an epizootic? The question is addressed through the discipline of epizootiology, i.e., the study of animal disease dynamics on a population level. The major factors influencing an epizootic can be divided into four basic components: (1) the pathogen population; (2) the host population; (3) transmission; and (4) the environment. Although the question pertaining to the causes of an epizootic is of great interest to all aspects of insect pathology, it is of particular interest to microbial control efforts. Microbial control can be defined as the use of entomopathogens (viruses, fungi, bacteria, protists, or nematodes) for pest suppression. This chapter presents a summary and analysis of epizootiological principles and the concepts of microbial control. The goal is to promote expanded studies in epizootiology, and foster research and implementation toward improved microbial control programs.

3.1. INTRODUCTION

Epizootiology is a central component of insect pathology that affects diverse aspects of the discipline. Epizootiology is defined as the study of animal disease dynamics on the basis of mass phenomena (Steinhaus, 1967; Fuxa and Tanada, 1987; Onstad *et al.*, 2006).

Thus, epizootiology deals with disease on a population level rather than an individual basis. The subject concerns both infectious and non-infectious diseases (though this chapter will focus primarily on the former) and affects all pathogen groups. Epizootiology deals with epizootic and enzootic levels of animal disease. An epizootic is defined as an outbreak of disease in which there is an unusually large number of cases, whereas an enzootic refers to a low level of disease that is constantly present in a population (Steinhaus, 1967; Onstad *et al.*, 2006).

The study of epizootiology crosses multiple interests within insect pathology. Epizootiology is the foundation for studying disease fluctuation in natural insect populations from an ecological perspective. However, the primary motivation of many insect pathologists is not the study of natural disease levels but the use of diseases to suppress insect pest populations, i.e., microbial control. Microbial control can be viewed as applied epizootiology with the goal of inducing an epizootic in the targeted insect population through manipulation. In contrast, many insect pathologists are motivated by reducing or eliminating the prevalence of disease in insect populations, e.g., in beneficial insects (e.g., pollinators) or insects in culture (see Chapter 12). In either case, understanding factors that cause epizootics is critical to the successful implementation of microbial control and equally important in developing methods to guard against epizootics.

Steinhaus (1949) first introduced terminology regarding epizootiology in relation to insect pathology, and made the connection between the principles of epizootiology and microbial control. Despite that introduction more than six decades ago and significant reviews on the principles of epizootiology since then (Tanada, 1963; Tanada and Kaya, 1993), including an entire book on the subject (Fuxa and Tanada, 1987), the importance of epizootiology in insect pathology and its application to microbial control has arguably not received the attention it deserves. It is hoped that this chapter will offer additional support for incorporating concepts of epizootiology into studies of insect pathology and microbial control.

Specifically, this chapter offers an updated examination of the principles of epizootiology and microbial control. Four primary areas that influence epizootiology (Tanada, 1963; Fuxa and Tanada, 1987) are discussed: (1) the pathogen population; (2) the host population; (3) transmission; and (4) the environment. The chapter then examines microbial control as applied epizootiology and discuss factors that influence success in microbial control as well as methods to enhance efficacy. Case studies and other specific examples are offered to illustrate points, and the chapter concludes with a synthesis of topics and suggestions for future research.

3.2. EPIZOOTIOLOGY: BASIC PRINCIPLES

A clear understanding of the basic principles of epizootiology is necessary to understand fully the science of causes and forms of disease. Given that the basic definition of an epizootic entails an unusually large number of cases of a disease in a host population, the question arises: what constitutes an unusually large number (Fuxa and Tanada, 1987)? To answer this question, one must establish what the long-term prevalence of the disease is in a space—time framework. Prevalence is defined as the total number of cases of a particular disease at a given time, in a given population (Onstad *et al.*, 2006). The prevalence of a causal agent at any point in time from any location can then be subjected to statistical analysis to establish when prevalence is outside the norm. While epizootics are sporadic in their occurrence and marked by a sudden change in prevalence, enzootics are present over very long periods and their prevalence in the host population varies little. With proper environmental conditions, host densities and a suitable entomopathogen (= insect pathogen) population, an enzootic disease can become epizootic. Steinhaus (1949) described this transition in disease prevalence in time from enzootic to epizootic levels and back again as an epizootic wave (Fig. 3.1). Steinhaus further divided the epizootic wave into the preepizootic, epizootic and postepizootic phases.

Two epidemiological terms often misused in insect pathogen epizootiology are prevalence and incidence. Prevalence, as defined above, is dependent on both the proportion of hosts inflicted with a disease and the duration of the infection. When one determines the prevalence of disease at a single point in time, infected individuals early and late in the progress of the infection are counted equally. This is a key point that differentiates disease prevalence from incidence. The incidence of a disease is the number of new cases of infection over a defined period (Tanada and Kaya, 1993). Because prevalence refers to all incidences of disease, new and old, at a given point in time, whereas incidence is specifically the new cases of infection over a defined duration, these terms must not be used interchangeably.

To further avoid confusion, the present discussion will rely on the following definitions of terms commonly used in insect pathology. The definitions of pathogenicity and virulence used are those proposed by Steinhaus and Martignoni (1970) and reiterated by Shapiro-Ilan *et al.* (2005a) and Onstad *et al.* (2006), which in large part have been adhered to in the insect pathology literature. To restate, pathogenicity is the quality or state of being pathogenic, and virulence is the disease-producing power of an organism, i.e., the degree of pathogenicity within a group or species. Pathogenicity is absolute, a disease-causing

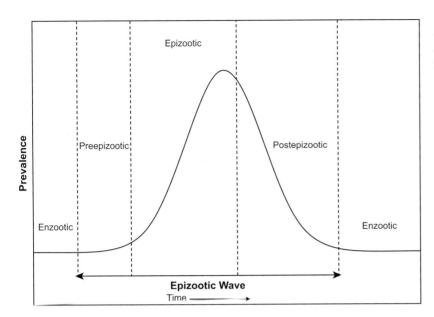

FIGURE 3.1 A curve showing the epizootic wave. The epizootic cycle is divided into the enzootic, preepizootic, epizootic, and postepizootic phases. *(Modified after Steinhaus, 1949, and Tanada and Kaya, 1993.)*

organism is pathogenic to a host or it is not, whereas virulence is a quantitative measure of the ability of a given amount of an agent to cause disease. Infectivity is the ability of an organism to cause an infection (Tanada and Kaya, 1993).

The four basic components of an epizootic — the pathogen population, the host population, transmission, and the environment — work alone or in combination in a manner that is conducive for the development of an epizootic or precludes an epizootic from taking place. This text will discuss how these components affect epizootic development individually and in combination. In the field, it is the sum effect of all of these factors that determines whether or not an epizootic takes place.

3.2.1. The Pathogen Population

Numerous properties associated with the pathogen population (e.g., pathogen density, dispersal, infectivity and latency, virulence, and genetics) play a large role in determining whether or not conditions are in place for an epizootic wave to be initiated. Foremost, the pathogen population density must be sufficient to initiate an epizootic. In addition, within the vicinity of the host, the pathogen population must also be able to disperse sufficiently, actively or passively, to reach the host, and once making contact, the pathogen must be able to infect; therefore, dispersal and infectivity are among the key factors of importance. Once reaching the host, the pathogen's level of virulence is of utmost importance. Furthermore, there are various genetic factors associated with pathogen strains that may influence pathogenicity and virulence. Each of the properties associated with the pathogen population alone or in combination with one or more of the others may have

a large impact on whether or not disease occurs within a host and subsequently spreads to other individuals in the population resulting in an epizootic. These pathogen properties, which vary widely among pathogen groups, are discussed below.

Pathogen Density

Of all of the factors of the pathogen population that influence an epizootic, none is more important than pathogen density, which includes the spatial distribution of the pathogen (Tanada and Fuxa, 1987; Onstad, 1993; D'Amico et al., 1996). Although pathogen density is fundamentally related to other intrinsic characteristics (e.g., pathogen reproductive rate and capacity to survive), it is an intrinsic character of a pathogen in and of itself. A pathogen with a high density and widespread spatial distribution in the field is inherently well suited to cause an epizootic. Susceptible hosts encountering a high pathogen density are more likely to come into contact with and become infected than those in an environment of low pathogen density and limited spatial distribution. However, pathogens with low densities in the field or limited spatial distribution can result in localized enzootics, and under favorable biological and environmental conditions can result in epizootics (Dwyer and Elkinton, 1993). High pathogen densities often occur after a widespread epizootic in which local host populations are depleted and propagule numbers maximized. This phenomenon is often seen with lepidopteran pests infected with a virus (Fuxa, 2004). After a severe epizootic, insect populations often fail to develop in the subsequent year, as neonate larvae are highly susceptible to infection and die before causing severe damage. The implementation of a microbial control program most often attempts to inundate an entire crop with high levels of a pathogen and initiate an epizootic. Timing of the pathogen's application in microbial control programs is critical so that the pathogen and the susceptible stage of the insect occur simultaneously in the environment.

Dispersal

The ability of a pathogen to disperse, either under its own power or passively, has implications on pathogen spatial distribution. Most pathogens have limited or no capacity to disperse on their own. In the aquatic environment, Oomycete (kingdom Chromista) and Chytridiomycete pathogens of mosquitoes and other aquatic Diptera produce zoospores which actively seek out and penetrate the cuticle of their host (Domnas, 1981; Sweeney, 1981; Andreadis, 1987). Entomopathogenic nematodes in the families Steinernematidae and Heterorhabditidae (see Chapter 11) are the best example of terrestrial entomopathogens capable of dispersing under their own power. Under proper soil conditions, nematodes are able to move through the soil to locate a host. Foraging strategies exhibited by entomopathogenic nematodes exist along a continuum from ambushers to cruisers. Ambushers use a sit-and-wait strategy; they usually stand on their tails (nictating) and wait until a host comes close before infecting. Cruisers actively seek out their hosts and cue into certain target volatiles (e.g., carbon dioxide) before contacting the host. Examples of nematodes that exhibit foraging behavior characteristic of ambushers include *Steinernema carpocapsae* and *S. scapterisci*. Those exhibiting behavior typical of cruisers include *Heterorhabditis bacteriophora*, *H. megidis*, and *S. glaseri*, and those with intermediate search behavior include *S. feltiae*, and *S. riobrave* (Campbell and Gaugler, 1997; Lewis, 2002; E. E. Lewis *et al.*, 2006). Other pathogens that use the soil as a reservoir (viruses, fungi, bacteria) have no capacity to disperse on their own. Rather, these pathogens rely on biotic and abiotic factors to distribute them in the environment (see Sections 3.2.2 and 3.2.3).

Infectivity and Latency

Infectivity is closely related to transmission, as it is the method of entry into a potential host. Routes of entry include oral ingestion (*per os*) and entry via the digestive tract, direct cuticular penetration, tracheal entry, entry via the reproduction system, or entry via the action of parasitoids. Pathogens with multiple routes of entry into their host are expected to result more easily in epizootics as disease transmission is facilitated (Tanada and Fuxa, 1987). Fungi and nematodes commonly use multiple routes of entry. Fungal entomopathogens most commonly gain entry to their host directly through the cuticle, but entry can also be gained through the alimentary canal (Wraight *et al.*, 2007). Entomopathogenic nematodes in the Steinernematidae primarily penetrate their host through the mouth,

anus, and spiracles, while members of the Heterorhabditidae use these routes of entry in addition to direct penetration of the cuticle. However, Koppenhöfer *et al.* (2007) demonstrated that *S. glaseri* and *S. scarabaei* directly penetrated scarab cuticles, albeit at lower rates than *H. zealandica* and *H. bacteriophora*, which demonstrated excellent cuticular penetration. *Steinernema feltiae* is also reported to penetrate directly the cuticle of crane flies (Peters and Ehlers, 1994). Furthermore, microsporidia (see Chapter 7) most often cause chronic infections and use multiple routes of entry via the egg, digestive tract, or passive transfer via the ovipositor of parasitoids (Reardon and Podgwaite, 1976; Solter and Becnel, 2007). Bacterial and viral infections are most often transmitted orally (Szewczyk *et al.*, 2006; Cory and Evans, 2007; Garczynski and Siegel, 2007) or are passively transferred via the ovipositor of parasitoids (Reardon and Podgwaite, 1976; Cossentine, 2009).

Some pathogen infections may be latent (i.e., a non-infective and non-replicative state) until triggered by some environmental stressor. However, the development of a visible infection after an environmental stressor is not proof in and of itself that a disease is latent, as would be the case for pathogens causing chronic infections (Tanada and Fuxa, 1987). Latency occurs in most pathogen groups. Molecular tools now allow for detailed studies of latency of nucleopolyhedrovirus (NPV) infections of lepidopterans that had been suspected for several decades. Persistent latent NPV infections occur in wild populations of cabbage moth, *Mamestra brassicae* (Burden *et al.*, 2003), and African armyworm, *Spodoptera exempta* (Cory and Evans, 2007; Vilaplana *et al.*, 2010). Covert baculoviral infections are present, and potentially frequent, in wild lepidopteran populations in which the virus can persist in the latent state (Kukan, 1999; Burden *et al.*, 2003). Overwintering European corn borer (*Ostrinia nubilalis*) larvae also harbor latent *Beauveria bassiana* infections that become active with the breaking of larval diapauses, resulting in fungal spores being produced and released into the environment in spring when insect hosts are abundant (Bruck and Lewis, 1999).

Virulence

Of the key pathogen properties discussed, the concepts of pathogenicity and virulence have been debated recently in the literature (Thomas and Elkinton, 2004; Shapiro-Ilan *et al.*, 2005a). The median lethal dose of a pathogen needed to kill 50% of the tested insects (LD_{50}) is a typical measure of virulence (Shapiro-Ilan *et al.*, 2005a). A more virulent pathogen requires fewer infective propagules to cause disease relative to a pathogen that is less virulent. The length of time from infection to 50% death of the host (median lethal time or LT_{50}) is also a common measure of virulence (Tanada and Kaya, 1993). A pathogen that infects and kills its host quickly is more virulent than those that are slower acting or result in long-term non-lethal chronic infection. Speed of replication in the host is also an important factor. Pathogens that replicate quickly are necessarily more virulent that those that replicate slowly. Bioassays can also be performed in the laboratory using a distinguishing dosage (based on prior work) of a pathogen with sufficient replication to indicate relative virulence of unique isolates (Shapiro-Ilan et al., 2003a; Bruck, 2004; Fisher et al., 2011). Pathogen virulence can be manipulated in the laboratory by a variety of approaches such as: (1) repeated passages through susceptible hosts; (2) identification of strains within a species of pathogens with differential virulence; and (3) introducing pathogens with other substances to enhance their invasive ability or virulence. These approaches are discussed in detail later in this chapter, in the context of microbial control programs (see Section 3.3.3).

The production of toxins and secondary metabolites is a common biological property of entomopathogens that is tied to pathogenicity and virulence. The production of toxins often results in reduced LD_{50} and LT_{50} values. Toxins and secondary metabolites may function in one or more of the following ways: direct toxicity to the host (thereby directly aiding or causing disease and death, immunosuppression (aiding in overcoming host defense; antibiotic activity), or to ward off competition from other pathogens or saprobes (Vey et al., 2001; Charnley, 2003). Secondary metabolites are commonly produced by bacteria and fungi to aid the pathogen before or in the course of invasion (Schnepf et al., 1998; Vey et al., 2001; Bravo et al., 2007).

Bacillus thuringiensis (Bt) is the most well-studied example of a pathogen that produces both exotoxin and endotoxins during the infection process (Schnepf et al., 1998). Bt produces a wide variety of toxins, many of which form the basis of transgenic plants producing insecticidal toxins (James, 2006; Sanchis and Bourguet, 2008). Fungal entomopathogens produce enzymes including chitinases, lipases, and proteases to aid in invasion of the integument (da Silva et al., 2005; Silva et al., 2005; Boldo et al., 2009). Overproduction of the cuticle-degrading protease (Pr1) in *Metarhizium anisopliae* increased virulence against the tobacco hornworm, *Manduca sexta* (St. Leger et al., 1996). In addition, fungal entomopathogens produce secondary metabolites that can have a wide range of effects (Charnley, 2003) (see Chapter 6). Destruxins are secondary metabolites produced by *M. anisopliae* (and other fungi) that could increase the rate of mortality of infected insects (Schrank and Vainstein, 2010). *Metarhizium anisopliae* isolates that produce higher quantities of destruxins are more virulent (Sree and Padmaja, 2008). The symbiotic bacteria *Xenorhabdus* and *Photorhabdus*, of *Steinernema* and *Heterorhabditis*, respectively, also produce toxins that benefit the nematodes in the infection process (Forst et al., 1997; Bowen et al., 1998; ffrench-Constant and Bowen, 1999). An analysis of the *Photorhabdus luminescens* genome identified more predicted toxin genes than any other bacteria in prior studies (Duchaud et al., 2003), which has led to increased interest in these symbiotic bacteria as a new source of genetic material for second generation transgenic plants. A large toxin gene from *P. luminescens* has been expressed in transgenic *Arabidopsis* plants and has been shown to convey resistance to *M. sexta* (Liu et al., 2003). Larvae of the beet armyworm (*Spodoptera exigua*), diamondback moth (*Plutella xylostella*), and black vine weevil (*Otiorhynchus sulcatus*), as well as nymphs of the desert locust, *Schistocerca gregaria*, were killed by both *Xenorhabdus nematophila* cells and cell secretions, indicating that toxic secretions were responsible for the observed lethal effects (Mahar et al., 2008) and suggesting alternative modes for insect pest management with these toxins. The genetics of pathogenicity in *Photorhabdus* appears to be very similar to that described for other enteric bacterial pathogens, indicating the universal nature of pathogenicity (Clarke, 2008).

Strain Effects and Genetics

A strain can be defined as a pure culture of a microorganism with relatively constant properties (Onstad et al., 2006). A pathogen isolate from a particular time, place, or substrate may constitute a strain if the population exhibits differences in some measurable biological property relative to other populations (or strains) of that species. The properties that differ among strains can affect the potential for epizootics. Pathogenicity is perhaps the most important biological trait used to differentiate strains and is generally quantified in the laboratory. Strains pathogenic to a variety of hosts are intuitively expected to result in more frequent and severe epizootics in the field as the number of potential hosts, and hence the amount of inoculum produced, is greater than that of an isolate with a limited host range. However, it is clearly desirable in many situations to have a pathogen with a host range limited to the target pest in order to minimize infections in non-target hosts (Goettel, 1995).

Virulence is the other most commonly characterized property between pathogen strains. Reduced time to kill, exit the host, and become available in the environment to infect subsequent hosts no doubt enhances the ability of a strain to induce an epizootic, at least in the short term. The long-term effects of releasing a more virulent strain in the field and its ability to produce subsequent epizootics are less clear as the host may respond with increased resistance, resulting in less severe epizootics and therefore reduced amounts of inoculum in the environment (Tanada and Fuxa, 1987).

Foreign pathogen strains that are introduced into a naïve host population can range from being non-infective to being more virulent than native strains, which have co-evolved with their host (Hajek *et al.*, 1995; Solter *et al.*, 1997). *Microsporidium* sp. from European populations of the gypsy moth, *Lymantria dispar*, exposed to non-target lepidopteran species indigenous to North America did not support optimal reproduction and it is unlikely that horizontal transmission within non-target populations in the field would result (Solter *et al.*, 1997). The reciprocal scenario can also occur. For example, the exotic ladybirds *Harmonia axyridis* and *Coccinella septempunctata* were not susceptible to endemic isolates of *B. bassiana* (Cottrell and Shapiro-Ilan, 2003, 2008).

Recent advances in molecular biology have shed considerable light on the genetics of insect pathogens and strain differentiation. Until recently, morphological characters were most often used to distinguish among species, which did not allow for differentiation of strains within a group of morphologically identical species. However, it is now possible not only to differentiate morphologically identical sister taxa (Rehner and Buckley, 2005; Bischoff *et al.*, 2006, 2009) but to also track individual pathogen strains in the field based on their unique genetic profile (Coates *et al.*, 2002a, b; Enkerli *et al.*, 2004; Wang *et al.*, 2004). Whole genomes of various insect pathogens have been sequenced, e.g., *Autographa californica* NPV (Ayres *et al.*, 1994), *P. luminescens* (Duchaud *et al.*, 2003), *Ascosphaera apis* (Qin *et al.*, 2006), *Pseudomonas entomophila* (Vodovar *et al.*, 2006), *B. thuringiensis* (Challacombe *et al.*, 2007), *M. robertsii* (as *M. anisopliae*), and *M. acridum* (Gao *et al.*, 2011). Genome sequencing allows for a better understanding of phylogenetic relationships and biological traits, and will invariably lead to an improved understanding of the multitude of factors unique to each pathogen.

3.2.2. The Host Population

The susceptibility of a host population to pathogen infection plays a critical role in determining whether or not an epizootic takes place. Inherent susceptibility of an insect to infection by a pathogen is genetically based. The more susceptible a host is to infection, the lower the pathogen dose necessary and subsequently the easier it is for infections to spread from one host to another. The level of host susceptibility is most commonly measured with an estimate of LD_{50}. The lower the LD_{50}, the fewer propagules are necessary to cause mortality.

Epizootics generally occur at high host densities (Andreadis, 1987; Watanabe, 1987; Onstad and Carruthers, 1990). High host density facilitates transmission of pathogens between infected and healthy hosts and results in large numbers of infective propagules being produced and released into the environment. Insects that are stressed are also generally more susceptible to pathogen infection. For example, nutritional stress due to host plant senescence increases the susceptibility of insects to NPV (Richter *et al.*, 1987). Crowding and environmental stressors such as high humidity also contribute to making hosts more susceptible to pathogen infection (Fuxa *et al.*, 1999). As a result, insect pathogens primarily act as density-dependent mortality factors infecting more hosts as the host population increases. Yet, epizootics can occur in situations where pathogens are widely distributed and host populations are low (Dwyer and Elkinton, 1993). Interactions between trophic levels can aid in pathogen transmission at low host population densities (Sait *et al.*, 1994; Cory, 2003).

Genetic Resistance

Host resistance (see Chapter 13) is one mechanism that can cause a substantial reduction in host susceptibility. The first example of the development of resistance to a pathogen outside the laboratory occurred in the Indian meal moth, *Plodia interpunctella*, exhibiting resistance to Bt (McGaughey, 1985). Moths isolated from treated grain bins were more resistant than those isolated from untreated bins, indicating that resistance developed quickly in the field. Rapid development of resistance to Bt in the laboratory demonstrates that many pests harbor natural genetic variation in susceptibility and have the potential to evolve resistance (Tabashnik *et al.*, 2003, 2006; Gould, 1998; US EPA, 1998).

The first reported resistance to Bt sprays in the field was observed in *P. xylostella* (Tabashnik *et al.*, 1990). The large-scale planting of crops engineered to produce the protein toxins (Bt crops) represents intense selection pressure for development of resistance to the gene product of an insect pathogen. First generation transgenic crops with only one Bt toxin were introduced in the mid-1990s targeting Lepidoptera. Bt cotton and Bt corn have been grown on more than 165 million ha worldwide (James, 2006). The frequency of resistance alleles has increased in some populations of *Helicoverpa zea*, but not in five other major Lepidoptera pests where large-scale Bt crops are grown successfully on a wide scale, suggesting that current refuge strategy has helped to delay resistance (Tabashnik *et al.*, 2008).

Insects develop resistance to pathogens as a result of repeated prolonged exposure through a process of selection. The most detailed work on the mode of resistance inheritance and mechanisms of resistance has been developed for prolonging the effectiveness of transgenic plants expressing Bt protein toxin. A simple model of resistance evolution is conferred by one gene with two alleles, *r* (resistance) and *s* (susceptibility), yielding three possible genotypes (Gould, 1998; Tabashnik and Carrière, 2007). Transgenic crops are designed to produce sufficient toxin to kill all heterozygous individuals, rendering resistance as functionally recessive (Gould *et al.*, 1995; Liu *et al.*, 2001; Tabashnik *et al.*, 2004). Because the *r* alleles are rare in the host populations (Gould *et al.*, 1997; Burd *et al.*, 2003; Wenes *et al.*, 2006), the extremely rare *rr* individuals surviving in a Bt crop are likely to mate with the abundant *ss* individuals from refuge areas resulting in *rs* progeny that are susceptible to the Bt crop (Gassmann *et al.*, 2009).

Behavioral Resistance

Hosts exhibit a wide range of behavioral responses to the presence of pathogen propagules in the environment before infection, in addition to behavioral responses after infection. The type of behavior response can vary widely. Many responses help the host either to avoid or to mitigate infection, while others aid the pathogen in subsequent dissemination of propagules in the environment.

There are many reports of host-mediated behavior reducing pathogen transmission among social insect colonies. Behavior observed includes grooming, nest cleaning, secretion of antibiotics, avoidance, removal of infected individuals, and colony relocation (Roy *et al.*, 2006). Grooming behavior has been documented among solitary and social insects (Siebeneicher *et al.*, 1992; Oi and Pereira, 1993). Grooming is common among termites and other social insects and can result in the increased spread of a pathogen within a colony (Kramm *et al.*, 1982) or conversely serve as an effective means of actively removing pathogen propagules attached to the cuticle (Oi and Pereira, 1993). Termites reared in groups physically removed 80% of *M. anisopliae* conidia from their nest mates and eliminated the conidia through the alimentary tract, while individually reared termites did not reduce their surface contamination (Yanagawa and Shimizu, 2007). Hygienic behavior goes beyond grooming to include nest cleaning, i.e., detecting and removing diseased and parasitized individuals from the nest (Wilson-Rich *et al.*, 2009), and enrichment of nest material with antimicrobial substances from the environment (Christe *et al.*, 2003). Synergy among termite defense mechanisms is apparently responsible for protecting colonies against epizootics (Chouvenc and Su, 2010). Red imported fire ants, *Solenopsis invicta*, bury nest mates infected with *B. bassiana* to reduce transmission (Pereira and Stimac, 1992). Ants also spray antimicrobial secretions from the gaster over the brood to reduce *B. bassiana* transmission (Oi and Pereira, 1993). Reduced genetic diversity of the ant *Cardiocondyla obscurior* reduces the colony's collective disease response by reducing the colony's ability to detect and react to the presence of fungal spores (Ugelvig *et al.*, 2010). Insects can also avoid areas of high pathogen densities when establishing new nest sites (Oi and Pereira, 1993). Mole crickets modify their behavior in response to *M. anisopliae* and *B. bassiana* in

a way that reduces their exposure to these fungi (Villani *et al.*, 2002; Thompson and Brandenburg, 2005).

Insects can also avoid or reduce the impact of infection by altering their physiology. For example, the brassy willow leaf beetle, *Phratora vitellinae*, constitutively releases volatile glandular secretions to inhibit a range of pathogens in its microclimate (Gross *et al.*, 2008). Thermoregulation is also a behavioral response in adaptation to pathogen infection. In the field, many locusts and grasshoppers are active thermoregulators capable of maintaining a preferred body temperature in excess of ambient in sunny conditions (Chappel and Whitam, 1990). A further increase above preferred body temperature following fungal infection, termed "behavioral fever", has been observed in a number of acridids (Blanford and Thomas, 1999) and has been shown to reduce fungal-induced mortality (Inglis *et al.*, 1997; Blanford *et al.*, 1998; Blanford and Thomas, 1999). Although thermoregulation can reduce the impact of *M. acridum* (as *M. anisopliae* var. *acridum*) infection, the insects have been reported to become more active and thus more susceptible to predation (Arthurs and Thomas, 2001).

Host Associations

Pathogens have different capacities for surviving in primary or secondary hosts. As such, pathogens capable of infecting more than one host in the environment are often better able to survive and persist (Goettel, 1995). The transmission and spread of a pathogen through the environment may rely not only on the infection of the primary host, but also on associations with alternative hosts, scavengers, parasitoids, predators, or other organisms. While in many cases these associations do not directly result in increased numbers of propagules in the environment, they do play a role in enhancing pathogen spread and persistence. To a large degree, the success, persistence, and spread of a pathogen in the environment are due to associations with organisms other than the primary host.

Primary pathogen survival and replication in the environment take place on primary hosts. Pathogens with long periods of infection that do not kill the host quickly are better suited to persist in the environment as they do not destroy the host population too quickly. For instance, infection of the chronic pathogen *Nosema pyrausta* in *O. nubilalis* populations is cyclic, with alternating enzootic and epizootic patterns over a period of several decades (Lewis *et al.*, 2009). To aid further in maintaining the cycle, prior to death, pathogen propagules are deposited and persist in fecal matter until they are fed on by additional host insects (Lewis and Cossentine, 1986; Vasconcelos, 1996). In another mechanism that enhances pathogen maintenance after the death of the primary host, pathogens are able to persist in the decomposing cadaver until environmental conditions are favorable for subsequent infection

(Koppenhöfer, 2007; Wraight *et al.*, 2007; Raymond *et al.*, 2010).

Secondary hosts (also called intermediate or alternate hosts) are host species in which immature, intermediate, or asexual stages of a pathogen occur (Onstad *et al.*, 2006). Secondary hosts can play an important role in pathogen population dynamics and survival in the field. A secondary host can serve an obligatory role in the host's life cycle, or it may be non-obligatory and be used solely to prolong the pathogen's survival until the primary host or target is infected. An early example of an entomopathogen using an obligatory secondary host for completion of its life cycle involves the transmission of the microsporidium *Amblyospora* sp. infecting mosquitoes to alternate copepod hosts (Andreadis, 1985; Sweeney *et al.*, 1985). It has since been shown that many species require two host generations to complete their life cycle, and members of at least four genera, *Amblyospora*, *Duboscqia*, *Hyalinocysta*, and *Parathelohania*, require obligatory development in a secondary copepod (Becnel and Andreadis, 1999).

The ability to infect a wide range of hosts beyond the primary target, none of which is obligatory, provides additional resources for pathogen survival. Fungal entomopathogens with wide host ranges are generally facultative pathogens, ubiquitous soil saprophytes and enzootic (Goettel, 1995). *Beauveria bassiana* is a well-studied example of a facultative pathogen with a broad host range, infecting over 700 species of arthropods from different orders (Li, 1988; Meyling *et al.*, 2009). The ability of *B. bassiana* and other facultative fungal entomopathogens to utilize various resources in the environment no doubt plays a large role in the persistence of these fungi in nearly all areas on Earth. In addition to secondary arthropod hosts, *B. bassiana* and *M. anisopliae* persist as endophytes (Wagner and Lewis, 2000; Vega, 2008) or in the rhizosphere of a wide variety of plants (Fisher *et al.*, 2011).

Carriers, or phoretic hosts, are insects or other organisms that are not susceptible to the pathogen, but play a role in dispersing the pathogen in the environment. Local dispersal of viruses by biotic agents has been reviewed by Fuxa (1989, 1991). Viral particles are dispersed via hemipterans (e.g., *Podisus maculiventris*), coleopterans (e.g., *Calleida decora*), hymenopterans (e.g., *Brachymeria ovata*), dipterans (e.g., *Archytas apicifer*), spiders (e.g., *Misumenops* sp.), and bird droppings (Fuxa *et al.*, 1993; Fuxa and Richter, 1994). Phoretic relationships have been reported between entomopathogenic nematodes and earthworms (Shapiro *et al.*, 1995; Campos-Herrera *et al.*, 2006), mites (Epsky *et al.*, 1988), and isopods (Eng *et al.*, 2005). The sap beetle, *Carpophilus freemani*, vectored *B. bassiana* mechanically and via their fecal material, resulting in *O. nubilalis* infection in the laboratory (Bruck and Lewis, 2002a). In the soil environment, collembolans vector fungal entomopathogens mechanically and via their

gut contents in numbers sufficient to cause host infection (Dromph, 2001, 2003). The vectoring of fungal spores by these small soil arthropods is believed to have an important impact on the epizootiology of soil-borne fungal entomopathogens in the field (Dromph, 2003).

3.2.3. Transmission

Transmission is the process by which a pathogen or parasite is passed from a source of infection to a new host (Anderson and May, 1981; Andreadis, 1987; Onstad and Carruthers, 1990). Insect pathogens have evolved a wide array of mechanisms to ensure their long-term survival and transmission to new hosts. A fundamental knowledge of the methods of transmission is key to understanding the dynamics of disease and the occurrence of disease epizootics. Direct transmission of a pathogen occurs when it is transmitted from an infected to a susceptible new host without the aid of another living organism, whereas indirect transmission relies on one or more secondary hosts or vectors to facilitate transmission (Tanada and Kaya, 1993). Insect pathogens are primarily transmitted either directly host to host, or from the infected host into the environment, where they are subsequently acquired by a susceptible host.

Methods of Transmission

Transmission can be vertical or horizontal. Horizontal transmission is the transfer of a pathogen from individual to individual, either host to host, or host to environment to host (Canning, 1982). Vertical transmission, also referred to as congenital, parental, or hereditary, is the direct transmission of a pathogen from parent to progeny (Fine, 1975; Andreadis, 1987). Horizontal transmission occurs widely in all pathogen groups and is the primary mode of transmission that leads to epizootics, particularly when host population densities are high. Infective propagules in the environment are predominantly transmitted directly to uninfected hosts leading to an epizootic. An increased pathogen titer within an individual may be caused by additional consumption of infected propagules (e.g., bacteria, viruses, protists) or by acquiring propagules externally (fungi), and by initial propagules completing a reproductive cycle and generating new ones.

Fungal pathogens are readily transmitted either host to host, or host to environment to host. Horizontal transmission of *M. acridum* (as *M. flavoviride*) is a key biological factor responsible for long-term persistence of the fungus in the environment (Thomas *et al.*, 1995). Host-to-host horizontal transmission of *B. bassiana* and *M. anisopliae* between tsetse flies, *Glossina morsitans* (Kaaya and Okech, 1990), the Mexican fruit fly, *Anastrepha ludens* (Toledo *et al.*, 2007), and the German cockroach, *Blatella germanica* (Quesada-Moraga *et al.*, 2004), among others, has been demonstrated. Fungal-treated and control tsetse flies mixed in cages for 32 days resulted in significant mortality due to fungal infection of untreated flies. In another example, direct and indirect transmission of *M. anisopliae* resulted in 40% infection of mites (*Psoroptes* spp.) responsible for mange in a wide range of vertebrate hosts (Brooks and Wall, 2005).

Horizontal transmission of bacterial and viral pathogens primarily occurs via the host-to-environment-to-host pathway. However, direct horizontal transfer from host to host does occur in the case of cannibalism in which an insect consumes infected conspecifics (Polis, 1981; Boots, 1998; Chapman *et al.*, 1999). Horizontal transmission of bacterial and viral pathogens plays an important role in determining the rate of spatial spread of a pathogen (Dwyer, 1992, 1994).

Microsporidia are a well-studied pathogen group for their proclivity to transmit horizontally within a generation of hosts. The rate of horizontal transmission is governed by the percentage of a population infected, time in an instar, virulence, and the tendency of the microsporidium to transmit vertically. Most investigations on the dynamics of horizontal transmission of *N. pyrausta* to *O. nubilalis* report very little to no horizontal transmission during the first generation of *O. nubilalis* (Siegel *et al.*, 1988; Onstad and Maddox, 1989; Solter *et al.*, 1990). Andreadis (1986) suggested that horizontal transmission is greater in the second generation owing to greater larval density and interplant movement of larvae. Lewis (1978) demonstrated interplant movement of second generation larvae and horizontal transmission of spores. Experimentally, *N. pyrausta* can be transferred from first to second generation larvae when spores deposited by first generation larvae are consumed by second generation larvae feeding on the same plants (Lewis and Cossentine, 1986). In an experiment conducted in a single location for six consecutive growing seasons, researchers found that horizontal and vertical transmission influenced both the intensity of the *N. pyrausta* infection in the *O. nubilalis* population and the percentage of the population infected (L. C. Lewis *et al.*, 2006). In a field study conducted over a 16-year period in two Nebraska counties, an increase in the percentage of the *O. nubilalis* population infected with *N. pyrausta* always followed periods of *O. nubilalis* population build-up, implicating insect density as a basis for horizontal transmission (Hill and Gary, 1979).

Entomopathogenic nematodes are exclusively horizontally transmitted. Infective juveniles (IJs) exit their host at the conclusion of the infection cycle, enter the environment (most often soil) and begin their quest in search of a new host. Recycling of nematodes as infections spread from host to host in the field decreases host populations and aids in nematode performance (Koppenhöfer, 2007). Nematode recycling is highly desirable in pest management programs as it not only provides additional control, but may also

allow the nematodes to persist to the next host generation. While recycling is common, the factors that influence its occurrence are not clearly understood, and often the level of recycling is low (Shapiro-Ilan *et al.*, 2006; Koppenhöfer, 2007). Most of the factors that influence nematode persistence, infectivity, and mobility are expected to play a role in nematode recycling.

Vertical transmission of a pathogen from parent to offspring is a key epizootiological factor for viruses and protists. In some instances, vertical transmission is the primary means by which a viral or microsporidian pathogen persists from one host generation to the next. In other cases, vertical transmission simply augments horizontal transmission, particularly when host populations are low. Vertical transmission occurs primarily through the female. The pathogen can be transmitted on the surface of the egg (transovum transmission) or within the egg via the ovary (transovarial transmission). The pathogen gains entry to the egg when the female reproductive organs (ovaries and accessory glands) are infected.

The mechanisms and epizootiological implications of vertical transmission of microsporidia are well studied. However, the intricate relationship between horizontal and vertical infection must be kept in mind. Horizontal infection of *O. nubilalis* larvae with *N. pyrausta* during the first, second, or third instar results in microsporidian spores being evident in ovarian tissue seven days after exposure. If exposed as a fourth or fifth instar, ovarian tissue is not infected until the pupal stage (Sajap and Lewis, 1988). The intensity of the infection during ovarian development governs whether or not there is destruction of germ cells and a subsequent reduction in oogenesis. This results in a fine line between reduction in fecundity because of damage to the ovary and that caused by transovarial transmission of the spore (Sajap and Lewis, 1988).

Many insect viruses are vertically transmitted in the environment. Understanding the relative contribution of vertical transmission and the conditions that maintain the pathogen is critical to understanding their ecology (Cory and Evans, 2007). Vertical transmission has been best studied in baculoviruses, but even within this group, the nature and role of vertical transmission on epizootiology are poorly understood (Cory and Meyers, 2003). Overt disease in the next generation can result from transovum and transovarial transmission (Kukan, 1999). Vertical transmission of NPV through the adult host is an adaptation for long-range environmental transport to new locations or for difficult-to-reach areas within a location (e.g., trees) (Fuxa, 2004). There is strong evidence of vertical transmission of NPV (Smirnoff, 1972; Olofson, 1989) and cytoplasmic polyhedrosis viruses (= cypoviruses) of Lepidoptera (Sikorowski *et al.*, 1973), indicating continuation of gut infections initiated in the larval stage. In these cases, transovum transmission as an external contaminant of the deposited eggs appears to be the most likely route of entry.

Modes of Dissemination

Dissemination is the ability of the pathogen to spread within a host population and throughout the environment (Tanada, 1963). There are four primary routes of pathogen dissemination in the field (Tanada, 1964; Andreadis, 1987), based on: (1) the pathogen's motility; (2) the behavior and movements of the primary host; (3) the behavior and movement of secondary hosts and non-host carriers; and (4) the actions of physical environmental factors.

Pathogen motility in general is very limited in terms of both the ability of pathogens to move under their own power and the distance they are able to travel in the environment. As such, their importance in the initiation or continuance of an epizootic is more profound when host densities are high and these mobile propagules are able to contact nearby hosts. Movement and activities of the primary host in the environment are important factors in pathogen dissemination. The direct mechanisms of pathogen dissemination (host to host) by the primary host include vertical transmission, cannibalism, and grooming. Indirect pathogen dissemination (host to environment to host) occurs via horizontal transmission by excretion of infective stages in feces and anal discharge (Fuxa *et al.*, 1998) and meconial discharges, disintegration of cadavers with infective pathogen propagules (Fuxa, 2004), and elimination of infected exuvia during molting. Perhaps the most well-cited example of primary host behavior that aids in the dissemination of disease is the treetop disease (also known as summit disease) of many lepidopterous and hymenopterous larvae infected with NPV (Andreadis, 1987), whereby infected larvae migrate to the tops of the plant or tree on which they are feeding and subsequently die attached to leaves by their prolegs (Fig. 3.2). Being positioned in this way ensures that the foliage below is showered with inoculum as the cadaver breaks down. This facilitates pathogen transmission to healthy larvae on the plant (Harper, 1958; Fuxa, 2004). A similar phenomenon occurs in aquatic systems when mosquito larvae and pupae killed by *Erynia aquatica* are buoyant and the conidiophores grow from the cadaver above water and are discharged onto mosquitoes nearby (Steinkraus and Kramer, 1989).

Wind and rain are the most important abiotic environmental factors responsible for pathogen dispersal. Their effect on the dispersal and subsequent epizootiology of plant pathogens has been well studied (Gregory *et al.*, 1959; Hunter and Kunimoto, 1974; Aylor, 1990). Wind was suspected in the dissemination up to 30 m of *Hyphantria cunea* NPV and *Neodiprion sertifer* NPV (Hukuhara, 1973). Rainfall distributes *L. dispar* NPV from branch to branch in a downward direction, increasing virus spread

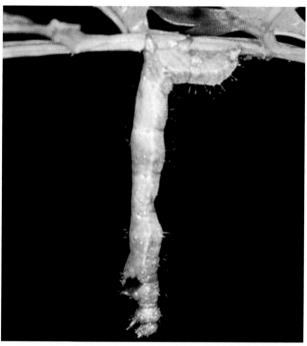

FIGURE 3.2 Soybean looper, *Thysanoplusia orichalcea*, infected with *Thysanoplusia orichalcea* nucleopolyhedrovirus. The image illustrates what is known as treetop disease or summit disease, whereby the host climbs to an exposed position (e.g., on vegetation) before death to facilitate dispersal of the pathogen. *(Courtesy of M. Shepard, G. R. Carner, and P. A. C. Ooi, Insects and their Natural Enemies Associated with Vegetables and Soybean in Southeast Asia, Bugwood.org, with permission.)*

and the likelihood of horizontal transmission (D'Amico and Elkinton, 1995). Rainfall is an efficient means of transfer from the fungal entomopathogen reservoir in the soil to the surface of whorl-stage corn plants (Bruck and Lewis, 2002b). Plants receiving rainfall had a mean of 8.8 colony-forming units (cfu) of *B. bassiana* per plant, while those not receiving rainfall had a mean of 0.03 cfu per plant. *Beauveria bassiana* conidia splash from the soil to corn leaves, where the pathogen can potentially come into contact with *O. nubilalis* larvae and *Diabrotica* spp. adults (Bruck and Lewis, 2002b). Prolonged rainfall can also transfer viral occlusion bodies from soil to the plant surface via rain splash (Fuxa and Richter, 2001).

3.2.4. The Environment

The role of environmental factors in disease dynamics in nature cannot be overstated. Epizootics of insect pathogens are heavily influenced by environmental factors (biotic and abiotic) and in many cases environmental factors are the most relevant in the epizootiological process (Ignoffo, 1992). The three primary factors contributing to the epizootics of disease — the host population, the pathogen population, and transmission — have already been discussed. Environmental factors may have direct or indirect impacts on any or all of these primary factors in the environment in which the host and pathogen occur (Benz, 1987). The result is a complex interaction between primary disease factors and the environment that impact all stages and processes of the disease cycle and cannot be separated.

Pathogens have varying levels of capacity to survive both abiotic and biotic factors in the environment. Many pathogens produce resistant forms designed to persist in the environment for extended periods in the absence of a host. The role of abiotic factors on pathogen persistence has been well studied in insect pathology. Because of the long-standing desire to develop microbial programs for a variety of soil, foliar, and aquatic pests, understanding and enhancing the ability of microbial agents to persist in the environment are key components to improving microbial control (see Section 3.3.3). The following text discusses environmental influences on epizootiology in aerial and aquatic environments, edaphic environments, and interactions among trophic levels.

Aerial and Aquatic Environments

Abiotic factors or the physical environment are the most well-studied factors affecting the ability of a pathogen to survive. Abiotic factors that are of particular importance for terrestrial applications are ultraviolet (UV) radiation, temperature, humidity, and moisture (Ignoffo, 1992). Of these, UV radiation affects all pathogen groups and is the most damaging (Fuxa, 1987). The exposure of Bt spores to UV radiation is believed to be largely responsible for its inactivation and low persistence in nature (Myasnik et al., 2001). Conidial exposure of multiple isolates of *Metarhizium* and *Beauveria* spp. indicated that the viability of all isolates dropped markedly with increasing exposure to UV radiation (Morley-Davies et al., 1995; Fargues et al., 1996; Braga et al., 2001a). Despite this intrinsic susceptibility, fungal species and strains differ significantly in their tolerance to UV radiation (Morley-Davies et al., 1995; Braga et al., 2001a).

Ambient temperature, humidity, and moisture can affect both pathogen persistence and infectivity (Reyes et al., 2004; Ebssa et al., 2004; Lacey et al., 2006a). In aquatic environments, extreme temperatures can inactivate microsporidian spores or cause them to germinate prematurely (Undeen and Vávra, 1997). Temperature and moisture extremes can be highly detrimental to nematode survival (Smits, 1996; Glazer, 2002; Koppenhöfer, 2007). Occluded viruses are tolerant to fairly wide temperature extremes, surviving cold storage at $-70°C$ and rapid heat inactivation at temperatures above $40°C$ (Cory and Evans, 2007; Lacey et al., 2008a). Fungal entomopathogens, in general, have a wide range of temperature tolerances, but optimal conditions for infection, growth and sporulation range from 20 to $30°C$ (Vidal and Fargues, 2007; Wraight et al., 2007). Most studies on the impact of temperature on fungal

pathogen efficacy to date focus on constant temperatures and not fluctuating temperatures as would be observed in the field (Inglis et al., 1997; Bruck, 2007). Temperatures below $15°C$ significantly slowed mycelial growth of *M. anisopliae in vitro* and *O. sulcatus* larval infection in soilless potting media (Bruck, 2007). *Otiorhynchus sulcatus* larval mortality due to *M. anisopliae* occurred at temperatures as low as $10°C$; however, the progression of the infection was significantly retarded (Bruck, 2007). The effect of temperature and moisture on fungal spore survival varies considerably among fungal species and strains (Hong et al., 1997). In addition to temperature, humidity, and moisture, pH, dissolved minerals, and inhibitors are important abiotic factors affecting microsporidium survival (Undeen, 1990). Shifts in pH cause spores of *N. algerae* to germinate (Undeen and Avery, 1988), and ammonia and calcium can inhibit spore germination of this microsporidian species as well (Undeen, 1978).

Plants, and more often than not particular areas of the plant architecture, provide a physical barrier from the adverse effects of abiotic factors and enhance the ability of pathogens to survive. The leaf collar of corn plants is a favorable habitat for *B. bassiana* and Bt, providing an environment with adequate moisture and protection from UV radiation (McGuire et al., 1994; Bruck and Lewis, 2002c; Lewis et al., 2002). Inglis et al. (1993) also found improved fungal survival in the protected environment of the alfalfa canopy. Plant leaf surface topography and leaf surface wax can influence conidial acquisition (Inyang et al., 1998; Duetting et al., 2003; Ugine et al., 2007) and leaf microclimate can influence conidial germination (Baverstock et al., 2005). Plant chemistry can also be detrimental to insect pathogens. For example, plant secondary chemicals sequestered by the brassy willow leaf beetle, *Phratora vitellinae*, reduce LD_{50} and LT_{50} values of *B. bassiana* and *M. anisopliae* (Gross et al., 2008) as well as germination and conidial production of *B. bassiana* infecting the whiteflies *Bemisia tabaci* and *B. argentifolii* (also known as *Bemisia tabaci* biotype B) (Poprawski and Jones, 2000; Santiago-Álvarez et al., 2006).

Insect pathogens have evolved a variety of adaptations to overcome environmental barriers. For example, there is clear evidence that the presence of an occlusion body increases virus survival in the field over free virions in the environment (Wood et al., 1994; Cory and Evans, 2007). In another example, various fungal entomopathogens have evolved the ability to grow endophytically with a wide variety of plants (Vega, 2008). By doing so, these fungi persist within the plant protected from sunlight.

Edaphic Environment

The soil is a natural pathogen reservoir. Pathogens are well suited to persist and survive in the soil, where they are protected from UV radiation and buffered from desiccation and rapid variations in moisture and temperature. *Bacillus thuringiensis* (Ohba and Aizawa, 1986; Martin and Travers, 1989; Meadows, 1993), fungal entomopathogens (Bidochka *et al.*, 1998; Klingen *et al.*, 2002; Bruck, 2004), nematodes (Chandler *et al.*, 1997; Rosa *et al.*, 2000; Hominick, 2002), and NPVs (Thompson *et al.*, 1981; Fuxa, 2004) are all well established as ubiquitous soil organisms.

The moisture content of soil can affect pathogen persistence and performance. *Invertebrate iridescent virus 6* loses activity in dry soil (6% moisture) in less than 24 h, while soil moistures between 17 and 37% did not influence persistence (Reyes *et al.*, 2004). Moisture is the most important factor influencing nematode performance, as IJs require a water film on the soil particles for effective movement. If the water layer becomes too thin or the interspaces between soil particles are too thick, nematode movement is restricted (Koppenhöfer *et al.*, 1995). Inactive IJs may persist for longer in dry soil, but infection will be impeded (Kaya, 1990).

Soil protects pathogens from the damaging effects of UV radiation once the pathogen enters the soil profile. However, nematode IJs applied to the soil surface are acutely sensitive to UV radiation. Exposure to short UV radiation (254 nm) and natural sunlight inhibited 95% of *Neoaplectana* (*Steinernema*) *carpocapsae* (Gaugler and Boush, 1978). Exposure of *S. kushidai* IJs to the UV-C portion of the spectrum resulted in 62 and 100% mortality after 20 s and more than 60 s exposure, respectively (Azusa and Tomoko, 1999). Exposure to simulated sunlight in the laboratory for a few hours, particularly to the UV-B portion of the spectrum, fully inactivates *Metarhizium* conidia (Fargues *et al.*, 1996; Braga *et al.*, 2001b, c) and delays the germination of surviving conidia (Braga *et al.*, 2001d).

The chemical and structural make-up of the soil affects the persistence and performance of pathogens. Soil parameters such as texture, organic matter, and electrical conductivity can be significant factors in entomopathogenic nematode persistence or efficacy (Kaya, 1990; Sturhan, 1999; Kaspi *et al.*, 2010). Entomopathogenic nematode survival and pathogenicity are generally greatest in soils with larger soil particles (sandy-loam), decreasing as soils transition to smaller soil particles (clay) (Kung *et al.*, 1990; Barbercheck and Kaya, 1991). Fertilization, which is reflected in conductivity and the concentration of soluble nutrients in the soil, has been shown to decrease the persistence of fungal entomopathogens, presumably owing to increased activity of antagonistic microbes (Lingg and Donaldson, 1981; Rosin *et al.*, 1996, 1997). Groden and Lockwood (1991) found that fungistasis levels for *B. bassiana* in soils increased exponentially with increases in soil pH from 6.1 to 7. *Beauveria bassiana* persisted best in peat that also had the highest pH (Vänninen *et al.*, 2000). Conversely, Rath *et al.* (1992) found no effect of electrical

conductivity or pH on the natural occurrence of *M. anisopliae* in Tasmania.

Biotic agents in the soil also play an important role in enhancing, distributing, synergizing, and competing with entomopathogens. The presence of plants and their roots in the soil is proving to be important in the overall biology and ecology of fungal entomopathogens. The rhizosphere has been shown to provide a favorable microhabitat for fungal survival in the soil (Hu and St. Leger, 2002; Bruck, 2010; Fisher *et al.*, 2011). During the six months following fungal application, the *M. anisopliae* titer in the bulk soil decreased from 10^5 propagules/g in the top 3 cm of soil to 10^3 propagules/g. However, fungal titers in the rhizosphere remained at 10^5 propagules/g six months after application, resulting in a 100:1 ratio in fungal densities between the rhizosphere and bulk soil (Hu and St. Leger, 2002). The *M. anisopliae* population in the rhizosphere of *Picea abies* was significantly higher than the population in the surrounding bulk soil (Bruck, 2005). Subsequent studies demonstrated variability within *M. anisopliae* isolates in rhizosphere competence between plants (Bruck, 2010).

Macroorganisms within the soil environment interact with pathogens on multiple levels. Earthworms can enhance the dispersal of *S. carpocapsae* (Shapiro *et al.*, 1993, 1995). *Lumbricus terrestris* increased the upward dispersal of *S. carpocapsae* and *S. feltiae* but had no impact on *S. glaseri* upward dispersal. Collembolans vector fungal entomopathogens (mechanically and via their gut contents) in numbers sufficient to cause host infection (Dromph, 2001, 2003). The vectoring of fungal spores by these small soil arthropods is believed to have an important impact on the epizootiology of soil-borne fungal entomopathogens in the field (Dromph, 2003).

In addition to positive biotic associations (e.g., phoresy), there are antagonistic biotic interactions that are fostered natural enemies of entomopathogens. For example, nematophagous mites and collembolans feed on *Steinernema* and *Heterorhabditis* spp. with some mite species able to complete their development on a diet consisting entirely of IJs (Epsky *et al.*, 1988; Cakmak *et al.*, 2010). Besides predators, nematodes (or their bacterial symbionts) have a suite of natural enemies as they are susceptible to infection by microorganisms including phages (Poinar *et al.*, 1989; Boemare *et al.*, 1993), protists (Poinar and Hess, 1988), and nematophagous fungi (Koppenhöfer *et al.*, 1996).

Interactions among Trophic Levels

Interactions between trophic levels (i.e., plants and insect pathogens) have been studied extensively (Cory and Ericsson, 2010). Some of the more interesting examples of tritrophic interactions between pathogens and their host involve the production of herbivore-induced plant volatiles, the cassava green mite, *Mononychellus tanajoa*, and its fungal pathogen *Neozygites tanajoae*. Green leaf volatiles inhibit the germination of *N. tanajoae* conidia, while herbivore-induced plant volatiles enhance conidiation (Hountondji *et al.*, 2005). Tritrophic interactions involving host plant volatiles have also been shown to occur below ground in experiments examining entomopathogenic nematodes and their attraction to the conifer *Thuja occidentalis* after root herbivory by *O. sulcatus* larvae (van Tol *et al.*, 2001). The induction of natural enemy attractants in response to root herbivores has also been identified in turnips (Neveu *et al.*, 2002), tulips (Aratchige *et al.*, 2004), and corn (Rasmann *et al.*, 2005). In addition, *O. sulcatus* larvae are attracted to the roots of *P. abies* grown in the presence of *M. anisopliae* conidia, indicating the operation of a previously undescribed tritrophic interaction (Kepler and Bruck, 2006).

3.2.5. Modeling Epizootics

A model is an idealized representation of reality that is used to understand a particular defined system. Given the complexity of factors that contribute to epizootiology, mathematical models are required for understanding and predicting the dynamics of insect disease in natural populations. Furthermore, models may be highly useful in understanding and predicting parameters required for success in microbial control applications.

Similar to many aspects in epizootiology of insect diseases, a significant amount of the conceptual basis for modeling arises from epidemiology of human diseases or diseases of other non-insect animals (Onstad and Carruthers, 1990). According to Hethcote (2000), deterministic epidemiological modeling started early in the twentieth century for studying the dynamics of measles epidemics (Hamer, 1906). Although much of today's focus on epidemiological modeling is on non-infectious disease (Hethcote, 2000), the continued importance of infectious diseases in developing countries, as well as emerging or reemerging diseases in developed countries, has sustained a substantial need for and interest in mathematical modeling of infectious diseases (Hethcote, 2000; Maines *et al.*, 2008; Temime *et al.*, 2008). Some basic concepts in epidemiology (as reviewed by Hethcote, 2000) that are carried in epizootiology include basic classes and flow patterns in epidemiological models such as SEIR, where S is susceptible individuals, E is those exposed, I those infected, and R recovered (immune individuals), as well as the threshold concept, i.e., that the density of susceptible individuals must exceed a critical value for an epidemic outbreak to occur.

General Concepts in Epizootiological Modeling

The basis for modeling epizootiology of insect diseases, and the ecological theory behind it, is derived primarily from Anderson and May's work in the early 1980s (1980, 1981, 1982), with substantial review and analyses by Brown (1987), Onstad and Carruthers (1990), Hesketh *et al.* (2010), and others. With particular relevance to applied microbial control, Brown (1987) discusses epizootiological modeling using an integrated pest management (IPM) system approach, i.e., taking a defined and delineated (host–pathogen) system and breaking it into mathematical components that adequately represent the whole. Typically, a systems model is generated by building a conceptual framework based on traditional empirical research, developing a mathematical representation of that framework, and finally devising a computer program that will implement the mathematical representation (Brown, 1987).

Fundamentally, host–pathogen models are based on the concept that the number of susceptible individuals in a population at a given time depends on the number of contacts between susceptible and infected individuals, and the transmission rate. A simple equation for this concept may be written as:

$$S_{t+1} = S_t - pS_tI_t$$

where S is the number of susceptible individuals, t is time, p is transmission efficiency (e.g., if $p = 0.05$ then 5% of possible contacts result in transmission), and I is the number of infected individuals.

Similarly, the number of infected individuals at a given time can be represented as:

$$I_{t+1} = I_t + pS_tI_t - mI_t$$

where m is the mortality rate, e.g., if the incubation period is four days then m is 0.25 (25% of infected individuals die each day); m can be considered virulence (power to cause disease/mortality).

The equations can be rewritten as simple differential equations, e.g., for the change in number of susceptible individuals over time with the intrinsic rate of increase (r) incorporated:

$$dS/dt = rS - pSI$$

Another basic concept in epizootiological models that can be derived from the above equations is that of the threshold (Brown, 1987; Onstad and Carruthers, 1990). Anderson and May (1980, 1981) defined the threshold as the host density required for the pathogen to persist within the population; the threshold is directly proportional to virulence and indirectly proportional to transmissibility. The threshold occurs when the number of susceptible individuals is equal to the mortality rate divided by transmission ($S_t = m/p$). If the susceptible population is greater than the threshold then disease prevalence can increase and if the susceptible population is less than the threshold then disease prevalence can only decrease. In other terms, Hesketh et al. (2010) discusses R_0 as the basic rate of pathogen increase; R_0 must be > 1 for the pathogen to persist and spread, and H_T is a critical threshold below which prevalence will decline and above which it will rise, i.e., when $R_0 = 1$.

Beyond the basic concepts described above, a systems model increases in its complexity as it incorporates additional factors in order to predict prevalence, e.g., impact of the pathogen on reproductive potential, vertical transmission, and impact of non-infective stages. The factors that can be considered for incorporation may be divided into three groups: (1) pathogen stages outside the host (dispersal, survival, and transmission mode); (2) factors at the host–pathogen interface (host stress, inoculums load, etc.); and (3) pathogen stages inside the host (incubation period; host feeding and aging rate). Certainly, not all potential factors can be incorporated into the model

Examples of Epizootiological Models

Examples of epizootiological models may be found across pathogen groups, e.g., viruses (Bianchi et al., 2002; Sun et al., 2006), bacteria (Zahiri et al., 2004, and with more emphasis in recent years on modeling resistance to Bt crops, e.g., Tabashnik et al., 2008; Onstad and Meinke, 2010), protists (Otterstatter and Thomson, 2008), fungi (Hesketh et al., 2010), microsporidia (Onstad and Maddox, 1990; Kelly et al., 2001), and nematodes (Stuart et al., 2006; Ram et al., 2008). Epizootiological models for insect pathogens have ranged from simple to complex and covered both natural and introduced pathogen populations.

In some cases, relatively simple models have provided benefits toward understanding and predicting epizootiology. For example, Milks et al. (2008) used multiple regression analysis to describe the prevalence of the microsporidium Thelohania solenopsae in populations of S. invicta. Several factors that appeared to impact T. solenopsae prevalence were defined, including the number of colonies at a site, precipitation, proximity to waterways, and habitat type. In another example, Edelstein et al. (2005) used a non-linear model to study the influence of temperature on the developmental rate of the fungus Nomuraea rileyi in larvae of the velvetbean caterpillar, Anticarsia gemmatalis. The estimated lower and upper thresholds of fungal vegetative development were observed to coincide with conditions during the natural epizootics in central Argentina. Although the model was developed under controlled conditions in the laboratory, some understanding of field dynamics and potential for microbial control was gained and opportunities for additional research and further model development were obtained.

Some more complex or comprehensive models than those described above have addressed inundative or inoculative approaches to microbial control of specific target pests. An example is depicted in the work of Bianchi et al. (2002), in which a model was constructed when applying Spodoptera exigua multicapsid nucleopolyhedrovirus (SeMNPV) in the greenhouse. Parameters included crop growth, insect development (growth rate and reproductive potential), spatial distribution of plants, virus inactivation within occlusion bodies (based on expected decay due to UV radiation), rate of infection by ingestion, rate of horizontal transmission, and environmental variables (e.g., temperature). The models were generally in close agreement with experimental data (Bianchi et al., 2002).

In another example directed toward predicting efficacy of microbial control applications, a model was developed to predict the potential of fungal entomopathogens to control mosquitoes and suppress malaria (Hancock et al., 2009). In reference to the mosquito–plasmodium dynamics, the concept was based on a simple model considering susceptible hosts, exposed individuals (those that carry malaria but cannot yet transmit the disease), and infected individuals. Exposure to a fungal entomopathogen and the probability of fungal infection upon exposure were also incorporated. When realistic assumptions were made about mosquito, fungus, and malaria biology and with moderate to low daily fungal infection probabilities, the model indicated the potential for substantial reductions in malaria transmission with fungal applications (Hancock et al., 2009).

In addition to models geared toward optimizing microbial control applications, several models have been developed for predicting the epizootiology of endemic pathogen populations. For example, substantial research has been directed toward the ecology and epizootiology of Entomophaga maimaiga in relation to a mortality factor of L. dispar (Weseloh et al., 1993; Hajek, 1999; Weseloh, 2002). Models to simulate fungal prevalence based on parameters including temperature, humidity, precipitation, fungal reservoirs, and L. dispar density have been developed and tested for validity (Weseloh, 2004). Models that sufficiently account for dispersal of conidia can provide a good fit to infection prevalence in forests (Weseloh, 2004).

Malakar et al. (1999a) demonstrated that models might also be used to simulate interactions between two pathogens. Specifically, a model to determine the impact of E. maimaiga on Lymantria dispar multiple nucleopolyhedrovirus (LdMNPV) in L. dispar populations was developed. Pathogen activity was determined in plots with and without artificial rain (the latter inducing higher fungal mortality in L. dispar). Despite the negative impact of dual infection on production of LdMNPV in a separate laboratory study (Malakar *et al.*, 1999b), the model for field interactions indicated that in moderate host populations, the fungus would not substantially affect virus infection, owing to temporal separation of the two pathogens' activities (Malakar *et al.*, 1999a). The fungus remained at low levels until the later instars, at which time *L. dispar* is less susceptible to LdMNPV infection (Malakar *et al.*, 1999a).

Models have also been used for predictive value in simulating the epizootiology of genetically modified pathogens. For example, Sun *et al.* (2006) developed a comprehensive model to simulate epizootics in wild-type or genetically modified, *Helicoverpa armigera* single nucleocapsid nucleopolyhedrovirus (HaSNPV) for the control of cotton bollworm, *Helicoverpa armigera*, in cotton. With multiple applications, both wild-type and genetically modified viruses were found to be capable of keeping the target pest under its economic injury level (EIL). The recombinant virus, which kills the host more quickly, provided better short-term protection than the wild-type virus. However, limited persistence of the recombinant virus in the host population resulted in reduced horizontal transmission and higher pupal survival at the end of the growing season. Thus, the simulation indicated that the fast acting genetically modified viruses may be more efficacious for short-term (inundative) control, whereas the wild-type may be superior for longer term (inoculative) approaches.

The modeling examples described above are generally directed at representation of whole insect—pathogen systems. There have also been some models that focus on simulating a specific portion of the epizootiological process. For example, models have been developed to explain group infection behavior in entomopathogenic nematodes (Fenton and Rands, 2004; Fushing *et al.*, 2008). Fushing *et al.*'s (2008) model infection was based on the concept of risk-sensitive foraging and a follow-the-leader behavior, i.e., individuals that are more prone to taking risk infect the host first, and subsequently the remaining infective population follows. In their analytical review articles, Onstad and Carruthers (1990) and Hesketh *et al.* (2010) recommend advancing the discipline of epizootiological modeling within insect pathology through better incorporation of spatial and temporal heterogeneity and their associated variables, improved merging of empirical and theoretical data, and greater cooperation among the empiricists and theorists. Additional models that focus on the components of epizootiology (such as infectivity, success within the host, and survival outside the host) may lead to superior models of the whole system once the components are coupled together.

3.3. MICROBIAL CONTROL (APPLIED EPIZOOTIOLOGY)

3.3.1. Basic Concepts in Microbial Control

Microbial control can be defined as the use of insect pathogens for pest suppression. Some uses include byproducts of the pathogenic organisms as well (e.g., toxins). Furthermore, some consider natural suppression of pests (i.e., without any human intervention) to be included in microbial control, but this chapter will just consider microbial control in relation to intentional manipulation of the targeted system. The goal of microbial control programs is to eliminate or reduce a pest population below an economically damaging level. The goal is achieved by causing disease prevalence in a targeted population that is sufficient to keep the pest below the EIL. The principles that underlie the ability to cause disease in targeted populations are guided by the principles that cause epizootics in natural insect populations. Therefore, microbial control can be considered applied epizootiology.

Like other insect pest management strategies, the decision to implement a microbial control tactic for pest suppression should be weighed against other options, including not taking any action. In most systems, chemical insecticides still predominate as the primary pest control tactic. Therefore, microbial control is often compared with the use of chemical insecticides. In general, relative to chemical insecticides, there are several disadvantages that vary in degree among the pathogen groups, including cost of production, lack of a wider host range, susceptibility to environmental degradation, and a longer time to kill the host (Fuxa, 1987; Tanada and Kaya, 1993; Lacey and Shapiro-Ilan, 2008). There are also a number of general advantages that are associated with microbial control relative to the use of chemical insecticides. Although not all pathogen groups hold these advantages (or the levels at which they hold them vary), the advantages of microbial control include a reduced potential for development of resistance in the target pest and safety to humans, other non-target organisms, and the environment; relative safety to non-targets also leads to a reduced chance of secondary pest outbreaks through conservation of natural enemies (Fuxa, 1987; Tanada and Kaya, 1993; Lacey and Shapiro-Ilan, 2008).

The safety of microbial control agents is largely based on their narrow host ranges, which allows for safe application with reduced potential for impact on beneficial organisms and secondary pest outbreaks. A narrow host range, however, can also be considered a disadvantage; for example, if the user wants to target several unrelated pests at once, then a broad-spectrum tactic may be more attractive. The host ranges of some pathogen groups are narrower than others, e.g. certain NPVs and granuloviruses (GVs) may infect within only one genus or species, whereas some, such as *Autographica californica* NPV, infect numerous species in different lepidopteran families (Cory and Evans, 2007). Microsporidia and various protists also tend to have narrow host ranges, whereas entomopathogenic nematodes and hypocrealean fungi have wider host ranges. Yet even within these groups, exceptions exist. For example, *S. scarabaei* and *S. scapterisci* are specific to white grubs and orthopterans (especially mole crickets), respectively (Shapiro-Ilan *et al.*, 2002a). Some bacteria have intermediate host ranges, e.g., Bt strains to various Lepidoptera, whereas *Paenibacillus popilliae* is specific to several subfamilies within Scarabaeidae (Garczynski and Siegel, 2007). *Bacillus sphaericus* is specific for several, but not all species of the Culicidae (Lacey, 2007).

A primary concern in pest management is protection of human health. In general, microbial control agents are safe to humans and other vertebrates, yet some exceptions exist, e.g., the bacterium *Serratia marcescens* can be an opportunistic pathogen in humans causing septicemia, formulations of *B. bassiana* can cause allergic reactions in humans (Westwood *et al.*, 2006), and in rare cases *Beauveria* sp. has been reported to be capable of causing infection in immunosuppressed humans (Henke *et al.*, 2002). Nonetheless, microbial control agents are overall considerably safer to humans than chemical insecticides, which have been documented to cause approximately 220,000 human deaths per year (Pimentel, 2008).

In addition to safety to humans, microbial control agents are generally considered to have little or no effect on other non-targets including beneficial insects. Even so, exceptions, in which entomopathogens have a negative impact on insect biocontrol agents, have been documented, e.g., a negative association between the microsporidium *N. pyrausta* and the parasitoid *Macrocentrus grandii* (Bruck and Lewis, 1999). In another example, entomopathogenic nematodes were observed to infect several hymenopteran parasitoids (Shannag and Capinera, 2000; Lacey *et al.*, 2003; Mbata and Shapiro-Ilan, 2010) or coleopteran predators (Shapiro-Ilan and Cottrell, 2005). However, several studies or analyses have indicated that the impact of entomopathogenic nematodes on field populations of insect natural enemies is negligible owing to various factors such as spatial or temporal dynamics that prevent negative impact between the control agents (Georgis *et al.*, 1991; Bathon, 1996; Koppenhöfer and Grewal, 2005).

Another primary advantage in the use of microbial control agents is the reduced potential for resistance development in the target pest. Resistance to entomopathogens, however, has been demonstrated in the laboratory,

e.g., resistance to *M. anisopliae* in termites (Rosengaus *et al.*, 1998; Traniello *et al.*, 2002). Resistance to certain entomopathogens has also been reported under field conditions, including field resistance to GV (Asser-Kaiser *et al.*, 2007), NPV (Fuxa *et al.*, 1988), Bt (McGaughey, 1985; Tabashnik *et al.*, 1990), and Bt crops (Tabashnik *et al.*, 2008). Host resistance is discussed in Chapter 13.

3.3.2. Factors Affecting Efficacy in Microbial Control

Beyond safety and possessing a low potential for resistance, some other factors affect the efficacy of microbial control agents. An ideal microbial control agent is likely to possess the following attributes: (1) a high level of virulence to the target pest; (2) ease of production and storage; and (3) the ability to persist in the environment. Factors that affect efficacy are discussed below and summarized by pathogen group in Table 3.1.

Production technology is critical to cost competitiveness and meeting market demands for a target pest. Production of some pathogen groups cannot be achieved outside an insect host (e.g., microsporidia) and thus relies on *in vivo* culture methods, which limit or prevent economy of scale. Insect virus production also relies on *in vivo* methods, and although culture methods have been streamlined by mechanization, attempts to advance mass production in cell culture have thus far been unsuccessful (Vail *et al.*, 1999; Szewczyk *et al.*, 2006).

Mass production has been greatly facilitated in entomopathogens that can be cultured in liquid or solid fermentation. Among bacteria, the successful bioinsecticides Bt and *B. sphaericus* are produced via submerged *in vitro* culture, whereas *P. popilliae* must be grown *in vivo* (Garczynski and Siegel, 2007). Hypocrealean conidia can be produced using a solid fermentation or a diphasic approach, whereas liquid culture can be used for production of blastospores, sclerotia, or in some cases, conidia (Feng *et al.*, 1994; Leland *et al.*, 2005; Jackson *et al.*, 2010) (see Chapter 6). Depending on the microbial control approach and level of persistence required, some culture methods for fungi may be more appropriate than others (Jackson *et al.*, 2010). Entomopathogenic nematodes are produced using solid or liquid fermentation as well as *in vivo* approaches. Most entomopathogenic nematodes are produced using *in vitro* liquid culture, which offers the greatest economy of scale relative to other approaches (Shapiro-Ilan and Gaugler, 2002; Ehlers and Shapiro-Ilan, 2005). In some studies or for certain species, the efficacy or quality of nematodes produced in liquid culture was found to be inferior to *in vivo* produced products (Gaugler and Georgis, 1991; Abu Hatab and Gaugler, 1999; Cottrell *et al.*, 2011), whereas in other cases no differences due to culture method were detected (Gaugler and Georgis, 1991;

TABLE 3.1 Positive and Negative Factors of Pathogen Groups that Affect Microbial Control Efficacy and Commercialization[a]

Pathogen Group	Positive	Negative
Virus (e.g., baculovirus)	Specificity High level of virulence Capable of persisting in the environment (soil) Can store at room temperature	Specificity Environmental sensitivity (UV) Slow acting for most baculoviruses Cost of production (*in vivo* only)
Bacteria (e.g., *Bacillus thuringiensis, B. sphaericus*)	Wide host range Ease of mass production Speed of kill Can store at room temperature	Environmental sensitivity (UV) For *Paenibacillus popilliae*: cost of production (*in vivo*) and slow acting
Microsporidia[b]	Persistent stages Specificity	Cost of production (*in vivo*) Most cause chronic infection and are slow acting Environmental sensitivity (UV)
Fungi (e.g., *Hypocreales*)	Wide host range Ease of mass production (*in vitro*) High virulence Can penetrate cuticle (infection through contact) Can store at room temperature	Environmental sensitivity (UV, relative humidity) Slow acting
Nematodes (*Steinernema* and *Heterorhabditis*)	Wide host range Speed of kill Ease of mass production (*in vivo* or *in vitro*) Mobility, can actively detect and seek or ambush host Little or no registration required	Cost of production (despite *in vitro* methodology) Environmental sensitivity (UV, desiccation) Storage requires refrigeration

UV = ultraviolet.
[a]The properties described pertain to a generalized view of each pathogen group; exceptions may exist in each category.
[b]Although microsporidia are classified as fungi they are treated separately here.

Shapiro and McCoy, 2000a). Despite advances in the production of microbial control agents, they are still generally more expensive than chemical insecticides, yet with the advent of certain "soft" chemicals (with a narrower host range) the gap in price has decreased.

Virulence clearly has a direct effect on microbial control efficacy. Pathogens, even at the strain level, can vary greatly in their ability to cause high levels of mortality in a specific host. Speed of kill or disease onset can also vary greatly among pathogens. Entomopathogenic nematodes are relatively fast acting, with the ability to kill the host within 24–48 h postinfection. In contrast, hypocrealean fungi and insect viruses (e.g., NPVs) are slower acting and can take a week or more to kill the host (Fuxa, 1987; Shapiro-Ilan *et al.*, 2004a). Some bacterial pathogens such as Bt are fast acting, killing their hosts within hours to a few days, whereas *P. popilliae* is slow, taking 20 days or more to kill their white grub hosts (Dutky, 1963; Fuxa, 1987). Some pathogens do not cause mortality directly; for example, although there are a few exceptions (e.g.,

Vairimorpha necatrix, which can rapidly cause mortality), most microsporidia cause chronic infections that may affect various aspects of host fitness such as reproductive capacity, longevity, or development without causing direct mortality (Solter and Becnel, 2007). A virulent and fast-acting pathogen is generally considered desirable and is required to suppress pests with a low EIL; however, slower, less virulent pathogens may control pests with higher EILs (Fuxa, 1987). In addition, a low level of virulence and long infection process can enhance the pathogen's persistence in the host population (Anderson, 1982; Fuxa, 1987). Fast-acting pathogens can be disadvantageous when targeting social insects, which may impair virulent pathogens through avoidance and removal of infected individuals (Rath, 2000; Wilson-Rich *et al.*, 2007, 2009). In contrast, chronic pathogens can have the advantage of infiltrating social insect colonies, including accessing the queen (Oi, 2006; Milks *et al.*, 2008).

Pathogen population density affects microbial control efficacy. A higher pathogen population density (as well as a higher host density) leads to increased host–pathogen contact. Thus, in susceptible hosts, it is assumed that a certain pathogen density is required to suppress the target pest below an EIL, and that lower densities will fail. There are many examples that demonstrate this dose–response relationship between pathogen and host in the laboratory (Lacey, 1997), e.g., using fungi (Hesketh *et al.*, 2008; Wraight *et al.*, 2010), nematodes (Power *et al.*, 2009), or viruses (Figueiredo *et al.*, 2009). Effects of pathogen application rate in field studies have been reported (Lacey and Kaya, 2007; Lacey and Shapiro-Ilan, 2008), e.g., using nematodes (McCoy *et al.*, 2002; Arthurs *et al.*, 2005; Chambers *et al.*, 2010) or fungi (Wraight and Ramos, 2002), whereas in other studies the effects of field application rates varied, were not detected, or were not deemed important relevant to other factors (e.g., strain or species effect) (Cappaert and Koppenhöfer, 2003; Grewal *et al.*, 2004; Dillon *et al.*, 2007). Nonetheless, although specific pest requirements may differ, minimum baseline application rates have been established that are expected to result in pathogen population densities that produce pest suppression, e.g., 10^{13}–10^{14} conidia/ha for certain hypocrealean fungi (Wraight and Carruthers, 1999; Jaronski, 2010), 10^{11}–10^{12} occlusion bodies/ha for NPVs (e.g., when applied for control of various lepidopteran pests) (Vail *et al.*, 1999), and 2.5×10^{9} IJs/ha for entomopathogenic nematodes (Shapiro-Ilan *et al.*, 2006).

At a minimum, a pathogen must persist in the environment long enough to infect the target host. A longer term persistence can have a positive impact on efficacy, e.g., by allowing the pathogen to persist when host population density is low. In some cases, superior environmental persistence may compensate for lower virulence (Shields *et al.*, 1999; Shapiro-Ilan *et al.*, 2002a). A variety of factors may limit pathogen persistence. As indicated in Section 3.2.4, all pathogen groups are susceptible to degradation by UV radiation (Fuxa, 1987). Thus, achieving persistence of above-ground entomopathogen applications is more challenging than in applications to the soil or other environments (e.g., greenhouses) that are more protected. Nonetheless, some microbial control agents such as Bt and certain viruses have been successful in suppressing pests above ground in the field (Lacey *et al.*, 2001).

In addition to susceptibility to UV radiation, entomopathogenic nematodes are highly sensitive to desiccation (Kaya and Gaugler, 1993; Shapiro-Ilan *et al.*, 2006) and fungal entomopathogens generally require high levels of relative humidity for germination. Yet, exceptions exist for the fungal requirement of high humidity (Wraight *et al.*, 2007); e.g., low humidity was reported to be beneficial for control of the lesser grain borer, *Rhyzopertha dominica*, with *B. bassiana* (Lord, 2005). Furthermore, high levels of moisture in the soil can be detrimental to fungi by enhancing environmental degradation, e.g., through increased antagonists in soil (Shapiro-Ilan *et al.*, 2004a; Jaronski, 2007; Wraight *et al.*, 2007) (see Section 3.2.4 for more details on environmental impact).

In nature and in field applications, persistence of entomopathogens in the soil tends to be greater than above ground. Some viruses such as NPVs can persist for several years or even decades in soil (Thompson *et al.*, 1981; England *et al.*, 1998). In contrast, the duration of pest control resulting from most entomopathogenic nematode applications is limited to two to eight weeks (Shapiro-Ilan *et al.*, 2006). Persistence of efficacy, however, can depend on host density, e.g., multiseason persistence of entomopathogenic nematodes was observed for suppression of white grubs with high population densities (Klein and Georgis, 1992). A number of other factors can affect entomopathogen persistence in soil, including pH, texture, aeration, antagonists, temperature, and use of amendments (fertilizers or pesticides) (Lacey *et al.*, 2001; Shapiro-Ilan *et al.*, 2006; Meyling and Eilenberg, 2007).

3.3.3. Improving Efficacy in Microbial Control

Based on the factors affecting microbial control efficacy described above, a number of avenues may be considered for improving the use of entomopathogens. Approaches to improving microbial control efficacy may be divided into several categories: (1) improving the entomopathogen; (2) improving production and application methods; and (3) improving the environment.

Improving the Entomopathogen

Achieving or improving efficacy in microbial control can rely on choosing the best entomopathogen for a particular system. The most suitable entomopathogen from a variety of candidates can be selected simply by screening existing species and strains that possess superior desired traits such as virulence and environmental tolerance. New entomopathogens can be discovered through surveys and screened in parallel to existing strains; such surveys have been conducted extensively for entomopathogenic nematodes (Shapiro-Ilan *et al.*, 2003a, 2008a; Campos-Herrera *et al.*, 2008) and fungi (Leland *et al.*, 2005; McGuire *et al.*, 2005; Lubeck *et al.*, 2008). The screening process is often accomplished by first narrowing down the number of candidates in laboratory comparisons; such comparisons have been made to find superior entomopathogen strains for numerous target pests such as the emerald ash borer, *Agrilus planipennis* (Castrillo *et al.*, 2010), aphids (Shapiro-Ilan *et al.*, 2008a), cowpea weevils, *Callosobruchus maculatus* (Cherry *et al.*, 2005), plum curculio, *Conotrachelus nenuphar* (Shapiro-Ilan *et al.*, 2008b), *Lygus* spp. (Liu *et al.*, 2002; Leland *et al.*, 2005),

O. sulcatus (Bruck, 2004), and mites (e.g., *Tetranychus cinnabarinus*) (Shi and Feng, 2004).

The importance of verifying laboratory efficacy in the field cannot be overemphasized. An entomopathogen that shows high virulence in the controlled environment of a laboratory could fail to suppress the target pest in the field owing to various biotic or abiotic factors that render the organism incompatible. A lack of understanding of the biological and ecological constraints required for pathogen persistence and proliferation in the environment is likely to lead to a discrepancy between laboratory and field efficacy (Hu and St. Leger, 2002; Bruck, 2005, 2010). Some examples of laboratory screening studies that selected entomopathogen strains or species that later proved successful in the field include *S. riobrave* and *H. indica* for control of the citrus weevil, *Diaprepes abbreviatus* (Duncan and McCoy, 1996; Shapiro *et al.*, 1999a; Shapiro and McCoy, 2000b), *S. riobrave* for control of *C. nenuphar* (Shapiro-Ilan *et al.*, 2002b, 2004b; Pereault *et al.*, 2009), hypocrealean fungi for suppression of the brown citrus aphid, *Toxoptera citricida* (Poprawski *et al.*, 1999), and *M. anisopliae* for control of *A. ludens* (Lezama-Gutiérrez *et al.*, 2000).

In contrast, in some cases a high level of laboratory virulence or efficacy has not been corroborated under field conditions. For instance, *S. feltiae* was highly virulent to *C. nenuphar* in the laboratory, but failed to control the pest in Georgia peach orchards, possibly because of unsuitable soil temperatures (Shapiro-Ilan *et al.*, 2004b). In another example, Leland *et al.* (2005) screened strains of *B. bassiana* isolated from *Lygus* spp. populations for *in vitro* conidia production, temperature growth optima, tolerance to UV radiation, and production of beauvericin and compared these to a commercial *B. bassiana* isolate (GHA). *Lygus* spp. isolates were orders of magnitude more virulent than the commercial isolate based on LC_{50} values. However, field-collected isolates that were superior to GHA in the laboratory (Leland *et al.*, 2005) did not provide significantly higher levels of control of *L. hesperus* infesting alfalfa (McGuire *et al.*, 2006). Indeed, strain selection based primarily on pathogenicity or mass production that ignores habitat preferences of the pathogen has often been unsuccessful (Hu and St. Leger, 2002; Bruck, 2005, 2010). A recent focus on pathogen ecology and habitat preferences when selecting strains for microbial control is expected to enhance the pathogen performance in the field and the frequency and magnitude of epizootics (Jaronski, 2007, 2010; Vega *et al.*, 2009).

If existing or newly discovered entomopathogen strains or species cannot achieve desired levels of microbial control efficacy, another option is to improve selected candidates through genetic approaches. Genetic improvement is directed toward enhancement of single or various beneficial traits, e.g., virulence, reproductive capacity, or environmental tolerance. Approaches may include molecular or non-molecular methods. One non-molecular method entails selection for desired traits. Selection for improved virulence can be obtained by passing the pathogen through a susceptible host (Steinhaus, 1949; Daoust and Roberts, 1982). Some examples of genetic selection for other traits include improvements in entomopathogenic nematode host-finding (Gaugler *et al.*, 1989a) and nematicide resistance (Glazer *et al.*, 1997). Directed selection can, however, have the shortcoming of inadvertently selecting for an inferior level of one trait while selecting for the targeted trait (Gaugler, 1987). For example, Gaugler *et al.* (1990) reported a loss in storage capacity in entomopathogenic nematodes that had been selected for improved host finding. In addition, *B. bassiana* and *M. brunneum* selected for fungicide resistance exhibited tradeoffs with other traits, e.g., reproductive capacity (Shapiro-Ilan *et al.*, 2011).

Another non-molecular approach to strain improvement is hybridization, e.g., the transfer of beneficial traits from one strain to another. Examples of hybridization for improved biocontrol include the development of superior environmental tolerance and or virulence in entomopathogenic nematodes (Shapiro *et al.*, 1997; Shapiro-Ilan *et al.*, 2005c) and improved virulence in protoplast fusion hybrids of *B. bassiana* (Couteaudier *et al.*, 1996). The two non-molecular approaches (selection and hybridization) have also been combined for the development of superior entomopathogenic nematode strains (Mukaka *et al.*, 2010).

Substantial progress has been made in using transgenic or other molecular approaches for improving microbial control agents. Thus, transgenic approaches have been used to increase the virulence in NPVs, e.g., through the addition of scorpion toxin (Stewart *et al.*, 1991; Harrison and Bonning, 2000) or insect hormones (Maeda, 1989; Chen *et al.*, 2000). *Beauveria bassiana* has been transformed for benomyl resistance (Sandhu *et al.*, 2001), and *H. bacteriophora* has been transformed for increased heat tolerance (Gaugler *et al.*, 1997). *Metarhizium anisopliae* has been genetically engineered to express the 70 amino acid *Androctonus australis* neurotoxin AaIT, a toxic insect-selective peptide (Zlotkin *et al.*, 2000), resulting in increased virulence against *M. sexta* larvae and *Aedes aegypti* adults (Wang and St. Leger, 2007). Genetic modification for improvement of natural Bt strains has been examined (Sansinenea *et al.*, 2010), and as an extension of microbial control research on Bt, various crops have been genetically modified with Bt toxins, resulting in widespread implementation (Tabashnik *et al.*, 2003, 2008).

One issue that can jeopardize strains with beneficial traits is repeated subculturing, resulting in attenuation due to genetic factors (e.g., inbreeding, drift, inadvertent selection) or non-genetic factors (e.g., disease) (Tanada and Kaya, 1993; Hopper *et al.*, 1993; Chaston *et al.*, 2011).

Serial *in vitro* transfer of *N. rileyi* quickly caused attenuation and reduced virulence to *A. gemmatalis* (Morrow *et al.*, 1989). Trait deterioration has also been observed during laboratory culturing of entomopathogenic nematodes (Shapiro *et al.*, 1996; Wang and Grewal, 2002; Bilgrami *et al.*, 2006). Commercial manufacturers of entomopathogens pay special attention to ensuring that repeated subculturing is avoided, to reduce the likelihood of attenuation of the microbial control agents that they produce. Nonetheless, even with these efforts detrimental trait changes have been observed; e.g., Bt from lepidopteran cadavers produced by DiPel (a commercial Bt formulation) infection was significantly less infective than Bt from cadavers produced by a strain recently isolated from the field, suggesting that the wild-type strain was more efficient at producing spores in the host (Naryanan, 2006). For some organisms (e.g., entomopathogenic nematodes), trait loss can be reduced through the development of selected inbred lines (Bai *et al.*, 2005; Chaston *et al.*, 2011).

The effectiveness of entomopathogen species or strain improvement for implementation in microbial control may also be hindered by the extremely high cost of commercial development and registration. For example, registration of new fungal entomopathogen strains requires the generation of human and environmental safety data, and in some countries, replicated verification of efficacy is also needed; a process that can require approximately 1–1.5 million US dollars (Jaronski, 2010). Registration of modified organisms can be even more costly. Thus, in many cases, private sector development of a new entomopathogen may only be considered if improvements in virulence, production efficiency or other attributes in the new organism are extreme relative to the existing products. Therefore, screening processes should generally include strains and species that are already commercially registered and available. The cost of developing new entomopathogen strains or species, however, is less for some entomopathogen groups than for others, e.g., in the USA endemic entomopathogenic nematodes are not regulated.

Improving Production and Application Methods

Improvements in entomopathogen production methods can lead to improved quality, improved fitness, and reduced costs. Lower costs of production can allow for increased application rates and cost competitiveness with other pest management strategies. *In vivo* production of entomopathogenic nematodes or viruses can be accomplished through enhanced mechanization, thereby reducing labor requirements (Gaugler *et al.*, 2002; Shapiro-Ilan *et al.*, 2002c), or through improved host diets, resulting in lower costs or improved pathogen quality (Shapiro-Ilan *et al.*, 2008c; Elvira *et al.*, 2010).

In vitro solid and liquid culture can be enhanced for entomopathogen production. Similar to *in vivo* approaches, solid fermentation of entomopathogenic nematodes can be enhanced through improved host nutrition and process automation (Gaugler and Han, 2002). Liquid production of entomopathogenic nematodes has been substantially improved through superior media development and elucidation of bioreactor conditions needed for optimum recovery and fecundity (Shapiro-Ilan and Gaugler, 2002; Ehlers and Shapiro-Ilan, 2005). *In vitro* approaches to fungal production can also be improved through media enhancement (Gao and Liu, 2010; Jaronski, 2010). Bt has been produced in submerged culture for over 40 years (Couch and Ross, 1980). One of the factors that has played a role in making Bt the most successful microbial control agent for insect pests (Lacey *et al.*, 2001) is the efficient and relatively inexpensive production methods that have been developed. Nonetheless, research on improving Bt production media continues, e.g., using waste materials instead of raw materials (Brar *et al.*, 2009; Zhuang *et al.*, 2011).

Application and delivery methods offer another opportunity to improve microbial control efficacy. Most entomopathogens can be applied using common agricultural equipment including various spray and irrigation systems. Standard application techniques and equipment for microbial control agents are reviewed in Lacey and Kaya (2007) and will not be covered in this chapter. Despite well-established procedures, equipment used for entomopathogen application can be improved, such as optimizing spray systems for enhanced pathogen survival and dispersion (Fife *et al.*, 2006; Shapiro-Ilan *et al.*, 2006). Other application parameters can also be further optimized for many host–pathogen systems, e.g., rate and timing of application.

In addition to optimization of application equipment or parameters, improved application techniques can be sought. For example, application of entomopathogenic nematodes in nematode-killed hosts has been considered (Jansson *et al.*, 1993; Shapiro and Glazer, 1996; Dolinski and Lacey, 2007). Advantages to the cadaver application approach relative to standard application in aqueous suspension have been reported, including superior nematode dispersal (Shapiro and Glazer, 1996), infectivity (Shapiro and Lewis, 1999), survival (Perez *et al.*, 2003), and efficacy (Shapiro-Ilan *et al.*, 2003b), whereas other studies did not detect a benefit in the cadaver approach (Bruck *et al.*, 2005). Methods to facilitate application of cadavers through formulation have been developed to protect cadavers from rupture and improving ease of handling (Fig. 3.3) (Shapiro-Ilan *et al.*, 2001, 2010a; Del Valle *et al.*, 2009). Yet commercial application of host cadavers with nematodes has been minimal, possibly because of costs and a need to develop mass-application

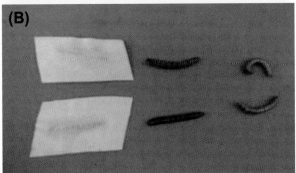

FIGURE 3.3 Formulated and non-formulated entomopathogenic nematode-killed host (i.e., cadavers). *Galleria mellonella* (A) and *Tenebrio molitor* (B) are, from right to left healthy, killed by *Heterorhabditis bacteriophora*, or coated with a starch–clay matrix (A) or masking tape (B). *[Photo credits: (A) P. Greb and (B) K. Halat, both with the US Department of Agriculture, Agricultural Research Service.]*

methods. Recently, nematodes applied as within the host cadavers were demonstrated to be effective and persistent when added to bags of potting media for subsequent distribution to target pest sites (Deol *et al.*, 2011).

Another application approach that may offer advantages in microbial control is autodissemination, i.e., the use of insects as natural dispersal organisms to spread pathogens (Vega *et al.*, 2007). This approach can be enhanced with a device that promotes contact with the pathogen (Vega *et al.*, 1995, 2007). In one example of leveraging movement of a non-host, the bumble bee, *Bombus impatiens*, was used to carry *B. bassiana* for control of the greenhouse whitefly, *Trialeurodes vaporariorum* (Kapongo *et al.*, 2008). Autodissemination has been used against a variety of pests, including *G. morsitans* (Kaaya and Okech, 1990; Maniania, 1998, 2002), the Japanese beetle, *Popillia japonica* (Klein and Lacey, 1999), *B. germanica* (Kaakeh *et al.*, 1996), house flies, *Musca domestica* (Renn *et al.*, 1999), the spruce bark beetle, *Ips typographus* (Kreutz *et al.*, 2004), and *P. xylostella* (Furlong and Pell, 2001). In other novel approaches to using fungal entomopathogens, endophytic relationships with the host plant or colonization of the rhizosphere might be utilized to develop low-cost microbial control strategies (Vega *et al.*, 2008; Bruck, 2010).

Successful microbial control applications can be facilitated through improved formulation. Research on improved entomopathogen formulations is directed toward ease of handling and enhanced persistence in the environment. Significant work has been undertaken to mitigate the effect of UV radiation on entomopathogens via improved formulation of microbial control products (Burges, 1998; Jackson *et al.*, 2010). Recent advances in formulation of entomopathogenic nematodes that have facilitated aboveground use, a major barrier to expanding use of this pathogen group, using mixtures with a surfactant and polymer (Schroer and Ehlers, 2005), postapplication protective covers using foam (Lacey *et al.*, 2010), and a sprayable gel thought to provide resistance to UV radiation and desiccation (Shapiro-Ilan *et al.*, 2010b). UV radiation-protecting formulations have also been developed for fungal entomopathogens (Behle *et al.*, 2011), and optical brighteners have been demonstrated to protect entomopathogenic viruses from UV radiation (Shapiro, 1992). Protective formulations have also substantially improved the persistence of Bt (Garczynski and Siegel, 2007). Bait formulations can enhance entomopathogen persistence and reduce the quantity of microbial agents required per unit area; e.g., baits have been developed for entomopathogenic nematodes (Grewal, 2002), fungi (Geden and Steinkraus, 2003), and Bt (Navon *et al.*, 1997), yet thus far, the market impact of baits has not been substantial relative to other formulations.

Another approach to improving application methodology and achieving higher levels of microbial control efficacy is to combine entomopathogens with each other or with other biotic agents. Although combinations of entomopathogens can result in synergistic levels of mortality, interactions may also be antagonistic or additive (Kreig, 1971; Koppenhöfer and Grewal, 2005). The nature of the interaction depends on several factors, including the specific combination, the host species, and the timing and rate of application (Kreig, 1971; Koppenhöfer and Grewal, 2005). Some examples of entomopathogen combinations that have been reported to be synergistic include Bt combined with entomopathogenic nematodes for scarab grub control (Koppenhöfer and Kaya, 1997), and entomopathogenic nematodes combined with *M. anisopliae* against the white grub *Hoplia philanthus* (Ansari *et al.*, 2006). Examples of antagonism among entomopathogens include the microsporidium *V. necatrix* combined with *Heliothis* NPV (except at very high concentrations of the microsporidium) (Fuxa, 1979), combinations of Bt and *Anagrapha falcifera* multiple NPV against three lepidopteran corn pests (*H. zea, O. nubilalis, Spodoptera frugiperda*) (Pingel and Lewis, 1999), and the bacterium *S. marcescens* combined with entomopathogenic nematodes when targeting the pecan weevil, *Curculio caryae* (Shapiro-Ilan *et al.*, 2004c).

Entomopathogens may also be combined with chemical agents to enhance microbial control efficacy (Benz, 1971). Similar to combination with biotic agents, combination with abiotic agents can vary and result in synergy, additivity, or antagonism, and the nature of the interaction can depend on various factors including application parameters such as timing and rate (Benz, 1971; Koppenhöfer and Grewal, 2005). Positive interactions include the addition of optical brighteners to NPVs. The brighteners not only provide UV protection (as mentioned above), but also have been shown to enhance the virulence of NPV (Shapiro and Argauer, 1995; Shapiro, 2000; Boughton et al., 2001). The combination of NPV and the neem compound azadirachtin reduced time to death of *Spodoptera litura* larvae (Nathan and Kalaivani, 2005) most likely owing to synergistic effect on gut enzymic activity (Nathan et al., 2005). Foliar applications of the chemical insecticide carbofuran to corn plants treated with *B. bassiana* reduced tunneling of *O. nubilalis* (Lewis et al., 1996), and the LC_{50} value for *N. pyrausta*-infected *O. nubilalis* larvae fed Bt was significantly lower than for uninfected larvae (Pierce et al., 2001).

The impact on efficacy varies depending on the specific chemical agent and target pest (Benz, 1971; Koppenhöfer and Grewal, 2005). Granulovirus infection of *S. litura* was synergistic with chlorpyrifos, additive with fenvalerate and endosulfan, and antagonistic with cartap hydrochloride (Subramanian et al., 2005). Imidacloprid was reported to be synergistic with nematodes used against white grubs (Koppenhöfer and Kaya, 1998; Koppenhöfer et al., 2000) or with *B. bassiana* against *D. abbreviatus* (Quintela and McCoy, 1998), but antagonistic when combined with *B. bassiana* against *B. argentifolii* (James and Elzen, 2001).

Entomopathogen efficacy may also be enhanced through combination with physical agents. For example, a synergist effect of diatomaceous earth combined with *B. bassiana* has been observed with a number of coleopteran stored grain pests (Lord, 2001; Akbar et al., 2004; Athanassiou and Steenberg, 2007) as well as with the microsporidium *T. solenopsae* infecting the red imported fire ant (Brinkman and Gardner, 2001). While the exact details of the interaction between diatomaceous earth and pathogens are currently unclear, it appears to involve a combination of increased availability of water and other nutrients, removal or mitigation of inhibitory materials, alteration of adhesive properties, and physical disruption of the cuticular barrier (Akbar et al., 2004).

Improving the Environment

Manipulation of the environment at the target site can increase microbial control efficacy through a variety of mechanisms such as decreasing exposure to harmful biotic or abiotic factors or enhancing entomopathogen reproduction, virulence, and exposure to the host. Thus, the persistence of efficacy of entomopathogenic nematodes can be enhanced through the addition of soil amendments such as mulch or crop residues (Shapiro et al., 1999b; Lacey et al., 2006b). Addition of compost also enhanced the persistence of *B. bassiana* (Rosin et al., 1996, 1997).

Various cultural practices can also affect entomopathogen efficacy within a cropping system. Tillage or the movement of cattle can enhance NPV efficacy by increasing the amount of virus on the host plant (Fuxa, 1987). Narrowing soybean rows to increase relative humidity enhanced the activity of *N. rileyi* (Sprenkel et al., 1979). Gaugler et al. (1989b) observed enhanced persistence of *B. bassiana* in soil that was tilled versus untilled. In contrast, detrimental effects of tillage on the efficacy or persistence of hypocrealean fungi, such as *B. bassiana*, have been observed in other studies (Sosa-Gomez and Moscardi, 1994; Hummel et al., 2002; Shapiro-Ilan et al., 2008d). These discrepancies emphasize the need to test environmental manipulation approaches in a variety of cropping systems as their impact can vary based on biotic and abiotic factors.

3.3.4. Approaches to Microbial Control

Similar to biological control approaches that are defined for the use of insect predators and parasitoids (and as categorized in Chapter 1), microbial control can be classified into four approaches: classical, inoculation, inundation, and conservation. In the classical approach (also termed "introduction and establishment" approach), entomopathogens are released into areas where they do not occur naturally in an attempt to control a targeted pest (or conceivably, a complex of pests). The target pest can be endemic or exotic to the area; if it is the latter, the introduced entomopathogen might ideally be from the putative center of origin of the invasive pest. The ultimate goal is the establishment of the pathogen or pathogens for the total or partial suppression of the pest on a long-term basis. Optimally, introduction and establishment of exotic entomopathogens result in the reduction of the pest below the EIL. The pathogen should have a narrow host range, preferably specific to the pest, and have little or no impact on beneficial organisms. Several entomopathogens have been used as classical biological control agents, including viruses, bacteria, fungi, and nematodes. Classical biological control can also include exotic pathogens that have been accidentally introduced or are pathogens of unknown origin.

Unlike classical microbial control, pathogens that are released using the inoculation or inundation approaches are not expected to become permanently established. In inoculative microbial control, some recycling of the entomopathogen is expected; thus, seasonal or in some cases multiyear pest suppression may occur before reapplication is required. The inundative approach can be likened to a "pesticidal" approach; it is intended for short-term pest suppression and little or no recycling is expected. This approach may require repeated applications depending on the number of generations of the targeted insect and the duration of the stages that are injurious.

Insect pathogens are ubiquitous in nature, but endemic pathogens are frequently insufficient to keep pests below the EIL. Many may cause epizootics in pest populations, but often when the host population density is high and after the EIL has been surpassed. However, there are also occurrences of natural epizootics that hold pests in check, allowing the delay or avoidance of pesticide applications (Steinkraus, 2007a). Conservation microbial control relies on conserving or enhancing the activity of entomopathogens that occur naturally in the pest's habitat. In this approach, pathogen species are not added directly to the system. Rather, they are conserved or enhanced through agricultural and environmental practices that favor their survival or efficacy. Such practices could include selective timing or application of chemical pesticides (e.g., proper timing or reduction of pesticides that are inimical to the pathogen), or environmental modification, such as increased irrigation to provide moisture for some pathogens (e.g., fungi, entomopathogenic nematodes), or reduced use of conventional plowing. Under optimal environmental conditions, many entomopathogens have the natural ability to cause disease at epizootic levels. Several examples of conservation of naturally occurring entomopathogens are presented by Steinkraus (2007a), Cory and Evans (2007), and Elkinton and Burand (2007).

Case Studies: Classical Biological Control

One of the most successful case studies on the introduction and establishment of a microbial control agent is the virus of the coconut rhinoceros beetle, *Oryctes rhinoceros* (Huger, 2005). In 1966, Huger described a non-occluded virus of *O. rhinoceros* from Malaysia that demonstrated potential for long-term control of the beetle, a serious pest of oil and coconut palm (Huger, 1966, 2005). The beetles are infected through oral contact with the virus and subsequently serve as reservoirs and disseminators. Although there are no external symptoms of the disease in adults, the virus shortens the insect's lifespan and reduces fecundity (Zelazny, 1973). Transmission to larvae occurs when virus-infected females defecate in breeding sites during oviposition (Zelazny, 1972, 1973). Infection of larvae is always lethal (Zelazny, 1972). Pheromone lures and other methods have been used to capture, infect, and release beetles to disseminate the virus further (Lomer, 1986; Young, 1986). Introduction of the virus in conjunction with cultural practices such as the removal of larval habitats (e.g., rotting palm logs and the like) has significantly reduced *O. rhinoceros* populations (Huger, 2005). One of the key factors responsible for success of the virus is the persistence of the virus in adult and larval habitats.

Adults also serve as a reservoir of the virus. However, a ban on burning of palm logs and the recent invasion of other islands have resulted in the resurgence of the beetle in some locations (Jackson *et al.*, 2005).

The European spruce sawfly, *Diprion hercyniae*, is an invasive pest of spruce in North America. An NPV found infecting *D. hercyniae* was accidentally introduced from Europe into Canada (Bird, 1955). When the NPV appeared in New Brunswick in 1938, disease spread throughout most of the area infested by *D. hercyniae* and by 1942 it was considered to be the major factor in the collapse of the outbreak (Bird and Elgee, 1957). Natural epizootics of the NPV have since kept populations of the sawfly under control in most locations. The virus was also introduced into a moderately infested area, near Sault Ste. Marie, where it became established and spread rapidly (Bird and Burk, 1961). Virus epizootics recurred each year and have prevented excessive increases in sawfly populations.

Another example of long-term control produced by an introduced viral pathogen is that of the NPV of the European pine sawfly, *Neodiprion sertifer*, which is an exotic pest of various pine varieties. The virus was isolated from *N. sertifer* in Canada by Bird and Whalen (1953). Isolates of the virus were also introduced from Sweden (Bird, 1953). Several field trials using aerial and mist blower applications of the imported virus produced mortalities exceeding 90%. Bird (1953) concluded that without treatment, defoliation would have been almost complete. These and other small-scale applications produced epizootics in *N. sertifer* populations that resulted in suppression of the pest below the EIL (Bird, 1950, 1953, 1955).

Introductions of fungal entomopathogens and microsporidia for classical biological control outnumber those of other entomopathogens. Among 136 programs using different groups of arthropod pathogens, 49% have introduced fungal pathogens (Hajek and Delalibera, 2010). The introduction of the fungus *E. maimaiga* for control of *L. dispar* is one of the most successful. *Lymantria dispar* was accidentally introduced into the north-east USA from Europe in the late 1860s and has steadily spread into other areas (Liebhold *et al.*, 1993). Although the fungal pathogen (concluded to be *E. maimaiga*) was obtained from gypsy moth in Japan and released at several locations in the Boston area in 1910−1911, no transmission was detected (Hajek, 1999). The first reports of epizootics caused by the fungus were published by Andreadis and Weseloh (1990) and Hajek *et al.* (1990). Since then, *E. maimaiga* has been observed in the north-eastern USA, producing dramatic epizootics in *L. dispar* (Elkinton and Burand, 2007). Entomophaga *maimaiga* is expanding its geographical range naturally but also with human assistance (Elkinton *et al.*, 1991; Hajek *et al.*, 1995). Smitley *et al.* (1995) reported on the distributions of resting spores of the fungus in Michigan; fungus-infected larvae were observed two years after inoculation. Epizootics in *L. dispar* induced by *E. maimaiga* resulted in up to 99% mortality and were documented at the same sites three years after inoculation. Resting spores survive under the bark of trees and surrounding soil where epizootics have occurred and are capable of producing disease in *L. dispar* for several years (Hajek *et al.*, 1998).

The Sirex woodwasp, *Sirex noctilio*, was accidentally introduced from Europe into New Zealand pine plantations (*Pinus* spp.). It is now one of the most important pests of pine in the southern hemisphere and threatens pine production on eight million ha in Australia, New Zealand, South America, and South Africa (Bedding and Iede, 2005; Carnegie *et al.*, 2005). In addition to damage due to feeding by larval *S. noctilio*, the ovipositing female introduces a tree pathogenic fungus, *Amylostereum areolatum*, which can result in death of the tree. It is the most important insect pest of *Pinus radiata* in Australia. Consequently, the first concerted use of biological control agents (parasitoids and nematodes) in an integrated program for control of the wasp was in Australia. The entomogenous nematode, *Deladenus* (= *Beddingia*) *siricidicola*, was imported from Hungary and, after preliminary trials in northern Tasmania, introduced into the state of Victoria in the early 1970s (Bedding and Akhurst, 1974). The nematode has two separate life cycles, one parasitic and the other free living (Bedding, 1967, 1972). Parasitism by the nematode sterilizes female wasps. Rearing of the infective stage of *D. siricidicola* on *A. areolatum* cultures enabled large-scale application to infested plantations (Bedding and Iede, 2005). Successful dispersal of the nematode to new sites was facilitated by infected females wasps ovipositing eggs filled with up to 200 juvenile nematodes (Bedding, 1972). The initial success of the program provided up to 100% infection in *S. noctilio* and a decline in wasp populations and tree damage in treated plantations (Bedding and Iede, 2005). However, subsequent outbreaks of the wasp in Australian pine plantations where the nematode had not been introduced resulted in severe tree damage (Haugen and Underdown, 1990). Subsequent introductions of *D. siricidicola* provided effective control of the wasp (Haugen and Underdown, 1990).

Following the introduction of *D. siricidicola* into Australia, this nematode has been introduced into New Zealand, South Africa, and Brazil. Although establishment and good control of *S. noctilio* have been reported in New Zealand (Zondag, 1979) and South Africa (Tribe and Cillie, 2004), variability in the establishment and success of the biological control agent has been reported in South Africa (Hurley *et al.*, 2008) and Brazil (Fenili *et al.*, 2000; Penteado *et al.*, 2008). Decreased parasitism has been ascribed to strain deterioration, incorrect inoculation procedures, climate, inability to adapt to different environments, and different populations of *S. noctilio* (Haugen

and Underdown, 1993; Bedding and Iede, 2005; Hurley *et al.*, 2008). Despite the variable successes, overall *D. siricidicola* is still regarded as the most important means of controlling *S. noctilio* (Bedding and Iede, 2005). In the absence of control agents, principally *D. siricidicola*, the wasp has the potential to cause 16—60 million US dollars of damage each year (Bedding and Iede, 2005).

Introduction and establishment of *S. scapterisci* in populations of invasive mole crickets, *Scapteriscus* spp., were reported by Hudson *et al.* (1988). The exotic *Scapteriscus* spp., a serious pest of lawn and turf, arrived in Florida from South America. The nematode was collected in Uruguay and introduced into Florida in small plot field tests (Hudson *et al.*, 1988). After demonstrating effective control and persistence, it was introduced into several locations around the state. *Scapteriscus scapterisci* has become established in most of these locations and is dispersing from sites where applications were made (Parkman *et al.*, 1993, 1996).

Case Studies: Inoculation

Klein (1992) and Klein *et al.* (2007) reported on the inoculation of *Paenibacillus* spp. into turf and lawn habitats. Inoculation produced localized epizootics and often resulted in persistence of the pathogen with periodic outbreaks of disease. The most widely used species is *P. popilliae* for control of *P. japonica* (Klein, 1992). The bacterium is an obligate pathogen of the beetle and must be ingested and subsequently gain access to the hemocoel. Mortality is due to bacteremia rather than toxemia (Garczynski and Siegel, 2007). Feeding or injecting spores into larvae is necessary for production of the spores that will be used for inoculative application. Production on artificial media has consistently failed to produce spores. *Paenibacillus popilliae* spores plus a carrier, such as talc, are typically applied to the surface of lawns (Klein *et al.*, 2007). Infected larvae turn white, a characteristic sign of the disease referred to as milky disease or milky spore. The spores may persist for years, and at unpredictable intervals cause epizootics. Naturally occurring epizootics of milky disease have been reported (Klein, 1992), as has failure of the bacterium to control *P. japonica* (Redmond and Potter, 1995).

Klein and Georgis (1992) demonstrated that applications of *S. carpocapsae* and *H. bacteriophora* for control of *P. japonica* produced an inoculative effect and resulted in 51% and 60% mortality, respectively, a month following application, and 90% and 96% mortality, respectively, the following spring. *Heterorhabditis bacteriophora*, but not *S. carpocapsae*, continued to control *P. japonica* larvae until the following autumn, resulting in up to 99% mortality. Their findings document reproduction of the nematodes in *P. japonica* larvae, survival in the field, and continuation of control for a longer period than previously demonstrated for inundative control of scarabs. For entomopathogenic nematodes, this level of recycling is an exception and may have been due to the high population density of *P. japonica* and favorable environmental conditions.

Case Studies: Inundation

Bacillus thuringiensis is by far the most widely used inundative microbial control agent for control of insect pests of annual and perennial crops, forests, and pests of humans and domestic animals (Beegle and Yamamoto, 1992; Gelernter and Lomer, 2000; Lacey *et al.*, 2001). A multitude of case histories is present in the literature (Beegle and Yamamoto, 1992). As with viruses, the active moieties of the bacterium, the so-called delta endotoxins, must be ingested in order to be larvicidal. Various toxins have been isolated from the delta-endotoxin, each of which has specific activity for a certain groups of insects (Garczynski and Siegel, 2007; Crickmore *et al.*, 2011). *Bacillus thuringiensis* subspecies *kurstaki* (Btk) has been commercially produced for several decades for control of lepidopteran pests (Beegle and Yamamoto, 1992). Typically, it is applied to pest populations using conventional ground and aircraft spray equipment (Hall and Menn, 1999). In general, the bacteria are applied at regular intervals owing to rapid degradation due to UV radiation. The host range and larvicidal activity of several varieties and toxins are presented by Garczynski and Siegel (2007) and Crickmore *et al.* (2011). In addition to UV radiation, a potential limiting factor is the development of resistance. Resistance to Btk by lepidopterans was reviewed by Shelton *et al.* (2007), who also present strategies to manage resistance to Bt toxins in various orders of insects.

The host range of *B. thuringiensis* subspecies *tenebrionis* (Btt) is considerably narrower than that of other commercially produced Bt subspecies. It was discovered in Germany by Langenbruch *et al.* (1985) and has been used principally against the Colorado potato beetle, *Leptinotarsa decemlineata*, in potato-growing regions worldwide. However, with the advent of a number of new pesticide chemistries for *L. decemlineata* control, commercial success of Btt has lagged considerably behind that of other Bt subspecies (Gelernter and Lomer, 2000).

After its discovery in Israel (Goldberg and Margalit, 1977), *B. thuringiensis* subspecies *israelensis* (Bti) was rapidly developed by the biopesticide industry (Margalit and Dean, 1985). Larval control of a multitude of mosquito species using Bti has been demonstrated worldwide (Lacey and Undeen, 1986; Lacey, 2007). Bti is applied using a variety of conventional aerial and ground spray equipment and formulations. Flowable concentrates, granules and slow-release formulations are used to target mosquito habitats. Certain mosquitoes such as *Anopheles* spp. are particularly difficult to target because they are water surface feeders and most Bti formulations rapidly sink from their feeding zone. The most economical and efficacious means of application of Bti to large mosquito habitats uses the Beecomist® ultralow volume (ULV) sprayer by air (Sandoski *et al.*, 1985). The 80 μm droplets generated by the ULV sprayer tend to float on the water surface and within the feeding zone of *Anopheles* spp.

Bti is also used for effective control of black fly larvae (Molloy, 1982; Lacey and Undeen, 1986; Merritt *et al.*, 1989). Its use in the Onchocerciasis Control Program in West Africa helped to control populations of *Simulium damnosum* that had become resistant to organophosphate and carbamate insecticides. Rapid development to insecticide resistance in non-resistant black fly populations was prevented by alternating Bti with chemical larvicides (Guillet *et al.*, 1990).

Another bacterium, *B. sphaericus*, has also been used for control of mosquito larvae, principally in organically enriched habitats (Lacey, 2007). Its larvicidal activity in these habitats is significantly prolonged relative to that of Bti. It is only effective for control of mosquitoes and, relative to Bti, it has a considerably narrower range of activity, primarily in the Culicidae. For example, it is very effective against *Culex quinquefasciatus*, an important vector of the filaroid nematodes that cause elephantiasis, but inactive against *Ae. aegypti*, the vector of dengue and yellow fever viruses. Another potential caveat to its use as an exclusive larvicide over longer periods is the development of extremely high levels of resistance (Rao *et al.*, 1995; Nielsen-LeRoux *et al.*, 2002; Mulla *et al.*, 2003).

Control of codling moth, *Cydia pomonella*, with its host-specific granulovirus (CpGV) is an excellent example of inundative control using an entomopathogenic virus (Fig. 3.4A). CpGV is one of the most widely used and successful baculoviruses. It is commercially produced in Europe and North America and used by both organic and conventional orchardists. Several locations where CpGV has been used worldwide are cited by Lacey *et al.* (2008a). Although CpGV can be used as a standalone means of control it must be applied at seven- to 14-day intervals to provide effective suppression, especially if there are multiple generations of the moth each year (Arthurs and Lacey, 2004; Arthurs *et al.*, 2005). The principal limiting factor of CpGV is UV radiation (Lacey *et al.*, 2008a). Another severely limiting factor is the development of CpGV resistance in some populations in Western Europe, where it has been used for 20 years or more as the principal means of *C. pomonella* control (Eberle and Jehle, 2006). An integrated approach where CpGV is used in conjunction with other means of control could forestall development of resistance and provide a broader range of control for *C. pomonella* and other orchard pest insects (Lacey and Shapiro-Ilan, 2008).

A uniquely successful inundative program using an NPV for control of *A. gemmatalis* in soybean is described by Moscardi (1999, 2007). Moscardi and colleagues

FIGURE 3.4 Microbial control applications in apple and pear orchards targeting codling moth, *Cydia pomonella*. (A) Application of codling moth granulovirus (CpGV) targeted for neonate larvae in an apple orchard using a typical sprayer for orchard applications. Very small droplets that provide general coverage of trees are applied using an airblast sprayer outfitted with small aperture nozzles and high pressure (1550 kPa) at a rate of $0.5-1.0 \times 10^{12}$ virus occlusion bodies in 1000 liters of water/ha. (B) The same type of sprayer modified for application of *Steinernema feltiae* in a spray stream that drenches the trees from the base into the scaffold branches for control of cocooned prepupae in a pear orchard. The sprayer is configured with large-aperture nozzles, low pressure (700 kPa), screens and swirl plates removed and delivered at a rate of 10^6 infective juveniles in 3740 liters of water/ha. The application procedures for CpGV in (A) and nematodes in (B) would be the same regardless of fruit variety. [*Photo credits: (A) S. Arthurs, (B) L. A. Lacey.*]

discovered, developed, and implemented the virus in a large-scale control program. At present, the NPV is used on approximately two million ha of soybeans in Brazil, representing the largest program worldwide for the use of an entomopathogen to control a pest in a single crop. Following the development of an efficient means of *in vivo* production (Moscardi *et al.*, 1997), farmer cooperatives were instructed on the methods for virus production and began producing the virus for use on their crops (Moscardi,

2007). Several other entomopathogenic GVs and NPVs are used for inundative control of lepidopterous pests (Hunter-Fujita *et al.* 1998; Moscardi, 1999; van Frankenhuyzen *et al.*, 2007). Some NPVs also provide inundative control of a limited number of hymenopteran forest pests (Moreau and Lucarotti, 2007).

Unlike virus and bacteria, fungi gain entry to the host through the integument (Hajek and St. Leger, 1994). This makes them especially valuable as microbial control agents for insects with piercing and sucking mouthparts such as aphids, whiteflies, psyllids, and other hemipterans (Wraight *et al.*, 2007, 2009; Goettel *et al.*, 2010). Many fungal species in the Hypocreales are pathogenic to a broad range of insect pests and are the main contenders among fungi for inundative control and commercial development. Of these, *B. bassiana*, *Isaria fumosorosea*, and *Metarhizium* spp. are the most widely used for insect and mite control (Alves, 1998; Goettel *et al.*, 2010). Although less commonly used, *Lecanicillium* spp. have been developed as effective means of inundative control for insect pests such as aphids and whiteflies in greenhouse crops.

A multitude of case studies of Hypocreales species used as inundatively applied microbial control agents is found throughout the literature. Field and greenhouse applications of Hypocreales are reported for control of whiteflies, aphids, psyllids, thrips, beetles, and other insect pests (Burges, 2007; Goettel *et al.*, 2010; Lacey *et al.*, 2011). For instance, successful control of *B. tabaci* and *B. argentifolii* with *B. bassiana* has been achieved in various places in the USA (Wraight *et al.*, 2000; Faria and Wraight, 2001; Lacey *et al.*, 2008b). One of the most successful large-scale inundative microbial control efforts using a fungus is the LUBILOSA program in West Africa (Moore, 2008) (see Chapter 6). Aerial application of an oil-based formulation of *M. acridum* (as *M. flavoviride*) provided effective control of locusts and grasshoppers despite hot, dry conditions (Lomer *et al.*, 1997, 1999; Bateman, 2004).

The predominant use of entomopathogenic nematodes (*Steinernema* and *Heterorhabditis* spp.) in microbial control is via inundation (Grewal *et al.*, 2005a; Georgis *et al.*, 2006). The inundative approach is taken as the nematodes usually do not recycle at levels that provide continual host suppression. Exceptions to this are presented above in the "Case Studies: Inoculation" and "Case Studies: Classical Biological Control" sections of this chapter. The nematodes are most effective in soil and cryptic habitats where the infective stage of the nematode will not desiccate before penetrating a host insect. One well-documented example of successful inundative application is the use of *S. riobrave* for control of *D. abbreviatus* (McCoy *et al.*, 2002, 2007; Stuart *et al.*, 2008), with high levels of mortality (e.g., 90% or greater) (Shapiro-Ilan *et al.*, 2002a, 2005b). The use of irrigation systems or herbicide boom sprayers for application of *S. riobrave* has been effective in delivering IJs into the zone below trees where larvae enter the soil. Other entomopathogenic nematode species attacking *D. abbreviatus*, with rates and percentage mortality, were summarized by Shapiro-Ilan *et al.* (2002a). Examples of nematode evaluation and efficacy in lawns and turf are presented by Klein and Georgis (1992), Grewal *et al.* (2005b), and Klein *et al.* (2007), and against several pest insects in orchard habitats (Fig. 3.4B) by Shapiro-Ilan *et al.* (2005b), Lacey *et al.* (2006a), and Lacey and Shapiro-Ilan (2008). Numerous other examples of insect control using entomopathogenic nematodes are provided by Grewal *et al.* (2005a) and Georgis *et al.* (2006).

Case Studies: Conservation

One of the most notable cases of conservation microbial control is the delay in insecticide application in order to promote epizootics caused by *Neozygites fresenii* in populations of the cotton aphid, *Aphis gossypii* (Steinkraus, 2007b; Abney *et al.*, 2008). Abney *et al.* (2008) showed that fungal epizootics caused by *N. fresenii* reduced aphid numbers below the EIL in 1999, 2000, and 2001, and occurred consistently in early to mid-July in their three-year study. The key factor in the success of this program is the prediction of epizootics caused by the fungus, and knowing when to advise cotton farmers to delay application (Hollingsworth *et al.*, 1995; Steinkraus *et al.*, 1996). The program was so successful that it expanded to include Louisiana and Mississippi (Steinkraus *et al.*, 1998).

Another example of the effect that agricultural practices can have on entomopathogen survival is presented by Hummel *et al.* (2002). Two tillage types (conventional plow and disk versus conservation strip tillage), two input approaches (chemical versus biologically based), and two cropping schedules (continuous versus rotation) were compared. A bait-trap bioassay (using *Galleria mellonella*) was used to monitor the abundance of *S. carpocapsae*, *H. bacteriophora*, *B. bassiana*, and *M. anisopliae* populations. Entomopathogens were significantly higher in conservation compared with conventional tillage systems. Pesticide use significantly reduced the detection of fungal entomopathogens. Ground cover (rye mulch and clover intercrop) that resulted in lower temperatures positively affected the abundance of *S. carpocapsae*. Hummel *et al.* (2002) concluded that although the type of tillage was the key factor affecting the abundance of entomopathogens, its benefit could be negated by pesticide use and ground cover that resulted in high temperatures.

3.4. FUTURE RESEARCH DIRECTIONS

Although substantial progress has been made in the discipline of epizootiology, additional advances and broader integration of epizootiology into other components of invertebrate pathology are needed. More research is needed in all aspects of epizootiology, including factors that impact epizootics and epizootiological modeling. Owing to the complex nature of epizootiology, multidisciplinary research is likely to be particularly fruitful. Towards this end, collaboration among specialists from various fields, including pathology, microbiology, physiology, and ecology (quantitative, behavioral, etc.), is to be encouraged. Incorporation of molecular tools in epizootiological studies related to microbial control will allow disease prevalence or history and pathogen movement to be tracked, and gene flow among pathogens and host populations to be studied. Furthermore, the discipline of epizootiology can be advanced through integration and emphasis in graduate and undergraduate curricula; epizootiology can be taught as a standalone course or incorporated into existing courses such as insect pathology, entomology, biological control, general biology, zoology, and ecology. Expansion and improved understanding in epizootiology will lead to enhanced microbial control efforts.

As illustrated in this chapter, there have been numerous successes in microbial control, yet there have also been many failures. The definition of success is debatable. For the purposes of this discussion, the criterion for success in microbial control can be considered commercial application of the entomopathogen as a suppressive agent on a significant scale. The key elements for achieving success can be boiled down to two factors: appropriate match of the entomopathogen to the target pest, and cost competitiveness (Shapiro-Ilan *et al.*, 2002a). To constitute an appropriate match, the entomopathogen must possess sufficient virulence, infection, dispersal capabilities, environmental tolerance, niche overlap (with the target), and other traits that facilitate an acceptable level of control. Although some attributes of the host—pathogen match can be estimated in the laboratory, the bottom line for determining an appropriate match is establishment of efficacy in the pest's habitat. Regardless of the level of efficacy achieved, if the microbial control application is not economically viable relative to other tactics then implementation is destined to fail. The economics of a putative control approach depends on a number of factors, including the value of the crop (a higher crop value being more favorable to microbial control approaches), the proportion of the target area that must be treated, and the ease of delivery or compatibility with existing practices.

If a microbial control tactic is not successful, elements for success might be reached using approaches discussed in the sections above, e.g., by improving production, formulation, or delivery mechanisms. Strain discovery and improvement techniques can also substantially enhance microbial control potential. Additional research is needed to develop new techniques or approaches for the improvement of microbial control agents and their application.

The greatest barrier to expanded implementation of microbial control is a lack of economic feasibility. Thus, research is required to facilitate the use of microbial control agents in a cost-effective manner. A larger emphasis on conservation approaches can augment microbial control usage without necessarily increasing the costs of crop production. In addition, the integration of microbial control tactics into more holistic IPM programs may allow for improved cost competitiveness. A microbial control agent that is incorporated into a multitactic approach may be more cost-effective than a standalone pest control solution, e.g., the microbial could be combined directly with other tactics (in a synergistic approach), or might account for partial suppression of the pest as part of a multiapplication or multistage approach.

Finally, to expand the success of microbial control, translational research must be promoted. Advances in applied microbial control can be made through greater linkage with basic research. Fundamental research in insect pathology, such as on host–pathogen relationships at the molecular and organismal level, ecological relationships, physiology, etc., is the foundation for microbial control and serves as a basis for the future growth and development of the discipline.

REFERENCES

Abney, M. R., Ruberson, J. R., Herzog, G. A., Kring, T. J., Steinkraus, D. C., & Roberts, P. M. (2008). Rise and fall of cotton aphid (Hemiptera: Aphididae) populations in southeastern cotton production systems. *J. Econ. Entomol., 101*, 23–35.

Abu Hatab, M., & Gaugler, R. (1999). Lipids of *in vivo* and *in vitro* cultured Heterorhabditis bacteriophora. *Biol. Control, 15*, 113–118.

Akbar, W., Lord, J. C., Nechols, J. R., & Howard, R. W. (2004). Diatomaceous earth increases the efficacy of *Beauveria bassiana* against *Tribolium castaneum* larvae and increases conidia attachment. *J. Econ. Entomol., 97*, 273–280.

Alves, S. B. (1998). Fungos entomopatogênicos. In *Controle Microbiano de Insetos*, (2nd ed.). (S.B. Alves, Ed.), (pp. 289–381). Bibioteca de Ciências Agrárias Luis de Queiroz, No. 4, Piracicaba.

Anderson, R. M., & May, R. M. (1980). Infectious diseases and population cycles of forest insects. *Science, 210*, 658–661.

Anderson, R. M., & May, R. M. (1981). The population dynamics of microparasites and their invertebrate hosts. *Philos. Trans. R. Soc. Lond., 291*, 451–524.

Anderson, R. M., & May, R. M. (1982). Theoretical basis for the use of pathogens as biological control agents of pest species. *Parasitology, 84*, 3–33.

Andreadis, T. G. (1985). Experimental transmission of a microsporidan pathogen from mosquitos to an alternate copepod host. *Proc. Natl. Acad. Sci. USA, 82*, 5574–5577.

Andreadis, T. G. (1986). Dissemination of *Nosema pyausta*, in feral populations of the European corn borer, *Ostrinia nubilalis. J. Invertebr. Pathol., 48*, 335–343.

Andreadis, T. G. (1987). Transmission. In J. R. Fuxa & Y. Tanada (Eds.), *Epizootiology of Insect Diseases* (pp. 159–176). New York: Wiley-Interscience.

Andreadis, T. G., & Weseloh, R. M. (1990). Discovery of *Entomophaga maimaiga* in North American gypsy moth, *Lymantria dispar. Proc. Natl. Acad. Sci. USA, 87*, 2461–2465.

Ansari, M. A., Shah, F. A., & Moens, T. M. (2006). Field trials against *Hoplia philanthus* (Coleoptera: Scarabaeidae) with a combination of an entomopathogenic nematode and the fungus *Metarhizium anisopliae* CLO 53. *Biol. Control, 39*, 453–459.

Aratchige, N. S., Lesna, I., & Sabelis, M. W. (2004). Below-ground plant parts emit herbivore-induced plant volatiles: olfactory responses of a predatory mite to tulip bulbs infested by rust mites. *Exp. Appl. Acarol., 33*, 21–30.

Arthurs, S. P., & Lacey, L. A. (2004). Field evaluation of commercial formulations of the codling moth granulovirus (CpGV): persistence of activity and success of seasonal applications against natural infestations in the Pacific Northwest. *Biol. Control, 31*, 388–397.

Arthurs, S., & Thomas, M. B. (2001). Behavioural changes in *Schistocerca gregaria* following infection with a fungal pathogen: implications for susceptibility to predation. *Ecol. Entomol., 26*, 227–234.

Arthurs, S. P., Lacey, L. A., & Fritts, R., Jr. (2005). Optimizing use of codling moth granulovirus: effects of application rate and spraying frequency on control of codling moth larvae in Pacific Northwest apple orchards. *J. Econ. Entomol., 98*, 1459–1468.

Asser-Kaiser, S., Fritsch, E., Undorf-Spahn, K., Kienzle, J., Eberle, K. E., Gund, N. A., Reineke, A., Zebitz, C. P., Heckel, D. G., Huber, J., & Jehle, J. A. (2007). Rapid emergence of baculovirus resistance in codling moth due to dominant, sex-linked inheritance. *Science, 317*, 1916–1918.

Athanassiou, C. G., & Steenburg, T. (2007). Insecticidal effect of *Beauveria bassiana* (Balsamo) Vuillemin (Ascomycota: Hypocreales) in combination with three diatomaceous earth formulations against *Sitophilus granarius* (L.) (Coleoptera: Curculionidae). *Biol. Control, 40*, 411–416.

Aylor, D. E. (1990). The role of intermittent wind in the dispersal of fungal pathogens. *Annu. Rev. Phytopathol., 28*, 73–92.

Ayres, M. D., Howard, S. C., Kuzio, J., Lopez-Ferber, M., & Possee, R. D. (1994). The complete DNA sequence of *Autographa californica* nuclear polyhedrosis virus. *Virology, 202*, 586–605.

Azusa, F., & Tomoko, Y. (1999). Effect of ultraviolet on insecticidal activity of the entomopathogenic nematode, *Steinernema kushidai*, sprayed on turfgrass. *J. Jpn. Soc. Turfgrass Sci., 27*, 147–151.

Bai, C., Shapiro-Ilan, D. I., Gaugler, R., & Hopper, K. R. (2005). Stabilization of beneficial traits in *Heterorhabditis bacteriophora* through creation of inbred lines. *Biol. Control, 32*, 220–227.

Barbercheck, M. E., & Kaya, H. K. (1991). Effect of host condition and soil texture on host finding by the entomogenous nematodes *Heterorhabditis bacteriophora* (Rhabditida: Heterorhabditidae) and *Steinernema carpocapsae* (Rhabditida: Steinernematidae). *Environ. Entomol., 20*, 582–589.

Bateman, R. (2004). Constraints and enabling technologies for mycopesticide development. *Outl. Pest Manag., 15*, 64–69.

Bathon, H. (1996). Impact of entomopathogenic nematodes on non-target hosts. *BioControl Sci. Technol., 6*, 421–434.

Baverstock, J., Alderson, P. G., & Pell, J. K. (2005). *Pandora neoaphidis* transmission and aphid foraging behaviour. *J. Invertebr. Pathol., 90*, 73–76.

Becnel, J. J., & Andreadis, T. G. (1999). Microsporida in insects. In M. Witter & L. M. Weiss (Eds.), *The Microsporidia and Micro-sporidiosis* (pp. 447–501). Washington, DC: ASM Press.

Bedding, R. A. (1967). Parasitic and free-living cycles in entomogenous nematodes of the genus, *Deladenus. Nature, 214*, 174–175.

Bedding, R. A. (1972). Biology of *Deladenus siricidicola* (Neo-tylenchidae), an entomophagous–mycetophagous nematode parasitic in Siricid woodwasps. *Nematologica, 18*, 482–493.

Bedding, R. A., & Akhurst, R. J. (1974). Use of the nematode *Deladenus siricidicola* in the biological control of *Sirex noctilio* in Australia. *J. Aust. Entomol. Soc., 13*, 129–135.

Bedding, R. A., & Iede, E. T. (2005). Application of *Beddingia sir-icidicola* for Sirex woodwasp control. In P. S. Grewal, R.-U. Ehlers & D. I. Shapiro-Ilan (Eds.), *Nematodes as Biological Control Agents* (pp. 385–399). Wallingford: CABI.

Beegle, C. C., & Yamamoto, T. (1992). History of *Bacillus thuringiensis* Berliner research and development. *Can. Entomol., 124*, 587–616.

Behle, R. W., Compton, D. L., Kenar, J. A., & Shapiro-Ilan, D. I. (2011). Improving formulations for biopesticides: enhanced UV protection for beneficial microbes. *J. ASTM Int., 8*, 137–157.

Benz, G. (1971). Synergism of micro-organisms and chemical insecti-cides. In H. D. Burgess & N. W. Hussey (Eds.), *Microbial Control of Insects and Mites* (pp. 327–355). London: Academic Press.

Benz, G. (1987). Environment. In J. R. Fuxa & Y. Tanada (Eds.), *Epizootiology of Insect Diseases* (pp. 177–199). New York: Wiley-Interscience.

Bianchi, F. J. J. A., Vlack, J. M., Rabbinge, R., & Van der Werf, W. (2002). Biological control of beet armyworm, *Spodoptera exigua*, with baculoviruses in greenhouses: development of a comprehensive process-based model. *Biol. Control, 23*, 35–46.

Bidochka, M. J., Kasperski, J. E., & Wild, G. A. M. (1998). Occurrence of the entomopathogenic fungi *Metarhizium anisopliae* and *Beauveria bassiana* in soils from temperate and near-northern habitats. *Can. J. Bot., 76*, 1198–1204.

Bilgrami, A. L., Gaugler, R., Shapiro-Ilan, D. I., & Adams, B. J. (2006). Source of trait deterioration in entomopathogenic nematodes *Heter-orhabditis bacteriophora* and *Steinernema carpocapsae* during *in vivo* culture. *Nematology, 8*, 397–409.

Bird, F. T. (1950). The dissemination and propagation of a virus disease affecting the European pine sawfly,. *Neodiprion sertifer (Geoff.). For. Insect Invest., 6*, 2–3.

Bird, F. T. (1953). The use of a virus disease in the biological control of the European pine sawfly,. *Neodiprion sertifer (Geoffr.). Can. Ento-mol., 85*, 437–446.

Bird, F. T. (1955). Virus diseases of sawflies. *Can. Entomol., 87*, 124–127.

Bird, F. T., & Burk, J. M. (1961). Artificially disseminated virus as a factor controlling the European spruce sawfly, *Diprion hercyniae* (Htg.) in the absence of introduced parasites. *Can. Entomol., 93*, 228–238.

Bird, F. T., & Elgee, D. E. (1957). A virus disease and introduced parasites as factors controlling the European spruce sawfly, *Diprion hercyniac* (Htg.), in central New Brunswick. *Can. Entomol., 89*, 371–378.

Bird, F. T., & Whalen, M. M. (1953). A virus disease of the European pine sawfly,. *Neodiprion sertifer (Geoffr.). Can. Entomol., 85*, 433–437.

Bischoff, J. F., Rehner, S. A., & Humber, R. A. (2006). *Metarhizium frigidum* sp. nov.: a cryptic species of *M. anisopliae* and a member of the *M. flavoviride* complex. *Mycologia, 98*, 737–745.

Bischoff, J. F., Rehner, S. A., & Humber, R. A. (2009). A multilocus phylogeny of the *Metarhizium anisopliae* lineage. *Mycologia, 101*, 512–530.

Blanford, S., & Thomas, M. B. (1999). Host thermal biology: the key to understanding insect–pathogen interactions and microbial pest control? *Agric. For. Entomol., 1*, 195–202.

Blanford, S., Thomas, M. B., & Langewald, J. (1998). Behavioural fever in the Senegalese grasshopper, *Oedaleus sengalensis*, and its impli-cations for biological control using pathogens. *Ecol. Entomol., 23*, 9–14.

Boemare, N. E., Boyer-Giglio, M.-H., Thaler, J.-O., & Akhurst, R. J. (1993). The phages and bacteriocins of *Xenorhabdus* sp., symbiont of the nematodes *Steinernema* spp. and *Heterorhabditis* spp. In R. Bedding, R. Akhurst & H. Kaya (Eds.), *Nematodes and the Biological Control of Insect Pests* (pp. 137–145). Victoria: CSIRO.

Boldo, J. T., Junges, A., Amaral, K. B., Staats, C. C., Vainstein, M. H., & Schrank, A. (2009). Endochitinase CH12 of the biocontrol fungus *Metarhizium anisopliae* affects its virulence toward the cotton strainer bug *Dysdercus peruvianus. Curr. Genet., 55*, 551–560.

Boots, M. (1998). Cannibalism and the stage-dependent transmission of a viral pathogen of the Indian meal moth. *Plodia interpunctella. Ecol. Entomol., 23*, 118–122.

Boughton, A. J., Lewis, L. C., & Bonning, B. C. (2001). Potential of *Agrotis ipsilon* nucleopolyhedrovirus for suppression of the black cutworm (Lepidoptera: Noctuidae) and effect of an optical brightener on virus efficacy. *J. Econ. Entomol., 94*, 1045–1052.

Bowen, D. J., Rocheleau, T. A., Blackburn, M., Andreev, O., Golubeva, E., Bhartia, R., & ffrench-Constant, R. H. (1998). Insec-ticidal toxins from the bacterium *Photorhabdus luminescens. Science, 280*, 2129–2132.

Braga, G. U. L., Flint, S. D., Miller, C. D., Anderson, A. J., & Roberts, D. W. (2001a). Both solar UVA and UVB radiation impair conidial curability and delay germination in the entomopathogenic fungus *Metarhizium anisopliae. Photochem. Photobiol. Sci., 74*, 734–739.

Braga, G. U. L., Flint, S. D., Messias, C. L., Anderson, A. J., & Roberts, D. W. (2001b). Effects of UVB irradiance on conidia and germinants of the entomopathogenic hyphomycete *Metarhizium anisopliae. Photochem. Photobiol., 73*, 140–146.

Braga, G. U. L., Flint, S. D., Messias, C. L., Anderson, A. J., & Roberts, D. W. (2001c). Effects of UV-B on conidia and germlings of the entomopathogenic hyphomycete *Metarhizium anisopliae. Mycol. Res., 105*, 874–882.

Braga, G. U. L., Flint, S. D., Miller, C. D., Anderson, A. J., & Roberts, D. W. (2001d). Variability in response to UV-B among species and strains of *Metarhizium* isolated from sites at latitudes from 61°N to 54°S. *J. Invertebr. Pathol., 78*, 98–108.

Brar, S., Verma, M., Tyagi, R. D., Valéro, J. R., & Surampalli, R. Y. (2009). Entomotoxicity, protease and chitinase activity of *Bacillus thuringiensis* fermented wastewater sludge with a high solids content. *Bioresour. Technol., 100*, 4317–4325.

Bravo, A., Gill, S. S., & Soberón, M. (2007). Mode of action of *Bacillus thuringiensis* Cry and Cyt toxins and their potential for insect control. *Toxicon., 49*, 423–435.

Brinkman, M. A., & Gardner, W. A. (2001). Use of diatomaceous earth and entomopathogen combinations against the red imported fire ant (Hymenoptera: Formicidae). *Fla. Entomol., 84*, 740–741.

Brooks, A., & Wall, R. (2005). Horizontal transmission of fungal infection by *Metarhizium anisopliae* in parasitic *Psoroptes* mites (Acari: Psoroptidae). *Biol. Control, 34*, 58−65.

Brown, G. C. (1987). Modeling. In J. R. Fuxa & Y. Tanada (Eds.), *Epizootiology of Insect Diseases* (pp. 43−68). New York: Wiley-Interscience.

Bruck, D. J. (2004). Natural occurrence of entomopathogens in Pacific Northwest nursery soils and their virulence to the black vine weevil, *Otiorhynchus sulcatus* (F.) (Coleoptera: Curculionidae). *Environ. Entomol., 33*, 1335−1343.

Bruck, D. J. (2005). Ecology of *Metarhizium anisopliae* in soilless potting media and the rhizosphere: implications for pest management. *Biol. Control, 32*, 155−163.

Bruck, D. J. (2007). Efficacy *of Metarhizium anisopliae* as a curative application for black vine weevil (*Otiorhynchus sulcatus*) infesting container-grown nursery crops. *J. Environ. Hort., 25*, 150−156.

Bruck, D. J. (2010). Fungal entomopathogens in the rhizosphere. *Bio Control, 55*, 103−112.

Bruck, D. J., & Lewis, L. C. (1999). *Ostrinia nubilalis* (Lepidoptera: Pyralidae) larval parasitism and infection with entomopathogens in cornfields with different border vegetation. *J. Agric. Urban Entomol., 16*, 255−272.

Bruck, D. J., & Lewis, L. C. (2002a). *Carpophilus freemani* (Coleoptera: Nitidulidae) as a vector of Beauveria bassiana. *J. Invertebr. Pathol., 80*, 188−190.

Bruck, D. J., & Lewis, L. C. (2002b). Rainfall and crop residue effects on soil dispersion and *Beauveria bassiana* spread to corn. *Appl. Soil. Ecol., 20*, 183−190.

Bruck, D. J., & Lewis, L. C. (2002c). Whorl and pollen-shed stage application of *Beauveria bassiana* for suppression of adult western corn rootworm. *Entomol. Exp. Appl., 103*, 161−169.

Bruck, D. J., Shapiro-Ilan, D. I., & Lewis, E. E. (2005). Evaluation of application technologies of entomopathogenic nematodes for control of the black vine weevil. *Otiorhynchus sulcatus. J. Econ. Entomol., 98*, 1884−1889.

Burd, A. D., Gould, F., Bradley, J. R., van Duyn, J. W., & Moar, W. J. (2003). Estimated frequency of nonrecessive Bt resistance genes in bollworm, *Helicoverpa zea* (Boddie) (Lepidoptera: Noctuidae), in eastern North Carolina. *J. Econ. Entomol., 96*, 137−142.

Burden, J. P., Nixon, C. P., Hodgkinson, A. E., Possee, R. P., Sait, S. M., King, L. A., & Hails, R. S. (2003). Covert infections as a mechanism for long-term persistence of baculoviruses. *Ecol. Lett., 6*, 524−531.

Burges, H. D. (Ed.). (1998). *Formulation of Microbial Biopesticides. Beneficial Microorganisms, Nematodes, and Seed Treatments*. Dordrecht: Kluwer Academic.

Burges, H. D. (2007). Techniques for testing microbials for control of arthropod pests in greenhouses. In L. A. Lacey & H. K. Kaya (Eds.), *Field Manual of Techniques in Invertebrate Pathology* (2nd ed.). (pp. 463−479) Dordrecht: Springer.

Cakmak, I., Ipek Ekmen, Z., Karagoz, M., Hazir, S., & Kaya, H. K. (2010). Development and reproduction of *Sancassania polyphyllae* (Acari: Acaridae) feeding on entomopathogenic nematodes and tissues of insect larvae. *Pedobiologia, 53*, 235−240.

Campbell, J. F., & Gaugler, R. (1997). Inter-specific variation in entomopathogenic nematode foraging strategy: dichotomy or variation along a continuum? *Fundam. Appl. Nematol., 20*, 393−398.

Campos-Herrera, R., Trigo, D., & Gutiérrez, C. (2006). Phoresy of the entomopathogenic nematode *Steinernema feltiae* by the earthworm *Eisenia fetida. J. Invertebr. Pathol., 92*, 50−54.

Campos-Herrera, R., Gómez-Ros, J. M., Escuer, M., Cuadra, L., Barrios, L., & Gutiérrez, C. (2008). Diversity, occurrence, and life characteristics of natural entomopathogenic nematode populations from La Rioja (northern Spain) under different agricultural management and their relationships with soil factors. *Soil Biol. Biochem., 40*, 1474−1484.

Canning, E. U. (1982). An evaluation of protozoal characteristics in relation to biological control of pests. *Parasitology, 84*, 119−149.

Cappaert, D. L., & Koppenhöfer, A. M. (2003). *Steinernema scarabaei*, an entomopathogenic nematode for control of the European chafer. *Biol. Control, 28*, 379−386.

Carnegie, A. J., Eldridge, R. H., & Waterson, D. G. (2005). History and management of sirex wood wasp in pine plantations in New South Wales. *Australia. N.Z.J. For. Sci., 35*, 3−24.

Castrillo, L. A., Bauer, L. S., Houping, L., Griggs, M. H., & Vandenberg, J. D. (2010). Characterization of *Beauveria bassiana* (Ascomycota: Hypocreales) isolates associated with *Agrilus planipennis* (Coleoptera: Buprestidae) populations in Michigan. *Biol. Control, 54*, 135−140.

Challacombe, J. F., Altherr, M. R., Xie, G., Bhotika, S. S., Brown, N., Bruce, D., Campbell, C. S., Campbell, M. L., Chen, J., Chertkov, O., Cleland, C., Dimitrijevic, M., Doggett, N. A., Fawcett, J. J., Glavina, T., Goodwin, L. A., Green, L. D., Han, C. S., Hill, K. K., Hitchcock, P., Jackson, P. J., Keim, P., Kewalramani, A. R., Longmire, J., Lucas, S., Malfatti, S., Martinez, D., McMurry, K., Meincke, L. J., Misra, M., Moseman, B. L., Mundt, M., Munk, A. C., Okinaka, R. T., Parson-Quintana, B., Reilly, L. P., Richardson, P., Robinson, D. L., Saunders, E., Tapia, R., Tesmer, J. G., Thayer, N., Thompson, L. S., Tice, H., Ticknor, L. O., Wills, P. L., Gilna, P., & Brettin, T. S. (2007). The complete genome sequence of *Bacillus thuringiensis* Al Hakam. *J. Bacteriol., 189*, 3680−3681.

Chambers, U., Bruck, D. J., Olsen, J., & Walton, V. M. (2010). Control of overwintering filbertworm (Lepidoptera: Tortricidae) larvae with *Steinernema carpocapsae. J. Econ. Entomol., 103*, 416−422.

Chandler, D., Hay, D., & Reid, A. P. (1997). Sampling and occurrence of entomopathogenic fungi and nematodes in UK soils. *Appl. Soil Ecol., 5*, 133−141.

Chapman, J. W., Williams, T., Escribiano, A., Caballero, P., Cave, R. D., & Goulson, D. (1999). Age-related cannibalism and horizontal transmission of a nuclear polyhedrosis virus in larval *Spodoptera frugiperda. Ecol. Entomol., 24*, 268−275.

Chappel, M. A., & Whitam, D. A. (1990). Grasshopper thermoregulation. In R. F. Chapman & A. Joern (Eds.), *Biology of Grasshoppers* (pp. 143−172). New York: Wiley Interscience.

Charnley, A. K. (2003). Fungal pathogens of insects: cuticle degrading enzymes and toxins. *Adv. Bot. Res., 40*, 241−321.

Chaston, J. M., Dillman, A. R., Shapiro-Ilan, D. I., Bilgrami, A. L., Gaugler, R., Hopper, K. R., & Adams, B. J. (2011). Outcrossing and crossbreeding recovers deteriorated traits in laboratory cultured *Steinernema carpocapsae* nematodes. *Int. J. Parasitol., 41*, 801−809.

Chen, X., Xiulian, S., Hu, Z., Li, M., Reilly, D. R. O., Zuidema, D., & Vlak, J. M. (2000). Genetic engineering of *Helicoverpa armigera* single-nucleocapsid nucleopolyhedrovirus as an improved pesticide. *J. Invertebr. Pathol., 76*, 140−146.

Cherry, A. J., Abalo, P., & Hell, K. (2005). A laboratory assessment of the potential of different strains of the entomopathogenic fungi

Beauveria bassiana (Balsamo) Vuillemin and *Metarhizium aniso-pliae* (Metschnikoff) to control *Callosobruchus maculatus* (F.) (Coleoptera: Bruchidae) in stored cowpea. *J. Stored Prod. Res., 41,* 295−309.

Chouvenc, T., & Su, N.-Y. (2010). Apparent synergy among defense mechanisms in subterranean termites (Rhinotermitidae) against epizootic events: limits and potential for biological control. *J. Econ. Entomol., 10,* 1327−1337.

Christe, P., Oppliger, A., Bancala, F., Castella, G., & Chapuisat, M. (2003). Evidence for collective medication in ants. *Ecol. Lett., 6,* 19−22.

Clarke, D. J. (2008). *Photorhabdus:* a model for the analysis of pathogenicity and mutualism. *Cell. Microbiol., 10,* 2159−2167.

Coates, B. S., Hellmich, R. L., & Lewis, L. C. (2002a). Allelic variation of a *Beauveria bassiana* (Ascomycota: Hypocreales) minisatellite is independent of host range and geographic origin. *Genome, 45,* 125−132.

Coates, B. S., Hellmich, R. L., & Lewis, L. C. (2002b). Nuclear small subunit rRNA group I intron variation among *Beauveria* spp. provide tools for strain identification and evidence of horizontal transfer. *Curr. Genet., 41,* 414−424.

Cory, J. S. (2003). Ecological impacts of virus insecticides: host range and non-target organisms. In H. Hokkanen & A. Hajek (Eds.), *Environmental Impacts of Microbial Insecticides* (pp. 73−91). Dordrecht: Kluwer Academic.

Cory, J. S., & Evans, H. F. (2007). Viruses. In L. A. Lacey & H. K. Kaya (Eds.), *Field Manual of Techniques in Invertebrate Pathology* (2nd ed.). (pp. 149−174) Dordrecht: Springer.

Cory, J. S., & Ericsson, J. D. (2010). Fungal entomopathogens in a tritrophic context. *BioControl, 55,* 75−88.

Cory, J. S., & Myers, J. H. (2003). The ecology and evolution of insect baculoviruses. *Annu. Rev. Ecol. Evol. Syst., 34,* 239−272.

Cossentine, J. E. (2009). The parasitoid factor in the virulence and spread of Lepidopteran baculoviruses. *Virol. Sin., 24,* 305−314.

Couch, T. L., & Ross, D. A. (1980). Production and utilization of *Bacillus thuringiensis. Biotechnol. Bioeng., 22,* 1297−1304.

Couteaudier, Y., Viaud, M., & Riba, G. (1996). Genetic nature, stability, and improved virulence of hybrids from protoplast fusion in *Beauveria. Microb. Ecol., 32,* 1−10.

Cottrell, T. E., & Shapiro-Ilan, D. I. (2003). Susceptibility of a native and an exotic lady beetle (Coleoptera: Coccinellidae) to *Beauveria bassiana. J. Invertebr. Pathol., 84,* 137−144.

Cottrell, T. E., & Shapiro-Ilan, D. I. (2008). Susceptibility of four species of North American Coccinellidae (Coleoptera) to *Beauveria bassiana. Eur. J. Entomol., 105,* 455−460.

Cottrell, T. E., Shapiro-Ilan, D. I., Horton, D. L., & Mizell, R. F. III (2011). Laboratory virulence and orchard efficacy of entomopathogenic nematodes toward the lesser peachtree borer (Lepidoptera: Sesiidae). *Environ. Entomol., 104,* 47−53.

Crickmore, N., Zeigler, D. R., Feitelson, J., Schnepf, E., Van Rie, J., Lereclus, D., Baum, J., & Dean, D. H. (2011). *Bacillus thuringiensis* toxin nomenclature. Available at: http://www.lifesci.sussex.ac.uk/Home/Neil_Crickmore/Bt/.

D'Amico, V., & Elkinton, J. S. (1995). Rainfall effects of transmission of gypsy moth (Lepidoptera Lymantriidae) nuclear polyhedrosis viruses. *Environ. Entomol., 24,* 1144−1149.

D'Amico, V., Elkinton, J. S., Dwyer, G., Burand, J. P., & Buonaccorsi, J. P. (1996). Virus transmission in gypsy moths is not a simple mass action process. *Ecology, 77,* 201−206.

Daoust, R. A., & Roberts, D. W. (1982). Virulence of natural and insect-passaged strains of *Metarhizium anisopliae* to mosquito larvae. *J. Invertebr. Pathol., 40,* 107−117.

Del Valle, E. E., Dolinksi, C., Barreto, E. L. S., & Souza, R. M. (2009). Effect of cadaver coatings on emergence and infectivity of the entomopathogenic nematode *Heterorhabditis baujardi* LPP7 (Rhabditida: Heterorhabditidae) and the removal of cadavers by ants. *Biol. Control, 50,* 21−24.

Deol, Y. S., Jagdale, G. B., Cañas, L., & Grewal, P. S. (2011). Delivery of entomopathogenic nematodes directly through commercial growing media via the inclusion of infected host cadavers: a novel approach. *Biol. Control, 58,* 60−67.

Dillon, A. B., Downes, M. J., Ward, D., & Griffin, C. T. (2007). Optimizing application of entomopathogenic nematodes to manage large pine weevil, *Hylobius abietis* L. (Coleoptera: Curculionidae) populations developing in pine stumps. *Pinus sylvestris. Biol. Control, 40,* 253−263.

Dolinski, C., & Lacey, L. A. (2007). Microbial control of arthropod pests of tropical tree fruits. *Neotrop. Entomol., 36,* 161−179.

Domnas, A. J. (1981). Biochemistry of *Lagenidium giganteum* infection in mosquito larvae. In E. W. Davidson (Ed.), *Pathogenesis of Invertebrate Microbial Diseases* (pp. 425−450). Totowa: Allanheld, Osmun.

Dromph, K. M. (2001). Dispersal of entomopathogenic hyphomycete fungi by collembolans. *Soil Biol. Biochem., 33,* 2047−2051.

Dromph, K. M. (2003). Collembolans as vectors of entomopathogenic fungi. *Pedobiologia, 47,* 245−256.

Duchaud, E., Rusniok, C., Frangeul, L., Buchrieser, C., Givaudan, A., Taourit, S., Bocs, S., Boursaux-Eude, C., Chandler, M., Charles, J. F., Dassa, E., Derose, R., Derzelle, S., Freyssinet, G., Gaudriault, S., Medigue, C., Lanois, A., Powell, K., Siguier, P., Vincent, R., Wingate, V., Zouine, M., Glaser, P., Boemare, N., Danchin, A., & Kunst, F. (2003). The genome sequence of the entomopathogenic bacterium *Photorhabdus luminescens. Nat. Biotechnol., 21,* 1307−1313.

Duetting, P. S., Ding, H., Neufeld, J., & Eigenbrode, S. D. (2003). Plant waxy bloom on peas affects infection of pea aphids by *Pandora neoaphidis. J. Invertebr. Pathol., 84,* 149−158.

Duncan, L. W., & McCoy, C. W. (1996). Vertical distribution in soil, persistence, and efficacy against citrus root weevil (Coleoptera: Curculionidae) of two species of entomogenous nematodes (Rhabditida: Steinernematidae; Heterorhabditidae). *Environ. Entomol., 25,* 174−178.

Dutky, S. R. (1963). In E. A. Steinhaus (Ed.), *Insect Pathology. Vol. 2. The milky diseases,* (pp. 75−115). New York: Academic Press.

Dwyer, G. (1992). On the spatial spread of insect viruses: theory and experiment. *Ecology, 73,* 479−494.

Dwyer, G. (1994). Density-dependence and spatial structure in the dynamics of insect pathogens. *Am. Nat., 143,* 533−562.

Dwyer, G., & Elkinton, J. S. (1993). Using simple models to predict virus epizootics in gypsy moth populations. *J. Anim. Ecol., 62,* 1−11.

Eberle, K. E., & Jehle, J. A. (2006). Field resistance of codling moth against *Cydia pomonella* granulovirus (CpGV) is autosomal and incompletely dominant inherited. *J. Invertebr. Pathol., 93,* 201−206.

Ebssa, L., Borgemeister, C., & Poehling, H.-M. (2004). Effectiveness of different species/strains of entomopathogenic nematodes for control of western flower thrips (*Frankliniella occidentalis*) at various concentrations, host densities, and temperatures. *Biol. Control, 29,* 145−154.

Edelstein, J. D., Trumper, E. V., & Lecuona, R. E. (2005). Temperature-dependent development of the entomopathogenic fungus *Nomuraea rileyi* (Farlow) Samson in *Anticarsia gemmatalis* (Hübner) larvae (Lepidoptera: Noctuidae). *Neotrop. Entomol., 34,* 593−599.

Ehlers, R.-U., & Shapiro-Ilan, D. I. (2005). Mass production. In P. S. Grewal, R.-U. Ehlers & D. I. Shapiro-Ilan (Eds.), *Nematodes as Biological Control Agents* (pp. 65−79). Wallingford: CABI.

Elkinton, J. S., & Burand, J. (2007). Assessing impact of naturally occurring pathogens of forest insects. In L. A. Lacey & H. K. Kaya (Eds.), *Field Manual of Techniques in Invertebrate Pathology* (2nd ed.). (pp. 283−296). Dordrecht: Springer.

Elkinton, J. S., Hajek, A. E., Boettner, G. H., & Simons, E. E. (1991). Distribution and apparent spread of *Entomophaga maimaiga* (Zygomycetes: Entomophthorales) in gypsy moth (Lepidoptera: Lymantriidae) populations in North America. *Environ. Entomol., 20,* 1601−1605.

Elvira, S., Gorría, N., Muñoz, D., Williams, T., & Caballero, P. (2010). A simplified low-cost diet for rearing *Spodoptera exigua* (Lepidoptera: Noctuidae) and its effect on *S. exigua* nucleopolyhedrovirus production. *J. Econ. Entomol., 103,* 17−24.

Eng, M. S., Preisser, E. L., & Strong, D. R. (2005). Phoresy of the entomopathogenic nematode *Heterorhabditis marelatus* by a non-host organism, the isopod *Porcellio scaber*. *J. Invertebr. Pathol., 88,* 173−176.

England, L. S., Holmes, S. B., & Trevors, J. T. (1998). Persistence of viruses and DNA in soil. *World J. Microbiol. Biotechnol., 14,* 163−169.

Enkerli, J., Widmer, F., & Keller, S. (2004). Long-term field persistence of *Beauveria brongniartii* strains applied as biocontrol agents against European cockchafer larvae in Switzerland. *Biol. Control, 29,* 115−123.

Epsky, N. D., Walter, D. E., & Capinera, J. L. (1988). Potential role of nematophagous microarthropods as biotic mortality factors of entomogenous nematodes (Rhabditida: Steinernematidae, Heterorhabditidae). *J. Econ. Entomol., 81,* 821−825.

Fargues, J., Goettel, M. S., Smits, N., Ouedraogo, A., Vidal, C., Lacey, L. A., Lomer, C. J., & Rougier, M. (1996). Variability in susceptibility to simulated sunlight of conidia among isolates of entomopathogenic Hyphomycetes. *Mycopathologia, 135,* 171−181.

Faria, M., & Wraight, S. P. (2001). Biological control of *Bemisia tabaci* with fungi. *Crop. Prot., 20,* 767−778.

Feng, M. G., Poprawski, T. J., & Khachatourians, G. G. (1994). Production, formulation and application of the entomopathogenic fungus *Beauveria bassiana* for insect control: current status. *Biocontrol Sci. Technol., 4,* 3−34.

Fenili, R., Mendes, C. J., Miquelluti, D. J., Mariano-da Silva, S., Xavier, Y., Ribas, H. S., & Furlan, G. (2000). *Deladenus siricidicola*, Bedding (Neotylenchidae) parasitism evaluation in adult *Sirex noctilio*, Fabricius, 1793 (Hymenoptera: Siricidae). *Rev. Bras. Biol., 60,* 683−687.

Fenton, A., & Rands, S. A. (2004). Optimal parasite infection strategies: a state-dependent approach. *Int. J. Parasitol., 34,* 813−821.

ffrench-Constant, R. H., & Bowen, D. J. (1999). *Photorhabdus* toxins: novel biological insecticides. *Curr. Opin. Microbiol., 2,* 284−288.

Fife, J. P., Ozkan, H. E., Derksen, R. C., & Grewal, P. S. (2006). Using computational fluid dynamics to predict damage of a biological pesticide during passage through a hydraulic nozzle. *Biosyst. Eng., 94,* 387−396.

Figueiredo, E., Muñoz, D., Murillo, R., Mexia, A., & Caballero, P. (2009). Diversity of Iberian nucleopolyhedrovirus wild-type isolates infecting *Helicoverpa armigera* (Lepidoptera: Noctuidae). *Biol. Control, 50,* 43−49.

Fine, P. E. M. (1975). Vectors and vertical transmission: an epidemiologic perspective. *Ann. N.Y. Acad. Sci., 266,* 173−194.

Fisher, J. J., Rehner, S. A., & Bruck, D. J. (2011). Diversity of rhizosphere associated entomopathogenic fungi of perennial herbs, shrubs and coniferous trees. *J. Invertebr. Pathol., 106,* 289−295.

Forst, S., Dowds, B., Boemare, N., & Stackebrandt, E. (1997). *Xenorhabdus* and *Photorhabdus* spp.: bugs that kill bugs. *Annu. Rev. Microbiol., 51,* 47−72.

van Frankenhuyzen, K., Reardon, R. C., & Dubois, N. R. (2007). Forest defoliators. In L. A. Lacey & H. K. Kaya (Eds.), *Field Manual of Techniques in Invertebrate Pathology* (2nd ed.). (pp. 481−504). Dordrecht: Springer.

Furlong, M. J., & Pell, J. K. (2001). Horizontal transmission of entomopathogenic fungi by the diamondback moth. *Biol. Control, 22,* 288−299.

Fushing, H., Zhu, L., Shapiro-Ilan, D. I., Campbell, J. F., & Lewis, E. E. (2008). State-space based mass event-history model. I: Many decision-making agents with one target. *Ann. Appl. Stat., 2,* 1503−1522.

Fuxa, J. A., Fuxa, J. R., & Richter, A. R. (1998). Host-insect survival time and disintegration in relation to population density and dispersion of recombinant and wild-type nucleopolyhedroviruses. *Biol. Control, 12,* 143−150.

Fuxa, J. R. (1979). Interactions of the microsporidium *Vairimorpha necatrix* with a bacterium, virus, and fungus in *Heliothis zea*. *J. Invertebr. Pathol., 33,* 316−323.

Fuxa, J. R. (1987). Ecological considerations for the use of entomopathogens in IPM. *Annu. Rev. Entomol., 32,* 225−251.

Fuxa, J. R. (1989). Fate of released entomopathogens with reference to risk assessment of genetically engineered microorganisms. *Bull. Entomol. Soc. Am., 35,* 12−25.

Fuxa, J. R. (1991). Release and transport of entomopathogenic microorganisms. In M. Levin & H. Strauss (Eds.), *Risk Assessment in Genetic Engineering. Environmental Release of Organisms* (pp. 83−113). New York: McGraw-Hill.

Fuxa, J. R. (2004). Ecology of insect nucleopolyhedroviruses. *Agric. Ecosyst. Environ., 103,* 27−43.

Fuxa, J. R., & Richter, A. R. (1994). Distance and rate of spread of *Anticarsia gemmatalis* (Lepidoptera: Noctuidae) nuclear polyhedrosis virus released into soybean. *Environ. Entomol., 23,* 1308−1316.

Fuxa, J. R., & Richter, A. R. (2001). Quantification of soil-to-plant transport of recombinant nucleopolyhedrovirus: effects of soil type and moisture, air currents, and precipitation. *Appl. Environ. Microbiol., 67,* 5166−5170.

Fuxa, J. R., & Tanada, Y. (1987). *Epizootiology of Insect Diseases.* New York: Wiley-Interscience.

Fuxa, J. R., Mitchell, F. L., & Richter, A. R. (1988). Resistance of *Spodoptera frugiperda* (Lep: Noctuidae) to a nuclear polyhedrosis virus in the field and laboratory. *Entomophaga, 33,* 55−63.

Fuxa, J. R., Richter, A. R., & Strother, M. S. (1993). Detection of *Anticarsia gemmatalis* nuclear polyhedrosis virus in predatory arthropods and parasitoids after viral release in Louisiana soybean. *J. Entomol. Sci., 28,* 51−60.

Fuxa, J. R., Sun, J.-Z., Weidner, E. H., & LaMotte, L. R. (1999). Stressors and rearing diseases of *Trichoplusia ni*: evidence of vertical transmission of NPV and CPV. *J. Invertebr. Pathol., 74,* 149−155.

Gao, L., & Liu, X. (2010). Sporulation of several biocontrol fungi as affected by carbon and nitrogen sources in a two-stage cultivation system. *J. Microbiol., 48,* 767−770.

Gao, Q., Kai, J., Ying, S.-H., Zhang, Y., Xiao, G., Shang, Y., Duan, Z., Hu, X., Xie, X.-Q., Zhou, G., Peng, G., Luo, Z., Huang, W., Wang, B., Fang, W., Wang, S., Zhong, Y., Ma, L.-J., St. Leger, R. J., Zhao, G.-P., Pei, Y., Feng, M.-G., Xia, Y., & Wang, C. (2011). Genome sequencing and comparative transcriptomics of the model and entomopathogenic fungi *Metarhizium anisopliae* and *M. acridum. PLoS Genet, 7.* e1001264.

Garczynski, S. F., & Siegel, J. P. (2007). Bacteria. In L. A. Lacey & H. K. Kaya (Eds.), *Field Manual of Techniques in Invertebrate Pathology* (2nd ed.). (pp. 175−197). Dordrecht: Springer.

Gassmann, A. J., Carrière, Y., & Tabashnik, B. E. (2009). Fitness costs of insect resistance to *Bacillus thuringiensis. Annu. Rev. Entomol., 54,* 147−163.

Gaugler, R. (1987). Entomogenous nematodes and their prospects for genetic improvement. In K. Maramorosch (Ed.), *Biotechnology in Invertebrate Pathology and Cell Culture* (pp. 457−484). San Diego: Academic Press.

Gaugler, R., & Boush, G. M. (1978). Effects of ultraviolet radiation and sunlight on the entomogenous nematode. *Neoaplectana carpocapsae. J. Invertebr. Pathol., 32,* 291−296.

Gaugler, R., & Georgis, R. (1991). Culture method and efficacy of entomopathogenic nematodes (Rhabditida: Steinernematidae and Heterorhabditidae). *Biol. Control, 1,* 269−274.

Gaugler, R., & Han, R. (2002). Production technology. In R. Gaugler (Ed.), *Entomopathogenic Nematology* (pp. 289−310). New York: CABI.

Gaugler, R., Campbell, J. F., & McGuire, T. R. (1989a). Selection for host-finding in *Steinernema feltiae. J. Invertebr. Pathol., 54,* 363−372.

Gaugler, R., Costa, S. D., & Lashomb, J. (1989b). Stability and efficacy of *Beauveria bassiana* soil inoculations. *Environ. Entomol., 18,* 412−417.

Gaugler, R., Campbell, J. F., & McGuire, T. R. (1990). Fitness of a genetically improved entomopathogenic nematode. *J. Invertebr. Pathol., 56,* 106−116.

Gaugler, R., Wilson, M., & Shearer, P. (1997). Field release and environmental fate of a transgenic entomopathogenic nematode. *Biol. Control, 9,* 75−80.

Gaugler, R., Brown, I., Shapiro-Ilan, D. I., & Atwa, A. (2002). Automated technology for *in vivo* mass production of entomopathogenic nematodes. *Biol. Control, 24,* 199−206.

Geden, C. J., & Steinkraus, D. C. (2003). Evaluation of three formulations of *Beauveria bassiana* for control of lesser mealworm and hide beetle in Georgia poultry houses. *J. Econ. Entomol., 96,* 1602−1607.

Gelernter, W. D., & Lomer, C. J. (2000). Success in biological control of above-ground insects by pathogens. In G. Gurr & S. Wratten (Eds.), *Biological Control: Measures of Success* (pp. 297−322). Dordrecht: Kluwer Academic.

Georgis, R., Kaya, H. K., & Gaugler, R. (1991). Effect of steinernematid and heterorhabditid nematodes (Rhabditida: Steinernematidae and Heterorhabditidae) on nontarget arthropods. *Environ. Entomol., 20,* 815−822.

Georgis, R., Koppenhöfer, A. M., Lacey, L. A., Bélair, G., Duncan, L. W., Grewal, P. S., Samish, M., Tan, L., Torr, P., & van Tol, R. W. H. M. (2006). Successes and failures in the use of parasitic nematodes for pest control. *Biol. Control, 38,* 103−123.

Glazer, I. (2002). Survival biology. In R. Gaugler (Ed.), *Entomopathogenic Nematology* (pp. 169−187). New York: CABI.

Glazer, I., Salame, L., & Segal, D. (1997). Genetic enhancement of nematicide resistance in entomopathogenic nematodes. *Biocontrol Sci. Technol., 7,* 499−512.

Goettel, M. S. (1995). The utility of bioassays in the risk assessment of entomopathogenic fungi. In *Biotechnology Risk Assessment: USEPA/USDA/Environment Canada/Agriculture and Agri-Food Canada. Risk Assessment Methodologies. Proceedings of the Biotechnology Risk Assessment Symposium, June 6−8, 1995. Pensacola, FL* (pp. 2−8). College Park: University of Maryland Biotechnology.

Goettel, M. S., Eilenberg, J., & Glare, T. R. (2010). Entomopathogenic fungi and their role in regulation of insect populations. In L. I. Gilbert & S. Gill (Eds.), *Insect Control: Biological and Synthetic Agents* (pp. 387−432). London: Academic Press.

Goldberg, L. J., & Margalit, J. (1977). A bacterial spore demonstrating rapid larvicidal activity against *Anopheles sergentii, Uranotaenia unguiculata, Culex univittatus, Aedes aegypti* and *Culex univitattus. Mosq. News, 37,* 355−358.

Gould, F. (1998). Sustainability of transgenic insecticidal cultivars: integrating pest genetics and ecology. *Annu. Rev. Entomol., 43,* 701−726.

Gould, F., Anderson, A., Reynolds, A., Bumgarner, L., & Moar, W. (1995). Selection and genetic analysis of a *Heliothis virescens* (Lepidoptera: Noctuidae) strain with high levels of resistance to *Bacillus thuringiensis* toxins. *J. Econ. Entomol., 88,* 1545−1559.

Gould, F., Anderson, A., Jones, A., Sumerford, D., Heckel, D. G., Lopez, J., Micinski, S., Leonard, R., & Laster, M. (1997). Initial frequency of alleles for resistance to *Bacillus thuringiensis* toxins in field populations of *Heliothis virescens. Proc. Natl. Acad. Sci. USA, 94,* 3519−3523.

Gregory, P. H., Guthrie, E. J., & Bunce, M. (1959). Experiments on splash dispersal of fungal spores. *J. Gen. Microbiol., 20,* 328−354.

Grewal, P. S. (2002). Formulation and application technology. In R. Gaugler (Ed.), *Entomopathogenic Nematology* (pp. 265−287). New York: CABI.

Grewal, P. S., Power, K. T., Grewal, S. K., Suggars, A., & Haupricht, S. (2004). Enhanced consistency of white grubs (Coleoptera: Scarabaeidae) with new strains of entomopathogenic nematodes. *Biol. Control, 30,* 73−82.

Grewal, P. S., Ehlers, R.-U. & Shapiro-Ilan, D. I. (Eds.). (2005a). *Nematodes as Biological Control Agents.* Wallingford: CABI.

Grewal, P. S., Koppenhöfer, A. M., & Choo, H. Y. (2005b). Lawn turfgrass and pasture applications. In P. S. Grewal, R.-U. Ehlers & D. I. Shapiro-Ilan (Eds.), *Nematodes as Biological Control Agents* (pp. 115−146). Wallingford: CABI.

Groden, E., & Lockwood, J. L. (1991). Effects of soil fungistasis on *Beauveria bassiana* and its relationship to disease incidence in the Colorado potato beetle *Leptinotarsa decemlineata*, in Michigan and Rhode Island soils. *J. Invertebr. Pathol., 57,* 7−16.

Gross, J., Schumacher, K., Schmidtberg, H., & Vilcinskas, A. (2008). Protected by fumigants: beetle perfumes in antimicrobial defense. *J. Chem. Ecol., 34,* 179−188.

Guillet, P., Kurtak, D. C., Philippon, B., & Meyer, R. (1990). Use of *Bacillus thuringiensis* for onchocerciasis control in West Africa. In H. de Barjac & D. Sutherland (Eds.), *Bacterial Control of Mosquitoes and Black Flies: Biochemistry, Genetics, and Applications of* Bacillus thuringiensis israelensis *and* Bacillus sphaericus (pp. 187–201). New Brunswick: Rutgers University Press.

Hajek, A. E. (1999). Pathology and epizootiology of *Entomophaga maimaiga* infections in forest Lepidoptera. *Microbiol. Mol. Biol. Rev., 63*, 814–835.

Hajek, A. E., & Delalibera, I., Jr. (2010). Fungal pathogens as classical biological control agents against arthropods. *BioControl, 55*, 147–158.

Hajek, A. E., & St. Leger, R. (1994). Interactions between fungal pathogens and insect hosts. *Annu. Rev. Entomol., 39*, 293–322.

Hajek, A. E., Humber, R. A., Elkinton, J. S., May, B., Walsh, S. R. A., & Silver, J. S. (1990). Allozyme and restriction fragment length polymorphism analyses confirm *Entomophaga maimaiga* responsible for 1989 epizootics in North America gypsy moth populations. *Proc. Natl. Acad. Sci. USA, 87*, 6979–6982.

Hajek, A. E., Butler, L., & Wheeler, M. M. (1995). Laboratory bioassays testing the host range of the gypsy moth fungal pathogen *Entomophaga maimaiga*. *Biol. Control, 5*, 530–544.

Hajek, A. E., Bauer, L., McManus, M. L., & Wheeler, M. M. (1998). Distribution of resting spores of the *Lymantria dispar* pathogen *Entomophaga maimaiga* in soil and on bark. *BioControl, 43*, 189–200.

Hall, F. R., & Menn, J. J. (1999). *Biopesticides: Use and Delivery*. New York: Humana Press.

Hamer, W. H. (1906). Epidemic disease in England. *Lancet., 1*, 733–739.

Hancock, P. A., Thomas, M. B., & Godfray, H. C. J. (2009). An age-structured model to evaluate the potential of novel malaria-control interventions: a case study of fungal biopesticide sprays. *Proc. R. Soc. B, 276*, 71–80.

Harper, A. M. (1958). Notes on behaviour of *Pemphigus betae* Doane (Homoptera: Aphididae) infected with *Entomophthora aphidis* Hoffm. *Can. Entomol., 90*, 439–440.

Harrison, R. L., & Bonning, B. C. (2000). Use of scorpion neurotoxins to improve the insecticidal activity of *Rachiplusia ou* multicapsid nucleopolyhedrovirus. *Biol. Control, 17*, 191–201.

Haugen, D. A., & Underdown, M. G. (1990). *Sirex noctilio* control program in response to the 1987 Green Triangle outbreak. *Aust. For., 53*, 33–40.

Haugen, D. A., & Underdown, M. G. (1993). Reduced parasitism of *Sirex noctilio* in radiata pines inoculated with the nematode *Beddingia siricidicola* during 1974–1989. *Aust. For., 56*, 45–48.

Henke, M. O., de Hoog, G. S., Gross, U., Zimmermann, G., Kraemer, D., & Weig, M. (2002). Human deep tissue infection with an entomopathogenic *Beauveria* species. *J. Clin. Microbiol., 40*, 2698–2702.

Hesketh, H., Alderson, P. G., Pye, B. J., & Pell, J. K. (2008). The development and multiple uses of a standardised bioassay method to select hypocrealean fungi for biological control of aphids. *Biol. Control, 46*, 242–255.

Hesketh, H., Roy, H. E., Eilenberg, J., Pell, J. K., & Hails, R. S. (2010). Challenges in modelling complexity of fungal entomopathogens in semi-natural populations of insects. *BioControl, 55*, 55–73.

Hethcote, H. W. (2000). The mathematics of infectious diseases. *SIAM Rev., 42*, 599–653.

Hill, R. E., & Gary, W. J. (1979). Effects of the microsporidium, *Nosema pyrausta*, on field populations of European corn borers in Nebraska. *Environ. Entomol., 8*, 91–95.

Hollingsworth, R. G., Steinkraus, D. C., & McNew, R. W. (1995). Sampling to predict fungal epizootics in cotton aphids (Homoptera: Aphididae). *Environ. Entomol., 24*, 1414–1421.

Hominick, W. M. (2002). Biogeography. In R. Gaugler (Ed.), *Entomopathogenic Nematology* (pp. 115–144). New York: CABI.

Hong, T. D., Ellis, R. H., & Moore, D. (1997). Development of a model to predict the effect of temperature and moisture on fungal spore longevity. *Ann. Bot., 79*, 121–128.

Hopper, K. R., Roush, R. T., & Powell, W. (1993). Management of genetics of biological-control introductions. *Annu. Rev. Entomol., 38*, 27–51.

Hountondji, F. C. C., Sabelis, M. W., Hanna, R., & Janssen, A. (2005). Herbivore-induced plant volatiles trigger sporulation in the entomopathogenic fungi: the case of *Neozygites tanajoae* infecting the cassava green mite. *J. Chem. Ecol., 31*, 1004–1021.

Hu, G., & St. Leger, R. J. (2002). Field studies using a recombinant mycoinsecticide (*Metarhizium anisopliae*) reveal that it is rhizosphere competent. *Appl. Environ. Microbiol., 68*, 6383–6387.

Hudson, W. G., Frank, J. H., & Castner, J. L. (1988). Biological control of *Scapteriscus* spp. mole crickets (Orthoptera: Gryllotalpidae) in Florida. *Bull. Entomol. Soc. Am., 34*, 192–198.

Huger, A. M. (1966). A virus disease of the Indian rhinoceros beetle *Oryctes rhinoceros* (Linnaeus), caused by a new type of insect virus, *Rhabdionvirus oryctes* gen. n., sp. n. *J. Invertebr. Pathol., 8*, 38–51.

Huger, A. M. (2005). The Oryctes virus: its detection, identification and implementation in biological control of the coconut palm rhinoceros beetle, *Oryctes rhinoceros* (Coleoptera: Scarabaeidae). *J. Invertebr. Pathol., 89*, 78–84.

Hukuhara, T. (1973). Further studies on the distribution of a nuclear polyhedrosis virus of the fall webworm, *Hyphantria cunea*, in soil. *J. Invertebr. Pathol., 22*, 345–350.

Hummel, R. L., Walgenbach, J. F., Barbercheck, M. E., Kennedy, G. G., Hoyt, G. D., & Arellano, C. (2002). Effects of production practices on soil-borne entomopathogens in Western North Carolina vegetable systems. *Environ. Entomol., 31*, 84–91.

Hunter, J. E., & Kunimoto, R. K. (1974). Dispersal of *Phytophthora palmivora* by wind-blown rain. *Phytopathology, 64*, 202–206.

Hunter-Fujita, F. R., Entwistle, P. F., Evans, H. F., & Crook, N. E. (1998). *Insect Viruses and Pest Management*. Chichester: John Wiley and Sons.

Hurley, B. P., Slippers, B., Croft, P. K., Hatting, H. J., van der Linde, M., Morris, A. R., Dyer, C., & Wingfield, M. J. (2008). Factors influencing parasitism of *Sirex noctilio* (Hymenoptera: Siricidae) by the nematode *Deladenus siricidicola* (Nematoda: Neotylenchidae) in summer rainfall areas of South Africa. *Biol. Control, 4*, 450–459.

Ignoffo, C. M. (1992). Environmental factors affecting persistence of entomopathogens. *Fla. Entomol., 75*, 516–525.

Inglis, G. D., Goettel, M. S., & Johnson, D. L. (1993). Persistence of the entomopathogenic fungus *Beauveria bassiana*, on phylloplanes of crested wheat grass and alfalfa. *Biol. Control, 3*, 258–270.

Inglis, G. D., Johnson, D. L., Cheng, K.-J., & Goettel, M. S. (1997). Use of pathogen combinations to overcome the constraints of temperature on entomopathogenic hyphomycetes against grasshoppers. *Biol. Control, 8*, 143–152.

Inyang, E. N., Butt, T. M., Ibrahim, L., Clark, S. J., Pye, B. J., Beckett, A., & Archer, S. (1998). The effect of plant growth and topography on the acquisition of conidia of the insect pathogen *Metarhizium anisopliae* by larvae of *Phaedon cochleariae*. *Mycol. Res., 102*, 1365–1374.

Jackson, T. A., Crawford, A. M., & Glare, T. R. (2005). *Oryctes* virus – time for a new look at a useful biocontrol agent. *J. Invertebr. Pathol., 89*, 91–94.

Jackson, M. A., Dunlap, C. A., & Jaronski, S. T. (2010). Ecological considerations in producing and formulating fungal entomopathogens for use in insect biocontrol. *BioControl, 55*, 129–145.

James, C. (2006). Global status of commercialized biotech/GM crops. *ISAAA Briefs, 35*, 1–9.

James, R., & Elzen, G. W. (2001). Antagonism between *Beauveria bassiana* and imidacloprid when combined for *Bemisia argentifolii* (Homoptera: Aleyrodidae) control. *J. Econ. Entomol., 94*, 357–361.

Jansson, R. K., Lecrone, S. H., & Gaugler, R. (1993). Field efficacy and persistence of entomopathogenic nematodes (Rhabditida: Steinernematidae, Heterorhabditidae) for control of sweetpotato weevil (Coleoptera: Apionidae) in southern Florida. *J. Econ. Entomol., 86*, 1055–1063.

Jaronski, S. T. (2007). Soil ecology of the entomopathogenic ascomycetes: a critical examination of what we (think) we know. In K. Maniana & S. Ekesi (Eds.), *Use of Entomopathogenic Fungi in Biological Pest Management* (pp. 91–144). Kerala: Research Signpost.

Jaronski, S. T. (2010). Ecological factors in the inundative use of fungal entomopathogens. *BioControl, 55*, 159–185.

Kaakeh, W., Reid, B. L., & Bennett, G. W. (1996). Horizontal transmission of the entomopathogenic fungus *Metarhizium anisopliae* (Imperfect Fungi: Hyphomycetes) and hydramethylnon among German cockroaches (Dictyoptera: Blattelolidae). *J. Entomol. Sci., 31*, 378–390.

Kaaya, G. P., & Okech, L. A. (1990). Horizontal transmission of mycotic infection in adult tsetse flies. *Glossina morsitans. Entomophaga, 35*, 589–600.

Kapongo, J. P., Shipp, L., Kevan, P., & Broadbent, B. (2008). Optimal concentration of *Beauveria bassiana* vectored by bumble bees in relation to pest and bee mortality in greenhouse tomato and sweet pepper. *BioControl, 53*, 797–812.

Kaspi, R., Ross, A., Hodson, A. K., Stevens, G. N., Kaya, H. K., & Lewis, E. E. (2010). Foraging efficacy of the entomopathogenic nematode *Steinernema riobrave* in different soil types from California citrus groves. *Appl. Soil Ecol., 45*, 243–253.

Kaya, H. K. (1990). Soil ecology. In R. Gaugler & H. K. Kaya (Eds.), *Entomopathogenic Nematodes in Biological Control* (pp. 93–115). Boca Raton: CRC Press.

Kaya, H. K., & Gaugler, R. (1993). Entomopathogenic nematodes. *Annu. Rev. Entomol., 38*, 181–206.

Kelly, A., Dunn, A. M., & Hatcher, M. J. (2001). Population dynamics of a vertically transmitted, parasitic sex ratio distorter and its amphipod host. *Oikos, 94*, 392–402.

Kepler, R. M., & Bruck, D. J. (2006). Examination of the interaction between the black vine weevil (Coleoptera: Curculionidae) and an entomopathogenic fungus reveals a new tritrophic interaction. *Environ. Entomol., 35*, 1021–1029.

Klein, M. G. (1992). Use of *Bacillus popilliae* in Japanese beetle control. In T. A. Jackson & T. R. Glare (Eds.), *Use of Pathogens in Scarab Pest Management* (pp. 179–189). Hampshire: Intercept.

Klein, M. G., & Georgis, R. (1992). Persistence of control of Japanese beetle (Coleoptera: Scarabaeidae) larvae with steinernematid and heterorhabditid nematodes. *J. Econ. Entomol., 85*, 727–730.

Klein, M. G., & Lacey, L. A. (1999). An attractant trap for autodissemination of entomopathogenic fungi into populations of the Japanese beetle, *Popillia japonica* (Coleoptera: Scarabaeidae). *Biocontrol Sci. Technol., 9*, 151–158.

Klein, M. G., Grewal, P. S., Jackson, T. A., & Koppenhöfer, A. M. (2007). Lawn, turf and grassland pests. In L. A. Lacey & H. K. Kaya (Eds.), *Field Manual of Techniques in Invertebrate Pathology* (2nd ed.). (pp. 655–675) Dordrecht: Springer.

Klingen, I., Eilenberg, J., & Meadow, R. (2002). Effects of farming system, field margins and bait insect on the occurrence of insect pathogenic fungi in soils. *Agric. Ecosyst. Environ., 91*, 191–198.

Koppenhöfer, A. M. (2007). Nematodes. In L. A. Lacey & H. K. Kaya (Eds.), *Field Manual of Techniques in Invertebrate Pathology* (2nd ed.). (pp. 249–264). Dordrecht: Springer.

Koppenhöfer, A. M., & Grewal, P. S. (2005). Compatibility and interactions with agrochemicals and other biocontrol agents. In P. S. Grewal, R.-U. Ehlers & D. I. Shapiro-Ilan (Eds.), *Nematodes as Biological Control Agents* (pp. 363–381). Wallingford: CABI.

Koppenhöfer, A. M., & Kaya, H. K. (1997). Additive and synergistic interactions between entomopathogenic nematodes and *Bacillus thuringiensis* for scarab grub control. *Biol. Control, 8*, 131–137.

Koppenhöfer, A. M., & Kaya, H. K. (1998). Synergism of imidacloprid and entomopathogenic nematodes: a novel approach to white grub control in turfgrass. *J. Econ. Entomol., 91*, 618–623.

Koppenhöfer, A. M., Kaya, H. K., & Taormino, S. P. (1995). Infectivity of entomopathogenic nematodes (Rhabditida: Steinernematidae) at different soil depths and moistures. *J. Invertebr. Pathol., 65*, 193–199.

Koppenhöfer, A. M., Jaffee, B. A., Muldoon, A. E., Strong, D. R., & Kaya, H. K. (1996). Effect of nematode-trapping fungi on an entomopathogenic nematode originating from the same field site in California. *J. Invertebr. Pathol., 68*, 246–252.

Koppenhöfer, A. M., Grewal, P. S., & Kaya, H. K. (2000). Synergism of imidacloprid and entomopathogenic nematodes against white grubs: the mechanism. *Entomol. Exp. Appl., 94*, 283–293.

Koppenhöfer, A. M., Grewal, P. S., & Fuzy, E. M. (2007). Differences in penetration routes and establishment rates of four entomopathogenic nematode species into four white grub species. *J. Invertebr. Pathol., 94*, 184–195.

Kreig, A. (1971). Interactions between pathogens. In H. D. Burgess & N. W. Hussey (Eds.), *Microbial Control of Insects and Mites* (pp. 459–468). London: Academic Press.

Kramm, K. R., West, D. F., & Rockenback, P. G. (1982). Termite pathogens: transfer of the entomopathogen *Metarhizium anisopliae* between *Reticulitermes* sp. termites. *J. Invertebr. Pathol., 40*, 1–6.

Kreutz, J., Zimmermann, G., & Vaupel, O. (2004). Horizontal transmission of the entomopathogenic fungus *Beauveria bassiana* among the spruce bark beetle, *Ips typographus* (Col., Scolytidae) in the laboratory and under field conditions. *Biocontrol Sci. Technol., 14*, 837–848.

Kukan, B. (1999). Vertical transmission of nucleopolyhedrovirus in insects. *J. Invertebr. Pathol., 74*, 103–111.

Kung, S.-P., Gaugler, R., & Kaya, H. K. (1990). Soil type and entomopathogenic nematode persistence. *J. Invertebr. Pathol., 55*, 401–406.

Lacey, L. A. (Ed.). (1997). *Manual of Techniques in Insect Pathology*. London: Academic Press.

Lacey, L. A. (2007). *Bacillus thuringiensis* serovariety *israelensis* and *Bacillus sphaericus* for mosquito control. In T. G. Floore (Ed.), *Biorational Control of Mosquitoes. Bull. Am. Mosq. Control Assoc., No*, 7 (pp. 133–163).

Lacey, L. A. & Kaya, H. K. (Eds.). (2007). *Field Manual of Techniques in Invertebrate Pathology* (2nd ed.). Dordrecht: Springer.

Lacey, L. A., & Shapiro-Ilan, D. I. (2008). Microbial control of insect pests in temperate orchard systems: potential for incorporation into IPM. *Annu. Rev. Entomol., 53*, 121–144.

Lacey, L. A., & Undeen, A. H. (1986). Microbial control of black flies and mosquitoes. *Annu. Rev. Entomol., 31*, 265–296.

Lacey, L. A., Frutos, R., Kaya, H. K., & Vail, P. (2001). Insect pathogens as biological control agents: do they have a future? *Biol. Control, 21*, 230–248.

Lacey, L. A., Unruh, T. R., & Headrick, H. L. (2003). Interactions of two idiobiont parasitoids (Hymenoptera: Ichneumonidae) of codling moth (Lepidoptera: Tortricidae) with entomopathogenic nematode *Steinernema carpocapsae* (Rhabditida: Steinernematidae). *J. Invertebr. Pathol., 83*, 230–239.

Lacey, L. A., Arthurs, S. P., Unruh, T. R., Headrick, H., & Fritts, R., Jr. (2006a). Entomopathogenic nematodes for control of codling moth (Lepidoptera: Tortricidae) in apple and pear orchards: effect of nematode species and seasonal temperatures, adjuvants, application equipment and post-application irrigation. *Biol. Control, 37*, 214–223.

Lacey, L. A., Arthurs, S. P., Granatstein, D., Headrick, H., & Fritts, R., Jr. (2006b). Use of entomopathogenic nematodes (Steinernematidae) in conjunction with mulches for control of codling moth (Lepidoptera: Tortricidae). *J. Entomol. Sci., 41*, 107–119.

Lacey, L. A., Thomson, D., Vincent, C., & Arthurs, S. P. (2008a). Codling moth granulovirus: a comprehensive review. *Biocontrol Sci. Technol., 18*, 639–663.

Lacey, L. A., Wraight, S. P., & Kirk, A. A. (2008b). Entomopathogenic fungi for control of *Bemisia* spp.: foreign exploration, research and implementation. In J. K. Gould, K. Hoelmer & J. Goolsby (Eds.), *Classical Biological Control of Bemisia tabaci in the USA: A Review of Interagency Research and Implementation* (pp. 33–69). Dordrecht: Springer.

Lacey, L. A., Shapiro-Ilan, D. I., & Glenn, G. M. (2010). Post-application of anti-desiccant agents improves efficacy of entomopathogenic nematodes in formulated host cadavers or aqueous suspension against diapausing codling moth larvae (Lepidoptera: Tortricidae). *Biocontrol Sci. Technol., 20*, 909–921.

Lacey, L. A., Liu, T.-X., Buchman, J. L., Munyaneza, J. E., Goolsby, J. A., & Horton, D. R. (2011). Entomopathogenic fungi (Hypocreales) for control of potato psyllid, *Bactericera cockerelli* (Šulc) (Hemiptera: Triozidae) in an area endemic for zebra chip disease of potato. *Biol. Control, 36*, 271–278.

Langenbruch, G. A., Krieg, A., Huger, A. M., & Schnetter, W. (1985). Erst Feldversuche zur Bekämpfung der Larven des Kartoffelkäfers (*Leptinotarsa decemlineata*) mit *Bacillus thuringiensis* var. *tenebrionis. Med. Fac. Land. Rij. Gent., 50*, 441–449.

Leland, J. E., McGuire, M. R., Grace, J. A., Jaronski, S. T., Ulloa, M., Young-Hoon, P., & Plattner, R. D. (2005). Strain selection of a fungal entomopathogen, *Beauveria bassiana*, for control of plant bugs (*Lygus* spp.) (Heteroptera: Miridae). *Biol. Control, 35*, 104–114.

Lewis, E. E. (2002). Behavioural ecology. In R. Gaugler (Ed.), *Entomopathogenic Nematology* (pp. 205–224). Wallingford: CABI.

Lewis, E. E., Campbell, J., Griffin, C., Kaya, H., & Peters, A. (2006). Behavioral ecology of entomopathogenic nematodes. *Biol. Control, 38*, 66–79.

Lewis, L. C. (1978). Migration of larvae of *Ostrinia nubilalis* (Lepidoptera: Pyralidae) infected with *Nosema pyrausta* (Microsporida: Nosematidae) and subsequent dissemination of this microsporiduim. *Can. Entomol., 110*, 897–900.

Lewis, L. C., & Cossentine, J. E. (1986). Season long intraplant epizootics of entomopathogens, *Beauveria bassiana* and *Nosema pyrausta*, in a corn agroecosystem. *Entomophaga, 31*, 363–369.

Lewis, L. C., Berry, E. C., Obrycki, J. J., & Bing, L. (1996). Aptness of insecticides (*Bacillus thuringiensis* and carbofuran) with endophytic *Beauveria bassiana*, in suppressing larval populations of the European corn borer. *Agric. Ecosyst. Environ., 15*, 27–34.

Lewis, L. C., Bruck, D. J., & Gunnarson, R. D. (2002). Measures of *Bacillus thuringiensis* persistence in the corn whorl. *J. Invertebr. Pathol., 80*, 69–71.

Lewis, L. C., Sumerford, D. V., Bing, L. A., & Gunnarson, R. D. (2006). Dynamics of *Nosema pyrausta* in natural populations of the European corn borer. *Ostrinia nubilalis: BioControl, 51*, 627–642.

Lewis, L. C., Bruck, D. J., Prasifka, J. R., & Raun, E. S. (2009). *Nosema pyrausta*: its biology, history, and potential role in landscape of transgenic insecticidal crops. *Biol. Control, 48*, 223–231.

Lezama-Gutiérrez, R., Trujillo-de la Luz, A., Molina-Ochoa, J., Rebolledo-Dominguez, O., Pescador, A. R., Lopez-Edwards, M., & Aluja, M. (2000). Virulence of *Metarhizium anisopliae* (Deuteromycotina: Hyphomycetes) on *Anastrepha ludens* (Diptera: Tephritidae): laboratory and field trials. *J. Econ. Entomol., 93*, 1080–1084.

Li, Z. (1988). In Y. Li, Z. Li, Z. Liang, J. Wu, Z. Wu & Q. Xu (Eds.), *Study and Application of Entomogenous Fungi in China. List on the insect hosts of* Beauveria bassiana, *Vol. 1* (pp. 241–255). Beijing: Academic Periodical Press.

Liebhold, A. M., Halverson, J. A., & Elmes, G. A. (1993). Gypsy moth invasion in North America: a quantitative analysis. *J. Gen. Virol., 74*, 513–520.

Lingg, A. J., & Donaldson, M. D. (1981). Biotic and abiotic factors affecting stability of *Beauveria bassiana* conidia in soil. *J. Invertebr. Pathol., 38*, 191–200.

Liu, D., Burton, S., Glancy, T., Li, Z. S., Hampton, R., Meade, T., & Merlo, D. J. (2003). Insect resistance conferred by 283-kDa *Photorhabdus luminescens* protein TcdA in *Arabidopsis thaliana. Nat. Biotechnol., 21*, 1222–1228.

Liu, H., Skinner, M., Parker, B. L., & Brownbridge, M. (2002). Pathogenicity of *Beauveria bassiana, Metarhizium anisopliae* (Deuteromycotina: Hyphomycetes) and other entomopathogenic fungi against *Lygus lineolaris* (Hemiptera: Miridae). *J. Econ. Entomol., 95*, 675–681.

Liu, Y.-B., Tabashnik, B. E., Meyer, S. K., Carrière, Y., & Bartlett, A. C. (2001). Genetics of pink bollworm resistance to *Bacillus thuringiensis* toxin Cry1Ac. *J. Econ. Entomol., 92*, 248–252.

Lomer, C. J. (1986). Release of Baculovirus oryctes into populations of *Oryctes monoceros* in the Seychelles. *J. Invertebr. Pathol., 47*, 237–246.

Lomer, C. J., Thomas, M. B., Godonou, I., Shah, P. A., Douro-Kpindou, O. K., & Langewald, J. (1997). Control of grasshoppers, particularly *Hieroglyphus daganensis*, in northern Benin using *Metarhizium flavoviride*. *Mem. Entomol. Soc. Can., 129*, 301–311.

Lomer, C. J., Bateman, R. P., Dent, D., De Groote, H., Douro-Kpindou, O. K., Kooyman, C., Langewald, J., Ouambama, Z., Peveling, R., & Thomas, M. (1999). Development of strategies for the incorporation of biological pesticides into the integrated management of locusts and grasshoppers. *Agric. For. Entomol., 1*, 71–88.

Lord, J. C. (2001). Desiccant dusts synergize the effect of *Beauveria bassiana* (Hyphomycetes: Moniliales) on stored-grain beetles. *J. Econ. Entomol., 94*, 367–372.

Lord, J. C. (2005). Low humidity, moderate temperature, and desiccant dust favor efficacy of *Beauveria bassiana* (Hyphomycetes: Moniliales) for the lesser grain borer, *Rhyzopertha dominica* (Coleoptera: Bruchidae). *Biol. Control, 34*, 180–186.

Lubeck, I. A., Arruda, W., Souza, B. K., Staniscuiaski, F., Carlini, C. R., Schrank, A., & Vainstein, M. H. (2008). Evaluation of *Metarhizium anisopliae* strains as potential biocontrol agents of the tick *Rhipicephalus (Boophilus) microplus* and the cotton stainer *Dysdercus peruvianus*. *Fungal Ecol., 1*, 78–88.

Maeda, S. (1989). Increased insecticidal effect by a recombinant baculovirus carrying a synthetic diuretic hormone gene. *Biochem. Biophys. Res. Commun., 165*, 1177–1183.

Mahar, A. N., Jan, N. K., Mahar, G. M., & Mahar, A. Q. (2008). Control of insects with entomopathogenic bacterium *Xenorhabdus nematophila* and its toxic secretions. *Int. J. Agric. Biol., 10*, 52–56.

Maines, T. R., Szretter, K. J., Perrone, L., Besler, J. A., Bright, R. A., Zeng, H., Tumpey, T. M., & Katz, J. M. (2008). Pathogenesis of emerging avian influenza viruses in mammals and the host innate immune response. *Immunol. Rev., 225*, 68–84.

Malakar, R., Elkinton, J. S., Carroll, S. D., & D'Amico, V. (1999a). Interactions between two gypsy moth (Lepidoptera: Lymantriidae) pathogens: Nucleopolyhedrovirus and *Entomophaga maimaiga* (Zygomycetes: Entomophthorales): field studies and a simulation model. *Biol. Control, 16*, 189–198.

Malakar, R., Elkinton, J., Hajek, A. E., & Burand, J. P. (1999b). Within-host interactions of *Lymantria dispar* (Lepidoptera: Lymantriidae) nucleopolyhedrosis virus and *Entomophaga maimaiga* (Zygomycetes: Entomophthorales). *J. Invertebr. Pathol., 73*, 91–100.

Maniania, N. K. (1998). A device for infecting adult tsetse flies, *Glossina* spp., with an entomopathogenic fungus in the field. *Biol. Control, 11*, 248–254.

Maniania, N. K. (2002). A low-cost contamination device for infecting adult tsetse flies, *Glossina* spp., with the entomopathogenic fungus *Metarhizium anisopliae* in the field. *Biocontrol Sci. Technol., 12*, 59–66.

Margalit, J., & Dean, D. (1985). The story of Bacillus thuringiensis var. *israelensis (Bti)*. *J. Am. Mosq. Control Assoc., 1*, 1–7.

Martin, P. A. W., & Travers, R. S. (1989). Worldwide abundance and distribution of *Bacillus thuringiensis* isolates. *Appl. Environ. Microbiol., 55*, 2437–2442.

Mbata, G. N., & Shapiro-Ilan, D. I. (2010). Compatibility of *Heterorhabditis indica* (Rhabditida: Heterorhabditidae) and *Hebrobracon hebetor* (Hymenoptera: Braconidae) for biological control of *Plodia interpunctella* (Lepidoptera: Pyralidae). *Biol. Control, 54*, 75–82.

McCoy, C. W., Stuart, R. J., Duncan, L. W., & Nguyen, K. (2002). Field efficacy of two commercial preparations of entomopathogenic nematodes against larvae of *Diaprepes abbreviatus* (Coleoptera: Curculionidae) in alfisol type soil. *Fla. Entomol., 85*, 537–544.

McCoy, C. W., Stuart, R. J., Shapiro-Ilan, D. I., & Duncan, L. W. (2007). Application and evaluation of entomopathogens for citrus pest control. In L. A. Lacey & H. K. Kaya (Eds.), *Field Manual of Techniques in Invertebrate Pathology* (2nd ed.). (pp. 567–581) Dordrecht: Springer.

McGaughey, W. H. (1985). Insect resistance to the biological insecticide *Bacillus thuringiensis*. *Science, 229*, 193–195.

McGuire, M. R., Shasha, B. S., Lewis, L. C., & Nelson, T. C. (1994). Residual activity of granular starch-encapsulated *Bacillus thuringiensis*. *J. Econ. Entomol., 87*, 631–637.

McGuire, M. R., Ulloa, M., Park, Y.-H., & Hudson, N. (2005). Biological and molecular characteristics of *Beauveria bassiana* isolates from California *Lygus hesperus* (Hemiptera: Miridae) populations. *Biol. Control, 33*, 307–314.

McGuire, M. R., Leland, J. E., Dara, S., Park, Y.-K., & Ulloa, M. (2006). Effect of different isolates of *Beauveria bassiana* on field populations of *Lygus hesperus*. *Biol. Control, 38*, 390–396.

Meadows, M. P. (1993). *Bacillus thuringiensis* in the environment: ecology and risk assessment. In P. F. Entwhistle, J. S. Corey, M. J. Bailey & S. Higgs (Eds.), Bacillus thuringiensis, *an Environmental Biopesticide: Theory and Practice* (pp. 193–220). Chichester: Wiley and Sons.

Merritt, R. W., Walker, E. D., Wilzbach, M. A., Cummins, K. W., & Morgan, W. T. (1989). A broad evaluation of B.t.i. for black fly (Diptera: Simuliidae) control in a Michigan river: efficacy, carry and nontarget effects on invertebrates and fish. *J. Am. Mosq. Control Assoc., 5*, 397–415.

Meyling, N. V., & Eilenberg, J. (2007). Ecology of the entomopathogenic fungi *Beauveria bassiana* and *Metarhizium anisopliae* in temperate agroecosystems: potential for conservation biological control. *Biol. Control, 43*, 145–155.

Meyling, N. V., Lubeck, M., Buckley, E. P., Eilenberg, J., & Rehner, S. A. (2009). Community composition, host range and genetic structure of the fungal entomopathogen *Beauveria* in adjoining agricultural and seminatural habitats. *Mol. Ecol., 18*, 1282–1293.

Milks, M. L., Fuxa, J. R., & Richter, A. R. (2008). Prevalence and impact of the microsporidium *Thelohania solenopsae* (Microsporidia) on wild populations of red imported fire ants, *Solenopsis invicta*, in Louisiana. *J. Invertebr. Pathol., 97*, 91–102.

Molloy, D. (1982). Biological control of black flies (Diptera: Simuliidae) with *Bacillus thuringiensis* var. *israelensis* (Serotype 14): a review with recommendations for laboratory and field protocol. *Misc. Pub. Entomol. Soc. Am., 12*, 1–30.

Moore, D. (2008). A plague on locusts – the LUBILOSA story. *Outl. Pest Manag., 19*, 14–17.

Moreau, G., & Lucarotti, C. J. (2007). A brief review of the past use of baculoviruses for the management of eruptive forest defoliators and recent developments on a sawfly virus in Canada. *For. Chron., 83*, 105–112.

Morley-Davies, J., Moore, D., & Prior, C. (1995). Screening of *Metarhizium* and *Beauveria* spp. conidia with exposure to simulated sunlight and a range of temperatures. *Mycol. Res., 100*, 31–38.

Morrow, B. J., Boucias, D. G., & Heath, M. A. (1989). Loss of virulence in an isolate of an entomopathogenic fungus, *Nomuraea rileyi*, after serial *in vitro* passage. *J. Econ. Entomol., 82*, 404–407.

Moscardi, F. (1999). Assessment of the application of baculoviruses for control of Lepidoptera. *Annu. Rev. Entomol., 44*, 257–289.

Moscardi, F. (2007). A nucleopolyhedrovirus for control of the velvet-bean caterpillar in Brazilian soybeans. In C. Vincent, M. S. Goettel & G. Lazarovits (Eds.), *Biological Control: A Global Perspective* (pp. 344–352). Wallingford: CABI.

Moscardi, F., Leite, L. G., & Zamataro, C. E. (1997). Produção do vírus de poliedrose nuclear de *Anticarsia gemmatalis* Hübner (Lepidoptera: Noctuidae): efeito da dose do patógeno, população e idade do hospedeiro. *An. Soc. Entomol. Bras., 26*, 121–132.

Mukaka, J., Strauch, O., Hoppe, C., & Ehlers, R.-U. (2010). Improvement of heat and desiccation tolerance in *Heterorhabditis bacteriophora* through cross-breeding of tolerant strains and successive genetic selection. *BioControl, 55*, 511–521.

Mulla, M. S., Thavara, U., Tawatsin, A., Chomposri, J., & Su, T. Y. (2003). Emergence of resistance and resistance management in field populations of tropical *Culex quinquefasciatus* to the microbial control agent *Bacillus sphaericus*. *J. Am. Mosq. Control Assoc., 19*, 39–46.

Myasnik, M., Manasherob, R., Ben-Dov, E., Zaritsky, A., Margalith, Y., & Barak, Z. (2001). Comparative sensitivity to UV-B radiation of two *Bacillus thuringiensis* subspecies and other *Bacillus* sp. *Curr. Microbiol., 43*, 140–143.

Naryanan, M. S. (2006). *Competitive ability and host exploitation in Bacillus thuringiensis.* MSc thesis. University of Oxford.

Nathan, S. S., & Kalaivani, K. (2005). Efficacy of nucleopolyhedrovirus and azadirachtin on *Spodoptera litura* Fabricius (Lepidoptera: Noctuidae). *Biol. Control, 34*, 93–98.

Nathan, S. S., Kalaivani, K., & Chung, P. G. (2005). The effects of azadirachtin and nucleopolyhedrovirus on midgut enzymatic profile of *Spodoptera litura* Fab. (Lepidoptera: Noctuidae). *Pestic. Biochem. Physiol., 83*, 46–57.

Navon, A., Keren, S., Levski, S., Grinstein, A., & Riven, Y. (1997). Granular feeding baits based on *Bacillus thuringiensis* products for the control of lepidopterous pests. *Phytoparasitica, 25* (Suppl.), 101S–110S.

Neveu, N., Grandgirard, J., Nenon, J. P., & Cortesero, A. M. (2002). Systemic release of herbivore-induced plant volatiles by turnips infested by concealed root-feeding larvae *Delia radicum*, L. *J. Chem. Ecol., 28*, 1717–1732.

Nielsen-LeRoux, C., Pasteur, N., Pretre, J., Charles, J.-F., Ben Sheikh, H., & Chevillon, C. (2002). High resistance to *Bacillus sphaericus* binary toxin in *Culex pipiens* (Diptera: Culicidae): the complex situation of west Mediterranean countries. *J. Med. Entomol., 39*, 729–735.

Ohba, M., & Aizawa, K. (1986). Distribution of *Bacillus thuringiensis* in soils of Japan. *J. Invertebr. Pathol., 47*, 277–282.

Oi, D. H. (2006). Effect of mono- and polygyne social forms on transmission and spread of a microsporidium in fire ant populations. *J. Invertebr. Pathol., 92*, 146–151.

Oi, D. H., & Pereira, R. M. (1993). Ant behaviour and microbial pathogens (Hymenoptera: Formicidae). *Fla. Entomol., 76*, 63–74.

Olofsson, E. (1989). Transmission of the nuclear polyhedrosis virus of the European pine sawfly from adult to offspring. *J. Invertebr. Pathol., 54*, 322–330.

Onstad, D. W. (1993). Thresholds and density dependence: the roles of pathogen and insect densities in disease dynamics. *Biol. Control, 3*, 353–356.

Onstad, D. W., & Carruthers, R. I. (1990). Epizootiological models of insect diseases. *Annu. Rev. Entomol., 35*, 399–419.

Onstad, D. W., & Maddox, J. V. (1989). Modeling the effects of the microsporidium, *Nosema pyrausta*, on the population dynamics of the insect. *Ostrinia nubilalis*. *J. Invertebr. Pathol., 53*, 410–421.

Onstad, D. W., & Maddox, J. V. (1990). Simulation model of *Tribolium confusum* and its pathogen. *Nosema whitei*. *Ecol. Model., 51*, 143–160.

Onstad, D. W., & Meinke, L. J. (2010). Modeling evolution of *Diabrotica virgifera virgifera* (Coleoptera: Chrysomelidae) to transgenic corn with two insecticidal traits. *J. Econ. Entomol., 103*, 849–860.

Onstad, D. W., Fuxa, J. R., Humber, R. A., Oestergaard, J., Shapiro-Ilan, D. I., Gouli, V. V., Anderson, R. S., Andreadis, T. G., & Lacey, L. A. (2006). An Abridged Glossary of Terms Used in Invertebrate Pathology. http://www.sipweb.org/glossary. Society for Invertebrate Pathology.

Otterstatter, M. C., & Thomson, J. D. (2008). Does pathogen spillover from commercially reared bumble bees threaten wild pollinators? *PLoS ONE, 3*, e2771.

Parkman, J. P., Frank, J. H., Nguyen, K. B., & Smart, G. C., Jr. (1993). Dispersal of *Steinernema scapterisci* (Rhabditida: Steinernematidae) after inoculative applications for mole cricket (Orthoptera: Gryllotalpidae) control in pastures. *Biol. Control, 3*, 226–232.

Parkman, J. P., Frank, J. H., Walker, T. J., & Schuster, D. J. (1996). Classical biological of *Scapteriscus* spp. (Orthoptera: Gryllotalpidae) in Florida. *Biol. Control, 25*, 1415–1420.

Penteado, S., do, R. C., Oliveira, de, E. B., & Iede, E. T. (2008). Utilização da amostragem sequencial para avaliar a eficiência do parasitismo de *Deladenus* (*Beddingia*) *siricidicola* (Nematoda: Neotylenchidae) em adultos de *Sirex noctilio* (Hymenoptera: Siricidae). *Ciên. Flor., 18*, 223–231.

Pereault, R. J., Whalon, M. E., & Alston, D. G. (2009). Field efficacy of entomopathogenic fungi and nematodes targeting caged last-instar plum curculio (Coleoptera: Curculionidae) in Michigan cherry and apple orchards. *Environ. Entomol., 38*, 1126–1134.

Pereira, R. M., & Stimac, J. L. (1992). Transmission of *Beauveria bassiana* within nests of *Solenopsis invicta* (Hymenoptera, Formicidae) in the laboratory. *Environ. Entomol., 21*, 1427–1432.

Perez, E. E., Lewis, E. E., & Shapiro-Ilan, D. I. (2003). Impact of host cadaver on survival and infectivity of entomopathogenic nematodes (Rhabditida: Steinernematidae and Heterorhabditidae) under desiccating conditions. *J. Invertebr. Pathol., 82*, 111–118.

Peters, A., & Ehlers, R.-U. (1994). Susceptibility of leatherjackets (*Tipula paludosa* and *T. oleracea*; Timulidae: Nematocera) to the entomopathogenic nematode *Steinernema feltiae*. *J. Invertebr. Pathol., 63*, 163–171.

Pierce, C. M. F., Solter, L. F., & Weinzierl, R. A. (2001). Interactions between *Nosema pyrausta* (Microsporidia: Nosematidae) and *Bacillus thuringiensis* subsp. *kurstaki* in the European corn borer (Lepidoptera: Pyralidae). *J. Econ. Entomol., 94*, 1361–1368.

Pimentel, D. (2008). Preface special issue: Conservation biological control. *Biol. Control, 45*, 171.

Pingel, R. L., & Lewis, L. C. (1999). Effect of *Bacillus thuringiensis*, *Anagrapha falcifera* multiple nucleopolyhedrovirus, and their mixture on three lepidopteran corn ear pests. *J. Econ. Entomol., 92*, 91–96.

Poinar, G. O., Jr., & Hess, R. (1988). In G. O. Poinar, Jr. & H.-B. Jansson (Eds.), *Diseases of Nematodes. Protozoan diseases, Vol. 1* (pp. 103–131). Boca Raton: CRC Press.

Poinar, G. O., Jr., Hess, R., Lanier, W., Kinney, S., & White, J. (1989). Preliminary observations of a bacteriophage infecting Xenorhabdus luminescens (Enterobacteriaceae). *Experientia, 45,* 191–192.

Polis, G. A. (1981). The evolution and dynamics of intraspecific predation. *Annu. Rev. Ecol. Syst., 12,* 225–251.

Poprawski, T. J., & Jones, W. J. (2000). Host plant effects on activity of the mitosporic fungi *Beauveria bassiana* and *Paecilomyces fumosoroseus* against two populations of *Bemisia* whiteflies (Homoptera: Aleyrodidae). *Mycopathologia, 151,* 11–20.

Poprawski, T. J., Parker, P. E., & Tsai, J. H. (1999). Laboratory and field evaluation of hyphomycete insect pathogenic fungi for control of brown citrus aphid (Homoptera: Aphididae). *Environ. Entomol., 28,* 315–321.

Power, K. T., An, R., & Grewal, P. S. (2009). Effectiveness of *Heterorhabditis bacteriophora* strain GPS11 applications targeted against different instars of the Japanese beetle *Popillia japonica*. *Biol. Control, 48,* 232–236.

Qin, X., Evans, J. D., Aronstein, K. A., Murray, K. D., & Weinstock, G. M. (2006). Genome sequences of the honey bee pathogens Paenibacillus larvae and Ascosphaera apis. *Insect Mol. Biol., 15,* 715–718.

Quesada-Moraga, E., Santos-Quirós, R., Valverde-García, P., & Santiago-Álvarez, C. (2004). Virulence, horizontal transmission, and sublethal reproductive effects of *Metarhizium anisopliae* (anamorphic fungi) on the German cockroach (Blattodea: Blattellidae). *J. Invertebr. Pathol., 87,* 51–58.

Quintela, E. D., & McCoy, C. W. (1998). Synergistic effect of imidacloprid and two entomopathogenic fungi on the behavior and survival of larvae of *Diaprepes abbreviatus* in soil. *J. Econ. Entomol., 91,* 110–122.

Ram, K., Preisser, E. L., Gruner, D. S., & Strong, D. R. (2008). Metapopulation dynamics override local limits on long-term parasite persistence. *Ecology, 89,* 3290–3297.

Rao, D. R., Mani, T. R., Rajendran, R., Joseph, A. S., Gajanana, A., & Reuben, R. (1995). Development of a high level of resistance to *Bacillus sphaericus* in a field population of *Culex quinquefasciatus* from Kochi, India. *J. Am. Mosq. Control Assoc., 11,* 1–5.

Rasmann, S., Köllner, T. G., Degenhardt, J., Hiltpold, I., Toepfer, S., Kuhlmann, U., Gershenzon, J., & Turlings, T. C. J. (2005). Recruitment of entomopathogenic nematodes by insect-damaged maize roots. *Nature, 434,* 732–737.

Rath, A. C. (2000). The use of entomopathogenic fungi for control of termites. *Biocontrol Sci. Technol., 10,* 563–581.

Rath, A. C., Koen, T. B., & Yip, H. Y. (1992). The influence of abiotic factors on the distribution and abundance of *Metarhizium anisopliae* in Tasmanian pasture soils. *Mycol. Res., 96,* 378–384.

Raymond, B., Johnston, P. R., Nielsen-LeRoux, C., Lereclus, D., & Crickmore, N. (2010). Bacillus thuringiensis: an impotent pathogen. *Trends Microbiol., 18,* 189–194.

Reardon, R., & Podgwaite, J. (1976). Disease–parasitoid relationships in natural populations of *Lymantria dispar* in the northeastern United States. *Entomophaga, 21,* 333–341.

Redmond, C., & Potter, D. A. (1995). Lack of efficacy of *in vivo* and putatively *in vitro* produced *Bacillus popilliae* against field populations of Japanese beetle (Coleoptera: Scarabaeidae) grubs in Kentucky. *J. Econ. Entomol., 88,* 846–854.

Rehner, S. A., & Buckley, E. (2005). A *Beauveria* phylogeny inferred from nuclear ITS and EF1-α sequences: evidence for cryptic diversification and links to *Coryceps* teleomorphs. *Mycologia, 97,* 84–98.

Renn, N., Bywater, A. F., & Barson, G. (1999). A bait formulated with *Metarhizium anisopliae* for the control of *Musca domestica* L. (Dipt., Muscidae) assessed in large-scale laboratory enclosures. *J. Appl. Entomol., 123,* 309–314.

Reyes, A., Christian, P., Valle, J., & Williams, T. (2004). Persistence of *Invertebrate iridescent virus 6* in soil. *BioControl, 49,* 433–440.

Richter, A. R., Fuxa, J. R., & Abdel-Fattah, M. (1987). Effect of host plant on the susceptibility of *Spodoptera frugiperda* (Lepidoptera: Noctuidae) to a nuclear polyhedrosis virus. *Environ. Entomol., 16,* 1004–1006.

Rosa, J. S., Bonifassi, E., Amaral, J., Lacey, L. A., Simões, N., & Laumond, C. (2000). Natural occurrence of entomopathogenic nematodes (*Rhabditida: Steinernema, Heterorhabditis*) in the Azores. *J. Nematol., 32,* 215–222.

Rosengaus, R. B., Maxmen, A. B., Coates, L. E., & Traniello, J. F. A. (1998). Disease resistance: a benefit of sociality in the dampwood termite *Zootermopsis angusticollis* (Isoptera: Termopsidae). *Behav. Ecol. Sociobiol., 44,* 125–134.

Rosin, F., Shapiro, D. I., & Lewis, L. C. (1996). Effects of fertilizers on the survival of *Beauveria bassiana*. *J. Invertebr. Pathol., 68,* 194–195.

Rosin, F., Shapiro, D. I., & Lewis, L. C. (1997). Erratum (in reference to Rosin et al. *J. Invertebr. Pathol.,* 68, 194–195). *J. Invertebr. Pathol., 69,* 84.

Roy, H. E., Steinkraus, D. C., Eilenberg, J., Hajek, A. E., & Pell, J. K. (2006). Bizarre interactions and endgames: entomopathogenic fungi and their arthropod hosts. *Annu. Rev. Entomol., 51,* 331–357.

Sait, S. M., Begon, M., & Thompson, D. J. (1994). The effects of a sublethal baculovirus infection in the Indian meal moth, *Plodia interpunctella*. *J. Anim. Ecol., 63,* 541–550.

Sajap, A. S., & Lewis, L. C. (1988). Histopathology of transovarial transmission of *Nosema pyrausta* in the European corn borer, *Ostrinia nubilalis*. *J. Invertebr. Pathol., 52,* 147–153.

Sanchis, V., & Bourguet., D. (2008). *Bacillus thuringiensis*: applications in agriculture and insect resistance management. A review. *Agron. Sustain. Dev., 28,* 11–20.

Sandhu, S. S., Unkles, S. E., Rajak, R. C., & Kinghorn, J. R. (2001). Generation of benomyl resistant *Beauveria bassiana* strains and their infectivity against *Helicoverpa armigera*. *Biocontrol Sci. Technol., 11,* 245–250.

Sandoski, C. A., Yates, M. W., Olson, J. K., & Meisch, M. V. (1985). Evaluation of Beecomist applied *Bacillus thuringiensis* (H-14) against *Anopheles quadrimaculatus* larvae. *J. Am. Mosq. Control Assoc., 1,* 316–319.

Sansinenea, E., Vázquez, C., & Ortiz, A. (2010). Genetic manipulation in *Bacillus thuringiensis* for strain improvement. *Biotechnol. Lett., 32,* 1549–1557.

Santiago-Álvarez, C., Maranhão, E. A., Maranhão, E., & Quesada-Moraga, E. (2006). Host plant influences pathogenicity of *Beauveria bassiana* to *Bemisia tabaci* and its sporulation on cadavers. *BioControl, 51,* 519–532.

Schnepf, E., Crickmore, N., Van Rie, J., Lereclus, D., Baum, J., Feitelson, J., Zeigler, D. R., & Dean, D. H. (1998). *Bacillus thuringiensis* and its pesticidal crystal proteins. *Microbiol. Mol. Biol. Rev., 62,* 775–806.

Schrank, A., & Vainstein, M. H. (2010). *Metarhizium anisopliae* enzymes and toxins. *Toxicon, 56,* 1267—1274.

Schroer, S., & Ehlers, R.-U. (2005). Foliar application of the entomopathogenic nematode *Steinernema carpocapsae* for biological control of diamondback moth larvae (*Plutella xylostella*). *Biol. Control, 33,* 81—86.

Shannag, H. K., & Capinera, J. C. (2000). Interference of *Steinernema carpocapsae* (Nematoda: Steinernematidae) with *Cardiochiles diaphaniae (Hymenoptera: Braconidae)*, a parasitoid of melonworm and pickleworm (Lepidoptera: Pyralidae). *Environ. Entomol., 29,* 612—617.

Shapiro, D. I., & Glazer, I. (1996). Comparison of entomopathogenic nematode dispersal from infected hosts versus aqueous suspension. *Environ. Entomol., 25,* 1455—1461.

Shapiro, D. I., & Lewis, E. E. (1999). Comparison of entomopathogenic nematode infectivity from infected hosts versus aqueous suspension. *Environ. Entomol., 28,* 907—911.

Shapiro, D. I., & McCoy, C. W. (2000a). Effects of culture method and formulation on the virulence of *Steinernema riobrave* (Rhabditida: Steinernematidae) to *Diaprepes abbreviatus* (Coleoptera: Curculionidae). *J. Nematol., 32,* 281—288.

Shapiro, D. I., & McCoy, C. W. (2000b). Virulence of entomopathogenic nematodes to *Diaprepes abbreviatus* (Coleoptera: Curculionidae) in the laboratory. *J. Econ. Entomol., 93,* 1090—1095.

Shapiro, D. I., Berry, E. C., & Lewis, L. C. (1993). Interactions between nematodes and earthworms: enhanced dispersal of *Steinernema carpocapsae*. *J. Nematol., 25,* 189—192.

Shapiro, D. I., Tylka, G. L., Berry, E. C., & Lewis, L. C. (1995). Effect of earthworms on the dispersal of *Steinernema* spp. *J. Nematol., 27,* 21—28.

Shapiro, D. I., Glazer, I., & Segal, D. (1996). Trait stability in and fitness of the heat tolerant entomopathogenic nematode *Heterorhabditis bacteriophora* IS5 strain. *Biol. Control, 6,* 238—244.

Shapiro, D. I., Glazer, I., & Segal, D. (1997). Genetic improvement of heat tolerance in *Heterorhabditis bacteriophora* through hybridization. *Biol. Control, 8,* 153—159.

Shapiro, D. I., Cate, J. R., Pena, J., Hunsberger, A., & McCoy, C. W. (1999a). Effects of temperature and host age on suppression of *Diaprepes abbreviatus* (Coleoptera: Curculionidae) by entomopathogenic nematodes. *J. Econ. Entomol., 92,* 1086—1092.

Shapiro, D. I., Obrycki, J. J., Lewis, L. C., & Jackson, J. J. (1999b). The effects of crop residue on the persistence of *Steinernema carpocapsae*. *J. Nematol., 31,* 517—519.

Shapiro, M. (1992). Use of optical brighteners as radiation protectants for gypsy moth (Lepidoptera: Lymantriidae) nuclear polyhedrosis virus. *J. Econ. Entomol., 85,* 1682—1686.

Shapiro, M. (2000). Enhancement in activity of homologous and heterologous baculoviruses infectious to beet armyworm (Lepidoptera: Noctuidae) by an optical brightener. *J. Econ. Entomol., 93,* 572—576.

Shapiro, M., & Argauer, R. (1995). Effects of pH, temperature, and ultraviolet radiation on the activity of an optical brightener as a viral enhancer for the gypsy moth (Lepidoptera: Lymantriidae) baculovirus. *J. Econ. Entomol., 88,* 1602—1606.

Shapiro-Ilan, D. I., & Cottrell, T. E. (2005). Susceptibility of lady beetles (Coleoptera: Coccinellidae) to entomopathogenic nematodes. *J. Invertebr. Pathol., 89,* 150—156.

Shapiro-Ilan, D. I., & Gaugler, R. (2002). Production technology for entomopathogenic nematodes and their bacterial symbionts. *J. Ind. Microbiol. Biotechnol., 28,* 137—146.

Shapiro-Ilan, D. I., Lewis, E. E., Behle, R. W., & McGuire, M. R. (2001). Formulation of entomopathogenic nematode-infected cadavers. *J. Invertebr. Pathol., 78,* 17—23.

Shapiro-Ilan, D. I., Gouge, D. H., & Koppenhöfer, A. M. (2002a). Factors affecting commercial success: case studies in cotton, turf and citrus. In R. Gaugler (Ed.), *Entomopathogenic Nematology.* (pp. 333—355). Wallingford: CABI.

Shapiro-Ilan, D. I., Mizell, R. F., III, & Campbell, J. F. (2002b). Susceptibility of the plum curculio, *Conotrachelus nenuphar*, to entomopathogenic nematodes. *J. Nematol., 34,* 246—249.

Shapiro-Ilan, D. I., Gaugler, R., Tedders, W. L., Brown, I., & Lewis, E. E. (2002c). Optimization of inoculation for *in vivo* production of entomopathogenic nematodes. *J. Nematol., 34,* 343—350.

Shapiro-Ilan, D. I., Gardner, W. A., Fuxa, J. R., Wood, B. W., Nguyen, K. B., Adams, B. J., Humber, R. A., & Hall, M. J. (2003a). Survey of entomopathogenic nematodes and fungi endemic to pecan orchards of the southeastern United States and their virulence to the pecan weevil (Coleoptera: Curculionidae). *Environ. Entomol., 32,* 187—195.

Shapiro-Ilan, D. I., Lewis, E. E., Tedders, W. L., & Son, Y. (2003b). Superior efficacy observed in entomopathogenic nematodes applied in infected-host cadavers compared with application in aqueous suspension. *J. Invertebr. Pathol., 83,* 270—272.

Shapiro-Ilan, D. I., Cottrell, T. E., & Gardner, W. A. (2004a). Trunk perimeter applications of *Beauveria bassiana* to suppress adult *Curculio caryae* (Coleoptera: Curculionidae). *J. Entomol. Sci., 39,* 337—349.

Shapiro-Ilan, D. I., Mizell, R. F., III, Cottrell, T. E., & Horton, D. L. (2004b). Measuring field efficacy of *Steinernema feltiae* and *Steinernema riobrave* for suppression of plum curculio, *Conotrachelus nenuphar*, larvae. *Biol. Control, 30,* 496—503.

Shapiro-Ilan, D. I., Jackson, M., Reilly, C. C., & Hotchkiss, M. W. (2004c). Effects of combining an entomopathogenic fungi or bacterium with entomopathogenic nematodes on mortality of *Curculio caryae* (Coleoptera: Curculionidae). *Biol. Control, 30,* 119—126.

Shapiro-Ilan, D. I., Fuxa, J. R., Lacey, L. A., Onstad, D. W., & Kaya, H. K. (2005a). Definitions of pathogenicity and virulence in invertebrate pathology. *J. Invertebr. Pathol., 88,* 1—7.

Shapiro-Ilan, D. I., Duncan, L. W., Lacey, L. A., & Han, R. (2005b). Orchard crops. In P. S. Grewal, R.-U. Ehlers & D. I. Shapiro-Ilan (Eds.), *Nematodes as Biological Control Agents* (pp. 215—229). Wallingford: CABI.

Shapiro-Ilan, D. I., Stuart, R. J., & McCoy, C. W. (2005c). Targeted improvement of *Steinernema carpocapsae* for control of the pecan weevil, *Curculio caryae* (Horn) (Coleoptera: Curculionidae) through hybridization and bacterial transfer. *Biol. Control, 34,* 215—221.

Shapiro-Ilan, D. I., Gouge, G. H., Piggott, S. J., & Patterson Fife, J. (2006). Application technology and environmental considerations for use of entomopathogenic nematodes in biological control. *Biol. Control, 38,* 124—133.

Shapiro-Ilan, D. I., Cottrell, T. E., Jackson, M. A., & Wood, B. W. (2008a). Virulence of Hypocreales fungi to pecan aphids (Hemiptera: Aphididae) in the laboratory. *J. Invertebr. Pathol., 99,* 312—317.

Shapiro-Ilan, D. I., Mizell, R. F., III, Cottrell, T. E., & Horton, D. L. (2008b). Control of plum curculio, *Conotrachelus nenuphar*, with entomopathogenic nematodes: effects of application timing, alternate host plant, and nematode strain. *Biol. Control, 44,* 207—215.

Shapiro-Ilan, D. I., Rojas, G. M., Morales-Ramos, J. A., Lewis, E. E., & Tedders, W. L. (2008c). Effects of host nutrition on virulence and fitness of entomopathogenic nematodes: lipid and protein based supplements in *Tenebrio molitor* diets. *J. Nematol., 40*, 13–19.

Shapiro-Ilan, D. I., Gardner, W. A., Cottrell, T. E., Behle, R. W., & Wood, B. W. (2008d). A comparison of application methods for suppressing the pecan weevil (Coleoptera: Curculionidae) with *Beauveria bassiana* under field conditions. *Environ. Entomol., 37*, 162–171.

Shapiro-Ilan, D. I., Morales, Ramos, J. A., Rojas, M. G., & Tedders, W. L. (2010a). Effects of a novel entomopathogenic nematode-infected host formulation on cadaver integrity, nematode yield, and suppression of *Diaprepes abbreviatus* and *Aethina tumida* under controlled conditions. *J. Invertebr. Pathol., 103*, 103–108.

Shapiro-Ilan, D. I., Cottrell, T. E., Mizell, R. F., III, Horton, D. L., Behle, R. W., & Dunlap, C. A. (2010b). Efficacy of *Steinernema carpocapsae* for control of the lesser peachtree borer, *Synanthedon pictipes*: improved aboveground suppression with a novel gel application. *Biol. Control, 54*, 23–28.

Shapiro-Ilan, D. I., Reilly, C. C., & Hotchkiss, M. W. (2011). Comparative impact of artificial selection for fungicide resistance on *Beauveria bassiana* and *Metarhizium brunneum*. *Environ. Entomol., 40*, 59–65.

Shelton, A. M., Wang, P., Zhao, J.-Z., & Roush, R. T. (2007). Resistance to insect pathogens and strategies to manage resistance: an update. In L. A. Lacey & H. K. Kaya (Eds.), *Field Manual of Techniques in Invertebrate Pathology* (2nd ed.). (pp. 793–811) Dordrecht: Springer.

Shi, W.-B., & Feng, M.-G. (2004). Lethal effect of *Beauveria bassiana, Metarhizium anisopliae*, and *Paecilomyces fumosoroseus* on the eggs of *Tetranychus cinnabarinus* (Acari: Tetranychidae) with a description of a mite egg bioassay system. *Biol. Control, 30*, 165–173.

Shields, E. J., Testa, A., Miller, J. M., & Flanders, K. L. (1999). Field efficacy and persistence of the entomopathogenic nematodes *Heterorhabditis bacteriophora* "Oswego" and *H. bacteriophora* "NC" on alfalfa snout beetle larvae (Coleoptera: Curculionidae). *Environ. Entomol., 28*, 128–136.

Siebeneicher, S. R., Vinson, S. B., & Kenerley, C. M. (1992). Infection of the red imported fire ant by *Beauveria bassiana* through various routes of exposure. *J. Invertebr. Pathol., 59*, 280–285.

Siegel, J. P., Maddox, J. V., & Ruesink, W. G. (1988). Seasonal progress of *Nosema pyrausta* in the European corn borer. *Ostrinia nubilalis*. *J. Invertebr. Pathol., 52*, 130–136.

Sikorowski, P. P., Andrews, G. L., & Broome, J. R. (1973). Transovum transmission of a cytoplasmic polyhedrosis virus of *Heliothis virescens* (Lepidoptera: Noctuidae). *J. Invertebr. Pathol., 21*, 41–45.

da Silva, M. V., Santi, L., Staats, C. C., da Costa, A. M., Colodel, E. M., Driemeier, D., Vainstein, M. H., & Schrank, A. (2005). Cuticle-induced endo/exoacting chitinase CHIT30 from *Metarhizium anisopliae* is encoded by an ortholog of the *chi3* gene. *Res. Microbiol., 156*, 382–392.

Silva, W. O. B., Mitidieri, S., Schrank, A., & Vainstein, M. (2005). Productions and extraction of an extracellular lipase from the entomopathogenic fungus *Metarhizium anisopliae*. *Process Biochem., 40*, 321–326.

Smirnoff, W. A. (1972). Promoting virus epizootics in populations of the swaine jack pine sawfly by infected adults. *BioScience, 22*, 662–663.

Smitley, D. R., Bauer, L. S., Hajek, A. E., Sapio, F. J., & Humber, R. A. (1995). Introduction and establishment of *Entomophaga maimaiga*, a fungal pathogen of gypsy moth (Lepidoptera: Lymantriidae) in Michigan. *Environ. Entomol., 24*, 1685–1695.

Smits, P. (1996). Post-application persistence of entomopathogenic nematodes. *Biocontrol Sci. Technol., 6*, 379–388.

Solter, L. F., & Becnel, J. J. (2007). Entomopathogenic microsporidia. In L. A. Lacey & H. K. Kaya (Eds.), *Field Manual of Techniques in Invertebrate Pathology* (2nd ed.). (pp. 199–221) Dordrecht: Springer.

Solter, L. F., Onstad, D. W., & Maddox, J. V. (1990). Timing of disease-influenced processes in the life cycle of *Ostrinia nubilalis* infected with *Nosema pyrausta*. *J. Invertebr. Pathol., 55*, 337–341.

Solter, L. F., Maddox, J. V., & McManus, M. L. (1997). Host specificity of microsporidia (Protista: Microspora) from European populations of *Lymantria dispar* (Lepidoptera: Lymantriidae) to indigenous North American Lepidoptera. *J. Invertebr. Pathol., 69*, 135–150.

Sosa-Gomez, D. R., & Moscardi, F. (1994). Effect of till and no-till soybean cultivation on dynamics of entomopathogenic fungi in the soil. *Fla. Entomol., 77*, 284–287.

Sprenkel, R. K., Brooks, W. M., Van Duyn, J. W., & Deitz, L. L. (1979). The effects of three cultural variables on the incidence of *Nomuraea rileyi*, phytophagous Lepidoptera, and their predators on soybeans. *Environ. Entomol., 8*, 334–339.

Sree, K. S., & Padmaja, V. (2008). Destruxin from *Metarhizium anisopliae* induces oxidative stress effecting larval mortality of the polyphagous pest *Spodoptera litura*. *J. Appl. Entomol., 132*, 68–78.

St. Leger, R. J., Joshi, L., Bidochka, M. J., & Roberts, D. W. (1996). Construction of an improved mycoinsecticide overexpressing a toxic protease. *Proc. Natl. Acad. Sci. USA, 93*, 6349–6354.

Steinhaus, E. A. (1949). *Principles of Insect Pathology*. New York: McGraw-Hill.

Steinhaus, E. A. (1967). *Principles of Insect Pathology* (2nd ed.). New York: Hafner.

Steinhaus, E. A., & Martignoni, M. E. (1970). An Abridged Glossary of Terms Used. In *Invertebrate Pathology, 2nd ed. Pacific Northwest Forest and Range Experimental Station*. USDA Forest Service.

Steinkraus, D. C. (2007a). Documentation of naturally-occurring pathogens and their impact in agroecosystems. In L. A. Lacey & H. K. Kaya (Eds.), *Field Manual of Techniques in Invertebrate Pathology* (2nd ed.). (pp. 267–281) Dordrecht: Springer.

Steinkraus, D. (2007b). Management of aphid populations in cotton through conservation: delaying insecticide spraying has its benefits. In C. Vincent, M. S. Goettel & G. Lazarovits (Eds.), *Biological Control: A Global Perspective* (pp. 383–391). Wallingford: CABI.

Steinkraus, D. C., & Kramer, J. P. (1989). Development of resting spores of *Erynia aquatica* (Zygomycetes, Entomophthoraceae) in *Aedes aegypti* (Diptera, Culicidae). *Environ. Entomol., 18*, 1147–1152.

Steinkraus, D. C., Hollingsworth, R. G., & Boys, G. O. (1996). Aerial spores of *Neozygites fresenii* (Entomophthorales: Neozygitaceae): density, periodicity, and potential role in cotton aphid (Homoptera: Aphididae) epizootics. *Environ. Entomol., 25*, 48–57.

Steinkraus, D. C., Boys, G. O., Bagwell, R. D., Johnson, D. R., Lorenz, G. M., Meyers, H., Layton, M. B., & O'Leary, P. F. (1998) *Expansion of extension-based aphid fungus sampling service to Louisiana and Mississippi*, 2. San Diego, CA, USA: Proc. Beltwide Cotton Conf. January 5–9, 1239–1242.

Stewart, L. M. D., Hirst, M., Ferber, M. L., Merryweather, A. T., Cayley, P. J., & Possee, R. D. (1991). Construction of an improved baculovirus insecticide containing an insect-specific toxin gene. *Nature, 352*, 85—88.

Stuart, R. J., Barbercheck, M. E., Grewal, P. S., Taylor, R. A. J., & Hoy, C. W. (2006). Population biology of entomopathogenic nematodes: concepts, issues, and models. *Biol. Control, 38*, 80—102.

Stuart, R. J., El-Borai, F. E., & Duncan, L. W. (2008). From augmentation to conservation of entomopathogenic nematodes: trophic cascades, habitat manipulation and enhanced biological control of *Diaprepes abbreviatus* root weevils in Florida citrus groves. *J. Nematol., 40*, 73—84.

Sturhan, D. (1999). Prevalence and habitat specificity of entomopathogenic nematodes in Germany. In R. L. Gwynn, P. H. Smits, C. Griffin, R.-U. Ehlers, N. Boemare & J.-P. Masson (Eds.), *Entomopathogenic Nematodes — Application and Persistence of Entomopathogenic Nematodes* (pp. 123—132). Luxembourg: European Commission.

Subramanian, S., Rabindra, R. J., Palaniswamy, S., Sathiah, N., & Rajasekaran, B. (2005). Impact of granulovirus infection on susceptibility *of Spodoptera litura* to insecticides. *Biol. Control, 33*, 165—172.

Sun, X., van der Werf, W., Banchi, F. J. J. A., Hu, Z., & Vlak, J. M. (2006). Modelling biological control with wild-type and genetically modified baculoviruses in the *Helicoverpa armigera*—cotton system. *Ecol. Model., 198*, 387—398.

Sweeney, A. W. (1981). Fungal pathogens of mosquito larvae. In E. W. Davidson (Ed.), *Pathogenesis of Invertebrate Microbial Diseases* (pp. 403—424). Totowa: Allanheld, Osmun.

Sweeney, A. W., Hazard, E. I., & Graham, M. F. (1985). Intermediate host for an *Amblyospora* sp. (Microspora) infecting the mosquito *Culex annulirostris*. *J. Invertebr. Pathol., 46*, 98—102.

Szewczyk, B., Hoyos-Carvajal, L., Paluszek, M., Skrzecz, I., & de Souza, M. L. (2006). Baculoviruses — re-emerging biopesticides. *Biotechnol. Adv., 24*, 143—160.

Tabashnik, B. E., & Carrière, Y. (2007). Evolution of insect resistance to transgenic crops. In K. J. Tilmon (Ed.), *The Evolutionary Biology of Herbivorous Insects: Specialization, Speciation, and Radiation* (pp. 267—279). Berkeley: University of California Press.

Tabashnik, B. E., Cushing, N. L., Finson, N., & Johnson, M. W. (1990). Field development of resistance to *Bacillus thuringiensis* in diamondback moth (Lepidoptera: Plutellidae). *J. Econ. Entomol., 83*, 1671—1676.

Tabashnik, B. E., Carrière, Y., Dennehy, J., Morin, S., Sisterson, M. S., Roush, R. T., Shelton, A. M., & Zhao, J.-Z. (2003). Insect resistance to transgenic Bt crops: lessons from the laboratory and field. *J. Econ. Entomol., 96*, 1031—1038.

Tabashnik, B. E., Gould, F., & Carrière, Y. (2004). Delaying evolution of insect resistance to transgenic crops by decreasing dominance and heritability. *J. Evol. Biol., 17*, 904—912.

Tabashnik, B. E., Fabrick, J. A., Henderson, S., Biggs, R. W., Yafuso, C. M., Nyboer, M. E., Manhardt, N. M., Coughlin, L. A., Carrière, Y., Dennehy, T. J., & Morin, S. (2006). DNA screening reveals pink bollworm resistance to Bt cotton remains rare after a decade of exposure. *J. Econ. Entomol., 99*, 1525—1530.

Tabashnik, B. E., Gassmann, A. J., Crowder, D. W., & Carrière, Y. (2008). Insect resistance to Bt crops: evidence versus theory. *Nat. Biotechnol., 26*, 199—202.

Tanada, Y. (1963). In E. A. Steinhaus (Ed.), *Insect Pathology: An Advanced Treatise. Epizootiology of infectious diseases, Vol. 2*. New York: Academic Press.

Tanada, Y. (1964). Epizootiology of insect diseases. In P. DeBach (Ed.), *Biological Control of Insect Pests and Weeds* (pp. 548—578). New York: Reinhold.

Tanada, Y., & Fuxa, J. R. (1987). The pathogen population. In J. R. Fuxa & Y. Tanada (Eds.), *Epizootiology of Insect Diseases* (pp. 71—112). New York: Wiley-Interscience.

Tanada, Y., & Kaya, H. K. (1993). *Insect Pathology*. San Diego: Academic Press.

Temime, L., Hejblum, G., Setbon, M., & Valleron, A. J. (2008). The rising impact of mathematical modelling in epidemiology: antibiotic resistance research as a case study. *Epidemiol. Infect., 136*, 289—298.

Thomas, M. B., Wood, S. N., & Lomer, C. J. (1995). Biological control of locusts and grasshoppers using a fungal pathogen: the importance of secondary cycling. *Proc. R. Soc. Lond. B, 259*, 265—270.

Thomas, S. R., & Elkinton, J. S. (2004). Pathogenicity and virulence. *J. Invertebr. Pathol., 85*, 146—151.

Thompson, C. G., Scott, D. W., & Wickman, B. E. (1981). Long-term persistence of the nuclear polyhedrosis virus of the Douglas-fir tussock moth, *Orgyia pseudotsugata* (Lepidoptera: Lymantriidae), in forest soil. *Environ. Entomol., 10*, 254—255.

Thompson, S. R., & Brandenburg, R. L. (2005). Tunneling responses of mole crickets (Orthoptera: Gryllotalpidae) to the entomopathogenic fungus, *Beauveria bassiana*. *Environ. Entomol., 34*, 140—147.

van Tol, R. W. H. M., Van Der Sommen, T. C., Boff, M. I. C., Van Bezooijen, J., Sabelis, M. W., & Smits, P. H. (2001). Plants protect their roots by alerting the enemies of grubs. *Ecol. Lett., 4*, 292—294.

Toledo, J., Campos, S. E., Flores, S., Liedo, P., Barrera, J. F., Villaseñor, A., & Montoya, P. (2007). Horizontal transmission of *Beauveria bassiana* in *Anastrepha ludens* (Diptera: Tephritidae) under laboratory and field cage conditions. *J. Econ. Entomol., 100*, 291—297.

Traniello, J. F. A., Rosengaus, R. B., & Savoie, K. (2002). The development of immunity in a social insect: evidence for the group facilitation of disease resistance. *Proc. Natl. Acad. Sci. USA, 99*, 6838—6842.

Tribe, G. D., & Cillie, J. J. (2004). The spread of *Sirex noctilio* Fabricius (Hymenoptera: Siricidae) in South African pine plantations and the introduction and establishment of its biological control agents. *Afr. Entomol., 12*, 9—17.

Ugelvig, L. V., Kronauer, D. J. C., Schrempf, A., Heinze, J., & Cremer, S. (2010). Rapid anti-pathogen response in ant societies relies on high genetic diversity. *Proc. R. Soc. B, 277*, 2821—2828.

Ugine, T. A., Wraight, S. P., & Sanderson, J. P. (2007). A tritrophic effect of host plant on susceptibility of western flower thrips to the entomopathogenic fungi *Beauveria bassiana*. *J. Invertebr. Pathol., 96*, 162—172.

Undeen, A. H. (1978). Spore-hatching processes in some *Nosema* species with particular reference to *Nosema algerae* Vavra and Undeen. *Misc. Publ. Entomol. Soc. Am., 11*, 29—49.

Undeen, A. H. (1990). A proposed mechanism for the germination of microsporidian (Protozoa, Microspora) spores. *J. Theor. Biol., 142*, 223—235.

Undeen, A. H., & Avery, S. W. (1988). Effect of anions on the germination of *Nosema algerae* (Microspora: Nosematidae) spores. *J. Invertebr. Pathol., 52*, 84—89.

Undeen, A. H., & Vavra, J. (1997). Research methods for entomopathogenic Protozoa. In L. A. Lacey (Ed.), *Manual of Techniques in Insect Pathology* (pp. 117–151). London: Academic Press.

US Environmental Protection Agency. (1998). *The Environmental Protection Agency's White Paper of Bt Plant-Pesticide Resistance Management (EPA Publication 739-S-98-001)*. Washington, DC, USA: Environmental Protection Agency. www.epa.gov/EPA-PEST/1998/January/Day-14/paper.pdf.

Vail, P. V., Hostetter, D. L., & Hoffman, D. F. (1999). Development of the multi-nucleocapsid nucleopolyhedroviruses (MNPVs) infectious to loopers (Lepidoptera: Noctuidae: Plusiinae) as microbial control agents. *Int. Pest Manag. Rev., 4*, 231–257.

Vänninen, I., Tyni-Juslin, J., & Hokkanen, H. (2000). Persistence of augmented *Metarhizium anisopliae* and *Beauveria bassiana* in Finnish agricultural soils. *BioControl, 45*, 201–222.

Vasconcelos, S. D. (1996). Alternative routes for the horizontal transmission of a nucleopolyhedrovirus. *J. Invertebr. Pathol., 68*, 269–274.

Vega, F. E. (2008). Insect pathology and fungal endophytes. *J. Invertebr. Pathol., 98*, 277–279.

Vega, F. E., Dowd, P. F., & Bartelt, R. J. (1995). Dissemination of microbial agents using an autoinoculating device and several insect species as vectors. *Biol. Control, 5*, 545–552.

Vega, F. E., Dowd, P. F., Lacey, L. A., Pell, J. K., Jackson, D. M., & Klein, M. G. (2007). Dissemination of beneficial microbial agents by insects. In L. A. Lacey & H. K. Kaya (Eds.), *Field Manual of Techniques in Invertebrate Pathology* (2nd ed.). (pp. 127–148). Dordrecht: Springer.

Vega, F. E., Posada, F., Aime, M. C., Pava-Ripoll, M., Infante, F., & Rehner, S. A. (2008). Entomopathogenic fungal endophytes. *Biol. Control, 46*, 72–82.

Vega, F. E., Goettel, M. S., Blackwell, M., Chandler, D., Jackson, M. A., Keller, S., Koike, M., Maniania, N. K., Monzón, A., Ownley, B. H., Pell, J. K., Rangel, D. E. N., & Roy, H. (2009). Fungal entomopathogens: new insights on their ecology. *Fungal Ecol., 2*, 149–159.

Vey, A., Hoagland, R. E., & Butt, T. M. (2001). Toxic metabolites of fungal biocontrol agents. In T. M. Butt, C. Jackson & N. Magan (Eds.), *Fungi as Biocontrol Agents* (pp. 311–334). New York: CABI.

Vidal, C., & Fargues, J. (2007). Climatic constraints for fungal biopesticides. In S. Ekesi & N. K. Maniania (Eds.), *Use of Entomopathogenic Fungi in Biological Pest Management* (pp. 39–55). Kerala: Research Signpost.

Vilaplana, L., Wilson, K., Redman, E. M., & Cory, J. S. (2010). Pathogen persistence in migratory insects: high levels of vertically-transmitted virus infection in field populations of the African armyworm. *Evol. Ecol., 24*, 147–160.

Villani, M. G., Allee, L. L., Preston-Wilsey, L., Consolie, Y. X., & Brandenburg, R. L. (2002). Use of radiography and tunnel castings for observing mole cricket (Orthoptera: Gryllotalpidae) behavior in soil. *Am. Entomol., 48*, 42–50.

Vodovar, N., Vallenet, D., Cruveiller, S., Rouy, Z., Barbe, V., Acosta, C., Cattolico, L., Jubin, C., Lajus, A., Segurens, B., Vacherie, B., Wincker, P., Weissenbach, J., Lemaitre, B., Medigue, C., & Boccard, F. (2006). Complete genome sequence of the entomopathogenic and metabolically versatile soil bacterium *Pseudomonas entomophila*. *Nat. Biotechnol., 24*, 673–679.

Wagner, B. L., & Lewis, L. C. (2000). Colonization of corn, *Zea mays*, by the entomopathogenic fungus *Beauveria bassiana*. *Appl. Environ. Microbiol., 66*, 3468–3473.

Wang, C., & St. Leger, R. J. (2007). A scorpion neurotoxin increases the potency of a fungal insecticide. *Nat. Biotechnol., 25*, 1455–1456.

Wang, C., Fan, M., Li, Z., & Butt, T. M. (2004). Molecular monitoring and evaluation of the application of the insect-pathogenic fungus *Beauveria bassiana* in southeast China. *J. Appl. Microbiol., 96*, 861–870.

Wang, X., & Grewal, P. S. (2002). Rapid genetic deterioration of environmental tolerance and reproductive potential of an entomopathogenic nematode during laboratory maintenance. *Biol. Control, 23*, 71–78.

Watanabe, H. (1987). The host population. In J. R. Fuxa & Y. Tanada (Eds.), *Epizootiology of Insect Diseases* (pp. 71–112). New York: Wiley-Interscience.

Wenes, A. L., Bourguet, D., Andow, D. A., Courtin, C., Carré, G., Lorme, P., Sanchez, L., & Augustin, S. (2006). Frequency and fitness cost of resistance to *Bacillus thuringiensis* in *Chrysomela tremulae* (Coleoptera: Chrysomelidae). *Heredity, 97*, 127–134.

Weseloh, R. M. (2002). Modeling the impact of the fungus, *Entomophaga maimaiga* Zygomycetes: Entomophthorales) on gypsy moth (Lepidoptera: Lymantriidae): Incorporating infection by conidia. *Environ. Entomol., 31*, 1071–1084.

Weseloh, R. M. (2004). Effect of conidial dispersal of the fungal pathogen *Entomophaga maimaiga* (Zygomycetes: Entomophthorales) on survival of its gypsy moth (Lepidoptera: Lymantriidae) host. *Biol. Control, 29*, 138–144.

Weseloh, R. M., Andreadis, T. G., & Onstad, D. W. (1993). Modeling the influence of rainfall and temperature on the phenology of infection of gypsy moth, *Lymantria dispar*, larvae by the fungus *Entomophaga maimaiga*. *Biol. Control, 3*, 311–318.

Westwood, G. S., Huang, S.-W., & Keyhani, N. O. (2006). Molecular and immunological characterization of allergens from the entomopathogenic fungus *Beauveria bassiana*. *Clin. Mol. Allergy., 4*, 12, Doi:10.1186/1476-7961-4-12.

Wilson-Rich, N., Stuart, R. J., & Rosengaus, R. B. (2007). Susceptibility and behavioral responses of the dampwood termite *Zootermopsis angusticollis* to the entomopathogenic nematode *Steinernema carpocapsae*. *J. Invertebr. Pathol., 95*, 17–25.

Wilson-Rich, N., Spivak, M., Fefferman, N. H., & Starks, T. S. (2009). Genetic, individual, and group facilitation of disease resistance in insect societies. *Annu. Rev. Entomol., 54*, 405–423.

Wood, H. A., Hughs, P. R., & Shelton, A. (1994). Field studies of the co-occlusion strategy with a genetically altered isolate of the *Autographa californica* nuclear polyhedrosis virus. *Environ. Entomol., 23*, 211–219.

Wraight, S. P., & Carruthers, R. I. (1999). Production, delivery and use of mycoinsecticides for control of insect pests of field crops. In F. R. Hall & J. J. Menn (Eds.), *Methods in Biotechnology: Vol. 5. Biopesticides: Use and Delivery* (pp. 233–269). Totowa: Humana Press.

Wraight, S. P., & Ramos, M. E. (2002). Application parameters affecting field efficacy of *Beauveria bassiana* foliar treatments against Colorado potato beetle *Leptinotarsa decemlineata*. *Biol. Control, 23*, 164–178.

Wraight, S. P., Carruthers, R. I., Jaronski, S. T., Bradley, C. A., Garza, C. J., & Galaini-Wraight, S. (2000). Evaluation of the entompathogenic fungi *Beauveria bassiana* and *Paecilomyces fumosoroseus* for microbial control of the silverleaf whitefly, *Bemisia argentifolii*. *Biol. Control, 17*, 203–217.

Wraight, S. P., Inglis, G. D., & Goettel, M. S. (2007). Fungi. In L. A. Lacey & H. K. Kaya (Eds.), *Field Manual of Techniques in Invertebrate Pathology* (2nd ed.). (pp. 223–248). Dordrecht: Springer.

Wraight, S. P., Lacey, L. A., Kabaluk, J. T., & Goettel, M. S. (2009). Potential for microbial biological control of coleopteran and hemipteran pests of potato. *Fruit Veg. Cer. Sci. Biotechnol., 3*, 25–38.

Wraight, S. P., Ramos, M. E., Avery, P. B., Jaronski, S. T., & Vandenberg, J. D. (2010). Comparative virulence of *Beauveria bassiana* isolates against lepidopteran pests of vegetable crops. *J. Invertebr. Pathol., 103*, 186–199.

Yanagawa, A., & Shimizu, S. (2007). Resistance of the termite, *Coptotermes formosanus* Shiraki to *Metarhizium anisopliae* due to grooming. *BioControl, 52*, 75–85.

Young, E. C. (1986). The rhinoceros beetle project: history and review of the research programme. *Agric. Ecosyst. Environ., 15*, 149–166.

Zahiri, N. S., Federici, B. A., & Mulla, M. S. (2004). Laboratory and simulated field evaluation of a new recombinant of *Bacillus thuringiensis* ssp. *israelensis* and *Bacillus sphaericus* against *Culex* mosquito larvae (Diptera: Culicidae). *J. Med. Entomol., 41*, 423–429.

Zelazny, B. (1972). Studies on *Rhabdionvirus rhinoceros*. I. Effects on larvae of *Oryctes rhinoceros* and inactivation of the virus. *J. Invertebr. Pathol., 20*, 235–241.

Zelazny, B. (1973). Studies on *Rhabdionvirus rhinoceros*. II. Effects on adults of *Oryctes rhinoceros*. *J. Invertebr. Pathol., 22*, 122–126.

Zhuang, L., Zhou, S., Wang, Y., Liu, Z., & Xu, R. (2011). Cost-effective production of *Bacillus thuringiensis* biopesticides by solid-state fermentation using wastewater sludge: effects of heavy metals. *Bioresour. Technol., 102*, 4820–4826.

Zlotkin, E., Fishman, Y., & Elazar, M. (2000). AaIT: from neurotoxin to insecticide. *Biochimie., 82*, 869–881.

Zondag, R. (1979). Control of *Sirex noctilio* F. with *Deladenus siricidicola* Bedding. Part II. Introductions and establishments in the South Island 1968–75. *N.Z.J. For. Sci., 9*, 68–76.

Chapter 4

Baculoviruses and Other Occluded Insect Viruses

Robert Harrison* and Kelli Hoover[†]

*United States Department of Agriculture, Agricultural Research Service, Beltsville, Maryland, USA, [†] Pennsylvania State University, Pennsylvania, USA

Chapter Outline

Insect Pathology. DOI: 10.1016/B978-0-12-384984-7.00004-X

Summary

Baculoviruses are among the most thoroughly studied insect pathogens. Members of *Baculoviridae* possess a large, circular double-stranded DNA genome contained within the enveloped nucleoprotein core of a rod-shaped virion. Baculovirus replication is distinguished by the production of two different virion phenotypes: the occlusion-derived virions, which occur within proteinaceous viral occlusions and initiate infection of the host, and budded virions, which spread infection to other cells and tissues within the host. Baculoviruses have been isolated exclusively from insects, and much of the initial interest in them stemmed from their potential role in insect pest management. The development of cell culture systems for growing baculoviruses and the application of molecular biology methodology to their study led to the development of baculoviruses as popular expression vectors and research tools with a profound expansion in knowledge and understanding of these viruses. This chapter summarizes the current state of baculovirology, covering: baculovirus classification, taxonomy, and phylogeny; basic characteristics of baculovirus virions and viral occlusions; features of two unrelated groups of occluded viruses (the entomopoxviruses and cypoviruses) and a related group of non-occluded viruses (the nudiviruses); mechanisms of infection, gene expression, and replication; their pathology, ecology, and use as biopesticides; and the potential new frontiers for study of this fascinating group of viruses.

4.1. INTRODUCTION

The first accounts of baculovirus pathology were associated with the silkworm, *Bombyx mori*, probably because of the economic significance of this species in silk production. Descriptions suggestive of aspects of nuclear polyhedrosis disease in silkworm larvae can be found in Spanish and Japanese manuals on silkworm rearing dating back to the sixteenth and eighteenth centuries, respectively, and in Marco Girolamo Vida's short poem "De Bombyce" ("Silkworm") in 1527 (Steinhaus, 1975). The first complete description of this disease, referred to as jaundice or grasserie, appeared in a treatise by P. H. Nysten in 1808 (Bergold, 1953; Steinhaus, 1975; Benz, 1986) (Chapter 2).

Examination of the tissues of diseased silkworms led to the identification of crystal-like structures called polyhedra (singular polyhedron) in the cell nuclei of silkworm larvae afflicted with jaundice (Cornalia, 1856; Maestri, 1856). Silkworm larvae that were fed polyhedra contracted jaundice, whereas jaundice did not occur if the polyhedra were removed or destroyed before inoculation (Bolle, 1906).

Polyhedra, referred to hereafter as occlusion bodies (OBs) or occlusions, were also identified in diseased larvae of other lepidopteran species, including larvae of the nun moth, *Lymantria monacha*, suffering from "wilt disease" (von Tubeuf, 1892). The cause of nuclear polyhedrosis was first described as a filterable virus in 1913 (Glaser and Chapman, 1913). A subsequent biochemical study indicated that the infectious particles of this virus were contained within the OBs (Komárek and Breindl, 1924). A polyhedrosis disease of a dipteran, the crane fly, *Tipula paludosa*, was later identified (Rennie, 1923), as was a nuclear polyhedrosis of a hymenopteran, the European spruce sawfly, *Gilpinia hercyniae*. The first granulosis disease, characterized by "granules" much smaller than OBs, was found in the cabbage butterfly, *Pieris brassicae* (Paillot, 1926).

The application of electron microscopy to the study of nuclear polyhedrosis and granulosis viruses (now called nucleopolyhedrovirus or NPV, and granulovirus or GV) allowed for the direct visualization of their structure and key moments in their life cycle. Electron micrographs showing bacilliform rod-shaped virions contained within OBs from diseased larvae of *B. mori* and the gypsy moth, *Lymantria dispar*, were first reported in 1947 (Bergold, 1947). The same characteristic rod-shaped virions were observed the following year in electron micrographs of a GV (Bergold, 1948). Electron microscopy subsequently enabled the visualization of structures formed by the virus in living cells (Hughes, 1953) and the paracrystalline lattice structure of the OBs (Morgan *et al.*, 1955).

A tremendous advance in the study of baculoviruses came with the development of *in vitro* cell culture systems that supported baculovirus infection and replication. The formation of OBs *in vitro* was first observed in 1917 in hemocytes harvested from infected larvae (Glaser, 1917), and the first report of a successful baculovirus infection of insect tissue cultures *in vitro* occurred in 1935 (Trager, 1935). Infection of an established cell line with an NPV was first reported in 1970 (Goodwin *et al.*, 1970). A baculovirus plaque assay was established shortly afterwards using an established cell line, enabling the isolation of clonal strains (Hink and Vail, 1973). Experiments with primary and established cell cultures led to the identification of two distinct baculovirus virion phenotypes, the budded virion (BV) and the occlusion-derived virion (ODV) (Vaughn and Faulkner, 1963; Henderson *et al.*, 1974; Summers and Volkman, 1976).

Early biochemical studies established the presence of DNA in virions of baculoviruses and determined some of the physical and chemical properties of baculovirus DNA molecules (Bergold, 1947; Shvedchikova *et al.*, 1969; Summers and Anderson, 1972). These DNA molecules were recognized as the genetic material of the virus, and the development of techniques for isolating pure preparations of viral DNA, digesting DNA with restriction endonucleases, and separating the digested DNA fragments on the basis of size by agarose gel electrophoresis allowed for the molecular genetic characterization and comparison of baculoviruses (Rohrmann and Beaudreau, 1977; Rohrmann et al., 1978). The application of protein electrophoresis techniques, transcription mapping methods, and DNA sequencing allowed for a description of baculovirus proteins and the identification and characterization of the genes that encoded them, beginning with the occlusion body matrix (polyhedrin) protein of the NPV from the alfalfa looper, *Autographa californica* (Hooft van Iddekinge et al., 1983; Smith et al., 1983c). Finally, the ability to initiate infections of cell culture with viral DNA alone by transfection (i.e., the deliberate introduction of nucleic acids into a cell) allowed for the development of methods to genetically modify baculovirus genomes (Burand et al., 1980; Carstens et al., 1980; Smith et al., 1983a). The procedure for producing recombinant baculovirus genomes was developed from a method for marker-rescuing baculovirus temperature-sensitive (*ts*) mutants that involved co-transfecting insect cells with *ts* mutant viral DNA and a restriction endonuclease fragment derived from wild-type DNA, followed by selection for, and plaque-purification of, virus clones in which the restriction fragment recombined with the *ts* DNA by homologous recombination and replaced the genomic segment containing the *ts* mutation (Miller, 1981). This procedure has allowed for the site-specific deletion of viral DNA and insertion of foreign DNA, which in turn has facilitated the analysis of gene function in baculoviruses and the development of baculoviruses as foreign gene expression vectors (Smith et al., 1983b; Pennock et al., 1984; Maeda et al., 1985). A modification of this method has been developed in which baculovirus genomes are grown and modified as a large plasmid in bacteria, a refinement that has significantly reduced the time and effort required to construct recombinant baculoviruses (Luckow et al., 1993; Westenberg et al., 2010).

Additional advances in the methodology of cellular and molecular biology, as well as further developments in insect tissue culture and the rearing of insect hosts of baculoviruses, have allowed for a wide range of questions about the structure and biology of baculoviruses to be asked and answered. Much of what is known about the basic virology of these viruses derives from studies on a multiple nucleopolyhedrovirus (MNPV) derived from *A. californica* (AcMNPV) (Vail et al., 1971) owing to the ease of working with this virus in cell culture, particularly with cell lines derived from the fall armyworm, *Spodoptera frugiperda*, and the cabbage looper, *Trichoplusia ni*. Other lepidopteran NPVs have also served as models for the study of baculovirus biology, including *Orgyia pseudotsugata* multiple nucleopolyhedrovirus (OpMNPV), *B. mori* nucleopolyhedrovirus (BmNPV), *Spodoptera exigua* multiple nucleopolyhedrovirus (SeMNPV), and *Helicoverpa armigera* single nucleopolyhedrovirus (HearSNPV). Less research has been carried out with GVs or with non-lepidopteran

NPVs, because of the lack of cell culture systems for growing these viruses.

The following sections describe the current state of knowledge about baculoviruses. Section 4.2.3 is devoted to three additional groups of arthropod viruses: the entomopoxviruses and cypoviruses, which also produce virion-containing OBs during infection, and the nudiviruses, which are closely related to baculoviruses. A discussion of the use of baculoviruses as gene expression vectors and the potential medical applications of baculoviruses can be found in other recent reviews (Jarvis, 2009; Krammer and Grabherr, 2010; Wang and Balasundaram, 2010).

4.2. CLASSIFICATION AND PHYLOGENY

The baculoviruses have undergone a series of reorganizations and changes in status and nomenclature. The current taxonomic organization of the baculoviruses reflects knowledge based on years of ultrastructural and biochemical characterization and the rapidly growing database of nucleotide sequence information.

4.2.1. General Characteristics

Morphology: Virions, Occlusion Bodies, and Genomes

Baculovirus virions occur in two structurally and biochemically distinct phenotypes, the ODV found embedded in viral occlusions, and the BV which is secreted from infected cells (Henderson et al., 1974; Summers and Volkman, 1976). Both types of virion consist of cylindrical nucleocapsids contained within a lipid envelope. The rod-like shape of the virions led to the name "baculovirus" for this family of viruses from the Latin word "baculum", which refers to a cane, walking stick, or staff (Vago et al., 1974).

The nucleocapsids range in size from 40 to 60 nanometer (nm) wide by 250 to 300 nm long (Adams and McClintock, 1991), and their length varies with size of the viral genome (Fraser, 1986; Kool et al., 1991; Ihalainen et al., 2010). The nucleocapsids are composed of a nucleoprotein core containing the viral genomic DNA packed into a protein capsid. The DNA in the nucleoprotein core is complexed with an arginine-rich basic protein (Tweeten et al., 1980). Ultrastructural studies and optical diffraction examination of electron micrographs indicate that the capsids consist of stacked rings of subunits arranged perpendicularly to the longitudinal axis of the nucleocapsid and terminate in a nipple-like cap on one end, and a base structure on the other end (Fig. 4.1A) (Beaton and Filshie, 1976; Burley et al., 1982; Fraser, 1986).

The envelope of a BV contains a single nucleocapsid. The BV envelope fits loosely around the nucleocapsid and is modified at one end by the occurrence of spikes or peplomers consisting of an envelope glycoprotein

FIGURE 4.1 Baculovirus virions. (A) Negatively stained nucleocapsids from virions of *Autographa californica* multiple nucleopolyhedrovirus (AcMNPV), with electron-translucent terminal cap structures (red boxes). (B, C) Budded virus virion phenotype, showing terminal arrays of envelope glycoprotein-containing spikes (red boxes). (D) Occlusion-derived virion phenotype of AcMNPV before incorporation into occlusion bodies, showing multiple nucleocapsids enveloped within each virion. Scale bars: 250 nm.

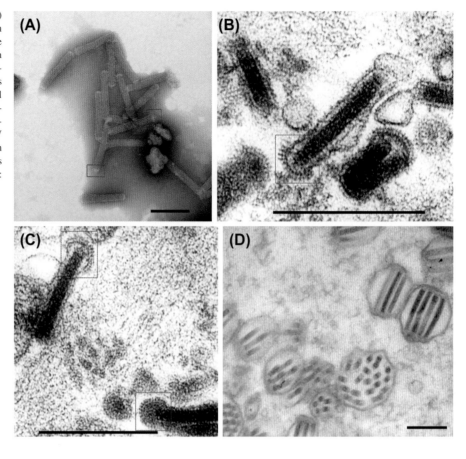

(Fig. 4.1B, C) (Summers and Volkman, 1976). ODV envelopes may contain one nucleocapsid or several nucleocapsids present in a parallel array (Fig. 4.1D) (Summers and Volkman, 1976; Adams and McClintock, 1991). The protein and lipid compositions of BV and ODV envelopes differ significantly (Braunagel and Summers, 1994).

The ODVs occur within paracrystalline viral occlusions (Fig. 4.2). Two distinct types of viral occlusion can be distinguished: the OBs, produced by lepidopteran, dipteran, and hymenopteran NPVs, and the granules, produced by the GVs. The OBs range in size from 0.4 to 2.5 μm in diameter, and can usually be easily seen under a light microscope (Fig. 4.2A). OBs are often roughly cuboidal (Fig. 4.2B), although dodecahedral, tetrahedral, globular, and irregular shapes have been described for some NPVs (Cheng *et al.*, 1998; Moser *et al.*, 2001; Shapiro *et al.*, 2004). The ODV can contain one or multiple nucleocapsids and are arranged randomly within the occlusion matrix (Fig. 4.2C, D). In contrast, granules are ovocylindrical in shape and much smaller in size, ranging from 120 to 350 nm in width by 300 to 500 nm in length (Fig. 4.2E) (Tanada and Hess, 1991). The granules contain only a single virion and are difficult to detect with a light microscope (Fig. 4.2F).

Both OBs and granules consist of a single viral protein that is produced in very high levels in infected cells. This protein is called polyhedrin or granulin, depending on whether it originates from OBs (NPVs) or granules (GVs). Electron microscopy and X-ray analysis indicate that the matrix formed by the occlusion proteins possesses a crystalline lattice structure that is only broken by the occluded ODV (Morgan *et al.*, 1955; Bergold, 1963). X-ray crystallographic analysis of wild-type and mutant AcMNPV OBs has revealed that the unit cell of the OB crystalline lattice consists of 24 copies of the polyhedrin protein subunit, arranged into eight tightly packed trimers (Ji *et al.*, 2010). The structure of the polyhedrin trimer resembles a cube with one corner missing. This corner corresponds to residues 32–48 in the AcMNPV polyhedrin amino acid sequence, the positions of which could not be placed in the final solved OB structure. The amino acid sequence in this region is poorly conserved among the polyhedrins of different NPVs, and it is thought to provide the flexibility required for the OB lattice unit cells to accommodate pleiomorphic ODV as the OB lattice is assembled. Mutations that prevent assembly of ODV into OB tend to group in this region (Ji *et al.*, 2010).

The baculovirus genome consists of a single circular, double-stranded DNA molecule which is covalently

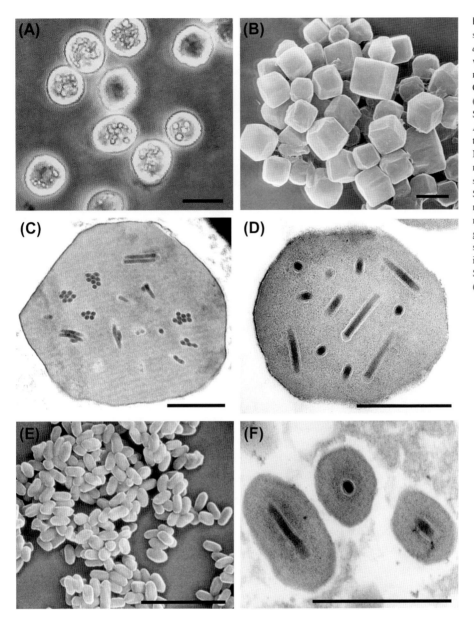

FIGURE 4.2 Baculovirus viral occlusions. (A) Light micrograph of *Spodoptera frugiperda* Sf9 cells infected with *Autographa californica* multiple nucleopolyhedrovirus (AcMNPV) strain C6, showing cuboidal occlusion bodies (OBs) within the infected cells. (B) Scanning electron micrograph of AcMNPV OBs. (C) Cross-section through an OB of a *Spodoptera eridania* NPV, with virions each containing multiple nucleocapsids. (D) Cross-section through an OB of *Trichoplusia ni* SNPV, with virions containing a single nucleocapsid each. (E) Scanning electron micrograph of *Spodoptera frugiperda* granulovirus (GV) granules. (F) Cross-section through granules of a GV infecting larvae of *Argyrotaenia velutinana*. Scale bars: (A), 20 μm; (B, E), 2 μm; (C, D, F), 0.5 μm.

closed and supercoiled (Summers and Anderson, 1972, 1973). The first baculovirus to have the nucleotide sequence of its genome completely determined was that of the C6 clone of AcMNPV (Ayres *et al.*, 1994). As of January 2011, the complete nucleotide sequences for 57 individual baculovirus genomes have been completed and deposited in GenBank (http://www.ncbi.nlm.nih.gov/genomes/GenomesGroup.cgi?taxid=10442).

When the AcMNPV genome was first sequenced, 154 open reading frames (ORFs) that were capable of encoding proteins of at least 50 amino acids and that had minimal (< 75 nucleotides (nt)) overlap with each other were designated as likely protein-encoding genes that were expressed during infection (Ayres *et al.*, 1994). This convention for potential baculovirus genes has been adopted for designating genes in novel baculovirus genome sequences. In addition, the 154 ORFs originally designated in AcMNPV were numbered consecutively. Since the first nucleotide of the AcMNPV sequence was set at the first nucleotide of a previously sequenced *Eco*R I restriction endonuclease fragment, the first gene in the AcMNPV genome was designated as the *ptp* gene that encodes a protein tyrosine/serine phosphatase (Kim and Weaver, 1993; Ayres *et al.*, 1994). This same convention was followed for the genome sequences of *Rachiplusia ou* MNPV and *Plutella xylostella* MNPV, viruses with genome sequences that were almost completely contiguous with that of AcMNPV (Harrison and Bonning, 2003; Harrison and Lynn, 2007). Somewhat different gene numbering schemes were adapted for genomes sequenced shortly after AcMNPV that exhibited differences in gene content and order (Ahrens *et al.*, 1997; Kuzio *et al.*, 1999). Currently, the convention for gene numbering that is followed sets the first nucleotide of the genome sequence as the adenine nucleotide of the first (initiation) codon of the polyhedrin or granulin gene, thus establishing the first gene as the polyhedrin or granulin gene.

The baculovirus genomes that have been sequenced to date vary in size over a roughly two-fold range, from 81,755 base pairs (bp) (*Neodiprion lecontei* NPV; NeleNPV) to 178,733 bp (*Xestia c-nigrum* GV; XecnGV),

with nucleotide distributions that range from 32.4% G + C (*Cryptophlebia leucotreta* GV) to 57.5% G + C (*L. dispar* MNPV). Baculovirus genomes are densely packed with potentially protein-encoding ORFs meeting the criteria described above. These genes are distributed on both strands, in both orientations (Fig. 4.3). The gene order and distribution do not follow a pattern but are conserved among related baculoviruses. Regions on the genomes not occupied by ORFs consist of intergenic spaces with promoter and untranslated regions associated with adjacent genes, and also of regions consisting of copies of repeated sequences known as homologous repeat regions or *hrs* (Fig. 4.3, yellow/black boxes).

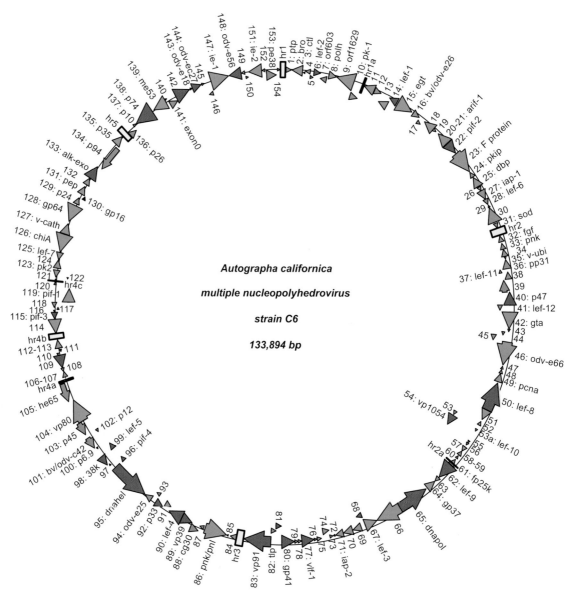

FIGURE 4.3 Circular map of the AcMNPV genome (C6 strain). Genes are presented as arrowheads or arrows, with the direction of the arrowhead indicating the orientation of the open reading frame. Blue arrows/arrowheads indicate core genes that are conserved throughout *Baculoviridae*. The locations for the nine homologous repeat regions (hrs) are indicated by black and yellow rectangles.

The numbers of genes on different baculovirus genomes vary in approximate proportion with the size of the genome. Hence, while 89 potentially protein-coding ORFs have been identified on the small genome of NeleNPV, the genome of XecnGV contains 181 ORFs. The NeleNPV genome, in spite of its small size, still contains intergenic regions not occupied by ORFs. Therefore, differences in genome size among different baculoviruses are mostly attributable to differences in the number of genes. A comparison of NeleNPV and XecnGV ORF composition suggests that the large size of XecnGV is due to the acquisition of many unique genes and copies of repeated genes (Hayakawa *et al.*, 1999; Lauzon *et al.*, 2004); the NeleNPV genome, in contrast, is missing many genes and has no repeated genes.

Genetic Definition of a Baculovirus: The Core Genes

Comparison of complete baculovirus genome sequences from lepidopteran, dipteran, and hymenopteran baculoviruses has identified a set of 30 core genes that are present in all baculoviruses (Table 4.1) (van Oers and Vlak, 2007). These core genes constitute the genetic definition of a baculovirus. Many of these genes encode proteins involved in basic functions such as infection, gene transcription, DNA replication, and progeny virion assembly and structure, while the functions of others are unclear.

Additional sets of genes are conserved within genera or large groups of related baculoviruses (e.g., lepidopteran NPVs) (Jehle *et al.*, 2006a). There are also genes that are present in only a small group of baculoviruses, and genes that are unique to a single baculovirus genome. The combination of core genes, genes characteristic of a baculovirus group, and genes unique to one or a few baculoviruses in a given baculovirus genome provides a genetic definition of that particular baculovirus. Such combinations are likely to influence the morphological and biological properties of individual baculoviruses.

4.2.2. Organization of *Baculoviridae*

Current Classification Scheme: Four Genera

Baculoviruses are placed within a single family, *Baculoviridae*, having the properties described above. A prior classification scheme placed all NPVs into a single genus (*Nucleopolyhedrovirus*), with granuloviruses placed into a second genus (*Granulovirus*). However, sequences from dipteran (mosquito) and hymenopteran (sawfly) NPVs indicated that nucleopolyhedroviruses are polyphyletic (Herniou *et al.*, 2004), which led to a proposed revision of *Baculoviridae* splitting the genus *Nucleopolyhedrovirus* into three genera consisting of lepidopteran, dipteran, and hymenopteran NPVs (Jehle *et al.*, 2006a). This revision has

been accepted by the International Committee on Taxonomy of Viruses (ICTV) and will appear in the Ninth Report of the ICTV (http://www.ictvonline.org/index.asp). On the basis of morphological, biological, and phylogenetic features, members of family *Baculoviridae* are now classified into the following four genera:

- *Alphabaculovirus*: This genus consists of NPVs that infect Lepidoptera. The ODV produced by members of this genus can contain one or many nucleocapsids per enveloped virion, a feature not found among members of the other genera. Most of the NPVs that have been described are found in this genus. The *Alphabaculovirus* type species is *Autographa californica multiple nucleopolyhedrovirus* (AcMNPV).
- *Betabaculovirus:* This genus consists of the GVs, all of which have been isolated from Lepidoptera. The type species for this genus is *Cydia pomonella granulovirus* (CpGV).
- *Gammabaculovirus*: This genus consists of NPVs from *Neodiprion* spp. of sawflies. The genomes of gammabaculoviruses that have been sequenced to date tend to be significantly smaller than those of other baculoviruses and do not contain any genes encoding a BV envelope fusion protein, something found in baculoviruses from the other genera. These viruses form infections that are restricted to the midgut of their hosts (Federici, 1997; Duffy *et al.*, 2007). The type species for this genus is *Neodiprion lecontei nucleopolyhedrovirus* (NeleNPV).
- *Deltabaculovirus*: This genus includes the NPV isolated from the mosquito, *Culex nigripalpus*. The viral occlusion matrix protein of this virus is significantly larger than those of the viruses from the other three genera, and appears to share no significant amino acid sequence identity with other polyhedrins and granulins (Perera *et al.*, 2006). The type species for *Deltabaculovirus* is *Culex nigripalpus nucleopolyhedrovirus* (CuniNPV).

The naming of baculoviruses has followed a binomial standard in which an NPV or GV isolate is named after the host from which it is identified (e.g., *Bombyx mori* NPV). This practice has led to some confusion about the identity of some baculoviruses due to the isolation of the same virus from different hosts, or the isolation of different viruses from the same host (Abdul Kadir and Payne, 1989; Harrison and Bonning, 1999; Jakubowska *et al.*, 2005). A criterion for distinguishing individual species of baculoviruses has been proposed which relies on pairwise nucleotide distances of baculovirus *lef-8*, *lef-9*, and polyhedrin (*polh*) gene sequences estimated using the Kimura-2-parameter model of nucleotide substitution (Jehle *et al.*, 2006b). The assumption underlying this criterion is that genetic differences between two viruses, particularly in

TABLE 4.1 Baculovirus Core Genes

AcMNPV ORF Number[b]	Gene Name	Category	Function/activity
ac6	*lef-2*; late expression factor-2	Transcription/replication	DNA primase accessory factor
ac14	*lef-1*; late expression factor-1	Transcription/replication	DNA primase
ac22	*pif-2*; *per os* infectivity factor-2	Virion structure/assembly	ODV envelope protein; required for oral infectivity
ac40	p47	Transcription/replication	RNA polymerase subunit
ac50	*lef-8*; late expression factor-8	Transcription/replication	RNA polymerase subunit
ac54	vp1054	Virion structure/assembly	Capsid protein; nucleocapsid assembly
ac62	*lef-9*; late expression factor-9	Transcription/replication	RNA polymerase subunit
ac65	*dnapol*; DNA polymerase	Transcription/replication	DNA polymerase
ac68	–	Virion structure/assembly	Nucleocapsid protein; polyhedron morphogenesis
ac77	*vlf-1*; very late factor-1	Transcription/replication; Virion structure/assembly	Hyperexpression of very late genes; virion structural protein; processing/packaging of viral genomes
ac80	gp41	Virion structure/assembly	Tegument protein
ac81	–	Auxiliary/unknown	Unknown
ac83	vp91	Virion structure/assembly	Capsid protein
ac89	*vp39*; major capsid protein	Virion structure/assembly	Major component of the capsid
ac90	*lef-4*; late expression factor-4	Transcription/replication	RNA polymerase subunit
ac92	p33	Auxiliary/unknown	FAD-linked sulfhydryl oxidase
ac95	*dnahel*; *p143*; DNA helicase	Transcription/replication	DNA helicase
ac96	*pif-4*; *per os* infectivity factor-4	Virion structure/assembly	ODV envelope protein; required for oral infectivity
ac98	38k	Virion structure/assembly	Nucleocapsid assembly
ac99	*lef-5*; late expression factor-5	Transcription/replication	Late gene transcription
ac100[a]	p6.9	Virion structure/assembly	DNA condensation; nucleocapsid assembly
ac109	–	Virion structure/assembly	Virion structural protein
ac115	*pif-3*; *per os* infectivity factor-3	Virion structure/assembly	ODV envelope protein; required for oral infectivity
ac119	*pif-1*; *per os* infectivity factor-1	Virion structure/assembly	ODV envelope protein; required for oral infectivity
ac133	*alk-exo*; *an*; alkaline nuclease	Transcription/replication	DNA recombination
ac138	p74	Virion structure/assembly	ODV envelope protein; required for oral infectivity
ac142	–	Virion structure/assembly	Virion assembly
ac143	odv-e18	Virion structure/assembly	ODV envelope protein; virion assembly
ac144	odv-ec27	Virion structure/assembly	Virion structural protein
ac148	odv-e56	Virion structure/assembly	ODV envelope protein; required for oral infectivity

[a]The p6.9 ORF in the currently available version of the NeabNPV genome sequence (GenBank accession no. NC_008252.1) is truncated at 20 codons.
[b]AcMNPV = Autographa californica *multiple nucleopolyhedrovirus*; ORF = open reading frame; ODV = occlusion-derived virus; FAD = flavin adenine dinucleotide.

functionally significant genes, are likely to correlate with phenotypic differences that are also relevant to the definition of a species. According to this criterion, if nucleotide distances between two baculoviruses estimated from single or concatenated alignments of partial *lef-8*, *lef-9*, and *polh* genes are less than 0.015 substitutions/site, the two baculoviruses are considered to be the same species. If pairwise nucleotide distances between two viruses at these loci are greater than 0.05 substitutions/site, the viruses are considered to be different species. If the nucleotide distances lie between 0.015 and 0.050 substitutions/site, additional data (from comparative bioassays, for example) are required to determine whether the two viruses being compared are the same species. Application of this criterion has helped to distinguish between baculovirus species and identify variants of previously identified species (Harrison and Popham, 2008; Mukawa and Goto, 2008).

Phylogeny and Evolution

Complete baculovirus genome sequences have been used to infer relationships and construct an evolutionary history of these viruses with phylogenomics (Herniou *et al.*, 2001, 2003; Herniou and Jehle, 2007). On the basis of whole phylogenetic analysis with the concatenated alignments of the core genes, alpha-, beta-, gamma-, and deltabaculoviruses form four distinct and major lineages (Fig. 4.4).

Many more sequences of individual genes and complete genomes have been determined for lepidopteran NPVs than for the other lineages, so a greater degree of detail about the relationships of these viruses can be observed. Phylogenetic analysis suggests that the lepidopteran NPVs occur in two groups, designated group I and group II (Fig. 4.4) (Zanotto *et al.*, 1993; Lange *et al.*, 2004; Herniou and Jehle, 2007). Two additional subgroups, clade Ia and clade Ib, are evident among group I viruses in analyses of large numbers of NPV isolates (Fig. 4.4) (Jehle *et al.*, 2006a; Herniou and Jehle, 2007), although not all group I NPVs occur in these clades. Analysis of currently available NPVs and GVs indicates that the group II NPVs and GVs generally are very divergent, with clades characterized by long branch lengths. The group I NPV clade, in contrast, is characterized by short branches and strong bootstrap support. The clade Ia NPVs, which include AcMNPV, are very closely related.

Observations of OBs in an ancient sandfly preserved in amber suggest that baculoviruses existed as early as 100 million years ago (Poinar and Poinar, 2005). The monophyletic nature of the alpha-, beta-, gamma-, and deltaculovirus clades, along with the presence of clusters of baculoviruses from the same insect families, indicates that baculoviruses have co-evolved with their insect hosts. Two hypotheses for baculovirus–host co-evolution are consistent with phylogenetic relationships of baculovirus genera: (1) baculoviruses originated with the first ancestral arthropods and co-evolved with their hosts as they diverged into the insect orders that are present today; or (2) baculoviruses originated with one or more already-established orders of arthropods and later cross-infected other insect orders very early during their evolution, thus forming different lineages that specialized in infecting the contemporary insect orders (Herniou *et al.*, 2004).

4.2.3. Other Occluded Viruses and Nudiviruses

Other viruses that share some of the features of baculoviruses are important pathogens of insects. These viruses include two groups of insect viruses unrelated to the *Baculoviridae*, the entomopoxviruses and the cypoviruses, which also package their virions in proteinaceous viral occlusions. In addition, some double-stranded DNA viruses that resemble baculoviruses but do not produce occlusions were previously classified in family *Baculoviridae*, but were subsequently removed. Except for the cypoviruses (see Chapter 5), these viruses are not covered extensively elsewhere in this book and are discussed here. In terms of the cypoviruses in this chapter, the emphasis is on those that infect Lepidoptera.

Occlusion of virions has been observed predominantly in viruses of insects, a feature that may be explainable in terms of the transient nature of insect populations and fluctuations in insect population density. Insect viruses that can persist outside the host for long intervals of time between opportunities for infection and replication would have a substantial selective advantage.

Entomopoxviruses

The entomopoxviruses (EPVs), first discovered in 1963 (Vago, 1963), are members of the family *Poxviridae*, which includes many significant DNA pathogens of vertebrates such as vaccinia (smallpox) virus. EPVs have been placed into their own subfamily, the *Entomopoxvirinae*. This subfamily has three insect-related genera: *Alphaentomopoxvirus*, found in Coleoptera; *Betaentomopoxvirus*, isolated from Orthoptera and Lepidoptera; and *Gammaentomopoxviruses*, isolated from Diptera.

The viral occlusions of EPVs, referred to as spheroids, are oval shaped and range in size from 5 to 20 μm (Fig. 4.5A). The major protein comprising the matrix of the spheroid, called spheroidin, is approximately 109–115 kilodaltons (kDa) (Hall and Moyer, 1991; Sanz *et al.*, 1994). The amino acid sequences determined for individual spheroidins share no significant degree of similarity with baculovirus polyhedrin or granulin sequences. The spheroids of some EPVs also contain a second spindle-shaped paracrystalline structure that is composed of a protein called fusolin (Dall *et al.*, 1993; Gauthier *et al.*, 1995). Fusolin

FIGURE 4.4 Phylogeny of 29 baculovirus core genes from the genome sequences of baculoviruses currently acknowledged as baculovirus species by the International Committee on Taxonomy of Viruses. The phylogenetic tree was inferred from the concatenated alignments of each gene's amino acid sequences by the neighbor-joining method. The four genera of *Baculoviridae* are distinguished by different colors, and subdivisions within *Alphabaculovirus* (groups I and II, clades Ia and Ib) are indicated. Virus taxa include AcMNPV, *Rachiplusia ou* MNPV (RoMNPV), *Bombyx mori* NPV (BmNPV), *Maruca vitrata* MNPV (MaviMNPV), *Epiphyas postvittana* NPV (EppoNPV), *Anticarsia gemmatalis* MNPV (AgMNPV), *Choristoneura fumiferana* DEF NPV (CfDEFNPV), *Antheraea pernyi* NPV (AnpeNPV), *Hyphantria cunea* NPV (HycuNPV), *Choristoneura fumiferana* MNPV (CfMNPV), *Spodoptera litura* MNPV-A and -B (SpltMNPV-A and -B), *Leucania separata* NPV (LeseNPV), *Helicoverpa armigera* SNPV (HearSNPV), *Adoxophyes orana* NPV (AdorNPV), *Adoxophyes honmai* NPV (AdhoNPV), *Orgyia leucostigma* NPV (OrleNPV), *Euproctis pseudoconspersa* NPV (EupsNPV), *Clanis bilineata* NPV (ClbiNPV), *Lymantria dispar* MNPV (LdMNPV), *Trichoplusia ni* SNPV (TnSNPV), *Chrysodeixis chalcites* NPV (ChchNPV), *Mamestra configurata* NPV-A and -B (MacoNPV-A and -B), *Agrotis segetum* NPV (AgseNPV), *Agrotis ipsilon* MNPV (AgipMNPV), *Spodoptera frugiperda* MNPV (SfMNPV), *Spodoptera exigua* MNPV (SeMNPV), *Pseudaletia unipuncta* GV (PsunGV), *Xestia c-nigrum* GV (XecnGV), *Plutella xylostella* GV (PlxyGV), *Adoxophyes orana* GV (AdorGV), *Phthorimaea operculella* GV (PhopGV), *Cydia pomonella* GV (CpGV), *Cryptophlebia leucotreta* GV (CrleGV), *Neodiprion sertifer* NPV (NeseNPV), *Neodiprion abietis* NPV (NeabNPV), *Neodiprion lecontei* NPV (NeleNPV), and *Culex nigripalpus* NPV (CuniNPV).

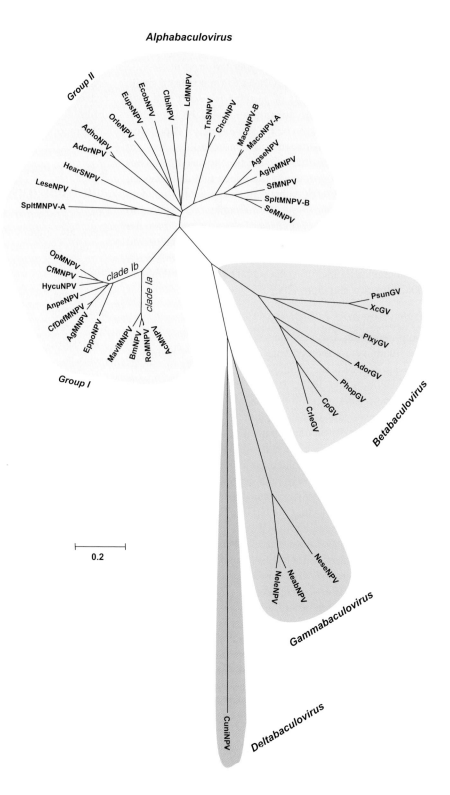

facilitates the infection of host midgut cells by disrupting the peritrophic matrix lining the gut (Mitsuhashi *et al.*, 2007).

Contained within EPV spheroids are virions that measure 230−250 × 250−350 nm and are either ovoid (alpha- and betaentomopoxviruses) or brick-shaped (gammaentomopoxviruses) (Fig. 4.5B). They consist of a DNA genome-containing core with one or two "lateral bodies" of unknown function. The core and lateral bodies are wrapped within one or two membranes, dependent on whether the virion is intracellular (occluded or

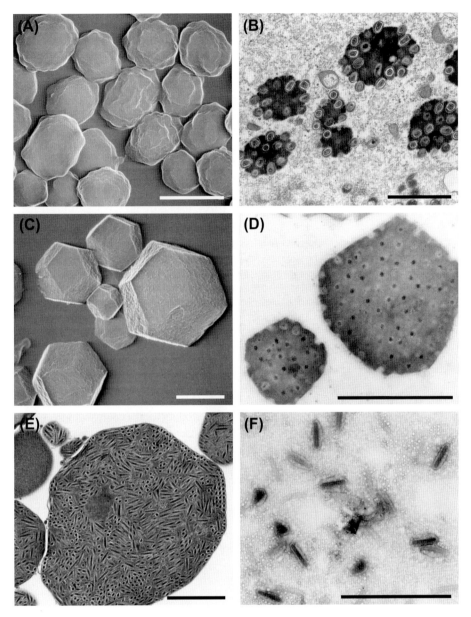

FIGURE 4.5 Viral occlusions and virions of other occluded insect viruses and nudiviruses. (A) Scanning electron micrograph of spheroids of an entomopoxvirus from the cutworm, *Euxoa auxiliaries*. (B) Cross-section through spheroids of *Amsacta moorei* entomopoxvirus. (C) Scanning electron micrograph of cypovirus occlusion bodies (OBs) from the pink bollworm, *Pectinophora gossypiella*. (D) Cross-section through cypovirus OBs in a larva of the tufted apple bud moth, *Platynota idaeusalis*. (E) Viral occlusion-like masses of virions of the nudivirus HzNV-2. (F) Virions of HzNV-2 after treatment of occlusion-like masses with sodium carbonate. Scale bars: 2 μm.

non-occluded) or extracellular. In larvae of the saltmarsh caterpillar, *Estigmene acrea*, infected with the betaentomopoxvirus *Amsacta moorei* EPV (abbreviated AmEPV or AMEV), the spheroids dissolve in the alkaline environment of the midgut, releasing the occluded virions (King *et al.*, 1998). The virions enter midgut cells by fusion between the viral membrane and the membranes of the midgut microvilli (Granados, 1973). The virions migrate to tissue layers adjacent to the gut, where limited replication takes place (King *et al.*, 1998). Infection spreads to the main site of replication in the fat body (Roberts and Granados, 1968) by way of infected hemocytes (Perera *et al.*, 2010a).

The EPV genome is a linear double-stranded molecule that ranges from 232 to 370 kilobase pairs (kbp). The genomes of AmEPV and the EPV from the migratory grasshopper, *Melanoplus sanguinipes*, have been fully sequenced (Afonso *et al.*, 1999; Bawden *et al.*, 2000). These genomes are 232 and 236 kbp, respectively, and have remarkably low G + C contents of 17−18%. The EPV genomes are bound by inverted terminal repeats characteristic of poxvirus genomes in general. However, only 49 (*M. sanguinipes* EPV) and 52 (AmEPV) of the more than 250 ORFs in their genomes are homologues of vertebrate poxvirus genes, and the order of these homologues is poorly conserved between the EPVs and the vertebrate poxviruses (Bawden *et al.*, 2000). Genome replication takes place in electron-dense areas referred to as viroplasms, followed by packaging into virion cores and maturation of virions (Granados and Roberts, 1970; Goodwin *et al.*, 1990; Marlow *et al.*, 1993). The mature virions are occluded into spheroids, with a proportion either leaving the cell to form extracellular virus, or remaining in the cytoplasm as non-occluded virus.

Symptoms of EPV infection vary from host to host and can take a relatively long time to manifest. Of the three EPV genera, members of *Betaentomopoxvirus* that infect lepidopteran hosts kill larvae in the shortest period, with symptoms appearing at 10−12 days followed by death in three to four weeks (Goodwin *et al.*, 1991; Nakai and Kunimi, 1998). Although EPVs have been isolated from significant grasshopper, beetle, and mosquito pests, they have held little appeal as potential biological control agents. However, there is ongoing interest in the development of EPVs as gene expression vectors (Perera *et al.*, 2010b).

Cypoviruses

A disease characterized by the occurrence of occlusions only in the cytoplasm of infected midgut cells was first characterized in *B. mori* larvae in 1934 (Ishimori, 1934). Viruses that cause cytoplasmic polyhedrosis have since been isolated from more than 250 species of insects (Hukuhara and Bonami, 1991) from the orders Lepidoptera, Diptera, and Hymenoptera. These viruses, now referred to as cypoviruses (see Chapter 5), belong to the genus *Cypovirus* within *Reoviridae*, a RNA family of viruses that infects a wide range of vertebrates, invertebrates, and plants. Unlike baculoviruses and entomopoxviruses, the cypoviruses (CPVs) possess genetic material consisting of RNA. The ICTV currently recognizes 16 different species of cypoviruses (CPV-1 to CPV-16), differentiated on the basis of electrophoretic mobilities of the 10 double-stranded RNA molecules that comprise their genomes, as well as nucleotide sequence and antigenic variations. An additional four species from mosquitoes, the winter moth, *Operophtera brumata*, and a black fly have been proposed (Green *et al.*, 2007).

Cypovirus occlusions vary in size from 0.1 to 10 μm and are composed of a polyhedrin protein that has no amino acid sequence similarity with the polyhedrins and granulins of baculoviruses (Fig. 4.5C) (Arella *et al.*, 1988). Contained within the occlusions are icosahedral non-enveloped virions of approximately 60−70 nm in diameter that differ from other reoviruses in having only a single shell surrounding the viral core, rather than two shells (Fig. 4.5D). The structure for CPV polyhedrin in occlusions was recently solved (Coulibaly *et al.*, 2007), and although both the NPV and CPV polyhedrins formed trimers within crystalline lattices characterized by the same-sized unit cells and space grouping, the three-dimensional structures of the two polyhedrins were not at all similar (Coulibaly *et al.*, 2009; Ji *et al.*, 2010).

Cypovirus infection occurs when larvae consume the occlusions, which dissolve in the midgut and release the occluded virions. The virions appear to enter the midgut cells by direct penetration of the cell membrane (Tan *et al.*, 2003). Transcription of the double-stranded RNA genomic segments probably takes place within the capsid and transcripts are capped and extruded from the capsid into the cytoplasm (Zhang *et al.*, 1999; Yu *et al.*, 2008). Progeny genomic segments and capsids are assembled into virions in the cytoplasm of infected cells, presumably by mechanisms shared with other reoviruses. Mature virions are incorporated into progeny occlusions.

Infections by CPVs generally do not lead to cell lysis (Belloncik *et al.*, 1996; Nagata *et al.*, 2003). Although virions have been observed in the hemolymph of infected larvae, infection does not proceed beyond the midgut (Miyajima and Kawase, 1968; Sikorowski *et al.*, 1971). OBs can be found in the feces of infected insects, presumably from infected epithelial cells shed into the midgut lumen. The outcome of a CPV infection can range from death of the infected host to pupation and emergence of a healthy adult, but infection most often results in larval developmental retardation, a decrease in adult fecundity, and the transmission of the disease to offspring (Hukuhara and Bonami, 1991; Belloncik and Mori, 1998).

The first microbial product for the control of insect pests to be registered in Japan was a cypovirus of the pine caterpillar, *Dendrolimus spectabilis* (Kunimi, 2007). There has otherwise been little interest in the use of CPVs for biological control. Researchers are currently developing and testing a system for incorporating heterologous proteins into *B. mori* CPV occlusions in a functional form for use on protein chips (Ijiri *et al.*, 2009).

Nudiviruses

At one point, family *Baculoviridae* consisted of one genus, *Baculovirus*, which was further divided into subgenus A (NPVs), subgenus B (GVs), and a subgenus C that contained non-occluded viruses (Bilimoria, 1986). Subgenus C viruses resembled baculoviruses in terms of replicating in the cell nucleus of their insect hosts and producing rod-shaped enveloped virions, but generally did not produce paracrystalline proteinaceous occlusions (Huger and Krieg, 1991). This group included the OrNV virus from the palm rhinoceros beetle, *Oryctes rhinoceros* (Huger, 1966), the HzNV-1 (or Hz-1) virus persistently infecting a cell line derived from *Helicoverpa zea* (Granados *et al.*, 1978), and a virus that caused mortality in a number of cricket colonies in Germany (Huger, 1985). Beginning with the Sixth Report of the International Committee on Taxonomy of Viruses (Murphy *et al.*, 1995), these viruses have been excluded from *Baculoviridae*. Since then, the genomes of these three viruses and another non-occluded virus, *Helicoverpa zea* virus HzNV-2 (also known as Hz-2 or gonad-specific virus; Raina and Adams, 1995), have been completely sequenced (Cheng *et al.*, 2002; Wang *et al.*, 2007b, 2008). The genomes range in size from 96,944 bp (for the cloned cricket *Gryllus bimaculatus* nudivirus GbNV) to 231,621 bp (for HzNV-2), with nucleotide distributions ranging from 28 to 42% G + C. The number, composition, and order of ORFs vary considerably among these four viruses, but there are 33 genes found in all four genomes (Wang and Jehle, 2009). Twenty of these 33 conserved genes are homologues of baculovirus core genes, and phylogenetic inference with a subset of these genes from baculoviruses and nudiviruses indicates that the four nudiviruses that have been sequenced are a monophyletic group (Wang *et al.*, 2007a; Wang and Jehle, 2009). On the basis of this observation, it has been proposed a new genus, *Nudivirus* (from the Latin *nudi-*, meaning bare, naked, or uncovered), be established to accommodate the non-occluded viruses (Wang *et al.*, 2007c).

Analysis of sequence data from an occluded virus of the shrimp, *Penaeus monodon* (Lightner and Redman, 1981), indicates that it is also a nudivirus. This virus, known as monodon baculovirus in the literature, encodes a poly-hedrin protein and produces viral occlusions, but the pol-yhedrin protein has no sequence similarity to any baculovirus polyhedrin (Chaivisuthangkura *et al.*, 2008). This property, along with observations of a polyhedrin homologue in nudivirus genomes (Wang and Jehle, 2009) and the occurrence of virus-containing bodies resembling viral occlusions in cells infected by OrNV and HzNV-2 (Fig. 4.5E) (Huger and Krieg, 1991; Raina *et al.*, 2000; Rallis and Burand, 2002), suggests that the absence of viral occlusions is not a reliable diagnostic feature of nudiviruses.

The nudiviruses differ considerably from each other in their structure and biology. While OrNV virions measure 120×220 nm (Payne *et al.*, 1977), the virions of HzNV-1 and HzNV-2 (Fig. 4.5F) are noticeably longer and thinner, measuring approximately 415×80 nm (Burand *et al.*, 1983; Hamm *et al.*, 1996). OrNV and GbNV can both be transmitted orally and establish lethal infections of larvae (Zelazny, 1972; Huger, 1985). OrNV infections of adult *O. rhinoceros* tend to be of a chronic nature and cause a cessation of feeding and reduction in oviposition (Zelazny, 1977). Infection and replication of OrNV occur in the midgut (Payne, 1974) and spread to the midgut and other tissues in adults (Huger, 1966; Huger and Krieg, 1991). GbNV replication is also observed in the fat body (Huger, 1985). In contrast, HzNV-1 was not able to estab-lish infections of either *H. zea* larvae or larvae of four other lepidopteran species by oral inoculation or intra-hemocoelomic injection (Granados *et al.*, 1978). HzNV-2, which possesses a genomic nucleotide sequence that is nearly identical to that of HzNV-1, establishes persistent infections characterized by atrophy of both male and female reproductive organs, resulting in sterility (Raina and Adams, 1995; Hamm *et al.*, 1996), along with increases in both the frequency of mating pheromone release and the quantity of pheromone released by infected females (Bur-and *et al.*, 2005). HzNV-2 is transmitted during mating, presumably through contact with a virus-containing waxy plug that forms in the vulva of infected female moths (Rallis and Burand, 2002; Burand *et al.*, 2004), and the effect of HzNV-2 infection on pheromone release is believed to promote transmission by increasing the frequency of mating with infected females.

The deployment of OrNV against *O. rhinoceros*, a pest of coconut palms, is hailed as a successful example of classical biological control of an insect pest (Huger, 2005) (Chapter 3). Progeny OrNV is defecated by infected adults, and an uninfected adult acquires the virus during mating, probably through oral contact with defecated infectious virus (Zelazny, 1976). To control beetle populations, virus inoculum from infected larval homogenates initially was applied at *O. rhinoceros* breeding sites (rotting palm coconut logs), but it was later discovered that the infected adults themselves served as reservoirs that could dissemi-nate the virus throughout the pest population, with subse-quent reductions in beetle populations and palm damage (Zelazny, 1977; Purrini, 1989; Zelazny *et al.*, 1992).

4.3. BACULOVIRUS INFECTION, REPLICATION, PATHOLOGY, AND TRANSMISSION

4.3.1. Primary Infection

Initial infection by baculoviruses generally occurs when the host larva consumes viral occlusions (polyhedra or gran-ules) in the environment while feeding. In addition to horizontal transmission, vertical transmission of baculovi-ruses occurs and is discussed in Section 4.3.5.

Solubilization of Occlusions

The lepidopteran midgut lumen is maintained at an alkaline pH by potassium pumps in the goblet cells of the midgut epithelium (Dow, 1984); this alkaline pH is thought to aid digestion and protect the midgut cells from toxic compounds (Berenbaum, 1980). The crystalline lattice of baculovirus occlusions dissolves rapidly in alkaline diges-tive juices of the midgut (Fig. 4.6). AcMNPV OBs, for example, are completely dissolved in 3 min upon exposure to *T. ni* larval digestive fluid (Pritchett *et al.*, 1982). The atomic structure of AcMNPV polyhedrin in viral occlu-sions suggests that OB dissolution is likely to be mediated by alkali-accelerated breaking of disulfide bonds of poly-hedrin trimers and the ionization of polyhedrin tyrosine side-chains at an alkaline pH (Ji *et al.*, 2010).

FIGURE 4.6 A model for baculovirus primary infection of the midgut and secondary infection of the tracheal system, generalized from studies on lepidopteran larvae infected with multiple nucleopolyhedrovirus (MNPV). Occlusions ingested by larvae dissolve in the alkaline environment of the host midgut lumen, liberating the occlusion-derived virions (ODVs) (1). These ODVs move through the peritrophic matrix (yellow) lining the midgut epithelium (2), in some cases with the assistance of matrix-degrading baculovirus-encoded proteases called enhancins. ODV bind to microvilli of midgut columnar epithelial cells, and nucleocapsids enter the microvilli after fusion between the ODV envelope and the cytoplasmic membrane (3). Nucleocapsids move through the cytoplasm via actin polymerization (red lines). Some nucleocapsids translocate to the nucleus (4), uncoat their DNA, and express early genes, including the genes encoding the major budded virion (BV) envelope glycoprotein (GP64 or F protein). Other nucleocapsids translocate to the basolateral domain of the cell and bud through the plasma membrane (5), forming BV with envelope glycoproteins expressed from the subpopulation of nucleocapsids that entered the nucleus of the cell. Expression of a viral fibroblast growth factor leads to turnover of the basement membrane surrounding tracheoblasts, which facilitates infection of the tracheoblasts with BV from midgut epithelial cells (6). Infection then spreads through tracheal epidermal cells to other tissues beyond the midgut sheath.

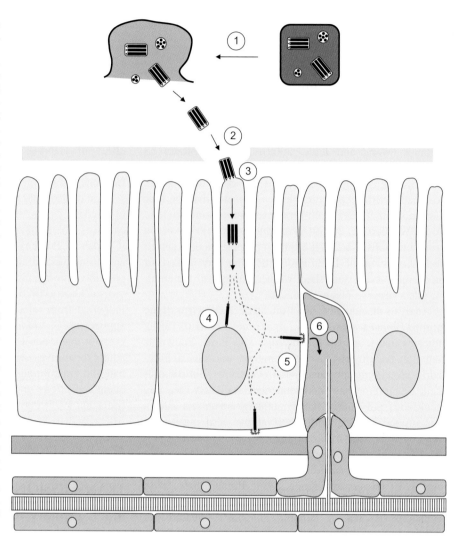

When OBs are dissolved *in vitro* with a sodium carbonate solution, a bag-like structure corresponding to the polyhedral envelope/calyx remains, and virions can be observed trapped within this structure (Harrap, 1972b; Hess and Falcon, 1978). In the larval midgut, host proteases are likely to digest the protein components of this structure and may also contribute to the dissolution of the polyhedrin matrix.

Penetration of the Peritrophic Matrix and Midgut Cell Entry

The midgut of most insects is lined by a mesh of fibers consisting of chitin fibrils linked to glycoproteins and proteoglycans, collectively called the peritrophic matrix (PM) (Hegedus *et al.*, 2009). Like the protective mucosal layer that lines the digestive tracts of mammals, the PM protects the midgut epithelium from abrasion by the insect diet and also restricts access of ingested insect pathogens and toxins to the midgut epithelium.

Enhancins (also referred to as synergistic factors (SFs) in the earlier literature) were the first baculovirus gene products identified that can surmount host defenses (Tanada *et al.*, 1973). They are found in the occlusions of most granuloviruses (Gallo *et al.*, 1991; Wang *et al.*, 1994; Roelvink *et al.*, 1995), and their discovery was serendipitous when it was found that a component of the *Pseudaletia unipuncta* granulovirus (PsunGV) increased the potency of PsunNPV in NPV/GV mixed infections (Tanada, 1959). Hashimoto *et al.* (1991) sequenced the enhancing factor from the *Trichoplusia ni* granulovirus (TnGV) granules that enhanced potency of AcMNPV. Sequencing of several NPV genomes has led to discovery of enhancin genes in some group II NPVs such as the *Lymantria dispar* multiple NPV (LdMNPV), the *Agrotis segetum* NPV (AgseNPV-A), and *Mamestra configurata* NPV (MacoNPV) (Bischoff and Slavicek, 1997; Li *et al.*, 2002b; Jakubowska *et al.*, 2006), but to date they have not been found in group I NPVs or baculoviruses that infect Hymenoptera or Diptera.

As metalloproteases (Lepore *et al.*, 1996), enhancins act by degrading insect intestinal mucin, a major glycoprotein of the PM, facilitating penetration of virions through this barrier to reach the midgut cells to initiate infection (Wang and Granados, 1997). However, earlier studies suggest that at least some enhancins also participate in entry of virions into midgut cells (Tanada *et al.*, 1975) by binding to specific sites on the cell membrane (Uchima *et al.*, 1988). This possibility is further supported by studies on the baculovirus that infects *L. dispar*. Immunoelectron microscopy indicated that LdMNPV has two enhancin proteins (E1 and E2) that are located in and traverse the ODV envelopes associated with the nucleocapsids (Slavicek and Popham, 2005). In addition, the N-termini of these enhancins extend out from the ODV envelope, placing them in a position to interact with the PM and midgut epithelial cells. In *L. dispar* larvae, removal of the PM using a stilbene optical brightener, which eliminated the enhancins' function of facilitating penetration of ODV through the PM, still resulted in a 14-fold lower potency of an LdMNPV mutant from which both enhancins had been deleted compared with wild-type virus containing intact enhancin genes (Hoover *et al.*, 2010). Taken together, these findings suggest that enhancins could facilitate viral entry into midgut cells.

After penetration through the PM, ODVs infect columnar epithelial cells in the midgut. Entry of ODVs into cells appears to occur by fusion between the host cell plasma membrane and the ODV envelope (Kawanishi *et al.*, 1972; Granados and Lawler, 1981). Fusion is preceded by binding of the ODV to the microvilli of the columnar cells by way of a specific, protease-sensitive receptor (Horton and Burand, 1993), and followed by the appearance of nucleocapsids in the microvilli (Figs. 4.6 and 4.7A) (Granados, 1978).

A group of ODV envelope proteins, known collectively as *per os* infectivity factors (PIFs), are required for efficient infection of the midgut. These proteins are encoded by core genes and include PIF-0 (P74, *ac138*), PIF-1 (*ac119*), PIF-2 (*ac22*), PIF-3 (*ac115*), PIF-4 (*ac96*), and PIF-5 (ODV-E56, *ac148*). Mutation or elimination of these genes severely impairs the oral infectivity of the resulting virus, but does not affect the infectivity of BVs (Kuzio *et al.*, 1989; Kikhno *et al.*, 2002; Pijlman *et al.*, 2003). PIF-0/P74, PIF-1, and PIF-2 are required for optimal levels of ODV binding and fusion with epithelial cell membranes (Haas-Stapleton *et al.*, 2004; Ohkawa *et al.*, 2005), but eliminating expression of PIF-3 and PIF-5/ODV-E56 does not affect ODV binding and fusion (Ohkawa *et al.*, 2005; Sparks *et al.*, 2011). These proteins, and possibly other ODV proteins, are presumably assembled into an oligomeric structure that mediates ODV binding and fusion, and a preliminary study of ODV protein–protein interactions as detected using a yeast two-hybrid screen has been published (Peng *et al.*, 2010). The *H. armigera* SNPV and AcMNPV P74 proteins and AcMNPV ODV-E56 bind to host proteins of 30, 35, and 97 kDa,

respectively, in BBMV prepared from host midguts (Yao *et al.*, 2004; W. Zhou *et al.*, 2005; Sparks *et al.*, 2011), but little is known about the ODV receptor(s).

Transport and Entry into the Nucleus

After virion entry into midgut cells, nucleocapsids travel through the cytoplasm to the nucleus, the site of baculovirus replication (Fig. 4.6). In a study of BV nucleocapsid motility, nucleocapsids were found to be propelled through the cytoplasm by actin polymerization and the formation of actin filaments, catalyzed by a cellular protein complex called Arp2/3, which is in turn activated by a protein, PP78/83 (*ac9*), located in the basal structure of the capsid (Russell *et al.*, 1997; Goley *et al.*, 2006; Ohkawa *et al.*, 2010). This mechanism is likely to be used by ODV nucleocapsids as well. Microscopic examination of nucleocapsids moving through the cytoplasm revealed that they are trailed by actin "comet tails" and frequently come into contact with the nuclear envelope (Ohkawa *et al.*, 2010). The gene encoding PP78/83 is not a core gene and is found only in alphabaculoviruses.

Nucleocapsids can be found in association with nuclear pores and also in the nucleoplasm very soon after infection, suggesting that nucleocapsids enter the nucleus through the nuclear pores (Summers, 1971; Granados and Lawler, 1981; Ohkawa *et al.*, 2010). This mechanism has been confirmed using inhibitors of transport through the nuclear pores, which cause a 70–78% decrease in the quantity of nucleocapsids found in the nucleus after infection (Ohkawa *et al.*, 2010).

4.3.2. Replication and Virion Assembly

Viral DNA Uncoating

Release and decondensation of DNA from a *T. ni* GV correlate with the activity of a capsid-associated kinase of unknown origin and phosphorylation of a homologue of the AcMNPV p6.9 protein (*ac100*), the major protein component of the nucleoprotein core found within the baculovirus capsid (Tweeten *et al.*, 1980; Wilson and Consigli, 1985). Chelation of zinc was found to trigger the release of DNA from TnGV nucleocapsids and to activate the capsid-associated kinase, suggesting that phosphorylation of p6.9 triggers DNA uncoating (Funk and Consigli, 1992, 1993). There is evidence that viral DNA is bound by host histones after uncoating (Wilson and Miller, 1986).

Temporally Regulated Viral Gene Expression

Baculovirus genes are expressed in a sequential pattern, with different genes expressed at different times after infection (Carstens *et al.*, 1979; Dobos and Cochran, 1980; Kelly and Lescott, 1981). Two broad temporal categories of genes, the early genes and the late genes, can be distinguished. Early genes are transcribed by host RNA

FIGURE 4.7 Baculovirus infection and replication. (A) Cross-section of midgut epithelial cell microvilli, with some microvilli containing *Helicoverpa zea* single nucleopolyhedrovirus (HzSNPV) nucleocapsids (red boxes). (B) Cell infected with HzSNPV showing nuclear hypertrophy and the formation of virogenic stroma (VS). The nuclear membrane (NM) is indicated. (C) Midgut of *Lymantria dispar* infected with a recombinant clone of *L. dispar* multiple nucleopolyhedrovirus (MNPV) engineered to express the marker gene β-galactosidase during infection. Blue staining indicates infection of the virus spreading to the tracheae (T) associated with the midgut. (D) Midgut epithelial cell infected with an MNPV, with envelopment of nucleocapsids and assembly of occlusion-derived virions (ODVs) into occlusion bodies (OBs) occurring in the nucleus. The NM and VS are indicated. (E) Cell infected with *Rachiplusia ou* MNPV, with mature OBs in the nucleus. Three fibrillar bodies in the nucleus and one fibrillar body in the cytoplasm are indicated (FB). (F) Cell infected with *Plodia interpuntella* granulovirus, with mature granules in a mixed nuclear—cytoplasmic environment formed after rupture of the nuclear membrane. Scale bars: (A) 0.5 μm, (B), (D—F) 2 μm, (C) 200 μm.

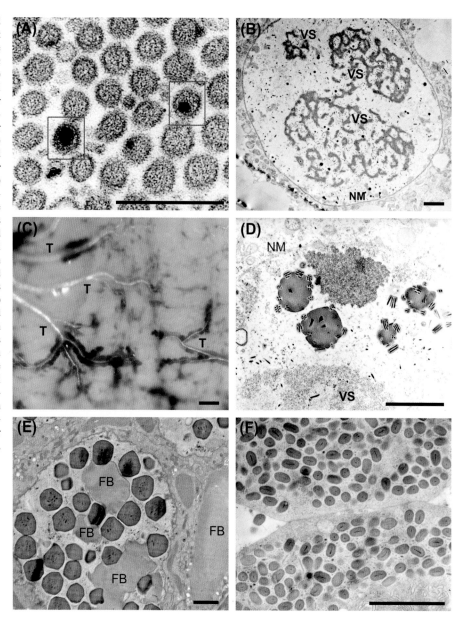

polymerase II before the onset of viral DNA replication, while late genes are transcribed by an RNA polymerase encoded by viral genes after viral DNA replication begins (Huh and Weaver, 1990; Hoopes and Rohrmann, 1991; Guarino *et al.*, 1998). Early genes can be further classified into immediate early genes and delayed early genes (Guarino and Summers, 1986a; Ross and Guarino, 1997). Immediate early genes require no viral gene products for expression, and generally encode transcriptional transactivators of viral genes. Delayed early genes require immediate early gene products for optimal expression.

The most extensively studied of the immediate early genes is *ie-1* (immediate early-1; *ac147*). In transient expression assays, the IE-1 protein can stimulate expression of a reporter gene under the control of baculovirus early gene promoters (Guarino and Summers, 1986a; Blissard and Rohrmann, 1991; Lu and Carstens, 1993). When an early promoter is linked to one of the homologous repeat (*hr*) regions of AcMNPV, the degree of transcriptional stimulation achieved is augmented several hundred-fold, indicating that the *hr* elements act as transcriptional enhancers (Guarino and Summers, 1986b; Guarino *et al.*, 1986). Stimulation of transcription by IE-1 is dependent on the assembly of IE-1 homodimers, and enhancer-dependent augmentation of transcriptional transactivation involves binding of IE-1 dimers to *hr* repeats (Rodems and Friesen, 1995; Rodems *et al.*, 1997). Transcription of the *ie-1* gene region early during AcMNPV infection also gives rise to a second immediate early transcript, *ie-0*, due to splicing of an additional exon to the 5′ end of the *ie-1* ORF (Chisholm and

Henner, 1988). This is the only instance of splicing that has been reported for any baculovirus gene. The protein encoded by this larger transcript, IE-0, consists of the entire IE-1 amino acid sequence and an additional 54 amino acids added to the N-terminus by the *ie-0* exon. IE-0 also is able to transactivate early genes (Kovacs *et al.*, 1991), but in AcMNPV it is not necessary to produce both IE-1 and IE-0 during infection to obtain a productive infection (Stewart *et al.*, 2005). In LdMNPV, only IE-0 is able to activate transcription (Pearson and Rohrmann, 1997). In addition to *ie-1*, the immediate early genes *ie-2* (*ac151*) and *pe38* (*ac153*) encode transcriptional transactivators (Lu and Carstens, 1993; Yoo and Guarino, 1994; Jiang *et al.*, 2006). AcMNPV clones from which expression of these genes has been eliminated are still able to infect and replicate in host cells, albeit with reductions in progeny virus production and infectivity (Prikhod'ko *et al.*, 1999; Milks *et al.*, 2003). In contrast, *ie-1* is essential for replication (Stewart *et al.*, 2005). The *ie-2* gene is found only in group I alphabaculoviruses, while *pe38* is found in group I alphabaculoviruses and four betabaculoviruses. Although the *ie-1* gene plays a crucial role in the AcMNPV life cycle, it is found only in alpha- and betabaculoviruses.

The delayed early genes encode proteins with a variety of functions, including proteins required for viral DNA synthesis and late gene expression. The promoter regions of several immediate and delayed early genes contain basal promoter elements of host cellular RNA polymerase II promoters, including TATA sequences (consensus TATAA) and initiator sites (consensus TCA(T/G)T), which interact with RNA polymerase II and basal transcription factors (Arnosti, 2003; Xing *et al.*, 2005). Deletion of these elements in baculovirus promoters eliminates expression (Blissard *et al.*, 1992; Guarino and Smith, 1992). Other promoter motifs have also been identified in baculovirus early gene promoters, and in some cases, have been demonstrated to function in transcription, such as GATA and CACGTG sequences in the *gp64* promoter (Kogan and Blissard, 1994).

Baculovirus late genes mostly encode structural components of virions and viral occlusions and are expressed through the action of proteins encoded by early genes. Baculovirus genes required for late gene expression have been identified using a transient expression assay that utilized a plasmid containing a reporter gene under the control of a late gene promoter, which was co-transfected into cultured insect cells with plasmids containing genes that were candidates for encoding proteins involved in late gene expression (Passarelli and Miller, 1993). This approach identified 19 late expression factors, or *lef*s, in the AcMNPV genome that were essential for, or stimulated, late gene promoter activation in the assay (Rapp *et al.*, 1998). Nine of these genes are directly or indirectly involved in viral DNA replication and may influence promoter activity indirectly by

fulfilling the requirement of late gene expression for DNA replication or increasing the copy number of the reporter gene plasmid template (Lu and Miller, 1997). IE-1 and IE-2 were also identified as *lef*s, possibly due to activating transcription of other *lef*s in the screen for late expression factors. The *p35* gene, which blocks host cell apoptosis (see Section 4.3.4), also functioned as a *lef*, probably by preventing apoptosis induced by IE-1 or the process of *lef*-stimulated DNA replication (Lu and Miller, 1995; Schultz *et al.*, 2009).

Four of the nineteen *lef*s — *lef-8*, *lef-9*, *p47*, and *lef-4* — are core genes that encode components of the baculovirus RNA polymerase (Guarino *et al.*, 1998). LEF-8 and LEF-9 are believed to catalyze RNA synthesis (Lu and Miller, 1994; Passarelli *et al.*, 1994), while LEF-4 confers a $5'$ 7-methylguanosine cap, a structure necessary for nuclear export, stability, and translation of eukaryotic messenger RNAs (Gross and Shuman, 1998; Jin *et al.*, 1998; Furuichi and Shatkin, 2000). Another core gene-encoded protein, LEF-5, stimulates initiation of transcription by the RNA polymerase complex (Guarino *et al.*, 2002). The baculovirus RNA polymerase initiates transcription from a consensus (A/T/G)TAAG sequence in late gene promoters (Rankin *et al.*, 1988; Xing *et al.*, 2005). Normally, transcripts produced by RNA polymerase II are processed by cleavage of the transcript upstream of the $3'$ end of the transcript near a consensus AAUAAA sequence, followed by addition of a polyadenylate tail (Zhao *et al.*, 1999). Baculovirus late gene transcripts terminate within an AT-rich region downstream of the ORF and are polyadenylated without cleavage (Jin and Guarino, 2000).

A subclass of baculovirus late genes, referred to as the very late genes, consists of two genes, *polh* (polyhedrin; *ac8*) and *p10* (*ac137*), which differ from late genes in terms of their timing and level of expression. Transcription of these two genes initiates approximately 6 h after initiation of transcription of other late genes (Thiem and Miller, 1990) and the transcripts and proteins encoded by these genes accumulate to enormous quantities, eventually accounting for much of the total mRNA and protein in infected cells (Rohel *et al.*, 1983; Smith *et al.*, 1983c). An additional expression factor called VLF-1 (very late expression factor-1), encoded by core gene *ac77*, was identified that augments transcription from the *polh* and *p10* promoters by interacting with an AT-rich "burst sequence" thought to account for the very high levels of transcription of these genes (Ooi *et al.*, 1989; Yang and Miller, 1999; Mistretta and Guarino, 2005).

Virogenic Stroma and Peristromal Region Formation, and DNA Replication

Within hours after cell infection, an electron-dense granular structure, the virogenic stroma, begins to form in the nuclei, surrounded by an electron-translucent region known as the ring zone or peristromal region (Figs. 4.7B and 4.8) (Xeros,

FIGURE 4.8 Model of secondary infection by budded virus (BV) and viral replication. BVs bind to a cell-surface receptor and are internalized through clathrin-mediated endocytosis (1). Endosomes containing BV are acidified, triggering fusion between the cellular membrane and the BV envelope (2), which releases the nucleocapsid into the cytoplasm. Nucleocapsids move through the cytoplasm via actin polymerization (red lines) and translocate to the nucleus (3). DNA is released in the nucleus and viral early genes are expressed. The nucleus swells and the virogenic stroma (VS) forms, accompanied by DNA replication and assembly of progeny nucleocapsids (4). Initially, progeny nucleocapsids exit the nucleus and bud from the plasma membrane to become BVs (5). Later during infection, nucleocapsids are enveloped within the peristromal region of the nucleus to form the occlusion-derived virions (ODV) (6). Polyhedrin crystallizes around ODVs to form the occlusion bodies (7). The surface of mature occlusion bodies is covered with the calyx, possibly with the involvement of fibrillar bodies (FB).

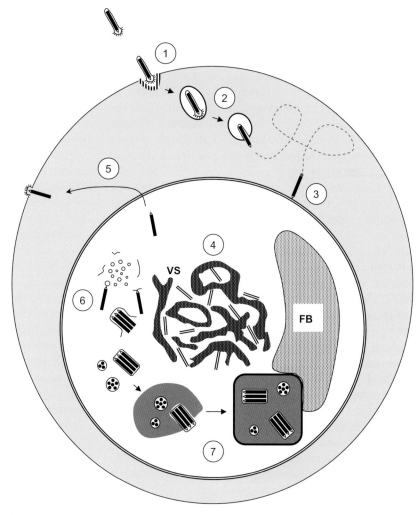

1956; Tanada and Hess, 1976; Granados, 1978). The virogenic stroma and peristromal region are two discrete subnuclear compartments in infected cells (Kawasaki *et al.*, 2004; Nagamine *et al.*, 2008). The virogenic stroma contains DNA (Young *et al.*, 1993) and is the site for viral DNA replication (Kawasaki *et al.*, 2004).

On the basis of plasmid replication studies in NPV- and GV-infected cells, *hr* regions have been identified as the origins (initiation sites) of DNA replication for baculovirus genomes (Pearson *et al.*, 1992; Kool *et al.*, 1993; Hilton and Winstanley, 2007). Non-*hr* origins of replication have also been identified (Pearson *et al.*, 1993) and plasmid DNA with no viral origin of replication is also replicated in baculovirus-infected cells (Wu *et al.*, 1999), which raises questions about the template specificity of baculovirus DNA replication enzymes.

Plasmid replication assays have also been used to identify baculovirus genes involved in DNA replication (Kool *et al.*, 1994; Lu and Miller, 1995). Baculoviruses encode their own DNA polymerase (*dnapol*), encoded by core gene *ac65*, which is capable of synthesizing deoxyribonucleotide polymers (Tomalski *et al.*, 1988; Hang and Guarino, 1999; Vanarsdall *et al.*, 2005). Baculoviruses also contain a gene encoding a DNA helicase (*dnahel*; *p143*), encoded by core gene *ac95*, which unwinds a DNA double helix so that new DNA strands can be synthesized using the individual parental strands as templates (Lu and Carstens, 1991; McDougal and Guarino, 2000). The *lef-1* core gene encodes a DNA primase, which synthesizes an RNA primer from which the DNA polymerase can initiate synthesis of a DNA strand (Mikhailov and Rohrmann, 2002). The *lef-2* core gene product interacts with LEF-1, but it is otherwise unclear what role it plays in DNA replication. The *lef-3* gene, which is not found in gamma- or deltabaculovirus genomes, encodes a single-stranded DNA binding protein that prevents the formation of double helices and other secondary structures by individual DNA strands (Hang *et al.*, 1995). IE-1 was also found to be essential for *hr*-initiated plasmid DNA replication, presumably due to its *hr*-binding activity or its ability to activate transcription of the other DNA replication genes. BmNPV IE-1, together with LEF-3, DNA helicase, and a BmNPV *hr* element, forms a structure in the nuclei of cells transfected with these elements that may be the precursor for the virogenic stroma (Nagamine *et al.*, 2006). Several other non-core genes appear to be involved in *hr*-initiated DNA synthesis (Lu and Miller, 1995; Gomi *et al.*, 1997; Milks *et al.*, 2003). Baculoviruses may also rely on additional host proteins, such as topoisomerase, for replication of their genomes (Rohrmann, 2008).

There is evidence to support a rolling circle mechanism of baculovirus genome replication in which concatamers of multiple genomes are synthesized and processed into individual circular molecules (Oppenheimer and Volkman, 1997; Wu *et al.*, 1999). Mutant AcMNPV clones missing the core gene encoding an alkaline nuclease (*ac133*; *alk-exo*) or a gene encoding a DNA binding protein (*dbp*; *ac25*) both produce subgenomic fragments (Okano *et al.*, 2007; Vanarsdall *et al.*, 2007). Both LEF-3 and DNA binding protein (DBP) have DNA unwinding and annealing activity (Mikhailov *et al.*, 2005, 2008) and LEF-3 also catalyzes DNA strand exchange (Mikhailov *et al.*, 2006). These activities are consistent with DNA recombination and, together with the mutant phenotypes of *dbp* and *alk-exo*, suggest that homologous recombination plays a role in DNA replication, possibly by joining subgenomic fragments into larger molecules that are then processed into circular genomes (Vanarsdall *et al.*, 2007; Rohrmann, 2008).

Virion Assembly and Budded Virion Egress

Electron microscopic observations on baculovirus-infected cells suggest that assembly of virion nucleocapsids occurs in the intrastromal spaces of the virogenic stroma (Fraser, 1986; Young *et al.*, 1993). The main structural component of the capsid is the VP39, a 39 kDa protein encoded by the core gene *ac89* (Pearson *et al.*, 1988; Thiem and Miller, 1989). Some minor capsid proteins also have been identified, including PP78/83 (Vialard and Richardson, 1993; Russell *et al.*, 1997); VLF-1, which localizes to one of the capsid ends (Vanarsdall *et al.*, 2006); VP1054 (*ac54*), encoded by a core gene and required for capsid assembly (Olszewski and Miller, 1997a); 38K (*ac98*), a core gene product that interacts with PP78/83 and VP39 (Wu *et al.*, 2008); a capsid-associated protein encoded by the core gene *ac142* (McCarthy *et al.*, 2008); and ODV-EC43 (*ac109*), a core gene product that occurs in both nucleocapsid and envelope protein fractions (Fang *et al.*, 2003).

The nucleoprotein core of the nucleocapsid consists of baculovirus DNA complexed with the p6.9 protein, which carries a substantial positive charge due to its high arginine content. This positive charge neutralizes the negatively charged DNA phosphodiester backbone and facilitates the condensation of the genome into the capsid structure (Kelly *et al.*, 1983). The *p6.9* gene is considered to be a core gene, although a full-length ORF for this gene is not present in the genome sequence currently available for the gamma-baculovirus NeabNPV (Duffy *et al.*, 2006). Electron microscopic observations indicate that capsids are preassembled then filled with the nucleoprotein core (Fraser, 1986). However, AcMNPV clones deleted for the genome processing genes DBP, alkaline nuclease, and VLF-1 also produce aberrant capsids, suggesting that capsid formation and genome replication are linked processes (Vanarsdall *et al.*, 2006, 2007; Okano *et al.*, 2007). Proper nucleocapsid assembly also requires the formation of actin filaments in the nucleus, a process that is mediated by the same Arp2/3-PP78/83 complex that is involved in cytoplasmic motility of parental nucleocapsids after infection (Volkman, 1988; Goley *et al.*, 2006).

After assembly in the nucleus, nucleocapsids exit the nucleus, travel to the plasma membrane, and bud through the plasma membrane, thus producing the budded virion phenotype (Figs. 4.6 and 4.8). Egress of assembled nucleocapsids from the nucleus to the cytoplasm requires GP41, a protein encoded by core gene *ac80* and found in the tegument between the envelope and capsid (Whitford and Faulkner, 1992; Olszewski and Miller, 1997b), as well as two additional proteins encoded by *ac66* and *ac141* (EXON0) (Fang *et al.*, 2007; Ke *et al.*, 2008). Electron micrographs of infected cells suggest that nucleocapsids may acquire a transient envelope from the nuclear envelope that is lost prior to budding (Granados and Lawler, 1981). Nucleocapsids migrate to regions of the host plasma membrane that are enriched for the BV envelope proteins GP64 and/or F protein (*ac23*/*ld130*) (Oomens and Blissard, 1999; Pearson *et al.*, 2001). These membrane regions possess properties of lipid rafts, which are highly structured, cholesterol-rich regions of the plasma membrane (Haines *et al.*, 2009). Filamentous actin is also found at areas of cytoplasm beneath the GP64-containing lipid rafts (Haines *et al.*, 2009). GP64 is required for virus budding in group I alphabaculoviruses (Oomens and Blissard, 1999). The AcMNPV BV is characterized by an enlarged terminus with spikes or peplomers thought to consist of homotrimers of GP64 (Summers and Volkman, 1976; Oomens *et al.*, 1995; Ihalainen *et al.*, 2010). A comprehensive mass spectrometric proteomic study of purified AcMNPV BV has identified several BV-associated viral proteins in addition to the ones described above, as well as 11 cellular proteins (R. Wang *et al.*, 2010).

Infectious BV has been detected in the hemolymph of infected larvae as early as 0.5 h postinfection, indicating that BV can be produced by some mechanism from the parental inoculum (Granados and Lawler, 1981). Recent studies provide convincing evidence that a multicapsid NPV such as AcMNPV can produce BV consisting of nucleocapsids derived from parental inoculum in midgut cells during primary infection of the midgut. The GP64 gene contains both early and late promoter elements, and consequently its biosynthesis begins well before biosynthesis of other viral structural proteins (Blissard and Rohrmann, 1989; Jarvis and Garcia, 1994). Assuming that the nucleocapsids from a multicapsid ODV are not physically linked together after entry into a midgut columnar cell, a subpopulation of nucleocapsids from a multicapsid ODV conceivably can enter the nucleus, uncoat, and express GP64 from an early promoter, while the remainder of the nucleocapsids can bud through the plasma membrane and acquire an envelope containing newly synthesized GP64 (Fig. 4.6) (Volkman, 1997). Consistent with this hypothesis are the observations that (1) an inoculum consisting of multiple-capsid AcMNPV ODV can infect other tissues more rapidly than density gradient-purified single-capsid AcMNPV ODV (Washburn et al., 1999), and (2) mutant AcMNPV in which early GP64 biosynthesis has been eliminated spreads infection to other tissues more slowly than wild-type AcMNPV (Washburn et al., 2003). This strategy of producing BV without undergoing the complete viral replication cycle may be a strategy evolved by MNPVs to overcome host developmental resistance (see Section 4.3.4) (Zhang et al., 2004).

4.3.3. Secondary Infection

For alphabaculoviruses and most betabaculoviruses, BV serves as a vehicle to spread infection to other tissues in the host. In the case of the gammabaculoviruses (hymenopteran NPVs), the betabaculovirus *Harrisina brillians* GV, and the deltabaculovirus CuniNPV, infection does not spread beyond the midgut (Bird and Whalen, 1953; Federici and Stern, 1990; Moser et al., 2001). Whereas BV has been observed in mosquito larvae infected with CuniNPV (Moser et al., 2001), the three gammabaculovirus genome sequences published to date do not contain genes for either GP64 or F protein, from which it has been inferred that these viruses may not produce BV (Duffy et al., 2006; Lauzon et al., 2006).

Route of Infection; Penetration of Basement Membranes

The insect midgut epithelium is surrounded by a sheath composed of basement membrane and smooth muscle. Basement membranes are layers of extracellular matrix composed of a network of fibrous proteins, particularly type IV collagen and laminin, which surround the tissues of all animals, providing structural support, a filtration function, and a surface for cell attachment, migration, and differentiation (Timpl and Brown, 1996). The size exclusion limit of the basement membranes of the lepidopteran *Calpodes ethlius* was measured at 15 nm (Reddy and Locke, 1990). It is unlikely that baculovirus virions, which measure approximately 40–60 nm wide by 250–300 nm long, can freely diffuse through a barrier with a 15 nm size exclusion limit, and progeny virions from baculovirus infection can be observed accumulating in spaces between the midgut epithelial layer and the surrounding basement membranes in electron micrographs (Hess and Falcon, 1987; Tang et al., 2007). Nevertheless, progeny BV traverse the basement membrane and infect other tissues (Granados and Lawler, 1981).

The insect tracheal system, which functions in respiration, is a network of tubes running inward from spiracles (cuticular pores) and terminating in all tissues, breaking up into finer branches along the way. The larger trunks of the tracheae consist of tubes of cuticle surrounded by epidermal cells, which in turn are surrounded by basement membrane. As the tracheae bifurcate into finer and finer branches, they penetrate the basement membranes surrounding the tissues they service. The lumens of the cuticular tubes become intracellular structures (tracheoles) contained within specialized epidermal cells called tracheoblasts, which are in intimate association with the cells of the tissues they service for gas exchange (Wigglesworth, 1977; Maina, 1989). The tracheoblasts that service larval midgut cells are targets of secondary infection in baculovirus-infected hosts, and systemic infection of larvae spreads from the tracheoblasts to epidermal cells in tracheal trunks and on to other tissues in the hemocoel (Figs. 4.6 and 4.7C) (Engelhard et al., 1994; Rahman and Gopinathan, 2004). Infection of other tissues is associated with the tracheae of those tissues, indicating that baculoviruses use the tracheal system as a conduit to bypass the basement membranes.

Lepidopteran baculoviruses encode and express a fibroblast growth factor homologue (VFGF; *ac32*) that triggers degradation of basement membrane surrounding the tracheae and possibly the branching and migration of tracheoblasts to sites of midgut cell infection (Detvisitsakun et al., 2005; Means and Passarelli, 2010). VFGF binds to a lepidopteran homologue of a fibroblast growth factor receptor (Katsuma et al., 2006a). This binding initiates a signal transduction pathway leading to the activation of a matrix metalloprotease, which in turn activates an effector caspase cysteine protease that degrades laminin and type IV collagen in tracheal basement membrane, but not in the midgut epithelial basement membrane (Means and Passarelli, 2010). Viruses from which the *vfgf* gene has been deleted take longer to kill host larvae than wild-type virus (Katsuma et al., 2006b; Detvisitsakun et al., 2007), and the spread of infection to hemocytes occurs more slowly (Katsuma *et al.*, 2008; Means and Passarelli, 2010), indicating that *vfgf* facilitates, but is not essential for, baculovirus systemic infection.

Tissue Tropism

In studies using recombinant AcMNPV clones that express marker genes, AcMNPV was found to infect both columnar and regenerative epithelial cells of the midgut simultaneously, although regenerative cells were not exposed to the midgut lumen (Keddie *et al.*, 1989; Flipsen *et al.*, 1995). Infection of regenerative cells was probably initiated by progeny BV produced from parental nucleocapsids, as described above (Zhang *et al.*, 2004). Although reporter gene expression indicative of infection was noted in midgut goblet cells, no replication of virus was found to occur in this cell type (Knebel-Morsdorf *et al.*, 1996). While OBs are produced in midgut cells infected with gamma- and deltabaculoviruses, occlusion formation generally is not observed in alphabaculovirus-infected midgut cells (Harrap and Robertson, 1968; Knebel-Morsdorf *et al.*, 1996). Alphabaculovirus infection is next observed in the tracheoblasts, tracheae, and muscle cells of the midgut sheath, at the same time or followed closely by the hemocytes (blood cells) (Keddie *et al.*, 1989; Engelhard *et al.*, 1994; Knebel-Morsdorf *et al.*, 1996). Infection spreads through the tracheal system and can be seen in fat body, epidermis, and glial cells (Keddie *et al.*, 1989; Knebel-Morsdorf *et al.*, 1996). Although reporter gene activity suggestive of infection has been reported in Malpighian tubules, salivary glands, and gonads, viral replication does not appear to occur in those tissues (Keddie *et al.*, 1989; Engelhard *et al.*, 1994; Knebel-Morsdorf *et al.*, 1996).

Federici (1997) classified the GVs on the basis of tissue tropism. Type 1 GVs, including GVs from *S. frugiperda* and *H. armigera*, appear to replicate solely in the fat body, with larvae living for longer than 20 days after infection (Hamm, 1968; Whitlock, 1974). Type 2 GVs, including the GV from *C. pomonella*, more closely resemble alphabaculoviruses in that they infect and replicate in multiple tissues, including the tracheae, epidermis, fat body, and Malpighian tubules (Tanada and Leutenegger, 1968), and kill the host within three to six days, depending on viral dose and host instar. The sole type 3 GV is that of *H. brillians*, whose replication is restricted to the midgut, and larval mortality takes approximately six days.

As mentioned above, gamma- and deltabaculoviruses both replicate solely in the midgut epithelium and produce viral occlusions in midgut cells. Infection by CuniNPV occurs in the posterior midgut and gastric caecae, but not in the anterior midgut of its mosquito host (Moser *et al.*, 2001; Andreadis *et al.*, 2003). The NPV of the crane fly, *T. paludosa*, infects and replicates in hemocytes (Smith, 1955),

but there are insufficient sequence data for this virus to determine whether it is in fact a deltabaculovirus.

Cell Entry

The BVs of baculoviruses enter cells by adsorptive endocytosis, in which BVs are internalized in endosomes, followed by acidification of the endosomes, fusion between the BV and endosomal membranes, and release of the nucleocapsid into the cytoplasm (Fig. 4.8) (Volkman and Goldsmith, 1985; IJkel *et al.*, 2000; Long *et al.*, 2006). Group I alphabaculovirus BV entry involves the interaction of GP64 trimers on the BV envelope surface with a receptor on the surface of cell plasma membranes (Hefferon *et al.*, 1999). The identity of the receptor is unknown. However, given that BVs of alphabaculoviruses infect the cells of many different tissues and are also able to enter mammalian cell lines, the receptor is likely to be a common component of cell membranes or a common post-translational modification of proteins (Wickham *et al.*, 1992; Tani *et al.*, 2001). GP64 is a class III penetrene, or fusion protein (Garry and Garry, 2008; Kadlec *et al.*, 2008) that can trigger membrane fusion at a low pH (Blissard and Wenz, 1992; Plonsky and Zimmerberg, 1996).

Group II alphabaculoviruses, betabaculoviruses, and deltabaculoviruses do not possess a gp64 orthologue. Rather, these viruses encode a functionally homologous protein, the F (fusion) protein (IJkel *et al.*, 2000; Pearson *et al.*, 2000). The F protein differs from GP64 in being a class I fusion protein (Garry and Garry, 2008) that is synthesized in a precursor form and activated by cleavage with a furin protease (Westenberg *et al.*, 2002). The AcMNPV *ac23* gene also encodes an F protein orthologue, but it does not appear to play a role in BV infection (Lung *et al.*, 2003). F protein amino acid sequences share striking similarity with the sequences of the envelope proteins encoded by insect errantivirus (retrotransposon) *env* genes, and it has been suggested that errantiviruses may have acquired *env* genes by a recombination event following integration of an errantivirus genome into a baculovirus genome (Malik *et al.*, 2000; Rohrmann and Karplus, 2001). While group II and GV F proteins are able to functionally substitute for GP64 in recombinant AcMNPV clones that are missing the *gp64* gene (Lung *et al.*, 2002; Yin *et al.*, 2008), GP64 compensates for the deletion of the F protein gene either partially or not at all (Westenberg and Vlak, 2008; M. Wang *et al.*, 2010). GP64 and F protein do not bind to the same receptor (Wickham *et al.*, 1992; Westenberg *et al.*, 2007).

Occlusion-Derived Virion Occlusion and Morphogenesis

Translocation to, and entry of BV nucleocapsids into the nucleus occur by the same mechanisms described above (see Section 4.3.1), as do gene expression and assembly of nucleocapsids. Unlike ODV-infected midgut cells, many of the tissues infected by BV allow for the morphogenesis and occlusion of ODVs.

Nucleocapsids destined to become ODVs are enveloped in the peristromal region of infected cell nuclei (Fig. 4.8) (Harrap, 1972a; Hughes, 1972). This intranuclear envelopment involves association of the nucleocapsids with vesicle-like membranous material, termed microvesicles, that appear to be the source of the ODV envelope (Summers and Arnott, 1969; Kawamoto et al., 1977; Fraser, 1986). The microvesicles are probably derived from the host cell inner nuclear membrane (Summers and Arnott, 1969; Tanada and Hess, 1976; Hong et al., 1994) and ODV envelope proteins can be found in both the microvesicles and the nuclear envelope (Hong et al., 1997; Beniya et al., 1998).

Some of the details of a pathway for the sorting of ODV envelope proteins to the inner nuclear membrane in AcMNPV-infected cells have been elucidated. A host cellular protein, importin-α-16, binds to an N-terminal amino acid sequence identified in the ODV envelope protein ODV-E66 (encoded by most lepidopteran baculoviruses) during its synthesis and insertion into the host endoplasmic reticulum (ER) (Saksena et al., 2006). This N-terminal sequence consists of a hydrophobic domain of approximately 18 amino acids followed by a positively charged amino acid within 4−8 residues of the hydrophobic domain, which can be found in other ODV envelope proteins (Braunagel et al., 2009). The AcMNPV proteins FP25K and BV/ODV-E26 associate with importin-α-16 and the ODV envelope protein at the ER membrane, and together these proteins mediate translocation of ODV envelope proteins from the ER to the inner nuclear membrane (Braunagel et al., 2004). Mutations in the gene encoding FP25K (ac61) occur during serial passage of lepidopteran baculoviruses through cell culture and result in a phenotype known as the "few polyhedra" (FP) phenotype (Fraser and Hink, 1982; Harrison and Summers, 1995). FP mutants exhibit altered plaque morphology and produce fewer viral occlusions and more BVs. Ultrastructural observations of FP mutant-infected cells reveal that intranuclear envelopment of nucleocapsids and occlusion of ODVs are both impaired. The features of the FP phenotype suggest that one or more proteins that are sorted to the inner nuclear membrane are required for proper intranuclear envelopment; intranuclear envelopment of nucleocapsids may also be required for the attenuation of nucleocapsid egress and BV production, as well as the production and maturation of ODVs.

The ODV envelopes are more densely packed with protein than BV envelopes (Braunagel and Summers, 1994) and contain a larger variety of proteins. In addition to the six per os infectivity factors (see Section 4.3.1), BV/ODV-E26, and ODV-E66 (Hong et al., 1994), the core gene product ODV-E18 (Braunagel et al., 1996) is found in ODV envelopes. Mass spectrometric proteomic studies of purified ODV from AcMNPV, CuniNPV, and HearSNPV have revealed the presence of additional proteins associated with ODVs, including DNA replication proteins that may not be a part of the virion structure but have been packaged into virions or co-purified with virions (Braunagel et al., 2003; Deng et al., 2007; Perera et al., 2007).

ODVs are occluded when polyhedrin or granulin condense around mature, enveloped virions to form a crystalline matrix (Figs. 4.7D and 4.8) (Arnott and Smith, 1968; Summers, 1971; Knudson and Harrap, 1976). The interaction between the matrix protein and the ODV surface is characterized by a degree of specificity, as T. ni GV granulin was not able to occlude AcMNPV ODV (Eason et al., 1998), but AcMNPV ODV were occluded to varying degrees of efficiency by the polyhedrins of the Thysanoplusia orichalcea SNPV (ThorSNPV), the Buzura suppressaria SNPV (BusuSNPV), and SeMNPV (Cheng et al., 1998; Hu et al., 1999). The occlusion formed by ThorSNPV polyhedrin had the same tetrahedral shape that native ThorSNPV OBs possess. When a single leucine codon in the ThorSNPV polh coding sequence was mutated, the occlusions formed were polyhedral instead of tetrahedral. Single amino acid substitutions in AcMNPV polyhedrin have also resulted in the formation of very large, cuboidal occlusions that often have few or no virions (Carstens et al., 1986; Jarvis et al., 1991). These observations indicate that the polyhedrin amino acid sequence determines the shape of the occlusions and interaction with virions may regulate the OB size.

Mature occlusions formed by lepidopteran NPVs possess an electron-dense polyhedron "envelope" or calyx around the outside of the occlusion (Harrap, 1972b). The calyx consists of carbohydrate and a phosphoprotein, PP34, encoded by ac131 and found in all lepidopteran NPVs (Minion et al., 1979; Whitt and Manning, 1988; Gombart et al., 1989). Electron-dense structures ("spacers") that resemble the calyx can be observed in infected cell nuclei during occlusion assembly in association with fibrous structures called fibrillar bodies (Summers and Arnott, 1969; MacKinnon et al., 1974). The fibrillar bodies are also associated with developing occlusion and can be found in the cytoplasm of infected cells (Figs. 4.7E and 4.8). Immunogold electron microscopy indicated that both the P10 protein (ac137) and PP34 co-localized with fibrillar bodies and electron-dense spacers (van der Wilk et al., 1987; Russell et al., 1991; Lee et al., 1996). These observations collectively suggested that the electron-dense spacers were precursors of the calyx and that the fibrillar bodies and P10 have a role in calyx synthesis and placement. AcMNPV clones in which p10 was fused with a marker gene did not produce fibrillar bodies, electron-dense spacers, or calyces (Vlak et al., 1988). However, some p10 mutants examined still produced electron-dense spacers, while others produced neither fibrillar bodies nor electron-dense spacers but still generated viral occlusions with a calyx (Williams *et al.*, 1989; van Oers *et al.*, 1993; Gross *et al.*, 1994). Hence, it is unclear what role P10 has in calyx formation or whether the electron-dense spacers are precursors of the calyx. Examination of occlusions with no calyx indicates that the calyx provides an enhanced degree of stability to occlusions and enables retention of virions near the polyhedron surface (Williams *et al.*, 1989; Gross *et al.*, 1994).

In GV-infected cells, the nuclear envelope breaks down and contents of the cytoplasm and nucleus mix together (Walker *et al.*, 1982; Federici and Stern, 1990; Winstanley and Crook, 1993). GV virion maturation and granule formation occur within this mixed cytoplasmic—nuclear environment (Fig. 4.7F). At the end of the replication cycle, host cell lysis, tissue liquefaction, and the weakening and rupturing of the cuticle lead to the release of mature progeny occlusions into the environment (see *Cytopathology, tissue liquefaction, and cuticular weakening* in Section 4.3.4, below).

4.3.4. Host Response and Pathology

Developmental Resistance Between and Within Instars

As the lepidopteran larval hosts of baculoviruses progress through development, they become increasingly less susceptible to infection, with doses of virus required to kill late-instar larvae being many orders of magnitude higher than those required to kill early-instar larvae (Stairs, 1965; Payne *et al.*, 1981; Li, 2005). This phenomenon has been termed developmental resistance. The degree of difference in susceptibility between early and late instars due to developmental resistance cannot be wholly explained by differences in larval mass (Boucias and Nordin, 1977).

Developmental resistance has also been observed to occur intrastadially, or during progression through an instar, with susceptibility tending to decrease after molting (David *et al.*, 1971; Teakle *et al.*, 1986; Washburn *et al.*, 1995). Infections of developmentally coordinated cohorts of *T. ni*, *Helicoverpa virescens*, and *L. dispar* larvae at different times after molting with recombinant baculoviruses carrying a marker gene revealed that larvae infected later after molting exhibited fewer infected midgut cells (Engelhard and Volkman, 1995; Kirkpatrick *et al.*, 1998; McNeil *et al.*, 2010a). Given prior observations of the clearance of infected columnar epithelial cells from infected larvae (Keddie *et al.*, 1989; Engelhard *et al.*, 1994), it has been suggested that intrastadial developmental resistance may be due to a higher rate of sloughing of infected cells from the midgut, which decreases the opportunities for progeny virus (or repackaged parental virus) to infect the tracheal network and establish systemic infection of the larva (Engelhard and Volkman, 1995). Apoptosis of infected midgut epithelial and tracheal cells may play a role in developmental resistance as well (see *Apoptosis* section below) (Washburn *et al.*, 1998; Dougherty *et al.*, 2006; McNeil *et al.*, 2010a).

Developmental resistance does not manifest itself in *T. ni* or *H. virescens* larvae that are injected intrahemocoelically with BVs, indicating that developmental resistance in these species is a midgut-mediated phenomenon. However, a systemic component of developmental resistance can be seen in *L. dispar* larvae injected with virus (Hoover *et al.*, 2002; Grove and Hoover, 2007). This systemic component is attributable to an enhanced immune response to virus-infected cells in larvae infected later after molting (see *Immune responses* section, below) (McNeil *et al.*, 2010b).

Apoptosis

Apoptosis is a type of programmed cell death, or cell death that results from the activation of genetic pathways within the cell rather than cell death caused by circumstances beyond the cell's control (e.g., injury or insult). Morphologically, apoptosis begins with the formation of protuberances on the plasma membrane ("blebbing") followed by condensation and disintegration of the nucleus, cleavage of chromosomal DNA into nucleosomal unit-length fragments, and fragmentation of cells into membrane-bound apoptotic bodies that undergo phagocytosis by other cells (Wyllie *et al.*, 1980, 1984). The process of apoptosis plays an important role in the removal of unneeded cells or cells with damaged DNA during development (Conradt, 2009). Apoptosis is also a means by which a virus host organism can block viral replication and thus thwart an infection, and viruses have evolved counter-measures to block apoptosis in infected cells (Benedict *et al.*, 2002). Baculoviruses were among the first viruses that were discovered to induce and block apoptosis. A mutant clone of AcMNPV, vAcAnh (or *annihilator*), was found to cause apoptosis of an *S. frugiperda* cell line, but not a *T. ni* cell line, upon infection (Clem *et al.*, 1991), while infection with wild-type AcMNPV did not cause apoptosis in either cell line. The mutation in vAcAnh associated with apoptosis was mapped to the AcMNPV *p35* gene (*ac135*) and the apoptosis induction phenotype could be reproduced in wild-type virus in which expression of *p35* had been disrupted by insertion of a marker gene in the *p35* ORF (Clem *et al.*, 1991). Cells induced to undergo apoptosis by infection with a *p35* mutant die within 24 h postinfection and produce greatly reduced yields of progeny virus (Clem and Miller, 1993). In addition to cell lines, infection of *S. frugiperda* larvae results in apoptosis of cells within larval tissues and yields of occluded virus from *p35* mutant-infected larvae are drastically reduced (Clem and Miller, 1993; Clarke and Clem, 2003).

Regulation of apoptosis occurs through the control of the activation of a class of cysteine proteases called caspases. Apoptotic stimuli trigger a pathway of events within the cell that leads to the activation of initiator caspases, which in turn activate effector caspases that are directly responsible for the destruction of cells during apoptosis. In AcMNPV-infected insect cells, the IE-1 protein is primarily responsible for inducing apoptosis by activating the apoptosis pathway directly or indirectly through its role in viral gene expression and replication (Prikhod'ko and Miller, 1996; Schultz et al., 2009). The protein encoded by the p35 gene binds to and irreversibly inhibits effector caspases (Bump et al., 1995; Bertin et al., 1996). Homologues of p35 are found in the genomes of a small number of group I and group II NPVs and the *Choristoneura occidentalis* GV (Escasa et al., 2006). Among these homologues are genes that encode the protein P49, a variant form of P35 which inhibits initiator caspases (Zoog et al., 2002). A second family of antiapoptotic genes unrelated to the p35/p49 genes, known as the *iap* (*inhibitor of apoptosis*) genes, was also originally discovered in baculovirus genomes (Crook et al., 1993; Birnbaum et al., 1994). While p35/p49 genes have a narrow distribution among baculoviruses, *iap* genes are found in almost all baculovirus genomes and in eukaryotic cellular genomes as well (Clem and Duckett, 1997). Different IAP proteins have been shown to have a wide array of different functions, including the inhibition of caspases (Rumble and Duckett, 2008). Research examining the mechanism of action of the *iap-3* gene of OpMNPV suggests that IAP-3 does not inhibit caspases, but blocks the apoptosis pathway at a point upstream of caspase activation (Wilkinson et al., 2004).

Infection by wild-type AcMNPV of larvae or primary cell cultures of semi- or non-permissive host species, including *Spodoptera litura* (Zhang et al., 2002), *L. dispar* (Dougherty et al., 2006), and *Anticarsia gemmatalis* (Chikhalya et al., 2009) also results in apoptosis of infected cells. The observation suggests that apoptosis and the ability of a baculovirus to shut down an apoptotic response are factors that can determine the host range of a baculovirus. This concept was supported by a recent study on host gene expression in different mosquito species infected with CuniNPV, in which the rapid expression of a proapoptotic gene in midgut cells of the mosquito *Aedes aegypti* infected with CuniNPV was correlated with the inability of CuniNPV to mount a productive infection of *Ae. aegypti* larvae (Liu et al., 2011). In contrast, delayed expression of the same gene was observed in larvae of the permissive mosquito species *Culex quinquefasciatus* upon infection.

Immune Responses

The insect immune system is composed of both humoral (cell-free) and cellular responses. The humoral immune response consists of the synthesis of antimicrobial peptides and other proteins such as lectins, lysozyme, and attacin, whereas the cellular immune response consists of the phagocytosis of foreign material by hemocytes; the process of nodulation, in which bacteria are trapped in a mesh of hemocytes; and encapsulation, in which pathogens or foreign matter too large to be phagocytosed are surrounded by layers of flattened hemocytes (Kanost et al., 2004; Jiravanichpaisal et al., 2006). Melanization, the formation and deposition of the pigment melanin, is an important part of the cellular immune response. The nodules and capsules formed by hemocytes are often melanized to form an inflexible, impenetrable capsule around the object or pathogen; the molecules produced during melanogenesis (including quinones and reactive oxygen species) are likely to have a role in killing hemocyte-entombed pathogens (Cerenius et al., 2008).

While humoral and cellular immune responses against bacteria, fungi, protists, nematodes, and parasitoids of insects are well documented, there have been comparatively few reports of immune responses against insect viruses. In infections by AcMNPV of *H. zea*, a host which is only semi-permissive for AcMNPV, infected tracheal epidermal cells associated with the midgut were found to be surrounded by melanized capsules of hemocytes (Washburn et al., 1996). This immune response does not occur in larvae of the fully permissive host *H. virescens*. While the hemocytes of *H. virescens* were productively infected by AcMNPV, the hemocytes of *H. zea* larvae were observed to be resistant to infection with this virus, suggesting that AcMNPV was able to shut down a cellular immune response to infection simply by mounting successful infections of the host's hemocytes (Andersons et al., 1990; Trudeau et al., 2001). Melanotic encapsulation of infected tracheae was also observed in AcMNPV infections of another semi-permissive host, *Manduca sexta* (Washburn et al., 2000). When *M. sexta* larvae were immunocompromised after parasitization by a parasitoid wasp or injection of its polydnavirus, a significantly greater number of larvae succumbed to AcMNPV infection. The signal triggering encapsulation of infected cells and tissues may consist of damage inflicted on, or alteration in the structure of, the surrounding basement membranes (Tang et al., 2007).

The enzyme phenoloxidase catalyzes key steps in a series of reactions leading to the production of melanin. Phenoloxidase is synthesized as an inactive zymogen, prophenoloxidase, that is activated in response to developmental and environmental cues by a cascade of trypsin-like serine proteases to produce melanin not only as part of the immune response, but also as part of wound healing and cuticular darkening and hardening (Cerenius et al., 2010). Larvae of the African armyworm, *Spodoptera exempta*, exhibit both increased resistance to baculovirus infection and a greater degree of cuticular melanization when reared at a high density and these properties correlated with higher hemolymph levels of phenoloxidase activity (Kunimi and Yamada, 1990; Reeson et al., 1998). Phenoloxidase activity was also higher in hemolymph from *L. dispar* larvae inoculated with LdMNPV 48 h after molting to fourth instar, compared with larvae inoculated immediately after molting (McNeil *et al.*, 2010b). In contrast, larvae that survived infection with *Plodia interpunctella* GV did not contain higher levels of phenoloxidase compared to larvae that succumbed to infection (Saejeng *et al.*, 2010). Cabbage looper larvae that exhibited reduced susceptibility to TnSNPV after feeding on broccoli also did not contain higher levels of phenoloxidase (Shikano *et al.*, 2010). These studies indicate that higher levels of phenoloxidase may not account for larval survival or for every case of resistance to baculovirus infection.

It is possible that higher levels of phenoloxidase activity may lead to a higher degree of melanotic encapsulation of baculovirus-infected cells in larvae. However, cell-free plasma from the hemolymph of *H. virescens* was found to have a virucidal effect on BV of HzSNPV, and this virucidal activity was eliminated by specific inhibitors of phenoloxidase, indicating that phenoloxidase by itself has the capacity to act as an antiviral humoral immune response (Popham *et al.*, 2004).

Behavioral and Developmental Effects

Pathogens and parasites often manipulate host behavior (Thomas *et al.*, 2005), and baculoviruses are no exception. The classic example of altered behavior by baculoviruses was first reported in 1891 and coined "Wipfelkrankheit" (treetop disease) (Hofmann, 1891). Infected larvae of *L. monacha* tended to migrate to and die from baculovirus infection at the tops of trees. This behavior has also been documented in *M. brassicae* infected with MbNPV (Goulson, 1997), *L. dispar* infected with LdMNPV (Murray and Elkinton, 1992), and a number of other infected insects. Healthy *L. dispar* are known to hide in bark crevices or climb down the tree to the soil during the day to avoid predation from birds, climbing back out on the leaves to feed at night, but infected *L. dispar* larvae are more likely to remain exposed on the upper surface of leaves during the day than healthy larvae (Murray and Elkinton, 1992). Similar behavior is induced in infected *M. brassicae* (Vasconcelos *et al.*, 1996). Infected larvae migrate to more external, exposed parts of the plant before death, while the majority of healthy larvae are found under the leaves. This behavior, together with the paler color of moribund larvae due to a body cavity filled with viral occlusions, is likely to render the larvae more conspicuous to predators, especially birds. Since occlusions remain infectious after passage through the gut of many predators, this process may benefit the virus by increasing the distribution of virus when the predators defecate (Fuxa *et al.*, 1993).

Elevation-seeking behavior clearly affects the spatial distribution of baculovirus inoculum. Once virus-killed insects liquefy on the top of plants, horizontal transmission is facilitated as OBs containing ODV runs down the plant or are dispersed by rainfall, contaminating the foliage and soil below (D'Amico and Elkinton, 1995). However, there may be a tradeoff between increased opportunities for horizontal transmission and inactivation of occlusions exposed to sunlight on the tops of leaves (Biever and Hostetter, 1985).

Although *L. dispar* die in elevated positions as early instars (usually on top of leaves), it is well known that baculovirus-infected older instars are most often found in bark crevices at various tree heights. This observation is consistent with a report by Raymond *et al.* (2005) showing that *O. brumata* NPV and MbNPV induced their hosts to die lower down on tree stems. This behavior resulted in better persistence of virus on tree stems relative to foliage, probably as the result of protection of virus reservoirs from sunlight. Despite low levels of virus contamination on plant stems, neonate larvae were able to acquire infections from these stems.

In addition to elevation-seeking behavior, many baculoviruses induce increased dispersal activity in their hosts. *Mamestra brassicae* infected with MbNPV moved markedly further from the plant they were first placed on than healthy larvae in field plots of cabbages and tended to die on the apex of leaves, until they were too moribund to move (Vasconcelos *et al.*, 1996; Goulson, 1997). In bioassays, leaves of cabbage plants from these plots revealed a linear decrease in inoculum from central to peripheral plants within the plots (Vasconcelos *et al.*, 1996), indicating that increased mobility facilitated dispersal of virus. Furthermore, exposure to rainfall increased the infectivity of leaves in bioassays (Goulson, 1997).

It has been speculated that baculovirus manipulation of host mobility benefits the virus by delivering a reservoir of inoculum to a location that enhances transmission to conspecifics. However, determining the evolutionary basis for these altered behaviors has proven difficult in the absence of a mechanistic explanation. In two recent studies, the genetic bases for baculovirus manipulation of increased host mobility were determined. Death at elevated positions in *L. dispar* is induced by the virus gene ecdysteroid UDP-glucosyltransferase (*egt*) (Hoover *et al.*, 2011). Deletion of *egt* from LdMNPV eliminated, and rescue of the gene restored, elevation-seeking behavior. The *egt* gene, which is conserved among all lepidopteran NPVs and some GVs, encodes an enzyme that inactivates the insect molting hormone 20-hydroxyecdysone (20E) by transferring a glucose or galactose from a nucleotide-sugar donor to a hydroxyl group on ecdysone (O'Reilly *et al.*, 1991, 1992). By inactivating 20E, EGT also disrupts hormonal regulation of several physiological processes. EGT blocks molting and pupation, preventing a period of feeding arrest that occurs before molting (O'Reilly and Miller, 1989; O'Reilly *et al.*, 1998; Wilson *et al.*, 2000). EGT also prolongs time to death in some hosts and larval instars (Slavicek *et al.*, 1999;

Cory and Myers, 2004; Sun *et al.*, 2004). In some cases (O'Reilly *et al.*, 1998; Cory and Myers, 2004), but not others (Georgievska *et al.*, 2010c), EGT increases the yield of progeny virus by producing larger larvae. The use of the same viral gene to manipulate host physiology and behavior to enhance pathogen fitness is indeed parsimonious.

The other host behavior reported to be induced by a baculovirus-coded gene is hypermobility, referred to by the authors as enhanced locomotor activity (ELA) (Kamita *et al.*, 2005). The *ptp* gene that encodes protein tyrosine phosphatase (PTP) in *B. mori* NPV was shown to induce ELA in silkworm larvae infected with BmNPV, and the distance the larvae dispersed was three-fold higher in response to light. Deletion of *ptp* eliminated the behavior, while reinsertion of the gene restored ELA. Whether this gene influences elevation-seeking behavior was not tested. In addition, *ptp* is not found in the genomes of group II NPVs.

Cytopathology, Tissue Liquefaction, and Cuticular Weakening

Baculovirus infection initially causes nuclear hypertrophy, a characteristic swelling of the nucleus (Fig. 4.7B) (Bergold, 1953; Faulkner and Henderson, 1972; Vail *et al.*, 1973). The progression of infected cells through the cell cycle is halted at either the G2/M or S phase, and cell division ceases (Braunagel *et al.*, 1998; Ikeda and Kobayashi, 1999). Microtubules within infected cells are rearranged and eventually depolymerized, with a concomitant rounding of infected cells (Volkman and Zaal, 1990), or form thick bundles that associate with cytoplasmic filaments composed of the baculovirus P10 protein (Carpentier *et al.*, 2008). Actin microfilaments in infected cells undergo rearrangement in the cytoplasm and microfilaments are formed in the nucleus (Charlton and Volkman, 1991). Nucleoli disappear and host cell chromatin is marginated along the inner nuclear membrane in correlation with formation of the virogenic stroma and peristromal region (Faulkner and Henderson, 1972; Tanada and Hess, 1976; Nagamine *et al.*, 2008). Expression of the host cell's nuclear genes is progressively shut down at the level of transcription (Ooi and Miller, 1988; Jarvis, 1993; Nobiron *et al.*, 2003). Eventually, after progeny occlusions are assembled in the cell nucleus, infected cells lyse (Goodwin *et al.*, 1970). Cell lysis appears to involve both disruption of the cytoplasmic membrane and disintegration of the nucleus; the P10 protein is specifically required for nuclear disintegration, possibly along with other viral factors (Williams *et al.*, 1989; van Oers *et al.*, 1994).

Lepidopteran larvae that die from baculovirus infection become flaccid owing to liquefaction of their internal anatomy (Fig. 4.9A). The cuticle eventually ruptures, releasing progeny viral occlusions into the environment (Fig. 4.9B). Lepidopteran NPVs and some GVs encode a chitinase and a cathepsin protease that are required for the

FIGURE 4.9 Gross pathology of baculovirus infection. (A) Cadaver of *Lymantria dispar* larva that was killed by NPV infection, attached to birch by its proleg crochets and exhibiting the flaccidity consistent with liquefaction of the internal anatomy. (B) NPV-killed *Spodoptera exigua* larva on foliage, exhibiting cuticular rupture and release of progeny occlusion-containing liquid. An apparently uninfected larva forages next to the cadaver. [*(A) Courtesy of Viatcheslav Martemyanov, Institute of Systematics and Ecology of Animals, Siberian Branch of Russian Academy of Science; (B) courtesy of Trevor Williams, Instituto de Ecología, AC.*]

characteristic wilting and dramatic "melting" of the host larva after death (Hawtin *et al.*, 1997; Kang *et al.*, 1998). The cathepsin protease, encoded by the *v-cath* gene (*ac127*), is synthesized as an inactive precursor and activated upon death of the cell (Hom *et al.*, 2002). It requires the baculovirus chitinase encoded by *chiA* (*ac126*) for processing and folding in the endoplasmic reticulum (Hom and Volkman, 2000). NPVs with mutations eliminating expression of either gene do not cause the tissue liquefaction and cuticular rupture characteristic of baculovirus-killed larvae (Ohkawa *et al.*, 1994; Slack *et al.*, 1995; Hawtin *et al.*, 1997).

4.3.5. Ecology

Transmission

The viral occlusions of baculoviruses are structures that have evolved specifically to transmit infection horizontally between individual host organisms, either within a population or between generations. OBs play a crucial role in epizootics or outbreaks of disease within a population of larval hosts of baculoviruses. Baculovirus occlusions capable of infecting larvae can persist for years and possibly decades in soil (Jaques, 1964, 1967; Thompson *et al.*, 1981). While there is little direct evidence in support of specific mechanisms for how baculovirus epizootics are initiated, the movement of occlusions from soil reservoirs to foliage by rain, wind, or other disturbances of the soil has been correlated with the occurrence of baculovirus epizootics (Young and Yearian, 1986; Olafsson, 1988; Fuxa and Richter, 2001). Initial infections can occur through consumption of occlusions from soil or other environmental reservoirs. Alternatively, infection can be initiated by consumption of egg shells contaminated with occlusions (Doane, 1969; Hamm and Young, 1974). Another proposed mechanism for epizootic initiation involves the conversion of asymptomatic, vertically transmitted baculovirus infections present within individuals in a population into lethal infections that lead to the production and release of OBs (Vilaplana *et al.*, 2010). This mechanism can account for epizootics under conditions where exposure of larvae to significant environmental reservoirs of occlusions is unlikely to occur.

After initial infection, subsequent horizontal transmission of baculovirus infection to other individuals in a population may occur through ingestion of occlusions from the liquefied cadavers or virus-laden feces of infected individuals, or from cannibalization of infected larvae (Vasconcelos, 1996). Infection rates in host populations can reach levels as high as 95% (Erlandson, 1990), with high levels of mortality and a decline in the host population (Doane, 1970).

The mass action assumption of pathogen transmission states that the degree of transmission within a host population increases in direct proportion to the density of the host and the density of the pathogen. This assumption has not been found to apply to transmission of baculoviruses (D'Amico *et al.*, 1996; Knell *et al.*, 1998). The efficiency of baculovirus transmission depends on (1) the frequency with which hosts encounter and ingest infectious virus, and (2) the likelihood that a productive infection results from the host–virus encounter. Neither the parental inoculum from the soil reservoir nor the inoculum generated from infected hosts is likely to be evenly distributed in the area where the host is foraging; rather, virus is more likely to occur in clumps as cadavers die and liquefy (Fuxa, 1982; Vasconcelos *et al.*, 1996; D'Amico *et al.*, 2005). Hence, above a particular host or virus density, the efficiency of transmission is likely to decrease. Feeding behavior also plays a role in whether hosts ingest or avoid the inoculum they encounter. *Lymantria dispar* larvae are able to detect and avoid conspecific cadavers on foliage, and this ability may

be heritable (Parker *et al.*, 2010). Variability in host innate susceptibility to infection occurs for a variety of reasons (see Section 4.3.4), which in turn affects the likelihood that ingested virus results in a fatal infection. In some cases, heterogeneity in the risk of viral infection as a function of host feeding behavior can equal that caused by effects of host susceptibility (Dwyer *et al.*, 2005).

Interactions between baculoviruses and their hosts do not always lead to epizootics and population decline (Cory and Myers, 2003). As described above, host feeding behavior and innate susceptibility can modify the impact of host density on the severity of epizootics (Dwyer *et al.*, 2000). Some ecosystems are not as consistently conducive to the development of epizootics as others and may allow for only a low prevalence of baculovirus infection (Fuxa, 2004). Also, at the level of the individual, baculovirus infection sometimes fails to result in mortality, with larvae surviving through to adulthood via a number of different mechanisms (see Section 4.3.4). Sublethal effects of infection in survivors of baculovirus infection include reduced adult longevity, mating and oviposition success, fertility, weight, fecundity and egg viability, and a slower rate of development (Rothman and Myers, 1996; Myers *et al.*, 2000; Sood *et al.*, 2010). In addition, surviving larvae and the pupae and adults developing from these larvae can carry infectious virus, and surviving adults can transmit the infection vertically to progeny (Shapiro and Robertson, 1987; Burden *et al.*, 2002; Fuxa *et al.*, 2002). Evidence of infection and replication in reproductive organs of surviving males and females, as well as virus-induced mortality in F_1 progeny of crosses involving surviving adults, points to both transovarial and venereal routes of vertical transmission (Burden *et al.*, 2002; Khurad *et al.*, 2004). In addition, some host species contain baculoviruses in a latent, seemingly asymptomatic form that can be activated by infection with a heterologous virus (Hughes *et al.*, 1993; Burden *et al.*, 2003; Cooper *et al.*, 2003) and possibly by other stressors such as host crowding (Fuxa *et al.*, 1999).

Host Resistance and Specificity

Variability in susceptibility of hosts to baculovirus infection has been documented both between and within populations, suggesting that baculovirus hosts contain selectable genetic potential for resistance to baculovirus infection (Briese and Mende, 1981; Dwyer *et al.*, 1997; Milks, 1997). Exposure to baculovirus can result in selection for virus resistance. Selection of resistance in laboratory colonies of *S. frugiperda* and *A. gemmatalis* has resulted in the development of low levels of resistance (approximately three- to 10-fold increases in resistance ratios) that were accompanied by fitness costs such as reduced egg production and lower egg hatch rates, but resistance was lost after selection pressure (e.g., exposure to virus) was halted (Fuxa and Richter, 1989, 1998; Milks and Myers, 2000). Significantly higher resistance ratios have been reported for other laboratory-selected colonies (Briese *et al.*, 1980; Briese and Mende, 1983; Abot *et al.*, 1996), and *T. ni* colonies have been generated that exhibit stable resistance with no fitness cost (Milks *et al.*, 2002). Populations of *C. pomonella* in Germany acquired resistance after repeated exposure to a CpGV isolate that was formulated and applied to control *C. pomonella* infestations in apple orchards, with resistance ratios ranging from 1000 to 10,000 × (Asser-Kaiser *et al.*, 2007). A laboratory colony developed from one of the resistant field colonies exhibited a stable resistance ratio of 100 × (Asser-Kaiser *et al.*, 2007).

Relatively little is known about the molecular genetic or physiological bases for baculovirus resistance. The resistance of *C. pomonella* populations to CpGV was attributed to a sex chromosome-linked dominant allele, but nothing else is currently known about this locus (Asser-Kaiser *et al.*, 2010). CpGV was able to enter the cells of resistant *C. pomonella* larvae, but CpGV DNA replication did not occur (Asser-Kaiser *et al.*, 2011). The *B. mori suppressor of profiling 2* (*Bmsop2*) gene is expressed at a higher level in a BmNPV-resistant silkworm moth strain than in more susceptible strains (Xu *et al.*, 2005), indicating that it may encode a BmNPV resistance factor. Resistance to baculovirus infection has been correlated in specific instances with a thicker peritrophic matrix (Levy *et al.*, 2007) and elevated levels of a midgut NADPH oxidoreductase (Selot *et al.*, 2007).

Cross-infectivity studies with baculoviruses have indicated that baculoviruses generally do not infect very many species beyond the genus or family of the original host (Gröner, 1986). GVs appear to be more specific in this regard than NPVs (Ignoffo, 1968). Some baculoviruses infect an unusually broad range of host species, including AcMNPV, *Anagrapha falcifera* MNPV (AfMNPV), *Mamestra brassicae* MNPV (MbMNPV), and *Xestia c-nigrum* GV (XecnGV). AcMNPV has been reported to infect up to 43 lepidopteran species from 11 families (Payne, 1986). The magnitude of this figure should be regarded with some skepticism, as the capacity of a baculovirus cross-infection to activate a latent baculovirus may contribute to reports of mortality caused by cross-infection with AcMNPV and spurious conclusions that AcMNPV can replicate in such species. Nevertheless, even with this qualification, AcMNPV and its variants have been isolated from many different host species (Smith and Summers, 1979; Yanase *et al.*, 2000; Jehle *et al.*, 2006b). AfMNPV has been reported to infect 31 species of Lepidoptera from 10 families (Hostetter and Puttler, 1991) and had also been isolated from the mint looper, *Rachiplusia ou* (Harrison and Bonning, 1999). Af/RoMNPV is closely related to

AcMNPV, although considered to be a different species (Federici and Hice, 1997; Jehle *et al.*, 2006b). MbMNPV was found to be infectious to 32 species from four families of Lepidoptera, a figure confirmed by analysis of the viral DNA from cadavers resulting from cross-infection (Doyle *et al.*, 1990). Variants of XecnGV have been isolated from nine additional species of Lepidoptera (Goto *et al.*, 1992; Lange *et al.*, 2004; Harrison and Popham, 2008).

Genetic Variation

Analysis of baculovirus DNA generally has indicated that a high degree of genetic variation exists in baculovirus isolates from the same population or from geographically distant host populations. Genetic variation can be found in a virus population from individual larvae; for example, 24 different genotypes of an NPV from a single larva of *Panolis flammea* have been identified (Cory *et al.*, 2005). However, relatively little genetic variation was identified in an isolate of *Agrotis ipsilon* MNPV (Harrison and Lynn, 2008; Harrison, 2009). Restriction endonuclease digestion and sequencing of DNA have revealed that baculovirus genetic diversity is attributable to both nucleotide substitutions and insertions and deletions (indels) of DNA (Lee and Miller, 1978; Durantel *et al.*, 1998; Li *et al.*, 2005). Indels noted within viral isolates and between baculovirus variants frequently occur near or within *baculovirus repeated orf* (*bro*) genes, *hrs*, and other repeated sequences, suggesting that such rearrangements are introduced through intramolecular and intermolecular homologous recombination (Li *et al.*, 2002a; de Jong *et al.*, 2005; Harrison and Popham, 2008). Different genotypes also differ in their biological activity against larval hosts (Stiles and Himmerich, 1998; Cory *et al.*, 2005).

Some naturally occurring genotypes in baculovirus isolates of SeMNPV and SfMNPV are defective with respect to initiating and establishing infection (Muñoz *et al.*, 1998; Simón *et al.*, 2004). The defective SeMNPV genotypes persist in a mixture of SeMNPV genotypes despite reducing the overall virulence of the mixture (Muñoz and Caballero, 2000). In principle, this pattern could be explained by simultaneous co-infection of host cells by virions containing the defective and non-defective genotypes, followed by co-envelopment and/or co-occlusion of progeny ODV containing two classes of genotypes. Experimental evidence that a parasitic genotype could maintain itself in this way was provided by two separate studies examining the persistence of an occlusion-negative AcMNPV recombinant clone and an AcMNPV FP mutant; both viruses exhibited ablated or reduced oral infectivity after co-inoculation with wild-type AcMNPV and serial passage through larvae (Bull *et al.*, 2001, 2003). Both the occlusion-negative clone and the AcMNPV FP mutant were able to maintain themselves as a stable proportion of the total NPV population during serial passage. In both studies, the rate of co-infection of cells in the infected larvae was estimated at 4.3 genomes/cell. In contrast to the situation with SeMNPV, defective genotypes of the Nicaraguan isolate of SfMNPV do not reduce the overall virulence of the isolate, but appear to be required for optimal activity against *S. frugiperda* larvae. The eight minority genotypes in the Nicaraguan isolate contained deletions ranging in size from 4.8 to 16.5 kb that were centered around the *egt* gene (Simón *et al.*, 2005). A 3:1 mixture of the complete, undeleted genotype and a genotype with a 16.4 kb deletion that removes the *pif-1* and *pif-2* genes is able to kill larvae with the same potency as the original field isolate and is approximately 2.5-fold more effective than the undeleted genotype alone (Lopez-Ferber *et al.*, 2003; Simón *et al.*, 2005). A similar trend was observed with single- and dual-genotype infections of the *P. flammea* NPV (Hodgson *et al.*, 2004). In addition, the SfNIC field isolate killed larvae more slowly and produced more progeny occlusions than the undeleted genotype by itself (Simón *et al.*, 2008). The advantages of the genotype mixture over the undeleted genotype by itself point to a selective basis for the maintenance of minority and/or defective genotypes in a field isolate, over and above the persistence of these genotypes due to co-infection and co-occlusion (Simón *et al.*, 2006).

The production and maintenance of genetic diversity in a baculovirus population create a reservoir of selectable variation that can be drawn upon to deal with genetically different and more resistant host populations, or even a host of a different species altogether. This notion is supported by experimental evidence showing changes in the ratio of genotypes of a PaflNPV isolate upon serial passage through different host species (Hitchman *et al.*, 2007).

Multitrophic Interactions

Plants can influence interactions between baculoviruses and their hosts in many ways, from key steps in pathogenesis to the ecology and evolution of the pathogen (Cory and Hoover, 2006). Several studies have documented marked differences in susceptibility of insects to baculoviruses when larvae feed on different plant species (Keating and Yendol, 1987; Ali *et al.*, 1998; Hoover *et al.*, 1998c), or when plants are damaged, which induces higher levels of plant secondary chemicals used in plant defense from herbivores (Hunter and Schultz, 1993; Hoover *et al.*, 1998b). For example, the potency (LD_{50}) of LdMNPV in *L. dispar* was over 50-fold higher for larvae fed on black oak *Quercus velutina* compared with big-tooth aspen *Populus grandidentata* (Keating *et al.*, 1988).

Both plant physical structure and phytochemicals impact host–baculovirus interactions by several different mechanisms (Duffey *et al.*, 1995). Plant architecture and

habitat can increase the persistence of the virus reservoir by providing shade that protects the virus from inactivation by ultraviolet (UV) light (Biever and Hostetter, 1985). Persistence of baculoviruses between seasons and spread/transmission from reservoirs on plants to new hosts have been shown to affect the frequency and severity of viral epizootics (Dwyer *et al.*, 2000). The infectivity of baculovirus on Sitka spruce, *Picea sitchensis*, and English oak, *Quercus robur*, to *O. brumata*, for example, was greater than virus on heather, *Calluna vulgaris*, grown in an open field without shade (Raymond *et al.*, 2005). In addition to inactivation by sunlight, glands on cotton and soybean plants produce cations that generate alkaline exudates (Elleman and Entwistle, 1985), which can inactivate baculoviruses on leaf surfaces (McLeod *et al.*, 1977; Young *et al.*, 1977), probably by premature release of virions from the occlusions (Elleman and Entwistle, 1985).

The palatability of foliage influences feeding behavior, which in turn affects the amount of virus ingested. Bixby and Potter (2010) reported that black cutworm, *Agrotis ipsilon*, allowed to feed *ad libitum* on ryegrass containing endophytes, and sprayed with AgipMNPV, ate less, thereby acquiring a lower dose of virus. These larvae were less likely to die from virus than larvae fed on the same species of ryegrass without endophytes, but this difference was absent when insects were given a single equivalent dose of virus on ryegrass with or without endophytes. Some endophytes contain alkaloids that confer resistance to several herbivores (Breen, 1994).

The influence of phytochemicals on viral pathogenesis occurs either through direct antagonism between plant chemicals and virus, or through indirect effects on insect physiology and susceptibility. Quinones formed by oxidation of phytochemicals such as phenolics can bind to and reduce the infectivity of baculovirus occlusions (Felton and Duffey, 1990). The influence of plants on mortality of *H. virescens* by AcMNPV was shown to act through plant effects on insect physiology. Larvae fed on several different plant species and inoculated with AcMNPV showed a direct inverse relationship between peritrophic matrix thickness of the larvae, which was plant mediated, and mortality by AcMNPV (Plymale *et al.*, 2008), indicating that plants can alter the effectiveness of the peritrophic matrix as a barrier to viral infection. Cotton-fed *H. virescens* are generally less susceptible to mortality by AcMNPV than larvae fed on iceberg lettuce or an artificial diet (Hoover *et al.*, 1998c). Subsequent studies showed that larvae fed on cotton had higher rates of sloughing of infected midgut cells than larvae fed on iceberg lettuce or an artificial diet (Hoover *et al.*, 2000). Premature sloughing of infected midgut cells can eliminate infected cells before the virus has the opportunity to spread systemically (Washburn *et al.*, 1998).

Host susceptibility to baculoviruses can be altered by plant quality (e.g., nutrients and allelochemicals). *Helicoverpa zea* showed reduced growth on tomato plants that were damaged by prior herbivory and these larvae were more susceptible to fatal infection by HzSNPV than larvae fed on undamaged plants (Ali *et al.*, 1998). Plant quality can also influence immune responses and disease resistance against baculoviruses. *Trichoplusia ni* experienced reduced growth, survival, and lower indicators of immune function when fed on cucumber compared with broccoli (Shikano *et al.*, 2010). In turn, susceptibility of *T. ni* to TnSNPV was significantly lower on the plant of higher quality for the larvae. Given that resistance against pathogens is costly (Wilson *et al.*, 2002), it follows that adequate nutrition can play a role in host susceptibility to baculoviruses. For example, *Spodoptera littoralis* challenged with baculovirus were allowed to choose among diets of different relative amounts of protein and carbohydrate (Lee *et al.*, 2005). Larvae that survived infection chose to eat more protein relative to carbohydrates compared with controls or larvae that died from the infection, suggesting that protein could compensate for the cost of resistance.

The effects of plants on the interactions between a baculovirus and its host go beyond host susceptibility to infection. Several parameters of viral fitness are influenced by plants, including time to death (Hoover *et al.*, 1998a, c; Raymond *et al.*, 2002), virus yield (Raymond *et al.*, 2002), and infectivity of progeny virus (Cory and Myers, 2004). The quantity of OBs of HzSNPV produced in *H. virescens* differed by up to 19-fold depending upon the plant species or plant tissue ingested by the larvae (Ali *et al.*, 1998), and the lethal infectivity of these occlusions differed by 11-fold among plant tissue types. Because these factors mediated by plants can affect viral fitness, it is possible that these interactions could influence virus evolution (Cory and Hoover, 2006). Populations of baculovirus isolates may adapt to the most locally abundant host plants of their larval hosts. Baculoviruses isolated from three geographically separate populations of Western tent caterpillar, *Malacosoma californicum pluviale*, that fed on different host plants were found to be most effective (faster speed of kill) against larvae fed on the host plant from which the virus was isolated (Cory and Myers, 2004).

4.4. USE OF BACULOVIRUSES FOR INSECT PEST CONTROL

Documented efforts to use baculoviruses to control insect pests date back to the nineteenth century (Huber, 1986). The benefits of baculoviruses as means to control pests stem from their narrow host specificity and environmental safety, their virulence towards insect pests, and their capacity to cause epizootics in the field. These features continue to drive efforts to develop baculovirus-based

pesticides, even as the actual use of baculoviruses in pest management has not met its potential.

4.4.1. Factors Involved in Controlling Pests with Baculoviruses

Host Range

The narrow host range of baculoviruses is a two-edged sword when it comes to their use as a biopesticide. Most baculoviruses efficiently infect very few, if any, species outside the species from which they were originally isolated, and baculovirus applications generally have no impact on beneficial predator and parasitoid species (Armenta *et al.*, 2003). As a consequence, secondary pest outbreaks and primary pest rebounds that are consequences of the use and abuse of chemical insecticides are avoided (Grzywacz *et al.*, 2010). The narrow host range also makes baculoviruses a good choice for controlling a pest when it is desired to conserve other non-target lepidopteran species (Reardon *et al.*, 2009). However, the baculovirus narrow host range makes it unappealing for situations where more than one pest species needs to be controlled. If additional pest species are not sufficiently controlled by other factors, then additional pesticides would be need to be purchased. The low sales potential of a pesticide that targets one or two pests is expected to discourage investment in development of baculovirus-based pesticides. As a consequence, the baculoviruses that have been registered and sold as pesticides are viruses that efficiently infect and kill pest species that by themselves are of great economic significance, such as *C. pomonella*, and not against pests that are part of a pest complex or that are pests only on an intermittent basis. Broad host-range baculoviruses such as AcMNPV would appear to offer a solution to the problems posed by host range, but while such viruses are able to infect a wide variety of species in laboratory bioassays, only a few of these species are susceptible at concentrations that translate to economically feasible application rates in the field (Black *et al.*, 1997).

Survival Time

Baculovirus-infected larvae live for days to weeks after infection and continue to feed and cause crop damage during this time. This issue has been resolved through the development of fast-killing recombinant baculoviruses (see Section 4.4.3), but recombinant baculoviruses currently are not in use in pest control programs. Early-instar lepidopteran larvae cause much less feeding damage and die more quickly from baculovirus infection than later instars; hence, for wild-type baculoviruses, the quantity of feeding damage can be reduced by timing the application of baculovirus so that neonate or early-instar larvae are infected (Harper, 1973; Bianchi *et al.*, 2000; Vasconcelos *et al.*, 2005).

Larger concentrations of baculovirus result in shorter larval survival times, so a high application rate will also reduce feeding damage (Harper, 1973; van Beek *et al.*, 1988a, b). It follows that a slow speed of kill is not a problem for crops that can tolerate some degree of defoliation without loss of yield or commercial value, such as soybean (Gazzoni and Moscardi, 1998).

Mass Production

For every baculovirus product that has been marketed or used on a large scale for control of insect pests to date, production of the virus has been carried out in infected larvae. At the most basic level, production in larvae has involved simply collecting cadavers of infected larvae from the field (Moscardi, 1989; Moreau *et al.*, 2005). It is possible that mass production of AgMNPV by field collection of larval cadavers may have selected for an AgMNPV genotype missing the viral cathepsin L and chitinase genes, as cadavers that have not liquefied are easier to collect (Slack *et al.*, 2004). When an artificial diet is available for the species used for amplifying a given baculovirus, mass production *in vivo* typically involves mass rearing and infection of larvae. A considerable amount of research has been necessary to determine the optimal parameters for baculovirus production, including diet composition, rearing conditions (containers used, temperature, humidity), and the age/developmental stage at which to infect larvae (Shapiro, 1986; Hunter-Fujita *et al.*, 1998). In addition, facilities, equipment, and processes for efficient infection of larvae, collection of cadavers, and isolation of viral occlusions have been established (Wood and Hughes, 1996; Black *et al.*, 1997; van Beek and Davis, 2007). Because of the idiosyncrasies of a given baculovirus and its host, conditions and procedures for mass production of baculoviruses are optimized for each virus−host pair (Biji *et al.*, 2006).

A considerable amount of research has also been carried out to develop means for the mass production of baculoviruses *in vitro* in insect cell culture. Mass production in cell culture avoids some of the problems encountered with mass production in larvae, such as the development of diseases in a production colony (McEwen and Hervey, 1960) and contamination of the final product with other microorganisms and potentially allergenic insect parts (Podgwaite *et al.*, 1983; Vail *et al.*, 1983; McKinley *et al.*, 1997). The use of insect cell culture requires less labor for production at a larger scale and allows for the production of baculoviruses that infect host species for which *in vivo* production is not feasible or desirable (e.g., species with small larvae, like *P. xylostella*, or species with urticacious hairs, like *L. dispar*). Production in cell culture at a large scale has become easier and less expensive with the development of media that support cell growth without the need for fetal bovine serum and of surfactants that protect cells from shear stresses generated during oxygenation and fermentation (Maiorella *et al.*, 1988; Murhammer and Goochee, 1988). As with *in vivo* production, the conditions for *in vitro* production of baculovirus require optimization. Cell lines capable of good growth in bioreactors and virus isolates able to produce large quantities of viral occlusions and high titers of BV are important for economical production of baculovirus in cell culture. The scaling up of baculovirus *in vitro* and the use of a continuous bioreactor system involve serial passage of BV stock, which can result in the accumulation of defective interfering particles (DIPs) and FP mutants with reduced or abrogated insecticidal activity (Fraser and Hink, 1982; Kool *et al.*, 1991). DIP and FP mutant accumulation can be delayed or avoided by using a low multiplicity of infection when scaling up virus, or using a virus isolate that is resistant to the formation and accumulation of FP mutants (Slavicek *et al.*, 2001; Giri *et al.*, 2010).

Formulation

After production, baculoviruses are prepared for application using methods that optimize the stability and biological activity of the product and facilitate its application. The materials and processes involved in this preparation or formulation have been the subject of much research. Baculoviruses are initially formulated for storage and distribution to the user and the chief goal of such formulation is stability during storage. Since baculoviruses are generally applied by spraying with conventional chemical pesticide application equipment, they are often formulated as liquid concentrates, sometimes with additional compounds to prevent aggregation of the viral occlusions or to facilitate dispersal of the occlusions before application. The baculovirus products sold by some commercial companies, for example, are in the form of liquid concentrates. Baculoviruses have also been formulated as "wettable" powders either as lyophilized occlusions (Dulmage *et al.*, 1970) or as occlusions microencapsulated in a carbohydrate, protein, or secondary metabolite polymer such as starch, gelatin, or lignin (Ignoffo and Batzer, 1971; Tamez-Guerra *et al.*, 2000). Some formulations have involved binding occlusions to a carrier particle to make a baculovirus product that is applied as dust or granules, with the carrier often serving as a bait (Montoya *et al.*, 1966; Bourner *et al.*, 1992; Boughton *et al.*, 2001).

If the baculovirus product is to be sprayed, additional materials may be added before application to form a tank mixture. Applying the baculovirus product in smaller droplets promotes a more even coverage of foliage, but the smaller droplets tend to shrink rapidly owing to evaporation, which necessitates the addition of compounds that prevent evaporation (Ignoffo *et al.*, 1976; Killick, 1990). Wetting agents to promote the adherence and spreading of droplets on plant surfaces can be added (Ignoffo and Montoya, 1966). Phagostimulants that promote consumption of baculovirus-contaminated foliage can be added as well (Ignoffo *et al.*, 1976; Luttrell *et al.*, 1983), although the effectiveness of a phagostimulant has been found to depend on the prior diet of the target population (Lasa *et al.*, 2009). Unformulated viral occlusions strongly adhere to plant surfaces, but a "sticker" reagent is sometimes added to promote adhesion of the formulated product and its adjuvants (David and Gardiner, 1966; Andrews *et al.*, 1975; Ignoffo *et al.*, 1997).

Perhaps the biggest factor affecting the efficacy of a baculovirus application is the rapid inactivation of baculovirus by UV light from solar radiation (Jaques, 1968; Young and Yearian, 1974; Ignoffo *et al.*, 1977). Compounds that reflect or absorb UV light have been tested and used in baculovirus formulations to prolong the persistence of baculovirus insecticidal activity. These compounds have included bright substances that reflect UV light such as titanium dioxide (Bull *et al.*, 1976) and compounds that absorb light in variable ranges of wavelengths that include the UV spectrum, such as carbon (Ignoffo and Batzer, 1971) and Congo red (Shapiro, 1989). Crude, unrefined, and unformulated baculovirus preparations have been found in some cases to exhibit a degree of persistence and performance in the field equal or superior to refined, formulated baculovirus applications, possibly owing to the absorbance of UV light by the insect parts in the crude preparation (Cherry *et al.*, 2000; Tadeu *et al.*, 2002). Microencapsulation of NPV occlusions in lignin has been found to confer resistance to inactivation by simulated sunlight (Tamez-Guerra *et al.*, 2000; Arthurs *et al.*, 2006).

Among the compounds tested for UV protection were the optical brighteners, a group of UV-absorbing compounds used as fluorescent brightening agents in a variety of industrial processes. In addition to conferring protection against inactivation by UV light, these compounds were found to augment the infectivity of baculoviruses, an effect documented as significant reductions in LC_{50} in bioassays (Shapiro, 1992; Shapiro and Robertson, 1992; Dougherty *et al.*, 1996). The degree of reductions in LC_{50} mediated by optical brighteners has depended on the specific virus—host pair, the optical brightener used, the larval instar tested, and the assay procedure, with values ranging from 2.6- to approximately 5731-fold (Vail *et al.*, 1996; Murillo *et al.*, 2003; Lasa *et al.*, 2007). Formulations of baculovirus with optical brighteners also resulted in significantly higher larval mortalities in greenhouse and field applications (Hamm *et al.*, 1994; Webb *et al.*, 1994; Lasa *et al.*, 2007) and at least one instance of a greater degree of reduction in feeding damage mediated by the baculovirus application formulated with an optical brightener (Webb *et al.*, 1996). However, not all field applications with optical brighteners recorded greater mortalities or reductions in feeding damage, perhaps due to a failure of the larvae to ingest sufficient quantities of the

brightener, rapid environmental degradation of the brightener itself, or a level of pest pressure that was insufficient to record a difference between treatments with or without brightener (Thorpe *et al.*, 1999; Vail *et al.*, 1999; Boughton *et al.*, 2001). Including brighteners in a microencapsulation formulation (i.e., encapsulating the brighteners along with the viral occlusions) did not augment insecticidal activity or improve persistence in the field, although a tank mix of microencapsulated virus that included the brightener did show both augmented activity and increased persistence (Tamez-Guerra *et al.*, 2000).

Optical brighteners were found to disrupt and/or prevent formation of the peritrophic matrix of larvae when fed continuously with diet containing the brighteners (Wang and Granados, 2000; Okuno *et al.*, 2003). The brighteners by themselves caused retardation of larval development, reduced larval and pupal weights, and mortality (Wang and Granados, 2000; Martínez *et al.*, 2004), probably due to the degradation of the peritrophic matrix and subsequent loss of protection of the midgut epithelium from chemical and mechanical insults that could compromise the absorption of nutrients. Optical brighteners have also been reported to prevent sloughing of, and reduce apoptosis by, baculovirus-infected midgut cells (Washburn *et al.*, 1998; Dougherty *et al.*, 2006). In spite of the benefits conferred by formulation with optical brighteners, it has been contended that the relatively high cost of brighteners, along with the requirement for addition of brighteners to a tank mixture as a percentage of the total volume instead of application at a given rate per unit area, has hindered their use in commercial baculovirus formulations (McGuire *et al.*, 2001).

4.4.2. Case Studies from the Field

Helicoverpa zea/Helicoverpa armigera *SNPVs*

Three lepidopteran species, the tobacco budworm, *H. virescens*, the corn earworm, *H. zea*, and the Old World bollworm, *H. armigera*, form a complex of highly polyphagous pests that chronically infest many economically important crops across the world, especially maize, cotton, sorghum, chickpea, and tomato (Fitt, 1989). *Helicoverpa virescens* and *H. zea* are major pests in the New World (particularly North America), and *H. armigera* is a major pest in the Old World (particularly Africa, India, Australia, and China). Reports of heliothine larvae exhibiting the classic "wilt" symptoms associated with nuclear polyhedrosis were published in the early twentieth century (Glaser and Chapman, 1915; Stahler, 1939). Early field tests indicated that heliothine NPVs had potential for controlling outbreaks of heliothine pests (Chamberlin and Dutky, 1958; Tanada and Reiner, 1962).

A SNPV from *H. zea* (HzSNPV) was isolated in 1961 from a cotton field in Texas, and evaluated as a potential insecticide for control of *H. virescens* and *H. zea* (Ignoffo,

1965; Ignoffo *et al.*, 1965). This virus isolate became the first baculovirus to be registered with the US Environmental Protection Agency for use as a pesticide and was the first baculovirus developed into a commercial product (Ignoffo, 1973, 1999). HzSNPV was produced and sold under the trade names Viron H (later changed to Elcar®) and Biotrol-VHZ®. With the availability of pyrethroid insecticides, sales of Elcar plummeted beginning in 1979 when pyrethroids became the insecticide of choice; subsequently, the manufacturer of Elcar halted production of all insect virus products in 1982.

Currently, the HzSNPV isolate in Elcar is marketed as Gemstar® in the USA. In addition to HzSNPV, SNPV isolates also have been identified from *H. armigera* (HearSNPV) that are the same baculovirus species as HzSNPV (Chen *et al.*, 2002; Jehle *et al.*, 2006b). HearSNPV isolates produced for *H. armigera* control are marketed under the trade names Helicovex® and VIVUS MAX in Europe and Australia, respectively (Buerger *et al.*, 2007). Of the 11 baculoviruses registered as commercial insecticides in China, HearSNPV is produced in the greatest quantities (Sun and Peng, 2007). HearSNPV is also commercially produced in India for control of *H. armigera* on a variety of crops (Cherry *et al.*, 2000) and has been field tested elsewhere in Asia and Africa (Parnell *et al.*, 1999; Moore *et al.*, 2004).

Heliothine SNPVs are produced *in vivo* and are generally formulated as liquid concentrates. There has been research to evaluate *in vitro* production of heliothine SNPVs in cell lines (Lynn and Shapiro, 1998; Chakraborty *et al.*, 1999; Ogembo *et al.*, 2008) but, like other NPVs, serial passage of these viruses through cell lines results in the accumulation of FP mutants with reduced oral infectivity (Chakraborty and Reid, 1999; Lua *et al.*, 2002).

Management of outbreaks of heliothine pests with pyrethroid insecticides has become increasingly ineffective and undesirable, partly due to the development of resistance to insecticides as a consequence of their excessive application (McCaffery, 1998). The development and deployment of transgenic corn and cotton expressing genes for Bt endotoxin proteins from *Bacillus thuringiensis* (Bt) have suppressed heliothine pests with significant reductions in the use of insecticides (Bravo *et al.*, 2007; James, 2009). However, evidence for the development of field-evolved resistance in populations of *H. zea* and *H. armigera* to the toxins in Bt crops has been reported (Tabashnik *et al.*, 2009; Liu *et al.*, 2010). It follows that baculoviruses ought not to be ignored as a potential approach to managing heliothine pests.

Lymantria dispar *MNPV*

Subspecies of the gypsy moth, *L. dispar*, are found throughout Europe, Russia, the northern two-thirds of China, the Korean peninsula, and Japan (Pogue and Schaefer, 2007). The European gypsy moth, *L. dispar*

dispar, was brought to the USA from France to hybridize with native silk-producing moths. In 1868 or 1869, gypsy moth adults were accidentally released from captivity and a *L. dispar* population was established in Massachusetts, USA, within 10 years. Today, the gypsy moth can be found from Nova Scotia south to North Carolina and west to Wisconsin, with occasional populations occurring elsewhere and is one of the most costly defoliating pests of forests. Gypsy moth larvae feed on over 300 species of plants, but prefer species of oak. Outbreaks of *L. dispar* are characterized by very dense populations of larvae that defoliate entire trees, causing tree mortality and millions of dollars of damage to forests and additional costs to control outbreaks.

A naturally occurring NPV (LdMNPV) was identified as a major mortality factor in dense populations of gypsy moth larvae, capable of causing periodic epizootics and population collapse (Reiff, 1911; Campbell and Podgwaite, 1972). (Since then, an entomophthorean fungus has been found to cause epizootics in gypsy moth populations in the USA and details of this fungus impact are covered in Chapter 6.) Efforts to control *L. dispar* with an NPV had begun in the former Soviet Union as early as 1959 (Lipa, 1998), while the US Forest Service initiated studies on the use of LdMNPV in 1963 (Rollinson *et al.*, 1965). This research led to the development of the products Virin-ENSh in the USSR and Gypchek® in the USA (Lipa, 1998; Reardon *et al.*, 2009), with an additional product, Disparvirus, developed from Gypchek for use in Canada (Zhang *et al.*, 2010).

Of these products, only Gypchek is still in production and even it is produced in limited supply by the US Forest Service, not by a commercial manufacturer. Although applications of Gypchek cause substantial reductions in gypsy moth populations and tree defoliation (e.g., Podgwaite *et al.*, 1992), the cost of production of Gypchek is prohibitively high owing to the necessity to produce the virus in larvae. Efforts to develop an *in vitro*, cell culture-based system for production of Gypchek were initially stymied by the propensity of LdMNPV BV stocks to accumulate FP mutants during passage in cell culture (Slavicek *et al.*, 1995). A LdMNPV clone, termed 122b, was isolated that did not accumulate FP mutants after 14 passages in cell culture (Slavicek *et al.*, 2001). In a field test in which *in vitro*-produced LdMNPV 122b and *in vivo*-produced Gypchek were applied at the same rate, larval mortality was significantly lower and defoliation significantly higher in 122b-treated plots than in Gypchek-treated plots. This result was likely to be due to the larger size of the Gypchek occlusions, which contained five-fold more volume than 122b occlusions and thus contained five-fold more ODV (Thorpe *et al.*, 1998).

In a specificity study, LdMNPV was found not to infect 46 non-target lepidopteran species (Barber *et al.*, 1993). In a long-term study on the impacts of LdMNPV and Bt application on non-target species, LdMNPV application was found to have little to no impact on non-target lepidopteran species, whereas caterpillar counts of many non-target species in Bt-treated plots showed significant reductions during the treatment period (Butler *et al.*, 2005). Thus, Gypchek is the preferred method of gypsy moth control in situations and areas where conservation of non-target species is important.

Anticarsia gemmatalis *MNPV*

The velvetbean caterpillar, *A. gemmatalis*, is the most serious pest of soybean in Brazil and the south-eastern USA (Panizzi and Corrêa-Ferreira, 1997). NPVs were isolated from *A. gemmatalis* in Brazil in 1977 (Allen and Knell, 1977; Carner and Turnipseed, 1977). Preliminary field tests with one of these *A. gemmatalis* MNPV (AgMNPV) isolates demonstrated that it was able to suppress *A. gemmatalis* populations in soybean (Moscardi *et al.*, 1981). AgMNPV was developed and deployed by the Brazilian Organization for Agricultural Research (EMBRAPA) as a microbial insecticide on soybean in Brazil over an increasingly large area during the 1980s and 1990s (Moscardi, 1999). Formulations of this virus are currently used to control *A. gemmatalis* on approximately 2 million ha annually in Brazil, which makes it the baculovirus that is used most widely for insect pest control (Oliveira *et al.*, 2006; Levy *et al.*, 2009).

AgMNPV for use against *A. gemmatalis* infestations is produced in the field, on soybean farms, by private companies. Virus-killed larvae (i.e., cadavers) are harvested and AgMNPV is extracted and prepared as a wettable powder for very low cost, such that it can be sold to the Brazilian farmer for less than a chemical insecticide (Moscardi, 1999). However, the amount of AgMNPV made this way varies from year to year and demand for AgMNPV often outstrips its supply (Moscardi, 1999). This has stimulated efforts to optimize production of AgMNPV in laboratory-reared larvae under controlled conditions where production levels are more predictable (Moscardi *et al.*, 1997). Research on growing AgMNPV *in vitro* has also been carried out (Rodas *et al.*, 2005; Castro *et al.*, 2006; Mengual Gomez *et al.*, 2010), but the accumulation of FP mutants in AgMNPV BV stocks during serial passage indicates that this will be a problem for AgMNPV as it is for LdMNPV (de Rezende *et al.*, 2009).

Neodiprion *spp. NPVs*

The larvae of many species of sawflies feed on pine trees. Among these sawflies, species of the genus *Neodiprion* are found throughout North America and Europe and are the most destructive defoliating pests of pine (Coppel and Benjamin, 1965; Haack and Mattson, 1993). During outbreaks of these pests, the larvae congregate in large numbers and feed on pine needles, features that make them an ideal target for control by baculoviruses.

The European pine sawfly, *Neodiprion sertifer*, is native to northern Europe and first appeared in North America in 1925. An NPV from a *N. sertifer* population in Sweden was first used successfully to control *N. sertifer* in Canada (Bird, 1953); this isolate was subsequently tested in several additional field trials in Europe and North America and used without registration (Cunningham, 1982). Although *N. sertifer* NPV (NeseNPV) products were developed in Finland, the UK, and the USA, attempts to register a NeseNPV product in Canada expired owing to lack of demand. The product registered in the USA, Neochek-S, reduced *N. sertifer* larval populations by 80–90% and defoliation by up to 96% in a field trial (Podgwaite *et al.*, 1984), but registration for Neochek-S was not renewed.

The redheaded pine sawfly, *Neodiprion lecontei*, is a New World species and a pest of red pine and jack pine plantations. An NPV from this species was identified and found to suppress *N. lecontei* populations in a field trial (Bird, 1961). This virus was developed into a product, Lecontvirus®, which was registered for use in Canada in 1983 and has been used routinely to suppress *N. lecontei* infestations of pine plantations. Lecontvirus is prepared from field-collected, NPV-killed *N. lecontei* larvae, which are lyophilized and ground up into a wettable powder (http://www.glfc.cfs.nrcan.gc.ca/Lecontvirus.pdf).

The balsam fir sawfly, *Neodiprion abietis*, is another native North American species that was a relatively minor pest on balsam fir in the Canadian provinces of Newfoundland and Labrador until the 1990s, when a severe outbreak resulted in defoliation of young and prethinned balsam fir stands (Moreau, 2006). To develop a control measure for this species, an NPV (NeabNPV) was isolated from *N. abietis* in 1997, amplified in the laboratory and field, and field-tested by aerial application to infested stands from 2000 to 2002 (Moreau *et al.*, 2005). Applications of NeabNPV to populations that were increasing or peaking in size were associated with a subsequent five- to 10-fold decrease in population density. After additional field trials that confirmed the efficacy of NeabNPV for biological control, it was registered with the name Abietev™ in 2006. Unlike other sawfly baculovirus products, Abietev is being commercially produced and is currently being applied to ameliorate the damage caused by the ongoing *N. abietis* outbreak in Newfoundland (Iqbal and MacLean, 2010).

Because infected sawfly larvae shed progeny virus in the frass, a considerable degree of horizontal transmission of sawfly baculovirus infection takes place within populations of gregariously feeding larvae. As a consequence, it has been possible to obtain satisfactory control of sawfly infestations with application rates that are 100–1000-fold lower in terms of viral occlusions/ha than are commonly used with lepidopteran pests (Lucarotti *et al.*, 2007).

Cydia pomonella GV

The codling moth, *C. pomonella*, is the most serious pest of apple orchards worldwide and is also a significant pest of other pome fruits and walnuts (Barnes, 1991). A GV was identified from *C. pomonella* cadavers (Tanada, 1964) and found to be a type 2 GV that could mount a lethal infection of neonate larvae with a single granule, killing larvae in three to four days at higher concentrations (Sheppard and Stairs, 1977). After field tests indicated that CpGV had promise as a control measure against the codling moth (Falcon *et al.*, 1968; Huber and Dickler, 1977), it was developed into a number of control products, including the European products Madex™, Granupom™, and Carpovirusine™, and the North American products Cyd-X™ and Virosoft™. The various CpGV products are produced in larvae and formulated as liquid concentrates. Although cell lines have been developed that support CpGV infection, the characteristics of these cell lines and the degree of progeny virus production in them were not conducive to the development of an *in vitro* production system for CpGV (Winstanley and Crook, 1993).

Neonate *C. pomonella* larvae move from the leaves where eggs were laid to fruit within two days and begin to bore into fruit shortly afterwards (Ballard *et al.*, 2000b). Even though CpGV kills larvae relatively quickly, some infected larvae still manage to initiate entry into fruit. The damage, or stings, caused by the larvae raises concerns among growers about using CpGV to control codling moth infestations. Thus, in addition to adjuvants that protect CpGV from solar inactivation, research has been carried out to identify adjuvants that increase the time spent on leaves, the likelihood of ingesting a lethal dose of virus, or the degree of virus ingested, by larvae prior to attacking the fruit. Both phagostimulants to stimulate feeding on leaves and a kairomone attractant to prolong larval wandering on leaves have been evaluated in the field, but results have either been inconsistent with respect to reducing the damage done to fruit or required levels of the adjuvant that are impractical (Ballard *et al.*, 2000a; Arthurs *et al.*, 2007).

CpGV is used to control codling moth on over 100,000 ha in Europe, both on organic orchards and in conjunction with standard control measures (Asser-Kaiser *et al.*, 2007; Eberle and Jehle, 2006). The original Mexican isolate described by Tanada in 1964, CpGV-M, had been used for all European CpGV products. The heavy usage of a single CpGV isolate is likely to have contributed to the occurrence of the relatively high, stable resistance to CpGV among populations of codling moth in France and Germany that led to control failures (see Section 4.3.5). Two different CpGV isolates, one from Iran and one from a commercial

virus collection, have been found to kill larvae from colonies established from CpGV-M-resistant populations with dosages that were orders of magnitude lower than CpGV-M (Eberle *et al.*, 2008; Berling *et al.*, 2009a). However, the capacity of these new isolates to control the CpGV-M-resistant populations in the field and provide an acceptable degree of crop protection remains unclear (Berling *et al.*, 2009b).

4.4.3. Recombinant Baculoviruses

The idea of genetically engineering baculoviruses with improved insecticidal activity occurred in the literature almost concurrently with the first publications on engineering AcMNPV to express foreign genes. In a review of entomopathogens, Miller *et al.* (1983) suggested that the gene for an insect-specific toxin could be introduced into a baculovirus genome that would reduce the survival time of the infected host when expressed during infection, thus making baculoviruses more attractive as a pest control agent to both producers and users of pest control products. The seminal paper to test this idea reported the production of a recombinant clone of AcMNPV that carried and expressed the gene for an insect-specific peptide neurotoxin from the venom of the scorpion *Buthus eupeus* (Carbonell *et al.*, 1988). The recombinant virus in this study did not produce active toxin or exhibit any improvement in insecticidal activity, but this negative result did nothing to discourage the flood of research that followed on improving the insecticidal properties of baculoviruses through the application of recombinant DNA technology.

Because the peptide toxins found in arthropod venoms often are small (≤ 70 amino acids), it has been relatively easy to determine their amino acid sequences and assemble coding sequences for these peptides from synthetic oligonucleotides for insertion into the baculovirus genome. Recombinant baculoviruses have been assembled that express peptide neurotoxins from the venoms of scorpions (Maeda *et al.*, 1991; Imai *et al.*, 2000), mites (Tomalski and Miller, 1992), spiders (Prikhod'ko *et al.*, 1996; Hughes *et al.*, 1997), sea anemones (Prikhod'ko *et al.*, 1996), and ants (Szolajska *et al.*, 2004). These toxins normally cause paralysis when injected into insects (e.g., Tomalski and Miller, 1991). When active toxin is produced, infection of larvae with recombinant viruses expressing these toxins results in cessation of feeding, paralysis, and a reduction in mean or median survival times of larvae by approximately 10−60% compared to wild-type (non-toxin-expressing) virus (Inceoglu *et al.*, 2006).

A considerable amount of research was carried out with the AaIT toxin from the North African scorpion, *Androctonus australis*, which was the first venom peptide to be isolated with specific toxicity towards insects (Zlotkin *et al.*, 1971, 2000). Like many of the insect-specific paralytic neurotoxins, AaIT binds to voltage-gated sodium channels in neurons and interrupts normal sodium conductance across neuronal membranes. Lepidopteran larvae are relatively insensitive to AaIT (Herrmann *et al.*, 1990). However, infection with baculoviruses carrying an AaIT coding sequence causes paralysis in the infected larvae with hemolymph levels of AaIT that are substantially lower than the quantities of AaIT required to cause paralysis by direct hemocoelic injection of the neurotoxin (McCutchen *et al.*, 1991). A histopathology study of *B. mori* larvae infected with a recombinant BmNPV clone expressing AaIT found that the virus was able to infect tracheal epidermal cells within the glial sheaths surrounding neuronal axons and express AaIT in those cells (Elazar *et al.*, 2001). Thus, baculoviruses "potentiate" the action of insect-selective neurotoxins by infecting cells in the vicinity of the axonal targets of the neurotoxins, continuously producing a supply of these toxins close to these targets. In addition, targets in the central nervous system, which normally are not accessed by toxin that has been injected into the hemocoel, are bound by toxin that is secreted from recombinant baculovirus-infected tracheal epidermal cells (Elazar *et al.*, 2001).

The virulence of baculoviruses has also been improved by engineering baculoviruses with genes for hormones and enzymes involved in development or maintenance of physiological homeostasis. Survival times of infected larvae were reduced when infected by viruses that over-expressed diuretic hormones (Maeda, 1989; Raina *et al.*, 2007), modified juvenile hormone esterases (Bonning *et al.*, 1995, 1999), and a pheromone biosynthesis activating neurohormone (Ma *et al.*, 1998) compared to infections with wild-type viruses. Viruses engineered to express chitinolytic and proteolytic enzymes also exhibited accelerated speeds of kill. Overexpression of a chitinase (Gopalakrishnan *et al.*, 1995), a GV enhancin (del Rincón-Castro and Ibarra, 2005), and a cathepsin cysteine protease (Harrison and Bonning, 2001) reduced survival times of infected larvae below the survival times achieved with wild-type viruses. Reductions in larval survival times have also been achieved with other heterologous genes, such a gene encoding the maize mitochondrial gene URF13 (Korth and Levings, 1993) and a segment of the human *c-myc* proto-oncogene inserted in an antisense orientation so that antisense RNA would be produced (Lee *et al.*, 1997).

The genes used to improve the insecticidal efficacy of baculoviruses were at first placed under control of the *polh* promoter. Although the *polh* promoter can mediate very high expression levels of cytoplasmic proteins, it is the last promoter to be activated during baculovirus infection. By the time that transcription from this promoter begins to reach optimal levels, the folding, processing, and trafficking of proteins intended to be secreted from infected

cells are impaired by the deleterious effects of infection (Jarvis and Summers, 1989; Ailor and Betenbaugh, 1999). Alternative promoters that are activated earlier during infection than the *polh* promoter have been evaluated to see whether earlier expression of insecticidal proteins further increases the improvement in efficacy achieved with insecticidal proteins. An improvement in speed of kill was observed when a synthetic promoter (P_{synXIV}) containing elements of both a late promoter (from *vp39*) and the *polh* promoter was used to drive mite toxin expression (Tomalski and Miller, 1992). The *p6.9* promoter, which can drive higher levels of gene expression than either the *polh* or *p10* promoters (Bonning *et al.*, 1994), was found to mediate faster speeds of kill than the P_{synXIV} and *p10* promoters when used to drive toxin expression in recombinant viruses (Lu *et al.*, 1996; Harrison and Bonning, 2000). Early viral promoters and host cellular promoters have also been used to drive insecticidal gene expression, with varying results.

The *Drosophila melanogaster* heat shock protein *hsp70* promoter and a modified cytomegalovirus promoter mediated faster speeds of kill against two different host species than the P_{synXIV} and *p10* promoters (Lu *et al.*, 1996; Tuan *et al.*, 2005). However, viruses using the early viral promoters of the *da26* and *p35* genes to drive insecticidal gene expression failed to kill larvae faster than viruses using the P_{synXIV}, *p6.9* or *p10* promoters (Lu *et al.*, 1996; Gershburg *et al.*, 1998). Two studies reported that an *hr5* enhancer/*ie-1* promoter combination used to drive insecticidal gene expression failed to mediate faster speeds of kill than *p6.9* and *p10* promoters (Jarvis *et al.*, 1996; Harrison and Bonning, 2001), but a subsequent study found that the *hr5/ie-1* combination did mediate faster speeds of kill than the *p10* promoter with some species and developmental stages (van Beek *et al.*, 2003).

Secreted proteins such as the insecticidal proteins used in the above studies bear an amino acid sequence at the N-terminus called the signal peptide that directs the protein into the endoplasmic reticulum, an obligate step in protein secretion. It was found that an insect signal peptide can drive a higher level of secretion of a heterologous, non-insect protein from baculovirus-infected cells than the native signal peptide of the protein being expressed (Tessier *et al.*, 1991). Consequently, different signal peptides were evaluated for their capacity to drive higher levels of peptide neurotoxin secretion and thus to mediate faster speeds of kill in recombinant baculoviruses (Lu *et al.*, 1996; van Beek *et al.*, 2003). Although some signal peptides promoted greater degrees of neurotoxin secretion and reductions in larval survival time, no generalization about the best signal peptide to use can be inferred from the data.

Mutation of the AcMNPV *egt* gene was found to reduce survival time of AcMNPV-infected larvae to a significant extent (O'Reilly and Miller, 1991; Treacy *et al.*, 1997). The *egt* gene is not required for baculovirus infection or replication at the molecular and cellular level, and thus mutants missing this gene can be generated easily in cell culture. Some strategies for producing baculovirus clones with increased insecticidal efficacy have involved replacing the *egt* gene with a gene for a peptide neurotoxin (Popham *et al.*, 1997; Sun *et al.*, 2009). In addition to *egt*, deletion of the AcMNPV *orf603* (*ac7*) gene was found to reduce survival time of infected *S. frugiperda* larvae, but not of *T. ni* larvae (Popham *et al.*, 1998).

In general, the recombinant baculoviruses engineered with insecticidal genes have not exhibited changes in the host dose—mortality response or host range (e.g., Ignoffo and Garcia, 1996; Harrison and Bonning, 2000). However, viral replication, measured as the quantity of progeny OBs produced, is reduced in larvae infected with fast-killing recombinant baculoviruses (O'Reilly and Miller, 1991; Kunimi *et al.*, 1996; Li *et al.*, 2007). This reduced yield of OBs from recombinant virus-killed larvae probably accounts in part for observations of fewer OBs on the leaves and in the soil of plants occupied by recombinant virus-infected larvae, compared to wild-type virus-infected larvae (Fuxa *et al.*, 1998, 2001). Significantly reduced horizontal transmission of recombinant baculoviruses expressing AaIT has been observed compared to wild-type virus, but this appears to be caused in large part by the tendency of recombinant virus-infected larvae to fall off the plant as a result of the effects of AaIT-mediated contractile paralysis (Cory *et al.*, 1994; Hoover *et al.*, 1995; M. Zhou *et al.*, 2005). An *egt*-deleted clone of HearSNPV also produces fewer OBs/larva than wild-type HearSNPV, but this reduction does not appear to affect transmission (Sun *et al.*, 2005; Georgievska *et al.*, 2010a,b).

Exposure, ingestion, or injection of fast-killing recombinant baculoviruses was not found to have an adverse effect on beneficial insects (Heinz *et al.*, 1995; Smith *et al.*, 2000; Boughton *et al.*, 2003). Field tests with recombinant HzSNPV and HearSNPV clones expressing genes for insecticidal proteins have demonstrated that these viruses provided degrees of plant protection greater than that provided by wild-type viruses and equivalent to that provided by applications of Bt and chemical insecticides (Smith *et al.*, 2000; Treacy *et al.*, 2000; Sun *et al.*, 2009). While providing viruses with insecticidal genes has successfully resolved efficacy issues caused by the baculovirus slow speed of kill, the reduced quantities of progeny OBs obtained with these viruses are a problem for large-scale production. One solution pursued by scientists in industry was to produce recombinant viruses in cell culture (Black *et al.*, 1997) since the insecticidal genes used generally have not had an effect on replication *in vitro*. Another industrial approach was to develop a system in which expression of the insecticidal gene in the recombinant baculovirus is repressed in the presence of tetracycline (van Beek and Davis, 2007). When larvae are provided with

tetracycline during production, the insecticidal gene is not expressed, and wild-type levels of OBs are produced upon infection. Despite successful research demonstrating the efficacy and safety of recombinant baculoviruses and resolving issues surrounding production, recombinant baculovirus-based insecticides have not been commercialized and are not used in any capacity outside experimental research in the USA.

4.5. FUTURE RESEARCH DIRECTIONS

The application of molecular biology to the study of baculoviruses, facilitated by the use of insect cell culture, has answered many questions about baculovirus infection, pathology, replication, and transmission that have arisen from earlier ultrastructural and field experiments and observations. Most of this research, however, has been carried out with lepidopteran baculoviruses, and specifically with AcMNPV. The recently published nucleotide sequences of mosquito and sawfly NPV genomes point to a genetic diversity among viruses of *Baculoviridae* that may render inaccurate many of the generalizations based on studies of lepidopteran NPVs. An exploration of this diversity will require more research on beta-, gamma-, and deltabaculoviruses.

Although great strides have been made in understanding the genetics and basic virology of baculoviruses, a detailed and comprehensive understanding of the host response to baculovirus infection remains to be elucidated. What host genes are upregulated or downregulated in response to baculovirus infection? Are these genes involved in previously studied antivirus processes, such as apoptosis? What additional genes are expressed to attenuate viral pathology? The decreasing cost of next-generation DNA sequencing technology, together with methods in bioinformatics, transcriptomics, proteomics, and RNAi, allow for the pursuit of detailed answers to these questions, and such studies have started to appear in the literature (Gatehouse *et al.*, 2009; Popham *et al.*, 2010; Salem *et al.*, 2011). More genome sequences of agronomically significant insect species need to be generated and made available and the genes involved in host response to baculovirus infection need to be determined by transcriptomic, proteomic, and RNAi approaches. The response of a host insect at the physiological, cellular, and molecular levels to baculovirus infection can significantly influence the timing and viral dose required to kill or incapacitate the infected host, as well as the level of progeny virus produced during infection and the performance of a baculovirus as a heterologous gene expression vector. A deeper insight into the individual genes and genetic pathways involved in host response may assist with the identification of baculovirus isolates ideal for the control of a given pest species and the specific conditions under which control with a baculovirus may best be achieved. Such insight may also lead to the optimization of the production of baculovirus occlusions, and of baculovirus vector-mediated expression of heterologous genes in insects and insect cell culture. Advances in these areas would also facilitate the development of baculoviruses as vectors for the production of commercially available vaccines and as treatments for cancer. Improvements may also occur in the mass production of baculovirus occlusions in cells or insects, which could reduce the price of baculovirus pesticidal formulations. Improved efficiency of production, better performance in the field, and an increased demand for environmentally safe, host-specific microbial pest control products may lead to a greater degree of commercialization and use of baculoviruses as pest control agents.

ACKNOWLEDGMENTS

Almost all of the electron micrographs in this chapter were selected from the collection of micrographs produced by the late Dr. Jean Adams during her career at the USDA-ARS in Beltsville, Maryland. We thank Chris Pooley of the USDA-ARS Soybean Genomics and Improvement Laboratory (Beltsville, MD) for scanning and processing Jean's micrographs. We also thank Charlie Murphy (also of the Soybean Genomics and Improvement Laboratory) for the electron micrograph in Fig. 4.1D, and Viatcheslav Martemyanov, John Podgwaite, and Trevor Williams for the photographs of baculovirus-infected larvae in Fig. 4.9. Dr. Harrison dedicates this chapter to the memory of his sister Kim, who passed away during its writing. Mention of trade names or commercial products in this publication is solely for the purpose of providing specific information and does not imply recommendation or endorsement by the US Department of Agriculture. USDA is an equal opportunity provider and employer.

REFERENCES

Abdul Kadir, H. B., & Payne, C. C. (1989). Relationship of a *Plutella xylostella* nuclear polyhedrosis virus (NPV) to NPV of *Galleria mellonella*. *J. Invertebr. Pathol., 53*, 113−115.

Abot, A. R., Moscardi, F., Fuxa, J. R., Sosa-Gomez, D. R., & Richter, A. R. (1996). Development of resistance by *Anticarsia gemmatalis* from Brazil and the United States to a nuclear polyhedrosis virus under laboratory selection pressure. *Biol. Control, 7*, 126−130.

Adams, J. R., & McClintock, J. T. (1991). Baculoviridae. Nuclear polyhedrosis viruses. In J. R. Adams & J. R. Bonami (Eds.), *Atlas of Invertebrate Viruses* (pp. 87−226). Boca Raton: CRC Press.

Afonso, C. L., Tulman, E. R., Lu, Z., Oma, E., Kutish, G. F., & Rock, D. L. (1999). The genome of *Melanoplus sanguinipes* entomopoxvirus. *J. Virol., 73*, 533−552.

Ahrens, C. H., Russell, R. L., Funk, C. J., Evans, J. T., Harwood, S. H., & Rohrmann, G. F. (1997). The sequence of the *Orgyia pseudotsugata* multinucleocapsid nuclear polyhedrosis virus genome. *Virology, 229*, 381−399.

Ailor, E., & Betenbaugh, M. J. (1999). Modifying secretion and posttranslational processing in insect cells. *Curr. Opin. Biotechnol., 10*, 142−145.

Ali, M., Felton, G., Meade, T., & Young, S. (1998). Influence of interspecific and intraspecific host plant variation on the susceptibility of heliothines to a baculovirus. *Biol. Control, 12*, 42−49.

Allen, G. E., & Knell, J. D. (1977). A nuclear polyhedrosis virus of *Anticarsia gemmatalis*: I. Ultrastructure, replication, and pathogenicity. *Fla. Entomol., 60*, 233−240.

Andersons, D., Gunne, H., Hellers, M., Johansson, H., & Steiner, H. (1990). Immune responses in *Trichoplusia ni* challenged with bacteria or baculoviruses. *Insect Biochem., 20*, 537−543.

Andreadis, T. G., Becnel, J. J., & White, S. E. (2003). Infectivity and pathogenicity of a novel baculovirus, CuniNPV from *Culex nigripalpus* (Diptera: Culicidae) for thirteen species and four genera of mosquitoes. *J. Med. Entomol., 40*, 512−517.

Andrews, G. L., Harris, F. A., Sikorowski, P. P., & McLaughlin, R. E. (1975). Evaluation of *Heliothis* nuclear polyhedrosis virus in a cottonseed oil bait for control of *Heliothis virescens* and *H. zea* on cotton. *J. Econ. Entomol., 68*, 87−90.

Arella, M., Lavallee, C., Belloncik, S., & Furuichi, Y. (1988). Molecular cloning and characterization of cytoplasmic polyhedrosis virus polyhedrin and a viable deletion mutant gene. *J. Virol., 62*, 211−217.

Armenta, R., Martínez, A. M., Chapman, J. W., Magallanes, R., Goulson, D., Caballero, P., Cave, R. D., Cisneros, J., Valle, J., Castillejos, V., Penagos, D. I., García, L. F., & Williams, T. (2003). Impact of a nucleopolyhedrovirus bioinsecticide and selected synthetic insecticides on the abundance of insect natural enemies on maize in southern Mexico. *J. Econ. Entomol., 96*, 649−661.

Arnosti, D. N. (2003). Analysis and function of transcriptional regulatory elements: insights from *Drosophila. Annu. Rev. Entomol., 48*, 579−602.

Arnott, H. J., & Smith, K. M. (1968). An ultrastructural study of the development of a granulosis virus in the cells of the moth *Plodia interpunctella. J. Ultrastruct. Res., 21*, 251−268.

Arthurs, S. P., Lacey, L. A., & Behle, R. W. (2006). Evaluation of spray-dried lignin-based formulations and adjuvants as solar protectants for the granulovirus of the codling moth, *Cydia pomonella* (L.). *J. Invertebr. Pathol., 93*, 88−95.

Arthurs, S. P., Hilton, R., Knight, A. L., & Lacey, L. A. (2007). Evaluation of the pear ester kairomone as a formulation additive for the granulovirus of codling moth (Lepidoptera: Tortricidae) in pome fruit. *J. Econ. Entomol., 100*, 702−709.

Asser-Kaiser, S., Fritsch, E., Undorf-Spahn, K., Kienzle, J., Eberle, K. E., Gund, N. A., Reineke, A., Zebitz, C. P., Heckel, D. G., Huber, J., & Jehle, J. A. (2007). Rapid emergence of baculovirus resistance in codling moth due to dominant, sex-linked inheritance. *Science, 317*, 1916−1918.

Asser-Kaiser, S., Heckel, D. G., & Jehle, J. A. (2010). Sex linkage of CpGV resistance in a heterogeneous field strain of the codling moth *Cydia pomonella* (L.). *J. Invertebr. Pathol., 103*, 59−64.

Asser-Kaiser, S., Radtke, P., El-Salamouny, S., Winstanley, D., & Jehle, J. A. (2011). Baculovirus resistance in codling moth (*Cydia pomonella* L.) caused by early block of virus replication. *Virology, 410*, 360−367.

Ayres, M. D., Howard, S. C., Kuzio, J., Lopez-Ferber, M., & Possee, R. D. (1994). The complete DNA sequence of *Autographa californica* nuclear polyhedrosis virus. *Virology, 202*, 586−605.

Ballard, J., Ellis, D. J., & Payne, C. C. (2000a). The role of formulation additives in increasing the potency of *Cydia pomonella* granulovirus for codling moth larvae, in laboratory and field experiments. *Biocontrol Sci. Technol., 10*, 627−640.

Ballard, J., Ellis, D. J., & Payne, C. C. (2000b). Uptake of granulovirus from the surface of apples and leaves by first instar larvae of the codling moth *Cydia pomonella* L. (Lepidoptera: Olethreutidae). *Biocontrol Sci. Technol., 10*, 617−625.

Barber, K. N., Kaupp, W. J., & Holmes, S. B. (1993). Specificity testing of the nuclear polyhedrosis virus of the gypsy moth, *Lymantria dispar* (L.) (Lepidoptera: Lymantriidae). *Can. Entomol., 125*, 1055−1066.

Barnes, M. M. (1991). Codling moth occurrence, host race formation, and damage. In L. P. S. Van der Geest & H. H. Evenhuis (Eds.), *Tortricid Pests: Their Biology, Natural Enemies and Control*, Vol. 5 (pp. 313−327). Amsterdam: Elsevier.

Bawden, A. L., Glassberg, K. J., Diggans, J., Shaw, R., Farmerie, W., & Moyer, R. W. (2000). Complete genomic sequence of the *Amsacta moorei* entomopoxvirus: analysis and comparison with other poxviruses. *Virology, 274*, 120−139.

Beaton, C. D., & Filshie, B. K. (1976). Comparative ultrastructural studies of insect granulosis and nuclear polyhedrosis viruses. *J. Gen. Virol., 31*, 151−161.

van Beek, N., & Davis, D. C. (2007). Baculovirus insecticide production in insect larvae. *Methods Mol. Biol., 388*, 367−378.

van Beek, N. A., Wood, H. A., Angellotti, J. E., & Hughes, P. R. (1988a). Rate of increase and critical amount of nuclear polyhedrosis virus in lepidopterous larvae estimated from survival time assay data with a birth−death model. *Arch. Virol., 100*, 51−60.

van Beek, N. A. M., Wood, H. A., & Hughes, P. R. (1988b). Quantitative aspects of nuclear polyhedrosis virus infections in lepidopterous larvae: the dose−survival time relationship. *J. Invertebr. Pathol., 51*, 58−63.

van Beek, N., Lu, A., Presnail, J., Davis, D., Greenamoyer, C., Joraski, K., Moore, L., Pierson, M., Herrmann, R., Flexner, L., Foster, J., Van, A., Wong, J., Jarvis, D., Hollingshaus, G., & McCutchen, B. (2003). Effect of signal sequence and promoter on the speed of action of a genetically modified *Autographa californica* nucleopolyhedrovirus expressing the scorpion toxin LqhIT2. *Biol. Control, 27*, 53−64.

Belloncik, S., & Mori, H. (1998). Cypoviruses. In L. K. Miller & L. A. Ball (Eds.), *The Insect Viruses* (pp. 337−370). New York: Plenum Press.

Belloncik, S., Liu, J., Su, D., & Arella, M. (1996). Identification and characterization of a new cypovirus, type 14, isolated from *Heliothis armigera*. *J. Invertebr. Pathol., 67*, 41−47.

Benedict, C. A., Norris, P. S., & Ware, C. F. (2002). To kill or be killed: viral evasion of apoptosis. *Nat. Immunol., 3*, 1013−1018.

Beniya, H., Braunagel, S. C., & Summers, M. D. (1998). *Autographa californica* nuclear polyhedrosis virus: subcellular localization and protein trafficking of BV/ODV-E26 to intranuclear membranes and viral envelopes. *Virology, 240*, 64−75.

Benz, G. A. (1986). Introduction: historical perspectives. In R. R. Granados & B. A. Federici (Eds.), *The Biology of Baculoviruses, Vol. I: Biological Properties and Molecular Biology* (pp. 2−27). Boca Raton: CRC Press.

Berenbaum, M. (1980). Adaptive significance of midgut pH in larval Lepidoptera. *Am. Nat., 115*, 138−146.

Bergold, G. H. (1947). Die Isolierung des Polyeder-Virus und die Natur der Polyeder. *Z. Naturforsch. Teil B, 2*, 122−143.

Bergold, G. H. (1948). Ueber die Kapselvirus-Krankheit. *Z. Naturforsch. Teil B, 3*, 338−342.

Bergold, G. H. (1963). The molecular structure of some insect virus inclusion bodies. *J. Ultrastruct. Res., 8*, 360−378.

Bergold, G. H. (1953). Insect viruses. *Adv. Virus Res., 1*, 91−139.

Berling, M., Blachere-Lopez, C., Soubabere, O., Lery, X., Bonhomme, A., Sauphanor, B., & Lopez-Ferber, M. (2009a). *Cydia pomonella* granulovirus genotypes overcome virus resistance in the codling moth and improve virus efficiency by selection against resistant hosts. *Appl. Environ. Microbiol., 75*, 925−930.

Berling, M., Rey, J.-B., Ondet, S.-J., Tallot, Y., Soubabère, O., & Bonhomme, A. (2009b). Field trials of CpGV virus isolates overcoming resistance to CpGV-M. *Virol. Sin., 24*, 470−477.

Bertin, J., Mendrysa, S. M., LaCount, D. J., Gaur, S., Krebs, J. F., Armstrong, R. C., Tomaselli, K. J., & Friesen, P. D. (1996). Apoptotic suppression by baculovirus P35 involves cleavage by and inhibition of a virus-induced CED-3/ICE-like protease. *J. Virol., 70*, 6251−6259.

Bianchi, F. J., Joosten, N. N., Vlak, J. M., & van der Werf, W. (2000). Greenhouse evaluation of dose− and time−mortality relationships of two nucleopolyhedroviruses for the control of beet armyworm, *Spodoptera exigua*, on chrysanthemum. *Biol. Control, 19*, 252−258.

Biever, K. D., & Hostetter, D. L. (1985). Field persistence of *Trichoplusia ni* (Lepidoptera: Noctuidae) single-embedded nuclear polyherdrosis virus on cabbage foliage. *Environ. Entomol., 14*, 579−581.

Biji, C. P., Sudheendrakumar, V. V., & Sajeev, T. V. (2006). Influence of virus inoculation method and host larval age on productivity of the nucleopolyhedrovirus of the teak defoliator. *Hyblaea puera (Cramer)*. *J. Virol. Methods, 133*, 100−104.

Bilimoria, S. L. (1986). Taxonomy and identification of baculoviruses. In R. R. Granados & B. A. Federici (Eds.), *The Biology of Baculoviruses, Vol. I: Biological Properties and Molecular Biology* (pp. 37−59). Boca Raton: CRC Press.

Bird, F. T. (1953). The use of a virus disease in the biological control of the European pine sawfly, *Neodiprion sertifer* (Geoffr.). *Can. Entomol., 85*, 437−446.

Bird, F. T. (1961). Transmission of some insect viruses with particular reference to ovarial transmission and its importance in the development of epizootics. *J. Insect Pathol., 3*, 352−380.

Bird, F. T., & Whalen, M. M. (1953). A virus disease of the European pine sawfly, *Neodiprion sertifer* (Geoffr.). *Can. Entomol., 85*, 433−437.

Birnbaum, M. J., Clem, R. J., & Miller, L. K. (1994). An apoptosis-inhibiting gene from a nuclear polyhedrosis virus encoding a polypeptide with Cys/His sequence motifs. *J. Virol., 68*, 2521−2528.

Bischoff, D. S., & Slavicek, J. M. (1997). Molecular analysis of an enhancin gene in the *Lymantria dispar* nuclear polyhedrosis virus. *J. Virol., 71*, 8133−8140.

Bixby, A. J., & Potter, D. A. (2010). Influence of endophyte (*Neotyphodium lolii*) infection of perennial ryegrass on susceptibility of the black cutworm (Lepidoptera: Noctuidae) to a baculovirus. *Biol. Control, 54*, 141−146.

Black, B. C., Brennan, L. A., Dierks, P. M., & Gard, I. E. (1997). Commercialization of baculoviral insecticides. In L. K. Miller (Ed.), *The Baculoviruses* (pp. 341−387). New York: Plenum Press.

Blissard, G. W., & Rohrmann, G. F. (1989). Location, sequence, transcriptional mapping, and temporal expression of the gp64 envelope glycoprotein gene of the *Orgyia pseudotsugata* multicapsid nuclear polyhedrosis virus. *Virology, 170*, 537−555.

Blissard, G. W., & Rohrmann, G. F. (1991). Baculovirus gp64 gene expression: analysis of sequences modulating early transcription and transactivation by IE1. *J. Virol., 65*, 5820−5827.

Blissard, G. W., & Wenz, J. R. (1992). Baculovirus gp64 envelope glycoprotein is sufficient to mediate pH-dependent membrane fusion. *J. Virol., 66*, 6829–6835.

Blissard, G. W., Kogan, P. H., Wei, R., & Rohrmann, G. F. (1992). A synthetic early promoter from a baculovirus: roles of the TATA box and conserved start site CAGT sequence in basal levels of transcription. *Virology, 190*, 783–793.

Bolle, J. (1906). Bericht über die Tätigkeit der k. k. landwirtschaftlich-chemischen Versuchsstation in Görz im Jahre 1905. *Z. Landwirtsch. Vers. Oesterr., 9*, 239–258.

Bonning, B. C., Roelvink, P. W., Vlak, J. M., Possee, R. D., & Hammock, B. D. (1994). Superior expression of juvenile hormone esterase and beta-galactosidase from the basic protein promoter of *Autographa californica* nuclear polyhedrosis virus compared to the p10 protein and polyhedrin promoters. *J. Gen. Virol., 75*, 1551–1556.

Bonning, B. C., Hoover, K., Booth, T. F., Duffey, S., & Hammock, B. D. (1995). Development of a recombinant baculovirus expressing a modified juvenile hormone esterase with potential for insect control. *Arch. Insect Biochem. Physiol., 30*, 177–194.

Bonning, B. C., Possee, R. D., & Hammock, B. D. (1999). Insecticidal efficacy of a recombinant baculovirus expressing JHE-KK, a modified juvenile hormone esterase. *J. Invertebr. Pathol., 73*, 234–236.

Boucias, D. G., & Nordin, G. L. (1977). Interinstar susceptibility of the fall webworm, *Hyphantria cunea*, to its nucleopolyhedrovirus and granulosis viruses. *J. Invertebr. Pathol., 30*, 68–75.

Boughton, A. J., Lewis, L. C., & Bonning, B. C. (2001). Potential of *Agrotis ipsilon* nucleopolyhedrovirus for suppression of the black cutworm (Lepidoptera: Noctuidae) and effect of an optical brightener on virus efficacy. *J. Econ. Entomol., 94*, 1045–1052.

Boughton, A. J., Obrycki, J. J., & Bonning, B. C. (2003). Effects of a protease-expressing recombinant baculovirus on nontarget insect predators of *Heliothis virescens*. *Biol. Control, 28*, 101–110.

Bourner, T. C., Vargas-Osuna, E., Williams, T., Santiago-Alvarez, C., & Cory, J. S. (1992). A comparison of the efficacy of nuclear polyhedrosis and granulosis viruses in spray and bait formulations for the control of *Agrotis segetum* (Lepidoptera: Noctuidae) in maize. *Biocontrol Sci. Technol., 2*, 315–326.

Braunagel, S. C., & Summers, M. D. (1994). *Autographa californica* nuclear polyhedrosis virus PDV and ECV viral envelopes and nucleocapsids: structural proteins, antigens, lipid and fatty acid profiles. *Virology, 202*, 315–328.

Braunagel, S. C., He, H., Ramamurthy, P., & Summers, M. D. (1996). Transcription, translation, and cellular localization of three *Autographa californica* nuclear polyhedrosis virus structural proteins: ODV-E18, ODV-E35, and ODV-EC27. *Virology, 222*, 100–114.

Braunagel, S. C., Parr, R., Belyavskyi, M., & Summers, M. D. (1998). *Autographa californica* nucleopolyhedrovirus infection results in Sf9 cell cycle arrest at G2/M phase. *Virology, 244*, 195–211.

Braunagel, S. C., Russell, W. K., Rosas-Acosta, G., Russell, D. H., & Summers, M. D. (2003). Determination of the protein composition of the occlusion-derived virus of *Autographa californica* nucleopolyhedrovirus. *Proc. Natl. Acad. Sci. USA, 100*, 9797–9802.

Braunagel, S. C., Williamson, S. T., Saksena, S., Zhong, Z., Russell, W. K., Russell, D. H., & Summers, M. D. (2004). Trafficking of ODV-E66 is mediated via a sorting motif and other viral proteins: facilitated trafficking to the inner nuclear membrane. *Proc. Natl. Acad. Sci. USA, 101*, 8372–8377.

Braunagel, S. C., Cox, V., & Summers, M. D. (2009). Baculovirus data suggest a common but multifaceted pathway for sorting proteins to the inner nuclear membrane. *J. Virol., 83*, 1280–1288.

Bravo, A., Gill, S. S., & Soberon, M. (2007). Mode of action of *Bacillus thuringiensis* Cry and Cyt toxins and their potential for insect control. *Toxicon, 49*, 423–435.

Breen, J. P. (1994). Acremonium endophyte interactions with enhanced plant-resistance to insects. *Annu. Rev. Entomol., 39*, 401–423.

Briese, D. T., & Mende, H. A. (1981). Differences in susceptibility to a granulosis virus between field populations of the potato moth, *Phthorimaea operculella* (Zeller) (Lepidoptera: Gelechiidae). *Bull. Entomol. Res., 71*, 11–18.

Briese, D. T., & Mende, H. A. (1983). Selection for increased resistance to a granulosis virus in the potato moth, *Phthorimaea operculella* (Zeller) (Lepidoptera: Gelechiidae). *Bull. Entomol. Res., 73*, 1–9.

Briese, D. T., Mende, H. A., Grace, T. D. C., & Geier, P. W. (1980). Resistance to a nuclear polyhedrosis virus in the light-brown apple moth *Epiphyas postvittana* (Lepidoptera: Tortricidae). *J. Invertebr. Pathol., 36*, 211–215.

Buerger, P., Hauxwell, C., & Murray, D. (2007). Nucleopolyhedrovirus introduction in Australia. *Virol. Sin., 22*, 173–179.

Bull, D. L., Ridgway, R. L., House, S., & Pryor, N. W. (1976). Improved formulations of the *Heliothis* nuclear polyhedrosis virus. *J. Econ. Entomol., 69*, 731–736.

Bull, J. C., Godfray, H. C., & O'Reilly, D. R. (2001). Persistence of an occlusion-negative recombinant nucleopolyhedrovirus in *Trichoplusia ni* indicates high multiplicity of cellular infection. *Appl. Environ. Microbiol., 67*, 5204–5209.

Bull, J. C., Godfray, H. C., & O'Reilly, D. R. (2003). A few-polyhedra mutant and wild-type nucleopolyhedrovirus remain as a stable polymorphism during serial coinfection in. *Trichoplusia ni. Appl. Environ. Microbiol., 69*, 2052–2057.

Bump, N. J., Hackett, M., Hugunin, M., Seshagiri, S., Brady, K., Chen, P., Ferenz, C., Franklin, S., Ghayur, T., Li, P., Licari, P., Mankovich, J., Shi, L., Greenberg, A. H., Miller, L. K., & Wong, W. W. (1995). Inhibition of ICE family proteases by baculovirus antiapoptotic protein p35. *Science, 269*, 1885–1888.

Burand, J. P., Summers, M. D., & Smith, G. E. (1980). Transfection with baculovirus DNA. *Virology, 101*, 286–290.

Burand, J. P., Stiles, B., & Wood, H. A. (1983). Structural and intracellular proteins of the nonoccluded baculovirus HZ-1. *J. Virol., 46*, 137–142.

Burand, J. P., Rallis, C. P., & Tan, W. (2004). Horizontal transmission of Hz-2V by virus infected *Helicoverpa zea* moths. *J. Invertebr. Pathol., 85*, 128–131.

Burand, J. P., Tan, W., Kim, W., Nojima, S., & Roelofs, W. (2005). Infection with the insect virus Hz-2v alters mating behavior and pheromone production in female *Helicoverpa zea* moths. *J. Insect Sci., 5*, 6.

Burden, J. P., Griffiths, C. M., Cory, J. S., Smith, P., & Sait, S. M. (2002). Vertical transmission of sublethal granulovirus infection in the Indian meal moth, *Plodia interpunctella*. *Mol. Ecol., 11*, 547–555.

Burden, J. P., Nixon, C. P., Hodgkinson, A. E., Possee, R. D., Sait, S. M., King, L. A., & Hails, R. S. (2003). Covert infections as a mechanism for long-term persistence of baculoviruses. *Ecol. Lett., 6*, 524–531.

Burley, S. K., Miller, A., Harrap, K. A., & Kelly, D. C. (1982). Structure of the *Baculovirus* nucleocapsid. *Virology, 120*, 433–440.

Butler, L., Cederbaum, S. B., Cooper, R. J., DeCecco, J. A., Gale, G. A., Hajek, A. E., Kondo, V., Marshall, M. R.,

Pauley, T. K., Raimondo, S., Rastall, K. E., Seidel, G. E., Strazanac, J. S., Sutton, W. V., Watson, M. B., Wheeler, M. M., & Williams, A. B. (2005). *Long term evaluation of the effects of Bacillus thuringiensis kurstaki, gypsy moth nucleopolyhedrosis virus product Gypchek®, and Entomophaga maimaiga on nontarget organisms in mixed broadleaf–pine forests in the central Appalachians. FHTET-2004-14.* Fort Collins: Forest Health Technology Enterprise Team.

Campbell, R. W., & Podgwaite, J. D. (1972). The disease complex of the gypsy moth. I. Major components. *J. Invertebr. Pathol., 18*, 101–107.

Carbonell, L. F., Hodge, M. R., Tomalski, M. D., & Miller, L. K. (1988). Synthesis of a gene coding for an insect-specific scorpion neurotoxin and attempts to express it using baculovirus vectors. *Gene, 73*, 409–418.

Carner, G. R., & Turnipseed, S. G. (1977). Potential of a nuclear polyhedrosis virus for the control of the velvetbean caterpillar in soybean. *J. Econ. Entomol., 70*, 608–610.

Carpentier, D. C., Griffiths, C. M., & King, L. A. (2008). The baculovirus P10 protein of *Autographa californica* nucleopolyhedrovirus forms two distinct cytoskeletal-like structures and associates with polyhedral occlusion bodies during infection. *Virology, 371*, 278–291.

Carstens, E. B., Tija, S. T., & Doerfler, W. (1979). Infection of *Spodoptera frugiperda* cells with *Autographa californica* nuclear polyhedrosis virus. I. Synthesis of intracellular proteins after virus infection. *Virology, 99*, 386–396.

Carstens, E. B., Tjia, S. T., & Doerfler, W. (1980). Infectious DNA from *Autographa californica* nuclear polyhedrosis virus. *Virology, 101*, 311–314.

Carstens, E. B., Krebs, A., & Gallerneault, C. E. (1986). Identification of an amino acid essential to the normal assembly of *Autographa californica* nuclear polyhedrosis virus polyhedra. *J. Virol., 58*, 684–688.

Castro, M. E. B., Ribeiro, Z. M. A., & Souza, M. L. (2006). Infectivity of *Anticarsia gemmatalis* nucleopolyhedrovirus to different insect cell lines: morphology, viral production, and protein synthesis. *Biol. Control, 36*, 299–304.

Cerenius, L., Lee, B. L., & Söderhäll, K. (2008). The proPO-system: pros and cons for its role in invertebrate immunity. *Trends Immunol., 29*, 263–271.

Cerenius, L., Kawabata, S., Lee, B. L., Nonaka, M., & Söderhäll, K. (2010). Proteolytic cascades and their involvement in invertebrate immunity. *Trends Biochem. Sci., 35*, 575–583.

Chaivisuthangkura, P., Tawilert, C., Tejangkura, T., Rukpratanporn, S., Longyant, S., Sithigorngul, W., & Sithigorngul, P. (2008). Molecular isolation and characterization of a novel occlusion body protein gene from *Penaeus monodon* nucleopolyhedrovirus. *Virology, 381*, 261–267.

Chakraborty, S., & Reid, S. (1999). Serial passage of a *Helicoverpa armigera* nucleopolyhedrovirus in *Helicoverpa zea* cell cultures. *J. Invertebr. Pathol., 73*, 303–308.

Chakraborty, S., Monsour, C., Teakle, R., & Reid, S. (1999). Yield, biological activity, and field performance of a wild-type *Helicoverpa* nucleopolyhedrovirus produced in *H. zea* cell cultures. *J. Invertebr. Pathol., 73*, 199–205.

Chamberlin, F. S., & Dutky, S. R. (1958). Tests of pathogens for the control of tobacco insects. *J. Econ. Entomol., 51*, 560.

Charlton, C. A., & Volkman, L. E. (1991). Sequential rearrangement and nuclear polymerization of actin in baculovirus-infected *Spodoptera frugiperda* cells. *J. Virol., 65*, 1219–1227.

Chen, X., Zhang, W. J., Wong, J., Chun, G., Lu, A., McCutchen, B. F., Presnail, J. K., Herrmann, R., Dolan, M., Tingey, S., Hu, Z. H., & Vlak, J. M. (2002). Comparative analysis of the complete genome sequences of *Helicoverpa zea* and *Helicoverpa armigera* single-nucleocapsid nucleopolyhedroviruses. *J. Gen. Virol., 83*, 673–684.

Cheng, C. H., Liu, S. M., Chow, T. Y., Hsiao, Y. Y., Wang, D. P., Huang, J. J., & Chen, H. H. (2002). Analysis of the complete genome sequence of the Hz-1 virus suggests that it is related to members of the Baculoviridae. *J. Virol., 76*, 9024–9034.

Cheng, X. W., Carner, G. R., & Fescemyer, H. W. (1998). Polyhedrin sequence determines the tetrahedral shape of occlusion bodies in *Thysanoplusia orichalcea* single-nucleocapsid nucleopolyhedrovirus. *J. Gen. Virol., 79*, 2549–2556.

Cherry, A. J., Rabindra, R. J., Parnell, M. A., Geetha, N., Kennedy, J. S., & Grzywacz, D. (2000). Field evaluation of *Helicoverpa armigera* nucleopolyhedrovirus formulations for control of the chickpea podborer, *H. armigera* (Hubn.), on chickpea (*Cicer arietinum* var. Shoba) in southern India. *Crop Prot., 19*, 51–60.

Chikhalya, A., Luu, D. D., Carrera, M., De La Cruz, A., Torres, M., Martinez, E. N., Chen, T., Stephens, K. D., & Haas-Stapleton, E. J. (2009). Pathogenesis of *Autographa californica* multiple nucleopolyhedrovirus in fifth-instar *Anticarsia gemmatalis* larvae. *J. Gen. Virol., 90*, 2023–2032.

Chisholm, G. E., & Henner, D. J. (1988). Multiple early transcripts and splicing of the *Autographa californica* nuclear polyhedrosis virus IE-1 gene. *J. Virol., 62*, 3193–3200.

Clarke, T. E., & Clem, R. J. (2003). *In vivo* induction of apoptosis correlating with reduced infectivity during baculovirus infection. *J. Virol., 77*, 2227–2232.

Clem, R. J., & Duckett, C. S. (1997). The iap genes: unique arbitrators of cell death. *Trends Cell. Biol., 7*, 337–339.

Clem, R. J., & Miller, L. K. (1993). Apoptosis reduces both the *in vitro* replication and the *in vivo* infectivity of a baculovirus. *J. Virol., 67*, 3730–3738.

Clem, R. J., Fechheimer, M., & Miller, L. K. (1991). Prevention of apoptosis by a baculovirus gene during infection of insect cells. *Science, 254*, 1388–1390.

Conradt, B. (2009). Genetic control of programmed cell death during animal development. *Annu. Rev. Genet., 43*, 493–523.

Cooper, D., Cory, J. S., Theilmann, D. A., & Myers, J. H. (2003). Nucleopolyhedroviruses of forest and western tent caterpillars: cross-infectivity and evidence for activation of latent virus in high-density field populations. *Ecol. Entomol., 28*, 41–50.

Coppel, H. C., & Benjamin, D. M. (1965). Bionomics of the nearctic pine-feeding diprionids. *Annu. Rev. Entomol., 10*, 69–96.

Cornalia, E. (1856). Monografia del bombice del gelso. *Mem. R. Istit. Lombardo Sci. Lett. Arte., 6*, 3–387.

Cory, J., & Hoover, K. (2006). Plant mediated effects in insect–pathogen interactions. *Trends Ecol. Evol., 21*, 278–286.

Cory, J. S., & Myers, J. H. (2003). The ecology and evolution of insect baculoviruses. *Annu. Rev. Ecol. Evol. Syst., 34*, 239–272.

Cory, J., & Myers, J. (2004). Adaptation in an insect host–plant pathogen interaction. *Ecol. Lett., 7*, 632–639.

Cory, J. S., Hirst, M. L., Williams, T., Hails, R. S., Goulson, D., Green, B. M., Carty, T. M., Possee, R. D., Cayley, P. J., & Bishop, D. H. L. (1994). Field trial of a genetically improved baculovirus insecticide. *Nature, 340*, 138–140.

Cory, J. S., Green, B. M., Paul, R. K., & Hunter-Fujita, F. (2005). Genotypic and phenotypic diversity of a baculovirus population within an individual insect host. *J. Invertebr. Pathol., 89*, 101–111.

Coulibaly, F., Chiu, E., Ikeda, K., Gutmann, S., Haebel, P. W., Schulze-Briese, C., Mori, H., & Metcalf, P. (2007). The molecular organization of cypovirus polyhedra. *Nature, 446*, 97–101.

Coulibaly, F., Chiu, E., Gutmann, S., Rajendran, C., Haebel, P. W., Ikeda, K., Mori, H., Ward, V. K., Schulze-Briese, C., & Metcalf, P. (2009). The atomic structure of baculovirus polyhedra reveals the independent emergence of infectious crystals in DNA and RNA viruses. *Proc. Natl. Acad. Sci. USA, 106*, 22205–22210.

Crook, N. E., Clem, R. J., & Miller, L. K. (1993). An apoptosis-inhibiting baculovirus gene with a zinc finger-like motif. *J. Virol., 67*, 2168–2174.

Cunningham, J. C. (1982). Field trials with baculoviruses: control of forest insect pests. In E. Kurstak (Ed.), *Microbial and Viral Pesticides* (pp. 335–386). New York: Marcel Dekker.

Dall, D., Sriskantha, A., Vera, A., Lai-Fook, J., & Symonds, T. (1993). A gene encoding a highly expressed spindle body protein of *Heliothis armigera* entomopoxvirus. *J. Gen. Virol., 74*, 1811–1818.

D'Amico, V., & Elkinton, J. S. (1995). Rainfall effects on transmission of gypsy moth (Lepidoptera: Lymantriidae) nuclear polyhedrosis virus. *Environ. Entomol., 24*, 1144–1149.

D'Amico, V., Elkinton, J. S., Dwyer, G., Burand, J. P., & Buonaccorsi, J. P. (1996). Virus transmission in gypsy moths is not a simple mass action process. *Ecology, 77*, 201–206.

D'Amico, V., Elkinton, J. S., Podgwaite, J. D., Buonaccorsi, J. P., & Dwyer, G. (2005). Pathogen clumping: an explanation for non-linear transmission of an insect virus. *Ecol. Entomol., 30*, 383–390.

David, W. A. L., & Gardiner, B. O. C. (1966). Persistence of a granulosis virus of *Pieris brassicae* on cabbage leaves. *J. Invertebr. Pathol., 8*, 180–183.

David, W. A., Clothier, S. E., Woolner, M., & Taylor, G. (1971). Bioassaying an insect virus on leaves. II. The influence of certain factors associated with the larvae and the leaves. *J. Invertebr. Pathol., 17*, 178–185.

Deng, F., Wang, R., Fang, M., Jiang, Y., Xu, X., Wang, H., Chen, X., Arif, B. M., Guo, L., & Hu, Z. (2007). Proteomics analysis of *Helicoverpa armigera* single nucleocapsid nucleopolyhedrovirus identified two new occlusion-derived virus-associated proteins, HA44 and HA100. *J. Virol., 81*, 9377–9385.

Detvisitsakun, C., Berretta, M. F., Lehiy, C., & Passarelli, A. L. (2005). Stimulation of cell motility by a viral fibroblast growth factor homolog: proposal for a role in viral pathogenesis. *Virology, 336*, 308–317.

Detvisitsakun, C., Cain, E. L., & Passarelli, A. L. (2007). The *Autographa californica* M nucleopolyhedrovirus fibroblast growth factor accelerates host mortality. *Virology, 365*, 70–78.

Doane, C. C. (1969). Trans-ovum transmission of a nuclear-polyhedrosis virus in the gypsy moth and the inducement of virus susceptibility. *J. Invertebr. Pathol., 14*, 199–210.

Doane, C. C. (1970). Primary pathogens and their role in the development of an epizootic in the gypsy moth. *J. Invertebr. Pathol., 15*, 21–33.

Dobos, P., & Cochran, M. A. (1980). Protein synthesis in cells infected by *Autographa californica* nuclear polyhedrosis virus (Ac-NPV): the effect of cytosine arabinoside. *Virology, 103*, 446–464.

Dougherty, E. M., Guthrie, K. P., & Shapiro, M. (1996). Optical brighteners provide baculovirus activity enhancement and UV radiation protection. *Biol. Control, 7*, 71–74.

Dougherty, E. M., Narang, N., Loeb, M., Lynn, D. E., & Shapiro, M. (2006). Fluorescent brightener inhibits apoptosis in baculovirus-infected gypsy moth larval midgut cells *in vitro*. *Biocontrol Sci. Technol., 16*, 157–168.

Dow, J. A. (1984). Extremely high pH in biological systems: a model for carbonate transport. *Am. J. Physiol., 246*, R633–R636.

Doyle, C. J., Hirst, M. L., Cory, J. S., & Entwistle, P. F. (1990). Risk assessment studies: detailed host range testing of wild-type cabbage moth, *Mamestra brassicae* (Lepidoptera: Noctuidae), nuclear polyhedrosis virus. *Appl. Environ. Microbiol., 56*, 2704–2710.

Duffey, S. S., Hoover, K., Bonning, B. C., & Hammock, B. D. (1995). The impact of host-plant on the efficacy of baculoviruses. In M. Roe & R. Kuhr (Eds.), *Reviews in Pesticide Toxicology* (pp. 137–275). Raleigh: CTI Toxicology Communications.

Duffy, S. P., Young, A. M., Morin, B., Lucarotti, C. J., Koop, B. F., & Levin, D. B. (2006). Sequence analysis and organization of the *Neodiprion abietis* nucleopolyhedrovirus genome. *J. Virol., 80*, 6952–6963.

Duffy, S. P., Becker, E. M., Whittome, B. H., Lucarotti, C. J., & Levin, D. B. (2007). *In vivo* replication kinetics and transcription patterns of the nucleopolyhedrovirus (NeabNPV) of the balsam fir sawfly, *Neodiprion abietis*. *J. Gen. Virol., 88*, 1945–1951.

Dulmage, H. T., Martinez, A. J., & Correa, J. A. (1970). Recovery of the nuclear polyhedrosis virus of the cabbage looper, *Trichoplusia ni*, by coprecipitation with lactose. *J. Invertebr. Pathol., 16*, 80–83.

Durantel, D., Croizier, L., Ayres, M. D., Croizier, G., Possee, R. D., & Lopez-Ferber, M. (1998). The pnk/pnl gene (ORF 86) of *Autographa californica* nucleopolyhedrovirus is a non-essential, immediate early gene. *J. Gen. Virol., 79*, 629–637.

Dwyer, G., Elkinton, J. S., & Buonaccorsi, J. P. (1997). Host heterogeneity in susceptibility and disease dynamics: tests of a mathematical model. *Am. Nat., 150*, 685–707.

Dwyer, G., Dushoff, J., Elkinton, J. S., & Levin, S. A. (2000). Pathogen-driven outbreaks in forest defoliators revisited: building models from experimental data. *Am. Nat., 156*, 105–120.

Dwyer, G., Firestone, J., & Stevens, E. (2005). Should models of disease dynamics in herbivorous insects include the effects of variability in host-plant foliage quality? *Am. Nat., 165*, 16–31.

Eason, J. E., Hice, R. H., Johnson, J. J., & Federici, B. A. (1998). Effects of substituting granulin or a granulin-polyhedrin chimera for polyhedrin on virion occlusion and polyhedral morphology in *Autographa californica* multinucleocapsid nuclear polyhedrosis virus. *J. Virol., 72*, 6237–6243.

Eberle, K. E., & Jehle, J. A. (2006). Field resistance of codling moth against *Cydia pomonella* granulovirus (CpGV) is autosomal and incompletely dominant inherited. *J. Invertebr. Pathol., 93*, 201–206.

Eberle, K. E., Asser-Kaiser, S., Sayed, S. M., Nguyen, H. T., & Jehle, J. A. (2008). Overcoming the resistance of codling moth against conventional *Cydia pomonella* granulovirus (CpGV-M) by a new isolate CpGV-I12. *J. Invertebr. Pathol., 98*, 293–298.

Elazar, M., Levi, R., & Zlotkin, E. (2001). Targeting of an expressed neurotoxin by its recombinant baculovirus. *J. Exp. Biol., 204*, 2637–2645.

Elleman, C. J., & Entwistle, P. F. (1985). Inactivation of nuclear polyhedrosis virus on cotton by substances produced by the cotton leaf glands. *Ann. Appl. Biol., 106*, 83–92.

Engelhard, E. K., & Volkman, L. E. (1995). Developmental resistance in fourth instar *Trichoplusia ni* orally inoculated with *Autographa californica* M nuclear polyhedrosis virus. *Virology, 209,* 384−389.

Engelhard, E. K., Kam-Morgan, L. N., Washburn, J. O., & Volkman, L. E. (1994). The insect tracheal system: a conduit for the systemic spread of *Autographa californica* M nuclear polyhedrosis virus. *Proc. Natl. Acad. Sci. USA, 91,* 3224−3227.

Erlandson, M. A. (1990). Biological and biochemical comparison of *Mamestra configurata* and *Mamestra brassicae* nuclear polyhedrosis virus isolates pathogenic for the bertha armyworm, *Mamestra configurata* (Lepidoptera: Noctuidae). *J. Invertebr. Pathol., 56,* 47−56.

Escasa, S. R., Lauzon, H. A. M., Mathur, A. C., Krell, P. J., & Arif, B. M. (2006). Sequence analysis of the *Choristoneura occidentalis* granulovirus genome. *J. Gen. Virol., 87,* 1917−1933.

Falcon, L. A., Kane, W. R., & Bethell, R. S. (1968). Preliminary evaluation of a granulosis virus for control of the codling moth. *J. Econ. Entomol., 61,* 1208−1213.

Fang, M., Wang, H., Yuan, L., Chen, X., Vlak, J. M., & Hu, Z. (2003). Open reading frame 94 of *Helicoverpa armigera* single nucleocapsid nucleopolyhedrovirus encodes a novel conserved occlusion-derived virion protein, ODV-EC43. *J. Gen. Virol., 84,* 3021−3027.

Fang, M., Dai, X., & Theilmann, D. A. (2007). *Autographa californica* multiple nucleopolyhedrovirus EXON0 (ORF141) is required for efficient egress of nucleocapsids from the nucleus. *J. Virol., 81,* 9859−9869.

Faulkner, P., & Henderson, J. F. (1972). Serial passage of a nuclear polyhedrosis disease virus of the cabbage looper (*Trichoplusia ni*) in a continuous tissue culture cell line. *Virology, 50,* 920−924.

Federici, B. A. (1997). Baculovirus pathogenesis. In L. K. Miller (Ed.), *The Baculoviruses* (pp. 33−59). New York: Plenum Press.

Federici, B. A., & Hice, R. H. (1997). Organization and molecular characterization of genes in the polyhedrin region of the *Anagrapha falcifera* multinucleocapsid NPV. *Arch. Virol., 142,* 333−348.

Federici, B. A., & Stern, V. M. (1990). Replication and occlusion of a granulosis virus in larval and adult midgut epithelium of the western grapeleaf skeletonizer, *Harrisina brillians. J. Invertebr. Pathol., 56,* 401−414.

Felton, G. W., & Duffey, S. S. (1990). Inactivation of a baculovirus by quinones formed in insect-damaged plant tissue. *J. Chem. Ecol., 16,* 1211−1236.

Fitt, G. P. (1989). The ecology of *Heliothis* species in relation to agroecosystems. *Annu. Rev. Entomol., 34,* 17−52.

Flipsen, J. T., Martens, J. W., van Oers, M. M., Vlak, J. M., & van Lent, J. W. (1995). Passage of *Autographa californica* nuclear polyhedrosis virus through the midgut epithelium of *Spodoptera exigua* larvae. *Virology, 208,* 328−335.

Fraser, M. J. (1986). Ultrastructural observations of virion maturation in *Autographa californica* nuclear polyhedrosis virus infected *Spodoptera frugiperda* cell cultures. *J. Ultrastruct. Mol. Struct. Res., 95,* 189−195.

Fraser, M. J., & Hink, W. F. (1982). The isolation and characterization of the MP and FP plaque variants of *Galleria mellonella* nuclear polyhedrosis virus. *Virology, 117,* 366−378.

Funk, C. J., & Consigli, R. A. (1992). Evidence for zinc binding by two structural proteins of *Plodia interpunctella* granulosis virus. *J. Virol., 66,* 3168−3171.

Funk, C. J., & Consigli, R. A. (1993). Phosphate cycling on the basic protein of *Plodia interpunctella* granulosis virus. *Virology, 193,* 396−402.

Furuichi, Y., & Shatkin, A. J. (2000). Viral and cellular mRNA capping: past and prospects. *Adv. Virus Res., 55,* 135−184.

Fuxa, J. R. (1982). Prevalence of viral infections in populations of fall armyworm, *Spodoptera frugiperda,* in southeastern Louisiana. *Environ. Entomol., 11,* 239−242.

Fuxa, J. R. (2004). Ecology of insect nucleopolyhedroviruses. *Agric. Ecosyst. Environ., 103,* 27−43.

Fuxa, J. R., & Richter, A. R. (1989). Reversion of resistance by *Spodoptera frugiperda* to nuclear polyhedrosis virus. *J. Invertebr. Pathol., 53,* 52−56.

Fuxa, J. R., & Richter, A. R. (1998). Repeated reversion of resistance to nucleopolyhedrovirus by anticarsia gemmatalis. *J. Invertebr. Pathol., 71,* 159−164.

Fuxa, J. R., & Richter, A. R. (2001). Quantification of soil-to-plant transport of recombinant nucleopolyhedrovirus: effects of soil type and moisture, air currents, and precipitation. *Appl. Environ. Microbiol., 67,* 5166−5170.

Fuxa, J. R., Richter, A. R., & Strother, M. S. (1993). Detection of *Anticarsia gemmatalis* nuclear polyhedrosis-virus in predatory arthropods and parasitoids after viral release in Louisiana soybean. *J. Entomol. Sci., 28,* 51−60.

Fuxa, J. A., Fuxa, J. R., & Richter, A. R. (1998). Host-insect survival time and disintegration in relation to population density and dispersion of recombinant and wild-type nucleopolyhedroviruses. *Biol. Control, 12,* 143−150.

Fuxa, J. R., Sun, J. Z., Weidner, E. H., & LaMotte, L. R. (1999). Stressors and rearing diseases of *Trichoplusia ni*: evidence of vertical transmission of NPV and CPV. *J. Invertebr. Pathol., 74,* 149−155.

Fuxa, J. R., Matter, M. M., Abdel-Rahman, A., Micinski, S., Richter, A. R., & Flexner, J. L. (2001). Persistence and distribution of wild-type and recombinant nucleopolyhedroviruses in soil. *Microb. Ecol., 41,* 222−231.

Fuxa, J. R., Richter, A. R., Ameen, A. O., & Hammock, B. D. (2002). Vertical transmission of TnSNPV, TnCPV, AcMNPV, and possibly recombinant NPV in *Trichoplusia ni. J. Invertebr. Pathol., 79,* 44−50.

Gallo, L. G., Corsaro, B. G., Hughes, P. R., & Granados, R. R. (1991). *In vivo* enhancement of baculovirus infection by the viral enhancing factor of a granulosis virus of the cabbage looper, *Trichoplusia ni* (Lepidoptera: Noctuidae). *J. Invertebr. Pathol., 58,* 203−210.

Garry, C. E., & Garry, R. F. (2008). Proteomics computational analyses suggest that baculovirus GP64 superfamily proteins are class III penetrenes. *Virol. J., 5,* 28.

Gatehouse, H. S., Poulton, J., Markwick, N. P., Gatehouse, L. N., Ward, V. K., Young, V. L., Luo, Z., Schaffer, R., & Christeller, J. T. (2009). Changes in gene expression in the permissive larval host lightbrown apple moth (*Epiphyas postvittana,* Tortricidae) in response to EppoNPV (Baculoviridae) infection. *Insect. Mol. Biol., 18,* 635−648.

Gauthier, L., Cousserans, F., Veyrunes, J. C., & Bergoin, M. (1995). The *Melolontha melolontha* entomopoxvirus (MmEPV) fusolin is related to the fusolins of lepidopteran EPVs and to the 37K baculovirus glycoprotein. *Virology, 208,* 427−436.

Gazzoni, D. L., & Moscardi, F. (1998). Effect of defoliation levels on recovery of leaf area, on yield and agronomic traits of soybeans. *Pesq. Agropec. Bras., 33,* 411−424.

Georgievska, L., De Vries, R. S., Gao, P., Sun, X., Cory, J. S., Vlak, J. M., & van der Werf, W. (2010a). Transmission of wild-type and recombinant HaSNPV among larvae of *Helicoverpa armigera* (Lepidoptera: Noctuidae) on cotton. *Environ. Entomol., 39,* 459−467.

Georgievska, L., Hoover, K., van der Werf, W., Munoz, D., Caballero, P., Cory, J. S., & Vlak, J. M. (2010b). Dose dependency of time to death in single and mixed infections with a wildtype and egt deletion strain of *Helicoverpa armigera* nucleopolyhedrovirus. *J. Invertebr. Pathol., 104*, 44−50.

Georgievska, L., Joosten, N., Hoover, K., Cory, J. S., Vlak, J. M., & van der Werf, W. (2010c). Effects of single and mixed infections with wild type and genetically modified *Helicoverpa armigera* nucleopolyhedrovirus on movement behaviour of cotton bollworm larvae. *Entomol. Exp. Appl., 135*, 56−67.

Gershburg, E., Stockholm, D., Froy, O., Rashi, S., Gurevitz, M., & Chejanovsky, N. (1998). Baculovirus-mediated expression of a scorpion depressant toxin improves the insecticidal efficacy achieved with excitatory toxins. *FEBS Lett., 422*, 132−136.

Giri, L., Li, H., Sandgren, D., Feiss, M. G., Roller, R., Bonning, B. C., & Murhammer, D. W. (2010). Removal of transposon target sites from the *Autographa californica* multiple nucleopolyhedrovirus *fp25k* gene delays, but does not prevent, accumulation of the few polyhedra phenotype. *J. Gen. Virol., 91*, 3053−3064.

Glaser, R. W. (1917). The growth of insect blood cells *in vitro*. *Psyche., 24*, 1−6.

Glaser, R. W., & Chapman, J. W. (1913). The wilt disease of gypsy caterpillars. *J. Econ. Entomol., 6*, 479−488.

Glaser, R. W., & Chapman, J. W. (1915). A preliminary list of insects which have wilt, with a comparative study of their polyhedra. *J. Econ. Entomol., 8*, 140−150.

Goley, E. D., Ohkawa, T., Mancuso, J., Woodruff, J. B., D'Alessio, J. A., Cande, W. Z., Volkman, L. E., & Welch, M. D. (2006). Dynamic nuclear actin assembly by Arp2/3 complex and a baculovirus WASP-like protein. *Science, 314*, 464−467.

Gombart, A. F., Pearson, M. N., Rohrmann, G. F., & Beaudreau, G. S. (1989). A baculovirus polyhedral envelope-associated protein: genetic location, nucleotide sequence, and immunocytochemical characterization. *Virology, 169*, 182−193.

Gomi, S., Zhou, C. E., Yih, W., Majima, K., & Maeda, S. (1997). Deletion analysis of four of eighteen late gene expression factor gene homologues of the baculovirus, BmNPV. *Virology, 230*, 35−47.

Goodwin, R. H., Vaughn, J. L., Adams, J. R., & Louloudes, S. J. (1970). Replication of a nuclear polyhedrosis virus in an established insect cell line. *J. Invertebr. Pathol., 16*, 284−288.

Goodwin, R. H., Adams, J. R., & Shapiro, M. (1990). Replication of the entomopoxvirus from *Amsacta moorei* in serum-free cultures of a gypsy moth cell line. *J. Invertebr. Pathol., 56*, 190−205.

Goodwin, R. H., Milner, R. J., & Beaton, C. D. (1991). Entomopoxvirinae. In J. R. Adams & J. R. Bonami (Eds.), *Atlas of Invertebrate Viruses*. Boca Raton: CRC Press.

Gopalakrishnan, K., Muthukrishnan, S., & Kramer, K. J. (1995). Baculovirus-mediated expression of a *Manduca sexta* chitinase gene: properties of the recombinant protein. *Insect Biochem. Mol. Biol., 25*, 255−265.

Goto, C., Minobe, Y., & Iizuka, T. (1992). Restriction endonuclease analysis and mapping of the genomes of granulosis viruses isolated from *Xestia c-nigrum* and five other noctuid species. *J. Gen. Virol., 73*, 1491−1497.

Goulson, D. (1997). *Wipfelkrankheit*: modification of host behaviour during baculoviral infection. *Oecologia, 109*, 219−228.

Granados, R. R. (1973). Entry of an insect poxvirus by fusion of the virus envelope with the host cell membrane. *Virology, 52*, 305−309.

Granados, R. R. (1978). Early events in the infection of *Heliothis zea* midgut cells by a baculovirus. *Virology, 90*, 170−174.

Granados, R. R., & Lawler, K. A. (1981). *In vivo* pathway of *Autographa californica* baculovirus invasion and infection. *Virology, 108*, 297−308.

Granados, R. R., & Roberts, D. W. (1970). Electron microscopy of a poxlike virus infecting an invertebrate host. *Virology, 40*, 230−243.

Granados, R. R., Nguyen, T., & Cato, B. (1978). An insect cell line persistently infected with a baculovirus-like particle. *Intervirology, 10*, 309−317.

Green, T. B., White, S., Rao, S., Mertens, P. P., Adler, P. H., & Becnel, J. J. (2007). Biological and molecular studies of a cypovirus from the black fly *Simulium ubiquitum* (Diptera: Simuliidae). *J. Invertebr. Pathol., 95*, 26−32.

Gröner, A. (1986). Specificity and safety of baculoviruses. In R. R. Granados & B. A. Federici (Eds.), *The Biology of Baculoviruses, Vol. II. Practical Application for Insect Control* (pp. 177−202). Boca Raton: CRC Press.

Gross, C. H., & Shuman, S. (1998). RNA 5′-triphosphatase, nucleoside triphosphatase, and guanylyltransferase activities of baculovirus LEF-4 protein. *J. Virol., 72*, 10020−10028.

Gross, C. H., Russell, R. L., & Rohrmann, G. F. (1994). *Orgyia pseudotsugata* baculovirus p10 and polyhedron envelope protein genes: analysis of their relative expression levels and role in polyhedron structure. *J. Gen. Virol., 75*, 1115−1123.

Grove, M. J., & Hoover, K. (2007). Intrastadial developmental resistance of third instar gypsy moths (*Lymantria dispar* L.) to *L. dispar* nucleopolyhedrovirus. *Biol. Control, 40*, 355−361.

Grzywacz, D., Rossbach, A., Rauf, A., Russell, D. A., Srinivasan, R., & Shelton, A. M. (2010). Current control methods for diamondback moth and other brassica insect pests and the prospects for improved management with lepidopteran-resistant Bt vegetable brassicas in Asia and Africa. *Crop Prot., 29*, 68−79.

Guarino, L. A., & Smith, M. (1992). Regulation of delayed-early gene transcription by dual TATA boxes. *J. Virol., 66*, 3733−3739.

Guarino, L. A., & Summers, M. D. (1986a). Functional mapping of a trans-activating gene required for expression of a baculovirus delayed-early gene. *J. Virol., 57*, 563−571.

Guarino, L. A., & Summers, M. D. (1986b). Interspersed homologous DNA of *Autographa californica* nuclear polyhedrosis virus enhances delayed-early gene expression. *J. Virol., 60*, 215−223.

Guarino, L. A., Gonzalez, M. A., & Summers, M. D. (1986). Complete sequence and enhancer function of the homologous DNA regions of *Autographa californica* nuclear polyhedrosis virus. *J. Virol., 60*, 224−229.

Guarino, L. A., Xu, B., Jin, J., & Dong, W. (1998). A virus-encoded RNA polymerase purified from baculovirus-infected cells. *J. Virol., 72*, 7985−7991.

Guarino, L. A., Dong, W., & Jin, J. (2002). *In vitro* activity of the baculovirus late expression factor LEF-5. *J. Virol., 76*, 12663−12675.

Haack, R. A., & Mattson, W. J. (1993). Life history patterns of North American tree-feeding sawflies. In M. R. Wagner & K. F. Raffa (Eds.), *Sawfly Life History Adaptations to Woody Plants* (pp. 503−545). San Diego: Academic Press.

Haas-Stapleton, E. J., Washburn, J. O., & Volkman, L. E. (2004). P74 mediates specific binding of *Autographa californica* M nucleopolyhedrovirus occlusion-derived virus to primary cellular targets in the midgut epithelia of *Heliothis virescens* Larvae. *J. Virol., 78*, 6786−6791.

Haines, F. J., Griffiths, C. M., Possee, R. D., Hawes, C. R., & King, L. A. (2009). Involvement of lipid rafts and cellular actin in AcMNPV GP64 distribution and virus budding. *Virol. Sin., 24*, 333–349.

Hall, R. L., & Moyer, R. W. (1991). Identification, cloning, and sequencing of a fragment of *Amsacta moorei* entomopoxvirus DNA containing the spheroidin gene and three vaccinia virus-related open reading frames. *J. Virol., 65*, 6516–6527.

Hamm, J. J. (1968). Comparative histopathology of a granulosis and a nuclear polyhedrosis of *Spodoptera frugiperda. J. Invertebr. Pathol., 10*, 320–326.

Hamm, J. J., & Young, J. R. (1974). Mode of transmission of nuclear-polyhedrosis virus to progeny of adult *Heliothis zea. J. Invertebr. Pathol., 24*, 70–81.

Hamm, J. J., Chandler, L. D., & Sumner, H. R. (1994). Field tests with a fluorescent brightener to enhance infectivity of fall armyworm (Lepidoptera: Noctuidae) nuclear polyhedrosis virus. *Fla. Entomol., 77*, 425–437.

Hamm, J. J., Carpenter, J. E., & Styer, E. L. (1996). Oviposition day effect on incidence of agonadal progeny of *Helicoverpa zea* (Lepidoptera: Noctuidae) infected with a virus. *Ann. Entomol. Soc. Am., 56*, 535–556.

Hang, X., & Guarino, L. A. (1999). Purification of *Autographa californica* nucleopolyhedrovirus DNA polymerase from infected insect cells. *J. Gen. Virol., 80*, 2519–2526.

Hang, X., Dong, W., & Guarino, L. A. (1995). The lef-3 gene of *Autographa californica* nuclear polyhedrosis virus encodes a single-stranded DNA-binding protein. *J. Virol., 69*, 3924–3928.

Harper, J. D. (1973). Food consumption by cabbage loopers infected with nuclear polyhedrosis virus. *J. Invertebr. Pathol., 21*, 191–197.

Harrap, K. A. (1972a). The structure of nuclear polyhedrosis viruses. III. Virus assembly. *Virology, 50*, 133–139.

Harrap, K. A. (1972b). The structure of nuclear polyhedrosis viruses. I. The inclusion body. *Virology, 50*, 114–123.

Harrap, K. A., & Robertson, J. S. (1968). A possible infection pathway in the development of a nuclear polyhedrosis virus. *J. Gen. Virol., 3*, 221–225.

Harrison, R. L. (2009). Genomic sequence analysis of the Illinois strain of the *Agrotis ipsilon* multiple nucleopolyhedrovirus. *Virus Genes, 38*, 155–170.

Harrison, R. L., & Bonning, B. C. (1999). The nucleopolyhedroviruses of *Rachiplusia ou* and *Anagrapha falcifera* are isolates of the same virus. *J. Gen. Virol., 80*, 2793–2798.

Harrison, R. L., & Bonning, B. C. (2000). Use of scorpion neurotoxins to improve the insecticidal activity of *Rachiplusia ou* multicapsid nucleopolyhedrovirus. *Biol. Control, 17*, 191–201.

Harrison, R. L., & Bonning, B. C. (2001). Use of proteases to improve the insecticidal activity of baculoviruses. *Biol. Control, 20*, 199–209.

Harrison, R. L., & Bonning, B. C. (2003). Comparative analysis of the genomes of *Rachiplusia ou* and *Autographa californica* multiple nucleopolyhedroviruses. *J. Gen. Virol., 84*, 1827–1842.

Harrison, R. L., & Lynn, D. E. (2007). Genomic sequence analysis of a nucleopolyhedrovirus isolated from the diamondback moth, *Plutella xylostella. Virus Genes, 35*, 857–873.

Harrison, R. L., & Lynn, D. E. (2008). New cell lines derived from the black cutworm, *Agrotis ipsilon*, that support replication of the *A. ipsilon* multiple nucleopolyhedrovirus and several group I nucleo-polyhedroviruses. *J. Invertebr. Pathol., 99*, 28–34.

Harrison, R. L., & Popham, H. J. (2008). Genomic sequence analysis of a granulovirus isolated from the Old World bollworm, *Helicoverpa armigera. Virus Genes, 36*, 565–581.

Harrison, R. L., & Summers, M. D. (1995). Mutations in the *Autographa californica* multinucleocapsid nuclear polyhedrosis virus 25 kDa protein gene result in reduced virion occlusion, altered intranuclear envelopment and enhanced virus production. *J. Gen. Virol., 76*, 1451–1459.

Hashimoto, Y., Corsaro, B. G., & Granados, R. R. (1991). Location and nucleotide sequence of the gene encoding the viral enhancing factor of the *Trichoplusia ni* granulosis virus. *J. Gen. Virol., 72*, 2645–2651.

Hawtin, R. E., Zarkowska, T., Arnold, K., Thomas, C. J., Gooday, G. W., King, L. A., Kuzio, J. A., & Possee, R. D. (1997). Liquefaction of *Autographa californica* nucleopolyhedrovirus-infected insects is dependent on the integrity of virus-encoded chitinase and cathepsin genes. *Virology, 238*, 243–253.

Hayakawa, T., Ko, R., Okano, K., Seong, S. I., Goto, C., & Maeda, S. (1999). Sequence analysis of the *Xestia c-nigrum* granulovirus genome. *Virology, 262*, 277–297.

Hefferon, K. L., Oomens, A. G., Monsma, S. A., Finnerty, C. M., & Blissard, G. W. (1999). Host cell receptor binding by baculovirus GP64 and kinetics of virion entry. *Virology, 258*, 455–468.

Hegedus, D., Erlandson, M., Gillott, C., & Toprak, U. (2009). New insights into peritrophic matrix synthesis, architecture, and function. *Annu. Rev. Entomol., 54*, 285–302.

Heinz, K. M., McCutchen, B. F., Herrmann, R., Parrella, M. P., & Hammock, B. D. (1995). Direct effects of recombinant nuclear polyhedrosis viruses on selected nontarget organisms. *J. Econ. Entomol., 88*, 259–264.

Henderson, J. F., Faulkner, P., & MacKinnon, E. A. (1974). Some biophysical properties of virus present in tissue cultures infected with the nuclear polyhedrosis virus of *Trichoplusia ni. J. Gen. Virol., 22*, 143–146.

Herniou, E. A., & Jehle, J. A. (2007). Baculovirus phylogeny and evolution. *Curr. Drug Targets, 8*, 1043–1050.

Herniou, E. A., Luque, T., Chen, X., Vlak, J. M., Winstanley, D., Cory, J. S., & O'Reilly, D. R. (2001). Use of whole genome sequence data to infer baculovirus phylogeny. *J. Virol., 75*, 8117–8126.

Herniou, E. A., Olszewski, J. A., Cory, J. S., & O'Reilly, D. R. (2003). The genome sequence and evolution of baculoviruses. *Annu. Rev. Entomol., 48*, 211–234.

Herniou, E. A., Olszewski, J. A., O'Reilly, D. R., & Cory, J. S. (2004). Ancient coevolution of baculoviruses and their insect hosts. *J. Virol., 78*, 3244–3251.

Herrmann, R., Fishman, L., & Zlotkin, E. (1990). The tolerance of lepidopterous larvae to an insect selective toxin. *Insect Biochem., 20*, 625–637.

Hess, R. T., & Falcon, L. A. (1978). Electron microscope observations of the membrane surrounding polyhedral inclusion bodies of insects. *Arch. Virol., 56*, 169–176.

Hess, R. T., & Falcon, L. A. (1987). Temporal events in the invasion of the codling moth, *Cydia pomonella*, by a granulosis virus. *J. Invertebr. Pathol., 50*, 85–105.

Hilton, S., & Winstanley, D. (2007). Identification and functional analysis of the origins of DNA replication in the *Cydia pomonella* granulovirus genome. *J. Gen. Virol., 88*, 1496–1504.

Hink, W. F., & Vail, P. V. (1973). A plaque assay for titration of alfalfa looper nuclear polyhedrosis virus in a cabbage looper (TN-368) cell line. *J. Invertebr. Pathol., 22*, 168–174.

Hitchman, R. B., Hodgson, D. J., King, L. A., Hails, R. S., Cory, J. S., & Possee, R. D. (2007). Host mediated selection of pathogen genotypes as a mechanism for the maintenance of baculovirus diversity in the field. *J. Invertebr. Pathol., 94,* 153—162.

Hodgson, D. J., Hitchman, R. B., Vanbergen, A. J., Hails, R. S., Possee, R. D., & Cory, J. S. (2004). Host ecology determines the relative fitness of virus genotypes in mixed-genotype nucleopolyhedrovirus infections. *J. Evol. Biol., 17,* 1018—1025.

Hofmann, O. (1891). *Die Schlaffsucht (Flacherie) der Nonne (*Liparis monacha*) nebst einem Anhang. Insektenötende Pilze mit besonderer Berücksichtigung der Nonne.* Frankfurt: P. Weber.

Hom, L. G., & Volkman, L. E. (2000). *Autographa californica* M nucleopolyhedrovirus chiA is required for processing of V-CATH. *Virology, 277,* 178—183.

Hom, L. G., Ohkawa, T., Trudeau, D., & Volkman, L. E. (2002). *Autographa californica* M nucleopolyhedrovirus proV-CATH is activated during infected cell death. *Virology, 296,* 212—218.

Hong, T., Braunagel, S. C., & Summers, M. D. (1994). Transcription, translation, and cellular localization of PDV-E66: a structural protein of the PDV envelope of *Autographa californica* nuclear polyhedrosis virus. *Virology, 204,* 210—222.

Hong, T., Summers, M. D., & Braunagel, S. C. (1997). N-terminal sequences from *Autographa californica* nuclear polyhedrosis virus envelope proteins ODV-E66 and ODV-E25 are sufficient to direct reporter proteins to the nuclear envelope, intranuclear microvesicles and the envelope of occlusion derived virus. *Proc. Natl. Acad. Sci. USA, 94,* 4050—4055.

Hooft van Iddekinge, B. J. L., Smith, G. E., & Summers, M. D. (1983). Nucleotide sequence of the polyhedrin gene of *Autographa californica* nuclear polyhedrosis virus. *Virology, 131,* 561—565.

Hoopes, R. R., Jr., & Rohrmann, G. F. (1991). *In vitro* transcription of baculovirus immediate early genes: accurate mRNA initiation by nuclear extracts from both insect and human cells. *Proc. Natl. Acad. Sci. USA, 88,* 4513—4517.

Hoover, K., Schultz, C. M., Lane, S. S., Bonning, B. C., Duffey, S. S., McCutchen, B. F., & Hammock, B. D. (1995). Reduction in damage to cotton plants by a recombinant baculovirus that knocks moribund larvae of *Heliothis virescens* off the plant. *Biol. Control, 5,* 419—426.

Hoover, K., Alaniz, S. A., Yee, J. L., Rocke, D. M., Hammock, B. D., & Duffey, S. S. (1998a). Influence of dietary protein and chlorogenic acid on disease of noctuid (Lepidoptera: Noctuidae) larvae infected with wild-type or recombinant baculoviruses. *Environ. Entomol., 27,* 1264—1272.

Hoover, K., Stout, M. J., Alaniz, S. A., Hammock, B. D., & Duffey, S. S. (1998b). Influence of induced plant defenses in cotton and tomato on efficacy of baculoviruses. *J. Chem. Ecol., 24,* 253—271.

Hoover, K., Yee, J. L., Schultz, C. M., Hammock, B. D., Rocke, D. M., & Duffey, S. S. (1998c). Effects of plant identity and chemical constituents on the efficacy of a baculovirus against *Heliothis virescens. J. Chem. Ecol., 24,* 221—252.

Hoover, K., Washburn, J. O., & Volkman, L. E. (2000). Midgut-based resistance of *Heliothis virescens* to baculovirus infection mediated by phytochemicals in cotton. *J. Insect Physiol., 46,* 999—1007.

Hoover, K., Grove, M. J., & Su, S. (2002). Systemic component to intrastadial developmental resistance in *Lymantria dispar* to its baculovirus. *Biol. Control, 25,* 92—98.

Hoover, K., Humphries, M. A., Gendron, A. R., & Slavicek, J. M. (2010). Impact of viral enhancin genes on potency of *Lymantria dispar* multiple nucleopolyhedrovirus in *L. dispar* following disruption of the peritrophic matrix. *J. Invertebr. Pathol., 104,* 150—152.

Hoover, K., Grove, M., Gardner, M., Hughes, D. P., McNeil, J., & Slavicek, J. (2011). A gene for an extended phenotype. *Science, 333,* 1401.

Horton, H. M., & Burand, J. P. (1993). Saturable attachment sites for polyhedron-derived baculovirus on insect cells and evidence for entry via direct membrane fusion. *J. Virol., 67,* 1860—1868.

Hostetter, D. L., & Puttler, B. (1991). A new broad spectrum nuclear polyhedrosis virus isolated from the celery looper, *Anagrapha falcifera* (Kirby) (Lepidoptera: Noctuidae). *Environ. Entomol., 20,* 1480—1488.

Hu, Z., Luijckx, T., van Dinten, L. C., van Oers, M. M., Hajos, J. P., Bianchi, F. J., van Lent, J. W., Zuidema, D., & Vlak, J. M. (1999). Specificity of polyhedrin in the generation of baculovirus occlusion bodies. *J. Gen. Virol., 80,* 1045—1053.

Huber, J. (1986). Use of baculoviruses in pest management programs. In R. R. Granados & B. A. Federici (Eds.), *The Biology of Baculoviruses. Vol. II. Practical Application for Insect Control* (pp. 181—202). Boca Raton: CRC Press.

Huber, J., & Dickler, E. (1977). Codling moth granulosis virus: its efficiency in the field in comparison with organophosphorus insecticides. *J. Econ. Entomol., 70,* 557—561.

Huger, A. M. (1966). A virus disease of the Indian rhinoceros beetle, *Oryctes rhinoceros* (Linnaeus), caused by a new type of insect virus, *Rhabdionvirus oryctes* gen. n., sp. n. *J. Invertebr. Pathol., 8,* 38—51.

Huger, A. M. (1985). A new virus disease of crickets (Orthoptera: Gryllidae) causing macronucleosis of fatbody. *J. Invertebr. Pathol., 45,* 108—111.

Huger, A. M. (2005). The *Oryctes* virus: its detection, identification, and implementation in biological control of the coconut palm rhinoceros beetle, *Oryctes rhinoceros* (Coleoptera: Scarabaeidae). *J. Invertebr. Pathol., 89,* 78—84.

Huger, A. M., & Krieg, A. (1991). Baculoviridae. Nonoccluded baculoviruses. In J. R. Adams & J. R. Bonami (Eds.), *Atlas of Invertebrate Viruses* (pp. 287—319). Boca Raton: CRC Press.

Hughes, D. S., Possee, R. D., & King, L. A. (1993). Activation and detection of a latent baculovirus resembling *Mamestra brassicae* nuclear polyhedrosis virus in *M. brassicae* insects. *Virology, 194,* 608—615.

Hughes, K. M. (1953). The development of an insect virus within cells of its host. *Hilgardia, 22,* 391—406.

Hughes, K. M. (1972). Fine structure and development of two polyhedrosis viruses. *J. Invertebr. Pathol., 19,* 198—207.

Hughes, P. R., Wood, H. A., Breen, J. P., Simpson, S. F., Duggan, A. J., & Dybas, J. A. (1997). Enhanced bioactivity of recombinant baculoviruses expressing insect-specific spider toxins in lepidopteran crop pests. *J. Invertebr. Pathol., 69,* 112—118.

Huh, N. E., & Weaver, R. F. (1990). Identifying the RNA polymerases that synthesize specific transcripts of the *Autographa californica* nuclear polyhedrosis virus. *J. Gen. Virol., 71,* 195—201.

Hukuhara, T., & Bonami, J. R. (1991). Reoviridae. In J. R. Adams & J. R. Bonami (Eds.), *Atlas of Invertebrate Viruses.* Boca Raton: CRC Press.

Hunter, M. D., & Schultz, J. C. (1993). Induced plant defenses breached? Phytochemical induction protects an herbivore from disease. *Oecologia, 94,* 195—203.

Hunter-Fujita, F. R., Entwistle, P. F., Evans, H. F., & Crook, N. E. (1998). Virus production. In F. R. Hunter-Fujita, P. F. Entwistle, H. F. Evans & N. E. Crook (Eds.), *Insect Viruses and Pest Management* (pp. 92—116). Chichester: John Wiley & Sons.

Ignoffo, C. M. (1965). The nuclear-polyhedrosis virus of *Heliothis zea* (Boddie) and *Heliothis virescens* (Fabricius). I. Virus propagation and its virulence. *J. Invertebr. Pathol., 7,* 209–216.

Ignoffo, C. M. (1968). Specificity of insect viruses. *Bull. Entomol. Soc. Am., 14,* 265–276.

Ignoffo, C. M. (1973). Development of a viral insecticide: concept to commercialization. *Exp. Parasitol., 33,* 380–406.

Ignoffo, C. M. (1999). The first viral pesticide: past, present, and future. *J. Ind. Microbiol. Biotechnol., 22,* 407–417.

Ignoffo, C. M., & Batzer, O. F. (1971). Microencapsulation and ultraviolet protectants to increase sunlight stability of an insect virus. *J. Econ. Entomol., 64,* 850–853.

Ignoffo, C. M., & Garcia, C. (1996). Rate of larval lysis and yield and activity of inclusion bodies harvested from *Trichoplusia ni* larvae fed a wild or recombinant strain of the nuclear polyhedrosis virus of *Autographa californica. J. Invertebr. Pathol., 68,* 196–198.

Ignoffo, C. M., & Montoya, E. L. (1966). The effects of chemical insecticides and insecticidal adjuvants of a *Heliothis* nuclear-polyhedrosis virus. *J. Invertebr. Pathol., 8,* 409–412.

Ignoffo, C. M., Chapman, A. J., & Martin, D. F. (1965). The nuclear-polyhedrosis virus of *Heliothis zea* (Boddie) and *Heliothis virescens* (Fabricius) III. Effectiveness of the virus against field populations of *Heliothis* on cotton, corn, and grain sorghum. *J. Invertebr. Pathol., 7,* 227–235.

Ignoffo, C. M., Hostetter, D. L., & Smith, D. B. (1976). Gustatory stimulant, sunlight protectant, evaporation retardant: three characteristics of a microbial insecticidal adjuvant. *J. Econ. Entomol., 69,* 207–210.

Ignoffo, C. M., Hostetter, D. L., Sikorowski, P. P., Sutter, G., & Brooks, W. M. (1977). Inactivation of representative species of entomopathogenic viruses, a bacterium, fungus, and protozoan by an ultraviolet light source. *Environ. Entomol., 6,* 411–415.

Ignoffo, C. M., Garcia, C., & Saathoff, S. G. (1997). Sunlight stability and rain-fastness of formulations of *Baculovirus heliothis. Environ. Entomol., 26,* 1470–1474.

Ihalainen, T. O., Laakkonen, J. P., Paloheimo, O., Ylä-Herttuala, S., Airenne, K. J., & Vihinen-Ranta, M. (2010). Morphological characterization of baculovirus *Autographa californica* multiple nucleopolyhedrovirus. *Virus Res., 148,* 71–74.

Ijiri, H., Coulibaly, F., Nishimura, G., Nakai, D., Chiu, E., Takenaka, C., Ikeda, K., Nakazawa, H., Hamada, N., Kotani, E., Metcalf, P., Kawamata, S., & Mori, H. (2009). Structure-based targeting of bioactive proteins into cypovirus polyhedra and application to immobilized cytokines for mammalian cell culture. *Biomaterials, 30,* 4297–4308.

IJkel, W. F., Westenberg, M., Goldbach, R. W., Blissard, G. W., Vlak, J. M., & Zuidema, D. (2000). A novel baculovirus envelope fusion protein with a proprotein convertase cleavage site. *Virology, 275,* 30–41.

Ikeda, M., & Kobayashi, M. (1999). Cell-cycle perturbation in Sf9 cells infected with *Autographa californica* nucleopolyhedrovirus. *Virology, 258,* 176–188.

Imai, N., Ali, S. E., El-Singaby, N. R., Iwanaga, M., Matsumoto, S., Iwabuchi, K., & Maeda, S. (2000). Insecticidal effects of a recombinant baculovirus expressing scorpion toxin LqhIT2. *J. Seric. Sci., 69,* 197–205.

Inceoglu, A. B., Kamita, S. G., & Hammock, B. D. (2006). Genetically modified baculoviruses: a historical overview and future outlook. *Adv. Virus Res., 68,* 323–360.

Iqbal, J., & MacLean, D. A. (2010). Prediction of balsam fir sawfly defoliation using a Bayesian network model. *Can. J. For. Res., 40,* 2322–2332.

Ishimori, N. (1934). Contribution à l'étude de la grasserie du ver à soie (*Bombyx mori*). *C.R. Soc. Biol., 116,* 1169–1170.

Jakubowska, A., van Oers, M. M., Ziemnicka, J., Lipa, J. J., & Vlak, J. M. (2005). Molecular characterization of *Agrotis segetum* nucleopolyhedrovirus from Poland. *J. Invertebr. Pathol., 90,* 64–68.

Jakubowska, A. K., Peters, S. A., Ziemnicka, J., Vlak, J. M., & van Oers, M. M. (2006). Genome sequence of an enhancin gene-rich nucleopolyhedrovirus (NPV) from *Agrotis segetum*: collinearity with *Spodoptera exigua* multiple NPV. *J. Gen. Virol., 87,* 537–551.

James, C. (2009). *Global status of commercialized biotech/GM crops: 2008.* In: *ISAAA Brief,* Vol. 39. Ithaca: ISAAA.

Jaques, R. P. (1964). The persistence of a nuclear-polyhedrosis virus in soil. *J. Invertebr. Pathol., 6,* 251–254.

Jaques, R. P. (1967). The persistence of a nuclear polyhedrosis virus in the habitat of the host insect, *Trichoplusia ni*: II. Polyhedra in soil. *Can. Entomol., 99,* 820–829.

Jaques, R. P. (1968). The inactivation of the nuclear polyhedrosis virus of *Trichoplusia ni* by gamma and ultraviolet radiation. *Can. J. Microbiol., 14,* 1161–1163.

Jarvis, D. L. (1993). Effects of baculovirus infection on IE1-mediated foreign gene expression in stably transformed insect cells. *J. Virol., 67,* 2583–2591.

Jarvis, D. L. (2009). Baculovirus–insect cell expression systems. *Methods Enzymol., 463,* 191–222.

Jarvis, D. L., & Garcia, A., Jr. (1994). Biosynthesis and processing of the *Autographa californica* nuclear polyhedrosis virus gp64 protein. *Virology, 205,* 300–313.

Jarvis, D. L., & Summers, M. D. (1989). Glycosylation and secretion of human tissue plasminogen activator in recombinant baculovirus-infected insect cells. *Mol. Cell. Biol., 9,* 214–223.

Jarvis, D. L., Bohlmeyer, D. A., & Garcia, J. A. (1991). Requirements for nuclear localization and supramolecular assembly of a baculovirus polyhedrin protein. *Virology, 185,* 795–810.

Jarvis, D. L., Reilly, L. M., Hoover, K., Hammock, B., & Guarino, L. A. (1996). Construction and characterization of immediate-early baculovirus insecticides. *Biol. Control, 7,* 228–235.

Jehle, J. A., Blissard, G. W., Bonning, B. C., Cory, J. S., Herniou, E. A., Rohrmann, G. F., Theilmann, D. A., Thiem, S. M., & Vlak, J. M. (2006a). On the classification and nomenclature of baculoviruses: a proposal for revision. *Arch. Virol., 151,* 1257–1266.

Jehle, J. A., Lange, M., Wang, H., Hu, Z., Wang, Y., & Hauschild, R. (2006b). Molecular identification and phylogenetic analysis of baculoviruses from Lepidoptera. *Virology, 346,* 180–193.

Ji, X., Sutton, G., Evans, G., Axford, D., Owen, R., & Stuart, D. I. (2010). How baculovirus polyhedra fit square pegs into round holes to robustly package viruses. *EMBO J., 29,* 505–514.

Jiang, S. S., Chang, I. S., Huang, L. W., Chen, P. C., Wen, C. C., Liu, S. C., Chien, L. C., Lin, C. Y., Hsiung, C. A., & Juang, J. L. (2006). Temporal transcription program of recombinant *Autographa californica* multiple nucleopolyhedrosis virus. *J. Virol., 80,* 8989–8999.

Jin, J., & Guarino, L. A. (2000). 3′-End formation of baculovirus late RNAs. *J. Virol., 74,* 8930–8937.

Jin, J., Dong, W., & Guarino, L. A. (1998). The LEF-4 subunit of baculovirus RNA polymerase has RNA 5′-triphosphatase and ATPase activities. *J. Virol., 72,* 10011–10019.

Jiravanichpaisal, P., Lee, B. L., & Söderhäll, K. (2006). Cell-mediated immunity in arthropods: hematopoiesis, coagulation, melanization and opsonization. *Immunobiology, 211*, 213−236.

de Jong, J. G., Lauzon, H. A., Dominy, C., Poloumienko, A., Carstens, E. B., Arif, B. M., & Krell, P. J. (2005). Analysis of the *Choristoneura fumiferana* nucleopolyhedrovirus genome. *J. Gen. Virol., 86*, 929−943.

Kadlec, J., Loureiro, S., Abrescia, N. G., Stuart, D. I., & Jones, I. M. (2008). The postfusion structure of baculovirus gp64 supports a unified view of viral fusion machines. *Nat. Struct. Mol. Biol., 15*, 1024−1030.

Kang, W., Tristem, M., Maeda, S., Crook, N. E., & O'Reilly, D. R. (1998). Identification and characterization of the *Cydia pomonella* granulovirus cathepsin and chitinase genes. *J. Gen. Virol., 79*, 2283−2292.

Kamita, S. G., Nagasaka, K., Chua, J. W., Shimada, T., Mita, K., Kobayashi, M., Maeda, S., & Hammock, B. D. (2005). A baculovirus-encoded protein tyrosine phosphatase gene induces enhanced locomotory activity in a lepidopteran host. *Proc. Natl. Acad. Sci. USA, 102*, 2584−2589.

Kanost, M. R., Jiang, H., & Yu, X. Q. (2004). Innate immune responses of a lepidopteran insect. *Manduca sexta. Immunol. Rev., 198*, 97−105.

Katsuma, S., Daimon, T., Mita, K., & Shimada, T. (2006a). Lepidopteran ortholog of Drosophila breathless is a receptor for the baculovirus fibroblast growth factor. *J. Virol., 80*, 5474−5481.

Katsuma, S., Horie, S., Daimon, T., Iwanaga, M., & Shimada, T. (2006b). *In vivo* and *in vitro* analyses of a *Bombyx mori* nucleopolyhedrovirus mutant lacking functional vfgf. *Virology, 355*, 62−70.

Katsuma, S., Horie, S., & Shimada, T. (2008). The fibroblast growth factor homolog of *Bombyx mori* nucleopolyhedrovirus enhances systemic virus propagation in *B. mori* larvae. *Virus Res., 137*, 80−85.

Kawamoto, F., Kumada, N., & Kobayashi, M. (1977). Envelopment of the nuclear polyhedrosis virus of the oriental tussock moth, *Euproctis subflava. Virology, 77*, 867−871.

Kawanishi, C. Y., Summers, M. D., Stoltz, D. B., & Arnott, H. J. (1972). Entry of an insect virus *in vivo* by fusion of viral envelope and microvillus membrane. *J. Invertebr. Pathol., 20*, 104−108.

Kawasaki, Y., Matsumoto, S., & Nagamine, T. (2004). Analysis of baculovirus IE1 in living cells: dynamics and spatial relationships to viral structural proteins. *J. Gen. Virol., 85*, 3575−3583.

Ke, J., Wang, J., Deng, R., & Wang, X. (2008). *Autographa californica* multiple nucleopolyhedrovirus ac66 is required for the efficient egress of nucleocapsids from the nucleus, general synthesis of pre-occluded virions and occlusion body formation. *Virology, 374*, 421−431.

Keating, S. T., & Yendol, W. G. (1987). Influence of selected host plants on gypsy moth (Lepidoptera: Lymantriidae) larval mortality caused by a baculovirus. *Environ. Entomol., 16*, 459−462.

Keating, S. T., Yendol, W. G., & Schultz, J. C. (1988). Relationship between susceptibility of gypsy moth larvae (Lepidoptera: Lymantriidae) to a baculovirus and host-plant constituents. *Environ. Entomol., 17*, 942−958.

Keddie, B. A., Aponte, G. W., & Volkman, L. E. (1989). The pathway of infection of *Autographa californica* nuclear polyhedrosis virus in an insect host. *Science, 243*, 1728−1730.

Kelly, D. C., & Lescott, T. (1981). Baculovirus replication: protein synthesis in *Spodoptera frugiperda* cells infected with *Trichoplusia ni* nuclear polyhedrosis virus. *Microbiologica, 4*, 35−37.

Kelly, D. C., Brown, D. A., Ayres, M. D., Allen, C. J., & Walker, I. O. (1983). Properties of the major nucleocapsid protein of *Heliothis zea* singly enveloped nuclear polyhedrosis virus. *J. Gen. Virol., 64*, 399−408.

Khurad, A. M., Mahulikar, A., Rathod, M. K., Rai, M. M., Kanginakudru, S., & Nagaraju, J. (2004). Vertical transmission of nucleopolyhedrovirus in the silkworm, *Bombyx mori* L. *J. Invertebr. Pathol., 87*, 8−15.

Kikhno, I., Gutierrez, S., Croizier, L., Croizier, G., & Ferber, M. L. (2002). Characterization of *pif*, a gene required for the per os infectivity of *Spodoptera littoralis* nucleopolyhedrovirus. *J. Gen. Virol., 83*, 3013−3022.

Killick, H. J. (1990). Influence of droplet size, solar ultraviolet light and protectants, and other factors on the efficacy of baculovirus sprays against *Panolis flammea* (Schiff.) (Lepidoptera: Noctuidae). *Crop Prot., 9*, 21−28.

Kim, D., & Weaver, R. F. (1993). Transcription mapping and functional analysis of the protein tyrosine/serine phosphatase (PTPase) gene of the *Autographa californica* nuclear polyhedrosis virus. *Virology, 195*, 587−595.

King, L. A., Wilkinson, N., Miller, D. P., & Marlow, S. A. (1998). Entopoxviruses. In L. K. Miller & L. A. Ball (Eds.), *The Insect Viruses* (pp. 1−29). New York: Plenum Press.

Kirkpatrick, B. A., Washburn, J. O., & Volkman, L. E. (1998). AcMNPV pathogenesis and developmental resistance in fifth instar Heliothis virescens. *J. Invertebr. Pathol., 72*, 63−72.

Knebel-Morsdorf, D., Flipsen, J. T., Roncarati, R., Jahnel, F., Kleefsman, A. W., & Vlak, J. M. (1996). Baculovirus infection of *Spodoptera exigua* larvae: *lacZ* expression driven by promoters of early genes *pe38* and *me53* in larval tissue. *J. Gen. Virol., 77*, 815−824.

Knell, R. J., Begon, M., & Thompson, D. J. (1998). Transmission of *Plodia interpunctella* granulosis virus does not conform to the mass action model. *J. Anim. Ecol., 67*, 592−599.

Knudson, D. L., & Harrap, K. A. (1976). Replication of a nuclear polyhedrosis virus in a continuous cell culture of *Spodoptera frugiperda*: microscopy study of the sequence of events of the virus infection. *J. Virol., 17*, 254−268.

Kogan, P. H., & Blissard, G. W. (1994). A baculovirus gp64 early promoter is activated by host transcription factor binding to CACGTG and GATA elements. *J. Virol., 68*, 813−822.

Komárek, J., & Breindl, V. (1924). Die Wipfel-Krankheit der Nonne und der Erreger derselben. *Z. Angew. Entomol., 10*, 99−162.

Kool, M., Voncken, J. W., van Lier, F. L., Tramper, J., & Vlak, J. M. (1991). Detection and analysis of *Autographa californica* nuclear polyhedrosis virus mutants with defective interfering properties. *Virology, 183*, 739−746.

Kool, M., van den Berg, P. M., Tramper, J., Goldbach, R. W., & Vlak, J. M. (1993). Location of two putative origins of DNA replication of *Autographa californica* nuclear polyhedrosis virus. *Virology, 192*, 94−101.

Kool, M., Ahrens, C. H., Goldbach, R. W., Rohrmann, G. F., & Vlak, J. M. (1994). Identification of genes involved in DNA replication of the *Autographa californica* baculovirus. *Proc. Natl. Acad. Sci. USA, 91*, 11212−11216.

Korth, K. L., & Levings, C. S., III (1993). Baculovirus expression of the maize mitochondrial protein URF13 confers insecticidal activity in cell cultures and larvae. *Proc. Natl. Acad. Sci. USA, 90*, 3388−3392.

Kovacs, G. R., Guarino, L. A., & Summers, M. D. (1991). Novel regulatory properties of the IE1 and IE0 transactivators encoded by the baculovirus *Autographa californica* multicapsid nuclear polyhedrosis virus. *J. Virol., 65*, 5281–5288.

Krammer, F., & Grabherr, R. (2010). Alternative influenza vaccines made by insect cells. *Trends Mol. Med., 16*, 313–320.

Kunimi, Y. (2007). Current status and prospects on microbial control in Japan. *J. Invertebr. Pathol., 95*, 181–186.

Kunimi, Y., & Yamada, E. (1990). Relationship of larval phase and susceptibility of the armyworm, *Pseudaletia separata* Walker (Lepidoptera: Noctuidae), to a nuclear polyhedrosis virus and a granulosis virus. *Appl. Entomol. Zool., 25*, 289–297.

Kunimi, Y., Fuxa, J. R., & Hammock, B. D. (1996). Comparison of wild type and genetically engineered nuclear polyhedrosis viruses of *Autographa californica* for mortality, virus replication and polyhedra production in *Trichoplusia ni* larvae. *Entomol. Exp. Appl., 81*, 251–257.

Kuzio, J., Jaques, R., & Faulkner, P. (1989). Identification of p74, a gene essential for virulence of baculovirus occlusion bodies. *Virology, 173*, 759–763.

Kuzio, J., Pearson, M. N., Harwood, S. H., Funk, C. J., Evans, J. T., Slavicek, J. M., & Rohrmann, G. F. (1999). Sequence and analysis of the genome of a baculovirus pathogenic for Lymantria dispar. *Virology, 253*, 17–34.

Lange, M., Wang, H., Zhihong, H., & Jehle, J. A. (2004). Towards a molecular identification and classification system of lepidopteran-specific baculoviruses. *Virology, 325*, 36–47.

Lasa, R., Ruiz-Portero, C., Alcázar, M. D., Belda, J. E., Williams, T., & Caballero, P. (2007). Efficacy of optical brightener formulations of *Spodoptera exigua* multiple nucleopolyhedrovirus (SeMNPV) as a biological insecticide in greenhouses in southern Spain. Biol. *Control, 40*, 89–96.

Lasa, R., Williams, T., & Caballero, P. (2009). The attractiveness of phagostimulant formulations of a nucleopolyhedrovirus-based insecticide depends on prior insect diet. *J. Pest. Sci., 82*, 247–250.

Lauzon, H. A., Lucarotti, C. J., Krell, P. J., Feng, Q., Retnakaran, A., & Arif, B. M. (2004). Sequence and organization of the *Neodiprion lecontei* nucleopolyhedrovirus genome. *J. Virol., 78*, 7023–7035.

Lauzon, H. A., Garcia-Maruniak, A., Zanotto, P. M., Clemente, J. C., Herniou, E. A., Lucarotti, C. J., Arif, B. M., & Maruniak, J. E. (2006). Genomic comparison of *Neodiprion sertifer* and *Neodiprion lecontei* nucleopolyhedroviruses and identification of potential hymenopteran baculovirus-specific open reading frames. *J. Gen. Virol., 87*, 1477–1489.

Lee, H. H., & Miller, L. K. (1978). Isolation of genotypic variants of *Autographa californica* nuclear polyhedrosis virus. *J. Virol., 27*, 754–767.

Lee, K. P., Cory, J. S., Wilson, K., Raubenheimer, D., & Simpsom, S. J. (2005). Flexible diet choice offsets protein costs of pathogen resistance in a caterpillar. *Proc. R. Soc. B., 273*, 823–829.

Lee, S. Y., Poloumienko, A., Belfry, S., Qu, X., Chen, W., MacAfee, N., Morin, B., Lucarotti, C., & Krause, M. (1996). A common pathway for p10 and calyx proteins in progressive stages of polyhedron envelope assembly in AcMNPV-infected *Spodoptera frugiperda* larvae. *Arch. Virol., 141*, 1247–1258.

Lee, S. Y., Qu, X., Chen, W., Poloumienko, A., MacAfee, N., Morin, B., Lucarotti, C., & Krause, M. (1997). Insecticidal activity of a recombinant baculovirus containing an antisense *c-myc* fragment. *J. Gen. Virol., 78*, 273–281.

Lepore, L. S., Roelvink, P. R., & Granados, R. R. (1996). Enhancin, the granulosis virus protein that facilitates nucleopolyhedrovirus (NPV) infections, is a metalloprotease. *J. Invertebr. Pathol., 68*, 131–140.

Levy, S. M., Falleiros, A. M., Moscardi, F., & Gregorio, E. A. (2007). Susceptibility/resistance of *Anticarsia gemmatalis* larvae to its nucleopolyhedrovirus (AgMNPV): structural study of the peritrophic membrane. *J. Invertebr. Pathol., 96*, 183–186.

Levy, S. M., Moscardi, F., Falleiros, A. M., Silva, R. J., & Gregorio, E. A. (2009). A morphometric study of the midgut in resistant and nonresistant *Anticarsia gemmatalis* (Hubner) (Lepidoptera: Noctuidae) larvae to its nucleopolyhedrovirus (AgMNPV). *J. Invertebr. Pathol., 101*, 17–22.

Li, H., Tang, H., Harrison, R. L., & Bonning, B. C. (2007). Impact of a basement membrane-degrading protease on dissemination and secondary infection of Autographa californica multiple nucleopolyhedrovirus in *Heliothis virescens* (Fabricus). *J. Gen. Virol., 88*, 1109–1119.

Li, L., Donly, C., Li, Q., Willis, L. G., Keddie, B. A., Erlandson, M. A., & Theilmann, D. A. (2002a). Identification and genomic analysis of a second species of nucleopolyhedrovirus isolated from *Mamestra configurata*. *Virology, 297*, 226–244.

Li, L., Donly, C., Li, Q., Willis, L. G., Theilmann, D. A., & Erlandson, M. A. (2002b). Sequence and organization of the *Mamestra configurata* nucleopolyhedrovirus genome. *Virology, 294*, 106–121.

Li, L., Li, Q., Willis, L. G., Erlandson, M., Theilmann, D. A., & Donly, C. (2005). Complete comparative genomic analysis of two field isolates of *Mamestra configurata* nucleopolyhedrovirus-A. *J. Gen. Virol., 86*, 91–105.

Li, S. Y. (2005). Virulence of a nucleopolyhedrovirus to *Neodiprion abietis* (Hymenoptera: Diprionidae). *J. Econ. Entomol., 98*, 1870–1875.

Lightner, D. V., & Redman, R. M. (1981). A baculovirus-caused disease of penaeid shrimp, *Penaeus monodon*. *J. Invertebr. Pathol., 38*, 299–302.

Lipa, J. J. (1998). Eastern Europe and the former Soviet Union. In F. R. Hunter-Fujita, P. F. Entwhistle, H. F. Evans & N. E. Crook (Eds.), *Insect Viruses and Pest Management* (pp. 216–231). Chichester: John Wiley & Sons.

Liu, B., Becnel, J. J., Zhang, Y., & Zhou, L. (2011). Induction of reaper ortholog mx in mosquito midgut cells following baculovirus infection. *Cell. Death Differ., 18*, 1337–1345, *in press*.

Liu, F., Xu, Z., Zhu, Y. C., Huang, F., Wang, Y., Li, H., Gao, C., Zhou, W., & Shen, J. (2010). Evidence of field-evolved resistance to Cry1Ac-expressing Bt cotton in *Helicoverpa armigera* (Lepidoptera: Noctuidae) in northern China. *Pest. Manag. Sci., 66*, 155–161.

Long, G., Pan, X., Kormelink, R., & Vlak, J. M. (2006). Functional entry of baculovirus into insect and mammalian cells is dependent on clathrin-mediated endocytosis. *J. Virol., 80*, 8830–8833.

López-Ferber, M., Simón, O., Williams, T., & Caballero, P. (2003). Defective or effective? Mutualistic interactions between virus genotypes. *Proc. Biol. Sci. B., 270*, 2249–2255.

Lu, A., & Carstens, E. B. (1991). Nucleotide sequence of a gene essential for viral DNA replication in the baculovirus *Autographa californica* nuclear polyhedrosis virus. *Virology, 181*, 336–347.

Lu, A., & Carstens, E. B. (1993). Immediate-early baculovirus genes transactivate the p143 gene promoter of *Autographa californica* nuclear polyhedrosis virus. *Virology, 195*, 710–718.

Lu, A., & Miller, L. K. (1994). Identification of three late expression factor genes within the 33.8- to 43.4-map-unit region of *Autographa californica* nuclear polyhedrosis virus. *J. Virol., 68*, 6710—6718.

Lu, A., & Miller, L. K. (1995). The roles of eighteen baculovirus late expression factor genes in transcription and DNA replication. *J. Virol., 69*, 975—982.

Lu, A., & Miller, L. K. (1997). Late and very late gene expression. In L. K. Miller (Ed.), *The Baculoviruses* (pp. 193—216). New York: Plenum Press.

Lu, A., Seshagiri, S., & Miller, L. K. (1996). Signal sequence and promoter effects on the efficacy of toxin-expressing baculoviruses as biopesticides. *Biol. Control, 7*, 320—332.

Lua, L. H., Pedrini, M. R., Reid, S., Robertson, A., & Tribe, D. E. (2002). Phenotypic and genotypic analysis of *Helicoverpa armigera* nucleopolyhedrovirus serially passaged in cell culture. *J. Gen. Virol., 83*, 945—955.

Lucarotti, C. J., Morin, B., Graham, R. I., & Lapointe, R. (2007). Production, application, and field performance of Abietiv™, the balsam fir sawfly nucleopolyhedrovirus. *Virol. Sin., 22*, 163—172.

Luckow, V. A., Lee, S. C., Barry, G. F., & Olins, P. O. (1993). Efficient generation of infectious recombinant baculoviruses by site-specific transposon-mediated insertion of foreign genes into a baculovirus genome propagated in *Escherichia coli. J. Virol., 67*, 4566—4579.

Lung, O., Westenberg, M., Vlak, J. M., Zuidema, D., & Blissard, G. W. (2002). Pseudotyping *Autographa californica* multicapsid nucleopolyhedrovirus (AcMNPV): F proteins from group II NPVs are functionally analogous to AcMNPV GP64. *J. Virol., 76*, 5729—5736.

Lung, O. Y., Cruz-Alvarez, M., & Blissard, G. W. (2003). Ac23, an envelope fusion protein homolog in the baculovirus *Autographa californica* multicapsid nucleopolyhedrovirus, is a viral pathogenicity factor. *J. Virol., 77*, 328—339.

Luttrell, R. G., Yearian, W. C., & Young, S. Y. (1983). Effect of spray adjuvants on *Heliothis zea* (Lepidoptera: Noctuidae) nuclear polyhedrosis virus efficacy. *J. Econ. Entomol., 76*, 162—167.

Lynn, D. E., & Shapiro, M. (1998). New cell lines from *Heliothis virescens*: characterization and susceptibility to baculoviruses. *J. Invertebr. Pathol., 72*, 276—280.

Ma, P. W., Davis, T. R., Wood, H. A., Knipple, D. C., & Roelofs, W. L. (1998). Baculovirus expression of an insect gene that encodes multiple neuropeptides. *Insect Biochem. Mol. Biol., 28*, 239—249.

MacKinnon, E. A., Henderson, J. F., Stoltz, D. B., & Faulkner, P. (1974). Morphogenesis of nuclear polyhedrosis virus under conditions of prolonged passage *in vitro. J. Ultrastruct. Res., 49*, 419—435.

Maeda, S. (1989). Increased insecticidal effect by a recombinant baculovirus carrying a synthetic diuretic hormone gene. *Biochem. Biophys. Res. Commun., 165*, 1177—1183.

Maeda, S., Kawai, T., Obinata, M., Fujiwara, H., Horiuchi, T., Saeki, Y., Sato, Y., & Furusawa, M. (1985). Production of human α-interferon in silkworm using a baculovirus vector. *Nature, 315*, 592—594.

Maeda, S., Volrath, S. L., Hanzlik, T. N., Harper, S. A., Majima, K., Maddox, D. W., Hammock, B. D., & Fowler, E. (1991). Insecticidal effects of an insect-specific neurotoxin expressed by a recombinant baculovirus. *Virology, 184*, 777—780.

Maestri, A. (1856). *Fragmmenti anatomici fisiologici e pathologici sul baco da seta* (Bombyx mori *Linn*). Pavia: Fratelli Fusi.

Maina, J. N. (1989). Scanning and transmission electron microscopic study of the tracheal air sac system in a grasshopper *Chrotogonus*

senegalensis (Kraus) − Orthoptera: Acrididae: Pyrgomorphinae. *Anat. Rec., 223*, 393—405.

Maiorella, B., Inlow, D., Shauger, A., & Harano, D. (1988). Large-scale insect cell culture for recombinant protein production. *Bio/Technology, 6*, 1406—1410.

Malik, H. S., Henikoff, S., & Eickbush, T. H. (2000). Poised for contagion: evolutionary origins of the infectious abilities of invertebrate retroviruses. *Genome Res., 10*, 1307—1318.

Marlow, S. A., Billam, L. J., Palmer, C. P., & King, L. A. (1993). Replication and morphogenesis of Amsacta moorei entomopoxvirus in cultured cells of *Estigmene acrea* (salt marsh caterpillar). *J. Gen. Virol., 74*, 1457—1461.

Martínez, A.-M., Caballero, P., & Williams, T. (2004). Effects of an optical brightener on the development, body weight and sex ratio of *Spodoptera frugiperda* (Lepidoptera: Noctuidae). *Biocontrol Sci. Technol., 14*, 193—200.

McCaffery, A. R. (1998). Resistance to insecticides in heliothine Lepidoptera: a global view. *Philos. Trans. R. Soc. Lond. B., 353*, 1735—1750.

McCarthy, C. B., Dai, X., Donly, C., & Theilmann, D. A. (2008). *Autographa californica* multiple nucleopolyhedrovirus ac142, a core gene that is essential for BV production and ODV envelopment. *Virology, 372*, 325—339.

McCutchen, B. F., Choudary, P. V., Crenshaw, R., Maddox, D., Kamita, S. G., Palekar, N., Volrath, S., Fowler, E., Hammock, B. D., & Maeda, S. (1991). Development of a recombinant baculovirus expressing an insect-selective neurotoxin: potential for pest control. *Bio/Technology, 9*, 848—852.

McDougal, V. V., & Guarino, L. A. (2000). The *Autographa californica* nuclear polyhedrosis virus *p143* gene encodes a DNA helicase. *J. Virol., 74*, 5273—5279.

McEwen, F. L., & Hervey, G. E. R. (1960). Mass-rearing the cabbage looper, *Trichoplusia ni*, with notes on its biology in the laboratory. *Ann. Entomol. Soc. Am., 53*, 229—234.

McGuire, M. R., Tamez-Guerra, P., Behle, R. W., & Streett, D. A. (2001). Comparative field stability of selected entomopathogenic virus formulations. *J. Econ. Entomol., 94*, 1037—1044.

McKinley, D., Jones, K. A., & Moawad, G. (1997). Microbial contamination in *Spodoptera littoralis* nuclear polyhedrosis virus produced in insects in Egypt. *J. Invertebr. Pathol., 69*, 151—156.

McLeod, P. J., Yearian, W. C., & Young, S. Y., III (1977). Inactivation of *Baculovirus heliothis* by ultraviolet irradiation, dew, and temperature. *J. Invertebr. Pathol., 30*, 237—241.

McNeil, J., Cox-Foster, D., Gardner, M., Slavicek, J., Thiem, S., & Hoover, K. (2010a). Pathogenesis of *Lymantria dispar* multiple nucleopolyhedrovirus in *L. dispar* and mechanisms of developmental resistance. *J. Gen. Virol., 91*, 1590—1600.

McNeil, J., Cox-Foster, D., Slavicek, J., & Hoover, K. (2010b). Contributions of immune responses to developmental resistance in *Lymantria dispar* challenged with baculovirus. *J. Insect Physiol., 56*, 1167—1177.

Means, J. C., & Passarelli, A. L. (2010). Viral fibroblast growth factor, matrix metalloproteases, and caspases are associated with enhancing systemic infection by baculoviruses. *Proc. Natl. Acad. Sci. USA, 107*, 9825—9830.

Mengual Gomez, D. L., Belaich, M. N., Rodriguez, V. A., & Ghiringhelli, P. D. (2010). Effects of fetal bovine serum deprivation in cell cultures on the production of *Anticarsia gemmatalis* multinucleopolyhedrovirus. *BMC Biotechnol., 10,* 68.

Mikhailov, V. S., & Rohrmann, G. F. (2002). Baculovirus replication factor LEF-1 is a DNA primase. *J. Virol., 76,* 2287–2297.

Mikhailov, V. S., Okano, K., & Rohrmann, G. F. (2005). The redox state of the baculovirus single-stranded DNA-binding protein LEF-3 regulates its DNA binding, unwinding, and annealing activities. *J. Biol. Chem., 280,* 29444–29453.

Mikhailov, V. S., Okano, K., & Rohrmann, G. F. (2006). Structural and functional analysis of the baculovirus single-stranded DNA-binding protein LEF-3. *Virology, 346,* 469–478.

Mikhailov, V. S., Vanarsdall, A. L., & Rohrmann, G. F. (2008). Isolation and characterization of the DNA-binding protein (DBP) of the *Autographa californica* multiple nucleopolyhedrovirus. *Virology, 370,* 415–429.

Milks, M. L. (1997). Comparative biology and susceptibility of cabbage looper (Lepidoptera: Noctuidae) lines to a nuclear polyhedrosis virus. *Environ. Entomol., 26,* 839–848.

Milks, M. L., & Myers, J. H. (2000). The development of larval resistance to a nucleopolyhedrovirus is not accompanied by an increased virulence in the virus. *Evol. Ecol., 14,* 645–664.

Milks, M. L., Myers, J. H., & Leptich, M. K. (2002). Costs and stability of cabbage looper resistance to a nucleopolyhedrovirus. *Evol. Ecol., 16,* 369–385.

Milks, M. L., Washburn, J. O., Willis, L. G., Volkman, L. E., & Theilmann, D. A. (2003). Deletion of pe38 attenuates AcMNPV genome replication, budded virus production, and virulence in Heliothis virescens. *Virology, 310,* 224–234.

Miller, L. K. (1981). Construction of a genetic map of the baculovirus *Autographa californica* nuclear polyhedrosis virus by marker rescue of temperature-sensitive mutants. *J. Virol., 39,* 973–976.

Miller, L. K., Lingg, A. J., & Bulla, L. A., Jr. (1983). Bacterial, viral, and fungal insecticides. *Science, 219,* 715–721.

Minion, F. C., Coons, L. B., & Broome, J. R. (1979). Characterization of the polyhedral envelope of the nuclear polyhedrosis virus of *Heliothis virescens. J. Invertebr. Pathol., 34,* 303–307.

Mistretta, T. A., & Guarino, L. A. (2005). Transcriptional activity of baculovirus very late factor 1. *J. Virol., 79,* 1958–1960.

Mitsuhashi, W., Kawakita, H., Murakami, R., Takemoto, Y., Saiki, T., Miyamoto, K., & Wada, S. (2007). Spindles of an entomopoxvirus facilitate its infection of the host insect by disrupting the peritrophic membrane. *J. Virol., 81,* 4235–4243.

Miyajima, S., & Kawase, S. (1968). Changes in virus-infectivity titer in the hemolymph and midgut during the course of a cytoplasmic polyhedrosis in the silkworm. *J. Invertebr. Pathol., 12,* 329–334.

Montoya, E. L., Ignoffo, C. M., & McGarr, R. L. (1966). A feeding stimulant to increase effectiveness of, and a field test with, a nuclear-polyhedrosis virus of *Heliothis. J. Invertebr. Pathol., 8,* 320–324.

Moore, S. D., Pittaway, T., Bouwer, G., & Fourie, J. G. (2004). Evaluation of *Helicoverpa armigera* nucleopolyhedrovirus (HearNPV) for control of *Helicoverpa armigera* (Lepidoptera: Noctuidae) on citrus in South Africa. *Biocontrol Sci. Technol., 14,* 239–250.

Moreau, G. (2006). Past and present outbreaks of the balsam fir sawfly in western Newfoundland: an analytical review. *Forest Ecol. Manag., 221,* 215–219.

Moreau, G., Lucarotti, C. J., Kettela, E. G., Thurston, G. S., Holmes, S., Weaver, C., Levin, D. B., & Morin, B. (2005). Aerial application of nucleopolyhedrovirus induces decline in increasing and peaking populations of *Neodiprion abietis. Biol. Control, 33,* 65–73.

Morgan, C., Bergold, G. H., Moore, D. H., & Rose, H. M. (1955). The macromolecular paracrystalline lattice of insect viral polyhedral bodies demonstrated in ultrathin sections examined in the electron microscope. *J. Biophys. Biochem. Cytol., 1,* 187–190.

Moscardi, F. (1989). Use of viruses for pest control in Brazil: the case of the nuclear polyhedrosis virus of the soybean caterpillar, *Anticarsia gemmatalis. Mem. Inst. Oswaldo Cruz., 84*(Suppl. 3), 51–56.

Moscardi, F. (1999). Assessment of the applications of baculoviruses for control of Lepidoptera. *Annu. Rev. Entomol., 44,* 257–289.

Moscardi, F., Allen, G. E., & Greene, G. L. (1981). Control of the velvetbean caterpillar by nuclear polyhedrosis virus and insecticides and impact of treatments on the natural incidence of the entomopathogenic fungus Nomuraea rileyi. *J. Econ. Entomol., 74,* 480–485.

Moscardi, F., Leite, L. G., & Zamataro, C. E. (1997). Production of nuclear polyhedrosis virus of *Anticarsia gemmatalis* Hübner (Lepidoptera: Noctuidae): effect of virus dosage, host density and age. *An. Soc. Entomol. Bras., 26,* 121–132.

Moser, B. A., Becnel, J. J., White, S. E., Afonso, C., Kutish, G., Shanker, S., & Almira, E. (2001). Morphological and molecular evidence that *Culex nigripalpus* baculovirus is an unusual member of the family Baculoviridae. *J. Gen. Virol., 82,* 283–297.

Mukawa, S., & Goto, C. (2008). *In vivo* characterization of two granuloviruses in larvae of *Mythimna separata* (Lepidoptera: Noctuidae). *J. Gen. Virol., 89,* 915–921.

Muñoz, D., & Caballero, P. (2000). Persistence and effects of parasitic genotypes in a mixed population of the *Spodoptera exigua* nucleopolyhedrovirus. *Biol. Control, 19,* 259–264.

Muñoz, D., Castillejo, J. I., & Caballero, P. (1998). Naturally occurring deletion mutants are parasitic genotypes in a wild-type nucleopolyhedrovirus population of Spodoptera exigua. *Appl. Environ. Microbiol., 64,* 4372–4377.

Murhammer, D. W., & Goochee, C. F. (1988). Scaleup of insect cell cultures: protective effects of pluronic F-68. *Bio/Technology, 6,* 1411–1418.

Murillo, R., Lasa, R., Goulson, D., Williams, T., Muñoz, D., & Caballero, P. (2003). Effect of tinopal LPW on the insecticidal properties and genetic stability of the nucleopolyhedrovirus of *Spodoptera exigua* (Lepidoptera: Noctuidae). *J. Econ. Entomol., 96,* 1668–1674.

Murphy, F. A., Fauquet, C. M., Bishop, D. H. L., Ghabrial, S. A., Jarvis, A. W., Martelli, G. P., Mayo, M. A. & Summers, M. D. (Eds.). (1995). *Virus Taxonomy: Sixth Report of the International Committee on Taxonomy of Viruses.* New York: Springer.

Murray, K. D., & Elkinton, J. S. (1992). Vertical-distribution of nuclear polyhedrosis virus-infected gypsy-moth (Lepidoptera, Lymantriidae) larvae and effects on sampling for estimation of disease prevalence. *J. Econ. Entomol., 85,* 1865–1872.

Myers, J. H., Malakar, R., & Cory, J. S. (2000). Sublethal nucleopolyhedrovirus infection effects on female pupal weight, egg mass size, and vertical transmission in gypsy moth (Lepidoptera: Lymantriidae). *Environ. Entomol., 29,* 1268–1272.

Nagamine, T., Kawasaki, Y., & Matsumoto, S. (2006). Induction of a subnuclear structure by the simultaneous expression of baculovirus proteins, IE1. *LEF3, and P143 in the presence of hr. Virology, 352,* 400–407.

Nagamine, T., Kawasaki, Y., Abe, A., & Matsumoto, S. (2008). Nuclear marginalization of host cell chromatin associated with expansion of two discrete virus-induced subnuclear compartments during baculovirus infection. *J. Virol., 82*, 6409−6418.

Nagata, M., Ohta, M., Kambara, M., & Aoki, F. (2003). Infection of *Bombyx mori* cypovirus 1 in the BmN cell line. *J. Insect Biotechnol. Sericol., 72*, 51−55.

Nakai, M., & Kunimi, Y. (1998). Effects of the timing of entomopoxvirus administration to the smaller tea tortrix, *Adoxophyes* sp. (Lepidoptera: Tortricidae) on the survival of the endoparasitoid, *Ascogaster reticulatus* (Hymenoptera: Braconidae). *Biol. Control, 13*, 63−69.

Nobiron, I., O'Reilly, D. R., & Olszewski, J. A. (2003). *Autographa californica* nucleopolyhedrovirus infection of *Spodoptera frugiperda* cells: a global analysis of host gene regulation during infection, using a differential display approach. *J. Gen. Virol., 84*, 3029−3039.

van Oers, M. M., & Vlak, J. M. (2007). Baculovirus genomics. *Curr. Drug Targets, 8*, 1051−1068.

van Oers, M. M., Flipsen, J. T., Reusken, C. B., Sliwinsky, E. L., Goldbach, R. W., & Vlak, J. M. (1993). Functional domains of the p10 protein of *Autographa californica* nuclear polyhedrosis virus. *J. Gen. Virol., 74*, 563−574.

van Oers, M. M., Flipsen, J. T., Reusken, C. B., & Vlak, J. M. (1994). Specificity of baculovirus p10 functions. *Virology, 200*, 513−523.

Ogembo, J. G., Chaeychomsri, S., Caoili, B. L., Ikeda, M., & Kobayashi, M. (2008). Susceptibility of newly established cell lines from *Helicoverpa armigera* to homologous and heterologous nucleopolyhedroviruses. *J. Insect Biotechnol. Sericol., 77*, 25−34.

Ohkawa, T., Majima, K., & Maeda, S. (1994). A cysteine protease encoded by the baculovirus *Bombyx mori* nuclear polyhedrosis virus. *J. Virol., 68*, 6619−6625.

Ohkawa, T., Washburn, J. O., Sitapara, R., Sid, E., & Volkman, L. E. (2005). Specific binding of *Autographa californica* M nucleopolyhedrovirus occlusion-derived virus to midgut cells of *Heliothis virescens* larvae is mediated by products of pif genes Ac119 and Ac022 but not by Ac115. *J. Virol., 79*, 15258−15264.

Ohkawa, T., Volkman, L. E., & Welch, M. D. (2010). Actin-based motility drives baculovirus transit to the nucleus and cell surface. *J. Cell Biol., 190*, 187−195.

Okano, K., Vanarsdall, A. L., & Rohrmann, G. F. (2007). A baculovirus alkaline nuclease knockout construct produces fragmented DNA and aberrant capsids. *Virology, 359*, 46−54.

Okuno, S., Takatsuka, J., Nakai, M., Ototake, S., Masui, A., & Kunimi, Y. (2003). Viral-enhancing activity of various stilbene-derived brighteners for a *Spodoptera litura* (Lepidoptera: Noctuidae) nucleopolyhedrovirus. *Biol. Control, 26*, 146−152.

Olafsson, E. (1988). Dispersal of the nuclear polyhedrosis virus of *Neodiprion sertifer* from soil to pine foliage with dust. *Entomol. Exp. Appl., 46*, 181−186.

Oliveira, J. V., Wolff, J. L., Garcia-Maruniak, A., Ribeiro, B. M., de Castro, M. E., de Souza, M. L., Moscardi, F., Maruniak, J. E., & Zanotto, P. M. (2006). Genome of the most widely used viral biopesticide: *Anticarsia gemmatalis* multiple nucleopolyhedrovirus. *J. Gen. Virol., 87*, 3233−3250.

Olszewski, J., & Miller, L. K. (1997a). Identification and characterization of a baculovirus structural protein, VP1054, required for nucleocapsid formation. *J. Virol., 71*, 5040−5050.

Olszewski, J., & Miller, L. K. (1997b). A role for baculovirus GP41 in budded virus production. *Virology, 233*, 292−301.

Ooi, B. G., & Miller, L. K. (1988). Regulation of host RNA levels during baculovirus infection. *Virology, 166*, 515−523.

Ooi, B. G., Rankin, C., & Miller, L. K. (1989). Downstream sequences augment transcription from the essential initiation site of a baculovirus polyhedrin gene. *J. Mol. Biol., 210*, 721−736.

Oomens, A. G., & Blissard, G. W. (1999). Requirement for GP64 to drive efficient budding of *Autographa californica* multicapsid nucleopolyhedrovirus. *Virology, 254*, 297−314.

Oomens, A. G., Monsma, S. A., & Blissard, G. W. (1995). The baculovirus GP64 envelope fusion protein: synthesis, oligomerization, and processing. *Virology, 209*, 592−603.

Oppenheimer, D. I., & Volkman, L. E. (1997). Evidence for rolling circle replication of *Autographa californica* M nucleopolyhedrovirus genomic DNA. *Arch. Virol., 142*, 2107−2113.

O'Reilly, D. R., & Miller, L. K. (1989). A baculovirus blocks insect molting by producing ecdysteroid UDP-glucosyl transferase. *Science, 245*, 1110−1112.

O'Reilly, D. R., & Miller, L. K. (1991). Improvement of a baculovirus pesticide by deletion of the *egt* gene. *Bio/Technology, 9*, 1086−1089.

O'Reilly, D. R., Howarth, O. W., Rees, H. H., & Miller, L. K. (1991). Structure of the ecdysone glucoside formed by a baculovirus ecdysteroid UDP-glucosyltransferase. *Insect Biochem., 21*, 795−801.

O'Reilly, D. R., Brown, M. R., & Miller, L. K. (1992). Alteration of ecdysteroid metabolism due to baculovirus infection of the fall armyworm *Spodoptera frugiperda*: host ecdysteroids are conjugated with galactose. *Insect Biochem. Mol. Biol., 22*, 313−320.

O'Reilly, D. R., Hails, R. S., & Kelly, T. J. (1998). The impact of host developmental status on baculovirus replication. *J. Invertebr. Pathol., 72*, 269−275.

Paillot, A. (1926). Sur une nouvelle maladie du noyau ou grasserie des chenilles de *P. brassicae* et un nouveau groupe de microorganismes parasites. *C.R. Acad. Sci., 182*, 180−182.

Panizzi, A. R., & Corrêa-Ferreira, B. S. (1997). Dynamics in the insect fauna adaptation to soybean in the tropics. *Trends Entomol., 1*, 71−88.

Parker, B. J., Elderd, B. D., & Dwyer, G. (2010). Host behaviour and exposure risk in an insect−8pathogen interaction. *J. Anim. Ecol., 79*, 863−870.

Parnell, M. A., King, W. J., Jones, K. A., Ketunuti, U., & Wetchakit, D. (1999). A comparison of motorised knapsack mistblower, medium volume application, and spinning disk, very low volume application, of *Helicoverpa armigera* nuclear polyhedrosis virus on cotton in Thailand. *Crop Prot., 18*, 259−265.

Passarelli, A. L., & Miller, L. K. (1993). Identification and characterization of *lef-1*, a baculovirus gene involved in late and very late gene expression. *J. Virol., 67*, 3481−3488.

Passarelli, A. L., Todd, J. W., & Miller, L. K. (1994). A baculovirus gene involved in late gene expression predicts a large polypeptide with a conserved motif of RNA polymerases. *J. Virol., 68*, 4673−4678.

Payne, C. C. (1974). The isolation and characterization of a virus from. *Oryctes rhinoceros. J. Gen. Virol., 25*, 105−116.

Payne, C. C. (1986). Insect pathogenic viruses as pest control agents. In J. M. Franz (Ed.), *Biological Plant and Health Protection* (pp. 183−200). Stuttgart: G. Fischer.

Payne, C. C., Compson, D., & de looze, S. M. (1977). Properties of the nucleocapsids of a virus isolated from *Oryctes rhinoceros. Virology, 77*, 269−280.

Payne, C. C., Tatchell, G. M., & Williams, C. F. (1981). The comparative susceptibilities of *Pieris brassae* and *P. rapae* to a granulosis virus from *P. brassicae*. *J. Invertebr. Pathol., 38*, 273−280.

Pearson, M. N., & Rohrmann, G. F. (1997). Splicing is required for transactivation by the immediate early gene 1 of the *Lymantria dispar* multinucleocapsid nuclear polyhedrosis virus. *Virology, 235*, 153−165.

Pearson, M. N., Russell, R. L., Rohrmann, G. F., & Beaudreau, G. S. (1988). p39, a major baculovirus structural protein: immunocytochemical characterization and genetic location. *Virology, 167*, 407−413.

Pearson, M., Bjornson, R., Pearson, G., & Rohrmann, G. (1992). The *Autographa californica* baculovirus genome: evidence for multiple replication origins. *Science, 257*, 1382−1384.

Pearson, M. N., Bjornson, R. M., Ahrens, C., & Rohrmann, G. F. (1993). Identification and characterization of a putative origin of DNA replication in the genome of a baculovirus pathogenic for *Orgyia pseudotsugata*. *Virology, 197*, 715−725.

Pearson, M. N., Groten, C., & Rohrmann, G. F. (2000). Identification of the *Lymantria dispar* nucleopolyhedrovirus envelope fusion protein provides evidence for a phylogenetic division of the Baculoviridae. *J. Virol., 74*, 6126−6131.

Pearson, M. N., Russell, R. L. Q., & Rohrmann, G. F. (2001). Characterization of a baculovirus-encoded protein that is associated with infected-cell membranes and budded virions. *Virology, 291*, 22−31.

Peng, K., Wu, M., Deng, F., Song, J., Dong, C., Wang, H., & Hu, Z. (2010). Identification of protein−protein interactions of the occlusion-derived virus-associated proteins of *Helicoverpa armigera* nucleopolyhedrovirus. *J. Gen. Virol., 91*, 659−670.

Pennock, G. D., Shoemaker, C., & Miller, L. K. (1984). Strong and regulated expression of *Escherichia coli* beta-galactosidase in insect cells with a baculovirus vector. *Mol. Cell. Biol., 4*, 399−406.

Perera, O. P., Valles, S. M., Green, T. B., White, S., Strong, C. A., & Becnel, J. J. (2006). Molecular analysis of an occlusion body protein from *Culex nigripalpus* nucleopolyhedrovirus (CuniNPV). *J. Invertebr. Pathol., 91*, 35−42.

Perera, O., Green, T. B., Stevens, S. M., Jr., White, S., & Becnel, J. J. (2007). Proteins associated with *Culex nigripalpus* nucleopolyhedrovirus occluded virions. *J. Virol., 81*, 4585−4590.

Perera, S., Li, Z., Pavlik, L., & Arif, B. (2010a). Entomopoxviruses. In S. Asgari & K. Johnson (Eds.), *Insect Virology* (pp. 83−102). Norfolk: Caister Academic Press.

Perera, S. C., Wong, P., Krell, P. J., & Arif, B. M. (2010b). Expression of heterologous genes in the Amsacta moorei entomopoxvirus. *J. Virol. Methods, 165*, 1−8.

Pijlman, G. P., Pruijssers, A. J., & Vlak, J. M. (2003). Identification of *pif-2*, a third conserved baculovirus gene required for per os infection of insects. *J. Gen. Virol., 84*, 2041−2049.

Plonsky, I., & Zimmerberg, J. (1996). The initial fusion pore induced by baculovirus GP64 is large and forms quickly. *J. Cell Biol., 135*, 1831−1839.

Plymale, R., Grove, M. J., Cox-Foster, D., Ostiguy, N., & Hoover, K. (2008). Plant-mediated alteration of the peritrophic matrix and baculovirus infection in lepidopteran larvae. *J. Insect Physiol., 54*, 737−749.

Podgwaite, J. D., Bruen, R. B., & Shapiro, M. (1983). Microorganisms associated with production lots of the nucleopolyhedrosis virus of the gypsy moth, *Lymantria dispar* [Lep.: Lymantriidae]. *Entomophaga., 28*, 9−15.

Podgwaite, J. D., Rush, P., Hall, D., & Walton, G. S. (1984). Efficacy of the *Neodiprion sertifer* (Hymenoptera: Diprionidae) nucleopolyhedrosis virus (baculovirus) product, Neochek-S. *J. Econ. Entomol., 77*, 525−528.

Podgwaite, J., Reardon, R., Walton, G., & Witcosky, J. (1992). Efficacy of aerially applied Gypchek against gypsy moth in the Appalachian highlands. *J. Entomol. Sci., 27*, 337−344.

Pogue, M. G., & Schaefer, P. W. (2007). *A Review of Selected Species of* Lymantria Hubner *[1819] (Lepidoptera: Noctuidae: Lymantriinae) from Subtropical and Temperate Regions of Asia, including the Descriptions of Three New Species, Some Potentially Invasive to North America*. FHTET-2006-07. Fort Collins: Forest Health Technology Enterprise Team.

Poinar, G. O., Jr., & Poinar, R. (2005). Fossil evidence of insect pathogens. *J. Invertebr. Pathol., 89*, 243−250.

Popham, H. J. R., Li, Y., & Miller, L. K. (1997). Genetic improvement of *Helicoverpa zea* nuclear polyhedrosis virus as a biopesticide. *Biol. Control, 10*, 83−91.

Popham, H. J. R., Pellock, B. J., Robson, M., Dierks, P. M., & Miller, L. K. (1998). Characterization of a variant of *Autographa californica* nuclear polyhedrosis virus with a nonfunctional ORF 603. *Biol. Control, 12*, 223−230.

Popham, H. J., Shelby, K. S., Brandt, S. L., & Coudron, T. A. (2004). Potent virucidal activity in larval *Heliothis virescens* plasma against *Helicoverpa zea* single capsid nucleopolyhedrovirus. *J. Gen. Virol., 85*, 2255−2261.

Popham, H. J., Grasela, J. J., Goodman, C. L., & McIntosh, A. H. (2010). Baculovirus infection influences host protein expression in two established insect cell lines. *J. Insect Physiol., 56*, 1237−1245.

Prikhod'ko, E. A., & Miller, L. K. (1996). Induction of apoptosis by baculovirus transactivator IE1. *J. Virol., 70*, 7116−7124.

Prikhod'ko, G. G., Robson, M., Warmke, J. W., Cohen, C. J., Smith, M. M., Wang, P., Warren, V., Kaczorowski, G., van der Ploeg, L. H. T., & Miller, L. K. (1996). Properties of three baculovirus-expressing genes that encode insect-selective toxins: μ-Aga-IV, As II, and Sh I. *Biol. Control, 7*, 236−244.

Prikhod'ko, E. A., Lu, A., Wilson, J. A., & Miller, L. K. (1999). *In vivo* and *in vitro* analysis of baculovirus ie-2 mutants. *J. Virol., 73*, 2460−2468.

Pritchett, D. W., Young, S. Y., & Yearian, W. C. (1982). Dissolution of *Autographa californica* nuclear polyhedrosis virus polyhedra by the digestive fluid of *Trichoplusia ni* (Lepidoptera: Noctuidae) larvae. *J. Invertebr. Pathol., 39*, 354−361.

Purrini, K. (1989). *Baculovirus oryctes* release into *Oryctes monoceros* population in Tanzania, with special reference to the interaction of virus isolates used in our laboratory infection experiments. *J. Invertebr. Pathol., 53*, 285−300.

Rahman, M. M., & Gopinathan, K. P. (2004). Systemic and *in vitro* infection process of *Bombyx mori* nucleopolyhedrovirus. *Virus Res., 101*, 109−118.

Raina, A. K., & Adams, J. R. (1995). Gonad-specific virus of corn earworm. *Nature, 374*, 770.

Raina, A. K., Adams, J. R., Lupiani, B., Lynn, D. E., Kim, W., Burand, J. P., & Dougherty, E. M. (2000). Further characterization of the gonad-specific virus of corn earworm, *Helicoverpa zea*. *J. Invertebr. Pathol., 76*, 6−12.

Raina, A. K., Vakharia, V. N., Leclerc, R. F., & Blackburn, M. B. (2007). Engineering a recombinant baculovirus with a peptide hormone gene and its effect on the corn earworm, *Helicoverpa zea. Biopestic. Int., 3*, 43–52.

Rallis, C. P., & Burand, J. P. (2002). Pathology and ultrastructure of Hz-2V infection in the agonadal female corn earworm, *Helicoverpa zea. J. Invertebr. Pathol., 81*, 33–44.

Rankin, C., Ooi, B. G., & Miller, L. K. (1988). Eight base pairs encompassing the transcriptional start point are the major determinant for baculovirus polyhedrin gene expression. *Gene, 70*, 39–49.

Rapp, J. C., Wilson, J. A., & Miller, L. K. (1998). Nineteen baculovirus open reading frames, including LEF-12, support late gene expression. *J. Virol., 72*, 10197–10206.

Raymond, B., Vanbergen, A., Pearce, I., Hartley, S., Cory, J., & Hails, R. (2002). Host plant species can influence the fitness of herbivore pathogens: the winter moth and its nucleopolyhedrovirus. *Oecologia, 131*, 533–541.

Raymond, B., Hartley, S., Cory, J., & Hails, R. (2005). The role of food plant and pathogen-induced behaviour in the persistence of a nucleopolyhedrovirus. *J. Invertebr. Pathol., 88*, 49–57.

Reardon, R. C., Podgwaite, J., & Zerillo, R. (2009). *Gypchek-Bioinsecticide for the Gypsy Moth.*. FHTET 2009-01 Fort Collins: Forest Health Technology Enterprise Team.

Reddy, J. T., & Locke, M. (1990). The size limited penetration of gold particles through insect basal laminae. *J. Insect Physiol., 36*, 397–407.

Reeson, A. F., Wilson, K., Gunn, A., Hails, R. S., & Goulson, D. (1998). Baculovirus resistance in the noctuid *Spodoptera exempta* is phenotypically plastic and responds to population density. *Proc. R. Soc. Lond. B., 265*, 1787–1791.

Reiff, W. (1911). *The "wilt disease" or "flacherie" of the gypsy moth.* Boston: Wright & Potter Printing Company.

Rennie, J. (1923). Polyhedral disease of *Tipula paludosa* (Meigen). *Proc. R. Phys. Soc. Edinb. A., 20*, 265–267.

de Rezende, S. H., Castro, M. E., & Souza, M. L. (2009). Accumulation of few-polyhedra mutants upon serial passage of *Anticarsia gemmatalis* multiple nucleopolyhedrovirus in cell culture. *J. Invertebr. Pathol., 100*, 153–159.

del Rincón-Castro, M. C., & Ibarra, J. E. (2005). Effect of a nucleopolyhedrovirus of *Autographa californica* expressing the enhancin gene of *Trichoplusia ni* granulovirus on *T. ni* larvae. *Biocontrol Sci. Technol., 15*, 701–710.

Roberts, D. W., & Granados, R. R. (1968). A poxlike virus from *Amsacta moorei* (Lepidoptera: Arctiidae). *J. Invertebr. Pathol., 12*, 141–143.

Rodas, V. M., Marques, F. H., Honda, M. T., Soares, D. M., Jorge, S. A., Antoniazzi, M. M., Medugno, C., Castro, M. E., Ribeiro, B. M., Souza, M. L., Tonso, A., & Pereira, C. A. (2005). Cell culture derived AgMNPV bioinsecticide: biological constraints and bioprocess Issues. *Cytotechnology, 48*, 27–39.

Rodems, S. M., & Friesen, P. D. (1995). Transcriptional enhancer activity of hr5 requires dual-palindrome half sites that mediate binding of a dimeric form of the baculovirus transregulator IE1. *J. Virol., 69*, 5368–5375.

Rodems, S. M., Pullen, S. S., & Friesen, P. D. (1997). DNA-dependent transregulation by IE1 of *Autographa californica* nuclear polyhedrosis virus: IE1 domains required for transactivation and DNA binding. *J. Virol., 71*, 9270–9277.

Roelvink, P. W., Corsaro, B. G., & Granados, R. R. (1995). Characterization of the *Helicoverpa armigera* and *Pseudaletia unipuncta* granulovirus enhancin genes. *J. Gen. Virol., 76*, 2693–2705.

Rohel, D. Z., Cochran, M. A., & Faulkner, P. (1983). Characterization of two abundant mRNAs of *Autographa californica* nuclear polyhedrosis virus present late in infection. *Virology, 124*, 357–365.

Rohrmann, G. F. (2008). DNA replication and genome processing. In *Baculovirus Molecular Biology*. Bethesda: National Library of Medicine, National Center for Biotechnology Information.

Rohrmann, G. F., & Beaudreau, G. S. (1977). Characterization of DNA from polyhedral inclusion bodies of the nucleopolyhedrosis single-rod virus pathogenic for *Orgyia pseudotsugata. Virology, 83*, 474–478.

Rohrmann, G. F., & Karplus, P. A. (2001). Relatedness of baculovirus and gypsy retrotransposon envelope proteins. *BMC Evol. Biol., 1*, 1.

Rohrmann, G. F., McParland, R. H., Martignoni, M. E., & Beaudreau, G. S. (1978). Genetic relatedness of two nucleopolyhedrosis viruses pathogenic for *Orgyia pseudotsugata. Virology, 84*, 213–217.

Rollinson, W. D., Lewis, F. B., & Waters, W. E. (1965). The successful use of a nuclear-polyhedrosis virus against the gypsy moth. *J. Invertebr. Pathol., 7*, 515–517.

Ross, L., & Guarino, L. A. (1997). Cycloheximide inhibition of delayed early gene expression in baculovirus-infected cells. *Virology, 232*, 105–113.

Rothman, L. D., & Myers, J. H. (1996). Debilitating effects of viral diseases on host lepidoptera. *J. Invertebr. Pathol., 67*, 1–10.

Rumble, J. M., & Duckett, C. S. (2008). Diverse functions within the IAP family. *J. Cell Sci., 121*, 3505–3507.

Russell, R. L., Pearson, M. N., & Rohrmann, G. F. (1991). Immunoelectron microscopic examination of *Orgyia pseudotsugata* multicapsid nuclear polyhedrosis virus-infected *Lymantria dispar* cells: time course and localization of major polyhedron-associated proteins. *J. Gen. Virol., 72*, 275–283.

Russell, R. L., Funk, C. J., & Rohrmann, G. F. (1997). Association of a baculovirus-encoded protein with the capsid basal region. *Virology, 227*, 142–152.

Saejeng, A., Tidbury, H., Siva-Jothy, M. T., & Boots, M. (2010). Examining the relationship between hemolymph phenoloxidase and resistance to a DNA virus, *Plodia interpunctella* granulosis virus (PiGV). *J. Insect Physiol., 56*, 1232–1236.

Saksena, S., Summers, M. D., Burks, J. K., Johnson, A. E., & Braunagel, S. C. (2006). Importin-alpha-16 is a translocon-associated protein involved in sorting membrane proteins to the nuclear envelope. *Nat. Struct. Mol. Biol., 13*, 500–508.

Salem, T. Z., Zhang, F., Xie, Y., & Thiem, S. M. (2011). Comprehensive analysis of host gene expression in *Autographa californica* nucleopolyhedrovirus-infected *Spodoptera frugiperda* cells. *Virology, 412*, 167–178.

Sanz, P., Veyrunes, J. C., Cousserans, F., & Bergoin, M. (1994). Cloning and sequencing of the spherulin gene, the occlusion body major polypeptide of the *Melolontha melolontha* entomopoxvirus (MmEPV). *Virology, 202*, 449–457.

Schultz, K. L., Wetter, J. A., Fiore, D. C., & Friesen, P. D. (2009). Transactivator IE1 is required for baculovirus early replication events that trigger apoptosis in permissive and nonpermissive cells. *J. Virol., 83*, 262–272.

Selot, R., Kumar, V., Shukla, S., Chandrakuntal, K., Brahmaraju, M., Dandin, S. B., Laloraya, M., & Kumar, P. G. (2007). Identification of a soluble NADPH oxidoreductase (BmNOX) with antiviral activities in the gut juice of *Bombyx mori. Biosci. Biotechnol. Biochem., 71*, 200–205.

Shapiro, A. M., Becnel, J. J., & White, S. E. (2004). A nucleopolyhedrovirus from *Uranotaenia sapphirina* (Diptera: Culicidae). *J. Invertebr. Pathol., 86*, 96−103.

Shapiro, M. (1986). *In vivo* production of baculoviruses. In R. R. Granados & B. A. Federici (Eds.), *The Biology of Baculoviruses, Vol. II: Practical Application for Insect Control* (pp. 31−61). Boca Raton: CRC Press.

Shapiro, M. (1989). Congo red as an ultraviolet protectant for the gypsy moth (Lepidoptera: Lymantriidae) nuclear polyhedrosis virus. *J. Econ. Entomol., 82*, 548−550.

Shapiro, M. (1992). Use of optical brighteners as radiation protectants for gypsy moth (Lepidoptera: Lymantriidae) nuclear polyhedrosis virus. *J. Econ. Entomol., 85*, 1682−1686.

Shapiro, M., & Robertson, J. L. (1987). Yield and activity of gypsy moth (Lepidoptera: Lymantriidae) nucleopolyhedrosis virus recovered from survivors of viral challenge. *J. Econ. Entomol., 80*, 901−905.

Shapiro, M., & Robertson, J. L. (1992). Enhancement of gypsy moth (Lepidoptera: Lymantriidae) baculovirus activity by optical brighteners. *J. Econ. Entomol., 85*, 1120−1124.

Sheppard, R. F., & Stairs, G. R. (1977). Dosage−mortality and time−mortality studies of a granulosis virus in a laboratory strain of the codling moth, *Laspeyresia pomonella*. *J. Invertebr. Pathol., 29*, 216−221.

Shikano, I., Ericsson, J. D., Cory, J. S., & Myers, J. H. (2010). Indirect plant-mediated effects on insect immunity and disease resistance in a tritrophic system. *Basic Appl. Ecol., 11*, 15−22.

Shvedchikova, N. G., Ulanov, V. P., & Tanasevich, L. M. (1969). Structure of the granulosis virus of Siberian silkworm *Dendrolinus sibiricus* Tschetw. *Molekulyarwaya Biologiya, 3*, 361−365.

Sikorowski, P. P., Andrews, G. L., & Broome, J. R. (1971). Presence of cytoplasmic polyhedrosis virus in the hemolymph of *Heliothis virescens* larvae and adults. *J. Invertebr. Pathol., 18*, 167−168.

Simón, O., Williams, T., López-Ferber, M., & Caballero, P. (2004). Genetic structure of a *Spodoptera frugiperda* nucleopolyhedrovirus population: high prevalence of deletion genotypes. *Appl. Environ. Microbiol., 70*, 5579−5588.

Simón, O., Williams, T., Lopez-Ferber, M., & Caballero, P. (2005). Functional importance of deletion mutant genotypes in an insect nucleopolyhedrovirus population. *Appl. Environ. Microbiol., 71*, 4254−4262.

Simón, O., Williams, T., Caballero, P., & López-Ferber, M. (2006). Dynamics of deletion genotypes in an experimental insect virus population. *Proc. R. Soc. B., 273*, 783−790.

Simón, O., Williams, T., López-Ferber, M., Taulemesse, J.-M., & Caballero, P. (2008). Population genetic structure determines speed of kill and occlusion body production in *Spodoptera frugiperda* multiple nucleopolyhedrovirus. *Biol. Control, 44*, 321−330.

Slack, J. M., Kuzio, J., & Faulkner, P. (1995). Characterization of *v-cath*, a cathepsin L-like proteinase expressed by the baculovirus *Autographa californica* multiple nuclear polyhedrosis virus. *J. Gen. Virol., 76*, 1091−1098.

Slack, J. M., Ribeiro, B. M., & de Souza, M. L. (2004). The *gp64* locus of *Anticarsia gemmatalis* multicapsid nucleopolyhedrovirus contains a 3′ repair exonuclease homologue and lacks *v-cath* and *ChiA* genes. *J. Gen. Virol., 85*, 211−219.

Slavicek, J. M., & Popham, H. J. R. (2005). The *Lymantria dispar* nucleopolyhedrovirus enhancins are components of occlusion-derived virus. *J. Virol., 79*, 10578−10588.

Slavicek, J. M., Hayes-Plazolles, N., & Kelly, M. E. (1995). Rapid formation of few polyhedra mutants of *Lymantria dispar* multi-nucleocapsid nuclear polyhedrosis virus during serial passage in cell culture. *Biol. Control, 5*, 251−261.

Slavicek, J. M., Popham, H. J. R., & Riegel, C. I. (1999). Deletion of the *Lymantria dispar* multicapsid nucleopolyhedrovirus ecdysteroid UDP-glucosyl transferase gene enhances viral killing speed in the last instar of the gypsy moth. *Biol. Control, 16*, 91−103.

Slavicek, J. M., Hayes-Plazolles, N., & Kelly, M. E. (2001). Identification of a *Lymantria dispar* nucleopolyhedrovirus isolate that does not accumulate few-polyhedra mutants during extended serial passage in cell culture. *Biol. Control, 22*, 159−168.

Smith, C. R., Heinz, K. M., Sansone, C. G., & Flexner, J. L. (2000). Impact of recombinant baculoviruses on target heliothines and nontarget predators in cotton. *Biol. Control, 19*, 201−214.

Smith, G. E., & Summers, M. D. (1979). Restriction maps of five *Autographa californica* MNPV variants, *Trichoplusia ni* MNPV, and *Galleria mellonella* MNPV DNAs with endonucleases SmaI, KpnI, BamHI, SacI, XhoI, and EcoRI. *J. Virol., 30*, 828−838.

Smith, G. E., Fraser, M. J., & Summers, M. D. (1983a). Molecular engineering of the *Autographa californica* nuclear polyhedrosis virus genome: deletion mutations within the polyhedrin gene. *J. Virol., 46*, 584−593.

Smith, G. E., Summers, M. D., & Fraser, M. J. (1983b). Production of human beta interferon in insect cells infected with a baculovirus expression vector. *Mol. Cell. Biol., 3*, 2156−2165.

Smith, G. E., Vlak, J. M., & Summers, M. D. (1983c). Physical analysis of *Autographa californica* nuclear polyhedrosis virus transcripts for poly-hedrin and a 10,000-molecular-weight protein. *J. Virol., 45*, 215−225.

Smith, K. A. (1955). Intranuclear changes in the polyhedrosis of *Tipula paludosa* (Diptera). *Nature, 176*, 255.

Sood, P., Mehta, P. K., Bhandari, K., & Prabhakar, C. S. (2010). Transmission and effect of sublethal infection of granulosis virus (PbGV) on *Pieris brassicae* Linn. (Pieridae: Lepidoptera). *J. Appl. Entomol., 134*, 774−780.

Sparks, W. O., Harrison, R. L., & Bonning, B. C. (2011). *Autographa californica* multiple nucleopolyhedrovirus ODV-E56 is a *per os* infectivity factor, but is not essential for binding and fusion of occlusion-derived virus to the host midgut. *Virology, 409*, 69−76.

Stahler, N. (1939). A disease of the corn ear worm, *Heliothis obsoleta* (F.). *J. Econ. Entomol., 32*, 151.

Stairs, G. R. (1965). Quantitative differences in susceptibility to nuclear-polyhedrosis virus among larval instars of the forest tent caterpillar, *Malacosoma disstria* (Hübner). *J. Invertebr. Pathol., 7*, 427−429.

Steinhaus, E. A. (1975). *Disease in a Minor Chord*. Columbus: Ohio State University Press.

Stewart, T. M., Huijskens, I., Willis, L. G., & Theilmann, D. A. (2005). The *Autographa californica* multiple nucleopolyhedrovirus *ie0- ie1* gene complex is essential for wild-type virus replication, but either IE0 or IE1 can support virus growth. *J. Virol., 79*, 4619−4629.

Stiles, B., & Himmerich, S. (1998). Autographa californica NPV isolates: restriction endonuclease analysis and comparative biological activity. *J. Invertebr. Pathol., 72*, 174−177.

Summers, M. D. (1971). Electron microscopic observations on granulosis virus entry, uncoating and replication processes during infection of the midgut cells of *Trichoplusia ni*. *J. Ultrastruct. Res., 35*, 606−625.

Summers, M. D., & Anderson, D. L. (1972). Granulosis virus deoxy-ribonucleic acid: a closed, double-stranded molecule. *J. Virol., 9*, 710−713.

Summers, M. D., & Anderson, D. L. (1973). Characterization of nuclear polyhedrosis virus DNAs. *J. Virol., 12*, 1336−1346.

Summers, M. D., & Arnott, H. J. (1969). Ultrastructural studies on inclusion formation and virus occlusion in nuclear polyhedrosis and granulosis virus-infected cells of *Trichoplusia ni* (Hubner). *J. Ultrastruct. Res., 28*, 462−480.

Summers, M. D., & Volkman, L. E. (1976). Comparison of biophysical and morphological properties of occluded and extracellular non-occluded baculovirus from *in vivo* and *in vitro* host systems. *J. Virol., 17*, 962−972.

Sun, X., & Peng, H. (2007). Recent advances in biological control of pest insects by using viruses in China. *Virol. Sin., 22*, 158−162.

Sun, X., Wua, D., Sun, X., Jin, L., Mab, Y., Bonning, B. C., Peng, H., & Hu, Z. (2009). Impact of *Helicoverpa armigera* nucleopolyhedroviruses expressing a cathepsin L-like protease on target and nontarget insect species on cotton. *Biol. Control, 49*, 77−83.

Sun, X. L., Wang, H. L., Sun, X. C., Chen, X. W., Peng, C. M., Pan, D. M., Jehle, J. A., van der Werf, W., Vlak, J. M., & Hu, Z. H. (2004). Biological activity and field efficacy of a genetically modified *Helicoverpa armigera* SNPV expressing an insect-selective toxin from a chimeric promoter. *Biol. Control, 29*, 124−137.

Sun, X. L., Sun, X. C., Bai, B. K., van der Werf, W., Vlak, J. M., & Hu, Z. H. (2005). Production of polyhedral inclusion bodies from *Helicoverpa armigera* larvae infected with wild-type and recombinant HaSNPV. *Biocontrol Sci. Technol., 15*, 353−366.

Szolajska, E., Poznanski, J., Ferber, M. L., Michalik, J., Gout, E., Fender, P., Bailly, I., Dublet, B., & Chroboczek, J. (2004). Poneratoxin, a neurotoxin from ant venom. Structure and expression in insect cells and construction of a bio-insecticide. *Eur. J. Biochem., 271*, 2127−2136.

Tabashnik, B. E., van Rensburg, J. B. J., & Carrière, Y. (2009). Field-evolved insect resistance to Bt crops: definition, theory, and data. *J. Econ. Entomol., 102*, 2011−2025.

Tadeu, M., Silva, B., & Moscardi, F. (2002). Field efficacy of the nucleopolyhedrovirus of *Anticarsia gemmatalis* Hübner (Lepidoptera: Noctuidae): effect of formulations, water pH, volume and time of application, and type of spray nozzle. *Neotrop. Entomol., 31*, 75−83.

Tamez-Guerra, P., McGuire, M. R., Behle, R. W., Hamm, J. J., Sumner, H. R., & Shasha, B. S. (2000). Sunlight persistence and rainfastness of spray-dried formulations of baculovirus isolated from *Anagrapha falcifera* (Lepidoptera: Noctuidae). *J. Econ. Entomol., 93*, 210−218.

Tan, Y.-R., Sun, J.-C., Lu, X.-Y., Su, D.-M., & Zhang, J.-Q. (2003). Entry of *Bombyx mori* cypovirus 1 into midgut cells *in vivo*. *J. Electron. Microsc., 52*, 485−489.

Tanada, Y. (1959). Synergism between two viruses of the armyworm, *Pseudaletia unipuncta* (Haworth) (Lepidoptera, Noctuidae). *J. Insect Pathol., 1*, 215−231.

Tanada, Y. (1964). A granulosis virus of the codling moth, *Carpocapsa pomonella* (Linnaeus) (Olethreutidae, Lepidoptera). *J. Insect Pathol., 6*, 378−380.

Tanada, Y., & Hess, R. T. (1976). Development of a nuclear polyhedrosis virus in midgut cells and penetration of the virus into the hemoeoel of the armyworm, *Pseudaletia unipuncta*. *J. Invertebr. Pathol., 28*, 67−76.

Tanada, Y., & Hess, R. T. (1991). Baculoviridae. Granulosis viruses. In J. R. Adams & J. R. Bonami (Eds.), *Atlas of Invertebrate Viruses* (pp. 227−257). Boca Raton: CRC Press.

Tanada, Y., & Leutenegger, R. (1968). Histopathology of a granulosis-virus disease of the codling moth, *Carpocapsa pomonella*. *J. Invertebr. Pathol., 10*, 39−47.

Tanada, Y., & Reiner, C. (1962). The use of pathogens in the control of the corn earworm *Heliothis zea* (Boddie). *J. Invertebr. Pathol., 4*, 139−154.

Tanada, Y., Himeno, M., & Omi, E. M. (1973). Isolation of a factor, from the capsule of a granulosis virus, synergistic for a nuclear-polyhedrosis virus of the armyworm. *J. Invertebr. Pathol., 21*, 31−40.

Tanada, Y., Hess, R. T., & Omi, E. M. (1975). Invasion of a nuclear polyhedrosis virus in midgut of the armyworm, *Pseudaletia unipuncta*, and the enhancement of a synergistic enzyme. *J. Invertebr. Pathol., 26*, 99−104.

Tang, H., Li, H., Lei, S. M., Harrison, R. L., & Bonning, B. C. (2007). Tissue specificity of a baculovirus-expressed, basement membrane-degrading protease in larvae of *Heliothis virescens*. *Tissue Cell, 39*, 431−443.

Tani, H., Nishijima, M., Ushijima, H., Miyamura, T., & Matsuura, Y. (2001). Characterization of cell-surface determinants important for baculovirus infection. *Virology, 279*, 343−353.

Teakle, R. E., Jensen, J. M., & Giles, J. E. (1986). Age-related susceptibility of *Heliothis punctiger* to a commercial formulation of nuclear polyhedrosis virus. *J. Invertebr. Pathol., 47*, 82−92.

Tessier, D. C., Thomas, D. Y., Khouri, H. E., Laliberte, F., & Vernet, T. (1991). Enhanced secretion from insect cells of a foreign protein fused to the honeybee melittin signal peptide. *Gene, 98*, 177−183.

Thiem, S. M., & Miller, L. K. (1989). Identification, sequence, and transcriptional mapping of the major capsid protein gene of the baculovirus *Autographa californica* nuclear polyhedrosis virus. *J. Virol., 63*, 2008−2018.

Thiem, S. M., & Miller, L. K. (1990). Differential gene expression mediated by late, very late and hybrid baculovirus promoters. *Gene, 91*, 87−94.

Thomas, F., Adamo, S., & Moore, J. (2005). Parasitic manipulation: where are we and where should we go? *Behav. Processes, 68*, 185−199.

Thompson, C. G., Scott, D. W., & Wickman, B. E. (1981). Long-term persistence of the nuclear polyhedrosis virus of the Douglas-fir tussock moth, *Orgyia pseudotsugata* (Lepidoptera: Lymantriidae), in forest soil. *Environ. Entomol., 10*, 254−255.

Thorpe, K. W., Podgwaite, J., Slavicek, J. M., & Webb, R. E. (1998). Gypsy moth (Lepidoptera: Lymantriidae) control with ground-based hydraulic applications of Gypchek, *in vitro*-produced virus, and *Bacillus thuringiensis*. *J. Econ. Entomol., 91*, 875−880.

Thorpe, K. W., Cook, S. P., Webb, R. E., Podgwaite, J. D., & Reardon, R. C. (1999). Aerial application of the viral enhancer Blankophor BBH with reduced rates of gypsy moth (Lepidoptera: Lymantriidae) nucleopolyhedrovirus. *Biol. Control, 16*, 209−216.

Timpl, R., & Brown, J. C. (1996). Supramolecular assembly of basement membranes. *Bioessays, 18*, 123−132.

Tomalski, M. D., & Miller, L. K. (1991). Insect paralysis by baculovirus-mediated expression of a mite neurotoxin gene. *Nature, 352*, 82−85.

Tomalski, M. D., & Miller, L. K. (1992). Expression of a paralytic neurotoxin gene to improve insect baculoviruses as biopesticides. *Nat. Biotechnol., 10*, 545−549.

Tomalski, M. D., Wu, J. G., & Miller, L. K. (1988). The location, sequence, transcription, and regulation of a baculovirus DNA polymerase gene. *Virology, 167*, 591−600.

Trager, W. (1935). Cultivation of the virus of grasserie in silkworm tissue cultures. *J. Exp. Med., 61,* 501–514.

Treacy, M. F., All, J. N., & Ghidiu, G. M. (1997). Effect of ecdysteroid UDP-glucosyltransferase gene deletion on efficacy of a baculovirus against *Heliothis virescens* and *Trichoplusia ni* (Lepidoptera: Noctuidae). *J. Econ. Entomol., 90,* 1207–1214.

Treacy, M. F., Rensner, P. E., & All, J. N. (2000). Comparative insecticidal properties of two nucleopolyhedrovirus vectors encoding a similar toxin gene chimer. *J. Econ. Entomol., 93,* 1096–1104.

Trudeau, D., Washburn, J. O., & Volkman, L. E. (2001). Central role of hemocytes in *Autographa californica* M nucleopolyhedrovirus pathogenesis in *Heliothis virescens* and. *Helicoverpa zea. J. Virol., 75,* 996–1003.

Tuan, S. J., Hou, R. F., Kao, S. S., Lee, C. F., & Chao, Y. C. (2005). Improved plant protective efficacy of a baculovirus using an early promoter to drive insect-specific neurotoxin expression. *Bot. Bull. Acad. Sin., 46,* 11–20.

von Tubeuf, C. (1892). Die Krankheiten der Nonne. *Forstl. Naturwiss. Z., 1,* 34–37.

Tweeten, K. A., Bulla, L. A., & Consigli, R. A. (1980). Characterization of an extremely basic protein derived from granulosis virus nucleocapsids. *J. Virol., 33,* 866–876.

Uchima, K., Harvey, J. P., Omi, E. M., & Tanada, Y. (1988). Binding sites on the midgut cell membrane of the synergistic factor of a granulosis virus of the armyworm (*Pseudaletia unipunctata*). *Insect Biochem., 18,* 645–650.

Vago, C. (1963). A new type of insect virus. *J. Insect Pathol., 5,* 275–276.

Vago, C., Aizawa, K., Ignoffo, C., Martignoni, M. E., Tarasevitch, L., & Tinsley, T. W. (1974). Present status of the nomenclature and classification of invertebrate viruses. *J. Invertebr. Pathol., 23,* 133–134.

Vail, P., Sutter, G., Jay, D., & Gough, D. (1971). Reciprocal infectivity of nuclear polyhedrosis viruses of the cabbage looper and alfalfa looper. *J. Invertebr. Pathol., 17,* 383–388.

Vail, P. V., Jay, D. L., & Hink, W. F. (1973). Replication and infectivity of the nuclear polyhedrosis virus of the alfalfa looper, *Autographa californica,* produced in cells grown *in vitro. J. Invertebr. Pathol., 22,* 231–237.

Vail, P. V., Morris, T. J., Collier, S. S., & Mackey, B. (1983). An RNA virus in *Autographa californica* nuclear polyhedrosis virus preparations: incidence and influence on baculovirus activity. *J. Invertebr. Pathol., 41,* 171–178.

Vail, P. V., Hoffmann, D. F., & Tebbets, J. S. (1996). Effects of a fluorescent brightener on the activity of *Anagrapha falcifera* (Lepidoptera: Noctuidae) nuclear polyhedrosis virus to four noctuid pests. *Biol. Control, 7,* 121–125.

Vail, P. V., Hoffmann, D. F., & Tebbets, J. S. (1999). Influence of fluorescent brighteners on the field activity of the celery looper nucleopolyhedrovirus. *Southwest. Entomol., 24,* 87–97.

Vanarsdall, A. L., Okano, K., & Rohrmann, G. F. (2005). Characterization of the replication of a baculovirus mutant lacking the DNA polymerase gene. *Virology, 331,* 175–180.

Vanarsdall, A. L., Okano, K., & Rohrmann, G. F. (2006). Characterization of the role of very late expression factor 1 in baculovirus capsid structure and DNA processing. *J. Virol., 80,* 1724–1733.

Vanarsdall, A. L., Mikhailov, V. S., & Rohrmann, G. F. (2007). Characterization of a baculovirus lacking the DBP (DNA-binding protein) gene. *Virology, 364,* 475–485.

Vasconcelos, S. D. (1996). Alternative routes for the horizontal transmission of a nucleopolyhedrovirus. *J. Invertebr. Pathol., 68,* 269–274.

Vasconcelos, S. D., Cory, J. S., Wilson, K. R., Sait, S. M., & Hails, R. S. (1996). Modified behavior in baculovirus-infected lepidopteran larvae and its impact on the spatial distribution of inoculum. *Biol. Control, 7,* 299–306.

Vasconcelos, S. D., Hails, R. S., Speight, M. R., & Cory, J. S. (2005). Differential crop damage by healthy and nucleopolyhedrovirus-infected *Mamestra brassicae* L. (Lepidoptera: Noctuidae) larvae: a field examination. *J. Invertebr. Pathol., 88,* 177–179.

Vaughn, J. L., & Faulkner, P. (1963). Susceptibility of an insect tissue culture to infection by virus preparations of the nuclear polyhedrosis of the silkworm (*Bombyx mori* L.). *Virology, 20,* 484–489.

Vialard, J. E., & Richardson, C. D. (1993). The 1,629-nucleotide open reading frame located downstream of the *Autographa californica* nuclear polyhedrosis virus polyhedrin gene encodes a nucleocapsid-associated phosphoprotein. *J. Virol., 67,* 5859–5866.

Vilaplana, L., Wilson, K., Redman, E. M., & Cory, J. S. (2010). Pathogen persistence in migratory insects: high levels of vertically transmitted virus infection in field populations of the African armyworm. *Evol. Ecol., 24,* 147–160.

Vlak, J. M., Klinkenberg, F. A., Zaal, K. J., Usmany, M., Klinge-Roode, E. C., Geervliet, J. B., Roosien, J., & van Lent, J. W. (1988). Functional studies on the p10 gene of *Autographa californica* nuclear polyhedrosis virus using a recombinant expressing a p10-beta-galactosidase fusion gene. *J. Gen. Virol., 69,* 765–776.

Volkman, L. E. (1988). *Autographa californica* MNPV nucleocapsid assembly: inhibition by cytochalasin D. *Virology, 163,* 547–553.

Volkman, L. E. (1997). Nucleopolyhedrovirus interactions with their insect hosts. *Adv. Virus. Res., 48,* 313–348.

Volkman, L. E., & Goldsmith, P. A. (1985). Mechanism of neutralization of budded *Autographa californica* nuclear polyhedrosis virus by a monoclonal antibody: inhibition of entry by adsorptive endocytosis. *Virology, 143,* 185–195.

Volkman, L. E., & Zaal, K. J. (1990). *Autographa californica* M nuclear polyhedrosis virus: microtubules and replication. *Virology, 175,* 292–302.

Walker, S., Kawanishi, C. Y., & Hamm, J. J. (1982). Cellular pathology of a granulosis virus infection. *J. Ultrastruct. Res., 80,* 163–177.

Wang, M., Yin, F., Shen, S., Tan, Y., Deng, F., Vlak, J. M., Hu, Z., & Wang, H. (2010). Partial functional rescue of *Helicoverpa armigera* single nucleocapsid nucleopolyhedrovirus infectivity by replacement of F protein with GP64 from *Autographa californica* multicapsid nucleopolyhedrovirus. *J. Virol., 84,* 11505–11514.

Wang, P., & Granados, R. R. (1997). An intestinal mucin is the target substrate for a baculovirus enhancin. *Proc. Natl. Acad. Sci. USA, 94,* 6977–6982.

Wang, P., & Granados, R. R. (2000). Calcofluor disrupts the midgut defense system in insects. *Insect Biochem. Mol. Biol., 30,* 135–143.

Wang, P., Hammer, D. A., & Granados, R. R. (1994). Interaction of *Trichoplusia ni* granulosis virus-encoded enhancin with the midgut epithelium and peritrophic membrane of four lepidopteran insects. *J. Gen. Virol., 75,* 1961–1967.

Wang, R., Deng, F., Hou, D., Zhao, Y., Guo, L., Wang, H., & Hu, Z. (2010). Proteomics of the *Autographa californica* nucleopolyhedrovirus budded virions. *J. Virol., 84,* 7233–7242.

Wang, S., & Balasundaram, G. (2010). Potential cancer gene therapy by baculoviral transduction. *Curr. Gene Ther., 10*, 214–225.

Wang, Y., & Jehle, J. A. (2009). Nudiviruses and other large, double-stranded circular DNA viruses of invertebrates: new insights on an old topic. *J. Invertebr. Pathol., 101*, 187–193.

Wang, Y., Burand, J. P., & Jehle, J. A. (2007a). Nudivirus genomics: diversity and classification. *Virol. Sin., 22*, 128–136.

Wang, Y., Kleespies, R. G., Huger, A. M., & Jehle, J. A. (2007b). The genome of *Gryllus bimaculatus* nudivirus indicates an ancient diversification of baculovirus-related nonoccluded nudiviruses of insects. *J. Virol., 81*, 5395–5406.

Wang, Y., van Oers, M. M., Crawford, A. M., Vlak, J. M., & Jehle, J. A. (2007c). Genomic analysis of Oryctes rhinoceros virus reveals genetic relatedness to *Heliothis zea* virus 1. *Arch. Virol., 152*, 519–531.

Wang, Y., Kleespies, R. G., Ramle, M. B., & Jehle, J. A. (2008). Sequencing of the large dsDNA genome of Oryctes rhinoceros nudivirus using multiple displacement amplification of nanogram amounts of virus DNA. *J. Virol. Methods, 152*, 106–108.

Washburn, J. O., Kirkpatrick, B. A., & Volkman, L. E. (1995). Comparative pathogenesis of *Autographa californica* M nuclear polyhedrosis virus in larvae of *Trichoplusia ni* and *Heliothis virescens*. *Virology, 209*, 561–568.

Washburn, J. O., Kirkpatrick, B. A., & Volkman, L. E. (1996). Insect protection against viruses. *Nature, 383*, 767.

Washburn, J. O., Kirkpatrick, B. A., Haas-Stapleton, E., & Volkman, L. E. (1998). Evidence that the stilbene-derived optical brightener M2R enhances *Autographa californica* M nucleopolyhedrovirus infection of *Trichoplusia ni* and *Helithosis virescens* by preventing sloughing of infected midgut epithelial cells. *Biol. Control, 11*, 58–69.

Washburn, J. O., Lyons, E. H., Haas-Stapleton, E. J., & Volkman, L. E. (1999). Multiple nucleocapsid packaging of *Autographa californica* nucleopolyhedrovirus accelerates the onset of systemic infection in *Trichoplusia ni*. *J. Virol., 73*, 411–416.

Washburn, J. O., Haas-Stapleton, E. J., Tan, F. F., Beckage, N. E., & Volkman, L. E. (2000). Co-infection of *Manduca sexta* larvae with polydnavirus from *Cotesia congregata* increases susceptibility to fatal infection by *Autographa californica* M nucleopolyhedrovirus. *J. Insect. Physiol., 46*, 179–190.

Washburn, J. O., Chan, E. Y., Volkman, L. E., Aumiller, J. J., & Jarvis, D. L. (2003). Early synthesis of budded virus envelope fusion protein GP64 enhances *Autographa californica* multicapsid nucleopolyhedrovirus virulence in orally infected. *Heliothis virescens*. *J. Virol., 77*, 280–290.

Webb, R. E., Shapiro, M., Podgwaite, J. D., Ridgway, R. L., Venables, L., White, G. B., Argauer, R. J., Cohen, D. L., Witcosky, J., Kester, K. M., & Thorpe, K. W. (1994). Effect of optical brighteners on the efficacy of gypsy moth (Lepidoptera: Lymantriidae) nuclear polyhedrosis virus in forest plots with high or low levels of natural virus. *J. Econ. Entomol., 87*, 134–143.

Webb, R. E., Dill, N. H., McLaughlin, J. M., Kershaw, L. S., Podgwaite, J. D., Cook, S. P., Thorpe, K. W., Farrar, J. R. R., Ridgway, R. L., Fuester, R. W., Shapiro, M., Argauer, R. J., Venables, L., & White, G. B. (1996). Blankophor BBH as an enhancer of nuclear polyhedrosis virus in arborist treatments against the gypsy moth (Lepidoptera: Lymantriidae). *J. Econ. Entomol., 89*, 957–962.

Westenberg, M., & Vlak, J. M. (2008). GP64 of group I nucleopolyhedroviruses cannot readily rescue infectivity of group II f-null nucleopolyhedroviruses. *J. Gen. Virol., 89*, 424–431.

Westenberg, M., Wang, H., WF, I. J., Goldbach, R. W., Vlak, J. M., & Zuidema, D. (2002). Furin is involved in baculovirus envelope fusion protein activation. *J. Virol., 76*, 178–184.

Westenberg, M., Uijtdewilligen, P., & Vlak, J. M. (2007). Baculovirus envelope fusion proteins F and GP64 exploit distinct receptors to gain entry into cultured insect cells. *J. Gen. Virol., 88*, 3302–3306.

Westenberg, M., Soedling, H. M., Mann, D. A., Nicholson, L. J., & Dolphin, C. T. (2010). Counter-selection recombineering of the baculovirus genome: a strategy for seamless modification of repeat-containing BACs. *Nucleic Acids Res., 38*, e166.

Whitford, M., & Faulkner, P. (1992). A structural polypeptide of the baculovirus *Autographa californica* nuclear polyhedrosis virus contains O-linked *N*-acetylglucosamine. *J. Virol., 66*, 3324–3329.

Whitlock, V. H. (1974). Symptomatology of two viruses infecting *Heliothis armigera*. *J. Invertebr. Pathol., 23*, 70–75.

Whitt, M. A., & Manning, J. S. (1988). A phosphorylated 34-kDa protein and a subpopulation of polyhedrin are thiol linked to the carbohydrate layer surrounding a baculovirus occlusion body. *Virology, 163*, 33–42.

Wickham, T. J., Shuler, M. L., Hammer, D. A., Granados, R. R., & Wood, H. A. (1992). Equilibrium and kinetic analysis of *Autographa californica* nuclear polyhedrosis virus attachment to different insect cell lines. *J. Gen. Virol., 73*, 3185–3194.

Wigglesworth, V. B. (1977). Structural changes in the epidermal cells of *Rhodnius* during tracheole capture. *J. Cell Sci., 26*, 161–174.

van der Wilk, F., van Lent, J. W. M., & Vlak, J. M. (1987). Immunogold detection of polyhedrin, p10 and virion antigens in *Autographa californica* nuclear polyhedrosis virus-infected *Spodoptera frugiperda* cells. *J. Gen. Virol., 68*, 2615–2623.

Wilkinson, J. C., Wilkinson, A. S., Scott, F. L., Csomos, R. A., Salvesen, G. S., & Duckett, C. S. (2004). Neutralization of Smac/Diablo by inhibitors of apoptosis (IAPs). A caspase-independent mechanism for apoptotic inhibition. *J. Biol. Chem., 279*, 51082–51090.

Williams, G. V., Rohel, D. Z., Kuzio, J., & Faulkner, P. (1989). A cytopathological investigation of *Autographa californica* nuclear polyhedrosis virus *p10* gene functions using insetion/deletion mutants. *J. Gen. Virol., 70*, 187–202.

Wilson, K., Thomas, M. B., Blanford, S., Doggett, M., Simpson, S. J., & Moore, S. L. (2002). Coping with crowds: density-dependent disease resistance in desert locusts. *Proc. Natl. Acad. Sci. USA, 99*, 5471–5475.

Wilson, K. R., O'Reilly, D. R., Hails, R. S., & Cory, J. S. (2000). Age-related effects of the *Autographa californica* multiple nucleopolyhedrovirus *egt* gene in the cabbage looper (*Trichoplusia ni*). *Biol. Control, 19*, 57–63.

Wilson, M. E., & Consigli, R. A. (1985). Functions of a protein kinase activity associated with purified capsids of the granulosis virus infecting *Plodia interpunctella*. *Virology, 143*, 526–535.

Wilson, M. E., & Miller, L. K. (1986). Changes in the nucleoprotein complexes of a baculovirus DNA during infection. *Virology, 151*, 315–328.

Winstanley, D., & Crook, N. E. (1993). Replication of *Cydia pomonella* granulosis virus in cell cultures. *J. Gen. Virol., 74*, 1599–1609.

Wood, H. A., & Hughes, P. R. (1996). Recombinant viral insecticides: delivery of environmentally safe and cost-effective products. *Entomophaga, 41*, 361−373.

Wu, W., Liang, H., Kan, J., Liu, C., Yuan, M., Liang, C., Yang, K., & Pang, Y. (2008). *Autographa californica* multiple nucleopolyhedrovirus 38K is a novel nucleocapsid protein that interacts with VP1054, VP39, VP80, and itself. *J. Virol., 82*, 12356−12364.

Wu, Y., Liu, G., & Carstens, E. B. (1999). Replication, integration, and packaging of plasmid DNA following cotransfection with baculovirus viral DNA. *J. Virol., 73*, 5473−5480.

Wyllie, A. H., Kerr, J. F., & Currie, A. R. (1980). Cell death: the significance of apoptosis. *Int. Rev. Cytol., 68*, 251−306.

Wyllie, A. H., Morris, R. G., Smith, A. L., & Dunlop, D. (1984). Chromatin cleavage in apoptosis: association with condensed chromatin morphology and dependence on macromolecular synthesis. *J. Pathol., 142*, 67−77.

Xeros, N. (1956). The virogenic stroma in nuclear and cytoplasmic polyhedroses. *Nature, 178*, 412−413.

Xing, K., Deng, R., Wang, J., Feng, J., Huang, M., & Wang, X. (2005). Analysis and prediction of baculovirus promoter sequences. *Virus Res., 113*, 64−71.

Xu, J. P., Chen, K. P., Yao, Q., Lin, M. H., Gao, G. T., & Zhao, Y. (2005). Identification and characterization of an NPV infection-related gene *Bmsop2* in *Bombyx mori* L. *J. Appl. Entomol., 129*, 425−431.

Yanase, T., Hashimoto, Y., & Kawarabata, T. (2000). Identification of insertion and deletion genes in *Autographa californica* nucleopolyhedrovirus variants isolated from *Galleria mellonella, Spodoptera exigua, Spodoptera litura* and *Xestia c-nigrum. Virus Genes, 21*, 167−177.

Yang, S., & Miller, L. K. (1999). Activation of baculovirus very late promoters by interaction with very late factor 1. *J. Virol., 73*, 3404−3409.

Yao, L., Zhou, W., Xu, H., Zheng, Y., & Qi, Y. (2004). The *Heliothis armigera* single nucleocapsid nucleopolyhedrovirus envelope protein P74 is required for infection of the host midgut. *Virus Res., 104*, 111−121.

Yin, F., Wang, M., Tan, Y., Deng, F., Vlak, J. M., Hu, Z., & Wang, H. (2008). A functional F analogue of *Autographa californica* nucleopolyhedrovirus GP64 from the *Agrotis segetum* granulovirus. *J. Virol., 82*, 8922−8926.

Yoo, S., & Guarino, L. A. (1994). The *Autographa californica* nuclear polyhedrosis virus *ie2* gene encodes a transcriptional regulator. *Virology, 202*, 746−753.

Young, J. C., MacKinnon, E. A., & Faulkner, P. (1993). The architecture of the virogenic stroma in isolated nuclei of *Spodoptera frugiperda* cells *in vitro* infected by *Autographa californica* nuclear polyhedrosis virus. *J. Struct. Biol., 110*, 141−153.

Young, S. Y., & Yearian, W. C. (1974). Persistence of *Heliothis* NPV on foliage of cotton, soybean, and tomato. *Environ. Entomol., 3*, 253−260.

Young, S. Y., & Yearian, W. C. (1986). Movement of a nuclear polyhedrosis virus from soil to soybean and transmission in *Anticarsia gemmatalis* (Hübner) (Lepidoptera: Noctuidae) populations on soybean. *Environ. Entomol., 15*, 573−580.

Young, S. Y., Yearian, W. C., & Kim, K. S. (1977). Effect of dew from cotton and soybean foliage on activity of *Heliothis* nuclear polyhedrosis virus. *J. Invertebr. Pathol., 29*, 105−111.

Yu, X., Jin, L., & Zhou, Z. H. (2008). 3.88 Å Structure of cytoplasmic polyhedrosis virus by cryo-electron microscopy. *Nature, 453*, 415−419.

Zanotto, P. M., Kessing, B. D., & Maruniak, J. E. (1993). Phylogenetic interrelationships among baculoviruses: evolutionary rates and host associations. *J. Invertebr. Pathol., 62*, 147−164.

Zelazny, B. (1972). Studies on *Rhabdionvirus oryctes* I. Effect on larvae of *Oryctes rhinoceros* and inactivation of the virus. *J. Invertebr. Pathol., 20*, 235−241.

Zelazny, B. (1976). Transmission of a baculovirus in populations of. *Oryctes rhinoceros. J. Invertebr. Pathol., 27*, 221−227.

Zelazny, B. (1977). *Oryctes rhinoceros* populations and behavior influenced by a bacuivirus. *J. Invertebr. Pathol., 29*, 210−215.

Zelazny, B., Lolong, A., & Pattang, B. (1992). *Oryctes rhinoceros* (Coleoptera: Scarabaeidae) populations suppressed by a baculovirus. *J. Invertebr. Pathol., 59*, 61−68.

Zhang, H., Zhang, J., Yu, X., Lu, X., Zhang, Q., Jakana, J., Chen, D. H., Zhang, X., & Zhou, Z. H. (1999). Visualization of protein−RNA interactions in cytoplasmic polyhedrosis virus. *J. Virol., 73*, 1624−1629.

Zhang, J., Lapointe, R., Thumbi, D., Morin, B., & Lucarotti, C. J. (2010). Molecular comparisons of alphabaculovirus-based products: Gypchek with Disparvirus (*Lymantria dispar*) and TM BioControl-1 with Virtuss (*Orgyia pseudotsugata*). *Can. Entomol., 142*, 546−556.

Zhang, J. H., Washburn, J. O., Jarvis, D. L., & Volkman, L. E. (2004). Autographa californica M nucleopolyhedrovirus early GP64 synthesis mitigates developmental resistance in orally infected noctuid hosts. *J. Gen. Virol., 85*, 833−842.

Zhang, P., Yang, K., Dai, X., Pang, Y., & Su, D. (2002). Infection of wild-type *Autographa californica* multicapsid nucleopolyhedrovirus induces *in vivo* apoptosis of *Spodoptera litura* larvae. *J. Gen. Virol., 83*, 3003−3011.

Zhao, J., Hyman, L., & Moore, C. (1999). Formation of mRNA 3′ ends in eukaryotes: mechanism, regulation, and interrelationships with other steps in mRNA synthesis. *Microbiol. Mol. Biol. Rev., 63*, 405−445.

Zhou, M., Sun, X., Vlak, J. M., Hu, Z., & van der Werf, W. (2005). Horizontal and vertical transmission of wild-type and recombinant *Helicoverpa armigera* single-nucleocapsid nucleopolyhedrovirus. *J. Invertebr. Pathol., 89*, 165−175.

Zhou, W., Yao, L., Xu, H., Yan, F., & Qi, Y. (2005). The function of envelope protein P74 from *Autographa californica* multiple nucleopolyhedrovirus in primary infection to host. *Virus Genes, 30*, 139−150.

Zlotkin, E., Rochat, H., Kopeyan, Miranda, F., & Lissitzky, S. (1971). Purification and properties of the insect toxin from the venom of the scorpion *Androctonus australis* Hector. *Biochimie., 53*, 1073−1078.

Zlotkin, E., Fishman, Y., & Elazar, M. (2000). AaIT: from neurotoxin to insecticide. *Biochimie, 82*, 869−881.

Zoog, S. J., Schiller, J. J., Wetter, J. A., Chejanovsky, N., & Friesen, P. D. (2002). Baculovirus apoptotic suppressor P49 is a substrate inhibitor of initiator caspases resistant to P35 *in vivo. EMBO J., 21*, 5130−5140.

Chapter 5

RNA Viruses Infecting Pest Insects

Yan Ping Chen,* James J. Becnel[†] and Steven M. Valles[†]
*United States Department of Agriculture, Agricultural Research Service, Beltsville, Maryland, USA, [†] United States Department of Agriculture, Agricultural Research Service, Gainesville, Florida, USA

Chapter Outline

Summary

Viruses that contain ribonucleic acid (RNA) as their genetic material are known as RNA viruses and may be double or single stranded based on the type of RNA they contain. Single-stranded RNA viruses can be further grouped into negative-sense or positive-sense viruses according to the polarity of their RNA. Further, RNA viruses can be categorized according to whether they have a lipid envelope, and whether they have single (non-fragmented) or multiple (fragmented) genomes. RNA viruses comprise a wide variety of infectious agents, many of which induce disease in plant, vertebrate, and invertebrate hosts. RNA viruses exhibit structural simplicity, reduced genome size, and prolific replication rates, making them ideal models for the study of various aspects of virus biology. Although they are the most prevalent group of viruses, RNA viruses infecting insect pests are currently limited to six families: *Nodaviridae* (genus *Alphanodavirus*), *Dicistroviridae* (genera *Cripavirus* and *Aparavirus*), *Flaviviridae* (genus *Flavivirus*), *Iflaviridae* (genus *Iflavirus*), *Tetraviridae* (genera *Betatetravirus* and *Omegatetravirus*), and *Reoviridae* (genus *Cypovirus*). Over the past two decades, significant advances have been made in elucidating the fundamental molecular biology and pathology of insect RNA viruses. The progress in the study of these viruses presents extraordinary research opportunities for a better understanding of the diseases they cause, and for subsequent prevention and treatment of such diseases. While RNA viruses that infect beneficial insects are discussed elsewhere in this book, this chapter focuses on RNA viruses infecting pest insects. It provides an overview of the current understanding of morphology, genome organization, phylogeny, natural history, transmission, and pathogenesis of these viruses. The prospects for future research and the potential of RNA viruses to be used as biological control agents are also discussed.

5.1. INTRODUCTION

In a summary of the status of arthropod ribonucleic acid (RNA) virus research, Christian and Scotti (1998) stated succinctly, "… in the insect virus field, the ultimate application of any work is in the development of novel control agents". However, these authors also lamented that RNA viruses would see limited application as microbial control agents until commercial profitability for their use was perceived and significant advances were made in production technology. Although the impact on beneficial arthropods (e.g., honey bees and shrimp) is a concern, insect control arguably remains the primary impetus for most research conducted on arthropod-infecting viruses. Despite the growing numbers of arthropod-infecting RNA viruses with fully sequenced genomes that are available in public databases, few have been characterized with respect to their use as insect control agents since Christian and Scotti's (1998) summary. However, significant economic losses caused by many insect pests continue to spur the research and development of RNA viruses as microbial control agents, albeit slowly.

The use of RNA viruses as microbial control agents presents unique challenges and concerns including the taxonomic similarity with human RNA viruses and the possible exchange of genetic material, a high genome mutation rate, and difficulties in producing large, stable quantities of virus that hamper their development and use in insect control. Taxonomic similarity or "homology" (Christian and Scotti, 1998) with human-infecting RNA viruses is a legitimate concern. Many significant human diseases (e.g., polio, the common cold, and hepatitis A, C, and E) are caused by RNA viral infections that are related, however distantly, to many of the insect-infecting RNA viruses. In fact, serological cross-reactivity between some picornaviruses and arthropod RNA viruses has been reported (Longworth et al., 1973; Moore et al., 1981; Tinsley et al., 1984). This apparent cross-reactivity could have actually stemmed from previous host exposure to the insect-infecting RNA viruses as many are ubiquitous in the environment (Culley et al., 2006; Djikeng et al., 2009).

Recombination among human- and arthropod-infecting positive-strand RNA viruses is a remote possibility because both viruses would have to be replicating in the same cellular environment. Host specificity of each virus would be likely to preclude this event from occurring. However, template switching, a mechanism by which the RNA-dependent RNA polymerase (RdRp) switches from one virus RNA template to another, could result in exchange of genetic material between viruses and lead to the evolution of variants with enhanced virulence and/or host ranges (Kirkegaard and Baltimore, 1986; Jarvis and Kirkegaard, 1992). Recombination by template switching appears to depend strongly on sequence homology, as evidenced by a higher prevalence of intratypic events compared with intertypic events (Kirkegaard and Baltimore, 1986). Nonetheless, "gene shuffling" (Dolja and Koonin, 1991) by this mechanism must be a consideration when intending to develop RNA viruses for insect control purposes. Indeed, phylogenetic analyses among vertebrate-infecting and even insect-infecting RNA viruses illustrate their diversification capacities (Lukashev, 2005; Cristina and Costa-Mattioli, 2007; de Miranda et al., 2010a). For instance, Acute bee paralysis virus (ABPV), Kashmir bee virus (KBV), and Israeli acute paralysis virus (IAPV), all insect-infecting RNA viruses, are considered a closely related virus complex with common origins (de Miranda et al., 2010a). More recently, Moore et al. (2011) have demonstrated recombination events between Deformed wing virus

(DWV) and Varroa destructor virus 1 (VDV-1). They suggested that the genome modularity of these viruses facilitated independent evolution, resulting in better adapted host transmissibility. Similarly, Terio et al. (2008) described a virus with intermediate features between DWV and Kakugo virus (KV; "kakugo" means "ready to attack" in Japanese).

It is well known that RNA viruses lack proofreading machinery and, as such, exhibit high mutation rates (Holland et al., 1982) which, coupled with high virus yields and short replication times (Domingo and Holland, 1997), lead to individual hosts infected with a diverse population of phylogenetically related genome variants, or quasispecies (Smith et al., 1997). In fact, RNA virus evolution can be more than a million-fold greater than that of their hosts (Holland et al., 1982). Thus, significant phenotypic changes could alter virulence and/or host range (Waters et al., 2007).

Despite these concerns, successful laboratory and field evaluations of RNA viruses for insect control have been reported. Drosophila C virus (DCV) and Cricket paralysis virus (CrPV) were evaluated against the olive fruit fly, Dacus oleae, and CrPV was shown to replicate in the fly host, disperse in fly feces, and cause up to 80% mortality 12 days after initial exposure (Manousis and Moore, 1987). CrPV (strain CrPV$_{brk}$) was also reported to be effective in killing adult Mediterranean fruit flies, Ceratitis capitata (Plus and Scotti, 1984). Similarly, high mortality rates (up to 100%) were observed in laboratory and field tests of newly described RNA viruses against two lepidopteran pests (Epicerura pergrisea and Latoia viridissima) in Côte d'Ivoire (Fédière et al., 1990; Kanga and Fediere, 1991), and successful field tests were conducted against Helicoverpa armigera using the Helicoverpa armigera stunt virus (HaSV) (Christian et al., 2005). HaSV was shown to be as effective as a commercial preparation of the baculovirus Helicoverpa zea single-nucleopolyhedrovirus (HzSNPV) on sorghum. Although these tests demonstrate successful use of RNA viruses in insect control, they also identify a significant limitation, namely, large-scale production of virus.

In all laboratory and field tests featured above, the source of the viruses was purified preparations from infected hosts, which involves an arduous and inefficient process. Unfortunately, insect host cell lines supporting viral production are only available for a handful of the viruses. Indeed, the absence of host cell lines has hampered biological investigation of most arthropod RNA viruses. As an alternative, infectious RNA virus transcripts have been produced successfully (Boyer and Haenni, 1994). Several insect-infecting RNA virus transcripts have been produced including Black queen cell virus (BQCV) and Rhopalosiphum padi virus (RhPV) (Benjeddou et al., 2002a, b; Boyapalle et al., 2008). Although infectious transcripts facilitate laboratory study of these viruses, they offer limited application as microbial control agents. However, *in vitro* production of an insect-infecting positive-strand RNA virus has been reported. Pal *et al.* (2007) successfully constructed a baculovirus that expressed the dicistrovirus, RhPV. The baculovirus-driven RhPV construct was infectious to its natural host, *R. padi. In vitro* production of insect-infecting positive-strand RNA viruses not only facilitates study of their biology, but also provides a means for large-scale production of these viruses for use as microbial control agents.

Many comprehensive reviews on insect RNA viruses have been published in recent years, including Hanzlik and Gordon (1997), Ball and Johnson (1998), Belloncik and Mori (1998), Christian and Scotti (1998), Gordon and Hanzlik (1998), Gordon and Waterhouse (2006), Bonning and Johnson (2010), Dorrington and Short (2010), Mori and Metcalf (2010), van Oers (2010), and Venter *et al.* (2010). Recent discoveries of new RNA viruses from pestiferous insects and advances in their production offer the development of novel methods of insect-specific control agents. While RNA viruses that infect beneficial insects are discussed in Chapter 12, this chapter focuses on the recent progress in understanding the molecular biology and pathology of RNA viruses infecting pest insects and their potential application for use as microbial control agents for integrated pest management.

5.2. CLASSIFICATION, PHYLOGENY, STRUCTURE, AND GENOME ORGANIZATION

RNA viruses are a group of evolutionarily related viruses with RNA as their genetic material. As with other viruses, the classification of RNA viruses is based mainly on phenotypic characters, including (1) virion properties (e.g., the structure of the capsid and the number of protein subunits); (2) ribonucleic acid composition (single- or double-stranded, polarity if single-stranded); (3) mechanisms of viral genome replication; (4) host organisms; and (5) pathology (Francki *et al.*, 1991). The International Committee for the Taxonomy of Viruses (ICTV) adheres to the Baltimore system (Baltimore, 1971) to classify RNA viruses into group III [viruses possessing double-stranded RNA (dsRNA) genomes], group IV [viruses possessing positive-sense single-stranded RNA (ssRNA) genomes], or group V (viruses possessing negative-sense ssRNA genomes). While the traditional classification scheme sorts RNA viruses into different groups according to their convergent similarities and differences in morphological and genomic features as well as associated replication strategies, it does not provide natural evolutionary relationships among the groups. Beginning in the 1990s, phylogenetic analysis of shared traits has been incorporated

as a tool to aid virus taxonomy. Molecular phylogeny aids classification boundaries and evolutionary relationships among viral taxa. The recent proliferation of available viral gene and genome sequences has facilitated and developed evolutionary relationships and taxonomy of viruses, and continues to do so.

Although genetic variation and microevolutionary processes such as mutation, natural selection, gene flow, and genetic drift that result in genetic changes of RNA viruses have received a great deal of attention, the origins of RNA viruses remain obscure because of the absence of a fossil record. Three mutually non-exclusive hypotheses (regressive, cellular origin, and primordial) have been proposed to explain the origin of RNA viruses (review in Morse, 1994). The regressive hypothesis postulates that the ancestry of RNA viruses might have originated from parasitized cellular organisms through successive reductions in genome size to become simple replicating molecules within host cells. The cellular origin hypothesis posits that RNA viruses were genetic elements that escaped from cellular genes and acquired a protective proteinaceous coat and the ability to replicate in new hosts. Finally, the primordial hypothesis (also referred to as the virus-first hypothesis) postulates that viruses evolved from precellular RNA life forms at the same time as the first cells on Earth and independent of cellular life for billions of years.

The lack of a clear understanding of the origin of RNA viruses is an impediment to direct phylogenetic analysis among groups. In addition, the highly divergent nature of RNA viruses limits the sequence-based method for inferring a meaningful phylogeny of the viruses. As a result, some attempts have been made to reveal evolutionary relationships based on the similarity of genome organization and secondary protein structures of RNA viruses (Sankoff, 2003). Despite the substantial differences in phenotypic characters and life cycles, all RNA viruses with no DNA stage encode an essential protein, RdRp, that is not found in cellular species. RdRps specifically recognize different origins of replication at the $3'$ termini of both (plus) and (minus) sense RNAs to direct replication of viral genomes (Koonin and Dolja, 1993). The catalytic center of the RdRps includes a ubiquitous $\alpha\beta$ palm subdomain structure comprised of conserved amino acid sequence motifs crucial for catalysis (Koonin and Dolja, 1993; Goldbach and deHaan, 1994; Gorbalenya *et al.*, 2002). Because of its unique and universal conservation, the palm subdomain of RdRps has been used widely for molecular taxonomy of the RNA viruses and for inferring their evolutionary relationships.

RNA viruses are subdivided into 40 families and infect a broad range of vertebrate and invertebrate hosts. The RNA viruses identified in insects are currently limited to five positive-stranded RNA virus families including *Nodaviridae* (genus *Alphanodavirus*), *Dicistroviridae* (genera *Cripavirus* and *Aparavirus*), *Flaviviridae* (genus *Flavivirus*), *Iflaviridae* (genus *Iflavirus*), and *Tetraviridae* (genera *Betatetravirus* and *Omegatetravirus*), and one dsRNA virus family, *Reoviridae* (genus *Cypovirus*).

5.2.1. Alphanodaviruses

The *Nodaviridae* contains two genera, *Alphanodavirus* that primarily infect insects and *Betanodavirus* that infects fish (Schneemann *et al.*, 2005). The family name is derived from the *Nodamura virus* (NoV), the type species of the *Alphanodavirus* genus, which was first discovered in mosquitoes near the Japanese village of Nodamura (now the city of Nodashi) (Scherer and Hurlbut, 1967). NoV is the only alphanodavirus that also multiplies in vertebrates. Members of the *Alphanodavirus* contain five well-characterized species including NoV, *Black beetle virus* (BBV), *Boolarra virus* (BoV), *Flock house virus* (FHV), and *Pariocoto virus* (PaV). The genus *Alphanodavirus* also contains some tentative species, including *Gypsy moth virus* (GMV), *Lymantria ninayi virus* (LNV), *Manawatu virus* (MwV), *New Zealand virus* (NZV), and *Drosophila line 1 virus* (DLV). A recently isolated nodavirus, *Wuhan nodavirus* (WhNV), from larvae of the imported cabbageworm, *Pieris rapae*, is the most distantly related of both the alphanodaviruses and betanodaviruses. Similarity of a coat protein of WhNV with homologous proteins of other nodaviruses is very low, suggesting that WhNV may represent a new genus in the family *Nodaviridae* (Liu *et al.*, 2006a, b).

Nodaviruses are classified into genera based on the shape, size, buoyant density of the particles, RNA genome organization, and host range. However, definitive classification is based on the nucleotide sequence of the coat protein gene. Identity of the coat protein gene sequences between the alphanodaviruses and betanodaviruses is low and does not exceed 17%. The sequence comparisons of the coat protein gene have been used not only to separate genera in the family but also to resolve the phylogenetic relationships at the level of genus (Nishizawa *et al.*, 1997; Valle *et al.*, 2001).

The X-ray structures of BBV, FHV, NoV, and PaV have been determined at high resolution and provided significant insights into the organization of the virion structure of RNA viruses (Johnson and Reddy, 1998; Banerjee *et al.*, 2010; Venter *et al.*, 2010). The *Alphanodavirus* capsid is not enveloped and is round with icosahedral symmetry on a T = 3 surface lattice (Fisher and Johnson, 1993; Tang *et al.*, 2001). The isometric capsid has a diameter of 29−30 nm and consists of 32 capsomers. Virions consist of 180 copies of the same capsid precursor, protein α, without any distinctive surface structure (Hosur *et al.*, 1987).

Studies of the structures and assembly of nodaviruses reveal extensive interactions between the viral RNA genome and the icosahedral capsid. The capsid contains 12 five-fold, 20 three-fold, and 30 two-fold rotation axes of symmetry. At the five-fold axes of the virion, the capsid proteins form a pentameric helical bundle that has been hypothesized to play a role in release of viral RNA into the host cells during the virus uncoating process. At the quasi-three-fold axis, three protomers form a prominent peak. At the icosahedral two-fold axes of the virion, C-terminal residues of the capsid protein contact with ordered genomic RNA forming double-helical segments (Schneemann and Marshall, 1998). The crystal structure of PaV (Tang *et al.*, 2001) reveals a dodecahedral cage of duplex RNA that accounts for 35% of the RNA genome inside the icosahedral virus capsid. The ordered portion of RNA is in close association with the basic N-terminal region of 60 subunits of the capsid protein and the capsid protein displays a remarkable complementarity to the geometry of the dodecahedral cage. However, the remaining 65% of the viral RNA lies in the interior of the particle and does not adopt icosahedral symmetry.

The genome is linear, positive sense, bipartite (composed of two segments) ssRNA. Both genomic segments contain a methylated cap at the 5′ terminus but lack poly (A) tails at the 3′ terminus and are co-encapsidated in the same viral particle (Newman and Brown, 1973, 1976) (Fig. 5.1A). The larger genomic segment, RNA1, such as for NoV, is 3204 nucleotides (nt) long and encodes protein A [112 kilodaltons (kDa)], a catalytic subunit of RdRp that is required for virus replication (Ball and Johnson, 1999; Johnson *et al.*, 2001). The distinguishing structural features of the RdRp include conserved canonical motifs in the C-terminal amino acid sequences and conserved elements of predicted secondary structures throughout (Johnson *et al.*, 2001). In contrast, the 5′ and 3′ untranslated regions (UTRs) of RNA1 segments do not exhibit common features of RNA sequence or secondary structure (Johnson *et al.*, 2001). The smaller genomic segment, RNA 2 contains about 1.4 kilobases (kb) and encodes the capsid protein precursor α (43 kDa) that is subsequently cleaved into capsid proteins β and γ. Except for PaV, which has an asparagine/serine cleavage site on the inside surface of the virion, the cleavage site asparagine/alanine is conserved among the alphanodaviruses. This cleavage site is required for acquisition of virion infectivity and stability. A subgenomic RNA (RNA 3), 3′ co-linear to RNA1, is also produced and encodes proteins B1 and B2. Protein B2 is a suppressor of host-mediated RNA silencing (Li *et al.*, 2002, 2004). Among members of genus *Alphanodavirus*, the B2 proteins share limited sequence homology. At present, the function of protein B1 has not been determined.

5.2.2. Dicistroviruses

Viruses comprising the *Dicistroviridae* were historically classified as "picorna-like" without succinct taxonomic

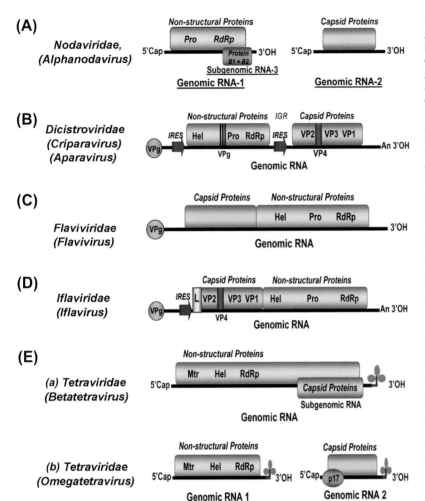

FIGURE 5.1 Schematic representation of the genome organization of positive-sense single-stranded insect RNA viruses. (A) Alphanodavirus genome organization showing bipartite genome composed of genomic segments RNA 1 and RNA 2 as well as a subgenomic RNA 3. Both genomic segments are capped at their 5′ end and not polyadenlyated at their 3′ end. (B) Dicistrovirus monopartite, dicistronic genome organization with two non-overlapping open reading frames (ORFs) separated by an intergenic untranslated region (IGR). Distinct internal ribosome entry sites (IRES) are located in the 5′ untranslated region and IGR. The 5′ end of the genome carries a covalently linked genome-linked virus protein (VPg) and the 3′ end of the genome is polyadenylated. (C) Flavivirus monopartite genome organization composed of a single RNA segment. The 5′ end of the genome carries a covalently linked genome-linked virus protein (VPg) and the 3′ end of the genome is not polyadenylated. (D) Iflavirus monopartite genome organization composed of a single RNA segment. The viral capsid proteins are characteristically preceded by a short leader protein (L). The 5′ end of the genome carries a covalently linked VPg and the 3′ end of the genome is polyadenylated. (E-a) Betatetravirus monopartite genome organization with two overlapping ORFs. Subgenomic RNA is generated from the 3′ -capsid region of the large genomic RNA. The genomic RNA is capped at the 5′ end and has a tRNA-like structure at the 3′ end. (E-b) Omegatetravirus bipartite genome organization composed of genomic segments RNA1 and RNA 2. The genomic RNAs are capped at their 5′ ends and have a tRNA-like structure at their 3′ ends.

placement (Christian and Scotti, 1998). However, virus discoveries within the past decade, complete genome sequencing, and extensive phylogenetic analyses have firmly defined the *Dicistroviridae*, which was officially adopted by the ICTV in 2002 (Mayo, 2002). The number of viruses comprising the *Dicistroviridae* has increased considerably since the discovery of ABPV (Bailey *et al.*, 1963). Currently, the *Dicistroviridae* is comprised of 13 viruses in two genera, *Cripavirus* (derived from *Cri*cket *pa*ralysis *virus*) and *Aparavirus* (derived from *A*cute bee *pa*ralysis *virus*) (Table 5.1).

The genera are distinct phylogenetically and exhibit unique characteristics; most notable are differences in the structure of the intergenic internal ribosome entry site (IRES) (Jan, 2006; Firth *et al.*, 2009; Jang *et al.*, 2009). The two IRES structures are classified as type I (cripaviruses) or type II (aparaviruses) depending on characteristic bulge sequences in the 5′ region (UGAUCU and UGC type I; UGGUUACCCAU and UAAGGCUU type II). In addition, the structure of the IRES elements in the two genera is also distinguished by the presence of an

additional stem loop in the 3′ region of the intergenic region (IGR) of the IRES of the aparaviruses but not in the cripaviruses. The IRES provides a mechanism for 7-methyl guanosine cap-independent translation, which results in more efficient production of capsid proteins (Jan and Sarnow, 2002). All of these viruses infect arthropods, with the majority (12) infecting insects (Bonning and Miller, 2010a,b); one virus, *Taura syndrome virus* (TSV), infects penaeid shrimps (Hasson *et al.*, 1995). Also, single-stranded and positive-sense RNA viruses consistent with characteristics of the *Dicistroviridae* have been reported recently in the mud crab, *Scylla serrata* (Zhang *et al.*, 2010a,b), and red imported fire ant, *Solenopsis invicta* (Valles and Hashimoto, 2009), but their taxonomic placement awaits further characterization. *Solenopsis invicta virus 3* (SINV-3) appears to cause significant fire ant colony mortality reminiscent of honey bee colony collapse disorder (CCD) (Fig. 5.2) and appears to be an excellent candidate for development as a microbial control agent against fire ants in the USA.

TABLE 5.1 Members of the *Dicistroviridae* and Their Hosts[a]

Virus (Acronym)	Host Status	Host Range[c]	Pathology and/or Manifestation Thereof[d]	In Vitro Propagation	Tissue Tropism[f]
Genus: Cripavirus					
Cricket paralysis virus (CrPV)	Pest (agriculture), beneficial	Lepidoptera (6:12:12) Orthoptera (1:3:4) Diptera (1:1:2) Hemiptera (1:2:2) Hymenoptera (2:2:2)	Paralysis of rear legs, paralysis, mortality	Cell culture	Epidermis, alimentary canal, ganglia
Aphid lethal paralysis virus (ALPV)	Pest (agriculture)	Hemiptera (2:7:8)	Paralysis, decreased fecundity, mortality	None available	Intestinal epithelium, alimentary canal, protocerebrum
Black queen cell virus (BQCV)	Beneficial	Hymenoptera (1:1:1) Mesostigmata[e] (1:1:1)	Queen mortality, black-colored cells	Infectious transcript	Alimentary canal
Drosophila C virus (DCV)	Pest (agriculture), beneficial	Diptera (1:1:16)	Mortality, dark larvae/pupae, increased fecundity, altered development time	Cell culture	Basal gut cells, gut contents
Himetobi P virus (HiPV)	Pest (agriculture)	Hemiptera (1:3:3)	Asymptomatic	None available	Midgut, hindgut, gut contents
Plautia stali intestine virus (PSIV)	Pest (agriculture)	Hemiptera (1:3:3)	Decreased lifespan, mortality	None available	Alimentary canal
Rhopalosiphum padi virus (RhPV)	Pest (agriculture)	Hemiptera (1:5:7)	Decreased fecundity, decreased lifespan	Cell culture, infectious transcript, expression system	Posterior midgut, hindgut
Triatoma virus (TrV)	Pest (human health)	Hemiptera (1:2:2)	Decreased fecundity, decreased lifespan, molting disruption, paralysis of rear legs, mortality	None available	Midgut, nervous tissue

Virus	Pest (agriculture)	Host[c]	Disease signs[d]	Cell line	Tissue tropism[f]
Homalodisca coagulata virus-1 (HoCV-1)	Pest (agriculture)	Hemiptera (1:2:3)	Mortality	None available	Midgut
Genus: *Aparavirus*[b]					
Acute bee paralysis virus (ABPV)	Beneficial	Hymenoptera (2:2:4) Mesostigmata	Paralysis, cuticular darkening, mortality	None available	Brain, hypopharyngeal glands, semen, fat body
Taura syndrome virus (TSV)	Beneficial	Decapoda[g] (1:2:10)	Mortality, lethargy, melanized cuticular lesions	None available[h]	Systemic
Kashmir bee virus (KBV)	Beneficial	Hymenoptera (3:3:4) Mesostigmata[e] (1:1:1)	Trembling, mortality	None available	Alimentary canal, epidermis, trachea, hemocytes, oenocytes
Solenopsis invicta virus-1 (SINV-1)	Pest (agriculture, human health, urban)	Hymenoptera (1:1:6)	Mortality	None available	Midgut, hindgut, gut contents
Israeli acute paralysis virus (IAPV)[b]	Beneficial	Hymenoptera (1:1:1)	Paralysis, cuticular darkening, mortality	None available	Undetermined

[a]Table modeled after Christian and Scotti (1993). Information acquired from Bailey and Gibbs (1964), Furgala and Lee (1966), Bailey and Milne (1969), Bailey and Woods (1974), D'Arcy et al. (1981a), Muscio et al. (1987), Gildow and D'Arcy (1990), Hatfill et al. (1990), Anderson (1991), Laubscher and von Wechmar (1992), Lautié-Harivel (1992), Suzuki et al. (1993), Hasson et al. (1995), Scotti et al. (1996), Christian and Scotti (1998), Johnson and Christian (1998), Nakashima et al. (1998), Muscio et al. (2000), Rozas-Dennis and Cazzaniga (2000), Benjeddou et al. (2002b), Rozas-Dennis et al. (2002), van Munster et al. (2002), Valles et al. (2004), Hunnicutt et al. (2006), Ban et al. (2007), Boyapalle et al. (2007), Chen and Siede (2007), Hashimoto and Valles (2007), Pal et al. (2007), Boyapalle et al. (2008), Marti et al. (2008), Ribière et al. (2008), Bonning and Miller (2010b), de Miranda et al. (2010a), and Kapun et al. (2010).

[b]Proposed additions pending ICTV approval.

[c]Taxonomic order and corresponding number of families, genera, and species (n:n:n) in which virus has been detected or isolated. After Bonning and Miller (2010b).

[d]Characteristic of many of these viruses is their prevalence in apparently healthy host populations as inapparent, asymptomatic infections. Therefore, "asymptomatic" could have been listed for all of the dicistroviruses. The pathological effects of infection listed are often observed after injection of virus particles into the host — an unnatural form of exposure. Natural events of disease progression are not completely characterized. Environmental (or other) factors may strongly influence the lethality of infection (Bailey, 1968; Chen and Siede, 2007; Bonning, 2009).

[e]Taxonomic order in the class Arachnida (Varroa destructor mite). Varroa destructor has been implicated as a mechanism of virus activation, vectoring, and host immune suppression. Whether these viruses actually replicate within the mites is currently unresolved.

[f]The tissue tropism under natural conditions is not known for many of the dicistroviruses. Some studies have noted virus-infected cells or tissues in insects exposed to the virus by injection into the hemocoel — an artificial method of virus acquisition. Thus, the natural tissue tropism of most of the viruses is largely unknown. There is a strong association of many dicistroviruses with the alimentary canal — a likely route of infection and dissemination (de Miranda et al., 2010a).

[g]Taxonomic order Decapoda; class Malacostraca.

[h]TSV was evaluated in human and monkey cell lines by two laboratories, with conflicting results. One laboratory reported that the cell lines supported replication of TSV (Audelo-del-Valle et al., 2003) while the other laboratory failed to support TSV on these cell lines (Pantoja et al., 2004).

FIGURE 5.2 (A) Purified preparations of *Solenopsis invicta virus 3* (SINV-3). Scale bar: 100 nm. *Solenopsis invicta* fire ant laboratory colony (B) two weeks and (C) eight weeks after infection with SINV-3. Brood and worker death were significant (> 90%) over the course of the infection.

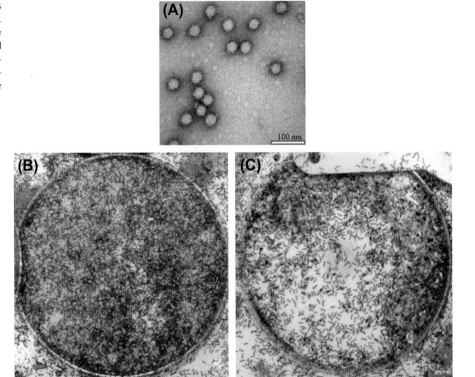

The number of viruses in the *Dicistroviridae* will undoubtedly continue to expand. In fact, many discoveries have been and will continue to be made with the metagenomics approach (Hunnicutt *et al.*, 2006; Valles *et al.*, 2008; Djikeng *et al.*, 2009). Large pyrosequencing projects of environmental samples have demonstrated that RNA viruses have a ubiquitous and highly prevalent presence in the environment (Culley *et al.*, 2006; Djikeng *et al.*, 2009).

Comprehensive reviews of the *Dicistroviridae* have been published recently, including molecular characterization, replication strategy (Bonning and Miller, 2010a, b; Bonning and Johnson, 2010), and taxonomic placement (Koonin *et al.*, 2008; Le Gall *et al.*, 2008; Chen *et al.*, 2011), which provide the most up-to-date information concerning this group. Furthermore, reviews concerned with dicistrovirus infections of social insects (honey bees and viral effects on the immune responses in ants) have been published recently (Ribière *et al.*, 2008; Schluns and Crozier, 2009; de Miranda *et al.*, 2010a). Thus, the reader is directed to these resources for specific information regarding the genome organization, molecular characteristics, virion structure, phylogenetics, host immune responses elicited by these viruses, and the relationship of these viruses with honey bees and ants.

Although outside the scope of this section, the reader must be aware of the fact that dicistroviruses also infect and cause considerable damage to beneficial arthropods. Several dicistroviruses are pathogenic to honey bees, *Apis*

mellifera (Bailey, 1968, 1982; Chen and Siede, 2007), one of which is IAPV, which has been found to be associated with honey bee CCD (Cox-Foster *et al.*, 2007). In addition, TSV, which infects penaeid shrimp species, regularly causes severe losses (50−100% mortality) to farm-raised shrimp (Hasson *et al.*, 1995; Bonami *et al.*, 1997; Mari *et al.*, 2002).

Among the viruses comprising the *Dicistroviridae*, nine infect pestiferous insects (Table 5.1) and, as a result, may find utility as microbial control agents (Christian *et al.*, 1992; Christian and Scotti, 1996). Although only a portion of the viruses in the *Dicistroviridae* infect pest insects, they are discussed collectively with beneficial insects to present a more complete understanding of the mechanisms of infection, replication, pathology, transmission, and host range of these viruses. Much additional research on these interesting viruses will be required to gain a complete understanding of their biology.

The dicistrovirus virions are roughly spherical with a particle diameter of approximately 30 nm and no envelope. The virions exhibit icosahedral, *pseudo* T = 3 symmetry and are composed of 60 protomers, each comprised of a single molecule of each of three structurally similar proteins, VP1, VP2, and VP3. Each capsid contains a beta-barrel arrangement of eight beta-sheets that make up a jelly roll structure. A smaller protein (< 5 kDa), VP4, is also present in the virions of some members and is located on the internal surface of the five-fold axis below VP1.

Virions contain a single molecule of infectious, linear, positive-sense, ssRNA of approximately 8500−10,000 nt. The genome is monopartite and dicistronic with two non-overlapping open reading frames (ORFs) of approximately 5500 and 2500 nt, respectively. A small genome-linked virus protein [viral protein genome (VPg)] is covalently attached to the 5′ end of the genome. The 3′ end of the viral RNA genome is polyadenylated (Fig. 5.1B).

The ORFs are separated by an intergenic UTR. The 5′-proximal ORF encodes a non-structural precursor which is auto-proteolytically cleaved into RNA helicase (Hel), cysteine protease with a chymotrypsin-like fold (Pro), and RdRp components. The VPg sequence is repeated in most dicistrovirus genomes in Hel-(VPg)n-Pro-RdRp module with the number of repeats being species dependent (Nakashima and Shibuya, 2006). The 3′-proximal ORF encodes the capsid proteins VP1, VP2, and VP3. A fourth smaller capsid protein (VP4) has also been reported in some species. In most species a protein precursor (VP0) is present which is cleaved to yield capsid proteins VP3 and VP4.

5.2.3. Flaviviruses

Flaviviridae is derived from the Latin word "flavus", meaning yellow, and referring to the *Yellow fever virus* transmitted by the mosquito *Aedes aegypti* which is the type species of the genus *Flavivirus*. Of the three genera in this family, *Flavivirus* is the only genus where only insects serve as hosts, and has historically been subdivided into three groups: the tick-borne flaviviruses (TBFV), the mosquito-borne flaviviruses (MBFV), and the flaviviruses with no known arthropod vector (NKV). Flaviviruses can also be categorized into different subgroups based on serological cross-reactivity or antigenic determinants, or into clusters, clades, and species according to molecular phylogeny, host species, vector, and associated diseases (Kuno *et al.*, 1998; Gaunt *et al.*, 2001; Gould *et al.*, 2001; Calisher and Gould, 2003). Recently, a group of insect-specific flaviviruses (ISV) has been isolated from mosquitoes that do not replicate in vertebrate hosts or cell lines but share traits with other members of the *Flavivirus*. The arthropod-borne viruses (arboviruses) within the genus *Flavivirus* that have vertebrate hosts are not considered insect pathogens, and therefore the focus of this section will be to review new information on the ISV.

The flaviviruses share remarkable similarities in virion properties, structure, and genome organization but vary in biological and antigenic properties (Kuhn *et al.*, 2002). Virions are spherical to pleomorphic with a diameter of 40−60 nm, and consist of an envelope and a nucleocapsid. The nucleocapsid exhibits icosahedral symmetry with a diameter of 25−30 nm. The surface projections of the envelope are made up of small spikes that are surrounded by a prominent fringe. The spikes are 6 nm long and constructed from two viral proteins: E (envelope) and M (membrane). Protein E is a glycosolated protein found on the outer surface of the lipid bilayered envelope and is the major antigenic determinant on virus particles. Protein E has been shown to be a viral hemagglutinin and the primary target for host neutralizing antibodies (Sanchez *et al.*, 2005; Stiasny *et al.*, 2007). Protein M spans the membrane and protects protein E as it is transported from the infected cell during replication.

The molecular mass of virions is 60×10^6 daltons (Da) and the virions sedimentation coefficient is 170−210 S_{20w} with a buoyant density in cesium chloride of 1.22−1.24 g/cm^3. Virions are stable in an alkaline environment of pH 8 under *in vitro* conditions but are sensitive to treatment with organic solvents and detergents. Virions are composed of 17% lipids by weight and they are derived from host cell membranes.

Flavivirus genomes consist of a linear, single-stranded, positive-sense, infectious RNA. In the case of type species *Yellow fever virus*, the complete genome is 10,233 nt long, which is translated as a single ORF into one large polypetitide of 3411 amino acids (Rice *et al.*, 1985). The structural genes encoding an envelope glycoprotein, a nucleoprotein, and a small membrane protein are located at the 5′ end of the genome. The 5′ terminus has a methylated nucleotide cap or a genome-linked protein (VPg). The non-structural proteins including serine protease, helicase, and RdRp are encoded at the 3′ end of the genome. The 3′ terminus has no poly (A) tail (Fig. 5.1C).

5.2.4. Iflaviruses

The *Iflaviridae* is a recently established virus family (Carstens and Ball, 2009) classified under the order *Picornavirales* (Le Gall *et al.*, 2008), and comprises six assigned species isolated from insects and a parasitic mite, all in the *Iflavirus* genus. *Iflavirus* is derived from the type species *Infectious flacherie virus* (IFV). Among the iflaviruses, four infect arthropods of pest status, including *Ectropis obliqua virus* (EoV), IFV, *Perina nuda virus* (PnV), and *Varroa destructor virus-1* (VDV-1). *Deformed wing virus* (DWV) and *Sacbrood virus* (SBV) infect honey bees. Two additional viruses, *Brevicoryne brassicae virus 1* (BrBV-1) and *Bee slow paralysis virus* (BsPV) (Ryabov, 2007; de Miranda *et al.*, 2010b), exhibit characteristics consistent with iflaviruses and are currently under consideration by the ICTV for inclusion in the *Iflaviridae*. A comprehensive review of the *Iflaviridae*, including taxonomy, genome organization, replication, pathology, and transmission, was published recently (van Oers, 2010).

Virions of iflaviruses are roughly spherical and exhibit icosahedral symmetry with a 26−30 nm diameter. They do not possess an envelope or distinctive surface structures.

Iflaviruses have a genome organization similar to viruses in the families, *Picornaviridae*, *Marnaviridae*, and *Secoviridae*, with a monopartite and monocistronic genome comprising one large, uninterrupted ORF. Virions contain one molecule of linear, infectious, positive-sense, ssRNA of 8800−10,100 nt. The genome contains a single large ORF encoding the capsid proteins at the 5′ end and the non-structural proteins at the 3′ end. The 5′ end of the genome is covalently linked to a small peptide, VPg, which plays an important role in RNA replication (Fig. 5.1D).

Phylogenetic analysis of the RdRp region shows that iflaviruses form a distinct monophyletic clade distantly related to the *Picornaviridae*, *Secoviridae*, and *Dicistroviridae* (Chen *et al.*, 2011). In addition, the position of the smallest capsid protein VP4 in iflaviruses is located in the second position of the capsid precursor coding region, a feature that differs from other viral families within the *Picornavirales*.

The replicases of iflaviruses resemble those of the dicistroviruses, picornaviruses, marnaviruses, and secoviruses by containing sequences with homology to an RNA helicase (Hel), a chymotrypsin-like 3C protease (Pro), and an RdRp in the 5′ to 3′ orientation. The capsid proteins, arranged in the order VP2−VP4−VP3−VP1, are preceded by a short leader protein (L). It is unclear whether there are any conserved RNA secondary structures in the 5′ and 3′ UTRs.

5.2.5. Tetraviruses

The *Tetraviridae* is a family of viruses whose host range is restricted to lepidopteran insects. The family name is derived from T = 4 icosahedral capsid symmetry that distinguishes the family from other non-enveloped viruses (Finch *et al.*, 1974). Since the description of the first tetravirus, *Nudaurelia β virus* (NβV), which caused an epizootic outbreak in larvae of the emperor pine moth, *Nudaurelia cytherea capensis*, four additional species have been discovered in infected larvae and named *Nudaurelia α, γ, δ,* and ε *virus* (Grace and Mercer, 1965; Hendry *et al.*, 1968). A species physically similar to NβV but different by possession of a second genomic RNA strand was isolated from *Nudaurelia* larvae and named *Nudaurelia ω virus* (NωV) (Hendry *et al.*, 1985). So far, there are 12 recognized and 11 unassigned viruses in the family. The tetraviruses are divided into two genera, *Betatetravirus* and *Omegatetravirus*, on the bases of the sequences and morphological features of their capsid proteins and the number of ssRNA segments that comprises their genomes. Of these assigned members, 10 species are grouped within the genus *Betatetravirus* and three species are classified within the genus *Omegatetravirus* (Agrawal and Johnson, 1992; Gordon *et al.*, 1995). NβV and NωV are the best characterized members of the family *Tetraviridae* and are the type species of the *Betatetravirus* and *Omegatetravirus* genera, respectively. While tetraviruses exhibit different serological relationship patterns, serological evaluations have not been the most reliable approach for distinguishing tetraviruses (Grace and Mercer, 1976; King and Moore, 1985; Hanzlik *et al.*, 1993; Yi *et al.*, 2005).

Virions consist of a round, unenveloped capsid, with quasi-icosahedral symmetry (T = 4) and a 40 nm diameter (Finch *et al.*, 1974; Agrawal and Johnson, 1992). The capsid shell is composed of 240 copies of two protein subunits of approximately 60 kDa and 8 kDa. Structural studies have revealed that betatetravirus capsids display three characteristic pits with a distinct groove on each side in high resolution electron micrographs (Olson *et al.*, 1990), a feature that is not visible in omegatetravirus capsids.

Currently, the complete genome sequences are available for four betatetraviruses including NβV, *Euprosterna elaeasa virus* (EeV), *Thosea asigna virus* (TaV), and *Providence virus* (PrV) (Gordon *et al.*, 1999; Pringle *et al.*, 1999; Walter *et al.*, 2010). Betatetraviruses, such as NβV, have monopartite, linear, positive-sense, ssRNA genomes of approximately 7000 nt (Fig. 5.1E-a). For omegatetraviruses, the capsid gene sequence is known for the type species, NωV, and the complete genome sequences for *Dendrolimus punctatus tetravirus* (DpTV) and HaSV have been determined (Hanzlik *et al.*, 1993, 1995; Yi *et al.*, 2005). Omegatetraviruses have two segmented, bipartite genomes that are linear, positive-sense, ssRNA. RNA segment 1 and RNA segment 2 are 5312−5492 nt and 2445−2490 nt, respectively. For both genera, the 5′ terminus of the RNA genome is capped and the 3′ terminus of the genome is not polyadenylated and may be folded into tRNA-like secondary structures (Gordon *et al.*, 1995; Hanzlik *et al.*, 1995; Zeddam *et al.*, 2010) (Fig. 5.1E-b).

5.2.6. Cypoviruses

The *Reoviridae* is a family of viruses that can infect an extremely broad range of hosts including vertebrates, invertebrates and plants. The name "*Reoviridae*" is derived from respiratory, enteric, orphan viruses. *Reoviridae* consists of 15 genera that were ratified by the ICTV to be grouped into two subfamilies *Spinareovirinae* and *Sedoreovirinae* based on the presence or absence of a "turret" protein on the inner capsid, respectively.

Cypovirus [= cytoplasmic polyhedrosis virus (CPV)] is a genus in the *Spinareovirinae* subfamily; viruses from this genus have been isolated from insects only, primarily in the Lepidoptera, Diptera, and Hymenoptera. Since the first report of occluded CPV in the silkworm, *Bombyx mori* (Ishimori, 1934), there have been more than 250 CPVs described (Hukuhara and Bonami, 1991). The classification of CPVs is based on the electrophoretic migration patterns of their dsRNA genome segments on polyacrylamide gels and the host species from which the viruses were originally isolated (Payne *et al.*, 1977; Shapiro *et al.*, 2005; Y. Li *et al.*, 2007). Among 20 CPVs that have been identified in the genus *Cypovirus* (Mori and Metcalf, 2010), 14 CPV types have been recognized by ICTV and two tentative species have also been assigned to the group. Type 1 *Cypovirus*, represented by *Bombyx mori CPV-1* (BmCPV-1), is the type species for the genus (Index of Viruses, http://www.ictvdb.org/ICTVdB/descindex.htm). The validity of this classification has been confirmed by cross-hybridization analyses of dsRNA, serological comparisons of cypovirus proteins, and sequence analysis of genome segments in the coding regions and terminal regions (Payne and Rivers, 1976; Mertens *et al.*, 2004a, b; Graham *et al.*, 2006).

CPVs are unique among the *Reoviridae* in that their capsid is composed of a single shell of protein and they produce polyhedrin protein that crystallizes in the cell cytoplasm. Virions are non-enveloped and are embedded within the polyhedrin matrix, forming a polyhedral occlusion body, within which the single or multiple enveloped virus particles become occluded (Hill *et al.*, 1999; Zhang *et al.*, 1999). The symmetry of polyhedral occlusion bodies is icosahedral, and the polyhedrin protein is arranged as a face-centered cubic lattice. The non-occluded viral particles have the shape of spheres with a diameter of approximately 60 nm. The capsid is round and exhibits an icosahedral symmetry. The capsid surface structure reveals a regular pattern with 12 turret-like spikes 20 nm in length and 15−23 nm in width at the five-fold axis positions. The capsomer arrangement is clearly visible in cryoelectron microscopy (Miura *et al.*,

1969). CPVs are morphologically similar to *Baculovirus* [nucleopolyhedroviruses (NPVs)], a genus in the *Baculoviridae*, but are distinct in that they have a multipartite dsRNA genome and replicate in the cytoplasm of the infected cells, whereas the baculoviruses have a circular dsDNA genome and replicate in the nucleus. There is no significant homology between cypovirus and baculovirus polyhedrins at the amino acid sequence level.

CPV genomes are multipartite, consisting of 10 segments of linear dsRNAs encoding 10−12 proteins. A copy of all segments is packaged within each single-shelled, icosahedral, turreted viral particle. The RNA segments range in size from 1 to 4.2 kb and the complete genome is 20,500−24,000 nt depending on the virus. Each genome segment contains a single ORF. The complete nucleotide sequence of dsRNA genomes has been determined for several cypoviruses including BmCPV-1 (Ikeda *et al.*, 1998; Hagiwara and Matsumoto, 2000; Hagiwara *et al.*, 2002), *Dendrolimus punctatus CPV-1* (Zhao *et al.*, 2003), *Helicoverpa armigera CPV-5* (Belloncik *et al.*, 1996; Li *et al.*, 2006; Tan *et al.*, 2008), and *Trichoplusia ni CPV-15* (Rao *et al.*, 2003). The availability of these cypovirus sequences has provided insight into the roles of each dsRNA segment in the process of the virus infection. The dsRNA segments 1, 2, 3, 4, 6, and 8, in the case of BmCPV-1 (strain 1), encode structural capsid proteins VP1 (148 kDa), VP2 (136 kDa), VP3 (140 kDa), VP4 (120 kDa), VP6 (64 kDa), and VP7 (31 kDa), respectively. The dsRNA segment 10 encodes a polyhedrin protein of 25−37 kDa. The dsRNA segments 5, 7, and 9 encode functional proteins RdRp, p101, p44, and NS5, respectively (McCrae and Mertens, 1983; Ikeda *et al.*, 1998; Hagiwara and Matsumoto, 2000) (Fig. 5.3).

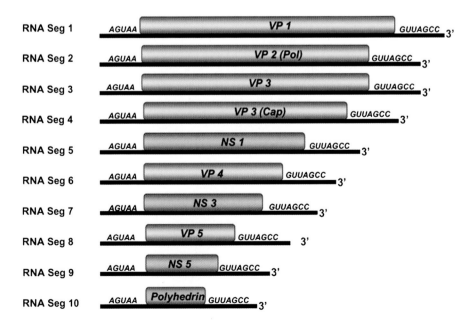

FIGURE 5.3 Schematic representation of cypovirus genomic structure (*Bombyx mori cypovirus-1*, BmCPV-1). The monomeric genome consists of 10 linear double-stranded RNA segments packaged as exactly one copy of each segment within a single virus particle. Segments 1, 2, 3, 4, 6, and 8 encode structural capsid proteins VP1, VP2, VP3, VP4, VP6, and VP7, respectively. Segments 5, 7, and 9 encode functional proteins. Segment 10 encodes a polyhedrin protein. The 5′-terminus and 3′-terminus of each segment have conserved nucleotide sequences, AGUAA and GUUAGCC, respectively. The 3′ end of genome segments has no poly(A) tract.

5.3. INFECTION, REPLICATION, PATHOLOGY, TRANSMISSION, AND HOST RANGE

5.3.1. Alphanodaviruses

Most knowledge acquired concerning alphanodavirus infection in insects has come from studies of NoV infection in wax moth larvae, *Galleria mellonella*. Experimental inoculation of wax moth larvae with NoV led to localized lesions in the cytoplasm of muscles, nerve, salivary gland, and molting glands (Garzon *et al.*, 1978). Larvae became paralyzed at four to six days postinoculation and died in seven to 14 days (Garzon *et al.*, 1978). Further studies showed that NoV infection resulted in impairment of the mitochondrial structure and function in infected larvae (Garzon *et al.*, 1990). It was also reported that NoV had lethal effects in honey bees (Bailey and Scott, 1973). While nodaviruses have been isolated from other insects including various grass grub species, scarab beetles, mosquitoes, and *Drosophila*, limited data are available for the pathogenicity of BBV, BoV, FHV, and PaV *in vivo*.

NoV was originally isolated from mosquitoes (*Culex tritaeniorhynchus*) in Nodamura, Japan (Scherer and Hurlbut, 1967). Pigs in the same locality tested positive for antibodies to NoV, suggesting that pigs may be a natural host of the virus (Scherer *et al.*, 1968). In addition to *Cx. tritaeniorhynchus*, infection of NoV has been identified in other mosquito species including *Ae. aegypti*, *Ae. albopictus*, and *Toxorhynchites amboinensis*, without causing apparent disease symptoms (Scherer and Hurlbut, 1967; Tesh, 1980). NoV is unique among alphanodaviruses in that it is considered pathogenic to both invertebrates and vertebrates. In addition to wax moth larvae, infectivity assays show that NoV could cause a lethal infection in adult honey bees (Bailey and Scott, 1973). The host range for NoV is not limited to insects; the virus could infect suckling mice and result in tissue abnormalities and, in later stages, paralysis and death (Scherer, 1968; Murphy *et al.*, 1970). The virus is transmissible to suckling mice by infected *Ae. aegypti*, and the multiplication of NoV has been observed in suckling mice and suckling hamsters (Scherer *et al.*, 1968). *In vitro* experiments show that NoV replicates in a wide range of cell lines including cultured mosquito cells from *Ae. pseudoscutellaris*, *Ae. aegypti*, and *Ae. albopictus*, and the hamster kidney cell, BHK21 (Ball *et al.*, 1992). When NoV genomic RNAs were introduced into cell cultures by transfection, they caused no cytopathic effect but the production of maturing progeny viral particles could be readily seen in many cell types in a wide range of species (Ball and Johnson, 1998). Despite the fact that its potential as a biological control agent remains to be demonstrated, NoV has provided an

excellent model system for studying RNA virus replication, virion structure, assembly, and factors that impact virus infection (Johnson, 2008).

FHV is one of the best studied members of the *Alphanodavirus*. It was originally isolated from the larvae of grass grubs, *Costelytra zealandica*, collected near the Flock House agricultural station in New Zealand, hence the name (Scotti *et al.*, 1983). Infection of FHV has been documented in Coleoptera, Lepidoptera, Diptera, and Hemiptera. When different genera of adult mosquitoes (*Aedes*, *Culex*, *Anopheles*, and *Armigeres*) were orally fed or injected with FHV, FHV antigen was detected in various tissues (fat bodies, salivary glands, and head tissue), and the infectious FHV was recovered from inoculated hosts (Dasgupta *et al.*, 2003). Active replication of FHV has been observed in major tissues of the tsetse fly, *Glossina morsitans morsitans* (Dasgupta *et al.*, 2007). In addition, FHV could be grown in mosquito cell cultures, *Drosophila* cell lines (Dasgupta *et al.*, 1994, 2003), and mammalian cells (Ball *et al.*, 1992). Replication assays have shown that FHV replicates well in cell line C6/36, derived from *Ae. albopictus*, and Schneider's DL cells, derived from *Drosophila melanogaster*.

Despite its insect origin, FHV can infect and multiply in hosts from different kingdoms, including plants and yeasts (Selling *et al.*, 1990; Price *et al.*, 1996; Santi *et al.*, 2006). Yeast cells have been shown to support complete replication of FHV and the transfection of yeast with FHV genomic RNA can lead to RNA replication and production of infectious FHV virions (Price *et al.*, 1996). The inoculation of plants with genomic RNA or intact virions of FHV can result in synthesis of progeny virions containing newly synthesized RNA and coat protein. The results indicated that the intracellular environment of plants can support FHV RNA replication, virion assembly, and movement within the plant (Selling *et al.*, 1990; Dasgupta *et al.*, 2001). As with NoV, FHV also possesses the smallest genome known for animal viruses and is able to replicate in a wide variety of cell lines. As a result, it has been a valuable model system for studying various aspects of virology, including virus replication, virion structure, assembly, and suppressors of RNA silencing to counteract host antiviral immunity (Kopek *et al.*, 2007; Johnson, 2008; Odegard *et al.*, 2010).

The early phase of virus infection is characterized by the absorption of viral particles to the surface of their target cells and the penetration of the virus genome into the host cells. However, the molecular processes associated with nodavirus attachment and entry are not yet well understood. The direct interactions between viral capsid proteins and specific receptors on the host cell cytoplasmic membrane remain unidentified. Although viral entry remains uncharacterized, a mechanism for translocation of alphanodavirus genomes into host cells has been proposed. It has been suggested that low endocytic pH and peptide-triggered disruption of the endosomal membrane are preludes to the nodavirus uncoating process. Low endocytic pH has been shown to promote the autoproteolytic cleavage of the capsid protein precursor to release lipophilic γ-peptides that have a membrane permeabilization function, ultimately leading to the release of viral genomic RNA into the host cell cytoplasm (Bong *et al.*, 1999; Odegard *et al.*, 2009). These studies also showed that the cleavage of coat protein in purified provirions of FHV was accompanied by a five- to eight-fold increase in specific infectivity (Odegard *et al.*, 2009).

Replication of nodaviruses takes place in the cytoplasm of infected cells. The bipartite genome carries information necessary for replication (RNA 1) of the viral genome and packaging (RNA 2) of the genetic material into the new virus (Friesen and Rueckert, 1981; Gallagher *et al.*, 1983; Johnson *et al.*, 2001; Li *et al.*, 2002). Both genomic segments, RNA 1 and RNA 2, are packaged in the same virion and are required for infection. Upon delivery of viral genomic RNA into the host cell's cytoplasm, RNA 1 functions as mRNA and is translated to produce protein A which contains the RdRp that catalyzes replication of the viral genome. A negative-sense viral RNA is synthesized using the genomic RNA as template. The negative-sense RNA is then used as a template to synthesize new genomic RNA and subgenomic RNA 3 that is derived from the 3′-proximal region of RNA 1. RNA 3 is not packaged into viral particles and transcription of the subgenomic RNA 3 produces protein B2 (Friesen and Rueckert, 1982). Protein B2 plays a vital role in coordinating the synthesis of genomic RNA 1 and RNA 2. The synthesis of protein A and protein B2 starts immediately after infection, reaches a maximum at 5 h and 6−10 h postinfection, respectively, and declines abruptly thereafter (Friesen *et al.*, 1980; Friesen and Rueckert, 1981). The replication of RNA 1 and RNA 2 represses synthesis of subgenomic RNA 3 ensuring the accumulation of an approximately equal ratio of genomic RNAs. Despite the presence of an equal molar ratio of RNA 1 and RNA 2 throughout infection, the translation of RNA 2 predominates and the major product produced during infection is capsid protein α. Translation of RNA 2 during infection yields abundant levels of the capsid precursor protein α. The assembled capsid protein α is subsequently cleaved into capsid protein β and protein γ during maturation of the particle, an essential reaction for infectivity (Schneemann *et al.*, 1992; Venter *et al.*, 2010).

A characteristic feature of alphanodaviruses is the RNA interference (RNAi) suppressive protein B2. It has been shown that protein B2 from FHV and NoV suppresses the host's RNAi response by binding dsRNA as well as siRNA in a sequence-independent manner (Chao *et al.*, 2005; Lu *et al.*, 2005; Sullivan and Ganem, 2005). The suppression of RNAi by protein B2 can therefore enhance virus replication and pathogenicity. Furthermore, the interactions between host RNAi antiviral defense and activities of viral RNAi suppressors can lead to a persistent viral infection. The persistent infection of alphanodaviruses has been described in several *Drosophila* cell lines and the cell lines that developed persistent infections were found to be resistant to superinfection (Dasgupta *et al.*, 1994; Czech *et al.*, 2008; Flynt *et al.*, 2009).

The composition and structure of genomes may have important implications for genetic variation of the viruses. As with other viruses that have multiple genome segments that are packaged in the same virion, alphanodaviruses readily undergo exchange of genetic information through reassortment of RNA genomic segments. Viruses have been generated by reassortment of RNA 1 and RNA 2 molecules from possible combinations of FHV, BBV, and BoV in cultured cells. However, the resulting progeny virions containing genome segments from NoV and FHV were not infectious (Gallagher, 1987).

Large numbers of insect and mammalian cell lines are susceptible to nodavirus replication (Ball and Johnson, 1998). The cell lines that support productive infection of nodaviruses include cells derived from mosquito species, *Ae. aegypti*, *Ae. albopictus*, *Ae. pseudoscutellaris*, and *T. amboinensis*, *Drosophila* DL1 and DL2 cells, and various mammalian cell lines such as BHK21 cells. A latent nodavirus closely related to FHV was also detected in a cabbage looper (*Trichoplusia ni*) cell line, BTI-TN-5B1-4 (Hi5) (T. C. Li *et al.*, 2007). While no characteristic cytopathic effect was observed in most infected cells, viruses including NoV, FHV, and BBV multiply well in cultured cells, as evidenced by the accumulation of large numbers of viral particles or crystalline arrays in membrane-bound structures inside cells and detection of viral antigen by enzyme-linked immunosorbent assay. In addition, viral replication can occur if genomic RNAs are introduced into a wide range of cells by transfection (Selling *et al.*, 1990; Ball, 1992; Ball *et al.*, 1992). The virulence and transmissibility of alphanodaviruses have not yet been fully elucidated.

5.3.2. Dicistroviruses

Consequences of infection by dicistroviruses vary somewhat depending upon the host and, most certainly, additional environmental circumstances. Some common characteristics of infection do appear to be emerging for these viruses (Table 5.1). First, and perhaps incongruously, many dicistroviruses are present in apparently normal, healthy arthropod populations as chronic−asymptomatic infections (Chen and Siede, 2007; Bonning, 2009; Bonning and Miller, 2010a). Under proper and, as yet, unknown conditions, these chronic−asymptomatic infections can rapidly become acute and lethal (Valles *et al.*, 2004;

de Miranda *et al.*, 2010a). Dicistrovirus infections of honey bees offer an excellent illustration of this conversion. It has long been thought that environmental factors (e.g., external stressors) induce changes in the virus' lethality to the host (Bailey, 1967). Dicistrovirus prevalence and distribution among honey bee populations can be normally quite high (Bailey, 1967; Tentcheva *et al.*, 2004; Berenyi *et al.*, 2006) and co-infections by multiple viruses are common (Evans, 2001; Chen *et al.*, 2004b). However, symptoms and/or mortalities are observed inconsistently (Chen and Siede, 2007; Cox-Foster *et al.*, 2007; Cox-Foster and vanEngelsdorp, 2009). The conditions necessary for the chronic—asymptomatic to acute—lethal conversion are not known currently and offer an interesting area for future research.

Pathology associated with dicistrovirus infection varies with virus and host. For example, the two most common and extreme manifestations of infection are no observable symptoms (i.e., asymptomatic) and mortality (Table 5.1). Virulence is strongly influenced by the mode of viral acquisition; injection of virus into the host hemocoel (an unnatural form of transmission) typically results in rapid mortality. *Varroa* ectoparasitic mite infestations are associated with a more lethal form of several of the viral infections in honey bees, which may stem from injection of virus particles during feeding by these mites. A brief description of the gross pathological effects of each of the dicistroviruses (ostensibly from the primary host) follows.

Cricket paralysis virus (CrPV) was first observed in laboratory colonies of the field cricket, *Teleogryllus oceanicus* (Reinganum *et al.*, 1970). Early- to mid-instar nymphs were most susceptible to the virus, with adults exhibiting a degree of resistance. The virus spreads rapidly among domesticated cricket colonies with nearly 100% mortality. Affected crickets exhibited uncoordinated behavior concomitant with paralysis of their hind legs, which was followed by death.

Aphid lethal paralysis virus (ALPV) infections of *R. padi* (Williamson *et al.*, 1988) produce symptoms similar to CrPV. Affected aphids stopped feeding and moved away from their food source. They exhibited uncoordinated movement and paralysis, followed by death.

Black queen cell virus (BQCV) was isolated from honey bee queen larvae and pupae sealed in their cells (Bailey and Woods, 1977). Infected bee larvae exhibited a yellow, tough, sac-like integument (Chen and Siede, 2007), and before death the larvae and the cell walls in which they are contained turn black. The prevalence of BQCV has been linked with the microsporidium *Nosema apis* (Bailey, 1981), and the virus may be vectored by *Varroa* mites (Bailey, 1976). Recent conflicting studies raise doubt about the role *Varroa* plays in disease transmission (Tentcheva *et al.*, 2004; Chantawannakul *et al.*, 2006).

Drosophila C virus (DCV) induces a cytopathic effect in cultured insect cells. In *D. melanogaster*, the virus is associated with changes in developmental time and fecundity and is the cause of occasional bouts of significant mortality among laboratory fly cultures (Christian and Scotti, 1998). Gomariz-Zilber *et al.* (1995) reported that fly death occurred once a threshold level of viral particles was reached.

Himetobi P virus (HiPV) was serendipitously discovered during purification of rice stripe virus, a rice disease vectored by the small brown ("himetobi" is Japanese for "small brown") planthopper, *Laodelphax striatellus* (Toriyama *et al.*, 1992). The virus causes a chronic, asymptomatic infection of the midgut and hindgut in the planthopper (Suzuki *et al.*, 1993).

Plautia stali intestine virus (PSIV) was discovered from diseased brown-winged green bugs, *Plautia stali* (Nakashima *et al.*, 1998). PSIV has been reported to be lethal after being fed to healthy bugs (Nakashima *et al.*, 1998). However, it appears that the virus is most commonly observed as a chronic—asymptomatic infection causing decreased aphid longevity and fecundity (D'Arcy *et al.*, 1981b).

Rhopalosiphum padi virus (RhPV) was discovered in laboratory colonies of *R. padi* and *Schizaphis graminum* (D'Arcy *et al.*, 1981b). RhPV was shown to decrease significantly the longevity and fecundity of infected aphids (D'Arcy *et al.*, 1981a) and to induce cytopathic changes in posterior midgut and hindgut cells that included loss of ribosomes and formation of intracellular vesicles (Gildow and D'Arcy, 1990). A cell line derived from the glassy-winged sharpshooter, *Homalodisca vitripennis* (formerly *H. coagulata*), appeared to be permissive to infection by transfection with RhPV RNA (Boyapalle *et al.*, 2007). Recognition of the potential use of RhPV as a microbial control agent has led to the development of an infectious transcript (Boyapalle *et al.*, 2008) and a baculovirus-driven clone expressed in lepidopteran cells (Pal *et al.*, 2007). The RhPV genome has been sequenced and characterized (Moon *et al.*, 1998).

Triatoma virus (TrV) was discovered in laboratory colonies of *Triatoma infestans*, an important insect vector of the protist, *Trypanosoma cruzi*, the causative agent of Chagas disease (François-Xavier *et al.*, 2010). TrV infection of *T. infestans* causes rear leg paralysis (Muscio *et al.*, 1987), inhibition of the molting process, decreased longevity and fecundity (Rozas-Dennis and Cazzaniga, 2000), and mortality (Muscio *et al.*, 1987).

Homalodisca coagulata virus 1 (HoCV-1) is the most recent addition to the *Dicistroviridae* (Carstens, 2010). Metagenomic analysis of the transcriptome of *H. vitripennis* (=*H. coagulata*) led to the discovery of this virus (Hunter *et al.*, 2006). Pathological effects of HoCV-1 infection have not been fully described, although host mortality was associated with the virus.

Acute bee paralysis virus (ABPV) was discovered from diseased honey bees (Bailey *et al.*, 1963). Infection of adult bees may be asymptomatic or result in paralysis, uncoordinated movement, and death (Bailey and Gibbs, 1964). Among dead and dying honey bees, ABPV was present in brain tissue in high quantity (Furgala and Lee, 1966; Bailey and Milne, 1969).

Taura syndrome virus (TSV) was discovered in Ecuadorian marine shrimp (*Penaeus vannamei*) and is considered one of the most economically damaging viral diseases of shrimp aquaculture (Jimenez, 1992; Dhar *et al.*, 2010). Mortality rates among farm-raised shrimp typically range between 50 and 90% (Bonami *et al.*, 1997). Preceding mortality, infected shrimp exhibit a range of symptoms including melanized lesions of the cuticle and cellular lesions characterized by hemocytic infiltration and melanization (Hasson *et al.*, 1995). In addition, epithelial necrosis has been observed within the stomach, appendages, gills, and cuticle (Dhar *et al.*, 2010).

Kashmir bee virus (KBV) was first known from diseased Asian honey bee (*Apis cerana*) populations in Kashmir, India (Bailey and Woods, 1977). The virus attacks all honey bee stages and persists as a chronic, asymptomatic infection (Hornitzky, 1982; Anderson and Gibbs, 1988; Anderson, 1991). In laboratory settings, the virus can be highly virulent, causing significant mortality (Chen and Siede, 2007). Disease symptoms have not been defined. Conversion from a chronic−asymptomatic to acute-lethal infection has been reported to occur in the presence of *Varroa* mites (Bailey *et al.*, 1979). KBV has been detected in the saliva of these mites, suggesting that they may vector the virus (Shen *et al.*, 2005a). Furthermore, the mites have been shown to suppress the immune response of honey bees, making them more vulnerable to KBV infection (Shen *et al.*, 2005b). KBV has been associated with honey bee CCD, but no causal relationship between the disorder and virus has been demonstrated (Todd *et al.*, 2007).

As with HoCV-1, *Solenopsis invicta virus 1* (SINV-1) was discovered by large-scale sequencing of the *S. invicta* transcriptome (Valles *et al.*, 2004, 2008). SINV-1 infects all developmental stages of *S. invicta* (Hashimoto and Valles, 2007; Hashimoto *et al.*, 2007). Laboratory colonies of *S. invicta* infected with SINV-1 may exhibit significant mortality in larval and adult stages. Large midden piles containing workers, larvae, and pupae are often observed in SINV-1-infected laboratory colonies. However, it appears that the majority of SINV-1 infections of fire ants are of the chronic−asymptomatic type. Mortality in the field has not been reported. Intercolony SINV-1 prevalence can be quite high (> 50%) and is temperature dependent (Valles *et al.*, 2007, 2010).

Israeli acute paralysis virus (IAPV) was isolated from honey bees in Israel (Maori *et al.*, 2007a). Honey bee workers infected with the virus exhibited shivering wings, followed by paralysis, and then death (typically occurring outside the hive) (Maori *et al.*, 2007b). IAPV has been linked to CCD, but no causal relationship between the disorder and virus has been established (Cox-Foster *et al.*, 2007). However, in recent studies, short interfering RNA targeting a portion of the IAPV genome provided some protection to the bees from the virus (Hunter *et al.*, 2010). As with KBV and ABPV, *Varroa* mites could serve as a vector of IAPV to uninfected members of the colony (Di Prisco *et al.*, 2011).

In order for the dicistroviruses to cause the pathologies illustrated above, viral pathogenesis must begin by host−virus contact and the virus has to contact the target tissue or cell and receptor to gain entry and initiate the infection. These events are often linked closely to host behavior. For many of the dicistroviruses, natural host exposure probably occurs through ingestion as these viruses are associated strongly with the alimentary canal (de Miranda *et al.*, 2010a). Indeed, 12 of the 14 dicistroviruses exhibit tissue tropism toward some part of the alimentary canal (Table 5.1). In all likelihood, multiple cell or tissue types support replication of the viruses during pathogenesis. The oral route of infection is well illustrated in social insect hosts, honey bees, and fire ants. BQCV and ABPV in honey bees were shown to be more prevalent in colonies during the spring and summer months (Bailey *et al.*, 1981). The seasonal phenology has been hypothesized to stem from the accumulation of honey bee fecal deposits containing BQCV and ABPV on the combs during the overwintering stage (Chen *et al.*, 2006). In the spring, housekeeping bees are thought to acquire and disseminate the viruses throughout the colony, most likely by trophallaxis. This mechanism of active virus spread (by cleaning) and increased colony growth during warmer periods are thought to facilitate virus−host (intracolony) infections.

A similar mode of viral acquisition and dissemination has been proposed for SINV-1 and its host, the red imported fire ant (Valles *et al.*, 2004; Oi and Valles, 2009). Intracolony and intercolony SINV-1 prevalence declines precipitously in the winter, only to rebound strongly in the spring and summer (Valles *et al.*, 2007, 2010). SINV-1 exhibits a strong tropism for cells of the midgut of *S. invicta* larvae and adult workers (Hashimoto and Valles, 2007). More precisely, SINV-1 is found primarily in fire ant midgut contents and epithelial cells [based on quantitative polymerase chain reaction (QPCR) for SINV-1 genome].

A model for colony dissemination of SINV-1 has emerged in which SINV-1 replicates in cells of the midgut/hindgut epithelial layer where the virus particles are somehow released into the gut lumen. From there they are available to leave the alimentary canal through the anus by defecation (workers) or oral cavity by trophallaxis (larvae and workers) (Oi and Valles, 2009). Thus, normal colony behaviors result in dissemination of viral particles throughout the colony. Furthermore, *S. invicta* larvae may be serving as reservoirs for SINV-1 for the entire colony. *Solenopsis invicta* larvae have a blind gut requiring feeding and waste elimination through the oral opening and mediated by worker ants (Petralia and Vinson, 1979). *Solenopsis invicta* larvae were reported to contain in excess of 1×10^8 SINV-1 particles, with a large proportion contained in the midgut contents (Hashimoto and Valles, 2007). Consequently, as *S. invicta* larvae are tended by workers, SINV-1 particles are passed to other members of the colony.

The mechanism by which SINV-1 midgut epithelial cells constantly produce viral particles and release them into the gut lumen without apparently harming the host is unclear. However, Cherry and Perrimon (2004) reported that DCV production and release from fat body cells occurred non-cytolytically without affecting the integrity of the infected cells. Non-cytolytic release of some related picornaviruses has also been demonstrated from other cell types (Tucker *et al.*, 1993) and TrV viral particles have been reported in midgut contents of the host, *T. infestans* (Muscio *et al.*, 1988), as well as RhPV in its host, *R. padi* (Gildow and D'Arcy, 1990).

Further evidence for the significant role of the alimentary canal in horizontal transmission of dicistroviruses is illustrated by the presence of viral particles in excreta of infected hosts. In fact, excreta has been reported to be an important source of viral inoculum for PSIV (Nakashima *et al.*, 1998), HiPV (Guy *et al.*, 1992), TrV (Muscio *et al.*, 2000), BQCV, ABPV (Bailey *et al.*, 1981), and KBV (Hung, 2000). In addition, host gut contents have been shown to contain high numbers of viral particles (or genome equivalents) for HiPV (Suzuki *et al.*, 1993), SINV-1 (Hashimoto and Valles, 2007), DCV (Lautié-Harivel, 1992), and TrV (Muscio *et al.*, 1987, 2000). Likewise, the alimentary canal (fecal—oral route) plays a prominent role in the infection process of many of the *Picornaviridae* (Racaniello, 2006).

Horizontal transmission of dicistrovirus infections also occurs orally among some plant-feeding insects, often with the plant host playing a key role in virus dissemination. During the feeding process by *R. padi*, RhPV particles are introduced into the plant host from salivary gland secretions (Ban *et al.*, 2007). These RhPV particles are transported throughout the plant's vascular system, which then effectively serves as a virus reservoir (Gildow and D'Arcy, 1988). RhPV-uninfected aphids that subsequently feed on the contaminated plant host can acquire the virus and become infected (Ban *et al.*, 2007). The plant does not serve as a host to RhPV because the virus does not replicate in the plant, but rather, serves only as a repository for RhPV. The duration for which RhPV particles are detected in the plant is transient and, therefore, unlikely to serve as a long-term reservoir of RhPV (Ban *et al.*, 2007). A similar host—plant—virus relationship was associated with

HoCV-1 and its host, *H. vitripennis* (= *H. coagulata*) (Hunnicutt *et al.*, 2008). Conversely, HiPV was not detected or translocated in plant host tissue (Guy *et al.*, 1992).

Horizontal dicistrovirus transmission may also occur by vectoring. The prevalence of KBV in honey bees has been reported to coincide with presence of the parasitic mite *V. destructor* (Brodsgaard *et al.*, 2000; Todd *et al.*, 2007). Honey bee pupae were later shown to contain higher levels of KBV (RNA and structural proteins) when *V. destructor* was present (Shen *et al.*, 2005b). A strong linear relationship between the number of mites present in a colony and levels of KBV viral RNA was also demonstrated. Later, KBV was detected in mites and their saliva, illustrating the importance of vectoring in disease transmission (Shen *et al.*, 2005a). However, in addition to physically vectoring viruses, mite infestations influence the ability of honey bees to mount an immune response. *Varroa destructor* was associated with a decrease in antimicrobial peptides and overall immune health in honey bees (Gregory *et al.*, 2005; Yang and Cox-Foster, 2005).

The potential for dicistroviruses being vectored was also examined in the imported fire ant, and one of its natural enemies (*Pseudacteon* spp. phorid flies) (Porter, 1998). During the oviposition process, adult female flies inject eggs into the thorax of worker ants (Porter, 1998). Because egg laying by female flies is an intrusive and repetitive action and flies develop within the ant host while consuming internal tissues, it was hypothesized that the *Pseudacteon* parasitoids that developed in SINV-1-infected ants may have been capable of harboring this virus and potentially vectoring it to other fire ant colonies (Valles and Porter, 2007). In addition, flies may acquire the virus during the oviposition process and mechanically transmit virus to subsequent worker ants. However, laboratory and field experiments showed that SINV-1 did not replicate in any of the phorid fly species examined (*P. obtusus*, *P. litoralis*, and *P. curvatus*). Oviposition by flies among SINV-1-infected and -uninfected fire ant workers showed that SINV-1 mechanical transmission was unlikely (Valles and Porter, 2007).

In addition to horizontal transmission, many dicistrovirus hosts are capable of vertically transmitting the virus to their progeny. Transmission may occur by direct zygote infection (transovarial) or by egg surface contamination (transovum). Vertical transmission via these routes has been reported for ALPV (Hatfill *et al.*, 1990; Laubscher and von Wechmar, 1992), PSIV (Nakashima *et al.*, 1998), RhPV (Ban *et al.*, 2007), TrV (Rozas-Dennis and Cazzaniga, 2000), and SINV-1 (Valles *et al.*, 2004). Conversely, HiPV does not appear to be transovarially transmitted (Guy *et al.*, 1992).

Once ingested, dicistroviruses must encounter a target cell/receptor and gain entry for their RNA genome. There is a dearth of information concerning this portion of dicistrovirus pathogenesis, but some interesting progress and conclusions concerned with cell entry and its relationship with mortality among infected hosts are emerging from investigations with DCV. Ingestion is likely a route of transmission for DCV; however, whether this method of acquisition leads to mortality is unclear (Gomariz-Zilber et al., 1995; Thomas-Orillard et al., 1995). Strains of *D. melanogaster* exhibit differential susceptibility to DCV by injection (Plus et al., 1978; Thomas-Orillard et al., 1995; Cherry and Perrimon, 2004). This strain-dependent susceptibility had a strong maternal connection and linkage to a gene(s) on chromosome III (Thomas-Orillard et al., 1995). Recent evidence with DCV suggests that the endocytotic pathway was essential for DCV cell entry, infection, and pathogenesis (Cherry and Perrimon, 2004). Cherry and Perrimon (2004) reported that DCV entry (and infection) into fat body cells of *D. melanogaster* was prevented by administration of drugs that inhibited endocytosis. They also showed that flies with mutations in certain genes of clathrin-mediated endocytosis (e.g., α-adaptin) were resistant to DCV infection and had decreased production of viral antigens *in vivo*. These experiments may explain why *D. melanogaster* susceptibility to DCV varies by strain (Cherry and Perrimon, 2004). Cell entry by endocytosis was also observed in RhPV (Gildow and D'Arcy, 1990). Genetic differences inherent to virus "strains" undoubtedly influence their virulence (Gomariz-Zilber et al., 1995; Valles and Strong, 2005; Palacios et al., 2008).

Temperature may play an important role in dicistrovirus pathogenesis. Honey bees infected with ABPV appeared healthy when maintained at 35°C, but died when held at 30°C (Bailey and Milne, 1969). Conversely, a positive correlation was associated between temperature and prevalence of ALPV (Laubscher and von Wechmar, 1992) and SINV-1 (Valles et al., 2007, 2010). Further indirect support implicating a temperature dependency includes increased viral prevalence during warmer seasons and in warmer regions (Plus et al., 1975). Few reports examining this temperature dependency at the molecular level are available. However, Cevallos and Sarnow (2010) showed that exposure to elevated temperatures greatly increased CrPV viral protein and RNA production, yet appeared to inhibit virion formation in *Drosophila* S2 cells. They concluded that temperature dependency of host susceptibility to virus infection in poikilotherms may be more pronounced than in warm-blooded animals (Cevallos and Sarnow, 2010). Thus, viral replication may be affected directly by temperature changes.

Host range is an important consideration when evaluating the potential usefulness of an organism as a microbial control agent. A narrow host range is ideal to maintain host specificity and minimize impact to non-target organisms. Table 5.1 provides a summary of the host range of the dicistroviruses. In addition to providing the total number of

species in which each virus has been detected or isolated, taxonomic placement of each of these hosts illustrates the breadth of range. *Dicistroviridae* is a rapidly expanding group and information about these viruses continues to be reported. As such, host range data for most of the dicistroviruses are quite limited. Further hampering elucidation of host range is the inability to mass produce pure virus in culture for use in host specificity tests. Thus, the actual host range of each of the viruses may be considerably wider than is known currently.

CrPV, has been isolated from, or detected in, 22 species of insects from five insect orders and 11 families, exhibiting one of the widest host ranges of any insect virus (Christian and Scotti, 1998). The pathology associated with CrPV infection in each of these potential hosts is not known. While CrPV exhibits an exceptionally wide host range, most of the dicistroviruses appear quite limited in the number of hosts they infect. Typically, infection is limited to a single taxonomic order or family. For example, DCV infects 16 species of flies, all in the genus *Drosophila* (Kapun et al., 2010). Similarly, extensive field surveys for SINV-1 among nearly 2000 arthropod specimens revealed a host range limited to several species in the genus *Solenopsis* (Valles et al., 2007).

The replication strategy of the *Dicistroviridae* has not been elucidated, but is likely to be very similar to the *Picornaviridae* (Bonning and Miller, 2010a, b). The viral coat plays a crucial role in the infection process. It must provide protection for the RNA genome from the severity of the environment (e.g., variable pH, proteases), yet be capable of releasing the genome under the appropriate conditions (Hogle and Racaniello, 2002).

Viral entry into the cell is likely to be mediated by a dicistrovirus receptor (Dvr) on the host's target cell (Fig. 5.4). Clathrin-mediated endocytosis is not involved in cell entry in some *Picornaviridae* (e.g., poliovirus), but has been shown to play an essential role in rhinovirus (DeTulleo and Kirchhausen, 1998) and DCV infection and pathogenesis (Cherry and Perrimon, 2004). Surface differences in viral coat protein 1 (VP1) suggest different receptor types for the *Picornaviridae* and *Dicistroviridae* (Tate et al., 1999). Dvr binding probably simultaneously initiates endocytosis and conformational changes in the dicistrovirus capsid, permitting release of the RNA genome into the cytoplasm (Fig. 5.4) (Rossmann et al., 2000). RhPV was reported to gain cell entry by endocytosis in *R. padi* (Gildow and D'Arcy, 1990). Once the viral genome is released into the host cell cytoplasm, it serves a dual purpose as transcript for polyprotein synthesis and template for production of complementary minus-strand copies of the genome that, in turn, serve as templates for production of the infectious plus strand genome. During this time, host cellular rearrangement occurs, including the formation of a variety of membranous vesicles (Gildow and D'Arcy,

FIGURE 5.4 Depiction of the general infection and replication processes of dicistroviruses in host cells as they are currently thought to occur. Replication occurs in the cytoplasm and no DNA stage is formed. Viral docking (1) and entry (2) into the cell (CM = cell membrane) are probably mediated by a dicistrovirus receptor (Dvr) on the host's target cell. Dvr binding is likely to initiate cellular endocytosis and conformational changes in the dicistrovirus capsid, permitting release of the RNA genome into the cytoplasm (Gildow and D'Arcy, 1990; Rossmann *et al.*, 2000). Cellular rearrangement (3) is induced after cell entry including the formation of a variety of membranous vesicles (Gildow and D'Arcy, 1990) likely to facilitate viral component trafficking and replication (Mackenzie, 2005). These vesicular structures (or replication complexes; Kopek *et al.*, 2007) are thought to increase replication efficiency by providing a scaffolding to concentrate viral components during replication and assembly (Wileman, 2006). Once the viral genome is released into the host cell cytoplasm, it serves a dual purpose as transcript for polyprotein synthesis (4) and template for production of complementary minus-strand copies of the genome (5) that, in turn, serve as template for production of the infectious plus strand genome. Viral assembly (6) occurs by encapsidation of a plus strand genome by the viral coat proteins. The mechanisms by which mature virions are released (7) from dicistrovirus-infected host cells are not known. However, among the *Picornaviridae*, this process has been reported to occur cytolytically and non-cytolytically (e.g., by exocytosis) (Tucker *et al.*, 1993; Agol, 2002).

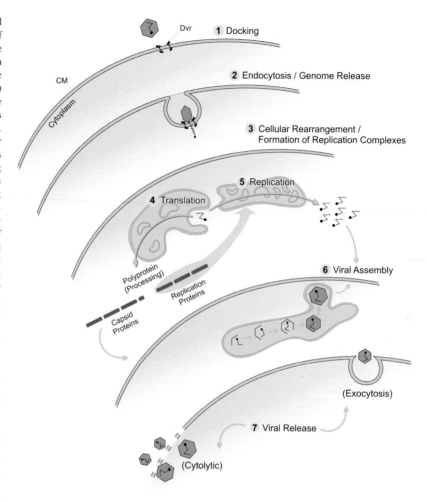

1990) likely to facilitate viral component trafficking and replication (Mackenzie, 2005). These vesicular structures are thought to increase replication efficiency by providing a scaffolding to concentrate viral components during replication and assembly (Wileman, 2006). Many of these processes may be regulated by phosphorylation events (Jakubiec and Jupin, 2007). Further usurpation of cellular processes includes the inhibition of host translation (Garrey *et al.*, 2010) while viral RNA cap-independent translation continues from the 5′ and intergenic IRES (Wilson *et al.*, 2000; Jan and Sarnow, 2002; Pestova and Hellen, 2003). Gildow and D'Arcy (1990) showed that midgut cells of *R. padi* infected with RhPV progressively lose ribosomes and other organelles. Recent evidence suggests that some picornaviral protein components interact directly with host nuclear proteins to shut down certain cellular processes (Weidman *et al.*, 2003). Viral assembly occurs by encapsidation of a plus strand genome by the viral proteins (Fig. 5.4). Mechanisms by which mature virions are released from dicistrovirus-infected host cells are not

known. However, among the *Picornaviridae*, this process has been reported to occur cytolytically and non-cytolytically (e.g., by exocytosis) (Tucker *et al.*, 1993; Agol, 2002). Perhaps the mechanism by which the virus is released from the host cell (either destructively or not) is related to the chronic—asymptomatic/acute—lethal characteristic exhibited by many of the dicistroviruses.

5.3.3. Flaviviruses

The genus *Flavivirus* contains viruses vectored by mosquitoes and ticks and includes well-known members such as dengue virus and West Nile virus as well as those with no known vectors (Thiel *et al.*, 2005). These arboviruses are not normally considered as insect pathogens and will not be covered in this chapter. However, there are several insect-specific flaviviruses (ISVs) that have been studied since the original isolation of cell fusing agent virus (CFAV) isolated from an *Ae. aegypti* cell line (Stollar and Thomas, 1975). CFAV remained the only reported insect-specific flavivirus until isolation of *Kamiti River virus* (KRV) from *Ae. mcintoshi* in Kenya (Crabtree *et al.*, 2003; Sang *et al.*, 2003). The nucleotide sequence of KRV was closely related to CFAV (Cook *et al.*, 2006). KRV can be grown in mosquito cell culture but not in vertebrate cells or mice. An isolate closely related to CFAV and KRV has been identified from *Culex* spp. mosquitoes in Japan (Hoshino *et al.*, 2007) and named *Culex flavivirus* (CxFV).

These ISVs share traits with other members of the *Flavivirus* genus in that they are single-stranded, positive-sense RNA viruses with icosahedral virions of approximately 50 nm and genomes of 10−11 kb encoding a single ORF that produces three structural proteins and seven non-structural proteins (Cammisa-Parks *et al.*, 1992; Crabtree *et al.*, 2003; Hoshino *et al.*, 2007). Genetic comparisons of the ISVs recognize three main groups (CFVA, KRV, and CxFV), which are considered to represent the CFVA complex within the genus *Flavivirus* (Kim *et al.*, 2009).

One finding of significance for this complex is the identification of sequences closely related to CFAV and KRV integrated into the genomes of laboratory and wild-type *Ae. aegypti* and *Ae. albopictus* as well as the C6/36 cell line (Crochu *et al.*, 2004). These authors have suggested that these sequences became integrated into these genomes as a result of infections by CFAV and/or KRV. If confirmed, this could have important evolutionary implications as it would represent a novel mechanism for horizontal gene transfer in eukaryotes (Kim *et al.*, 2009).

Natural infections with a member of the CFAV complex may influence the susceptibility and competence of the mosquito host to arboviruses such as dengue and West Nile virus (Farfan-Ale *et al.*, 2009; Kim *et al.*, 2009). The susceptibility and refractoriness of mosquitoes to arboviruses are generally considered to be controlled by genetic factors, but the widespread occurrence of CFAV complex members in natural mosquito populations (Farfan-Ale *et al.*, 2009; Kim *et al.*, 2009; Pabbaraju *et al.*, 2009) suggests that superinfection is another factor that should be investigated.

Specific studies on the replication strategy for ISVs have not been undertaken but is likely to be similar to other members of the group (Thiel *et al.*, 2005). Cell entry is by receptor-mediated endocytosis with transcription and replication occurring in the cytoplasm. In C6/36 cells, CxFV virions were shown to accumulate in the endoplasmic reticulum (Hoshino *et al.*, 2007; Kim *et al.*, 2009).

Cytopathic effects for these CFAV complex viruses have only been observed in cell culture. CFAV was originally isolated from an *Ae. aegypti* cell line and recognized by its ability to cause cell fusion and CPE in an *Ae. albopictus* cell line (Stollar and Thomas, 1975). Cell fusion did not occur until 60 h after inoculation and was not extensive until after 72 h. Although CFAV caused no cytopathic effects or cell fusion in the *Ae. aegypti* cell line, the titer of

CFAV increased. KRV also caused prominent cytopathic effects in C6/36 (*Ae. albopictus*) cells (Sang *et al.*, 2003). CxFV did not cause severe cytopathic effects in C6/36 or AeA1-2 cells (*Ae. albopictus*), indicating that ISVs can replicate avirulently in mosquito cells (Hoshino *et al.*, 2007).

The inability of ISVs to infect and replicate in vertebrate cells implies that there is no vertebrate reservoir and other mechanisms for spread and persistence are likely. Some authors have suggested that ISVs may be vertically transmitted in nature because KRV was isolated from larvae and pupae of *Ae. mcintoshi* (Sang *et al.*, 2003), and CxFV has been isolated from asymptomatic adult males and females (Hoshino *et al.*, 2007). If adult mosquitoes infected with an ISV can obtain a blood meal and develop eggs, it is possible that the virus could be transmitted to progeny. Laboratory studies with KRV in *Ae. aegypti* have confirmed that this virus is horizontally and vertically transmitted (Lutomiah *et al.*, 2007). KRV is horizontally transmitted to larvae when virions are ingested and resulted in infection rates of between 74 and 96% in male and female adults, respectively (Lutomiah *et al.*, 2007). In addition, adults became infected when fed KRV via blood meals using an artificial membrane system. Female adults infected either as larvae or as adults did vertically transmit KRV to both male and female progeny but at relatively low rates of between 2.8 and 4.5% (Lutomiah *et al.*, 2007). It is therefore likely that horizontal and vertical transmissions are important mechanisms for the persistence and spread of ISVs within and between mosquito species.

Evidence thus far suggests that the CFAV complex of flaviviruses is widely distributed in *Aedes* and *Culex* populations from many different regions of the world. The original CFAV isolate is only known from cell culture but a closely related strain isolated from Puerto Rico (CFAV Culebra strain) was found in a variety of mosquito species including *Ae. aegypti*, *Ae. albopictus*, and two *Culex* species (Cook *et al.*, 2006). KRV has thus far only been isolated from natural populations of *Ae. mcintoshi* from Kenya (Sang *et al.*, 2003) and transmitted in the laboratory to *Ae. aegypti* (Lutomiah *et al.*, 2007). Isolates of the CxFV group have been found in *Cx. pipiens*, *Cx. tritaeniorhynchus*, and *Cx. quinquefasciatus* in Japan and Indonesia (Hoshino *et al.*, 2007), *Cx. quinquefasciatus* in Guatemala (Morales-Betoulle *et al.*, 2008) and Mexico (Farfan-Ale *et al.*, 2009), *Cx. quinquefasciatus* and *Cx. restuans* in the USA and Trinidad (Kim *et al.*, 2009), and *Cx. tarsalis* from Canada (Pabbaraju *et al.*, 2009).

5.3.4. Iflaviruses

Iflaviruses have been isolated from arthropods including insect species in the Lepidoptera and Hymenoptera, and arachnid species in the Mesostigmata (Table 5.2). Many of

TABLE 5.2 Members of the *Iflaviridae* and Their Hosts[a]

Virus (Acronym)	Host Status	Host Range[d]	Pathology and/or Manifestation Thereof[e]	In Vitro Propagation	Tissue Tropism[g]
Genus: *Iflavirus*					
Deformed wing virus (DWV)	Beneficial	Hymenoptera (2:2:5) Mesostigmata[f] (2:2:2)	Malformed wings/appendages, mortality, learning deficits, paralysis	None available	Brain, wings, thorax, legs, hemolymph, gut epithelial cells, sperm
Kakugo virus (KV)[b]	Beneficial	Hymenoptera (1:1:1) Mesostigmata[f] (1:1:1)	Altered behavior (aggressiveness)	None available	Brain
Ectropis obliqua virus (EoV)	Pest (agriculture)	Lepidoptera (1:1:1)	Granulovirus infection, mortality	None available	Undetermined
Infectious flacherie virus (IFV)	Beneficial, pest (agriculture)	Lepidoptera (2:2:2)	Lethal diarrhea	None available	Midgut epithelial cells
Perina nuda virus (PnV)	Pest (agriculture)	Lepidoptera (1:1:1)	Delayed development	Cell culture	Undetermined
Sacbrood virus SBV	Beneficial	Hymenoptera (1:1:2) Mesostigmata[f] (1:1:1)	Failure to pupate, ecdysial fluid accumulation, larval color change (pale to yellow)	None available	Fat body, undetermined
Varroa destructor virus 1 (VDV-1)	Pest (agriculture)	Mesostigmata[f] (1:1:1)	None reported	None available	Gastric caecum
Bee slow paralysis virus (BsPV)[c]	Beneficial	Hymenoptera (1:1:1) Mesostigmata[f] (1:1:1)	Paralysis of rear legs	None available	Systemic
Brevicoryne brassicae virus 1 (BrBV-1)[c]	Pest (agriculture)	Hemiptera (1:1:1)	None reported	None available	Undetermined

[a]Information acquired from Bailey et al. (1964), Lee and Furgala (1967), Inoue and Ayuzama (1972), Choi (1989), Bailey and Woods (1974), Bowen-Walker et al. (1999), Wang et al. (1999, 2004a, b), Wu et al. (2002), Ongus et al. (2004), Fujiyuki et al. (2005, 2006, 2009), Fievet et al. (2006), Chen and Siede (2007), Iqbal and Mueller (2007), Ryabov (2007), Zhang et al. (2007), de Miranda and Fries (2008), Terio et al. (2008), Shah et al. (2009), and de Miranda et al. (2010b).

[b]Classification of KV is currently unresolved. KV may simply be a geographical variant of DWV (Lanzi et al., 2006).

[c]Proposed additions pending ICTV approval.

[d]Taxonomic order (class Insecta or Arachnida) and corresponding number of families, genera, and species in which virus has been detected or isolated.

[e]Characteristic of many of these viruses is their prevalence in apparently healthy host populations as inapparent, asymptomatic infections. The pathological effects of infection listed are often observed after injection of virus particles into the host — an unnatural form of exposure. Environmental factors may strongly influence the lethality of infection (Bailey, 1968; Chen and Siede, 2007).

[f]Taxonomic order in the class Arachnida (includes primarily Varroa destructor, but Tropilaelaps mercedesae has also been identified as a possible vector of DWV (Dainat et al., 2009; Forsgren et al., 2009; Varroa destructor has been implicated as a mechanism of virus activation, vectoring and host immune suppression (Shen et al., 2005a; Yang and Cox-Foster, 2005). DWV has been reported to replicate within V. destructor (Yue and Genersch, 2005; Ongus et al., 2006). Whether DWV actually replicates within T. mercedesae is currently unresolved. Although a close association between KV, SBV, BsPV, and Varroa mites has been established, whether these viruses actually replicate within the mites is currently unresolved.

[g]The tissue tropism under natural conditions is not known for many of the iflaviruses. Some studies have noted virus-infected cells or tissues in insects exposed to the virus by injection into the hemocoel — an artificial method of acquisition. Thus, the natural tissue tropism of most of the viruses is largely unknown.

the iflaviruses infect beneficial insects, including the honey bee and silkworm. As in the *Dicistroviridae*, chronic—asymptomatic infections are also observed for many of the iflaviruses with conversion to acute—lethal infections as a result of exposure of the host to stress (Bowen-Walker *et al.*, 1999). Pathological symptoms are typically exhibited above certain threshold levels of iflavirus concentrations (Bowen-Walker *et al.*, 1999). As with dicistroviruses, the environmental or host changes that ostensibly initiate viral replication and subsequent exhibition of symptoms are unknown. However, among honey bee colonies, parasitic mite infestations (especially *Varroa* sp.) feature prominently in the horizontal transmission and activation of the iflaviruses (Lanzi *et al.*, 2006).

Deformed wing virus (DWV) and *Kakugo virus* (KV) are considered together because they exhibit nucleotide identities commensurate with variability levels typically observed among isolates (or strains) of the same virus (Lanzi *et al.*, 2006). Arguments for the possibility that these two viruses are simply geographical variants of the same virus have been posited (Lanzi *et al.*, 2006).

DWV has been implicated as an etiological agent responsible for declines of honey bee colonies (Highfield *et al.*, 2009; de Miranda and Genersch, 2010). Several symptoms have been reported to be associated with DWV infection of honey bees including deformed wings, altered abdominal shape, paralysis, reduced life span, and learning deficits (Lanzi *et al.*, 2006; Iqbal and Mueller, 2007; de Miranda and Genersch, 2010). DWV has been detected in a number of developmental stages including eggs, larvae, pupae, and workers (Chen *et al.*, 2004a, 2005).

KV was discovered while investigators were conducting differential display analyses of brain tissue to identify gene expression patterns in aggressive honey bees (Fujiyuki *et al.*, 2004). KV infection of honey bee workers is associated with the exhibition of aggressive behavior; however, a causal link has not been established. KV may systemically infect honey bees, but the brain appears to be the primary tissue infected (Fujiyuki *et al.*, 2009). Genome nucleotide sequences of different isolates of KV exhibited up to 99% similarity (Fujiyuki *et al.*, 2006; Berenyi *et al.*, 2007). Comparisons between genome regions of KV and DWV likewise exhibited approximately 97% nucleotide sequence similarity, suggesting a strong phylogenetic relationship between the viruses (Fujiyuki *et al.*, 2006). Indeed, this relationship was further established by detection of virus sequences from honey bees with intermediate characteristics between KV and DWV (Terio *et al.*, 2008). Complicating this relationship further is the high nucleotide sequence similarity of VDV-1 with DWV and KV (Fujiyuki *et al.*, 2006; Lanzi *et al.*, 2006). However, DWV does not appear to induce aggressive behavior in honey bees as does KV (Rortais *et al.*, 2006), and extensive phylogenetic analysis among globally collected DWV, KV, and VDV-1

samples supports genetic segregation of these viruses (Berenyi *et al.*, 2007). Discovery, molecular characterization, and verification of replication of VDV-1 in *V. destructor* have been reported (Ongus *et al.*, 2004, 2006), but symptoms associated with, or caused by VDV-1 infection have not been defined. Further investigation will be required to establish the relationship among these viruses.

Ectropis obliqua virus (EoV) was discovered from tea loopers, *Ectropis obliqua*, in China, co-infected with a granulovirus (Wang *et al.*, 2004b). The virus is associated with a lethal granulovirus infection of larvae, but specific symptoms of infection have not been defined. The EoV genome has been sequenced and characterized (Wang *et al.*, 2004a).

Infectious flacherie virus (IFV) is considered an etiological agent responsible, perhaps partially, for infectious flacherie disease of the silkworm (Aizawa *et al.*, 1964; Choi *et al.*, 1989). IFV is often found in conjunction with a densovirus (Tanada and Kaya, 1993). The disease is characterized by lethal diarrhea in silkworm larvae, resulting in significant economic losses to the silk industry. IFV has also been shown to infect *Glyphodes pyloalis*, a pyralid pest of mulberry (Watanabe *et al.*, 1988). The IFV genome has been sequenced and phylogenetic relationships evaluated (Isawa *et al.*, 1998).

Perina nuda virus (PnV) infects larvae of the *Ficus* transparent wing moth, *Perina nuda*, and possibly causes symptoms similar to IFV (Wang *et al.*, 1999). PnV appears to produce an antagonistic effect on *P. nuda* NPV infections in the moth (Wang *et al.*, 1999). The complete genome sequence of PnV has been elucidated (Wu *et al.*, 2002), but symptoms for the disease have yet to be described.

Sacbrood virus (SBV) infects workers and larvae of honey bees. SBV infections of workers are primarily asymptomatic, possibly reducing the lifespan of those infected (Bailey and Fernando, 1972). Two-day-old larvae appear to be most vulnerable to infection. Larvae infected with SBV become pale yellow, the larval cuticle becomes leathery, and a large amount of SBV-containing fluid accumulates between the thickened integument and the body, which prevents the larva from successfully pupating (Bailey *et al.*, 1964; Chen and Siede, 2007). Dead larvae become dark and brittle and serve as a source of further virus inocula for the colony. SBV infection is strongly associated with *Varroa* mite infestations (Shen *et al.*, 2005a; Chantawannakul *et al.*, 2006). The genome of SBV was sequenced by Ghosh *et al.* (1999).

Bee slow paralysis virus (BsPV) is not a formal member of the *Iflaviridae*. It does possess genome characteristics consistent with iflaviruses and is under consideration for inclusion by the ICTV (de Miranda *et al.*, 2010b). The virus was first described by Bailey and Woods (1974) from infected honey bees. Injection of BsPV into honey bees resulted in paralysis of the hind legs and subsequent mortality (Bailey and Woods, 1974). Oral transmission is likely to occur and *Varroa* mites are thought to play an important role in virus dissemination throughout honey bee colonies (Santillán-Galicia *et al.*, 2010). The BsPV genome was published by de Miranda *et al.* (2010b).

Brevicoryne brassicae virus 1 (BrBV-1) also exhibits characteristics consistent with iflaviruses and is under consideration for inclusion in the *Iflaviridae* by the ICTV (Ryabov, 2007). BrBV-1 was discovered from the cabbage aphid, *B. brassicae*, by employing a method to amplify viral nucleic acids from a small number of aphids (Ryabov, 2007). The pathology of BrBV-1 has not been described. An undescribed parasitoid wasp of *B. brassicae* appears to support replication of BrBV-1 (E. Ryabov, pers. comm.).

The tissue tropism of iflaviruses is not as consistent among the group as has been shown for dicistroviruses (Table 5.2). Although the fecal—oral route of transmission appears to play an important role in the acquisition and dissemination of many of the iflaviruses (Yue and Genersch, 2005), the alimentary canal is not featured as prominently as an infected tissue. At present, the tissue tropism for PnV, EoV, and BrBV-1 has not been determined. KV has been shown to be localized in the brains of worker honey bees (Fujiyuki *et al.*, 2005, 2009; Shah *et al.*, 2009), whereas DWV and BsPV were found in nearly all honey bee tissues (Fievet *et al.*, 2006; de Miranda *et al.*, 2010b). IFV and VDV-1 were found localized to the midgut and gastric caecum, respectively (Inoue and Ayuzawa, 1972; Inoue, 1974; Zhang *et al.*, 2007).

Among the viruses in *Iflaviridae*, transmission routes and consequences of infection by differing routes are probably best characterized and illustrated for DWV (de Miranda and Genersch, 2010). In iflaviruses infecting honey bees (DWV, SBV, KV), ectoparasitic mites are important vectors and appear to play a crucial role in their apparent virulence. Two mites, *Tropilaelaps mercedesae* and *V. destructor*, have been reported to serve as vectors of DWV (Yue and Genersch, 2005; Dainat *et al.*, 2009; Mockel *et al.*, 2011). The majority of reports examining the possible replication of DWV in the mites concluded that they do support the replication of the virus (Yue and Genersch, 2005; Dainat *et al.*, 2009; Gisder *et al.*, 2009). However, Santillán-Galicia *et al.* (2008) reported that DWV was found in the midgut and feces of *V. destructor* but did not replicate. It appears that the presence of DWV in honey bee colonies devoid of parasitic mite infestations represents a persistent, covert (chronic—asymptomatic) infection (de Miranda and Genersch, 2010). In contrast, when DWV and *Varroa* mites occur simultaneously, overt symptoms develop, including pupal death and emergence of worker bees with deformed wings (de Miranda and Genersch, 2010). Although not as well characterized, *Varroa* mite infestations appear to influence expression of

disease progression of SBV similarly (Tentcheva *et al.*, 2004; Shen *et al.*, 2005a; Chantawannakul *et al.*, 2006). Horizontal transmission of DWV may also occur by acquisition of the virus by nurse bees through cannibalization of infected pupae and subsequent dispersal by trophallaxis and other behaviors (Mockel *et al.*, 2011). DWV has also been shown to be vertically transmitted to progeny by infection of eggs (from DWV-infected queens) or sperm (from DWV-infected drones) (Yue *et al.*, 2007; de Miranda and Fries, 2008).

Disease progression and severity are influenced in PnV and IFV by co-infections with other viruses. As stated earlier, IFV is often found in conjunction with a densovirus (Tanada and Kaya, 1993) and PnV is found with *P. nuda* NPV (Wang *et al.*, 1999). Thus, "flacherie" disease is thought to be caused by mixed viral infections in each respective host.

The host range of the iflaviruses appears fairly restricted (Table 5.2). However, this observation may be simply a function of the limited number of studies examining the host range of these viruses. DWV, KV, and SBV may replicate in ectoparasitic mites of honey bees. Aside from this developing relationship, only DWF and IFV have been reported infecting multiple hosts (Watanabe *et al.*, 1988; Genersch *et al.*, 2006).

Very little is known about virus replication in *Iflaviridae*, but the close phylogenetic relationship with other positive-strand RNA viruses suggests a similar mode of replication (Fig. 5.4). Iflavirus entry into the cell is likely to be mediated by a receptor on the host's target cell. Following receptor binding, the genome enters the cytoplasm of the host cell, where it serves a dual purpose as a transcript for polyprotein synthesis and a template for production of complementary minus-strand copies of the genome. The minus strand, in turn, serves as a template for production of the infectious plus strand genome. Plus- and minus-strand genome copies are synthesized by the viral-encoded RdRp. In EoV, RNA synthesis occurs in a primer- and poly-(A)-dependent manner *in vitro* (Lin *et al.*, 2010). Viral assembly occurs by encapsidation of a plus-strand genome by the viral proteins. The mechanisms by which mature virions are released from iflavirus-infected host cells are not known. Unique IRES structures that enhance translation have been identified in the 5′ UTRs of VDV-1 (Ongus *et al.*, 2006), EoV (Lu *et al.*, 2006, 2007), and PnV (Wu *et al.*, 2007).

5.3.5. Tetraviruses

Tetravirus replication occurs in the foregut and midgut but is mostly seen in the midgut cells of infected larvae. Studies of HaSV infection in larvae of *H. armigera* have yielded important information on tetravirus-induced pathology (Bawden *et al.*, 1999; Brooks *et al.*, 2002). Like other tetraviruses, HaSV generally targets the midgut tissues of the larval lepidopteran host. When HsSV was injected directly into the hemocoel of larvae, viral replication could be observed only in the midgut; no virus was detected in other tissues (Brooks *et al.*, 2002). The outcome of HaSV infection varies from an inapparent to acute lethal infection depending on the age of host. When the virus was introduced into the first three larval instars, infection was manifested by cessation of feeding, stunted growth, and sometimes the eventual death of the host. Larval death was usually accompanied by discoloration, flaccidness, and subsequent liquefaction of the internal organs. Similar signs and symptoms were also reported in NβV-, AeV-, and Dp-V infected larvae (Grace and Mercer, 1965; Tripconey, 1970; Greenwood and Moore, 1984). In contrast, infection of HaSV in late-stage larvae did not lead to overt pathology and no apparent signs and symptoms of disease were found in infected hosts (Christian *et al.*, 2001). The sloughing of infected midgut cells followed by regeneration of stem cells has been suggested as an adaptive immune mechanism that may be involved in host resistance to the effective infection of the viruses (Gordon and Waterhouse, 2006).

Study of *Tetravirus* replication has been hindered by the lack of an efficient virus culture system. Previous studies on the tetraviruses have mainly been focused on the molecular characterization of their virion structure, genome organization, and capsid protein assembly. In contrast, little is known about virus—host interactions involved in viral replication.

Tetraviruses replicate in the cytoplasm of infected cells. The monopartite genome of betatetraviruses such as NβV contains two ORFs with the 5' proximal ORF encoding the multidomain RNA replicase of 204 kDa and the 3' proximal ORF encoding capsid precursor (VCAP) of 66.4 kDa. Both ORFs overlap by 1517 bp. Virions also encapsidate a 2.5 kb subgenomic RNA, hybridized to the 3'-capsid region of the larger genomic RNA. The betatetraviruses utilize the subgenomic RNA for translation of the capsid precursor protein (Gordon *et al.*, 1999). For TaV and EeV (Pringle *et al.*, 1999; Zeddam *et al.*, 2010), there is a 2A-like processing site in the N-terminal of the VCAP. The activity of 2A sites leads to production of a small ORF encoding a protein of 17 kDa (p17) and a large ORF encoding the capsid protein precursor of 65 kDa. For PrV, there are two 2A-like processing sites near the N-terminal of VCAP. The activity of these 2A-like sites results in the production of two ORFs encoding small peptides of 7 kDa (p7) and 8 kDa (p8), along with a large ORF encoding capsid protein precursor of 68 kDa. However, the 2A-like site is absent in NβV.

The omegatetraviruses possess a bipartite genome that is segmented into two segments, RNA 1 and RNA 2. RNA 1 (5.2−5.5 kb) contains one ORF encoding the replicase and three small ORFs encoding proteins p11, p15, and p8

(Gordon *et al.*, 1995; du Plessis *et al.*, 2005; Yi *et al.*, 2005). In HaSV, all three small ORFs overlap with the 3' end of the replicase, while in DpTV, only ORFs encoding p11 and P15 overlap with the replicase. The bicistronic RNA 2 (2.4−2.5 kb) contains two ORFs encoding proteins of molecular weight approximately 17 kDa (p17) and the second encoding a capsid precursor (70 kDa). The ORF encoding p17 partially overlaps with the ORF of the capsid protein precursor. Both genomic RNAs are encapsidated within a single particle. The function of p17 is unclear.

For both betatetraviruses and omegatetraviruses, the viral particles are assembled from a single capsid protein precursor that is autoproteolytically cleaved at its C-terminus during maturation, leading to production of the β- and γ-peptides of approximately 60 kDa and 8 kDa, respectively (Agrawal and Johnson, 1995). The cleavage of the capsid protein precursor results in the stabilization of the mature viral particle. This stabilization appears to be important for viral infectivity at the early phase of infection. The exact functions of small peptides released by autoproteolysis at the 2A-like processing sites remain largely unknown, but may be involved in regulatory functions in the virus life cycle.

While capsid proteins appear to be conserved among members of the family, the sequences of tetravirus replicases have diverged widely and cluster into three distinct groups. The first group includes betatetravirus NβV, omegatetraviruses HaSV and DpTV, and belongs to supergroup III (alphavirus-like) cluster by having highly conserved enzymic domains organization, methyltransferase (Mtr)—helicase (Hel)—RdRp. The second group includes the betatetraviruses TaV and EeV and has some similarities with supergroup I (picorna-like) viruses by lacking both Mtr and Hel domains in their replicases. The RdRp of PrV appears to form a third phylogenetic group and clusters with members of the families *Tombusviridae* and *Umbraviridae*, representing a divergent lineage in the family *Tetraviridae* (Dorrington and Short, 2010).

A systematic survey of approximately 1000 diseased insects from different orders showed that tetraviruses have a very narrow host range that is restricted to the Lepidoptera, mainly from Saturniidae, Limacodidae, and Noctuidae (Greenwood and Moore, 1982). Efforts of artificial inoculation with a high titer of tetraviruses showed that the viruses failed to replicate in animals other than lepidopteran larvae (Kalmakoff and McMillan, 1985).

HaSV was originally isolated from larvae of *H. armigera*, a pest of many important cultivated crops. In addition, HaSV has been shown to infect other species in the Heliothinae including *H. zea*, *H. virescens*, and *H. punctigera* populations in the USA and Australia (Christian *et al.*, 2001; Brooks *et al.*, 2002).

Despite extensive efforts to explore a wide range of cell types for tetravirus replication other than PrV, no tetravirus has yet displayed the ability to grow in cell culture. PrV was originally isolated from an infected cell line (MG8) derived from the midgut tissue of the corn earworm, *H. zea* (Pringle *et al.*, 2003). The virus from persistently PrV-infected MG8 cells also grew well in *H. zea* fat body cells. The infectivity of PrV in a cell line (Se-1), derived from beet armyworm embryos, *Spodoptera exigua* (Pringle *et al.*, 2003), represents the first case of tetravirus that is able to replicate in culture cells and will facilitate study of tetravirus molecular biology.

The successful infection of HaSV in plant protoplasts was achieved when plasmids carrying full-length cDNA copies corresponding to the HaSV genome segments of RNA 1, RNA 2, and P71 were used for transfection (Gordon *et al.*, 2001). The HaSV also grew to high titer in laboratory-reared *H. armigera*.

In NωV, HaSV, and TaV, virus-like particles (VLPs) can be expressed from capsid protein in *Spodoptera frugiperda* cell lines (Sf9 and Sf21) using a baculovirus expression system and in *Saccharomyces cerevisiae* cells (Tomasicchio *et al.*, 2007). The VLPs revealed similar morphological features to the wild-type virions and underwent an assembly-dependent autocatalytic cleavage of the capsid protein precursor. The VLPs were also resistant to protease treatment. Furthermore, mRNAs for the coat proteins were encapsulated in the VLPs, suggesting that at least one of the signals required for encapsulation of viral RNAs resides within the coat protein ORF.

The infection of tetraviruses is normally initiated by larval ingestion of virus-contaminated food. Horizontal transmission of NβV to the emperor gum moth, *Antherea eucalypti*, by this oral route has been demonstrated experimentally (Hanzlik *et al.*, 2005). Vertical transmission of tetravirus from parent to progeny has been proposed and supported by detection of TnV in adults that emerged from TnV-inoculated larvae, and of DpV in larvae that emerged from eggs laid by DpV-infected females (Morris *et al.*, 1979; Greenwood and Moore, 1981).

5.3.6. Cypoviruses

There are currently 20 cypovirus types that have been recognized or proposed, with all but two types isolated from Lepidoptera: CPV-17 from mosquitoes and CPV-20 from a black fly (Mertens *et al.*, 2004a, b: Mori and Metcalf, 2010). CPVs from Lepidoptera have been extensively reviewed (Hukuhara and Bonami, 1991; Belloncik and Mori, 1998; Mori and Metcalf, 2010). This section will review new information on CPVs from Diptera.

Cypoviruses from Diptera have been reported only from mosquitoes (Culicidae) and black flies (Simuliidae), with the possible exception of a single report from a sand fly (Ceratopogonidae) (Dubitskii, 1978). Historically referred to as cytoplasmic polyhedrosis viruses (CPVs) owing to

similarities to viruses found in Lepidoptera, they are the most common viruses reported from mosquito and black fly larvae. The larvae of 20 different species of mosquitoes representing seven different genera, and of at least 10 species of North American black flies have been reported infected with CPVs (Federici, 1985; Adler *et al.*, 2004). These viruses are characterized by producing chronic infections in the cytoplasm of larval midgut epithelial cells and the presence of polyhedra (occlusion bodies) (Becnel and White, 2007). While most commonly reported from larvae, there have been four reports of CPVs from the midgut epithelium of adult mosquitoes (Davies *et al.*, 1971; Bird *et al.*, 1972; Shapiro *et al.*, 2005).

CPVs in Diptera remain unclassified by the ICTV because of the lack of any molecular evidence that would confirm their relationship to any classified group. Molecular studies were conducted on CPVs from two mosquitoes and one black fly species comparing the migration patterns of their genome segments and the sequence of segment 10 (codes for viral polyhedrin protein) with other known cypoviruses. This series of studies proposed a new type 17 (species) UsCPV-17 and CrCPV-17 from mosquitoes *Uranotaenia sapphirina* and *Cx. restuans*, and type 20, SuCPV-20, from a black fly, *Simulium ubiquitum* (Shapiro *et al.*, 2005; Green *et al.*, 2006, 2007). No further studies have been conducted on CPVs from Diptera.

Cypovirus infections in mosquitoes and black flies are initiated upon ingestion of the occlusion bodies by the larval stages. The highly alkaline condition of the larval gut facilitates release of the virions that invade midgut cells of the gastric caeca and the posterior regions of the midgut (Figs. 5.5A and 5.6A). Replication is restricted to the cytoplasm, where virions are assembled within the virogenic stroma (Fig. 5.5B) and occluded by the deposition of a crystalline polyhedron matrix around individual particles (Fig. 5.6C). The polyhedrin protein appears to be arranged as a face-centered crystalline lattice (Fig. 5.5C). Most occlusion bodies in mosquitoes and black flies (Fig. 5.6C) are large and irregularly shaped, and contain multiple virions (Clark *et al.*, 1969; Clark and Fukuda, 1971; Andreadis, 1981, 1986). In contrast, a few species have occlusion bodies that contain only one virion (Fig. 5.5B and C) and are cuboidal in shape (Anthony *et al.*, 1973; Shapiro *et al.*, 2005).

Cypoviruses in Diptera are most commonly observed in the gastric caeca and posterior portions of the intestine of larvae (Fig. 5.5A) but can also be found in the anterior and posterior regions of the adult midgut from individuals infected as larvae (Shapiro *et al.*, 2005; Becnel and White, 2007). The accumulation of occlusion bodies in the cytoplasm of infected cells (Figs. 5.5B and 5.6B) produces a characteristic porcelain-white or iridescent color depending on the arrangement of the occlusion bodies (Becnel and White, 2007). Cells infected with a CPV are

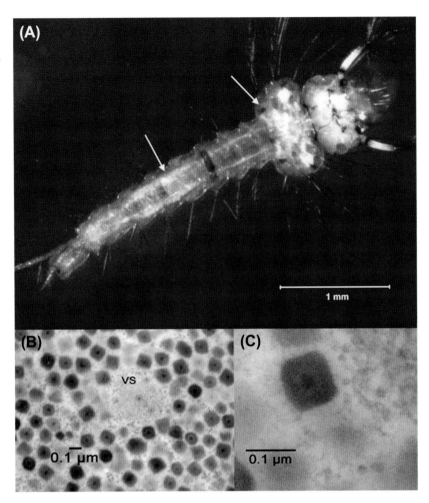

FIGURE 5.5 Cypovirus in mosquitoes. (A) *Culex erraticus* infected with UsCPV-17. Arrows indicate infected cells in the posterior midgut and gastric caeca. (B) Group of singly enveloped occlusion bodies of UsCPV-17; vs = virogenic stroma. (C) Occlusion body of UsCPV-17 containing a single virion and demonstrating the face-centered crystalline lattice.

irregular in shape and the nuclei are compressed and almost undetectable (Becnel and White, 2007). CPVs in Diptera are chronic and benign, with most infected larvae developing normally and emerging into adults with a persistent infection (Federici, 1985; Andreadis, 1986; Shapiro *et al.*, 2005).

Horizontal transmission to larvae and vertical transmission by adults have been documented for cypoviruses in mosquitoes (Becnel, 2006). Ingestion of occlusion bodies by larvae has been shown to initiate infections in the midgut (Clark and Fukuda, 1971) but at relatively low levels of less than 17% (Federici, 1973; Andreadis, 1986). Because divalent cations were found to influence transmission of a baculovirus in mosquitoes (Becnel *et al.*, 2001), Shapiro *et al.* (2005) conducted bioassays with UsCPV against *Ae. aegypti* larvae in the presence of magnesium and calcium. Exposure of UsCPV to *Ae. aegypti* in deionized water did not result in infections but with magnesium present (10 mM) infection levels averaged 34%. The addition of 10 mM calcium to this mixture reduced infection levels to 0.7% (Shapiro *et al.*, 2005). A subsequent study with CrCPV further verified the influence of divalent cations on

per os transmission, where the addition of magnesium resulted in 30% infection levels of *Cx. quinquefasciatus* larvae and the addition of calcium reduced the infection levels to less than 2% (Green *et al.*, 2006). The role of divalent cations in mediating CPV transmission in black flies is unknown. Studies with rotovirus have shown that in the extracellular medium, calcium stabilizes the structure of the viral capsid (Ruiz *et al.*, 2000). In addition, during cell entry, low cytoplasmic calcium concentrations induce the solubilization of the outer protein layer of the capsid, enhancing entry (Ruiz *et al.*, 2000). Whether similar mechanisms are involved in CPV transmission requires further study.

Vertical transmission was first documented for a CPV from *Ae. solicitans*, where four of 33 wild adults produced infected larvae but at a very low rate of 0.03% (Clark and Fukuda, 1971). One other study on vertical transmission showed progeny infection rates can be significantly higher, where six of 32 field-collected rafts of *Cx. restuans* produced CPV-infected larvae with an average of 10.5% of the progeny infected (Andreadis, 1986). The mechanisms and role and significance of vertical transmission

FIGURE 5.6 Cypovirus in a black fly. (A) Larva of *Simulium ubiquitum* infected with SuCPV-20. Arrows indicate infected cells in the posterior midgut and gastric caeca. (B) Midgut epithelial cells with SuCPV-20 occlusion bodies dispersed in the cytoplasm (arrow); mv = microvilli; bm = basement membrane; n = nucleus. (C) Mature occlusion body with angular sides containing multiple virions, some in the process of becoming occluded.

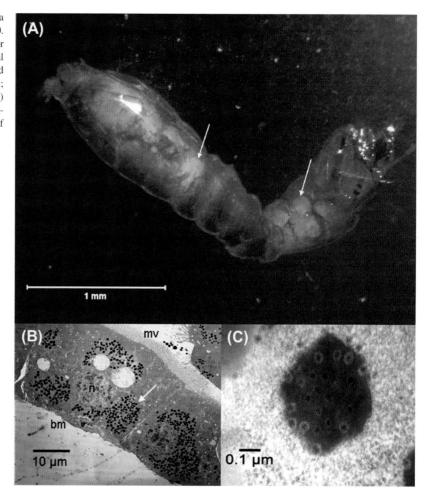

for mosquito CPVs are unknown but infected larvae can result in adults with both patent and covert infections (Shapiro *et al.*, 2005). Thirteen adults of *Ae. aegypti* from larvae infected with UsCPV resulted in four that showed patently infected midguts and nine with no apparent infections. However, when these nine adults were fed individually to healthy *Ae. aegypti* larvae, three induced typical UsCPV infections (Shapiro *et al.*, 2005). It is likely that vertical transmission of CPVs in mosquitoes plays a more important role in virus maintenance than previously recognized.

Several studies have examined the host range of various CPVs from Diptera, with the majority of the information from mosquito isolates. A CPV isolated from *Cx. salinarius* was transmitted to larvae of *Cx. territans* and *Culiseta inornada* but not *Ae. sollicitans* (Clark *et al.*, 1969). An isolate from *Ae. sollicitans* was transmitted to *Ae. taeniorhynchus* and *Psorophora ferox* but not *Cx. salinarius*. Andreadis (1986) found that a CPV isolate from *Cx. restuans* collected in Connecticut was host specific as it was not infectious to *Cx. pipiens* or *Cx. territans*. On the other hand, the isolate CrCPV from Florida was infectious to *Cx.*

quinquefasciatus but only in the presence of magnesium. Transmission experiments with UsCPV conducted with magnesium incorporated into the assays revealed a relatively broad host range including *Cx. quinquefasciatus*, *Ae. aegypti*, *Ae. triseriatus*, *Uranotaenia lowii*, and *An. albimanus*, although it was not transmitted to *An. quadrimaculatus* or the lepidopteran host, *H. zea*. In addition, *Cx. erraticus* and *An. crucians* may be hosts for UsCPV since they were found together with infected *U. sapphirina* in the field and exhibited similar infection characteristics, including the iridescence (Shapiro *et al.*, 2005). Additional assays with other CPV isolates from mosquitoes incorporating magnesium into the exposure media will help to determine the relative specificity of these viruses for the their mosquito hosts.

The host specificity of CPVs from black flies is poorly known, primarily because of difficulties conducting laboratory assays with black fly larvae. However, attempts to infect three species of mosquito larvae (*Cx. quinquefasciatus*, *An. quadrimaculatus*, and *An. albimanus*) with the black fly isolate, SuCPV, in the presence of magnesium were unsuccessful (Green *et al.*, 2007).

5.4. FUTURE RESEARCH DIRECTIONS

The structural simplicity, reduced genome size, and prolific replication rates of RNA viruses make them very well suited for the study of various aspects of virus biology and attractive candidates as microbial control agents against insect pests. In the past decade, there have been numerous advances in RNA virus research that have provided a wealth of important information relevant to our general understanding of viral disease pathogenicity. Many RNA viruses that infect insects have spurred interest in developing them as environmentally friendly microbial control agents for combating insect pests of agriculture as well as pests of medical importance. Furthermore, the exploitation of insect RNA viruses as protein expression vectors has proved useful for developing genetically engineered crop plants with pest resistance. Despite research advances and immense promise, however, key challenges relevant to development of RNA virus-based microbial control agents and expression vectors remain to be addressed and offer important directions for future research.

RNA viruses have often been identified as mortality factors of insect populations in nature. Of insect RNA viruses, tetraviruses have attracted particular interest because they exclusively infect insects within the order Lepidoptera, which contains many of the world's most serious pests. The high degree of host specificity and the ability to cause dramatic declines in host populations make tetraviruses excellent candidates for use as microbial control agents of insect pests. Successful field applications of tetraviruses, including *Darna trima virus* (DtV), NβV, and HaSV, for pest insect control were conducted by spraying aqueous suspensions of cadavers of the virus-killed larvae (Desmier *et al.*, 1988; Philippe *et al.*, 1997; Wood, 2002). Despite the proven ability of tetraviruses as microbial control agents, the problem of developing an *in vitro* system for mass production for commercial applications is still a major obstacle. With the exception of PrV, which was found to persistently infect a *H. zea* midgut cell line, tetraviruses are unable to replicate in any established cell lines. Their unusually narrow host range and tissue tropism raise interesting questions regarding replication biology of the viruses, including: (1) What host factors are involved in various steps of infections such as viral entry, replication and release? (2) What virus factors are responsible for virulence mechanisms and counteract defensive responses of the host? (3) What environmental factors impact host competence and disease expression? and (4) What intrinsic and extrinsic factors influence symptoms exhibited by the host; specifically, why are some viral infections present as asymptomatic or chronic infections and what conditions are necessary for conversion to an acute−lethal infection? Answers to these questions will not only clarify the molecular mechanisms of the viral infection process, but also provide critical information about host range, tissue tropism, and pathogenicity, which will facilitate the development and application of viral-based biopesticides.

In response to viral infections, host organisms have evolved effective defense mechanisms such as RNAi in which the host targets destruction of viral RNA by production of small interfering dsRNAs. In response, some viruses have evolved strategies to subvert this defense by encoding proteins that suppress host immunity (or RNAi mechanisms), thereby promoting their own replication. FHV was the first insect RNA virus that was found to encode a protein that functions as a suppressor of host-mediated RNAi. The induction and suppression of RNAi-mediated antiviral silencing are conserved features of nodaviruses. RNAi suppressors were also identified in ORF 1 of two dicistroviruses, CrPV and DCV. Expression of these proteins in *Drosophila* cells suppressed dsRNA-induced silencing and increased virus accumulation (Wang *et al.*, 2006). From an insect pest control perspective, one important future research direction would be to determine whether such immunosuppressive proteins are also present in the genomes of insect-infecting RNA viruses (other than nodaviruses and dicistroviruses). Specifically, RNAi suppressor proteins could be produced by gene cloning methods and used for pest control purposes to suppress insect host immunity and thereby enhance the pathogenic and virulence capacities of these RNA viruses.

The expression of pure VLPs from capsid proteins of insect RNA viruses in *Escherichia coli*, yeast, plant protoplasts, mammalian cells, and insect cell lines using baculovirus-driven systems is an emerging area that will facilitate research in the area of structure virology. Although they are replication incompetent and non-infectious, VLPs possess surface structure properties identical to the infectious virus and therefore provide useful platforms for obtaining high-resolution virus structure information by cryoelectron microscopy and X-ray crystallography. Furthermore, assembly of VLPs packed together with genes of interest has proven to be an effective RNA vector for foreign gene delivery. VLPs also have potential biomedical applications, as they are attractive candidates for the development of vaccines to prevent virus infection. While structural studies of some members of the *Nodaviridae, Tetraviridae, Dicistroviridae*, and Cypoviruses have yielded comprehensive information on the structure of viral capsids at an atomic level (Banerjee *et al.*, 2010), the high-resolution structures of many RNA viruses are yet to be determined. Future examination of VLP structures will lead to an enhanced understanding of virus maturation, assembly, and pathogenesis of insect RNA viruses and their eukaryotic cellular counterparts. The successful assembly of infectious HaSV particles in plant cells suggests that engineering plants to express RNA viruses specific to phytophagous insect pests has a promising future in controlling insect pests.

Cypoviruses produce a remarkably stable protein matrix, polyhedrin, that shields viral particles from the external environment. Virions embedded in the occlusion bodies are capable of surviving in unfavorable environmental conditions such as dehydration, freezing, and enzymic degradation for extended periods (Coulibaly *et al.*, 2007). Occlusion bodies, when ingested by a susceptible host, serve as the mechanism for the virions to infect the host. The infectious virions are released from the occlusion bodies in the alkaline midgut environment, allowing the infection process to proceed. The structural and functional properties of the occlusion bodies suggest that they can be used as multifunctional nanocontainers for the creation of effective systems for protein delivery. Advances in recombinant DNA techniques permit incorporation of different types of proteins into the occlusion bodies by co-expressing polyhedrin and target proteins (Ikeda *et al.*, 2001), offering potential in assisting the development and application of microbial control agents.

Although significant knowledge gaps and limitations in production of RNA viruses exist, the potential for exploiting these fascinating entities as safe, specific, environmentally friendly alternatives to chemical pesticides remains promising.

REFERENCES

Adler, P. H., Currie, D. C., & Wood, D. M. (2004). *The Black Flies (Simuliidae) of North America*. Ithaca: Cornell University Press.

Agol, V. I. (2002). Picornavirus genome, an overview. In B. L. Semler & E. Wimmer (Eds.), *Molecular Biology of Picornaviruses* (pp. 127–148). Washington DC: ASM Press.

Agrawal, D. K., & Johnson, J. E. (1992). Sequence and analysis of the capsid protein of Nudaurelia-capensis omega-virus, an insect virus with T = 4 icosahedral symmetry. *Virology, 190*, 806–814.

Agrawal, D. K., & Johnson, J. E. (1995). Assembly of the T = 4 *Nudaurelia capensis* omega virus capsid protein, post-translational cleavage, and specific encapsidation of its mRNA in a baculovirus expression system. *Virology, 207*, 89–97.

Aizawa, K., Furuta, Y., Kurata, K., & Sato, F. (1964). On the etiologic agent of the infectious flacherie of the silkworm, *Bombyx mori* (Linneaus). *Bull. Seric. Exp. Station, Tokyo, 19*, 223–240.

Anderson, D. L. (1991). Kashmir bee virus — a relatively harmless virus of honey bee colonies. *Am. Bee J., 131*, 767–770.

Anderson, D. L., & Gibbs, A. J. (1988). Inapparent virus infections and their interactions in pupae of the honey bee (*Apis mellifera* Linnaeus) in Australia. *J. Gen. Virol., 69*, 1617–1625.

Andreadis, T. G. (1981). A new cytoplasmic polyhedrosis virus from the salt-marsh mosquito, *Aedes cantator* (Diptera, Culicidae). *J. Invertebr. Pathol., 37*, 160–167.

Andreadis, T. G. (1986). Characterization of a cytoplasmic polyhedrosis virus affecting the mosquito *Culex restuans*. *J. Invertebr. Pathol., 47*, 194–202.

Anthony, D. W., Hazard, E. I., & Crosby, S. W. (1973). A virus disease in Anopheles quadrimaculatus. *J. Invertebr. Pathol., 22*, 1–5.

Audelo-del-Valle, J., Clement-Mellado, O., Magana-Hernandez, A., Flisser, A., Montiel-Aguirre, F., & Briseno-Garcia, B. (2003). Infection of cultured human and monkey cell lines with extract of penaeid shrimp infected with Taura syndrome virus. *Emerg. Infect. Dis., 9*, 265–266.

Bailey, L. (1967). The incidence of virus diseases in the honey bee. *Ann. Appl. Biol., 60*, 43–48.

Bailey, L. (1968). Honey bee pathology. *Annu. Rev. Entomol., 13*, 191–212.

Bailey, L. (1976). Viruses attacking the honey bee. *Adv. Virus Res., 20*, 271–304.

Bailey, L. (1981). *Honey Bee Pathology*. London: Academic Press.

Bailey, L. (1982). Viruses of honeybees. *Bee World, 63*, 165–173.

Bailey, L., & Fernando, E. F. W. (1972). Effects of sacbrood virus on adult honey bees. *Ann. Appl. Biol., 72*, 27–35.

Bailey, L., & Gibbs, A. J. (1964). Acute infection of bees with paralysis virus. *J. Invertebr. Pathol., 6*, 395–407.

Bailey, L., & Milne, R. G. (1969). The multiplication regions and interaction of acute and chronic bee-paralysis viruses in adult honey bees. *J. Gen. Virol., 4*, 9–14.

Bailey, L., & Scott, H. A. (1973). The pathogenicity of nodamura virus for insects. *Nature, 241*, 545.

Bailey, L., & Woods, R. D. (1974). Three previously undescribed viruses from the honey bee. *J. Gen. Virol., 25*, 175–186.

Bailey, L., & Woods, R. D. (1977). Two more small RNA viruses from honey bees and further observations on sacbrood and acute bee-paralysis viruses. *J. Gen. Virol., 37*, 175–182.

Bailey, L., Gibbs, A. J., & Woods, R. D. (1963). Two viruses from adult honey bees (*Apis mellifera* Linnaeus). *Virology, 21*, 390–395.

Bailey, L., Gibbs, A. J., & Woods, R. D. (1964). Sacbrood virus of the larval honey bee (*Apis mellifera* Linnaeus). *Virology, 23*, 425–429.

Bailey, L., Carpenter, J. M., & Woods, R. D. (1979). Egypt bee virus and Australian isolates of Kashmir bee virus. *J. Gen. Virol., 43*, 641–647.

Bailey, L., Ball, B. V., & Perry, J. N. (1981). The prevalence of viruses of honey bees in Britain. *Ann. Appl. Biol., 103*, 13–20.

Ball, L. A. (1992). Cellular expression of a functional nodavirus RNA replicon from vaccinia virus vectors. *J. Virol., 66*, 2335–2345.

Ball, L. A., & Johnson, K. L. (1998). Nodaviruses of insects. In L. K. Miller & L. A. Ball (Eds.), *The Insect Viruses* (pp. 225–267). New York: Plenum Publishing Corporation.

Ball, L. A., & Johnson, K. L. (1999). Reverse genetics of nodaviruses. *Adv. Virus Res., 53*, 229–244.

Ball, L. A., Amann, J. M., & Garrett, B. K. (1992). Replication of nodamura virus after transfection of viral RNA into mammalian cells in culture. *Virology, 66*, 2326–2334.

Baltimore, D. (1971). Expression of animal virus genomes. *Bacteriol. Rev., 35*, 235–241.

Ban, L., Didon, A., Jonsson, L. M. V., Glinwood, R., & Delp, G. (2007). An improved detection method for the *Rhopalosiphum padi virus* (RhPV) allows monitoring of its presence in aphids and movement within plants. *J. Virol. Methods, 142*, 136–142.

Banerjee, M., Speir, J. A., & Johnson, J. E. (2010). Structural comparison of insect RNA viruses. In S. Asgari & K. Johnson (Eds.), *Insect Virology* (pp. 327–346). Norfolk: Caister Academic Press.

Bawden, A. L., Gordon, K. H. J., & Hanzlik, T. N. (1999). The specificity of *Helicoverpa armigera* stunt virus infectivity. *J. Invertebr. Pathol., 74*, 156−163.

Becnel, J. J. (2006). Transmission of viruses to mosquito larvae mediated by divalent cations. *J. Invertebr. Pathol., 92*, 141−145.

Becnel, J. J., & White, S. E. (2007). Mosquito pathogenic viruses − the last 20 years. *J. Am. Mosq. Control Assoc., 23*(Suppl. 2), 36−49.

Becnel, J. J., White, S. E., Moser, B. A., Fukuda, T., Rotstein, M. J., Undeen, A. H., & Cockburn, A. (2001). Epizootiology and transmission of a newly discovered baculovirus from the mosquitoes *Culex nigripalpus* and *Culex quinquefasciatus*. *J. Gen. Virol., 82*, 275−282.

Belloncik, S., & Mori, H. (1998). Cypoviruses. In L. K. Miller & L. A. Ball (Eds.), *The Insect Viruses* (pp. 337−369). New York: Plenum Publishing Corporation.

Belloncik, S., Liu, J., Su, D., & Arella, M. (1996). Identification and characterization of a new Cypovirus, type 14, isolated from *Heliothis armigera*. *J. Invertebr. Pathol., 67*, 41−47.

Benjeddou, M., Leat, N., Allsopp, M., & Davison, S. (2002a). Development of infectious transcripts and genome manipulation of *Black queen-cell virus* of honey bees. *J. Gen. Virol., 83*, 3139−3146.

Benjeddou, M., Leat, N., & Davison, S. (2002b). *Black queen-cell virus* RNA is infectious in honey bee pupae. *J. Invertebr. Pathol., 81*, 205−206.

Berenyi, O., Bakonyi, T., Derakhshifar, I., Koglberger, H., & Nowotny, N. (2006). Occurrence of six honeybee viruses in diseased Austrian apiaries. *Appl. Environ. Microbiol., 72*, 2414−2420.

Berenyi, O., Bakonyi, T., Derakhshifar, I., Koglberger, H., Topolska, G., Ritter, W., Pechhacker, H., & Nowotny, N. (2007). Phylogenetic analysis of deformed wing virus genotypes from diverse geographic origins indicates recent global distribution of the virus. *Appl. Environ. Microbiol., 73*, 3605−3611.

Bird, R. G., Draper, C. C., & Ellis, D. S. (1972). A cytoplasmic polyhedrosis virus in midgut cells of *Anopheles stephensi* and in the sporogonic stages of *Plasmodium berghei yoelii*. *Bull. World Health Organ., 46*, 337−343.

Bong, D. T., Steinern, C., Janshoff, A., Johnson, J. E., & Ghadiri, M. R. (1999). A highly membrane-active peptide in Flock House virus, implications for the mechanism of nodavirus infection. *Chem. Biol., 6*, 473−481.

Bonami, J., Hasson, K., Mari, J., Poulos, B., & Lightner, D. (1997). Taura syndrome of marine penaeid shrimp: characterization of the viral agent. *J. Gen. Virol., 78*, 313−319.

Bonning, B. C. (2009). The *Dicistroviridae*: an emerging family of invertebrate viruses. *Virol. Sin., 24*, 415−427.

Bonning, B. C., & Johnson, K. N. (2010). Dicistroviruses. In S. Asgari & K. Johnson (Eds.), *Insect Virology* (pp. 201−228). Norfolk: Caister Academic Press.

Bonning, B. C., & Miller, W. A. (2010a). Dicistroviruses. In S. Asgari & K. Johnson (Eds.), *Insect Virology* (pp. 201−229). Norfolk: Caister Academic Press.

Bonning, B. C., & Miller, W. A. (2010b). Dicistroviruses. *Annu. Rev. Entomol., 55*, 129−150.

Bowen-Walker, P. L., Martin, S. J., & Gunn, A. (1999). The transmission of deformed wing virus between honeybees (*Apis mellifera* L.) by the ectoparasitic mite *Varroa jacobsoni* Oud. *J. Invertebr. Pathol., 73*, 101−106.

Boyapalle, S., Pal, N., Miller, W. A., & Bonning, B. C. (2007). A glassy-winged sharpshooter cell line supports replication of *Rhopalosiphum padi* virus (Dicistroviridae). *J. Invertebr. Pathol., 94*, 130−139.

Boyapalle, S., Beckett, R. J., Pal, N., Miller, W. A., & Bonning, B. C. (2008). Infectious genomic RNA of *Rhopalosiphum padi* virus transcribed *in vitro* from a full-length cDNA clone. *Virology, 375*, 401−411.

Boyer, J. C., & Haenni, A. L. (1994). Infectious transcripts and cDNA clones of RNA viruses. *Virology, 198*, 415−426.

Brodsgaard, C. J., Ritter, W., Hansen, H., & Brodsgaard, H. F. (2000). Interactions among *Varroa jacobsoni* mites, acute paralysis virus, and *Paenibacillus larvae* larvae and their influence on mortality of larval honeybees in vitro. *Apidologie, 31*, 543−554.

Brooks, E. M., Karl, H. J., Gordon, K. H. J., Dorrian, S. J., Hines, E. R., & Hanzlik, T. N. (2002). Infection of its lepidopteran host by the *Helicoverpa armigera* stunt virus (*Tetraviridae*). *J. Invertebr. Pathol., 80*, 97−111.

Calisher, C. H., & Gould, E. A. (2003). Taxonomy of the virus family *Flaviridae*. *Adv. Virus Res., 59*, 1−19.

Cammisa-Parks, H., Cisar, L. A., Kane, A., & Stollar, V. (1992). The complete nucleotide sequence of cell fusing agent (CFA), homology between the nonstructural proteins encoded by CFA and the nonstructural proteins encoded by arthropod-borne flaviviruses. *Virology, 189*, 511−524.

Carstens, E. (2010). Ratification vote on taxonomic proposals to the International Committee on Taxonomy of Viruses (2009). *Arch. Virol., 155*, 133−146.

Carstens, E., & Ball, L. (2009). Ratification vote on taxonomic proposals to the International Committee on Taxonomy of Viruses (2008). *Arch. Virol., 154*, 1181−1188.

Cevallos, R. C., & Sarnow, P. (2010). Temperature protects insect cells from infection by cricket paralysis virus. *J. Virol., 84*, 1652−1655.

Chantawannakul, P., Ward, L., Boonham, N., & Brown, M. (2006). A scientific note on the detection of honeybee viruses using real-time PCR (TaqMan) in *Varroa* mites collected from a Thai honeybee (*Apis mellifera*) apiary. *J. Invertebr. Pathol., 91*, 69−73.

Chao, J. A., Lee, J. H., Chapados, B. R., Debler, E. W., Schneemann, A., & Williamson, J. R. (2005). Dual modes of RNA-silencing suppression by Flock House virus protein B2. *Nat. Struct. Mol. Biol., 12*, 952−957.

Chen, Y. P., & Siede, R. (2007). Honey bee viruses. *Adv. Virus Res., 70*, 33−80.

Chen, Y. P., Smith, I. B., Collins, A. M., Pettis, J. S., & Feldlaufer, M. F. (2004a). Detection of deformed wing virus infection in honey bees, *Apis mellifera* L., in the United States. *Am. Bee J., 144*, 557−559.

Chen, Y. P., Zhao, Y., Hammond, J., Hsu, H. T., Evans, J., & Feldlaufer, M. (2004b). Multiple virus infections in the honey bee and genome divergence of honey bee viruses. *J. Invertebr. Pathol., 87*, 84−93.

Chen, Y. P., Higgins, J. A., & Feldlaufer, M. F. (2005). Quantitative real-time reverse transcription−PCR analysis of deformed wing virus infection in the honeybee (*Apis mellifera* L.). *Appl. Environ. Microbiol., 71*, 436−441.

Chen, Y. P., Pettis, J. S., Collins, A., & Feldlaufer, M. F. (2006). Prevalence and transmission of honeybee viruses. *Appl. Environ. Microbiol., 72*, 606−611.

Chen, Y. P., Nakashima, N., Christian, P., Bonning, B. C., Valles, S. M., & Lightner, D. V. (2011). Dicistroviridae. In A. M. King (Ed.), *Virus Taxonomy, Ninth Report of the ICTV*. London: Elsevier. (in press).

Cherry, S., & Perrimon, N. (2004). Entry is a rate-limiting step for viral infection in a *Drosophila melanogaster* model of pathogenesis. *Nat. Immunol., 5*, 81−87.

Choi, H. K., Kobayashi, M., & Kawase, S. (1989). Changes in infectious flacherie virus-specific polypeptides and translatable mRNA in the midgut of the silkworm, *Bombyx mori*, during larval molt. *J. Invertebr. Pathol., 53*, 128−131.

Christian, P. D., & Scotti, P. D. (1996). Biopesticides from small RNA viruses on insects: aspects of their *in vitro* production. In K. Maramorosch & M. J. Loeb (Eds.), *Invertebrate Cell Culture: Looking Toward the Twenty First Century* (pp. 73−81). San Francisco: Proceedings of the IX International Conference on Invertebrate Cell Culture. Society for In Vitro Biology.

Christian, P. D., & Scotti, P. D. (1998). The picorna-like viruses of insects. In L. Miller & A. Ball (Eds.), *The Insect Viruses* (pp. 301−336). New York: Plenum Press.

Christian, P. D., Hanzlik, T. N., Dall, D. J., & Gordon, K. (1992). Insect viruses: new strategies for pest control. In J. G. Oakeshott & M. J. Whitten (Eds.), *Molecular Approaches to Fundamental and Applied Entomology* (pp. 128−163). New York: Springer.

Christian, P. D., Dorrian, S. J., Gordon, K. H. J., Terry, N., & Hanzlik, T. N. (2001). Pathology and properties of the tetravirus *Helicoverpa armigera* stunt virus. *Biol. Control, 20*, 65−75.

Christian, P. D., Murray, D., Powell, R., Hopkinson, J., Gibb, N. N., & Hanzlik, T. N. (2005). Effective control of a field population of *Helicoverpa armigera* by using the small RNA virus *Helicoverpa armigera* stunt virus (*Tetraviridae*: Omegatetravirus). *J. Econ. Entomol., 98*, 1839−1847.

Clark, T. B., & Fukuda, T. (1971). Field and laboratory observations of two viral diseases in *Aedes sollicitans* (Walker) in southwestern Louisiana. *Mosq. News, 31*, 193−199.

Clark, T. B., Chapman, H. C., & Fukuda, T. (1969). Nuclear-polyhedrosis and cytoplasmic-polyhedrosis virus infections in Louisiana mosquitoes. *J. Invertebr. Pathol., 14*, 284−286.

Cook, S., Bennett, S. N., Holmes, E. C., de Chesse, R., Moureau, G., & de Lamballerie, X. (2006). Isolation of a new strain of the flavivirus cell fusing agent virus in a natural mosquito population from Puerto Rico. *J. Gen. Virol., 87*, 735−748.

Coulibaly, F., Chiu, E., Ikeda, K., Gutmann, S., Haebel, P. W., Schulze-Briese, C., Mori, H., & Metcalf, P. (2007). The molecular organization of cypovirus polyhedra. *Nature, 446*, 97−101.

Cox-Foster, D., & vanEngelsdorp, D. (2009). Saving the honeybee. *Sci. Am., 300*, 40−47.

Cox-Foster, D. L., Conlan, S., Holmes, E. C., Palacios, G., Evans, J. D., Moran, N. A., Quan, P., Briese, T., Hornig, M., Geiser, M., Martinson, V., VanEngelsdorp, D., Kalkstein, A. L., Drysdale, A., Hui, J., Ahai, J., Cui, L., Hutchinson, S. K., Simons, J. F., Egholm, M., Pettis, J., & Lipkin, W. I. (2007). A metagenomic survey of microbes in honey bee colony collapse disorder. *Science, 318*, 283−287.

Crabtree, M. B., Sang, R. C., Stollar, V., Dunster, L. M., & Miller, B. R. (2003). Genetic and phenotypic characterization of the newly described insect flavivirus, Kamiti River virus. *Arch. Virol., 148*, 1095−1118.

Cristina, J., & Costa-Mattioli, M. (2007). Genetic variability and molecular evolution of hepatitis A virus. *Virus Res., 127*, 151−157.

Crochu, S., Cook, S., Attoui, H., Charrel, R. N., De Chesse, R., Belhouchet, M., Lemasson, J. J., de Micco, P., & de Lamballerie, X. (2004). Sequences of flavivirus-related RNA viruses persist in DNA form integrated in the genome of *Aedes spp.* mosquitoes. *J. Gen. Virol., 85*, 1971−1980.

Culley, A. I., Lang, A. S., & Suttle, C. A. (2006). Metagenomic analysis of coastal RNA virus communities. *Science, 312*, 1795−1798.

Czech, B., Malone, C. D., Zhou, R., Stark, A., Schlingeheyde, C., Dus, M., Perrimon, N., Kellis, M., Wohlschlegel, J. A., Sachidanandam, R., Hannon, G. J., & Brennecke, J. (2008). An endogenous small interfering RNA pathway in *Drosophila*. *Nature, 453*, 798−802.

Dainat, B., Ken, T., Berthoud, H., & Neumann, P. (2009). The ectoparasitic mite *Tropilaelaps mercedesae* (Acari, Laelapidae) as a vector of honeybee viruses. *Insect. Soc., 56*, 40−43.

D'Arcy, C. J., Burnett, P. A., & Hewings, A. D. (1981a). Detection, biological effects, and transmission of a virus of the aphid *Rhopalosiphum padi*. *Virology, 114*, 268−272.

D'Arcy, C. J., Burnett, P. A., Hewings, A. D., & Goodman, R. M. (1981b). Purification and characterization of a virus from the aphid *Rhopalosiphum padi*. *Virology, 112*, 346−349.

Dasgupta, R., Selling, B., & Rueckert, R. (1994). Flock house virus, a simple model for studying persistent infection in cultured *Drosophila* cells. *Arch. Virol.,* (Suppl. 9), 121−132.

Dasgupta, R., Garcia, B. H., III, & Goodman, R. M. (2001). Systemic spread of an RNA insect virus in plants expressing plant viral movement protein genes. *Proc. Natl. Acad. Sci. USA, 98*, 4910−4915.

Dasgupta, R., Cheng, L.-L., Bartholomay, L. C., & Christensen, B. M. (2003). Flock house virus replicates and expresses green fluorescent protein in mosquitoes. *J. Gen. Virol., 84*, 1789−1797.

Dasgupta, R., Free, H. M., Zietlow, S. L., Paskewitz, S. M., Aksoy, S., Shi, L., Fuchs, J., Hu, C., & Christensen, B. M. (2007). Replication of flock house virus in three genera of medically important insects. *J. Med. Entomol., 44*, 102−110.

Davies, E. E., Howells, R. E., & Venters, D. (1971). Microbial infections associated with plasmodial development in *Anopheles stephensi*. *Ann. Trop. Med. Parisitol., 65*, 403−408.

Desmier de Chenon, R., Mariau, D., Monsarrat, P., Fédière, G., & Sipayung, A. (1988). Research into entomopathogenic agents of viral origin in leaf-eating Lepidoptera of the oil palm and coconut. *Oléagineux, 43*, 107−117.

DeTulleo, L., & Kirchhausen, T. (1998). The clathrin endocytic pathway in viral infection. *EMBO J., 17*, 4585−4593.

Dhar, A. K., Lakshman, D. K., Amundsen, K., Robles-Sikisaka, R., Kaizer, K. N., Roy, S., Hasson, K. W., & Allnutt, F. C. T. (2010). Characterization of a Taura syndrome virus isolate originating from the 2004 Texas epizootic in cultured shrimp. *Arch. Virol., 155*, 315−327.

Di Prisco, G., Pennacchio, F., Caprio, E., Boncristiani, H. F., Jr., Evans, J. D., & Chen, Y. (2011). *Varroa destructor* is an effective vector of Israeli acute paralysis virus in the honeybee. *Apis mellifera*. *J. Gen. Virol., 92*, 151−155.

Djikeng, A., Kuzmickas, R., Anderson, N. G., & Spiro, D. J. (2009). Metagenomic analysis of RNA viruses in a fresh water lake. *PLoS ONE, 4*, e7264.

Dolja, V. V., & Koonin, E. V. (1991). Phylogeny of capsid proteins of small icosahedral RNA plant viruses. *J. Gen. Virol., 72,* 1481–1486.

Domingo, E., & Holland, J. J. (1997). RNA virus mutations and fitness for survival. *Annu. Rev. Microbiol., 51,* 151–178.

Dorrington, R. A., & Short, J. R. (2010). In S. Asgari & K. Johnson (Eds.), *Insect Virology* (pp. 283–305). Norfolk: Caister Academic Press.

Dubitskii, A. M. (1978). *Biological Control of Bloodsucking Flies in the USSR.* Alma-Ata, Kazakhstan: Nauka. (In Russian.).

Evans, J. D. (2001). Genetic evidence for coinfection of honey bees by acute bee paralysis and Kashmir bee viruses. *J. Invertebr. Pathol., 78,* 189–193.

Farfan-Ale, J. A., Loroño-Pino, M. A., Garcia-Rejon, J. E., Hovav, E., Powers, A. M., Lin, M., Dorman, K. S., Platt, K. B., Bartholomay, L. C., Soto, V., Beaty, B. J., Lanciotti, R. S., & Blitvich, B. J. (2009). Detection of RNA from a novel West Nile-like virus and high prevalence of an insect-specific flavivirus in mosquitoes in the Yucatan Peninsula of Mexico. *Am. J. Trop. Med. Hyg., 80,* 85–95.

Federici, B. A. (1973). Preliminary studies of cytoplasmic polyhedrosis virus of *Aedes taeniorhynchus. 5th Intern. Colloq. Insect. Pathol. Microbiol. Control, 1,* 34, (Abstract).

Federici, B. A. (1985). Viral pathogens of mosquito larvae. *Bull. Am. Mosq. Control Assoc., 6,* 62–74.

Fédière, G., Philippe, R., Veyrunes, J., & Monsarrat, P. (1990). Biological control of the oil palm pest *Latoia viridissima* (Lepidoptera, Limacodidae) in Cote D'Ivoire, by a new Picornavirus. *Entomophaga, 35,* 347–354.

Fievet, J., Tentcheva, D., Gauthier, L., de Miranda, J., Cousserans, F., Colin, M. E., & Bergoin, M. (2006). Localization of deformed wing virus infection in queen and drone *Apis mellifera* L. *Virol. J., 3,* 16.

Finch, J. T., Crowther, R. A., Hendry, D. A., & Struthers, J. K. (1974). The structure of *Nudaurelia capensis* β virus: the first example of a capsid with icosahedral surface symmetry T = 4. *J. Gen. Virol., 24,* 191–200.

Firth, A. E., Wang, Q. S., Jan, E., & Atkins, J. F. (2009). Bioinformatic evidence for a stem-loop structure 5′-adjacent to the IGR-IRES and for an overlapping gene in the bee paralysis dicistroviruses. *Virol. J., 6,* 193.

Fisher, A. J., & Johnson, J. E. (1993). Ordered duplex RNA controls capsid architecture in an icosahedral animal virus. *Nature, 361,* 176–179.

Flynt, A., Liu, N., Martin, R., & Lai, E. C. (2009). Dicing of viral replication intermediates during silencing of latent *Drosophila* viruses. *Proc. Natl. Acad. Sci. USA, 106,* 5270–5275.

Forsgren, E., de Miranda, J. R., Isaksson, M., Wei, S., & Fries, I. (2009). Deformed wing virus associated with *Tropilaelaps mercedesae* infesting European honey bees (*Apis mellifera*). *Exp. Appl. Acarol., 47,* 87–97.

Francki, R. I. B., Fauquet, C. M., Knudson, D. L., & Brown, F. (1991). Classification and nomenclature of viruses. *Fifth Report of the International Committee on Taxonomy of Viruses.* In: *Arch. Virol.,* (Suppl. 2). Springer, New York.

François-Xavier, L., Guillaume Le, L., Hector, F., Michel, D., Luc, P., Laurent, B., & Gilles, P. (2010). Chagas disease: changes in knowledge and management. *Lancet Infect. Dis., 10,* 556–570.

Friesen, P. D., & Rueckert, R. R. (1981). Synthesis of black beetle virus proteins in cultured *Drosophila* cells: differential expression of RNAs 1 and 2. *J. Virol., 37,* 876–886.

Friesen, P. D., & Rueckert, R. R. (1982). Black beetle virus: messenger for protein B is a subgenomic viral RNA. *J. Virol., 42,* 986–995.

Friesen, P., Scotti, P., Longworth, J., & Rueckert, R. (1980). Black beetle virus, propagation in *Drosophila* line 1 cells and an infection-resistant subline carrying endogenous black beetle virus-related particles. *J. Virol., 35,* 741–747.

Fujiyuki, T., Takeuchi, H., Ono, M., Ohka, S., Sasaki, T., Nomoto, A., & Kubo, T. (2004). Novel insect picorna-like virus identified in the brains of aggressive worker honeybees. *J. Virol., 78,* 1093–1100.

Fujiyuki, T., Takeuchi, H., Ono, M., Ohka, S., Sasaki, T., Nomoto, A., & Kubo, T. (2005). Kakugo virus from brains of aggressive worker honeybees. *Adv. Virus Res., 65,* 1–27.

Fujiyuki, T., Ohka, S., Takeuchi, H., Ono, M., Nomoto, A., & Kubo, T. (2006). Prevalence and phylogeny of Kakugo virus, a novel insect picorna-like virus that infects the honeybee (*Apis mellifera* L.), under various colony conditions. *J. Virol., 80,* 11528–11538.

Fujiyuki, T., Matsuzaka, E., Nakaoka, T., Takeuchi, H., Wakamoto, A., Ohka, S., Sekimizu, K., Nomoto, A., & Kubo, T. (2009). Distribution of Kakugo virus and its effects on the gene expression profile in the brain of the worker honeybee *Apis mellifera* L. *J. Virol., 83,* 1560–1568.

Furgala, B., & Lee, P. E. (1966). Acute bee paralysis virus, a cytoplasmic insect virus. *Virology, 29,* 346–348.

Gallagher, T. M. (1987). *Synthesis and assembly of nodaviruses. PhD thesis.* Madison: University of Wisconsin.

Gallagher, T. M., Friesen, P. D., & Rueckert, R. R. (1983). Autonomous replication and expression of RNA 1 from black beetle virus. *J. Virol., 46,* 481–489.

Garrey, J. L., Lee, Y. Y., Au, H. H., Bushell, M., & Jan, E. (2010). Host and viral translational mechanisms during cricket paralysis virus infection. *J. Virol., 84,* 1124–1138.

Garzon, S., Charpentier, G., & Kurstak, E. (1978). Morphogenesis of the Nodamura virus in the larvae of the Lepidopteran *Galleria mellonella* (L.). *Arch. Virol., 56,* 61–76.

Garzon, S., Strykowski, H., & Charpentier, G. (1990). Implication of mitochondria in the replication of Nodamura virus in larvae of the Lepidopteran *Galleria mellonella* (L.) and in suckling mice. *Arch. Virol., 113,* 165–176.

Gaunt, M. W., Sall, A. A., de Lamballerie, X., Falconar, A. K., Dzhivanian, T. I., & Gould, E. A. (2001). Phylogenetic relationships of flaviviruses correlate with their epidemiology, disease association and biogeography. *J. Gen. Virol., 82,* 1867–1876.

Generschh, E., Yue, C., Fries, I., & de Miranda, J. R. (2006). Detection of deformed wing virus, a honey bee viral pathogen, in bumble bees (*Bombus terrestris* and *Bombus pascuorum*) with wing deformities. *J. Invertebr. Pathol., 91,* 61–63.

Ghosh, R. C., Ball, B. V., Willcocks, M. M., & Carter, M. J. (1999). The nucleotide sequence of sacbrood virus of the honey bee: an insect picorna-like virus. *J. Gen. Virol., 80,* 1541–1549.

Gildow, F. E., & D'Arcy, A. (1988). Barley and oats as reservoirs for an aphid virus and influence on barley yellow dwarf virus transmission. *Phytopathology, 78,* 811–816.

Gildow, F. E., & D'Arcy, C. J. (1990). Cytopathology and experimental host range of *Rhopalosiphum padi* virus, a small isometric RNA virus infecting cereal grain aphids. *J. Invertebr. Pathol., 55,* 245–257.

Gisder, S., Aumeier, P., & Genersch, E. (2009). Deformed wing virus: replication and viral load in mites (*Varroa destructor*). *J. Gen. Virol., 90.* 463–457.

Goldbach, R., & de Hann, P. (1994). RNA viral supergroups and evolution of RNA viruses. In S. Morse (Ed.), *The Evolutionary Biology of Viruses* (pp. 105–119). New York: Raven Press.

Gomariz-Zilber, E., Poras, M., & Thomas-Orillard, M. (1995). *Drosophila* C virus: experimental study of infectious yields and underlying pathology in *Drosophila melanogaster* laboratory populations. *J. Invertebr. Pathol., 65*, 243–247.

Gorbalenya, A. E., Pringle, F. M., Zeddam, J.-L., Luke, B. T., Cameron, C. E., Kalmakoff, J., Hanzlik, T. N., Gordon, K. H. J., & Ward, V. K. (2002). The palm subdomain-based active site is internally permuted in viral RNA-dependent RNA polymerases of an ancient lineage. *J. Mol. Biol., 324*, 47–62.

Gordon, K. H. J., & Hanzik, T. N. (1998). Tetraviruses. In L. K. Miller & L. A. Ball (Eds.), *The Insect Viruses* (pp. 269–299). New York: Plenum Press.

Gordon, K. H. J., & Waterhouse, P. M. (2006). Small RNA viruses of insects: expression in plants and RNA silencing. *Adv. Virus Res., 68*, 459–502.

Gordon, K. H. J., Johnson, K. N., & Hanzlik, T. N. (1995). The larger genomic RNA of *Helicoverpa armigera* stunt tetravirus encodes the viral RNA polymerase and has a novel 39-terminal tRNA-like structure. *Virology, 208*, 84–98.

Gordon, K. H. J., Williams, M. R., Hendry, D. A., & Hanzlik, T. N. (1999). Sequence of the genomic RNA of *Nudaurelia* beta virus (*Tetraviridae*) defines a novel virus genome organization. *Virology, 258*, 42–53.

Gordon, K. H. J., Williams, M. R., Baker, J. S., Gibson, J. M., Bawden, A. L., Millgate, A. G., Larkin, P. J., & Hanzlik, T. N. (2001). Replication-independent assembly of and insect virus (*Tetraviridae*) in plant cells. *Virology, 288*, 36–50.

Gould, E. A., de Lamballerie, X., Zanotto, P. M., & Holmes, E. C. (2001). Evolution, epidemiology, and dispersal of flaviviruses revealed by molecular phylogenies. *Adv. Virus Res., 57*, 71–103.

Grace, T. D. C., & Mercer, E. H. (1965). A new virus of the saturniid *Antherea eucalypti* (Scott). *J. Invertebr. Pathol., 7*, 241–244.

Grace, T. D. C., & Mercer, E. H. (1976). Serological relations between twelve small RNA viruses of insects. *J. Gen. Virol., 31*, 131–134.

Graham, R. I., Rao, S., Possee, R. D., Sait, S. M., Peter, P. C., Mertens, P. P. C., & Hails, R. S. (2006). Detection and characterisation of three novel species of reovirus (*Reoviridae*), isolated from geographically separate populations of the winter moth *Operophtera brumata* (Lepidoptera, Geometridae) on Orkney. *J. Invertebr. Pathol., 91*, 79–87.

Green, T. B., Shapiro, A. M., White, S. E., Rao, R., Mertens, P. P. C., Carner, G., & Becnel, J. J. (2006). Molecular and genomic characterization of *Culex restuans* cypovirus. *J. Invertebr. Pathol., 91*, 27–34.

Green, T. B., White, S. E., Rao, R., Mertens, P. P. C., Adler, P. H., & Becnel, J. J. (2007). Biological and molecular studies of a cypovirus from the blackfly *Simulium ubiquitum*. *J. Invertebr. Pathol., 95*, 26–32.

Greenwood, L. K., & Moore, N. F. (1981). A single protein Nudaurelia p like virus of the pale tussock moth, *Dasychira pudibunda*. *J. Invertebr. Pathol., 38*, 305–306.

Greenwood, L. K., & Moore, N. F. (1982). The Nudaurelia a group of small RNA-containing viruses of insects, serological identification of several new isolates. *J. Invertebr. Pathol., 39*, 407–409.

Greenwood, L. K., & Moore, N. F. (1984). Determination of the location of an infection in *Trichoplusia ni* larvae by a small RNA-containing virus using enzyme-linked immunosorbent assay and electron microscopy. *Microbiologica, 7*, 97–102.

Gregory, P. G., Evans, J. D., Rinderer, T., & de Guzman, L. (2005). Conditional immune-gene suppression of honeybees parasitized by *Varroa* mites. *J. Insect. Sci., 5*, 7.

Guy, P. L., Toriyama, S., & Fuji, S. (1992). Occurrence of a picorna-like virus in planthopper species and its transmission in *Laodelphax striatellus*. *J. Invertebr. Pathol., 59*, 161–164.

Hagiwara, K., & Matsumoto, T. (2000). Determination of the nucleotide sequence of *Bombyx mori* cytoplasmic polyhedrosis virus segment 9 and its expression in BmN4 cells. *J. Gen. Virol., 81*, 1143–1147.

Hagiwara, K., Rao, S., Scott, W., & Carner, G. R. (2002). Nucleotide sequences of segments 1, 3, and 4 of the genome of *Bombyx mori* cypovirus 1 encoding putative capsid proteins VP1, VP3, and VP4, respectively. *J. Gen. Virol., 83*, 1477–1482.

Hanzlik, T. N., & Gordon, K. H. J. (1997). The Tetraviridae. *Adv. Virus Res., 48*, 101–168.

Hanzlik, T. N., Dorrian, S. J., Gordon, K. H. J., & Christian, P. D. (1993). A novel small RNA virus isolated from the cotton bollworm, *Helicoverpa armigera*. *J. Gen. Virol., 74*, 1805–1810.

Hanzlik, T. N., Dorrian, S. J., Johnson, K. N., Brooks, E. M., & Gordon, K. H. J. (1995). Sequence of RNA2 of the *Helicoverpa armigera* stunt virus (Tetraviridae) and bacterial expression of its genes. *J. Gen. Virol., 76*, 799–811.

Hanzlik, T. N., Gordon, K. H. J., Gorbalenya, A. E., Hendry, D. A., Pringle, F. M., Ward, V. K., & Zeddam, J. L. (2005). Family Tetraviridae. In C. M. Hanzlik, M. A. Mayo, J. Maniloff, U. Desselberger & L. A. Ball (Eds.), *Virus Taxonomy, Eighth Report of the International Committee on Taxonomy of Viruses* (pp. 877–883). San Diego: Elsevier Academic Press.

Hashimoto, Y., & Valles, S. M. (2007). *Solenopsis invicta* virus-1 tissue tropism and intra-colony infection rate in the red imported fire ant: a quantitative PCR-based study. *J. Invertebr. Pathol., 96*, 156–161.

Hashimoto, Y., Valles, S. M., & Strong, C. A. (2007). Detection and quantitation of *Solenopsis invicta* virus in fire ants by real-time PCR. *J. Virol. Methods, 140*, 132–139.

Hasson, K. W., Lightner, D. V., Poulos, B. T., Redman, R. M., White, B. L., Brock, J. A., & Bonami, J. R. (1995). Taura syndrome in *Penaeus vannamei*: demonstration of a viral etiology. *Dis. Aquat. Organ., 23*, 115–126.

Hatfill, S. J., Williamson, C., Kirby, R., & von Wechmar, M. B. (1990). Identification and localization of aphid lethal paralysis virus particles in thin tissue sections of the *Rhopalosiphum padi* aphid by in situ nucleic acid hybridization. *J. Invertebr. Pathol., 55*, 265–271.

Hendry, D. A., Becker, M. F., & van Regenmortel, M. V. H. (1968). A non inclusion virus of the pine emperor moth *Nudaurelia cytherea capensis* Stoll. *S. Afr. Med. J., 42*, 117.

Hendry, D., Hodgson, V., Clark, R., & Newman, J. (1985). Small RNA viruses co-infecting the pine emperor moth (*Nudaurelia cytherea capensis*). *J. Gen. Virol., 66*, 627–632.

Highfield, A. C., El Nagar, A., Mackinder, L. C., Noel, L. M., Hall, M. J., Martin, S. J., & Schroeder, D. C. (2009). Deformed wing virus implicated in overwintering honeybee colony losses. *Appl. Environ. Microbiol., 75*, 7212–7220.

Hill, C. L., Booth, T. F., Venkataram Prasad, B. V., Grimes, J. M., Mertens, P. P. C., Sutton, G. C., & Stuart, D. I. (1999). The structure of a cypovirus and the functional organization of dsRNA viruses. *Nat. Struct. Mol. Biol., 6,* 565–568.

Hogle, J. M., & Racaniello, V. R. (2002). Poliovirus receptors and cell entry. In B. L. Semler & E. Wimmer (Eds.), *Molecular Biology of Picornaviruses* (pp. 71–83). Washington, DC: ASM Press.

Holland, J. J., Spindler, K., Horodyski, F., Grabau, E., Nichol, S., & VandePol, S. (1982). Rapid evolution of RNA genomes. *Science, 215,* 1577–1585.

Hornitzky, M. (1982). Bee diseases research. *Aus. Beekeeper, 84,* 7–10.

Hoshino, K., Isawa, H., Tsuda, Y., Yano, K., Sasaki, T., Yuda, M., Takasaki, T., Kobayashi, M., & Sawabe, K. (2007). Genetic characterization of a new insect flavivirus isolated from *Culex pipiens* mosquito in Japan. *Virology, 359,* 405–414.

Hosur, M. V., Schmidt, T., Tucker, R. C., Johnson, J. E., Gallagher, T. M., Selling, B. H., & Rueckert, R. R. (1987). Structure of an insect virus at 3.0 A resolution. *Proteins, 2,* 167–176.

Hukuhara, T., & Bonami, J. R. (1991). Reoviridae. In J. R. Adams & J. R. Bonami (Eds.), *Atlas of Invertebrate Viruses* (pp. 393–434). Boca Raton: CRC Press.

Hung, A. C. F. (2000). PCR detection of Kashmir bee virus in honey bee excreta. *J. Api. Res., 39,* 103–106.

Hunnicutt, L. E., Hunter, W. B., Cave, R. D., Powell, C. A., & Mozoruk, J. J. (2006). Genome sequence and molecular characterization of *Homalodisca coagulata* virus-1, a novel virus discovered in the glassy-winged sharpshooter (Hemiptera: Cicadellidae). *Virology, 350,* 67–78.

Hunnicutt, L. E., Mozoruk, J., Hunter, W. B., Crosslin, J. M., Cave, R. D., & Powell, C. A. (2008). Prevalence and natural host range of *Homalodisca coagulata* virus-1 (HoCV-1). *Arch. Virol., 153,* 61–67.

Hunter, W., Ellis, J., Vanengelsdorp, D., Hayes, J., Westervelt, D., Glick, E., Williams, M., Sela, I., Maori, E., Pettis, J., Cox-Foster, D., & Paldi, N. (2010). Large-scale field application of RNAi technology reducing Israeli acute paralysis virus disease in honey bees (*Apis mellifera,* Hymenoptera: Apidae). *PLoS Path, 6,* e1001160.

Hunter, W. B., Katsar, C. S., & Chaparro, J. X. (2006). Molecular analysis of capsid protein of *Homalodisca coagulata* virus-1, a new leafhopper-infecting virus from the glassy-winged sharpshooter, *Homalodisca coagulata. J. Insect Sci., 6,* 1–10.

Ikeda, K., Nakazawa, H., Alain, R., Belloncik, S., & Mori, H. (1998). Characterizations of natural and induced polyhedrin gene mutants of *Bombyx mori* cytoplasmic polyhedrosis viruses. *Arch. Virol., 143,* 241–248.

Ikeda, K., Nakazawa, H., Winkler, S., Kotani, K., Yagi, H., Nakanishi, K., Miyajima, S., Kobayashi, J., & Mori, H. (2001). Molecular characterization of *Bombyx mori* cytoplasmic polyhedrosis virus genome segment 4. *J. Virol., 75,* 988–995.

Inoue, H. (1974). Multiplication of an infectious-flacherie virus in the resistant and susceptible strains of the silkworm, *Bombyx mori. J. Sericult. Sci. Jpn., 43,* 318–324.

Inoue, H., & Ayuzawa, C. (1972). Studies on the infectious flacherie of the silkworm, *Bombyx mori. J. Sericult. Sci. Jpn., 41,* 345–348.

Iqbal, J., & Mueller, U. (2007). Virus infection causes specific learning deficits in honeybee foragers. *Proc. R. Soc. B., 274,* 1517–1521.

Isawa, H., Asano, S., Sahara, K., Iizuka, T., & Bando, H. (1998). Analysis of genetic information of an insect picorna-like virus, infectious flacherie virus of silkworm: evidence for evolutionary relationships among insect, mammalian and plant picorna (-like) viruses. *Arch. Virol., 143,* 127–143.

Ishimori, N. (1934). Contribution à l'étude de la Grasserie du ver à soie. *C.R. Seances Soc. Biol. Fil., 116,* 1169–1174.

Jakubiec, A., & Jupin, I. (2007). Regulation of positive-strand RNA virus replication: the emerging role of phosphorylation. *Virus Res., 129,* 73–79.

Jan, E. (2006). Divergent IRES elements in invertebrates. *Virus Res., 119,* 16–28.

Jan, E., & Sarnow, P. (2002). Factorless ribosome assembly on the internal ribosome entry site of cricket paralysis virus. *J. Mol. Biol., 324,* 889–902.

Jang, C. J., Lo, M. C., & Jan, E. (2009). Conserved element of the dicistrovirus IGR IRES that mimics an E-site tRNA/ribosome interaction mediates multiple functions. *J. Mol. Biol., 387,* 42–58.

Jarvis, T. C., & Kirkegaard, K. (1992). Poliovirus RNA recombination: mechanistic studies in the absence of selection. *EMBO J., 11,* 3135–3145.

Jimenez, R. (1992). Síndrome de Taura. *Acuacultura del Ecuador., 1,* 1–16.

Johnson, J. E. (2008). Multidisciplinary studies of viruses: the role of structure in shaping the questions and answers. *J. Struct. Biol., 163,* 246–253.

Johnson, J. E., & Reddy, V. (1998). Structural studies of nodaviruses and tetraviruses. In L. K. Miller & L. A. Ball (Eds.), *The Insect Viruses* (pp. 171–223). New York: Plenum Press.

Johnson, K., & Christian, P. (1998). The novel genome organization of the insect picorna-like virus *Drosophila* C virus suggests this virus belongs to a previously undescribed virus family. *J. Gen. Virol., 79,* 191–203.

Johnson, K. N., Johnson, K. L., Dasgupta, R., Gratschb, T., & Ball, L. A. (2001). Comparisons among the larger genome segments of six nodaviruses and their encoded RNA replicases. *J. Gen. Virol., 82,* 1855–1866.

Kalmakoff, J., & McMillan, N. A. J. (1985). Research study on viruses isolated from *Darna catenatus* and *Setothosea asigna. FAO Report.* INS/86/014.

Kanga, L., & Fediere, G. (1991). Towards integrated control of *Epicerura pergrisea* (Lepidoptera: Notodontidae), defoliator of *Terminalia ivorensis* and *T. superba,* in the Côte d'Ivoire. *For. Ecol. Manag., 39,* 73–79.

Kapun, M., Nolte, V., Flatt, T., & Schlatterer, C. (2010). Host range and specificity of the Drosophila C virus. *PLoS ONE, 5,* e12421.

Kim, D. Y., Guzman, H., Bueno, R., Jr., Dennett, J. A., Auguste, A. J., Carrington, C. V. F., Popov, V. L., Weaver, S. C., Beasley, D. W. C., & Tesh, R. B. (2009). Characterization of *Culex flavivirus* (Flaviviridae) strains isolated from mosquitoes in the United States and Trinidad. *Virology, 386,* 154–159.

King, L. A., & Moore, N. F. (1985). The RNAs of two viruses of the Nudaurelia β family share little homology and have no terminal poly(A) tracts. *FEMS Microbiol. Lett., 26,* 41–43.

Kirkegaard, K., & Baltimore, D. (1986). The mechanism of RNA recombination in poliovirus. *Cell, 47,* 433–443.

Koonin, E. V., & Dolja, V. V. (1993). Evolution and taxonomy of positive strand RNA viruses, implications of comparative analysis of amino acid sequences. *Crit. Rev. Biochem. Mol. Biol., 28,* 375–430.

Koonin, E. V., Wolf, Y. I., Nagasaki, K., & Dolja, V. V. (2008). The big bang of picorna-like virus evolution antedates the radiation of eukaryotic supergroups. *Nat. Rev. Microbiol., 6,* 925–939.

Kopek, B. G., Perkins, G., Miller, D. J., Ellisman, M. H., & Ahlquist, P. (2007). Three-dimensional analysis of a viral RNA replication complex reveals a virus-induced mini-organelle. *PLoS Biol., 5*, e220.

Kuhn, R. J., Zhang, W., Rossmann, M. G., Pletnev, S. V., Corver, J., Lenches, E., Jones, C. T., Mukhopadhyay, S., Chipman, P. R., Strauss, E. G., Baker, T. S., & Strauss, J. H. (2002). Structure of dengue virus: implications for flavivirus organization, maturation, and fusion. *Cell, 8*, 717−725.

Kuno, G., Chang, G. J., Tsuchiya, K. R., Karabatsos, N., & Cropp, C. B. (1998). Phylogeny of the genus *Flavivirus. J. Virol., 72*, 73−83.

Lanzi, G., de Miranda, J. R., Boniotti, M. B., Cameron, C. E., Lavazza, A., Capucci, L., Camazine, S. M., & Rossi, C. (2006). Molecular and biological characterization of deformed wing virus of honeybees (*Apis mellifera* L.). *J. Virol., 80*, 4998−5009.

Laubscher, J. M., & von Wechmar, M. B. (1992). Influence of aphid lethal paralysis virus and *Rhopalosiphum padi* virus on aphid biology at different temperatures. *J. Invertebr. Pathol., 60*, 134−140.

Lautié-Harivel, N. (1992). *Drosophila* C virus cycle during the development of two *Drosophila melanogaster* strains (Charolles and Champetières) after larval contamination by food. *Biol. Cell, 76*, 151−157.

Le Gall, O., Christian, P., Fauquet, C. M., King, A. M., Knowles, N. J., Nakashima, N., Stanway, G., & Gorbalenya, A. E. (2008). Picornavirales, a proposed order of positive-sense single-stranded RNA viruses with a pseudo-T = 3 virion architecture. *Arch. Virol., 153*, 715−727.

Lee, P. E., & Furgala, B. (1967). Viruslike particles in adult honey bees (*Apis mellifera* Linnaeus) following injection with sacbrood virus. *Virology, 32*, 11−17.

Li, H., Li, X. W., & Ding, S. W. (2002). Induction and suppression of RNA silencing by an animal virus. *Science, 296*, 1319−1321.

Li, T. C., Scotti, P. D., Miyamura, T., & Takeda, N. (2007). Latent infection of a new Alphanodavirus in an insect cell line. *J. Virol., 81*, 10890−10896.

Li, W.-X., Li, H., Lu, R., Li, F., Dus, M., Atkinson, P., Brydon, E. W. A., Johnson, K. L., García-Sastre, A., Ball, L. A., Palese, P., & Ding, S.-W. (2004). Interferon antagonist proteins of influenza and vaccinia viruses are suppressors of RNA silencing. *Proc. Natl. Acad. Sci. USA, 101*, 1350−1355.

Li, Y., Li, T., Li, Y., Chen, W., Zhang, J., & Hu, Y. (2006). Identification and genome characterization of *Heliothis armigera* cypovirus types 5 and 14 and *Heliothis assulta* cypovirus type 14. *J. Gen. Virol., 87*, 387−394.

Li, Y., Zhang, J., Li, Y., Tan, L., Wuguo Chen, W., Haishan Luo, H., & Hu, Y. (2007). Phylogenetic analysis of *Heliothis armigera* cytoplasmic polyhedrosis virus type 14 and a series of dwarf segments found in the genome. *J. Gen. Virol., 88*, 991−997.

Lin, M., Ye, S., Xiong, Y., Cai, D., Zhang, J., & Hu, Y. (2010). Expression and characterization of RNA-dependent RNA polymerase of *Ectropis obliqua* virus. *BMB Rep., 43*, 284−290.

Liu, C., Zhang, J., Wang, J., Lu, J., Chen, W., Cai, D., & Hu, Y. (2006a). Sequence analysis of coat protein gene of Wuhan nodavirus isolated from insect. *Virus Res., 121*, 17−22.

Liu, C., Zhang, J., Yi, F., Wang, J., Wang, X., Jiang, H., Xu, J., & Hu, Y. (2006b). Isolation and RNA1 nucleotide sequence determination of a new insect nodavirus from *Pieris rapae* larvae in Wuhan city, China. *Virus Res., 120*, 28−35.

Longworth, J. F., Robertson, J. S., Tinsley, T. W., Rowlands, D. J., & Brown, F. (1973). Reactions between an insect picornavirus and naturally occurring IgM antibodies in several mammalian species. *Nature, 242*, 314−316.

Lu, J., Zhang, J., Wang, X., Jiang, H., Liu, C., & Hu, Y. (2006). *In vitro* and *in vivo* identification of structural and sequence elements in the 5′ untranslated region of *Ectropis obliqua* picorna-like virus required for internal initiation. *J. Gen. Virol., 87*, 3667−3677.

Lu, J., Hu, Y., Hu, L., Zong, S., Cai, D., Wang, J., Yu, H., & Zhang, J. (2007). *Ectropis obliqua* picorna-like virus IRES-driven internal initiation of translation in cell systems derived from different origins. *J. Gen. Virol., 88*, 2834−2838.

Lu, R., Maduro, M., Li, F., Li, H. W., Broitman-Maduro, G., Li, W. X., & Ding, S. W. (2005). Animal virus replication and RNAi-mediated antiviral silencing in *Caenorhabditis elegans. Nature, 436*, 1040−1043.

Lukashev, A. N. (2005). Role of recombination in evolution of enteroviruses. *Rev. Med. Virol., 15*, 157−167.

Lutomiah, J. L. L., Mwandawiro, C., Magambo, J., & Sang, R. C. (2007). Infection and vertical transmission of Kamiti river virus in laboratory bred *Aedes aegypti* mosquitoes. *J. Insect Sci., 55*, 1−7.

Mackenzie, J. (2005). Wrapping things up about virus RNA replication. *Traffic, 6*, 967−977.

Manousis, T., & Moore, N. F. (1987). Cricket paralysis virus, a potential control agent for the olive fruit fly, *Dacus oleae* Gmel. *Appl. Environ. Microbiol., 53*, 142−148.

Maori, E., Lavi, S., Mozes-Koch, R., Gantman, Y., Peretz, Y., Edelbaum, O., Tanne, E., & Sela, I. (2007a). Isolation and characterization of Israeli acute paralysis virus, a dicistrovirus affecting honeybees in Israel: evidence for diversity due to intra- and inter-species recombination. *J. Gen. Virol., 88*, 3428−3438.

Maori, E., Tanne, E., & Sela, I. (2007b). Reciprocal sequence exchange between non-retro viruses and hosts leading to the appearance of new host phenotypes. *Virology, 362*, 342−349.

Mari, J., Poulos, B. T., Lightner, D. V., & Bonami, J.-R. (2002). Shrimp Taura syndrome virus: genomic characterization and similarity with members of the genus cricket paralysis-like viruses. *J. Gen. Virol., 83*, 915−926.

Marti, G. A., Gonzalez, E. T., Garcia, J. J., Viguera, A. R., Guerin, D. M., & Echeverria, M. G. (2008). AC-ELISA and RT-PCR assays for the diagnosis of triatoma virus (TrV) in triatomines (Hemiptera: Reduviidae) species. *Arch. Virol., 153*, 1427−1432.

Mayo, M. A. (2002). Virus taxonomy − Houston 2002. *Arch. Virol., 147*, 1071−1076.

McCrae, M. A., & Mertens, P. P. C. (1983). *In vitro* translation of the genome and RNA coding assignments for cytoplasmic polyhedrosis type 1. In R. W. Compans & D. H. L. Bishop (Eds.), *Double-stranded RNA Viruses* (pp. 35−42). Amsterdam: Elsevier.

Mertens, P. P. C., Rao, S., & Zhou, H. (2004a). Cypovirus. In C. M. Fauquet, M. A. Mayo, J. Maniloff, U. Desselberger & L. A. Ball (Eds.), *Virus Taxonomy, Eighth Report of the International Committee on Taxonomy of Viruses* (pp. 22−33). London: Elsevier/ Academic Press.

Mertens, P. P. C., Rao, S., & Zhou, H. (2004b). Cypovirus, Reoviridae. In C. M. Fauquet, M. A. Mayo, J. Maniloff, U. Desselberger & L. A. Ball (Eds.), *Virus Taxonomy, Eighth Report of the International Committee on Taxonomy of Viruses* (pp. 522−533). London: Elsevier/ Academic Press.

de Miranda, J. R., & Fries, I. (2008). Venereal and vertical transmission of deformed wing virus in honeybees (*Apis mellifera* L.). *J. Invertebr. Pathol., 98*, 184−189.

de Miranda, J. R., & Genersch, E. (2010). Deformed wing virus. *J. Invertebr. Pathol., 103*(Suppl. 1), S48−S61.

de Miranda, J. R., Cordoni, G., & Budge, G. (2010a). The acute bee paralysis virus−Kashmir bee virus−Israeli acute paralysis virus complex. *J. Invertebr. Pathol., 103*, S30−S47.

de Miranda, J. R., Dainat, B., Locke, B., Cordoni, G., Berthoud, H., Gauthier, L., Neumann, P., Budge, G. E., Ball, B. V., & Stoltz, D. B. (2010b). Genetic characterization of slow bee paralysis virus of the honeybee (*Apis mellifera* L.). *J. Gen. Virol., 91*, 2524−2530.

Miura, K., Fujii-Kawata, I., Iwata, H., & Kawase, S. (1969). Electron-microscopic observation of a cytoplasmic-polyhedrosis virus from the silkworm. *J. Invertebr. Pathol., 14*, 262−265.

Mockel, N., Gisder, S., & Genersch, E. (2011). Horizontal transmission of deformed wing virus: pathological consequences in adult bees (*Apis mellifera*) depend on the transmission route. *J. Gen. Virol., 92*, 370−377.

Moon, J. S., Domier, L. L., McCoppin, N. K., D'Arcy, C. J., & Jin, H. (1998). Nucleotide sequence analysis shows that *Rhopalosiphum padi* virus is a member of a novel group of insect-infecting RNA viruses. *Virology, 243*, 54−65.

Moore, J., Jironkin, A., Chandler, D., Burroughs, N., Evans, D. J., & Ryabov, E. V. (2011). Recombinants between deformed wing virus and *Varroa destructor* virus-1 may prevail in *Varroa destructor*-infested honeybee colonies. *J. Gen. Virol., 92*, 156−161.

Moore, N. F., McKnight, L., & Tinsley, T. W. (1981). Occurrence of antibodies against insect virus proteins in mammals: simple model to differentiate between passive exposure and active virus growth. *Infect. Immunol., 31*, 825−827.

Morales-Betoulle, M. E., Monzón-Pineda, M. L., Sosa, S. M., Panella, N., López, B. M. R., Cordón-Rosales, C., Komar, N., Powers, A., & Johnson, B. W. (2008). *Culex* flavivirus isolates from mosquitoes in Guatemala. *J. Med. Entomol., 45*, 1187−1190.

Mori, H., & Metcalf, P. (2010). Cypoviruses. In S. Asgari & K. Johnson (Eds.), *Insect Virology* (pp. 307−323). Norfolk: Caister Academic Press.

Morris, T. J., Hess, R. T., & Pinnock, D. E. (1979). Physicochemical characterization of a small RNA virus associated with baculovirus infection in. *Trichoplusia ni. Intervirology, 11*, 238−247.

Morse, S. S. (1994). Evolution of genetic exchange in RNA viruses. In S. Morse (Ed.), *The Evolutionary Biology of Viruses* (pp. 1−28). New York: Raven Press.

van Munster, M., Dullemans, A. M., Verbeek, M., van den Heuvel, J. F. J. M., Clérivet, A., & van Der Wilk, F. (2002). Sequence analysis and genomic organization of Aphid lethal paralysis virus: a new member of the family *Dicistroviridae*. *J. Gen. Virol., 83*, 3131−3138.

Murphy, F. A., Scherer, W. F., Harrison, A. K., Dunne, H. W., & Gary, G. W., Jr. (1970). Characterization of Nodamura virus, an arthropod transmissible picornavirus. *Virology, 40*, 1008−1021.

Muscio, O. A., LaTorre, J. L., & Scodeller, E. A. (1987). Small non-occluded viruses from triatomine bug *Triatoma infestans* (Hemiptera: Reduviidae). *J. Invertebr. Pathol., 49*, 218−220.

Muscio, O. A., La Torre, J. L., & Scodeller, E. A. (1988). Characterization of *Triatoma* virus, a picorna-like virus isolated from the triatomine bug *Triatoma infestans*. *J. Gen. Virol., 69*, 2929−2934.

Muscio, O., Bonder, M. A., La Torre, J. L., & Scodeller, E. A. (2000). Horizontal transmission of *Triatoma* virus through the fecal−oral route in *Triatoma infestans* (Hemiptera: Triatomidae). *J. Med. Entomol., 37*, 271−275.

Nakashima, N., & Shibuya, N. (2006). Multiple coding sequences for the genome-linked virus protein (VPg) in dicistroviruses. *J. Invertebr. Pathol., 92*, 100−104.

Nakashima, N., Sasaki, J., Tsuda, K., Yasunaga, C., & Noda, H. (1998). Properties of a new picorna-like virus of the brown-winged green bug, *Plautia stali. J. Invertebr. Pathol., 71*, 151−158.

Newman, J. F. E., & Brown, F. (1973). Evidence for a divided genome in Nodamura virus, an arthropod-borne picornavirus. *Gen. Virol., 21*, 371−384.

Newman, J. F. E., & Brown, F. (1976). Absence of poly (A) from the infective RNA of Nodamura virus. *J. Gen. Virol., 30*, 137−140.

Nishizawa, T., Furuhashi, M., Nagai, T., Nakai, T., & Muroga, K. (1997). Genomic classification of fish nodaviruses by molecular phylogenetic analysis of the coat protein gene. *Appl. Environ. Microbiol., 63*, 1633−1636.

Odegard, A., Banerjee, M., & Johnson, J. E. (2010). Flock house virus: a model system for understanding non-enveloped virus entry and membrane penetration. *Curr. Top. Microbiol. Immunol., 343*, 1−22.

Odegard, A. L., Kwan, M. H., Walukiewicz, H. E., Banerjee, M., Schneemann, A., & Johnson, J. E. (2009). Low endocytic pH and capsid protein autocleavage are critical components of flock house virus cell entry. *J. Virol., 83*, 8628−8637.

van Oers, M. M. (2010). Genomics and biology of iflaviruses. In S. Asgari & K. Johnson (Eds.), *Insect Virology* (pp. 231−250). Norfolk: Caister Academic Press.

Oi, D. H., & Valles, S. M. (2009). Fire ant control with entomopathogens in the USA. In A. E. Hajek, T. R. Glare & M. O'Callaghan (Eds.), *Use of Microbes for Control and Eradication of Invasive Arthropods* (pp. 237−258). New York: Springer Science.

Olson, N. H., Baker, T. S., Johnson, J. E., & Hendry, D. A. (1990). The three-dimensional structure of frozen-hydrated *Nudaurelia capensis* β virus, a T = 4 insect virus. *J. Struct. Biol., 105*, 111−122.

Ongus, J. R., Peters, D., Bonmatin, J. M., Bengsch, E., Vlak, J. M., & van Oers, M. M. (2004). Complete sequence of a picorna-like virus of the genus Iflavirus replicating in the mite *Varroa destructor*. *J. Gen. Virol., 85*, 3747−3755.

Ongus, J. R., Roode, E. C., Pleij, C. W., Vlak, J. M., & van Oers, M. M. (2006). The 5' non-translated region of *Varroa destructor* virus 1 (genus Iflavirus): structure prediction and IRES activity in *Lymantria dispar* cells. *J. Gen. Virol., 87*, 3397−3407.

Pabbaraju, K., Ho, K. C. Y., Wong, S., Fox, J. D., Kaplen, B., Tyler, S., Drebot, M., & Tilley, P. A. G. (2009). Surveillance of mosquito-borne viruses in Alberta using reverse transcription polymerase chain reaction with generic primers. *J. Med. Entomol., 46*, 640−648.

Pal, N., Boyapalle, S., Beckett, R., Miller, W. A., & Bonning, B. C. (2007). A baculovirus-expressed dicistrovirus that is infectious to aphids. *J. Virol., 81*, 9339−9345.

Palacios, G., Hui, J., Quan, P. L., Kalkstein, A., Honkavuori, K. S., Bussetti, A. V., Conlan, S., Evans, J., Chen, Y. P., vanEngelsdorp, D., Efrat, H., Pettis, J., Cox-Foster, D., Holmes, E. C., Briese, T., & Lipkin, W. I. (2008). Genetic analysis of Israeli acute paralysis virus: distinct clusters are circulating in the United States. *J. Virol., 82*, 6209−6217.

Pantoja, C. R., Navarro, S. A., Naranjo, J., Lightner, D. V., & Gerba, C. P. (2004). Nonsusceptibility of primate cells to Taura syndrome virus. *Emerg. Infect. Dis., 10*, 2106−2112.

Payne, C. C., & Rivers, C. F. (1976). A provisional classification of cytoplasmic polyhedrosis viruses based on the sizes of the RNA genome segments,. *J. Gen. Virol., 33*, 71−85.

Payne, C. C., Piasecka-Serafin, M., & Pilley, B. (1977). The properties of two recent isolates of cytoplasmic polyhedrosis viruses. *Intervirology, 8*, 155−163.

Pestova, T. V., & Hellen, C. U. (2003). Translation elongation after assembly of ribosomes on the cricket paralysis virus internal ribosomal entry site without initiation factors or initiator tRNA. *Genes Dev., 17*, 181−186.

Petralia, R. S., & Vinson, S. B. (1979). Developmental morphology of larvae and eggs of the imported fire ant, *Solenopsis invicta. Ann. Entomol. Soc. Am., 72*, 472−484.

Philippe, R., Veyrunes, J.-C., Mariau, D., & Bergoin, M. (1997). Biological control using entomopathogenic viruses. Application to oil palm and coconut pests. *Plantations, recherche, développement, 1*, 39−45.

du Plessis, L., Hendry, D. A., Dorrington, R. A., Hanzlik, T. N., Johnson, J. E., & Appel, M. (2005). Revised RNA2 sequence of the tetravirus, *Nudaurelia capensis* ω virus (NωV). *Arch. Virol., 150*, 2397−2402.

Plus, N., & Scotti, P. D. (1984). The biological properties of eight different isolates of cricket paralysis virus. *Ann. Inst. Pasteur Vir., 135*, 257−268.

Plus, N., Crozier, G., Jousset, F. X., & David, J. (1975). Picornaviruses of laboratory and wild *Drosophila melanogaster*: geographical distribution and serotypic composition. *Ann. Microbiol., 126A*, 107−121.

Plus, N., Croizier, G., Reinganum, C., & Scotti, P. D. (1978). Cricket paralysis virus and *Drosophila C virus*: serological analysis and comparison of capsid polypeptides and host range. *J. Invertebr. Pathol., 31*, 296−302.

Porter, S. D. (1998). Biology and behavior of *Pseudacteon* decapitating flies (Diptera: Phoridae) that parasitize *Solenopsis* fire ants (Hymenoptera: Formicidae). *Fla. Entomol., 81*, 292−309.

Price, B. D., Rueckert, R. R., & Ahlquist, P. (1996). Complete replication of an animal virus and maintenance of expression vectors derived from it in *Saccharomyces cerevisiae. Proc. Natl. Acad. Sci. USA, 93*, 9465−9470.

Pringle, F. M., Gordon, K. H. J., Hanzlik, T. N., Kalmakoff, J., Scotti, P. D., & Ward, V. K. (1999). A novel capsid expression strategy for *Thosea asigna* virus, a member of the Tetraviridae. *J. Gen. Virol., 80*, 1855−1863.

Pringle, F. M., Johnson, K. N., Goodman, C. L., McIntosh, A. H., & Ball, L. A. (2003). Providence virus, a new member of the Tetraviridae that infects cultured insect cells. *Virology, 306*, 359−370.

Racaniello, V. R. (2006). One hundred years of poliovirus pathogenesis. *Virology, 344*, 9−16.

Rao, S., Carner, G. R., Scott, S. W., Omura, T., & Hagiwara, K. (2003). Comparison of the amino acid sequences of RNA-dependent RNA polymerases of cypoviruses in the family Reoviridae. *Arch. Virol., 148*, 209−219.

Reinganum, C., O'Loughlin, G. T., & Hogan, T. W. (1970). A nonoccluded virus of the field crickets *Teleogryllus oceanicus* and *T. commodus* (Orthoptera: Gryllidae). *J. Invertebr. Pathol., 16*, 214−220.

Ribière, M., Ball, B., & Aubert, M. F. A. (2008). Natural history and geographic distribution of honey bee viruses. In M. Aubert, B. Ball, I. Fries, R. Moritz, N. Milani & I. Bernardinelli (Eds.), *Virology and the Honey Bee* (pp. 15−84). Brussels: EC Publications.

Rice, C. M., Lenches, E. M., Eddy, S. R., Shin, S. J., Sheets, R. L., & Strauss, J. H. (1985). Nucleotide sequence of yellow fever virus, implications for flavivirus gene expression and evolution. *Science, 229*, 726−733.

Rortais, A., Tentcheva, D., Papachristoforou, A., Gauthier, L., Arnold, G., Colin, M. E., & Bergoin, M. (2006). Deformed wing virus is not related to honey bees' aggressiveness. *Virol. J., 3*, 61.

Rossmann, M. G., Bella, J., Kolatkar, P. R., He, Y., Wimmer, E., Kuhn, R. J., & Baker, T. S. (2000). Cell recognition and entry by rhino- and enteroviruses. *Virology, 269*, 239−247.

Rozas-Dennis, G. S., & Cazzaniga, N. J. (2000). Effects of *Triatoma* virus (TrV) on the fecundity and moulting of *Triatoma infestans* (Hemiptera: Reduviidae). *Ann. Trop. Med. Parasitol., 94*, 633−641.

Rozas-Dennis, G. S., Cazzaniga, N. J., & Guerin, D. M. (2002). *Triatoma patagonica* (Hemiptera, Reduviidae), a new host for *Triatoma* virus. *Mem. Inst. Oswaldo Cruz., 97*, 427−429.

Ruiz, M. C., Cohen, F., & Michelangeli, F. (2000). Role of Ca^{2+} in the replication and pathogenesis of rotavirus and other viral infections. *Cell Calcium, 28*, 137−149.

Ryabov, E. V. (2007). A novel virus isolated from the aphid *Brevicoryne brassicae* with similarity to Hymenoptera picorna-like viruses. *J. Gen. Virol., 88*, 2590−2595.

Sanchez, M. D., Pierson, T. C., McAllister, D., Hanna, S. L., Puffer, B. A., Valentine, L. E., Murtadha, M. M., Hoxie, J. A., & Doms, R. W. (2005). Characterization of neutralizing antibodies to West Nile virus. *Virology, 336*, 70−82.

Sang, R. C., Gichogo, A., Gachoya, J., Dunster, M. D., Ofula, V., Hunt, A. R., Crabtree, M. B., Miller, B. R., & Dunster, L. M. (2003). Isolation of a new flavivirus related to cell fusing agent virus (CFAV) from field-collected flood-water *Aedes* mosquitoes sampled from a dambo in central Kenya. *Arch. Virol., 148*, 1085−1093.

Sankoff, D. (2003). Rearrangements and chromosomal evolution. *Curr. Opin. Genet. Dev., 13*, 583−587.

Santi, L., Huang, Z., & Mason, H. (2006). Virus-like particles production in green plants. *Methods, 40*, 66−76.

Santillán-Galicia, M. T., Carzaniga, R., Ball, B. V., & Alderson, P. G. (2008). Immunolocalization of deformed wing virus particles within the mite *Varroa destructor. J. Gen. Virol., 89*, 1685−1689.

Santillán-Galicia, M. T., Ball, B. V., Clark, S. J., & Alderson, P. G. (2010). Transmission of deformed wing virus and slow paralysis virus to adult bees (*Apis mellifera* L.) by *Varroa destructor. J. Apic. Res., 49*, 141−148.

Scherer, W. F. (1968). Variable results of sodium deoxycholate tests of Nodamura virus, an ether and chloroform resistant arbovirus. *Proc. Soc. Exp. Biol. Med., 129*, 194−199.

Scherer, W. F., & Hurlbut, H. S. (1967). Nodamura virus from Japan, a new and unusual arbovirus resistant to diethyl ether and chloroform. *Am. J. Epidemiol., 86*, 271−285.

Scherer, W. F., Verna, J. E., & Richter, G. W. (1968). Nodamura virus, an ether- and chloroform-resistant arbovirus from Japan. Physical and biological properties, with ecologic observations. *Am. J. Trop. Med. Hyg., 17*, 120−128.

Schluns, H., & Crozier, R. H. (2009). Molecular and chemical immune defenses in ants (Hymenoptera: Formicidae). *Myrmecol. News, 12,* 237−249.

Schneemann, A., & Marshall, D. (1998). Specific encapsidation of Nodavirus RNAs is mediated through the C terminus of capsid precursor protein alpha. *J. Virol., 72,* 8738−8746.

Schneemann, A., Zhang, W., Gallagher, T. M., & Rueckert, R. R. (1992). Maturation cleavage required for infectivity of a nodavirus. *J. Virol., 66,* 6728−6734.

Schneemann, A., Ball, L. A., Delsert, C., Johnson, J. E., & Nishizawa, T. (2005). Family Nodaviridae. In C. M. Fauquet, M. A. Mayo, J. Maniloff, U. Desselberger & L. A. Ball (Eds.), *Virus Taxonomy, Eighth Report of the International Committee on Taxonomy of Viruses* (pp. 865−872). San Diego: Elsevier/Academic Press.

Scotti, P. D., Dearing, S., & Mossop, D. W. (1983). Flock house virus: a nodavirus isolated from *Costelytra zealandica* (White) (Coleoptera: Scarabaeidae). *Arch. Virol., 75,* 181−189.

Scotti, P. D., Hoefakker, P., & Dearing, S. (1996). The production of cricket paralysis virus in suspension cultures of insect cell lines. *J. Invertebr. Pathol., 68,* 109−112.

Selling, B. H., Allison, R. F., & Kaesberg, P. (1990). Genomic RNA of an insect virus directs synthesis of infectious virions in plants. *Proc. Natl. Acad. Sci. USA, 87,* 434−438.

Shah, K. S., Evans, E. C., & Pizzorno, M. C. (2009). Localization of deformed wing virus (DWV) in the brains of the honeybee, *Apis mellifera* Linnaeus. *Virol. J., 6,* 182.

Shapiro, A., Green, T., Rao, S., White, S., Carner, G., Mertens, P. P. C., & Becnel, J. J. (2005). Morphological and molecular characterization of a cypovirus (*Reoviridae*) from the mosquito *Uranotaenia sapphirina* (Diptera, Culicidae). *J. Virol., 79,* 9430−9438.

Shen, M., Cui, L., Ostiguy, N., & Cox-Foster, D. (2005a). Intricate transmission routes and interactions between picorna-like viruses (Kashmir bee virus and sacbrood virus) with the honeybee host and the parasitic varroa mite. *J. Gen. Virol., 86,* 2281−2289.

Shen, M., Yang, X., Cox-Foster, D., & Cui, L. (2005b). The role of varroa mites in infections of Kashmir bee virus (KBV) and deformed wing virus (DWV) in honey bees. *Virology, 342,* 141−149.

Smith, D. B., McAllister, J., Casino, C., & Simmonds, P. (1997). Virus "quasispecies": making a mountain out of a molehill? *J. Gen. Virol., 78,* 1511−1519.

Stiasny, K., Kossl, C., Lepault, J., Rey, F. A., & Heinz, F. X. (2007). Characterization of a structural intermediate of flavivirus membrane fusion. *PLoS Pathog., 3,* e20.

Stollar, V., & Thomas, V. L. (1975). An agent in the *Aedes aegypti* cell line (Peleg) which causes fusion of *Aedes albopictus* cells. *Virology, 64,* 367−377.

Sullivan, C. S., & Ganem, D. A. (2005). Virus-encoded inhibitor that blocks RNA interference in mammalian cells. *J. Virol., 79,* 7371−7379.

Suzuki, Y., Toriyama, S., Matsuda, I., & Kojima, M. (1993). Detection of a picorna-like virus, Himetobi P virus, in organs and tissues of *Laodelphax striatellus* by immunogold labeling and enzyme-linked immunosorbent assay. *J. Invertebr. Pathol., 62,* 99−104.

Tan, L., Zhang, J., Li, Y., Li, Y., Jiang, H., Cao, X., & Hu, Y. (2008). The complete nucleotide sequence of the type 5 *Helicoverpa armigera* cytoplasmic polyhedrosis virus genome. *Virus Genes, 36,* 587−593.

Tanada, Y., & Kaya, H. K. (1993). *Insect Pathology.* San Diego: Academic Press.

Tang, L., Johnson, K. N., Ball, L. A., Lin, T., Yeager, M., & Johnson, J. E. (2001). The structure of Pariacoto virus reveals a dodecahedral cage of duplex RNA. *Nat. Struct. Biol., 8,* 77−83.

Tate, J., Liljas, L., Scotti, P., Christian, P., Lin, T., & Johnson, J. E. (1999). The crystal structure of cricket paralysis virus: the first view of a new virus family. *Nat. Struct. Biol., 6,* 765−774.

Tentcheva, D., Gauthier, L., Zappulla, N., Dainat, B., Cousserans, F., Colin, M. E., & Bergoin, M. (2004). Prevalence and seasonal variations of six bee viruses in *Apis mellifera* L. and *Varroa destructor* mite populations in France. *Appl. Environ. Microbiol., 70,* 7185−7191.

Terio, V., Martella, V., Camero, M., Decaro, N., Testini, G., Bonerba, E., Tantillo, G., & Buonavoglia, C. (2008). Detection of a honeybee iflavirus with intermediate characteristics between kakugo virus and deformed wing virus. *New Microbiol., 31,* 439−444.

Tesh, R. B. (1980). Infectivity and pathogenicity of Nodamura virus for mosquitoes. *J. Gen. Virol., 48,* 177−182.

Thiel, H. J., Collett, M. S., Gould, E. A., Heinz, F. X., Houghton, M., Meyers, G., Purcell, R. H., & Rice, C. M. (2005). Family *Flaviviridae*. In C. M. Fauquet, M. A. Mayo, J. Maniloff, U. Desselberger & L. A. Ball (Eds.), *Virus Taxonomy. Classification and Nomenclature of Viruses. Eighth Report of the International Committee on the Taxonomy of Viruses* (pp. 981−999). San Diego: Elsevier/Academic Press.

Thomas-Orillard, M., Bernard, J., & Cusset, G. (1995). *Drosophila*-host genetic control of susceptibility to *Drosophila* C virus. *Genetics, 140,* 1289−1295.

Tinsley, T. W., MacCallum, F. O., Robertson, J. S., & Brown, F. (1984). Relationship of encephalomyocarditis virus to cricket paralysis virus of insects. *Intervirology, 21,* 181−186.

Todd, J. H., De Miranda, J. R., & Ball, B. V. (2007). Incidence and molecular characterization of viruses found in dying New Zealand honey bee (*Apis mellifera*) colonies infested with Varroa destructor. *Apidologie, 38,* 354−367.

Tomasicchio, M., Venter, P. A., Gordon, K. H. J., Hanzlik, T. N., & Dorrington, R. A. (2007). Induction of apoptosis in *Saccharomyces cerevisiae* results in the spontaneous maturation of tetravirus procapsids *in vivo*. *J. Gen. Virol., 88,* 1576−1582.

Toriyama, S., Guy, P. L., Fuji, S.-I., & Takahashi, M. (1992). Characterization of a new picorna-like virus, himetobi P virus, in planthoppers. *J. Gen. Virol., 73,* 1021−1023.

Tripconey, D. (1970). Studies on a non-occluded virus of the pine tree emperor moth. *J. Invertebr. Pathol., 15,* 268−275.

Tucker, S. P., Thornton, C. L., Wimmer, E., & Compans, R. W. (1993). Vectorial release of poliovirus from polarized human intestinal epithelial cells. *J. Virol., 67,* 4274−4282.

Valle, L. D., Negrisolo, E., Patarnello, P., Zanella, L., Maltese, C., Bovo, G., & Colombo, L. (2001). Sequence comparison and phylogenetic analysis of fish nodaviruses based on the coat protein gene. *Arch. Virol., 146,* 1125−1137.

Valles, S. M., & Hashimoto, Y. (2009). Isolation and characterization of *Solenopsis invicta* virus 3, a new positive-strand RNA virus infecting the red imported fire ant. *Solenopsis invicta. Virology, 388,* 354−361.

Valles, S. M., & Porter, S. D. (2007). *Pseudacteon* decapitating flies: potential vectors of a fire ant virus? *Fla. Entomol., 90,* 268−270.

Valles, S. M., & Strong, C. A. (2005). *Solenopsis invicta* virus-1A (SINV-1A): distinct species or genotype of SINV-1? *J. Invertebr. Pathol., 88,* 232−237.

Valles, S. M., Strong, C. A., Dang, P. M., Hunter, W. B., Pereira, R. M., Oi, D. H., Shapiro, A. M., & Williams, D. F. (2004). A picorna-like virus from the red imported fire ant, *Solenopsis invicta*: initial discovery, genome sequence, and characterization. *Virology, 328*, 151–157.

Valles, S. M., Strong, C. A., Oi, D. H., Porter, S. D., Pereira, R. M., Vander Meer, R. K., Hashimoto, Y., Hooper-Bùi, L. M., Sánchez-Arroyo, H., Davis, T., Karpakakunjaram, V., Vail, K. M., Fudd Graham, L. C., Briano, J. A., Calcaterra, L. A., Gilbert, L. E., Ward, R., Ward, K., Oliver, J. B., Taniguchi, G., & Thompson, D. C. (2007). Phenology, distribution, and host specificity of *Solenopsis invicta* virus-1. *J. Invertebr. Pathol., 96*, 18–27.

Valles, S. M., Strong, C. A., Hunter, W. B., Dang, P. M., Pereira, R. M., Oi, D. H., & Williams, D. F. (2008). Expressed sequence tags from the red imported fire ant, *Solenopsis invicta*: annotation and utilization for discovery of viruses. *J. Invertebr. Pathol., 99*, 74–81.

Valles, S. M., Oi, D. H., & Porter, S. D. (2010). Seasonal variation and the co-occurrence of four pathogens and a group of parasites among monogyne and polygyne fire ant colonies. *Biol. Control, 54*, 342–348.

Venter, P. A., Jovel, J., & Schneemann, A. (2010). Nodaviruses. In S. Asgari & K. Johnson (Eds.), *Insect Virology* (pp. 251–282). Norfolk: Caister Academic Press.

Walter, C. T., Pringle, F. M., Nakayinga, R., de Felipe, P., Ryan, M. D., Ball, L. A., & Dorrington, R. A. (2010). Genome organization and translation products of Providence virus: insight into a unique tetra-virus. *J. Gen. Virol., 91*, 2826–2835.

Wang, C. H., Wu, C. Y., & Lo, C. F. (1999). A new picorna-like virus, PnPV, isolated from ficus transparent wing moth, *Perina nuda* (Fabricius). *J. Invertebr. Pathol., 74*, 62–68.

Wang, X., Zhang, J., Lu, J., Yi, F., Liu, C., & Hu, Y. (2004a). Sequence analysis and genomic organization of a new insect picorna-like virus, *Ectropis obliqua* picorna-like virus, isolated from *Ectropis obliqua*. *J. Gen. Virol., 85*, 1145–1151.

Wang, X. C., Zhang, J. M., Liu, C. F., & Hu, Y. Y. (2004b). A new insect picornavirus isolated from *Ectropis obliqua*. *Virol. Sin., 19*, 39–43.

Wang, X. H., Aliyari, R., Li, W. X., Li, H. W., Kim, K., Carthew, R., Atkinson, P., & Ding, S. W. (2006). RNA interference directs innate immunity against viruses in adult *Drosophila*. *Science, 312*, 452–454.

Watanabe, H., Kurihara, Y., Wang, Y.-X., & Shimizu, T. (1988). Mulberry pyralid, *Glyphodes pyloalis*: habitual host of nonoccluded viruses pathogenic to the silkworm, *Bombyx mori*. *J. Invertebr. Pathol., 52*, 401–408.

Waters, A., Coughlan, S., & Hall, W. W. (2007). Characterisation of a novel recombination event in the norovirus polymerase gene. *Virology, 363*, 11–14.

Weidman, M. K., Sharma, R., Raychaudhuri, S., Kundu, P., Tsai, W., & Dasgupta, A. (2003). The interaction of cytoplasmic RNA viruses with the nucleus. *Virus Res., 95*, 75–85.

Wileman, T. (2006). Aggresomes and autophagy generate sites for virus replication. *Science, 312*, 875–878.

Williamson, C., Rybicki, E. P., Kasdorf, G. G. F., & Von Wechmar, M. B. (1988). Characterization of a new picorna-like virus isolated from aphids. *J. Gen. Virol., 69*, 787–795.

Wilson, J. E., Powell, M. J., Hoover, S. E., & Sarnow, P. (2000). Naturally occurring dicistronic cricket paralysis virus RNA is regulated by two internal ribosome entry sites. *Mol. Cell. Biol., 20*, 4990–4999.

Wood, B. J. (2002). Pest control in Malaysia's perennial crops, a half century perspective tracking the pathway to integrated pest management. *Integr. Pest Manag. Rev, 7*, 173–190.

Wu, C. Y., Lo, C. F., Huang, C. J., Yu, H. T., & Wang, C. H. (2002). The complete genome sequence of *Perina nuda* picorna-like virus, an insect-infecting RNA virus with a genome organization similar to that of the mammalian picornaviruses. *Virology, 294*, 312–323.

Wu, T. Y., Wu, C. Y., Chen, Y. J., Chen, C. Y., & Wang, C. H. (2007). The 5′ untranslated region of *Perina nuda* virus (PnV) possesses a strong internal translation activity in baculovirus-infected insect cells. *FEBS Lett., 581*, 3120–3126.

Yang, X., & Cox-Foster, D. L. (2005). Impact of an ectoparasite on the immunity and pathology of an invertebrate: evidence for host immunosuppression and viral amplification. *Proc. Natl. Acad. Sci. USA, 102*, 7470–7475.

Yi, F., Zhang, J., Yu, H., Liu, C., Wang, J., & Hu, Y. (2005). Isolation and identification of a new tetravirus from *Dendrolimus punctatus* larvae collected from Yunnan province, China. *J. Gen. Virol., 86*, 789–796.

Yue, C., & Genersch, E. (2005). RT-PCR analysis of deformed wing virus in honeybees (*Apis mellifera*) and mites (*Varroa destructor*). *J. Gen. Virol., 86*, 3419–3424.

Yue, C., Schroder, M., Gisder, S., & Genersch, E. (2007). Vertical-transmission routes for deformed wing virus of honeybees (*Apis mellifera*). *J. Gen. Virol., 88*, 2329–2336.

Zeddam, J. L., Gordon, K. H. J., Lauber, C., Felipe Alves, C. A., Luke, B. T., Hanzlik, T. N., Ward, V. K., & Gorbalenya, A. E. (2010). *Euprosterna elaeasa* virus genome sequence and evolution of the Tetraviridae family: emergence of bipartite genomes and conservation of the VPg signal with the dsRNA Birnaviridae family. *Virology, 397*, 145–154.

Zhang, H., Zhang, J., Yu, X., Lu, X., Zhang, Q., Jakana, J., Chen, D. H., Zhang, X., & Zhou, Z. H. (1999). Visualization of protein-RNA interactions in cytoplasmic polyhedrosis virus. *J. Virol., 73*, 1624–1629.

Zhang, Q., Ongus, J. R., Boot, W. J., Calis, J., Bonmatin, J. M., Bengsch, E., & Peters, D. (2007). Detection and localisation of picorna-like virus particles in tissues of *Varroa destructor*, an ecto-parasite of the honey bee, *Apis mellifera*. *J. Invertebr. Pathol., 96*, 97–105.

Zhang, R., He, J., Su, H., Dong, C., Guo, Z., Ou, Y., Deng, X., & Weng, S. (2010a). Identification of the structural proteins of VP1 and VP2 of a novel mud crab dicistrovirus. *J. Virol. Methods, 171*, 323–328.

Zhang, R., He, J., Su, H., Dong, C., Guo, Z., & Weng, S. (2010b). Monoclonal antibodies produced against VP3 of a novel mud crab dicistrovirus. *Hybridoma (Larchmt), 29*, 437–440.

Zhao, S. L., Liang, C. Y., Hong, J. J., Xu, H. G., & Peng, H. Y. (2003). Molecular characterization of segments 7–10 of *Dendrolimus punctatus* cytoplasmic polyhedrosis virus provides the complete genome. *Virus Res., 94*, 17–23.

Chapter 6

Fungal Entomopathogens

Fernando E. Vega*, Nicolai V. Meyling†, Janet Jennifer Luangsa-ard** and Meredith Blackwell‡

*United States Department of Agriculture, Agricultural Research Service, Beltsville, Maryland, USA, †University of Copenhagen, Frederiksberg, Denmark, **BIOTEC, Khlong Luang, Pathumthani, Thailand, ‡Louisiana State University, Baton Rouge, Louisiana, USA

Chapter Outline

SUMMARY

Fungal entomopathogens are important biological control agents worldwide and have been the subject of intense research for more than 100 years. They exhibit both sexual and asexual reproduction and produce a variety of infective propagules. Their mode of action against insects involves attachment of the spore to the insect cuticle followed by germination, cuticle penetration, and internal dissemination throughout the insect. During this process, which may involve the production of secondary metabolites, the internal organs of the insect are eventually degraded. Environmental factors such as ultraviolet light, temperature, and humidity

Insect Pathology. DOI: 10.1016/B978-0-12-384984-7.00006-3

can influence the effectiveness of fungal entomopathogens in the field. Phylogenetic studies have resulted in a better understanding of associations with other fungi as well as a new classification scheme. Ecological studies have revealed fascinating aspects related to their host range, distribution, abundance, and trophic interactions. The development of fungal entomopathogens as effective biological control agents requires knowledge of bioassay methods, as well as production, formulation, and application methodologies. Some important case studies involve the gypsy moth in the USA, locusts and grasshoppers in Africa, and spittlebugs in Brazil. This chapter focuses on the biology, classification and phylogeny, ecology, and use of fungal entomopathogens as biological control agents.

6.1. INTRODUCTION

Members of the kingdom Fungi are common in terrestrial and aquatic environments throughout the world. Terrestrial fungi have been reported as pathogens or parasites of humans, animals, and plants, as plant endophytes, as symbionts of arthropods and roots of plants, and as components of the soil microbiota, among others (Alexopoulos *et al.*, 1996; Watanabe, 2010).

Based on molecular clock dating methods used to assess fungal evolution, fungal presence on Earth varies widely depending on the fossils used for calibration of the clock, and current estimates propose that fungi evolved approximately 0.5–1.5 billion years ago (Wang *et al.*, 1999; Heckman *et al.*, 2001; Lücking *et al.*, 2009; Berbee and Taylor, 2010). Fossil records from the Rhynie chert in northern Scotland confirm the presence of a fungal community (Ascomycota, Chytridiomycota, and Glomeromycota) in the Early Devonian [400 million years ago (mya)] (Taylor *et al.*, 2005; Taylor and Berbee, 2006; Berbee and Taylor, 2010). The earliest evidence of fungal entomopathogens, however, comes later from three specimens preserved in amber. A scale insect attacked by an *Ophiocordyceps*-like anamorph named *Paleoophiocordyceps coccophagus* is the oldest specimen from Myanmar (Burmese) amber from the Early Cretaceous (100–110 mya) (Sung *et al.*, 2008). The other two examples are preserved in amber from the Dominican Republic from the Oligocene (20–30 mya), and include what appears to be an *Entomophthora* species attacking a termite (Poinar and Thomas, 1982) and an ant infected with *Beauveria bassiana* (Poinar and Thomas, 1984).

Of the estimated 1.5–5.1 million species of fungi in the world (Hawksworth, 1991; O'Brien *et al.*, 2005; Hibbett *et al.*, 2011), approximately 100,000 have been described (Blackwell, 2011). Of these, approximately 750–1000 are fungal entomopathogens placed in over 100 genera (Roberts and Humber, 1981; McCoy *et al.*, 1988; St. Leger and Wang, 2010). However, based on the number of cryptic species revealed by recent molecular phylogeny studies (Rehner, 2009), it is evident that these

estimates are low. Fungal entomopathogens thus constitute the largest number of taxa that are insect pathogens (Ignoffo, 1973). de Faria and Wraight (2007) identified 171 fungal-based products used as biocontrol agents since the 1960s, most of them based on *B. bassiana*, *Beauveria brongniartii*, *Metarhizium anisopliae*, and *Isaria fumosorosea*.

Even though many terms have been used to refer to fungi that infect and kill insects, e.g., insect-pathogenic fungi, entomopathogenic fungi, entomogenous fungi, or entomophthorous fungi, the term "fungal entomopathogens" will be used throughout this chapter. In addition, the term "isolate" will be used for a "pure culture of microorganism obtained from some natural substrate" to differentiate from a "strain", which is "a population of homogeneous organisms possessing a set of defined characteristics" (Onstad *et al.*, 2006). Finally, some fungal pathogens of spiders have been included to illustrate their phylogenetic relatedness to fungal entomopathogens. In addition, spiders are commonly found in the same habitats as insects and, generally speaking, fall under the purview of entomologists, even though spiders (Class Arachnida) are not insects (Class Insecta).

A note on taxonomy: In recent taxonomic revisions, *Paecilomyces fumosoroseus* and *P. farinosus* have been reclassified as *Isaria fumosorosea* and *I. farinosa*, respectively (Luangsa-ard *et al.*, 2004, 2005). In this chapter, the new taxonomy is used even when the authors have reported these species studied under their former names. However, the former *Metarhizium anisopliae* and *Verticillium lecanii* have been separated into numerous species (Zare and Gams, 2001; Bischoff *et al.*, 2009). Consequently, citations of previous studies that involve *M. anisopliae* refer to *M. anisopliae sensu lato*. The former *Verticillium lecanii* was reclassified into several species in the new genus *Lecanicillium*, which includes *L. lecanii* (Zare and Gams, 2001). Therefore, unless it is known that the specific isolates in question were reclassified according to the new nomenclature, it is erroneous to call *L. lecanii* anything that was formerly referred to as *V. lecanii* (Goettel *et al.*, 2008). Accordingly, a species that was formerly described as *V. lecanii* should be referred to as *Lecanicillium* spp. unless it is known that the specific isolate referred to has been reclassified.

6.2. CLASSIFICATION AND PHYLOGENY

6.2.1. Classification of Fungi and Fungus-like Organisms

Since the first DNA sequencing studies began more than 20 years ago (Alexopoulos *et al.*, 1996), the phylogenetic classification of fungi and fungus-like organisms has changed dramatically (Figs. 6.1 and 6.2) (Hibbett *et al.*,

2007; Beakes and Sekimoto, 2009; Bhattacharya *et al.*, 2009). DNA provides a large number of characters that overcome the problems encountered by mycologists, such as few phenotypic traits, absence of common morphological characters, and convergent characters. Even the earliest phylogenetic studies using single gene sequences answered several long-standing questions about which organisms actually are fungi. For example, the phylogenetic analyses showed that organisms with a single posterior whiplash flagellum at some stage in their life cycles are fungi, although most fungi lack a flagellum. Within the monophyletic Fungi the early diverging fungal groups are not all well resolved, but they appear to be more diverse than previously recognized (Hibbett *et al.*, 2007). These lineages contain zoosporic fungi, i.e., those with flagellated spores, and zygosporic fungi, i.e., those producing zygospores (a type of resting spore) in sexual reproduction (Fig. 6.1). Microsporidia, intracellular pathogens of a variety of insects, as well as other invertebrates and vertebrates, are not an early diverging lineage in the tree of life as they were considered in the past, but rather they are a lineage diverging early near the base of the fungal clade (Lee *et al.*, 2010) (Chapter 7).

Oomycetes and the slime molds have been excluded from Fungi. The oomycetes, including the insect pathogen

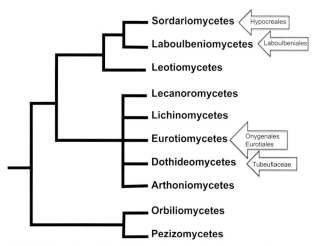

FIGURE 6.2 About 55 orders are placed in the 10 classes of the ascomycete subphylum Pezizomycotina (Schoch *et al.*, 2009). The classes and orders containing entomopathogens mentioned in the text are indicated by the arrows. For example, the Sordariomycetes contains about 13 orders, one of which is the Hypocreales, the order with the largest number of entomopathogenic species. Some members of the Pezizomycotina remain unclassified. (Tree based on Schoch *et al.*, 2009.)

Lagenidium, are placed among other anteriorly biflagellated organisms in the stramenopiles (Beakes and Sekimoto, 2009; Green, 2011). Two types of flagella are present in most members of this group. One flagellum with mastigonemes (hollow hairs) is projected forward to act as a rudder, and a trailing smooth whiplash flagellum propels the cell forward. Some of the stramenopiles have chloroplasts with chlorophylls *a* and *c* (Beakes and Sekimoto, 2009; Green, 2011). The slime mold groups previously considered to be fungi have been placed in Amoebozoa and Heterolobosa (Baldauf *et al.*, 2000; Steenkamp *et al.*, 2006). The arthropod gut symbionts, Eccrinales and Amoebidiales, previously considered to be Trichomycetes (Zygomycota), are members of Mesomycetozoea (Cafaro, 2005).

Fungi with a yeast growth form were particularly difficult to circumscribe and classify because they possess few morphological characters. For this reason, yeast systematists relied on physiology to separate species. Although useful in sorting different yeast strains, the physiological characters were not useful for establishing a phylogenetic classification because they are too variable among closely related species. An example of the shortcomings of using morphology and physiology to classify organisms with a yeast growth form was the combining of disparate taxa such as *Schizosaccharomyces* and *Saitoella* in a group with what we now define as Saccharomycotina (Fig. 6.1). In the past, it was often not possible to match sexual and asexual states of fungi unless a patient observer linked both states in pure culture. Other insect-associated fungi such as asexual yeast-like obligate symbionts of hemipteran planthoppers have been placed among their closest relatives in the *Ophiocordyceps*

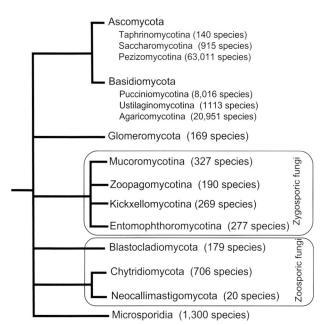

FIGURE 6.1 Diagrammatic representation of relationships of fungal taxa and approximate number of species in each group (Kirk *et al.*, 2008). Zoosporic and zygosporic fungi are more diverse than previously recognized on the basis of morphological traits, and they are not monophyletic; however, not all branches of the tree have been resolved. Evidence from multilocus sequencing and genomics reveals that microsporidians are fungi (Lee *et al.*, 2010). (Tree based on Hibbett *et al.*, 2007, and Blackwell, 2011.)

clade, and the toxic black molds, species of *Stachybotrys*, have been discovered to be members of the Hypocreales. Both of these placements of asexual organisms with their relatives would not have been possible without the use of DNA sequence comparisons. Convergence is a common feature of insect-associated fungi, and convergent traits are likely to have been selected in order to facilitate fungal entomopathogen dispersal and insect infection. The characters include the raising of spores on a pedestal, formation of sticky spores, and production of chemical attractants. For example, the erect stroma (pl., stromata; a hard, compact macroscopic somatic structure that develops from hypha and often bears spore-producing structures) that emerges from the insect cadaver is characteristic of all species previously placed in *Cordyceps*, apparently a convergent feature that raises the ascospores for release and dispersal. This trait occurs among entomopathogens now classified in three fungal families (Clavicipitaceae, Cordycipitaceae, and Ophiocordycipitaceae) spread over at least six genera. Beetle-associated species of *Ophiostoma* and *Ceratocystis* also have long-necked perithecia with sticky ascospores that rise to the tip for easy contact with insects (Alexopoulos *et al.*, 1996; Sung *et al.*, 2007).

An important undertaking for mycologists and insect pathologists is the use of phylogenetic and genomic information for guidance in determining robust morphological characters (Humber, 1997; Sung *et al.*, 2007), tracking production of secondary metabolites (Geiser *et al.*, 2000; Panaccione, 2005; Chang *et al.*, 2006; Frisvad *et al.*, 2008), and determination of host and substrate affiliations (Spatafora *et al.*, 2007). The application of functional genomics also is essential to the development of efficient insect biological control by fungi. The genome of *Metarhizium robertsii* with upregulation of large number of genes in the presence of plants and of insects reflects the presence of specialist genes required for the associations of a fungus with a broad array of organisms (St. Leger *et al.*, 2011). The larger number of secondary metabolites and extracellular enzymes is reflected in the genome of the broad host spectrum fungus compared to the host-specific species *Metarhizium acridum*. The genomes of both *M. robertsii* and *M. acridum*, however, have more secreted proteins belonging to certain gene families (i.e., trypsin) that could be involved in host colonization than the necrotrophic plant pathogens with which they were compared (Gao *et al.*, 2011; St. Leger *et al.*, 2011).

6.2.2. An Overview of Fungal Entomopathogens in a Phylogenetic Context

As a result of molecular phylogenetic studies, many higher level (kingdom, phylum) taxonomic concepts have changed, but the new concepts are better resolved and are expected to be somewhat stable into the future. For example, the use of molecular techniques and phylogenetic analyses has resulted in a better understanding of genetic relatedness that allows for linking asexual (anamorphic) stages of fungi to their sexual (teleomorphic) stages. This has led to the abandonment of the terms Deuteromycota, deuteromycetes, hyphomycetes, and Fungi Imperfecti (Taylor, 1995; Blackwell *et al.*, 2006), in which many fungal entomopathogens had been traditionally placed, and their reclassification in the Ascomycota, one of the two phyla in the fungal subkingdom Dikarya, the other being the Basidiomycota (Fig. 6.1). Phylogenetic classifications are important because they often are predictive of common traits inherited by descendants of a common ancestor (Hibbett *et al.*, 2007). The same methods with different marker genes are used at lower taxonomic levels as well, and the reclassification of *Cordyceps* (see below) serves as an example of the reclassification that results from the use of molecular methods (Spatafora *et al.*, 2007; Sung *et al.*, 2007).

Synopses of groups that are associated with insects (Madelin, 1966; Benjamin *et al.*, 2004; Humber, 2008), usually as pathogens, follow, with emphasis on lineages in which true entomopathogenic organisms occur. It should be kept in mind, however, that careful assessment is needed to determine whether a fungus growing on a dead insect is a true pathogen (i.e., the one that killed the dead host) or just a hyperparasite or a saprophyte.

Stramenopiles (Oomycetes)

For many years oomycetes were considered to be fungi on the basis of their filamentous morphology, heterotrophic nutrition, and similar habitats (Dick, 2001). A number of basic differences was disclosed as new technologies developed. For example, the cell walls of oomycetes contain cellulose rather than chitin, the anterior heterokont flagellation differs from the single posterior whiplash flagellum, and the site of meiosis in the life cycles occurs during gamete formation (gametic meiosis) rather than in spore formation (sporic meiosis) in most fungi. Finally, phylogenetic analyses of rDNA and other genes have made it clear that oomycetes are relatives of other organisms with heterokont flagellation, including brown algae, diatoms, and other organisms with chlorophylls *a* and *c* (Beakes and Sekimoto, 2009; Bhattacharya *et al.*, 2009).

Although water molds (saprolegnialean oomycetes) are repeatedly isolated from the shed exoskeletons (exuviae) of aquatic arthropods (Dick, 1970), fewer oomycetes are actually pathogens of insects. The true entomopathogens include close relatives of certain mammalian, crayfish, and plant pathogens. The best studied of these include *Lagenidium giganteum* and *Aphanomyces laevis*, both of which parasitize mosquito larvae (Kerwin and Petersen, 1997; Patwardhan *et al.*, 2005). Species of *Pythium* and *Leptolegnia* also have been reported as pathogens to mosquito larvae (Phillips *et al.*, 2008). Several products containing *L. giganteum* mycelia and oospores are registered in the USA for use as mosquito larvicides. However, the use of extracts of *L. giganteum* secreted in culture media has been suggested as more effective than release of the living organism, because the extracts are more effective over a broad range of temperatures (Vyas *et al.*, 2007). Another genus, *Leptolegnia*, contains two species (*L. caudata* and *L. chapmanii*) that are pathogenic to insects. While most species of *Pythium* are pathogens of vascular plants, fungi, and algae, there are species that are moderately to highly pathogenic to mosquitoes, e.g., *Pythium flevoense* (Washburn *et al.*, 1988).

Phylum Microsporidia

First considered to be early diverging intracellular parasites and most often grouped with the Protozoa, microsporidia now are considered to be fungi (Fig. 6.1) (Hirt *et al.*, 1999; Keeling and Fast, 2002; Lee *et al.*, 2010) and are obligate pathogens of animals. The most common hosts are fish and arthropods, including many insects. Microsporidia have been studied with the intent of using them for biological control of hosts such as the European corn borer, other caterpillars, locusts, grasshoppers, and mosquitoes. For example, *Paranosema locustae* is marketed as a biological control agent of grasshoppers (Lomer *et al.*, 2001). The Microsporidia are discussed in depth in Chapter 7.

Zoosporic Fungi

Phylum Chytridiomycota

All flagellated fungi were once included in the Chytridiomycota, but they have been split into three phyla (Blastocladiomycota, Chytridiomycota, and Neocallimastigomycota) (Fig. 6.1) based on phylogenetic analyses using multilocus approaches (James *et al.*, 2006a). In addition, flagellated species in the *Rozella* clade and *Olpidium* fall outside these phyla and their classification is uncertain, referred to as *incertae sedis*. Previous listings of entomopathogenic chytrids included fungi that have been reclassified in Blastocladiomycota (see below). *Myiophagus* species, members of the Chytridiomycota, occur on scale insects and have not been the object of extensive study. Phylum Neocallimastigomycota contains anaerobic fungi in the rumen (James *et al.*, 2006a; Hibbett *et al.*, 2007).

Phylum Blastocladiomycota

Blastocladiomycetes (Fig. 6.1) are unusual among all other fungi because their life cycle is characterized by meiosis during formation of spores in thick-walled meiosporangia resulting in an alternation between haploid and diploid generations, the type of life cycle described for plants

(James *et al.*, 2006b). Examples of genera included in the Blastocladiales are *Coelomycidium*, *Catenaria*, and *Coelomomyces*. Species of *Catenaria* are known primarily as pathogens of nematodes, but some species also infect small flies. *Coelomycidium* species are known from scale insects, beetle larvae, and dipteran pupae (Tanada and Kaya, 1993). Species of *Coelomomyces* are obligate pathogens requiring two aquatic hosts, mosquito larvae and crustaceans such as copepods, in different stages of their life cycle (Kerwin and Petersen, 1997). Furthermore, species of *Coelomomyces* are unique among all fungi because they lack hyphal as well as gametangial walls. Zoospores are produced in thick-walled meiosporangia in the diploid sporothallus within the mosquito host and gametes are produced by the haploid gametothallus within the copepod host. There are no mitosporangia (also called thin-walled sporangia) to propagate the sporophyte in the absence of the copepod host as there are in some blastocladialean fungi; the second host, therefore, is required for mosquito infection (Whisler *et al.*, 1975; Gleason *et al.*, 2010).

Zygosporic Fungi

Current molecular phylogenetic studies (White *et al.*, 2006a; Hibbett *et al.*, 2007) place the former members of the Zygomycetes in four unaffiliated subphyla (Mucoromycotina, Zoopagomycotina, Kickxellomycotina, and Entomophthoromycotina) (Fig. 6.1). An excellent website for the zygosporic fungi (as Zygomycota) provides information on the genera and their biology (Benny, 2009). Members of the Mucoromycotina include Mucorales and Mortierellales, which usually are saprobic in soil and dung or weak parasites of plants and animals. A few species such as *Sporodiniella umbellata* are probably weak pathogens of insects (Benny, 2009); none, however, is reported as a primary pathogen of healthy insects (Madelin, 1966). Zoopagomycotina and Kickxellomycotina form a clade along with the Harpellales (White *et al.*, 2006a). This clade, sometimes referred to as the DKH clade, contains the orders Dimargaritales (primarily haustorial parasites of Mucorales), Kickxellales (saprobes and non-haustorial parasites of fungi), and two orders (Harpellales and Asellariales) of arthropod gut fungi previously placed in the Trichomycetes (Lichtwardt *et al.*, 2001). Members of the clade have septate hyphae with distinctive septal structures in common, characters that were recognized to link them before molecular studies had been conducted (O'Donnell *et al.*, 1998; White *et al.*, 2006a). Members of the Harpellales and Asellariales are considered to be gut commensals, but the holdfast of one widespread species, *Smittium morbosum*, penetrates through the gut lining and kills mosquito larvae by inhibiting ecdysis (Sweeney, 1981). In addition, several members of the Harpellales parasitize adult black flies (Simuliidae) by filling their ovaries with fungal cysts, so that the flies disperse the fungus when they attempt to lay their eggs (Undeen and Nolan, 1977; White *et al.*, 2006b).

Members of Zoopagomycotina have been divided into several families, with the Zoopagaceae restricted to predaceous forms that trap hosts with adhesives and produce a restricted haustorium; and the Cochlonemataceae, which are parasites of rotifers, amoebae, rhizopods, and other fungi (Benny, 2009). Although few members of the Zoopagales have been targeted in molecular phylogenetic studies, several fungal and invertebrate parasites in Cochlonemataceae were been grouped together (White *et al.*, 2006a). Only one member of the Zoopagales, *Zoophagus insidians*, a parasite of rotifers included in an analysis, grouped with Harpellales, but without statistical support (White *et al.*, 2006a).

The subphylum Entomophthoromycotina (Fig. 6.1) contains another lineage of zygosporic fungi, the Entomophthorales (Fig. 6.3) (White *et al.*, 2006b; Hibbett *et al.*, 2007; Humber, 2008). Many entomopathogens are found in the group (Keller, 2007), but a more inclusive phylogenetic study is needed to access the evolutionary trends within the subphylum. The monophyly of members of genera (*Conidiobolus* and *Entomophthora*) in two families sampled, however, is supported by a common major sterol, 24-methyl cholesterol, while most other zygosporic fungi have ergosterol as the major membrane sterol (Weete *et al.*, 2010). Although most mycologists consider members of the Entomophthoromycotina to be entomopathogens, there is a variety of other nutritional modes among the taxa. For example, *Completoria* in the monotypic Completoriaceae is a parasite of fern gametophytes that has not been seen since 1895 when it was isolated from a university greenhouse (Benny, 2009). Members of Ancylistaceae, placed in the oomycetes until their forcibly ejected spores were discovered, include parasites of desmids, nematodes, and tardigrades but also *Conidiobolus* with some virulent insect and vertebrate pathogens. The Meristacraceae comprise genera once placed in Ancylistaceae and have a similar range of invertebrate hosts (Benny, 2009). The placement of the Basidiobolaceae remains problematic although it is usually considered to be a member of the Entomophthoromycotina. Molecular studies and morphological characters such as large nuclei, however, set these species apart from other members of the subphylum. Species of *Basidiobolus* range from saprobes to insect and vertebrate pathogens. The entomopathogens among the zygosporic fungi are classified in the Entomophthoraceae (Fig. 6.3A–E, G, H) and Neozygitaceae (Fig. 6.3F). Some of the species included in these families produce secondary capilliconidia that are effectively dispersed by insects. The Entomophthorales includes the genera *Entomophthora* (Fig. 6.3E, G), *Pandora* (Fig. 6.3C, H), *Zoophthora* (Fig. 6.3D), and segregate genera. A segregate genus is one that is split off from an existing genus. For example, *Entomophaga* is a segregate genus of *Entomophthora* because some species formerly placed in *Entomophthora* are now placed in *Entomophaga*. In addition to species-level specificity, some genera are restricted to certain hosts such as *Strongwellsea* (Fig. 6.3A, B) on flies and *Massospora* on periodical cicadas, with emergences of the cicadas often followed by *Massospora* epizootics (Benny, 2009).

Phylum Ascomycota

Ascomycota are classified in three subphyla (Fig. 6.1). Two of the subphyla do not have pathogenic associations with insects. The Taphrinomycotina are saprobic or parasitic on plants and vertebrates (Hibbett *et al.*, 2007). Many ascomycete yeasts (Saccharomycotina) are closely associated with insects for dispersal, and the fungi provide vitamins, enzymes, and other resources for the insect hosts (Vega and Dowd, 2005). Yeasts, however, are not pathogens or even parasites of insects (Kurtzman *et al.*, 2011). Several different lineages of Pezizomycotina (Figs. 6.1 and 6.2) have insect-associated members (Benjamin *et al.*, 2004; Humber, 2008; Blackwell, 2010).

Pezizomycotina (Fig. 6.2) are the most numerous and the most morphologically and ecologically complex of the ascomycetes (Schoch *et al.*, 2009). The life cycles often have two states, anamorph (asexual state) and teleomorph (sexual state), and members of the Hypocreales provide good examples of the phenomenon (Figs. 6.4–6.8). In some cases, the sexual state may never or only rarely be produced. Also, more than one morphologically distinct asexual state may occur in the same organism, and in this case, each is referred to as a synanamorph. Under the new classification, anamorphs often are useful in providing taxonomic information. For example, the anamorphs linked to *Torrubiella* (Fig. 6.6B–D), a genus whose primary host range consists of spiders and scale insects, include *Akanthomyces* species (Figs. 6.4I and 6.6E, F, H, K) having a broader host range than *Gibellula* species (Figs. 6.4F and 6.6J), which are restricted to spiders (Hodge, 2003; Johnson *et al.*, 2009). Among insect-associated fungi, asexual states (Figs. 6.4A–C, F–I, L, M, 6.5B, C, E, F, I, 6.6E–K, M, and 6.7G, I, K, L) often precede the production of sexual states (Figs. 6.4D, E, J, K, 6.5A, D, G, H, J, K, 6.6A–D, L, and 6.7A–F, H, J) (Sung *et al.*, 2007).

Conidia (asexual spores) that form on insect cadavers can be produced directly on conidiophores as in *Metarhizium* species on adult insects (Fig. 6.4A, B) and larvae. In some species of *Isaria* and *Akanthomyces*, conidia are formed on synnemata (Fig. 6.4I), structures formed by fusion of groups of individual conidiophores. Among the Hypocreales, many of the asexual and sexual states occur separately in time, and it is convenient to refer to the morphs by different generic names (Table 6.1). Occasionally, both

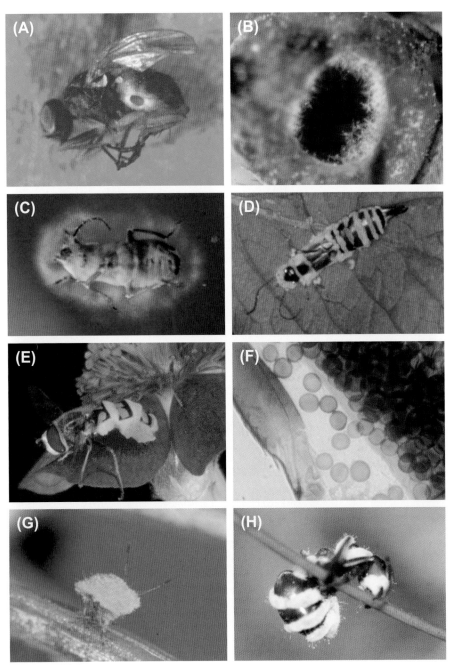

FIGURE 6.3 Insects infected with species of Entomophthorales. (A) The fungus *Strongwellsea* sp. creates a characteristic hole in the abdomen of the fly host *Coenosia testacea*. (B) Abdominal hole of the fly *Paradelia intersecta* created by *Strongwellsea* sp. Infective spores are actively ejected through the hole of the still living host. (C) Soldier beetle *Rhagonycha fulva* killed by the entomophthoralean fungus *Pandora lipai*. Spores are actively ejected from the fungus-infected cadaver and a halo of spores is visible around the dead host. (D) Earwig *Forficula* sp. infected by *Zoophthora forficulae*. The fungus emerges through the thin cuticular intersegmental parts of the host body. (E) Hoverfly infected by *Entomophthora syrphii*. The host dies in an elevated location (summit disease) to maximize dispersal of fungal spores. (F) Resting spores produced by *Neozygites parvispora* in the thrips host *Limothrips dentricornis*. (G) Green spruce aphid *Elatobium abietinum* infected by *Entomophthora planchoniana*. (H) Wood ant *Formica* sp. killed by *Pandora formicae*. Before the host dies, the fungus manipulates it to crawl to an elevated point in the vegetation and lock its jars around the grass straw. [*Photo credits: (A−D, G) J. Eilenberg; (E) H. Philipsen; (F) C. Nielsen; (H) J. Małagocka.*]

asexual and sexual states are present on one specimen at the same time, as in the case of *Hypocrella calendulina* (Fig. 6.4D, E) and its *Aschersonia calendulina* anamorph on scale insects as well as other species (Luangsa-ard *et al.*, 2007, 2008, 2010). Both asexual and sexual spores of insect-associated fungi usually are produced on raised structures that enhance the opportunity for dispersal to a new host. The production of asexual spores may occur in a stroma, with development of the asci and ascospores occurring at a later time. Perithecia may be immersed in a stroma, as in *Cordyceps militaris* (Fig. 6.8), *C. ninchukispora* (Fig. 6.6L), *Ophiocordyceps nutans* (Fig. 6.7E), *Elaphocordyceps paradoxa* (Fig. 6.7F), and other species.

Pezizomycotina includes some of the poorly known entomopathogens such as members of the Tubeufiaceae (Dothideomycetes *incertae sedis*). Species in the Tubeufiaceae have a broad nutritional range, with many species saprobic, parasitic on fungi and leaves, or lichen-forming with algae (Kodsueb *et al.*, 2006). Species of one genus, *Podonectria*, are pathogens of scale insects, and for many

FIGURE 6.4 Microscopic features commonly used in identification of entomopathogens. A–C, F–I, L, and M are structures from asexual states (anamorphs). Other figures of asci and ascospores are of sexual states (teleomorphs). (A) Phialides of *Metarhizium anisopliae*. (B) Conidiophores and conidia of *Metarhizium cylindrosporae*. (C) Conidiogenous cells and conidia of *Aschersonia coffeae*. (D) Mature and developing asci of *Hypocrella calendulina*. (E) Whole ascospores of *H. calendulina*. (F) Conidiophores of *Gibellula pulchra*. (G) Conidiophores of *Isaria* sp. (H) Conidiogenous structures of *Beauveria* sp. (I) Synnema of *Akanthomyces* sp. (J) Asci of *Ophiocordyceps communis*. (K) Ascospores of *O. communis*. (L) Conidiogenous cells of *Hirsutella* sp. (M) Conidiogenous cells of *Hymenostilbe ventricosa*. [*Photo credits: (A, B, G, H, J, K, M) J. Jennifer Luangsa-ard; (C−E) S. Mongkolsamrit; (F, I, L) K. Tasanathai.*]

years these fungi were mistakenly placed in the Hypocreales (Rossman, 1978). Other entomopathogens include members of the Eurotiomycetes, including species of *Ascosphaera* (Onygenales), and members of the Eurotiales, including certain species of *Aspergillus* and *Penicillium* and their relatives (Bailey, 1981; Sosa-Gómez *et al.*, 2010). Species of *Ascosphaera* are obligate parasites of larval honey bees (*Apis mellifera*) (see Chapter 12). The honey bee chalkbrood disease is caused by the fungus *Ascosphaera apis* (Geiser *et al.*, 2006).

The Laboulbeniomycetes (Fig. 6.2) is an interesting group of insect parasites, although they are not usually considered to be pathogens. The 2000 known species of the order Laboulbeniales are obligate haustorial ectoparasites of insects and a few other arthropods (Weir and Blackwell, 2005). Two-celled ascospores germinate on the host surface

to produce a haustorium from an attached cell and the outer cell divides produce a perithecium with determinate growth. No mycelium is produced and the lack of an asexual state in the order is unique among ascomycetes. Beetles and flies are the major hosts for members of the Laboulbeniales. A second order, Pyxidiophorales, appears to be mostly mycoparasitic, and is also associated with arthropod hosts. These fungi are mycelial and usually have prominent asexual states preceding the development of the sexual state. Insects and their phoretic mite associates are effective dispersers of the Pyxidiophorales (Weir and Blackwell, 2005).

Hypocreales, members of the class Sordariomycetes, are the best known entomopathogens among the ascomycetes (Fig. 6.2). Recent phylogenetic analyses have led to a dramatic revision of our understanding of relationships

FIGURE 6.5 Family Clavicipitaceae. B, C, E, F, and I are asexual states (anamorphs). Other figures are of sexual states (teleomorphs). (A) *Cordyceps* aff. *martialis* on lepidopteran larva. (B) *Metarhizium flavoviride* on cicada. (C) *Metarhizium* sp. on cockroach; D–H and J–K on scale insects. (D) *Moelleriella reineckiana*. (E) *Aschersonia marginata*. (F) *Aschersonia coffeae*. (G) *Orbiocrella petchii*. (H) *Hypocrella calendulina*. (I) Different stages of colonization of a cicada by *Metarhizium cylindrosporae*. (J) *Hypocrella* sp. (K) *Conoideocrella tenuis*. [*Photo credits: (A–C, I) R. Ridkaew; (D–H, J, K) S. Mongkolsamrit.*]

and host switching among the entomopathogenic taxa of Hypocreales (Spatafora *et al.*, 2007; Sung *et al.*, 2007). The former Clavicipitaceae has been divided into three monophyletic families: Clavicipitaceae, Cordycipitaceae, and Ophiocordycipitaceae. An excellent website provides detailed information about the revision of the entomopathogens in these groups (Spatafora and *Cordyceps* Working Group, 2011). Clavicipitaceae (Fig. 6.5) includes, among others, a number of grass endophytes (*Balansia*, *Claviceps*, *Epichloë*), and the well-known entomopathogens *Aschersonia*, *Hypocrella*, *Regiocrella*, and *Metarhizium*. A new genus, *Metacordyceps*, includes former *Cordyceps* species related to the grass endophytes as well as anamorphs and teleomorphs related to *Metarhizium* (Sung *et al.*, 2007). The family Cordycipitaceae (Fig. 6.6) contains *C. militaris* (Fig. 6.8), the type of the

genus, and includes most of the former *Cordyceps* species characterized by brightly colored, fleshy stromata. Entomopathogenic asexual states include species of *Beauveria*, *Isaria*, *Lecanicillium*, and other genera. Most members of Ophiocordycipitaceae (Fig. 6.7) can be distinguished morphologically from the other subphyla because they often have dark stromata and mature ascospores. In addition to *Ophiocordyceps*, species of the genus *Elaphocordyceps* parasitize not only insects but also *Elaphomyces*, a mycorrhizal ascomycete symbiont of trees. Other unnamed species in the family are unusual because they are obligate yeast-like symbionts of planthoppers (Suh *et al.*, 2001).

Host preference is not specific to a family or genus, although there are exceptions. There are trends, however, and spider pathogens are mostly found within the Cordycipitaceae,

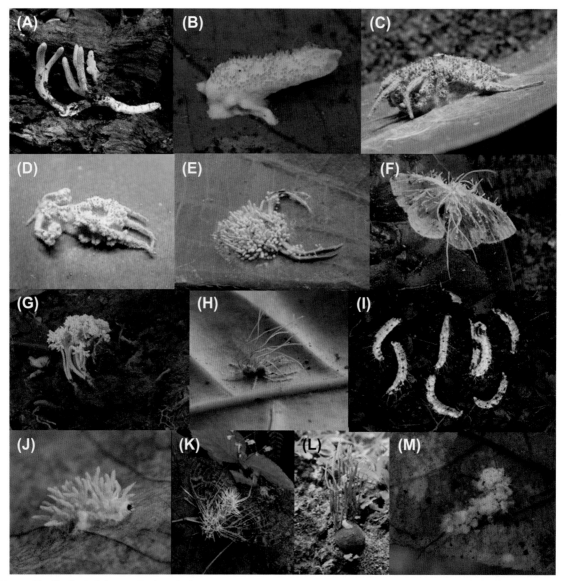

FIGURE 6.6 Family Cordycipitaceae. E—K, and M are asexual states (anamorphs). Other figures are of sexual states (teleomorphs). (A) *Cordyceps bassiana* on coleopteran larva. (B) *Torrubiella hemipterigena* on planthopper. C—E, H, J on spiders. (C) *Torrubiella* aff. *neofusiformis*. (D) *Torrubiella* aff. *corniformis*. (E) *Akanthomyces koratensis*. (F) *Akanthomyces aculeatus*. (G) *Isaria tenuipes*. (H) *Akanthomyces aranearum*. (I) *Beauveria bassiana*. (J) *Gibellula pulchra*. (K) *Akanthomyces pistillariiformis* on adult moth. (L) *Cordyceps ninchukispora* on lepidopteran pupa. (M) *Isaria farinosa* on lepidopteran larva. [*Photo credits: (A, B, F—I, K, M) P. Aphiphunya; (C) R. Ridkaew; (D, J) K. Tasanathai; (E) T. Chohmee; (L) P. Srikitikulchai.*]

while scale entomopathogens are common in the Clavicipitaceae. Pathogens of ants, termites, or dipterans, and endosymbionts of planthoppers are often found in the Ophiocordycipitaceae (Evans and Samson, 1982, 1984; Sung *et al.*, 2007). Because of the broad range of very different lifestyles and diversity of hosts, some of these fungi have been the objects of studies of host jumping (Nikoh and Fukatsu, 2000; Suh *et al.*, 2001; Spatafora *et al.*, 2007).

Phylum Basidiomycota

Basidiomycetes often are closely associated with insects as their sole nutritional resource and as a habitat, and many insects fertilize and disperse these fungi. Insect associations occur in the three subphyla of basidiomycetes (Pucciniomycotina, Ustilaginomycotina, and Agaricomycotina), but parasites are found only in the Septobasidiales of the Pucciniomycetes, a group that includes the plant-pathogenic rust fungi (Aime *et al.*, 2006). The genus *Septobasidium* and four additional genera, *Auriculoscypha*, *Uredinella*, *Coccidiodictyon*, and *Ordonia*, are specialized parasites of scale insects (Couch, 1938; Henk and Vilgalys, 2007) that seldom kill their hosts but use them as providers of nutrients while keeping them alive (Humber, 2008).

FIGURE 6.7 Family Ophiocordycipitaceae. G, I, K, and L are asexual states (anamorphs). Other figures are of sexual states (teleomorphs). (A) *Ophiocordyceps halabalaensis* on *Camponotus gigas*; arrow indicates the stroma with the anamorphic form where the conidiogenous structures and the conidia are for the *Hirsutella* anamorph. (B) *Ophiocordyceps sphecocephala* on wasp. (C) *Ophiocordyceps myrmecophila* on formicine ant. (D) *Ophiocordyceps dipterigena*; arrow indicates *Hymenostilbe* anamorph. (E) *Ophiocordyceps nutans* on pentatomid bug. (F) *Elaphocordyceps paradoxa* on cicada nymph. (G) *Hymenostilbe* state of *O. brunneipunctata*. (H) *O. acicularis* on lepidopteran larva. (I) *Hirsutella nivea* on hopper. (J) *Ophiocordyceps blattae* on cockroach. (K) *Hirsutella saussurei* on wasp. (L) *Paecilomyces lilacinus* on cydnid bug. [*Photo credits: (A) R. Ridkaew; (B, C, E–I, K, L) P. Aphiphunya; (D) P. Srikitikulchai; (J) J. Punya.*]

6.3. BIOLOGY

Fungi are heterotrophic, eukaryotic microorganisms; as such, they are unable to fix carbon and therefore have to absorb the organic compounds used as primary sources of energy from other organisms. They accomplish this as saprobes (i.e., by consuming dead organic matter), or as parasites of animals, plants, and other fungi. In addition, fungi can have mutualistic associations with algae in the form of lichens, or with plant roots in the form of endomycorrhiza or ectomycorrhiza. The presence of fungi as plant endophytes is well known, and is being studied as a possible biological control strategy against insects (see Section 6.4.1).

Fungi typically have tubular compartmentalized branched bodies referred to as hyphae (sing., hypha). A mass of hyphae is collectively known as the mycelium

(pl., mycelia). A different form of fungal growth is exhibited by yeasts, unicellular organisms that divide asexually by budding. Depending upon environmental conditions, many fungal entomopathogens can exhibit both growth forms (i.e., hyphal and yeast like), in which case they are referred to as dimorphic. A specialized fungal form, a sclerotium, is a hard, frequently spherical structure consisting of a mass of sterile hyphae considered to be resistant to adverse environmental conditions (Alexopoulos *et al.*, 1996). Sclerotia have been reported for *Cordyceps*, *Hirsutella*, and *Synnematium* (= *Hirsutella*) (Speare, 1920; Evans and Samson, 1982). The formation of small sclerotia, termed microsclerotia, was reported for *M. anisopliae* grown in liquid culture. The microsclerotia were defined as

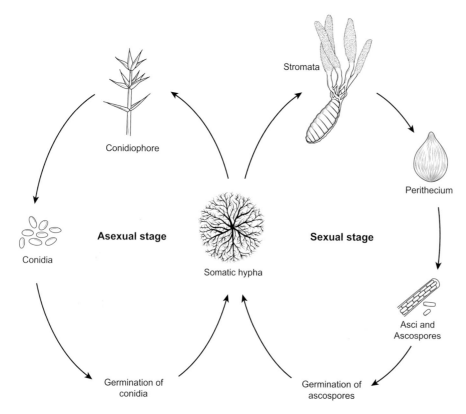

Conidiophore

Stromata

Asexual stage

Sexual stage

Conidia

Somatic hypha

Perithecium

Germination of
conidia

Germination of
ascospores

Asci and
Ascospores

FIGURE 6.8 Life cycle of *Cordyceps militaris* (Cordycipitaceae) and its *Lecanicillium* anamorph. The infected lepidopteran larva producing the sexual state in a stromata with perithecia, asci, and ascospores. The ascospores are forcibly discharged from the asci. Successful dispersal results in contact with other larvae. Ascospores germinate and produce hyphae. The anamorph, characterized by the production of conidiophores bearing conidia, is often formed on the host before development of the sexual state. Either ascospores or conidia may infect other larvae.

"undifferentiated, melanized, compact hyphal aggregates" which upon rehydration, enter sporogenic germination, i.e., produce masses of conidia directly on the surface of the microsclerotia. Laboratory bioassays revealed that microsclerotia incorporated into soil can cause high mortality of sugar beet root maggots (Jackson and Jaronski, 2009).

6.3.1. Reproduction

Fungi can reproduce sexually or asexually, and throughout the scientific literature, various terms have been used to describe these two types of reproduction: sexual versus asexual, teleomorphic versus anamorphic, perfect versus imperfect, and meiosporic versus mitosporic.

Sexual reproduction involves the fusion of two nuclei followed by meiosis, thus allowing for the new combination of genes, a possibility that usually does not occur in asexual reproduction. Asexual reproduction occurs through mycelial fragmentation or the production of reproductive spores known as conidia (sing., conidium). Conidia are the result of mitotic division and are formed directly on hyphal strands or from conidiogenous cells often at the ends of hyphae. Parasexual recombination involves mitotic crossing over, whereby hyphae with two different types of nuclei (i.e., heterokaryons) will fuse and produce a recombinant type without the involvement of sexual

reproduction. The parasexual cycle occurs in fungi growing in culture, but it is not clear how often the process occurs in natural populations (Alexopoulos *et al.*, 1996; Read *et al.*, 2010). The shape and arrangement of conidia and conidiophores (the hypha where conidiogenous cells are formed) are useful in identification of fungal entomopathogens (Humber, 1997), although recent molecular evidence shows that many traditionally accepted species are assemblages of cryptic taxa (see Section 6.4.1).

6.3.2. Types of Infective Propagules and Cell Wall Surface Properties

Fungal entomopathogens can produce different forms of infective propagules, including aerial conidia, submerged conidia, and blastospores (Jackson *et al.*, 2010). Aerial conidia are produced externally on infected insects or on the surface of solid substrates. Submerged conidia are produced in nutrient-limited culture and have been reported in *B. bassiana* (Thomas *et al.*, 1987; Hegedus *et al.*, 1990; Cho *et al.*, 2006), *Hirsutella thompsonii* (van Winkelhoff and McCoy, 1984), *Metarhizium flavoviride* (Jenkins and Prior, 1993), and *M. anisopliae* (Kassa *et al.*, 2004). In nutrient-rich liquid culture and in the insect hemocoel, some fungal entomopathogens such as *B. bassiana* develop what are called blastospores (Bidochka *et al.*, 1987),

TABLE 6.1 Alphabetical Listing of the Asexual (Anamorphs) and Sexual (Teleomorphs) Genera in Revised Hypocrealean Families

Anamorphs	Teleomorphs
Clavicipitaceae	
Aschersonia[a] (Figs. 6.4C, 6.5E, F)	*Aciculosporium*
Ephelis	*Atkinsonella*
Metarhizium[a] (Figs. 6.4A, B, 6.5I)	*Balansia*
Neotyphodium	*Conoideocrella*[a] (Fig. 6.5K)
Nomuraea[a]	*Claviceps*
Pochonia	*Epichloë*
Sphacelia	*Heteroepichloë*
	Hypocrella[a] (Figs. 6.4D, E, 6.5H, J)
	Metacordyceps[a]
	Moelleriella[a] (Fig. 6.5D)
	Myriogenospora
	Neoclaviceps
	Orbiocrella[a] (Fig. 6.5G)
	Parepichloë
	Regiocrella[a]
	Shimizuomyces
Cordycipitaceae	
Akanthomyces[a] (Figs. 6.4I, 6.6E, F, H, K)	*Ascopolyporus*
Beauveria[a] (Figs. 6.4H, 6.6I)	*Cordyceps*[a] (Figs. 6.6A, L, 6.8)
Engyodontium[a]	*Hyperdermium*
Gibellula[a] (Figs. 6.4F, 6.6J)	*Torrubiella*[a] (Fig. 6.6B–D)
Isaria[a] (Figs. 6.4G, 6.6G, M)	
Lecanicillium[a]	
Microhilum[a]	
Simplicillium[a]	
Ophiocordycipitaceae	
Haptocillium	*Elaphocordyceps*[a] (Fig. 6.7F)
Harposporium	*Ophiocordyceps*[a] (Figs. 6.4J, K, 6.7A–E, H, J)
Hirsutella[a] (Figs. 6.4L, 6.7I, K)	
Hymenostilbe[a] (Figs. 6.4M, 6.7G)	
Paraisaria[a]	
Paecilomyces[a] (Fig. 6.7L)	
Syngliocladium[a]	
Tolypocladium[a]	

Many of these taxa were previously placed in a single family. The major reassessment of the entomopathogens and their relatives has resulted in a new understanding of the interkingdom host shifts that involve host jumps among plants, insects, and fungi within the Hypocreales (Spatafora *et al.*, 2007). Not all of the species in the Hypocreales have been included in the taxonomic revision (Sung *et al.*, 2007).
[a]*Entomopathogens.*
Source: Sung *et al.* (2007), Spatafora *et al.* (2011).

a vegetative fungal cell that forms by budding from hyphae. Blastospores formed in liquid culture are referred to as *in vitro* blastospores and those formed in the hemocoel are called *in vivo* blastospores or hyphal bodies. This distinction should be made owing to differences between them (Pendland *et al.*, 1993; Lewis *et al.*, 2009; Wanchoo *et al.*, 2009). Some fungal entomopathogens, e.g., *B. bassiana* (Alves *et al.*, 2002), *I. fumosorosea* (de la Torre and Cárdenas-Cota, 1996), and *M. anisopliae* (Liu *et al.*, 2010), exhibit microcycle conidiation, i.e., the production of conidia directly from germinated spores without the involvement of hyphal growth (Hanlin, 1994).

The cell wall surface properties of different developmental stages of *B. bassiana* have been studied in more detail than for any other fungal entomopathogen. The cell surface appears dynamic, with different cell types displaying variations in surface carbohydrate epitopes that have lectin-binding properties. In addition, aerial conidia are characterized by the presence of a proteinaceous rodlet layer believed to consist mostly of hydrophobins (Holder and Keyhani, 2005; Wanchoo *et al.*, 2009; Zhang *et al.*, 2011). These amphipathic proteins (i.e., they exhibit both hydrophobic and hydrophilic properties) are also present on submerged conidia, but they do not appear to form the characteristic rodlet layer found on aerial conidia (Bidochka *et al.*, 1995). Aerial conidia are highly hydrophobic, in contrast to the hydrophilic submerged conidia and blastospores (Holder *et al.*, 2007). A rodlet layer has also been reported on the conidia of *Nomuraea rileyi*, *I. fumosorosea*, and *M. anisopliae* (Boucias *et al.*, 1988; Boucias and Pendland, 1991). Jeffs *et al.* (1999) reported that aerial conidia of *B. bassiana*, *B. brongniartii*, *M. anisopliae*, and *I. farinosa* exhibiting a rugose surface had different levels of hydrophobicity, while aerial conidia of *Beauveria densa*, *Tolypocladium cylindrosporum*, *T. nivea*, and *Lecanicillium* spp. had smooth surfaces and were hydrophilic. Aerial conidia of *N. rileyi* and *I. fumosorosea* are also hydrophobic (Boucias and Pendland, 1991), in contrast to blastospores of *I. fumosorosea*, which are hydrophilic (Dunlap *et al.*, 2005). The conidia of fungal entomopathogens of various genera, including *Aschersonia*, *Conidiobolus*, *Culicinomyces*, *Entomophaga*, *Hirsutella*, *Lecanicillium*, and *Neozygites*, produce mucilage that confers hydrophilic properties (Boucias and Pendland, 1991).

Cell wall surface carbohydrates have been shown to influence pathogenesis through a possible effect on the host insect immune system recognition. Pendland *et al.* (1993) reported that *in vivo* blastospores of *B. bassiana* collected from hemolymph of the beet armyworm (*Spodoptera exigua*), after originally injecting the insect with *in vitro* blastospores, lacked a galactomannan coat, an important antigenic component. Loss of or changes to the galactomannan coat have also been reported to possibly affect immune system recognition of *in vitro* blastospores of

I. farinosa (Pendland and Boucias, 1993), and for galactose in *in vitro* blastospores of *N. rileyi* (Pendland and Boucias, 1992). In contrast, Wanchoo *et al.* (2009) reported that very few differences exist between carbohydrate epitopes of aerial and submerged *B. bassiana* conidia.

The type of infective propagule could also influence insect mortality. For example, blastospores of *B. bassiana* can cause higher and earlier mortality of the tobacco budworm (*Heliothis virescens*), when either applied topically or injected, in comparison to submerged or aerial conidia (Holder *et al.*, 2007). Similarly, Hall (1979) reported that blastospores of *Lecanicillium* spp. were twice as virulent as conidia against the aphid *Macrosiphoniella sanborni*, but owing to the larger size of blastospores, they were only 0.6 times as virulent on a spore volume basis. In contrast to these studies, Vandenberg *et al.* (1998) reported no difference in Russian wheat aphid (*Diuraphis noxia*) mortality when aerial conidia or blastospores of *I. fumosorosea* were used, and Jenkins and Thomas (1996) found no difference in virulence between aerial and submerged conidia of *M. acridum* (as *M. flavoviride*) used against the desert locust, *Schistocerca gregaria*.

As discussed below, cell wall surface properties also play a critical role on spore adhesion to the cuticle of an insect (Boucias *et al.*, 1988; Boucias and Pendland, 1991; Holder and Keyhani, 2005). In addition, knowledge of cell surface properties is essential for developing formulations of fungal entomopathogens. For example, infective propagules with hydrophilic surfaces would be inherently difficult to formulate in oils.

6.3.3. The Infection Process

A generic model for the infection process described below is shown in Fig. 6.9.

Attachment to the Cuticle

Because fungi are heterotrophic organisms, they have to absorb organic compounds produced by other organisms as their primary sources of energy. For infection to be successful, fungal entomopathogens must therefore penetrate the insect cuticle, a polymer network composed of chitin (a polysaccharide) embedded in a protein matrix (see *Cuticle Penetration*, below).

Holder and Keyhani (2005) examined the adhesion of aerial conidia, submerged conidia, and blastospores of *B. bassiana* to hydrophobic and hydrophilic surfaces. Their results show that (1) aerial conidia bind well to hydrophobic surfaces but weakly to hydrophilic surfaces; (2) submerged conidia will bind to both hydrophobic and hydrophilic surfaces; and (3) *in vitro* blastospores will bind strongly to hydrophilic surfaces and weakly to hydrophobic surfaces. They proposed that knowledge concerning the surface properties of various infectious cell types

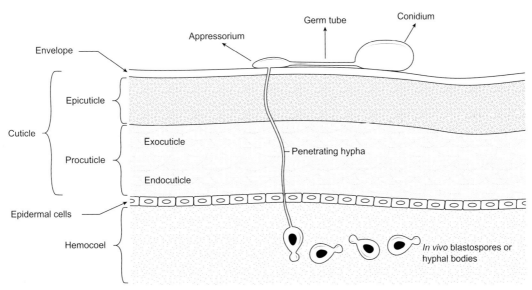

FIGURE 6.9 Generic model showing the infection process for fungal entomopathogens. A conidium lands on the insect cuticle, where it germinates to form a germ tube and appressorium. A penetrating hypha breaches the cuticular layers, reaching the hemocoel where *in vivo* blastospores or hyphal bodies are formed. Fungal growth continues and moves throughout the hemocoel, invading organs, disrupting metabolic processes, and possibly producing toxic metabolites, eventually causing death.

can be exploited to improve the efficacy of fungal entomopathogens, and particularly, in the development of formulations.

In addition to the initial passive adherence of conidia due to hydrophobic interactions (Holder and Keyhani, 2005), Wang and St. Leger (2007a) reported that *M. anisopliae* conidia adhesion to insect cuticle is mediated by an adhesin-like protein, known as MAD1. A different adhesin-like protein, MAD2, is involved in *M. anisopliae* adhesion to plant surfaces, and also appears to be related to the rhizosphere competency of *M. anisopliae* (Wang and St. Leger, 2007a; Pava-Ripoll *et al.*, 2011). MAD1 was also shown to be involved in conidial germination and formation of blastospores (Wang and St. Leger, 2007a).

Entomopathogens in the Entomophthorales produce conidia that are evolutionarily derived from sporangia. Conidia of many entomophthoralean fungi are forcibly discharged (Papierok and Hajek, 1997). Some taxa, e.g., species of *Neozygites* and *Zoophthora*, also produce elongated capillioconidia that develop at the end of long, slender capilliconidiophores. These propagules become easily detached when touched by an insect and adhere to the cuticle by an adhesive droplet that exudes from the capilliconidium tip. They may be important for infection in some species (Glare *et al.*, 1985). As mentioned in Section 6.3.2, some genera of fungal entomopathogens produce conidia with a mucilaginous coat that serves to attach it to a substrate (Boucias and Pendland, 1991). Conidia of *Entomophthora* have a mucilaginous coat within two cell wall layers that is released upon impact with a substrate, resulting in adhesion (Eilenberg *et al.*, 1986).

In plant-pathogenic fungi, the mucilage associated with the spore protects against desiccation and toxic plant chemicals produced in response to infection (Nicholson and Moraes, 1980; Nicholson *et al.*, 1989). To the authors' knowledge, it is not yet known whether such protection occurs with spores of fungal entomopathogens that produce mucilage, although it has been shown that plant allelochemicals can have inhibitor effects on fungal entomopathogens. For example, Vega *et al.* (1997) examined the *in vitro* effects of seven plant allelochemicals on blastospore germination of *I. fumosorosea* and hypothesized that sequestration of these compounds in the insect cuticle or hemolymph might reduce fungal effectiveness. Lacey and Mercadier (1998) also reported inhibitory effects on germination of conidia and blastospores of *I. fumosorosea* when exposed to allelochemicals.

Spore Germination

A critical event following attachment of the spore to the cuticle is germination, defined as the formation of a germ tube, with or without the subsequent formation of an appresorium (pl., appresoria). In addition to temperature and humidity (Gottlieb, 1978), nutritional, chemical, and physical factors can influence germination. Smith and Grula (1981) showed that in order for *B. bassiana* to germinate and form a germ tube, carbon and nitrogen sources are required. Some components of the insect cuticle could have inhibitory (Smith and Grula, 1982; Latgé *et al.*, 1987; Sosa-Gomez *et al.*, 1997; Priyatno and Ibrahim, 2002) or stimulatory (Notini *et al.*, 1944; Veen, 1968) effects on conidia, and fungi or bacteria present on the cuticle might also inhibit germination (Schabel, 1976).

If the host is in an immature stage, molting could reduce the effectiveness of the fungal entomopathogen, in part owing to the shedding of conidia attached to the molted cuticle (Fargues and Vey, 1974; Vey and Fargues, 1977; Luz et al., 2003). Fast germination would thus be advantageous for conidia to reduce exposure not only to adverse external environmental factors, but also to adaptations in the host reducing fungal infection success (discussed below). Vega et al. (1999) reported that blastospores of I. fumosorosea germinate more quickly on the cuticle of an insect than aerial conidia, and in some cases, faster germination has been shown to correlate with virulence (Milner et al., 1991).

As for physical factors, the adhesion of conidia to the cuticle can be region specific, e.g., to intersegmental membranes (Butt et al., 1995; Hajek and Eastburn, 2003). Working with M. anisopliae and four different elaterids, McCauley et al. (1968) found higher infection through intersegmental folds and ascribed this to: (1) conidia within folds being protected in contrast to conidia on the smooth, exposed sclerites; (2) microclimate within the folds being more conducive to higher humidity, which enhanced germination; and (3) the folds lacking the sclerotinous protein found on the cuticle, thereby making it easier to penetrate. Adhesion of conidia can also be influenced by cuticular topography, e.g., conidia can collect in areas containing a large number of setae and spines (Sosa-Gomez et al., 1997).

Once a conidium has attached to the cuticle, an appressorium might form at the end of the germ tube. The appressorium is the infection structure from which penetration into the host occurs through a penetration or infection peg. The pressure exerted by the appressorium has been measured for some fungal plant pathogens (Howard and Valent, 1996; Bechinger et al., 1999) but not for fungal entomopathogens. Some Entomophthorales (e.g., *Conidiobolus obscurus*, *Pandora neoaphidis*, *P. blunckii*, *Entomophthora planchoniana*, and *Batkoa apiculata*) do not form appresoria and can penetrate the cuticle directly from the germ tube (Hajek et al., 2002). Other fungal entomopathogens (e.g., *M. anisopliae*, *I. farinosa*, *Entomophaga aulicae*, *C. obscurus*, *Neozygites fresenii*, *Erynia radicans*, and *N. rileyi*) produce mucilage during appresorial and germ tube formation; this exocellular mucilage is believed to aid in adhesion to the cuticle (Boucias and Pendland, 1991).

In addition to invading the insect via cuticular areas and intersegmental membranes, fungal entomopathogens have been reported to invade insects through sense organs and spiracles, and *per os* (i.e., though the mouth) (Charnley, 1984; St. Leger, 1991). For *Ascosphaera* and *Culicinomyces*, the normal mode of infection is *per os* (Humber, 2009). For excellent reviews on the fate of microorganisms in the gut see Dillon and Charnley (1991) and Bignell (1994).

Cuticle Penetration

The insect cuticle is a non-living polymer network composed of chitin and up to 70% proteins (Clarkson and Charnley, 1996). The cuticle is divided in three sections (Fig. 6.9) (Locke, 2001; Klowden, 2007). The outermost layer is called the envelope (also known as the cuticulin layer) and is 10–30 nm in thickness. Underneath the envelope is the epicuticle (0.5–2 μm thick), followed by the procuticle (hundreds of times wider than the envelope and epicuticle), which is where chitin and proteins form a matrix. The procuticle can be further divided into an inner layer called the endocuticle (10–200 μm thick), and an outer layer known as the exocuticle. Cuticular proteins can vary across anatomical areas of the insect as well as with the developmental stage of the insect (Willis et al., 2005). The epidermal cells are found at the base of the procuticle, and the hemocoel is underneath the epidermal cells.

The cuticle must be breached by mechanical pressure (McCauley et al., 1968; Zacharuk, 1970) and via the production of cuticle-degrading enzymes for the fungus to obtain nutrition and eventually to colonize the insect. Wang and St. Leger (2007b) have implicated a *Metarhizium* perilipin-like protein (MPL1) in the generation of appresorial turgor pressure required to breach the cuticle. The enzymes involved in cuticle degradation include proteases, chitinases, and lipases, needed to degrade proteins, chitin, and lipids, respectively. For reviews of cuticle degradation by fungal entomopathogens see Charnley (2003), Pedrini et al. (2007), and Schrank and Vainstein (2010).

Once the fungus has gained access to the hemocoel, it starts utilizing available nutrients for growth and reproduction. It has been hypothesized that fungal isolates that grow profusely in the hemolymph kill the insect by consuming the host nutrients and physically damaging the tissues, thereby disrupting the host's physiology. This is in contrast to those that exhibit sparse vegetative growth, instead relying on secondary metabolite production (see below) to kill their host. These fungal strategies, which are based on *M. anisopliae* studies, have been referred to as "growth strategy" and "toxin strategy", respectively (Charnley, 1984; Clarkson and Charnley, 1996; Kershaw et al., 1999). Eventually, hyphae emerge from within the insect and sporulate, producing aerial conidia.

Insect Responses to Infection

Insect behavior can alter the success of fungal entomopathogen infection. In theory, social insects are at greater risk of infection by pathogens because many individuals live close together and are often closely related. However, social insects have evolved a range of mechanisms to defend against pathogens (Cremer *et al.*, 2007). For example, termites (Boucias *et al.*, 1996; Rosengaus *et al.*, 1998; Yanagawa *et al.*, 2008) and ants (Kermarrec *et al.*, 1986; Jaccoud *et al.*, 1999) exhibit mutual grooming behavior to remove conidia from themselves and from colony members. Furthermore, the metapleural glands of ants produce antibiotic secretions that inactivate fungal entomopathogens (Hughes *et al.*, 2002; Yek and Mueller, 2011). Cremer *et al.* (2007) list a range of mechanisms by which social insects can combat pathogen establishment. Nielsen *et al.* (2010) showed that aphid-tending ants remove sporulating aphid corpses infected with *P. neoaphidis* and will also groom aphids and remove conidia from the aphid cuticle, possibly to protect their mutualistic honeydew-producing resource.

Some non-social insects have been shown to exhibit density-dependent prophylaxis, i.e., increased resistance to pathogens when reared under high-density conditions (Wilson and Reeson, 1998). *Spodoptera littoralis* (Wilson *et al.*, 2001) and *Tenebrio molitor* (Barnes and Siva-Jothy, 2010) have shown increased resistance to *B. bassiana* and *M. anisopliae*, respectively, when reared in crowded conditions. This increased resistance appears to be related to cuticular melanization, i.e., the formation of the pigment melanin by polymerization of phenolic compounds (Charnley, 1984; Jacobson, 2000). In contrast, the increased resistance against *M. acridum* in *S. gregaria* reared in crowded conditions appears to be related to increased antimicrobial activity (Wilson *et al.*, 2002),

Once the fungal entomopathogen reaches the hemolymph, a range of immune responses can be initiated (see Chapter 13). Some are general antimicrobial responses while some are specifically targeting the invading fungus (see Rolff and Reynolds, 2009, for reviews of insect immune mechanisms, including evolutionary and ecological implications). Fungal units can be cellularly phagocytosed or encapsulated (Charnley, 1984; Gillespie *et al.*, 1997). *Metarhizium anisopliae* and *B. bassiana* are capable of avoiding encapsulation in the hemocoel, and this adaptation has been hypothesized to be a consequence of these fungi being facultative entomopathogens in soil environments where they can survive encapsulation by soil amoeboid predators (Bidochka *et al.*, 2010). A fascinating response to fungal infections is the concept of behavioral fever (Roy *et al.*, 2006), whereby infected insects (e.g., locusts, flies) modify their behavior (e.g., basking in the sun) to increase body temperature, with an adverse effect on the fungal entomopathogen present in the hemocoel (Carruthers *et al.*, 1992; Kalsbeek *et al.*, 2001; Elliot *et al.*, 2002).

6.3.4. Secondary Metabolites

As part of their biological activity, organisms produce chemical compounds that are considered to be essential for normal growth and development, such as amino acids, nucleotides, proteins, and carbohydrates. These are referred to as primary metabolites. Any other compound that is not essential for growth and development is referred to as a secondary metabolite. (For a review on the origin and use of the term "secondary metabolites", see Bennett and Bentley, 1989.) Some fungal secondary metabolites have medicinal properties in humans (e.g., penicillin, cephalosporin, cyclosporine, statins), while others, known as mycotoxins, are toxic (e.g., ergot alkaloids, aflatoxins, ochratoxins) (Keller *et al.*, 2005). Even though fungal entomopathogens produce many secondary metabolites, for the most part, the role of the metabolite in pathogenesis remains unclear (Molnár *et al.*, 2010). The detection of a secondary metabolite in culture does not necessarily imply that it is being produced in the insect or that it plays a role in pathogenicity.

Beauveria spp. have been reported to produce several different cyclodepsipeptides (peptide lactones). Cyclodepsipeptides can act as ionophore antibiotics owing to their selective interactions with potassium or sodium ions, thus altering the permeability of cell membranes (Ngoka *et al.*, 1999). The most studied *B. bassiana*-produced cyclodepsipeptide is beauvericin, first isolated from liquid culture (Hamill *et al.*, 1969). Beauvericin is toxic to fall armyworm (*Spodoptera frugiperda*) cell lines (Calo *et al.*, 2003; Fornelli *et al.*, 2004) and has a negative effect on the ultrastructure of mosquito (*Culex pipiens*) larvae (Žižka and Weiser, 1993). Xu *et al.* (2008) reported that "… beauvericin plays a highly significant but not indispensable role in virulence."

Other cyclodepsipeptides produced by *Beauveria* spp. include beauverolides (Elsworth and Grove, 1977; Frappier *et al.*, 1978), bassianolides (Suzuki *et al.*, 1977), beauveriolides (Mochizuki *et al.*, 1993), and bassiatin (Kagamizono *et al.*, 1995). Bassianolide gene disruption experiments by Xu *et al.* (2009) confirmed that this metabolite serves as a significant factor in virulence when the bassianolide non-producer was tested in bioassays involving dipping of various lepidopteran species (*Galleria mellonella*, *S. exigua*, or *Helicoverpa zea*) in conidial suspensions. Bassianolide has been shown to be toxic to the silkworm (*Bombyx mori*) when injected into larvae or incorporated in the diet (Suzuki *et al.*, 1977; Kanaoka *et al.*, 1978), or when injected into *H. zea* (Champlin and Grula, 1979). Kwon *et al.* (2000) reported the detection of bassianolide and beauvericin in *B. mori* larvae that had been killed by *B. bassiana*. Beauveriolides I and II have been reported to have moderate insecticidal activity against *Callosobrochus chinensis* and *Spodoptera litura* (Mochizuki *et al.*, 1993).

Beauveria spp. are not the only fungi that produce these secondary metabolites. For example, bassianolide is also produced by *Verticillium lecanii* (= *Lecanicillium* spp.)

(Suzuki *et al.*, 1977), beauverolides are produced by *I. fumosorosea* (Jegorov *et al.*, 1994), and beauvericin is produced by *Cordyceps*, *Fusarium*, and *Isaria* (Logrieco *et al.*, 1998; Luangsa-ard *et al.*, 2009).

Other secondary metabolites produced by *Beauveria* include three non-peptide pigments, oosporein (a red dibenzoquinone) (Vining *et al.*, 1962) and the yellow 2-pyridones tenellin and bassianin (El-Basyouni *et al.*, 1968; Wat *et al.*, 1977). Eley *et al.* (2007) conducted tenellin knockout experiments that demonstrated the non-involvement of tenellin in pathogenesis. Takahashi *et al.* (1998a,b) reported two pyridone alkaloids from *B. bassiana*: pyridovericin and pyridomacrolidin, while Quesada-Moraga and Vey (2004) reported the isolation of bassiacridin from a *B. bassiana*-infected locust, followed by tests that demonstrated its toxicity to locusts after injection into their hemocoel.

Destruxins are cyclodepsipeptides produced by various fungi and were first isolated from *Oospora destructor* (= *Metarhizium anisopliae*) (Kodaira, 1961, 1962) and later on from an *Aschersonia* sp. (Krasnoff *et al.*, 1996). Some of the 38 different destruxins or destruxin analogues have been shown to be insecticidal (Pedras *et al.*, 2002; Schrank and Vainstein, 2010), apparently by opening calcium channels in insect muscle membranes (Samuels *et al.*, 1988a). Samuels *et al.* (1988b) tested four *M. anisopliae* isolates against larvae of *Manduca sexta*; of these, three were pathogenic, but detection of destruxin A in the hemolymph was positive only for the most virulent isolate. For this isolate, growth was sparse in the hemolymph, compared to the two other pathogenic isolates, which exhibited a proliferation of hyphal bodies (see also Kershaw *et al.*, 1999).

Metarhizium spp. also produce cytochalasins (Aldridge and Turner, 1969), myroridins (Kondo *et al.*, 1980), swainsoine (Patrick *et al.*, 1993), viridoxin (Gupta *et al.*, 1993), helvolic acid (Espada and Dreyfuss, 1997), 12-hydroxy-ovalicin (Kuboki *et al.*, 1999), hydroxyfungerins (Uchida *et al.*, 2005), fusarin analogues (Krasnoff *et al.*, 2006), serinocyclins (Krasnoff *et al.*, 2007), aurovertins (Azumi *et al.*, 2008), macrolides (Kozone *et al.*, 2009), and tyrosine betaine (Carollo *et al.*, 2010).

Other metabolites produced by fungal entomopathogens include the trichotechane tenuipesine A, produced by *I. tenuipes* (as *Paecilomyces tenuipes*) (Kikuchi *et al.*, 2004); the pyridone akanthomycin, produced by *Akanthomyces gracilis* (Wagenaar *et al.*, 2002); cordyanhydrides, produced by *Cordyceps pseudomilitaris* (Isaka *et al.*, 2000); cordycepin, produced by various *Cordyceps* species (Cunningham *et al.*, 1950; Ling *et al.*, 2002); cyclosporin, produced by *Tolypocladium inflatum* (Aarnio and Agathos, 1989); and hopanoids, produced by *Aschersonia aleyrodis* (van Eijk *et al.*, 1986). *Hirsutella thompsonii* has been reported to produce phomalactone (Krasnoff and Gupta,

1994) as well as the protein hirsutellin (Mazet and Vey, 1995; Liu *et al.*, 1995). Phomalactone was shown to be toxic to the tephritid flies *Rhagoletis pomonella* and *Ceratitis capitata*, and to inhibit conidial germination of *B. bassiana*, *Tolypocladium geodes*, *T. cylindrosporum*, and *M. anisopliae* (Krasnoff and Gupta, 1994). In laboratory bioassays, hirsutellin was shown to be toxic to larvae of the greater wax moth, *G. mellonella*, and the common fruit fly, *Drosophila melanogaster* (Vey *et al.*, 1993), and to the citrus rust mite, *Phyllocoptruta oleivora* (Omoto and McCoy, 1998).

This coverage of secondary metabolites is by no means comprehensive and is meant to illustrate the diversity of secondary metabolites produced by fungal entomopathogens. For excellent reviews the reader is encouraged to consult Strasser *et al.* (2000), Vey *et al.* (2001), Isaka *et al.* (2005), Frisvad *et al.* (2008), Molnár *et al.* (2010), and Rohlfs and Churchill (2011).

6.3.5. Environmental Factors Influencing Stability

Ultraviolet Light

A major constraint to the survival of conidia in the epigeal habitat is exposure to solar radiation. The solar spectrum contains electromagnetic radiation at different wavelengths (measured in nanometers, nm). Ultraviolet light (UV) occurs at three different spectra: UVC (100−280 nm), UVB (280−315 nm), and UVA (315−400 nm). The visible range occurs at 380−780 nm. UVC is blocked by ozone and does not reach the Earth (Blumthaler, 1993). Conidia are very susceptible to UVB. The detrimental damage caused by UV light is due to photoreactions of nucleic acids, proteins, lipids, and membranes (Tevini, 1993). Sublethal exposure to UV radiation can cause physiological or genetic alterations that reduce virulence, e.g., reduced and delayed germination (Hunt *et al.*, 1994; Braga *et al.*, 2001a).

The pigmentation of conidia often influences susceptibility to solar radiation, with the most pigmented conidia usually being more tolerant to UV radiation (Ignoffo and Garcia, 1992; Fargues *et al.*, 1996; Braga *et al.*, 2006). Artificial methods have been studied to protect conidia from UV radiation. These involve the use of sunscreens that act as solar blockers by reflecting radiation (e.g., optical brighteners such as Tinopal) or as UVA/UVB absorbents (e.g., dyes such as Congo red). Inglis *et al.* (1995) tested 21 sunscreens in oil or water formulations in both the laboratory and the field and reported that some sunscreens increased *B. bassiana* survival. In contrast, Shah *et al.* (1998) reported that *M. acridum* (as *M. flavoviride*) conidia treated with oxybenzone as a sunscreen and used against grasshoppers did not exhibit increased efficacy in the field.

The natural pigments contained in vegetable oils used in formulations can also protect conidia by blocking solar radiation (Braga *et al.*, 2001b). Similarly, sublethal stresses (nutritional, osmotic, etc.) can enhance conidial tolerance to solar radiation (Rangel *et al.*, 2008). An extensive list of formulation additives, including sunscreens, has been published by Bernhard *et al.* (1998).

Temperature

One major issue encountered while assessing the effects of temperature on fungal entomopathogens is that most studies are based on constant temperatures in the laboratory, which do not mimic natural conditions. In addition, the environmental conditions usually measured in the field might not reflect those encountered in the microenvironment inhabited by the insect and fungal entomopathogens (Benz, 1987).

Nevertheless, various studies show that optimal ambient temperatures for vegetative growth of fungal entomopathogens vary widely across and within species. For example, working with 37 isolates of *I. fumosorosea*, Vidal *et al.* (1997) reported optimal temperatures of 20–30°C, whereas optimal temperatures for 65 isolates of *B. bassiana* were between 25 and 28°C, although the optimal temperatures for some isolates were 20 and 30°C (Fargues *et al.*, 1997). Roberts and Campbell (1977) published an extensive table indicating that the optimal temperatures for germination, growth, and sporulation are between 20 and 30°C. In contrast, entomophthoralean fungi are generally considered to have temperature optima for growth at 20°C or below (Hajek, 1997), although some species such as *Neozygites tanajoae* and *N. floridana* are found in tropical regions.

Some fungal entomopathogens have been shown to remain infective at low temperatures. Thus, Doberski (1981) assessed infection of elm bark beetles inoculated with *B. bassiana* or *I. farinosa* and incubated at 2, 6, 10, 15, and 20°C, and reported positive infection even at 2°C. These results imply that infection can occur at temperatures at which host insects are inactive.

Various genera of fungal entomopathogens have been isolated in Antarctica, e.g., *B. bassiana* (Mahaney *et al.*, 2001), *Neozygites* cf. *acaridis* (Bridge and Worland, 2004), and *Conidiobolus antarcticus* (Tosi *et al.*, 2004). Several undescribed species of *Strongwellsea* were found as infections in *Spilogona* spp. flies in high Arctic Greenland (Eilenberg, 2002), while *M. anisopliae* and *B. bassiana* have been isolated from sub-Antarctic soil in Macquarie Island, and *M. anisopliae* was able to germinate at 2.5°C (Roddam and Rath, 1997). De Croos and Bidochka (1999) recovered 26 isolates of *M. anisopliae* in Ontario, Canada, and at least seven of these could germinate and grow at 8°C.

These data indicate that temperature is a variable for which generalizations are difficult to make. One important fact related to temperature is that most fungal entomopathogens will not grow at human body temperature (37°C), a positive aspect in risk assessment during the registration process (Butt, 2002), although some tropical isolates of *M. anisopliae* and *M. flavoviride* have been shown to grow at 37°C and above (Rangel *et al.*, 2005; Zimmermann, 2007b). Human infection with *Beauveria* has been reported in immunocompromised individuals (Henke *et al.*, 2002; Tucker *et al.*, 2004). An isolate of *B. bassiana* from a liver lesion in a patient receiving immunosuppressant therapy could not grow at 37°C but was highly virulent to the Colorado potato beetle, *Leptinotarsa decemlineata* (Henke *et al.*, 2002).

Zimmermann (1982) showed that a correlation exists between median lethal temperatures for *M. anisopliae* conidia and humidity, with mean lethal temperature increasing as humidity decreases. Temperature might have a stronger negative effect on fungal entomopathogen infection of insects close to the soil surface, owing to the higher temperatures (up to 60–65°C) that have been recorded on the soil surface (Arthurs *et al.*, 2001; Rangel *et al.*, 2005).

Humidity

The level of humidity can have an effect on the germination of conidia and on the subsequent fungal growth and sporulation on the insect cadaver (Jaronski, 2010). Roberts and Campbell (1977) mention several studies in which it was shown that high humidity was needed for successful infection, and others in which it was not needed; this discrepancy has been ascribed to the microclimate surrounding the conidium on the insect cuticle (Ferron, 1977; Benz, 1987; Inglis *et al.*, 2001). For vegetative growth, Ferron (1977) demonstrated that mycelial development of *B. bassiana* on the insect cadaver required 92% humidity or higher, whereas Benz (1987) presented several examples of varying humidity levels resulting in mycelial growth and sporulation.

6.4. ECOLOGY

Ecology is the study of the patterns of distribution and abundance of organisms, and of how interactions among and between organisms and their environment influence the observed patterns (Begon *et al.*, 1990). Ecology is therefore implicitly integrated in many other disciplines of biology, including the study of biological control, which is a subdiscipline of applied ecology. The ecology of fungal entomopathogens was reviewed by Hajek (1997). Major advances made since then have been reviewed by Vega *et al.* (2009) and by Roy *et al.* (2010).

As discussed in Section 6.3.5, environmental factors have an impact on fungal entomopathogens in various ways and are likely to influence patterns of distribution and abundance. This section will focus on the host range of fungal entomopathogens and on how fungal entomopathogens can interact directly and indirectly with other organisms in complex trophic settings. These topics are highly relevant for the fundamental understanding of fungal entomopathogen ecology as well as for future development of sustainable approaches to biological control. Examples will be focused on anamorphic taxa of Hypocreales because most research in agricultural ecosystems has been based on this group. However, ecological studies on a few taxa within Entomophthorales have also been conducted and will be included in this section.

6.4.1. Host Range of Fungal Entomopathogens

Knowledge of the host range of fungal entomopathogens is important for several reasons. For fundamental ecological studies, it is relevant to know which hosts a fungus infects and the degree of virulence. Host distribution determines which host population the fungus must rely on for resources. In applied ecology (e.g., biological control), it is relevant to determine the host range of a fungal biological control agent, including both target and non-target hosts. However, determining host range is not straightforward and requires an in-depth understanding of both the fungus and the insect host environment.

Narrow and Broad Host Ranges: Specialist and Generalist Entomopathogens

Fungal entomopathogens include a variety of fungal taxa. Functionally, they also span a great variation of host specificity from extreme specialists with narrow host ranges to generalists with very broad host ranges. It is impossible to generalize as to which fungal taxa are specialist and generalist entomopathogens. However, considering the two most diverse orders of fungal entomopathogens, the Entomophthorales and Hypocreales, the former consists primarily of fungi with narrow host ranges whereas the latter exhibits a variety of host ranges from narrow to very broad. A fungal entomopathogen with a narrow host range is usually considered to be highly virulent and obligate (Goettel, 1995). This means that only a few spores are necessary to infect hosts, and that the fungus depends fully on infecting the host for obtaining nutrients. Furthermore, fungal entomopathogens with narrow host ranges usually produce specialized survival structures such as resting spores to survive periods when their host is not present. Finally, they often develop epizootics in their host populations when conditions are suitable. In contrast, fungal entomopathogens with broad host ranges are often considered facultative entomopathogens, i.e., they can survive on nutrients that are not

obtained from the living host. The ability to extract nutrients from *in vitro* media could be interpreted as an ability of the fungus to survive as a saprobe in the soil from non-host-derived nutrients. However, fungal entomopathogens are poor competitors compared to opportunistic microorganisms (Hajek, 1997; Meyling and Eilenberg, 2007). The fungal entomopathogens with broad host range are able to survive in their non-entomopathogenic part of the life cycle, the saprobic phase, in which they exploit the resources of their killed host (i.e., cadaver) (Vega *et al.*, 2009). As discussed below, some fungal taxa do appear to exploit nutrients that are not arthropod derived and they may therefore have other ecological roles, besides being entomopathogenic (Goettel *et al.*, 2008; Vega *et al.*, 2009). The fungal entomopathogens with broad host ranges generally exhibit relatively low virulence (Goettel, 1995). For long-term survival without a living host, they produce no specific survival structures, but their infective propagules can persist for prolonged periods in the environment. In epidemiological (epizootiological) terms, these fungi persist within their host populations at an endemic (or enzootic) level.

Hajek and Butler (2000) have defined two types of host range: ecological and physiological. The ecological host range is used for hosts that are infected by the fungus in nature, in contrast to the physiological host range, which is used for hosts that can be infected in the laboratory. For biological control studies, it is usually the physiological host that is in focus as many studies aim to find the most virulent isolate of a fungus towards a particular pest. However, infections induced in the laboratory are not necessarily reproducible in the field. For example, host ranges of *Entomophaga maimaiga* within Lepidoptera were tested against 78 species from 10 superfamilies in laboratory experiments (Hajek, 1999). About one-third of the lepidopteran species could be infected to produce sporulating cadavers, and in particular *Malacosoma disstria* showed the highest level of successful infection by *E. maimaiga*. Yet, when surveys were conducted in the field during an epizootic of *E. maimaiga* in the target host, the gypsy moth (*Lymantria dispar*), only one out of 318 *M. disstria* specimens, was infected by the fungus. Among 52 lepidopteran species collected, just two individuals from each of two species other than *L. dispar* were infected by *E. maimaiga* (Hajek, 1999). Presumably, only *L. dispar* is spatiotemporally sympatric with *E. maimaiga* in the forest habitat and by showing a diurnal migration pattern from the tree canopy to the forest floor, the caterpillars are exposed to the fungal inoculum. Therefore, caution should be taken in extrapolating laboratory data to field situations when evaluating target and non-target effects of fungal entomopathogens for biological control (Hajek and Butler, 2000; Jaronski *et al.*, 2003). As mentioned before, in laboratory experiments conditions for infection are usually optimized;

such conditions are generally not prevalent in the habitat of the host, thereby further reducing the likelihood of pathogen establishment. Thorough insights into the ecology and life history of the host and knowledge of the distribution of the fungus in the habitat are also crucial for assessment of ecological host ranges.

Some pathogens in the Entomophthorales have extremely narrow ecological host ranges although only a few have been studied in detail. Thus, epizootics caused by a particular fungus can be observed in only a single host species even when closely related species occur simultaneously in the same habitat (see Hajek and Butler, 2000, for examples). Using DNA-based characterization, it has been shown that within *Entomophthora muscae sensu stricto*, specific clades of the fungus infect one particular host species in the field (Jensen *et al.*, 2001). Likewise, specific host-adapted clades can be found within individual species of the genus *Strongwellsea* (J. Eilenberg and A. B. Jensen, pers. comm.). In contrast, entomophthoralean fungi infecting aphids appear to have broader host ranges, with the same fungal isolate infecting several aphid species (Ekesi *et al.*, 2005).

The ability of fungal entomopathogens to infect different host species in the field can be exploited in conservation biological control strategies. Conservation biological control is defined as the "Modification of the environment or existing practices to protect and enhance specific natural enemies or other organisms to reduce the effect of pests" (Eilenberg *et al.*, 2001). Specifically, if a fungal entomopathogen infects several hosts, it can be harbored in reservoir species in the vicinity of crops and then potentially be dispersed to the target pest (Meyling and Eilenberg, 2007), thus regulating the pest through the ecological process of apparent competition (Meyling and Hajek, 2010). Based on a field survey conducted in an English meadow locality, *P. neoaphidis* was found to infect up to 11 aphid species in one year, and during two years *P. neoaphidis* was the most common among five aphid pathogenic entomophthoralean fungi (van Veen *et al.*, 2008). The ability of *P. neoaphidis* to infect several host species in aphid communities has led to investigations aimed at infecting target aphids by movement of the pathogen from other aphid species occurring in non-crop areas (Ekesi *et al.*, 2005). Ecological studies of *P. neoaphidis* have been performed for development of conservation biological control of aphids (Pell *et al.*, 2010).

Cryptic Species and the Use of Molecular Methods for Ecological Studies

Fungal entomopathogens in the Hypocreales are generally considered to have broad host ranges. For instance, Goettel *et al.* (1990a) and Zimmermann (2007a, b, 2008) provide extensive host lists for many species of fungal entomopathogens. More than 700 arthropod host species have been recorded for *B. bassiana* (Li, 1988, cited in Zimmermann, 2007a) and more than four decades ago, over 200 insect host species were recorded to be infected by *M. anisopliae* (Veen, 1968, cited in Zimmermann, 2007b). The list compiled by Zimmermann (2008) for hosts susceptible to *I. farinosa* and *I. fumosorosea* shows a relatively high representation of species from Lepidoptera (42/89 and 36/81, respectively), whereas the list for *M. anisopliae* (Zimmermann, 2007b) has a relatively high representation of species from Coleoptera (134/204). This could indicate variability in host specificity but could also be the product of associations of differential habitat requirements of fungus and host (see below).

The compiled host lists of Goettel *et al.* (1990a) and Zimmermann (2007a, b) are based on identification of morphological characters such as conidia size, shape, and color. Recent molecular studies of *Beauveria* and *Metarhizium* have shown that morphological characters often are insufficient in determining species identity of these taxa as several cryptic species exist, many with a cosmopolitan distribution (Rehner and Buckley, 2005; Bischoff *et al.*, 2006, 2009). The host range of each "real" fungal species may be disentangled only by an identification based on molecular characters, as the fungi identified by morphological characters may consist of individual cryptic species showing different host associations. However, although fungal isolates that morphologically resemble *B. bassiana* can be any of five species (*B. bassiana*, *B. pseudobassiana*, *B. varroae*, *B. kipukae*, or *B. australis*) (Rehner *et al.*, 2011), the most widespread of these species can be found infecting different insects hosts from several orders (Rehner and Buckley, 2005; Ghikas *et al.*, 2010). Even *B. varroae*, whose species epithet may indicate host specificity towards varroa mites, has also been found infecting coleopteran hosts (Rehner *et al.*, 2011). Some of the newly defined species of *Beauveria* are poorly sampled and their host ranges can therefore not be conclusively determined, but *B. brongniartii* and *B. caledonica* appear to be associated with hosts within the Coleoptera (Zimmermann, 2007a; Reay *et al.*, 2008).

Isolates identified as *M. anisopliae* based on morphological characters may also belong to any of several species molecularly identified by Bischoff *et al.* (2009). Similar conidial morphology was demonstrated for *M. anisopliae*, *M. brunneum*, *M. lepidiotae*, *M. pingshaense*, and *M. robertsii*. Thus, studies reporting the host range of *M. anisopliae* may include any of these newly erected species as well as other species showing very similar or overlapping morphological characters, including *M. guizhouense* and *M. majus* (Bischoff *et al.*, 2009). Within *Metarhizium*, *M. acridum* (*M. anisopliae* var. *acridum sensu* Driver *et al.*, 2000; elevated to species level by

Bischoff *et al.*, 2009) appears to be the most host-specific species, primarily infecting locusts and grasshoppers.

With these new explicit definitions of fungal entities, future ecological studies related to *Beauveria* and *Metarhizium* should implement molecular markers. These markers will provide the scientific community with common tools to determine explicitly the collected isolates, making comparisons among studies possible. In particular, sequencing of the intron region of elongation factor 1-alpha (EF1-α) will provide good resolution for identification of *Metarhizium* spp. (Bischoff *et al.*, 2006, 2009). In addition to EF1-α, sequencing the Bloc region (Rehner *et al.*, 2006, 2011; Meyling *et al.*, 2009) will resolve clades within *Beauveria* spp. The molecular methods even allow for further phylogenetic resolution within individual species (e.g., *B. bassiana*), with clades identities comparable to a framework defined and presented by Rehner *et al.* (2006) and Meyling *et al.* (2009).

Application of the explicitly defined phylogenetic species concept to *Beauveria* isolates collected from latent infections of insect hosts living sympatrically within the same hedgerow habitat showed that the two most commonly found clades of *B. bassiana sensu stricto* (Eu_1 and Eu_5) infected seven and six different host species, respectively (Meyling *et al.*, 2009). These hosts belonged to three insect orders: Hemiptera, Diptera, and Coleoptera. The remaining clades were represented by only a few samples and their host ranges could not be evaluated from this study, but it can be concluded that even when focusing on individual clades of *B. bassiana*, broad ecological host ranges are evident.

Host Range Beyond Insects

Several studies have shown that the host range of fungal entomopathogens can extend beyond insects. For example, *B. bassiana* (reviewed by Ownley *et al.*, 2010), *Lecanicillium* spp. (reviewed by Goettel *et al.*, 2008), and *M. anisopliae* (Kang *et al.*, 1996) can also act as antagonists of plant pathogens via antibiosis or mycoparasitism.

Above-ground propagules of *B. bassiana* can be isolated from phylloplanes of different plant species (Meyling and Eilenberg, 2006a). Moreover, most plants harbor fungi asymptomatically within their tissues. These fungi are called endophytes and are comprised mainly of members of the Ascomycota, of which many are in the Hypocreales (Vega *et al.*, 2008). Fungal entomopathogens have also been included in this spectrum of endophytes (Vega, 2008). Endophytic activity of *B. bassiana* was first demonstrated in corn (*Zea mays*) (Bing and Lewis, 1991, 1992) and the host plant range of endophytic *B. bassiana* has since expanded (Vega, 2008; Vega *et al.*, 2008). To date, there is no evidence that the presence of *B. bassiana* as an endophyte has fitness benefits for the plant.

6.4.2. Distribution and Abundance of Fungal Entomopathogens

Fungal entomopathogens are globally distributed in almost all terrestrial ecosystems. Diversity is at its highest in the tropical forests, but fungal entomopathogens are also found in extreme habitats such as in the high Arctic tundra (Eilenberg, 2002) and Antarctica (Bridge and Worland, 2004; Tosi *et al.*, 2004). In general, the trend of global distribution patterns indicates that fungal entomopathogens in the Entomophthorales mostly occur in temperate climates, decreasing in abundance towards the subtropics and tropics (Hajek, 1997), although *N. tanajoae* is a significant mite pathogenic fungus in tropical regions (Delalibera *et al.*, 2004). The sexual stages of the Hypocreales are mostly found in tropical climates while the asexual stages are found in both tropical and temperate climates. Bidochka and Small (2005) hypothesized that anamorphic lineages of *Metarhizium* had dispersed from Southeast Asia, where the teleomorphs (*Metacordyceps*) presumably are found. From this center, most anamorphs have become cosmopolitan, although the highest *Metarhizium* diversity remains in the Asian region (Bidochka and Small, 2005). It has further been hypothesized that the anamorphs such as *Metarhizium* (and presumably other genera of fungal entomopathogens in the Hypocreales) colonized managed ecosystems such as farmland around the world as generalist entomopathogens reproducing asexually (Bidochka and Small, 2005). However, teleomorphs do occur in colder climates, although they are often restricted to pristine and mature ecosystems, e.g., *Ophiocordyceps sinensis* in the Tibetan plateau (Weckerle *et al.*, 2010) and several *Cordyceps* species in North America (Mains, 1958; Sung and Spatafora, 2004).

Fungal entomopathogens are often not readily visible, making the assessment of their distribution in the environment very challenging. Emerging fungal structures from a host cadaver are reliable signs of the presence of a fungus, but often hosts are small and die in an inconspicuous place such as underground, in aquatic habitats, or under bark. Exceptions are some taxa belonging to Entomophthorales that cause the host to die at an elevated point on vegetation to improve conidial dispersal (also known as summit disease), ensuring cadaver attachment to the substrate by producing rhizoids (Roy *et al.*, 2006). Similarly, some ascomycete taxa, usually the teleomorphic stages, produce conspicuous stroma emerging from their hosts (see Section 6.2.1).

The anamorphic stages of many ascomycetous fungal entomopathogens occur widely in most terrestrial ecosystems. Yet, their fungal structures are subtle in appearance and their often minute, mycosed arthropod hosts are inconspicuously located. Besides cadavers, microscopic fungal spores can be located outside their hosts in several ecosystem compartments. These compartments include predominantly the soil environment, which generally provides a stable habitat for fungal populations by buffering population fluctuations and protecting against detrimental abiotic conditions (van der Putten *et al.*, 2001).

Distribution Patterns of Generalist Fungal Entomopathogens

Specific methods are often necessary for the indirect detection of the presence of fungal entomopathogens in the soil environment. The most widely used is the insect bait method, originally developed for isolation of entomopathogenic nematodes, but adapted to the isolation of fungal entomopathogens from soil samples (Zimmermann, 1986). The principle of forcing a susceptible insect through a substrate potentially containing pathogenic microorganisms is intriguing and has become widespread as a standard method in surveys of fungal entomopathogen communities of soils. The main advantage of using the insect bait method on soil samples is that this method selectively isolates fungal entomopathogens that are biologically active. However, different host insects used as bait may not become infected by the same fungal entomopathogens (Klingen *et al.*, 2002; Goble *et al.*, 2010). This suggests that using more than one insect host as bait will result in a better assessment of fungal entomopathogen diversity in soils. Incubation temperature can affect which fungal taxa will dominate the fungi isolated from soil samples (Mietkiewski and Tkaczuk, 1998).

The insect bait method has been termed semi-quantitative at best (Jaronski, 2007) or merely qualitative (Scheepmaker and Butt, 2010). This means that extrapolations to the level of fungal inoculum in the soil samples based on the insect bait method are not possible as the information gathered from each sample is qualitative. Plating dilution series of soil suspensions on selective agar media will provide quantitative information of densities of fungal inoculum in individual samples. Furthermore, growth of a fungal colony on artificial media does not prove Koch's postulate of pathogenicity (see Chapter 1), and Koch's test should be carried out. Densities of naturally occurring fungal entomopathogens in soils based on dilution plating techniques have been reviewed by Scheepmaker and Butt (2010). Densities of *M. anisopliae* and *B. bassiana* varied greatly, but means were in the range of 10^2-10^3 colony forming units (cfu)/g soil (Scheepmaker and Butt, 2010). Keller *et al.* (2003) used both methods to evaluate communities of fungal entomopathogens in soil habitats in Switzerland and concluded that baiting soil samples with *G. mellonella* tended to be more sensitive for detection of fungal entomopathogens than using selective agar media. The threshold level of inoculum for initiation of infections is not clear, but presumably, high-density

areas are most likely to be responsible for successful infections. Such high-density patches are recorded in studies using dilution series plating (Scheepmaker and Butt, 2010).

In spite of its methodological limitations and the reservations raised about the ecological interpretations that can be made, the insect bait method, particularly using *G. mellonella*, is a valuable tool for analyses of distributions of fungal entomopathogens in soil samples. The method has provided much information about distributional patterns of fungal entomopathogens over the past 15–20 years, primarily in temperate regions. Although the method may not be truly quantitative, emerging patterns of abundance can be deduced from the studies. A recently developed method, involving the isolation of fungal entomopathogens from suspended-soil arthropods, revealed a higher number of fungal entomopathogen species in the suspended-soil arthropods than in ground-soil arthropods (Kurihara *et al.*, 2008).

Selected results from surveys of naturally occurring taxa of fungal entomopathogens in soil habitats are presented in Table 6.2. In all surveys, soil samples were baited with *G. mellonella* and samples were taken at several different sites within the specific geographical regions. Each site was defined as being either subjected to cultivation, mostly agricultural fields, or a less disturbed "natural" habitat type, mostly hedgerows or forested areas. In a study in China, however, Sun *et al.* (2008) defined their "cultivated" category as heavily cultivated cereal and corn fields, while "natural" habitats were represented by less cultivated orchards. The trend in these studies indicates that there are differences in the entomopathogenic fungal communities of cultivated and "natural" habitats within the geographical regions. In general, *B. bassiana* and *I. fumosorosea* (when the latter was found) were isolated more frequently from "natural" than from cultivated soils. Although not consistently so, *M. anisopliae* appears to be most frequent in soils of cultivated habitats. This observation has been interpreted as *M. anisopliae* populations showing more resilience to disturbance, which is a characteristic of cultivated habitats (Bidochka *et al.*, 1998; Meyling and Eilenberg, 2007). However, it is not always possible to predict the compositions of the entomopathogenic fungal community at a particular locality within a specific region based on distribution and abundance patterns from regional sampling of several localities.

In Denmark, the regional survey conducted by Steenberg (1995) indicates that *M. anisopliae* dominates the fungal entomopathogen community in cultivated soil habitats (Table 6.2). However, a subsequent soil sampling of a single agroecosystem in Denmark and isolation of fungal entomopathogens by baiting samples with *G. mellonella* revealed that *M. anisopliae* was practically absent from this locality, which was instead dominated by

TABLE 6.2 Frequency of Occurrence (%) of Three Selected Taxa of Anamorphic Fungal Entomopathogens (Hypocreales) from Soil Samples at Sites Defined by Habitat

Country (latitude, average)	Fungus	Cultivated	"Natural"	Reference
Finland	*B. bassiana*	5.6	28.1	Vänninen (1996)
(62° N)	*M. anisopliae*	14.9	24.2	
	I. fumosorosea	0.5	1.7	
Denmark	*B. bassiana*	38.5	52.9	Steenberg (1995)
(55° N)	*M. anisopliae*	51.3	7.8	
	I. fumosorosea	2.6	9.8	
UK	*B. bassiana*	1.0	7.7	Chandler *et al.* (1997)
(52° N)	*M. anisopliae*	1.0	1.3	
	I. fumosorosea	0.0	3.3	
Canada	*B. bassiana*	35	65	Bidochka *et al.* (1998)
(45° N)	*M. anisopliae*	63	36	
	I. fumosorosea	N/A	N/A	
Spain	*B. bassiana*	34	53	Quesada-Moraga *et al.* (2007)
(40° N)	*M. anisopliae*	10	4	
	I. fumosorosea	N/A	N/A	
China	*B. bassiana*	27.4	86.3	Sun *et al.* (2008)
(40° N)	*M. anisopliae*	60.0	26.4	
	I. fumosorosea	15.6	37.5	
South Africa	*B. bassiana*	13.0	8.3	Goble *et al.* (2010)
(33° S)	*M. anisopliae*	3.7	4.2	
	I. fumosorosea	N/A	N/A	

B. bassiana = *Beauveria bassiana*; *M. anisopliae* = *Metarhizium anisopliae*; *I. fumosorosea* = *Isaria fumosorosea*; N/A = not applicable.

B. bassiana (Meyling and Eilenberg, 2006b). A similar sampling and isolation approach at another locality in Denmark showed clear dominance of *M. anisopliae* in agricultural field soils (Meyling *et al.*, 2011). It is possible that the agroecosystem investigated by Meyling and Eilenberg (2006b) harbored an atypical soil-borne entomopathogenic fungal community, but it remains unclear why *M. anisopliae* was not found in the agricultural field. Variation in environmental factors could explain some of the distribution patterns reported. As discussed earlier, fungal entomopathogens are affected by abiotic factors and fungal taxa are affected differently by these factors (Zimmermann, 2007a, b, 2008; Jaronski, 2007; Scheepmaker and Butt, 2010). Thus, variations in soil type and other characteristics are likely to affect fungal entomopathogen

distribution patterns. However, Bidochka *et al.* (2001) suggested that abiotic factors prevalent in agricultural fields, such as high UV radiation and temporally high temperatures, were selecting for fungal genotypes able to be resilient to these conditions. Indeed, a specific clade of *M. anisopliae* was isolated most frequently from agricultural soils in Canada, and isolates from this clade were more tolerant to UV radiation and high temperatures than isolates from another clade that was more frequent in forest soils (Bidochka *et al.*, 2001). Selected isolates from the two clades were recently molecularly identified as *M. robertsii* and *M. brunneum*, respectively (Bischoff *et al.*, 2009).

Bidochka and Small (2005) suggested two main processes for the global distribution and abundance patterns of *Metarhizium* spp. In temperate regions, the fungal genotypes are associated with habitats defined by specific abiotic environmental factors that select for the genotypes present, while in tropical and subtropical regions fungal genotypes are associated with specific hosts (Bidochka and Small, 2005). While there is some evidence supporting the first process of habitat selection, it remains to be determined whether *Metarhizium* spp. in general are associated with certain hosts in tropical regions, especially when species are defined within the novel framework of Bischoff *et al.* (2009).

The habitat selection hypothesis was further expanded by Bidochka *et al.* (2002) to include *B. bassiana*. At the locality in Denmark mentioned above where the fungal community in the agricultural soil was dominated by *B. bassiana* (Meyling and Eilenberg, 2006b), the isolates were shown to belong to a single clade (*B. bassiana* Eu_1) while *B. bassiana* isolates from the bordering hedgerow were placed among several different clades (Meyling *et al.*, 2009). In Canada, isolates from certain clades of *B. bassiana* from agricultural fields were shown to be more tolerant to UV radiation and high temperatures than isolates from other clades that were recovered in forest habitats (Bidochka *et al.*, 2002). Whether the particular abiotic factors of agricultural fields select for fungi with specific tolerances remains to be elucidated, but it is clear from surveys of the fungal entomopathogen communities of soils (some included in Table 6.2) that differences are evident between habitat types. For example, it will be relevant to focus on why *I. fumosorosea* is almost absent from agricultural field soils while it is frequently isolated from hedgerows. Is this pattern reflecting particular abiotic or biotic factors of the "natural" habitat that are different from the cultivated ones?

Linking Below- and Above-ground Distribution of Fungal Entomopathogens

Even though fungal entomopathogens can be isolated from soils in almost any terrestrial ecosystem, their ecological significance in soils is not clear. One obvious possibility is that the soil environment acts as a reservoir for dormant infective units awaiting a susceptible host, or that it serves as a sink from which infective propagules cycle between below- and above-ground environments. It is also possible that fungal entomopathogens have alternative ecological roles in the soil, other than being entomopathogens.

Although it is well known that morphologically similar fungi, e.g., *B. bassiana sensu lato*, can be found both in the soil and as infections in insects above ground, the distributions must be studied within the same spatial locality to ensure the simultaneous occurrence in the two ecosystem compartments. Ormond *et al.* (2010) documented that *B. bassiana* could be isolated from soil as well as plant surfaces within a forest ecosystem in the UK. Similarly,

Meyling *et al.* (2009) isolated *B. bassiana* from soil and plant surfaces, and as latent infections in various insect hosts within a single hedgerow in Denmark. These studies indicate that *B. bassiana* cycles between below- and above-ground compartments within individual ecosystems in temperate regions of Europe. Indeed, the additional characterization of *B. bassiana* isolates by DNA sequencing showed that some individual *B. bassiana* clades were found in soil, on plants, and as infections in hosts at the same site (Meyling *et al.*, 2009). *Beauveria bassiana*, *I. farinosa*, and *I. fumosorosea* can be isolated from soils, and these taxa also occur as infections in arthropod hosts above ground at similar localities. However, detailed population studies on distribution patterns of *Isaria* spp. have not yet been conducted.

In tropical forests, *M. anisopliae* can be readily isolated from the soil environment by baiting soil samples with *G. mellonella* and the yellow mealworm, *T. molitor* (Hughes *et al.*, 2004). In temperate agroecosystems, *M. anisopliae* is also widespread in the soil environment (see Table 6.2) but is absent or rare above ground. Although *M. anisopliae* can occur as infections in arthropod hosts, the prevalence of natural infections has rarely been quantified. In a unique study, Yaginuma (2007) collected 10,411 mycosed arthropod cadavers from an apple orchard in Japan over a six-year period. The fungal entomopathogen community was dominated by *B. bassiana* (88% of all cadavers) and in most years, cadaver numbers peaked in September. Low prevalences of *M. anisopliae* (1% of all cadavers) were documented, thus supporting the statement that *M. anisopliae* rarely infects above-ground hosts in temperate regions. However, it could be that *M. anisopliae* was rare in the particular ecosystem sampled, and therefore, the results might simply reflect the general absence of the fungus at that site. To explore this, simultaneous assessments of different environmental compartments, both below and above ground, must be conducted. An evaluation of fungal entomopathogenic communities below and above ground at the same locations was conducted by Meyling *et al.* (2011). Soils of agricultural field plots were baited with *G. mellonella* and above-ground distribution and abundance of fungal entomopathogens were assessed by searching the ground surface for mycosed cadavers in the same plots. Hosts succumbing to infections by generalist entomopathogens such as *B. bassiana*, *M. anisopliae*, and *I. farinosa* are likely to fall to the ground upon death. Two years of surveys showed that *B. bassiana*, *I. farinosa*, and *M. flavoviride* occurred both in the soil and as infections in various hosts above ground, with *B. bassiana* being the dominating fungus among mycosed cadavers. In contrast, *M. anisopliae* was not found above ground even though it was the dominating fungus in the soil environment in the same plots (Meyling *et al.*, 2011). These observations indicate that although *M. anisopliae* occurs widely in temperate agroecosystems, it is likely to be restricted to the below-ground compartments and, therefore, soil-dwelling arthropods are potential hosts for this fungus.

Are Fungal Entomopathogens Only Associated with Arthropods?

So far, the ecological role of *M. anisopliae* in the soil remains elusive. Are fungal propagules isolated from the soil environment dormant and inactive conidia, mycelia derived from saprobic exploitation of insect cadavers, or is *M. anisopliae* obtaining nutrients from substrates of non-arthropod origin? Recent advances in the unraveling of *M. anisopliae* as a plant-associated fungus in the rhizosphere have questioned whether *M. anisopliae* should be strictly considered to be an entomopathogen (Hu and St. Leger, 2002; Vega *et al.*, 2009; Bruck, 2010).

Hu and St. Leger (2002) showed that experimentally applied *M. anisopliae* propagules did not decline in the soil immediately surrounding the rhizosphere of cabbage plants, while fungal densities in the non-rhizosphere bulk soil declined over a period of six months. Likewise, Bruck (2005) demonstrated that densities of applied inoculum of *M. anisopliae* were higher in growth media close to roots of Norway spruce (*Picea abies*) than in the bulk medium. These observations indicate that *M. anisopliae* survives better in the rhizosphere than in the soil. Whether this observation merely reflected delayed population decline or continued population growth was resolved when Bruck (2010) observed increased densities of *M. anisopliae* in rhizospheres of *P. abies*. However, increases in densities were not observed for *Taxus baccata*, indicating that rhizosphere competence of *M. anisopliae* is not related to all host plants (Bruck, 2010). In addition, Fang and St. Leger (2010) found that densities of *M. robertsii* increased from approximately 10^3 to 3×10^4 cfu/g soil in the rhizosphere of grass roots over three months.

Although the ability for rhizosphere competences has been demonstrated, it must be established whether rhizosphere colonization occurs naturally (Vega *et al.*, 2009) and whether some plants are more likely to be associated with *M. anisopliae* than others. In addition, given the high diversity within the *M. anisopliae* complex (Bischoff *et al.*, 2009), it is possible that not all species exhibit rhizosphere competence. Recently, Fisher *et al.* (2011) sampled roots from four different plants in the USA. Roots with adhering soil were incubated with larvae of *G. mellonella* in a modified insect bait approach. DNA extracted from fungal entomopathogens emerging from mycosed larvae were sequenced for EF1-α (Bischoff *et al.*, 2009) if the fungi resembled *M. anisopliae*, and for Bloc (Rehner *et al.*, 2006) if fungal morphology indicated *Beauveria* spp. Based on this molecular diagnosis, three different species of the *Metarhizium* complex were identified in the

rhizosphere: *M. brunneum*, *M. guizhouense*, and *M. robertsii*. Moreover, *M. brunneum* was significantly associated with strawberries and blueberries, whereas *M. guizhouense* and *M. robertsii* were associated with roots of coniferous trees (Fisher *et al.*, 2011). The natural occurrence of nine clades of *Beauveria* spp. was documented, including *B. bassiana*, *B. brongniartii*, and *B. pseudobassiana*, although their frequencies were very low and no associations of clades with particular host plants were evident (Fisher *et al.*, 2011). Thus, the isolation of *Beauveria* spp. is likely to be interpreted as simply reflecting the occurrence in the surrounding soil environment. Fisher *et al.* (2011) also occasionally isolated *M. flavoviride* var. *pemphigi* from root samples. This fungus is known to infect aphids in the genus *Pemphigus* (Driver *et al.*, 2000). Some aphid species in this genus live on roots of various plants (Heie, 1980); therefore, the presence of *M. flavoviride* var. *pemphigi* may be coinciding with the occurrence of this aphid host. The study by Fisher *et al.* (2011) has partly addressed one of the critical questions related to plant–fungus–insect association in the soil environment posed by Vega *et al.* (2009). Nevertheless, there is still a long way to go to unravel the ecological role of fungal entomopathogens in the soil and whether the plant benefits from the presence of the fungi in the rhizosphere.

A study by Kabaluk and Ericsson (2007) provides evidence that plants do indeed benefit from fungal entomopathogens: corn seeds treated with conidia of *M. anisopliae* before planting resulted in higher stand density, compared to untreated controls. Nevertheless, the mechanisms for the observed beneficial effects remain unknown and could be caused by increased pest mortality of root-feeding insects or by growth-promoting plant–fungus interactions, or both. Specific predictions of the potential effects of such interactions are discussed below.

Distribution of Specialist Fungal Entomopathogens

Fungi in the order Entomophthorales develop spectacular above-ground epizootics in host populations (Eilenberg and Pell, 2007; Keller and Wegensteiner, 2007). These epizootics are often temporally limited, as they principally occur when abiotic conditions are conducive to fungal development (van Veen *et al.*, 2008). However, during most of the season, the entomophthoralean fungi remain inconspicuous. As previously mentioned, most species in the Entomophthorales have a narrow host range and some are known to produce survival structures such as resting spores in hosts for dormancy in the soil during unfavorable conditions (Eilenberg, 2002). In Europe, much research has focused on the ecology of entomophthoralean aphid pathogenic fungi since these fungi have great potential for the control of aphids (Shah and Pell, 2003; Ekesi *et al.*, 2005).

As these fungi are considered obligate entomopathogens of aphids, they depend on a limited range of above-ground hosts that are available only during parts of the season. Possibly, the fungi can survive as continuous enzootic infections within aphid populations, as hyphal bodies in aphid cadavers, or as latent infections in overwintering aphids on their primary host plant (Eilenberg and Pell, 2007). The latter hypothesis was investigated by Nielsen and Steenberg (2004), who collected bird cherry aphids (*Rhopalosiphum padi*) on their primary host, *Prunus padus*, in autumn and spring over a four-year period. The aphids were then incubated for fungal infections. Among 3272 aphids collected in spring, 218 were infected with five different entomophthoralean species. However, none of 322 fundatrices (i.e., viviparous, parthenogenic females from overwintering eggs) collected in spring was infected and among 4860 parthenogenic offspring, only two were infected (with *C. obscurus* and *P. neoaphidis*) (Nielsen and Steenberg, 2004). Thus, fungal infections in aphid populations, at least *R. padi*, are unlikely to be initiated from latent infections in overwintering hosts. Instead, fungal inoculum can be expected to survive in the environment of the aphid populations.

Reservoirs of *P. neoaphidis* and other aphid-pathogenic entomophthoralean fungi should be expected in the soil, but surveys using *G. mellonella* as a bait insect are not effective in isolating the aphid-specific pathogens. Moreover, specific *in vitro* cultivation methods from soils are not available for fungi in the Entomophthorales. Nielsen *et al.* (2003) succeeded in isolating both *P. neoaphidis* and another aphid-pathogenic fungus, *C. obscurus*, by baiting soil samples collected in spring with aphids. Both fungi could be isolated from the soil below the primary host and from agricultural fields (Nielsen *et al.*, 2003). Accordingly, the few infections of parthenogenic offspring on the primary host could be caused by fungal inoculum in the soil. However, for *P. neoaphidis*, higher levels of inoculum could be recovered from soil in agricultural fields and, therefore, initiations of epizootics could be based on inoculum surviving in this habitat. Nielsen *et al.* (2003) further showed that *P. neoaphidis* remained infectious in soil for at least 96 days at 5°C. Field margin soils are also considered important reservoirs for *P. neoaphidis*; in these areas, the fungus predominantly infects nettle aphids (*Microlophium carnosum*) (van Veen *et al.*, 2008). Thus, field margin habitats contain viable inoculum as assessed by aphid baits (Baverstock *et al.*, 2008b; Fournier *et al.*, 2010a).

Recently, molecular techniques have been developed for the specific identification and detection of *P. neoaphidis* (Fournier *et al.*, 2010b). Furthermore, these methods allow for detection of *P. neoaphidis* in both aphids and soil samples and further quantification of inoculum level by quantitative polymerase chain reaction (PCR) (Fournier *et al.*, 2010a). In addition, separation of genotypes can provide valuable new information on population genetics and succession of fungal isolates during the season. Fournier *et al.* (2010a) showed in field experiments that an introduced isolate of *P. neoaphidis* did not survive in the soil during winter and that infections in spring were initiated by inoculum with a different genotype. *Entomophaga maimaiga* has also been isolated from soil by selectively baiting with the main host, *L. dispar* (Hajek and Siegert, 2004), and quantitative PCR methods targeting *E. maimaiga* have provided specific tools for ecological studies (Castrillo *et al.*, 2007). Clearly, the application of molecular techniques will greatly increase our understanding of the ecology of fungal entomopathogens.

The traditional method to assess for the presence of fungal entomopathogens in soils is based on indirect methods, either cultivation (on bait insect or *in vitro* media) or by molecular diagnosis. In contrast, observations of mycosed cadavers in the field are direct assessments of the entomopathogenic role of the fungi. However, this remains a snapshot of recent fungal mortality in host populations and is only one of several ways of assessing the prevalence of fungal entomopathogens in host populations (Hesketh *et al.*, 2010). Even so, quantitative studies of mycosed cadavers are rare. Most include entomophthoralean fungi in pest populations within agricultural systems during epizootic development (Klingen *et al.*, 2000; Jensen *et al.*, 2008; Díaz *et al.*, 2010) but have also been conducted in more natural systems (van Veen *et al.*, 2008). These studies can provide information on the actual contribution of fungal entomopathogens to the mortality of the host population.

In tropical ecosystems, a great diversity of fungal entomopathogens abounds, including both anamorphs and teleomorphs, and these are usually diagnosed by observing mycosed cadavers. However, there are few quantitative studies focusing on the distribution and abundance of these fungi in the tropics. A fascinating fungus–host association has recently been studied in detail in Thailand rain forests, involving the teleomorph *Ophiocordyceps unilateralis* infecting the ant *Camponotus leonardi*. Infected hosts die after attaching themselves, by biting into leaf veins, to the underside of leaves close to the forest floor (Andersen *et al.*, 2009; Pontoppidan *et al.*, 2009). Ants succumb to fungus infections in aggregations, so-called graveyards, with up to 26 infected ants/m^2 (Pontoppidan *et al.*, 2009). The fungus induces the ants to attach to leaves at approximately 25 cm above ground level in a north-west direction, a location with a microclimate of high relative humidity and lower temperatures than elsewhere in the forest (Andersen *et al.*, 2009). This location is presumably conducible to spore release and dispersal. The host ant, *C. leonardi*, was confined to the forest canopy far from sporulating cadavers.

6.4.3. Trophic Interactions Involving Fungal Entomopathogens

Up to this point, this chapter has discussed bitrophic interactions between the fungal entomopathogens and their arthropod hosts by focusing on the host range of the fungi. In addition, the plant–fungus interaction, an area that will result in a better understanding of fungal entomopathogen ecology, has been mentioned. The bitrophic interaction involving two organisms is often the focus of insect pathologists, particularly when they are interested in the use of fungal entomopathogens as potential biological control agents. However, in both natural and managed ecosystems, host and pathogen species are embedded in ecological webs of interactions with many other organisms. It is therefore relevant to look at the more complex context in which fungal entomopathogens function as members of ecological communities. Such a view will improve our understanding of how the fungi may affect organisms other than their hosts, and how these other organisms may affect the fungi. Moreover, such knowledge will provide a better basis for future development of an ecological approach to biological control.

Community Modules Including Arthropod– Fungus Interactions: Tritrophic Context

Communities are assemblages of populations of different species that co-occur and interact in the same habitat or area. Traditionally, the interactions are considered to be direct, meaning that two species engage in a direct confrontation. These could be trophic interactions (e.g., one consumes the other) or competitive interactions (e.g., interference competition). Species also interact with each other indirectly when a third party mediates a specific interaction. For example, if two species consume a common resource but never encounter each other and thus do not interact directly, they can have an indirect interaction with each other by one species reducing the amount of the resource that is available to the other species (exploitative competition). In this case, the indirect interaction is mediated by the shared resource. Trophic interactions within a community can be visualized by constructing food webs linking consumers and resources by solid lines, in which species that consume others are placed at higher trophic levels than their resources. Emerging indirect interactions can then be shown by broken lines, increasing the complexity of the food web. For simplicity, the complex architecture of the whole community with all interactions can be broken into community modules (*sensu* Holt, 1997). Attention can then be paid to particular community modules of interest (Holt, 1997; Hatcher *et al.*, 2006; Holt and Dobson, 2006). Community modules consist of a few species that interact as "multi-species extensions of basic pair-wise interactions" (Holt and Hochberg, 2001). By adding more species or trophic levels, more complex modules embedded in a food web can be considered. This section will present examples from selected community modules, including direct and indirect interactions among species in the modules, and discuss their relevance for fungal entomopathogen ecology. Principally, the discussion will concern fungal entomopathogens as natural enemies of arthropod hosts and the ecological implications of such interactions, but will also include the plant–fungus associations mentioned earlier and the potential related ecological implications of these.

A simple community module could be a food chain embracing three trophic levels (Fig. 6.10). In this food chain, an intermediate species H (herbivore) feeds on a resource, a plant (P), and H is infected by a fungal entomopathogen (FE). Each species in this chain reduces the abundance of the species of the trophic level immediately below through consumption. This leads to direct positive effects on higher trophic levels by nutrient gain and direct negative effects on lower trophic levels (solid lines in Fig. 6.10A). In addition, the fungal entomopathogen in the top level affects the lowest level indirectly by limiting the effect of the herbivore on the plant. Thus, FE has an indirect positive effect on P (broken line in Fig. 6.10A). The effect is indirect because it is mediated by a third species, H. This effect leads to a trophic cascade, i.e., the effects of trophic links are cascading along the food chain. Achieving and enhancing indirect effects through trophic cascades are the overall objectives of traditional biological control, when the abundance or impact of a natural enemy of a pest is increased to reduce the density and/or impact of the pest on a resource. However, studies of direct effects of fungal entomopathogens on their host populations (host–pathogen interactions) rarely include data on whether this interaction has an indirect effect on the resource of the host, although protecting the host resource is the overall aim. Most studies are limited to the host–pathogen interaction alone, although some studies include the effect on the fungal entomopathogen on the resource, e.g., plant productivity.

In a potato production system in the USA, Lacey *et al.* (2011) tested the biological control effects of two commercially available fungal entomopathogens, *M. anisopliae* F52® and *I. fumosorosea* Pfr 97®, on the potato psyllid, *Bactericera cockerelli*. Psyllids can transmit a bacterium causing zebra chip disease in the tubers. Inundative introductions of the fungal entomopathogens significantly reduced numbers of psyllid eggs and nymphs compared to controls (direct effects), and decreased plant damage and increased the tuber yield (indirect effects). In fact, this system consists of four species, not just three, as the bacterium is also infecting the potato plants. When vectoring the bacterium, the psyllids have a negative indirect effect on the potatoes by increasing the abundance and distribution of the disease-causing organism. Increasing the

(A)   **(B)**

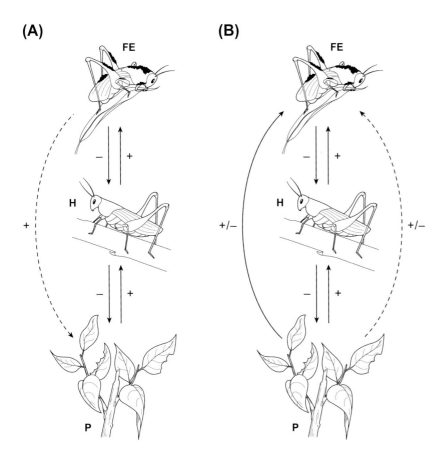

FIGURE 6.10 Three-species food chains indicating direct and indirect effects. The basal species, a plant (P), is consumed by a herbivorous arthropod (H), creating a positive direct effect on H and a negative direct effect on P. Likewise, a fungal entomopathogen (FE) infects H, yielding a positive direct effect for itself while H suffers. By consuming H, FE also has a positive indirect effect on P (broken line in A) by reducing the direct negative effect of H on P. H is therefore mediating the effect of FE on P. In B, the plant P can have a direct effect on the fungal entomopathogen FE (solid line) if plant substances or microclimatic conditions on the plant alter fungus performance, either positively (+) or negatively (−). P can also have an indirect effect on FE by altering the susceptibility of the herbivore to fungal infections. This effect can also be either positive or negative depending on whether the plant increases or decreases the susceptibility of H.

abundance of fungal entomopathogens by inundative applications also reduced the zebra chip disease symptoms (Lacey *et al.*, 2011). Therefore, the fungi also had a negative indirect effect on the bacterium, again mediated by the reduction in abundance of the psyllids. Taking this approach, the biological control system can also be viewed explicitly from the principles of community ecology (see Meyling and Hajek, 2010). Few studies have documented trophic cascades caused by fungal entomopathogens from the Entomophthorales. Fournier *et al.* (2010a) showed in field experiments that lucerne plant survival was higher in treatments combining aphids and the fungus *P. neoaphidis* than when only aphids were present on plants. Thus, this case also indicates indirect effects of *P. neoaphidis* on plant growth. Future considerations of the host–pathogen interaction as embedded in community modules as described above will provide more ecological insight into the indirect effects of fungal entomopathogens on host resources, mediated by the host.

Effects of Interactions Among Plants and Fungal Entomopathogens

Direct and indirect effects between plants and fungal entomopathogens can be expected, particularly in light of

the recent literature dealing with fungal entomopathogens as fungal endophytes (Vega, 2008; Vega *et al.*, 2008). Cory and Ericsson (2010) discussed in detail how plants might interact with fungal entomopathogens in a tritrophic context, specifically focusing on effects of the plants on the fungi. It is known that fungal propagules of *B. bassiana* occur on leaf surfaces (Meyling and Eilenberg, 2006a), and the plant could potentially have direct effects on the fungus by altering conidial survival or germination on phylloplanes. It can be hypothesized that chemical compounds on phylloplanes could enhance or hamper germination of conidia which, depending on the presence of susceptible hosts, would have positive or negative direct effects. For example, leaf waxes have been shown to affect germination abilities of *M. anisopliae* (Inyang *et al.*, 1999) and *P. neoaphidis* (Duetting *et al.*, 2003). Leaf topology may also affect fungal inoculum on phylloplanes by altering microclimatic conditions. Potentially, variation in leaf topology could protect against UV radiation, enhancing inoculum persistence. Furthermore, plants may have indirect effects on fungal entomopathogens if the plant alters the susceptibility of herbivores, making them more (or less) vulnerable to infection (Cory and Ericsson, 2010). Increased susceptibility could be expected if plant defense compounds reduce the immune response of the herbivore, if the cuticle becomes more penetrable to fungal hyphae, or if the plant induces behavioral changes in the herbivore making it more likely to encounter fungal inoculum on the plant (Cory and Ericsson, 2010). Such an effect would then be an indirect positive effect of the plant on the fungus (Fig. 6.10B). In contrast, sequestered plant compounds obtained by the herbivore (see Section 6.3.3) could also be hypothesized to increase defenses against infection; thus, the indirect effect would be negative (Fig. 6.10B). See Cory and Ericsson (2010) for examples and discussions of these effects.

An intriguing hypothesis is that plants should be able to manipulate entomopathogens and exploit them as bodyguards to protect against arthropod herbivores (Elliot *et al.*, 2000). Plants are known to emit volatiles as response to attacks by insect herbivores, which aid parasitoids to be recruited to plants harboring their hosts (Turlings *et al.*, 1990). Below ground, corn plants can recruit entomopathogenic nematodes by emission of compounds from the roots when attacked by coleopteran larvae (Rasmann *et al.*, 2005). Cory and Ericsson (2010) discuss this aspect from the perspective of fungal entomopathogens and suggest that the plant should either support populations of the fungus or enhance the infection efficacy of the fungus to manipulate the fungi as bodyguards. One major difference between fungal entomopathogens and other natural enemies mentioned above is that the fungi do not have their own means of locomotion and cannot be recruited from

a distance. Rather, they should be maintained as a "standing defense." As the potentially attacking herbivores are unknown to the plant, Cory and Ericsson (2010) argue that fungal entomopathogens with relatively broad host ranges should be best suited as candidates for becoming bodyguards. However, the applicability of the bodyguard hypothesis to fungal entomopathogens is yet to be resolved.

As described above, many plants harbor endophytic fungi in their tissues, including taxa of known entomopathogens. These fungi occur naturally in many plants and experimental infections can be induced (Vega *et al.*, 2008). The benefits of this association to the endophytic fungus and to the plant remain to be elucidated. It can be speculated that the plant has a direct positive effect on the fungus by providing nutrients and a suitable habitat for survival (Fig. 6.11). However, the mode of action of the endophytic fungal entomopathogen appears to involve the production of chemicals that are detrimental to a herbivore (Vega *et al.*, 2008; Cory and Ericsson, 2010). Disregarding the mechanism, if the fungus aids the plant by reducing consumption by the herbivore, the positive effect of the fungus on the plant will be indirect (Fig. 6.11), as reported for several other fungal endophytes (Saikkonen *et al.*, 2010). The negative effect of the fungus on the herbivore will be direct (Fig. 6.11). Unless the endophyte has specific growth-related benefits to the plant (e.g., providing nutrients), then the fungus has no direct positive effect on the plant. A negative direct effect would be the fungus acting as a plant

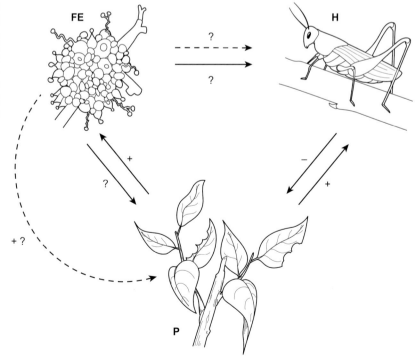

FIGURE 6.11 Direct (solid lines) and indirect (broken lines) effects among members in a community module including a plant (P), a herbivorous arthropod (H), and a plant-associated fungal entomopathogen (FE). Plant association of the fungal entomopathogen can be either as endophyte or rhizosphere associated (see text for details). Positive and negative effects are indicated by plus and minus, respectively. Unresolved effects are indicated by question marks.

endoparasite. Other fungal entomopathogens can be regarded as growth-promoting symbionts of plants, as suggested by Vega *et al.* (2009), in which case their positive effect on the plant would be direct.

Similar ecological effects to those presented in Fig. 6.11 can be considered for the described rhizosphere competence shown for *Metarhizium* spp. It is still unknown whether the fungus has benefits for the plants. Although Kabaluk and Ericsson (2007) showed that corn plants increased stand density when plants were grown from seeds treated with *M. anisopliae* conidia, the specific mechanism was not investigated. Evidently, the fungus benefits by residing in the rhizosphere, based on increased titer densities over time (Bruck, 2010; Fang and St. Leger, 2010); thus, a direct positive effect of the plant on the fungus is likely. St. Leger (2008) further suggests that *M. anisopliae* may be manipulated as a biofertilizer. Whether the fungus has positive indirect effect by infecting below-ground herbivores remains to be determined.

Fungal Entomopathogens as Part of Natural Enemy Communities

As previously discussed, fungal entomopathogens should not be considered as the sole contributor to the outcome of a host–pathogen interaction. Instead, fungal entomopathogens should be considered as integral members of natural enemy communities, with all the potential interactions this includes. The natural enemies of an organism include all organisms in the trophic level above it, which obtain nutrients by consuming the organism, partly or entirely. Natural enemies of arthropods include pathogens, parasites, predators, and parasitoids. Pathogens and parasites are sometimes merged or kept separate according to the context in which they are studied. Fungal entomopathogens can broadly be included in the definition of parasites as "symbionts that cause harm to another organism, the host, which the parasites utilize as habitat" (Raffel *et al.*, 2008). Pathogens and parasites can collectively be defined as "organisms that feed on a host individual, usually living on or in it and often causing harm but not immediate death" (Hatcher *et al.*, 2006). The term microparasites has been applied to pathogenic microbes in epidemiological studies (Anderson and May, 1981), in contrast to macroparasites or true parasites. In epidemiology, microparasites or pathogens are typically intensity independent, implying that a single infection event can lead to high within-host reproduction of the pathogen, resulting in characteristic host pathology. In contrast, macroparasites are intensity dependent, as their impact on host individuals is mostly dependent on increasing infection events, i.e., the number of infectious units entering the host (Lafferty *et al.*, 2008). Nevertheless, interactions among natural enemies within and between categories are likely to affect the outcome of

the host–pathogen interactions studied in insect pathology and also of biological control. Considering fungal entomopathogens as players in a multitrophic theater may improve our understanding and predictions of their impact as natural enemies.

When considering multiple natural enemies for biological control, a range of outcomes can be predicted based on the types of enemies, their life history and spatiotemporal occurrence, and their interactions. Straub *et al.* (2008) discussed the potential impacts on biological control of increasing natural enemy diversity based on ecological theory. The main predictions are that the increase of natural enemy richness can lead to three main outcomes: (1) decreased biological control as a consequence of conflicting effects such as intraguild predation (Polis and Holt, 1992; Rosenheim *et al.*, 1995); (2) neutral biological control effects due to functional redundancies or counteracting positive and negative effects; and (3) improved biological control when natural enemies supplement each other's effects by complementarity through resource partitioning and/or facilitation (Straub *et al.*, 2008). However, the outcomes may not be straightforward to predict from isolated studies of the natural enemies and may be highly context dependent and rely on species traits. Thus, future studies should focus on integrating members of natural enemy communities, including fungal entomopathogens, to improve our knowledge of the effects of multitrophic interactions on biological control.

It has been suggested that predation can reduce the prevalence of a pathogen in the host population (Packer *et al.*, 2003), and fungal entomopathogens may be directly or indirectly negatively affected by predators of a common prey or host. A presumably negative indirect effect of predators on a fungal entomopathogen was demonstrated by Laws *et al.* (2009) while studying grasshoppers in field cages. The grasshoppers could be infected by the entomophthoralean fungus *Entomophaga grylli* or consumed by wolf spiders. By manipulating spider presence and time in the cages it was shown that the presence of spiders decreased the prevalence of *E. grylli* in host populations compared to cages where spiders were absent, but only when spiders were added to the cages at an early stage. The observations were independent of grasshopper density. The presence of wolf spiders negatively affected the fungus, possibly by altering the behavior of the grasshoppers, causing them to be less likely to become infected than if predators were not around. Alternatively, predators could feed selectively on infected prey, thus becoming intraguild predators (Polis *et al.*, 1989; Polis and Holt, 1992). In the latter case, the negative effect of the spider on *E. grylli* would be direct.

Intraguild predation is also seen when the sevenspotted lady beetle, *Coccinella septempunctata*, consumes aphids infected by *P. neoaphidis* (Roy *et al.*, 1998). The lady beetle itself does not become infected by the fungus (Roy et al., 2001) and, therefore, the direct effect should be expected to be negative on the fungus. However, the lady beetle also vectors *P. neoaphidis* conidia to new hosts (Roy et al., 2001), thus potentially increasing pathogen transmission for the benefit of the fungus. Furthermore, other herbivorous insects that are not attacked by the aphid's natural enemies may also impact *P. neoaphidis* dispersal (Baverstock et al., 2008a); thus, the outcomes of species interactions are not straightforward to predict. Studies on these types of interactions are important to develop ecologically based biological control strategies.

In a field experiment, Ramirez and Snyder (2009) studied the effect of increasing natural enemy diversity on the biological control effect on *L. decemlineata*. The natural enemies included predatory beetles and predatory bugs inhabiting above-ground parts of potato plants, below-ground entomopathogenic nematodes, and the fungal entomopathogen *B. bassiana*. The natural enemy communities were manipulated by increasing enemy species and combining species among the enemies. Increased enemy diversity reduced herbivore numbers and thereby indirectly increased plant biomass (Ramirez and Snyder, 2009), in accordance with what was previously discussed for Fig. 6.10A. However, herbivore suppression was highest when the two enemy species were represented by a predator and a pathogen, as opposed to two predators or two pathogens. Predators and pathogens complemented each other in exploitation of the prey/host. In addition, the system proved to be a case of resource partitioning as the enhanced combination effect was realized when *L. decemlineata* larvae moved to the soil for pupation. Thus, the two types of natural enemies showed niche complementarity and no conflicting effects as the predators did not become infected by the entomopathogens below ground. The predators actually further enhanced the effects of entomopathogens on *L. decemlineata* since their presence weakened the immune response of the beetle larvae (Ramirez and Snyder, 2009). Accordingly, the predators facilitated the function of the pathogens and the two types of natural enemies acted synergistically.

Studying the same potato production system, Crowder et al. (2010) demonstrated that it is not simply the number of natural enemies species present and their total abundance that affect biological pest control, but rather the distribution of individuals (or densities) among the species of natural enemies in the community. In fact, the more evenly distributed individuals are, the better the suppression of the prey/host population. Crowder et al. (2010) experimentally manipulated evenness among natural enemy communities, but kept species number and total numbers (or densities) of natural enemies constant. The natural enemy communities consisted of predatory insects above ground, and entomopathogenic nematodes and fungi

(*B. bassiana*) below ground. Increasing evenness of predators and pathogens (that is, no particular species is most frequent) caused decreased numbers of *L. decemlineata* individuals, which resulted in higher potato plant biomass (Crowder et al., 2010). Niche complementarity by resource partitioning and facilitation is also dependent on the relative abundance of natural enemies within the community. As mentioned, the outcome of multitrophic interactions is context dependent and the potato production system may be a particularly well-suited system to demonstrate these ecological effects since, during their life cycles, the prey and host occupy separate habitats, which harbor different natural enemies. However, the general principles based on ecological theory were proven to work in the system, thus highlighting the insights of ecology being implemented in a biological control context. Future biological control strategies, including those involving fungal entomopathogens, should be based on such ecological insights by considering the entire natural enemy community, not the effects of the entomopathogens in isolation.

6.5. USE OF FUNGAL ENTOMOPATHOGENS AS BIOLOGICAL CONTROL AGENTS

Hajek et al. (2005) compiled a catalogue of pathogens and nematodes that had been released in classical biological control programs against insects and mites. The catalogue was summarized in Hajek et al. (2007a) and further updated with specific information on fungal pathogens (Hajek and Delalibera, 2010). Out of 136 classical biological control programs involving arthropod pathogens, 59 (43%) dealt with 20 different species of fungal entomopathogens and of these, 19 (32%) resulted in establishment. *Metarhizium anisopliae* was the most commonly introduced fungal entomopathogen (13 introductions), followed by *E. maimaiga* (seven introductions). One problem with introduction studies is that sometimes they are not conducted for long periods after the introduction, and establishment cannot be assessed. For example, Hajek et al. (2007a) considered this to be the case for introductions occurring after 1999. An excellent protocol on how properly to introduce and document fungal entomopathogen establishment has been published by Hajek et al. (2007b).

6.5.1. Pathogenicity, Virulence, and Bioassays

Pathogenicity has been defined as the ability to cause disease (qualitative measure) while virulence is the degree of pathogenicity (quantitative measure) (Watson and Brandly, 1949; Steinhaus and Martignoni, 1970;

Shapiro-Ilan *et al.*, 2005). Thus, pathogenicity is by definition a general term, and it encompasses virulence. The virulence of a fungal entomopathogen can be measured by conducting dose–response bioassays in the laboratory to estimate LD_{50}, LC_{50}, or LT_{50}, i.e., the median lethal dose, concentration, or time, respectively, needed to kill 50% of the tested insects (Marcus and Eaves, 2000). The difference between dose and concentration is that a dose is more precise and involves an exact quantity (e.g., conidia per insect), while concentration involves a relative content (e.g., mean number of conidia per milliliter of suspension). Some virulence factors include speed of replication within an insect, and enzyme and toxic metabolite production (Tanada and Fuxa, 1987). For critical assessments of the terms pathogenicity and virulence see Thomas and Elkinton (2004) and Shapiro-Ilan *et al.* (2005). The evolution of entomopathogenicity in fungi has been reviewed by Humber (2008).

Laboratory bioassays to compare virulence among fungal isolates are usually conducted under optimal conditions for fungal growth (e.g., high humidity and constant temperatures and photoperiods), which are obviously very different from the environmental conditions that will be encountered in the field (Butt and Goettel, 2000). Hence, the bioassays might reveal differences in virulence among isolates, but these differences might not be expressed under field conditions (see Section 6.4.1). Nevertheless, laboratory bioassays are traditionally used to select isolates for subsequent field trials and studies related to mass production, formulation, and storage properties, among others. If the results merit it, the isolate might then be a good candidate for registration and commercialization. A good example of this sequence of events is the *M. acridum* strain (IMI 330189) used in the LUBILOSA project (Douthwaite *et al.*, 2001) (see Section 6.5.3). In contrast, soil containing resting spores of *E. maimaiga* was directly introduced from infested sites in Massachusetts and New York to sites in Maryland, Pennsylvania, Virginia, and West Virginia as a biological control agent against gypsy moths without conducting preliminary laboratory or field bioassays, with very successful results (Hajek *et al.*, 1996). For a review on different types of bioassays, including spray, immersion, dusting, direct deposition, soil inoculation, bait, forcibly discharged conidia, etc., the reader is referred to Butt and Goettel (2000).

One problem encountered when running bioassays is that repeated *in vitro* culturing of the fungal entomopathogen may result in attenuation of virulence (Butt and Goettel, 2000; Butt *et al.*, 2006). This issue is sometimes resolved by passing the fungus through an insect host, but the exact mechanism resulting in attenuated virulence has not been elucidated, although recent research is moving in that direction. For example, working with *B. brongniartii*, Loesch *et al.* (2010) have implicated the carbon utilization

pattern as a factor affecting virulence after repeated *in vitro* passage. Hussain *et al.* (2010) compared the volatile organic compounds (VOCs) produced by *B. bassiana* and *M. anisopliae* grown on potato dextrose agar with the VOCs produced when the fungus was growing on termites, and reported significant differences in some components; how these differences might influence virulence remains unclear. More significantly, they found that insect passage reduced virulence. Finally, Shah *et al.* (2007) reported differences in properties of *M. anisopliae* conidia after repeated *in vitro* culturing, including changes in hydrophobicity and a reduction in Pr1, a cuticle-degrading protease bound to the cell wall.

To avoid problems with attenuated virulence, it is essential to establish a single-spore stock culture that can be used for extended periods in bioassays and other research projects. The establishment of a single-spore stock culture eliminates possible contaminants and avoids the possibility that the fungal entomopathogen that was attacking an insect might consist of different populations of the same species and might not be genetically uniform. Basically, once the fungus has sporulated on an insect, a small sample is taken with a sterile needle or inoculating loop and transferred to a Petri plate containing a non-selective medium, in order to establish a clean culture of the fungus. Once the fungus in this clean culture sporulates, a small sample of conidia is placed in a tube containing water and a wetting agent. This is followed by several stepwise dilutions (1/10, 1/100, 1/1000, and 1/10000) and, based on visual observations of each dilution in a hemacytometer or microscope slide, it can then be decided from which one an aliquot is taken and spread on an agar that provides extreme clarity (e.g., Difco™ Noble agar) for visualization of the conidia under a stereoscope. This procedure will guarantee that single conidia will be distributed throughout the media. Using a stereoscope, it is then possible to transfer a single germinating conidium to a Petri plate and allow it to grow until it sporulates and covers the entire area. This method will provide a pure culture of the fungus. From this plate, a stock culture is stored, either by freeze drying and storing harvested conidia, or by placing small sections of the mycelial mat in 10% glycerol and storing at −80°C. For other methods of obtaining single genotypes, see Goettel and Inglis (1997).

6.5.2. Production, Formulation, and Application

The production of fungal entomopathogens for use as biological control agents against insects requires the formation of an infective form such as a conidium or blastospore. These spore forms can be mass produced using solid-substrate fermentation (also known as diphasic fermentation or solid-state fermentation) or liquid culture (also known as submerged culture or submerged fermentation). These methods make available the basic requirements for fungal growth and nutrition, i.e., adequate pH, temperature, nitrogen, carbohydrates, water, and oxygen and carbon dioxide exchange.

Solid-substrate culture involves inoculating a sterile agar-based medium (e.g., potato dextrose agar) or natural substrate (e.g., barley, rice, wheat) with the fungal isolate and allowing it to sporulate. Conidia are then harvested by scraping off the mycelial mat with a spatula and washing it with sterile water or a buffer, and finally sieving to remove mycelial fragments. Liquid culture involves the fungal inoculation of a liquid medium placed in a rotary shaker, followed by harvesting of the fungal spore using filtration or a continuous centrifugation system. A two-stage system, combining liquid and solid culture, is known as solid-substrate fermentation. This method involves the production of spores in liquid media until they have reached their exponential growth (previously determined based on growth curves), followed by inoculation of a solid substrate such as barley or rice, which will absorb the liquid medium (Grace and Jaronski, 2005). Solid-substrate fermentation takes seven to 10 days for *Beauveria* and up to 14 days for *Metarhizium*, after which a drying and harvesting protocol must be followed (Grace and Jaronski, 2005).

One of the major challenges for the success of a fungal entomopathogen in the field is the development of an adequate formulation. The formulation process must keep several objectives in mind. Foremost is the stabilization of the organism to allow prolonged storage. In addition, the formulation must be suitable for handling and delivery in the field, while protecting the organism from adverse environmental factors (Jones and Burges, 1998). Formulations can be liquid suspensions or dry solids.

Liquid formulations are normally water or oil based, and emulsions. If the carrier for conidia is water, then a wetting agent (also known as a surfactant) must be used. These are used to reduce surface tension, thereby allowing for dispersion of conidia. When a wetting agent is used, it is important to assess its possible toxicity to the spore (Burges, 1998). An emulsion is one liquid dispersed in another, e.g., oil in water. An inverted emulsion would be water dispersed in oil.

Vegetable- and petroleum-derived oils are used as spray carriers for conidia, owing to their hydrophobic properties. Formulating in oil is essential for ultralow-volume (ULV) applications, which produce droplets of 50–100 μm diameter. Low volumes have the advantage of reducing the weight that needs to be carried in the field and are advantageous in areas where water is not readily available. Enhanced infectivity has been reported when conidia of *M. anisopliae* (Bateman *et al.*, 1993) or *B. bassiana* (Prior *et al.*, 1988) were formulated in oil rather than water. This increased infectivity could be a result of wider spread and stronger adhesion of the oil to the cuticle, possibly reaching intersegmental membranes, in contrast to the water formulation. For instance, Inglis *et al.* (1996) fed grasshopper nymphs lettuce and bran preinoculated with *B. bassiana* formulated in either oil or water. More nymphs died of mycosis after they had ingested the substrates inoculated with conidia in oil than in water, regardless of the food type. When nymphs ingested lettuce disks treated with a fluorescent dye in either oil or water, more dye was observed on the nymphs in the oil treatment, demonstrating that oil facilitates surface contamination of the host, even when conidia are used as baits.

Solid formulations include wettable powders, dusts, granules, and baits (Jones and Burges, 1998). For reviews dealing with production and formulation of fungal entomopathogens see Burges (1998) and Wraight *et al.* (2001). For a review of spray application of mycopesticide formulations see Bateman and Chapple (2001).

6.5.3. Some Important Case Studies

Gypsy Moth in the USA

The Frenchman Léopold Trouvelot (1827–1895) was an amateur entomologist interested in silk production. In the late 1860s, he imported gypsy moth eggs from France to Massachusetts, USA, in an attempt to hybridize them with American silkworms and develop a line resistant to a microsporidian disease prevalent in colonies of silk-producing moths (Liebhold *et al.*, 1989). In 1869, some gypsy moths escaped from his colony, and since then, outbreaks have occurred on a cyclic basis spanning one to three years followed by low population levels for a few years, followed by another outbreak in infested areas. The gypsy moth defoliates hardwood forests in the USA and is currently present over most of the north-east, as far south as North Carolina, and west into Wisconsin. Because of the lack of natural enemies of the insect in the USA, in 1909 some gypsy moths infected with the "gypsy fungus" (now believed to be *E. maimaiga*) (Soper *et al.*, 1988) were brought to the USA from Japan, and released in the Boston area in 1910 and 1911, but did not become established (Speare and Colley, 1912). In 1985 and 1986, an *E. maimaiga* isolate originally collected in Japan in 1984 was released in New York and Virginia, but apparently, did not become established (Hajek *et al.*, 1995). In 1989, gypsy moths infected with *E. maimaiga* were discovered in Connecticut (Andreadis and Weseloh, 1990), the first report of this fungus in North America. In 1991 and 1992, *E. maimaiga* was introduced in Maryland, Pennsylvania, Virginia, and West Virginia, and demonstrated "spectacular spread" (Hajek *et al.*, 1996). The origin of the 1989 isolate remains unknown, although several hypotheses have been suggested (Hajek *et al.*, 1995; Nielsen *et al.*, 2005; Hajek, 2007). *Entomophaga maimaiga* is considered to be "the most important enemy of gypsy moth" (Nielsen *et al.*, 2005) and has caused epizootics that have significantly reduced gypsy moth defoliation. These epizootics are not consistent, most likely owing to environmental factors (Siegert *et al.*, 2009). In addition to the fungus, a nucleo-polyhedrovirus causes natural epizootics in gypsy moth populations, and a registered viral product is available for limited application in some infested states in the USA (Chapter 4).

Locusts and Grasshoppers

Africa has had a long history of plant devastation caused by swarms of locusts. The affected area extends for hundreds of millions of hectares, from West Africa to India. This problem led to the creation of an international consortium headed by CABI in the UK in 1989, to deal with the problem in an environmentally friendly manner. The consortium was known as LUBILOSA, an acronym for "LUtte BIologique contre les LOcustes et les SAuteriaux" and translated into English as "Biological Control of Locusts and Grasshoppers." The consortium developed and registered a *Metarhizium* strain (IMI 330189; presented in the literature as *M. anisopliae* var. *acridum* as well as *M. flavoviride* and recently classified as *M. acridum*) (Bischoff *et al.*, 2009), marketed in South Africa under the trade name Green Muscle® and also produced in Senegal. The conidia are formulated in oil and applied as a ULV spray at rates of $1.2-3.7 \times 10^{12}$ conidia/ha (Langewald and Kooyman, 2007). LUBILOSA lasted for 12 years at a cost of $15 million dollars (Douthwaite *et al.*, 2001), and resulted in the publication of more than 100 refereed journal articles. The project led to important developments in mass production, formulation, and application (Lomer *et al.*, 2001), as well as the elucidation of interesting aspects of the biology of infected insects, e.g., the finding that infected grasshoppers were more prone to predation than healthy individuals and were less voracious feeders (Thomas *et al.*, 1997, 1998), and the discovery of behavioral fever in grasshoppers, whereby, as discussed above, infected insects would bask and thermoregulate their temperatures (up to 42°C) in an attempt to adversely affect the development of the fungus (Blanford *et al.*, 1998). Another significant finding was a study conducted in the Republic of Niger, comparing the mortality of Sahelian grasshoppers after applications of *M. acridum* or the organophosphate insecticide fenitrothion (Langewald *et al.*, 1999). Even though the fungal application took 16 days to cause 93% mortality, in contrast to fenitrothion, which caused 90% mortality shortly after application, grasshopper immigration to the fenitrothion plots resulted in populations reaching the control levels within 16 days. In addition, *M. acridum*

remained infectious for up to three weeks after spraying. The project also resulted in the development and commercialization of a machine (MycoHarvester) used to harvest fungal spores from the solid substrate (e.g., milled rice) on which they are grown. LUBILOSA serves as a case study for the development of a biological control strategy in extremely difficult environmental conditions and extending over an enormous area, with myriad political and economical issues to consider. For reviews of the LUBILOSA project see Lomer *et al.* (2001), Langewald and Kooyman (2007), and Moore (2008).

Spittlebugs in Brazil

The most extensive example of use of a fungal entomopathogen for pest management is the use of *M. anisopliae* in Brazil for biological control of several species of spittlebugs (Cercopidae) in sugarcane and pastures. In 2011, approximately 1.5 million ha were being treated with *M. anisopliae*, mostly to control *Mahanarva posticata* in sugarcane (M. de Faria, pers. commun.). Mass production, formulation, and application of *M. anisopliae* for this project have been described by Mendonça (1992) and Alves (1998). The topic has been reviewed by de Faria and Magalhães (2001), Alves *et al.* (2003), and Li *et al.* (2010).

6.5.4. Commercial Products

de Faria and Wraight (2007) compiled a comprehensive list of mycoinsecticides and mycoacaricides that have been used throughout the world since the 1960s. In total, 171 products were identified, based on 12 species, mostly *B. bassiana* (34%), *M. anisopliae* (34%), *I. fumosorosea* (6%), and *B. brongniartii* (4%). Of the 171 products, 26 were no longer in use at the time of publication. The regional distribution of these products was as follows: South America 42.7%, North America 20.5%, Europe and Asia 12.3% each, Central America 7%, Africa 2.9%, and Oceania 2.3%. Only one of the products listed is based on an Entomophthorales (*Conidiobolus thromboides*). The availability of registered products and regulatory issues involving the use of microbial pesticides in Kenya, China, India, South Korea, the European Union, Ukraine, Russia, Moldova, Argentina, Brazil, Cuba, Canada, the USA, Australia, and New Zealand are discussed by Kabaluk *et al.* (2010). Ravensberg (2011) has written an excellent book dealing with development and commercialization of entomopathogens.

6.5.5. Genetic Modification of Fungal Entomopathogens

Several studies have reported genetic modifications of fungal entomopathogens. For instance, *M. anisopliae* has been genetically modified for: (1) resistance to the fungicide benomyl (Bernier *et al.*, 1989; Goettel *et al.*, 1990b) and the herbicides bialaphos and glufosinate ammonium (Inglis *et al.*, 2000); (2) increased speed of kill via protease over expression (St. Leger *et al.*, 1996); (3) chitinase overproduction (Screen *et al.*, 2001); and (4) expression of a scorpion toxin gene (Wang and St. Leger, 2007c; Pava-Ripoll *et al.*, 2008). *Beauveria bassiana* has been modified for chitinase overproduction (Fang *et al.*, 2005; Fan *et al.*, 2007) and for expression of: a cuticle-degrading protease from *M. anisopliae* (Góngora, 2004; Lu *et al.*, 2008), a scorpion neurotoxin (Lu *et al.*, 2008), *Bacillus thuringiensis* vegetative insecticidal proteins (Qin *et al.*, 2010), a hybrid protease (Fan *et al.*, 2010), and a chitinase gene from *O. unilateralis* (Chantasingh *et al.*, 2011). Finally, protoplast fusion has been used to hybridize different species of *Lecanicillium* to expand the host range (Goettel *et al.*, 2008).

One of the most innovative uses of genetic modification of a fungal entomopathogen involves the production of recombinant *M. anisopliae* strains with various modes of action against the malaria parasite, *Plasmodium falciparum*, in the hemolymph of mosquitoes (Fang *et al.*, 2011). The transgenic fungus will still allow the mosquitoes to reproduce, thus presumably avoiding selection for resistance in the vector, which could be a problem if the fungus killed the mosquitoes too quickly (Fang *et al.*, 2011). Such a delivery mechanism could also be used in the future against other vector-borne diseases of human importance such as filariasis, dengue fever, and trypanosomiasis (Fang *et al.*, 2011). Another very interesting finding combining fungal entomopathogens and mosquitoes involves the report by Howard *et al.* (2010) of reduced blood feeding in *Culex quinquefasciatus* after exposure to *B. bassiana*.

In 2011, the genome for *M. anisopliae* strain ARSEF 23 (which has been reclassified by Bischoff *et al.*, 2009, as *M. robertsii*) and *M. acridum* was published (Gao *et al.*, 2011). This constitutes the second and third completed fungal entomopathogen genome, after *Ascosphaera apis*, the causative agent for chalkbrood disease in bees (Qin *et al.*, 2006). The genome for *M. robertsii* and *M. acridum* was 39 and 38 Mb (1 mega base pair = 1,000,000 bp), respectively, and *M. robertsii* was predicted to have 10,582 protein coding genes, in contrast to 9,849 in *M. acridum* (Gao *et al.*, 2011). The genome sequencing effort will result in an intimate understanding of host−pathogen interactions including factors related to pathogenicity, metabolite biosynthesis, and the identification and exploitation of novel fungal virulence genes. In addition, it should provide insights on host specificity and fungal lifestyle and nutritional modes (e.g., saprobic, endophytic). For a review of transformation methods for fungal entomopathogens see St. Leger and Wang (2009, 2010) and Fan *et al.* (2011).

6.6. FUTURE RESEARCH DIRECTIONS

Generally speaking, our understanding of fungal entomopathogens has progressed from basic observations of insects killed by fungi to a concerted effort aimed at understanding how to use these organisms as biological control agents. This journey has included the development of classical taxonomical and classification approaches, focused studies aimed at understanding the specific mechanisms of pathogenicity, production and formulation methodologies and technologies, and molecular studies that have teased apart phylogenetic relationships.

One area of paramount importance that is still in its infancy is the ecology of fungal entomopathogens. Even though there have been some significant developments in this area in recent years (Vega *et al.*, 2009; Roy *et al.*, 2010), in-depth knowledge of the most common genera of fungal entomopathogens should reveal important insights that may lead us to understand how better to exploit these fungi in an ecological approach to biological control.

Vega *et al.* (2009) have presented several research areas that should be pursued to gain a better understanding of fungal entomopathogens. Still, there are many other areas in need of research. For example, what will be the effect of climate change on fungal entomopathogens? Increased in-depth knowledge on spore surface properties, not only of *M. anisopliae* and *B. bassiana*, but also of other species, should result in new insights on how to preserve, store, formulate, and apply fungal entomopathogens in the field. Similarly, there should be a concerted effort to sequence the genomes of genera of fungal entomopathogens. This effort will reveal important information on genes, proteins, virulence, metabolite biosynthesis, and genetic relationships, among other things. Finally, in order to increase the chances of successfully using fungal entomopathogens in the field, it is necessary to develop novel delivery methods. The current technologies have been based on the chemical insecticide paradigm. Several attempts throughout the world have examined the possible use of fungal entomopathogens as fungal endophytes. This is another area that deserves more in-depth analysis and consideration.

ACKNOWLEDGMENTS

Special thanks to M. S. Goettel, W. S. Higgins, and H. K. Kaya for their thorough review of the chapter. For reviewing parts of this chapter or providing information we thank M. de Faria, R. Humber, M. A. Jackson, S. Jaronski, N. O. Keyhani, R. Kleespies, D. Moore, D. E. N. Rangel, and A. Simpkins.

REFERENCES

Aarnio, T. H., & Agathos, S. N. (1989). Production of extracellular enzymes and cyclosporin by *Tolypocladium inflatum* and morphologically related fungi. *Biotechnol. Lett.*, *11*, 759−764.

Aime, M. C., Matheny, P. B., Henk, D. A., Frieders, E. M., Nilsson, R. H., Piepenbring, M., McLaughlin, D. J., Szabo, L. J., Begerow, D., Sampaio, J. P., Bauer, R., Weiß, M., Oberwinkler, F., & Hibbett, D. (2006). An overview of the higher level classification of Pucciniomycotina based on combined analyses of nuclear large and small subunit rDNA sequences. *Mycologia, 98,* 896−905.

Aldridge, D. C., & Turner, W. B. (1969). Structures of cytochalasin C and cytochalasin D from *Metarhizium anisopliae. J. Chem. Soc., C 6,* 923−928.

Alexopoulos, C. J., Mims, C. W., & Blackwell, M. (1996). *Introductory Mycology.* New York: John Wiley & Sons.

Alves, S. B. (1998). Fungos entomopatogênicos. In S. B. Alves (Ed.), *Controle Microbiano de Insetos* (2nd ed.) (pp. 289−381) Piracicaba: Bibioteca de Ciências Agrárias Luis de Queiroz, No. 4.

Alves, S. B., Rossi, L. S., Lopes, R. B., Tamai, M. A., & Pereira, R. M. (2002). *Beauveria bassiana* yeast phase on agar medium and its pathogenicity against *Diatraea saccharalis* (Lepidoptera: Crambidae) and *Tetranychus urticae* (Acari: Tetranychidae). *J. Invertebr. Pathol., 81,* 70−77.

Alves, S. B., Pereira, R. M., Lopes, R. B., & Tamai, M. A. (2003). Use of entomopathogenic fungi in Latin America. In R. K. Upadhyay (Ed.), *Advances in Microbial Control of Insect Pests* (pp. 193−211). New York: Kluwer Academic/Plenum Publishers.

Andersen, S. B., Gerritsma, S., Yusah, K. M., Mayntz, D., Hywel-Jones, N. L., Billen, J., Boomsma, J. J., & Hughes, D. P. (2009). The life of a dead ant: the expression of an adaptive extended phenotype. *Am. Nat., 174,* 424−433.

Anderson, R. M., & May, R. M. (1981). The population dynamics of microparasites and their invertebrate hosts. *Philos. Trans. R. Soc. Lond. B, 291,* 451−524.

Andreadis, T. G., & Weseloh, R. M. (1990). Discovery of *Entomophaga maimaiga* in North American gypsy moth, *Lymantria dispar. Proc. Natl. Acad. Sci. USA, 87,* 2461−2465.

Arthurs, S. P., Thomas, M. B., & Lawton, J. L. (2001). Seasonal patterns of persistence and infectivity of *Metarhizium anisopliae* var. *acridum* in grasshopper cadavers in the Sahel. *Entomol. Exp. Appl., 100,* 69−76.

Azumi, M., Ishidoh, K.-I., Kinoshita, H., Nihira, T., Ihara, F., Fujita, T., & Igarashi, Y. (2008). Aurovertins F−H from the entomopathogenic fungus *Metarhizium anisopliae. J. Nat. Prod., 71,* 278−280.

Bailey, L. (1981). *Honey Bee Pathology.* London: Academic Press.

Baldauf, S. L., Roger, A. J., Wenk-Sierfert, I., & Doolittle, W. F. (2000). A kingdom-level phylogeny of eukaryotes based on combined protein data. *Science, 290,* 972−977.

Barnes, A. I., & Siva-Jothy, M. T. (2010). Density-dependent prophylaxis in the mealworm beetle *Tenebrio molitor* L. (Coleoptera: Tenebrionidae): cuticular melanization is an indicator of investment in immunity. *Proc. R. Soc. Lond. B, 267,* 177−182.

Bateman, R., & Chapple, A. (2001). The spray application of mycopesticide formulations. In T. M. Butt, C. Jackson & N. Magan (Eds.), *Fungi as Biocontrol Agents* (pp. 289−309). Wallingford: CABI.

Bateman, R. P., Carey, M., Moore, D., & Prior, C. (1993). The enhanced infectivity of *Metarhizium flavoviride* in oil formulations to desert locusts at low humidities. *Ann. Appl. Biol., 122,* 145−152.

Baverstock, J., Baverstock, K. E., Clark, S. J., & Pell, J. K. (2008a). Transmission of *Pandora neoaphidis* in the presence of co-occurring arthropods. *J. Invertebr. Pathol., 98,* 356−359.

Baverstock, J., Clark, S. J., & Pell, J. K. (2008b). Effect of seasonal abiotic conditions and field margin habitat on the activity of *Pandora neoaphidis* inoculum on soil. *J. Invertebr. Pathol., 97,* 282−290.

Beakes, G. W., & Sekimoto, S. (2009). The evolutionary phylogeny of Oomycetes − insights gained from studies of holocarpic parasites of algae and invertebrates. In K. Lamour & S. Kamoun (Eds.), *Oomycete Genetics and Genomics. Diversity, Interactions and Research Tools* (pp. 1−24). New York: John Wiley & Sons.

Bechinger, C., Giebel, K.-F., Schnell, M., Leiderer, P., Deising, H. B., & Bastmeyer, M. (1999). Optical measurements of invasive forces exerted by appresoria of a plant pathogenic fungus. *Science, 285,* 1896−1899.

Begon, M., Harper, J. L., & Townsend, C. R. (1990). *Ecology: Individuals, Populations and Communities* (2nd ed.). Boston: Blackwell Scientific Publications.

Benjamin, R. K., Blackwell, M., Chapela, I. H., Humber, R. A., Jones, K. G., Klepzig, K. D., Lichtwardt, R. W., Malloch, D., Noda, H., Roeper, R. A., Spatafora, J. W., & Weir, A. (2004). The search for diversity of insects and other arthropod associated fungi. In G. M. Mueller, G. F. Bills & M. S. Foster (Eds.), *Biodiversity of Fungi: Inventory and Monitoring Methods* (pp. 395−433). New York: Elsevier Academic Press.

Bennett, J. W., & Bentley, R. (1989). What's in a name? − Microbial secondary metabolism. In: *Advances in Applied Microbiology,* Vol. 34 (pp. 1−28) San Diego: Academic Press.

Benny, G. L. (2009). http://www.Zygomycetes.org.

Benz, G. (1987). Environment. In J. R. Fuxa & Y. Tanada (Eds.), *Epizootiology of Insect Diseases* (pp. 177−214). New York: Wiley-Interscience.

Berbee, M. L., & Taylor, J. W. (2010). Dating the molecular clock in fungi − how close are we? *Fungal Biol. Rev., 24,* 1−16.

Bernhard, K., Holloway, P. J., & Burges, H. D. (1998). Appendix I: A catalogue of formulation additives: function, nomenclature, properties, and suppliers. In H. D. Burges (Ed.), *Formulation of Microbial Biopesticides* (pp. 333−365). Dordrecht: Kluwer Academic Publishers.

Bernier, L., Cooper, R. M., Charnley, A. K., & Clarkson, J. M. (1989). Transformation of the entomopathogenic fungus *Metarhizium anisopliae* to benomyl resistance. *FEMS Microbiol. Lett., 60,* 261−266.

Bhattacharya, D., Yoon, H. S., Hedges, S. B., & Hackett, D. (2009). Eukaryotes (Eukaryota). In S. B. Hedges & S. Kumar (Eds.), *The Timetree of Life* (pp. 116−120). New York: Oxford University Press.

Bidochka, M. J., & Small, C. L. (2005). Phylogeograpy of *Metarhizium,* an insect pathogenic fungus. In F. E. Vega & M. Blackwell (Eds.), *Insect−Fungal Associations: Ecology and Evolution* (pp. 28−50). New York: Oxford University Press.

Bidochka, M. J., Pfeifer, T. A., & Khachatourians, G. G. (1987). Development of the entomopathogenic fungus *Beauveria bassiana* in liquid cultures. *Mycopathologia, 99,* 77−83.

Bidochka, M. J., St. Leger, R. J., Joshi, L., & Roberts, D. W. (1995). The rodlet layer from aerial and submerged conidia of the entomopathogenic fungus *Beauveria bassiana* contains hydrophobins. *Mycol. Res., 99,* 403−406.

Bidochka, M. J., Kasperski, J. E., & Wild, G. A. M. (1998). Occurrence of the entomopathogenic fungi *Metarhizium anisopliae* and *Beauveria bassiana* in soils from temperate and near-northern habitats. *Can. J. Bot., 76,* 1198−1204.

Bidochka, M. J., Kamp, A. M., Lavender, T. M., Dekoning, J., & De Croos, J. N. A. (2001). Habitat association in two genetic groups of the insect-pathogenic fungus *Metarhizium anisopliae:* uncovering cryptic species? *Appl. Environ. Microbiol., 67,* 1335−1342.

Bidochka, M. J., Menzies, F. V., & Kamp, A. M. (2002). Genetic groups of the insect-pathogenic fungus *Beauveria bassiana* are associated with habitat and thermal growth preferences. *Arch. Microbiol., 178,* 531–537.

Bidochka, M. J., Clark, D. C., Lewis, M. W., & Keyhani, N. O. (2010). Could insect phagocytic avoidance by entomogenous fungi have evolved via selection against soil amoeboid predators? *Microbiology, 156,* 2164–2171.

Bignell, D. E. (1994). The arthropod gut as an environment for microorganisms. In J. M. Anderson, A. D. M. Rayner & D. W. H. Walton (Eds.), *Invertebrate—Microbial Interactions* (pp. 205–227). Cambridge: Cambridge University Press.

Bing, L. A., & Lewis, L. C. (1991). Suppression of *Ostrinia nubilalis* (Hubner) (Lepidoptera, Pyralidae) by endophytic *Beauveria bassiana* (Balsamo) Vuillemin. *Environ. Entomol., 20,* 1207–1211.

Bing, L. A., & Lewis, L. C. (1992). Endophytic *Beauveria bassiana* (Balsamo) Vuillemin in corn: the influence of the plant growth stage and *Ostria nubialis* (Hübner). *Biocontrol Sci. Technol., 2,* 39–47.

Bischoff, J. F., Rehner, S. A., & Humber, R. A. (2006). *Metarhizium frigidum* sp. nov.: a cryptic species of *M. anisopliae* and a member of the *M. flavoviride* complex. *Mycologia, 98,* 737–745.

Bischoff, J. F., Rehner, S. A., & Humber, R. A. (2009). A multilocus phylogeny of the *Metarhizium anisopliae* lineage. *Mycologia, 101,* 512–530.

Blackwell, M. (2010). Fungal evolution and taxonomy. *BioControl, 55,* 7–16.

Blackwell, M. (2011). The fungi: 1, 2, 3, … 5.1 million species? *Am. J. Bot., 98,* 426–438.

Blackwell, M., Hibbett, D. S., Taylor, J. W., & Spatafora, J. W. (2006). Research coordination networks: a phylogeny for the kingdom Fungi (Deep Hypha). *Mycologia, 98,* 829–837.

Blanford, S., Thomas, M. B., & Langewald, J. (1998). Behavioural fever in the Senegalese grasshopper, *Oedaleus senegalensis*, and its implications for biological control using pathogens. *Ecol. Entomol., 23,* 9–14.

Blumthaler, M. (1993). Solar UV measurements. In M. Tevini (Ed.), *UV-B Radiation and Ozone Depletion: Effects on Humans, Animals, Plants, Microorganisms, and Materials* (pp. 71–94). Boca Raton: Lewis Publishers.

Boucias, D. G., & Pendland, J. C. (1991). Attachment of mycopathogens to cuticle. The initial event of mycoses in arthropod hosts. In G. T. Cole & H. C. Hoch (Eds.), *The Fungal Spore and Disease Initiation in Plants and Animals* (pp. 101–127). New York: Plenum Press.

Boucias, D. G., Pendland, J. C., & Latge, J. P. (1988). Nonspecific factors involved in attachment of entomopathogenic Deuteromycetes to host insect cuticle. *Appl. Environ. Microbiol., 54,* 1795–1805.

Boucias, D. G., Stokes, C., Storey, G., & Pendland, J. C. (1996). The effects of imidacloprid on the termite *Reticulitermes flavipes* and its interaction with the mycopathogen *Beauveria bassiana*. *Pflanzen-schutz-Nachrichten Bayer, 49,* 103–144.

Braga, G. U. L., Flint, S. D., Messias, C. L., Anderson, A. J., & Roberts, D. W. (2001a). Effects of UV-B irradiance on conidia and germinants of the entomopathogenic Hyphomycete *Metarhizium anisopliae*: a study of reciprocity and recovery. *Photochem. Photo-biol., 73,* 140–146.

Braga, G. U. L., Flint, S. D., Messias, C. L., Anderson, A. J., & Roberts, D. W. (2001b). Effects of UV-B on conidia and germlings of the entomopathogenic hyphomycete *Metarhizium anisopliae*. *Mycol. Res., 105,* 874–882.

Braga, G. U. L., Rangel, D. E. N., Flint, S. D., Anderson, A. J., & Roberts, D. W. (2006). Conidial pigmentation is important to tolerance against solar-simulated radiation in the entomopathogenic fungus *Metarhizium anisopliae*. *Photochem. Photobiol., 82,* 418–422.

Bridge, P. D., & Worland, M. R. (2004). First report of an entomophthoralean fungus on an arthropod host in Antarctica. *Polar Biol., 27,* 190–192.

Bruck, D. J. (2005). Ecology of *Metarhizium anisopliae* in soilless potting media and the rhizosphere: implications for pest management. *Biol. Control, 32,* 155–163.

Bruck, D. J. (2010). Fungal entomopathogens in the rhizosphere. *Biocontrol, 55,* 103–112.

Burges, H. D. (1998). Formulation of mycoinsecticides. In H. D. Burges (Ed.), *Formulation of Microbial Biopesticides* (pp. 131–185). Dordrecht: Kluwer Academic Publishers.

Butt, T. M. (2002). Use of entomogenous fungi for the control of insect pests. In F. Kempken (Ed.), *The Mycota XI, Agricultural Applications* (pp. 111–134). Berlin: Springer.

Butt, T. M., & Goettel, M. S. (2000). Bioassays of entomogenous fungi. In A. Navon & K. R. S. Ascher (Eds.), *Bioassays of Entomopathogenic Microbes and Nematodes* (pp. 141–195). Wallingford: CAB Publishing.

Butt, T. M., Ibrahim, L., Clark, S. J., & Beckett, A. (1995). The germination behaviour of *Metarhizium anisopliae* on the surface of aphid and flea beetle cuticles. *Mycol. Res., 99,* 945–950.

Butt, T. M., Wang, C., Shah, F. A., & Hall, R. (2006). Degeneration of entomogenous fungi. In J. Eilenberg & H. M. T. Hokkanen (Eds.), *An Ecological and Societal Approach to Biological Control* (pp. 213–226). Dordrecht: Springer.

Cafaro, M. J. (2005). Eccrinales (Trichomycetes) are not fungi, but a clade of protists at the early divergence of animals and fungi. *Mol. Phylogenet. Evol., 35,* 21–34.

Calo, L., Fornelli, F., Nenna, S., Tursi, A., Caiaffa, M. F., & Macchia, L. (2003). Beauvericin cytotoxicity to the invertebrate cell line SF-9. *J. Appl. Genet., 44,* 515–520.

Carollo, C. A., Calil, A. L. A., Schiave, L. A., Guaratini, T., Roberts, D. W., Lopes, N. P., & Braga, G. U. L. (2010). Fungal tyrosine betaine, a novel secondary metabolite from conidia of entomopathogenic *Metarhizium* spp. fungi. *Fungal Biol., 114,* 473–480.

Carruthers, R. L., Larkin, T. S., Firstencel, H., & Feng, Z. (1992). Influence of thermal ecology on the mycosis of a rangeland grasshopper. *Ecology, 73,* 190–204.

Castrillo, L. A., Thomsen, L., Juneja, P., & Hajek, A. E. (2007). Detection and quantification of *Entomophaga maimaiga* resting spores in forest soil using real-time PCR. *Mycol. Res., 111,* 324–331.

Champlin, F. R., & Grula, E. A. (1979). Noninvolvement of beauvericin in the entomopathogenicity of *Beauveria bassiana*. *Appl. Environ. Microbiol., 37,* 1122–1125.

Chandler, D., Hay, D., & Reid, A. P. (1997). Sampling and occurrence of entomopathogenic fungi and nematodes in UK soils. *Appl. Soil Ecol., 5,* 133–141.

Chang, P.-K., Ehrlich, K. C., & Hua, S.-S. T (2006). Cladal relatedness among *Aspergillus oryzae* isolates and *Aspergillus flavus* S and L morphotype isolates. *Int. J. Food Microbiol., 108,* 172–177.

Chantasingh, D., Kitikhun, S., Eurwilaichitr, L., Uengwetwanit, T., & Pootanakit, K. (2011). Functional expression in *Beauveria bassiana* of a chitinase gene from *Ophiocordyceps unilateralis*, an ant-pathogenic fungus. *Biocontrol Sci. Technol., 21*, 677–686.

Charnley, A. K. (1984). Physiological aspects of destructive pathogenesis in insects by fungi: a speculative review. In J. M. Anderson, A. D. M. Rayner & D. W. H. Walton (Eds.), *Invertebrate–Microbial Interactions* (pp. 229–270). Cambridge: Cambridge University Press.

Charnley, A. K. (2003). Fungal pathogens of insects: cuticle degrading enzymes and toxins. *Adv. Bot. Res., 40*, 241–321.

Cho, E.-M., Liu, L., Farmerie, W., & Keyhani, N. O. (2006). EST analysis of cDNA libraries from the entomopathogenic fungus *Beauveria* (*Cordyceps*) *bassiana*. I. Evidence for state-specific gene expression in aerial conidia, *in vitro* blastospores, and submerged conidia. *Microbiology, 152*, 2843–2854.

Clarkson, J. M., & Charnley, A. K. (1996). New insights into the mechanisms of fungal insect pathogenesis in insects. *Trends Microbiol., 4*, 197–203.

Cory, J. S., & Ericsson, J. D. (2010). Fungal entomopathogens in a tritrophic context. *BioControl, 55*, 75–88.

Couch, J. N. (1938). *The genus Septobasidium*. Chapel Hill: University of North Carolina Press.

Cremer, S., Armitage, S. A. O., & Schmid-Hempel, P. (2007). Social immunity. *Curr. Biol., 17*, R693–R702.

Crowder, D. W., Northfield, T. D., Strand, M. R., & Snyder, W. E. (2010). Organic agriculture promotes evenness and natural pest control. *Nature, 466*, 109–112.

Cunningham, K. G., Manson, W., Spring, F. S., & Hutchinson, S. A. (1950). Cordycepin, a metabolic product isolated from cultures of *Cordyceps militaris* (Linn.) Link. *Nature, 166*, 949–952.

De Croos, J. N. A., & Bidochka, M. J. (1999). Effects of low temperature on growth parameters in the entomopathogenic fungus *Metarhizium anisopliae*. *Can. J. Microbiol., 45*, 1055–1061.

Delalibera, I., Jr., Hajek, A. E., & Humber, R. A. (2004). *Neozygites tanajoae* sp. nov., a pathogen of the cassava green mite. *Mycologia, 96*, 1002–1009.

Díaz, B. M., Legarrea, S., Marcos-García, M. A., & Fereres, A. (2010). The spatio-temporal relationships among aphids, the entomophthoran fungus, *Pandora neoaphidis*, and aphidophagous hoverflies in outdoor lettuce. *Biol. Control, 53*, 304–311.

Dick, M. W. (1970). Saprolegniaceae on insect exuviae. *Trans. Br. Mycol. Soc., 55*, 449–458.

Dick, M. W. (2001). *Straminipilous Fungi: Systematics of the Peronosporomycetes Including Accounts of the Marine Straminipilous Protists, the Plasmodiophorids and Similar Organisms*. Dordrecht: Kluwer Academic Publishers.

Dillon, R. J., & Charnley, A. K. (1991). The fate of fungal spores in the insect gut. In G. T. Cole & H. C. Hoch (Eds.), *The Fungal Spore and Disease Initiation in Plants and Animals* (pp. 129–156). New York: Plenum Press.

Doberski, J. W. (1981). Comparative laboratory studies on three fungal pathogens of the elm bark beetle *Scolytus scolytus*: effect of temperature and humidity on infection by *Beauveria bassiana, Metarhizium anisopliae*, and *Paecilomyces farinosus*. *J. Invertebr. Pathol., 37*, 195–200.

Douthwaite, B., Langewald, J., & Harris, J. (2001). *Development and Commercialization of the Green Muscle Biopesticide*. Ibadan: International Institute of Tropical Agriculture.

Driver, F., Milner, R. J., & Trueman, J. W. H. (2000). A taxonomic revision of *Metarhizium* based on a phylogenetic analysis of rDNA sequence data. *Mycol. Res., 104*, 134–150.

Duetting, P. S., Ding, H. J., Neufeld, J., & Eigenbrode, S. D. (2003). Plant waxy bloom on peas affects infection of pea aphids by *Pandora neoaphidis*. *J. Invertebr. Pathol., 84*, 149–158.

Dunlap, C. A., Biresaw, G., & Jackson, M. A. (2005). Hydrophobic and electrostatic cell surface properties of blastospores of the entomopathogenic fungus *Paecilomyces fumosoroseus*. *Colloids Surf. B: Biointerfaces, 46*, 261–266.

van Eijk, G. W., Roeijmans, H. J., & Seykens, D. (1986). Hopanoids from the entomogenous fungus *Aschersonia aleyrodis*. *Tetrahedron Lett., 27*, 2533–2534.

Eilenberg, J. (2002). *Biology of fungi from the order Entomophthorales*. DSc thesis. Denmark: Royal Veterinary and Agricultural University.

Eilenberg, J., & Pell, J. K. (2007). Ecology. *Arthropod Pathogenic Entomophthorales: Biology, Ecology, Identification* (pp. 7–26). Brussels: Office des Publications Officielles des Communautés Européennes.

Eilenberg, J., Bresciani, J., & Latgé, J.-P. (1986). Ultrastructural studies of primary spore formation and discharge in the genus *Entomophthora*. *J. Invertebr. Pathol., 48*, 318–324.

Eilenberg, J., Hajek, A. E., & Lomer, C. (2001). Suggestions for unifying the terminology in biological control. *BioControl, 46*, 387–400.

Ekesi, S., Shah, P. A., Clark, S. J., & Pell, J. K. (2005). Conservation biological control with the fungal pathogen *Pandora neoaphidis*: implications of aphid species, host plant and predator foraging. *Agric. For. Entomol., 7*, 21–30.

El-Basyouni, S. H., Brewer, D., & Vining, L. C. (1968). Pigments of the genus *Beauveria*. *Can. J. Bot., 46*, 441–448.

Eley, K. L., Halo, L. M., Song, Z., Powles, H., Cox, R. J., Bailey, A. M., Lazarus, C. M., & Simpson, T. J. (2007). Biosynthesis of the 2-pyridone tenellin in the insect pathogenic fungus *Beauveria bassiana*. *Chembiochem, 8*, 289–297.

Elliot, S. L., Sabelis, M. W., Janssen, A., van der Geest, L. P. S., Beerling, E. A. M., & Fransen, J. (2000). Can plants use entomopathogens as bodyguards? *Ecol. Lett., 3*, 228–235.

Elliot, S. L., Blanford, S., & Thomas, M. B. (2002). Host–pathogen interactions in a varying environment: temperature, behavioural fever and fitness. *Proc. R. Soc. Lond. B, 269*, 1599–1607.

Ellsworth, J. F., & Grove, J. F. (1977). Cyclodepsipeptides from *Beauveria bassiana* Bals. Part 1. Beauverolides H and I. *J. Chem. Soc. Perkin, 1*(1), 270–273.

Espada, A., & Dreyfuss, M. M. (1997). Effect of the cyclopeptolide 90–215 on the production of destruxins and helvolic acid by *Metarhizium anisopliae*. *J. Ind. Microbiol. Biotechnol., 19*, 7–11.

Evans, H. C., & Samson, R. A. (1982). *Cordyceps* species and their anamorphs pathogenic on ants (Formicidae) in tropical forest ecosystems. I. The *Cephalotes* (Myrmicinae) complex. *Trans. Br. Mycol. Soc., 79*, 431–453.

Evans, H. C., & Samson, R. A. (1984). *Cordyceps* species and their anamorphs pathogenic on ants (Formicidae) in tropical forest ecosystems. II. The *Camponotus* (Formicinae) complex. *Trans. Br. Mycol. Soc., 82*, 127–150.

Fan, Y., Fang, W., Guo, S., Pei, X., Zhang, Y., Xiao, Y., Li, D., Jin, K., Bidochka, M. J., & Pei, Y. (2007). Increased insect virulence in *Beauveria bassiana* strains overexpressing an engineered chitinase. *Appl. Environ. Microbiol., 73*, 295–302.

Fan, Y., Pei, X., Guo, S., Zhang, Y., Luo, Z., Liao, X., & Pei, Y. (2010). Increased virulence using engineered protease-chitin binding domain hybrid expressed in the entomopathogenic fungus *Beauveria bassiana*. *Microb. Pathog., 49*, 376–380.

Fan, Y., Zhang, S., Kruer, N., & Keyhani, N. O. (2011). High-throughput insertion mutagenesis and functional screening in the entomopathogenic fungus *Beauveria bassiana*. *J. Invertebr. Pathol., 106*, 274–279.

Fang, W. G., & St. Leger, R. J. (2010). *Mrt*, a gene unique to fungi, encodes an oligosaccharide transporter and facilitates rhizosphere competency in *Metarhizium robertsii*. *Plant Physiol., 154*, 1549–1557.

Fang, W., Leng, B., Xiao, Y., Jin, K., Ma, J., Fan, Y., Feng, J., Yang, X., Zhang, Y., & Pei, Y. (2005). Cloning of *Beauveria bassiana* chitinase gene *Bbchit1* and its application to improve fungal strain virulence. *Appl. Environ. Microbiol., 71*, 363–370.

Fang, W., Vega-Rodríguez, J., Ghosh, A. K., Jacobs-Lorena, M., Kang, A., & St. Leger, R. J. (2011). Development of transgenic fungi that kill human malaria parasites in mosquitoes. *Science, 331*, 1074–1077.

Fargues, J., & Vey, A. (1974). Modalités d'infection des larves de *Leptinotarsa decemlineata* par *Beauveria bassiana* au cours de la mue. *Entomophaga, 19*, 311–323.

Fargues, J., Goettel, M. S., Smits, N., Ouedraogo, A., Vidal, C., Lacey, L. A., Lomer, C. J., & Rougier, M. (1996). Variability in susceptibility to simulated sunlight of conidia among isolates of entomopathogenic Hyphomycetes. *Mycopathologia, 135*, 171–181.

Fargues, J., Goettel, M. S., Smits, N., Ouedraogo, A., & Rougier, M. (1997). Effect of temperature on vegetative growth of *Beauveria bassiana* isolates from different origins. *Mycologia, 89*, 383–392.

de Faria, M. R., & Magalhães, B. P. (2001). O uso de fungos entomopatogênicos no Brasil. *Biotecnologia Ciência & Desenvolvimiento, 22*, 18–21.

de Faria, M. R., & Wraight, S. P. (2007). Mycoinsecticides and mycoacaricides: a comprehensive list with worldwide coverage and international classification of formulation types. *Biol. Control, 43*, 237–256.

Ferron, P. (1977). Influence of relative humidity on the development of fungal infection caused by *Beauveria bassiana* [Fungi Imperfecti, Moniliales] in imagines of *Acanthoscelides obtectus* [Col.: Bruchidae]. *Entomophaga, 22*, 393–396.

Fisher, J. J., Rehner, S. A., & Bruck, D. J. (2011). Diversity of rhizosphere associated entomopathogenic fungi of perennial herbs, shrubs and coniferous trees. *J. Invertebr. Pathol., 2*, 289–295.

Fornelli, F., Minervini, F., & Logrieco, A. (2004). Cytotoxicity of fungal metabolites to lepidopteran (*Spodoptera frugiperda*) cell line (SF-9). *J. Invertebr. Pathol., 85*, 74–79.

Fournier, A., Widmer, F., & Enkerli, J. (2010a). Assessing winter-survival of *Pandora neoaphidis* in soil with bioassays and molecular approaches. *Biol. Control, 54*, 126–134.

Fournier, A., Widmer, F., & Enkerli, J. (2010b). Development of a single-nucleotide polymorphism (SNP) assay for genotyping of *Pandora neoaphidis*. *Fungal Biol., 114*, 498–506.

Frappier, F., Pais, M., Elsworth, J. F., & Grover, J. F. (1978). Revised structure of beauvellide, a cyclodepsipeptide from *Beauveria tenella*. *Phytochemistry, 17*, 545–546.

Frisvad, J. C., Andersen, B., & Thrane, U. (2008). The use of secondary metabolite profiling in chemotaxonomy of filamentous fungi. *Mycol. Res., 112*, 231–240.

Gao, Q., Jin, K., Ying, S.-H., Zhang, Y., Xiao, G., Shang, Y., Duan, Z., Hu, X., Xi, X.-Q., Zhou, G., Peng, G., Luo, Z., Huang, W., Wang, B., Fang, W., Wang, S., Zhong, Y., Ma, L.-J., St. Leger, R. J., Zhao, G.-P., Pei, Y., Feng, M.-G., Xia, Y., & Wang, C. (2011). Genome sequencing and comparative transcriptomics of the model entomopathogenic fungi *Metarhizium anisopliae* and *M. acridum*. *PLoS Genet., 7*, e1001264.

Geiser, D. M., Dorner, J. W., Horn, B. W., & Taylor, J. W. (2000). The phylogenetics of mycotoxin and sclerotium production in *Aspergillus flavus* and *Aspergillus oryzae*. *Fungal Genet. Biol., 31*, 169–179.

Geiser, D. M., Gueidan, C., Miadlikowska, J., Lutzoni, F., Kauff, F., Hofstetter, V., Fraker, E., Schoch, C. L., Tibell, L., Untereiner, W. A., & Aptroot, A. (2006). Eurotiomycetes: Eurotiomycetidae and Chaetothyriomycetidae. *Mycologia, 98*, 1053–1064.

Ghikas, D. V., Kouvelis, V. N., & Typas, M. A. (2010). Phylogenetic and biogeographic implications inferred by mitochondrial intergenic region analyses and ITS1-5.8S-ITS2 of the entomopathogenic fungi *Beauveria bassiana* and *B. brongniartii*. *BMC Microbiol., 10*, 174.

Gillespie, J. P., Kanost, M. R., & Trenczek, T. (1997). Biological mediators of insect immunity. *Annu. Rev. Entomol., 42*, 611–643.

Glare, T. R., Chilvers, G. A., & Milner, R. J. (1985). Capilliconidia as infective spores in *Zoophthora phalloides* (Entomophthorales). *Trans. Br. Mycol. Soc., 85*, 463–470.

Gleason, F. H., Marano, A. V., Johnson, P., & Martin, W. W. (2010). Blastocladian parasites of invertebrates. *Fungal Biol. Rev., 24*, 56–67.

Goble, T. A., Dames, J. F., Hill, M. P., & Moore, S. D. (2010). The effects of farming system, habitat type and bait type on the isolation of entomopathogenic fungi from citrus soils in the Eastern Cape Province, South Africa. *BioControl, 55*, 399–412.

Goettel, M. S. (1995). The utility of bioassays in the risk assessment of entomopathogenic fungi. *Biotechnology Risk Assessment: USEPA/USDA/Environment Canada/Agriculture and Agri-Food Canada. Risk Assessment Methodologies* (pp. 2–8). College Park: University of Maryland Biotechnology. Proceedings of the Biotechnology Risk Assessment Symposium, June 6–8, 1995. Pensacola, FL.

Goettel, M. S., & Inglis, G. D. (1997). Fungi: Hyphomycetes. In L. A. Lacey (Ed.), *Manual of Techniques in Insect Pathology* (pp. 213–249). San Diego: Academic Press.

Goettel, M. S., Poprawski, T. J., Vandenberg, J. D., Li, Z., & Roberts, D. W. (1990a). Safety to nontarget invertebrates of fungal biocontrol agents. In M. Laird, L. A. Lacey & E. W. Davidson (Eds.), *Safety of Microbial Insecticides* (pp. 209–231). Boca Raton: CRC Press.

Goettel, M. S., St. Leger, R. J., Bhairi, S., Jung, M. K., Oakley, B. R., Roberts, D. W., & Staples, R. C. (1990b). Pathogenicity and growth of *Metarhizium anisopliae* stably transformed to benomyl resistance. *Curr. Genet., 17*, 129–132.

Goettel, M. S., Koike, M., Kim, J. J., Aiuchi, D., Shinya, R., & Brodeur, J. (2008). Potential of *Lecanicillium* spp. for management of insects, nematodes and plant diseases. *J. Invertebr. Pathol., 98*, 256–261.

Góngora, B., C. E (2004). Transformación de *Beauveria bassiana* cepa Bb9112 con los genes de la proteína verde fluorescente y la proteasa pr1A de *Metarhizum anisopliae*. *Rev. Colombiana Entomol., 30*, 15–21.

Gottlieb, D. (1978). *The Germination of Fungus Spores*. Durham: Meadowfield Press.

Grace, J., & Jaronski, S. (2005). *The Joy of Zen and the Art of Fermentation or the Tao of Fungi. Solid Substrate Fermentation Workshop*. Sydney: USDA Northern Plains Agricultural Research Laboratory.

Green, B. R. (2011). Chloroplast genomes of photosynthetic eukaryotes. *Plant J., 66*, 34–44.

Gupta, S., Krasnoff, S. B., Renwick, J. A. A., Roberts, D. W., Steiner, J. R., & Clardy, J. (1993). Viridoxins A and B: novel toxins from the fungus *Metarhizium flavoviride. J. Org. Chem., 58*, 1062–1067.

Hajek, A. E. (1997). Ecology of terrestrial fungal entomopathogens. *Adv. Microb. Ecol., 15*, 193–249.

Hajek, A. E. (1999). Pathology and epizootiology of *Entomophaga maimaiga* infections in forest lepidoptera. *Microbiol. Mol. Biol. Rev., 63*, 814–835.

Hajek, A. E. (2007). Introduction of a fungus into North America for control of gypsy moth. In C. Vincent, M. S. Goettel & G. Lazarovits (Eds.), *Biological Control: A Global Perspective* (pp. 53–62). Wallingford: CABI.

Hajek, A. E., & Butler, L. (2000). Predicting the host range of entomopathogenic fungi. In P. A. Follett & J. J. Duan (Eds.), *Nontarget Effects of Biological Control* (pp. 263–276). Boston: Kluwer Academic Publishers.

Hajek, A. E., & Delalibera, I., Jr. (2010). Fungal pathogens as classical biological control agents against arthropods. *BioControl, 55*, 147–158.

Hajek, A. E., & Eastburn, C. C. (2003). Attachment of germination of *Entomophaga maimaiga* conidia on host and non-host larval cuticle. *J. Invertebr. Pathol., 82*, 12–22.

Hajek, A. E., & Siegert, N. W. (2004). Using bioassays to estimate abundance of *Entomophaga maimaiga* resting spores in soil. *J. Invertebr. Pathol., 86*, 61–64.

Hajek, A. E., Humber, R. A., & Elkington, J. S. (1995). Mysterious origin of *Entomophaga maimaiga* in North America. *Am. Entomol., 41*, 31–42.

Hajek, A. E., Elkinton, J. S., & Witcosky, J. J. (1996). Introduction and spread of the fungal pathogen *Entomophaga maimaiga* (Zygomycetes: Entomophthorales) along the leading edge of gypsy moth (Lepidoptera: Lymantriidae) spread. *Environ. Entomol., 25*, 1235–1247.

Hajek, A. E., Filotas, M. J., & Ewing, D. C. (2002). Formation of appressoria by two species of lepidopteran-pathogenic Entomophthorales. *Can. J. Bot., 80*, 220–225.

Hajek, A. E., McManus, M. L., & Delalibera, I., Jr. (2005). *Catalogue of introductions of pathogens and nematodes for classical biological control of insects and mites*. USDA. Forest Service FHTET-2005-05.

Hajek, A. E., McManus, M. L., & Delalibera, I., Jr. (2007a). A review of introductions of pathogens and nematodes for classical biological control of insects and mites. *Biol. Control, 41*, 1–13.

Hajek, A. E., Delalibera, I., Jr., & McManus, M. L. (2007b). Introduction of exotic pathogens and documentation of their establishment and impact. In L. A. Lacey & H. K. Kaya (Eds.), *Field Manual of Techniques in Invertebrate Pathology* (2nd ed.) (pp. 299–325). Dordrecht: Springer.

Hall, R. A. (1979). Pathogenicity of *Verticillium lecanii* conidia and blastospores against the aphid *Macrosiphoniella sanborni. Entomophaga, 12*, 191–198.

Hamill, R. L., Higgens, C. E., Boaz, H. E., & Gorman, M. (1969). The structure of beauvericin, a new depsipeptide antibiotic toxic to *Artemia salina. Tetrahedron Lett., 49*, 4255–4258.

Hanlin, R. T. (1994). Microcycle conidiation – a review. *Mycoscience, 35*, 113–123.

Hatcher, M. J., Dick, J. T. A., & Dunn, A. M. (2006). How parasites affect interactions between competitors and predators. *Ecol. Lett., 9*, 1253–1271.

Hawksworth, D. L. (1991). The fungal dimension of biodiversity: magnitude, significance, and conservation. *Mycol. Res., 95*, 641–655.

Heckman, D. S., Geiser, D. M., Eidell, B. R., Stauffer, R. L., Kardos, N. L., & Blair Hedges, S. (2001). Molecular evidence for the early colonization of land by fungi and plants. *Science, 293*, 1129–1133.

Hegedus, D. D., Bidochka, M. J., & Khachatourians, G. G. (1990). *Beauveria bassiana* submerged conidia production in a defined medium containing chitin, two hexosamines or glucose. *Appl. Microbiol. Biotechnol., 33*, 641–647.

Heie, O. E. (1980). *The Aphidoidea (Hemiptera) of Fennoscandia and Denmark*. Denmark: Scandinavian Science Press.

Henk, D. A., & Vilgalys, R. (2007). Molecular phylogeny suggests a single origin of insect symbiosis in the Pucciniomyces with support for some relationships within the genus *Septobasidium. Am. J. Bot., 94*, 1515–1526.

Henke, M. O., de Hoog, G. S., Gross, U., Zimmermann, G., Kraemer, D., & Weig, M. (2002). Human deep tissue infection with an entomopathogenic *Beauveria* species. *J. Clin. Microbiol., 40*, 2698–2702.

Hesketh, H., Roy, H. E., Eilenberg, J., Pell, J. K., & Hails, R. S. (2010). Challenges in modelling complexity of fungal entomopathogens in semi-natural populations of insects. *BioControl, 55*, 55–73.

Hibbett, D. S., Binder, M., Bischoff, J. F., Blackwell, M., Cannon, P. F., Eriksson, O. E., Huhndorf, S., James, T., Kirk, P. M., Lücking, R., Lumbsch, H. T., Lutzoni, F., Matheny, P. B., McLaughlin, D. J., Powell, M. J., Redhead, S., Schoch, C. L., Spatafora, J. W., Stalpers, J. A., Vilgalys, R., Aime, M. C., Aptroot, A., Bauer, R., Begerow, D., Benny, G. L., Castlebury, L. A., Crous, P. W., Dai, Y. C., Gams, W., Geiser, D. M., Griffith, G. W., Gueidan, C., Hawksworth, D. L., Hestmark, G., Hosaka, K., Humber, R. A., Hyde, K. D., Ironside, J. E., Kõljalg, U., Jyrtznabm, C. P., Larsson, K. H., Lichtwardt, R., Longcore, J., Miadlikowska, J., Miller, A., Moncalvo, J. M., Mozley-Standridge, S., Oberwinkler, F., Parmasto, E., Reeb, V., Rogers, J. D., Roux, C., Ryvarden, L., Sampaio, J. P., Schüßler, A., Sugiyama, J., Thorn, R. G., Tibell, L., Untereiner, W. A., Walker, C., Wang, Z., Weir, A., Weiß, M., White, M. M., Winka, K., Yao, Y. J., & Zhang, N. (2007). A higher-level phylogenetic classification of the Fungi. *Mycol. Res., 111*, 509–547.

Hibbett, D. S., Ohman, A., Glotzer, D., Nuhn, M., Kirk, P., & Nilsson, R. H. (2011). Progress in molecular and morphological taxon discovery in *Fungi* and options for formal classification of environmental sequences. *Fungal Biol. Rev., 25*, 38–47.

Hirt, R. P., Logsdon, J. M., Jr., Healy, B., Dorey, M. W., Doolittle, W. F., & Embley, T. M. (1999). Microsporidia are related to Fungi: evidence from the largest subunit of RNA polymerase II and other proteins. *Proc. Natl. Acad. Sci. USA, 96*, 580–585.

Hodge, K. T. (2003). Clavicipitaceous anamorphs. In J. F. White, Jr., C. W. Bacon, N. L. Hywel-Jones & J. W. Spatafora (Eds.), *Clavicipitalean Fungi: Evolutionary Biology, Chemistry, Biocontrol, and Cultural Impacts* (pp. 75–123). New York: Marcel Dekker.

Holder, D. J., & Keyhani, N. O. (2005). Adhesion of the entomopathogenic fungus *Beauveria* (*Cordyceps*) *bassiana* to substrates. *Appl. Environ. Microbiol., 71*, 5260–5266.

Holder, D. J., Kirkland, B. H., Lewis, M. W., & Keyhani, N. O. (2007). Surface characteristics of the entomopathogenic fungus *Beauveria* (*Cordyceps*) *bassiana. Microbiology, 153*, 3448–3457.

Holt, R. D. (1997). Community modules. In A. C. Gange & V. K. Brown (Eds.), *Multitrophic Interactions in Terrestrial Systems* (pp. 333–350). London: Blackwell Science.

Holt, R. D., & Dobson, A. P. (2006). Extending the principles of community ecology to address the epidemiology of host–pathogen systems. In S. K. Collinge & C. Ray (Eds.), *Disease Ecology – Community Structure and Pathogen Dynamics* (pp. 6–27). New York: Oxford University Press.

Holt, R. D., & Hochberg, M. E. (2001). Indirect interactions, community modules and biological control: a theoretical perspective. In E. Wajnberg, J. K. Scott & P. C. Quimby (Eds.), *Evaluating Indirect Effects of Biological Control* (pp. 13–38). Wallingford: CABI.

Howard, A. F. V., N'Guessan, R., Koenraadt, C. J. M., Asidi, A., Farenhorst, M., Agokbéto, M., Thomas, M. B., Knols, B. G. J., & Takken, W. (2010). The entomopathogenic fungus *Beauveria bassiana* reduces instantaneous blood feeding in wild multi-insecticide-resistant *Culex quinquefasciatus* mosquitoes in Benin, West Africa. *Parasit. Vectors, 3,* 87.

Howard, R. J., & Valent, B. (1996). Breaking and entering: host penetration by the fungal rice blast pathogen *Magnaporthe grisea. Annu. Rev. Microbiol., 50,* 491–512.

Hu, G., & St. Leger, R. J. (2002). Field studies using a recombinant mycoinsecticide (*Metarhizium anisopliae*) reveal that it is rhizosphere competent. *Appl. Environ. Microbiol., 68,* 6383–6387.

Hughes, W. O. H., Eilenberg, J., & Boomsma, J. J. (2002). Trade-offs in group living: transmission and disease resistance in leaf-cutting ants. *Proc. R. Soc. Lond. B, 269,* 1811–1819.

Hughes, W. O. H., Thomsen, L., Eilenberg, J., & Boomsma, J. J. (2004). Diversity of entomopathogenic fungi near leaf-cutting ant nests in a neotropical forest, with particular reference to *Metarhizium anisopliae* var. *anisopliae. J. Invertebr. Pathol., 85,* 46–53.

Humber, R. A. (1997). Fungi: identification. In L. A. Lacey (Ed.), *Manual of Techniques in Insect Pathology* (pp. 153–185). San Diego: Academic Press.

Humber, R. A. (2008). Evolution of entomopathogenicity in fungi. *J. Invertebr. Pathol., 98,* 262–266.

Humber, R. A. (2009). Entomogenous fungi. In M. Schaechter (Ed.) (3rd ed.). *Encyclopedia of Microbiology,* Vol. 3 (pp. 443–456) San Diego: Academic Press.

Hunt, T. R., Moore, D., Higgins, P. M., & Prior, C. (1994). Effect of sunscreens, irradiance and resting periods on the germination of *Metarhizium flavoviride* conidia. *Entomophaga, 39,* 313–322.

Hussain, A., Tian, M.-Y., He, Y.-R., & Lei, Y.-Y. (2010). Differential fluctuation in virulence and VOC profiles among different cultures of entomopathogenic fungi. *J. Invertebr. Pathol., 104,* 166–171.

Ignoffo, C. M. (1973). Effects of entomopathogens on vertebrates. *Ann. N.Y. Acad. Sci., 217,* 141–172.

Ignoffo, C. M., & Garcia, C. (1992). Influence of conidial color on inactivation of several entomogenous fungi (Hyphomycetes) by simulated sunlight. *Environ. Entomol., 21,* 913–917.

Inglis, G. D., Goettel, M. S., & Johnson, D. L. (1995). Influence of ultraviolet light protectants on persistence of the entomopathogenic fungus, *Beauveria bassiana. Biol. Control, 5,* 581–590.

Inglis, G. D., Johnson, D. L., & Goettel, M. S. (1996). Effect of bait substrate and formulation on infection of grasshopper nymphs by *Beauveria bassiana. Biocontrol Sci. Technol., 6,* 35–50.

Inglis, P. W., Aragão, F. J. L., Frazão, H., Magalhães, B. P., & Valadares-Inglis, M. C. (2000). Biolistic co-transformation of *Metarhizium anisopliae* var. *acridum* strain GC423 with green fluorescent protein and resistance to glufosinate ammonium. *FEMS Microbiol. Lett., 191,* 249–254.

Inglis, G. D., Goettel, M. S., Butt, T. M., & Strasser, H. (2001). Use of hyphomycetous fungi for managing insect pests. In T. M. Butt, C. Jackson & N. Magan (Eds.), *Fungi as Biocontrol Agents: Progress, Problems and Potential* (pp. 23–69). Wallingford: CABI.

Inyang, E. N., Butt, T. M., Beckett, A., & Archer, S. (1999). The effect of crucifer epicuticular waxes and leaf extracts on the germination and virulence of *Metarhizium anisopliae* conidia. *Mycol. Res., 103,* 419–426.

Isaka, M., Tantichaoren, M., & Thebtaranonth, Y. (2000). Cordyanhydrides A and B. Two unique anhydrides from the insect pathogenic fungus *Cordyceps pseudomilitaris* BCC 1620. *Tetrahedron Lett., 41,* 1657–1660.

Isaka, M., Kittakoop, P., Kirtikara, K., Hywel-Jones, N. L., & Thebtaranonth, Y. (2005). Bioactive substances from insect pathogenic fungi. *Acc. Chem. Res., 38,* 813–823.

Jaccoud, D. B., Hughes, W. O. H., & Jackson, C. W. (1999). The epizootiology of a *Metarhizium* infection in mini-nests of the leaf-cutting ant *Atta sexdens rubropilosa. Entomol. Exp. Appl., 93,* 51–61.

Jackson, M. A., & Jaronski, S. T. (2009). Production of microsclerotia of the fungal entomopathogen *Metarhizium anisopliae* and their potential for use as a biocontrol agent for soil-inhabiting insects. *Mycol. Res., 113,* 842–850.

Jackson, M. A., Dunlap, C. A., & Jaronski, S. T. (2010). Ecological considerations in producing and formulating fungal entomopathogens for use in insect biocontrol. *BioControl, 55,* 129–145.

Jacobson, E. S. (2000). Pathogenic roles for fungal melanins. *Clin. Microbiol. Rev., 13,* 708–717.

James, T. Y., Letcher, P. M., Longcore, J. E., Mozley-Standridge, S. E., Porter, D., Powell, M. J., Griffith, G. W., & Vilgalys, R. (2006a). A molecular phylogeny of the flagellated fungi (Chytridiomycota) and description of a new phylum (Blastocladiomycota). *Mycologia, 98,* 860–871.

James, T. Y., Kauff, F., Schoch, C. L., Matheny, P. B., Hofstetter, V., Cox, C. J., Celio, G., Gueidan, C., Fraker, E., Miadlikowska, J., Lumbsch, H. T., Rauhut, A., Reeb, V., Arnold, A. E., Amtoft, A., Stajich, J. E., Hosaka, K., Sung, G. H., Johnson, D., O'Rourke, B., Crockett, M., Binder, M., Curtis, J. M., Slot, J. C., Wang, Z., Wilson, A. W., Schussler, A., Longcore, J. E., O'Donnell, K., Mozley-Standridge, S., Porter, D., Letcher, P. M., Powell, M. J., Taylor, J. W., White, M. M., Griffith, G. W., Davies, D. R., Humber, R. A., Morton, J. B., Sugiyama, J., Rossman, A. Y., Rogers, J. D., Pfister, D. H., Hewitt, D., Hansen, K., Hambleton, S., Shoemaker, R. A., Kohlmeyer, J., Volkmann-Kohlmeyer, B., Spotts, R. A., Serdani, M., Crous, P. W., Hughes, K. W., Matsuura, K., Langer, E., Langer, G., Untereiner, W. A., Lucking, R., Budel, B., Geiser, D. M., Aptroot, A., Diederich, P., Schmitt, I., Schultz, M., Yahr, R., Hibbett, D. S., Lutzoni, F., McLaughlin, D. J., Spatafora, J. W., & Vilgalys, R. (2006b). Reconstructing the early evolution of Fungi using a six-gene phylogeny. *Nature, 443,* 818–822.

Jaronski, S. T. (2007). Soil ecology of the entomopathogenic Ascomycetes: a critical examination of what we (think) we know. In S. Ekesi & N. K. Maniania (Eds.), *Use of Entomopathogenic Fungi in Biological Pest Management* (pp. 91–143). Kerala: Research Signpost.

Jaronski, S. T. (2010). Ecological factors in the inundative use of fungal entomopathogens. *BioControl, 55,* 159–185.

Jaronski, S. T., Goettel, M. S., & Lomer, C. L. (2003). Regulatory requirements for ecotoxicological assessments of microbial insecticides — how relevant are they? In H. M. T. Hokkanen & A. E. Hajek (Eds.), *Environmental Impacts of Microbial Insecticides* (pp. 237—260) Dordrecht: Kluwer Academic Publishers.

Jeffs, L. B., Xavier, I. J., Matai, R. E., & Khachatourians, G. G. (1999). Relationships between fungal spore morphologies and surface properties for entomopathogenic members of the genera *Beauveria, Metarhizium, Paecilomyces, Tolypocladium*, and *Verticillium. Can. J. Microbiol., 45*, 936—948.

Jegorov, A., Sedmera, P., Matha, V., Simek, P., Zaradnícková, H., Landa, Z., & Eyal, J. (1994). Beauverolides L and La from *Beauveria tenella* and *Paecilomyces fumosoroseus. Phytochemistry, 37*, 1301—1303.

Jenkins, N. E., & Prior, C. (1993). Growth and formation of true conidia by *Metarhizium flavoviride* in a simple liquid medium. *Mycol. Res., 97*, 1489—1494.

Jenkins, N. E., & Thomas, M. B. (1996). Effect of formulation and application method on the efficacy of aerial and submerged conidia of *Metarhizium flavoviride* for locust and grasshopper control. *Pestic. Sci., 46*, 299—306.

Jensen, A. B., Thomsen, L., & Eilenberg, J. (2001). Intraspecific variation and host specificity of *Entomophthora muscae sensu stricto* isolates revealed by random amplified polymorphic DNA, universal primed PCR, PCR-restriction fragment length polymorphism, and conidial morphology. *J. Invertebr. Pathol., 78*, 251—259.

Jensen, A. B., Hansen, L. M., & Eilenberg, J. (2008). Grain aphid population structure: no effect of fungal infections in a 2-year field study in Denmark. *Agric. For. Entomol., 10*, 279—290.

Johnson, D., Sung, G.-H., Hywell-Jones, N. L., Luangsa-ard, J. J., Bischoff, J. F., Kepler, R. M., & Spatafora, J. W. (2009). Systematics and evolution of the genus *Torrubiella* (Hypocreales, Ascomycota). *Mycol. Res., 113*, 279—289.

Jones, K. A., & Burges, H. D. (1998). Technology of formulation and application. In H. D. Burges (Ed.), *Formulation of Microbial Biopesticides* (pp. 2—30). Dordrecht: Kluwer Academic Publishers.

Kabaluk, J. T., & Ericsson, J. D. (2007). *Metarhizium anisopliae* seed treatment increases yield of field corn when applied for wireworm control. *Agron. J., 99*, 1377—1381.

Kabaluk, J. T., Svircev, A. M., Goettel, M. S. & Woo, S. G. (Eds.). (2010). *The Use and Regulation of Microbial Pesticides in Representative Jurisdictions Worldwide*. IOBC Global. Available online: www. IOBC-Global.org.

Kagamizono, T., Nishino, E., Matsumoto, K., Kawashima, A., Kishimoto, M., Sakai, N., He, B.-M., Chen, Z.-X., Adachi, T., Morimoto, S., & Hanada, K. (1995). Bassiatin, a new platelet aggregation inhibitor produced by *Beauveria bassiana* K-717. *J. Antibiot., 48*, 1407—1412.

Kalsbeek, V., Mullens, B. A., & Jespersen, J. B. (2001). Field studies of *Entomophthora* (Zygomycetes: Entomophthorales)-induced behavioral fever in *Musca domestica* (Diptera: Muscidae) *in Denmark. Biol. Control, 21*, 264—273.

Kanaoka, M., Isogai, A., Murakoshi, S., Ichinoe, M., Suzuki, A., & Tamura, S. (1978). Bassianolide, a new insecticidal cyclodepsipeptide from *Beauveria bassiana* and *Lecanicillium lecanii. Agric. Biol. Chem., 42*, 629—635.

Kang, S. C., Bark, Y. G., Lee, D. G., & Kim, Y. H. (1996). Antifungal activities of *Metarhizium anisopliae* against *Fusarium oxysporum, Botrytis cinerea*, and *Alternaria solani. Korean J. Mycol., 24*, 49—55.

Kassa, A., Stephan, D., Vidal, S., & Zimmermann, G. (2004). Production and processing of *Metarhizium anisopliae* var. *acridum* submerged

conidia for locust and grasshopper control. *Mycol. Res., 108*, 93—100.

Keeling, P. J., & Fast, N. M. (2002). Microsporidia: biology and evolution of highly reduced intracellular parasites. *Annu. Rev. Microbiol., 56*, 93—116.

Keller, N. P., Turner, G., & Bennett, J. W. (2005). Fungal secondary metabolism — from biochemistry to genomics. *Nat. Rev. Microbiol., 3*, 937—947.

Keller, S. (Ed.). (2007). *Arthropod-Pathogenic Entomophthorales: Biology, Ecology, Identification*. Luxembourg: Office for Official Publications of the European Communities.

Keller, S., & Wegensteiner, R. (2007). Introduction. In S. Keller (Ed.), *Arthropod-Pathogenic Entomophthorales: Biology, Ecology, Identification* (pp. 1—7). Luxembourg: Office for Official Publications of the European Communities.

Keller, S., Kessler, P., & Schweizer, C. (2003). Distribution of insect pathogenic soil fungi in Switzerland with special reference to *Beauveria brongniartii* and *Metarhizium anisopliae. BioControl, 48*, 307—319.

Kermarrec, A., Febvay, G., & Decharme, M. (1986). Protection of leaf-cutting ants from biohazards: is there a future for microbiological control? In C. S. Lofgren & R. K. VanderMeer (Eds.), *Fire Ants and Leaf-Cutting Ants: Biology and Management* (pp. 339—356) Boulder: Westview Press.

Kershaw, M. J., Moorhouse, E. R., Bateman, R., Reynolds, S. E., & Charnley, A. K. (1999). The role of destruxins in the pathogenicity of *Metarhizium anisopliae* for three species of insects. *J. Invertebr. Pathol., 74*, 213—223.

Kerwin, J. L., & Petersen, E. E. (1997). Fungi: Oomycetes and Chytridiomycetes. In L. A. Lacey (Ed.), *Manual of Techniques in Insect Pathology* (pp. 251—268). New York: Academic Press.

Kikuchi, H., Miyagawa, Y., Nakamura, K., Sahashi, Y., Inatomi, S., & Oshima, Y. (2004). A novel carbon skeletal trichothecane, tenuipesine A, isolated from an entomopathogenic fungus, *Paecilomyces tenuipes. Org. Lett., 6*, 4531—4533.

Kirk, P. M., Cannon, P. F., Minter, D. W., & Stalpers, J. A. (2008). *Dictionary of the Fungi* (10th ed.). Wallingford: CABI.

Klingen, I., Meadow, R., & Eilenberg, J. (2000). Prevalence of fungal infections in adult *Delia radicum* and *Delia floralis* captured on the edge of a cabbage field. *Entomol. Exp. Appl., 97*, 265—274.

Klingen, I., Eilenberg, J., & Meadow., R. (2002). Effects of farming system, field margins and bait insect on the occurrence of insect pathogenic fungi in soils. *Agric. Ecosyst. Environ., 91*, 191—198.

Klowden, M. J. (2007). *Physiological Systems in Insects* (2nd ed.). San Diego: Academic Press.

Kodaira, Y. (1961). Toxic substances to insects, produced by *Aspergillus ochraceus* and *Oospora destructor. Agric. Biol. Chem., 25*, 261—262.

Kodaira, Y. (1962). Studies on the new toxic substances to insects, destruxin A and B, produced by *Oospora destructor*. Part I. Isolation and purification of destruxin A and B. *Agric. Biol. Chem., 26*, 36—42.

Kodsueb, R., Jeewon, R., Vijaykrishna, D., McKenzie, E. H. C., Lumyong, P., Lumyong, S., & Hyde, K. D. (2006). Systematic revision of Tubeufiaceae based on morphological and molecular data. *Fungal Divers., 21*, 105—130.

Kondo, S., Meguriya, N., Mogi, H., Aota, T., Miura, K., Fujii, T., Hayashi, I., Makino, K., Yamamoto, M., & Nakajima, N. (1980). K-582, a new peptide antibiotic. *I. J. Antibiot., 33*, 533—542.

Kozone, I., Ueda, J.-Y., Watanabe, M., Nogami, S., Nagai, A., Inaba, S., Ohya, Y., Tagaki, M., & Shin-ya, K. (2009). Novel 24-membered macrolides, JBIR-19 and -20 isolated from *Metarhizium* sp. fE61. *J. Antibiot., 62*, 159–162.

Krasnoff, S. B., & Gupta, S. (1994). Identification of the antibiotic phomalactone from the entomopathogenic fungus *Hirsutella thompsonii* var. *synnematosa*. *J. Chem. Ecol., 20*, 293–302.

Krasnoff, S. B., Gibson, D. M., Belofsky, G. N., Gloer, K. B., & Gloer, J. B. (1996). New destruxins from the entomopathogenic fungus *Aschersonia* sp. *J. Nat. Prod., 59*, 485–489.

Krasnoff, S. B., Sommers, C. H., Moon, S.-Y., Donzelli, B. G. G., Vandenberg, J. D., Churchill, A. C. L., & Gibson, D. M. (2006). Production of mutagenic metabolites by *Metarhizium anisopliae*. *J. Agric. Food Chem., 54*, 7083–7088.

Krasnoff, S. B., Keresztes, I., Gillilan, R. E., Szebenyi, D. M. E., Donzelli, B. G. G., Churchill, A. C. L., & Gibson, D. M. (2007). Serinocyclins A and B, cyclic heptapeptides from *Metarhizium anisopliae*. *J. Nat. Prod., 70*, 1919–1924.

Kuboki, H., Tsuchida, T., Wakazono, K., Isshiki, K., Kumagai, H., & Yoshioka, T. (1999). Mer-f3, 12-hydroxy-ovalicin, produced by *Metarhizium* sp. f3. *J. Antibiot., 52*, 590–593.

Kurihara, Y., Sukarno, N., Ilyas, M., Yuniarti, E., Mangunwardoyo, W., Saraswati, R., Park, J.-Y., Inaba, S., Widyastuti, Y., & Ando, K. (2008). Entomopathogenic fungi isolated from suspended-soil-inhabiting arthropods in East Kalimantan, Indonesia. *Mycoscience, 49*, 241–249.

Kurtzman, C. P., Fell, J. W., & Boekhout, T. (2011). *The Yeasts: A Taxonomic Study* (5th ed.). Amsterdam: Elsevier.

Kwon, H. C., Bang, E. J., Choi, S. U., Lee, W. C., Cho, S. Y., Jung, I. Y., Kim, S. Y., & Lee, K. R. (2000). Cytotoxic cyclodepsipeptides of *Bombycis corpus* 101A. *Yakhak Hoechi, 44*, 115–118.

Lacey, L. A., & Mercadier, G. (1998). The effect of selected allelochemicals on germination of conidia and blastospores and mycelial growth of the entomopathogenic fungus, *Paecilomyces fumosoroseus* (Deuteromycotina: Hyphomycetes). *Mycopathologia, 142*, 17–25.

Lacey, L. A., Liu, T. X., Buchman, J. L., Munyaneza, J. E., Goolsby, J. A., & Horton, D. R. (2011). Entomopathogenic fungi (Hypocreales) for control of potato psyllid, *Bactericera cockerelli* (Sulc) (Hemiptera: Triozidae) in an area endemic for zebra chip disease of potato. *Biol. Control, 56*, 271–278.

Lafferty, K. D., Allesina, S., Arim, M., Briggs, C. J., De Leo, G., Dobson, A. P., Dunne, J. A., Johnson, P. T. J., Kuris, A. M., Marcogliese, D. J., Martinez, N. D., Memmott, J., Marquet, P. A., McLaughlin, J. P., Mordecai, E. A., Pascual, M., Poulin, R., & Thieltges, D. W. (2008). Parasites in food webs: the ultimate missing links. *Ecol. Lett., 11*, 533–546.

Langewald, J., & Kooyman, C. (2007). Green Muscle™, a fungal biopesticide for control of grasshoppers and locusts in Africa. In C. Vincent, M. S. Goettel & G. Lazarovits (Eds.), *Biological Control: A Global Perspective* (pp. 311–318). Wallingford: CABI.

Langewald, J., Ouambama, Z., Mamadou, A., Peveling, R., Stolz, I., Bateman, R., Attignon, S., Blanford, S., Arthurs, S., & Lomer, C. (1999). Comparison of organophosphate insecticide with a mycoinsecticide for the control of *Oedaleus senegalensis* (Orthoptera: Acrididae) and other Sahelian grasshoppers at an operational scale. *Biocontrol Sci. Technol., 9*, 199–214.

Latgé, J.-P., Sampedro, L., Brey, P., & Diaquin, M. (1987). Aggressiveness of *Conidiobolus obscurus* against the pea aphid: influence of cuticular extracts on ballistospore germination of aggressive and non-aggressive strains. *J. Gen. Microbiol., 133*, 1987–1997.

Laws, A. N., Frauendorf, T. C., Gómez, J. E., & Algaze, I. M. (2009). Predators mediate the effects of a fungal pathogen on prey: an experiment with grasshoppers, wolf spiders, and fungal pathogens. *Ecol. Entomol., 34*, 702–708.

Lee, S. C., Corradi, N., Doan, S., Dietrich, F. S., Keeling, P. J., & Heitman, J. (2010). Evolution of the sex-related locus and genomic features shared in Microsporidia and Fungi. *PLoS ONE 5* e10539.

Lewis, M. W., Robalino, I. V., & Keyhani, N. O. (2009). Uptake of fluorescent probe FM4-64 by hyphae and haemolymph-derived *in vivo* hyphal bodies of the entomopathogenic fungus *Beauveria bassiana*. *Microbiology, 155*, 3110–3120.

Li, Z. Z. (1988). List of the insect hosts of *Beauveria bassiana*. In *Study and Application of Entomogenous Fungi in China*, Vol. 1 (pp. 241–255). Beijing: Academic Periodical Press.

Li, Z., Alves, S. B., Roberts, D. W., Fan, M., Delalibera, I., Jr., Tang, J., Lopes, R. B., Faria, M., & Rangel, D. E. N. (2010). Biological control of insects in Brazil and China: history, current programs and reasons for their successes using entomopathogenic fungi. *Biocontrol Sci. Technol., 20*, 117–136.

Lichtwardt, R. W., Cafaro, M. J., & White, M. M. (2001). *The Trichomycetes: Fungal Associates of Arthropods*. Available online: http://www.nhm.ku.edu/~fungi/Monograph/Text/Mono.htm.

Liebhold, A., Mastro, V., & Schaefer, P. W. (1989). Learning from the legacy of Léopold Trouvelot. *Bull. Entomol. Soc. Am., 35*, 20–22.

Ling, J. Y., Sun, Y. J., Zhang, H., Lv, P., & Zhang, C. K. (2002). Measurement of cordycepin and adenosine in stroma of *Cordyceps* sp. by capillary zone electrophoresis (CZE). *J. Biosci. Bioeng., 94*, 371–374.

Liu, J., Cao, Y., & Xia, Y. (2010). *Mmc*, a gene involved in microcycle conidiation of the entomopathogenic fungus *Metarhizium* anisopliae. *J. Invertebr. Pathol., 105*, 132–138.

Liu, W.-Z., Boucias, D. G., & McCoy, C. W. (1995). Extraction and characterization of the insecticidal toxin hirsutellin A produced by *Hirsutella thompsonii* var. *thompsonii*. *Exp. Mycol., 19*, 254–262.

Locke, M. (2001). The Wigglesworth Lecture: Insects for studying fundamental problems in biology. *J. Insect Physiol., 47*, 495–507.

Loesch, A., Hutwimmer, S., & Strasser, H. (2010). Carbon utilization pattern as a potential quality control criterion for virulence of *Beauveria brongniartii*. *J. Invertebr. Pathol., 104*, 58–65.

Logrieco, A., Moretti, A., Castella, G., Kostecki, M., Golinski, P., Ritieni, A., & Chelkowski, J. (1998). Beauvericin production by *Fusarium* species. *Appl. Environ. Microbiol., 64*, 3084–3088.

Lomer, C. J., Bateman, R. P., Johnson, D. L., Langewald, J., & Thomas, M. (2001). Biological control of locusts and grasshoppers. *Annu. Rev. Entomol., 46*, 667–702.

Lu, D., Pava-Ripoll, M., Li, Z., & Wang, C. (2008). Insecticidal evaluation of *Beauveria bassiana* engineered to express a scorpion neurotoxin and a cuticle degrading protease. *Appl. Microbiol. Biotechnol., 81*, 515–522.

Luangsa-ard, J. J., Hywel-Jones, N. L., & Samson, R. A. (2004). The polyphyletic nature of *Paecilomyces sensu lato* based on 18S-generated rDNA phylogeny. *Mycologia, 96*, 773–780.

Luangsa-ard, J. J., Hywel-Jones, N. L., Manoch, L., & Samson, R. A. (2005). On the relationships of *Paecilomyces* sect. *Isarioidea* species. *Mycol. Res., 109*, 581–589.

Luangsa-ard, J. J., Tasanathai, K., Mongkolsamrit, S., & Hywel-Jones, N. L. (2007). *Atlas of Invertebrate-Pathogenic Fungi of*

Thailand, Vol. 1. Thailand: National Center for Genetic Engineering and Biotechnology.

Luangsa-ard, J. J., Tasanathai, K., Mongkolsamrit, S., & Hywel-Jones, N. L. (2008). *Atlas of Invertebrate-Pathogenic Fungi of Thailand*, Vol. 2. Thailand: National Center for Genetic Engineering and Biotechnology.

Luangsa-ard, J. J., Berkaew, P., Ridkaew, R., Hywel-Jones, N. L., & Isaka, M. (2009). A beauvericin hot spot in the genus *Isaria*. *Mycol. Res., 13*, 1389–1395.

Luangsa-ard, J. J., Tasanathai, K., Mongkolsamrit, S., & Hywel-Jones, N. L. (2010). *Atlas of Invertebrate-Pathogenic Fungi of Thailand*, Vol. 3. Thailand: National Center for Genetic Engineering and Biotechnology.

Lücking, R., Huhndorf, S., Pfister, D. H., Plata, E. R., & Lumbsch, H. T. (2009). Fungi evolved right on track. *Mycologia, 101*, 810–822.

Luz, C., Fargues, J., & Romaña, C. (2003). Influence of starvation and blood meal-induced moult on the susceptibility of nymphs of *Rhodnius prolixus* Stål (Hem., Triatominae) to *Beauveria bassiana* (Bals.) Vuill. infection. *J. Appl. Entomol., 127*, 153–156.

Madelin, M. F. (1966). Fungal parasites of insects. *Annu. Rev. Entomol., 11*, 423–448.

Mahaney, W. C., Dohm, J. M., Baker, V. R., Newsom, H. E., Malloch, D., Hancock, R. G. V., Campbell, I., Sheppard, D., & Milner, M. W. (2001). Morphogenesis of Antarctic paleosols: Martian analogue. *Icarus, 154*, 113–130.

Mains, E. B. (1958). North American entomogenous species of *Cordyceps*. *Mycologia, 50*, 169–222.

Marcus, R., & Eaves, D. M. (2000). Statistical and computational analysis of bioassay data. In A. Navon & K. R. S. Ascher (Eds.), *Bioassays of Entomopathogenic Microbes and Nematodes* (pp. 249–293). Wallingford: CAB Publishing.

Mazet, I., & Vey, A. (1995). Hirsutellin A, a toxic protein produced *in vitro* by *Hirsutella thompsonii*. *Microbiology, 141*, 1343–1348.

McCauley, V. J. E., Zacharuk, R. Y., & Tinline, R. D. (1968). Histopathology of green muscardine in larvae of four species of Elateridae (Coleoptera). *J. Invertebr. Pathol., 12*, 444–459.

McCoy, C. W., Samson, R. A., & Boucias, D. G. (1988). In C. M. Ignoffo (Ed.), *CRC Handbook of Natural Pesticides. Entomogenous fungi*, Vol. 5 (Part A) (pp. 151–236). Boca Raton: CRC Press.

Mendonça, A. F. (1992). Mass production, application and formulation of *Metarhizium anisopliae* for control of sugarcane froghopper, *Mahanarva posticata*, in Brazil. In C. J. Lomer & C. Prior (Eds.), *Biological Control of Locusts and Grasshoppers* (pp. 239–244). Wallingford: CABI.

Meyling, N. V., & Eilenberg, J. (2006a). Isolation and characterisation of *Beauveria bassiana* isolates from phylloplanes of hedgerow vegetation. *Mycol. Res., 110*, 188–195.

Meyling, N. V., & Eilenberg, J. (2006b). Occurrence and distribution of soil borne entomopathogenic fungi within a single organic agroecosystem. *Agric. Ecosyst. Environ., 113*, 336–341.

Meyling, N. V., & Eilenberg, J. (2007). Ecology of the entomopathogenic fungi *Beauveria bassiana* and *Metarhizium anisopliae* in temperate agroecosystems: potential for conservation biological control. *Biol. Control, 43*, 145–155.

Meyling, N. V., & Hajek, A. E. (2010). Principles from community and metapopulation ecology: application to fungal entomopathogens. *BioControl, 55*, 39–54.

Meyling, N. V., Lubeck, M., Buckley, E. P., Eilenberg, J., & Rehner, S. A. (2009). Community composition, host range and genetic structure of the fungal entomopathogen *Beauveria* in adjoining agricultural and seminatural habitats. *Mol. Ecol., 18*, 1282–1293.

Meyling, N. V., Thorup-Kristensen, K., & Eilenberg, J. (2011). Below- and aboveground abundance and distribution of fungal entomopathogens in experimental conventional and organic cropping systems. *Biol. Control, 59*, 180–186.

Mietkiewski, R., & Tkaczuk, C. (1998). The spectrum and frequency of entomopathogenic fungi in litter, forest soil and arable soil. *IOBC wprs Bull., 21*, 41–44.

Milner, R. J., Huppatz, R. J., & Swaris, S. C. (1991). A new method for assessment of germination of *Metarhizium anisopliae* conidia. *J. Invertebr. Pathol., 57*, 121–123.

Mochizuki, K., Ohmori, K., Tamura, H., Shizuri, Y., Nishiyama, S., Miyoshi, W., & Yamamura, S. (1993). The structures of bioactive cyclodepsipeptides, beauveriolides I and II, metabolites of entomopathogenic fungi *Beauveria* sp. *Bull Chem. Soc. Jpn, 66*, 3041–3046.

Molnár, I., Gibson, D. M., & Krasnoff, S. B. (2010). Secondary metabolites from entomopathogenic Hypocrealean fungi. *Nat. Prod. Rep., 27*, 1241–1275.

Moore, D. (2008). A plague on locusts – the Lubilosa story. *Outlooks Pest Manag., 19*, 14–17.

Ngoka, L. C. M., Gross, M. L., & Toogood, P. L. (1999). Sodium-directed selective cleavage of lactones: a method for structure determination of cyclodepsipeptides. *Int. J. Mass Spectrom, 182/183*, 289–298.

Nicholson, R. L., & Moraes, W. B. C. (1980). Survival of *Colletotrichum graminicola*: importance of the spore matrix. *Phytopathology, 70*, 255–261.

Nicholson, R. L., Hipskind, J., & Hanau, R. M. (1989). Protection against phenol toxicity by the spore mucilage of *Colletotrichum graminicola*, an aid to secondary spread. *Physiol. Mol. Plant Pathol., 35*, 243–252.

Nielsen, C., & Steenberg, T. (2004). Entomophthoralean fungi infecting the bird cherry-oat aphid, *Rhopalosiphum padi*, feeding on its winter host bird cherry, *Prunus padus*. *J. Invertebr. Pathol., 87*, 70–73.

Nielsen, C., Hajek, A. E., Humber, R. A., Bresciani, J., & Eilenberg, J. (2003). Soil as an environment for winter survival of aphid-pathogenic Entomophthorales. *Biol. Control, 28*, 92–100.

Nielsen, C., Milgroom, M. G., & Hajek, A. E. (2005). Genetic diversity in the gypsy moth fungal pathogen *Entomophaga maimaiga* from founder populations in North America and source populations in Asia. *Mycol. Res., 109*, 941–950.

Nielsen, C., Agrawal, A. A., & Hajek, A. E. (2010). Ants defend aphids against lethal disease. *Biol. Lett., 6*, 205–208.

Nikoh, N., & Fukatsu, T. (2000). Interkingdom host jumping underground: phylogenetic analysis of entomoparasitic fungi of the genus *Cordyceps*. *Mol. Biol. Evol., 17*, 629–638.

Notini, G., Mathlein, R., & Lihnell, D. (1944). Green mycosis caused by *Metarhizium anisopliae* (Metsch.) Sorok. I. Green mycoses as a biological means of insect control. II. Physiological investigations on the green mycosis fungus. *Rev. Appl. Mycol., 25*, 161–162.

O'Brien, H. E., Parrent, J. L., Jackson, J. A., Moncalvo, J.-M., & Vilgalys, R. (2005). Fungal community analysis by large-scale sequencing of environmental samples. *Appl. Environ. Microbiol., 71*, 5544–5550.

O'Donnell, K., Cigelnik, E., & Benny, G. L. (1998). Phylogenetic relationships among the Harpellales and Kickxellales. *Mycologia, 90*, 624–639.

Omoto, C., & McCoy, C. W. (1998). Toxicity of purified fungal toxin hirsutellin A to the citrus rust mite *Phyllocoptruta oleivora* (Ash.). *J. Invertebr. Pathol., 72*, 319–322.

Onstad, D. W., Fuxa, J. R., Humber, R. A., Oestergaard, J., Shapiro-Ilan, D. I., Gouli, V. V., Anderson, R. S., Andreadis, T. G., & Lacey, L. A. (2006). *An Abridged Glossary of Terms Used in Invertebrate Pathology* (3rd ed.). Society for Invertebrate Pathology. Available online: http://www.sipweb.org/glossary.

Ormond, E. L., Thomas, A. P. M., Pugh, P. J. A., Pell, J. K., & Roy, H. E. (2010). A fungal pathogen in time and space: the population dynamics of *Beauveria bassiana* in a conifer forest. *FEMS Microbiol. Ecol., 74*, 146–154.

Ownley, B. H., Gwinn, K. D., & Vega, F. E. (2010). Endophytic fungal entomopathogens with activity against plant pathogens: ecology and evolution. *BioControl, 55*, 113–128.

Packer, C., Holt, R. D., Hudson, P. J., Lafferty, K. D., & Dobson, A. P. (2003). Keeping the herds healthy and alert: implications of predator control for infectious disease. *Ecol. Lett., 6*, 797–802.

Panaccione, D. G. (2005). Origins and significance of ergot alkaloid diversity in fungi. *FEMS Microbiol. Lett., 25*, 9–17.

Papierok, B., & Hajek, A. E. (1997). Fungi: Entomophthorales. In L. A. Lacey (Ed.), *Manual of Techniques in Insect Pathology* (pp. 187–212). London: Academic Press.

Patrick, M., Adlard, M. W., & Keshavarz, T. (1993). Production of an indolizidine alkaloid, swainsonine by the filamentous fungus, *Metarhizium anisopliae. Biotechnol. Lett., 15*, 997–1000.

Patwardhan, A., Gandhe, R., Ghole, V., & Mourya, D. (2005). Larvicidal activity of the fungus *Aphanomyces* (Oomycetes: Saprolegniales) against *Culex quinquefasciatus. J. Commun. Dis., 37*, 269–274.

Pava-Ripoll, M., Posada, F. J., Momem, B., Wang, C., & St. Leger, R. (2008). Increased pathogenicity against coffee berry borer, *Hypothenemus hampei* (Coleoptera: Curculionidae) by *Metarhizium anisopliae* expressing the scorpion toxin (AaIT) gene. *J. Invertebr. Pathol., 99*, 220–226.

Pava-Ripoll, M., Angelini, C., Fang, W., Wang, S., Posada, F. J., & St. Leger, R. (2011). The rhizosphere-competent entomopathogen *Metarhizium anisopliae* expresses a specific subset of genes in plant root exudate. *Microbiology, 157*, 47–55.

Pedras, M. S. C., Zaharia, L. I., & Ward, D. E. (2002). The destruxins: synthesis, biosynthesis, biotransformation, and biological activity. *Phytochemistry, 59*, 579–596.

Pedrini, N., Crespo, R., & Juárez, M. P. (2007). Biochemistry of insect epicuticle degradation by entomopathogenic fungi. *Comp. Biochem. Physiol. C, 146*, 124–137.

Pell, J. K., Hannam, J. J., & Steinkraus, D. C. (2010). Conservation biological control using fungal entomopathogens. *BioControl, 55*, 187–198.

Pendland, J. C., & Boucias, D. G. (1992). Ultrastructural localization of carbohydrate in cell walls of the entomogenous hyphomycete *Nomuraea rileyi. Can. J. Microbiol., 38*, 377–386.

Pendland, J. C., & Boucias, D. G. (1993). Variations in the ability of galactose and mannose-specific lectins to bind to cell wall surfaces during growth of the insect pathogenic fungus *Paecilomyces farinosus. Eur. J. Cell Biol., 60*, 322–330.

Pendland, J. C., Hung, S.-Y., & Boucias, D. G. (1993). Evasion of host defense by *in vivo*-produced protoplast-like cells of the insect mycopathogen *Beauveria bassiana. J. Bacteriol., 175*, 5962–5969.

Phillips, A. J., Anderson, V. L., Robertson, E. J., Secombes, C. J., & van West, P. (2008). New insights into animal pathogenic oomycetes. *Trends Microbiol., 16*, 13–19.

Poinar, G. O., Jr., & Thomas, G. M. (1982). An entomophthoralean fungus from Dominican amber. *Mycologia, 74*, 332–334.

Poinar, G. O., Jr., & Thomas, G. M. (1984). A fossil entomogenous fungus from Dominican amber. *Experientia, 40*, 578–579.

Polis, G. A., & Holt, R. D. (1992). Intraguild predation – the dynamics of complex trophic interactions. *Trends Ecol. Evol., 7*, 151–154.

Polis, G. A., Myers, C. A., & Holt, R. D. (1989). The ecology and evolution of intraguild predation: potential competitors that eat each other. *Annu. Rev. Ecol. Syst., 20*, 297–330.

Pontoppidan, M. B., Himaman, W., Hywel-Jones, N. L., Boomsma, J. J., & Hughes, D. P. (2009). Graveyards on the move: the spatio-temporal distribution of dead *Ophiocordyceps*-infected ants. *PLoS ONE 4* e4835.

Prior, C., Jollands, P., & le Patourel, G. (1988). Infectivity of oil and water formulations of *Beauveria bassiana* (Deuteromycotina: Hyphomycetes) to the cocoa weevil pest *Pantorhytes plutus* (Coleoptera: Curculionidae). *J. Invertebr. Pathol., 52*, 66–72.

Priyatno, T. P., & Ibrahim, Y. B. (2002). Free fatty acids on the integument of the striped flea beetle, *Phyllotreta striolata* F., and their effects on conidial germination of the entomopathogenic fungi *Metarhizium anisopliae, Beauveria bassiana* and *Paecilomyces fumosoroseus. Pertanika J. Trop. Agric. Sci., 25*, 115–120.

van der Putten, W. H., Vet, L. E. M., Harvey, J. A., & Wickers, F. L. (2001). Linking above- and belowground multitrophic interactions of plants, herbivores, pathogens, and their antagonists. *Trends Ecol. Evol., 16*, 547–554.

Qin, X., Evans, J. D., Aronstein, K. A., Murray, K. D., & Weinstock, G. M. (2006). Genome sequences of the honey bee pathogens *Paenibacillus larvae* and *Ascosphaera apis. Insect Mol. Biol., 15*, 715–718.

Qin, Y., Ying, S.-H., Chen, Y., Shen, Z.-C., & Feng, M.-G. (2010). Integration of insecticidal protein Vip3Aa1 into *Beauveria bassiana* enhances fungal virulence to *Spodoptera litura* larvae by cuticle and *per os* infection. *Appl. Environ. Microbiol., 76*, 4611–4618.

Quesada-Moraga, E., & Vey, A. (2004). Bassiacridin, a protein toxic for locusts secreted by the entomopathogenic fungus *Beauveria bassiana. Mycol. Res., 108*, 441–452.

Quesada-Moraga, E., Navas-Cortés, J. A., Maranhao, E. A. A., Ortiz-Urquiza, A., & Santiago-Alvarez, C. (2007). Factors affecting the occurrence and distribution of entomopathogenic fungi in natural and cultivated soils. *Mycol. Res., 111*, 947–966.

Raffel, T. R., Martin, L. B., & Rohr, J. R. (2008). Parasites as predators: unifying natural enemy ecology. *Trends Ecol. Evol., 23*, 610–618.

Ramirez, R. A., & Snyder, W. E. (2009). Scared sick? Predator-pathogen facilitation enhances exploitation of a shared resource. *Ecology, 90*, 2832–2839.

Rangel, D. E. N., Braga, G. U. L., Anderson, A. J., & Roberts, D. W. (2005). Variability in conidial thermotolerance of *Metarhizium anisopliae* isolates from different geographic regions. *J. Invertebr. Pathol., 88*, 116–125.

Rangel, D. E. N., Anderson, A. J., & Roberts, D. W. (2008). Evaluating physical and nutritional stress during mycelial growth as inducers of tolerance to heat and UV-B radiation in *Metarhizium anisopliae* conidia. *Mycol. Res., 112*, 1362–1372.

Rasmann, S., Kollner, T. G., Degenhardt, J., Hiltpold, I., Toepfer, S., Kuhlmann, U., Gershenzon, J., & Turlings, T. C. J. (2005). Recruitment of entomopathogenic nematodes by insect-damaged maize roots. *Nature, 434*, 732–737.

Ravensberg, W. J. (2011). *A Roadmap to the Successful Development and Commercialization of Microbial Pest Control Products for Control of*

Arthropods. In: *Progress in Biological Control*, Vol. 10. Dordrecht: Springer.

Read, N. D., Fleißner, A., Roca, M. G., & Glass, N. L. (2010). Hyphal fusion. In K. A. Borkovich & D. J. Ebbole (Eds.), *Cellular and Molecular Biology of Filamentous Fungi* (pp. 260–273). Washington, DC: American Society for Microbiology Press.

Reay, S. D., Brownbridge, M., Cummings, N. J., Nelson, T. L., Souffre, B., Lignon, C., & Glare, T. R. (2008). Isolation and characterization of *Beauveria* spp. associated with exotic bark beetles in New Zealand *Pinus radiata* plantation forests. *Biol. Control, 46*, 484–494.

Rehner, S. A. (2009). Molecular systematics of entomopathogenic fungi. In S. P. Stock, J. Vandenberg, I. Glazer & N. Boemare (Eds.), *Insect Pathogens: Molecular Approaches and Techniques* (pp. 145–165). Wallingford: CABI.

Rehner, S. A., & Buckley, E. P. (2005). A *Beauveria* phylogeny inferred from nuclear ITS and EF1-alpha sequences: evidence for cryptic diversification and links to *Cordyceps* teleomorphs. *Mycologia, 97*, 84–98.

Rehner, S. A., Posada, F., Buckley, E. P., Infante, F., Castillo, A., & Vega, F. E. (2006). Phylogenetic origins of African and neotropical *Beauveria bassiana s.l.* pathogens of the coffee berry borer, *Hypothenemus hampei*. *J. Invertebr. Pathol., 93*, 11–21.

Rehner, S. A., Minnis, A. M., Sung, G.-H., Luangsa-ard, J. J., Devotto, L., & Humber, R. A. (2011). Phylogeny and systematics of the anamorphic, entomopathogenic genus *Beauveria*. *Mycologia, 103*, 1055–1073.

Roberts, D. W., & Campbell, A. S. (1977). Stability of entomopathogenic fungi. *Misc. Publ. Entomol. Soc. Am., 10*, 19–76.

Roberts, D. W., & Humber, R. A. (1981). Entomogenous fungi. In G. T. Cole & B. Kendrick (Eds.), *Biology of Conidial Fungi*, Vol. 2 (pp. 201–236). New York: Academic Press.

Roddam, L. F., & Rath, A. C. (1997). Isolation and characterization of *Metarhizium anisopliae* and *Beauveria bassiana* from subantarctic Macquarie Island. *J. Invertebr. Pathol., 69*, 285–288.

Rohlfs, M., & Churchill, A. C. L. (2011). Fungal secondary metabolites as modulators of interactions with insects and other arthropods. *Fungal Gen. Ecol., 48*, 23–34.

Rolff, J. & Reynolds, S. E. (Eds.). (2009). *Insect Infection and Immunity: Evolution, Ecology and Mechanisms*. Oxford: Oxford University Press.

Rosengaus, R. B., Maxmen, A. B., Coates, L. E., & Traniello, J. F. A. (1998). Disease resistance: a benefit of sociality in the dampwood termite *Zootermopsis angusticollis* (Isoptera: Termopsidae). *Behav. Ecol. Sociobiol., 44*, 125–134.

Rosenheim, J. A., Kaya, H. K., Ehler, L. E., Marois, J. J., & Jaffee, B. A. (1995). Intraguild predation among biological-control agents: theory and evidence. *Biol. Control, 5*, 303–335.

Rossman, A. Y. (1978). *Podonectria*, a genus in the Pleosporales on scale insects. *Mycotaxon, 7*, 163–182.

Roy, H. E., Pell, J. K., Clark, S. J., & Alderson, P. G. (1998). Implications of predator foraging on aphid pathogen dynamics. *J. Invertebr. Pathol., 71*, 236–247.

Roy, H. E., Pell, J. K., & Alderson, P. G. (2001). Targeted dispersal of the aphid pathogenic fungus *Erynia neoaphidis* by the aphid predator *Coccinella septempunctata*. *Biocontrol Sci. Technol., 11*, 99–110.

Roy, H. E., Steinkraus, D., Eilenberg, J., Hajek, A. E., & Pell, J. K. (2006). Bizarre interactions and endgames: entomopathogenic fungi and their arthropod hosts. *Annu. Rev. Entomol., 51*, 331–357.

Roy, H. E., Vega, F. E., Chandler, D., Goettel, M. S., Pell, J. K. & Wajnberg, E. (Eds.). (2010). *The Ecology of Fungal Entomopathogens*. Dordrecht: Springer.

Saikkonen, K., Saari, S., & Helander, M. (2010). Defensive mutualism between plants and endophytic fungi? *Fungal Divers., 41*, 101–113.

Samuels, R. I., Reynolds, S. E., & Charnley, A. K. (1988a). Calcium channel activation of insect muscle by destruxins, insecticidal compounds produced by the entomopathogenic fungus *Metarhizium anisopliae*. *Comp. Biochem. Physiol. C, 90*, 403–412.

Samuels, R. I., Charnley, A. K., & Reynolds, S. E. (1988b). The role of destruxins in the pathogenicity of 3 strains of *Metarhizium anisopliae* for the tobacco hornworm *Manduca sexta*. *Mycopathologia, 104*, 51–58.

Schabel, H. G. (1976). Green muscardine disease of *Hylobius pales* (Herbst) (Coleoptera: Curculionidae). *Z. Ang. Ent., 81*, 413–421.

Scheepmaker, J. W. A., & Butt, T. M. (2010). Natural and released inoculum levels of entomopathogenic fungal biocontrol agents in soil in relation to risk assessment and in accordance with EU regulations. *Biocontrol Sci. Technol., 20*, 503–552.

Schoch, C. L., Sung, G.-H., López-Giráldez, F., Towsend, J. P., Miadlikowska, J., Hofstetter, V., Robbertse, B., Matheny, P. B., Kauff, F., Wang, Z., Gueidan, C., Andrie, R. M., Trippe, K., Ciufetti, L. M., Wynns, A., Fraker, E., Hodkinson, B. P., Bonito, G., Groenewald, J. Z., Arzanlou, M., de Hoog, G. S., Crous, P. W., Hewitt, D., Pfister, D. H., Peterson, K., Gryzenhout, M., Wingfield, M. J., Aptroot, A., Suh, S.-O., Blackwell, M., Hillis, D. M., Griffith, G. W., Castlebury, L. A., Rossman, A. Y., Lumbsch, H. T., Lücking, R., Büdel, B., Rauhut, A., Diederich, P., Ertz, D., Geiser, D. M., Hosaka, K., Inderbitzin, P., Kohlmeyer, J., Volkmann-Kohlmeyer, B., Mostert, L., O'Donnell, K., Sipman, H., Rogers, J. D., Shoemaker, R. A., Sugiyama, J., Summerbell, R. C., Untereiner, W., Johnston, P. R., Stenroos, S., Zuccaro, A., Dyer, P. S., Crittenden, P. D., Cole, M. S., Hansen, K., Trappe, J. M., Yahr, R., Lutzoni, F., & Spatafora, J. W. (2009). The Ascomycota tree of life: a phylum-wide phylogeny clarifies the origin and evolution of fundamental reproductive and ecological traits. *Syst. Biol., 58*, 224–239.

Schrank, A., & Vainstein, M. H. (2010). *Metarhizium anisopliae* enzymes and toxins. *Toxicon, 56*, 1267–1274.

Screen, S. E., Hu, G., & St. Leger, R. J. (2001). Transformants of *Metarhizium anisopliae* sf. *anisopliae* overexpressing chitinase from *Metarhizium anisopliae* sf. *acridum* show early induction of native chitinase but are not altered in pathogenicity to *Manduca sexta*. *J. Invertebr. Pathol., 78*, 260–266.

Shah, F. A., Allen, N., Wright, C. J., & Butt, T. M. (2007). Repeated *in vitro* subculturing alters spore surface properties and virulence of *Metarhizium anisopliae*. *FEMS Microbiol. Lett., 276*, 60–66.

Shah, P. A., & Pell, J. K. (2003). Entomopathogenic fungi as biological control agents. *Appl. Microbiol. Biotechnol., 61*, 413–423.

Shah, P. A., Douro-Kpindou, O.-K., Sidibe, A., Daffe, C. O., van der Pauuw, H., & Lomer, C. J. (1998). Effects of the sunscreen oxybenzone on field efficacy and persistence of *Metarhizium flavoviride* conidia against *Kraussella amabile* (Orthoptera: Acrididae) in Mali, West Africa. *Biocontrol Sci. Technol., 8*, 357–364.

Shapiro-Ilan, D. I., Fuxa, J. R., Lacey, L. A., Onstad, D. W., & Kaya, H. K. (2005). Definition of pathogenicity and virulence in invertebrate pathology. *J. Invertebr. Pathol., 88*, 1–7.

Siegert, N. W., McCullough, D. G., Venette, R. C., Hajek, A. E., & Andresen, J. A. (2009). Assessing the climatic potential for epizootics of the gypsy moth fungal pathogens *Entomophaga maimaiga* in the North Central United States. *Can. J. For. Res., 39,* 1958–1970.

Smith, R. J., & Grula, E. A. (1981). Nutritional requirements for conidial germination and hyphal growth of *Beauveria bassiana. J. Invertebr. Pathol., 37,* 222–230.

Smith, R. J., & Grula, E. A. (1982). Toxic components on the larval surface of the corn earworm (*Heliothis zea*) and their effects on germination and growth of *Beauveria bassiana. J. Invertebr. Pathol., 39,* 15–22.

Soper, R. S., Shimazu, M., Humber, R. A., Ramos, M. E., & Hajek, A. E. (1988). Isolation and characterization of *Entomophaga maimaiga* sp. nov., a fungal pathogen of gypsy moth, *Lymantria dispar,* from Japan. *J. Invertebr. Pathol., 51,* 229–241.

Sosa-Gomez, D. R., Boucias, D. G., & Nation, J. L. (1997). Attachment of *Metarhizium anisopliae* to the southern green stink bug *Nezara viridula* cuticle and fungistatic effect of cuticular lipids and aldehydes. *J. Invertebr. Pathol., 69,* 31–39.

Sosa-Gómez, D. R., Lastra, C. C., & Humber, R. A. (2010). An overview of arthropod associated fungi from Argentina and Brazil. *Mycopathologia, 170,* 61–76.

Spatafora, J. W., and *Cordyceps* Working Group. (2011). *Cordyceps.US.* An Electronic Monograph of *Cordyceps* and Related Fungi. Available online: http://cordyceps.us/.

Spatafora, J. W., Sung, G.-H., Sung, J.-M., Hywel-Jones, N., & White, J. F., Jr. (2007). Phylogenetic evidence for an animal pathogen origin of ergot and the grass endophytes. *Mol. Ecol., 16,* 1701–1711.

Speare, A. T. (1920). On certain entomogenous fungi. *Mycologia, 12,* 62–76.

Speare, A. T., & Colley, R. H. (1912). *The Artificial Use of the Brown-tail Fungus in Massachusetts with Practical Suggestions for Private Experiment, and a Brief Note on a Fungous Disease of the Gypsy Caterpillar.* Boston: Wight & Potter Printing Company.

St. Leger, R. J. (1991). Integument as a barrier to microbial infections. In K. Binnington & A. Retnakaran (Eds.), *Physiology of the Insect Epidermis* (pp. 284–306). Australia: CSIRO.

St. Leger, R. J. (2008). Studies on adaptations of *Metarhizium anisopliae* to life in the soil. *J. Invertebr. Pathol., 98,* 271–276.

St. Leger, R. J., & Wang, C. (2009). Entomopathogenic fungi and the genomics era. In S. P. Stock, J. Vandenberg, I. Glazer & N. Boemare (Eds.), *Insect Pathogens: Molecular Approaches and Techniques* (pp. 365–400). Wallingford: CABI.

St. Leger, R. J., & Wang, C. (2010). Genetic engineering of fungal biocontrol agents to achieve efficacy against insect pests. *Appl. Microbiol. Biotechnol., 85,* 901–907.

St. Leger, R. J., Joshi, L., Bidochka, M. J., & Roberts, D. W. (1996). Construction of an improved mycoinsecticide overexpressing a toxic protease. *Proc. Natl. Acad. Sci. USA, 93,* 6349–6354.

St. Leger, R. J., Wang, C., & Fang, W. (2011). New perspectives on insect pathogens. *Fungal Biol. Rev., 25,* 84–88.

Steenberg, T. (1995). *Natural occurrence of Beauveria bassiana (Bals.) Vuill. with focus on infectivity to Sitona species and other insects in lucerne.* Denmark: Royal Veterinary and Agricultural University. PhD thesis.

Steenkamp, E. T., Wright, J., & Baldauf, S. L. (2006). The protistan origins of animals and fungi. *Mol. Biol. Evol., 23,* 93–106.

Steinhaus, E. A., & Martignoni, M. E. (1970). *An Abridged Glossary of Terms Used in Invertebrate Pathology* (2nd ed.). Pacific Northwest Forest and Range Experiment Station, Portland, Oregon: USDA Forest Service.

Strasser, H., Vey, A., & Butt, T. M. (2000). Are there any risks in using entomopathogenic fungi for pest control, with particular reference to the bioactive metabolites of *Metarhizium, Tolypocladium,* and *Beauveria* species? *Biocontrol Sci. Technol., 10,* 717–735.

Straub, C. S., Finke, D. L., & Snyder, W. E. (2008). Are the conservation of natural enemy biodiversity and biological control compatible goals? *Biol. Control, 45,* 225–237.

Suh, S.-O., Noda, H., & Blackwell, M. (2001). Insect symbiosis: derivation of yeast-like endosymbionts within an entomopathogenic lineage. *Mol. Biol. Evol., 18,* 995–1000.

Sun, B.-D., Yu, H.-Y., Chen, A. J., & Liu, X.-Z. (2008). Insect-associated fungi in soils of field crops and orchards. *Crop Prot., 27,* 1421–1426.

Sung, G.-H., & Spatafora, J. W. (2004). *Cordyceps cardinallis* sp. nov., a new species of *Cordyceps* with an east Asian-eastern North American distribution. *Mycologia, 96,* 658–666.

Sung, G.-H., Hywel-Jones, N. L., Sung, J.-M., Luangsa-ard, J. J., Srestha, B., & Spatafora, J. W. (2007). Phylogenetic classification of *Cordyceps* and the clavicipitaceous fungi. *Stud. Mycol., 57,* 5–59.

Sung, G.-H., Poinar, G. O., Jr., & Spatafora, J. W. (2008). The oldest fossil evidence of animal parasitism by fungi supports a Cretaceous diversification of fungal–arthropod symbioses. *Mol. Phylogenet. Evol., 49,* 49–502.

Suzuki, A., Kanaoka, M., Isogai, A., Murakoshi, S., Ichinoe, M., & Tamura, S. (1977). Bassianolide, a new insecticidal cyclodepsipeptide from *Beauveria bassiana* and *Verticillium lecanii. Tetrahedron Lett., 25,* 2167–2170.

Sweeney, A. W. (1981). An undescribed species of *Smittium* (Trichomycetes) pathogenic to mosquito larvae in Australia. *Trans. Br. Mycol. Soc., 77,* 55–60.

Takahashi, S., Karinuma, N., Uchida, K., Hashimoto, R., Yanagisawa, T., & Nakagawa, A. (1998a). Pyridovericin and pyridomacrolidin: novel metabolites from entomopathogenic fungi, *Beauveria bassiana. J. Antibiot., 51,* 596–598.

Takahashi, S., Uchida, K., Karinuma, N., Hashimoto, R., Yanagisawa, T., & Nakagawa, A. (1998b). The structures of pyridovericin and pyridomacrolidin, new metabolites from the entomopathogenic fungus, *Beauveria bassiana. J. Antibiot., 51,* 1051–1054.

Tanada, Y., & Fuxa, J. R. (1987). The pathogen population. In J. R. Fuxa & Y. Tanada (Eds.), *Epizootiology of Insect Diseases* (pp. 113–157). New York: Wiley Interscience.

Tanada, Y., & Kaya, H. K. (1993). *Insect Pathology.* San Diego: Academic Press.

Taylor, J. W. (1995). Making the Deuteromycota redundant: a practical integration of mitosporic and meiosporic fungi. *Can. J. Bot., 73S,* S754–S759.

Taylor, J. W., & Berbee, M. L. (2006). Dating divergences in the fungal tree of life: review and new analysis. *Mycologia, 98,* 838–849.

Taylor, T. N., Haas, H., Kerp, H., Krings, M., & Hanlin, R. T. (2005). Perithecial ascomycetes from the 400 million year old Rhynie chert: an example of ancestral polymorphism. *Mycologia, 97,* 269–285.

Tevini, M. (1993). Molecular biological effects of ultraviolet radiation. In M. Tevini (Ed.), *UV-B Radiation and Ozone Depletion: Effects on Humans, Animals, Plants, Microorganisms, and Materials* (pp. 1–15). Boca Raton: Lewis Publishers.

Thomas, K. C., Khachatourians, G. G., & Ingledew, W. M. (1987). Production and properties of *Beauveria bassiana* conidia cultivated in submerged culture. *Can. J. Microbiol., 33,* 12–20.

Thomas, M. B., Blanford, S., & Lomer, C. J. (1997). Reduction of feeding by the variegated grasshopper, *Zonocerus variegatus*, following infection by the fungal pathogen, *Metarhizium flavoviride*. *Biocontrol Sci. Technol., 7*, 327–334.

Thomas, M. B., Blanford, S., Gbongboui, C., & Lomer, C. J. (1998). Experimental studies to evaluate spray applications of a mycoinsecticide against the rice grasshopper, *Hieroglyphus daganensis*, in northern Benin. *Entomol. Exp. Appl., 87*, 93–102.

Thomas, S. R., & Elkinton, J. S. (2004). Pathogenicity and virulence. *J. Invertebr. Pathol., 85*, 146–151.

de la Torre, M., & Cárdenas-Cota, H. M. (1996). Production of *Paecilomyces fumosoroseus* conidia in submerged culture. *Entomophaga, 41*, 443–453.

Tosi, S., Caretta, G., & Humber, R. A. (2004). *Conidiobolus antarcticus*, a new species from continental Antarctica. *Mycotaxon, 90*, 343–347.

Tucker, D. L., Beresford, C. H., Sigler, L., & Rogers, K. (2004). Disseminated *Beauveria bassiana* infection in a patient with acute lymphoblastic leukemia. *J. Clin. Microbiol., 42*, 5412–5414.

Turlings, T. C. J., Tumlinson, J. H., & Lewis, W. J. (1990). Exploitation of herbivore-induced plant odors by host-seeking parasitic wasps. *Science, 250*, 1251–1253.

Uchida, R., Imasato, R., Yamaguchi, Y., Masuma, R., Shiomi, K., Tomoda, H., & Omura, S. (2005). New insecticidal antibiotics, hydroxyfungerins A and B, produced by *Metarhizium* sp. FK1-1079. *J. Antibiot., 58*, 804–809.

Undeen, A. H., & Nolan, R. A. (1977). Ovarian infection and fungal spore oviposition in the blackfly *Prosimulium mixtum*. *J. Invertebr. Pathol., 30*, 97–98.

Vandenberg, J. D., Jackson, M. A., & Lacey, L. A. (1998). Relative efficacy of blastospores and aerial conidia of *Paecilomyces fumosoroseus* against the Russian wheat aphid. *J. Invertebr. Pathol., 72*, 181–183.

Vänninen, I. (1996). Distribution and occurrence of four entomopathogenic fungi in Finland: effect of geographical location, habitat type and soil type. *Mycol. Res., 100*, 93–101.

van Veen, F. J. F., Mueller, C. B., Pell, J. K., & Godfray, H. C. J. (2008). Food web structure of three guilds of natural enemies: predators, parasitoids and pathogens of aphids. *J. Anim. Ecol., 77*, 191–200.

Veen, K. H. (1968). Recherches sur la maladie, due à *Metarhizium anisopliae* chez le criquet pèlerin. *Mededel. Landbouwhogeschool Wageningen, 68*, 1–77.

Vega, F. E. (2008). Insect pathology and fungal endophytes. *J. Invertebr. Pathol., 98*, 277–279.

Vega, F. E., & Dowd, P. F. (2005). The role of yeasts as insect endosymbionts. In F. E. Vega & M. Blackwell (Eds.), *Insect–Fungal Associations: Ecology and Evolution* (pp. 211–243). New York: Oxford University Press.

Vega, F. E., Dowd, P. F., McGuire, M. R., Jackson, M. A., & Nelsen, T. C. (1997). *In vitro* effects of secondary plant compounds on germination of blastospores of the entomopathogenic fungus *Paecilomyces fumosoroseus* (Deuteromycotina: Hyphomycetes). *J. Invertebr. Pathol., 70*, 209–213.

Vega, F. E., Jackson, M. A., & McGuire, M. R. (1999). Germination of conidia and blastospores of *Paecilomyces fumosoroseus* on the cuticle of the silverleaf whitefly, *Bemisia argentifolii*. *Mycopathologia, 147*, 33–35.

Vega, F. E., Posada, F., Aime, M. C., Pava-Ripoll, M., Infante, F., & Rehner, S. A. (2008). Entomopathogenic fungal endophytes. *Biol. Control, 46*, 72–82.

Vega, F. E., Goettel, M. S., Blackwell, M., Chandler, D., Jackson, M. A., Keller, S., Koike, M., Maniania, N. K., Monzón, A., Ownley, B. H., Pell, J. K., Rangel, D. E. N., & Roy, H. E. (2009). Fungal entomopathogens: new insights on their ecology. *Fungal Ecol., 2*, 149–159.

Vey, A., & Fargues, J. (1977). Histological and ultrastructural studies of *Beauveria bassiana* infection in *Leptinotarsa decemlineata* larvae during ecdysis. *J. Invertebr. Pathol., 30*, 207–215.

Vey, A., Quiot, J. M., Mazet, I., & McCoy, C. W. (1993). Toxicity and pathology of crude broth filtrate produced by *Hirsutella thompsonii* var. *thompsonii* in shake culture. *J. Invertebr. Pathol., 61*, 131–137.

Vey, A., Hoagland, R. E., & Butt, T. M. (2001). Toxic metabolites of fungal biocontrol agents. In T. M. Butt, C. Jackson & N. Magan (Eds.), *Fungi as Biocontrol Agents* (pp. 311–346). Wallingford: CABI Publishing.

Vidal, C., Fargues, J., & Lacey, L. A. (1997). Intraspecific variability of *Paecilomyces fumosoroseus*: effect of temperature on vegetative growth. *J. Invertebr. Pathol., 70*, 18–26.

Vining, L. C., Kelleher, W. J., & Schwarting, A. E. (1962). Oosporein production by a strain of *Beauveria bassiana* originally identified as *Amanita muscaria*. *Can. J. Microbiol., 8*, 931–933.

Vyas, N., Dua, K. K., & Prakash, S. (2007). Efficacy of *Lagenidium giganteum* metabolites on mosquito larvae with reference to nontarget organisms. *Parasitol. Res., 101*, 385–390.

Wagenaar, M. W., Gibson, D. M., & Clardy, J. (2002). Akanthomycin, a new antibiotic pyridone from the entomopathogenic fungus *Akanthomyces gracilis*. *Org. Lett., 4*, 671–673.

Wanchoo, A., Lewis, M. W., & Keyhani, N. O. (2009). Lectin mapping reveals stage-specific display of surface carbohydrates in *in vitro* and haemolymph derived cells of the entomopathogenic fungus *Beauveria bassiana*. *Microbiology, 155*, 3121–3133.

Wang, C., & St. Leger, R. J. (2007a). The MAD1 adhesin of *Metarhizium anisopliae* links adhesion with blastospore production and virulence to insects, and the MAD2 adhesin enables attachment to plants. *Eukaryot. Cell, 6*, 808–816.

Wang, C., & St. Leger, R. J. (2007b). The *Metarhizium anisopliae* perilipin homolog MPL1 regulates lipid metabolism, appresorial turgor pressure, and virulence. *J. Biol. Chem., 282*, 21110–21115.

Wang, C., & St. Leger, R. J. (2007c). A scorpion neurotoxin increases the potency of a fungal insecticide. *Nat. Biotechnol., 25*, 1455–1456.

Wang, D. Y.-C., Kumar, S., & Blair Hedges, S. (1999). Divergence time estimates for the early history of animal phyla and the origin of plants, animals and fungi. *Proc. R. Soc. Lond. B, 266*, 163–171.

Washburn, J. O., Egerter, D. E., Anderson, J. R., & Saunders, G. A. (1988). Density reduction in larval mosquito (Diptera: Culicidae) populations by interactions between a parasitic ciliate (Ciliophora: Tetrahymenidae) and an opportunistic fungal (Oomycetes: Pythiaceae) parasite. *J. Med. Entomol., 25*, 307–314.

Wat, C.-H., McInnes, A. G., Smith, D. G., Wright, J. L. C., & Vining, L. C. (1977). The yellow pigments of *Beauveria* species. Structures of tenellin and bassianin. *Can. J. Chem., 55*, 4090–4098.

Watanabe, T. (2010). *Pictorial Atlas of Soil and Seed Fungi. Morphologies of Cultured Fungi and Key to Species* (3rd ed.). Boca Raton: CRC Press.

Watson, D. W., & Brandly, C. A. (1949). Virulence and pathogenicity. *Annu. Rev. Microbiol., 3*, 195–220.

Weckerle, C. S., Yang, Y., Huber, F. K., & Li, Q. (2010). People, money, and protected areas: the collection of the caterpillar mushroom

Ophiocordyceps sinensis in the Baima Xueshan Nature Reserve, Southwest China. *Biodiv. Conserv., 19,* 2685–2698.

Weete, J. D., Abril, M., & Blackwell, M. (2010). Phylogenetic distribution of fungal sterols. *PLoS ONE 5* e10899.

Weir, A., & Blackwell, M. (2005). Fungal biotrophic parasites of insects and other arthropods. In F. E. Vega & M. Blackwell (Eds.), *Insect–Fungal Associations: Ecology and Evolution* (pp. 119–145). New York: Oxford University Press.

Whisler, H. C., Zebold, S. L., & Shemanchuk, J. A. (1975). Life history of *Coelomomyces psorophorae*. *Proc. Natl. Acad. Sci. USA, 72,* 693–696.

White, M. M., James, T. Y., O'Donnell, K., Cafaro, M. J., Tanabe, Y., & Sugiyama, J. (2006a). Phylogeny of the Zygomycota based on nuclear ribosomal sequence data. *Mycologia, 98,* 872–884.

White, M. M., Lichtwardt, R. W., & Colbo, M. H. (2006b). Confirmation and identification of parasitic stages of obligate endobionts (Harpellales) in blackflies (Simuliidae) by means of rRNA sequence data. *Mycol. Res., 110,* 1070–1079.

Willis, J. H., Iconomidou, V. A., Smith, R. F., & Hamodrakas, S. J. (2005). Cuticular proteins. In L. I. Gilbert, K. Iatrou & S. S. Gill (Eds.), *Comprehensive Molecular Insect Science*, Vol. 4 (pp. 79–109). Oxford: Elsevier.

Wilson, K., & Reeson, A. F. (1998). Density-dependent prophylaxis: evidence from Lepidoptera–baculovirus interactions? *Ecol. Entomol., 23,* 100–101.

Wilson, K., Cotter, S. C., Reeson, A. F., & Pell, J. K. (2001). Melanism and disease resistance in insects. *Ecol. Lett., 4,* 637–649.

Wilson, K., Thomas, M. B., Blanford, S., Doggett, M., Simpson, S. J., & Moore, S. L. (2002). Coping with crowds: density-dependent disease resistance in desert locusts. *Proc. Natl. Acad. Sci. USA, 99,* 5471–5475.

van Winkelhoff, A. J., & McCoy, C. W. (1984). Conidiation of *Hirsutella thompsonii* var. *synnematosa* in submerged culture. *J. Invertebr. Pathol., 43,* 59–68.

Wraight, S. P., Jackson, M. A., & de Kock, S. L. (2001). Production, stabilization and formulation of fungal biocontrol agents. In T. M. Butt, C. Jackson & N. Magan (Eds.), *Fungi as Biocontrol Agents* (pp. 253–287). Wallingford: CABI Publishing.

Xu, Y., Orozco, R., Wijeratne, E. M. K., Gunatilaka, A. A. L., Stock, S. P., & Molnár, I. (2008). Biosynthesis of the cyclooligomer depsipeptide beauvericin, a virulence factor of the entomopathogenic fungus *Beauveria bassiana. Chem. Biol., 15,* 898–907.

Xu, Y., Orozco, R., Wijeratne, E. M. K., Espinosa-Artiles, P., Gunatilaka, A. A. L., Stock, S. P., & Molnár, I. (2009). Biosynthesis of the cyclooligomer depsipeptide bassianolide, an insecticidal virulence factor of *Beauveria bassiana. Fungal Genet. Biol., 46,* 353–364.

Yanagawa, A., Yokohari, F., & Shimizu, S. (2008). Defense mechanism of the termite, *Coptotermes formosanus* Shiraki, to entomopathogenic fungi. *J. Invertebr. Pathol., 97,* 165–170.

Yaginuma, K. (2007). Seasonal occurrence of entomopathogenic fungi in apple orchard not sprayed with insecticides. *Jpn. J. Appl. Entomol. Zool., 51,* 213–220.

Yek, S.-H., & Mueller, U. G. (2011). The metapleural gland of ants. *Biol. Rev., 91,* 201–224.

Zacharuk, R. Y. (1970). Fine structure of the fungus *Metarhizium anisopliae* infecting three species of larval Elateridae (Coleoptera). III. Penetration of the host integument. *J. Invertebr. Pathol., 15,* 372–396.

Zare, R., & Gams, W. (2001). A revision of *Verticillium* section *Prostrata.* IV. The genera *Lecanicillium* and *Simplicillium* gen. nov. *Nova Hedwigia, 73,* 1–50.

Zhang, S., Kim, B., Xia, Y.-Y., & Keyhani, N. O. (2011). Two hydrophobins are involved in fungal spore coat rodlet layer assembly and each play distinct roles in surface interactions, development and pathogenesis in the entomopathogenic fungus, *Beauveria bassiana. Mol. Microbiol., 80,* 811–826.

Zimmermann, G. (1982). Effect of high temperatures and artificial sunlight on the viability of conidia of *Metarhizium anisopliae. J. Invertebr. Pathol., 40,* 36–40.

Zimmermann, G. (1986). The *Galleria* bait method for detection of entomopathogenic fungi in soil. *J. Appl. Entomol., 102,* 213–215.

Zimmermann, G. (2007a). Review on safety of the entomopathogenic fungi *Beauveria bassiana* and *Beauveria brongniartii. Biocontrol Sci. Technol., 17,* 553–596.

Zimmermann, G. (2007b). Review of safety of the entomopathogenic fungus *Metarhizium anisopliae. Biocontrol Sci. Technol., 17,* 879–920.

Zimmermann, G. (2008). The entomopathogenic fungi *Isaria farinosa* (formerly *Paecilomyces farinosus*) and the *Isaria fumosorosea* species complex (formerly *Paecilomyces fumosoroseus*): biology, ecology and use in biological control. *Biocontrol Sci. Technol., 18,* 865–901.

Žižka, J., & Weiser, J. (1993). Effect of beauvericin, a toxic metabolite of *Beauveria bassiana*, on the ultrastructure of *Culex pipiens autogenicus* larvae. *Cytobios, 75,* 13–19.

Chapter 7

Microsporidian Entomopathogens

Leellen F. Solter,* James J. Becnel[†] and David H. Oi[†]

*Illinois Natural History Survey, University of Illinois, Champaign, Illinois, USA, [†] United States Department of Agriculture, Agricultural Research Service, Gainesville, Florida, USA

Chapter Outline

SUMMARY

Microsporidia, pathogenic protists related to the Fungi, are considered to be primary pathogens of many aquatic and terrestrial insect species and have important roles in insect population dynamics, managed insect disease, and biological control of insect pests. Hosts are infected when spores are ingested and/or by transmission via the eggs. When ingested, spores germinate in a unique fashion: a polar tube that is coiled within the spore rapidly everts and punctures the host midgut cells, injecting the spore contents into the cell cytoplasm. Mitochondria and Golgi bodies are lacking in these obligate intracellular pathogens, and energy is evidently extracted from host cells via direct uptake of ATP. Effects on the host are typically chronic; therefore, the use of microsporidia in biological control programs focuses on inoculative introductions, augmentative release, and conservation biology. This chapter reviews the biology, ecology, pathology, and classification of microsporidia with examples of several long-term research efforts to manipulate these pathogens for the suppression of insect pests.

7.1. INTRODUCTION

"Enigmatic", in the words of Maddox and Sprenkel (1978) and echoed by succeeding generations of researchers (Canning *et al.*, 1985; Bigliardi, 2001; Susko *et al.*, 2004;

Insect Pathology. DOI: 10.1016/B978-0-12-384984-7.00007-5

James *et al.*, 2006), aptly describes the group of fascinating but elusive pathogens comprising the phylum Microsporidia. This particular label remains a not-so-subtle indication of the difficulties encountered in studies of these organisms and the still perplexing phylogenetic relationships at all taxonomic levels. Nevertheless, it is important to consider the role of the microsporidia in natural systems and in biological control programs because they are nearly ubiquitous pathogens of insects and other arthropods. Most of the more than 1300 described species in approximately 186 genera are pathogens of invertebrate animals, with insects being type hosts of nearly half of described genera (Becnel and Andreadis, 1999). Close inspection of most insect taxa yields new species (Brooks, 1974). All microsporidia are obligate intracellular pathogens of protists and animals, including both warm-blooded and cold-blooded vertebrates. They do not reproduce as free-living organisms and they are not plant pathogens.

Microsporidia are important primary pathogens of both pest and beneficial insects. The first known microsporidium, *Nosema bombycis*, was described from the silkworm, *Bombyx mori*, by Nägeli (1857), who considered the pathogen to be a yeast. Symptoms caused by *N. bombycis* infection, such as dark spots on the larval integument, were first noted by Jean Louis Armand de Quatrefages in the early 1800s and named "pébrine" (pepper) disease. Louis Pasteur's studies of pébrine disease (Pasteur, 1870) provided methods of prevention and control and were credited with saving the silkworm industry in France (see Chapters 2 and 12).

The silkworm was the first beneficial insect noted to be devastated by microsporidia, but several microsporidian species have serious impacts on other managed insects that are reared in high-density colonies, as well as on natural populations of beneficial insects (Chapter 12). Two microsporidian species, *Nosema apis* and *N. ceranae*, are pathogens of honey bees; *N. ceranae* has been implicated as a contributing factor in the decline of managed bees (reviewed by Paxton, 2010), although effects of the disease reported in the literature are highly variable. *Nosema bombi* is a Holarctic generalist pathogen of bumble bees (*Bombus* spp.). High prevalence levels in some *Bombus* species that are apparently in decline have led to questions about the possible introduction of virulent strains and the potential role these pathogens play in permanently reducing *Bombus* spp. range and population numbers in North America (Thorp and Shepherd, 2005; Cameron *et al.*, 2011).

Microsporidia are also important as regulatory factors in populations of insect pests. Hajek *et al.* (2005) listed five microsporidian species that have been introduced as biological control agents globally. They include *Paranosema* (*Nosema*) *locustae* used as a microbial insecticide for grasshoppers and established in Argentina in several orthopteran species (see Section 7.4.4); *Vavraia* (*Pleistophora*) *culicis* introduced into *Culex* mosquito populations in Nigeria, where establishment was not confirmed; *Nosema pyrausta*, a pathogen of the European corn borer, *Ostrinia nubilalis*, introduced from Iowa as an augmentative control in Illinois *O. nubilalis* populations, and persisting in the Illinois populations; and the European gypsy moth (*Lymantria dispar*) pathogens *Nosema portugal* and *Endoreticulatus* sp. (*E. schubergi*, = *Vavraia* sp., = *pleistophora schubergi* introduced into an isolated gypsy moth population in Maryland. Persistence of *N. portugal* was confirmed for one year (see Section 7.4.6). While probably lacking utility as microbial insecticides, these typically chronic pathogens clearly have a role as major components of the natural enemy complex of many insect species (Kohler and Wiley, 1992; Maddox, 1994; Lewis *et al.*, 2009) and warrant continued study for their use in classical, augmentative and conservation biological control programs.

7.2. CLASSIFICATION AND PHYLOGENY

Only in the past 15 years have microsporidian genes been analyzed to the extent that placement of this eukaryotic pathogen group in the Protozoa by Balbiani (1882) was seriously questioned. Recent molecular studies placed the phylum Microsporidia within the kingdom Fungi (Keeling and Doolittle, 1996; Hirt *et al.*, 1999; Bouzat *et al.*, 2000; James *et al.*, 2006; Hibbett *et al.*, 2007), with evidence, albeit controversial, that the group is related to the entomopathogenic Zygomycetes (Keeling, 2003; Lee *et al.*, 2008; Corradi and Slamovits, 2010). Koestler and Ebersberger (2011) point out that although Microsporidia and Zygomycetes alone among the fungi share a three-gene cluster encoding a sugar transporter, a specific transcription factor, and RNA helicase, clustering of these genes appears to be ancestral in eukaryotes and is, therefore, not phylogenetically informative. Questions about the phylogenetic placement of microsporidia thus remain without definitive answers. Although the predominance of evidence places microsporidia within the fungal lineage (Corradi and Keeling, 2009), their morphology, biology, and host interactions are unique. In this chapter, the microsporidia are treated as a separate group of insect pathogens.

The first modern classification for the microsporidia was proposed by Tuzet *et al.* (1971) and since that time, the classification of the microsporidia at all taxonomic levels has been controversial (reviews by Sprague, 1977a; Sprague *et al.*, 1992; Tanada and Kaya, 1993; Franzen, 2008). The classical morphological approach (based on nuclear arrangement, spore shape, structure and size, nature of vesicles, etc.) and molecular-based methods have produced conflicting interpretations (Vossbrinck and Debrunner-Vossbrinck, 2005; Larsson, 2005). A phylogenetic tree was created using rDNA data from 125 species of microsporidia from vertebrate and invertebrate hosts (Vossbrinck and Debrunner-Vossbrinck, 2005). Five major clades were identified, and three major groups based on host habitat (freshwater, marine, and terrestrial) were suggested as being more consistent with evolutionary relationships than previous higher level classifications based on morphology and life cycle characteristics. Nevertheless, there were numerous examples of species that did not fit into the expected grouping, such as some true *Nosema* species (type host from Lepidoptera) reported from aquatic crustaceans.

Like habitat, host taxon is not always a reliable basis for microsporidian classification. Examples include the genus *Encephalitozoon*, for which all known species are pathogens of vertebrate animals with the exception of one species isolated from *Romalea microptera*, a lubber grasshopper (Lange *et al.*, 2009). Ribosomal RNA sequences from a crayfish microsporidium placed it in the genus *Vairimorpha* (Moodie *et al.*, 2003), previously described only from Lepidoptera and two other insect orders. In other genera, similar morphological characters among species that were isolated from unrelated host species have probably obscured true phylogenetic relationships. According to Vossbrinck and Debrunner-Vossbrinck (2005), "Molecular phylogenetic analysis has revealed that genera such as *Nosema*, *Vairimorpha*, *Amblyospora*, *Thelohania* and *Pleistophora* are polyphyletic in origin and efforts are being made to reclassify species unrelated to the type species." There are many microsporidian species that will need to be reassigned to new genera and families. Determining the critical cytological and genetic features is the next challenge in establishing a sound taxonomic system for the microsporidia.

7.2.1. Overview of Microsporidian Entomopathogens in a Phylogenetic Context

Insects are represented as type hosts of 90 microsporidian genera, nearly half of all those described (Table 7.1). Microsporidia isolated from insects have also been assigned to genera with type species from hosts in other taxonomic classes, and microsporidia from non-insect hosts have been assigned to genera with an insect type species. In both cases, additional information, primarily at the molecular level, will determine whether the generic assignments are valid with respect to the type species, and will perhaps clarify genera that may be restricted to a particular host group and those that have a broad host range. Examples of two microsporidian clades that represent a large portion of insect microsporidia are presented here.

Amblyospora/Parathelohania *Clade*

The *Amblyospora/Parathelohania* clade contains more than 122 of the approximately 150 species of microsporidia that have been described from mosquito hosts (*Amblyospora* more than 100 species, *Parathelohania* about 22 species), with at least 21 other genera reported from mosquitoes but none with more than two species (Andreadis, 2007). The *Amblyospora/Parathelohania* clade is therefore a very important and common group of microsporidia notable for a number of reasons. First, it is the most widely observed group of microsporidian pathogens in larval mosquitoes because the infections are easily recognized by the presence of large white cysts filled with spores located in the fat body. Second, *Amblyospora* and *Parathelohania* species are representative of polymorphic microsporidia (more than one sporulation sequence) and were the first species for which an intermediate copepod host was found to be involved in the life cycle (Becnel and Andreadis, 1999; Andreadis, 2007). The complex life cycles include alternations of haploid and diploid cell states (which usually involves meiosis) as well as horizontal and vertical (transovarial) transmission involving larval and adult mosquitoes and copepods. Members of this clade are usually host specific, with *Amblyospora* spp. mainly restricted to *Aedes/Culex* mosquitoes and *Parathelohania* spp. to *Anopheles* spp.

Small subunit ribosomal DNA sequences from a large number of species in the *Amblyospora/Parathelohania* clade have permitted an in-depth analysis of the phylogenetic relationships among the microsporidia, as well as the relationships to the host groups. The *Amblyospora/Parathelohania* and *Amblyospora/Parathelohania*-like genera form a very strongly supported clade, demonstrating the close relationship among these polymorphic microsporidia in mosquitoes (Vossbrinck *et al.*, 2004; Vossbrinck and Debrunner-Vossbrinck, 2005). Analysis has also shown a possible evolutionary correlation between *Aedes* and *Culex* hosts with *Amblyospora* spp. and between *Anopheles* hosts with *Parathelohania* spp. (Baker *et al.*, 1998; Vossbrinck *et al.*, 2004). Not all species within this clade involve an intermediate host (notably *Edhazardia aedis* and *Culicospora magna*) but otherwise have complex life cycles that include both horizontal and vertical transmission. Studies of species within the *Amblyospora/Parathelohania* clade have made significant contributions to an overall understanding of microsporidian biology relative to developmental sequences and complete life cycles. As additional genomic data are generated on this and other groups of microsporidia, a better understanding should emerge of phylogenetic relationships, as well as information on sexuality, host range determinants, and evasion of host immune response, among many other basic biological features.

TABLE 7.1 Genera of microsporidian type species isolated from insects; listed by insect host taxonomic order.

Insect Order	Microsporidian Genus	Insect Order	Microsporidian Genus
Collembola	Auraspora	Diptera	Tricornia
Diptera	Aedispora		Tubilinosema
	Amblyospora		Vavraia
	Andreanna		Weiseria
	Anisofilariata		Pegmatheca
	Bohuslavia		Pernicivesicula
	Campanulospora		Pilosporella
	Caudospora		Polydispyrenia
	Chapmanium		Ringueletium
	Coccospora		Scipionospora
	Crepidulospora		Semenovaia
	Crispospora		Senoma
	Cristulospora		Simuliospora
	Culicospora		Spherospora
	Culicosporella		Spiroglugea
	Cylindrospora		Striatospora
	Dimeiospora		Systenostrema
	Edhazardia		Tabanispora
	Evlachovaia		Toxoglugea
	Flabelliforma		Toxospora
	Golbergia		Trichoctosporea
	Hazardia	Coleoptera	Anncaliia
	Helmichia		Cannngia
	Hessea		Chytridiopsis
	Hirsutusporos		Endoreticulatus
	Hyalinocysta		Ovavesicula
	Intrapredatorus	Ephemeroptera	Mitoplistophora
	Janacekia		Pankovaia
	Krishtalia		Stempellia
	Merocinta		Telomyxa
	Napamichum		Trichoduboscqia
	Neoperezia	Hemiptera	Becnelia
	Octosporea	Hymenoptera	Antonospora
	Octotetraspora		Burenella
	Parapleistophora		Kneallhazia
	Parastempellia	Isoptera	Duboscqia
	Parathelohania		

TABLE 7.1 Genera of microsporidian type species isolated from insects; listed by insect host taxonomic order—cont'd

Insect Order	Microsporidian Genus	Insect Order	Microsporidian Genera Totals
Lepidoptera	*Cystosporogenes*	Collembola	1
	Larssoniella	Diptera	57
	Nosema	Coleoptera	5
	Orthosomella	Ephemeroptera	5
	Vairimorpha	Hemiptera	1
Odonata	*Nudispora*	Hymenoptera	3
	Resiomeria	Isoptera	1
Orthoptera	*Heterovesicula*	Lepidoptera	5
	Johenrea	Odonata	2
	Liebermannia	Orthoptera	4
	Paranosema	Siphonaptera	2
Siphonaptera	*Nolleria*	Thysanura	1
	Pulicispora	Trichoptera	3
Thysanura	*Buxtehudea*		
Trichoptera	*Episeptum*	**Total**	**90**
	Issia		
	Tardivesicula		

Nosema/Vairimorpha *Clade*

Microsporidia in the *Nosema/Vairimorpha* clade are common and important pathogens of insects with terrestrial life cycles and are particularly well represented in Lepidoptera. The type microsporidian species, *Nosema bombycis*, is currently described (formally and informally) on the basis of having one mature infective spore type (*N. bombycis* is now known to produce a second spore type, an internally infective spore; see Section 7.2.2), no pansporoblastic membrane, diplokaryotic nuclei throughout the life cycle, and one sporont giving rise to two mature spores by binary fission. This relatively simple life cycle, or recognizable stages of it, resulted in well over 200 species descriptions and an additional 50 descriptions to the genus level (Sprague, 1977a). The genus *Vairimorpha*, several species of which were considered to be mixed infections of *Nosema* and *Thelohania* species (e.-g., *Nosema lymantriae* + *Thelohania similis* in the gypsy moth; *Nosema necatrix* + *Thelohania diazoma* in noctuid species), was recognized in the 1970s to be a polymorphic taxon with *Nosema*-like dikaryotic spores and *Thelohania*-like monokaryotic octospores (Maddox and Sprenkel, 1978), and a new genus was erected (Pilley, 1976).

Molecular data subsequently identified the close relationship of the monomorphic *Nosema* and polymorphic *Vairimorpha* (Baker *et al.*, 1994; Vossbrinck and Debrunner-Vossbrinck, 2005). The octospores are ancestral sexual meiospores that have been lost multiple times within the *Nosema/Vairimorpha* clade (Ironside, 2007), leaving approximately 10 *Vairimorpha*-type species represented within most closely related species groups of the *Nosema* clade. In addition to loss of meiospores, loss of diplokarya may have occurred in the *Nosema/Vairimorpha* clade. *Oligosporidium occidentalis*, a monokaryotic (or haplokarotic) species isolated from predatory mites (Becnel *et al.*, 2002), is a sister species to *N. bombi* (Vossbrinck and Debrunner-Vossbrinck, 2005).

Species in the *Nosema/Vairimorpha* group are primarily pathogens of Lepidoptera but have also been isolated from hosts in other insect orders. For example, species from Hymenoptera include *Nosema vespula* (wasps), *N. ceranae* and *N. apis* (honey bees), *N. bombi* (bumble bees), and *Vairimorpha invictae* (fire ants), as well as from other classes of arthropods including *Nosema granulosis* (copepods) and *V. cheracis* (crayfish). The genus *Nosema* has, however, long been recognized as a "catch-all" group for microsporidia with similar life cycles, and a large number of species has recently been transferred to other genera based on molecular characters. Additional genetic information should lead to a new characterization for the genus *Nosema* or possible new genera for the larger clade.

Species descriptions for the phylum Microsporidia are currently governed by the International Code of Zoological Nomenclature. Microsporidian taxonomists, mycologists, and other researchers agree that, although the Fungi are classified under the Code of Botanical Nomenclature, the Microsporidia should remain with the zoological code (Weiss, 2005; Redhead *et al.*, 2009).

7.2.2. General Characteristics of Microsporidia

Microsporidia, as obligate intracellular pathogens, utilize host tissues for reproductive energy and development. Organelles that are typically found in eukaryotic organisms, including peroxisomes, vesicular Golgi membranes, and mitochondria, are lacking in microsporidia. Thin Golgi-like tubules transport proteins from the endoplasmic reticulum, but no Golgi vesicles form (Beznoussenko *et al.*, 2007). Vestigial mitochondria are present in the form of mitosomes that may retain some capability for metabolic import (Williams and Keeling, 2005; Burri *et al.*, 2006). At a minimum, the microsporidia produce two or three reproductive vegetative stages (described below in Sections 7.2.2 and 7.3.1), and a mature, environmentally resistant, infective spore or "environmental spore" (Maddox *et al.*, 1999), although species exist for which no presporulation stages have been observed (Larsson, 1993). Additional characteristics that are detailed in this chapter include: (1) spore types, of which there may be one to four per species; (2) tissue tropism that ranges from utilization of one or a few host tissues to systemic infections; (3) effects on infected hosts, which vary from apparently benign to highly virulent; (4) transmission, which most typically occurs via oral ingestion of infective spores, infected female to offspring, or by both mechanisms; and (5) host specificity, ranging from highly host specific species to generalists, usually within an insect order; however, parasitoids and intermediate hosts in other orders or classes may also become infected.

Morphology

Environmental spores are readily observed in infected tissues with a light microscope ($\geq 250\times$) and are distinguished by a relatively featureless and smooth rounded surface (Fig. 7.1A) that results in birefringence under phase-contrast microscopy (Tiner, 1988). Mature spores possess a dense spore wall composed of a proteinaceous exospore and alpha-chitin and protein endospore layers (Vávra and Larsson, 1999; Y. Xu *et al.*, 2006), and are evenly bright in appearance. The endospore in aquatic species tends to be thinner than that of terrestrial species [compare transmission electron micrographs of mature microsporidian spores from aquatic hosts (Terry *et al.*, 1999; Micieli *et al.*, 2003; Nylund *et al.*, 2010) with those isolated from terrestrial hosts (Vávra *et al.*, 2006; Wang *et al.*, 2009; Sokolova *et al.*, 2010)], and the distal vacuole or spore contents may be vaguely visible. The general spore shapes, frequently oval, long oval, or egg, spindle, or tear shaped, are consistent within each sporulation sequence for a species, but there is much overlap among species and shape alone cannot be used for species identification. Spore length, also relatively consistent within a species (usually within 0.5 μm), ranges from 1.5 to 10 μm, with 2–6 μm being most common in insect hosts. Giemsa-stained environmental spores (Fig. 7.1B) appear white with blue-tinted spore walls and generally show a characteristic blue stain on the surface. Spores appearing gray under phase-contrast microscopy may be either immature or inviable (Fig. 7.1A) and, if recently germinated, may appear to be a "shell" with a gray interior and strongly visible spore walls (Fig. 7.1C). Some microsporidian species produce a "primary" spore, an internally infective stage that possesses a thin endospore and organelles that appear to be less well developed. Typical of the genus *Nosema* as well as some other genera, the primary spores are less refringent than environmental spores and may appear slightly more rounded in shape with a visible polar vacuole (Fig. 7.1C). Vegetative (reproducing) forms are difficult to detect under light microscopy, but when infections are intense, they may be observed as small (typically 2–10 μm), round cells with smooth cytoplasm and visible nuclei (Fig. 7.1D). Reproduction in some species entails a series of nuclear divisions that produce vegetative cells with multiple nuclei called plasmodia. These cells may be round (Fig. 7.2) or ribbon like.

The intracellular features of microsporidia are clearly discernible only in high-magnification transmission electron micrographs. Each environmental spore contains either a single nucleus (monokaryon) or a double nucleus (diplokaryon), depending on species. Some species possess both spore types over the course of the life cycle; for example, the terrestrial *Vairimorpha* species produce diplokaryotic spores and monokaryotic meiospores that are formed in a vesicle of eight spores (Moore and Brooks, 1992; Vávra *et al.*, 2006). The aquatic *Amblyospora* and *Edhazardia* species also produce both diplokaryotic and monokaryotic spores (Becnel, 1992a, 1994) (Fig. 7.3). The diplokaryotic form in those species that possess this configuration is considered to be genetically diploid. Vegetative forms may differ in number of nuclei from one to many, depending on the stage of development; whether monokaryotic or dikaryotic or developing as plasmodia

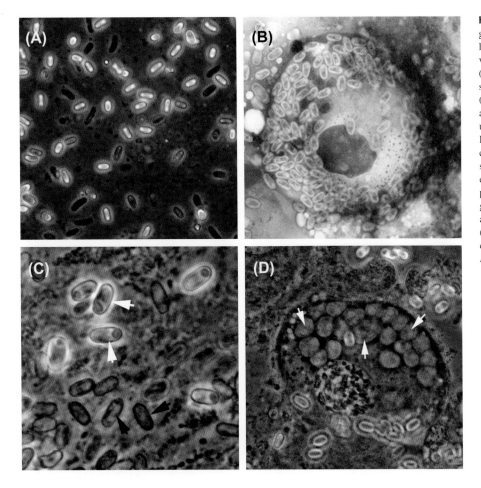

FIGURE 7.1 Phase-contrast micrographs of life cycle stages of a *Nosema*-like microsporidium isolated from black vine weevil, *Otiorynchus sulcatus.* (A) Environmental (brightly refractive) spores and immature (dark gray) spores (1000 ×); mature microsporidian spores are typically smooth and birefringent under phase-contrast microscopy. Immature spores lack a developed endospore and appear gray. (B) Giemsa stained environmental spores in a midgut epithelial cell (500 ×). (C) Group of primary spores (white arrows) and germinated primary spores (black arrows) in midgut epithelia (1000 ×). (D) Vegetative forms (arrows), in midgut epithelial cell (1000 ×). *(Photos by L. F. Solter.)*

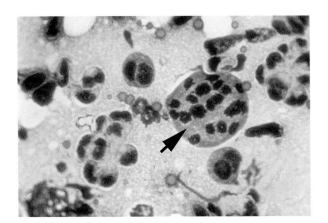

FIGURE 7.2 Giemsa stain of *Culicosporella lunata* vegetative plasmodium (black arrow). *(Photo by J. J. Becnel.)*

(Fig. 7.2); and whether division is by binary fission or multiple fission/budding.

Microsporidia possess uniquely identifying ultrastructural features (Fig. 7.4). A plate or anchoring disk at the anterior end of the spore attaches a specialized organelle, the polar filament, that coils around the interior of the

spore. This flagellum-like filament is everted to become the polar tube through which the spore contents are injected into a host cell (Frixione *et al.,* 1992). Polar tubes occasionally can be observed under light microscopy immediately after the spores have germinated. This structure is unique to the Microsporidia and defines the taxon (Sprague, 1977a). Layers of membranes, the polaroplast, at the anterior end of the spore appear to be involved in the germination process and may serve as the plasmalemma of the new vegetative stage (germ) infecting the host cell (Weidner *et al.,* 1984). The posterior of the spore is occupied by a posterior vacuole that appears to be involved in spore germination (Findley *et al.,* 2005). The fine structure of microsporidia is presented in a detailed review by Vávra and Larsson (1999), and the reader is referred to this excellent text for more detailed information.

Genetic Characters

Information about the microsporidian genome is changing rapidly with increasingly sensitive and affordable technology. The first sequenced genome, *Encephalitozoon cuniculi* (Katinka *et al.,* 2001), provided information about one of the

FIGURE 7.3 Transmission electron micrographs of (A) *Culicosporella lunata* diplokaryon; (B) a meiotic division in progress that will terminate in two cells, each monokaryotic. DN = diplokaryotic nucleus; N = nucleus. *(Photos by J. J. Becnel.)*

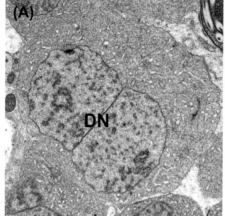
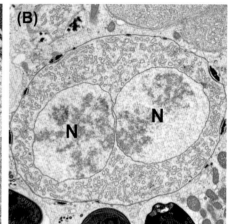

FIGURE 7.4 Schematic (left) and transmission electron micrograph (right) of an environmental spore. AD = anchoring disk; PP = polaroplast; PF = polar filament; N = nucleus; PV = posterior vacuole. *(Drawing and photo by J. J. Becnel.)*

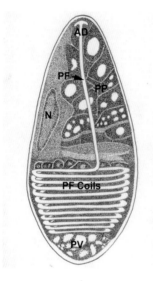
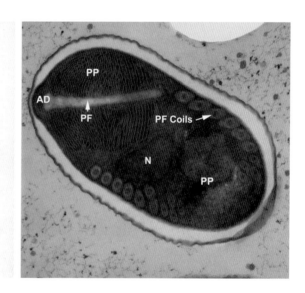

most reduced and compact genomes known in eukaryotes at 2.9 million base pairs (Mbp) (Keeling *et al.*, 2005), and was notable for short intergenic sequences, only 2000 protein-encoding genes, and its lack of repeats and transposable elements (Keeling, 2009). With more genomes currently sequenced and at least partially analyzed, the genome architecture has been shown to vary considerably across the phylum. Insect pathogenic species apparently possess numerous transposable elements and genome sizes that range from about 7 to 20 Mbp (J. Xu *et al.*, 2006; Williams *et al.*, 2008b; Keeling, 2009; Texier *et al.*, 2010). One grasshopper species, *Encephalitozoon romaleae*, and its human pathogen congener, *Encephalitozoon hellem*, were shown to possess protein-encoding gene sequences of insect origin in the genome (Selman *et al.*, 2011).

Microsporidian species are currently described based on a combination of morphology, host species, and the small subunit ribosomal RNA gene (SSU rRNA) sequence of approximately 1200 base pairs (Weiss and Vossbrinck, 1999). The gene is, however, highly conserved and may vary little among closely related species (Pieniazek *et al.*, 1996; Pomport-Castillon *et al.*, 1997; Vávra *et al.*, 2006). The internal transcribed spacer (ITS) region, with flanking portions of the SSU and the large subunit (LSU), is also used to evaluate close taxonomic relationships, but polymorphisms in the ITS region as well as in the SSU of some species can obscure analysis. Other genes such as alpha-tubulin, beta-tubulin, and heat shock protein (HSP-70) have been used for taxonomic purposes, but are also highly conserved.

7.3. LIFE HISTORY

While there are unifying characteristics of infection by microsporidia, life cycles range from the very simple, involving oral transmission and reproduction of vegetative forms in host tissues to form infective environmentally resistant spores, to complex cycles that entail production of two to four spore types and involve intermediate hosts. Pathologies range from relatively slight, allowing the host to develop and reproduce, to severe, resulting in early mortality in a large percentage of infected hosts. Tissue tropism and transmission mechanisms also vary among species.

7.3.1. Infection and Replication

Microsporidia typically infect a susceptible host when spores are ingested during feeding. The current understanding is that the spores are activated by constituents of the host gut environment, possibly pH, ions, or a combination of these, as well as other conditions that the pathogen recognizes as "host" (reviewed by Keohane and Weiss, 1998). Although some species are probably activated to germinate by direct contact with host cell membranes (Magaud *et al.*, 1997; Xu *et al.*, 2003; Hayman *et al.*, 2005; Southern *et al.*, 2006), most studies evaluated human pathogens in the genus *Encephalitozoon* that do not need to breach a peritrophic membrane in the host alimentary track. Germination of the entomopathogen *N. ceranae* occurred with no contact between spores and host cells in tissue culture (Gisder *et al.*, 2011). During the germination process, the cell swells with water, causing the polar filament to evert, rapidly extend from the spore and puncture a host cell. The spore contents, including the nucleus, membranes, and other cellular constituents, move through the polar tube and are injected into the cytoplasm of the host cell.

Obligate parasitism has resulted in the loss of functioning mitochondria (Williams *et al.*, 2002; Williams and Keeling, 2005) and recent research suggests that microsporidia probably import adenosine triphosphate (ATP) directly from host cells (Weidner and Trager, 1973; Bonafonte *et al.*, 2001; Williams *et al.*, 2008a; Keeling *et al.*, 2010) as an energy source. The type of nutrititive resources taken from the host cell environment is not specifically known, but the *E. cuniculi* genome encodes several transporters that may be involved with absorption of sugars and ions (Katinka *et al.*, 2001), consistent with carbohydrate depletion in gypsy moth larvae observed by Hoch *et al.* (2002).

Vegetative division begins as early as 30 min postinoculation (Takvorian *et al.*, 2005) by binary fission (mitotic merogony), or by production of many nuclei without cytokinesis to form multinucleate plasmodia, depending on species. Plasmodia undergo multiple fission to produce daughter cells that can continue in merogony or enter sporogony. Sporulation, sporogony followed by spore morphogenesis, is the process of spore formation from a late vegetative form called the sporont, which then produces sporoblasts that mature into infective spores. Budding in the manner of yeasts has been observed in only one species, *Chytridiopsis typographi*, a pathogen of the bark beetle *Ips typographus* (Tonka *et al.*, 2010). Movement within the host tissues from cell to cell is not well understood, but some species form primary spores that germinate within the tissues, apparently to infect adjacent cells and tissues. This cycle occurs in several microsporidian genera including those infecting terrestrial insect hosts (Iwano and Ishihara, 1989; Fries *et al.*, 1992; Solter and Maddox, 1998a) and aquatic insect hosts (Johnson *et al.*, 1997). Phagocytosis by host cells and subsequent germination of spores held within the phagosomes may also be involved (Takvorian *et al.*, 2005).

Reproduction in most species of microsporidia was assumed to be clonal (evolution from ancient asexual organisms), but recent genetic studies suggest that sexual genetic exchange may occur during vegetative reproduction (Sagastume, 2011). In addition, spore forms that are evidently the end-products of meiotic divisions are found in some species, e.g., the terrestrial *Vairimorpha* species and aquatic *Amblyospora* species, while other closely related species have no obvious sexual forms, suggesting that microsporidia evolved from sexual ancestors but have lost sexuality multiple times (Ironside, 2007).

7.3.2. Pathology

Microsporidian infections tend to be chronic in nature, but effects range from nearly benign to relatively virulent, with infections causing death of the host, albeit often slowly, even when dosages are low. Typical effects are sublethal and include extended larval development period, sluggishness, lower pupal weight, and lower fecundity if the host survives to eclose and mate. Adult life span is frequently reduced and immature insects that acquire infections may not survive later molts or pupation.

Microsporidia may invade one or a few tissues, typically the alimentary tract or fat body, or may cause systemic infections. Tissue tropism is a species-specific character, varying among microsporidian genera and even species within genera. All known *Endoreticulatus* spp. are pathogens of the alimentary tract (Brooks *et al.*, 1988), while species in the *Nosema/Vairimorpha* clade may be restricted to the midgut (e.g., *N. apis* and *N. ceranae* in

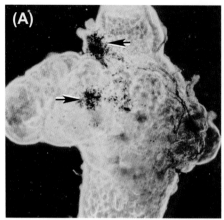

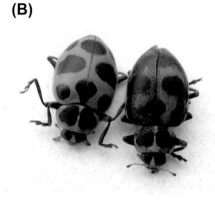

honey bees), primarily fat body pathogens (*Vairimorpha* spp.) or systemic (*N. bombycis* and closely related "true *Nosema*" species).

Tissue trophism of a well-studied microsporidian species in mosquitoes involves sequential infection of specific tissues, with initial infections in the larval alimentary tract followed by infection of larval oenocyes that are carried into the adult mosquito. These oenocytes invade the ovaries of adult females where binucleate spores are formed that germinate and infect the developing eggs. Sexual dimorphism in the progeny of some mosquito species results in fat body infections in male larvae causing death and release of meiospores, while infections in female larvae are benign and produce infected adults to complete the life cycle. The meiospores infect a copepod intermediate host, which produces spores that are infectious to mosquito larvae upon ingestion (Becnel and Andreadis, 1999; Andreadis, 2007).

Microsporidia that are systemic tend to produce fewer spores per unit of infected tissue and are, therefore, often less virulent than species that specifically target fat body tissues. Lower virulence may allow the host to complete its life cycle and, for many species, ensures transmission of the pathogen to the next host generation via the egg or embryo (Becnel and Johnson, 2000; Dunn *et al.*, 2001; Andreadis, 2005; Haine *et al.*, 2005). Fat body pathogens, for example, the lepidopteran *Vairimorpha* species, may result in nearly 100% mortality of larvae at any infective dosage (Goertz and Hoch, 2008a). The fate of immature insects infected with many other species of microsporidia may depend on number and viability of spores consumed, age and stage at exposure, and additional external stress. Insects infected as embryos often have high mortality rates (Andreadis, 1986; Han and Watanabe, 1988)

Microsporidia primarily develop in the cytoplasm of host cells, but several species infecting fish mature in

host cell nuclei (Lom and Dykova, 2002; Nylund *et al.*, 2010), as does a recently described species, *Enterospora canceri*, isolated from an arthropod, the brown crab (Stentiford *et al.*, 2007). Reports of development in host cell nuclei of insects are rare (Sprague *et al.*, 1992; Becnel and Andreadis, 1999). Vegetative reproduction, but not spore formation of *N. portugal*, was observed in the nuclei of silk gland cells of the gypsy moth host (Maddox *et al.*, 1999), and vegetative forms of the type species of several microsporidian genera develop in close contact with the host cell nucleus, sometimes in deep indentations of the nuclear membranes (Sprague *et al.*, 1992).

Many microsporidian species cause hyperplasia and/ or hypertrophy of the nuclei and cytoplasm of infected cells; xenomas, fusion or extension of hypertrophic infected cells to form tumors or cysts, are rare but do occur in insects (Becnel and Andreadis, 1999). Infections may become so severe that the cells of target tissues are filled with spores, interfering with normal cellular function and often resulting in death of the host. A common immune response by insects to microsporidian infection is melanization of infected cells (Fig. 7.5A). Dark melanized areas can sometimes be observed through the epidermis of the host (Fig. 7.5B) and may result in a spotted or mottled appearance called pébrine (as discussed earlier). The only other observable symptom is a puffy appearance and light color of some immature and aquatic insects with patent fat body infections. Patent infections in tissues, however, may be easily observed when insects are dissected.

7.3.3. Transmission

Most microsporidian species are horizontally transmitted to susceptible hosts when environmental spores are disseminated in the feces or regurgitated matter of infected hosts, possibly via silk (Jeffords *et al.*, 1987), and from decomposing infected hosts (Goertz and Hoch, 2011). Transmission by parasitoids, either by infected female wasps or by mechanical contamination when a wasp first oviposits in an infected host, then in an uninfected host, has been shown in laboratory studies (Own and Brooks, 1986; Siegel *et al.*, 1986a), but it is not known whether parasitoid transmission is important to the overall dynamics of microsporidian disease in insect populations. Horizontal transmission is likely whenever infections occur in gut tissues and Malpighian tubules; infections limited to the fat body require cannibalism, the death and decomposition of the host, or possibly vectoring by ovipositing parasitoids.

Many species of microsporidia, in addition to being horizontally transmitted, are vertically transmitted to the offspring of infected female hosts within or on the surface of the eggs. The egg chorion may be contaminated with infective spores as the egg is oviposited, or the embryos become infected during development. Both types of vertical transmission are considered to be transovum transmission; the latter special case is termed transovarial transmission. Venereal transmission via infected males is hypothesized to occur but does not appear to be common or important in disseminating the pathogen, even when male gonads are infected (Solter *et al.*, 1991; Patil *et al.*, 2002; Goertz and Hoch, 2011). For a detailed overview of transmission mechanisms see Becnel and Andreadis (1999).

The generation time, from inoculation to production of environmental spores in a newly infected host, may be as little as four days, although there may be a period of latency before spores are released from infected cells into the feces (Siegel *et al.*, 1988; Goertz *et al.*, 2007). Some species appear to have longer generation times and some may require a signal to form mature spores; for example, a host blood meal is required for the mosquito pathogen, *Amblyospora campbelli*, to sporulate (Dickson and Barr, 1990).

7.3.4. Environmental Persistence

The mature, infective microsporidian spore is the only life stage that is sufficiently environmentally resistant to survive outside the cells of living hosts. Although many terrestrial microsporidia can be stored for over 30 years in liquid nitrogen (Maddox and Solter, 1996), in general, infective spores from terrestrial hosts that are reasonably protected from ultraviolet (UV) radiation and other degradation factors, including other microbes, can probably survive for one month to one year under normal environmental conditions (Maddox, 1973, 1977; Brooks, 1980, 1988; Goertz and Hoch, 2008b). Some species of microsporidia isolated from aquatic hosts

have survived for 10 years in sterile water suspensions under refrigeration in the laboratory (Oshima, 1964; Undeen and Vávra, 1997), but spores from such hosts cannot be frozen, cannot tolerate desiccation, and are also subject to environmental degradation in the aquatic environment (Becnel and Johnson, 2000). Microsporidia have adapted to the presence and absence of hosts due to seasonality or population density cycles with strategies such as maintenance in other related host species (Lange and Azzaro, 2008) and maintenance in alternate or intermediate hosts (Micieli *et al.*, 2009). *Edhazardia aedis* infections in the host *Aedes aegypti* may be nearly benign in some individuals, favoring vertical transmission, and patent in other individuals, serving to inoculate the local environment with spores for horizontal transmission among mosquito larvae (Koella *et al.*, 1998). Survival of microsporidia in diapausing host stages, such as *N. pyrausta* infections in overwintering fifth instar *O. nubilalis* (Andreadis, 1986; Siegel *et al.*, 1988) and *N. portugal* in transovarially infected eggs (Maddox *et al.*, 1999), is probably common. Although some hosts may die during diapause owing to additional physiological stress caused by infection (Andreadis, 1986), the pathogen is protected from environmental degradation in surviving hosts, ensuring survival until the next generation of hosts is available.

7.3.5. Life Cycles

The life cycles of several microsporidian species from terrestrial hosts and aquatic hosts were covered in detail by Becnel and Andreadis (1999), and are distinctive for both similarities and differences among species. A generalized life cycle for the genus *Nosema* is briefly described here, as well as life cycles for two species from mosquito hosts. Each of these has similar counterparts in other genera, but these descriptions do not cover the full range of possible sequences in microsporidian development.

Life Cycles of Microsporidia in the Nosema/ Vairimorpha *Clade*

With the exception of the *Vairimorpha*-type microsporidia that appear in the *Nosema* clade, all *Nosema* species isolated from terrestrial hosts follow a similar sequence of sporulation events (Fig. 7.6A), with the major difference among species being tissue specificity. Some *Nosema* species, for example, those in the "true" *Nosema* group isolated from Lepidoptera, are systemic pathogens that, when orally ingested, first infect the midgut tissues, then invade most other tissues, or are systemic in the embryo if transovarially transmitted.

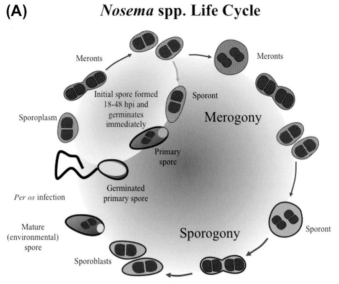

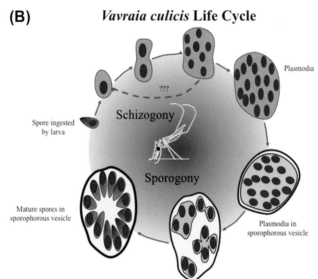

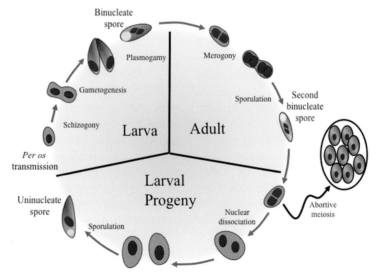

FIGURE 7.6 Life cycles of (A) *Nosema* spp. (h.p.i. = hours postinfection); (B) *Vavraia culicis*; (C) *Edhazardia aedis*. *(Drawings by S. E. White; courtesy of Folia Parasitologica, Institute of Parasitology.)*

Some *Nosema* species are restricted to the midgut tissues, notably, *N. apis* and *N. ceranae*, pathogens of honey bees. The gypsy moth *Nosema* species, *N. lymantriae* and *N. portugal*, initiate infection in the midgut epithelial cells, then primarily target the silk glands and fat body, with reproduction also occurring in the Malpighian tubules, gonads, and nerve tissues.

When *Nosema* spores are ingested, the infection cycle begins when germination occurs in the midgut lumen of the host and the sporoplasm, extruded through the polar tube, is injected into the cytoplasm of the midgut epithelial cells. The sporoplasm grows in size to become a meront, then the nucleus divides, followed by binary fission of the cell. For some species, this first division occurs between 30 mins and 12 h of infection. Monokaryotic meronts have been occasionally reported for a brief period shortly after initial infection (Maddox et al., 1999; Vávra et al., 2006), but it is not known whether this is an artifact of sectioning or an ephemeral phase that occurs in some species. All other stages, including those in merogonic cycles, are binucleate. Whether or not multiple merogonic divisions occur in the first cycle is probably a species character, but between 18 and 48 h postinfection, sporonts form from meronts and divide once to become two sporoblasts, each of which matures into a primary spore (described in Section 7.2.2). Primary spores germinate spontaneously, presumably infecting adjacent cells and tissues. Merogony is repeated in this second cycle and in ensuing proliferative cycles with, ultimately, development into sporonts that will divide to form thick-walled environmental spores. If the midgut tissues or Malpighian tubules are infected, environmental spores are shed in the feces of the host into the environment where they are available for ingestion by susceptible hosts.

Many *Nosema* species that produce systemic infections are also transovarially transmitted to the next host generation. These species infect the gonads of the host, and either spores or vegetative forms are incorporated into the egg during oogenesis. *Nosema granulosis*, a pathogen of the crustacean *Gamerus duebeni*, primarily infects the gonads and is apparently only transovarially transmitted. This species also feminizes its host (Terry et al., 1999).

Life cycles of the *Vairimorpha* species involve a primary spore cycle in the midgut tissues and two secondary sporulation sequences in the same host: a *Nosema*-type sequence producing binucleate spores, and a sequence in which sporonts divide to form eight mononucleated spores (meiospores or, more commonly, "octospores") within a sporophorous (parasite-derived) membrane (Vávra et al., 2006). With one exception, all known *Vairimorpha* species are primarily fat body pathogens and both spore types are produced in this tissue. Only

the crayfish pathogen *V. cheracis* invades the muscle tissues of the host (Moodie et al., 2003).

Life Cycle of *Vavraia culicis*

Vavraia culicis is a multisporous (multiple spores produced from plasmodia) microsporidian pathogen of *Culex pipiens* (Weiser, 1947) with all stages uninucleate (Fig. 7.6B). One type of environmental spore is produced. A sporoplasm has not been described for this species; the earliest described stage is a schizont with two or more nuclei. The plasmalemma of the schizont becomes coated with a thin, electron-dense, two-layer substance that divides when the plasmodium divides. Inside the plasmodia, nuclear divisions produce up to 16 nuclei before plasmotomy. Before sporogony, divisions occur by plasmotomy and multiple fission. Sporogony generates plasmodia that commonly produce eight, 16, or 32 nuclei. The surface layer separates from the plasmodium when sporogony is initiated, producing a sporophorous vesicle inside which lobate plasmodial stages divide to produce eight, 16, or 32 uninucleate spores. Recent evaluation of the ultrastructure and life cycles of various *Vavraia* isolates from different hosts suggests that the isolates represent a closely related species group with similar morphology and life cycles (Vávra and Becnel, 2007).

Life Cycle of *Edhazardia aedis*

Edhazardia aedis, a pathogen of *Ae. aegypti*, is an example of a microsporidium with a complex life cycle, involving four spore types (Fig. 7.6C), but lacking an intermediate host that is typical for the *Amblyospora* species. It is horizontally transmitted via ingestion of spores as well as vertically transmitted.

Germination of mature lanceolate-shaped spores in the midgut lumen of larvae results in infection of the epithelial cells of the gastric caeca where the microsporidium asexually produces uninucleate, pyriform gametes that have a distinctive double-membraned papilla on the plasmalemma. Plasmogamy produces diplokaryotic stages that develop into small binucleate spores (a form of primary spores) that germinate and disseminate *E. aedis* to other host tissues; the oenocytes in the host hemolymph appear to be a primary target tissue. Most lightly to moderately infected mosquito larvae survive the early infection sequences to develop to the adult stage. In adult mosquitoes, *E. aedis* again multiplies asexually in the oenocytes. The oenocytes move through the hemocoel, coming into proximity with the ovaries in the female mosquitoes. After the female mosquito takes a blood meal, a second binucleate spore, the transovarial spore, is formed and infects the ovaries of the host. This larger, oblong spore infects the filial generation, but fecundity and longevity are reduced in infected adult females (Becnel et al., 1995), as is success at blood feeding (Koella and Agnew, 1997). *Edhazardia aedis* infects the fat body tissues of the transovarially infected larval host and undergoes a third merogony followed by two types of division. Meiotic division occurs but is abortive, rarely producing meiospores, similar to the meiotic division that produces abortive octospores in the lepidopteran pathogen *Vairimorpha imperfecta* (Canning et al., 2000). A second type of division by nuclear dissociation forms two haploid cells, each of which undergoes sporogony to form large numbers of uninucleate spores. The transovarially infected host dies, releasing infective spores in the aquatic environment that are ingested by susceptible conspecific larvae to complete the cycle. Although infective in the laboratory to a number of mosquito species, *E. aedis* can only complete the full life cycle in *Ae. aegypti* (Becnel and Johnson, 1993; Andreadis, 1994).

7.3.6. Epizootiology and Host Population Effects

Microsporidia infecting insects at the population level are frequently host density dependent (Thomson, 1960; Andreadis, 1984; Streett and Woods, 1993; Briano et al., 1995a). With the exception of laboratory storage, there are no data to suggest that the infective spores survive longer than a year in the environment (Becnel and Andreadis, 1999), so a build-up of long-lived infective units, similar to the resting spores of entomophthoralean fungi (Weseloh and Andreadis, 2002), is unlikely. Enzootic prevalence of microsporidian infection varies depending on specific pathogen–host interactions, from occasional or continuous presence at low levels to widely fluctuating levels that increase greatly when host density is high, the latter suggesting a relatively important role in the population dynamics of the host.

Because immature insects infected with microsporidia often develop slowly, they are potentially vulnerable to biotic and abiotic mortality factors for a longer period than are healthy insects (Maddox et al., 1998). Impacts on the host population caused by delayed development and other sublethal effects that are typical of microsporidian infections, particularly reduced fecundity and, frequently, transovum or transovarially infected offspring, may be evidenced in the generation following an epizootic rather than in the generation experiencing peak prevalence (Maddox et al., 1998; Régnière and Nealis, 2008). This type of host population response has been documented, for example, in spruce budworm infected with *Nosema fumiferanae* (details in Section 7.4.6), the mosquito *Ochlerotatus cantator* infected with

Amblyospora connecticus (Andreadis, 2005), gypsy moth infected with *Vairimorpha disparis* (Pilarska et al., 1998), and second generation *O. nubilalis* infected with *N. pyrausta* (Hill and Gary, 1979). Some microsporidian species do, however, directly affect the host generation in which prevalence increases occur. Thus, first generation *O. nubilalis* declined as prevalence increased in a two-year study in Connecticut (Andreadis, 1984) and, notably, high density populations of the caddisfly, *Glossosoma nigrior*, appeared to collapse almost completely as microsporidian infection prevalence increased in one season (Kohler and Wiley, 1992).

Seasonal epizootics are common for some insect–microsporidian interactions, including *O. nubilalis* and *N. pyrausta* (Hill and Gary, 1979) and several species of microsporidia and their mosquito hosts (Andreadis, 2007). For populations of other insect species, microsporidian infection may be occasional and rarely epizootic, for example, *N. lymantriae* in gypsy moth populations (Pilarska et al., 1998) and *N. bombi* in populations of *Bombus impatiens* (Cameron et al., 2011). In these latter interactions, microsporidia may never be a major cause of mortality but probably serve as components of the natural enemy complex of their hosts that regulate population densities.

7.3.7. Host Specificity

Microsporidia as a group have been variously reported to be relatively host specific or, primarily based on laboratory host range studies (Sprague et al., 1977b), to have a generally broad host range. Indeed, some microsporidian species do infect multiple insect species. *Nosema bombi*, for example, was reported to infect 22 bumble bee species in the USA (Cordes, 2010) and at least eight bumble bee species in western Europe (Tay et al., 2005), *P. locustae* is known to infect large numbers of grasshopper species (Lange, 2005), and *Cystosporogenes* sp. (probably *Cystosporogenes operophterae*) was recovered from 21 species in eight lepidpoteran families collected in two small research sites in Slovakia (Solter et al., 2010). There are, in addition, records of hymenopteran parasitoids acquiring microsporidian infections from their lepidopteran hosts (Cossentine and Lewis, 1987; Futerman et al., 2006) and of a microsporidian pathogen of mosquitoes infecting humans (Becnel and Andreadis, 1999). Nevertheless, few of the many microsporidian entomopathogens reviewed by Sprague (1977b) have been isolated in field populations of putative host species that were found to be susceptible in laboratory bioassays.

Several laboratory evaluations have shown that infections in non-target hosts are often atypical and suboptimal and, while some non-target insects fed high dosages of spores may develop patent infections, these infections often are not horizontally and/or vertically transmitted to conspecific susceptible individuals (Andreadis, 1989a, 1994; Solter and Maddox, 1998b; Solter et al., 2005). Some microsporidian species with apparently broad host ranges may actually represent a complex of biotypes, each of which only readily infects a restricted number of hosts. For instance, *N. bombi* infecting bumble bees may represent multiple strains that are each relatively host specific (Schmid-Hempel and Loosli, 1998; Larsson, 2007).

Infections in human hosts by the mosquito pathogen *Anncaliia* (= *Nosema*, = *Brachiola*) *algerae* raised the first concerns about safety of entomopathogenic microsporidia, and *A. algerae* arguably has the broadest known microsporidian host range. Originally isolated from the mosquito *Anopheles stephensi* (Vávra and Undeen, 1970), it has been grown at 37°C in mammalian/human cell cultures (Moura et al., 1999; Lowman et al., 2000), in warm water fish cell culture (Monaghan et al., 2011), and in the extremities of mice (Undeen and Maddox, 1973; Undeen and Alger, 1976). It has also been isolated from corneal epithelial cells (Visvesvara et al., 1999), muscle tissue (Coyle et al., 2004), and vocal cords (Cali et al., 2010) of human patients. Although reported human cases are extremely rare, *A. algerae* is considered to be an emergent disease organism, particularly of concern for humans who are immunocompromised (Cali et al., 2005). Two other entomopathogenic microsporidia are of some concern, although no vertebrate infections have been recorded thus far. *Vavraia culicis* is genetically closely related to *Trachaepleistophora homminis*, a species described from human patients but capable of infecting mosquitoes (Weidner et al., 1999), and *E. romaleae*, isolated from lubber grasshoppers, is the only known species in the genus *Encephalitozoon* that is not a vertebrate pathogen (Lange et al., 2009).

Despite these concerns, with the exception of *A. algerae*, there is no evidence that other entomopathogenic microsporidia are pathogenic to vertebrate animals. The safety issues that are important to consider for their use in biological control programs primarily involve the potential susceptibility of non-target native insects to introduced pathogens (Solter and Becnel, 2003). Because microsporidian species vary widely in breadth of host range, both field and laboratory studies are needed to evaluate the potential for a species introduced as a classical biological control agent to "jump" to a non-target host. The physiological (laboratory) host range is important to consider should microsporidia be used as microbial insecticides because inundative field rates may be similar to the typically high dosages used in laboratory bioassays.

Microsporidia that are relatively host specific and are inoculatively released in augmentative or classical biological programs are far less likely to affect non-target species, particularly if release does not involve inundative methods such as spraying (Solter et al., 2010). Host specificity studies were conducted for several of the microsporidia covered in the following case histories and are mentioned in the context of the various biological control programs.

7.4. BIOLOGICAL CONTROL PROGRAMS: CASE HISTORIES

This section presents several long-term studies of microsporidia as potential biological control agents. Although not every attempt has been successful, the information gathered is a resource for increased understanding of microsporidian−host interactions and provides a framework for future considerations of the role of microsporidia in biological control programs.

7.4.1. Use of Microsporidia in Biological Control Programs

Microsporidia are best suited for use as natural controls, introduced as classical biological control agents, and in augmentative and conservation biological control programs (Lacey et al., 2001). They have not been found to be suitable as inundatively released microbial insecticides for two primary reasons. First, even at relatively high application rates, infections are typically chronic rather than acute, with the major effects on the host being reduced lifespan and fecundity. While these effects have been documented to suppress populations of insect pests, microsporidia are not sufficiently fast acting, nor do they typically produce the high mortality rates expected for chemical or microbial insecticides (Lacey et al., 2001; Solter and Becnel, 2007). Second, microsporidia cannot be inexpensively cultured in artificial media or fermentation tanks. They require living cells to reproduce, therefore, *in vivo* production requiring mass rearing of hosts or cell culture is needed for mass production. Both methods are costly and labor intensive. Most species of microsporidia are difficult to culture successfully or produce fewer spores in cell culture than in the host (Khurad et al., 1991; Kurtti et al., 1994). Despite 50 years of research on microsporidia of insect pests, only one species, *P. locustae*, a pathogen of a suite of grasshopper species, has been registered by the US Environmental Protection Agency (USEPA) for use as a microbial insecticide.

There are, nevertheless, numerous situations where adequate control of established pests can be achieved by manipulating their naturally occurring pathogens.

FIGURE 7.7 (A) *Edhazardia aedes* infecting fat body tissues (arrows) of mosquito host; (B) *E. aedes* environmental spores; and (C) germinating spore with everted polar filament. *(Photos by J. J. Becnel.)*

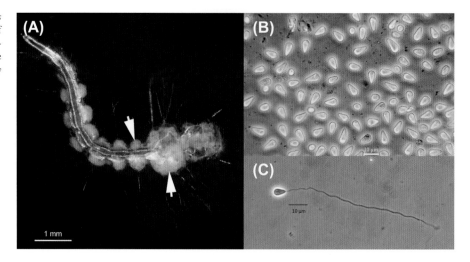

Outbreaks of native pests in economically important areas, and accidental or intentional introduction of non-native plants and animals, have created serious global issues concerning habitat damage and economic loss. The development of methods to suppress outbreak and invasive species while avoiding collateral damage to populations of native species is critical for many systems. Microsporidia are present in insects as naturally occurring primary pathogens. Although they do not function well as microbial insecticides, their more subtle effects can have important impacts on host population dynamics and, in the context of the natural enemy complex of a pest species, they may be important regulators of outbreak species (Anderson and May, 1980).

This section includes descriptions of several long-term research efforts to evaluate the role and potential of microsporidia as biological control agents for several major pests in a variety of managed and natural habitats. These laboratory and field studies include general biology and taxonomy of the pathogens, efficacy testing, host density dependence and persistence in host populations, and evaluation of host specificity.

7.4.2. Aquatic Diptera

Microsporidia are generally the most commonly found pathogens in natural populations of aquatic Diptera but usually occur at very low prevalence levels. The Culicidae have been the focus of much of the research on microsporidia in aquatic Diptera because of their medical importance and microbial control potential. Microsporidia are also common in the Simuliidae, Chironomidae, and Ceratopogonidae.

Microsporidia were seriously considered as biological control agents for mosquitoes because of their ability to cause larval epizootics, continuously cycle within a host

population, and spread to new habitats. The idea of utilizing these natural enemies of mosquitoes as manipulative control agents was perhaps first raised by Kudo (1921), who suggested that larval sites might be contaminated with microsporidian-infected mosquito tissues. Unaware at that time of the mechanism of transovarial transmission, he further suggested that infected adults could distribute the parasite to new sites that "may escape our watchful eye" by dying during oviposition.

The first microsporidia evaluated for mosquito control were those with relatively simple life cycles that were directly infectious to the host. The initial attempt to introduce a microsporidian parasite was presumably with *V. culicis* (Fig. 7.7) against *Cx. pipiens fatigans* on the South Pacific island of Nauru (Reynolds, 1972). The second attempt was with *A. algerae* against *Anopheles albimanus* in Panama (Anthony *et al.*, 1978). Both of these studies demonstrated that microsporidian parasites could be introduced into natural larval sites and result in considerable infection that reduced adult longevity and fecundity; they did not, however, cause significant larval mortality or persist and spread at levels considered important for population reduction. Because of public health and safety concerns (see Section 7.3.7 and reviews by Becnel *et al.*, 2005, and Andreadis, 2007), *A. algerae* and *V. culicis* can no longer be considered as viable biological control agents.

Two discoveries led to a renewed interest in microsporidia as microbial control agents of mosquitoes during the 1980s and 1990s. The first discovery was the involvement of an intermediate copepod host in the life cycle of *Amblyospora* (Sweeney *et al.*, 1985) and *Parathelohania* (Avery, 1989; Avery and Undeen, 1990), documenting for the first time the mechanism for horizontal transmission in these genera. The second was the identification of a new genus of microsporidia, *Edhazardia*, that is both horizontally and vertically transmitted but does not require an intermediate host (Hembree, 1979; Becnel *et al.*, 1989).

Much of the information concerning the life cycles and evaluations of polymorphic microsporidia has been derived from studies on *Amblyospora dyxenoides* (Sweeney *et al.*, 1988; Sweeney and Becnel, 1991), *A. connecticus* (Andreadis, 1988a), and *E. aedis* (Becnel *et al.*, 1989), but little recent work has been conducted. This discussion will focus on *E. aedis*, a species that does not require an intermediate host, and *A. connecticus*, an example of a species that requires a copepod intermediate host.

Edhazardia aedis

Edhazardia aedis, a pathogen of *Ae. aegypti*, was isolated in Thailand (Hembree, 1979). This pathogen has a complex life cycle involving both horizontal and vertical transmission affecting two successive generations of the host (Hembree, 1982; Hembree and Ryan, 1982). Usually, one sporulation sequence occurs in the adult female (infected orally as a larva) and results in the formation of binucleate spores. These spores are involved in vertical transmission of *E. aedis* to the subsequent generation via infected eggs. In infected progeny, there are two sporulation sequences in larval fat body but, owing to the abortion of one of the sequences, only one viable spore type is produced (Becnel *et al.*, 1989). Larval death results in the release of uninucleate spores that are responsible for horizontal transmission when ingested by larvae. This developmental sequence leads to the formation of binucleate spores in the adult to complete the cycle.

There are two known deviations from this parental host–filial host alternation that may play important roles in maintenance of *E. aedis* under natural circumstances (Becnel *et al.*, 1989). In these instances, the parasite completes its development through repeated cycles of horizontal transmission or, alternatively, by repeated cycles of vertical transmission.

Edhazardia aedis was transmitted to its natural host, *Ae. aegypti*, and to eight alternate hosts in the laboratory: *Ae. albopictus*, *Ae. triseriatus*, *Ae. taeniorhynchus*, *Ae. atropalpus*, *Ae. vexans*, *Anopheles quadrimaculatus*, *Orthopodomyia signifera*, and *Toxorhynchites rutilus rutilus* (Becnel and Johnson, 1993; Andreadis, 1994). Thirteen other mosquito species were not susceptible to *E. aedis*, including all species of *Culex* tested. In all susceptible hosts, the microsporidium underwent normal development but transovarial transmission was successful only in *Ae. aegypti*. Therefore, while a variety of mosquito species representing diverse genera is physiologically susceptible to *E. aedis per os*, the pathogen is specific for *Ae. aegypti*. Common non-target aquatic organisms were not susceptible to infection by *E. aedis* and, hence, there was no mortality due to *E. aedis* (Becnel, 1992b).

Edhazardia aedis causes larval death as a result of the highly efficient mechanism of vertical transmission, with approximately 95% of the progeny infected (Hembree, 1982; Hembree and Ryan, 1982; Becnel *et al.*, 1995). Spores released from these infected progeny are infectious to all instars of *Ae. aegypti* and result in infected adults. The influence of the microsporidium *E. aedis* on the survival and reproduction of its mosquito host *Ae. aegypti* was studied in the laboratory (Becnel *et al.*, 1995). Survival, fecundity, egg hatch, and percentage emergence for four gonotrophic cycles were compared for control and infected mosquitoes. Infected females oviposited 70% fewer eggs and percentage hatch was lower than for control females. Emergence in progeny of infected female *Ae. aegypti* was significantly less than for control mosquitoes in all gonotrophic cycles. The reproductive capacity (R_0) for control and infected adults was 168 and four, respectively, representing a 98% decrease. A semi-field study that involved the inoculative and inundative release of *E. aedis* produced encouraging results (Becnel and Johnson, 2000). Limitations to incorporating *E. aedis* into an integrated control program are the high costs involved in production in host mosquitoes and methods to store the fragile spore stage. The only possible field application of this microsporidian parasite would be as part of a classical biological control program to establish *E. aedis* in naïve host populations for long-term control (Becnel, 1990).

Ecological and epizootiological studies of *E. aedis* in natural populations of *Ae. aegypti* have not been completed. Optimism regarding the role of *E. aedis* as part of a program to control *Ae. aegypti* is based on a number of desirable traits determined in laboratory and semi-field studies. These studies have demonstrated that both the vertical and horizontal components of the life cycle of *E. aedis* are highly efficient in providing the means for the parasite to become established, persist, and spread in populations of *Ae. aegypti*. The profound effect on reproductive capacity of infected adults suggest the *E. aedis* can have a strong influence on host population dynamics (Becnel *et al.*, 1995). Both inoculative and inundative release strategies were evaluated using *E. aedis* against a semi-natural population of *Ae. aegypti* (Becnel and Johnson, 2000). Inoculative release resulted in dispersal of *E. aedis* to all containers within the study site over a 20-week period. Inundative release eliminated the population of *Ae. aegypti* within 11 weeks of introduction. Good persistence is expected in release sites owing to life cycle flexibility with dissemination to other mosquito-inhabiting sites by means of vertical transmission. Survival during dry periods occurs within the mosquito eggs, where the pathogen can survive for the life of the egg (Becnel *et al.*, 1989). In addition, an obligatory intermediate host is not required for horizontal transmission.

These desirable traits of *E. aedis*, including host specificity, support the belief that this pathogen can play an important role as a classical biocontrol agent in developing strategies to control *Ae. aegypti*. Ideally, *E. aedis* would be introduced via inoculative releases (rather than inundative releases), lessening the need for mass production. For this method to be successful, the pathogen must become permanently established or augmented seasonally to maintain acceptable levels of control.

Amblyospora connecticus

Amblyospora connecticus is a pathogen of the brown saltmarsh mosquito, *Aedes cantator*, from Connecticut, USA. A characteristic of most known species of *Amblyospora* is the requirement of two generations of the mosquito host and an obligate sequence in a copepod intermediate host to complete the life cycle (Andreadis, 1988a, 1990b).

Like *E. aedis*, the life cycle of *A. connecticus* involves both horizontal and vertical transmission in the mosquito host. In the infected adults, binucleate spores formed after a blood meal are responsible for transovarial transmission to progeny. Some female and most male progeny from infected adults develop fatal fat body infections that produce a distinct spore type, the meiospores. Released meiospores are infectious *per os* to the copepod intermediate host, *Acanthocyclops vernalis*. A sporulation sequence in the intermediate host ends with the production of uninucleate spores that are infectious *per os* to a new generation of the mosquito host. In larval mosquitoes, a developmental sequence leads to the formation of binucleate spores in adults to complete the cycle.

Amblyospora connecticus was experimentally transmitted to four alternate mosquito hosts; *Ae. atropalpus*, *Ae. epacticus*, *Ae. sierrensis*, and *Ae. triseriatus* (Andreadis, 1989a). Binucleate spores were formed only in *Ae. epacticus* and none of the alternate hosts was able to transmit the parasite vertically to progeny. Therefore, *A. connecticus* was unable to complete its life cycle in any of the alternate hosts, demonstrating a high degree of specificity for *Ae. cantator*. Fifteen other mosquito species belonging to the genera *Aedes*, *Anopheles*, *Culex*, *Culiseta*, and *Psorophora* were not susceptible.

Seasonal epizootics of *A. connecticus* in *Ae. cantator* larvae occur each fall (autumn; October–November) with infection rates of up to 100%, and some sites in Connecticut had an overall prevalence of 98% (Andreadis, 1983). These lethal infections are a result of the synchronous hatch of transovarially infected eggs and serve to produce meiospores that infect the intermediate host, *A. vernalis*. A small sample of field-collected infected *Ae. cantator* were examined to determine whether infection with *A. connecticus* is detrimental to reproductive success. Of 195 females examined, four were found to be infected with *A. connecticus*. There was no significant difference in fecundity or the percentage of eggs that hatched between healthy and infected females (Andreadis, 1983). It was concluded that *A. connecticus* does not detrimentally affect the reproductive success of *Ae. cantator*.

Life-cycle Based Management Strategies

Knowledge about the dynamics of a microsporidia–mosquito system can be utilized as part of a mosquito management project and is demonstrated using the extensive ecological and epizootiological data collected for *A. connecticus* (Andreadis, 1988b, 1990a). The mosquito host, *Ae. cantator*, is a typical multivoltine, saltmarsh species. It overwinters in the egg stage and may produce up to four generations a year in the north-eastern USA. Eggs hatch in spring (March) and the first adults emerge by May. Subsequent generations appear periodically as the saltmarsh pools are flooded by rain and/or high tides. The copepod intermediate host, *A. vernalis*, is a common species that has one to two generations a year in the northeastern USA and overwinters as a diapausing fourth or fifth stage copepodid. It is abundant during the spring and fall but aestivates during the summer months.

In a three-year study on natural ecology, it was demonstrated that there is a well-defined seasonality to the transmission cycles and epizootics involving *A. connecticus* (Andreadis, 1988b). Larval epizootics as a result of transovarial transmission occur each fall and routinely produce infection levels of 80–100%. These epizootics in *Ae. cantator* result in larval death and release of meiospores which coincides with the fall appearance of the intermediate host, *A. vernalis*. Copepods become infected by ingesting meiospores but development of *A. connecticus* is arrested until the following spring when winter dormancy ends. Sporulation of *A. connecticus* and subsequent death of the copepod coincides with the hatch of *Ae. cantator* in the spring when horizontal transmission to the mosquito host occurs. Infected female adult mosquitoes produced at this time lay infected eggs throughout the summer. Infected eggs hatch synchronously in the fall, causing epizootics, but few hatch during the summer, even if flooded.

The two major events identified as critical to the maintenance of *A. connecticus* are (1) the fall epizootic responsible for significant larval mortality of *Ae. cantator* and, concurrently, the infection of the copepod intermediate host to provide for overwintering of the parasite; and (2) the spring epizootic, when spores formed in the copepod are responsible for horizontal transmission to larval *Ae. cantator* and result in infected adults. Awareness of these critical windows in the maintenance of *A. connecticus* suggests a control strategy that precludes larviciding these habitats during the spring and fall as this would be counterproductive and disrupt the natural balance of the disease cycle. Larviciding these habitats during the summer is suggested by life cycle data that demonstrate an absence of infected mosquitoes as well as copepods during the summer months. This strategy would also target the healthy part of the mosquito population that had escaped infection by *A. connecticus*. In addition, *A. connecticus* was successfully introduced and maintained in a field population of *Ae. cantator* via infected *A. vernalis* (Andreadis, 1989b). This demonstrates that augmentation can be used as part of a classical biological control project to introduce microsporidia with complex life cycles involving an intermediate host.

Microsporidia for mosquito management has been proposed as one component of the natural complex of regulatory factors utilized for control. This approach recognizes that eradication of the target mosquito is an unrealistic goal but with a combination of physical, cultural, chemical, and biological control methods, mosquito vectors and pests can be regulated.

7.4.3. Lepidopteran Pests in Row Crop Systems

Classical or conservation biological control of plant pests should be appropriate for row crop systems, including those ranging from small truck farms to large-scale grain acreage. Most arthropod herbivore populations are, indeed, limited to a certain extent by their complex of natural enemies, even in crop monoculture, and many seed and fruit crops can tolerate some pest damage to foliage without damage to the product. Stem damage causes weakness and reduced yield (Bode and Calvin, 1990), however, and few markets will bear significant damage to fruit or other edible plant parts. The economics involved with pest control in these high value and ecologically artificial systems demand fast, inexpensive, and nearly complete control at a level that most natural enemies cannot achieve. Nevertheless, natural enemies, including microsporidia, can be factored into treatment decisions that benefit the producer, consumer, and environment. Although microsporidia have been isolated from a number of field crop pests, only two species, *Nosema pyrausta* in *Ostrinia nubilalis*, and *Vairimorpha necatrix* isolated from several noctuid species, have been extensively studied to evaluate host–pathogen interactions and effects on host population dynamics. *Vairimorpha necatrix* is a virulent and promising pathogen that can produce mortality in the range of a nucleopolyhedrovirus and may have some potential for use in greenhouse production (Down *et al.*, 2004). Unfortunately, *V. necatrix* has been disappointing in field trials because it is not persistent at levels that are effective for the control of targeted noctuid pests (Brooks, 1980). In contrast, *N. pyrausta* is particularly important as a persistent and primary pathogen of *O. nubilalis*.

Although *O. nubilalis* larvae have a wide host range that includes a variety of weeds, grasses, and vegetable crops with stems or fruit sufficiently large enough to support boring, it has, since its introduction to the USA in the early 1900s, become the most serious pest of corn in North America. The life cycle includes one to three generations annually; in the Midwest, there are typically two. The larvae first feed on leaves of the corn host, causing minor damage, then move to leaf whorls and tassels before boring into the stalk to complete their development. Damage includes broken stalks (lodging) and ear shanks, damaged ears (particularly in popcorn and sweet corn), and decreased yield. Losses in the late 1990s were reported to exceed $1 billion/year in the USA alone (CSREES, 2008). Treatments for *O. nubilalis* have included introduced parasitoids and all permitted pesticides from arsenicals and multiple synthetic compounds to microbial insecticides and, since the late 1990s, transgenic corn hybrids expressing *Bacillus thuringiensis kurstaki* (Btk) Cry proteins. *Ostrinia nubilalis* is host to a small number of pathogens, with *Beauveria bassiana* and *N. pyrausta* having the most important effects on population dynamics of the host. Nearly 80 years of research on *N. pyrausta* were recently reviewed by Lewis *et al.* (2009), and the reader is directed to this history for details. Here, the importance of this microsporidian pathogen in *O. nubilalis* population dynamics is briefly described for the two-generation cycle.

Nosema pyrausta is a sister species to the "true" *Nosema* group (Vossbrinck and Debrunner-Vossbrinck, 2005), closely related to *N. bombycis* and, like other members of this microsporidian clade infecting Lepidoptera, infections are chronic and systemic. *Nosema pyrausta* is transmitted both horizontally and vertically, bridging the two (or three) annual generations of the host. Infected overwintering fifth instar larvae have lower survival rates than uninfected larvae (Siegel *et al.*, 1986b), but those that survive are generally able to eclose, mate, and oviposit. Larvae that are transovarially infected contaminate the leaf whorls and newly bored tunnels with spore-laden frass and decomposing infected hosts. Mortality is high in first generation transovarially infected larvae and, thus, prevalence of the disease often declines during the first generation (Siegel *et al.*, 1986b), but the second generation is infected by both disseminated spores and transovarial transmission. This generation enters fifth instar diapause, serving as inoculum for the following spring generation.

Nosema pyrausta prevalence in *O. nubilalis* populations is density dependent (Hill and Gary, 1979; Andreadis, 1986), and enzootic levels are high relative to many other microsporidia–host interactions. Before the use of transgenic corn, it was not unusual to find populations with prevalence levels that fluctuated between 5% and 60% or higher (Hill and Gary, 1979; Andreadis, 1986; Lewis *et al.*, 2006). The well-documented effects of *N. pyrausta* infections on the *O. nubilalis* host are typical for chronic microsporidian infections. In addition to high mortality rates in overwintering larvae and transovarially infected larvae, adult fecundity is generally reduced (reviewed by Lewis *et al.*, 2009), larval growth rates and development time are slowed (Lewis *et al.*, 1983; Solter *et al.*, 1990), and flight behavior is negatively affected (Dorhout *et al.*, 2011). The effects of *N. pyrausta* are exacerbated (additive effects) by exposure to chemical insecticides (reviewed by Lewis *et al.*, 2009) and Btk (Pierce *et al.*, 2001; Lopez *et al.*, 2010). Evaluations of *O. nubilalis* populations 10 years after the first and, currently, expanded use of transgenic corn hybrids have determined that the use of Bt hybrid corn has not eliminated *N. pyrausta* in *O. nubilalis* populations and it is recommended that care should be taken in the management of Bt corn to conserve this pathogen in *O. nubilalis* integrated pest management (IPM) programs (Lopez *et al.*, 2010).

7.4.4. Grasshoppers and *Paranosema locustae*

The extensive western grasslands of the USA are a region grazed by livestock and plagued by irregular outbreaks of grasshoppers. The pest grasshoppers fall within four main subfamilies in the family Acrididae, with about 10−15 species being major pests (Capinera and Sechrist, 1982; Pfadt, 2002). A diversity of developmental times, foraging preferences, aggregation and migration patterns, and other biological and behavioral traits is represented by the numerous species. Integrating these variable traits with habitat and abiotic factors such as moisture and temperature results in population outbreaks with highly variable and unpredictable intensity and frequency (Lockwood and Lockwood, 2008). In addition, this system is unique in that the spatial scale of the outbreak can be vast, encompassing thousands or even millions of hectares. Forage devastation can occur quickly in localized but dispersed foci (Lockwood *et al.*, 2001). Treatments are most efficacious when applied within a limited time-frame that targets third instar nymphs to suppress grasshopper populations before substantial forage is consumed (Hewitt and Onsager, 1983). Similarly, grasshoppers plague other rangeland ecosystems of the world including northern China (Shi *et al.*, 2009) and the Pampas in Argentina (Lange and Cigliano, 2005).

Given the significant destruction of crops and rangeland forage by grasshopper outbreaks (Hewitt and Onsager, 1983; Joern and Gaines, 1990; Lockwood *et al.*, 2002), grasshoppers have long been the target of intense control programs utilizing various insecticide sprays and baits (Latchininsky and VanDyke, 2006). Insecticide sprays and baits can quickly kill grasshoppers (Quinn *et al.*, 1989), with the latter having less detrimental effects on non-target organisms (Quinn *et al.*, 1990; Peach *et al.*, 1994; McEwen *et al.*, 2000; Foster *et al.*, 2001). With applications of either formulation, the usual objective is to immediately quell the destructive

feeding by grasshoppers in that season. It is within this paradigm that the entomopathogen *Paranosema locustae* became the first and, currently, the only microsporidium to be commercially produced and registered for grasshopper control in rangelands (USEPA, 1992; Lockwood *et al.*, 1999).

Paranosema locustae was originally described in 1953, from colonies of African migratory locusts, *Locusta migratoria mirgratoroides*, being reared in England. It was initially named *Nosema locustae* (Canning, 1953), but it has since been placed in a new genus *Paranosema* (Sokolova *et al.*, 2003). Subsequently, the microsporidium was transferred to the genus *Antonospora* (Slamovits *et al.*, 2004) but the change was refuted by Sokolova *et al.* (2005a). In addition to *L. migratoria*, *P. locustae* was identified from several species of North American grasshoppers (Canning, 1962a) and is now known to have an unusually broad host range for a microsporidium, infecting 121 species of Orthoptera (Lange, 2005). The mature infective spores are ellipsoidal in shape, 3.5−5.5 μm in length by 1.5−3.5 μm in width, with a thick spore wall (Canning, 1953, 1962a, b).

Paranosema locustae primarily infects the adipocytes of fat body tissue, resulting in the disruption of the host's metabolism and energy storage. Other tissues can also be infected in heavily infected hosts. Similar to other microsporidian infections, *P. locustae* causes chronic debilitation, which is associated with reduced feeding, development, and fecundity, in addition to increased mortality rates (Canning, 1962b; Henry and Oma, 1974; Ewen and Mukerji, 1980; Johnson and Pavlikova, 1986). Sublethal infections in locusts were associated with a shift from the gregarious form to the less damaging solitary phase (Fu *et al.*, 2010). In extensive infections, the fat body is greatly hypertrophied with spores and acquires an opaque, cream coloration that can progress to a pink and eventually a dull red color.

Horizontal transmission of *P. locustae* is very efficient. Ingestion of vegetation inoculated or contaminated with spores, and cannibalism and necrophagy of infected hosts serve as routes of transmission (Canning, 1962b; Henry and Oma, 1981; Ewen and Mukerji, 1980). The third instar nymph is the stage most susceptible to acquiring infection. Infection at this stage development causes the highest initial mortality within a population; yet, a high prevalence of infection is maintained in survivors (Canning, 1962b; Henry *et al.*, 1973). In addition, *P. locustae* can be transmitted transovarially, with infected nymphs potentially serving as inoculum for horizontal transmission via cannibalism (Raina *et al.*, 1995). Observations of vertical transmission in hatchlings from field-collected eggs, however, revealed low levels of infection that were considered inadequate to establish infections in the following year (Ewen and Mukerji, 1980).

Early assessments considered the virulence of *P. locustae* to be insufficient to protect crops despite high nymphal mortality and efficient transmission (Canning, 1962b). However, initial field studies with *P. locustae* as an augmentative biological control were encouraging (Lockwood *et al.*, 1999). Locust populations were inoculated by applying spores to wheat bran that grasshoppers ingested, resulting in a peak infection prevalence of 43% and population reductions of some species. There was also evidence of reduced fecundity in infected females (Henry, 1971a). The efficient horizontal transmission and moderate virulence in adult hosts enabled large quantities of spores to be produced *in vivo*. This afforded the opportunity to further develop and evaluate *P. locustae* bait (Henry *et al.*, 1973; Ewen and Mukerji, 1980; Henry and Oma, 1981). The broad host range of *P. locustae* was considered by some to be a desirable attribute, as many species of grasshoppers damage crops and different developmental phenologies among species provide a continuous availability of susceptible hosts to maintain infections (Henry, 1971b), but it has also been noted that the variable susceptibilities among species would contribute to inconsistent efficacy (Lockwood *et al.*, 1999). Another advantage is that spores can be stored by freezing, but while storage capability made conducting trials with *P. locustae* logistically easier, freezing was reported to be a hindrance (Henry and Oma, 1974; Lockwood *et al.*, 1999). *Paranosema locustae* was registered as *Nosema locustae* (this species name is still recorded on EPA labels) and produced commercially in the USA beginning in the 1980s by a variety of companies and marketed under trade names such as Nolo Bait™, Semaspore™, and Grasshopper Attack™ (Lockwood *et al.*, 1999). Nolo Bait is still produced by a company in Colorado, and Semaspore is produced by a company in Montana. The USEPA (1992) produces a factsheet with further details on *P. locustae* bait products. The efficacy of *P. locustae* bait has been variable in numerous field studies with typical population reductions of 30% accompanied by 20–40% infection prevalence among survivors (Johnson, 1997). This level of control, however, would be perceived inadequate by end-users when compared to 70–95% control with insecticides (Vaughn *et al.*, 1991).

The construal of *P. locustae* as a microbial insecticide brought inappropriate and unfulfilled expectations of fast and thorough suppression of grasshopper populations and damage. This shrouded the original concept of using *P. locustae* as an augmentative biological control agent to improve long-term suppression of grasshoppers (Henry and Oma, 1981; Lange and Cigliano, 2005). Attempts to increase infection prevalence and population reductions of grasshoppers with inundative applications of *N. locustae* have not been successful. For example, two successive annual applications of the commercially produced *P. locustae* bait (Nolo Bait) did not sufficiently increase

prevalence among or the severity of infection in infected grasshoppers or result in significant population impacts, nor were sublethal effects of reduced fecundity and feeding apparent. These results were for both years of the applications and the subsequent third season and resulted in the conclusion that crop protection with *P. locustae* alone would be difficult to detect (Johnson and Dolinski, 1997). Infections of some grasshopper species reached 35%, however, which suggested a value for including *P. locustae* as a biological control component of an IPM program or in areas that can tolerate some levels of damage (Johnson and Dolinski, 1997; Vaughn *et al.*, 1991). The inconsistent efficacy of *P. locustae* bait relative to chemical insecticides, the high cost of *in vivo* spore production, and limited long-term spore storage have made marketing commercial formulations difficult in the USA and internationally. Even extensive government-subsidized applications of *P. locustae* in China were perceived to be inadequate by growers, with grasshopper reductions of 60% or less and slow mortality times (Lockwood *et al.*, 1999).

Pathogen enzootics and epizootics are generally thought to be among the many factors that influence grasshopper populations. The degree to which pathogens play a role in diminishing the frequency and severity of grasshopper outbreaks is difficult to ascertain (Joern and Gaines, 1990). *Paranosema locustae* was introduced into naïve grasshopper populations in several locations in Argentina in 1978–1982 using inocula from North America. Formal monitoring of the fate or impact of the introductions was not reported until the pathogen was detected in 1991 in the western Pampas region. *Paranosema locustae* continued to be observed over a span of 11 consecutive years. The sustained infections suggested efficient horizontal transmission and even a possible persistent decline in population abundance and frequency of grasshopper outbreaks (Lange and Cigliano, 2005). Similar conclusions were reported from China where a single application of *P. locustae* bait resulted in variable population declines and 10-year persistence. Indeed, *P. locustae* has been applied extensively in China for over 20 years and is thought to be useful for grasshopper and locust management when populations are moderate (Shi *et al.*, 2009). In both Argentina and China, the use of *P. locustae* is an example of "neoclassical" biological control, where an exotic organism is introduced to control a native pest (Lockwood, 1993). The establishment of an exotic biocontrol agent elicits concern regarding its impact on non-target organisms, shifts in species assemblages, and other components of the rangeland. While shifts in grasshopper species have been reported in Argentina, they are considered to be part of the inevitable changes to the Pampas as agroecosystems become prevalent (Lange and Cigliano, 2005). With the exception of China, the application of *P. locustae* bait for grasshopper suppression has dwindled. However,

P. locustae bait was reregistered by the USEPA and is sold in the USA mainly as an organic treatment for grasshoppers.

7.4.5. Fire Ants in Urban Landscapes

Fire ants are invasive stinging insects native to South America that were inadvertently introduced into the USA in the early 1930s. They currently infest the southern USA and parts of California and, since 2000, infestations have established in Australia, China, and Mexico (Ascunce *et al.*, 2011; Sanchez-Peña *et al.*, 2009). Fire ants have become ubiquitous within urban, agricultural, and natural landscapes. They are most noted for the painful, burning stings they inflict on humans, pets, and livestock. A conservative estimate is that 1% of people stung in the USA are potentially allergic to the venom and are at risk for anaphylaxis (Triplett, 1976). In addition, fire ants can damage crops, and the dominance of these ants in natural landscapes has reduced biodiversity (Wojcik *et al.*, 2001).

Despite the availability of insecticides that can effectively suppress fire ant populations, it is often logistically impractical or cost prohibitive to treat many infested sites. Untreated or unmanaged infested areas are key sources of reinfestations in areas cleared of fire ants or are the source of new infestations. Using entomopathogens as biological control agents is considered to be one of the few viable methods of sustainable control for invasive ants.

Kneallhazia solenopsae

Extensive efforts have been made to utilize the microsporidium *Kneallhazia solenopsae* for the biological control of fire ants. This pathogen was first reported from *Solenopsis invicta* in 1974 and described in 1977 as *Thelohania solenopsae*, but was subsequently placed in a new genus, *Kneallhazia* (Allen and Buren, 1974; Knell *et al.*, 1977; Sokolova and Fuxa, 2008).

Four spore types have been reported from *K. solenopsae* (Table 7.2). These are as follows: (1) Octets of pyriform, uninucleate, meiospores within sporophorous vesicles (octospores) are the most abundant and are typically found in the worker and reproductive caste of the adult ants. They are concentrated, though not exclusively, in the fat body tissues. (2) Diplokaryotic, *Nosema*-like, free spores (not enclosed within a vesicle) are also found in adult ants of both castes, and occasionally in larvae and pupae. These spores are ovoid and slightly larger than individual meiospores and are much less abundant. (3) Exclusively found in fourth instars and pupae, the diplokaryotic primary spores are ovoid with a distinctive posterior vacuole. (4) Megaspores are elongated ovals and the largest of the spores. They are diplokaryotic and, while most frequently observed in the adult reproductive caste, they are also present in adult worker ants, fourth instar larvae, and pupae.

Although the functions of the various spores of *K. solenopsae* are not definitively known, the biology and a hypothetical life cycle have been developed (Sokolova and Fuxa, 2008). In general, latent infections are found in the early immature stages (eggs, first to third instars) of fire ants. Mature spores were not observed by light microscopy in these stages, but infections were detected by polymerase chain reaction (Briano *et al.*, 1996; Valles *et al.*, 2002; Sokolova and Fuxa, 2008). It was proposed that infections in the early immature stages precede the proliferation of primary spores in fourth instars and pupae. Primary spores initiate the rapid distribution of infection within the host and serve as the initial source of the *Nosema*-like spores and megaspores. *Nosema*-like spores are thought to be involved in the early stage of pathogenesis, autoinfecting adipocytes. This infection sequence switches to megaspore and intensive meiospore production, which results in a conglomerate of spores enclosed within the expanded membrane of the now dysfunctional adipocyte, forming cysts, or sporocytosacs (Sokolova *et al.*, 2005b) (Fig. 7.8). While meiospores are by far the most copious spore type, their function is not understood. Megaspores are regularly observed in the cells of muscle, tracheoles, and fat bodies associated with ovaries and testes of fire ant reproductives. Germinated megaspores have also been observed in the ovaries of inseminated queens (Sokolova and Fuxa, 2008). Coupling these observations with the detection of vegetative stages in eggs (Briano *et al.*, 1996) provides evidence for megaspore involvement in the transovarial transmission of *K. solenopsae*.

Infections of *K. solenopsae* have been initiated in fire ant colonies by introducing live, infected fourth instar larvae and pupae (Williams *et al.*, 1999; Oi *et al.*, 2001); however, the mechanism of horizontal transmission between individual ants is not known. *Per os* inoculation of suspensions of the primary, *Nosema*-like, and meiospore types have not resulted in infections (Shapiro *et al.*, 2003). Adult *S. invicta* workers and virgin female reproductives possess an efficient filtering system that traps and removes solid particles from ingested liquids. They are able to filter particles 0.5−0.75 μm or larger in diameter (Glancey *et al.*, 1981; Petti, 1998; Oi, unpubl. data). Thus, *K. solenopsae* spores theoretically should not be ingested by adult *S. invicta* since all known spore types are considerably larger than 0.75 μm (Table 7.2). However, fourth instar larvae have ingested maximum particle sizes of 45.8 μm (Glancey *et al.*, 1981). While larvae are capable of ingesting spores and become infected, how adult queens become infected is yet to be determined (Chen *et al.*, 2004; Sokolova and Fuxa, 2008).

The ability to initiate colony infections by introducing live, *K. solenopsae*-infected brood has facilitated the evaluation of this pathogen as a biological control agent. Laboratory inoculations have resulted in brood reductions

TABLE 7.2 *Kneallhazia solenopsae* Spore Types and Size, Primary Fire Ant Stages and Castes, and Spore Images and Descriptions

Spore Type	Size (μm)	Primary Fire Ant Stages (Occasional Host)	Spore Image	Spore Description
Uninucleate meiospore (octospore)	3.3 × 1.95[a]	Adult workers, queens, alates (pupae)[b c]		Eight meiospores in sporophorous vesicle
Binucleate free or *Nosema*-like spore	4.9 × 1.85[a]	Adult workers, female reproductives (larvae, pupae)[c]		Free spore (right) and octet of meiospores
Diplokaryotic primary spore	4.5 × 2.3[d]	Pupae (larvae)[e]		Primary spores, with large posterior vacuole
Binucleate megaspore	6.2 × 3.6[c]	Queens, alate females, workers, larvae, pupae[e]		Primary spore and larger megaspore

[a]*Knell* et al. *(1977);*
[b]*Briano* et al. *(1996);*
[c]*Sokolova* et al. *(2004);*
[d]*Shapiro* et al. *(2003);*
[e]*Sokolova* et al. *(2010).*

FIGURE 7.8 Cyst (arrow) visible though the cuticle of a red imported fire ant infected with *Kneallhazia* (*Thelohania*) *solenopsae*. (*Photo courtesy USDA-ARS*)

of 88−100% and the demise of *S. invicta* colonies in 6 to 12 months. Queens from infected colonies were debilitated, exhibiting reductions in weight and oviposition (Fig. 7.9), as well as dying prematurely (Williams *et al.*, 1999; Oi and Williams, 2002). Reductions also have been documented in *K. solenopsae*-infected fire ant populations in the field. In Argentina, for example, reductions as high as 83% have occurred in the black imported fire ant, *Solenopsis richteri*, and reductions of 63% have been reported for *S. invicta* in the USA (Briano *et al.*, 1995b; Oi and Williams 2002). Reductions were generally due to decreases in colony size, less brood and fewer workers per nest, and population levels fluctuated (Cook, 2002; Oi and Williams, 2002; Fuxa *et al.*, 2005a). The slow demise of laboratory colonies suggested that field declines could be masked by the immigration or founding of new colonies.

FIGURE 7.9 *Solenopsis invicta* queen infected with *Kneallhazia solenopsae* (right) is smaller than the uninfected queen (left). (*Photo courtesy USDA-ARS*)

Inoculations of K. solenopsae in fire ant populations and surveys for natural infections in the USA revealed that the host's social form and associated behaviors have a profound effect on the prevalence of the pathogen. This microsporidium is found mainly in polygyne *S. invicta*, which is one of the two social forms of fire ants. Monogyne colonies have a single queen per colony and are territorial, thus they fight with other *S. invicta* colonies. In contrast, polygyne colonies are not territorial, and queens, workers, and brood are moved between colonies. The polygynous characteristics extend the persistence and spread of *K. solenopsae* among polygyne colonies (Oi *et al.*, 2004; Fuxa *et al.*, 2005b). The persistence of *K. solenopsae* in polygyne colonies may partly be attributed to the prolonged survivorship of infected colonies afforded by having multiple queens per colony (Williams *et al.*, 1999; Oi and Williams, 2002). Queen infections within a colony are not simultaneous; infection rates of queens from field-collected polygyne colonies ranged from 25 to 75% per *S. invicta* colony (Oi and Williams, 2002). Asynchrony of queen infections is likely to result in a slower demise of colonies because declines in brood production and queen death are staggered. In addition, unlike monogyne colonies, new queens can be adopted by polygyne colonies (Glancey and Lofgren, 1988; Vander Meer and Porter, 2001), providing another source of queens that can prolong the persistence of infected colonies. Polygyny also facilitates the spread of infections given that intercolony movement of infected brood and queens can initiate new colony infections. In contrast, infections in monogyne populations are infrequent in the USA, and when infections do occur they usually dissipate (Fuxa *et al.*, 2005a, b; Milks *et al.*, 2008). In Argentina, *K. solenopsae* infections were found to be nearly equally present in both social forms of *S. invicta*. Perhaps an intermediate host or a more pathogenic strain of *K. solenopsae* contributed to the observed difference in South America (Valles and Briano, 2004).

As indicated, the chronic debilitation of *K. solenopsae*-infected queens suppresses their reproductive capacity. The pathogen also impairs colony founding by infected queens, possibly because of reduced lipid reserves (Cook *et al.*, 2003; Oi and Williams, 2003; Overton *et al.*, 2006; Preston *et al.*, 2007). The abundance of uninfected colonies and the persistent survivorship of infected polygyne colonies, however, often diminish the perceived impact the pathogen is having on field populations. When an area is cleared of fire ants using insecticides, it is normally reinfested within months by colonies migrating from the surrounding area and colonies established by queens that land after mating flights (Collins *et al.*, 1992). Establishment of biological control agents such as *K. solenopsae* in untreated, or unmanaged, landscapes has been used to slow the reinfestation of landscapes cleared of fire ants. Reinfestations have been delayed by over a year where *K. solenopsae* and fire ant parasitic flies were released and established (Oi *et al.*, 2008). The classical biological control approach of establishing a natural enemy and relying on natural proliferation and spread is potentially the most sustainable method of suppressing well-established invasive species like fire ants which dominate diverse swaths of natural, agricultural, and urban habitats. Of the natural enemies used against fire ants, *K. solenopsae* has the most documented colony-level impacts. To date, *K. solenopsae* has not been found on other continents more recently invaded by fire ants (Yang *et al.*, 2010).

Regional epizootics in both social forms are probably needed to maintain fire ant population suppression. That monogyne fire ants are also commonly infected in South America may signal the possibility of the pathogen being vectored. Several species of fire ant parasitic phorid flies in the genus *Pseudacteon* have been introduced from South America and established in the USA. *Kneallhazia solenopsae* has been detected in at least three species of these flies that develop in infected fire ants; however, there is not yet evidence that the flies can vector the microsporidium (Oi *et al.*, 2009; Oi, unpubl. data).

Kneallhazia solenopsae was reported to be host specific to fire ant species in the *Solenopsis saevissima* species group based on field surveys and laboratory inoculations (Oi and Valles, 2009). However, infections recently have been confirmed in the tropical fire ant, *Solenopsis geminata*, and the *S. geminata* × *Solenopsis xyloni* hybrid, which are in the *Solenopsis geminata* species group. Analysis of 16S ribosomal RNA gene sequences in *K. solenopsae* revealed 22 haplotypes that grouped into two distinct clades. The molecular diversity and expanded host range suggest the presence of variants of *K. solenopsae* with different pathogenicities and host specificity (Ascunce *et al.*, 2010), which may perhaps include different affinities between the fire ant social forms. With the discovery of *K. solenopsae* in the USA, quarantine restrictions were eased and spurred the aforementioned interest and study of its biology and use as a biocontrol agent.

Vairimorpha invictae

Vairimorpha invictae is another microsporidian pathogen of fire ants from South America that exhibits potential as a microbial biocontrol agent. It was apparently first observed in an infected colony from Brazil a few years after the observation of *K. solenopsae* (Jouvenaz *et al.*, 1980). It has unicellular meiospores that develop in groups of eight within sporophorous vesicles and, in addition, has binucleate, free (not vesicle bound) spores (Jouvenaz and Ellis, 1986). Free spores are bacilliform and develop before the meiospores. They are present as mature spores in larvae,

pupae, and adult fire ants. Meiospores are ovoid with a slight narrowing at the anterior end. They begin development in pupae, but mature spores typically are seen in adult ants. Infections were found in all castes of *S. invicta*. Both spore types are larger than *K. solenopsae* spores and have an amber coloration internally under phase-contrast microscopy of unstained specimens (Jouvenaz and Ellis, 1986; Briano and Williams, 2002) (Table 7.3).

Vairimorpha invictae develops in fat bodies located throughout the ant, transforming infected cells into large sacs filled with both spore types (Jouvenaz and Ellis, 1986). Vegetative stages were observed in all immature stages, but they were most frequently seen in fourth instars and pupae. The presence of either vegetative stages or spores in queens and eggs is rare, suggesting inconsistent transovarial transmission (Briano and Williams, 2002). *Vairimorpha invictae* infections can be initiated in colonies with the introduction of live, infected brood. However, unlike *K. solenopsae*, dead *V. invictae*-infected adults have been used successfully as inocula to infect colonies (Oi *et al.*, 2005). The spore type(s) and mechanism involved in intracolony and intercolony infection are not yet known. Infection by isolated spores has not been accomplished, but larvae can become infected when reared to the pupal stage by infected adult workers (Jouvenaz and Ellis, 1986; Oi *et al.*, 2010).

The pathogenicity of *V. invictae* was indicated in laboratory studies where there was a higher prevalence of infection among dead workers than in live workers from diseased *S. invicta* colonies. In addition, mortality occured sooner among naturally infected, starved adult workers than uninfected starved workers (Briano and Williams, 2002). Significant reductions of more than 80% in colony growth were documented after infecting small laboratory colonies of *S. invicta* with live infected brood or dead infected adults (Oi *et al.*, 2005). In Argentina, declines of 69% in field populations of *S. invicta* were associated with natural *V. invictae* infections and also with *K. solenopsae* infections in the same colony (Briano, 2005). Colony declines were faster in the laboratory when infections of the microsporidia occurred together (Williams *et al.*, 2003). The prevalence of dual infections in field colonies has been reported to be about 2% among various sites in South America. The prevalence of *K. solenopsae* and *V. invictae* in field colonies was reported to be 8−13% and 2−10%, respectively, in South American surveys (Briano *et al.*, 1995b, 2006; Briano and Williams, 2002).

The persistence of *V. invictae* field infections appears to be more sporadic, with wide and abrupt fluctuations in prevalence that typically include periods of no infection, whereas infections of *K. solenopsae* were sustained and maintained at fluctuating levels. Peaks of prevalence for the two microsporidian pathogens generally did not coincide, resulting in periods of constant and high disease pressure.

TABLE 7.3 *Vairimorpha invictae* Spore Types and Size, Primary Fire Ant Stages and Castes, Spore Images, and Descriptions

Spore Type	Size (μm)	Primary Host Stages and Castes (Occasional Host)	Spore Image[c]	Image Descriptions
Uninucleate meiospores	6.3 × 4.2[a]	Adults of all castes (pupae)[b]		Meiospores without sporophorus vesicle
Binucleate free spores	11.2 × 3.1[a]	Adults of all castes, pupae, larvae[b]		Bacilliform free spores and ovoid meiospores

[a]*Jouvenaz and Ellis (1986);*
[b]*Briano and Williams (2002).*
[c]*Images by J. Briano.*

Using a classical biological control approach with these microsporidia against fire ants would be beneficial for long-term population suppression (Briano *et al.*, 2006). Host specificity testing and field surveys are also supportive of *V. invictae* as a biological control agent. This microsporidium only infects fire ants in the *Solenopsis saevissima* species group, which includes only the invasive *S. invicta* and *S. richteri* in the USA (Briano *et al.*, 2002; Porter *et al.*, 2007; Oi *et al.*, 2010).

The co-occurrence of microsporidia and other natural enemies (Valles *et al.*, 2010) could provide enough diversity in pathogenicity, transmission, and persistence to achieve the goal of self-sustaining suppression of the ubiquitous, invasive fire ant. There are documented laboratory colony declines and population-level field reductions with *K. solenopsae* and *V. invictae*, but sustained reductions perceivable by the public are not consistent in the USA, with only *K. solenopsae* present. *Vairimorpha invictae* is not yet approved for release in the USA, and additional challenges of mass production and efficient inoculation

protocols must be developed and implemented. A major biological hurdle is to understand the mechanism(s) of transmission to facilitate consistent infections of one or preferably both microsporidia into monogyne populations in the USA. This would limit reinfestations and slow the further spread of fire ants. Despite the challenges, classical biological control using microsporidia represents a key route toward successfully combating well-established invasive pest ants.

7.4.6. Control of Forest Insect Pests

Compared to most agricultural settings, forests are highly complex ecosystems that, whether old growth or secondary, are vestiges of natural systems with the potential to be harmed by broad-scale chemical inputs. The level of risk to non-target invertebrate forest species posed by chemical insecticides, including chitinase inhibitors such as diflubenzeron that have non-selective toxicity to insects (Eisler, 2000), is controversial (Perry *et al.*, 1997; Beck *et al.*,

2004), and some government entities have severely limited their use, particularly in environmentally sensitive areas. Development of methods to suppress outbreaks and invasive species while avoiding collateral damage to populations of native species is critical for natural systems. When microsporidia are important, host-specific components of the natural enemy complex of exotic forest pests, consideration for introduction as classical biological control agents or augmentative introduction is warranted. Naturally occurring microsporidian pathogens of several forest pests have been studied for their impacts on host populations, and several species of microsporidia infecting the gypsy moth have been extensively evaluated for classical and augmentative introduction for biological control of this serious defoliator.

Impacts of Naturally Occurring Microsporidia on Forest Pests

Microsporidia have been isolated from a number of economically important forest insect species, and it is likely that this pathogen group is far more common than published reports indicate. Solter *et al.* (2010), for example, recovered microsporidia from 23 lepidopteran species collected in two sites in Slovakia (approximately 2000 m^2 total area), although one generalist pathogen, a *Cystosporogenes* sp., represented the majority of the infections (see Section 7.3.7). Microsporidia were also recovered from three species of curculionid bark beetles collected in a one-season survey in Bulgaria (Takov *et al.*, 2011). Microsporidian epizootics are seldom as obvious as those of viruses and fungi, and the role of microsporidia in the population dynamics of their hosts is less well documented. However, a few species are known to be important mortality factors, particularly in species of forest Lepidoptera. Characterizations and descriptions of microsporidia infecting terrestrial forest insect species typically have been based on light microscopy and host relationships, but more detailed documentation that includes biology, molecular data, and epizootiology is available for a few species infecting important pests (Table 7.4). Although acute mortality caused by microsporidiosis is not common, the sublethal effects detailed in this chapter may strongly affect populations of many forest insects (Gaugler and Brooks, 1975; Wilson, 1980), and several species have been evaluated for biological control programs.

Spruce Budworm Microsporidia

Four species of microsporidia have been reported infecting the spruce budworm, *Choristoneura fumiferana*: *Cystosporogenes* sp., *Endoreticulatus* (*Pleistophora*) *schubergi*, *N. fumiferanae* and *Thelohania* sp. (van Frankenhuyzen *et al.*, 2004). Most of these microsporidia are encountered infrequently, but *N. fumiferanae*, a species

closely related to the microsporidian type species, *N. bombycis* (Kyei-Poku *et al.*, 2008), is common and often occurs at high prevalence levels, increasing in spruce budworm populations as larvae mature over the summer months (Wilson, 1973) and over the years that host populations persist at high densities (Thomson, 1960; Wilson, 1973). In one spruce budworm population, prevalence increased from 36% to 81% over a five-year period (Thomson, 1960) and, 17 years later, increased from 13% to 69% over a six-year period in the same site (Wilson, 1977). The spruce budworm population declined in the years following the peak prevalence (Thomson, 1960).

A fine-scale model of spruce budworm population dynamics that included *N. fumiferanae* infection as an independent variable showed that the microsporidium increased mortality of spring-emerging larvae (Régnière and Nealis, 2008). The model was corroborated in studies showing that transovarial transmission of *N. fumiferanae* by intensely infected spruce budworm females in older outbreak populations usually produces 100% infected offspring, resulting in reduced survival during dormancy, delayed emergence from hibernacula and dispersal, reduced establishment on feeding sites and, based on prevalence studies, high mortality between the period of dispersal and establishment in feeding sites (van Frankenhuyzen *et al.*, 2007). In a study of western spruce budworm, *Choristoneura occidentalis*, males infected with a closely related *Nosema* sp. (Kyei-Poku *et al.*, 2008) were less likely to fly upwind to a pheromone source, suggesting that infection reduces mating success in infected populations (Sweeney and McLean, 1987).

Both *N. fumiferanae* and another microsporidium, *E. schubergi*, were evaluated in the laboratory and field as potential microbial insecticides. Laboratory studies conducted to evaluate transmission and effects of the two species showed reduced pupal weight, fecundity, and adult longevity (Thomson, 1958; Wilson, 1984) in orally infected hosts, as well as high larval mortality for F1 larvae transovarially infected by *N. fumiferanae* (Bauer and Nordin, 1989a). Bauer and Nordin (1989b) determined that both LC$_{50}$ and LT$_{50}$ of *B. thuringiensis* treatments were lower for infected spruce budworm than for uninfected hosts, but LC$_{50}$ values observed in larvae from multiple spruce budworm populations were not affected when natural prevalence of *N. fumiferanae* was not augmented in the laboratory (van Frankenhuyzen *et al.*, 1995).

Sprayed on trees to increase prevalence in host populations, levels of *N. fumiferanae* remained higher in treated sites than in untreated sites for several years until levels of natural infections rose to similar levels, suggesting that augmentation successfully advanced epizootics by two to three years (Wilson and Kaupp, 1976). *Endoreticulatus schubergi* persisted at low levels; it was hypothesized that

TABLE 7.4 Partial List of Microsporidia Infecting Forest Insect Pests

Host Species	Microsporidian Species	References[a]
Agrilus anxius	*Cystosporogenes* sp.	Kyei-Poku *et al.* (2011)
Archips cerasivoranus	*Endoreticulatus* (*Pleistophora*) *schubergi*	Wilson and Burke (1978)
Choristoneura conflictana	*Nosema thomsoni*	Wilson and Burke (1971)
Choristoneura fumiferana	*Cystosporogenes* sp. (*legeri?*)	van Frankenhuyzen *et al.* (2004)
	Endoreticulatus (*Pleistophora*) *schubergi*	Wilson (1975)
	Nosema fumiferanae	Thompson (1955)
	Thelohania sp.	Wilson (1975)
Dendroctonus species	*Nosema dendroctoni*	Weiser (1970)
	Chytridiopsis typographi	Knell and Allen (1978)
	Unikaryon minutum	
Euproctis chrysorrhoea	*Nosema chrysorrhoeae*	Hyliš *et al.* (2006)
	Nosema kovacevici	Purrini and Weiser (1975)
	Endoreticulatus sp.	Purrini (1975)
Hyphantria cunea	*Endoreticulatus schubergi hyphantriae*	Weiser (1961)
	Nosema sp. (*bombycis*-type)	Solter and Maddox (1998b)
	Vairimorpha sp.	
Ips spp.	*Chytridiopsis typographi*	Purrini and Weiser (1985)
	Larssoniella duplicati	Weiser *et al.* (2006)
	Nosema typographi	Weiser *et al.* (1997)
	Unikaryon montanum	Wegensteiner *et al.* (1996)
Lymantria dispar	*Endoreticulatus schubergi*	McManus and Solter (2003)
	Nosema lymantriae	
	Nosema portugal	
	Nosema serbica	
	Vairimorpha disparis	
Malacosoma americanum	*Nosema* sp. (*bombycis*-type)	Weiser and Veber (1975)
Malacosoma disstria	*Nosema disstriae*	Thomson (1959)
Operophtera brumata	*Cystosporogenes operophterae*	Canning *et al.* (1983)
	Nosema wistmansi	
	Orthosoma operophterae	
Pristiphora erichsoni	*Thelohania pristiphorae*	Smirnoff (1967)
Tomicus piniperda	*Caningia tomici*	Kohlmayr *et al.* (2003)
Tortrix viridana	*Nosema tortricis*	Franz and Huger (1971)

[a]*References are species descriptions or research reports.*

low prevalence was due to mortality in the year of treatment, but no mortality studies were conducted.

Microsporidian Pathogens of Gypsy Moth

The gypsy moth, a defoliating outbreak species that was introduced to North America in the late 1860s, is a generalist feeder with preference for oaks (*Quercus* spp.) and aspens (*Populus* spp.) in North America (Leonard, 1981). Although the adult female does not fly, natural spread by ballooning neonate larvae and, more importantly, inadvertent human transport of egg masses (Allen *et al.*, 1993) have served to advance the movement of the pest from the site of introduction in Massachusetts to the current western leading edge in Wisconsin and north-eastern Illinois and south to Virginia, North Carolina, and Kentucky.

Nearly a century of attempts to control the gypsy moth resulted in a federal government, private industry, and state agency partnership in the USA to produce a comprehensive insect pest management effort. The "Slow the Spread of Gypsy Moth" (STS) program, established in 1999 and spearheaded by the USDA Forest Service, uses intensive monitoring and control, primarily with the microbial insecticide *B. thuringiensis*, to eradicate isolated populations that occur in advance of the leading edge of invasion, as well as treatment of outbreak populations at the leading edge and elsewhere in the infested zones. A cornerstone of the program is the use and manipulation of naturally occurring pathogens to suppress populations in the area of establishment. Solter and Hajek (2009) reviewed the research on several microsporidian species isolated from gypsy moth populations, as well as studies of other pathogens, parasites, and predators that have been manipulated as biological control agents of the gypsy moth. There follows a brief summary of the program and the studies leading up to release of microsporidia as augmentative and classical biological control agents.

Naturally occurring pathogens have been extensively incorporated in the gypsy moth STS program; two of them, nucleopolyhedrovirus, *Ld*MNPV (Chapter 4) and the fungal pathogen *Entomophaga maimaiga* (Chapter 6), have potential impacts on introductions of microsporidia. *Ld*MNPV was probably accidentally introduced into North America with the pest and is currently formulated by the USDA Forest Service as a host-specific microbial pesticide, Gypcheck® (Podgwaite, 1999). Produced in limited quantities, it is a particularly useful biopesticide in environmentally sensitive areas (Reardon *et al.*, 2009). The introduction from Asia, probably Japan (Nielsen *et al.*, 2005), of the host-specific *E. maimaiga* is an inadvertent success story. Whether the introduction was purposeful or accidental has not been determined (Solter and Hajek, 2009), but *E. maimaiga* is now well established in the USA and is currently present and frequently causes epizootics in gypsy moth populations on the leading edge of invasion (Villidieu and van Frankenhuyzen, 2004; L. F. Solter, unpubl. data). *Entomophaga maimaiga* has been introduced in eastern Europe and is now established in Bulgaria and Macedonia (Pilarska *et al.*, 2000, 2007).

Although several microsporidian species are commonly present in gypsy moth populations in Europe, evidently none was present in the gypsy moth population that first invaded US forests in the late 1860s; they have never been recovered from North American gypsy moth field populations (Campbell and Podgwaite, 1971; Podgwaite, 1981; Andreadis *et al.*, 1983; Jeffords *et al.*, 1989). There are seven species descriptions for microsporidia pathogenic to gypsy moth in the literature (Solter and Hajek, 2009). Of those, five species are currently recognized: *E. schubergi*, *N. lymantriae*, *V. disparis* (= *Thelohania disparis*, = *Thelohania similis* + *N. lymantriae*), *Nosema serbica* (known only from stained slides), and *N. portugal*. Vávra *et al.* (2006) disentangled the early species descriptions in the literature and recharacterized *V. disparis*; however, it is still not clear whether the named species in the *Nosema* group are biotypes of the same species or should retain species status.

Microsporidia were first reported to be important factors in gypsy moth population declines in central and eastern Europe, including Serbia (Sidor, 1976) and the Ukraine (Zelinskaya, 1980), where the gypsy moth is native. Particularly noted was the precipitous decline of populations with high prevalence of both microsporidia and *Ld*MNPV. Indeed, synergy between the virus and *N. portugal*, based on mortality rate, was shown for mixed infections in laboratory bioassays, particularly when *N. portugal* was orally inoculated before inoculation with the virus (Bauer *et al.*, 1998).

The first efforts to manipulate the pathogens for gypsy moth control were augmentative field experiments with *N. lymantriae* in Slovakia. A 79% increase in mortality was recorded in the site treated with *N. lymantriae* foliar sprays by the time adults eclosed and numbers of progeny were reduced by 90% (Weiser and Novotny, 1987). Novotny (1988) reported that artificial declines in gypsy moth populations were nearly equally precipitous for plots treated with *Ld*MNPV and those treated with *N. lymantriae*. In a 1986 study in the USA, two species, *N. portugal* and *E. schubergi*, were released in isolated 4 ha (10 acre) woodlots in Maryland by attaching gypsy moth egg masses soaked in spore suspensions to oak trees. *Endoreticulatus schubergi* was not recovered the following year, but *N. portugal* overwintered and was horizontally transmitted during the 1987 season (Jeffords *et al.*, 1989), indicating that it could become established. *Nosema portugal* apparently did not persist; it was not recovered in a 1989 field collection (M. Jeffords, LFS, unpubl. data). A second attempt in Michigan to release *N. portugal* via spore-saturated egg masses failed, perhaps owing to loss of viability resulting from overlong exposure of the spores to environmental conditions before hatch (Solter and Becnel, 2007).

Beginning in the mid-1990s, laboratory and field experiments were focused on determining mechanisms of transmission, and on safety to non-target organisms. As with other microsporidian species, transmission is related to the tissue specificity of the pathogen. *Endoreticulatus schubergi*, strictly a midgut parasite, is spread via infective spores in the feces of the host. In contrast, *V. disparis*, primarily a fat body pathogen, has little egress from the host except when infections are very intense and persist for over two weeks, in which case, the Malpighian tubules become infected and some spores are present in the feces. More typically, *V. disparis* is transmitted by degradation of dead infected hosts in the environment (Goertz and Hoch, 2008a, b). *Nosema lymantriae* and *N. portugal* are more systemic, maturing in the silk glands (Fig. 7.10), fat body tissues, Malpighian tubules, and gonads. Environmental spores were found in the silk of *N. portugal* and it was hypothesized that the pathogen could be transmitted on silk trails (Jeffords *et al.*, 1987), but Goertz *et al.* (2007) found little evidence that the silk served as a conduit for infective *N. lymantriae* spores among susceptible larvae. *Nosema lymantriae*, like *V. disparis*, is probably spread via decomposing infected hosts, but, after a latency period, is also transmitted via the feces at an increasing rate over the course of infection (Goertz and Hoch, 2008a), and is transovarially transmitted from infected female to the offspring (Goertz and Hoch, 2008b). Gonads of the gypsy moth host are also infected by *V. disparis*, but this species is a more virulent pathogen and kills its host before eclosion, no matter the dosage or age at inoculation (Goertz and Hoch, 2008b). Field cage experiments quantified horizontal transmission of *N. lymantriae* and also demonstrated the latency period for this microsporidium (Hoch *et al.*, 2008).

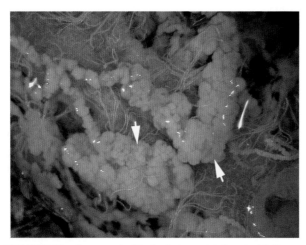

FIGURE 7.10 Gypsy moth silk glands (arrows) infected with *Nosema lymantriae*. (*Photo by G. Hoch.*)

Test larvae acquired infections nearly three weeks after exposure to newly inoculated treated larvae.

Solter *et al.* (1997) tested 50 species of North American forest Lepidoptera for susceptibility to *N. portugal*, two isolates of *N. lymantriae*, *V. disparis*, and *E. schubergi*. All isolates were infective to some of the lepidopteran species orally inoculated in the laboratory, and *E. schubergi* infections were often patent, but atypical *V. disparis* and *N. lymantriae* infections observed in many non-target hosts suggested that the pathogens would not reproduce optimally and would not be transmitted. Using gypsy moth as a model non-target host, Solter and Maddox (1998b) determined that non-target infections (microsporidia of native North American forest Lepidoptera infective to gypsy moth) were not horizontally transmitted among gypsy moth larvae. In addition, field studies in Bulgaria and Slovakia found no Lepidoptera sympatric with gypsy moth to be infected with naturally occurring gypsy moth microsporidia (Solter *et al.*, 2000, 2010). Inundative sprays of *V. disparis* in Slovakia (Solter *et al.*, 2010) caused apparently "dead-end" *V. disparis* infections in limited numbers of several non-target species; some infections were atypical, and no microsporidia were found infecting the susceptible species in subsequent years. Only one non-target infection was found where *N. lymantriae* was sprayed. Based on host specificity studies, US regulatory agencies permitted releases of Bulgarian isolates of *V. disparis* and *N. lymantriae* in 2008 and 2010. Studies of releases into rising natural gypsy moth populations in Illinois and in naïve Bulgarian populations are ongoing (L. F. Solter, D. K. Pilarska, and A. Linde, unpubl. data).

7.4.7. Microsporidia Infecting Biological Control Agents

The introduction of biological control agents to a new system is a complicated process encompassing a range of issues including host specificity, mass production, establishment in the field, and efficacy of control of the target pest. Another potential layer of complexity includes accidental importation of natural enemies of the agent as "contaminants" (Goettel and Inglis, 2006). While the more easily observed predators and parasites of a putative biological control agent usually can be eliminated during a quarantine period, pathogens may be much more difficult to detect, particularly if prevalence is initially low and effects are chronic, all typical for microsporidian infections. Unless obvious signs of disease occur in quarantined insects, pathogens arriving with their hosts may only begin to cause noticeable problems when mass rearing operations are established. High density and stresses on mass-reared insects can exacerbate disease transmission, resulting in unusually high prevalence and mortality rates (Goodwin, 1984), and infection can result in the loss of entire colonies

(Bhat *et al.*, 2009; Solter *et al.*, in press). A pathogenic organism released into the environment by way of infected biological control agents can negatively impact the release effort owing to debilitation and excessive mortality of the agent. There is also a risk that an exotic pathogen infecting the biological control agent can invade populations of closely related hosts in the field.

Goettel and Inglis (2006) suggested that the risk posed by pathogens of invertebrate biological control agents is greater for invertebrates collected from the field for direct release against a pest than for mass-reared agents that have been observed and screened for pathogens, and that exotic invertebrates intended for release as classical biological control agents may pose a higher risk than indigenous agents that are used in inundative biological control programs. Nevertheless, inundative releases of mass-reared ubiquitous (augmentative) agents may be problematic if epizootics occur during mass production and populations of conspecific naturally occurring predators and parasitoids are inadvertently exposed to the pathogen, as described for local bumble bee populations where commercially reared bees are used as pollinators (Colla *et al.*, 2006; Otterstatter and Thomson, 2008).

Goeden and Louda (1976) and Bjørnson and Schütte (2003) presented detailed discussions about the effects of microsporidia and other pathogens infecting a variety of insects reared for use in biological control programs. A *Nosema* sp., probably *Nosema tyriae* (Canning *et al.*, 1999), for example, was cited as significantly reducing the effectiveness of the arctiid *Tyria jacobaeae* in a biological control program for tansy ragwort (Harris *et al.*, 1971; Hawkes, 1973), and the pyralid *Nosema cactoblastis* compromised field releases and insectary rearing of the cactus moth *Cactoblastis cactorum* in control programs for prickly pear cactus (Pettey, 1947). Microsporidian disease reduced the fecundity and predation activity of predatory mites, including *Amyblyseius* spp. (Van Der Geest, 2000; Bjørnson and Schütte, 2003) and *Phytoseiulus* spp. (Bjørnson and Keddie, 1999). In addition to these reports, Solter *et al.* (in press) recovered five microsporidian species from four species of coccinellid and derodontid beetles being mass reared for biological control of the hemlock woolly adelgid, *Adelges tsugae* (Reardon and Onken, 2004). Three of the species caused high mortality in the colonies and resulted in loss of mass-reared insects for the release effort. One species, isolated from the coccinellid *Sasajiscymnus tsugae*, produced heavy infections and mortality in three other species of predatory beetles being reared for the control effort, raising concerns that high infection prevalence in *S. tsugae* released in or near sites where other beetle species were released could compromise those release efforts.

The eradication of all pathogens is probably impossible for any insect rearing situation, particularly those that do not have obvious symptoms or cause acute mortality and are, thus, difficult to detect. The *A. tsugae* study and those previously reported (Bjørnson and Schütte, 2003; Goettel and Inglis, 2006), however, highlight the importance of determining whether deleterious infectious diseases are present in putative biological control agents and eliminating pathogens before full-scale mass production as initiated.

7.5. FUTURE RESEARCH DIRECTIONS

Microsporidia are important as regulators of natural insect populations and have potential as classical and augmentative biological control agents of insect pests. Species infecting beneficial and economically important insects are also potentially harmful if accidentally introduced to new habitats or to managed insect populations and colonies. Key areas of basic research needed include increased understanding of host−parasite interactions such as mechanisms of infection, transmission, and host specificity. Tissue specificity, host immune response, and evasion of immune response should be elucidated at the cellular and molecular levels, and model systems designed to better understand the ecology of microsporidian parasites from different host and habitat types.

Genome sequences of a diversity of microsporidian entomopathogens, together with those of their hosts, are essential to understanding gene function and interactions at the protein level, as well as contributing to a better understanding of phylogenetic relationships within and between species with invertebrate and vertebrate hosts. Taxonomic data, including identity of strains, will become increasingly important information to collect for pathogens under consideration for use in biological control programs.

The effects on host fitness caused by multiple parasitism, infection of individual hosts and host populations by multiple pathogen and parasite species, is also an important area of study. Unexpected and confounding results of introducing biological control agents may be explained and failures avoided if the antagonistic, additive, or synergetic effects of multiple parasitism are known and accounted for. In addition, multiple pathogens and parasites can cause major problems with economically important managed insects; declines in honey bee populations, for example, have been hypothesized to be related to high loads of multiple pathogen and parasite species in high-density, heavily managed apiaries.

Several extensive efforts to use microsporidia as classical biological control agents in a variety of habitats have been undertaken. They appear to be most important in systems where other compatible natural enemies are present, where transmission potential is high, where suppression rather than eradication is the goal, and where immediate control of pests is not required. Efficacy in aquatic systems is confounded by the frequent necessity for fast-acting control measures, inconsistent persistence of the pathogens and hosts, and concerns about pathogens of mosquitoes, which take blood meals from humans and other vertebrate animals. Evidence that some mosquito-infecting microsporidia or closely related species can or do infect humans, albeit infrequently, has resulted in eliminating some species from consideration as biological control agents. Nevertheless, augmentative releases of some microsporidian species into naïve host populations is a viable method of increasing the load of natural enemies for long-term control of pest aquatic insects.

The situation is similar for microsporidia of terrestrial pests; for short-term control needed in most agricultural settings, microsporidia are not sufficiently virulent to act as microbial insecticides. Their best use is as augmentative or classical biological control agents in less intense settings, e.g., forests, rangelands, orchards, and prairies. *Paranosema locustae* has arguably had the most significant impact by controlling grasshoppers in the rangeland ecosystem (Lange, 2005; Shi *et al.*, 2009). High virulence and abundant, easily dispersed spores all contribute to its success, and it is still sold in North America as a "garden use" microbial insecticide. The inherent chronic effects caused by most microsporidian infections, however, have prevented declarations of completely successful biological control using these unique pathogens. In general, completely successful biological control requires a complex of diverse agents, and certainly microsporidia can and do play a part in suppression of many insect species. To enhance the utilization and impact of microsporidia, future efforts to characterize microsporidia and their interactions with their hosts at all levels will aid in the discovery of exceptionally adapted strains or cryptic species for biological control.

REFERENCES

Allen, G. E., & Buren, W. F. (1974). Microsporidian and fungal diseases of *Solenopsis invicta* Buren in Brazil. *J.N.Y. Entomol. Soc., 82*, 125−130.

Allen, J. C., Foltz, J. L., Dixon, W. N., Liebhold, A. M., Colbert, J. J., Regniere, J., Gray, D. R., Wilder, J. W., & Christie, I. (1993). Will the gypsy moth become a pest in Florida? *Fla. Entomol., 76*, 102−113.

Anderson, R. M., & May, R. M. (1980). Infectious diseases and population cycles of forest insects. *Science, 210*, 658−661.

Andreadis, T. G. (1983). Life cycle and epizootiology of *Amblyospora* sp. (Microspora: Amblyosporidae) in the mosquito, *Aedes cantator*. *J. Protozool., 30*, 509−518.

Andreadis, T. G. (1984). Epizootiology of *Nosema pyrausta* in field populations of the European corn borer (Lepidoptera: Pyralidae). *Environ. Entomol., 13*, 882−887.

Andreadis, T. G. (1986). Dissemination of *Nosema pyrausta* in feral population of the European corn borer *Ostrinis nublilis*. *J. Invertebr. Pathol., 48*, 335−343.

Andreadis, T. G. (1988a). *Amblyospora connecticus* sp. nov. (Microsporida: Amblyosporidae): horizontal transmission studies in the mosquito *Aedes cantator* and formal description. *J. Invertebr. Pathol., 52*, 90−101.

Andreadis, T. G. (1988b). Comparative susceptibility of the copepod *Acanthocyclops vernalis* to a microsporidian parasite, *Amblyospora connecticus* from the mosquito, *Aedes cantator*. *J. Invertebr. Pathol., 52*, 73−77.

Andreadis, T. G. (1989a). Host specificity of *Amblyospora connecticus*, a polymorphic microsporidian parasite of *Aedes cantator*. *J. Med. Entomol., 26*, 140−145.

Andreadis, T. G. (1989b). Infection of a field population of *Aedes cantator* with a polymorphic microsporidium, *Amblyospora connecticus* via release of the intermediate copepod host, *Acanthocyclops vernalis*. *J. Am. Mosq. Control Assoc., 5*, 81−85.

Andreadis, T. G. (1990a). Epizootiology of *Amblyospora connecticus* (Microsporida) in field populations of the saltmarsh mosquito, *Aedes cantator*, and the cyclopoid copepod, *Acanthocyclops vernalis*. *J. Protozool., 37*, 174−182.

Andreadis, T. G. (1990b). Polymorphic microsporidia of mosquitoes: potential for biological control. In R. E. Baker & P. E. Dunn (Eds.), *New Directions in Biological Control: Alternatives for Suppressing Agricultural Pests and Diseases* (pp. 177−188). New York: Alan R. Liss.

Andreadis, T. G. (1994). Host range tests with *Edhazardia aedis* (Microsporida: Culicosporidae) against northern Nearctic mosquitoes. *J. Invertebr. Pathol., 64*, 46−51.

Andreadis, T. G. (2005). Evolutionary strategies and adaptations for survival between mosquito−parasitic microsporidia and their intermediate copepod hosts: a comparative examination of *Amblyospora connectius* and *Hyalinocysta chapmani* (Microsporidia: Amblyosporidae). *Folia Parasitol., 52*, 23−35.

Andreadis, T. G. (2007). Microsporidian parasites of mosquitoes. *J. Am. Mosq. Control Assoc., 23*, 3−29.

Andreadis, T. G., Dubois, N. R., Moore, R. E. B., Anderson, J. F., & Lewis, F. B. (1983). Single applications of high concentrations of *Bacillus thuringiensis* for control of gypsy moth (Lepidoptera: Lymantriidae) populations and their impact on parasitism and disease. *J. Econ. Entomol., 76*, 1417−1422.

Anthony, D. W., Savage, K. E., Hazard, E. I., Avery, S. W., Boston, M. D., & Oldacre, S. W. (1978). Field tests with *Nosema algerae* Vávra and Undeen (Microsporida, Nosematidae) against *Anopheles albimanus* Wiedemann in Panama. *Misc. Publ. Entomol. Soc. Am., 11*, 17−28.

Ascunce, M. S., Valles, S. M., Oi, D. H., Shoemaker, D., Plowes, R., Gilbert, L., LeBrun, E. G., Sánchez-Arroyo, H., & Sanchez-Peña, S. (2010). Molecular diversity of the microsporidium *Kneallhazia solenopsae* reveals an expanded host range among fire ants in North America. *J. Invertebr. Pathol., 105*, 279−288.

Ascunce, M. S., Yang, C. C., Oakey, J., Calcaterra, L., Wu, W. J., Shih, C. J., Goudet, J., Ross, K. G., & Shoemaker, D. (2011). Global invasion history of the fire ant, *Solenopsis invicta*. *Science, 331*, 1066−1068.

Avery, S. W. (1989). Horizontal transmission of *Parathelohania obesa* (Protozoa: Microsporidia) to *Anopheles quadrimaculatus* (Diptera: Culicidae). *J. Invertebr. Pathol., 53*, 424−426.

Avery, S. W., & Undeen, A. H. (1990). Horizontal transmission of *Parathelohania anophelis* to the copepod *Microcyclops varicans*, and the mosquito, *Anopheles quadrimaculatus. J. Invertebr. Pathol., 56*, 98–105.

Baker, M. D., Vossbrinck, C. R., Maddox, J. V., & Undeen, A. H. (1994). Phylogenetic relationships among *Vairimorpha* and *Nosema* species (Microspora) based on ribosomal RNA sequence data. *J. Invertebr. Pathol., 64*, 100–106.

Baker, M. D., Vossbrinck, C. R., Becnel, J. J., & Andreadis, T. G. (1998). Phylogeny of *Amblyospora* (Microsporida: Amblyosporidae) and related genera based on small subunit ribosomal DNA data: a possible example of host parasite cospeciation. *J. Invertebr. Pathol., 71*, 199–206.

Balbiani, G. (1882). Sur les microsporidies ou psorospermies des articulés. *C.R. Acad. Sci., 95*, 1168–1171.

Bauer, L. S., & Nordin, G. L. (1989a). Effect of *Nosema fumiferanae* (Microsporida) on fecundity fertility and progeny performance of *Choristoneura fumiferana* (Lepidoptera: Tortricidae). *Environ. Entomol., 18*, 261–265.

Bauer, L. S., & Nordin, G. L. (1989b). Response of spruce budworm (Lepidoptera: Tortricidae infected with *Nosema fumiferanae* (Microsporida) to *Bacillus thuringiensis* treatments. *Environ. Entomol., 18*, 816–821.

Bauer, L. S., Miller, D. L., Maddox, J. V., & McManus, M. L. (1998). Interactions between a *Nosema* sp. (Microspora: Nosematidae) and nuclear polyhedrosis virus infecting the gypsy moth, *Lymantria dispar* (Lepidoptera: Lymantriidae). *J. Invertebr. Pathol., 74*, 147–153.

Beck, L., Rombke, J., Ruf, A., Prinzing, A., & Woas, S. (2004). Effects of diflubenzuron and *Bacillus thuringiensis* var. *kurstaki* toxin on soil invertebrates of a mixed deciduous forest in the Upper Rhine Valley, Germany. *Eur. J. Soil Biol., 40*, 55–62.

Becnel, J. J. (1990). *Edhazardia aedis* (Microsporida: Amblyosporidae) as a biocontrol agent of *Aedes aegypti* (Diptera: Culicidae). *Proc. 5th Int. Colloq. Invertebr. Pathol. Microb. Control*, (pp. 56–59) Australia: Adelaide.

Becnel, J. J. (1992a). Horizontal transmission and subsequent development of *Amblyospora californica* (Microsporida: Amblyosporidae) in the intermediate and definitive hosts. *Dis. Aquat. Organ., 13*, 17–28.

Becnel, J. J. (1992b). Safety of *Edhazardia aedis* (Microspora, Amblyosporidae) for nontarget aquatic organisms. *J. Am. Mosq. Control, 8*, 256–260.

Becnel, J. J. (1994). Life cycles and host–parasite relationships of microsporidia in culicine mosquitoes. *Folia Parasitol., 41*, 91–96.

Becnel, J. J., & Andreadis, T. G. (1999). Microsporidia in insects. In M. Wittner & L. M. Weiss (Eds.), *The Microsporidia and Microsporidiosis* (pp. 447–501). Washington, DC: American Society for Microbiology Press.

Becnel, J. J., & Johnson, M. A. (1993). Mosquito host range and specificity of *Edhazardia aedis* (Microspora: Culicosporidae). *J. Am. Mosq. Control Assoc., 9*, 269–274.

Becnel, J. J., & Johnson, M. A. (2000). Impact of *Edhazardia aedis* (Microsporidia: Culicosporidae) on a seminatural population of *Aedes aegypti* (Diptera: Culicidae). *Biol. Control, 18*, 39–48.

Becnel, J. J., Sprague, V., Fukuda, T., & Hazard, E. I. (1989). Development of *Edhazardia aedis* (Kudo, 1930) n. g., n. comb. (Microsporida: Amblyosporidae) in the mosquito *Aedes aegypti* (L.) (Diptera: Culicidae). *J. Protozool., 36*, 119–130.

Becnel, J. J., Garcia, J. J., & Johnson, M. A. (1995). *Edhazardia aedis* (Microspora: Culicosporidae) effects on the reproductive capacity of *Aedes aegypti* (Diptera: Culicidae). *J. Med. Entomol., 32*, 549–553.

Becnel, J. J., Jeyaprakash, A., Hoy, M. A., & Shapiro, A. (2002). Morphological and molecular characterization of a new microsporidian species from the predatory mite *Metaseiulus occidentalis* (Nesbitt) (Acari, Phytoseiidae). *J. Invertebr. Pathol., 79*, 163–172.

Becnel, J. J., White, S. E., & Shapiro, A. M. (2005). Review of microsporidia–mosquito relationships: from the simple to the complex. *Folia Parasitol., 52*, 41–50.

Beznoussenko, G. V., Dolgikh, V. V., Seliverstova, E. V., Semenov, P. B., Tokarev, Y. S., Trucco, A., Micaroni, M., Di Giandomenico, D., Auinger, P., Senderskiy, I. V., Skarlato, S. O., Snigirevskaya, E. S., Komissarchik, Y. Y., Pavelka, M., De Matteis, M. A., Luini, A., Sokolova, Y. Y., & Mironov, A. A. (2007). Analogs of the Golgi complex in microsporidia: structure and avesicular mechanisms of function. *J. Cell Sci., 120*, 1288–1298.

Bhat, S. A., Bashir, I., & Kamili, A. S. (2009). Microsporidiosis of silkworm, *Bombyx mori* (Lepidoptera: Bombycidae): a review. *Afr. J. Agric. Res., 4*, 1519–1523.

Bigliardi, E. (2001). Microsporidia, enigmatic parasites. *Ital. J. Zool., 68*, 263–271.

Bjørnson, S., & Keddie, B. A. (1999). Effects of *Microsporidium phytoseiuli* (Microsporidia) on the performance of the predatory mite, *Phytoseiulus persimilis* (Acari: Phytoseiidae). *Biol. Control, 15*, 153–161.

Bjørnson, S., & Schütte, C. (2003). Pathogens of mass-produced natural enemies and pollinators. In J. C. Van Lenteren (Ed.), *Quality Control and Production of Biological Control Agents, Theory and Testing Procedures* (pp. 133–165). Wallingford: CABI.

Bode, W. M., & Calvin, D. D. (1990). Yield–loss relationships and economic injury levels for European corn borer (Lepidoptera: Pyralidae) populations infesting Pennsylvania field corn. *J. Econ. Entomol., 83*, 1595–1603.

Bonafonte, M. T., Stewart, J., & Mead, J. R. (2001). Identification of two putative ATP-cassette genes in *Encephalitozoon intestinalis. Int. J. Parasitol., 31*, 1681–1685.

Bouzat, J. L., McNeil, L. K., Robertson, H. M., Solter, L. F., Nixon, J., Beever, J. E., Gaskins, H. R., Olsen, G., Subramaniam, S., Sogin, M. L., & Lewin, J. A. (2000). Phylogenomic analysis of the alpha proteasome gene family from early diverging eukaryotes. *J. Mol. Evol., 51*, 532–543.

Briano, J. A. (2005). Long-term studies of the red imported fire ant, *Solenopsis invicta*, infected with the microsporidia *Vairimorpha invictae* and *Thelohania solenopsae* in Argentina. *Environ. Entomol., 34*, 124–132.

Briano, J. A., & Williams, D. F. (2002). Natural occurrence and laboratory studies of the fire ant pathogen *Vairimorpha invictae* (Microsporida: Burenellidae) in Argentina. Environ. *Entomol., 31*, 887–894.

Briano, J. A., Patterson, R. S., & Cordo, H. A. (1995a). Long-term studies of the black imported fire ant (Hymenoptera: Formicidae) infected with a microsporidium. *Environ. Entomol., 24*, 1328–1332.

Briano, J. A., Jouvenaz, D. P., Wojcik, D. P., Cordo, H. A., & Patterson, R. S. (1995b). Protozoan and fungal diseases in *Solenopsis richteri* and *S. quinquecuspis* (Hymenoptera: Formicidae), in Buenos Aires province, Argentina. *Fla. Entomol., 78*, 531–537.

Briano, J. A., Patterson, R. S., Becnel, J. J., & Cordo, H. A. (1996). The black imported fire ant, *Solenopsis richteri*, infected with *Thelohania solenopsae*: intracolonial prevalence of infection and evidence for transovarial transmission. *J. Invertebr. Pathol., 67*, 178–179.

Briano, J. A., Williams, D. F., Oi, D. H., & Davis, L. R., Jr. (2002). Field host range of the fire ant pathogens *Thelohania solenopsae*

(Microsporida: Thelohaniidae) and *Vairimorpha invictae* (Microsporida: Burenellidae) in South America. *Biol. Control, 24,* 98–102.

Briano, J., Calcaterra, L., Vander Meer, R. K., Valles, S. M., & Livore, J. (2006). New survey for the fire ant microsporidia *Vairimorpha invictae* and *Thelohania solenopsae* in southern South America, with observations on their field persistence and prevalence of dual infections. *Environ. Entomol., 35,* 1358–1365.

Brooks, W. M. (1974). Protozoan infections. In G. E. Cantwell (Ed.), *Insect Diseases,* Vol. 1 (pp. 237–300). New York: Marcel Dekker.

Brooks, W. M. (1980). Production and efficacy of protozoa. *Biotechnol. Bioeng., 22,* 1415–1440.

Brooks, W. M. (1988). Entomogenous protozoa. In C. M. Ignoffo (Ed.), *CRC Handbook of Natural Pesticides. Microbial Insecticides, Part A: Entomogenous Protozoa and Fungi* (pp. 1–149). Boca Raton: CRC Press.

Brooks, W. M., Becnel, J. J., & Kennedy, G. G. (1988). Establishment of *Endoreticulatus* N.G. for *Pleistophora fidelis* (Hostounský & Weiser, 1975) (Microsporida: Pleistophoridae) based on the ultrastructure of a microsporidium in the Colorado potato beetle, *Leptinotarsa decemlineata* (Say) (Coleoptera: Chrysomelidae). *J. Eukaryot. Microbiol., 35,* 481–488.

Burri, L., Williams, B. A. P., Bursac, D., Lithgow, T., & Keeling, P. J. (2006). Microsporidian mitosomes retain elements of the general mitochondrial targeting system. *Proc. Natl. Acad. Sci. USA, 103,* 15916–15920.

Cali, A., Weiss, L. M., & Takvorian, P. M. (2005). An analysis of microsporidian genus *Brachiola,* with comparisons of human and insect isolates of *Brachiola algerae. J. Eukaryot. Microbiol., 51,* 678–685.

Cali, A., Neafie, R., Weiss, L. M., Ghosh, K., Vergara, R. B., Gupta, R., & Takvorian, P. M. (2010). Human vocal cord infection with the microsporidium *Anncaliia algerae. J. Eukaryot. Microbiol., 57,* 562–567.

Cameron, S. A., Lozier, J. D., Strange, J. P., Koch, J. B., Cordes, N., Solter, L. F., & Griswold, T. L. (2011). Recent widespread decline of some North American bumble bees: current status and causal factors. *Proc. Natl. Acad. Sci. USA, 108,* 662–667.

Campbell, R. W., & Podgwaite, J. D. (1971). The disease complex of the gypsy moth. I. Major components. *J. Invertebr. Pathol., 18,* 101–107.

Canning, E. U. (1953). A new microsporidian, *Nosema locustae* n.sp., from the fat body of the African migratory locust, *Locusta migratoria migratorioides* R. & F. *Parasitology, 43,* 287–290.

Canning, E. U. (1962a). The life cycle of *Nosema locustae* Canning in *Locusta migratoria migratorioides* (Reiche and Fairmaire) and its infectivity to other hosts. *J. Invertebr. Pathol., 4,* 237–247.

Canning, E. U. (1962b). The pathogenicity of *Nosema locustae* Canning. *J. Invertebr. Pathol., 4,* 248–256.

Canning, E. U., Wigley, P. J., & Barker, R. J. (1983). The taxonomy of three species of microsporidia (Protozoa: Microspora) from an oakwood population of winter moths *Operophtera brumata* (L.) (Lepidoptera: Geometridae). *Syst. Parasitol., 5,* 147–159.

Canning, E. U., Barker, R. J., Nicholas, J. P., & Page, A. M. (1985). The ultrastructure of three microsporidia from winter moth, *Operophtera brumata* (L.), and the establishment of a new genus *Cystosporogenes* n.g. for *Pleistophora operophterae* (Canning, 1960). *Syst. Parasitol., 7,* 213–225.

Canning, E. U., Curry, A., Cheney, S. A., Lafranchi-Tristem, N. J., Iwano, H., & Ishihara, R. (1999). *Nosema tyriae* n.sp and *Nosema* sp.,

microsporidian parasites of cinnabar moth *Tyria jacobaeae. J. Invertebr. Pathol., 74,* 29–38.

Canning, E. U., Curry, A., Cheney, S., Lafranchi-Tristem, N. J., & Haque, M. A. (2000). *Vairimorpha imperfecta* n. sp., a microsporidian exhibiting an abortive octosporous sporogony in *Plutella xylostella* L. (Lepidoptera: Yponomeutidae). *Parasitology, 119,* 273–286.

Capinera, J. L., & Seachrist, T. S. (1982). Grasshoppers (Acrididae) of Colorado: identification, biology, and management. *Colorado State Univ. Exp. Sta. Bull.* 584S.

Chen, J. S. C., Snowden, K., Mitchell, F., Sokolova, J., Fuxa, J., & Vinson, S. B. (2004). Sources of spores for the possible horizontal transmission of *Thelohania solenopsae* (Microspora: Thelohaniidae) in the red imported fire ants, *Solenopsis invicta. J. Invertebr. Pathol., 85,* 139–145.

Colla, S. R., Otterstatter, M. C., Gegear, R. J., & Thomson, J. D. (2006). Plight of the bumble bee: pathogen spillover from commercial to wild populations. *Biol. Conserv., 129,* 461–467.

Collins, H. L., Callcott, A. M., Lockley, T. C., & Ladner, A. (1992). Seasonal trends in effectiveness of hydramethylnon (AMDRO) and fenoxycarb (LOGIC) for control of red imported fire ants (Hymenoptera: Formicidae). *J. Econ. Entomol., 85,* 2131–2137.

Cook, T. J. (2002). Studies of naturally occurring *Thelohania solenopsae* (Microsporida: Thelohaniidae) infection in red imported fire ants, *Solenopsis invicta* (Hymenoptera: Formicidae). *Environ. Entomol., 31,* 1091–1096.

Cook, T. J., Lowery, M. B., Frey, T. N., Rowe, K. E., & Lynch, L. R. (2003). Effect of *Thelohania solenopsae* (Microsporida: Thelohaniidae) on weight and reproductive status of polygynous red imported fire ant, *Solenopsis invicta* (Hymenoptera: Formicidae), alates. *J. Invertebr. Pathol., 82,* 201–203.

Cordes, N. (2010). *The role of pathogens in the decline of North American bumble bees with a focus on the microsporidium Nosema bombi.* MS thesis. Urbana-Champaign: University of Illinois.

Corradi, N., & Keeling, P. J. (2009). Microsporidia: a journey through radical taxonomic revisions. *Fungal Biol. Rev., 23,* 1–8.

Corradi, N., & Slamovits, C. H. (2010). The intriguing nature of microsporidian genomes. *Brief. Funct. Genomics, 10,* 115–124.

Cossentine, J. E., & Lewis, L. C. (1987). Development of *Macrocentrus grandii* Goidanich within microsporidian-infected *Ostrinia nubilalis* Hübner host larvae. *Can. J. Zool., 65,* 2532–2535.

Coyle, C. M., Weiss, L. M., Rhodes, L. V., III, Cali, A., Takvorian, P. M., Brown, D. F., Visvesvara, G. S., Xiao, L., Naktin, J., Young, E., Gareca, M., Colasante, G., & Wittner, M. (2004). Fatal myositis due to the microsporidian *Brachiola algerae,* a mosquito pathogen. *N. Engl. J. Med., 351,* 42–47.

CSREES. (2008). *Bt corn & European corn borer. Long term success through resistance management.* University of Minnesota Extension Publication WW-07055.

Dickson, D. L., & Barr, A. R. (1990). Development of *Amblyospora campbelli* (Microsporidia: Amblyosporidae) in the mosquito *Culiseta incidens* Thomson. *J. Protozool., 37,* 71–77.

Dorhout, D. L., Sappington, T. W., Lewis, L. C., & Rice, M. E. (2011). Flight behaviour of European corn borer infected with *Nosema pyrausta. J. Appl. Entomol., 135,* 25–37.

Down, R. E., Bell, H. A., Kirkbride, A. E., & Edwards, J. P. (2004). The pathogenicity of *Vairimorpha necatrix* (Microspora: Microsporidia) against the tomato moth, *Lacanobia oleracea* (Lepidoptera:

Noctuidae) and its potential use for the control of lepidopteran glasshouse pests. *Pest Manag. Sci., 60,* 755−764.

Dunn, A. M., Terry, R., & Smith, J. E. (2001). Transovarial transmission in the microsporidia. *Adv. Parasitol., 48,* 57−100.

Eisler, R. (2000). *Handbook of Chemical Risk Assessment: Health Hazards to Humans, Plants, and Animals, Vol. 2, Organics.* Boca Raton: Lewis Publishers, CRC Press.

Ewen, A. B., & Mukerji, M. K. (1980). Evaluation of *Nosema locustae* (Microsporida) as a control agent of grasshopper populations in Saskatchewan. *J. Invertebr. Pathol., 35,* 295−303.

Findley, A. M., Weidner, E. H., Carman, K. R., Xu, Z., & Goodvar, J. S. (2005). Role of the posterior vacuole in *Spraguea lophii* (Microsporidia) spore hatching. *Folia Parasitol., 52,* 111−117.

Foster, R. N., Reuter, K. C., Winks, K., Reule, T. E., & Richard, R. D. (2001). *Biological control of leafy spurge,* Euphorbia esula *L.: impacts of eight rangeland grasshopper insecticide treatments on* Aphthona lacertosa *(Rosh.) and* A. nigriscutis *Foudras (Coleoptera: Chrysomelidae) in western N. Dakota, Final Report.* USDA-APHIS-PPQ, Center Plant Health Sci. and Technol. Phoenix: Decision Support and Pest Manag. Systems Lab.

van Frankenhuyzen, K., Nystrom, C. W., & Tabashnik, B. E. (1995). Variation in tolerance to *Bacillus thuringiensis* among and within populations of the spruce budworm (Lepidoptera: Tortricidae) in Ontario. *J. Econ. Entomol., 88,* 97−105.

van Frankenhuyzen, K., Ebling, P., McCron, B., Ladd, T., Gauthier, D., & Vossbrinck, C. (2004). Occurrence of *Cystosporogenes* sp. (Protozoa, Microsporidia) in a multi-species insect production facility and its elimination from a colony of the eastern spruce budworm, *Choristoneura fumiferana* (Clem.) (Lepidoptera: Tortricidae). *J. Invertebr. Pathol., 87,* 16−28.

van Frankenhuyzen, K., Nystrom, C., & Liu, Y. (2007). Vertical transmission of *Nosema fumiferanae* (Microsporidia: Nosematidae) and consequences for distribution, post-diapause emergence and dispersal of second-instar larvae of the spruce budworm, *Choristoneura fumiferana* (Clem.) (Lepidoptera: Tortricidae). *J. Invertebr. Pathol., 96,* 173−182.

Franz, J. M., & Huger, A. M. (1971). Microsporidia causing the collapse of an outbreak of the green trotrix (*Tortrix viridana* L.) in Germany. *Proc. 4th Int. Colloq. Insect Pathol.,* 48−53.

Franzen, C. (2008). Microsporidia: a review of 150 years of research. *Open Parasitol. J., 2,* 1−34.

Fries, I., Granados, R. R., & Morse, R. A. (1992). Intracellular germination of spores of *Nosema apis. Z. Apidol., 23,* 61−70.

Frixione, E., Ruiz, L., Santillán, M., de Vargas, L. V., Tejero, J. M., & Undeen, A. H. (1992). Dynamics of polar filament discharge and sporoplasm expulsion by microsporidian spores. *Cell Motil. Cytoskeleton, 22,* 38−50.

Fu, X. J., Hunter, D. M., & Shi, W. P. (2010). Effect of *Paranosema* (*Nosema*) *locustae* (Microsporidia) on morphological phase transformation of *Locusta migratoria manilensis* (Orthoptera: Acrididae). *Biocontrol Sci. Technol., 20,* 683−693.

Futerman, P. H., Layen, S. J., Kotzen, M. L., Franzen, C., Kraaijeveld, A. R., & Godfray, H. C. (2006). Fitness effects and transmission routes of a microsporidian parasite infecting *Drosophila* and its parasitoids. *Parasitology, 132,* 479−492.

Fuxa, J. R., Milks, M. L., Sokolova, Y. Y., & Richter, A. R. (2005a). Interaction of an entomopathogen with an insect social form: an epizootic of *Thelohania solenopsae* (Microsporidia) in a population

of the red imported fire ant, *Solenopsis invicta. J. Invertebr. Pathol., 88,* 79−82.

Fuxa, J. R., Sokolova, Y. Y., Milks, M. L., Richter, A. R., Williams, D. F., & Oi, D. H. (2005b). Prevalence, spread, and effects of the microsporidium *Thelohania solenopsae* released into populations with different social forms of the red imported fire ant (Hymenoptera: Formicidae). *Environ. Entomol., 34,* 1139−1149.

Gaugler, R. R., & Brooks, W. M. (1975). Sublethal effects of infection by *Nosema heliothidis* in the corn earworm, *Heliothis zea. J. Invertebr. Pathol., 26,* 57−63.

Gisder, S., Möckel, N., Linde, A., & Genersch, E. (2011). A cell culture model for *Nosema ceranae* and *Nosema apis* allows new insights into the life cycle of these important honey bee-pathogenic microsporidia. *Environ. Microbiol., 13,* 404−413.

Glancey, B. M., & Lofgren, C. S. (1988). Adoption of newly-mated queens: a mechanism for proliferation and perpetuation of polygynous red imported fire ants, *Solenopsis invicta* Buren. *Fla. Entomol., 71,* 581−587.

Glancey, B. M., Vander Meer, R. K., Glover, A., Lofgren, C. S., & Vinson, S. B. (1981). Filtration of microparticles from liquids ingested by the red imported fire ant, Buren. *Insectes Soc., 28,* 395−401.

Goeden, R. D., & Louda, S. M. (1976). Biotic interference with insects imported for weed control. *Annu. Rev. Entomol., 21,* 325−342.

Goertz, D., & Hoch, G. (2008a). Horizontal transmission pathways of terrestrial microsporidia: A quantitative comparison of three pathogens infecting different organs in *Lymantria dispar* L. (Lep: Lymantriidae) larvae. *Biol. Control, 44,* 196−206.

Goertz, D., & Hoch, G. (2008b). Vertical transmission and overwintering of microsporidia in the gypsy moth, *Lymantria dispar. J. Invertebr. Pathol., 99,* 43−48.

Goertz, D., & Hoch, G. (2011). Modeling horizontal transmission of microsporidia infecting gypsy moth, *Lymantria dispar* (L.), larvae. *Biol. Control, 56,* 263−270.

Goertz, D., Solter, L. F., & Linde, A. (2007). Horizontal and vertical transmission of a *Nosema* sp. (Microsporidia) from *Lymantria dispar* (L.) (Lepidoptera: Lymantriidae). *J. Invertebr. Pathol., 95,* 9−16.

Goettel, M. S., & Inglis, G. D. (2006). Methods for assessment of contaminants of invertebrate biological control agents and associated risks. In F. Bigler, D. Babendreier & U. Kuhlmann (Eds.), *Environmental Impact of Invertebrates for Biological Control of Arthropods, Methods and Risk Assessment* (pp. 145−165). Wallingford: CABI.

Goodwin, R. H. (1984). Recognition and diagnosis of diseases in insectaries and the effects of disease agents on insect biology. In E. G. King & N. C. Leppla (Eds.), *Advances and Challenges in Insect Rearing* (pp. 96−130). New Orleans: US Department of Agriculture, Agricultural Research Service.

Haine, E. R., Bouchansaud, K., & Rigaud, T. (2005). Conflict between parasites with different transmission strategies infecting an amphipod host. *Proc. R. Soc. B., 272,* 2505−2510.

Hajek, A. E., McManus, M. L., & Delalibera, I., Jr. (2005). *Catalogue of introductions of pathogens and nematodes for classical biological control of insects and mites.* USDA. Forest Service FHTET-2005-05.

Han, M. S., & Watanabe, H. (1988). Transovarial transmission of two microsporidia in the silkworm, *Bombyx mori,* and disease occurrence in the progeny population. *J. Invertebr. Pathol., 51,* 41−45.

Harris, P., Wilkinson, A. T. S., Neary, M. E., & Thompson, L. S. (1971). *Senecio jacobaeae* L., tansy ragwort (Compositae). *Biological*

Control Programs Against Insects and Weeds in Canada 1959–1968. Slough: Commonwealth Agricultural Bureaux. (pp. 97-104). Tech. Comm. No. 4.

Hawkes, R. B. (1973). Natural mortality of cinnabar moth in California. *Ann. Entomol. Soc. Am., 66*, 137–146.

Hayman, J. R., Southern, T. R., & Nash, T. E. (2005). Role of sulfated glycans in adherence of the microsporidian *Encephalitozoon intestinalis* to host cells *in vitro*. *Infect. Immun., 73*, 841–848.

Hembree, S. C. (1979). Preliminary report of some mosquito pathogens from Thailand. *Mosq. News, 39*, 575–582.

Hembree, S. C. (1982). Dose–response studies of a new species of *per os* and vertically transmittable microsporidian pathogens of *Aedes aegypti* from Thailand. *Mosq. News, 42*, 55–61.

Hembree, S. C., & Ryan, J. R. (1982). Observations on the vertical transmission of a new microsporidian pathogen of *Aedes aegypti* from Thailand. *Mosq. News, 42*, 49–54.

Henry, J. E. (1971a). Experimental application of *Nosema locustae* for control of grasshoppers. *J. Invertebr. Pathol., 18*, 389–394.

Henry, J. E. (1971b). Epizootiology of infections by *Nosema locustae* Canning (Microsporida: Nosematidae) in grasshoppers. *Acrida, 1*, 111–120.

Henry, J. E., & Oma, E. A. (1974). Effect of prolonged storage of spores on field applications of *Nosema locustae* (Microsporida-Nosematidae) against grasshoppers. *J. Invertebr. Pathol., 23*, 371–377.

Henry, J. E., & Oma, E. A. (1981). Pest control by *Nosema locustae*, a pathogen of grasshoppers and crickets. In H. D. Burges (Ed.), *Microbial control of pests and plant diseases 1970–1980* (pp. 573–585). London: Academic Press.

Henry, J. E., Tiahrt, K., & Oma, E. A. (1973). Importance of timing, spore concentrations, and levels of spore carrier in applications of *Nosema locustae* (Microsporida: Nosematidae) for control of grasshoppers. *J. Invertebr. Pathol., 21*, 263–272.

Hewitt, G. B., & Onsager, J. A. (1983). Control of grasshoppers on rangeland in the United States – a perspective. *J. Range Manag., 36*, 202–207.

Hibbett, D. S., Binder, M., Bischoff, J. F., Blackwell, M., Cannon, P. F., Eriksson, O. E., Huhndorf, S., James, T., Kirk, P. M., Lucking, R., Lumbsch, H. T., Lutzoni, F., Matheny, P. B., McLaughlin, D. J., Powell, M. J., Redhead, S., Schoch, C. L., Spatafora, J. W., Stalpers, J. A., Vilgalys, R., Aime, M. C., Aptroot, A., Bauer, R., Begerow, D., Benny, G. L., Castlebury, L. A., Crous, P. W., Dai, Y.-C., Gams, W., Geiser, D. M., Griffith, G. W., Gueidan, C., Hawksworth, D. L., Hestmark, G., Hosaka, K., Humber, R. A., Hyde, K. D., Ironside, J. E., Koljalg, U., Kurtzman, C. P., Larsson, K.-H., Lichtwardt, R., Longcore, J., Miadlikowska, J., Miller, A., Moncalvo, J.-M., Mozley-Standridge, S., Oberwinkler, F., Parmasto, E., Reeb, V., Rogers, J. D., Roux, C., Ryvarden, L., Sampaio, J. P., Schüßler, A., Sugiyama, J., Thorn, R. G., Tibell, L., Untereiner, W. A., Walker, C., Wang, Z., Weir, A., Weiss, M., White, M. M., Winka, K., Yao, Y.-J., & Zhang, N. (2007). A higher-level phylogenetic classification of the Fungi. *Mycol. Res., 111*, 509–547.

Hill, R. E., & Gary, W. J. (1979). Effects of the microsporidium, *Nosema pyrausta*, on field populations of European corn borer in Nebraska. *Environ. Entomol., 8*, 91–95.

Hirt, R. P., Logsdon, J. M., Jr., Healy, B., Dorey, M. W., Dolittle, W. F., & Embley, T. M. (1999). Microsporidia are related to Fungi: evidence from the largest subunit of RNA polymerase II and other proteins. *Proc. Natl. Acad. Sci. USA, 96*, 580–585.

Hoch, G., Schafellner, C., Henn, M. W., & Schopf, A. (2002). Alterations in carbohydrate and fatty acid levels of *Lymantria dispar* larvae caused by a microsporidian infection and potential adverse effects on a co-occurring endoparasitoid, *Glyptapanteles liparidis*. *Arch. Insect Biochem. Physiol., 50*, 109–120.

Hoch, G., D'Amico, V. D., Solter, L. F., Zubrik, M., & McManus, M. L. (2008). Quantifying horizontal transmission of a microsporidian pathogen of the gypsy moth, *Lymantria dispar* (Lep., Lymantriidae) in field cage studies. *J. Invertebr. Pathol., 99*, 146–150.

Hyliš, M., Pilarska, D. K., Oborník, M., Vávra, J., Solter, L. F., Weiser, J., Linde, A., & McManus, M. L. (2006). *Nosema chrysorrhoeae* n. sp. (Microsporidia), isolated from browntail moth (*Euproctis chrysorrhoea* L.) (Lepidoptera, Lymantriidae) in Bulgaria: characterization and phylogenetic relationships. *J. Invertebr. Pathol., 91*, 105–114.

Ironside, J. E. (2007). Multiple losses of sex within a single genus of microsporidia. *BMC Evol. Biol., 7*, 48.

Iwano, H., & Ishihara, R. (1989). Intracellular germination of spores of a *Nosema* sp. immediately after their formation in cultured cell. *J. Invertebr. Pathol., 54*, 125–127.

James, T. Y., Kauff, F., Schoch, C. L., Matheny, P. B., Hofstetter, V., Cox, C. J., Celio, G., Gueidan, C., Fraker, E., Miadlikowska, J., Lumbsch, H. T., Rauhut, A., Reeb, V., Arnold, A. E., Amtoft, A., Stajich, J. E., Hosaka, K., Sung, G.-H., Johnson, D., O'Rourke, B., Crockett, M., Binder, M., Curtis, J. M., Slot, J. C., Wang, Z., Wilson, A. W., Schuszler, A., Longcore, J. E., O'Donnell, K., Mozley-Standridge, S., Porter, D., Letcher, P. M., Powell, M. J., Taylor, J. W., White, M. M., Griffith, G. W., Davies, D. R., Humber, R. A., Morton, J. B., Sugiyama, J., Rossman, A. Y., Rogers, J. D., Pfister, D. H., Hewitt, D., Hansen, K., Hambleton, S., Shoemaker, R. A., Kohlmeyer, J., Volkmann-Kohlmeyer, B., Spotts, R. A., Serdani, M., Crous, P. W., Hughes, K. W., Matsuura, K., Langer, E., Langer, G., Untereiner, W. A., Lucking, R., Budel, B., Geiser, D. M., Aptroot, A., Diederich, P., Schmitt, I., Schultz, M., Yahr, R., Hibbett, D. S., Lutzoni, F., McLaughlin, D. J., Spatafora, J. W., & Vilgalys, R. (2006). Reconstructing the early evolution of Fungi using a six-gene phylogeny. *Nature, 443*, 818–822.

Jeffords, M. R., Maddox, J. V., & O'Hayer, K. W. (1987). Microsporidian spores in gypsy moth larval silk: a possible route of horizontal transmission. *J. Invertebr. Pathol., 49*, 332–333.

Jeffords, M. R., Maddox, J. V., McManus, M. E., Webb, R. E., & Wieber, A. (1989). Evaluation of the overwintering success of two European microsporidia inoculatively released into gypsy moth populations in Maryland. *J. Invertebr. Pathol., 53*, 235–240.

Joern, A., & Gaines, S. B. (1990). Population dynamics and regulation in grasshoppers. In R. F. Chapman & A. Joern (Eds.), *Biology of Grasshoppers* (pp. 415–482). New York: Wiley.

Johnson, D. L. (1997). Nosematidae and other protozoa as agents for the control of grasshoppers and locusts: current status and prospects. *Mem. Entomol. Soc. Can., 171*, 375–389.

Johnson, D. L., & Dolinski, M. G. (1997). Attempts to increase the incidence and severity of infection of grasshoppers with the entomopathogen *Nosema locustae* Canning (Microsporida: Nosematidae) by repeated field application. *Mem. Entomol. Soc. Can., 171*, 391–400.

Johnson, D. L., & Pavlikova, E. (1986). Reduction of consumption by grasshoppers (Orthoptera: Acrididae) infected with *Nosema locustae* Canning (Microsporida: Nosematidae). *J. Invertebr. Pathol., 48*, 232–238.

Johnson, M. A., Becnel, J. J., & Undeen, A. H. (1997). A new sporulation sequence in *Edhazardia aedis* (Microsporidia: Culicosporidae), a parasite of the mosquito *Aedes aegypti* (Diptera: Culicidae). *J. Invertebr. Pathol., 70*, 69–75.

Jouvenaz, D. P., & Ellis, E. A. (1986). *Vairimorpha invictae* n. sp. (Microspora: Microsporida), a parasite of the red imported fire ant, *Solenopsis invicta* Buren (Hymenoptera: Formicidae). *J. Protozool., 33*, 457–461.

Jouvenaz, D. P., Banks, W. A., & Atwood, J. D. (1980). Incidence of pathogens in fire ants, *Solenopsis* spp., in Brazil. *Fla. Entomol., 63*, 345–346.

Katinka, M. D., Duprat, S., Cornillot, E., Méténier, G., Thomarat, F., Prensier, G., Barbe, V., Peyretaillade, E., Brottier, P., Wincker, P., Delbac, F., El Alaoui, H., Peyret, P., Saurin, W., Gouy, M., Weissenbach, J., & Vivarès, C. P. (2001). Genome sequence and gene compaction of the eukaryote parasite *Encephalitozoon cuniculi. Nature, 414*, 450–453.

Keeling, P. J. (2003). Congruent evidence from α-tubulin and β-tubulin gene phylogenies for a zygomycete origin of microsporidia. *Fungal Genet. Biol., 38*, 298–309.

Keeling, P. (2009). Five questions about microsporidia. *PLoS Pathog., 5*, e1000489.

Keeling, P. J., & Doolittle, W. F. (1996). Alpha-tubulin from early-diverging eukaryotic lineages and the evolution of the tubulin family. *Mol. Biol. Evol., 13*, 1297–1305.

Keeling, P. J., Fast, N. M., Law, J. S., Williams, B. A. P., & Slamovits, C. H. (2005). Comparative genomics of microsporidia. *Folia Parasitol., 52*, 8–14.

Keeling, P. J., Corradi, N., Morrison, H. G., Haag, K. L., Ebert, D., Weiss, L. M., Akiyoshi, D. E., & Tzipori, S. (2010). The reduced genome of the parasitic microsporidian *Enterocytozoon bieneusi* lacks genes for core carbon metabolism. *Genome Biol. Evol., 2*, 304–309.

Keohane, E. M., & Weiss, L. M. (1998). Characterization and function of the microsporidian polar tube: a review. *Folia Parasitol., 45*, 117–127.

Khurad, A. M., Raina, S. K., & Pandharipande, T. N. (1991). *In vitro* propagation of *Nosema locustae* using fat body cell line derived from *Mythimna convecta* (Lepidoptera: Noctuidae). *J. Protozool., 38*, 91S–93S.

Knell, J. D., & Allen, G. E. (1978). Morphology and ultrastructure of *Unikaryon minutum* (Microsporidia: Protozoa), a parasite of the southern pine beetle, *Dendroctonus frontalis. Acta Protozool., 17*, 271–278.

Knell, J. D., Allen, G. E., & Hazard, E. I. (1977). Light and electron microscope study of *Thelohania solenopsae* n. sp. (Microsporida: Protozoa) in the red imported fire ant, *Solenopsis invicta. J. Invertebr. Pathol., 29*, 192–200.

Koella, J. C., & Agnew, P. (1997). Blood-feeding success of the mosquito *Aedes aegypti* depends on the transmission route of its parasite *Edhazardia aedis. Oikos, 78*, 311–316.

Koella, J. C., Agnew, P., & Michalakis, Y. (1998). Coevolutionary interactions between host life histories and parasite life cycles. *Parasitology, 116*, S47–S56.

Koestler, T., & Ebersberger, I. (2011). Zygomycetes, microsporidia, and the evolutionary ancestry of sex determination. *Genome Biol. Evol., 3*, 186–194.

Kohler, S. L., & Wiley, M. J. (1992). Parasite-induced collapse of populations of a dominant grazer in Michigan streams. *Oikos, 65*, 443–449.

Kohlmayr, B., Weiser, J., Wegensteiner, R., Händel, U., & Zizka., Z. (2003). Infection of *Tomicus piniperda* (Col., Scolytidae) with *Canningia tomici* sp. n. (Microsporidia, Unikaryonidae). *Anz. Schädlingskunde, 76*, 65–73.

Kudo, R. (1921). Studies on microsporidia with special reference to those parasitic in mosquitoes. *J. Morphol., 35*, 153–193.

Kurtti, T. J., Ross, S. E., Liu, Y., & Munderloh, U. G. (1994). *In vitro* developmental biology and spore production in *Nosema furnacalis* (Microspora, Nosematidae). *J. Invertebr. Pathol., 63*, 188–196.

Kyei-Poku, G., Gauthier, D., & van Frankenhuyzen, K. (2008). Molecular data and phylogeny of *Nosema* infecting lepidopteran forest defoliators in the genera *Choristoneura* and *Malacosoma. J. Eukaryot. Microbiol., 55*, 51–58.

Kyei-Poku, G., Gauthier, D., Schwarz, R., & van Frankenhuyzen, K. (2011). Morphology, molecular characteristics and prevalence of a *Cystosporogenes* species (Microsporidia) isolated from *Agrilus anxius* (Coleoptera: Buprestidae). *J. Invertebr. Pathol., 107*, 1–10.

Lacey, L. A., Frutos, R., Kaya, H. K., & Vail, P. (2001). Insect pathogens as biological control agents: do they have a future? *Biol. Control, 21*, 230–248.

Lange, C. E. (2005). The host and geographical range of the grasshopper pathogen *Paranosema* (*Nosema*) *locustae* revisited. *J. Orthoptera Res., 14*, 137–141.

Lange, C. E., & Azzaro, F. G. (2008). New case of long-term persistence of *Paranosema locustae* (Microsporidia) in melanopline grasshoppers (Orthoptera: Acrididae: Melanoplinae) of Argentina. *J. Invertebr. Pathol., 99*, 357–359.

Lange, C. E., & Cigliano, M. M. (2005). Overview and perspectives on the introduction and establishment of the grasshopper (Orthoptera: Acridoidea) biocontrol agent *Paranosema locustae* (Canning) (Microsporidia) in the western pampas of Argentina. *Vedalia, 12*, 61–84.

Lange, C. E., Johny, S., Baker, M. D., Whitman, D. W., & Solter, L. F. (2009). A new *Encephalitozoon* species (Microsporidia) isolated from the lubber grasshopper, *Romalea microptera* (Beauvois) (Orthoptera: Romaleidae). *J. Parasitol., 95*, 976–986.

Larsson, J. I. R. (1993). Description of *Chytridiopsis trichopterae*, new species (Microspora, Chytridiopsidae), a microsporidian parasite of the caddis fly *Polycentropus flavomaculatus* (Trichoptera, Polycentropodidae), with comments on relationships between the families Chytridiopsidae and Metchnikovellidae. *J. Eukaryot. Microbiol., 40*, 37–48.

Larsson, J. I. R. (2005). Molecular versus morphological approach to microsporidian classification. *Folia Parasitol., 52*, 143–144.

Larsson, J. I. R. (2007). Cytological variation and pathogenicity of the bumble bee parasite *Nosema bombi* (Microspora, Nosematidae). *J. Invertebr. Pathol., 94*, 1–11.

Latchininsky, A. V., & VanDyke, K. A. (2006). Grasshopper and locust control with poisoned baits: a renaissance of the old strategy? *Outlooks Pest Manag., 17*, 105–111.

Lee, S. C., Corradi, N., Byrnes, E. J., III, Torres-Martinez, S., Dietrich, F. S., Keeling, P. J., & Heitman, J. (2008). Microsporidia evolved from ancestral sexual fungi. *Curr. Biol., 18*, 1675–1679.

Leonard, D. E. (1981). Bioecology of the gypsy moth. In D. C. Doane & M. L. McManus (Eds.), *The Gypsy Moth: Research toward Integrated Pest Management* (pp. 9–29). USDA Forest Service, Tech. Bull, 1584.

Lewis, L. C., Cossentine, J. E., & Gunnarson, R. D. (1983). Impact of two microsporidia, *Nosema pyrausta* and *Vairimorpha necatrix*, in *Nosema pyrausta* infected European corn borer (*Ostrinia nubilalis*) larvae. *Can. J. Zool., 61,* 915–921.

Lewis, L. C., Sumerford, D. V., Bing, L. A., & Gunnarson, R. D. (2006). Dynamics of *Nosema pyrausta* in natural populations of the European corn borer, *Ostrinia nubilalis*: a six year study. *BioControl, 51,* 627–642.

Lewis, L. C., Bruck, D. J., Prasifka, J. R., & Raun, E. S. (2009). *Nosema pyrausta*: its biology, history, and potential role in a landscape of transgenic insecticidal crops. *Biol. Control, 48,* 223–231.

Lockwood, D. R. (1993). Environmental issues involved in biological control of rangeland grasshoppers (Orthoptera: Acrididae) with exotic agents. *Environ. Entomol., 22,* 503–518.

Lockwood, D. R., & Lockwood, J. A. (2008). Grasshopper population ecology: catastrophe, criticality, and critique. *Ecol. Soc., 13,* 34.

Lockwood, J. A., Bomar, C. R., & Ewen, A. B. (1999). The history of biological control with *Nosema locustae*: lessons for locust management. *Insect Sci. Applic., 19,* 333–350.

Lockwood, J. A., Showler, A. T., & Latchininsky, A. V. (2001). Can we make locust and grasshopper management sustainable? *J. Orthoptera Res., 10,* 315–329.

Lockwood, J. A., Anderson-Sprecher, R., & Schell, S. P. (2002). When less is more: optimization of reduced agent-area treatments (RAATs) for management of rangeland grasshoppers. *Crop Prot., 21,* 551–562.

Lom, J., & Dykova, I. (2002). Ultrastructure of *Nucleospora secunda* n. sp. (Microsporidia), parasite of enterocytes of *Nothobranchius rubripinnis*. *Eur. J. Protistol., 38,* 19–27.

Lopez, M. D., Sumerford, D. V., & Lewis, L. C. (2010). *Nosema pyrausta* and Cry1Ab-incorporated diet led to decreased survival and developmental delays in European corn borer. *Entomol. Exp. Appl., 134,* 146–153.

Lowman, P. M., Takvorian, P. M., & Cali, A. (2000). The effects of elevated temperature and various time–temperature combinations on the development of *Brachiola* (*Nosema*) *algerae* n. comb, in mammalian cell culture. *J. Eukaryot. Microbiol., 47,* 221–234.

Maddox, J. V. (1973). The persistence of the microsporidia in the environment. *Misc. Publ. Entomol. Soc. Am., 9,* 99–104.

Maddox, J. V. (1977). Stability of entomopathogenic Protozoa. *Misc. Publ. Entomol. Soc. Am., 10,* 3–18.

Maddox, J. V. (1994). Insect pathogens as biological control agents. In R. L. Metcalf & W. H. Luckmann (Eds.), *Introduction to Insect Pest Management* (3rd ed.). (pp. 199–244) New York: John Wiley & Sons.

Maddox, J. V., & Solter, L. F. (1996). Long-term storage of infective microsporidian spores in liquid nitrogen. *J. Eukaryot. Microbiol., 43,* 221–225.

Maddox, J. V., & Sprenkel, R. K. (1978). Some enigmatic microsporidia of the genus *Nosema*. *Misc. Publ. Entomol. Soc. Am., 11,* 65–84.

Maddox, J. V., McManus, M. L., & Solter, L. F. (1998). Microsporidia affecting forest Lepidoptera. *Proceedings: Population Dynamics, Impacts and Integrated Management of Forest Defoliating Insects.* USDA Forest Service General Technical Report. NE-247.

Maddox, J. V., Baker, M., Jeffords, M. R., Kuras, M., Linde, A., McManus, M., Solter, L., Vávra, J., & Vossbrinck, C. (1999). *Nosema portugal* n sp., isolated from gypsy moths (*Lymantria dispar* L.) collected in Portugal. *J. Invertebr. Pathol., 73,* 1–14.

Magaud, A., Achbarou, A., & Desportes-Livage, I. (1997). Cell invasion by the microsporidium *Encephalitozoon intestinalis*. *J. Eukaryot. Microbiol., 44,* 81S.

McEwen, L. C., Althouse, C. M., & Peterson, B. E. (2000). Direct and indirect effects of grasshopper integrated pest management chemicals and biologicals on nontarget animal life. *Grasshopper Integrated Pest Management User Handbook.* Washington DC: USDA-APHIS.

McManus, M. L., & Solter, L. (2003). Microsporidian pathogens in European gypsy moth populations. *Proceedings: Ecology, Survey, and Management of Forest Insects* (pp. 44–51) USDA Forest Service, Northeast Research Station Gen. Tech. Rep., NE-311.

Micieli, M. V., Garcia, J. J., & Becnel, J. J. (2003). Life cycle and epizootiology of *Amblyospora ferocis* (Microspora: Amblyosporidae) in the mosquito *Psorophora ferox* (Diptera: Culicidae). *Folia Parasitol., 50,* 171–175.

Micieli, M. V., Garcia, J. J., & Andreadis, T. G. (2009). Factors affecting horizontal transmission of the microsporidium *Amblyospora albifasciati* to its intermediate copepod host *Mesocyclops annulatus*. *J. Invertebr. Pathol., 101,* 228–233.

Milks, M. L., Fuxa, J. R., & Richter, A. R. (2008). Prevalence and impact of the microsporidium *Thelohania solenopsae* (Microsporidia) on wild populations of red imported fire ants, *Solenopsis invicta*, in Louisiana. *J. Invertebr. Pathol., 97,* 91–102.

Monaghan, S. R., Rumney, R. L., Nguyen, Vo, T. K., Bols, N. C., & Lee, L. E. J. (2011). *In vitro* growth of microsporidia *Anncaliia algerae* in cell lines from warm water fish. *In Vitro Cell. Dev. Biol. Anim., 47,* 104–113.

Moodie, E. G., Le Jambre, L. F., & Katz, M. E. (2003). Ultrastructural characteristics and small subunit ribosomal DNA sequence of *Vairimorpha cheracis* sp. nov. (Microspora: Burenellidae), a parasite of the Australian yabby, *Cherax destructor* (Decapoda: Parastacidae). *J. Invertebr. Pathol., 84,* 198–213.

Moore, C. B., & Brooks, W. M. (1992). An ultrastructural study of *Vairimorpha necatrix* (Microspora: Microsporidia) with particular reference to episporontal inclusions during octosporogony. *J. Protozool., 39,* 392–398.

Moura, H., da Silva, A., Moura, I., Schwartz, D. A., Leitch, G., Wallace, S., Pieniazek, N. J., Wirtz, R. A., & Visvesvara, G. S. (1999). Characterization of *Nosema algerae* isolates after continuous cultivation in mammalian cells at 37°C. *J. Eukaryot. Microbiol., 46,* 14S–16S.

Nägeli, C. (1857). Über die neue Krankheit der Seidenraupe und verwandte Organismen. *Bot. Zeitung., 15,* 760–761.

Nielsen, C., Milgroom, M. G., & Hajek, A. E. (2005). Genetic diversity in the gypsy moth fungal pathogen *Entomophaga maimaiga* from founder populations in North America and source populations in Asia. *Mycol. Res., 109,* 941–950.

Novotny, J. (1988). The use of nucleopolyhedrosis virus NPV and microsporidia in the control of the gypsy moth *Lymantria dispar* L. *Folia Parasitol., 35,* 199–208.

Nylund, S., Nylund, A., Watanabe, K., Arnesen, C. E., & Karlsbakk, E. (2010). *Paranucleospora theridion* n. gen., n. sp. (Microsporidia, Enterocytozoonidae) with a life cycle in the salmon louse (*Lepeophtheirus salmonis*, Copepoda) and Atlantic salmon (*Salmo salar*). *J. Eukaryot. Microbiol., 57,* 95–114.

Oi, D. H., & Williams, D. F. (2002). Impact of *Thelohania solenopsae* (Microsporidia: Thelohaniidae) on polygyne colonies of red imported fire ants (Hymenoptera: Formicidae). *J. Econ. Entomol., 95*, 558−562.

Oi, D. H., & Williams, D. F. (2003). *Thelohania solenopsae* (Microsporidia: Thelohaniidae) infection in reproductives of red imported fire ants (Hymenoptera: Formicidae) and its implication for intercolony transmission. *Environ. Entomol., 32*, 1171−1176.

Oi, D. H., & Valles, S. M. (2009). Fire ant control with entomopathogens in the USA. In A. E. Hajek, T. R. Glare & M. O'Callaghan (Eds.), *Use of Microbes for Control and Eradication of Invasive Arthropods* (pp. 237−257). New York: Springer.

Oi, D. H., Becnel, J. J., & Williams, D. F. (2001). Evidence of intracolony transmission of *Thelohania solenopsae* (Microsporidia: Thelohaniidae) in red imported fire ants (Hymenoptera: Formicidae) and the first report of spores from pupae. *J. Invertebr. Pathol., 78*, 128−134.

Oi, D. H., Valles, S. M., & Pereira, R. M. (2004). Prevalence of *Thelohania solenopsae* (Microsporidia: Thelohaniidae) infection in monogyne and polygyne red imported fire ants (Hymenoptera: Formicidae). *Environ. Entomol., 33*, 340−345.

Oi, D. H., Briano, J. A., Valles, S. M., & Williams, D. F. (2005). Transmission of *Vairimorpha invictae* (Microsporidia: Burenellidae) infections between red imported fire ant (Hymenoptera: Formicidae) colonies. *J. Invertebr. Pathol., 88*, 108−115.

Oi, D. H., Williams, D. F., Pereira, R. M., Horton, P. M., Davis, T. S., Hyder, A. H., Bolton, H. T., Zeichner, B. C., Porter, S. D., Hoch, A. L., Boswell, M. L., & Williams, G. (2008). Combining biological and chemical controls for the management of red imported fire ants (Hymenoptera: Formicidae). *Am. Entomol., 54*, 46−55.

Oi, D. H., Porter, S. D., Valles, S. M., Briano, J. A., & Calcaterra, L. A. (2009). *Pseudacteon* decapitating flies (Diptera: Phoridae): are they potential vectors of the fire ant pathogens *Kneallhazia* (= *Thelohania*) *solenopsae* (Microsporidia: Thelohaniidae) and *Vairimorpha invictae* (Microsporidia: Burenellidae)? *Biol. Control, 48*, 310−315.

Oi, D. H., Valles, S. M., & Briano, J. A. (2010). Laboratory host specificity testing of the fire ant microsporidian pathogen *Vairimorpha invictae* (Microsporidia: Burenellidae). *Biol. Control, 53*, 331−336.

Oshima, K. (1964). Stimulative or inhibitive substance to evaginate the filament of *Nosema bombycis*. I. The case of artificial buffer solution. *Jpn. J. Zool., 14*, 209−229.

Otterstatter, M. C., & Thomson, J. D. (2008). Does pathogen spillover from commercially reared bumble bees threaten wild pollinators? *PLoS ONE, 3*, e2771.

Overton, K., Rao, A., Vinson, S. B., & Gold, R. E. (2006). Mating flight initiation and nutritional status (protein and lipid) of *Solenopsis invicta* (Hymenoptera: Formicidae) alates infected with *Thelohania solenopsae* (Microsporidia: Thelohaniidae). *Ann. Entomol. Soc. Am., 99*, 524−529.

Own, O. S., & Brooks, W. M. (1986). Interactions of the parasite *Pediobius foveolatus* (Hymenoptera: Eulophidae) with two *Nosema* spp. microsporidia (Nosematidae) of the Mexican bean beetle *Epilachna varivestis* (Coleoptera: Coccinellidae). *Environ. Entomol., 15*, 32−39.

Pasteur, L. (1870). *Études sur la Maladie des Vers à Soie*. Paris: Tome I et II. Gauthier Villars.

Patil, C. S., Jyothi, N. B., & Dass, C. M. S. (2002). Role of *Nosema bombycis* infected male silk moths in the venereal transmission of pebrine disease in *Bombyx mori* (Lep., Bombycidae). *J. Appl. Entomol., 126*, 563−566.

Paxton, R. J. (2010). Does infection by *Nosema ceranae* cause colony collapse disorder in honey bees (*Apis mellifera*)? *J. Apicult. Res., 49*, 80−84.

Peach, M. L., Alston, D. G., & Tepedino, V. J. (1994). Bees and bran bait; is carbaryl bran bait lethal to alfalfa leafcutting bee (Hymenoptera: Megachilidae) adults or larvae? *J. Econ. Entomol., 87*, 311−317.

Perry, W. B., Christianson, T. A., & Perry, S. A. (1997). Response of soil and leaf litter microarthropods to forest application of diflubenzuron. *Ecotoxicology, 6*, 87−99.

Pettey, F. W. (1947). The biological control of prickly pears in South Africa. *Union S. Africa Dept. Agric. Sci. Bull., 271*.

Petti, J. M. (1998). *The structure and function of the buccal tube filter in workers of the ant species* Solenopsis invicta, Camponotus floridanus, *and* Monomorium pharaonis. MS thesis. Gainesville: University of Florida.

Pfadt, R. E. (2002). Field Guide to the Common Western Grasshoppers, 3rd ed. *Wyoming Agric. Exp. Sta. Bull. 912*.

Pieniazek, N. J., da Silva, A. J., Slemenda, S. B., Visvesvara, G. S., Kurti, T. J., & Yasunaga, C. (1996). *Nosema trichoplusiae* is a synonym of *Nosema bombycis* based on the sequence of the small subunit ribosomal RNA coding region. *J. Invertebr. Pathol., 67*, 316−317.

Pierce, C. M. F., Solter, L. F., & Weinzierl, R. A. (2001). Interactions between *Nosema pyrausta* (Microsporida: Nosematidae) and *Bacillus thuringiensis* subsp. *kurstaki* in the European corn borer (Lepidoptera: Pyralidae). *J. Econ. Entomol., 94*, 1361−1368.

Pilarska, D. K., Solter, L. F., Maddox, J. V., & McManus, M. L. (1998). Microsporidia from gypsy moth (*Lymantria dispar* L.) populations in central and western Bulgaria. *Acta Zool. Bulgar., 50*, 109−113.

Pilarska, D., McManus, M., Hajek, A. E., Herard, F., Vega, F. E., Pilarski, P., & Markova, G. (2000). Introduction of the entomopathogenic fungus *Entomophaga maimaiga* Hum., Shim & Sop. (Zygomycetes: Entomophthorales) to a *Lymantria dispar* (L.) population in Bulgaria. *J. Pest Sci., 73*, 125−126.

Pilarska, D., Georgiev, G., McManus, M., Mirchev, P., Pilarski, P., & Linde, A. (2007). *Entomophaga maimaiga* − an effective introduced pathogen of the gypsy moth (*Lymantria dispar* L.) in Bulgaria. *Proc. Int. Conf. Alien Arthropods in South East Europe − Crossroad of Three Continents*, (pp. 37−43). Sofia: Bulgaria.

Pilley, B. M. (1976). New genus, *Vairimorpha* (Protozoa: Microsporida), for *Nosema necatrix* Kramer 1965: pathogenicity and life cycle in *Spodoptera exempta* (Lepidoptera: Noctuidae). *J. Invertebr. Pathol., 28*, 177−183.

Podgwaite, J. D. (1981). Natural disease within dense gypsy moth populations. In C. C. Doane & M. L. McManus (Eds.), *The Gypsy Moth: Research Toward Integrated Pest Management* (pp. 125−134). US Dept. Agric. Tech. Bull. No, 1584.

Podgwaite, J. D. (1999). Gypchek: biological insecticide for the gypsy moth. *J. For., 97*, 16−19.

Pomport-Castillon, C., Romestand, B., & De Jonchkheere, J. F. (1997). Identification and phylogenetic relationships of microsporidia by riboprinting. *J. Eukaryot. Microbiol., 44*, 540−544.

Porter, S. D., Valles, S. M., Davis, T. S., Briano, J. A., Calcaterra, L. A., Oi, D. H., & Jenkins, R. A. (2007). Host specificity of the microsporidian pathogen *Vairimorpha invictae* at five field sites with infected *Solenopsis invicta* fire ant colonies in northern Argentina. *Fla. Entomol., 90*, 447−452.

Preston, C. A., Fritz, G. N., & Vander Meer, R. K. (2007). Prevalence of *Thelohania solenopsae* infected *Solenopsis invicta* newly mated queens within areas of differing social form distributions. *J. Invertebr. Pathol., 94*, 119−124.

Purrini, K. (1975). On the distribution of microorganisms infecting the larvae of *Euproctis chrysorrhoea* L. (Lep., Lymantriidae) near Kosova, Yugoslavia. *Anz. Schädlingskde. Pflanzenschutz. Umweltschutz, 48*, 182−183.

Purrini, K., & Weiser, J. (1975). Natural enemies of *Euproctis chrysorrhoea* in orchards in Yugoslavia. *Anz. Schädlingskde. Pflanzenschutz, Umweltschutz, 48*, 11−12.

Purrini, K., & Weiser, J. (1985). Ultrastructural study of the microsporidian *Chytridiopsis typographi* (Chytridiopsida: Microspora) infecting the bark beetle, *Ips typographus* (Scolytidae: Coleoptera), with new data on spore dimorphism. *J. Invertebr. Pathol., 45*, 66−74.

Quinn, M. A., Kepner, R. L., Walgenbach, D. D., Foster, R. N., Bohls, R. A., Pooler, P. D., Reuter, K. C., & Swain, J. L. (1989). Immediate and 2nd-year effects of insecticide spray and bait treatments on populations of rangeland grasshoppers. *Can. Entomol., 121*, 589−602.

Quinn, M. A., Kepner, R. L., Walgenbach, D. D., Foster, R. N., Bohls, R. A., Pooler, P. D., Reuter, K. C., & Swain, J. L. (1990). Effect of habitat and perturbation on populations and community structure of darkling beetles (Coleoptera: Tenebrionidae) on mixed-grass rangeland. *Environ. Entomol., 19*, 1746−1755.

Raina, S. K., Das, S., Rai, M. M., & Khurad, A. M. (1995). Transovarial transmission of *Nosema locustae* (Microsporida: Nosematidae) in the migratory locust *Locusta migratoria migratorioides*. *Parasitol. Res., 81*, 38−44.

Reardon, R., & Onken, B. (2004). *Biological control of hemlock woolly adelgid*. USDA. Forest Service FHTET-2004-04.

Reardon, R. C., Podgwaite, J., & Zerillo, R. (2009). *Gypchek − biopesticide for the gypsy moth*. USDA. Forest Service FHTET-2009-1.

Redhead, S. A., Kirk, P., Keeling, P. J., & Weiss, L. M. (2009). Proposals to exclude the phylum Microsporidia from the Code. *Taxon, 58*, 10−11.

Régnière, J., & Nealis, V. G. (2008). The fine-scale population dynamics of spruce budworm: survival of early instars related to forest condition. *Ecol. Entomol., 33*, 362−373.

Reynolds, D. G. (1972). Experimental introduction of a microsporidian into a wild population of *Culex pipiens fatigans* Wied. *Bull. Org. Mond. Santé., 46*, 807−812.

Sagastume, S., del Águila, C., Martín-Hernández, R., Higes, M., & Henriques-Gil, N. (2011). Polymorphism and recombination for rDNA in the putatively asexual microsporidian *Nosema ceranae*, a pathogen of honeybees. *Environ. Microbiol., 13*, 84−95.

Sanchez-Peña, S. R., Chacón-Cardosa, M. C., & Resendez-Perez, D. (2009). Identification of fire ants (Hymenoptera: Formicidae) from northeastern Mexico with morphology and molecular markers. *Fla. Entomol., 92*, 107−115.

Schmid-Hempel, P., & Loosli, R. (1998). A contribution to the knowledge of *Nosema* infections in bumble bees, *Bombus* spp. *Apidologie, 29*, 525−535.

Selman, M., Pombert, J.-F., Solter, L., Farinelli, L., Weiss, L. M., Keeling, P. J., & Corradi, N. (2011). Acquisition of an animal gene by microsporidian intracellular parasites. *Curr. Biol., 21*, R576−R577.

Shapiro, A. M., Becnel, J. J., Oi, D. H., & Williams, D. F. (2003). Ultrastructural characterization and further transmission studies of *Thelohania solenopsae* from *Solenopsis invicta* pupae. *J. Invertebr. Pathol., 83*, 177−180.

Shi, W. P., Wang, Y. Y., Lv, F., Guo, C., & Cheng, X. (2009). Persistence of *Paranosema* (*Nosema*) *locustae* (Microsporidia: Nosematidae)

among grasshopper (Orthoptera: Acrididae) populations in the inner Mongolia rangeland, China. *BioControl, 54*, 77−84.

Sidor, C. (1976). Oboljenja Izazvana microorganizmima kod nekih Lymantriidae u Jugoslavifi I Njihov Znacaj za entomofaunu. (Diseases provoked with microorganisms by some Limantriidae in Yugoslavia and their importance for entomofauna.). *Arh. boil. nauka Beograd, 28*, 127−137.

Siegel, J. P., Maddox, J. V., & Ruesink, W. G. (1986a). Impact of *Nosema pyrausta* on a braconid, *Macrocentrus grandii*, in Central Illinois. *J. Invertebr. Pathol., 47*, 271−276.

Siegel, J. P., Maddox, J. V., & Ruesink, W. G. (1986b). Lethal and sublethal effects of *Nosema pyrausta* on the European corn borer (*Ostrinia nubilalis*) in Central Illinois. *J. Invertebr. Pathol., 48*, 167−173.

Siegel, J. P., Maddox, J. V., & Ruesink, W. G. (1988). Seasonal progress of *Nosema pyrausta* in the European corn borer, *Ostrinia nubilalis*. *J. Invertebr. Pathol., 52*, 130−136.

Slamovits, C. H., Williams, B. A. P., & Keeling, P. J. (2004). Transfer of *Nosema locustae* (Microsporidia) to *Antonospora locustae* n. comb. based on molecular and ultrastructural data. *J. Eukaryot. Microbiol., 51*, 207−213.

Smirnoff, W. A. (1967). Diseases of the larch sawfly, *Pristiphora erichsonii*, in Quebec. *J. Invertebr. Pathol., 10*, 417−424.

Sokolova, Y. Y., & Fuxa, J. R. (2008). Biology and life-cycle of the microsporidium *Kneallhazia solenopsae* Knell Allan Hazard 1977 gen. n., comb. n., from the fire ant *Solenopsis invicta*. *Parasitology, 135*, 903−929.

Sokolova, Y. Y., Dolgikh, V. V., Morzhina, E. V., Nassonova, E. S., Issi, I. V., Terry, R. S., Ironside, J. E., Smith, J. E., & Vossbrinck, C. R. (2003). Establishment of the new genus *Paranosema* based on the ultrastructure and molecular phylogeny of the type species *Paranosema grylli* Gen. Nov., Comb. Nov. (Sokolova, Selezniov, Dolgikh, Issi 1994), from the cricket *Gryllus bimaculatus* Deg. *J. Invertebr. Pathol., 84*, 159−172.

Sokolova, Y. Y., McNally, L. R., Fuxa, J. R., & Vinson, S. B. (2004). Spore morphotypes of *Thelohania solenopsae* (microsporidia) described microscopically and confirmed by PCR of individual spores microdissected from smears by position ablative laser microbeam microscopy. *Microbiol., 150*, 1261−1270.

Sokolova, Y. Y., Issi, I. V., Morzhina, E. V., Tokarev, Y. S., & Vossbrinck, C. R. (2005a). Ultrastructural analysis supports transferring *Nosema whitei* Weiser 1953 to the genus *Paranosema* and creation [of] a new combination, *Paranosema whitei*. *J. Invertebr. Pathol., 90*, 122−126.

Sokolova, Y. Y., Fuxa, J. R., & Borkhsenious, O. N. (2005b). The nature of *Thelohania solenopsae* (Microsporidia) cysts in abdomens of red imported fire ants, *Solenopsis invicta*. *J. Invertebr. Pathol., 90*, 24−31.

Sokolova, Y. Y., Sokolov, I. M., & Carlton, C. E. (2010). New microsporidia parasitizing bark lice (Insecta: Psocoptera). *J. Invertebr. Pathol., 104*, 186−194.

Solter, L. F., & Becnel, J. J. (2003). Environmental safety of microsporidia. In H. M. T. Hokkanen & A. E. Hajek (Eds.), *Environmental Impacts of Microbial Insecticides: Need and Methods for Risk Assessment* (pp. 93−118). New York: Kluwer Academic Publishers (Springer).

Solter, L. F., & Becnel, J. J. (2007). Entomopathogenic microsporidia. In L. Lacey & H. K. Kaya (Eds.), *Field Manual of*

Techniques in Invertebrate Pathology (2nd ed.). (pp. 199−221) Dordrecht: Springer.

Solter, L. F., & Hajek, A. E. (2009). Control of the gypsy moth, *Lymantria dispar*, in North America since 1878. In A. E. Hajek, T. R. Glare & M. O'Callaghan (Eds.), *Use of Microbes for Control and Eradication of Invasive Arthropods* (pp. 181−212). Dordrecht: Springer.

Solter, L. F., & Maddox, J. V. (1998a). Timing of an early sporulation sequence of microsporidia in the genus *Vairimorpha* (Microsporidia: Burenllidae). *J. Invertebr. Pathol., 72*, 323−329.

Solter, L. F., & Maddox, J. V. (1998b). Physiological host specificity of microsporidia as an indicator of ecological host specificity. *J. Invertebr. Pathol., 71*, 207−216.

Solter, L. F., Onstad, D. W., & Maddox, J. V. (1990). Timing of disease-influenced processes in the life cycle of *Ostrinia nubilalis* infected with *Nosema pyrausta*. *J. Invertebr. Pathol., 55*, 337−341.

Solter, L. F., Maddox, J. V., & Onstad, D. W. (1991). Transmission of *Nosema pyrausta* in adult European corn borers. *J. Invertebr. Pathol., 57*, 220−226.

Solter, L. F., Maddox, J. V., & McManus, M. L. (1997). Host specificity of microsporidia (Protista: Microspora) from European populations of *Lymantria dispar* (Lepidoptera: Lymantriidae) to indigenous North American Lepidoptera. *J. Invertebr. Pathol., 69*, 135−150.

Solter, L. F., Pilarska, D. K., & Vossbrinck, C. R. (2000). Host specificity of microsporidia pathogenic to forest Lepidoptera. *Biol. Control, 19*, 48−56.

Solter, L. F., Maddox, J. V., & Vossbrinck, C. R. (2005). Physiological host specificity: a model using the European corn borer, *Ostrinia nubilalis* (Hübner) (Lepidoptera: Crambidae) and microsporidia of row crop and other stalk-boring hosts. *J. Invertebr. Pathol., 90*, 127−130.

Solter, L. F., Pilarska, D. K., McManus, M. L., Zubrik, M., Patocka, J., Huang, W.-H., & Novotny, J. (2010). Host specificity of microsporidia pathogenic to the gypsy moth, *Lymantria dispar* (L.): field studies in Slovakia. *J. Invertebr. Pathol., 105*, 1−10.

Solter, L. F., Huang, W.-F., & Onken, B. (2011). Microsporidian disease in predatory beetles. *Implementation and Status of Biological Control of Hemlock Woolly Adelgid*. Forest Service FHTET: USDA. (in press).

Southern, T. R., Jolly, C. E., Lester, M. E., & Hayman, J. R. (2006). Identification of a microsporidia protein potentially involved in spore adherence to host cells. *J. Eukaryot. Microbiol., 53* (Suppl. 1), S68−S69.

Sprague, V. (1977a). Classification and phylogeny of the microsporidia. In L. A. Bulla, Jr. & T. C. Cheng (Eds.), *Comparative Pathobiology, Vol. 2. Systematics of the Microsporidia* (pp. 1−30). New York: Plenum Press.

Sprague, V. (1977b). Annotated list of species of microsporidia. In L. A. Bulla, Jr. & T. C. Cheng (Eds.), *Comparative Pathobiology, Vol. 2. Systematics of the Microsporidia* (pp. 31−510). New York: Plenum Press.

Sprague, V., Becnel, J. J., & Hazard, E. I. (1992). Taxonomy of Phylum Microspora. *Crit. Rev. Microbiol., 18*, 285−395.

Stentiford, G. D., Bateman, K. S., Longshaw, M., & Feist, S. W. (2007). *Enterospora canceri* n. gen., n. sp., intranuclear within the hepatopancreatocytes of the European edible crab *Cancer pagurus*. *Dis. Aquat. Organ., 75*, 61−72.

Streett, D. A., & Woods, S. A. (1993). Epizootiology of a *Nosema* sp. (Microsporida: Nosematidae) infecting the grasshopper *Chorthippus*

curtipennis (Harris) (Orthoptera: Acrididae). *Can. Entomol., 125*, 457−461.

Susko, E., Inagaki, Y., & Roger, A. J. (2004). On inconsistency of the neighbor-joining, least squares, and minimum evolution estimation when substitution processes are incorrectly modeled. *Mol. Biol. Evol., 21*, 1629−1642.

Sweeney, A. W., & Becnel, J. J. (1991). Potential of microsporidia for the biological control of mosquitoes. *Parasitol. Today, 7*, 217−220.

Sweeney, A. W., Hazard, E. I., & Graham, M. F. (1985). Intermediate host for an *Amblyospora* sp. (Microspora) infecting the mosquito *Culex annulirostris*. *J. Invertebr. Pathol., 46*, 98−102.

Sweeney, A. W., Graham, M. F., & Hazard, E. I. (1988). Life cycle of *Amblyospora dyxenoides* sp. nov. in the mosquito *Culex annulirostris* and the copepod *Mesocyclops albicans*. *J. Invertebr. Pathol., 51*, 46−57.

Sweeney, J. D., & McLean, J. A. (1987). Effect of sublethal infection levels of *Nosema* sp. on the pheromone-mediated behavior of the western spruce budworm *Choristoneura occidentalis* Freeman (Lepidoptera: Tortricidae). *Can. Entomol., 119*, 587−594.

Takov, D., Doychev, D., Linde, A., Draganova, S., & Pilarska, D. (2011). Pathogens of bark beetles (Coleoptera: Curculionidae) in Bulgarian forests. *Phytoparasitica, 39*, 343−352.

Takvorian, P. M., Weiss, L. M., & Cali, A. (2005). The early events of *Brachiola algerae* (Microsporidia) infection: spore germination, sporoplasm structure, and development within host cells. *Folia Parasitol., 52*, 118−129.

Tanada, Y., & Kaya, H. K. (1993). *Insect Pathology*. San Diego: Academic Press.

Tay, W. T., O'Mahony, E. M., & Paxton, R. J. (2005). Complete rRNA gene sequences reveal that the microsporidium *Nosema bombi* infects diverse bumblebee (*Bombus* spp.) hosts and contains multiple polymorphic sites. *J. Eukaryot. Microbiol., 52*, 505−513.

Terry, R. S., Smith, J. E., Bouchon, D., Rigaud, T., Duncanson, P., Sharpe, R. G., & Dunn, A. M. (1999). Ultrastructural characterisation and molecular taxonomic identification of *Nosema granulosis* n. sp., a transovarially transmitted feminising (TTF) microsporidium. *J. Eukaryot. Microbiol., 46*, 492−499.

Texier, C., Vidau, C., Viguès, B., El Alaoui, H., & Delbac, F. (2010). Microsporidia: a model for minimal parasite−host interactions. *Curr. Opin. Microbiol., 13*, 443−449.

Thomson, H. M. (1955). *Perezia fumiferanae* n. sp., a new species of microsporidia from the spruce budworm *Choristoneura fumiferana* (Clem.). *J. Parasitol., 41*, 416−423.

Thomson, H. M. (1958). Some aspects of the epidemiology of a microsporidian parasite of the spruce budworm, *Choristoneura fumiferana* (Lepidoptera: Tortricidae). *Can J. Zool., 53*, 1799−1802.

Thomson, H. M. (1959). A microsporidian parasite of the forest tent caterpillar *Malacosoma disstria* Hbn. *Can. J. Zool., 37*, 217−221.

Thomson, H. M. (1960). The possible control of a budworm infestation by a microsporidian disease. *Can. Dept. Agric. Bi-Mon. Prog. Rep., 16*, 1.

Thorp, R. W., & Shepherd, M. D. (2005). Subgenus *Bombus* Latreille, 1802 (Apidae: Apinae: Bombini). In M. Shepherd, D. M. Vaughan & S. H. Black (Eds.), *Red List of Pollinator Insects of North America*. Available online: http://www.xerces.org/pollinator-redlist/.

Tiner, J. D. (1988). Birefringent spores differentiate *Encephalitozoon* and other microsporidia from Coccidia. *Vet. Pathol., 25*, 227−230.

Tonka, T., Weiser, J., Jr., & Weiser, J. (2010). Budding: a new stage in the development of *Chytridiopsis typographi* (Zygomycetes: Microsporidia). *J. Invertebr. Pathol., 104*, 17−22.

Triplett, R. F. (1976). The imported fire ant: health hazard or nuisance? *South. Med. J., 69,* 258–259.

Tuzet, O., Maurand, J., Fize, A., Michel, R., & Fenwick, B. (1971). Proposition d'un nouveau cadre systématique pour les genres de Microsporidies. *C.R. Acad. Sci., 272,* 1268–1271.

Undeen, A. H., & Alger, N. E. (1976). *Nosema algerae*: infection of the white mouse by a mosquito parasite. *Exp. Parasitol., 40,* 86–88.

Undeen, A. H., & Maddox, J. V. (1973). The infection of non-mosquito hosts by injection with spores of the Microsporidian *Nosema algerae*. *J. Invertebr. Pathol., 22,* 258–265.

Undeen, A. H., & Vávra, J. (1997). Research methods for entomopathogenic Protozoa. In L. A. Lacey (Ed.), *Manual of Techniques in Insect Pathology* (pp. 117–151). London: Academic Press.

USEPA. (1992). *Reregistration Eligibility Document (RED) facts: Nosema locustae. US Environmental Protection Agency, EPA-738-F-792–011.* Sept. 1992. http://www.epa.gov/oppbppd1/biopesticides/ingredients/factsheets/factsheet_117001.htm.

Valles, S. M., & Briano, J. A. (2004). Presence of *Thelohania solenopsae* and *Vairimorpha invictae* in South American populations of *Solenopsis invicta. Fla. Entomol., 87,* 625–627.

Valles, S. M., Oi, D. H., Perera, O. P., & Williams, D. F. (2002). Detection of *Thelohania solenopsae* (Microsporidia: Thelohaniidae) in *Solenopsis invicta* (Hymenoptera: Formicidae) by multiplex PCR. *J. Invertebr. Pathol., 81,* 196–201.

Valles, S. M., Oi, D. H., & Porter, S. D. (2010). Seasonal variation and the co-occurrence of four pathogens and a group of parasites among monogyne and polygyne fire ant colonies. *Biol. Control, 54,* 342–348.

Van der Geest, L. P. S., Elliot, S. L., Breeuwer, J. A. J., & Beerling, E. A. M. (2000). Diseases of mites. *Exp. Appl. Acarol., 24,* 497–560.

Vander Meer, R. K., & Porter, S. D. (2001). Fate of newly mated queens introduced into monogyne and polygyne *Solenopsis invicta* (Hymenoptera: Formicidae) colonies. *Ann. Entomol. Soc. Am., 94,* 289–297.

Vaughn, J. L., Brooks, W. M., Capinera, J. L., Couch, T. L., & Maddox, J. V. (1991). I.4 Utility of *Nosema locustae* in the suppression of rangeland grasshoppers. In Grasshopper IPM User Handbook, Issued 1996–2000. In G. L. Cunningham & M. W. Sampson (Eds.), *USDA-APHIS Tech. Bull. No.,* 1809.

Vávra, J., & Becnel, J. J. (2007). *Vavraia culicis* (Weiser, 1947) Weiser, 1977 revisited: cytological characterisation of a *Vavraia culicis*-like microsporidium isolated from mosquitoes in Florida and the establishment of *Vavraia culicis floridensis* subsp. n. *Folia Parasitol., 54,* 259–271.

Vávra, J., & Larsson, J. I. R. (1999). Structure of the microsporidia. In M. Wittner & L. M. Weiss (Eds.), *The Microsporidia and Microsporidiosis* (pp. 7–84). Washington, DC: American Society for Microbiology Press.

Vávra, J., & Undeen, A. H. (1970). *Nosema algerae* n. sp. (Cnidospora, Microsporida), a pathogen in a laboratory colony of *Anopheles stephensi* Liston (Diptera: Culicidae). *J. Protozool., 17,* 240–249.

Vávra, J., Hyliš, M., Vossbrinck, C. R., Pilarska, D. K., Linde, A., Weiser, J., McManus, M. L., Hoch, G., & Solter, L. F. (2006). *Vairimorpha disparis* n. comb. (Microsporidia: Burenellidae): a redescription and taxonomic revision of *Thelohania disparis* Timofejeva 1956, a microsporidian parasite of the gypsy moth *Lymantria dispar* (L.) (Lepidoptera: Lymantriidae). *J. Eukaryot. Microbiol., 53,* 292–304.

Villidieu, Y., & van Frankenhuyzen, K. (2004). Epizootic occurrence of *Entomophaga maimaiga* at the leading edge of an expanding population of the gypsy moth (Lepidoptera: Lymantriidae) in north-central Ohio. *Can. Entomol., 136,* 875–878.

Visvesvara, G. S., Belloso, M., Moura, H., Da Silva, A., Moura, I., Leitch, G., Schwartz, D. A., Chevez-Barrios, P., Wallace, S., Pieniazek, N. J., & Goosey, J. (1999). Isolation of *Nosema algerae* from the cornea of an immunocompetent patient. *J. Eukaryot. Microbiol., 46,* 10S.

Vossbrinck, C. R., & Debrunner-Vossbrinck, B. A. (2005). Molecular phylogeny of the Microsporidia: ecological, ultrastructural and taxonomic considerations. *Folia Parasitol., 52,* 131–142.

Vossbrinck, C. R., Andreadis, T. G., Vávra, J., & Becnel, J. J. (2004). Molecular phylogeny and evolution of mosquito parasitic Microsporidia (Microsporidia: Amblyosporidae). *J. Eukaryot. Microbiol., 51,* 88–95.

Wang, C. Y., Huang, W. F., Tsai, Y. C., Solter, L. F., & Wang, C. H. (2009). A new species, *Vairimorpha ocinarae* n. sp., isolated from *Ocinara lida* Moore (Lepidoptera: Bombycidae) in Taiwan. *J. Invertebr. Pathol., 100,* 68–78.

Wegensteiner, R., Weiser, J., & Führer, E. (1996). Observations on the occurrence of pathogens in the bark beetle *Ips typographus* L. (Col., Scolytidae). *J. Appl. Entomol., 120,* 199–204.

Weidner, E., & Trager, W. (1973). Adenosine triphosphate in the extracellular survival of an intracellular parasite (*Nosema michaelis*) Microsporidia. *J. Cell Biol., 57,* 586–591.

Weidner, E., Byrd, W., Scaroborough, A., Pleshinger, J., & Sibley, D. (1984). Microsporidian spore discharge and the transfer of polaroplast organelle membrane into plasma membrane. *J. Protozool., 31,* 195–198.

Weidner, E., Canning., E. U., Rutledge, C. R., & Meek, C. L. (1999). Mosquito (Diptera: Culicidae) host compatibility and vector competency for the human myositic parasite *Trachipleistophora hominis* (Phylum Microspora). *J. Med. Entomol., 36,* 522–525.

Weiser, J. (1947). Klíč kurčování Mikrosporidií (A key for the determination of microsporidia). *Acta Soc. Sci. Nat. Moravicae., 18,* 1–64.

Weiser, J. (1961). Die mikiosporidien als parasiten der insekte. *Monogr. Angew. Entomol. No.,* 17.

Weiser, J. (1970). Three new pathogens of the Douglas fir beetle, *Dendroctonus pseudotsugae: Nosema dendroctoni* n. sp., *Ophryocystic dendroctoni* n. sp., and *Chytridiopsis typographi* n. comb. *J. Invertebr. Pathol., 16,* 436–441.

Weiser, J., & Novotny, J. (1987). Field application of *Nosema lymantriae* against the gypsy moth, *Lymantria dispar* L. *J. Appl. Entomol., 104,* 58–62.

Weiser, J., & Veber, J. (1975). Die mikrosporidie *Thelohania hyphantriae* Weiser des Weissen barenspinner (*Hyphantria cunea*) und anderer mitghiden seiner luaconose. *Z. Angew Ent., 40,* 55–70.

Weiser, J., Wegensteiner, R., & Žižka, Z. (1997). Ultrastructures of *Nosema typographi* Weiser 1955 (Microspora: Nosematidae) of the bark beetle *Ips typographus* L. (Coleoptera; Scolytidae). *J. Invertebr. Pathol., 70,* 156–160.

Weiser, J., Holuša, J., & Žižka, Z. (2006). *Larssoniella duplicati* n.sp. (Microsporidia, Unikaryonidae), a newly described pathogen infecting the double-spined spruce bark beetle, *Ips duplicatus* (Coleoptera, Scolytidae) in the Czech Republic. *J. Pest Sci., 79,* 127–135.

Weiss, L. M. (2005). The first united workshop on Microsporidia from invertebrate and vertebrate hosts. *Folia Parasitol., 52,* 1–7.

Weiss, L. M., & Vossbrinck, C. R. (1999). Molecular biology, molecular phylogeny, and molecular diagnostic approaches to the microsporidia. In M. Mittner & L. M. Weiss (Eds.), *The Microsporidia and Microsporidiosis* (pp. 129−171). Washington, DC: American Society for Microbiology Press.

Weseloh, R., & Andreadis, T. G. (2002). Detecting the titer in forest soils of spores of the gypsy moth (Lepidoptera: Lymantriidae) fungal pathogen, *Entomophaga maimaiga* (Zygomycetes: Entomophthorales). *Can. Entomol., 134*, 269−279.

Williams, B. A. P., & Keeling, P. J. (2005). Microsporidian mitochondrial proteins: expression in *Antonospora locustae* spores and identification of genes coding for two further proteins. *J. Eukaryot. Microbiol., 52*, 271−276.

Williams, B. A. P., Hirt, R. P., Licocq, J. M., & Embley, M. (2002). A mitochondrial remnant in the microsporidian *Trachipleistophora hominis. Nature, 418*, 865−869.

Williams, B. A. P., Haferkamp, I., & Keeling, P. J. (2008a). An ADP/ATP-specific mitochondrial carrier protein in the microsporidian *Antonospora locustae. J. Mol. Biol., 375*, 1249−1257.

Williams, B. A. P., Lee, R. C. H., Becnel, J. J., Weiss, L. M., Fast, N. M., & Keeling, P. J. (2008b). Genome sequence surveys of *Brachiola algerae* and *Edhazardia aedis* reveal microsporidia with low gene densities. *BMC Genomics, 9*, 200.

Williams, D. F., Oi, D. H., & Knue, G. J. (1999). Infection of red imported fire ant (Hymenoptera: Formicidae) colonies with the entomopathogen *Thelohania solenopsae* (Microsporidia: Thelohaniidae). *J. Econ. Entomol., 92*, 830−836.

Williams, D. F., Oi, D. H., Porter, S. D., Pereira, R. M., & Briano, J. A. (2003). Biological control of imported fire ants (Hymenoptera: Formicidae). *Am. Entomol., 49*, 150−163.

Wilson, G. G. (1973). Incidence of microsporidia in a field population of spruce budworm. *Environ. Can. Bi-Mon. Res., 29*, 35−36.

Wilson, G. G. (1975). Occurrence of *Thelohania* sp. and *Pleistophora* sp. (Microsporida: Nosematidae) in *Choristoneura fumiferana* (Lepidoptera: Tortricidae). *Can. J. Zool., 53*, 1799−1802.

Wilson, G. G. (1977). The effects of feeding microsporidian (*Nosema fumiferana*) spores to naturally infected spruce budworm (*Choristoneura fumiferana*). *Can. J. Zool., 55*, 249−250.

Wilson, G. G. (1980). Effects of *Nosema fumiferanae* (Microsporida) on rearing stock of spruce budworm, *Choristoneura fumiferana* (Lepidoptera: Tortricidae). *Proc. Entomol. Soc. Ontario., 111*, 115−116.

Wilson, G. G. (1984). The transmission and effects of *Nosema fumiferanae* and *Pleistophora schubergi* (Microsporida) on *Choristoneura fumiferana* (Lepidoptera: Tortricidae). *Proc. Entomol. Soc. Ontario., 115*, 71−75.

Wilson, G. G., & Burke, J. M. (1971). *Nosema thomsoni* n. sp., a microsporidian from *Choristoneura conflictana* (Lepidoptera: Tortricidae). *Can. J. Zool., 49*, 786−788.

Wilson, G. G., & Burke, J. M. (1978). Microsporidian parasites of *Archips cerasivoranus* (Fitch) in the District of Algoma, Ontario. *Proc. Entomol. Soc. Ontario., 109*, 84−85.

Wilson, G. G., & Kaupp, W. J. (1976). Application of *Nosema fumiferanae* and *Pleistophera schubergi* (microsporidia) against the spruce budworm in Ontario, 1976. *Can. For. Serv., Sault Ste. Marie Ing. Rep. IP-X11.*

Wojcik, D. P., Allen, C. R., Brenner, R. J., Forys, E. A., Jouvenaz, D. P., & Lutz, R. S. (2001). Red imported fire ants: impact on biodiversity. *Am. Entomol., 47*, 16−23.

Xu, J., Pan, G., Fang, L., Li, J., Tian, X., Li, T., Zhou, Z., & Xiang, Z. (2006). The varying microsporidian genome: existence of long-terminal repeat retrotransposon in domesticated silkworm parasite *Nosema bombycis. Int. J. Parasitol., 36*, 1049−1056.

Xu, Y., Takvorian, P., Cali, A., & Weiss, L. M. (2003). Lectin binding of the major polar tube protein (PTP1) and its role in invasion. *J. Eukaryot. Microbiol., 50* (Suppl), 600−601.

Xu, Y., Takvorian, P., Cali, A., Wang, F., Zhang, H., Orr, G., & Weiss, L. M. (2006). Identification of a new spore wall protein from *Encephalitozoon cuniculi. Infect. Immun., 74*, 239−247.

Yang, C.-C., Yu, Y.-C., Valles, S. M., Oi, D. H., Chen, Y.-C., Shoemaker, D., Wu, W.-J., & Shih, C.-J. (2010). Loss of microbial (pathogen) infections associated with recent invasions of the red imported fire ant *Solenopsis invicta. Biol. Invasions, 12*, 3307−3318.

Zelinskaya, L. M. (1980). Role of microsporidia in the abundance dynamics of the gypsy moth (*Porthetria dispar*) in forest plantings along the lower Dnepr River (Ukrainian Republic, USSR). *Vestnik Zoology (Zool. Bull.), 1*, 57−62.

Chapter 8

Bacterial Entomopathogens

Juan Luis Jurat-Fuentes* and Trevor A. Jackson[†]

*University of Tennessee, Knoxville, Tennessee, USA, [†]Ag Research Lincoln Research Centre, Canterbury, New Zealand

Chapter Outline

Summary

Bacterial entomopathogens and/or their toxins must be ingested and enter the alimentary tract of insects where they multiply or are activated to initiate disease. Released bacterial toxins and other virulence factors (enzymes) target the midgut cells to disrupt the epithelial barrier and break through to the main body cavity. Bacterial proliferation in the hemocoel leads to septicemia that kills the infected host. Both Gram-positive and Gram-negative bacteria will kill their insect hosts, yet most microbial products for insect control are based on Gram-positive, spore-forming bacteria in the genus Bacillus. From this group, Bacillus thuringiensis (Bt)

Insect Pathology. DOI: 10.1016/B978-0-12-384984-7.00008-7

has been the most successful microbial pesticide to date and has dominated microbial control of insect pests. The crystal (*cry*) and vegetative (*vip*) toxin genes from Bt have been cloned and transformed into plants to develop transgenic Bt crops, which have revolutionized pest control and the agricultural landscapes. These Bt crops are protected from insect attack by constitutive production of the Bt toxins. Alternative entomopathogenic bacteria developed commercially include the Gram-positive *Bacillus sphaericus* and *Paenibacillus popilliae*, and Gram-negative bacteria in the genus *Serratia*. Current goals of research on bacterial pesticides include the search for novel pathogens and toxins, and the development of modified toxins to increase efficacy and extend the activity range of these microbial insect control technologies. This chapter reviews bacterial entomopathogens and discusses their taxonomy, genetics, and pathology, and the implications for insect control now and in the future.

8.1. INTRODUCTION

As with other living organisms, insects are intimately associated with bacteria at all stages of their lives. Insect eggs are surrounded by bacterial films and, on emergence from the egg, neonate larvae ingest bacteria from their surroundings and are covered with adherent organisms. Distinct bacterial communities develop within the insect gut through larval development and may occupy distinct bacteriomes and aid digestion. Bacteria that are carried within the body can produce pheromones in the adult insects, and after death, the cadavers are biodegraded by decomposing bacteria. During 250 million years of insect evolution, insects and bacteria have evolved complex relationships ranging from commensalism and symbiosis to pathogenesis. Bacterial insect pathogens were comprehensively discussed in Chapters 4 and 5 of the previous edition of this book (Tanada and Kaya, 1993) and are the focus of this chapter, with emphasis on those species that are currently in use, or under development, as agents for microbial control of pests.

Bacteria are defined within the Prokaryotae as unicellular microorganisms lacking a nuclear membrane to separate the genetic material from the cytoplasm and other intracellular membrane-enclosed organelles. Bacteria can be isolated from virtually any environment, and this diversity is reflected in varied metabolic strategies to use sunlight (phototrophs), or inorganic (lytotrophs) or organic material (organotrophs) for energy. Typically from one to a few micrometers in width, bacteria also display diverse morphologies, including spherical (cocci), rod-shaped (bacilli), and spiral (spirochaetes) shapes. Proliferation is usually through binary fission, an asexual process in which daughter cells are clonal copies of the mother cell.

Genetic variation in these organisms originates from mutation and selection or acquisition of genetic material from the environment (transformation), bacteriophages (transduction), or other bacteria (conjugation). These processes, combined with the rapid generation time of bacteria, result in a high level of variation and wide ranges of functionality among bacterial strains. The single, supercoiled bacterial chromosome (genophore) is organized as a covalently closed double-stranded DNA molecule lacking chromatin. Bacterial genes do not contain introns and are organized in functional units called operons that contain a number of genes involved in a specific pathway or process. Operons are controlled by a common promoter and terminator, allowing for fast expression of functionally related genes in response to environmental cues. Horizontal gene transfer of mobile genome fragments known as genomic islands also increases genetic variation among bacterial populations. Once stabilized in the bacterial genophore, genomic islands containing genes encoding diverse pathogenic factors (toxins and enzymes) can determine pathogenicity or increase virulence. Pathogenicity islands that allow production of entomocidal toxins or symbiotic factors related to pathogenicity in insects have been reported in diverse bacteria, including *Photorhabdus luminescens* (Waterfield *et al.*, 2002), *Xenorhabdus nematophila* (Brown *et al.*, 2004), *Serratia* spp. (Dodd *et al.*, 2006), *Yersinia* spp. (Fuchs *et al.*, 2008), and *Pseudomonas aeruginosa* (Kim *et al.*, 2008).

Genetic information is also sometimes stored in a number of plasmids: small and usually circular DNA molecules that are self-replicating and that can be transferred between bacteria through conjugation. These plasmids usually contain genes that are crucial for specific functions, including pathogenicity. For instance, most crystal toxins from *Bacillus thuringiensis* (Bt) are located in plasmids (Held *et al.*, 1982), which can be exchanged between diverse strains and *Bacillus* spp. (González *et al.*, 1982). The pathogenicity island of *Serratia* spp. containing *tc* genes is also located on a plasmid (Dodd *et al.*, 2006). Genetic information can also be stored in prophages, DNA from bacteriophages that is inserted into the bacterial chromosome or plasmid through transduction, which can also confer phenotypes conducive to pathogenicity.

Bacteria and insects have co-evolved a wide range of complex relationships from commensalism to parasitism or pathogenesis over more than 250 million years. The specific host—bacteria relationships that have developed are the outcome of dynamic co-evolution underpinned by genetic diversity and driven by selection pressure. While opportunistic bacterial infections occur, often as a result of injury or stress, most consistent bacterial entomopathogens have obligate or facultative relationships with their hosts. This chapter will concentrate on the latter species owing to their relevance for application and commercialization in insect control.

Obligate bacterial pathogens complete their life cycles within the insect host, for example *Paenibacillus* spp. infecting bees or scarabaeid beetles. In comparison, facultative pathogenic bacteria can also grow in the environment outside the host and some species in the genera *Bacillus* and *Serratia* fit within this category. Occasional pathogenesis is caused by potential or opportunistic pathogens, as defined by Bucher (1973), which are able to multiply within the insect hemolymph but are dependent on stress or other factors that weaken the insect to enable them to cross the gut barrier to multiply and cause disease. Successful bacterial pathogens are able to enter the host's tissues and overwhelm or avoid the insects' defenses to proliferate and multiply causing disease, usually assisted by production of pathogenic factors such as toxins and enzymes. In the final stages of disease, bacterial pathogens overwhelm and kill the host, after which the bacterial progeny must egress to infect a new host. In some cases, as in *Bacillus* spp., sporulation occurs producing a life stage resistant to adverse environmental conditions. This strategy is highly advantageous when considering the typically limited temporal availability of suitable insect hosts, and usually passive dispersal of these entomopathogenic bacteria.

Bacterial insecticidal sprays and transgenic crops expressing bacterial toxins have been the most commercially successful microbial pesticides to date. The success of most bacterial insecticides is due to ease and cost-effectiveness of mass production, specificity, persistence in the environment, and environmental safety. Cessation of feeding and rapid death induced by bacterial pathogens reduce loss of foliage, which usually allows the insect population to be maintained below economic thresholds. In comparison to chemical pesticides, bacterial insecticides have a narrow spectrum of activity, are more sensitive to environmental degradation, and may have a lower specific activity. The earliest documented use of a bacterial entomopathogen was application of a Gram-negative bacterium isolated from diseased grasshoppers (*Schistocerca americana*) during an epizootic (disease outbreak). Artificially cultured bacteria were applied for biological control of this insect, with variable results (d'Herelle, 1911, 1912). The bacterium *Bacillus* (now *Paenibacillus*) *popilliae* was the first bacterial entomopathogen to be used in a major insect control program to suppress larvae of the Japanese beetle, *Popillia japonica* (Klein and Jackson, 1992).

Undoubtedly, the best characterized and most widely used bacterium in microbial control is Bt. The historical development of Bt-based insecticides has been reviewed in detail by several authors (Burges, 2001; Federici, 2005; Sanchis, 2011). While products based on the Bt subsp. *israelensis* (against mosquitoes and black fly larvae) and *kurstaki* (against lepidopteran larvae) have been commercially successful, products based on Bt subsp. *tenebrionis* to control coleopteran larvae were not competitive against chemical alternatives, and thus their commercial success has been limited (Gelernter, 2004). The usually narrow

spectrum of toxicity and short field persistence have also hindered adoption of Bt microbial pesticides. Availability of cloned toxin genes from Bt isolates opened the possibility of expressing these toxins in heterologous systems to improve delivery and performance. Advances in plant molecular biology advocated that Bt toxin genes would be useful candidates for plant transformation in order to meet the need for increased stability, optimized delivery, and efficient control of tunneling lepidopteran larvae. This interest prompted the development of transgenic plants transformed to express Cry or Vip toxins. Several plant species have been transformed to express diverse Bt toxins to control lepidopteran and coleopteran pests. Increased yields, efficient pest control, and reductions in the use of chemical pesticide applications have greatly contributed to increasing adoption rates since the introduction of these transgenic crops in the market (Betz *et al.*, 2000; Kumar *et al.*, 2008; Naranjo, 2009).

In contrast to Gram-positive bacteria, there have been few attempts to commercialize bacteria from the Enterobacteriaceae, principally because of their lack of a resistance spore and instability in storage and delivery. An exception has been provided by a strain of the bacterium *Serratia entomophila* commercialized as a biocontrol agent for the New Zealand grass grub, *Costelytra zealandica* (Jackson *et al.*, 1992). A flowable granular formulation (Bioshield™) with long shelf-life in ambient conditions (Johnson *et al.*, 2001) overcomes the perceived limitations of this bacterial group. While there are several other candidates from the Enterobacteriaceae group with potential for microbial control, none has, as yet, been developed into a commercial product.

8.2. CLASSIFICATION AND PHYLOGENY

Two main groups of prokaryotic microbes are recognized based on the 16S ribosomal RNA sequence: Archaea, containing bacteria that share DNA replication, transcription, and translation features with eukaryotes; and Eubacteria or true bacteria. Entomopathogenic bacteria are classified within Eubacteria. This group contains three major divisions based on the presence or structure of the cell walls: bacteria with a Gram-negative type cell wall (Gracilicutes), Gram-positive type cell wall (Firmicutes), and Eubacteria lacking a cell wall (Tenericutes). Since there are representatives of all these three categories among entomopathogenic bacteria, species presented in this chapter have been organized according to this division.

The basic taxonomic group, bacterial species, has been difficult to resolve. Bacterial strains are grouped into species based on sharing of certain distinguishing phenotypes of ecological importance and overall genomic similarity (Stackebrandt *et al.*, 2002). A bacterial strain is defined as the descendants of a single isolation in pure culture and is usually derived from successive culture from an initial single colony. In practice, more than 70% DNA–DNA hybridization values are considered evidence of a single species (Wayne *et al.*, 1987). Although the DNA hybridization method remains the most commonly used technique, new analytical and sequencing methods have been indicated as desirable complements (Stackebrandt *et al.*, 2002). Within Eubacteria, current classification is mostly based on polyphasic (consensus) taxonomy, including analysis of the nucleotide sequence of the ribosomal small-subunit RNA (16S rDNA), DNA–DNA hybridization, as well as phenotypic, genotypic, and phylogenetic data (Vandamme *et al.*, 1996; Brenner *et al.*, 2005). For this chapter, "Bergey's Taxonomic Outlines" (Ludwig *et al.*, 2006, 2008a, b), the "List of Prokaryotic Names with Standing in Nomenclature" (Euzéby, 1997), and "Bergey's Manual of Determinative Bacteriology, 9th ed." (Holt, 1994), have been selected as references for systematics and nomenclature of bacteria.

8.2.1. General Characteristics

Bacteria exist as small individual cells, with most species surrounded by a rigid cell wall that functions to contain and maintain the cell shape. The cell wall also serves as an anchor for molecules, organelles for motility (flagella), and adhesion structures such as fimbriae, which may have a relevant role in pathogenicity. Cell walls of Gram-positive bacteria are composed of cross-linked peptidoglycan, whereas Gram-negative bacteria present a more complex structure composed of a periplasmic space containing a thin peptidoglycan layer and lipoproteins, and an outer membrane containing lipopolysaccharide (LPS). During host invasion bacterial LPS is recognized by host pattern recognition proteins, activating the host immune response. On contrast, bacteria in the class Mollicutes (Mycoplasmas and Spiroplasmas), which include some entomopathogenic species, do not possess a rigid cell wall.

Bacterial growth in closed systems follows three main phases: lag, exponential, and stationary, which are also usually observed during the disease process. In the lag phase, vegetative cells adjust to new environmental conditions, such as a new host or a microenvironment within the host. Once established in a susceptible host, favorable conditions initiate rapid proliferation by binary fission producing an exponential or log phase of growth. For entomopathogenic bacteria this log growth phase occurs in the host hemocoel, leading to septicemia and ultimately death of the insect host. Growth rate will slow owing to accumulation of toxic secondary metabolites, nutrient depletion, or adverse environmental conditions, and bacteria enter a stationary phase of growth that can be accompanied by sporulation. Bacterial spores do not exhibit metabolic activity and represent a highly resistant resting phase against adverse environments, allowing survival over astonishingly long periods. Spores of *Bacillus sphaericus* recovered from the digestive system of bees fossilized in amber 25–40 million years ago were able to germinate and grow under favorable conditions (Cano and Borucki, 1995). While vegetative cells age and are sensitive to environmental stresses leading to cell death, significant survival can be achieved in stasis conditions (Johnson *et al.*, 2001). Among the spore formers, sensing mechanisms detect diverse environmental cues to activate or repress spore germination (Foster and Johnstone, 1990).

In some cases, isolation and identification of disease-causing bacteria are complicated by lack of growth in artificial media, overgrowth by opportunistic microbes during disease, and lack of selective media. Species in the genus *Rickettsia* are highly fastidious and require living cells for propagation, whereas those in the genus *Mycoplasma* display an elongated lag phase, resulting in very slow growth and hindering propagation in the laboratory. Pathogenicity depends on virulence of the specific bacterial strain and on stress conditions in the host, which can result in variable virulence among strains of the same pathogen. Although many bacterial species were initially considered for their potential in insect control (Steinhaus, 1957, 1963), some of them have not been further considered for development of microbial control agents because they can also cause disease in humans.

8.2.2. Classification of Bacterial Entomopathogens

Classification of entomopathogenic bacteria into groups on the basis of their pathogenicity presents difficulties, as most bacteria are facultative pathogens and present virulence variations depending on the host environment and the specific strain. Bacterial populations, including pathogenic bacteria, are essentially clonal, implying that exchange of DNA in nature is infrequent. Bacterial species develop as clonal lines that are diverging slowly through genetic variations but are essentially identical within a lineage. Many bacterial species displaying pathogenicity exist as a number of distinct lineages, or clones, of which only a few contain the genetic information that encodes pathogenicity (Selander *et al.*, 1987). For example, genetic diversity among strains of *B. sphaericus* explained variation in toxicity against mosquito larvae (Krych *et al.*, 1980). In some cases, non-pathogenic and disease-causing clonal lineages have been named and classified as diverse species based on their pathogenic niche (Lan and Reeves, 2001). To resolve this issue, a bacterial species concept based on members sharing a core genome (housekeeping genes) was proposed (Dykhuizen and Green, 1991), in which species incorrectly described as diverse are named as clones within a bacterial species, based on sequence similarity (Lan and

Reeves, 2000, 2001). Advances in DNA sequencing technology with the concomitant increase in available bacterial genomic data, together with increased characterization of bacterial population genetics, are expected to allow for a more accurate and useful definition of bacterial species that will increase understanding of the entomopathogens. Entomopathogenic bacteria may be found in all three main bacterial Phyla (Fig. 8.1). However, when focusing on bacteria with demonstrated use in microbial insect control, species of interest occur in the following taxonomic families: Bacillaceae, Paenibacillaceae, Enterobacteriaceae, and Neisseriaceae.

Entomopathogenic bacteria in the genus *Bacillus* are the most widely used microbial insect control agents. Gram-negative bacteria that cause disease in insects also have potential as insect control agents, but their relationship to opportunistic pathogens of vertebrates initially limited development. Exceptions include the *Photorhabdus* and *Xenorhabdus* bacteria vectored by entomopathogenic nematodes (Chapter 11), some aspects of which are covered in this chapter, and species in the genus *Serratia*. More recently, Gram-negative bacteria from the genera *Chromobacterium*, *Yersinia*, and *Pseudomonas* have come under consideration as candidates for microbial insect control. Interest in cell wall-less bacteria in the class Mollicutes (mostly genera *Spiroplasma*) has been widespread because of their potential to infect vertebrates and plants, and although they are commonly found in the digestive system of diverse insects, their role as pathogens and their potential for insect control has not been explored adequately. Bacteria in the Rickettsiaceae and Coxiellaceae families are highly fastidious, obligate, intracellular parasites and are among the most common parasitic microbes, impacting on the insect host through cytoplasmic incompatibility or male-killing phenotypes during infection. The most common Rickettsiaceae genus infecting insects is *Wolbachia*, which is discussed in Chapter 9.

8.3. INFECTION, REPLICATION, PATHOLOGY, AND TRANSMISSION

8.3.1. Portals of Entry

In healthy insects, bacteria are mostly localized in the posterior section (hindgut) of the digestive tract, where they perform digestive functions that can be vital for their host. Bacteria are unable to penetrate the insect cuticle and can only invade the hemocoel after the gut epithelial barrier is

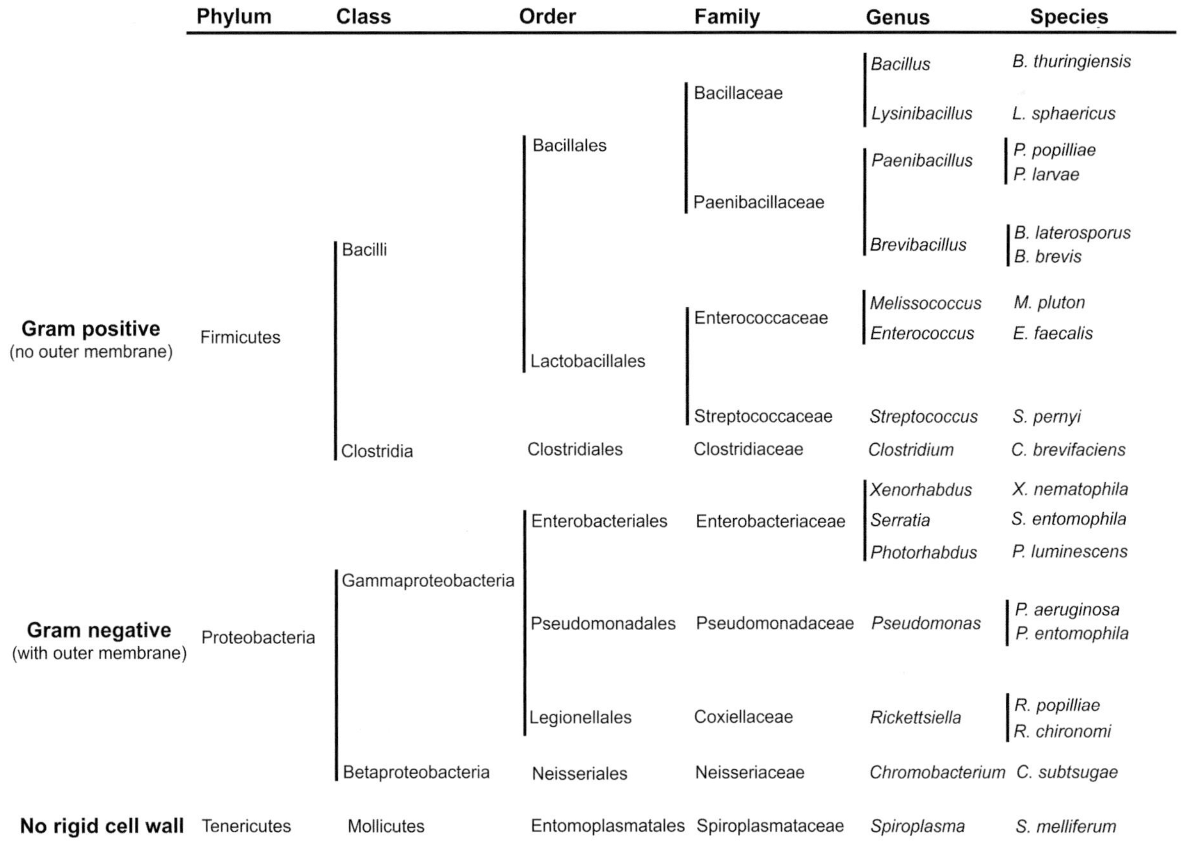

Phylum	Class	Order	Family	Genus	Species
		Bacillales	Bacillaceae	*Bacillus*	*B. thuringiensis*
				Lysinibacillus	*L. sphaericus*
	Bacilli		Paenibacillaceae	*Paenibacillus*	*P. popilliae* *P. larvae*
				Brevibacillus	*B. laterosporus* *B. brevis*
Gram positive (no outer membrane)	Firmicutes	Lactobacillales	Enterococcaceae	*Melissococcus* *Enterococcus*	*M. pluton* *E. faecalis*
			Streptococcaceae	*Streptococcus*	*S. pernyi*
	Clostridia	Clostridiales	Clostridiaceae	*Clostridium*	*C. brevifaciens*
	Gammaproteobacteria	Enterobacteriales	Enterobacteriaceae	*Xenorhabdus* *Serratia* *Photorhabdus*	*X. nematophila* *S. entomophila* *P. luminescens*
Gram negative (with outer membrane)	Proteobacteria	Pseudomonadales	Pseudomonadaceae	*Pseudomonas*	*P. aeruginosa* *P. entomophila*
		Legionellales	Coxiellaceae	*Rickettsiella*	*R. popilliae* *R. chironomi*
	Betaproteobacteria	Neisseriales	Neisseriaceae	*Chromobacterium*	*C. subtsugae*
No rigid cell wall	Tenericutes	Mollicutes	Entomoplasmatales	Spiroplasmataceae	*Spiroplasma*	*S. melliferum*

FIGURE 8.1 Classification of bacterial entomopathogens. Some specific examples with reported pathogenic activity against insects are included for each group.

compromised. Thus, the primary route of bacterial entry is the oral cavity (*per os*) during feeding, although infection is also possible through compromised integument or trachea, and the egg.

For successful entry through the oral cavity, the bacterial pathogens have to overcome a series of defensive mechanisms controlled by the host. Defense against pathogens starts at the insect mouthparts. For instance, before entering cotton bollworm (*Helicoverpa zea*) larvae, bacterial pathogens are exposed to reactive oxygen and other antibacterial compounds in the larval saliva, which may reduce infectivity (Musser *et al.*, 2005). In adult honey bees and ant species, a comb-like structure in the proventriculus helps to restrict passage of microbes in food (Sturtevant and Revell, 1953; Glancy *et al.*, 1981). After ingestion, bacteria are generally moved within the food bolus by peristalsis through the cuticle-lined stomodaeum (foregut) to reach the mesenteron (midgut). Attachment to the foregut cuticle of larvae of the scarab *C. zealandica* by the bacterium *S. entomophila* appears to be a precursor to amber disease (Wilson *et al.*, 1992). Many bacteria are unable to multiply in the physicochemical conditions (e.g., pH, ionic strength, redox potential) prevailing in the gut of healthy insects, but for some pathogens, these conditions are necessary for activation of bacterial pathogenesis. Thus, production of Sep toxins by *S. entomophila* is induced by midgut conditions (Hurst *et al.*, 2007b) and crystalline inclusions of Bt are solubilized by the alkaline pH in the midgut of lepidopteran larvae.

The peritrophic matrix is an important mechanical barrier to bacterial infection in the midgut. This thin chitinous layer secreted by midgut cells protects the epithelium from contact with microbials and abrasion and exposure to allelochemicals and also contains antimicrobial compounds (Brandt *et al.*, 1978; Barbehenn, 2001). Proteins and glycosylated moieties on the peritrophic matrix of some lepidopteran larvae have been proposed to contribute to reduced susceptibility to Bt by acting as a sink for the Cry toxins synthesized by this bacterium (Rees *et al.*, 2009). To overcome this barrier, several bacteria secrete chitinases during infection, which degrade the peritrophic matrix and allow access of bacteria and toxins to the ectoperitrophic space (Wiwat *et al.*, 2000; Thamthiankul *et al.*, 2001).

After traversing the peritrophic matrix, bacterial toxins and enzymes (lecithinase, phospholipase C, proteases) act on the midgut cells, disrupt the epithelial barrier and allow bacteria invasion of the hemocoel, resulting in septicemia. The bacterium *Xenorhabdus bovienii* produces a lecithinase important for virulence in wax moth, *Galleria mellonella*, larvae (Pinyon *et al.*, 1996). Another related bacterium, *X. nematophila*, also produces a lecithinase that does not display entomotoxic effect but has been proposed to be involved in lipid metabolism and progeny production

of its nematode host (Thaler *et al.*, 1998; Richards and Goodrich-Blair, 2010). Phospholipase C from *Bacillus* spp. is toxic to lepidopteran larvae (Lysenko, 1972) and cultured ovarian cells of the cabbage looper moth, *Trichoplusia ni* (Ikezawa *et al.*, 1989). Enolase produced by several spore-forming *Bacillus* species (Delvecchio *et al.*, 2006) has been reported to be a relevant virulence factor in American foulbrood (AFB) disease of honey bees (*Apis mellifera*) caused by *Paenibacillus larvae* (Antúnez *et al.*, 2011). Apart from mechanical and physicochemical conditions, the gut flora can also limit success of bacterial pathogens. Symbiotic bacteria in the gut may produce antimicrobial compounds to prevent successful colonization of the host by pathogens (Yoshiyama and Kimura, 2009). For example, the common bacterium *Streptococcus faecalis* in the guts of wax moth larvae has been reported to secrete antibacterial peptides that inhibit growth of alternative bacteria, including entomopathogens (Jarosz, 1979). However, the same bacterium has been described as pathogenic to larvae of the beet armyworm, *Spodoptera exigua* (Youngjin *et al.*, 2002).

The insect integument is normally contaminated by bacteria from the surrounding environment, but provides an effective barrier to bacterial colonization and invasion owing to its hydrophobicity and presence of antibacterial secretions (Asano, 2006). Entry of bacteria into the hemocoel after injury of the integument initiates a prophenyloxidase cascade producing sclerotization (Schachter *et al.*, 2007). As well as helping to seal wounds, this activation of phenol-oxidase enzymes produces reactive quinones with antimicrobial properties. Apart from the action of cuticle-degrading proteases from fungal entomopathogens (see Chapter 6), the insect integument can be damaged by crowding, parasites, parasitoids, or predators. For instance, the parasitic varroa mite damages the cuticle of the honey bee during feeding, which can lead to infection by *Melisococcus pluton*, the causative agent of European foulbrood (Kanbar and Engels, 2003), and other facultative bacterial entomopathogens (Glinsky and Jarosz, 1992). Injuries and stress caused by crowding conditions favor infection by diverse bacteria in the bark beetle, *Scolytis multistriatus* (Doane, 1960), and *Bacillus* spp. epizootics in the Mediterranean flour moth, *Anagasta kuehniella* (Flanders and Hall, 1965). There are few examples of reported entry and bacterial pathogenesis in the insect hemocoel through active degradation of the integument. Despite early reports of *Micrococcus nigrofasciens* invading larvae of the June beetle (*Phyllophaga* spp.) through degradation of connective integument (Northrup, 1914), larval mortality is also observed after injection, indicating that cuticle degradation is not necessary for infectivity (Poprawski and Yule, 1990). Some oil-degrading bacteria, such as *Pseudomonas* spp. and *Rhodococcus* spp., have been reported to cause mortality in the green peach aphid (*Myzus persicae*) through secretion of biosurfactants that degrade cuticular membranes (Kim *et al.*, 2010). In another example, production of chitinase and phospholipase A by the bacterium *Aeromonas punctata* was implicated in the development of cuticular and gut epithelium lesions conducive to bacterial entry into the hemocoel of *Anopheles annulipes* mosquito larvae (Kalucy and Daniel, 1972).

Pathogenic bacteria can also infect insects in the egg life stage by colonizing the surface or egg interior, usually resulting in failed hatching. Egg masses of the non-biting midge (*Chironomus* spp.) infected with *Vibrio cholerae* are degraded by proteases secreted by the bacteria (Halpern *et al.*, 2007). The most common bacteria in non-hatching egg masses of the European corn borer, *Ostrinia nubilalis*, were *Bacillus* spp. (Lynch *et al.*, 1976b), including *B. megaterium* and Bt subsp. *kurstaki*, which are pathogenic to *O. nubilalis* egg and larval stages, respectively (Lynch *et al.*, 1976a). Intimate association of the bacterium *P. aeruginosa* with eggs of the grasshopper *Schistocerca gregaria* results in normal egg hatch but mortality of the infected nymphs occurs before they reach the third instar (Bucher and Stephens, 1957).

8.3.2. Pathologies, Symptoms, and Factors Influencing Host Susceptibility

Insect disease normally originates from invading bacterial pathogens, although host stress or sickness can lead to infection by opportunistic bacteria. Normally, the pathogenic growth of saprophytic bacteria in the insect is prevented by the host immune system, limitation of nutrients, and competing bacteria. For instance, metabolism of secondary plant chemicals by gut microbiota in the desert locust, *S. gregaria*, produces phenolic compounds that contribute to host defense against pathogens (Dillon and Charnley, 2002). Infection by various pathogens can overcome the host's immune status and allow replication, resulting in disease. Entry of bacteria into an insect can potentially result in three distinct outcomes depending on the immune status of the host and virulence of the bacteria. Toxemia occurs when the bacteria remain confined to the host gut lumen but produce toxins that are disseminated into the hemolymph. When bacteria invade the hemolymph, they can cause bacteremia if they do not produce toxins or harmful factors, or septicemia when the bacteria multiply and produce toxins that lead to host death. By definition, bacteremia is common among symbiotic relationships but rare among entomopathogenic bacteria, with the exception of the intracellular pathogenic bacteria in the genera *Rickettsiella* and *Rhabdochlamydia*, which present very complex infection cycles with diverse structural forms (Huger and Krieg, 1967; Leclerque and Kleespies, 2008). Milky disease, caused by *Paenibacillus* spp. infection of

scarab larvae, also has the characteristics of a bacteremia as the invading bacteria multiply in the hemocoel for an extended period before death of the host.

After bacterial entry and establishment of disease, external symptoms in the infected host may vary depending on the host, bacterium, or stage of pathogenesis. Typical symptoms indicating the onset of bacteriosis include cessation of feeding, paralysis, diarrhea, or vomiting. Cessation of feeding after ingestion of bacterial pathogens has been described as a means of limiting exposure to toxins but may also increase immune function (Adamo *et al.*, 2010). The first reaction to ingestion of *Yersinia entomophaga* Tc toxins by scarab larvae is a violent vomiting and diarrhea leading to a brief amber appearance of the infected larvae due to clearance of food and enzymes from the midgut before death of the infected insect (Hurst *et al.*, 2011c). Similarly, an amber appearance resulting from a voided midgut is characteristic during the long chronic infection by *S. entomophila* in *C. zealandica* larvae (Jackson *et al.*, 1993). These host responses appear to represent defense mechanisms in an attempt to prevent establishment of infective foci and may be effective with less pathogenic microbes. Thus, cessation of feeding in larvae of the hornworm *Manduca sexta* after infection with *Serratia marcescens* has been reported to enhance the host immune response (Adamo *et al.*, 2007).

After traversing the gut epithelial barrier, bacteria proliferate in the hemocoel, producing bacteremia or septicemia. The high numbers of bacterial cells present in the hemolymph at this stage, and the concomitant tissue necrosis by the action of bacterial toxins and pathogenic factors, typically result in color and consistency changes in the insect host. Insects killed by bacteria usually darken in color and become soft and flaccid, although the integument initially remains intact. In some instances color change is due to the high bacterial loads of chromatic or refractive cells. For example, larvae of the Japanese beetle become red after infection with prodigiosin pigment-producing strains of the bacterium *S. marcescens*. Infection of the same larval species with *P. popilliae* results in whitish abdominal coloration caused by milkiness of the hemolymph, owing to the high number of refractile bacterial spores present. Body color change related to differential light reflection in infected host cells has also been noted in larvae of the midge *Chironomus thummi thummi* infected with the intracellular pathogen *Rickettsiella chironomi* (Götz, 1972). Rickettsial infections can also be recognized as white to blue phenotypes in scarab larvae (Jackson, 1992), which can sometimes be confused with milky disease. These visual changes are more noticeable in the early stages of hemimetabolous insects, such as larvae of Diptera, some Coleoptera, and Hymenoptera, because of their near-transparent integuments.

Bacterial infection may also change host body shape. Infection of the cockroach *Blatta orientalis* with an intracellular bacterium in the genus *Rhabdochlamydia* results in a characteristic swelling of the abdomen (Radek, 2000). As infection advances, cadavers may emit a putrid odor from the multiplying bacteria but, over time, the cadavers shrink, dry, and harden.

Conditions leading to stress, including prolonged exertion, deficient nutrition, unfavorable temperature and humidity, or pre-existing infection, activate responses that directly affect insect immune function and susceptibility to bacterial disease. Diverse acute stressors in the cricket *Gryllus texensis* reduce immune function and increase susceptibility to the opportunistic pathogen *S. marcescens* (Adamo and Parsons, 2006). Similarly, thermal stress promoted septicemia in larvae of the parasitoid wasp *Biosteres longicaudatus* and the Caribbean fruit fly *Anastrepha suspensa* (Greany *et al.*, 1977). Environmental stress conditions have been reported to have a significant role in the success of microbial control programs against the spruce budworm, *Choristoneura fumiferana*, using Bt sprays (Bauce *et al.*, 2006). However, stressors may have different effects on host susceptibility to bacteria. For example, while crowding conditions have been reported to promote bacterial infection (Doane, 1960), they also promote prophylactic behavior in the desert locust (*S. gregaria*) to reduce infectivity and support production of antimicrobials (Wilson *et al.*, 2002). This difference may represent adaptations in outbreak insect species to reduce risks of infection. Parasitization of the paper wasp, *Polistes dominulus*, by the strepsipteran *Xenos vesparum* results in reduced susceptibility to infection by *Staphylococcus aureus* (Manfredini *et al.*, 2010). In the lady beetle, *Hippodamia convergens*, deficient nutrition decreases susceptibility to infection by the bacterium *Pseudomonas fluorescens*, but when combined with high temperature had the opposite effect (James and Lighthart, 1992). Previous infection by a nucleopolyhedrovirus in larvae of the cabbage looper was reported to increase susceptibility to infection by Bt (McVay *et al.*, 1977). This synergistic effect increases options for biological control programs against insect pests with low susceptibility to individual pathogens (Brousseau *et al.*, 1998; Koppenhöfer *et al.*, 1999; Farrar *et al.*, 2004). However, there are also reports of antagonistic effects when using mixtures of entomopathogens (Ameen *et al.*, 1998; Pingel and Lewis, 1999).

8.3.3. Host Response to Infection

The insect immune system has been extensively reviewed (Vilmos and Kurucz, 1998; Rolff and Reynolds, 2009; Eleftherianos *et al.*, 2010) and includes both cellular and humoral responses. The initial response to infection occurs in the midgut epithelium. Upregulation of antimicrobial

genes is detected after ingestion of *Pseudomonas entomophila* by *Drosophila melanogaster* larvae (Vodovar *et al.*, 2005). Infection by Bt in larvae of the spruce budworm, *C. fumiferana*, leads to changes in expression levels of metabolic and stress-related gene products (Meunier *et al.*, 2006). Damage to the gut epithelium by Bt toxins has been described to activate a defensive gut response involving cell sloughing and tissue regeneration to overcome injury (Lacey and Federici, 1979; Delello *et al.*, 1984; Chiang *et al.*, 1986b).

After traversing the gut epithelium barrier, invading bacteria face phagocytosis from hemocytes circulating within the hemocoel, which will rapidly contact and engulf the pathogenic bacteria. The humoral response, which may take several hours from infection, is activated by bacterial invasion and initiated by pattern recognition proteins that are synthesized by hemocytes, fat body, and epidermal cells. These proteins recognize conserved bacterial surface determinants, LPS for Gram-negatives and peptidoglycans for Gram-positives, and initiate cascading chemical reactions leading to synthesis and secretion of antibacterial peptides from the fat body. There seems to be a crucial cross-talk between pathways controlling energy storage, metabolism, and defense against pathogens, as reported for infection of honey bee larvae with the bacterium *P. larvae* (Chan *et al.*, 2009).

8.4. GRAM-POSITIVE ENTOMOPATHOGENS: PHYLUM FIRMICUTES, CLASS BACILLI, ORDER BACILLALES

The order Bacillales includes families containing bacteria phenotypically defined as endospore-forming, Gram-positive, rods and cocci that are often arranged in chains. Under adverse environmental conditions, they undergo sporulation to form one oval-shaped spore per cell. In some members of the genus *Bacillus* (family Bacillaceae), one or more parasporal bodies are also formed during sporulation and remain in the sporangium until lysis. The most important entomopathogenic bacteria within this order are found in the genera *Bacillus* and *Paenibacillus* within the families Bacillaceae and Paenibacillaceae, respectively, and represent the most successful microbial insect control agents to date.

8.4.1. Family Bacillaceae, Genus *Bacillus*

This genus contains catalase-positive bacteria with peritrichous flagella and rod-shaped cells that may be arranged in chains (Fig. 8.2). Under adverse conditions they undergo sporulation, which is not repressed by air, to form an oval-shaped endospore that in some species may be accompanied by parasporal bodies. While information on

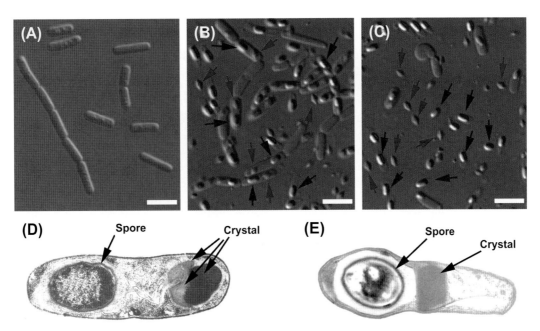

FIGURE 8.2   Different stages of a *Bacillus thuringiensis* subsp. *kurstaki* strain HD-73 culture documented through differential interference contrast (DIC) microscopy. Scale bars: 5 μm. During early stages (A) individual vegetative cells which may form long chains are observed. These cells are motile through the use of their flagella. As nutrients are depleted in the medium, *B. thuringiensis* cells undergo sporulation (B). At the late stages during the sporulation process both the spore (blue arrows) and bipyramidal parasporal crystal containing the Cry1Ac toxin (red arrows) can be visualized at this time within the sporulating cell. During later culture stages (C) free spores (blue arrows) and crystals (red arrows) are observed. (D) Spore and parasporal bodies of *B. thuringiensis* var. *israelensis*. Note the three different inclusion bodies (Crystal) representing different toxins. (E) Electron micrograph of a sporulated *Bacillus sphaericus* cell displaying the characteristic spherical spore located in a terminal position within the swollen sporangium. *[(A–C) Photos courtesy of Dr. John Dunlap, Advanced Microscopy and Imaging Center, University of Tennessee. (D) Courtesy of Federici et al., 1990, © Rutgers University Press. (E) Courtesy of Dr. Jean-François Charles, Institut Pasteur, Paris).]*

several reported *Bacillus* entomopathogenic species is included here, emphasis will be placed on *Bacillus thuringiensis* (and as introduced earlier, the commonly used abbreviation is "Bt") as the most important entomopathogenic species for insect control. This bacterium is part of the *B. cereus* group, which also contains *B. mycoides*, *B. weihenstephanensis*, *B. pseudomycoides*, and two vertebrate pathogens: *B. cereus*, causing food poisoning, and *B. anthracis*, which can cause anthrax. Data from diverse phylogenetic studies support the classification of *B. cereus* (*sensu stricto*), Bt, and *B. anthracis* as subspecies within a single species designated as *B. cereus sensu lato* (Daffonchio et al., 2000; Helgason et al., 2000; Bavykin et al., 2004). Plasmids containing host-specific genes explain differences among pathogenic phenotypes within the *B. cereus sensu lato* group (Rasko et al., 2005). Some authors have proposed that the group represents a collection of clones in the process of becoming diverse species, with speciation being driven by the presence of specific plasmids conferring advantageous phenotypes in specific natural environments (Priest and Dewar, 2000). Several observations support this model, including the low number of chromosomal differences among *Bacillus cereus sensu lato* species (Helgason et al., 2000), the highly clonal structure of the most specialized pathogens in the group

(Gaviria Rivera and Priest, 2003), and the transfer of plasmids between Bt strains infecting insects to create novel insecticidal toxin combinations (Jarrett and Stephenson, 1990). This hypothesis has been tested by introduction of insect hosts into an environment that increases the proportion of crystalliferous *Bacillus* spp. strains, resulting in an environment dominated by the most insecticidal Bt genotype (Raymond et al., 2010). In contrast, accelerated saprophytic growth is favored when the bacterium lacks plasmids containing insecticidal toxins.

The high degree of genetic similarity (Helgason et al., 2000) and potential for genetic exchange among species in the *B. cereus sensu lato* group (Ruhfel et al., 1984), the production of *B. cereus* toxins by Bt isolates (Abdel-Hameed and Landen, 1994; Ankolekar et al., 2009), and the highly similar genotypes of *B. cereus* and Bt isolates from food (Tourasse et al., 2011), have fueled concerns over the safety of Bt strains in pesticidal products. Early reports documented that the main difference between entomopathogenic Bt isolates and *B. cereus* or *B. anthracis* was the presence of crystalline inclusions in the Bt sporangium, which are composed of toxins responsible for insecticidal activity (Krieg, 1970). Further differences between subspecies in the *B. cereus sensu lato* group have been reported at the genomic and proteomic levels (Radnedge *et al.*, 2003; Gohar *et al.*, 2005). In contrast to both Bt and *B. cereus*, *B. anthracis* lacks a functional PlcR regulon, a series of virulence genes regulated by a single transcriptional activator that are critical for pathogenicity (Agaisse *et al.*, 1999; Salamitou *et al.*, 2000), resulting in the absence of specific secreted virulence proteins in *B. anthracis* compared to *B. cereus* or Bt (Gohar *et al.*, 2005). Analysis of genetic relatedness among natural populations of Bt and *B. cereus* suggests that populations of the same *B. cereus sensu lato* subspecies from diverse areas show higher genetic similarity than that between sympatric populations of different subspecies (Vilas-Boas *et al.*, 2002), indicating that gene flow in natural environments is more common among strains of the same *B. cereus sensu lato* subspecies. Although *B. cereus* and Bt isolates carrying plasmids related to B. anthracis plasmids pXO1 and pXO2 have been reported (Pannucci *et al.*, 2002a, b), these plasmids did not contain the anthrax toxin, regulator genes, or other virulence factors necessary to cause disease in vertebrates (Kolstø *et al.*, 2009). Furthermore, exchange of pathogenic plasmids between *B. anthracis* and Bt is hampered by *B. anthracis* existing mostly as dormant spores and rarely being found as vegetative cells in soil or insects, which are the preferred sites for genetic exchange between vegetative Bt cells (Yuan *et al.*, 2007). The genes encoding *B. cereus* toxins causing food-borne disease have been found in some insecticidal Bt strains (Frederiksen *et al.*, 2006), and expression of diarrheal-type toxins occurs during septicemia in *C. fumiferana* larvae infected with Bt subsp. *kurstaki* HD-1 (Kyei-Poku *et al.*, 2007). The safety of Bt-based pesticides is discussed in Subsection 8.4.1, *Safety of Bacillus thuringiensis pesticides and crops.*

Bacillus cereus (sensu stricto)

Most strains of *Bacillus cereus* (*sensu stricto*) consistently associated with insects are saprophytic or symbiotic bacteria occupying the digestive system. While genetically highly similar genetically to Bt, *B. cereus* does not produce parasporal crystalline toxins, which limits its virulence against insect hosts and favors saprophytic growth. Although it is most abundant in soil, *B. cereus* has wide distribution and is commonly detected among the gut flora of diverse insect species. Isolates of *B. cereus* have been reported to cause natural or induced infections in diverse insects, including larvae of scarab beetles (Selvakumar *et al.*, 2007); the flour beetle, *Tribolium castaneum* (Kumari and Neelgund, 1985); the spruce budworm, *C. fumiferana* (Strongman *et al.*, 1997); anopheline mosquitoes (Chatterjee *et al.*, 2010); *T. ni* (Wai Nam *et al.*, 1975); and adults of the tsetse fly, *Glossina morsitans* (Kaaya and Darji, 1989). The bacterium can produce virulence factors to limit iron acquisition by the host insect (Harvie and Ellar, 2005), evade host

phagocytic cells (Guillemet *et al.*, 2010), degrade host tissue (Lysenko, 1974), or inhibit growth of commensal gut flora (Abi Khattar *et al.*, 2009). Production of some of these virulence factors is correlated with pathogenicity, and most of them are controlled by the PlcR regulon (Gohar *et al.*, 2008). The combined action of these diverse virulence factors inhibits growth of other bacteria or clearance of *B. cereus* from the gut, typically resulting in accumulation of the bacterium and septicemia. In the cockroach, *Leucophaea maderae*, production of phospholipase C correlated with *B. cereus* pathogenicity (Rahmet-Alla and Rowley, 1989), whereas production of antibiotics by *B. cereus* in the gut of larvae of diamondback moth, *Plutella xylostella*, synergized Bt pathogenicity (Raymond *et al.*, 2008b).

While insecticidal toxins have been detected on the spore coat of *B. cereus* isolates, their activity is much lower than homologous toxins from Bt spores (Somerville and Pockett, 1975). Vegetative insecticidal proteins (or Vip toxins) were originally identified in a strain of *B. cereus* pathogenic to larvae of western and northern corn rootworms (*Diabrotica* spp.) (Warren *et al.*, 1996). The Vip toxins from *B. cereus* resemble the binary toxins from *Clostridium* and consist of a toxin subunit that binds to the membrane of target cells (Vip1 toxin) and a second toxin subunit displaying actin-ADP-ribosylating activity (Vip2 toxin) to inhibit actin polymerization and kill the host cell (Han *et al.*, 1999). Both Vip1 and Vip 2 toxins are necessary for toxicity, and their potential toxicity to non-target organisms or plants during expression limits their application for insect control. More recently, expression of the Vip2 toxin subunit in plants as an inactive proenzyme that is activated once ingested by the rootworm has been proposed as a strategy to produce transgenic plants for insect control (Jucovic *et al.*, 2008). Thermostable (Ohba *et al.*, 1981c) and non-proteinaceous (Perchat *et al.*, 2005) exotoxins have also been detected in strains of *B. cereus*. The non-proteinaceous exotoxins are highly active against the cotton boll weevil, *Anthonomus grandis*, while less activity was detected against the cotton leafworm, *Spodoptera littoralis*, or the black bean aphid, *Aphis fabae* (Perchat *et al.*, 2005). Vegetative cells of *B. cereus* can also obtain toxin-encoding plasmids from Bt cells in infected lepidopteran larvae, resulting in *B. cereus* cells producing crystalline insecticidal toxins (Yuan *et al.*, 2007).

Development and commercialization of *B. cereus* for microbial insect control have not been considered owing to the presence of toxins that induce diarrhea or emesis (vomiting) during food-borne illnesses in vertebrates. However, the use of *B. cereus* as host for expression of toxins targeting mosquito larvae was proposed because of its amenability to transformation and retention in the mosquito gut (Luxananil *et al.*, 2001). Transgenic plants expressing Vip toxins from *B. cereus* have also been produced (Jucovic *et al.*, 2008), although the safety of these products will need careful consideration before commercialization.

Bacillus thuringiensis: *Characteristics and Classification*

The history of the discovery and development of Bt as commercial pesticide has been extensively reviewed by several authors (Beegle and Yamamoto, 1992; Burges, 2001; Sanchis, 2011), and only key events are presented in this section. The bacterium was originally isolated by Shigetane Ishiwata in 1901 from diseased larvae of the silkworm (*Bombyx mori*) (Ishiwata, 1901), but his description of the Sottokin or "sudden death bacillus" was published in Japanese and did not become known among contemporary insect pathologists. Ishiwata already realized that toxins were involved in the pathogenic process, as larvae became rapidly paralyzed before multiplication of the bacillus (Ishiwata, 1905). However, it was Ernst Berliner who formally described and named the bacterium after isolation from diseased larvae of the Mediterranean flour moth, *A. kuehniella*, collected in the German state of Thuringia (Berliner, 1915). Later work with the Ishiwata isolate demonstrated that toxicity was associated with a protein present in sporulated cultures and not in vegetative cells (Aoki and Chigasaki, 1916). Berliner had noted the presence of a parasporal body or crystalline inclusion, which was later demonstrated to be proteinaceous in nature and soluble in alkaline solutions (Hannay and Fitz-James, 1955). The inclusions were found to be soluble in the alkaline digestive fluids of lepidopteran larvae and responsible for toxicity (Angus, 1954). Production of these parasporal inclusion bodies is the only phenotypic trait unique to Bt and this feature is used to differentiate this bacterium from other *Bacillus* species (Vilas-Bôas *et al.*, 2007). González et al. (1981) reported strong evidence for a correlation between loss of crystalline toxin production and specific plasmids. In the same year, Schnepf and Whiteley reported the cloning and expression of one of the genes encoding crystal proteins from Bt subsp. *kurstaki* HD1 (Schnepf and Whiteley, 1981). As expected, the heterologously expressed crystal protein displayed toxicity against *M. sexta* larvae.

Diverse biochemical, morphological, and antigenic methods have been used to describe and classify new Bt isolates. Early efforts concentrated on serological and biochemical methods, including morphological and biochemical typing (Heimpel and Angus, 1958), esterase patterns of vegetative cells (Norris, 1964), and serotyping of the H flagellar antigen in vegetative cells (de Barjac and Bonnefoi, 1968). Availability of serotyping services at the Pasteur Institute greatly contributed to wide acceptance of the H-serotyping method to classify Bt isolates and to assign them a subspecific name and serovariety type.

Initially, a gross correlation between classifications of Bt isolates based on biochemical characters and H serotyping with pathotypes was noted (Dulmage *et al.*, 1981), but this relation became more complicated as the number of Bt isolates with novel (and in some cases wide range) activities increased drastically (de Barjac and Frachon, 1990). Existence of strains that agglutinate spontaneously under conditions used for serotyping, and cross-reactivity of *B. cereus* strains with Bt H serotypes were also recognized as emerging limitations with the H-serotyping classification method (Lecadet *et al.*, 1999). Phage typing based on different sensitivity of bacterial strains to infection with diverse phages (Ackermann *et al.*, 1995), typing of ribosomal RNA genes (Priest *et al.*, 1994; Akhurst *et al.*, 1997), conservation of repetitive extragenic palindromic sequences (Reyes-Ramírez and Ibarra, 2005), and DNA hybridization and random amplified polymorphism analysis (Hansen *et al.*, 1998) have been proposed as valid methods to complement H-serotyping, which despite its limitations remains the most common and accepted method for classification of Bt isolates. A total of 82 Bt serovars, or more commonly used "subspecies," was described in the most recent revision of the H-antigen classification (Lecadet *et al.*, 1999). This list included 69 antigenic groups and 13 subgroups. Serovar *mogi* isolated from fallen leaves and displaying pathogenicity against mosquito larvae was recently isolated and described (Roh *et al.*, 2009a), resulting in the current list of 85 Bt serotypes presented in Table 8.1.

While pathogenicity cannot be considered a reliable taxonomic criterion, the distinction of Bt strains according to their pathogenicity (pathotype) is useful for practical purposes considering that the primary interest in Bt is its ability to kill insects. Initially, all Bt isolates identified displayed pathogenicity against larvae of Lepidoptera (pathotype A). The isolation of Bt subsp. *israelensis* by Goldberg and Margalit (1977) represented the first report of a Bt isolate pathogenic against larvae of Diptera (pathotype B). Later, isolates displaying both pathotypes A and B were described (Krieg *et al.*, 1968). A fourth pathotype was described with the isolation of Bt subsp. *tenebrionis* pathogenic against larvae of Coleoptera (Krieg *et al.*, 1983). Since then, Bt strains pathogenic against species of Blattaria (Quesada-Moraga *et al.*, 2004), Hemiptera (Lima *et al.*, 1994), Hymenoptera (Garcia-Robles *et al.*, 2001), Isoptera (de Castilhos-Fortes *et al.*, 2002), Orthoptera (Quesada-Moraga *et al.*, 2004), and nematodes (Bottjer *et al.*, 1985) have been reported. The ability to discriminate among Bt strains based on pathotype is greatly challenged by the production of multiple toxins with diverse specificities by individual isolates and the increasing number of pathotypes described. Thus, most Bt serovars include strains that are pathogenic against insects in diverse taxonomic orders. For some crystalliferous Bt strains no insect

TABLE 8.1 Current List of *Bacillus thuringiensis* Serovars (= Subspecies)

H Antigen	Serovar	Abbreviation	First Valid Description[a]
1	*thuringiensis*	THU	Heimpel and Angus (1958)
2	*finitimus*	FIN	Heimpel and Angus (1958)
3a:3c	*alesti*	ALE	Heimpel and Angus (1958)
3a:3b:3c	*kurstaki*	KUR	de Barjac and Lemille (1970)
3a:3b:3d	*mogi*	MOG	Roh *et al.* (2009a)
3a:3d	*sumiyoshiensis*	SUM	Ohba and Aizawa (1989)
3a:3d:3e	*fukuokaensis*	FUK	Ohba and Aizawa (1989)
4a:4b	*sotto*	SOT	Heimpel and Angus (1958)
4a:4c	*kenyae*	KEN	Bonnefoi and de Barjac (1963)
5a:5b	*galleriae*	GAL	de Barjac and Bonnefoi (1962)
5a:5c	*canadensis*	CAN	de Barjac and Bonnefoi (1972)
6	*entomocidus*	ENT	Heimpel and Angus (1958)
7	*aizawai*	AIZ	Bonnefoi and de Barjac (1963)
8a:8b	*morrisoni*	MOR	Bonnefoi and de Barjac (1963)
8a:8c	*ostriniae*	OST	Ren *et al.* (1975)
8b:8d	*nigeriensis*	NIG	Weiser and Prasertphon (1984)
9	*tolworthi*	TOL	de Barjac and Bonnefoi (1968)
10a:10b	*darmstadiensis*	DAR	de Barjac and Bonnefoi (1968)
10a:10c	*londrina*	LON	Arantes *et al.* (unpubl.)
11a:11b	*toumanoffi*	TOU	Krieg (1970)
11a:11c	*kyushuensis*	KYU	Ohba and Aizawa (1979)
12	*thompsoni*	THO	de Barjac and Thompson (1970)
13	*pakistani*	PAK	de Barjac *et al.* (1977)
14	*israelensis*	ISR	de Barjac (1978)
15	*dakota*	DAK	DeLucca *et al.* (1979)
16	*indiana*	IND	DeLucca *et al.* (1979)
17	*tohokuensis*	TOH	Ohba *et al.* (1981a)
18a:18b	*kumamotoensis*	KUM	Ohba *et al.* (1981b)
18a:18c	*yosoo*	YOS	H. H. Lee *et al.* (1995)
19	*tochigiensis*	TOC	Ohba *et al.* (1981b)
20a:20b	*yunnanensis*	YUN	Yu *et al.* (1984)
20a:20c	*pondicheriensis*	PON	Rajagopalan *et al.* (unpubl.)
21	*colmeri*	COL	DeLucca *et al.* (1984)
22	*shandongiensis*	SHA	Ying *et al.* (1986)
23	*japonensis*	JAP	Ohba and Aizawa (1986)
24a:24b	*neoleonensis*	NEO	Rodriguez-Padilla *et al.* (1990)
24a:24c	*novosibirsk*	NOV	Burtseva *et al.* (1995)

TABLE 8.1 Current List of *Bacillus thuringiensis* Serovars (= Subspecies)—cont'd

H Antigen	Serovar	Abbreviation	First Valid Description[a]
25	*coreanensis*	COR	Lee *et al.* (1994)
26	*silo*	SIL	de Barjac and Lecadet (unpubl.)
27	*mexicanensis*	MEX	Rodriguez-Padilla *et al.* (unpubl.)
28a:28b	*monterrey*	MON	Rodriguez-Padilla *et al.* (unpubl.)
28a:28c	*jegathesan*	JEG	Seleena *et al.* (1995)
29	*amagiensis*	AMA	Ohba (unpubl.)
30	*medellin*	MED	Orduz *et al.* (1992)
31	*toguchini*	TOG	Khodyrev (1990)
32	*cameroun*	CAM	Juárez-Pérez *et al.* (1994)
33	*leesis*	LEE	Lee *et al.* (1994)
34	*konkukian*	KON	Lee *et al.* (1994)
35	*seoulensis*	SEO	H. H. Lee *et al.* (1995)
36	*malaysiensis*	MAL	Ho (unpubl.)
37	*andaluciensis*	AND	Quesada-Moraga *et al.* (2004)
38	*oswaldocruzi*	OSW	Rabinovitch *et al.* (1995)
39	*brasiliensis*	BRA	Rabinovitch *et al.* (1995)
40	*huazhongensis*	HUA	Dai *et al.* (1996)
41	*sooncheon*	SOO	H. H. Lee *et al.* (1995)
42	*jinghongiensis*	JIN	Li *et al.* (1999)
43	*guiyangiensis*	GUI	Li *et al.* (1999)
44	*higo*	HIG	Ohba *et al.* (1995)
45	*roskildiensis*	ROS	Hinrinschen and Hansen (unpubl.)
46	*chanpaisis*	CHA	Chanpaisaeng *et al.* (1996)
47	*wratislaviensis*	WRA	Lonc *et al.* (1997)
48	*balearica*	BAL	Iriarte *et al.* (2000)
49	*muju*	MUJ	Park *et al.* (unpubl.)
50	*navarrensis*[b]	NAV	Iriarte *et al.* (2000)
51	*xiaguangiensis*	XIA	Yan (unpubl.)
52	*kim*	KIM	Kim *et al.* (unpubl.)
53	*asturiensis*	AST	Quesada-Moraga *et al.* (2004)
54	*poloniensis*	POL	Damgaard *et al.* (unpubl.)
55	*palmanyolensis*	PAL	Quesada-Moraga *et al.* (2004)
56	*rongseni*	RON	Li *et al.* (1999)
57	*pirenaica*	PIR	Porcar *et al.* (1999)
58	*argentinensis*	ARG	Campos-Dias *et al.* (unpubl.)
59	*iberica*	IBE	Porcar *et al.* (1999)
60	*pingluonsis*	PIN	Li *et al.* (1999)

(Continued)

TABLE 8.1 Current List of *Bacillus thuringiensis* Serovars (= Subspecies)—cont'd

H Antigen	Serovar	Abbreviation	First Valid Description[a]
61	*sylvestriensis*	SYL	Damgaard (unpubl.)
62	*zhaodongensis*	ZHA	Li *et al.* (1999)
63	*bolivia*[b]	BOL	Ferrandis *et al.* (1999)
64	*azorensis*	AZO	Santiago-Alvarez *et al.* (unpubl.)
65	*pulsiensis*	PUL	Khalique and Khalique (unpubl.)
66	*graciosensis*	GRA	Santiago-Alvarez *et al.* (unpubl.)
67	*vazensis*[b]	VAZ	Santiago-Alvarez *et al.* (unpubl.)
68	*thailandensis*	THA	Chanpaisaeng *et al.* (unpubl.)
69	*pahangi*	PAH	Seleena *et al.* (unpubl.)
70	*sinensis*	SIN	Li *et al.* (2000)
71	*jordanica*	JOR	Khyami-Horani *et al.* (2003)

Kindly supplied by Dr. J. Charles, Institute Pasteur, Paris, France.
The first valid description for each serovar is also presented.
[a]*Taken and updated from a previous review (Lecadet* et al., *1999).*
[b]*These serovars have been reported to represent strains of* Bacillus weihenstephanensis *(Soufiane and Côté, 2010).*

host has been identified. Since toxicity is correlated with production of toxins, identification of the toxins produced by a Bt strain may be used to predict the range of activity more accurately. However, Bt toxins are highly specific and a single Bt strain or crystal toxin may only be active against a limited number of insect species, even within a genus. For instance, Bt subsp. *tolworthi* is highly pathogenic to larvae of the fall armyworm, *Spodoptera frugiperda* (Hernandez, 1988), but not to the closely related oriental leafworm moth, *Spodoptera litura* (Amonkar *et al.*, 1985). Recently, phenotypic classification of a wide collection of Bt isolates indicated the existence of specific combinations of traits that correlate with host range (Martin *et al.*, 2010). For instance, production of urease, which has been shown to be important for effective reproduction and passage of Bt in larvae of the gypsy moth, *Lymantria dispar* (Martin *et al.*, 2009), was found in strains producing bipyramidal (diamond-like) crystals, which are usually active against lepidopteran larvae. The reader is referred to comprehensive reviews on invertebrate species susceptible to Bt strains (Glare and O'Callaghan, 2000) or purified Bt crystal toxins (van Frankenhuyzen, 2009).

Bacillus thuringiensis *Ecology, Biology, and Infection*

Even though Bt has been traditionally considered a ubiquitous soil microorganism, it has been isolated from multiple environments worldwide (Martin and Travers, 1989; Bernhard *et al.*, 1997; Chaufaux *et al.*, 1997), except

in Antarctica (Wasano *et al.*, 1999). Several extensive Bt screening projects have reported that the most prolific environment for isolation of Bt strains is dust and materials associated with grain stores and silos (Bernhard *et al.*, 1997; Chaufaux *et al.*, 1997; Iriarte *et al.*, 1998). In these screenings, slightly more than half of the Bt isolates were pathogenic to selected species of Lepidoptera, Coleoptera, or Diptera. The most abundant Bt serovar in any area varies with geographical region and environment (Martin and Travers, 1989). The most common shapes of crystalline inclusions in natural isolates are bipyramidal and spherical.

The synthesis of the parasporal crystalline inclusions represents a large commitment of metabolic resources, advocating that production of these crystals should confer an evolutionary advantage over acrystalliferous bacteria. The specific ecological niche occupied by Bt in the environment and the evolutionary advantage of producing crystalline inclusions are still a matter of debate, with little conclusive evidence to determine whether the bacterium is principally an insect pathogen or exists as an environmental bacterium with a facultative ability to infect insects. Bactericidal activity, which would favor Bt over competitive gut microflora, has only been reported for the cytocidal-type crystal toxins (Cahan *et al.*, 2008). Some crystal toxins remain attached to the Bt spore and their binding to the insect gut cells enhances spore germination, which may reflect a selective advantage over alternative bacteria colonizing this environment (Du and Nickerson, 1996b). Supported by observations that Bt naturally resides in the digestive system of terrestrial arthropods (Hansen and Salamitou, 2000), some authors have proposed a two-phase life cycle that includes endosymbiotic and infective phases for Bt and other bacteria in the *B. cereus sensu lato* group (Jensen *et al.*, 2003; Swiecicka, 2008). According to this hypothesis, acquisition of Bt may confer advantages for certain hosts, such as protection against pathogens through production of antibiotics and bacteriocins or enhanced degradation of organic materials by Bt-secreted enzymes.

Although Bt has been traditionally considered a soil bacterium, it is not capable of multiplying in soil or water at levels that would allow favorable competition with other bacteria (Yara *et al.*, 1997; Furlaneto *et al.*, 2000), and insects are considered the optimal site for Bt multiplication and exchange of genetic material (Jarrett and Stephenson, 1990). In fact, all the commercially relevant Bt subspecies used in biopesticides were originally isolated from dead insects, including *kurstaki* (Kurstak, 1964), *israelensis* (Goldberg and Margalit, 1977), and *tenebrionis* (Krieg *et al.*, 1983). It is likely, therefore, that the soil acts as a reservoir for bacterial spores rather than as a site of multiplication, especially as most known Bt hosts are foliar feeders that would preferentially come into contact with soil bacteria during the non-feeding pupal stage. In fact, higher proportions of entomopathogenic serovars are identified from phylloplane compared to soil samples, denoting an evolutionary advantage for pathogenic Bt on foliar surfaces and Bt as a phylloplane epiphyte (Smith and Couche, 1991). For instance, mulberry leaves were proposed as a source of the bacterium in silkworm insectaries (Ohba, 1996). Damgaard et al. (1997) reported a much higher prevalence of Bt strains pathogenic against common lepidopteran cabbage pests in isolates from the cabbage phylloplane compared to soil and other samples. The presence of Bt on plant foliage also explains the common isolation of the bacterium from feces of herbivorous animals (Lee *et al.*, 2002; Maheswaran *et al.*, 2010). However, Bt presence on foliage would be transient, whereas spores can bioaccumulate and remain viable for long periods in soil, from where they can be passively dispersed to the lower leaves of plants by rain splash (Pedersen *et al.*, 1995) or during plant emergence (Bizzarri and Bishop, 2008). Despite described metabolic activity and genetic transfer on the phylloplane (Bizzarri and Bishop, 2007, 2008), some authors argue that compared to common epiphytic bacteria, Bt does not efficiently colonize or grow extensively on plant surfaces (Maduell *et al.*, 2008). While the level of survival of Bt on the phylloplane remains under discussion, crystalliferous strains of Bt have also been reported as endophytes (Goryluk *et al.*, 2009; Monnerat *et al.*, 2009). When these Bt endophytes were inoculated in cotton and cabbage plants, mortality of leaf-feeding *S. frugiperda* and *P. xylostella* larvae was observed.

A generalized Bt infective cycle commences with the ingestion of spores and/or crystalline inclusions by a susceptible host. After ingestion, cessation of feeding and paralysis follows in the host, which can be reversible in some species (Angus and Heimpel, 1959). Disease becomes irreversible with increased hemolymph pH after toxins increase the permeability of the midgut epithelial barrier (Fast and Angus, 1965). Host symptoms during Bt infection allow definition of three major host types (Heimpel and Angus, 1959; Chiang *et al.*, 1986a). Type I and II hosts are characterized by a fast onset of gut paralysis continuing in type I to a general paralysis correlating with an increase in the pH and potassium concentration in the hemolymph (Angus and Heimpel, 1959), which are not observed for type II hosts. Paralysis of the digestive system in type I larvae is limited to the midgut region (Nishiitsutsuji-Uwo and Endo, 1980), and examples include the lepidopterans *B. mori*, *Protoparce* (= *Manduca*) *quinquemaculata* (Heimpel and Angus, 1959), *Philosamia ricini* (Pendleton, 1970), and *T. ni*; and the dipterans *Simulium vittatum* (Lacey and Federici, 1979) and the yellow fever mosquito, *Aedes aegypti*. Representatives of type IIA hosts, having midgut pH higher than 9.0, are larvae of *Corcyra cephalonica*, and *M. sexta* (Chiang *et al.*, 1986a), while *Pieris rapae*, *L. dispar*, and *Ephestia cautella* are examples of type IIB hosts. While both type I and II hosts are susceptible to crystal toxin alone, type III hosts do not develop paralysis and require the presence of spores for mortality. Larvae of *A. kuehniella* (Heimpel and Angus, 1959) and *G. mellonella* (Chiang *et al.*, 1986a) are examples of the type III group.

The symptoms and events leading to mortality of type I larvae have been thoroughly described (Heimpel and Angus, 1959; Nishiitsutsuji-Uwo and Endo, 1980). After ingestion of the crystal, the larvae appear normal for the first 30 min, although cessation of feeding is evident within 15 min post-ingestion and corresponds to gut paralysis and a rise in hemolymph pH and potassium concentration, which continue to increase until larval death. Midgut paralysis and anorexia prevent passage of the ingested bacteria through the digestive system. Between 30 and 120 min post-ingestion, larvae appear sluggish, start vomiting, and produce bead-like frass and diarrhea. This period corresponds to processing of crystal toxins by the host digestive fluids to active toxins, which target the gut cells and compromise the gut epithelial barrier. At this time (approximately 2 h post-ingestion), infected larvae are very sluggish and display only reflex movements, leading to complete paralysis and death at 3 h postingestion. However, at low toxin concentrations or in larvae less susceptible to the toxins, changes in the physicochemical conditions during lysis of the gut epithelium result in spore germination and secretion of virulence factors by vegetative cells. Binding of activated toxin to midgut receptors has been reported to act as a spore germination cue (Du and Nickerson, 1996b). Virulence factors secreted include chitinases, proteases, phospholipases, and iron-sequestering systems to degrade host tissues and inhibit growth of competing bacteria. While Bt is capable of growing in the alkaline pH of the midgut, the majority of growth occurs in the nutrient-rich and neutral pH environment of the hemolymph (Raymond *et al.*, 2008b). Effective Bt growth in the hemolymph has been detected even when the insect is not susceptible (Johnston and Crickmore, 2009), indicative that in this case factors limiting mortality are related to the intoxication process. Multiplication of vegetative cells in the hemocoel leads to septicemia and ultimately insect death.

The low mortality after antibiotic pretreatment of hosts led some authors (Broderick *et al.*, 2006, 2009) to propose that Bt may not be a true pathogen and that lethal septicemia, at least for some insect species, may be caused by saprophytic bacteria in the gut flora. However, subsequent investigations questioned the methodology used and verified that undesired effects of the antibiotic treatment (Johnston and Crickmore, 2009; Raymond *et al.*, 2009) and differences in gut flora composition or bioassay methodology (van Frankenhuyzen *et al.*, 2010) explained the unexpected results. Dependence of Bt pathogenicity on alternate gut bacteria is also unlikely given the observations of gut flora inhibiting Bt pathogenicity (Takatsuka and Kunimi, 2000; Hernández-Martínez *et al.*, 2010) and enhancement of Bt virulence by antimicrobials (Broderick *et al.*, 2000). Positive selection of crystal toxin residues involved in specificity (Wu *et al.*, 2007a, b) indicates that production of these toxins is evolutionarily advantageous to Bt and support its role as a *bona fide* pathogen. Selection of the most pathogenic Bt genotypes after the introduction of suitable insect hosts into an environment (Raymond *et al.*, 2010) confirms that production of toxins and concurrent pathogenicity in Bt are favorably selected.

After host death and Bt multiplication, sporulation is initiated by nutrient exhaustion and the concomitant reduction in intracellular guanosine triphosphate (GTP) concentration (Starzak and Bajpai, 1991). Information on the genetic regulation of Bt sporulation has been derived from *Bacillus subtilis* (reviewed by Stragier and Losick, 1996) and ultrastructural observations of Bt sporulation (Ribier and Lecadet, 1973; Bechtel and Bulla, 1976). Gene transcription during sporulation is temporally and spatially regulated by a cascade of transcription factors in the RNA polymerase sigma factor family, which bind the RNA polymerase to direct transcription from sporulation-specific promoters. Environmental cues, such as nutrient depletion, cell density, or DNA damage, activate a regulatory cascade resulting in activation of the sigma factor σ^A, during the vegetative phase. The σ^H factor participates in formation of the polar septum separating the sporulating cell (sporangium) into a large (mother cell) and small (forespore) compartments. The σ^E factor is active from septum formation to the formation of the spore cortex, and contributes to engulfment of the forespore by the mother cell. After engulfment, σ^K factors become activated in the mother cell and σ^G in the forespore compartment. Eventually, the forespore becomes a protoplast within the mother cell compartment, which describes Bt spores as endospores. Expression of most crystal toxin genes is directed by σ^E/σ^K type factors, and synthesis of parasporal bodies is localized to the mother cell compartment. After 8–12 h of spore development and maturation, the mother cell undergoes programmed cell death and liberates the mature spore into the environment.

Although Bt spores are very persistent in the environment, Bt has low rate of transmission between hosts. No evidence of vertical transmission has been found (Raymond *et al.*, 2008a), while effective horizontal transmission has been difficult to demonstrate experimentally (Takatsuka and Kunimi, 1998). The low levels of effective horizontal transmission help to explain why Bt epizootics are rare. Low levels of natural Bt in the environment may support enzootic disease (Burges, 1973; Damgaard *et al.*, 1997), but epizootics are usually related to external conditions such as the crowding commonly present in insect-rearing facilities. For instance, an outbreak in a *B. mori* rearing facility led to the discovery of Bt (Ishiwata, 1901). Multiplication of the bacterium in insect carcasses varies depending on the insect species (Suzuki *et al.*, 2004; Raymond *et al.*, 2008a) as well as virulence of the particular Bt strain (Prasertphon *et al.*, 1973), indicating that favorable conditions may be present in only some cases. The ability of Bt strains to survive repeated passages through larvae of *L. dispar* was reported to be directly correlated to urease production, probably by allowing more efficient degradation of the insect host cadaver (Martin *et al.*, 2009). However, although both Bt subsp. *kurstaki* and subsp. *aizawai* used in commercial Bt products are urease positive (Martin *et al.*, 2010), secondary infections in spray applications are uncommon (Smith and Barry, 1998). In contrast, persistent activity against mosquito larvae is found in applications of the urease-negative Bt subsp. *israelensis* (Tilquin *et al.*, 2008; Martin *et al.*, 2010), probably the result of specific environmental conditions involved. Thus, it is possible that conditions in rearing facilities and in selected environments such as stored grain mills may promote horizontal transmission. Diverse factors, including ultraviolet (UV) light degradation (Pozsgay *et al.*, 1987), plant species (Pinnock *et al.*, 1975), and soil microbes (West *et al.*, 1984), can directly affect viability of spores or degrade crystal toxins under field conditions, hindering the onset of natural epizootics. In fact, insecticidal activity on leaves after application of Bt sprays is short lived, with a half-life of one to three days (Ignoffo *et al.*, 1974; Pinnock *et al.*, 1974;

Beegle *et al.*, 1981). Yet, field epizootics of Bt disease have been recorded in *Ephestia* (= *Anagasta*) *kuehniella*, *E. elutella*, and *Plutella maculipennis* (= *P. xylostella*) in an area of Yugoslavia not exposed to Bt sprays (Vanková and Purrini, 1979), the Siberian silkworm (*Dendrolimus sibiricus*) (Talalaev, 1956), and in the Loreyi leafworm (*Mythimna loreyi*) in corn fields in Spain (Porcar and Caballero, 2000). A field epizootic in the mosquito *Culex pipiens* led to the isolation of Bt subsp. *israelensis* (Margalit and Dean, 1985). However, reported epizootics of disease are rare considering the widespread distribution of the bacterium.

The response to Bt infection and ability to recover has been mostly studied in lepidopteran hosts. Larvae of *Heliothis virescens* exposed to sublethal doses of Bt subsp. *kurstaki* HD-1 over a short period were able to recover from infection (Dulmage *et al.*, 1978), substantiating the existence of a defensive mechanism. In larvae of *C. cephalonica* (Chiang *et al.*, 1986b) and *M. sexta* (Spies and Spence, 1985), midgut cells damaged by Bt intoxication were extruded into the midgut lumen and replaced by newly differentiated cells. Since replacement of dying mature cells allows preservation of an effective midgut epithelial barrier, this defensive response would prevent invasion of the hemocoel and septicemia. Even with recovery from Bt infection, sublethal effects are often observed, including delayed development, reduction in larval weight and size, and increased number of larval instars (Nyouki *et al.*, 1996; Pedersen *et al.*, 1997; Flores *et al.*, 2004). Larvae surviving infection usually produce smaller pupae that may fail to develop or emerge as adults. Sublethal effects in adults developed from treated larvae are usually detected as reductions in fecundity or egg viability. Larvae that recover from Bt infection are also more susceptible to a second infective episode (Moreau and Bauce, 2003).

The entomopathogenic feature of Bt is dependent on the expression of virulence factors, including toxins, lipases, proteases, and chitinases. A nomenclature based on the Greek alphabet was proposed to designate the diverse toxins produced by the bacterium (Heimpel, 1967) to include α- and β-exotoxins, and γ-endotoxins. Exotoxins are secreted in the environment, while the endotoxins are in protein crystals.

General Characteristics of Bacillus thuringiensis Crystal Toxins

Production of the parasporal body or crystal is the defining feature of Bt. The crystals are proteinaceous in nature and are composed of millions of Cry (from "crystal") or Cyt ("cytolytic") toxin molecules. Production of the crystal(s) is generally concomitant with sporulation, although expression of the crystal toxin in Bt subsp. *tenebrionis* was demonstrated also to occur during vegetative growth

(Sekar, 1988). The shape of the crystal is variable depending on the toxins present in the crystal and growth conditions, with diverse crystal morphologies sometimes occurring concurrently. Among natural isolates, the most common morphologies are bipyramidal and circular (round) (Martin and Travers, 1989; Bernhard *et al.*, 1997). Cysteine residues located at the C terminus have been described to cross-link through disulfide bonds (Bietlot *et al.*, 1990), making this region crucial in assembly and stability of the crystal (Baum and Malvar, 1995). The specific protein composition of the crystal influences solubility in the midgut of the insect host, which is crucial for toxicity (Aronson, 1994, 1995). Synthesis of the crystal represents an enormous metabolic investment, as it may make up to 20–30% of the dry cell weight in the sporulated cell. These high levels of expression and its temporal limitation to the stationary phase of growth are controlled at the transcriptional, post-transcriptional, and post-translational levels.

Vegetative Bt cells have a strong ability to harbor DNA plasmids, sometimes in excess of the DNA amounts found in the chromosome (Zhong *et al.*, 2011). There is usually more than one *cry* gene and plasmid per strain (Kronstad *et al.*, 1983), although the presence of plasmids does not always translate in production of Cry proteins (González and Carlton, 1980). The *cry* and *cyt* genes are mostly located in plasmids of 30 to more than 200 megadaltons (MDa) in size (González *et al.*, 1981), and acrystalliferous strains can be generated using a plasmid curing technique involving heat shock (Ward and Ellar, 1983). Copies of *cry* genes can also be located in, and expressed from, the bacterial chromosome (Kronstad *et al.*, 1983; Jarrett, 1985). Different *cry* genes in a strain can direct the synthesis of related proteins that are stored in a single crystal or production of unique Cry proteins that form separate crystals of distinct shape. The number and size of plasmids in each strain are highly variable and although a characteristic plasmid pattern may be identified for each serovar, differences may be common between strains in the same serovar (Reyes-Ramírez and Ibarra, 2008). One of the most commercially relevant strains, Bt subsp. *kurstaki* HD-1, harbors the *cry1Aa*, *cry1Ac*, *cry2A*, and a cryptic *cry2B* gene in a single plasmid of approximately 100 MDa, while the *cry1Ab* gene is located on an unstable 44 MDa plasmid (Kronstad and Whiteley, 1986). In comparison, Bt subsp. *israelensis* contains the *cry4A*, *cry4B*, *cry10A*, *cry11A*, *cyt1A*, and *cyt2A* genes on a single 75 MDa plasmid called pBtoxis (Berry *et al.*, 2002).

Plasmids or chromosomal regions containing *cry* genes can be shared between Bt strains during conjugation. This phenomenon has been reported to occur in diverse environments, most importantly soil (Vilas-Bôas *et al.*, 2000), the phylloplane (Bizzarri and Bishop, 2008), and insect hosts (Jarrett and Stephenson, 1990). Highest rates of conjugation occur in the susceptible insect host during the latter stages of the pathogenic process, probably owing to the high number of vegetative cells present (Vilas-Bôas et al., 1998; Thomas et al., 2002), which limits horizontal transfer to other bacteria. Specific large plasmids (pXO11, pXO12, pXO13, pXO14, pXO15, and pXO16) were initially described to be responsible for plasmid mobilization, because strains harboring any of these plasmids had donor capability and strains cured of them were infertile (Battisti et al., 1985; Reddy et al., 1987), but subsequent work has identified self-transmissible plasmids (Wiwat et al., 1990; Wilcks et al., 1999). This conjugative process is not limited to exchanges between Bt strains, and transfer of genetic material has been reported between strains of Bt and other bacteria (González et al., 1982; Klier et al., 1983; Battisti et al., 1985). Horizontal cry gene transfer usually results in crystalliferous recipients with insecticidal activity (Hu et al., 2004, 2005). This process has been hypothesized to be vital in the evolution of Bt populations, allowing for novel combinations of toxins (Jarrett and Stephenson, 1990). In the natural environment, horizontal transfer of cry genes to other bacteria may result in new genes through recombination events or new phenotypes and possibly ecotypes in the recipient species.

Since the cloning of the first cry gene by Schnepf and Whiteley (Schnepf and Whiteley, 1981), more than 400 cry toxin genes have been cloned and sequenced (Crickmore et al., 2011). Usually, cry genes are monocistronic but in some cases they are part of a cluster of genes under the control of a single promoter (operon). For example, the cry40 and cry34 genes of Bt subsp. thompsoni are sequentially located in an operon with the promoter region located upstream of the cry40 gene (Brown, 1993). The cry34 and cry35 genes encoding a binary toxin complex (Ellis et al., 2002) are also sequentially organized in an operon (Schnepf et al., 2005). The cry2A gene of Bt subsp. kurstaki is one of the most well-characterized operons and is found as the most distal in a three-gene operon (Widner and Whiteley, 1990). The first open reading frame (orf1) in the operon encodes a protein of unknown function (Delattre et al., 1999) that is not necessary for production of the crystal or toxicity (Crickmore and Ellar, 1992). The second gene in the operon (orf2) encodes a 29 kilodalton (kDa) cytosolic chaperone that interacts directly with Cry2A protoxin and is needed for production of Cry2A inclusions in acrystalliferous strains (Crickmore and Ellar, 1992; Staples et al., 2001). Similarly, the cry11A gene from Bt subsp. israelensis is organized in an operon containing a gene encoding a 20 kDa chaperone (Dervyn et al., 1995). Functional analysis of the 29 kDa and 20 kDa helper proteins in the same system suggests that the 29 kDa protein stabilizes and organizes the protoxin molecules during crystal formation, whereas the 20 kDa protein acts as a chaperone to enhance protoxin synthesis (Ge et al.,

1998). The 20 kDa protein has been widely used to enhance expression of wild-type and mutant toxins in Bt strains (Wu and Federici, 1995; Rang et al., 1996; Shi et al., 2006), and Bin toxin from B. sphaericus (Park et al., 2007a). The 29 kDa protein was not needed for production of Cry2A crystals in insect cell cultures, indicating the existence of proteins performing a similar function in these cells (Lima et al., 2008). Lack of orf1 and accumulation of mutations in the orf2 region upstream of the cry2B gene result in a cryptic (not expressed) gene (Hodgman et al., 1993), and a similar phenomenon may explain the existence of other cryptic genes in Bt strains (Masson et al., 1998; Shisa et al., 2002).

The cry1A gene is the most well-characterized example of a sporulation-dependent cry gene. Detailed mapping of the structural organization of cry genes has shown that they are flanked by multiple copies of inverted repetitive DNA sequences (ISs) typical of transposon boundaries (Kronstad and Whiteley, 1984; Mahillon et al., 1985), or structurally associated with transposons (Lereclus et al., 1986). Expression of the cry gene is independent of the IS sequences, and they may assist in horizontal movement of cryptic cry genes (Hodgman et al., 1993). Six main different ISs (IS231, IS232, IS240, ISBt1, ISBt2, and ISBth) and their variations, and two transposons (Tn4430 and Tn5401) have been reported to be structurally associated with cry genes. These nucleotide sequences encode putative translocases that would allow them to move or transpose to new positions in plasmids or in the cell chromosome by transposition or recombination events mediated by these sequences (Hallet et al., 1991). Instability of the cry1Ab gene is well documented, as approximately 30% of Bt subsp. kurstaki HD-1 colonies lack this gene, while the cry1Aa and cry1Ac genes are common to all colonies (Aronson, 1994). Transfer of a fragment of the cry1Ab gene from Bt subsp. kurstaki to B. mycoides in soil was facilitated by IS231 (Donnarumma et al., 2010). Instability of plasmids and cry genes can also result in spontaneous loss of cry genes during culture (Roy et al., 1987; Sarrafzadeh et al., 2007; Bizzarri et al., 2008). Loss of the pBtoxis plasmid in Bt subsp. israelensis under field conditions resulted in acrystalliferous strains and reduced potency against mosquito larvae (Kamdar and Jayaraman, 1983).

Regulation of Cry Gene Expression

The diverse molecular mechanisms contributing to the high levels of protein synthesis and accumulation necessary for formation of protein crystals in Bt have been thoroughly reviewed (Agaisse and Lereclus, 1995; Baum and Malvar, 1995). The highly variable number of cry gene copies among Bt strains (Kronstad et al., 1983) was suggestive of a potential contribution to increased production of Cry proteins. Increased cry3A gene copy number in a mutant strain of Bt subsp. tenebrionis resulted in increased toxin production (Adams *et al.*, 1994). Increased *cry3A* gene copy number also correlated with higher toxin production in acrystalliferous Bt (Arantes and Lereclus, 1991). Arantes and Lereclus (1991) reported that the level of *cry* gene expression is not exclusively related to the vector copy number and proposed the existence of putative active repressors when the *cry* gene is present at a low copy number. Increased gene copy number was hypothesized to result in titration of this repressor and increased expression, as reported for production of Cry1A toxins in *B. megaterium* (Shivakumar *et al.*, 1989). However, expression levels are limited and a plateau in the direct relationship between *cry* gene copy number and expression was detected at 15 copies per chromosome (Arantes and Lereclus, 1991). Limited resources for *cry* gene expression within the cell explain increased levels of expression of the *cry1Aa*, *cry1Ac*, and *cry2A* genes in Bt subsp. *kurstaki* strains missing the *cry1Ab* gene (Aronson, 1994). Introduction of a single *cry2A* toxin gene in an acrystalliferous Bt strain resulted in about five-fold higher levels of Cry2A toxin production compared to the parental Bt strain containing *cry1A* and *cry2A* genes (Baum *et al.*, 1990). Restricted levels of *cry* gene expression also explain increased expression of chromosomal *cry* genes in Bt strains cured of plasmids containing these genes (Driss *et al.*, 2011).

Expression of most *cry* genes takes place during the stationary or sporulation phase, although in the case of the *cry3Aa* gene expression occurs during the late-exponential growth phase (Sekar, 1988). The *cry* gene transcription promoters are very strong (Agaisse and Lereclus, 1995). In the *cry1A* gene, two overlapping transcription promoters, BtI and BtII, are sequentially active during sporulation (Wong *et al.*, 1983). The BtI promoter is active during early to mid-sporulation, when the BtII is activated until late sporulation. Regions that are unique to each *cry* gene and are located upstream of the BtI and BtII promoter sequences explain differences in levels of expression of the same *cry* gene among Bt subspecies (Cheng *et al.*, 1999). Transcription from the BtI promoter is initiated by an RNA polymerase containing an alternative sigma factor, termed σ^{35} (Brown and Whiteley, 1988), while a different sigma factor, σ^{28}, initiates transcription from BtII (Brown and Whiteley, 1990). Based on high sequence identity and rescue of null mutants, the σ^{35} and σ^{28} factors were suggested as functionally equivalent homologues to the σ^E and σ^K sigma factors from *B. subtilis*, respectively (Adams *et al.*, 1991). The fact that the RNA polymerases containing the σ^{35} and σ^{28} factors were able to direct transcription from the *cry2A*, *cyt1A*, and *cry1B* gene promoters *in vitro* (Brown and Whiteley, 1988, 1990) supports the idea that factors similar to σ^{35}/σ^E and σ^{28}/σ^K drive transcription of other *cry* genes. Promoters similar to those recognized by these sigma factors have been reported for characterized

sporulation-specific *cry* genes, including *cry1B* (Brizzard *et al.*, 1991), *cry2A* (Widner and Whiteley, 1989), *cry4A* (Yoshisue *et al.*, 1993), *cry11A* (Dervyn *et al.*, 1995), or *cry40* and *cry34* (Brown, 1993). Transcription of the *cry1A* gene in a negative σ^K mutant strain was similar to levels in the wild type, signifying that high levels of toxin production can be sustained by transcription from the BtI promoter alone (Bravo *et al.*, 1996). RNA polymerases containing the σ^H factor can initiate low levels of *cry* gene transcription during early sporulation (Poncet *et al.*, 1997; Pérez-García *et al.*, 2010).

Non-sporulation-dependent expression has been demonstrated for the *cry3A* gene in Bt subsp. *tenebrionis* (Sekar, 1988). The promoter of this gene is expressed weakly during the late exponential phase of growth before the start of the stationary phase that marks the initiation of expression of the *cry1A*-type genes (Agaisse and Lereclus, 1995). Two active transcriptional start regions separated by an intermediate region were identified upstream of the *cry3A* and *cry3B* genes (Donovan *et al.*, 1992; de Souza *et al.*, 1993). The promoter regions in both of these transcriptional start sites are different from the BtI and BtII promoters and are very similar to promoters recognized by the σ^A factor in vegetative cells (Agaisse and Lereclus, 1994a). As expected from these observations, mutations affecting sporulation-dependent sigma factors do not disrupt expression of the *cry3A* gene (Agaisse and Lereclus, 1994b).

Stability of the mRNA is vital to obtain high levels of crystal proteins. The mRNAs produced from *cry* genes display a half-life about three times longer than the average bacterial mRNA, indicating the existence of efficient mRNA stabilizers. A Shine−Dalgarno (SD) sequence designated STAB-SD at the $5'$ untranslated region of the processed mRNA from the *cry3A* gene acts as a post-transcriptional mRNA stabilizer (Agaisse and Lereclus, 1996). The stabilization mechanism involves binding of a ribosomal subunit to STAB-SD to block exonucleolytic progression of RNase J1 (Mathy *et al.*, 2007). In contrast, *cry1A* mRNA stability is provided by a large inverted repeat situated on the *cry1A* transcription terminator located immediately after the toxin open reading frame (ORF) (Wong and Chang, 1986). This region forms stable stem-loop structures that inhibit mRNA degradation (Ramírez-Prado *et al.*, 2006).

Structure and Classification of Crystal Toxins

Correct folding and packaging of Cry proteins into crystals contributes towards prevention of proteolytic degradation, and resolves the osmotic challenge represented by the elevated Cry protein concentrations in the cytosol of the mother cell during sporulation. Roughly, two main types of Cry proteins are described based on the mass of their protoxin form (Fig. 8.3). The first comprises proteins of 130−140 kDa in mass sharing a highly conserved

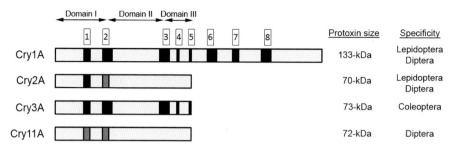

FIGURE 8.3 Schematic alignment of the four main Cry protoxin groups from *Bacillus thuringiensis*. Each group is defined by a representative protoxin family (Cry1A, Cry2A, Cry3A, and Cry11A) with a typical protoxin size as indicated. Numbered blocks represent conserved amino acid sequence blocks among Cry proteins. These blocks are colored black or gray to represent high and low sequence homology among protoxin types, respectively. The relative location of the three structural domains of Cry toxins is indicated. The positions of the conserved blocks and toxin domains, as well as the length of the protoxins, are approximate and not drawn to scale. *(Adapted and modified using information from published reviews: Baum and Malvar, 1995; Kumar et al., 1996; Schnepf et al., 1998.)*

C terminus containing 15−17 cysteine residues, which is necessary for formation of intermolecular disulfide bonds during crystal formation (Bietlot *et al.*, 1990; Yu *et al.*, 2002). Examples of protoxins in this group include Cry1, Cry4A, and Cry4B proteins. These protoxins can be cloned and expressed in diverse heterologous systems to form biologically active inclusions, confirming that they do not require the Bt environment for crystal production. The second group of protoxins includes proteins of 70−75 kDa in mass, such as Cry2A, Cry3A, or Cry11A, which do not contain the C-terminal half and are structurally similar to the N-terminal half of the protoxins in the 130−140 kDa toxin group. In the smaller protoxins, crystallization involves helper proteins that are part of the *cry* gene operon, which requires cloning of the whole operon for expression and production of crystals in recombinant strains and heterologous systems (Moar *et al.*, 1994). As an exception, these helper proteins have not been described for the *cry3A*

gene, and in this case crystallization seems to depend on four intermolecular salt bridges (Li *et al.*, 1991).

The shape of the crystal usually correlates with its protoxin composition. Thus, Cry1 crystals usually have a bipyramidal shape, Cry2 are cuboidal, Cry3A flat rectangular, Cry3B irregular, Cry4 spherical, and Cry11A rhomboidal (Schnepf *et al.*, 1998). Although crystals containing diverse Cry proteins are commonly described, this feature seems to be mostly limited to protoxins in the 130−140 kDa group. Within one of these crystals, the amounts of each of the diverse Cry1 protoxins are controlled at the transcriptional and translational levels. For instance, differences in transcription rates and translation account for more than 20-fold lower amounts of Cry1Da compared to Cry1Ab in crystals from Bt subsp. *aizawai* (Chang *et al.*, 2001).

The first three-dimensional structure for a Cry protein was reported by Li et al. (1991) using X-ray

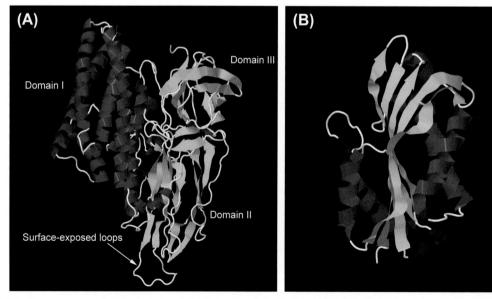

FIGURE 8.4 Cartoon representing the three-dimensional structure of the Cry1Aa (A) and Cyt2Ba (B) toxin monomers from *Bacillus thuringiensis*. α-Helices are colored magenta and β-sheets are presented in yellow. The Cry1Aa structure is composed of three structural domains as indicated. Protruding loops in domain II involved in specificity of toxin binding are also indicated. The Cyt2Ba crystal structure is composed of a single domain of α/β architecture with a β-sheet surrounded by two α-helical layers representing a cytolysin fold.

crystallography to resolve the structure of Cry3A (Fig. 8.4). Since then, the structures of Cry1Aa (Grochulski *et al.*, 1995), Cry1Ac complexed to *N*-acetyl galactosamine (Derbyshire *et al.*, 2001), Cry2Aa (Morse *et al.*, 2001), Cry3Bb1 (Galitsky *et al.*, 2001), Cry4Ba (Boonserm *et al.*, 2005), Cry4Aa (Boonserm *et al.*, 2006a), and Cry8Ea1 (Guo *et al.*, 2009a) have been resolved. Additional Cry toxin structures have been modeled using the resolved structures as a template (Xia *et al.*, 2008). While these proteins display low sequence identity and diverse insect specificities among them, their three-dimensional structures are highly conserved, containing the three structural domains originally described for Cry3A (Li *et al.*, 1991). Domain I consists of seven antiparallel α-helices in a bundle with helix 5 in the center surrounded by the other helices. The outer helices are amphipathic, with their hydrophobic side facing the central helix and the hydrophilic residues facing the solvent. Both domains II and III consist of antiparallel β-sheets, but they are arranged in a "Greek key" topology in domain II and a "jelly roll" topology in the case of domain III. Conservation of the structural domains advocates involvement in crucial steps in the mode of action of the toxin and indicates that there is a similarity in the mode of action of the different toxins.

Early attempts at Cry toxin classification were based on specificity and gene sequence similarities (Höfte and Whitely, 1989). Under this approach, CryI toxins were active against lepidopteran larvae, CryII had dual specificity against lepidopteran and dipteran larvae, CryIII were pathogenic to coleopteran larvae, and CryIV specific for dipteran larvae. Subgroups within these primary ranks were assigned according to amino acid sequence identity. Discrepancies between insecticidal specificity and sequence similarity as the number of described Cry toxins increased led to problems with this nomenclature. For instance, toxins in the CryI group were found to display activity against dipteran (Smith *et al.*, 1996) or coleopteran (Bradley *et al.*, 1995) targets. To resolve these problems, a Bt Toxin Nomenclature Committee was formed in 1993 and proposed a revised nomenclature procedure to overcome limitations of the previous method (Crickmore *et al.*, 1998). The revised nomenclature system is based on amino acid sequence identity and is currently used to name and classify new crystal toxins. Toxin names are assigned according to the degree of evolutionary divergence from other toxins based on phylogenetic trees, which allows for grouping of closely related toxins. Under the revised nomenclature, Roman numerals have been substituted for Arabic numerals for the first rank, assigned for toxins displaying at least 45% sequence identity. The secondary rank is represented by a capitalized letter which groups toxins displaying up to 75% amino acid sequence identity, and the tertiary rank is represented by a lower case letter identifying toxins displaying 75−95% sequence identity. The quaternary rank,

distinguishing toxins that are more than 95% identical, is optional and assigned to independently sequenced toxin genes, which means that toxins that differ only in their quaternary ranks may be identical. Currently, 229 holotype crystal toxins have been described, which are grouped in 68 Cry and three Cyt families at the primary rank (Crickmore *et al.*, 2011). This classification also includes families of highly related crystal proteins produced by bacteria other than Bt: Cry16 and Cry17 from *Clostridium bifermentans* (Barloy *et al.*, 1996, 1998), Cry18 from *P. popilliae* (Zhang *et al.*, 1997), Cry43 from *Paenibacillus lentimorbus* (Yokoyama *et al.*, 2004) and *P. popilliae* (Asano *et al.*, 2008), and the binary Cry48/Cry49 toxin produced by *B. sphaericus* (Jones *et al.*, 2007). Readers are directed to the Bt Toxin Nomenclature Committee website (http://www. lifesci.sussex.ac.uk/home/Neil_Crickmore/Bt/) for toxin dendrograms, more detailed information on the nomenclature procedure, and updated toxin lists.

Current knowledge on the specificity of Cry toxins is limited to the range of insect species tested in bioassays and the definition of activity. Cry toxicity has been reported for species in six taxonomic orders: Lepidoptera, Coleoptera, Diptera, Hymenoptera, Hemiptera, and Blattaria (van Frankenhuyzen, 2009). Activity to insects in more than one taxonomic order has been described for six of the 68 Cry families. For example, the Cry2Aa toxin has been reported as toxic against species in three taxonomic orders: *M. sexta* (Lepidoptera) and *Ae. aegypti* (Diptera) larvae (Widner and Whiteley, 1989), and the potato aphid *Macrosiphum euphorbiae* (Hemiptera) (Walters and English, 1995). In addition to activity against insects, some Cry toxins have been described as active against nematodes or mammalian tumor cells (van Frankenhuyzen, 2009). Readers are directed to the Bt Specificity Database (http://www.glfc. cfs.nrcan.gc.ca/bacillus/BtSearch.cfm) for detailed information on bioassays and the range of activity for specific Cry toxins. A simplified list of general specificities for the most well-characterized Cry toxin families is presented in Table 8.2.

Despite the observed low overall sequence identity among Cry protoxins, five blocks of amino acids (1−5) located at the N-terminal halves of the protoxins (Fig. 8.3) are highly conserved among diverse Cry toxins (Höfte and Whitely, 1989; Schnepf *et al.*, 1998). The Cry2A, Cry11A, and Cry18A protoxins only show significant homology to the sequence of block 1, with sequences in block 2 appearing as a truncated variant of the same block in other Cry protoxins. No significant homologous sequences to blocks 4 and 5 are detected in these three protoxins (Schnepf *et al.*, 1998). Amplification of the conserved blocks by polymerase chain reaction (PCR) has been reported to detect modified interblock regions or variations and identify new toxins (Ben-Dov *et al.*, 2001; Noguera and Ibarra, 2010). The five domains are distributed along the

TABLE 8.2 Specificity of Selected Crystal Toxins

Crystal Toxin Family	Order[a]	Examples
Cry1A	L	*Manduca sexta, Heliothis virescens, Trichoplusia ni*
Cry1B	C, D, L	*Anthonomus grandis, Musca domestica, Plutella xylostella*
Cry1C	D, L	*Aedes aegypti, Spodoptera exigua*
Cry1F	L	*Spodoptera* spp.*, Helicoverpa armigera, Trichoplusia ni*
Cry1G	L	*Agrotis ipsilon, Ostrinia nubilalis, Trichoplusia ni*
Cry1I	C	*Leptinotarsa decemlineata*
Cry2A	D, H, L	*Aedes aegypti, Macrosiphum euphorbiae, Pectinophora gossypiella*
Cry3A	C, H, Hy	*Tenebrio molitor, Macrosiphum euphorbiae, Solenopsis invicta*
Cry3B	C	*Diabrotica virgifera*
Cry4	D	*Anopheles gambiae, Aedes aegypti, Culex quinquefasciatus*
Cry5	Hy, Nem	*Diprion pini, Meloidogyne hapia*
Cry6A	Nem	*Pristionchus pacificus, Acrobeloides* spp.*, Pratylenchus* spp.
Cry7A	C	*Leptinotarsa decemlineata*
Cry7B	L	*Plutella xylostella*
Cry8A	C	*Leptinotarsa decemlineata*
Cry8B	C	*Diabrotica virgifera*
Cry8D	C, L	*Popillia japonica*
Cry9	L	*Phthorimaea operculella, Bombyx mori, Agrotis segetum*
Cry11B	D	*Aedes aegypti, Anopheles stephensi, Culex pipiens*
Cry12A	Nem	*Pratylenchus* spp.
Cry18A	C	*Melolontha melolontha*
Cry19A	D	*Anopheles stephensi, Culex pipiens*
Cry20A	D	*Aedes aegypti*
Cry22A	C, L	*Anthonomus grandis, Plutella xylostella*
Cry23A	C	*Tribolium castaneum*
Cry24C	D	*Aedes aegypti*
Cry32A	L	*Plutella xylostella*
Cry32B	D	*Aedes aegypti*
Cry34/Cry35	C	*Diabrotica virgifera*
Cry36A	C	*Diabrotica virgifera*
Cry37A	C	*Popillia japonica*
Cry39A	D	*Anopheles stephensi, Culex pipiens*
Cry43	C	*Anomala cuprea*
Cry47A	D	*Lucilia cuprina*
Cry48A/Cry49A	D	*Culex quinquefasciatus*
Cry51A	L	*Bombyx mori*
Cry55A	C, Nem	*Phyllotreta cruciferae, Meloidogyne hapia*

Examples of susceptible species were obtained from the *Bacillus thuringiensis* (Bt) toxin specificity relational database (van Frankenhuyzen and Nystrom, 2011) and published reviews (van Frankenhuyzen, 2009).
[a]*C = Coleoptera; D = Diptera; H = Hemiptera; Hy = Hymenoptera; L = Lepidoptera; Nem = toxic to nematode species.*

toxic core sequence, while three additional conserved sequence blocks (6–8) are detected at the C-terminal halves of the 130–140 kDa group of protoxins. The highly conserved block 1 in domain I includes the central helix (α5), while block 2 contains the C-terminal half of helix α6 and all of α7 from domain I plus the first β-strand of domain II. Domain substitution and mutagenesis within block 2 suggests that it is important to toxin stability and crystallization (Park and Federici, 2004). Next in the sequence (from N to C terminus) is block 3, which includes the last β-strand of domain II and the N-terminal segment of the first β-strand of domain III. Block 4 contains the second β-strand of domain III, and the highly conserved block 5 is located at the C terminus of domain III. Both blocks 4 and 5 were proposed to be involved in Cry toxin recognition of biotin-containing proteins (Du and Nickerson, 1996a), although the relevance of this binding to toxicity has not been further investigated.

Crystal Toxin Structure–Function

The functional role of the three structural Cry toxin domains has been methodically studied using comparisons with other bacterial toxins and mutant Cry toxins using diverse techniques. Based on observations that most of the amphipathic helices of domain I are long enough to span a hydrophobic cellular membrane and that domain I has similarities to pore-forming domains in alternative bacterial toxins, it was hypothesized that domain I was responsible for pore formation in Cry toxins (Li *et al.*, 1991). This hypothesis was supported by data from site-directed mutagenesis of residues in domain I, which affected toxicity but did not disturb interactions of the toxin with the insect midgut brush border membrane (Ahmad and Ellar, 1990; Wu and Aronson, 1992). Two main models for Cry toxin pore formation have been proposed and Cry toxin mode of action is discussed in the subsection below, entitled *Crystal Intoxication Process.*

Domain II consists of three antiparallel β-sheets in a "Greek key" conformation forming a β-prism, which is connected to domain I through hydrogen bonding and salt bridges that are highly conserved among Cry toxins. Interactions between domains II and III are through hydrophobic interactions and are connected by a short linker (Grochulski *et al.*, 1995). Domain II is the most variable of the toxin domains, with diversity in β-strand lengths and hypervariable protruding loop regions detected among diverse Cry toxins (Fig. 8.4). This high degree of variability in domain II may result from positive selection in regions determining specificity among Cry toxins (Wu *et al.*, 2007b). Domain II shares structural similarities with other carbohydrate-binding proteins containing β-prism folds (Pigott and Ellar, 2007). Clear structural folding similarities have been observed between domain II and vitelline protein from hen's eggs (Shimizu *et al.*, 1994), and

the plant lectins jacalin (Sankaranarayanan *et al.*, 1996) and KM+ (Rosa *et al.*, 1999) from *Artocarpus* sp., and the *Maclura pomifera* agglutinin (Lee *et al.*, 1989). This structural similarity advocates that domain II may bind to carbohydrates, which has not yet been demonstrated. The apex loop regions in domain II are similar to carbohydrate-binding sites of lectins and antigen-binding sites of immunoglobulins (Li *et al.*, 1991), substantiating their participation in binding to receptors. Studies using mutant Cry toxins in loops of domain II demonstrate the participation of these loops in binding specificity (Dean *et al.*, 1996; Smedley and Ellar, 1996). Toxins with high sequence similarity in domain II share at least some binding sites on the host midgut cells (Jurat-Fuentes and Adang, 2001; Hernández and Ferré, 2005), which can be used to predict potential cross-resistance patterns resulting from alterations in shared toxin binding sites (Tabashnik *et al.*, 1996). Substitution of domain II loops by antibody complementarity determining regions (CDRs) demonstrated that although some degree of sequence diversity is tolerated, some combinations are detrimental to toxin stability and toxicity (Pigott *et al.*, 2008). Directed mutagenesis of Cry1A toxins denotes that amino acid residues in domain II loop 3 are involved in initial interactions between the toxin and the midgut membrane (Rajamohan *et al.*, 1996c; Pacheco *et al.*, 2009a), while residues in loop 2 are important for irreversible toxin binding (Rajamohan *et al.*, 1996b). Similar observations have been reported for domain II loops of mosquitocidal toxins, such as Cry19Aa (Roh *et al.*, 2009b), Cry11Aa (Fernández *et al.*, 2005), or Cry4Aa (Howlader *et al.*, 2009). Diverse domain II loops may determine specificity to diverse insects, as binding of Cry1C to midgut proteins from *Ae. aegypti* depends on loop 2 while loop 3 is important for binding and toxicity in *S. littoralis* larvae (Abdul-Rauf and Ellar, 1999). Toxicity against larvae of *Culex* spp. mosquitoes could be transferred from Cry4Aa to Cry4Ba by swapping of domain II loops between these toxins (Abdullah *et al.*, 2003).

Domain III contains conserved blocks 3–4–5 and it is structured as a β-sandwich of two antiparallel β-sheets compressed in a "jelly roll" topology (Li *et al.*, 1991). Two long loops extend from near the end of the β-sheets to interact through hydrophobic interactions with the C-terminal portion of helices α6 and α7 of domain I (Grochulski *et al.*, 1995). Domain III is more conserved among Cry toxins than domain II, with loops being the region displaying the highest diversity. Of particular relevance is a unique loop extension creating a lectin binding pocket in domain III of Cry1Ac, which is not observed in related Cry1A toxins (Derbyshire *et al.*, 2001). The overall structure of domain III is similar to carbohydrate-binding domains of microbial glycoside hydrolases, suggestive of binding to carbohydrate residues on midgut proteins (Pigott and Ellar, 2007). Similarities to regions in alternative bacterial toxins that are involved in binding to receptors or maintaining toxin stability have also been noted (de Maagd et al., 2003). There are multiple reports confirming a vital role for domain III in determining Cry toxin binding specificity. The lectin pocket in domain III of Cry1Ac is specifically recognized by N-acetylgalactosamine (GalNAc) and is involved in binding to midgut proteins (Burton et al., 1999) and toxin insertion on the membrane (Pardo-López et al., 2006). Transfer of toxicity against Spodoptera spp. larvae from Cry1C to Cry1Ac or Cry1E was reported to involve swapping of domain III (Bosch et al., 1994; Ayra et al., 2008). Swapping of domain III of Cry1Ab into Cry3Aa resulted in toxicity against larvae of the western corn rootworm (Diabrotica virgifera), which are naturally not susceptible to Cry3A (Walters et al., 2010). The fact that this Cry3A/Cry1Ab hybrid toxin does not share binding sites with Cry3Aa toxin on midgut brush border membrane vesicles (BBMVs) from D. virgifera larvae supports the idea that domain III of Cry1Ab contains binding specificity determinants. Domain III of Cry1Ac has been reported to dictate specificity in diverse lepidopteran larvae including L. dispar (M. K. Lee et al., 1995), H. virescens and T. ni (Ge et al., 1991), and M. sexta (de Maagd et al., 1999a). In addition, structural and pore formation roles have been proposed for domain III. For instance, mutations in block 4 of domain III in Cry1Aa result in mutant toxins that form faulty channels in BBMVs from target insects (Chen et al., 1993; Wolfersberger et al., 1996). Substitution of residues in conserved block 5 of Cry4A resulted in higher susceptibility to protease digestion, confirming a structural role (Nishimoto et al., 1994).

Midgut Proteins Interacting with Crystal Toxins and Toxin Receptors

The described proteins and glycoconjugates from the midgut brush border membrane interacting with activated Cry toxins and their role in intoxication have been reviewed by Pigott and Ellar (2007). Aminopeptidases (APNs) were the first identified Cry toxin binding proteins in lepidopteran models (Knight et al., 1994; Vadlamudi et al., 1995), and have also been reported to interact with the mosquitocidal Cry11Ba toxin (Abdullah et al., 2006; R. Zhang et al., 2008). These enzymes are tethered to the midgut cell by a glycosyl-phosphatidylinositol (GPI) anchor (Garczynski and Adang, 1995), are naturally involved in protein digestion, and are ubiquitous to the midgut brush border membrane of insects (Terra and Ferreira, 1994). Interactions between Cry1 toxins and midgut APNs have been mapped to regions of domain III in the toxins including the unique lectin pocket of domain III in Cry1Ac binding to APN from M. sexta (Masson et al., 1995), L. dispar (Jenkins et al., 2000), and

H. virescens larvae (Luo et al., 1997). As an exception, binding of Cry1Ac to an alternative 106 kDa APN in H. virescens is not GalNAc mediated (Banks et al., 2001), although this binding does not result in toxicity (Banks et al., 2003). In contrast to carbohydrate-mediated binding in Cry1Ac, direct protein–protein interactions have been proposed to mediate Cry1Aa domain II or III binding to APN from B. mori larvae (Jenkins and Dean, 2001; Atsumi et al., 2005). In the case of Cry1Ab, domains II and III are both involved in binding to APN from M. sexta (Gómez et al., 2006; Pacheco et al., 2009a). Evidence verifying the functional receptor role of APN in Cry1 intoxication is limited to genetic knockdown of APN in S. litura resulting in lower susceptibility to Cry1C (Rajagopal et al., 2002) and correlation between reduced expression of APNs and resistance to Cry1 toxins in Diatraea saccharalis (Yang et al., 2010) and S. exigua (Herrero et al., 2005). Reports of Cry toxin binding to APNs being not conducive to toxicity substantiate the existence of alternative receptors (Jenkins et al., 1999; Lee et al., 2000).

Cadherin-like proteins are widely accepted as functional Cry toxin receptors. Contrary to the generally observed localization for cadherin proteins to regions of cell–cell interaction, cadherin proteins binding Cry toxins localize mostly to the brush border membrane of midgut cells (Chen et al., 2005; Aimanova et al., 2006). Based on their controlled expression during larval development, a role in midgut epithelial organization was proposed (Midboe et al., 2003), but the specific physiological role of these Cry-binding cadherins is unknown. The first Cry toxin-binding cadherin was identified and cloned from M. sexta larvae (Vadlamudi et al., 1993, 1995). This cadherin was named BT-R$_1$ (Bt receptor 1), and was shown to bind Cry1Aa, Cry1Ab, and Cry1Ac toxins specifically and with high affinity when expressed in insect cell cultures (Keeton and Bulla, 1997; Hua et al., 2004). Cultured insect cells expressing BT-R$_1$ were susceptible to low levels of Cry1A toxins, substantiating this cadherin as a functional receptor for Cry1A toxins in M. sexta. Functional evidence of a Cry1A receptor role has also been reported for cadherin proteins from B. mori (Nagamatsu et al., 1999), O. nubilalis (Flannagan et al., 2005), and H. virescens (Jurat-Fuentes and Adang, 2006). Evidence of the importance of cadherin proteins for toxicity comes from reports of resistance to Cry1Ac in laboratory strains of Pectinophora gossypiella (Morin et al., 2003), H. armigera (Xu et al., 2005; Zhao et al., 2010), and H. virescens (Gahan et al., 2001; Jurat-Fuentes et al., 2004). More recently, a cadherin protein from the mealworm beetle, T. molitor, has been reported as the first functional Cry3Aa toxin receptor in Coleoptera (Fabrick et al., 2009). Although an additional Cry toxin-binding cadherin has been reported from larvae of D. virgifera (Sayed et al., 2007), its interactions with Cry toxins or receptor functionality have not been reported. Cadherin-like proteins have also been reported as binding proteins in *Anopheles gambiae* larvae for Cry4Ba toxin, and in *Ae. aegypti* larvae for Cry4Ba (Bayyareddy *et al.*, 2009), Cry11Aa (Chen *et al.*, 2009), and Cry11Ba (Likitvivatanavong *et al.*, 2011) toxins. Although the functional Cry receptor role of these dipteran cadherins has not been established, reduced expression of a cadherin-like protein was reported in a strain of *Ae. aegypti* resistant to Bt subsp. *israelensis* (Bonin *et al.*, 2009).

Membrane-bound midgut alkaline phosphatases represent the third major group of Cry-toxin binding proteins that have been proposed as functional Cry toxin receptors. Interactions between midgut alkaline phosphatase (ALP) from *H. virescens* or *M. sexta* larvae and Cry1Ac toxin were described to result in reduced ALP activity (English and Readdy, 1989; Sangadala *et al.*, 1994). Binding of Cry toxins to ALP in BBMVs from *H. virescens* (Krishnamoorthy *et al.*, 2007), *M. sexta* (McNall and Adang, 2003), and *Ae. aegypti* (Bayyareddy *et al.*, 2009) larvae has been demonstrated through proteomic analyses. Binding of Cry1Ac to midgut ALP from *H. virescens* and *H. armigera* larvae involves interactions with GalNAc (Jurat-Fuentes and Adang, 2004; Ning *et al.*, 2010), indicating interactions with domain III of the toxin. Recently, interactions between Cry1Ab and ALP from *M. sexta* have been described as crucial to toxin binding and insertion on the membrane (Arenas *et al.*, 2010). Associations between reduced ALP expression levels and resistance to Cry toxins in strains of *H. virescens*, *H. armigera*, and *S. frugiperda* (Jurat-Fuentes *et al.*, 2011) further support a relevant role for ALP in toxicity to Lepidoptera. The mosquitocidal Cry11Aa and Cry11Ba toxins have been reported to bind to ALP from BBMV of *Ae. aegypti* and *An. gambiae* larvae, respectively. In both cases, any toxicity could be reduced by heterologous ALP regions predicted to interact with the toxin, confirming a vital role for ALP in the intoxication process (Fernandez *et al.*, 2006; Hua *et al.*, 2009). In the case of Cry11Aa, interactions with ALP were described to involve regions of domains II and III with two distinct sites on the *Ae. aegypti* ALP (Fernandez *et al.*, 2009). Functional expression data have been presented to support the ALP from *Ae. aegypti* as a Cry4Ba toxin receptor (Dechklar *et al.*, 2011). Receptor functionality of a recently described ALP protein from the cotton boll weevil (*A. grandis*) binding Cry1Ba toxin (Martins *et al.*, 2010) needs to be established.

Other molecules have been reported to interact with Cry toxins, including glycolipids from *M. sexta*, a glycoconjugate from *L. dispar*, a high molecular weight protein from *B. mori*, an ABC-type transporter from *H. virescens*, an ADAM metalloprotease in *Leptinotarsa decemlineata*, and an amylase in *Anopheles albimanus*. Studies on resistance mechanisms to the nematocidal Cry5Ba toxin in

Caenorhabditis elegans indicated that alterations in a glycosylation pathway were responsible for altered glycolipids and reduced toxin binding in resistant nematodes (Griffitts *et al.*, 2001). While insect-extracted glycolipids have been reported to bind Cry1 toxins (Dennis *et al.*, 1986; Griffitts *et al.*, 2005), the relevance of this interaction to toxicity has not established.

High-affinity binding of a 270 kDa glycoconjugate purified from *L. dispar* was specific to Cry1Aa, Cry1Ab, and Cry1Ba toxins (Valaitis *et al.*, 2001), but the biological significance of this binding is not known. Another high molecular weight molecule, a BBMV protein of 252 kDa (P252), was identified as binding Cry1A toxins in a carbohydrate-independent manner (Hossain *et al.*, 2004). This protein appears to exist as a 985 kDa oligomer and has been described as a member of the polycalin family of proteins (Pandian *et al.*, 2008). This oligomer interacts with Cry1A toxins to form a complex that retains activity against *B. mori* larvae (Pandian *et al.*, 2010). Although the existence of similar proteins in *Ae. aegypti* has been reported (Pandian *et al.*, 2010), the potential role for P252 in Cry intoxication needs further study.

Using a map-based cloning approach, an ABC transporter gene involved in resistance to Cry1Ac in *H. virescens* has recently been described (Gahan *et al.*, 2010). A natural genetic knockout of this gene in resistant *H. virescens* larvae correlates with lack of Cry1Ab and Cry1Ac binding. While this observation may suggest direct interactions between this ABC transporter and Cry1A toxins, direct testing of toxin binding to the ABC transporter proteins and its significance for toxicity need to be established.

An ADAM metalloprotease localized to the midgut brush border membrane of *L. decemlineata* larvae has been proposed as a Cry3Aa receptor (Ochoa-Campuzano *et al.*, 2007). A peptide containing a region of domain II loop 1 of Cry3Aa prevented interactions with the protease and greatly reduced pore formation, supporting a functional role for the metalloprotease—toxin interaction. However, blocking of Cry3Aa proteolysis also correlated with increased pore formation (Rausell *et al.*, 2007), which may indicate that diverse metalloproteases on the BBMV of *L. decemlineata* interact with Cry3Aa toxin. Another enzyme, an α-amylase from *An. albimanus*, was shown to bind Cry4Ba and Cry11Aa toxins *in vitro* (Fernandez-Luna *et al.*, 2010), although the functional role of this enzyme as a receptor has not been demonstrated.

Crystal Intoxication Process

The mode of action of Cry toxins has been mostly characterized using lepidopteran larvae as models (detailed in Fig. 8.5), and has been reviewed recently (Ibrahim *et al.*, 2010; Soberón *et al.*, 2010). Although Cry1 toxins were reported to be active in intrahemocoelic injections in

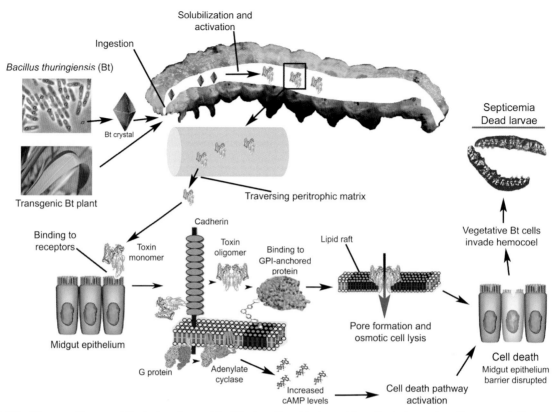

FIGURE 8.5 Diagram of the current Cry toxin action models. After ingestion of toxin crystals or Cry toxins produced by transgenic Bt plants, the Cry proteins are solubilized and processed to a toxin core which, after traversing the peritrophic matrix, binds to receptors on the brush border membrane of the midgut epithelial cells. Toxin binding may activate intracellular cell death pathways or facilitate further toxin processing resulting in oligomerization. Toxin oligomers bind to selected proteins tethered to the cell membrane by a glycosyl-phosphatidylinositol (GPI) anchor and insert in the membrane to form pores that lead to osmotic cell lysis. Disruption of the midgut epithelium barrier facilitates invasion of the hemocoel, conducive to septicemia and death of the insect.

L. dispar and *N. bullata* larvae (Cerstiaens *et al.*, 2001), it is widely accepted that the major role of Cry toxins is to disrupt the midgut epithelium barrier to facilitate bacterial invasion of the hemocoel. Upon ingestion of the toxin crystal by a susceptible insect, the alkaline pH and reducing conditions found in the midgut of lepidopteran larvae allow for solubilization of the protoxins. Solubilization of the protoxin molecules in the crystal renders them available to proteolysis (activation) to yield an active toxin core that is mostly resistant to further proteolysis (Bietlot *et al.*, 1989). The N-terminal half of the protoxin is tightly associated with 20 kilobase (kb) fragments of DNA (Bietlot *et al.*, 1993), which is also observed when expressing Cry toxins in *Escherichia coli*. These DNA fragments contain Cry toxin genes (Xia *et al.*, 2005) and have been reported to stabilize the protoxin form and assure sequential proteolysis during activation (Clairmont *et al.*, 1998). Interactions between the Cry8Ea1 protoxin and these DNA fragments have been recently reported to increase hydrophobicity and promote toxin insertion in phospholipid monolayers (Guo *et al.*, 2011), although the functional significance of these

toxin–DNA interactions for *in vivo* toxicity needs further characterization.

Proteases in the insect midgut fluids complete the processing of the protoxin to an active toxin core, although endogenous Bt proteases also participate in toxin activation (Chestukhina *et al.*, 1980; Haider and Ellar, 1987). These endogenous proteases are produced during sporulation and remain associated with the toxin crystals (Bulla *et al.*, 1977). Reduced protease production results in defective activation and lower activity against target insects (Bibilos and Andrews, 1988; Kumar and Venkateswerlu, 1998a), but deletion of protease genes in Bt subsp. *kurstaki* results in increased production of Cry1B protein (Tan and Donovan, 2000). Differences in Bt proteases may contribute to dictate specificity by matching protease activities found in target insects. For instance, while Bt subsp. *tenebrionis* active against coleopteran larvae produces metalloproteases, subsp. *israelensis* active against dipteran larvae mostly produces serine proteases (Reddy *et al.*, 1998). Some reports demonstrate that, at least in some cases, activation by endogenous or insect proteases results in a highly lethal toxin form that is not produced by exogenous proteases

(Kumar and Venkateswerlu, 1998b; Rausell *et al.*, 2004), and that the digestive fluids may have a relevant role in toxin pore formation (Brunet *et al.*, 2010b). However, it is generally considered that larval midgut extracts or commercial proteases can be used *in vitro* to activate Cry toxins without affecting specificity (Andrews *et al.*, 1985; Knowles *et al.*, 1986). Under *in vivo* conditions and since reduced rates of toxin activation result in lower toxicity (Oppert, 1999), endogenous Bt proteases may have a vital role in accelerating toxin activation (Bibilos and Andrews, 1988; Kumar and Venkateswerlu, 1998a).

In the case of the 130–140 kDa protoxins, proteolysis includes sequential removal of about 500 amino acids from the C terminus and a short (29–43 amino acid long) N-terminal peptide. For the Cry4A toxin, activation generates 20 and 45 kDa protease-resistant fragments, which remain associated to form a 60 kDa complex toxic to larvae of the mosquito *Cx. pipiens* (Yamagiwa *et al.*, 1999). Similarly, Cry8Da toxicity against Japanese beetle larvae is dependent on nicking of a region in the N-terminal section that remains associated with the toxin (Yamaguchi *et al.*, 2010). For the smaller (67–75 kDa) Cry protoxins, activation involves removal of the N-terminal peptide, which varies in size depending on the protoxin, from 28 amino acids in Cry11A to 58 in Cry3Aa.

Activated toxins traverse the peritrophic matrix before reaching the midgut epithelium. While the size of the matrix pores would allow passage of Cry toxins (Adang and Spence, 1983), the chitinous matrix contains glycoproteins (Adang and Spence, 1982) that may bind to Cry toxins and reduce their passage. Cry1 toxins have been reported to bind extensively to peritrophic matrices of diverse lepidopteran larvae *in vivo* and *in vitro* (Bravo *et al.*, 1992; Rees *et al.*, 2009). This toxin retention process has been hypothesized to explain low susceptibility of *B. mori* larvae to Cry1Ac (Hayakawa *et al.*, 2004). In contrast, retention of Cry11Bb was not observed in peritrophic matrix of mosquito larvae (Ruiz *et al.*, 2004), suggesting that Cry toxin retention may be specific for some toxins displaying higher affinity for binding to glycoproteins. Disruption of the peritrophic matrix with chitinases from Bt (Sampson and Gooday, 1998), *Bacillus licheniformis* (Thamthiankul *et al.*, 2004), *S. marcescens* (Regev *et al.*, 1996), *M. sexta* (Kramer and Muthukrishnan, 1997), and *T. ni* granulovirus (Granados *et al.*, 2001) has been reported to enhance the activity of Cry toxins.

After traversing the peritrophic matrix, Cry toxins bind to the brush border membrane of the midgut cells (Bravo *et al.*, 1992). BBMVs have been traditionally used as an *in vitro* biologically relevant model system to characterize interactions between Cry toxins and the midgut epithelial membrane. Two binding components, reversible and irreversible, are displayed by Cry toxins on BBMVs from lepidopteran larvae. Specific binding is necessary but not

sufficient for Cry toxicity, since heterologously expressed domains II and III of Cry4Ba and Cry1Ab bound reversibly to BBMV from *Ae. aegypti* (Moonsom *et al.*, 2007) and *M. sexta* (Flores *et al.*, 1997) larvae, respectively, but were not toxic to larvae. Reversible toxin binding has been proposed to concentrate Cry toxins on the brush border membrane to facilitate irreversible binding (Pacheco *et al.*, 2009a). The irreversible binding component is considered synonymous with toxin insertion on the membrane and is directly correlated to toxicity (Liang *et al.*, 1995). Irreversible binding is mostly dependent on domain I, as mutations in this region of the toxin result in reduced irreversible binding and toxicity (Chen *et al.*, 1995; Hussain *et al.*, 1996). However, loop 2 in domain II of Cry1Ab has also been reported to participate in irreversible binding to BBMV from *M. sexta* (Rajamohan *et al.*, 1995) but not *H. virescens* (Rajamohan *et al.*, 1996b) larvae, implying that residues in this loop may perform different roles in diverse insects.

Monomeric Cry1A toxins bind with high affinity (K_d between 1 and 5 nM) to cadherin proteins on BBMVs from larvae of *M. sexta* (Vadlamudi *et al.*, 1995), *B. mori* (Ihara *et al.*, 1998), and *H. virescens* (Xie *et al.*, 2005). In contrast, affinity of monomeric Cry1A toxins for APN is in the 32–110 nM range (Valaitis *et al.*, 1997; Cooper *et al.*, 1998; Jenkins and Dean, 2001), confirming that the toxin monomers preferentially interact with cadherin-like proteins. Cry1Ac and Cry3Aa protoxins bind specifically to cadherin from *P. gossypiella* and *T. molitor*, respectively, indicating that the activated toxin form may not be necessary for binding and that activation may occur after receptor binding (Fabrick and Tabashnik, 2007; Fabrick *et al.*, 2009). The cadherin repeat ectodomain most proximal to the cell membrane has been identified as responsible for Cry toxin binding and toxicity in all characterized cadherins (Pigott and Ellar, 2007; Park *et al.*, 2009; Likitvivatanavong *et al.*, 2011). The events following binding of Cry toxins to cadherin have been best characterized using the *M. sexta* cadherin (BT-R$_1$) and Cry1A toxins as a model. The "oligomers and pore formation" model (Soberón *et al.*, 2009) supports the theory that binding of Cry1Ab toxin to BT-R$_1$ results in a conformational change facilitating cleavage of helix α1 from domain I, which results in formation of a prepore toxin oligomer of approximately 250 kDa in size (Gómez *et al.*, 2002). This oligomer displays high affinity for binding to oligosaccharides on APN and ALP proteins (Pardo-López *et al.*, 2006). Toxin binding and concentration of APN and ALP on specific membrane regions of unique lipid composition called "lipid rafts" leads to oligomer insertion (Zhuang *et al.*, 2002) and formation of a toxin pore that leads to cell death by osmotic shock. In an alternative "cell death pathway" model (Ibrahim *et al.*, 2010), binding of monomeric Cry1Ab toxin to BT-R$_1$

activates intracellular oncotic cell pathways leading to enterocyte death (X. Zhang *et al.*, 2006).

Several observations support the "oligomer and pore formation" model. Oligomer formation can be induced *in vitro* by incubating Cry toxins with cadherin peptides containing the toxin binding region (Fabrick *et al.*, 2009; Pacheco *et al.*, 2009b; Peng *et al.*, 2010b). As predicted by the model, enhanced production of Cry toxin oligomers results in increased toxicity against lepidopteran (Pacheco *et al.*, 2009b) and coleopteran (Gao *et al.*, 2011) larvae. Oligomeric, but not monomeric, Cry1 structures permeated *M. sexta* BBMVs (Muñoz-Garay *et al.*, 2006), confirming the importance of these structures for pore formation. Mutant Cry1A toxins in helix α3 of domain I bind with high affinity to BBMVs but are unable to form oligomers or pores, resulting in lack of toxicity (Jiménez-Juárez *et al.*, 2007). Lipid rafts have also been shown to be involved in Cry intoxication (Avisar *et al.*, 2005). Undoubtedly, the strongest evidence supporting the "oligomer and pore formation" model was obtained from modified (Mod) Cry1Ab and Cry1Ac toxins lacking helix α1 from domain I. These Cry1AbMod and Cry1AcMod toxins formed oligomers spontaneously in the absence of cadherin, and these oligomers restored susceptibility in insects that were resistant to Cry1A toxins by decreased expression of cadherin genes (Soberón *et al.*, 2007). However, several studies have demonstrated that formation of Cry toxin oligomeric structures and functional pores occurs in model membranes in the absence of cadherin proteins (Vié *et al.*, 2001), or in solution depending on pH and temperature conditions (Walters *et al.*, 1994; Güereca and Bravo, 1999). Some of these oligomeric structures are not toxic to target insects (Guo *et al.*, 2009b), and at least one monomeric Cry1A toxin mutant unable to form oligomers has been reported to display enhanced toxicity (Aronson *et al.*, 1999).

In contrast, the alternative "cell death pathway" model proposed by the group of Bulla (X. Zhang *et al.*, 2006) is supported by evidence obtained from insect cell cultures expressing the BT-R$_1$ cadherin. In these cells, binding of monomeric Cry1Ab activates a magnesium-dependent signaling pathway associated with cell death (Zhang *et al.*, 2005). Data from experiments testing Cry1Ab cytotoxicity in the presence of G protein antagonists and inhibitors of adenylate cyclase (AC) and protein kinase A (PKA) support the idea that monomeric Cry1Ab toxin binding to BT-R$_1$ activates a pathway involving AC/PKA signaling (X. Zhang *et al.*, 2006). The membrane blebbing and cell swelling observed after exposure to Cry1Ab and before cell death suggest similarities to oncotic cell death pathways. Although binding of Cry1Ab to cultured cells expressing BT-R$_1$ and subsequent cell death correlate with increased cyclic adenosine monophosphate (cAMP) production, activators of AC or cAMP analogues did not result in cell death, in agreement with previous reports of increased intracellular cAMP levels after Cry intoxication as an indirect effect related to toxin disruption of the membrane rather than a direct cause of cytotoxicity (Knowles and Farndale, 1988). There are examples of bacterial pore-forming toxins inducing host cell death through oncosis (Dacheux *et al.*, 2001; Zhou *et al.*, 2009), which may suggest that the "oligomer and pore formation" and "cell death pathway" models are not exclusive and occur simultaneously during Cry-induced midgut cell death. Further work to establish causal connections among pore formation, intracellular signaling, and cytotoxicity would be necessary to determine the specific sequence of events resulting in midgut cell death.

Some Cry toxins have been demonstrated to function as a binary complex. For instance, the Cry34 protein (14 kDa) is active against larvae of the southern corn rootworm, *Diabrotica undecimpunctata* (Herman *et al.*, 2002), but toxicity against the closely related species *D. virgifera* is observed only when the Cry35 (44 kDa) protein is present, advocating a binary toxin complex (Ellis *et al.*, 2002). The Cry35 protein seems to depend on processing to a 40 kDa form to be active (Masson *et al.*, 2004). Neither of these proteins shares relevant homology with other Cry proteins. Members in the Cry34 family share sequence similarity with proteins predicted to be involved in intracellular signaling (Schnepf *et al.*, 2005), while Cry35 proteins are similar to the BinA and BinB mosquitocidal toxins of *B. sphaericus* (Ellis *et al.*, 2002). The Cry35 proteins also contain a predicted lectin fold, indicating a potential role for this protein in binding to glycoproteins (Schnepf *et al.*, 2005). The specific receptors recognized by the binary toxins have not been reported to date, although calcein efflux studies indicate that Cry34 and Cry35 can form ionic channels in liposomes and lipid bilayers independently or in combination, confirming that these proteins are pore-forming toxins (Masson *et al.*, 2004).

Two main models for Cry toxin pore formation have been proposed (Knowles, 1994). Based on similarity between Cry toxins and the colicin A toxin, Hodgman and Ellar (1990) proposed the "penknife" model, in which the hydrophobic α5 and α6 helices opened in a penknife fashion to insert into the membrane while the other helices in domain I lay flat on the membrane surface of the host cell or remained associated with the receptor. Li *et al.* (1991) proposed an alternative "umbrella" model derived from comparisons with pore-forming mechanisms described for bacterial toxins containing helical bundles similar to domain I. According to this second model, the hydrophobic helical hairpin represented by α4 and the central α5 helix initiated pore formation. Insertion of the helical hairpin presented by the umbrella model during pore formation is supported by observations that alterations in helix α4 or α5 render Cry toxins inactive, while mutations in alternative helices from domain I do not result in significant effects on toxin activity (Wu and Aronson, 1992; Aronson *et al.*, 1995; Kumar and Aronson, 1999). Insertion of the helical hairpin is dependent on a major conformational change in the toxin molecule upon binding to receptors on the cell membrane (Li *et al.*, 2001), during which domain I moves away from domains II and III through the hinge region between domains I and II (Schwartz *et al.*, 1997a). Furthermore, insertion of the hairpin requires that the α3 and α6 separate from helices α4 and α5, respectively, which has been proposed to occur through breaking of conserved interhelical bonds in the insect digestive fluids (Seale, 2005). The crucial role for the α5 and α7 helices during pore formation helps to explain the high degree of sequence conservation in blocks 1 and 2 encompassing these helices. Once the α4 and α5 hairpin inserts in the membrane, association with other toxin molecules results in formation of a tetrameric ion channel with a diameter of approximately 6 Å (Schwartz *et al.*, 1997a; Gerber and Shai, 2000). Oligomerization seems to depend on α5, while the passage of ions through the channel is controlled by charged amino acids on the side of the α4 helix that is exposed to the channel lumen (Kumar and Aronson, 1999; Masson *et al.*, 1999). However, the proposed umbrella or penknife models do not account for all the experimental observations. For example, helix α7 has been reported to appear buried in the membrane and to be involved in ion transport through the Cry toxin ion channel (Alcantara *et al.*, 2001), while α2 has been found to participate in channel permeability (Arnold *et al.*, 2001). Based on compilation of data from experiments using Cry toxin mutants, Dean et al. (1996) proposed a model in which both domains I and III inserted in the membrane, which is supported by protease protection assays with BBMVs from *M. sexta* (Aronson *et al.*, 1999; Aronson, 2000). Furthermore, cysteine mutants spanning the three domains of Cry1A toxins were detected within the membrane of BBMVs from *M. sexta*, supporting insertion of the whole Cry toxin molecule into the membrane (Nair and Dean, 2008). More recently and based on the topology of Cry1Aa inserted in planar lipid bilayers, an alternative model has been proposed to include domain I unfolding and transit to the internal leaflet of the membrane bilayer followed by insertion of the α3–α4 hairpin into the membrane (Groulx *et al.*, 2010). Pore formation in this model results from oligomerization of inserted toxins.

The minimum Cry toxin oligomer and pore structure has been proposed to be a tetramer, based on molecular size estimates (Vié *et al.*, 2001; Gómez *et al.*, 2002; Puntheeranurak *et al.*, 2005). However, posterior analysis of Cry4Ba inserted in lipid membranes using electron crystallography has indicated that pores formed by oligomers present a trimeric organization (Ounjai *et al.*, 2007), which is also supported by available mutagenesis (Torres *et al.*, 2008) and molecular modeling (Taveecharoenkool *et al.*, 2010) data. Cry toxin pores present selectivity for cations, although the pore properties may be greatly modified by midgut epithelium components (Schwartz *et al.*, 1997b; Peyronnet *et al.*, 2001). The inserted toxin remains associated with membrane receptors (Fortier *et al.*, 2007), which may also modulate channel function. While pH affects pore assembly, the channel properties are not affected by changes in pH conditions once the pores have formed (Vachon *et al.*, 2006). This effect of pH on the pores has been inferred to result from electrostatic interactions between the toxin and the membrane during insertion (Brunet *et al.*, 2010a). Estimations of the Cry toxin pore radius also vary depending on the conditions used for testing. For example, while a 1.0–1.3 nm radius was described for Cry1C pores in lipid bilayers (Peyronnet *et al.*, 2002) and Sf9 cultured cells (Villalon *et al.*, 1998) under neutral pH conditions (pH 6.5–8.0), a 1.1–2.6 nm radius was reported for Cry1Ac in *M. sexta* BBMV under alkaline (pH 9.8) conditions (Carroll and Ellar, 1997). The range of ionic selectivity and pore sizes reported may reflect the adaptability of Cry toxins to diverse functional environments (Schnepf *et al.*, 1998).

Extensive midgut cell death after intoxication with Cry toxins results in disruption of the midgut epithelial barrier, change in midgut conditions to promote spore germination, and invasion of the host hemocoel by Bt vegetative cells. While composition of the gut microbiota may influence the outcome of Bt intoxication (van Frankenhuyzen *et al.*, 2010), it is widely accepted that under normal conditions vegetative Bt cells are responsible for septicemia and insect death (Johnston and Crickmore, 2009; Raymond *et al.*, 2009).

Cytolytic Toxins: Description, Regulation, and Classification

Early comparisons of crystal proteins from diverse Bt subspecies detected a low molecular weight protein (28 kDa) specific to Bt subsp. *israelensis* (Tyrell *et al.*, 1981). Unlike lepidopteran-specific crystal proteins, this 28 kDa protein was soluble in alkaline buffer in the absence of reducing agents (Thomas and Ellar, 1983a). This protein induced rapid rounding up and swelling followed by membrane blebbing and cell lysis in mammalian and insect cell cultures, was hemolytic to erythrocytes and toxic when injected in mice, but was not active against larvae of the lepidopteran *Pieris brassicae* (Thomas and Ellar, 1983a). Further bioassays demonstrated activity of this protein against mosquito larvae, although at lower levels than the native Bt subsp. *israelensis* crystal (Yamamoto *et al.*, 1983; Davidson and Yamamoto, 1984; Chilcott and Ellar, 1988). The 28 kDa protein was also observed to synergize mosquitocidal activity of proteins from the Bt subsp. *israelensis* crystal (Wu and Chang, 1985; Chilcott and Ellar, 1988). Based on lack of sequence homology or structural similarities with *cry* genes, Höfte and Whiteley proposed the name *cytA* for the gene encoding the 28 kDa cytolytic crystal protein (Höfte and Whitely, 1989). In the revised crystal toxin nomenclature (Crickmore *et al.*, 1998), the *cytA* gene was renamed *cyt1A*. While originally *cyt* genes were thought to be limited to mosquitocidal Bt strains, these genes have also been found in strains targeting lepidopteran or coleopteran insects (Guerchicoff *et al.*, 1997). Currently, three *cyt* toxin gene families have been defined (*cyt1*, *cyt2*, and *cyt3*), which include 11 toxin holotypes in the current crystal toxin nomenclature (Crickmore *et al.*, 2011).

Expression of the *cyt1A* gene is developmentally regulated as for *cry* genes, with transcription occurring during early stages of sporulation (Waalwijck *et al.*, 1985), advocating similar regulatory expression mechanisms. Production of inclusions containing the Cyt1A protein in *E. coli* was found to depend on the presence of a 20 kDa helper protein encoded by short region upstream of the *cyt1A* gene (McLean and Whiteley, 1987). This helper protein interacts with Cyt1A to stabilize it against proteolysis (Visick and Whiteley, 1991). In contrast, Cyt2A protein is stabilized by a secondary structure at the C terminus and its production is not dependent on the helper protein (Koni and Ellar, 1993, 1994). Both Cyt1A and Cyt2A are expressed as protoxins and are processed at the same sites in the N and C termini to a protease-resistant toxin core of 25 kDa and 23 kDa, respectively (Koni and Ellar, 1994), by endogenous or insect proteases (Nisnevitch *et al.*, 2006). Solubilized Cyt2A toxin appears to form dimers in solution, which are released to monomers during processing to the toxin form (Li *et al.*, 1996). Both Cyt1A protoxin and toxin can permeate unilamellar lipid vesicles, although the toxin form is two to three times more effective (Butko *et al.*, 1996), probably owing to tighter binding to the lipids (Butko *et al.*, 1997).

Structure—Function Relationship in Cytolytic Toxins

The three-dimensional structures of the Cyt2A protoxin (Li *et al.*, 1996) and the Cyt2Ba toxin (Cohen *et al.*, 2008) were resolved using crystallography (Fig. 8.4), detecting high structural similarity. The Cyt structural model includes two α-helix hairpins encompassing α-helices A—B and C—D, flanking a β-sheet core containing seven β-strands (1—7). Sequence alignments of Cyt toxins reveal the existence of highly conserved blocks located to helix A, the loop connecting helix D and α4, β-strands 5 and 6, and strand 6a (Butko, 2003). Mutations of residues on the helix surfaces did not have an effect on activity while mutations on the β-sheet resulted in decreased toxicity, denoting a role for the β-sheet structure in binding and pore formation. Three of the β-strands (5, 6, and 7) are long enough to span the cell membrane. Based on these structural observations, a model was proposed in which

a conformational change in Cyt2A results in movement of the helices to expose the hydrophobic face of the β-sheet for insertion on the cell membrane (Li *et al.*, 1996).

The site of Cyt intoxication is the brush border membrane of the midgut cells (Ravoahangimalala *et al.*, 1993). Cyt toxins display high affinity for membrane lipids containing unsaturated acyl chains (Thomas and Ellar, 1983b; Gill *et al.*, 1987), which explains specificity to Diptera as their brush border membrane contains higher proportions of these unsaturated fatty acids (Li *et al.*, 1996). Homology of Cyt2Ba to the *Erwinia* virulence factor (Evf) allowed for identification of a putative phospholipid binding site pocket formed by residues in the αC—αD hairpin and the β4—β5—β6 strands (Rigden, 2009). Two non-exclusive mechanisms, pore formation and detergent-like membrane disruption, have been proposed to explain the interactions between Cyt1A and cell membranes that result in cytolysis (Butko, 2003). Based on conductivity changes induced by Cyt1A on planar bilayers, Knowles et al. (1989) proposed that the Cyt1A toxin acts by forming transmembrane ionic channels. In this model, upon binding to the membrane a conformational change in the toxin results in the helix hairpins moving away from the β-sheet and facilitates insertion of three long β-strands (β5, β6, and β7) in the membrane (Promdonkoy and Ellar, 2005), which is followed by oligomerization to form a β-barrel pore (Li *et al.*, 1996; Du *et al.*, 1999). Once the pore has been formed, osmotic pressure induces entry of ions and accompanying water through the pore, ultimately resulting in cell swelling and lysis (Knowles and Ellar, 1987). However, Cyt pore formation results in leakage of molecules of diverse size, which would not be expected with a pore of defined size (Butko *et al.*, 1996). Furthermore, large Cyt toxin aggregates, rather than smaller oligomers capable of forming pores, are observed during membrane permeation (Rodriguez-Almazan *et al.*, 2011). Data from diverse biophysical analyses support a substantial structural change upon Cyt toxin binding to the lipid membrane and advocates the localization of the toxin to the membrane surface (Butko *et al.*, 1997). Based on this evidence, Butko et al. (1996) proposed an alternative model in which large Cyt1A aggregates remain on the membrane surface to cause detergent-like defects in lipid packing that result in leakage of intracellular molecules. This model explains why cells and membranes affected by Cyt toxins either remain intact or leak their contents (Butko *et al.*, 1996). Large Cyt aggregates of diverse size are recovered from lipid membranes, and treatment of lipid vesicles with Cyt1A leads to fragmentation of vesicles into smaller forms, bolstering the detergent-like model (Manceva *et al.*, 2005).

Synergistic effects between Cyt and other bacterial toxins have frequently been reported. A combination of Cyt1A with Cry4 toxins overcame resistance (Wirth *et al.*,

1997) and delayed evolution of resistance to Cry11Aa (Wirth *et al.*, 2005b) in *Culex quinquefasciatus*. Synergism was not limited to mosquitocidal Bt toxins, as Cyt1A overcame resistance to Bin toxins in strains of *Cx. pipiens* and *Cx. quinquefasciatus* (Thiéry *et al.*, 1998), and synergized Mtx1 toxicity to *Cx. quinquefasciatus* (B. Zhang *et al.*, 2006). Furthermore, both Cyt1A and Cyt2A conferred susceptibility to *B. sphaericus* in naturally insensitive *Ae. aegypti* larvae (Wirth *et al.*, 2001). The synergistic effect between Cyt1A and Cry toxins is due to Cyt1A functioning as a membrane-bound receptor for the Cry toxins (Pérez *et al.*, 2005; Cantón *et al.*, 2010). Binding of Cry11Aa to membrane-bound Cyt1A resulted in formation of a Cry11Aa prepore oligomer necessary for pore formation (Pérez *et al.*, 2007). However, in contrast to reports of synergism between Cyt1A and Cry3A in the cottonwood leaf beetle (*Chrysomela scripta*) (Federici and Bauer, 1998), no binding of Cry3A to membrane-bound Cyt1A was observed (Pérez *et al.*, 2005), indicating dependence on host factors for synergism.

Non-Crystal Toxins: Vegetative Insecticidal Proteins

Bt synthesizes insecticidal toxins that are expressed during the logarithmic phase of growth and secreted to the medium. The important role of these vegetative insecticidal protein (Vip) toxins in the overall insecticidal activity displayed by Bt strains has been established (Donovan *et al.*, 2001). Because of their activity and unique events in their mode of action, Vip proteins are considered as relevant alternatives to Cry toxin for insect pest control. There are currently 29 holotype Vip toxins classified in four families (Vip1, Vip2, Vip3, and Vip4) (Crickmore *et al.*, 2011). The Vip1 and Vip2 families represent a binary toxin with activity against *D. virgifera* and other coleopteran larvae (Warren, 1997; Shi *et al.*, 2004). The *vip1A* and *vip2A* genes are arranged in a single operon, and high levels of expression are detected from before sporulation until after the spore stage. Both Vip1A and Vip2A are synthesized as preproteins containing an N-terminal signal peptide that is cleaved during secretion into the medium. Vip1A (66 kDa) is involved in recognition of membrane receptors, which is followed by formation of oligomers and insertion in the membrane to create pores that facilitate translocation of Vip2A (45 kDa) to the cell cytoplasm (Leuber *et al.*, 2006). Once in the cytoplasm, Vip2A blocks actin polymerization through ADP-ribosyltransferase activity, leading to loss of the actin cytoskeleton and eventual cell death (Han *et al.*, 1999). Thus, expression of both toxins is necessary for toxicity against coleopteran larvae (Warren, 1997).

In contrast to these binary toxins, Vip3 proteins are holotype toxins that are secreted without N-terminal processing (Estruch *et al.*, 1996; Bhalla *et al.*, 2005). Vip3A toxins are active against larvae of economically relevant lepidopteran pests that present low susceptibility to Cry toxins, such as *S. frugiperda*, *S. exigua*, or *H. zea*, substantiating their use for pest control (Estruch *et al.*, 1996; Donovan *et al.*, 2001). Expression of the *vip3A* gene is first detected during the logarithmic phase and remains active during the entry to stationary phase and sporulation (Estruch *et al.*, 1996). Expression of Vip3A as parasporal crystals in the Bt mother cell can be achieved using the sporulation-dependent BtI and BtII promoter regions and a 5′ STAB-SD sequence from *cry* toxin genes (Arora *et al.*, 2003b; R. Song *et al.*, 2008). After ingestion of the Vip3A toxins, susceptible larvae present symptoms typically observed for Cry toxins targeting the gut epithelium, including cessation of feeding and gut paralysis, although mortality is delayed. Histopathological effects include vacuolization of the cytoplasm of midgut cells, destruction of the brush border membrane, and cell lysis (Lee *et al.*, 2003; Abdelkefi-Mesrati *et al.*, 2011). While the Vip3A toxins (approximately 90 kDa) are processed to a 63 kDa active toxin core in susceptible and tolerant larvae, specific binding to midgut cells is observed only in susceptible insects (Yu *et al.*, 1997; Lee *et al.*, 2003). Vip3A toxin recognized 80 kDa and 100 kDa molecules in BBMVs from *M. sexta* (Lee *et al.*, 2003), 55 kDa and 100 kDa proteins in midgut extracts from *S. littoralis* (Abdelkefi-Mesrati *et al.*, 2011), and a 65 kDa protein in midgut extracts from *Prays oleae* larvae (Abdelkefi-Mesrati *et al.*, 2009). The Vip3A binding proteins in lepidopteran BBMV are not recognized by Cry1A toxins (Lee *et al.*, 2006; Sena *et al.*, 2009). Upon binding, the activated Vip3A toxin, but not the protoxin form, initiates pores in midgut cells and artificial lipid membranes (Lee *et al.*, 2003) that disrupt midgut cells in target larvae (Yu *et al.*, 1997), substantiating pore formation as the mode of action for this toxin. However, recently the ribosomal S2 protein was identified as a functional Vip3A receptor in Sf21 cells (Singh *et al.*, 2010), suggestive of activity by pore formation and activation of intracellular pathways resulting in cell lysis. The precise mechanism of action of Vip3A toxins needs further characterization to describe the steps conducive to death of the target midgut cells.

Exotoxins

McConnell and Richards were the first to report the production by Bt strains of thermostable insecticidal toxins that were active by injection into the hemocoel (McConnell and Richards, 1959). In contrast to the crystal endotoxins, these heat-stable vegetative-stage toxins are secreted, which led to their description as β-exotoxins. Based on further characterization of the β-exotoxin molecule as an ATP analogue, the name thuringiensin was designated as most appropriate (Sebesta *et al.*, 1981). Two types of thuringiensin (I and II) have been reported, but most research has been focused on type I. Type I thuringiensin has been traditionally considered a nucleotide analogue that inhibits transcription by blocking RNA polymerization (Sebesta and Horská, 1968, 1970), and adenine nucleoside oligosaccharide structure has been proposed based on the presence of glycosylation enzymes among the thuringiensin gene cluster (X. Y. Liu *et al.*, 2010). In comparison, type II thuringiensin is considered an analogue of uracil (Levinson *et al.*, 1990). Production of thuringiensin occurs during the vegetative growth phase, independent of spore and crystal formation. Since Bt itself is susceptible to thuringiensin (Klier *et al.*, 1974), it is synthesized as an inactive (non-phosphorylated) precursor that is later activated during secretion (X. Y. Liu *et al.*, 2010). Synthesis of thuringiensin is a strain-specific property and cannot be predicted by serotyping alone. Bt serotypes reported to produce thuringiensin include 1, 3a:3b, 3a:3b:3c, 4a:4b, 4a:4c, 5a5b, 7, 8a:8b, 9, 10, 11a:11b, 12, 14, 16, and 18a:18b (Ohba *et al.*, 1981c; Sebesta *et al.*, 1981; Hernández *et al.*, 2003). Production of thuringiensin is linked to the presence of plasmids harboring *cry* genes (Levinson *et al.*, 1990), and the genetic determinants for thuringiensin and Cry toxin production are contained in the same plasmid (Ozawa and Iwahana, 1986). High-level production of type I thuringiensin is strongly associated with the presence of the *cry1B* and *vip2* genes (Espinasse *et al.*, 2002; Dong *et al.*, 2007). The *thu* cluster containing the genes involved in thuringiensin production has been cloned (X. Y. Liu *et al.*, 2010). The plasmid harboring this cluster contains the *cry1Ba* gene and all genes needed for thuringiensin production, activation, and secretion.

Thuringiensin was reported as a feeding deterrent for diverse lepidopteran larvae (Mohd-Salleh and Lewis, 1982), explaining potentiation of Bt subsp. *kurstaki* toxicity against *S. exigua* larvae (Moar *et al.*, 1986). Although particularly active against dipteran larvae, thuringiensin is non-specific and has been reported to be active against diverse insects (Krieg and Langerbruch, 1981). Preparations containing thuringiensin as an active component were commercialized and effectively used for controlling house fly (Carlberg, 1986) and *L. decemlineata* (Sebesta *et al.*, 1981; Jaques and Laing, 1988). Despite control successes, safety concerns led the World Health Organization to ban public use of thuringiensin in bacterial larvicides (World Health Organization, 1999). Accordingly, much effort has been spent in the development of techniques to detect and quantify thuringiensin production in Bt cultures during preparation of insecticidal products. Chromatographic purification (Oehler *et al.*, 1982) detects the presence of type I thuringiensin, while insect bioassays are still required to detect the presence of type II thuringiensin (Hernández *et al.*, 2001).

A proteinaceous exotoxin, α-exotoxin, was described to be secreted by Bt during the growth phase and to be sensitive to heat or trypsin treatment (Fluer *et al.*, 1981). This toxin was active *per os* against lepidopteran larvae (Krieg, 1971) and had acute toxicity against mice and other vertebrates, which led to its definition as a "mouse factor."

Other Bacillus thuringiensis *Virulence Factors*

Vegetative Bt cells in the hemocoel can induce insect death by septicemia (Zhang *et al.*, 1993; Fedhila *et al.*, 2004), indicating that apart from toxins, additional factors contribute to virulence. Phospholipases secreted by bacteria promote host cell membrane disruption through catalytic activity against sphingomyelin, phosphatidylinositol, or phosphatidylcholine. Deficient production of phospholipase C in Bt mutants results in highly reduced toxicity against target insect larvae (Zhang *et al.*, 1993), confirming its relevance for entomopathogenicity. This enzyme targets proteins tethered to the cell membrane by phosphatidylinositol anchors (Ikezawa *et al.*, 1985), resulting in reduced cell growth, cell swelling, and deterioration of organelles (Ikezawa *et al.*, 1989). Phosphatidylinositol-specific phospholipase C in Bt is encoded by the *plcA* gene, and transcription is activated at the beginning of the stationary phase by the PlcR regulon (Lereclus *et al.*, 1996). Deletion of the *plcR* gene in Bt directly affects hemolytic activity and virulence (Salamitou *et al.*, 2000).

Secretion of chitinases and proteases by vegetative Bt cells aids in degradation of the peritrophic matrix (Regev *et al.*, 1996) and facilitates access of bacteria and toxins to the midgut epithelium. Manipulation of the level of chitinases in commercial or laboratory Bt preparations has shown a direct effect on pathogenicity (Smirnoff, 1971, 1973; Sampson and Gooday, 1998). Introduction of chitinase genes in Bt increases virulence against target insect hosts (Lertcanawanichakul *et al.*, 2004; Ding *et al.*, 2008). Low activity enhancement observed in some cases may be due to low levels of chitinase expression (Tantimavanich *et al.*, 1997), reduced sporulation ability and crystal toxin production (Sirichotpakorn *et al.*, 2001), or proteolytic degradation of the expressed chitinase (Driss *et al.*, 2007). Chromosomal integration of chitinase genes into Bt has been shown to overcome these limitations and to enhance toxicity (Thamthiankul *et al.*, 2004). Chitinase genes in Bt are located in the bacterial chromosome, and chitinase genes have been cloned, expressed, and characterized from a number of strains. Production of the enzyme is dependent on pH conditions (Liu *et al.*, 2002) and can be induced by addition of chitin to the media (Guttmann and Ellar, 2000), which has been used to increase chitinase production and enhance the activity of Bt products (Rojas-Avelizapa *et al.*, 1999; Vu *et al.*, 2009). Constitutively expressed chitinases cloned from Bt strains HD-1 and HD-73 were reported to function within a broad pH range, with optimal activity at pH 6.5 (Arora *et al.*,

2003a; Barboza-Corona *et al.*, 2008). When expressed heterologously and purified, the exochitinase from strain HD-1 was shown to potentiate the insecticidal effect of Vip toxins against larvae of *S. litura* (Arora *et al.*, 2003a). In contrast, chitinase from Bt subsp. *colmeri* did not enhance Bt toxicity against *S. exigua* or *H. armigera* larvae (D. Liu *et al.*, 2010), indicating that only some chitinases may be relevant to entomopathogenicity.

Proteolytic enzymes are produced by Bt throughout most of its life cycle, but proteases participating in virulence are produced at the onset of the stationary phase. Proteins extracted from the outermost layer surrounding the Bt spore, the exosporium, were described as toxic to *P. brassicae* larvae (Scherrer and Somerville, 1977). A Bt metalloprotease termed immune inhibitor A (InhA) has been shown to hydrolyze specific insect antibacterial proteins (Dalhammar and Steiner, 1984). Production of InhA is common to diverse Bt strains and is controlled by sporulation genes (Grandvalet *et al.*, 2001), except in the case of production of InhA2, which is regulated by PlcR (Fedhila *et al.*, 2003). Since InhA remains attached to the exosporium (Charlton *et al.*, 1999), which is the first bacterial layer to make contact with the host tissues, InhA and related proteases can participate in degradation of insect defensive proteins and the peritrophic matrix, explaining the variable Bt resistance to host defense systems. The InhA2 protease has been shown to be essential, although not sufficient, for Bt virulence through oral infection (Fedhila *et al.*, 2003). Injection of InhA protease has been reported to be toxic to *D. melanogaster* adults (Sidén *et al.*, 1979) and pupae of *Hyalophora cecropia* (Edlund *et al.*, 1976) or *T. ni* larvae (Lovgren *et al.*, 1990), substantiating a direct role for pathogenicity in the hemocoel. The S-layer proteins (or SLPs) of Bt have been proposed to promote conjugation (Wiwat *et al.*, 1995) and toxicity (Peña *et al.*, 2006).

Host Range

The list of insect species susceptible to Bt has greatly increased since its discovery owing to continuous isolation of novel strains with unique properties and activity range. Traditionally, Bt strains were sorted into pathotypes with activity against species of Lepidoptera, Coleoptera, or Diptera. Currently, Bt strains have been identified that are active against species of Blattaria (Quesada-Moraga *et al.*, 2004), Hemiptera (Lima *et al.*, 1994), Hymenoptera (Garcia-Robles *et al.*, 2001), Isoptera (de Castilhos-Fortes *et al.*, 2002), and Orthoptera (L. Song *et al.*, 2008). Pathogenicity to alternative invertebrate groups, such as nematodes (Bottjer *et al.*, 1985) or ticks (Fernández-Ruvalcaba *et al.*, 2010), has been also reported for Bt strains.

The range of activity of a novel Bt strain is difficult to predict, as it is determined by the combined effect of the specific toxins produced and virulence factors expressed.

Susceptibility to a specific toxin in an insect is determined by the correct processing of the toxin in the insect gut and the presence of functional toxin receptors on the gut epithelium. Since the number and type of toxin genes harbored by different strains are highly variable, different toxicities and activity range can be observed, even for strains belonging to the same serovar. However, activity is highly variable when testing strains against species within the same insect family and even genus (Peacock *et al.*, 1998). For example, while larvae of *H. virescens* are highly susceptible to Cry1Ac toxin, closely related *H. zea* larvae are more than 200 times less susceptible (Garczynski *et al.*, 1991), which results in differential effectivity in field control of these pests by Bt products (Luttrell *et al.*, 1998).

Combining Bt toxins can result in synergism, antagonism, or no effect (Tabashnik, 1992). A classic example of synergism between Cry toxins is the highly increased toxicity against mosquitoes observed when combining toxins produced by Bt subsp. *israelensis* (Angsuthanasombat *et al.*, 1992; Crickmore *et al.*, 1995). However, no synergism between subsp. *israelensis* toxins was detected against larvae of *Chironomus tepperi* (Hughes *et al.*, 2005), signifying that synergism is dependent not only on toxin combinations, but also on the insect host. For instance, while mixtures of Cry1Ac and Cry2Ab are synergistic against larvae of *Helicoverpa armigera*, they are additive against *Earias insulana* larvae (Ibargutxi *et al.*, 2008). Synergism is also observed for Cry toxins from diverse subspecies, as found for Cry4Ba from subsp. *israelensis* and Cry2A from subsp. *kurstaki* against *Cx. pipiens* larvae (Zghal *et al.*, 2005). Cyt toxins have been the most commonly reported toxins displaying synergism when combined with other toxins against mosquito larvae (Wu *et al.*, 1994; Promdonkoy *et al.*, 2005). Moreover, addition of Cyt1A overcomes high levels of resistance to Cry toxins in strains of *Cx. quinquefasciatus* (Wirth *et al.*, 1997) or *P. xylostella* (Sayyed *et al.*, 2001), suggesting commonalities in the synergism mechanism. In both dipteran and lepidopteran hosts, Cyt synergism of Cry toxicity correlates with increased binding (Cantón *et al.*, 2010; Sharma *et al.*, 2010), which has been proposed to result from membrane-inserted Cyt toxins serving as additional receptors for Cry toxins (Pérez *et al.*, 2005).

Antagonism between diverse Bt subspecies or toxins has also been reported. In *H. zea* larvae, commercial products based on subsp. *aizawai* and subsp. *kurstaki* were antagonistic (Ameen *et al.*, 1998). Although Cry1Aa and Cry1Ac toxins acted synergistically against larvae of *L. dispar*, combinations of Cry1Aa and Cry1Ab displayed a clear antagonistic effect (Lee *et al.*, 1996). Similarly, combinations of Cyt1A and Cry1Ac were antagonistic in toxicity against *T. ni* larvae (del Rincon-Castro *et al.*, 1999).

The presence of viable spores has been reported to enhance toxicity. Addition of toxin-free spore suspensions enhanced toxicity of crystals from subsp. *kurstaki* HD-1 against larvae of *P. xylostella* (Miyasono *et al.*, 1994) or Cry1Ab and Cry1C toxicity against *Plodia interpunctella* larvae (Johnson and McGaughey, 1996). In larvae of *L. dispar* this enhancing effect was also observed when using spores of other *Bacillus* spp., and spores alone were not toxic (Dubois and Dean, 1995). While spore coat proteins may actively contribute to enhanced Cry toxicity (Johnson *et al.*, 1998), it is generally considered that the synergistic effect of spores corresponds with germination septicemia caused by the vegetative cells.

Improvement of Bacillus thuringiensis *Activity*

Since specificity in Bt strains is principally dictated by the amounts and types of crystal and/or Vip toxins, Bt strains can be genetically manipulated to express novel toxin combinations or increased toxin levels with the objective of obtaining higher activity or extended host range properties. While some of the strategies used to increase insecticidal activity of Bt will be introduced here, the reader is directed to comprehensive reviews for further details (Kaur, 2000; Pardo-López *et al.*, 2009). Both non-recombinant and recombinant genetic manipulations have been used to develop improved Bt strains for commercialization. Loss or "curing" and transfer of plasmids encoding insecticidal toxins are examples of non-recombinant approaches, which generate novel strains that are considered non-genetically engineered, thus easing registration. Curing of plasmids results in increased expression of virulence factors and toxins located in the remaining genetic material (Driss *et al.*, 2011), which can result in augmented insecticidal potency (Carlton and Gawron-Burke, 1993). Conjugal mating allows transfer of compatible plasmids containing toxin genes selected for their potency or specificity, to generate improved strains with increased activity range, as when combining the *cry3A* and *cry1Ac* genes in a product targeting both lepidopteran and coleopteran pests (Carlton and Gawron-Burke, 1993). Conjugative transfer, however, is limited by plasmid incompatibility, the presence of undesirable genes in transmissible plasmids, and loss or unintended transfer of plasmids.

The development of transformation protocols (Bone and Ellar, 1989) and the availability of Bt shuttle vectors with multiple cloning sites (Baum *et al.*, 1990; Arantes and Lereclus, 1991) have greatly advanced the potential to produce improved strains. The use of vectors containing Bt replication origins and promoter sequences to drive expression of the introduced toxin gene (Gamel and Piot, 1992; Sanchis *et al.*, 1996), and site-specific recombination systems (Baum *et al.*, 1996), have allowed elimination of foreign DNA sequences in the generated strains to result in a non-transgenic product. The use of Bt transcript stability

sequences (STAB-SD) and the strong sporulation-dependent *cyt1A* promoter (the so-called *cyt1*AP/STAB combination) in expression vectors results in increased toxin production, although toxin yields and toxicity in the generated strains vary with the toxin synthesized (Park *et al.*, 1999, 2000, 2001). Integration vectors have been used to introduce *cry* genes by homologous recombination into resident plasmids facilitated by the transposon-like structures flanking *cry* genes (Lereclus *et al.*, 1992; Yue *et al.*, 2005a, b). Several recombinant Bt strains expressing combinations of selected toxins for enhanced toxicity or activity range have been successfully developed and commercialized (Baum, 1998). However, it has been well established that increased production of toxin can result in reduced sporulation rates (Adams *et al.*, 1994; Malvar *et al.*, 1994), which may have a detrimental effect on commercial production and field persistence.

Engineering of improved Cry toxins has usually been directed at enhancing effectiveness of a specific step in their mode of action. For example, faster *in vivo* activation of a modified Cry3A toxin induced by introduction of a chymotrypsin cleavage site resulted in three-fold higher toxicity against *D. virgifera* larvae (Walters *et al.*, 2008). A triple mutant (N372A, A282G, and L283S) in loop 2 of Cry1Ab domain II displayed 36-fold increased toxicity against *L. dispar* larvae, which correlated with increased irreversible toxin binding (Rajamohan *et al.*, 1996a). Since domains II and III are involved in binding specificity, mutations in these regions usually have species-dependent effects on toxicity. Thus, swapping of these domains between different Cry toxins results in extended binding and activity range (de Maagd *et al.*, 1999b; Naimov *et al.*, 2001; Walters *et al.*, 2010). In a more targeted strategy, toxicity against a specific host can be introduced through substitutions of shorter regions in domain II, as demonstrated during introduction of activity against *Culex* spp. mosquitoes in Cry4Ba by substitution of domain II loops with Cry4A (Abdullah *et al.*, 2003). The modified Cry1A toxins (Mod toxins) represent another example of a guided strategy to improve toxin activity. These toxins are capable of forming functional pores on cell membranes without previous interaction with the midgut receptors involved in oligomerization, which allows them to overcome resistance due to alterations of cadherin receptor genes (Soberón *et al.*, 2007).

Formulation, Delivery Systems, and Enhancers

Products based on Bt have been the most widely used biological insecticides, with sales accounting for about 80% of the biopesticides sold worldwide (Whalon and Wingerd, 2003). Formulations of Bt for insect control are manufactured using a standard fermentation batch process, including growing through a vegetative phase, a sporulation phase that commences upon depletion of nutrients in the medium, and a final phase in which spores and crystals are released from the sporangia. After this phase, fermentation solids are concentrated and dried or mixed with inert ingredients before packaging. Conditions used during Bt culturing can have a relevant effect on activity. For instance, production of crystal toxins decreases under limited oxygen conditions (Avignone-Rossa *et al.*, 1992) and is also sensitive to nutrient amounts in the media (Perani and Bishop, 2000). Diverse low-cost media derived from wastewater (Dang Vu *et al.*, 2009), fruit juice (Prabakaran *et al.*, 2008), or poultry waste products (Poopathi and Abidha, 2008) have been reported as alternatives for cost-effective production of Bt biopesticides. Because of the steps in the production process, Bt products contain carry-over materials from the fermentation: spores, crystals, cellular debris, and some vegetative cells (Glare and O'Callaghan, 2000). The majority of strains used in pesticide production are naturally occurring and include subspecies *kurstaki*, *thuringiensis*, *morrisoni*, and *aizawai*, used against lepidopteran larvae; subsp. *israelensis* active against mosquito and black fly larvae; subsp. *tenebrionis* toxic to coleopteran adults and larvae; and subsp. *japonensis* strain *buibui* active against soil-dwelling coleopteran larvae (Copping and Menn, 2000; Glare and O'Callaghan, 2000).

The majority of Bt products commercialized in the USA are used in specialty crops, representing a small niche (1−2%) of total insecticide sales (Nester *et al.*, 2002). The limited use of Bt products is due, in part, to the introduction of transgenic crops expressing *cry* or *vip* toxin genes, which has greatly reduced the use of Bt microbial products in the large commodity crops such as corn or cotton (Walker *et al.*, 2003). Several types of Bt formulation have been developed to target insects in different environments (Burges and Jones, 1998). Additives to increase coverage of sprayed materials and persistence, as well as insect phagostimulants have been included during manufacturing to increase exposure and intake of toxins and spores. The specific formulation type used, whether dry dust, granule, water-dispersible solid, or liquid solution, may also be important in obtaining high toxicity. For instance, the size of ingested particles was reported to greatly influence Bt toxicity against mosquito larvae (Ben-Dov *et al.*, 2003).

Factors limiting efficacy of Bt products include short persistence and residual activity due to environmental degradation, and poor control of tunneling or root-feeding pests (Walker *et al.*, 2003). These limitations led to the interest in discovering alternative delivery systems for more persistent/direct delivery and increased toxicity. Inhibition of late sporulation events (Sanchis *et al.*, 1999) or enhanced melanin production (Saxena *et al.*, 2002; J. T. Zhang *et al.*, 2008) has been reported to extend toxicity by protecting toxin crystals from UV degradation. In the case of mosquitocidal Bt products, delivery and persistence in

the larval feeding water areas are crucial for efficient control. Entrapment of Bt subsp. *israelensis* spores in calcium alginate microcapsules results in improvement of persistence in adverse environmental conditions (Elcin, 1995a). An alternative strategy to increase persistence in mosquito-feeding zones is to transform bacteria that are more persistent in these environments, such as *B. sphaericus* (Bar *et al.*, 1998) or photosynthetic *Anabaena* spp. (Manasherob *et al.*, 2003), with *cry* toxin genes from subsp. *israelensis*. Encapsulation of Bt subsp. *israelensis* spores and crystals in cells of the protist *Tetrahymena pyriformis* allows targeted delivery to the water surface and results in faster time to death and increased toxicity against mosquito larvae (Manasherob *et al.*, 1994, 1996). In the case of terrestrial insects, encapsulation of Bt toxins in non-pathogenic *P. fluorescens* cells that were killed before release resulted in increased resistance to environmental degradation and toxicity (Gaertner *et al.*, 1993). Epiphytic and endophytic bacteria have also been transformed to express *cry* genes to increase persistence of activity against leaf-feeding or boring insects (Obukowicz *et al.*, 1986; Tomasino *et al.*, 1995; Alberghini *et al.*, 2006), although this strategy was faced with considerable opposition related to the release of living recombinant organisms. Similarly, bacteria forming root nodules transformed to produce Cry toxins are also effective in controlling root-feeding larvae (Nambiar *et al.*, 1990; Skot *et al.*, 1990; Bezdicek *et al.*, 1994). Alternative entomopathogens, such as *Beauveria bassiana* producing Vip3Aa toxin, can also be used as delivery systems (Qin *et al.*, 2010).

Undoubtedly, the most successful delivery system developed to date for Bt toxins is the introduction of *cry* and/or *vip* genes into plants to produce the so-called "Bt crops", which are resistant to insect damage through endogenous production of the Bt toxins. Genes encoding Cry or Vip toxins have been inserted into diverse crop plants, including cotton, maize, potato, tobacco, rice, broccoli, lettuce, walnut, apple, alfalfa, soybean, and eggplant (aubergine). Since a detailed history of the development of Bt crops is outside the scope of this chapter, the reader is referred to published reviews on the subject for detailed information (Peferoen, 1997; Sharma *et al.*, 2003; Kumar *et al.*, 2008). Initial problems with low levels of toxin expression in Bt crops were resolved by modification of the *cry* genes to reduce potentially deleterious sequences and optimize expression in plants (Perlak *et al.*, 1991). Transgenic Bt crops present several advantages over Bt microbial products, including easier application, higher efficacy, longer pest control persistence, higher resistance to environmental degradation or washing off, and better cost-effectiveness (Walker *et al.*, 2003). Potential disadvantages of Bt crops include the generation of longer lasting residues, potential gene transfer to closely related species, and higher risk of development of insect resistance.

Diverse compounds, organisms, and proteins have been reported to synergize Bt entomopathogenicity. Burges and Jones (1998) presented multiple chemical compounds that caused varying degrees of synergism when applied with Bt products. Chemical pesticides have also been reported to enhance the toxicity of Bt (Trisyono and Whalon, 1999; Yu *et al.*, 2005; Singh *et al.*, 2007). Mycotoxins from *Metarhizium anisopliae* (destruxins) synergized toxicity of Bt against *C. fumiferana* (Brousseau *et al.*, 1998) and *S. exigua* (Rizwan-Ul-Haq *et al.*, 2009). The pigment prodigiosin produced by *S. marcescens* was reported to synergize Cry1C toxicity against *S. litura* but not in *P. xylostella* larvae (Asano *et al.*, 1999), although the mechanisms responsible for synergism and specificity are not known. Some antibiotics, such as zwittermicin A from *B. cereus*, have also been shown to synergize entomopathogenic properties of Bt (Broderick *et al.*, 2000). Gossypol, a natural phenolic compound from cotton, has been demonstrated to synergize Cry1Ac toxicity against *H. zea* larvae and increase fitness costs of Cry1Ac-resistant *H. zea* larvae on transgenic Bt cotton (Anilkumar *et al.*, 2009).

Microbes, such as epiphytic bacteria or *E. coli*, synergize toxicity of Cry1Aa and Cry1Ac against larvae of *L. dispar* (Dubois and Dean, 1995), which may relate to their potential involvement in septicemia during pathogenesis. Synergism between *Xenorhabdus* spp. or *Photorhabdus* spp. bacteria and Bt in *S. exigua* larvae results from disruption of the midgut epithelium by Cry toxins to facilitate bacterial growth in the hemolymph (Jung and Kim, 2006). Alternative entomopathogens usually present an additive effect in combination with Bt, although synergism has been reported for some baculoviruses (Marzban *et al.*, 2009) and the fungus *B. bassiana* (Wraight and Ramos, 2005).

Inhibitors of serine proteases, the most common proteases in insect digestive fluids, have been documented to potentiate Cry toxicity in lepidopteran, coleopteran, and dipteran hosts (MacIntosh *et al.*, 1990). While the specific mechanism involved in this enhancement has not been characterized, it is plausible that the protease inhibitors reduce levels of toxin degradation in the insect gut. The addition of chitinases to Bt preparations has also been reported to result in increased activity levels (see Other *Bacillus thuringiensis* Virulence Factors, above). Expression of *cry* genes as fusions with spider neurotoxins (Xia *et al.*, 2009a), the galactose-binding domain of the non-toxic ricin B-chain (Mehlo *et al.*, 2005), a protease from *B. bassiana* (Xia *et al.*, 2009b), and a fragment of a Cry1Ac receptor cadherin (Peng *et al.*, 2010a), have been reported to increase toxicity. The synergism of Cry1A toxicity by cadherin fragments was initially observed by Chen et al. (2007) using a peptide containing a Cry1Ab toxin-binding region from the *M. sexta* cadherin. Alternative fragments derived from the same cadherin synergize Cry1Ac and Cry1C toxicity against diverse lepidopteran larvae (Abdullah *et al.*, 2009). Similar effects have been reported for adherin fragments derived from *An. gambiae* (Hua *et al.*, 2008), *D. virgifera* (Park *et al.*, 2009), and *T. molitor* (Gao *et al.*, 2011). The enhancement of Cry toxicity by these cadherin peptides has been proposed to result from the peptides increasing the formation of toxin oligomers (Fabrick *et al.*, 2009; Pacheco *et al.*, 2009b).

Safety of Bacillus thuringiensis *Pesticides and Crops*

Concerns over the safety of Bt pesticidal technologies are based on health risks, adverse environmental effects, and potential gene flow from Bt crops. In terms of potential health risks of Bt pesticides to animals, dosages equivalent to field level exposure rarely result in toxicity or detrimental effects (reviewed in Glare and O'Callaghan, 2000; Siegel, 2001). Similarly, compiled data from *in vivo* safety evaluations clearly demonstrate that transgenic Bt plants are toxicologically and nutritively equivalent to non-transgenic plants (Shimada *et al.*, 2008). When analyzed for potential allergenicity, Cry toxins have been found to lack peptide sequences found in allergens (Randhawa *et al.*, 2011). Investigations of reported allergic reactions to Bt products concluded that the bacteria were not the causative agent (McClintock *et al.*, 1995). While human infections with Bt can be considered extremely rare considering the level of use, several cases have been reported (Glare and O'Callaghan, 2000). Laboratory and field studies with commercialized Bt pesticides demonstrate that extremely rare human intoxication events can be related to accidental exposure or infection of immunocompromised individuals (Siegel, 2001), confirming the safety of Bt pesticides when used appropriately. While Bt isolates from food can produce enterotoxins (Damgaard *et al.*, 1996), there are no demonstrated cases of food poisoning directly linked to Bt. This observation is explained by Bt products containing toxins that do not endure the highly acidic conditions found in the mammalian digestive system, and spores that would need to germinate and replicate in the food to produce enterotoxins. Thus, although Bt was isolated in feces of greenhouse workers after spraying, no adverse gastrointestinal symptoms were detected (Jensen *et al.*, 2002). As a solution to reduce the potential risks associated with enterotoxin production, mutant strains can be developed that are incapable of producing enterotoxins but maintain insecticidal properties (Klimowicz *et al.*, 2010). A potential source of toxicity to mammals is the non-specific β-exotoxin produced by some Bt strains (see section above).

Although Bt is a ubiquitous naturally occurring bacterium, extensive treatment programs with Bt pesticides result in episodes of increased levels of spores and toxins in the environment, which may have detrimental effects on non-target species. Owing to their high specificity, Bt pesticides are considered a safer alternative for non-target impacts compared to the wide-range toxicity of chemical pesticides. Reports from bioassays with non-target organisms (O'Callaghan et al., 2005) have shown that Bt products can be considered the safest commercially viable pesticides, only inferior to more specific pathogens that may not be commercially viable (Glare and O'Callaghan, 2000). Laboratory and greenhouse studies with Bt crops have revealed potential detrimental effects caused by prey factors rather than by Bt toxin activity (Romeis et al., 2004; Chen et al., 2008). In field studies, no differences in abundance and activity of parasitoids and predators were found when comparing transgenic Bt and non-Bt crops. Moreover, Bt crops have been designated as an optimal tool in integrated pest management programs (Romeis et al., 2006), since the efficacy of Bt crops leads to reductions in insecticide usage. However, as a note of caution, environmental factors affecting Bt crops may synergize or alter specificity (Then, 2009), potentially altering the impact on non-target species.

Bt product and Bt crop residues containing crystal toxins become associated with soil particles, enhancing their environmental persistence (Fu et al., 2009). Yet, there is no evidence of impact of residues from Bt corn, Bt cotton, or sprayed Bt formulations on environmental decomposers (Addison and Holmes, 1995; Liu et al., 2009; Zeilinger et al., 2010). Furthermore, multiple-year use of Bt crops had no effect on functional rhizosphere bacterial populations (Icoz et al., 2008; Hu et al., 2009). Another potential route for environmental effects is the leaching of Bt formulations or Bt crop residues into water streams. While no direct effects have been detected, a recent report described harmful effects on breeding birds by suppression of nematocera prey after Bt subsp. *israelensis* application (Poulin et al., 2010), indicating that further research is necessary to establish potential indirect detrimental effects on non-target species.

Hybridization between Bt-transgenic plants and their non-transformed relatives can lead to unintended reduced herbivory and increased fitness of the hybrid plants. However, stable introgression of toxin genes from Bt plants into the genome of conventional plants is complicated and would require several hybrid generations to become established in a plant population (Stewart et al., 2003). Research has established the possibility of Bt transgene escape through pollen or seed-mediated gene flow (Heuberger et al., 2010). Diverse containment strategies have been developed to avoid unintentional transgene escape from Bt-transgenic plants, including targeting of the transgene to the chloroplast genome, use of a mitigating gene linked to the transgene, and transgene excision from the pollen genome (Daniell, 2002). The genetic and ecological consequences of gene flow from Bt crops to wild flora are complex and depend on several variables, such as the specific *cry* or *vip* transgene expressed, its insertion site, the density of plants and insect hosts, and other ecological factors (Felber et al., 2007).

8.4.2. Family Bacillaceae, Genus *Lysinibacillus*

Lysinibacillus sphaericus is a highly heterogeneous species synonymous with *Bacillus sphaericus* (Ahmed et al., 2007) that contains both saprophytic and pathogenic strains. The bacterium is strictly aerobic and lacks genes encoding enzymes and transport systems to use sugars as a carbon source (Hu et al., 2008), so that amino and organic acids are used instead. Another defining feature of this bacterium is the production of a spherical spore that is located in a terminal position within the swollen sporangium (Fig. 8.2E). Phylogenetic analyses using 16S rDNA sequences and phenotypic characters substantiated the existence of seven clusters within *B. sphaericus sensu lato*, encompassing *B. sphaericus* (*sensu stricto*), *Bacillus fusiformis*, and four possible new species (Nakamura, 2000). Strains of *B. sphaericus* are well known as dipteran pathogens, and all of the pathogenic and some non-pathogenic strains align with cluster group 1, which is closely related to the *B. fusiformis* group and not to *B. sphaericus sensu stricto*. Ribosomal-RNA gene restriction patterns (Demuro et al., 1992) and DNA fingerprinting (Woodburn et al., 1995) demonstrated a high degree of genetic homogeneity among the mosquitocidal strains and supported their description as a new species. Thus, *B. sphaericus* has been reclassified in a new genus and renamed *Lysinibacillus sphaericus* (Ahmed et al., 2007). While this proposal has been accepted at the taxonomy level, the *B. sphaericus* synonym has remained in common use in the literature and will be used in this chapter to retain consistency with earlier reports. Flagellar H-serotyping (de Barjac et al., 1980) was used to classify *B. sphaericus* strains, defining 49 serotypes, of which nine (H1, H2, H3, H5, H6, H9, H25, H26, and H48) contain mosquitocidal strains. However, it has been demonstrated that serotyping is not a good predictor for mosquitocidal activity or the presence of toxin genes (Priest et al., 1997). The genome of *B. sphaericus* strain C3-41 has been sequenced, and consists of a circular chromosome and a large two-copy plasmid named pBsph (Hu et al., 2008). Virulence genes in C3-41 are widely distributed throughout the chromosome rather than as part of pathogenic islands.

The bacterium can be readily isolated from soil and aquatic habitats using methods to isolate spore-forming bacteria (Guerineau et al., 1991). High larvicidal activity in

B. sphaericus strains 1593 and 2297 isolated from diseased *Culex* spp. mosquito larvae (Wickremesinghe and Mendis, 1980), and strain 2362 from adult black fly, *Simulium damnosum* (Weiser, 1984), increased interest in the potential use of *B. sphaericus* as a larvicide (Singer, 1981). Strains 2362 and C3-41 are the most widely used as active ingredients in commercial *B. sphaericus* products (Park *et al.*, 2010). The isolation of new strains with slightly higher toxicity than 2362 against *Cx. quinquefasciatus* (Park *et al.*, 2007b) supports the prospect of isolating novel strains with higher activity.

Susceptibility to *B. sphaericus* is highly variable depending on the strain and mosquito species (Wraight *et al.*, 1987), and is due to production of toxins (Davidson *et al.*, 1975). A generalized decreasing order of susceptibility is *Culex*, *Anopheles*, *Mansonia*, and *Aedes* (Yap, 1990). The marked difference in susceptibility between *Culex* and *Aedes* mosquitoes is due to differences in toxin binding to gut epithelial cells (Aly *et al.*, 1989; Davidson, 1989). Feeding rates (Ramoska and Hopkins, 1981), sunlight and pH (Mulligan *et al.*, 1980), and microbiological aspects (Yousten, 1984) greatly influence mosquito susceptibility to *B. sphaericus*. In general, first instar larvae appear to be the most susceptible, with lower susceptibility observed in later instars (Wraight *et al.*, 1987).

Initial pathology of *B. sphaericus* can be observed in the larval midgut. Partial inhibition of feeding in infected *Cx. pipiens* larvae was observed within 10 min post-ingestion, with feeding stopping between 2 and 5 h, and death occurring within two days (Singer, 1981). Histological studies revealed that the bacteria are confined within the peritrophic matrix and are digested by the host, possibly contributing to the release of toxins (Davidson *et al.*, 1975; Davidson, 1979). The toxins appear to bind carbohydrate residues on midgut receptors (Davidson, 1988), entering and accumulating in the midgut cells of mosquito larvae. Disease progression has been monitored. From 5 to 15 min after exposure, hypertrophy and apocrine secretions in the gut epithelium were observed (Oliveira *et al.*, 2009). About 30 min after intoxication, midgut cells appeared swollen and separated from each other at their bases (Singh and Gill, 1988). Swollen cells contained numerous cytolysosomes, which increased in number and size until epithelial cell lysis by 36 h. After midgut disruption, neural and skeletal tissue cells are also damaged, which explains the observed cessation of gut peristalsis and paralysis between 24 and 48 h (Davidson *et al.*, 1975). After host death, the spores germinate and invade the cadaver, vegetative cells grow, and the cycle ends with sporulation amplifying spore numbers in the environment (Charles and Nicolas, 1986). At lower bacterial loads, mortality is delayed, but long-term effects on development of the population are observed (Lacey *et al.*, 1987). Long residual activity, less tendency of the spore to settle in the sediment (Yousten *et al.*, 1992), and bacterial recycling in larval cadavers (Davidson *et al.*, 1984) contribute to the characteristic longer environmental persistence of *B. sphaericus* compared to Bt subsp. *israelensis* (Lacey and Undeen, 1986).

Binary Toxin

The main virulence factors of *B. sphaericus* are toxins that target the midgut epithelium (reviewed in Porter *et al.*, 1993; Park *et al.*, 2010). The mosquitocidal properties of *B. sphaericus* are produced by the action of binary (Bin) and toxin mosquitocidal (Mtx) toxins, although some strains also produce an additional binary toxin (Cry48 and Cry49) on sporulation. The binary toxin (Bin) is observed as a single parasporal crystal during late stages of sporulation. This toxin is composed of equimolar amounts of two proteins of 42 kDa (BinA or P42) and 51 kDa (BinB or P51), which are necessary for maximum toxicity (Broadwell *et al.*, 1990; Nicolas *et al.*, 1993). The two toxins have low overall sequence similarity, but share several regions of identity that are essential for toxicity (Clark and Baumann, 1991). The Bin toxins among *B. sphaericus* strains used in field applications share high sequence identity, except in small regions that allow classification into four groups. Bin1 is found in strain IAB59, the most active and commonly used strains (2362, 1593, and C3−41) produce Bin2, Bin3 is found in strain 2297, and Bin4 is found in strain LP-1G (Priest *et al.*, 1997; Humphreys and Berry, 1998). Both BinA and BinB have been shown to be proteolytically processed to smaller proteins (Broadwell *et al.*, 1990; Clark and Baumann, 1990). Proteolysis of BinA was slower and generated a 40 kDa protein, while the BinB protein is rapidly converted to a stable protein of approximately 43 kDa. The BinB protein acts as binding domain for BinA, which is internalized in the midgut cell and exerts its toxic properties (Oei *et al.*, 1992; Berry *et al.*, 1993). In solution, Bin toxin exists as a heterotetramer consisting of two molecules each of BinA and BinB (Smith *et al.*, 2005) connected through disulfide bonds (Promdonkoy *et al.*, 2008; Boonyos *et al.*, 2010). The receptor for BinB in *Cx. pipiens* is a putative 60 kDa α-glucosidase (named Cpm1) that is tethered to the midgut brush border membrane by a GPI anchor (Silva-Filha *et al.*, 1999). Comparisons between Cpm1 orthologues indicate that a few sequence differences may be responsible for specificity (Opota *et al.*, 2008; Ferreira *et al.*, 2010). Studies with model lipid monolayers indicate that only BinB is involved in membrane insertion, and that the membrane-inserted residues in BinB alone or in the BinA−BinB complex adopt a β-sheet conformation, confirming the contention that the putative Bin toxin pores consist of β-barrels (Boonserm *et al.*, 2006b). However, the precise mechanism by which the toxin kills the target cells is still unclear. A Bin toxin-resistant *Cx. quinquefasciatus* cell line displayed toxin binding and insertion but no susceptibility, attesting that toxicity is due to post-internalization processes (Schroeder *et al.*, 1989). Once the Bin toxin is internalized, the host cell starts to form lysosomes, which is conducive to autophagy of the host cell (Opota *et al.*, 2011). Although key steps have been identified (Fig. 8.6), further work is needed to develop a more complete model of the Bin intoxication process.

The *binA* and *binB* genes are highly conserved among *B. sphaericus* strains and are contained in an operon of approximately 35 kb that is located both in the chromosome and in the pBsph plasmid (Hu *et al.*, 2008). This operon contains a total of 21 predicted coding regions, including chitin-binding proteins, transposases, and insertion elements, indicating the phage infection or transposon origins for the operon (Hu *et al.*, 2008). Potential transposition would explain why *bin* genes are not present in all *B. sphaericus sensu lato* strains and the high similarity among *bin* genes in diverse strains. The specific mechanisms controlling synthesis and Bin crystal size have yet to be elucidated.

Studies on BinB deletion derivatives indicated that the N-terminal region is required for binding to the larval gut (Singkhamanan *et al.*, 2010), while the C-terminal region is responsible for interacting with BinA (Oei *et al.*, 1992).

Amino acid positions 92–100 of BinA participate in formation of the BinA–BinB complex (Yuan *et al.*, 2001) and toxin activity (Sanitt *et al.*, 2008), and are crucial for toxicity against mosquito larvae (Berry *et al.*, 1993). The C-terminal half of BinA interacts with BinB in formation of the toxin complex (Limpanawat *et al.*, 2009).

Crystal Toxins of Bacillus sphaericus

Analysis of *B. sphaericus* strains able to overcome resistance to Bin toxin in *Culex* mosquitoes revealed the presence of a new binary toxin (Jones *et al.*, 2007). One of the subunits is related to the crystal (Cry) toxins of Bt and has been classified as Cry48Aa. Mosquitocidal activity for Cry48Aa is only detected when combined with a second protein, classified as Cry49Aa, which displays relatedness to both Bin toxin and Cry35/Cry36 of Bt. While purified Cry48Aa/Cry49Aa is highly active against *Cx. quinquefasciatus* larvae, production of the toxins in *B. sphaericus* resulted in suboptimal mosquitocidal activity, probably due to low levels of Cry48Aa production. The major cytopathological effects observed in susceptible and Bin-resistant *Cx. quinquefasciatus* larvae after Cry48Aa/Cry49Aa intoxication include intense mitochondrial vacuolation, breakdown of endoplasmic reticulum, vacuolization, and microvillus disruption (de Melo *et al.*, 2009). Similar

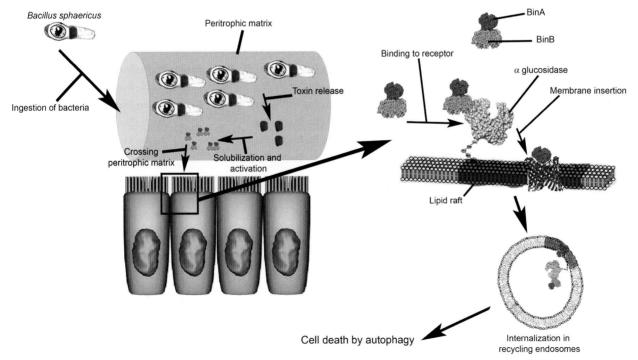

FIGURE 8.6   Sequential steps described for the mode of action of binary (Bin) toxin from *Bacillus sphaericus*. Only steps for which experimental evidence is available in the literature are included. After ingestion of the bacteria by a susceptible mosquito larvae, the Bin toxin is solubilized and processed to an active form, and the BinA–BinB complex binds to its described receptor (α-glucosidase) through interactions with BinB. The Bin complex can then insert in the cell membrane through oligomerization and formation of a β-barrel structure. The inserted toxin and/or Bin toxin bound to α glucosidase is internalized in the cell and locates in recycling endosomes. While autophagy is observed as the cell death mechanism, the specific events involved in the activation of this process after Bin toxin internalization are not known.

effects are observed for a mixture of Bin/Cry11Aa toxins, advocating that these effects result from the combined action of Cry-like and Bin-like toxins. The Cry48Aa/Cry49Aa toxin is specific to *Cx. quinquefasciatus* larvae, and is not toxic to *Aedes* and *Anopheles* mosquitoes, lepidopteran, or coleopteran insects (Jones *et al.*, 2008).

Mosquitocidal Toxins

Several strains of *B. sphaericus* produce the so-called mosquitocidal toxins (Mtx) during the vegetative stage (reviewed in Carpusca *et al.*, 2006; Park *et al.*, 2010). These toxins are soluble and do not form crystalline inclusions. Three Mtx proteins have been isolated from the low-activity strain SSII-1: Mtx1 of about 100 kDa in size (Thanabalu *et al.*, 1991), Mtx2 of 31.8 kDa (Thanabalu and Porter, 1996), and Mtx3 of 35.8 kDa (Liu *et al.*, 1996a). The *mtx1* gene promoter shows homology to the consensus sequence for the σ^{55} vegetative promoter of *B. subtilis* (Thanabalu *et al.*, 1991), and the *mtx2* and *mtx1* genes are closely located and separated by a region containing insertion elements (Hu *et al.*, 2008), denoting a relevant role for mobile genetic elements in *mtx* gene evolution. Sequence variability between Mtx toxins from diverse strains affects mosquitocidal activity and host range (Chan *et al.*, 1996). Both Mtx2 and Mtx3 display low homology to Mtx1 but have high homology to the ε-toxin of *Clostridium perfringens*, indicating that they may act as pore-forming toxins. Both Mtx1 and Mtx2 synergize the activity of *B. sphaericus* or Bt strains against susceptible and resistant *Cx. quinquefasciatus* larvae (Wirth *et al.*, 2007). Low activity detected for wild-type strains producing Mtx1 correlates with proteolytic degradation of the toxins (Thanabalu and Porter, 1995). Notably, the Mtx1 toxin induced higher cumulative larval and preadult mortalities on *Cx. quinquefasciatus* compared to Bin toxins (Wei *et al.*, 2006).

The 100 kDa Mtx1 protein displays similarity to motifs found in bacterial toxins that exert their action through ADP ribosylation (Porter *et al.*, 1993). The protein is produced as a single chain containing an N-terminal signal peptide that is cleaved in the mature protein to yield an N-terminal 27 kDa and a C-terminal 70 kDa fragment (Thanabalu *et al.*, 1992a). The 27 kDa peptide displays ADP-ribosyltransferase activity (Thanabalu *et al.*, 1993). In the case of Mtx2, the C-terminal half of the protein is required for proper folding and toxicity (Phannachet *et al.*, 2010). In solution, the 70 kDa fragment from Mtx1 binds and acts as a potent inhibitor of the ADP-ribosyltransferase activity of the 27 kDa fragment (Carpusca *et al.*, 2004). This 70 kDa fragment displays sequence similarity with the lectin-like binding component (ricin B) of the plant toxin ricin (Hazes and Read, 1995), suggesting involvement in binding to receptors and specificity. Loss of cell shape and cluster formation in *Cx. quinquefasciatus* cell cultures has been attributed to the ricin-like binding of the 70 kDa

fragment (Thanabalu *et al.*, 1993). After binding to unidentified receptors, the toxin is internalized into the target cells, where exposure to low pH in endosomes may lead to translocation of the catalytic (27 kDa) domain into the cytosol. Although Mtx1 ADP-ribosylates numerous eukaryotic proteins (Schirmer *et al.*, 2002), the relevant *in vivo* substrate(s) for Mtx toxins has not been elucidated.

Other Toxins in Bacillus sphaericus

A strain of *B. sphaericus* isolated from larvae of the ant lion, *Myrmeleon borfe*, was reported to produce a novel insecticidal toxin of 53 kDa in size named sphaericolysin. This toxin is unrelated to Bin or Mtx toxins, and is active against the German cockroach *Blattela germanica* (Nishiwaki *et al.*, 2007). The sphaericolysin gene displays high similarity with cereolysin O and other members of the cholesterol-dependent cytolysin family (Hu *et al.*, 2008). Thus, sphaericolysin is hemolytic and cholesterol abolishes its insecticidal activity (Nishiwaki *et al.*, 2007). Production of sphaericolysin is not limited to insecticidal *B. sphaericus* strains (From *et al.*, 2008).

Formulation, Improvement, and Safety

Although they display a more narrow spectrum of activity, products based on *B. sphaericus* overcome some of the limitations observed for field use of Bt subsp. *israelensis*, including environmental persistence, recycling in the environment, and activity in polluted waters (Mulligan *et al.*, 1980; Silapanuntakul *et al.*, 1983). Factors affecting persistence of *B. sphaericus* include recycling, inactivation by solar radiation or temperature, water quality, and larval feeding behavior (Lacey, 1990). Successful recycling has been documented in laboratory (Davidson *et al.*, 1975) and field (Davidson *et al.*, 1984) trials, and correlates with the presence of intact larval cadavers (Becker *et al.*, 1995). *Bacillus sphaericus* spores are less sensitive to inactivation by UV radiation than other *Bacillus* spp. spores owing to high concentrations of small acid-soluble proteins and DNA repair systems (Myasnik *et al.*, 2001). Encapsulation of spores in calcium alginate microcapsules increases resistance to UV radiation and high temperature (Elcin, 1995b). While larvicidal activity in clear water has been documented to persist for as long as nine months (Silapanuntakul *et al.*, 1983), shorter persistence has been described for polluted waters (Mulla *et al.*, 1984). Lower persistence has been proposed to result from faster spore settling (Mulligan *et al.*, 1980), excess particulate matter preventing ingestion by the larvae (Davidson *et al.*, 1984), and loss of spore toxicity (Skovmand and Guillet, 2000).

Stock suspensions of *B. sphaericus* products are very stable in storage (Thiery and Hamon, 1998). Owing to its inability to use sugars, amino acids are used as a carbon source when culturing *B. sphaericus*. Growth in media containing digested proteins (Fridlender *et al.*, 1989) or corn steep liquor (Sasaki *et al.*, 1998) as a source of amino acids and vitamins enhances sporulation and toxin production. Diverse formulations of *B. sphaericus* have been successfully used in laboratory trials and in the field for control of mosquitoes, including liquid concentrates, dry or wettable powders, granules, briquettes, and encapsulated forms. Further development of improved formulations and delivery systems is expected to have a great impact on the efficiency of *B. sphaericus* as a larvicide. For instance, a floating slow-release granule formulation was reported to increase persistence in demonstration trials (Skovmand *et al.*, 2009), and water-soluble pouches applied as prehatch treatment yielded long-term control of *Cx. quinquefasciatus* larvae (Su, 2008). Transgenic aquatic or mosquito gut microorganisms producing *B. sphaericus* toxins have been proposed as alternative delivery systems to improve activity and increase persistence (Thanabalu *et al.*, 1992b; Liu *et al.*, 1996b). Delivery of the *B. sphaericus* toxins in these bacteria has additional potential advantages, such as resistance to UV radiation, lack of toxin-degrading proteases, and low production costs.

Highly active strains *B. sphaericus* 2362 (active component in the commercial product VectoLex), C3-41 (mostly used in China and Southeast Asia), 1593, and 2297, have been evaluated in the field against all *Culex* and selected species of *Aedes*, *Psorophora*, and *Anopheles* mosquitoes in different countries. A strategy to increase activity, expand range, and delay resistance evolution has been the development of recombinant strains expressing mosquitocidal toxins. The pathology of *B. sphaericus* toxins is very different from that of Cry toxins, implying the possibility of combining these toxins in recombinant strains or alternative delivery methods for increased toxicity and expanded activity range. For instance, mosquito gut bacteria transformed to produce the Cry4B and Bin toxins are pathogenic against both *Aedes* and *Culex* larvae (Tanapongpipat *et al.*, 2003). Transferring the Bin toxin operon to Bt subsp. *israelensis* conferred enhanced activity against *Culex* mosquitoes (Park *et al.*, 2003). In addition, Cyt toxins synergize Bin and Mtx toxins (Wirth *et al.*, 2004; B. Zhang *et al.*, 2006), and help to delay the evolution of resistance in laboratory selection experiments (Wirth *et al.*, 2005a). Stable toxin expression in these strains could be achieved through insertion of the toxin genes in the *B. sphaericus* chromosome (Bar *et al.*, 1998). An alternative Bt virulence factor expressed in *B. sphaericus* is chiAC chitinase, which results in a recombinant strain more than 4,000-fold more active against Bin-resistant *Cx. quinquefasciatus* larvae (Cai *et al.*, 2007).

The safety of *B. sphaericus* against non-target invertebrates and vertebrates has been well documented (McClintock *et al.*, 1995). The high specificity of *B. sphaericus* for *Culex* and *Psorophora* species limits potential adverse effects on non-target aquatic species, including mosquito predators, other invertebrates, and fish (Aly and Mulla, 1987; Lacey and Mulla, 1990). While no direct negative impacts on aquatic fauna have been observed (Singer, 1981), there are no reported studies on potential deleterious indirect effects of *B. sphaericus* on invertebrate and vertebrate predators by prey reduction. Data from tests with vertebrate models support the safety of *B. sphaericus* for use in environments where human exposure may occur (Siegel and Shadduck, 1990; McClintock *et al.*, 1995). There are no reports on illness or death induced by exposure to diverse *B. sphaericus* strains, including commercially used isolates. Worst case exposure scenarios suggested the possibility of tissue lesions, but found no evidence of bacterial multiplication (Siegel and Shadduck, 1990).

8.4.3. Family Paenibacillaceae

Genus Paenibacillus

Milky disease, caused by proliferation of refractile bacilli through the hemolymph of scarab larvae, was first recognized in the 1930s as a disease of the Japanese beetle, *P. japonica*, that had invaded the USA (Dutky, 1940). As the name suggests, the infected larvae become milky as the bacterium grows throughout the hemolymph. Dutky (1940) described two forms of the infective bacteria: *B. popilliae*, containing a distinctive parasporal body within the sporangium, and *Bacillus lentimorbus*, lacking this feature. Milky disease has subsequently been found infecting larvae of the Scarabaeidae from a wide range of species dispersed through most continents and many islands (Klein and Jackson, 1992). Many strains of milky disease-causing bacteria are characterized by a distinctive resting stage with a spore and parasporal body contained within a thick sporangium, giving the cells a footprint-like appearance when visualized by light microscopy (Fig. 8.7). Strains can be differentiated by morphotype (Milner, 1981a) based on presence and shape of the parasporal body, with some isolates containing multiple bodies (Kaya *et al.*, 1992). Distinctive morphotypes associated with particular insect hosts confirm a high degree of strain specificity and a long co-evolution with an ancient order of insects (Morón Rios, 2004). The milky disease bacteria appear highly adapted to their specific niche. They are obligate pathogens, only found outside their hosts as resistant spores in the soil, and have become specialized pathogens utilizing the host's body to the maximum to produce large numbers of persistent spores.

Spores are ingested by scarabaeid larvae as they feed on roots and organic matter in the soil. The spores germinate in the high pH and enzyme-rich conditions of the scarab midgut (Jackson *et al.*, 2004). Once they have emerged from the sporangium, the vegetative rods penetrate the

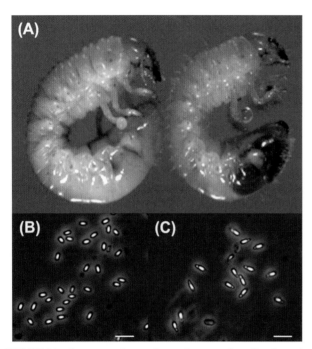

FIGURE 8.7 Symptomatology and causative agents for milky disease in larvae of *Popillia japonica*. (A) Live larvae of *P. japonica* with milky disease (left) and healthy (right) (*courtesy of Michael Klein*). Note milky appearance of hemolymph exuding from cut leg of the diseased larva. (B) Hemolymph sample of milky spore type B lacking a parasporal body. (C) Milky spore type A with spore and parasporal body. Scale bar: 5 μm.

midgut cells by phagocytosis (Splittstoesser *et al.*, 1973) and multiply initially in the regenerative nidi on the luminal side of the basal membrane before invasion of the hemocoel, where multiplication continues (Kawanishi *et al.*, 1978). Vegetative cells proliferate within the hemolymph but cause limited or no toxemia, allowing the larvae to remain active and continue feeding through a long period of infection. Vegetative cells multiply as the disease develops but with a proportion of the cells undergoing sporulation in waves throughout the disease process. Thus, a milky disease infection typically consists of vegetative rods and spores, with the latter predominating in the later stages of infection when the larva takes on an ivory-white appearance, becomes moribund, and finally dies (Klein, 1981). Once infected, the larvae rarely molt, there is little or no sign of a melanization response to infection and death of the larva is finally attributed to depletion of nutrients and fat body reserves (Sharpe and Detroy, 1979) as well as toxin action (Dutky, 1963).

By the end of the infection period, the hemocoel is filled with up to 20 billion spores (Dutky, 1940), which are released into the soil on death of the infected larva. Once in the soil, the spores persist for long periods, surviving until a new population of susceptible larvae is present on the same site. Spores germinate after ingestion by the feeding larvae, but high doses of spores are needed to induce

infections (Milner, 1981a). Sporangia applied *per os* in the laboratory have generally produced erratic results with variable levels of infection but indications of a high degree of strain/host specificity (Milner, 1981a; Franken *et al.*, 1996). Variability in infection rate can be reduced by direct injection of vegetative cells into the insect's hemocoel and this method appears to overcome strain specificity as single strains can produce high levels of infection across a wide range of scarab species (Milner, 1981b) and different strains will multiply in hosts resistant to *per os* infection (Franken *et al.*, 1996). This is evidence that strain/host specificity is determined by the ability of the bacterium to cross the gut wall barrier before multiplication in the favorable conditions of the hemolymph.

Inconsistencies in response to *per os* infection may also be due to the difficulty of obtaining consistent spore germination. Heat activation before nutrient enrichment (Stahly and Klein, 1992), or application of pressure followed by suspension in midgut fluids (Krieger *et al.*, 1996), induces germination. Once germinated, vegetative cells grow poorly in regular bacteriological media, and a complex medium (St. Julian *et al.*, 1963) has been widely used for multiplication of vegetative cells. Although the growth of the bacterium has been studied extensively, as yet, no satisfactory method has been developed to mass produce spores *in vitro* (Stahly and Klein, 1992).

The general inability of the milky disease bacteria to grow and sporulate on standard media has challenged taxonomic definition, and initial classification based on morphotype has been overtaken by molecular genetics methods. In the earliest studies, Rippere *et al.* (1998) examined a bank of 34 strains of milky disease causing bacteria and differentiated two major groupings on the basis of DNA similarity and rapid amplification of polymorphic DNA (RAPD) analysis, broadly corresponding to type strains of *B. popilliae* and *B. lentimorbus*, but the analysis denoted a wide variability across geographical regions and showed that the accepted species differentiation based on presence of the parasporal body was not valid. Vancomycin resistance appeared to be a uniform trait for the *B. popilliae* group isolated from the USA, but phylogenetic analysis of 16S rRNA gene sequences (Pettersson *et al.*, 1999) showed that the milky disease-causing Bacillus strains formed a well-defined cluster within the genus *Paenibacillus* and the authors argued for reclassification of the milky disease-causing species as *P. popilliae* and *P. lentimorbus*. The revised classification has been widely accepted, with the species initially differentiated on the basis of gene sequence and vancomycin resistance. The latter property was incorporated into a selective medium for isolation of *P. popilliae* from soil and biopesticide samples (Stahly and Klein, 1992). However, vancomycin resistance in *P. popilliae* has proven inconsistent as subsequent studies indicated that the resistance gene was absent from Central American strains of *P. popilliae* (Harrison *et al.*, 2000). Thus, neither the presence of the paraspore nor vancomycin resistance has proven to be a consistent differentiating characteristic between species.

Genetic characterization of strains has proven more conclusive, as Macdonald and Kalmakoff (1995) showed by differentiating New Zealand isolates of *P. popilliae* from the US biocontrol strain by pulsed-field gel electrophoresis (PFGE). When PGFE was applied to strains from North and South America (Correa and Yousten, 2001), regional similarities in banding patterns were found but unique variation indicated that the technique is best used for monitoring of individual isolates. Variation according to geographical region of isolation was further found in a study of milky disease bacteria from the USA using restriction fragment length polymorphism (RFLP) and 16S rDNA sequence comparison (Dingman, 2009), which supports the contention of localized evolution and specificity based on morphological and pathological data.

The role of the paraspore remains contentious. Resemblance to the toxin-bearing parasporal crystals of Bt and identification of a gene (*cryBP1*) with homology to *cry2Aa* (Zhang *et al.*, 1997) have suggested a toxic function. However, the similarity between milky disease caused by paraspore-bearing *P. popilliae* and by paraspore-free *P. lentimorbus* suggests that any toxic effect is limited.

The ecology of milky disease bacteria is defined by their long co-evolution with scarab beetles. Typically, these insects inhabit grasslands and forests and often colonize disturbed areas where populations will grow steadily in dense patches over several years before declining to low levels through biotic or environmental factors (Jackson and Klein, 2006). To use the sporadic resource of shifting beetle populations, the milky disease-causing strains of *Paenibacillus* have evolved to maximize multiplication in their host and persistence in the beetle's environment. Under favorable conditions, epizootics of milky disease will break out and contribute another wave of spores to the soil. Evidence of the success of this strategy comes from the wide distribution of scarab beetle/milky disease associations and the specificity of these interactions.

Another example of an entomopathogen in the genus *Paenibacillus* is the bacterium originally described as *Bacillus larvae*, the causative agent of AFB in honey bees (White, 1907). The bacterium was later reclassified based on 16S rRNA sequence analysis as *Paenibacillus larvae* (Ash *et al.*, 1993). Polyphasic taxonomic examinations have revealed that *Paenibacillus pulvifaciens*, the causative agent of powdery scale disease in honey bees (Katznelson, 1950), is a variant within the *P. larvae* species (Genersch *et al.*, 2006). The bacterium is catalase negative, facultative anaerobic, and pleomorphic, presenting diverse morphologies during its life cycle. AFB disease is the most important and common disease of the honey bee, and as

such, the bacterium and AFB disease are further discussed in more detail in Chapter 12.

Genus Brevibacillus

Bacillus laterosporus is a spore-forming bacterium characterized by production of a distinctive canoe-shaped lamellar parasporal inclusion that remains adjacent to the spore after lysis of the sporangium. This species has been reclassified into the genus *Brevibacillus* on the basis of 16S rRNA gene sequence analyses (Shida *et al.*, 1996). Insecticidal activity was first shown against mosquitoes (Favret and Yousten, 1985; Rivers *et al.*, 1991) but activity has also been established against Coleoptera and other organisms (Singer, 1996; de Oliveira *et al.*, 2004).

Some *B. laterosporus* strains have crystalline inclusions that are released during the lysis of the sporangium (Smirnova *et al.*, 1996). Crystal-producing strains 921 and 615 were toxic to *Ae. aegypti* and *Anopheles stephensi* at levels similar to some Bt strains, but were less toxic than *B. sphaericus* towards *C. pipiens* (Orlova *et al.*, 1998). Toxicity among these strains was associated with parasporal crystals containing a single protein of 130 kDa that was toxic to *Aedes* larvae (Zubasheva *et al.*, 2010). However, activity against dipteran larvae in an acrystalliferous strain was associated with proteins within the canoe-shaped parasporal body (Ruiu *et al.*, 2007a). Genetic analysis shows high levels of similarity among isolates, demonstrating a low level of genetic polymorphism within this species (de Oliveira *et al.*, 2004). Successful laboratory and field tests (Ruiu *et al.*, 2008), and lack of negative effects on parasitoids (Ruiu *et al.*, 2007b), demonstrate that *B. laterosporus* could be used in integrated pest management (IPM) approaches to fly control.

8.4.4. Other Entomopathogenic Bacteria in the Order Bacillales

While most research has been focused on Bt and species displaying high pathogenicity and virulence against pest insects, there are other examples of strains of bacteria in the order Bacillales, especially in the genus *Bacillus*, pathogenic to insects. The type species for this genus, *B. subtilis sensu lato*, forms part of a group of closely related bacteria commonly isolated from soil that produce oval spores that do not distend the mother cell. Members of this group have been reported as pathogenic to the bhindi leaf roller *Sylepta derogate* (Jacob *et al.*, 1982) and black fly larvae (*Simulium* spp.) (Reeves and Nayduch, 2002).

Bacillus alvei, which was later classified as *Paenibacillus alvei* based on 16S rRNA sequence similarity (Ash *et al.*, 1993), is commonly isolated from diseased honey bee larvae, which at one time contributed to the mistaken belief that it was the causative agent for European foulbrood. This bacterium can cause symptoms in honey bee larvae that resemble infection with *P. larvae* during AFB disease. However, unlike *P. larvae*, the distribution of *P. alvei* is not limited to honey bees, and the bacterium can be isolated from diverse environments. Its presence in the digestive tract of honey bees and larvae of the wax moth (*G. mellonella*), together with the synthesis of bioactive products by *P. alvei*, has led to the assumption that the bacterium has a role in insect nutrition (Gilliam, 1985, 1997). In contrast to this symbiotic relationship, strains of *P. alvei* have been reported as pathogenic to larvae of mosquito species in the genera *Culex*, *Anopheles*, and *Aedes* (Balaraman *et al.*, 1979).

The bacterium *Bacillus bombysepticus* was first isolated from the cadavers of diseased silkworm larvae (Hartman, 1931). This bacterium produces parasporal crystals and is highly pathogenic to silkworm larvae. The disease is characterized by the appearance of a darkening area on the thorax or abdomen of the larvae that later expands to the whole body (Huang *et al.*, 2009). Although the pathogenesis of *B. bombysepticus* is poorly characterized, the observations of toxicity by ingestion, production of crystalline inclusions, and invasion of the hemocoel denote similarities with Bt.

The soil bacterium *Bacillus firmus* was reported to be pathogenic against larvae of the narcissus moth, *Eligma narcissus* (Varma and Mohamed Ali, 1986). Dead larvae become black and soft, symptomatic of septicemia, and discharged gut contents contain the rod-shaped bacterium. Although this bacterium has been successfully developed and commercialized for use in control of plant-parasitic nematodes (Giannakou *et al.*, 2004), it has not been further studied for use in microbial insect control programs.

Recently, a "flacherie" of silkworm larvae reared in an artificial diet containing chloramphenicol was reported to be due to infection with *Enterococcus mundtii* (family Lactobacillaceae) (Cappellozza *et al.*, 2011). Infected larvae displayed symptoms of retarded growth, cessation of feeding, flaccidity, loss of body luster, sluggishness, and dysentery. The bacteria persisted at low levels on an artificial diet or on mulberry leaves, but could be detected at relevant loads in frass and on the surface of eggs from infected moths (Cappellozza *et al.*, 2011).

8.5. GRAM-NEGATIVE BACTERIA

8.5.1. General Introduction to the Group

Gram-negative bacteria historically have been separated from Gram-positive bacteria by their inability to absorb crystal violet dye. This definition has proven useful in diagnostics and defines bacterial genera in Prokaryote groups IV and V, which are often associated with insects (Lysenko, 1963; Boucias and Pendland, 1998). As a group, Gram-negative bacteria are characterized by the absence of spores and can occur in rods and cocciform shapes as aerobic and facultatively anaerobic species. Gram-negative bacteria are widely distributed and abundant in the environment, and thus it is not surprising that they can always be isolated from dead or diseased insects. In the early development of insect pathology, researchers devoted considerable effort to the isolation and diagnosis of Gram-negative bacteria associated with disease (Bucher, 1981), but studies of these bacteria lost favor owing to their instability, through lack of a spore, and their inconsistency in producing pathogenic effects and insect death. This led to their definition as "potential pathogens" (Bucher, 1963) or "opportunistic pathogens" (Boucias and Pendland, 1998), as although isolated from dead insects they show inconsistent performance in bioassays. However, Gram-negative bacteria have been found causing field epizootics (Trought *et al.*, 1982) and are a major cause of collapse of laboratory insect colonies unless effective hygiene is instituted (Sikorowski and Lawrence, 1994). Despite the general opportunistic associations, some highly specific relationships have developed between Gram-negative bacteria and insects. These vary from intracellular symbiotic associations such as those between *Serratia symbiotica* and aphids (Moran *et al.*, 2005), through the colonization of specific structures in the insect gut (Zhang and Jackson, 2008), to elements of disease discussed in this chapter.

Potential pathogens as defined by Bucher (1963) are highly pathogenic once within the hemocoel but have limited ability to penetrate the cuticular insect membranes and initiate disease. To be infective, potential pathogens must be vectored into the insect hemocoel. Nematodes are used to vector bacterial symbionts (*Photorhabdus* and *Xenorhabdus*) between insects hosts (see Chapter 11), but bacterial potential pathogens can also be vectored by parasitoids and may cause significant "unexpected" mortality among host insects (Jackson and McNeill, 1998). The use of parasitoid wasps to vector *S. marcescens* has been used as a method to monitor parasitoid behavior and has been considered as a potential delivery system for the pathogen (McNeill, 2000). Once in the hemocoel, Gram-negative potential pathogens are able to avoid the host's immune defense systems and/or produce toxins that can enable them to overwhelm the host, causing death through generalized septicemia (Tan *et al.*, 2006).

Gram-negative bacteria are well known for their ability to form natural transconjugants and have been widely used for genetic transformation. Genetic research on disparate groups of Enterobacteriaceae has shown the presence of common genes constituting a "toxin complex" (Tc) found among the genera *Serratia*, *Photorhabdus*, *Xenorhabdus*, *Pseudomonas*, *Yersinia*, and *Paenibacillus* (ffrench-Constant and Waterfield, 2006). Genes for the Tc have been found both on plasmids (Dodd *et al.*, 2006) and within the bacterial chromosome (Fuchs *et al.*, 2008).

The Tc are high molecular weight, multisubunit proteinaceous insecticidal toxins, derived from a range of bacteria, but having similar structures. ffrench-Constant and Waterfield (2006) defined a classification of toxin complexes and their genes based on three components (ABC) common to the different toxins. Under the classification, each mature complex is inferred to have three components, A, B, and C, and each gene name is correlated to its anticipated component within the complex. The A component provides the toxic fragment which is potentiated by activity of the B and C components. Despite a common genetic homology, bacteria containing elements of the toxin complex appear to cause quite different pathologies. While most Tc-containing bacteria are potential pathogens with a common ability to overwhelm the insect defenses once in the hemocoel, their methods of entry and pathologies are distinct. *Photorhabdus* and *Xenorhabdus*, once thought to be pathogenic only when mediated by nematode hosts, are now known to contain strains pathogenic *per os* at least among early instar larvae (Blackburn *et al.*, 2005). Isolates of *Yersinia* have proven to be highly pathogenic and capable of causing rapid septicemia (Hurst *et al.*, 2011c), while isolates of *S. entomophila* cause chronic disease through colonization of the gut for a long period before invasion of the hemocoel (Jackson *et al.*, 2001).

8.5.2. Family Enterobacteriaceae

Serratia *spp.*

The genus *Serratia* comprises 10 species which are ubiquitous in nature and commonly found in soil and water (Grimont and Grimont, 2006). Identification of infection in insects was facilitated by the characteristic red pigmentation of several strains of *S. marcescens* and *S. plymuthica*. A detailed report on insect associated *Serratia* spp. was produced by Grimont et al. (1979), which indicated that the bacteria isolated from insects were frequently from specific strains and biotypes.

Serratia marcescens is known to associate with and colonize the digestive tract of a broad range of insects, but can be found as a potential or facultative pathogen (Bucher, 1963) with a lethal dose that kills 50% of a test insect population (LD$_{50}$) of just a few cells per insect once in the hemocoel (Slatten and Larson, 1967; Podgwaite and Cosenza, 1976; Tan *et al.*, 2006). Some insects, however, are susceptible to *S. marcescens* strains through the oral route, such as tsetse flies, *Glossina* spp. (Poinar *et al.*, 1979); blow fly, *Lucilia sericata* (O'Callaghan *et al.*, 1996); and May beetles, *Melolontha melolontha* (Jackson and Zimmermann, 1996). Although high levels of pathogenicity could be achieved in some experiments, variability of response and concerns about relatedness to human pathogens have limited research on *S. marcescens* as a microbial pesticide.

A novel species, *S. entomophila*, causing amber disease in the larvae of the New Zealand grass grub, *C. zealandica*, was identified following recognition of an unusual condition among larvae associated with population decline in field studies (Trought *et al.*, 1982). The disease was caused by specific isolates of *S. entomophila* and *Serratia proteamaculans* (Stucki *et al.*, 1984; Grimont *et al.*, 1988) which were ingested by the larvae while feeding on grass roots and organic matter in the soil. The LD$_{50}$ for *S. entomophila* strain 154 against grass grub larvae was calculated to be $2-4 \times 10^4$ cells/larva (Jackson *et al.*, 2001). Ingestion of the bacteria has a major effect on the appearance of the infected larva (Fig. 8.8). Within one to three days of ingesting amber disease-causing bacteria, *C. zealandica* larvae cease feeding and the levels of the major digestive enzymes, trypsin and chymotrypsin, decrease dramatically in the midgut (Jackson, 1995; Jackson *et al.*, 2004). The midgut, which is normally dark, rapidly clears of organic matter and digestive enzymes, leaving larvae with a translucent amber coloration characteristic of the disease (Jackson *et al.*, 1993). Diseased larvae may remain in this active but non-feeding amber state for a period of several months, during which the fat bodies are autoconsumed and tissues weaken, until bacteria finally invade the hemocoel, causing death through septicemia (Jackson *et al.*, 1993, 2001). After ingestion, *S. entomophila* bacteria colonize the insect gut and adhere to the foregut cuticle (Wilson *et al.*, 1992; Jackson *et al.*, 1993). However, most *S. entomophila* appear to grow in association with particulate matter throughout the gut (Hurst and Jackson, 2002) and reach a peak of approximately 1×10^6 cells/larva, with the majority present in the hindgut (Jackson *et al.*, 2001), before invasion of the hemolymph and growth on the larva after death. The characteristic gut clearance and reduction of enzyme titer substantiated transcriptional down-regulation of the serine protease enzymes responsible for digestion in *C. zealandica* larvae (Marshall *et al.*, 2008). In contrast, protein levels and rates of protein synthesis were found to increase in the midgut of diseased insects, including large increases in soluble forms of both actin and tubulin, together with concurrent decreases in the levels of polymeric actin-associated proteins (Gatehouse *et al.*, 2008). These observations suggest that *Serratia* toxins act to cause degradation of the cytoskeletal network and prevent secretion of midgut digestive proteinases as both the actin cytoskeleton and microtubules are involved in exocytosis.

Serratia *Virulence Factors: Sep and Tc Toxins*

The occurrence of pathogenic and non-pathogenic forms among field-collected isolates of two species, *S. entomophila* and *S. proteamaculans*, was resolved by identification of a specific plasmid associated with the pathogenic

FIGURE 8.8 (A, B) Amber disease of *Costelytra zealandica*. (A) Phenotype — clear gut and amber appearance of the thorax and midgut anterior to the dark hindgut. (B) Disease process. 1. Ingestion of bacteria. 2. Colonization of particulate matter and cuticular surfaces. 3. Release of Afp and Sep. 4. Cessation of feeding and gut clearance. 5. Blocking of enzyme release in the cell. 6. Fat body auto-consumption. 7. Transit of the midgut lamina. 8. Septicemia and death. (C, D) *Yersinia entomophaga* infection of *C. zealandica*. (C) Phenotype — Soft cadaver with septicemia after bacterial invasion of the hemocoel. (D) Disease process. 1. Ingestion of bacteria. 2. Constitutive release of Tc toxin. 3. Vomiting and purging of the gut. 4. Degradation of the midgut epithelium and invasion of the hemocoel. 5. Septicemia and death.

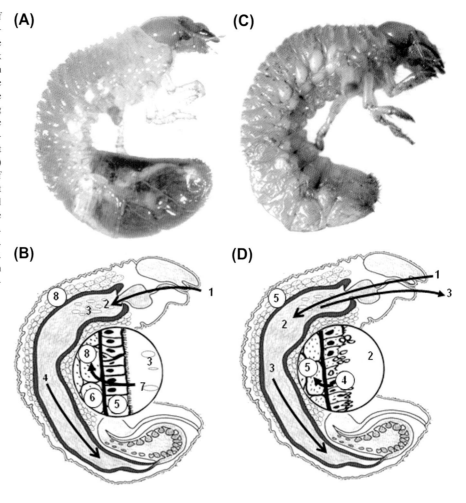

isolates (Glare *et al.*, 1993; Grkovic *et al.*, 1995). Genes causing amber disease were located on the 155 kb amber disease-associated plasmid (pADAP) (Hurst and Glare, 2002) and arranged into two clusters: a virulence-associated gene cluster, termed *sep* (*S. entomophila* pathogenicity), and an antifeeding gene cluster, termed *afp* (antifeeding prophage). The sep-virulence cluster is a Tc (see above) comprised of three genes, *sepA*, *sepB*, and *sepC*, which must all be expressed to produce the gut clearance and amber coloration of the diseased larvae (Hurst *et al.*, 2000). Sep proteins are induced in the grass grub gut after ingestion but can be induced artificially with an arabinose promoter allowing *in vitro* production (Hurst *et al.*, 2007b). The induced proteins also produced disease symptoms in grass grub larvae but had no effect against other scarab species (Hurst *et al.*, 2007b), indicating that specificity of the Sep toxin is intrinsic to *C. zealandica* rather than a consequence of induction in the insect gut.

The *afp* gene cluster is comprised of 18 ORFs forming part of a virus-like prophage (Hurst *et al.*, 2007a), and exerted strong antifeeding activity and often led to rapid

death when cloned and expressed in either *S. entomophila* or *E. coli*. After induction and purification, the expressed Afp could be visualized by electron microscopy and was found to resemble a phage-tail like bacteriocin with extended and contracted forms (Hurst *et al.*, 2007a). At greater resolution the prophage consists of an inner tube speculated to house the toxin molecule, surrounded by an outer helical sheath (Sen *et al.*, 2010). A natural isolate, *S. proteamaculans* strain 143, produces a weak form of amber disease and lacks an *afp* orthologue on the pADAP type plasmid (Hurst *et al.*, 2011a).

Thus, amber disease in *C. zealandica* requires both Sep and Afp virulence factors for observed pathogenesis. After ingestion, the release of Afp toxins causes a rapid cessation of feeding by the infected larva. Induction and release of Sep toxins lead to an expulsion of the gut contents through the hindgut and anus. Colonization of cuticular membranes of the foregut and hindgut ensures continued, constitutive production of Sep and Afp toxins. Maintenance of the non-feeding status of the infected larvae ensures that pathogenic bacteria are not purged from the insect and can colonize the alimentary tract before weakening of the gut membranes and bacterial penetration of the hemocoel, allowing colonization of the cadaver (Fig. 8.8).

Host Range and Ecology

Amber disease is unusual as it appears to be confined to a single species in an isolated environment, *C. zealandica* in New Zealand. While similar symptoms, cessation of feeding and gut clearance, have been observed in other insects, isolation of other similar disease-causing bacteria has remained elusive. The *S. entomophila* bacterium is a relatively rare member of the genus, although isolates have been recovered from Europe (Grimont *et al.*, 1988) and Mexico (Nuñez-Valdez *et al.*, 2008). The pADAP plasmid has only been found in New Zealand bacterial isolates and, despite considerable testing against New Zealand and international insect pests, no other insects have been found to be susceptible to the pADAP-bearing strains (Jackson, 2003).

Serratia spp. are ubiquitous in the environment, are present in most soils, and form a significant proportion of the bacteria found in average soils. The study of *Serratia* spp. in the environment has been facilitated by development of a semi-selective caprylate-thallous agar (CTA) (Starr *et al.*, 1976; O'Callaghan and Jackson, 1993). Species have been defined on the basis of phenotypic analysis and DNA homology (Grimont and Grimont, 1978), serology (Allardyce *et al.*, 1991), phage typing (O'Callaghan *et al.*, 1997), or DNA fingerprinting (Claus *et al.*, 1995). Identification of plasmid-bearing, disease-causing strains has been determined by bioassay (Stucki *et al.*, 1984) and DNA probes (Jackson *et al.*, 1997) or qPCR (Monk, 2010) targeting the plasmid genes. *Serratia* spp. form a significant proportion of the estimated 10^5-10^8 bacteria/g found in soil (Metting, 1993). In New Zealand, *S. entomophila* and *S. proteamaculans* are found at levels of $1 \times 10^3-10^5$ bacteria/g of pasture soil (O'Callaghan *et al.*, 1999). In new soils and pastures free from grass grub, most isolates are non-pathogenic. In these conditions, the univoltine grass grub populations will increase steadily from year to year, but as grass grub build up in the soil and amber disease enters the population, the presence of pathogenic isolates in the soil will increase until, following amber disease epizootics, pathogenic isolates of *Serratia* predominate. The presence of disease in a grass grub population will be an indicator of population decline (Jackson *et al.*, 1999). The impact of natural epizootics of amber disease on grass grub populations stimulated the development of a selected strain as a commercial biocontrol (Jackson *et al.*, 1992).

Although "amber" conditions have been observed in other insects, homologues of amber disease caused by similar bacteria have yet to be found. The putative *S. entomophila* isolate from scarab larvae in Mexico (Mor4.1) caused cessation of feeding without gut clearance and results indicated the involvement of a proteinaceous toxin (Nuñez-Valdez *et al.*, 2008). This specific isolate was active against scarab larvae in the *Phyllophaga* and *Anomala* genera, but showed no activity against *S. frugiperda*, while other *Serratia* strains have shown high lepidopteran activity. A strain of *S. marcescens* with high virulence against *H. zea* was found by Farrar *et al.* (2001) and a strain pathogenic to *P. xylostella* was found by Jeong *et al.* (2010), implying that there may be a wide range of activities remaining to be found from within this genus.

Yersinia *spp.*

The genus *Yersinia* is best known through colonization of the alimentary tract of the rat flea, *Xenopsylla cheopis*, by *Yersinia pestis*, the cause of bubonic plague (Jarrett *et al.*, 2004). Gene sequencing has revealed that homologues of the *tc* genes are common among *Yersinia* isolates (Fuchs *et al.*, 2008) but they express high variability in terms of insecticidal activity. Greatest oral lethality to *M. sexta* larvae was obtained with *Yersinia* strains containing the Tc pathogenicity island. Bresolin et al. (2006) found that low temperature was critical for induction of *tc* gene transcription and pathogenicity of whole cell extracts of *Yersinia enterocolitica* towards *M. sexta*.

Most recent research on the insecticidal activity of *Yersinia* spp. has focused on their possible role in virulence when administered orally to neonate larvae or by hemocoelic injection. The results suggest that these strains are only weakly pathogenic. For example, Champion *et al.* (2009) showed that 10^6 colony-forming units (cfu) of *Yersinia pseudotuberculosis* IP32953 were required to cause 53% mortality when injected into *G. mellonella* larvae. Furthermore, no assays have been reported showing success against damaging stages of agricultural pests. It was, therefore, surprising when a novel bacterium, *Y. entomophaga*, isolated from an infected larva of the New Zealand grass grub, *C. zealandica*, was found to have a wide insect host range and shown to be highly pathogenic to a range of important pest species (Hurst *et al.*, 2011b). The main disease determinant of *Y. entomophaga* MH96 is a 32 kb pathogenicity island that encodes an insecticidal Tc comprising the TcA, TcB, and TcC components and two chitinase proteins that form a composite Tc molecule (Hurst *et al.*, 2011c). As with other *Yersinia* spp., induction of *Y. entomophaga* MH96 Tc proteins is inhibited at high temperatures (greater than 25°C). While the bacterium was isolated from a scarabaeid larva, it has been shown to be pathogenic to a wide range of pest species including coleopteran, lepidopteran, and orthopteran species (McNeill and Hurst, 2008; Hurst *et al.*, 2011c). When ingested by larvae of *C. zealandica*, the bacterium causes cessation of feeding, followed by regurgitation and clearing of the gut contents causing a brief period of coloration resembling amber disease (Fig. 8.8). After ingestion, *Y. entomophaga* rapidly causes a degradation of the gut epithelial membranes and invades the hemocoel, causing septicemia and death. The LD_{50} for *Y. entomophaga* towards *C. zealandica* larvae has been calculated as 50 ng of purified Tc protein, and for larvae of the diamondback moth, *P. xylostella*, 30 ng of purified Tc protein (Hurst *et al.*, 2011c). *Yersinia entomophaga* can be differentiated from other *Yersinia* spp. by 16S rRNA gene sequence and DNA–DNA hybridization (Hurst *et al.*, 2011b) but the highly pathogenic *Y. entomophaga* MH96 appears to be rare in the environment as it has only been isolated on one occasion from New Zealand. Thus, the pathology of *Y. entomophaga* towards its hosts is simpler than that described previously for *S. entomophila*. In brief, bacteria are ingested from the soil or leaf surfaces and continue constitutive expression of the Tc toxin. Despite attempts by the host to purge the toxic bacteria through regurgitation and diarrhea, the toxin breaks down the midgut epithelial cells and basal membrane, allowing bacteria to enter the hemocoel, multiply and produce a septicemia, killing the insect host (Fig. 8.8).

8.5.3. Family Pseudomonadaceae: *Pseudomonas* spp.

Bacteria from the family Pseudomonadaceae are strictly aerobic, Gram-negative rods with polar flagellae providing motility. They are widely distributed in the environment and can be frequently isolated from dead and diseased insects. Strains of *P. aeruginosa* have been found to display the characteristics of potential pathogens (Bucher, 1963). In grasshoppers, the oral LD_{50} was calculated to be between 8 and 29×10^3 cells/insect, while that from intrahemocoelic inoculation was from 10 to 20 cells/insect (Bucher and Stephens, 1957). *Pseudomonas* strains have been found to be pathogenic to a range of insect species in laboratory tests (Kreig, 1987), but Bucher (1960) commented that *P. aeruginosa* had never been recorded to cause epizootics in insect populations in the field and was only destructive under laboratory conditions, especially at high temperatures and humidities.

Recently, a pseudomonad showing specific insecticidal activity was isolated from soil. The bacterium, *P. entomophila*, is highly pathogenic to both larvae and adults of the common vinegar fly, *D. melanogaster* (Vodovar *et al.*, 2005). Genomic analysis showed that the bacterium relies on a number of potential virulence factors, including toxins, proteases, putative hemolysins, hydrogen cyanide, and novel secondary metabolites, which are regulated by a two-component system, GacS/GacA, that can kill insects (Vodovar *et al.*, 2006). *Pseudomonas entomophila* encodes copies of the C and B elements of the Tc (see above), but lacks the A component (ffrench-Constant and Waterfield, 2006). Further analysis indicates that *P. entomophila* secretes a strong hemolytic activity linked to production of a new cyclic lipopeptide (Vallet-Gely *et al.*, 2010).

8.5.4. Family Coxiellaceae: *Rickettsiella* spp.

The pathogenic rickettsiae are Gram-negative, obligate intracellular pathogens with typical bacterial cell walls and no flagella. Entomopathogens belong to the genus *Rickettsiella*, which has been reassigned from the alphaproteobacteria order Rickettsiales to the gammaproteobacterial order Legionalles on the basis of 16S rRNA sequence analysis (Roux *et al.*, 1997; Cordaux *et al.*, 2007) and whole-genome analysis (Leclerque, 2008). Currently, the genus comprises three widely recognized entomopathogenic species, and their pathotypes, *Rickettsiella popilliae*, *R. grylli*, and *R. chironomi*, have been described (Dutky and Gooden, 1952; Huger and Krieg, 1967; Federici, 1980). Other entomopathogenic species in the genus are usually synonymized with the type species, *R. popilliae*. All species are highly fastidious intracellular pathogens and typically target the fat body and hemolymph cells of the host. The infective cells are typically small, dense rods ingested during feeding which traverse the midgut epithelium and enter the hemocoel where they gain entry to host cells through endocytosis. Once within the cell, pleiomorphic forms develop within the cytoplasmic vacuoles, varying from bacteria-like secondary cells to large, round rickettsogenic stroma (Fig. 8.9) (Huger and Krieg, 1967). As the disease develops, characteristic protein crystals form and cells revert to small rickettsia. Eventually, infected cells undergo lysis, releasing masses of rickettsia and crystals into the hemolymph and producing the white to blue coloration of infected larvae. Greenish blue discoloration of the fat body is typically observed in infection of Japanese beetle (*P. japonica*) larvae by *R. popilliae* cells (Dutky and Gooden, 1952). Rickettsial diseases of insects are usually chronic and the infected host will lose vigor over time, with affected larvae developing external symptoms weeks after inoculation but remaining normal for more than a month. Before death, the larvae become sluggish and cease feeding.

A behavioral response known as "behavioral fever" and consisting of the diseased insect actively seeking a high-temperature habitat to suppress bacterial development (Louis *et al.*, 1986) has been suggested to be specific to *R. grylli* infection in crickets (Adamo, 1998). Concern has been raised over potential inflammation and infection induced by entomopathogenic *Rickettsiella* in vertebrates (Delmas and Timon-David, 1985), so care should be taken when working with these organisms.

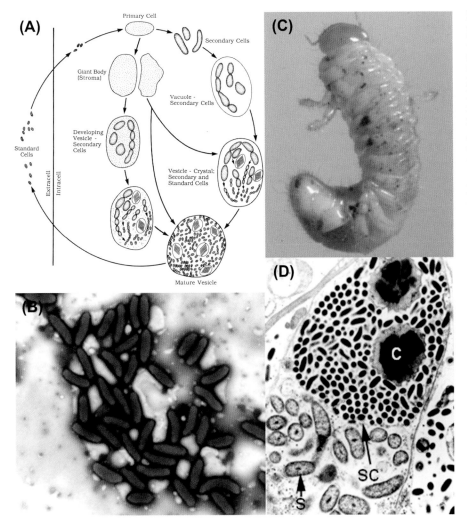

FIGURE 8.9 (A) Life cycle of an insect pathogenic rickettsia as proposed by A. M. Huger and A. Krieg. (B) Electron micrograph of *Rickettsiella* bacteria isolated from the fat body of *Pyronota setosa*. (C) Larva of *P. setosa* showing milky translucent symptoms and subcuticular melanized spots typical of *Rickettsiella* infection. (D) A thin section of *Phyllophaga* sp. tissue showing crystalline inclusion bodies (c), large secondary cells with arrow pointing to a cell (s) and standard cells of *Rickettsiella* with arrow pointing to a cluster of cells (sc). *[(A) From Tanada and Kaya, 1993. (B) Courtesy of R. G. Kleespies and S. D. Marshall. (D) Courtesy of T. G. Andreadis Press.]*

8.5.5. New Pathologies from Other Genera

Not surprisingly, examination of other non-spore formers has revealed a number of new pathologies. The intracellular entomopathogen *Coxiella burnetii* has been reported to infect dipteran larvae and adults (Padbidri *et al.*, 1982; Hucko, 1984) and has also been observed to persist in coleopteran larvae (Rehacek, 1979). However, this organism is also pathogenic to vertebrates and can cause pneumonia through inhalation. While the pigmented *Chromobacterium* spp. are generally soil- and water-associated organisms (Gillis and De Ley, 2006), isolate HM5-1 of a new species, *Chromobacterium subtsugae* (Betaproteobacteria), has been found with pathogenicity towards larvae of the Colorado potato beetle, *L. decemlineata*, and other insects (Martin *et al.*, 2007a). When incorporated into diet, *C. subtsugae* HM5-1 produced mortality in second instar *L. decemlineata* larvae, with an LC$_{50}$ of 6.4×10^7 cells/diet pellet. The bacterium has also been found to be toxic to stink bugs and rootworms (Martin *et al.*, 2007b) and whiteflies. Bacterial pathogenicity appears to be due to insecticidal toxins, which are present in the supernatants and filtrates.

8.5.6. Safety and Registration of Non-spore-forming Bacteria

Several species of non-spore-forming bacteria are recognized as human pathogens, which raises questions over their safety and use as agents for microbial control. Concern is raised when genera and species of insect pathogenic bacteria proposed for microbial control are also noted as human pathogens. Thus, *S. marcescens*, *P. aeruginosa*, and *Yersinia* spp. have all been isolated human infections and septicemia, especially in hospital-acquired, nosocomial infections. Although species and biotypes of insect pathogens may be different from those of the hospital-colonizing pathogens, safety-testing protocols have been developed as a part of product registrations (e.g., http://www.epa.gov/pesticides/biopesticides/)

Safety testing and registration of *S. entomophila* was reviewed by Jackson (2003). In summary, the bacterium has been registered for sale in New Zealand after required safety testing was carried out. Tier 1 testing (high dose challenge) was carried out against small mammals and no detrimental effects were noted following application by oral, intraperitoneal, dermal, and ocular routes. Environmental safety was assessed by direct challenge of closely related scarabs and beneficial organisms. Sheep and chickens were tested as representative vertebrates likely to be exposed to the bacterium. No infectivity was demonstrated for any non-target animal tested, indicating high specificity of *S. entomophila* to grass grub. The bacterium has been used in New Zealand for 15 years with no indications of safety problems or unexpected environmental effects from widespread application. Despite widespread testing, no other insect species have been shown to be susceptible to the plasmid-bearing strains of *S. entomophila*.

8.6. OTHER POTENTIALLY PATHOGENIC BACTERIA: INFECTION BY MOLLICUTES

Bacteria of Class Mollicutes are characterized by a lack of a cell wall, the low G + C content of their genomic DNA, and the small size of their genomes. They include organisms in the genera *Phytoplasma* [previously referred to as mycoplasma-like organisms (MLOs)] and *Spiroplasma* associated with insects but principally known as plant pathogens which must be vectored by an insect host (Garnier *et al.*, 2001). In plants, phytoplasmas and spiroplasmas induce yellowing diseases and stunting which have been recognized in monocotyledon and dicotyledon hosts. However, the fragile cells cannot survive in the environment and must be transmitted between plants by phloem-feeding homopterans including leafhoppers, planthoppers, and psyllids. For successful transmission to the next plant host, the microorganisms must first infect the vector and then be introduced into the phloem of a new host during feeding. This requires survival in the insect gut, movement across the epithelium, and transit through the hemocoel before colonization of the salivary gland and introduction into the phloem of a new plant during feeding. The movement of *Spiroplasma kunkelii* across the midgut epithelium of the leafhopper *Dalbulus maidis* has been described in detail by Ozbek *et al.* (2003). The spiroplasmas appear to move through the midgut epithelial cells without causing damage to the nuclei or other cell organelles. Spiroplasmas accumulated in vesicles on the basal side of the epithelial cells and degraded the basal lamina, creating openings for entry into the hemocoel. Heavy infection was found in the Malpighian tubule

epithelial cells without apparent damage but, in contrast, cellular lysis and necrosis were observed in muscle cells. A similar pathway was observed for *Spiroplasma citri* infecting the beet leafhopper, *Circulifer tenellus*, with spiroplasmas encountered in membranous pockets in the basal lamina of the salivary glands and disruption of muscle cells (Kwon *et al.*, 1999). Such cellular disruption indicates a detrimental effect on the insect host, but only minor effects on leafhopper longevity and vigor have been noted in infected hosts (Garnier et al., 2001).

Given the limited impact of spiroplasma and mycoplasma on host vigor, it is not surprising that these organisms were not recognized in insects until the 1960s and few interactions have been fully characterized. Molecular screening using PCR is revealing the widespread occurrence of phytoplasmas and spiroplasmas in a wide range of arthropod species (Enigl and Schausberger, 2007) but the pathological effects are not well understood. Infection by spiroplasma in *D. melanogaster* has a "male-killing" effect, leading to mortality of male progeny at the embryonic or larval stages and female dominance in subsequent fly generations (Kageyama *et al.*, 2007).

8.7. FIELD USE: EXAMPLES, SUCCESS, AND CONSTRAINTS

Bacterial biopesticides currently form the major proportion of the growing international microbial pesticide market and have become widely available, with approximately 225 microbial pesticides manufactured in the 30 member countries of the Organization for Economic Cooperation and Development (OECD) (Kabaluk and Gazdik, 2005). The biopesticide market is estimated to be growing at a rate of 10% per year (Bailey *et al.*, 2010), so it is important to review the current status of bacterial biopesticides and define factors that will limit or expand this market in the future.

The first commercialized bacterial insecticide was based on *P. popilliae* and targeted larvae of the Japanese beetle, *P. japonica*, presenting a very narrow host range, compatibility with alternative control methods, and persistence. Although these pesticides were expensive to produce and the narrow host range limited wider usage, this bacterium is still produced commercially and used for control of the Japanese beetle. Undoubtedly, the most important bacterial pesticide commercialized to date is Bt, with an estimated global market of more than $110 million per year, representing 80% of all biopesticides (Whalon and Wingerd, 2003). Despite constant increases in Bt crop adoption, this technology is best suited for high-volume crops (corn, cotton, and rice), while specialty crops and organic agriculture remain the main markets for Bt formulations, particularly in IPM programs. The most important limitation of these bacterial bioinsecticides has been their typically lower efficiency compared to synthetic pesticides. A typical example was the withdrawal from the market of formulations based on subsp. *tenebrionis* for control of *L. decemlineata* owing to market competition with highly effective imidacloprid-based products. Although the potential use of this strain in the organic market is not relevant enough to grant commercial production, development of resistance to imidacloprid may renew interest in *tenebrionis*-based pesticides.

Non-target effects of most synthetic pesticides and health-related issues related to their accumulation in the environment have also been important drivers for the growing interest in microbial (bacterial) pesticides in recent years. Although in some crop/pest situations Bt insecticides may be competitive with synthetic pesticides, the utility of preparations based on natural isolates is limited by their narrow activity spectra. Advances in the genetics of the Bt toxin genes and the ability to develop recombinant strains expressing toxins of interest have greatly influenced the development of new and improved products, and several recombinant Bt strains have been successfully commercialized (Baum, 1998). For example, strain EG2424 (active ingredient in Foil®) contains plasmids encoding toxins active against lepidopteran and coleopteran larvae for an expanded range of activity (Sanchis, 2011). An alternative has been the development of recombinant strains that produce crystal toxin chimeras with enhanced activity against key pests. For instance, strain EG7826 produces a Cry1Ac/Cry1Fa chimera with superior toxicity to *S. frugiperda* larvae compared to the parental wild-type toxins (Baum, 1998). Further success of such strategies to improve the usefulness of Bt insecticides depends on the availability of suitable toxin genes with new specificities or higher activity.

The most relevant uses of Bt pesticides in agriculture are in the control of diverse lepidopteran larvae in high-value and specialty crops, including apples, cabbage, beans, peppers, tomatoes, and strawberries (Walker *et al.*, 2003). Examples of activity include *P. xylostella* and *P. rapae* in cabbage (Endersby *et al.*, 1992), *H. virescens*, *Pseudoplusia includens*, *S. exigua*, and *H. zea* in cotton and soybeans (Luttrell *et al.*, 1998), and *D. saccharalis* in sugarcane (Rosas-García, 2006). Since Bt pesticides do not have tolerance levels, they may be applied shortly before harvest as a clean-up spray for lepidopteran larvae. Furthermore, Bt products have been designated as efficient components of IPM strategies (Lacey and Shapiro-Ilan, 2008; Reddy, 2011). These products are also very important in organic agriculture, where sometimes Bt pesticides are the only registered product to control specific pests (Collier *et al.*, 2001). Outside the agricultural environment, Bt sprays have been successfully used to control relevant lepidopteran pest species in forests, including the forest tent caterpillar, *Malacosoma disstria* (Wallner, 1971), gypsy

moth, *L. dispar*, and spruce budworm, *Choristoneura* spp. (Skyler *et al.*, 1990). While successful, these area-wide applications have been reported to potentially affect non-target Lepidoptera (Boulton *et al.*, 2007). Pesticides based on subsp. *israelensis* are very important components of programs to control mosquito and black fly vectors of serious diseases such as malaria and onchocerciasis (reviewed in Mittal, 2003; Lacey, 2007). The use of these Bt products is complemented by pesticides based on *B. sphaericus* in integrated control programs of disease vectors (Tchicaya *et al.*, 2009), including rotational application with deltamethrin for the control of *Cx. quinquefasciatus*, a vector of filariasis (Gayathri *et al.*, 2004). The lower stability and probability of evolution of resistance for subsp. *israelensis* are complemented by the higher residual activity and increased resistance probability observed for *B. sphaericus* pesticides. It is expected that the combined use of slow-release and sunlight-protected formulations and genetically engineered strains will greatly improve mosquito control.

Important limitations of Bt formulations, namely low field stability and lack of control of boring pests, have been addressed with the development of Bt crop technologies. As well as effective control of tunneling, cryptic, and root-feeding larvae, Bt crops bring additional important benefits, such as reduction in the amounts of chemical pesticides applied and reducing levels of mycotoxins produced by *Fusarium* spp. fungi in corn (Munkvold *et al.*, 1997; Folcher *et al.*, 2010). The adoption of Bt crops has been growing steadily since their introduction in the market in 1996 to account for about 30 million ha worldwide (James, 2010). Transgenic Bt cotton against larvae of the bollworm/budworm complex (*H. virescens*, *H. zea*, and *H. armigera*), and Bt corn targeting stalk-boring pests such as the European corn borer, *O. nubilalis*, have greatly dominated the transgenic Bt plant market worldwide. While Bt potato was very efficient in controlling *L. decemlineata* larvae, concerns relating to human consumption of transgenic potatoes led to the withdrawal of this product from the market (Kaniewski and Thomas, 2004).

The increased adoption rate of Bt crops has been very pronounced in the past decade in emerging and developing countries, especially China, India, Brazil, and Argentina. For instance, while the adoption rate of Bt cotton is currently estimated at 63% in the USA, it has reached 86% in India and 75% in Myanmar (James, 2010). The use of this transgenic technology in the USA has resulted in the highly effective control of targeted pests, including the European and south-western corn borers, tobacco budworm, and pink bollworm, while reducing reliance on conventional chemical pesticides (Betz *et al.*, 2000). Area-wide suppression of European corn borer populations by planting Bt corn also results in indirect benefits to non-Bt growers (Hutchison *et al.*, 2010). In China, planting of Bt cotton has been reported to account for excellent control of *H. armigera* bollworms in transgenic cotton and other host crops (Wu *et al.*, 2008). However, the reduction in insecticide use in Bt cotton has been reported to result in significant increases in mirid bug populations in China (Lu *et al.*, 2010). High Bt crop adoption rates also represent increased selection pressure for the evolution of resistance to Bt crops, one of the main areas of concern with the increasing use of this technology. While evolution of resistance to Bt crops threatens the future use of this technology as well as of Bt formulations, second generation Bt crops expressing multiple Bt toxins with distinct modes of action are expected to control resistance to single-toxin varieties and delay the onset of resistance evolution (Zhao *et al.*, 2003). The recent development of Bt rice using public sector resources and its approval for commercial use in China is expected to greatly affect the worldwide adoption of Bt crops and the benefits associated with this technology.

Since *S. entomophila* has a single susceptible host, the New Zealand grass grub, biopesticide products based on this bacterium are limited to use in New Zealand. Despite this limitation, it has been a model for the development of novel microbial controls and it fulfills an important role within the country (Jackson, 2007). Early "proof of concept" experiments relied on the application of *S. entomophila* by drenching the soil with the bacterium suspended in high volumes of water (Jackson *et al.*, 1986), which was clearly impractical and exposed the bacteria to UV radiation and desiccation stress. A solution was developed by applying a bacterial suspension directly into the soil using a modified seed drill, a method that was adopted commercially for the biocontrol product Invade™ between 1990 and 2002. Limitations associated with the need for high volumes of water and specialized equipment stimulated the development of a dry granular formulation of the bacteria, Bioshield™ (Johnson *et al.*, 2001), which was used in subsequent years (Townsend *et al.*, 2004). As a novel organism, *S. entomophila* was subject to full safety testing and registration (Jackson *et al.*, 1992) and has been marketed by a number of different New Zealand companies. The success of *S. entomophila* has been due to its ability to establish in the soil and initiate a cycle of disease in the target pest population after which the pathogenic bacteria will persist in the soil. It has also been facilitated by the ability to turn the bacteria into a stable product through fermentation and microbial formulation (Johnson *et al.*, 2001; Visnovsky *et al.*, 2008).

8.8. FUTURE RESEARCH DIRECTIONS

The continuous success and increased commercialization of bacterial products for insect control depend on the development of products with increased efficacy and/or lower production costs, wider range of activity, and enhanced persistence. There are several programs aimed at identifying novel bacterial isolates and toxins with enhanced activity or novel activity range. Molecular screening can enhance the speed of discovery of variants of known toxins and virulence factors, but robust screening systems using live insects will still be required to identify novel activities. As well as the isolation and characterization of new isolates, directed evolution and genetic engineering are expected to generate novel toxins with enhanced activity and/or expanded activity range. In directed evolution, multiple toxin variants are generated through high-throughput and error-prone genetic methods and screened for activity against a selected insect (Ishikawa *et al.*, 2007). Research in these areas should contribute improved toxins for expression in transgenic crops or introduction in recombinant strains. Ultimately, the need for improved bacterial insecticides may lead to generation of recombinant Bt expressing foreign insecticidal genes, which would require extensive characterization of potential genetic transfer processes for registration and commercialization. While registration of Bt products has been facilitated by the long history of safe use of this bacterium, available data have been obtained from a limited number of strains producing a small number of toxins (Baum, 1998), which would warrant further research on the safety of new toxins or Bt strains. Future prospects for the use of mosquitocidal formulations against disease vectors would also benefit from novel toxins but would depend on increasing persistence of the product in the larval feeding zone and the use of recombinant strains or alternative delivery systems producing optimized combinations of toxins. A growing area of interest in recent years has been the development of Cry toxin enhancers based on insect toxin receptor proteins. While these enhancers have been demonstrated to effectively increase toxicity and in some cases expand the range of activity, their use in agricultural systems will require further investigation. If successful, these enhancers can facilitate and expand insect control, as well as contribute to delaying the evolution of resistance.

Evolution of resistance is currently considered one of the main concerns with the use of Bt transgenic Bt crops. Owing to the use of similar toxins, this issue is highly relevant for Bt formulations. While much is known on mechanisms of resistance to Cry toxins involving alterations of receptors in lepidopteran and dipteran systems, there are no available data on mechanisms of resistance to Cry toxins in coleopteran insects. With new transgenic Bt crops targeting coleopteran larvae, characterization of the Cry toxin mode of action and potential resistance mechanisms in these insects will be crucial to maintain the utility of this technology. There is also a lack of information on the receptor and mechanisms of resistance to other Bt toxins, such as Vip3 toxin, that are being introduced in new

Bt crop varieties. Alteration of Cry1A receptors in resistant *H. virescens* larvae did not affect susceptibility to Vip3A toxin (Jackson *et al.*, 2007), implying that cross-resistance between these toxins would entail a more general mechanism. In this regard, the introduction of Bt crops producing multiple toxins that recognize diverse receptors would increase pressure for the evolution of mechanisms of resistance to diverse Bt toxins. The characterization of mechanisms involved in resistance to diverse Cry toxins in laboratory-selected insect strains should provide critical information to design strategies to preserve the effective use of Bt-based insecticidal technologies.

REFERENCES

Abdel-Hameed, A., & Landen, R. (1994). Studies on *Bacillus thuringiensis* strains isolated from Swedish soils: insect toxicity and production of *B. cereus*-diarrhoeal-type enterotoxin. *World J. Microb. Biotechnol., 10*, 406–409.

Abdelkefi-Mesrati, L., Rouis, S., Sellami, S., & Jaoua, S. (2009). *Prays oleae* midgut putative receptor of *Bacillus thuringiensis* vegetative insecticidal protein Vip3LB differs from that of Cry1Ac toxin. *Mol. Biotechnol., 43*, 15–19.

Abdelkefi-Mesrati, L., Boukedi, H., Dammak-Karray, M., Sellami-Boudawara, T., Jaoua, S., & Tounsi, S. (2011). Study of the *Bacillus thuringiensis* Vip3Aa16 histopathological effects and determination of its putative binding proteins in the midgut of *Spodoptera littoralis*. *J. Invertebr. Pathol., 106*, 250–254.

Abdul-Rauf, M., & Ellar, D. J. (1999). Mutations of loop 2 and loop 3 residues in domain II of *Bacillus thuringiensis* Cry1C delta-endotoxin affect insecticidal specificity and initial binding to *Spodoptera littoralis* and *Aedes aegypti* midgut membranes. *Curr. Microbiol., 39*, 94–98.

Abdullah, M. A., Alzate, O., Mohammad, M., McNall, R. J., Adang, M. J., & Dean, D. H. (2003). Introduction of *Culex* toxicity into *Bacillus thuringiensis* Cry4Ba by protein engineering. *Appl. Environ. Microbiol., 69*, 5343–5353.

Abdullah, M. A., Valaitis, A. P., & Dean, D. H. (2006). Identification of a *Bacillus thuringiensis* Cry11Ba toxin-binding aminopeptidase from the mosquito, *Anopheles quadrimaculatus*. *BMC Biochem., 7*, 16.

Abdullah, M. A., Moussa, S., Taylor, M. D., & Adang, M. J. (2009). *Manduca sexta* (Lepidoptera: Sphingidae) cadherin fragments function as synergists for Cry1A and Cry1C *Bacillus thuringiensis* toxins against noctuid moths *Helicoverpa zea*, *Agrotis ipsilon* and *Spodoptera exigua*. *Pest Manag. Sci., 65*, 1097–1103.

Abi Khattar, Z., Rejasse, A., Destoumieux-Garzón, D., Escoubas, J. M., Sanchis, V., Lereclus, D., Givaudan, A., Kallassy, M., Nielsen-Leroux, C., & Gaudriault, S. (2009). The *dlt* operon of *Bacillus cereus* is required for resistance to cationic antimicrobial peptides and for virulence in insects. *J. Bacteriol., 191*, 7063–7073.

Ackermann, H. W., Azizbekyan, R. R., Bernier, R. L., de Barjac, H., Saindoux, S., Valero, J. R., & Yu, M. X. (1995). Phage typing of *Bacillus subtilis* and *B. thuringiensis*. *Res. Microbiol., 146*, 643–657.

Adamo, S. A. (1998). The specificity of behavioral fever in the cricket *Acheta domesticus*. *J. Parasitol., 84*, 529–533.

Adamo, S. A., & Parsons, N. M. (2006). The emergency life-history stage and immunity in the cricket, *Gryllus texensis*. *Anim. Behav., 72*, 235–244.

Adamo, S. A., Fidler, T. L., & Forestell, C. A. (2007). Illness-induced anorexia and its possible function in the caterpillar, *Manduca sexta*. *Brain Behav. Immun., 21*, 292–300.

Adamo, S. A., Bartlett, A., Le, J., Spencer, N., & Sullivan, K. (2010). Illness-induced anorexia may reduce trade-offs between digestion and immune function. *Anim. Behav., 79*, 3–10.

Adams, L. F., Brown, K. L., & Whiteley, H. R. (1991). Molecular cloning and characterization of two genes encoding sigma factors that direct transcription from a *Bacillus thuringiensis* crystal protein gene promoter. *J. Bacteriol., 173*, 3846–3854.

Adams, L. F., Mathewes, S., O'Hara, P., Petersen, A., & Gurtler, H. (1994). Elucidation of the mechanism of CryIIIA overproduction in a mutagenized strain of *Bacillus thuringiensis* var. *tenebrionis*. *Mol. Microbiol., 14*, 381–389.

Adang, M. J., & Spence, K. D. (1982). Biochemical comparison of the peritrophic membrane of the lepidopteran *Orgyia pseudotsugata* and *Manduca sexta*. *Comp. Biochem. Physiol. B, 73*, 645–649.

Adang, M. J., & Spence, K. D. (1983). Permeability of the peritrophic membrane of the Douglas fir tussock moth (*Orgyia pseudotsugata*). *Comp. Biochem. Physiol. A, 75*, 233–238.

Addison, J. A., & Holmes, S. B. (1995). Effect of two commercial formulations of *Bacillus thuringiensis* subsp. *kurstaki* (Dipel 8L and Dipel 8AF) on the Collembolan species *Folsomia candida* in a soil microcosm study. *Bull. Environ. Contam. Toxicol., 55*, 771–778.

Agaisse, H., & Lereclus, D. (1994a). Structural and functional analysis of the promoter region involved in full expression of the cryIIIA toxin gene of *Bacillus thuringiensis*. *Mol. Microbiol., 13*, 97–107.

Agaisse, H., & Lereclus, D. (1994b). Expression in *Bacillus subtilis* of the *Bacillus thuringiensis* cryIIIA toxin gene is not dependent on a sporulation-specific sigma factor and is increased in a *spo0A* mutant. *J. Bacteriol., 176*, 4734–4741.

Agaisse, H., & Lereclus, D. (1995). How does *Bacillus thuringiensis* produce so much insecticidal crystal protein? *J. Bacteriol., 177*, 6027–6032.

Agaisse, H., & Lereclus, D. (1996). STAB-SD: a Shine-Dalgarno sequence in the 5′ untranslated region is a determinant of mRNA stability. *Mol. Microbiol., 20*, 633–643.

Agaisse, H., Gomlnet, M., Økstad, O. A., Kolstø, A. B., & Lereclus, D. (1999). PlcR is a pleiotropic regulator of extracellular virulence factor gene expression in *Bacillus thuringiensis*. *Mol. Microbiol., 32*, 1043–1053.

Ahmad, W., & Ellar, D. J. (1990). Directed mutagenesis of selected regions of a *Bacillus thuringiensis* entomocidal protein. *FEMS Microbiol. Lett., 56*, 97–104.

Ahmed, I., Yokota, A., Yamazoe, A., & Fujiwara, T. (2007). Proposal of *Lysinibacillus boronitolerans* gen. nov. sp. nov., and transfer of *Bacillus fusiformis* to *Lysinibacillus fusiformis* comb. nov. and *Bacillus sphaericus* to *Lysinibacillus sphaericus* comb. nov. *Int. J. Syst. Evol. Microbiol., 57*, 1117–1125.

Aimanova, K. G., Zhuang, M., & Gill, S. S. (2006). Expression of Cry1Ac cadherin receptors in insect midgut and cell lines. *J. Invertebr. Pathol., 92*, 178–187.

Akhurst, R. J., Lyness, E. W., Zhang, Q. Y., Cooper, D. J., & Pinnock, D. E. (1997). A 16S rRNA gene oligonucleotide probe for identification of *Bacillus thuringiensis* isolates from sheep fleece. *J. Invertebr. Pathol., 69*, 24–30.

Alberghini, S., Filippini, R., Shevelev, A. B., Squartini, A., & Battisti, A. (2006). Extended plant protection by an epiphytic *Pseudomonas* sp.

derivative carrying the *cry9Aa* gene from *Bacillus thuringiensis galleriae* against the pine processionary moth *Thaumetopoea pityocampa. Biocontrol Sci. Technol., 16*, 709−715.

Alcantara, E. P., Alzate, O., Lee, M. K., Curtiss, A., & Dean, D. H. (2001). Role of α-helix seven of *Bacillus thuringiensis* Cry1Ab δ-endotoxin in membrane insertion, structural stability, and ion channel activity. *Biochemistry, 40*, 2540−2547.

Allardyce, R. A., Keenan, J. I., O'Callaghan, M., & Jackson, T. A. (1991). Serological identification of *Serratia entomophila*, a bacterial pathogen of the New Zealand grass grub (*Costelytra zealandica. J. Invertebr. Pathol., 57*, 250−254.

Aly, C., & Mulla, M. S. (1987). Effect of two microbial insecticides on aquatic predators of mosquitoes. *J. Appl. Entomol., 103*, 113−118.

Aly, C., Mulla, M. S., & Federici, B. A. (1989). Ingestion, dissolution, and proteolysis of the *Bacillus sphaericus* toxin by mosquito larvae. *J. Invertebr. Pathol., 53*, 12−20.

Ameen, A. O., Fuxa, J. R., & Richter, A. R. (1998). Antagonism between formulations of different *Bacillus thuringiensis* subspecies in *Heliothis virescens* and *Helicoverpa zea* (Lepidoptera: Noctuidae). *J. Entomol. Sci., 33*, 129−135.

Amonkar, S. V., Kulkarni, U., & Anand, A. (1985). Comparative toxicity of *Bacillus thuringiensis* subspecies to *Spodoptera litura* (F.). *Curr. Sci., 54*, 475−478.

Andrews, R. E. J., Bibilos, M. M., & Bulla, L. A. J. (1985). Protease activation of the entomocidal protoxin of *Bacillus thuringiensis* ssp. *kurstaki. Appl. Environ. Microbiol., 50*, 737−742.

Angsuthanasombat, C., Crickmore, N., & Ellar, D. J. (1992). Comparison of *Bacillus thuringiensis* subsp. *israelensis* CryIVA and CryIVB cloned toxins reveals synergism *in vivo. FEMS Microbiol. Lett., 73*, 63−68.

Angus, T. A. (1954). A bacterial toxin paralyzing silkworm larvae. *Nature, 173*, 545−546.

Angus, T. A., & Heimpel, A. M. (1959). Inhibiton of feeding and blood pH changes in lepidopterous larvae infected with crystal-forming bacteria. *Can. Entomol., 91*, 352−358.

Anilkumar, K. J., Sivasupramaniam, S., Head, G., Orth, R., Van Santen, E., & Moar, W. J. (2009). Synergistic interactions between Cry1Ac and natural cotton defenses limit survival of Cry1Ac-resistant *Helicoverpa zea* (Lepidoptera: Noctuidae) on Bt cotton. *J. Chem. Ecol., 35*, 785−795.

Ankolekar, C., Rahmati, T., & Labbé, R. G. (2009). Detection of toxigenic *Bacillus cereus* and *Bacillus thuringiensis* spores in US rice. *Int. J. Food Microbiol., 128*, 460−466.

Antúnez, K., Anido, M., Arredondo, D., Evans, J. D., & Zunino, P. (2011). *Paenibacillus larvae* enolase as a virulence factor in honeybee larvae infection. *Vet. Microbiol., 147*, 83−89.

Aoki, K., & Chigasaki, Y. (1916). Uber die pathogenitat der sog. Sottobacillen (Ishiwata) bei seidenraupen. *Bull. Imperial Sericult. Exp. Sta., 1*, 97−139.

Arantes, O., & Lereclus, D. (1991). Construction of cloning vectors for *Bacillus thuringiensis. Gene., 108*, 115−119.

Arenas, I., Bravo, A., Soberón, M., & Gómez, I. (2010). Role of alkaline phosphatase from *Manduca sexta* in the mechanism of action of *Bacillus thuringiensis* Cry1Ab toxin. *J. Biol. Chem., 285*, 12497−12503.

Arnold, S., Curtiss, A., Dean, D. H., & Alzate, O. (2001). The role of a proline-induced broken-helix motif in alpha-helix 2 of *Bacillus thuringiensis* delta-endotoxins. *FEBS Lett., 490*, 70−74.

Aronson, A. I. (1994). Flexibility in the protoxin composition of *Bacillus thuringiensis. FEMS Microbiol. Lett., 117*, 21−27.

Aronson, A. (1995). The protoxin composition of *Bacillus thuringiensis* insecticidal inclusions affects solubility and toxicity. *Appl. Environ. Microbiol., 61*, 4057−4060.

Aronson, A. (2000). Incorporation of protease K into larval insect membrane vesicles does not result in disruption of integrity or function of the pore-forming *Bacillus thuringiensis* delta-endotoxin. *Appl. Environ. Microbiol., 66*, 4568−4570.

Aronson, A. I., Wu, D., & Zhang, C. (1995). Mutagenesis of specificity and toxicity regions of a *Bacillus thuringiensis* protoxin gene. *J. Bacteriol., 177*, 4059−4065.

Aronson, A. I., Geng, C., & Wu, L. (1999). Aggregation of *Bacillus thuringiensis* Cry1A toxins upon binding to target insect larval midgut vesicles. *Appl. Environ. Microbiol., 65*, 2503−2507.

Arora, N., Ahmad, T., Rajagopal, R., & Bhatnagar, R. K. (2003a). A constitutively expressed 36 kDa exochitinase from *Bacillus thuringiensis* HD-1. *Biochem. Biophys. Res. Commun., 307*, 620−625.

Arora, N., Selvapandiyan, A., Agrawal, N., & Bhatnagar, R. K. (2003b). Relocating expression of vegetative insecticidal protein into mother cell of *Bacillus thuringiensis. Biochem. Biophys. Res. Commun., 310*, 158−162.

Asano, S., Ogiwara, K., Nakagawa, Y., Suzuki, K., Hori, H., & Watanabe, T. (1999). Prodigiosin produced by *Serratia marcescens* enhances the insecticidal activity of *Bacillus thuringiensis* delta-endotoxin (Cry1C) against common cutworm, *Spodoptera litura. J. Pestic. Sci., 24*, 381−385.

Asano, S., Nozawa, M., & Bando, H. (2008). *Recombinant organisms producing insect toxins and methods for constructing same.* Washington, DC: US Patent No. 7,364,728.

Asano, T. (2006). Insect cuticle as the first line of defense reaction. *Comp. Biochem. Physiol. B, 145*. 404−404.

Ash, C., Priest, F. G., & Collins, M. D. (1993). Molecular identification of rRNA group 3 bacilli (Ash, Farrow, Wallbanks and Collins) using a PCR probe test. *Antonie Van Leeuwenhoek, 64*, 253−260.

Atsumi, S., Mizuno, E., Hara, H., Nakanishi, K., Kitami, M., Miura, N., Tabunoki, H., Watanabe, A., & Sato, R. (2005). Location of the *Bombyx mori* aminopeptidase N type 1 binding site on *Bacillus thuringiensis* Cry1Aa toxin. *Appl. Environ. Microbiol., 71*, 3966−3977.

Avignone-Rossa, C., Arcas, J., & Mignone, C. (1992). *Bacillus thuringiensis* growth, sporulation and δ-endotoxin production in oxygen limited and non-limited cultures. *World J. Microbiol. Biotechnol., 8*, 301−304.

Avisar, D., Segal, M., Sneh, B., & Zilberstein, A. (2005). Cell-cycle-dependent resistance to *Bacillus thuringiensis* Cry1C toxin in Sf9 cells. *J. Cell Sci., 118*, 3163−3170.

Ayra, C., Rodríguez, L., Trujillo, D., Fernández, Y., Ponce, M., Morán, I., & Téllez, P. (2008). Broadening the target host range of the insecticidal Cry1Ac1 toxin from *Bacillus thuringiensis* by biotechnological means. *Biotecnol. Apl., 25*, 270−275.

Bailey, K. L., Boyetchko, S. M., & Langle, T. (2010). Social and economic drivers shaping the future of biological control: a Canadian perspective on the factors affecting the development and use of microbial biopesticides. *Biol. Control, 52*, 221−229.

Balaraman, K., Bheema Rao, U. S., & Rajagopalan, P. K. (1979). Bacterial pathogens of mosquito larvae − *Bacillus alvei* (Cheshire and Cheyene) and *Bacillus brevis* (Migula) — isolated in Pondicherry. *Indian J. Med. Res., 70*, 615–619.

Banks, D. J., Jurat-Fuentes, J. L., Dean, D. H., & Adang, M. J. (2001). *Bacillus thuringiensis* Cry1Ac and Cry1Fa delta-endotoxin binding to a novel 110 kDa aminopeptidase in *Heliothis virescens* is not *N*-acetylgalactosamine mediated. *Insect Biochem. Mol. Biol., 31*, 909–918.

Banks, D. J., Hua, G., & Adang, M. J. (2003). Cloning of a *Heliothis virescens* 110 kDa aminopeptidase N and expression in *Drosophila* S2 cells. *Insect Biochem. Mol. Biol., 33*, 499–508.

Bar, E., Sandler, N., Makayoto, M., & Keynan, A. (1998). Expression of chromosomally inserted *Bacillus thuringiensis israelensis* toxin genes in *Bacillus sphaericus. J. Invertebr. Pathol., 72*, 206–213.

Barbehenn, R. V. (2001). Roles of peritrophic membranes in protecting herbivorous insects from ingested plant allelochemicals. *Arch. Insect Biochem. Physiol., 47*, 86–99.

Barboza-Corona, J. E., Reyes-Rios, D. M., Salcedo-Hernández, R., & Bideshi, D. K. (2008). Molecular and biochemical characterization of an endochitinase (ChiA-HD73) from *Bacillus thuringiensis* subsp. *kurstaki* HD-73. *Mol. Biotechnol., 39*, 29–37.

de Barjac, H. (1978). Une nouvelle variété de *Bacillus thuringiensis* très toxique pour les moustiques: *B. thuringiensis* var. *israeliensis* sérotype H14. *C.R. Acad. Sci. Ser. D, 286*, 797–800.

de Barjac, H., & Bonnefoi, A. (1962). Essai de classification biochimique et sérologique de 24 souches de *Bacillus* du type *B. thuringiensis*. *Entomophaga, 7*, 5–31.

de Barjac, H., & Bonnefoi, A. (1968). A classification of strains of *Bacillus thuringiensis* Berliner with a key to their differentiation. *J. Invertebr. Pathol., 11*, 335–347.

de Barjac, H., & Bonnefoi, A. (1972). Presence of H-antigenic subfactors in serotype 5 of *Bacillus thuringiensis* var. *canadiensis. J. Invertebr. Pathol., 20*, 212–213.

de Barjac, H., & Frachon, E. (1990). Classification of *Bacillus thuringiensis* strains. *Entomophaga, 35*, 233–240.

de Barjac, H., & Lemille, F. (1970). Presence of antigenic subfactors in serotype 3 of *Bacillus thuringiensis. J. Invertebr. Pathol., 15*, 139–140.

de Barjac, H., & Thompson, J. V. (1970). A new serotype of *Bacillus thuringiensis* var. *thompsoni* (serotype 12). *J. Invertebr. Pathol., 15*, 141–144.

de Barjac, H., Cosmao Dumanoir, V., Shaik, R., & Viviani, G. (1977). *Bacillus thuringiensis* var. *pakistani*: une nouvelle sous-espèce correspondant au serotype 13. *C.R. Acad. Sci. Ser. D, 284*, 2051–2053.

de Barjac, H. D., Véron, M., & Cosmaodumanoir, V. (1980). Biochemical and serological characterization of *Bacillus sphaericus* strains, pathogenic or non-pathogenic for mosquitos. *Ann. Microbiol., 131*, 191–201.

Barloy, F., Delécluse, A., Nicolas, L., & Lecadet, M. M. (1996). Cloning and expression of the first anaerobic toxin gene from *Clostridium bifermentans* subsp. *malaysia*, encoding a new mosquitocidal protein with homologies to *Bacillus thuringiensis* delta-endotoxins. *J. Bacteriol., 178*, 3099–3105.

Barloy, F., Lecadet, M. M., & Delécluse, A. (1998). Cloning and sequencing of three new putative toxin genes from *Clostridium bifermentans* CH18. *Gene., 211*, 293–299.

Battisti, L., Green, B. D., & Thorne, C. B. (1985). Mating system for transfer of plasmids among *Bacillus anthracis, Bacillus cereus*, and *Bacillus thuringiensis. J. Bacteriol., 162*, 543–550.

Bauce, E., Carisey, N., & Dupont, A. (2006). Carry over effects of the entomopathogen *Bacillus thuringiensis* ssp. *kurstaki* on *Choristoneura fumiferana* (Lepidoptera: Tortricidae) progeny under various stressful environmental conditions. *Agric. For. Entomol., 8*, 63–76.

Baum, J. (1998). Transgenic *Bacillus thuringiensis. Phytoprotection, 79*, 127–130.

Baum, J. A., & Malvar, T. (1995). Regulation of insecticidal crystal protein production in *Bacillus thuringiensis. Mol. Microbiol., 18*, 1–12.

Baum, J. A., Coyle, D. M., Gilbert, M. P., Jany, C. S., & Gawron-Burke, C. (1990). Novel cloning vectors for *Bacillus thuringiensis. Appl. Environ. Microbiol., 56*, 3420–3428.

Baum, J. A., Kakefuda, M., & Gawron-Burke, C. (1996). Engineering *Bacillus thuringiensis* bioinsecticides with an indigenous site-specific recombination system. *Appl. Environ. Microbiol., 62*, 4367–4373.

Bavykin, S. G., Lysov, Y. P., Zakhariev, V., Kelly, J. J., Jackman, J., Stahl, D. A., & Cherni, A. (2004). Use of 16S rRNA, 23S rRNA, and gyrB gene sequence analysis to determine phylogenetic relationships of *Bacillus cereus* group microorganisms. *J. Clin. Microbiol., 42*, 3711–3730.

Bayyareddy, K., Andacht, T. M., Abdullah, M. A., & Adang, M. J. (2009). Proteomic identification of *Bacillus thuringiensis* subsp. *israelensis* toxin Cry4Ba binding proteins in midgut membranes from *Aedes* (*Stegomyia*) *aegypti* Linnaeus (Diptera, Culicidae) larvae. *Insect Biochem. Mol. Biol., 39*, 279–286.

Bechtel, D. B., & Bulla, L. A., Jr. (1976). Electron microscope study of sporulation and parasporal crystal formation in *Bacillus thuringiensis. J. Bacteriol., 127*, 1472–1481.

Becker, N., Zgomba, M., Petric, D., Beck, M., & Ludwig, M. (1995). Role of larval cadavers in recycling processes of *Bacillus sphaericus. J. Am. Mosq. Control Assoc., 11*, 329–334.

Beegle, C. C., & Yamamoto, T. (1992). History of *Bacillus thuringiensis* Berliner research and development. *Can. Entomol., 124*, 587–616.

Beegle, C. C., Dulmage, H. T., Wolfenbarger, D. A., & Martinez, E. (1981). Persistence of *Bacillus thuringiensis* Berliner insecticidal activity on cotton foliage. *Environ. Entomol., 10*, 400–401.

Ben-Dov, E., Manasherob, R., Zaritsky, A., Barak, Z., & Margalith, Y. (2001). PCR analysis of *cry7* genes in *Bacillus thuringiensis* by the five conserved blocks of toxins. *Curr. Microbiol., 42*, 96–99.

Ben-Dov, E., Saxena, D., Wang, Q., Manasherob, R., Boussiba, S., & Zaritsky, A. (2003). Ingested particles reduce susceptibility of insect larvae to *Bacillus thuringiensis. J. Appl. Entomol., 127*, 146–152.

Berliner, E. (1915). Uber die schlaffsucht der mehlmottenraupe (*Ephestia kuhniella*, Zell.) und ihren erreger *B. thuringiensis* n. sp. *Z. Angew. Entom., 2*, 29–56.

Bernhard, K., Jarrett, P., Meadows, M., Butt, J., Ellis, D. J., Roberts, G. M., Pauli, S., Rodgers, P., & Burges, H. D. (1997). Natural isolates of *Bacillus thuringiensis*: worldwide distribution, characterization, and activity against insect pests. *J. Invertebr. Pathol., 70*, 59–68.

Berry, C., Hindley, J., Ehrhardt, A. F., Grounds, T., de Souza, I., & Davidson, E. W. (1993). Genetic determinants of host ranges of *Bacillus sphaericus* mosquito larvicidal toxins. *J. Bacteriol., 175*, 510–518.

Berry, C., O'Neil, S., Ben-Dov, E., Jones, A. F., Murphy, L., Quail, M. A., Holden, M. T., Harris, D., Zaritsky, A., & Parkhill, J. (2002). Complete sequence and organization of pBtoxis, the toxin-coding plasmid of *Bacillus thuringiensis* subsp. *israelensis*. *Appl. Environ. Microbiol., 68*, 5082—5095.

Betz, F. S., Hammond, B. G., & Fuchs, R. L. (2000). Safety and advantages of *Bacillus thuringiensis*-protected plants to control insect pests. *Regul. Toxicol. Pharmacol., 32*, 156—173.

Bezdicek, D. F., Quinn, M. A., Forse, L., Heron, D., & Kahn, M. L. (1994). Insecticidal activity and competitiveness of *Rhizobium* spp. containing the *Bacillus thuringiensis* subsp. *tenebrionis* delta-endotoxin gene *cryIII* in legume modules. *Soil Biol. Biochem., 26*, 1637—1646.

Bhalla, R., Dalal, M., Panguluri, S. K., Jagadish, B., Mandaokar, A. D., Singh, A. K., & Kumar, P. A. (2005). Isolation, characterization and expression of a novel vegetative insecticidal protein gene of *Bacillus thuringiensis*. *FEMS Microbiol. Lett., 243*, 467—472.

Bibilos, M., & Andrews, R. E., Jr. (1988). Inhibition of *Bacillus thuringiensis* proteases and their effects on crystal toxin proteins and cell-free translations. *Can. J. Microbiol., 34*, 740—747.

Bietlot, H. P., Carey, P. R., Choma, C., Kaplan, H., Lessard, T., & Pozsgay, M. (1989). Facile preparation and characterization of the toxin from *Bacillus thuringiensis* var. *kurstaki*. *Biochem. J, 260*, 87—91.

Bietlot, H. P., Vishnubhatla, I., Carey, P. R., Pozsgay, M., & Kaplan, H. (1990). Characterization of the cysteine residues and disulphide linkages in the protein crystal of *Bacillus thuringiensis*. *Biochem. J, 267*, 309—315.

Bietlot, H. P., Schernthaner, J. P., Milne, R. E., Clairmont, F. R., Bhella, R. S., & Kaplan, H. (1993). Evidence that the CryIA crystal protein from *Bacillus thuringiensis* is associated with DNA. *J. Biol. Chem., 268*, 8240—8245.

Bizzarri, M. F., & Bishop, A. H. (2007). Recovery of *Bacillus thuringiensis* in vegetative form from the phylloplane of clover (*Trifolium hybridum*) during a growing season. *J. Invertebr. Pathol., 94*, 38—47.

Bizzarri, M. F., & Bishop, A. H. (2008). The ecology of *Bacillus thuringiensis* on the phylloplane: colonization from soil, plasmid transfer, and interaction with larvae of *Pieris brassicae*. *Microb. Ecol., 56*, 133—139.

Bizzarri, M. F., Bishop, A. H., Dinsdale, A., & Logan, N. A. (2008). Changes in the properties of *Bacillus thuringiensis* after prolonged culture in a rich medium. *J. Appl. Microbiol., 104*, 60—69.

Blackburn, M. B., Domek, J. M., Gelman, D. B., & Hu, J. S. (2005). The broadly insecticidal *Photorhabdus luminescens* toxin complex a (Tca): activity against the Colorado potato beetle, *Leptinotarsa decemlineata*, and sweet potato whitefly, *Bemisia tabaci*. *J. Insect Sci., 5*, 32.

Bone, E. J., & Ellar, D. J. (1989). Transformation of *Bacillus thuringiensis* by electroporation. *FEMS Microbiol. Lett., 49*, 171—177.

Bonin, A., Paris, M., Tetreau, G., David, J. P., & Després, L. (2009). Candidate genes revealed by a genome scan for mosquito resistance to a bacterial insecticide: sequence and gene expression variations. *BMC Genomics., 10*, 551—562.

Bonnefoi, A., & de Barjac, H. (1963). Classification des souches du groupe *Bacillus thuringiensis* par la détermination de l'antigène flagellaire. *Entomophaga., 8*, 223—229.

Boonserm, P., Davis, P., Ellar, D. J., & Li, J. (2005). Crystal structure of the mosquito-larvicidal toxin Cry4Ba and its biological implications. *J. Mol. Biol., 348*, 363—382.

Boonserm, P., Mo, M., Angsuthanasombat, C., & Lescar, J. (2006a). Structure of the functional form of the mosquito larvicidal Cry4Aa

toxin from *Bacillus thuringiensis* at a 2.8-angstrom resolution. *J. Bacteriol., 188*, 3391—3401.

Boonserm, P., Moonsom, S., Boonchoy, C., Promdonkoy, B., Parthasarathy, K., & Torres, J. (2006b). Association of the components of the binary toxin from *Bacillus sphaericus* in solution and with model lipid bilayers. *Biochem. Biophys. Res. Commun., 342*, 1273—1278.

Boonyos, P., Soonsanga, S., Boonserm, P., & Promdonkoy, B. (2010). Role of cysteine at positions 67, 161 and 241 of a *Bacillus sphaericus* binary toxin BinB. *BMB Rep., 43*, 23—28.

Bosch, D., Schipper, B., van der Kleij, H., de Maagd, R. A., & Stiekema, W. J. (1994). Recombinant *Bacillus thuringiensis* crystal proteins with new properties: possibilities for resistance management. *Bio. Technology, 12*, 915—918.

Bottjer, K. P., Bone, L. W., & Gill, S. S. (1985). Nematoda: susceptibility of the egg to *Bacillus thuringiensis* toxins. *Exp. Parasitol., 60*, 239—244.

Boucias, D. G., & Pendland, J. C. (1998). *Principles of Insect Pathology*. Boston: Kluwer Academic Publishers.

Boulton, T. J., Otvos, I. S., Halwas, K. L., & Rohlfs, D. A. (2007). Recovery of nontarget Lepidoptera on Vancouver Island, Canada: one and four years after a gypsy moth eradication program. *Environ. Toxicol. Chem., 26*, 738—748.

Bradley, D., Harkey, M. A., Kim, M. K., Biever, K. D., & Bauer, L. S. (1995). The insecticidal CryIB crystal protein of *Bacillus thuringiensis* ssp. *thuringiensis* has dual specificity to coleopteran and lepidopteran larvae. *J. Invertebr. Pathol., 65*, 162—173.

Brandt, C. R., Adang, M. J., & Spence, K. D. (1978). The peritrophic membrane: ultrastructural analysis and function as a mechanical barrier to microbial infection in *Orgyia pseudotsugata*. *J. Invertebr. Pathol., 32*, 12—24.

Bravo, A., Hendrickx, K., Jansens, S., & Peferoen, M. (1992). Immunocytochemical analysis of specific binding of *Bacillus thuringiensis* insecticidal crystal proteins to lepidopteran and coleopteran midgut membranes. *J. Invertebr. Pathol., 60*, 247—253.

Bravo, A., Agaisse, H., Salamitou, S., & Lereclus, D. (1996). Analysis of cryIAa expression in sigE and sigK mutants of *Bacillus thuringiensis*. *Mol. Gen. Genet., 250*, 734—741.

Brenner, D. J., Staley, J. T., & Krieg, N. R. (2005). Classification of procaryotic organisms and the concept of bacterial speciation. In D. J. Brenner, N. R. Krieg & J. T. Staley (Eds.). (2nd ed.). *Bergey's Manual of Systematic Bacteriology*, (Vol. 2) (pp. 15—20). Boston: Springer, The Proteobacteria.

Bresolin, G., Neuhaus, K., Scherer, S., & Fuchs, T. M. (2006). Transcriptional activity of long-term adaptation of *Yersinia enterocolitica* to low-temperature growth. *J. Bacteriol., 188*, 2945—2958.

Brizzard, B. L., Schnepf, H. E., & Kronstad, J. W. (1991). Expression of the *cryIB* crystal protein gene of *Bacillus thuringiensis*. *Mol. Gen. Genet., 231*, 59—64.

Broadwell, A. H., Baumann, L., & Baumann, P. (1990). The 42-kilodalton and 51-kilodalton mosquitocidal proteins of *Bacillus sphaericus* 2362 — construction of recombinants with enhanced expression and *in vivo* studies of processing and toxicity. *J. Bacteriol., 172*, 2217—2223.

Broderick, N. A., Goodman, R. M., Raffa, K. F., & Handelsman, J. (2000). Synergy between zwittermicin A and *Bacillus thuringiensis* subsp *kurstaki* against gypsy moth (Lepidoptera: Lymantriidae). *Environ. Entomol., 29*, 101—107.

Broderick, N. A., Raffa, K. F., & Handelsman, J. (2006). Midgut bacteria required for *Bacillus thuringiensis* insecticidal activity. *Proc. Natl. Acad. Sci. USA, 103*, 15196–15199.

Broderick, N. A., Robinson, C. J., McMahon, M. D., Holt, J., Handelsman, J., & Raffa, K. F. (2009). Contributions of gut bacteria to *Bacillus thuringiensis*-induced mortality vary across a range of Lepidoptera. *BMC Biol., 7*, 11.

Brousseau, C., Charpentier, G., & Belloncik, S. (1998). Effects of *Bacillus thuringiensis* and destruxins (*Metarhizium anisopliae* mycotoxins) combinations on spruce budworm (Lepidoptera: Tortricidae). *J. Invertebr. Pathol., 72*, 262–268.

Brown, K. L. (1993). Transcriptional regulation of the *Bacillus thuringiensis* subsp. *thompsoni* crystal protein gene operon. *J. Bacteriol., 175*, 7951–7957.

Brown, K. L., & Whiteley, H. R. (1988). Isolation of a *Bacillus thuringiensis* RNA polymerase capable of transcribing crystal protein genes. *Proc. Natl. Acad. Sci. USA, 85*, 4166–4170.

Brown, K. L., & Whiteley, H. R. (1990). Isolation of the second *Bacillus thuringiensis* RNA polymerase that transcribes from a crystal protein gene promoter. *J. Bacteriol., 172*, 6682–6688.

Brown, S. E., Cao, A. T., Hines, E. R., Akhurst, R. J., & East, P. D. (2004). A novel secreted protein toxin from the insect pathogenic bacterium *Xenorhabdus nematophila*. *J. Biol. Chem., 279*, 14595–14601.

Brunet, J. F., Vachon, V., Juteau, M., Van Rie, J., Larouche, G., Vincent, C., Schwartz, J. L., & Laprade, R. (2010a). Pore-forming properties of the *Bacillus thuringiensis* toxin Cry9Ca in *Manduca sexta* brush border membrane vesicles. *Biochim. Biophys. Acta., 1798*, 1111–1118.

Brunet, J. F., Vachon, V., Marsolais, M., Van Rie, J., Schwartz, J. L., & Laprade, R. (2010b). Midgut juice components affect pore formation by the *Bacillus thuringiensis* insecticidal toxin Cry9Ca. *J. Invertebr. Pathol., 104*, 203–208.

Bucher, G. E. (1960). Potential bacterial pathogens of insects and their characteristics. *J. Insect Pathol., 2*, 172–195.

Bucher, G. E. (1963). Nonsporulating bacterial pathogens. In E. A. Steinhaus (Ed.), *Insect Pathology, An Advanced Treatise*, (Vol. 2) (pp. 117–147). New York: Academic Press.

Bucher, G. E. (1973). Definition and identification of insect pathogens. *Ann. N.Y. Acad. Sci., 217*, 8–17.

Bucher, G. E. (1981). Identification of bacteria found in insects. In H. D. Burges (Ed.), *Microbial Control of Pests and Plant Diseases 1970–1980* (pp. 7–33). London: Academic Press.

Bucher, G. E., & Stephens, J. M. (1957). A disease of grasshoppers caused by the bacterium *Pseudomonas aeruginosa* (Schroeter) Migula. *Can. J. Microbiol., 3*, 611–625.

Bulla, L. A., Jr., Kramer, K. J., & Davidson, L. I. (1977). Characterization of the entomocidal parasporal crystal of *Bacillus thuringiensis*. *J. Bacteriol., 130*, 375–383.

Burges, H. D. (1973). Enzootic diseases of insects. *Ann. N.Y. Acad. Sci., 217*, 31–49.

Burges, H. D. (2001). *Bacillus thuringiensis* in pest control. *Pestic. Outlook., 12*, 90–98.

Burges, H. D., & Jones, K. A. (1998). Formulation of bacteria, viruses and protozoa to control insects. In H. D. Burges (Ed.), *Formulation of Microbial Biopesticides: Beneficial Microorganisms, Nematodes and Seed Treatments* (pp. 33–127). Dordrecht: Kluwer Academic Publishers.

Burton, S. L., Ellar, D. J., Li, J., & Derbyshire, D. J. (1999). *N*-acetylgalactosamine on the putative insect receptor aminopeptidase N is recognised by a site on the domain III lectin-like fold of a *Bacillus thuringiensis* insecticidal toxin. *J. Mol. Biol., 287*, 1011–1022.

Burtseva, L. I., Burlak, V. A., Kalmikova, G. V., de Barjac, H., & Lecadet, M. M. (1995). *Bacillus thuringiensis novosibirsk* (serovar H24a24c), a new subspecies from the West Siberian plain. *J. Invertebr. Pathol., 66*, 92–93.

Butko, P. (2003). Cytolytic toxin Cyt1A and its mechanism of membrane damage: data and hypotheses. *Appl. Environ. Microbiol., 69*, 2415–2422.

Butko, P., Huang, F., Pusztai-Carey, M., & Surewicz, W. K. (1996). Membrane permeabilization induced by cytolytic delta-endotoxin CytA from *Bacillus thuringiensis* var. *israelensis*. *Biochemistry, 35*, 11355–11360.

Butko, P., Huang, F., Pusztai-Carey, M., & Surewicz, W. K. (1997). Interaction of the delta-endotoxin CytA from *Bacillus thuringiensis* var. *israelensis* with lipid membranes. *Biochemistry, 36*, 12862–12868.

Cahan, R., Friman, H., & Nitzan, Y. (2008). Antibacterial activity of Cyt1Aa from *Bacillus thuringiensis* subsp. *israelensis*. *Microbiology, 154*, 3529–3536.

Cai, Y., Yan, J., Hu, X., Han, B., & Yuan, Z. (2007). Improving the insecticidal activity against resistant *Culex quinquefasciatus* mosquitoes by expression of chitinase gene *chiAC* in *Bacillus sphaericus*. *Appl. Environ. Microbiol., 73*, 7744–7746.

Cano, R. J., & Borucki, M. K. (1995). Revival and identification of bacterial spores in 25- to 40-million-year-old Dominican amber. *Science, 268*, 1060–1064.

Cantón, P. E., Reyes, E. Z., Escudero, I. R., Bravo, A., & Soberón, M. (2010). Binding of *Bacillus thuringiensis* subsp. *israelensis* Cry4Ba to Cyt1Aa has an important role in synergism. *Peptides, 32*, 595–600.

Cappellozza, S., Saviane, A., Tettamanti, G., Squadrin, M., Vendramin, E., Paolucci, P., Franzetti, E., & Squartini, A. (2011). Identification of *Enterococcus mundtii* as a pathogenic agent involved in the "flacherie" disease in *Bombyx mori* L. larvae reared on artificial diet. *J. Invertebr. Pathol., 106*, 386–393.

Carlberg, G. (1986). *Bacillus thuringiensis* and microbial control of flies. *MIRCEN J. Appl. Microbiol. Biotechnol., 2*, 267–274.

Carlton, B. C., & Gawron-Burke, C. (1993). Genetic improvement of *Bacillus thuringiensis* for bioinsecticide development. In L. Kim (Ed.), *Advanced Engineered Pesticides* (pp. 43–61). New York: Marcel Dekker.

Carpusca, I., Schirmer, J., & Aktories, K. (2004). Two-site autoinhibition of the ADP-ribosylating mosquitocidal toxin (MTX) from *Bacillus sphaericus* by its 70-kDa ricin-like binding domain. *Biochemistry, 43*, 12009–12019.

Carpusca, I., Jank, T., & Aktories, K. (2006). *Bacillus sphaericus* mosquitocidal toxin (MTX) and pierisin: the enigmatic offspring from the family of ADP-ribosyltransferases. *Mol. Microbiol., 62*, 621–630.

Carroll, J., & Ellar, D. J. (1997). Analysis of the large aqueous pores produced by a *Bacillus thuringiensis* protein insecticide in *Manduca sexta* midgut-brush-border-membrane vesicles. *Eur. J. Biochem., 245*, 797–804.

de Castilhos-Fortes, R., Matsumura, A. T. S., Diehl, E., & Fiuza, L. M. (2002). Susceptibility of *Nasutitermes ehrhardti* (Isoptera:

Termitidae) to *Bacillus thuringiensis* subspecies. *Braz. J. Microbiol., 33*, 219—222.

Cerstiaens, A., Verleyen, P., Van Rie, J., Van Kerkhove, E., Schwartz, J. L., Laprade, R., De Loof, A., & Schoofs, L. (2001). Effect of *Bacillus thuringiensis* Cry1 toxins in insect hemolymph and their neurotoxicity in brain cells of *Lymantria dispar. Appl. Environ. Microbiol., 67*, 3923—3927.

Champion, O. L., Cooper, I. A. M., James, S. L., Ford, D., Karlyshev, A., Wren, B. W., Duffield, M., Oyston, P. C. F., & Titball, R. W. (2009). *Galleria mellonella* as an alternative infection model for *Yersinia pseudotuberculosis. Microbiology, 155*, 1516—1522.

Chan, Q. W. T., Melathopoulos, A. P., Pernal, S. F., & Foster, L. J. (2009). The innate immune and systemic response in honey bees to a bacterial pathogen, *Paenibacillus larvae. BMC Genomics, 10*, 387.

Chan, S. W., Thanabalu, T., Wee, B. Y., & Porter, A. G. (1996). Unusual amino acid determinants of host range in the Mtx2 family of mosquitocidal toxins. *J. Biol. Chem., 271*, 14183—14187.

Chang, L., Grant, R., & Aronson, A. (2001). Regulation of the packaging of *Bacillus thuringiensis* delta-endotoxins into inclusions. *Appl. Environ. Microbiol., 67*, 5032—5036.

Chanpaisaeng, J., Chaowanadisai, L., Jarupat, D., & Therragol, G. (1996). Characterization of a new serovar (H46), *Bacillus thuringiensis* subsp. *chanpaisis*, isolated from soil in Thailand. In *Proceedings of The Second Pacific Rim on Biotechnology of* Bacillus thuringiensis *and Its Impact to the Environment* (pp. 539—548). Thailand: Chiangmai.

Charles, J. F., & Nicolas, L. (1986). Recycling of *Bacillus sphaericus* 2362 in mosquito larvae: a laboratory study. *Ann. Inst. Pasteur Microbiol., 137*, 101—111.

Charlton, S., Moir, A. J. G., Baillie, L., & Moir, A. (1999). Characterization of the exposporium of *Bacillus cereus. J. Appl. Microbiol., 87*, 241—245.

Chatterjee, S., Ghosh, T. S., & Das, S. (2010). Virulence of *Bacillus cereus* as natural facultative pathogen of *Anopheles subpictus* Grassi (Diptera: Culicidae) larvae in submerged rice-fields and shallow ponds. *Afr. J. Biotechnol., 9*, 6983—6987.

Chaufaux, J., Marchal, M., Gilois, N., Jehanno, I., & Buisson, C. (1997). Research on natural strains of *Bacillus thuringiensis* in different biotopes throughout the world. *Can. J. Microbiol., 43*, 337—343.

Chen, J., Hua, G., Jurat-Fuentes, J. L., Abdullah, M. A., & Adang, M. J. (2007). Synergism of *Bacillus thuringiensis* toxins by a fragment of a toxin-binding cadherin. *Proc. Natl. Acad. Sci. USA, 104*, 13901—13906.

Chen, J., Brown, M. R., Hua, G., & Adang, M. J. (2005). Comparison of the localization of *Bacillus thuringiensis* Cry1A delta-endotoxins and their binding proteins in larval midgut of tobacco hornworm, *Manduca sexta. Cell Tissue Res., 321*, 123—129.

Chen, J., Aimanova, K. G., Fernandez, L. E., Bravo, A., Soberon, M., & Gill, S. S. (2009). *Aedes aegypti* cadherin serves as a putative receptor of the Cry11Aa toxin from *Bacillus thuringiensis* subsp. *israelensis. Biochem. J., 424*, 191—200.

Chen, M., Zhao, J. Z., Collins, H. L., Earle, E. D., Cao, J., & Shelton, A. M. (2008). A critical assessment of the effects of Bt transgenic plants on parasitoids. *PLoS ONE, 3*, e2284.

Chen, X. J., Lee, M. K., & Dean, D. H. (1993). Site-directed mutations in a highly conserved region of *Bacillus thuringiensis* δ-endotoxin affect inhibition of short circuit current across *Bombyx mori* midguts. *Proc. Natl. Acad. Sci. USA, 90*, 9041—9045.

Chen, X. J., Curtiss, A., Alcantara, E., & Dean, D. H. (1995). Mutations in domain I of *Bacillus thuringiensis* δ-endotoxin CryIAb reduce the irreversible binding of toxin to *Manduca sexta* brush border membrane vesicles. *J. Biol. Chem., 270*, 6412—6419.

Cheng, P., Wu, L., Ziniu, Y., & Aronson, A. (1999). Subspecies-dependent regulation of *Bacillus thuringiensis* protoxin genes. *Appl. Environ. Microbiol., 65*, 1849—1853.

Chestukhina, G. G., Zalunin, I. A., Kostina, L. I., Kotova, T. S., Kattrukha, S. P., & Stepanov, V. M. (1980). Crystal-forming proteins of *Bacillus thuringiensis*. The limited hydrolysis by endogoneous proteinases as a cause of their apparent multiplicity. *Biochem. J., 187*, 457—465.

Chiang, A. S., Yen, D. F., & Peng, W. K. (1986a). Mode of action of *Bacillus thuringiensis* to different types of hosts: in midgut cellular defense reaction and gut fluid pH changes of infected rice moth larvae aspects. *Plant Prot. Bull. (Taiwan R.O.C.), 28*, 179—189.

Chiang, A. S., Yen, D. F., & Peng, W. K. (1986b). Defense reaction of midgut epithelial cells in the rice moth larva (*Corcyra cephalonica*) infected with *Bacillus thuringiensis. J. Invertebr. Pathol., 47*, 333—339.

Chilcott, C. N., & Ellar, D. J. (1988). Comparative toxicity of *Bacillus thuringiensis* var. *israelensis* crystal proteins *in vivo* and *in vitro. J. Gen. Microbiol., 134*, 2551—2558.

Clairmont, F. R., Milne, R. E., Pham, V. T., Carrière, M. B., & Kaplan, H. (1998). Role of DNA in the activation of the Cry1A insecticidal crystal protein from *Bacillus thuringiensis. J. Biol. Chem., 273*, 9292—9296.

Clark, M. A., & Baumann, P. (1990). Deletion analysis of the 51-kilodalton protein of the *Bacillus sphaericus* 2362 binary mosquitocidal toxin — construction of derivatives equivalent to the larva-processed toxin. *J. Bacteriol., 172*, 6759—6763.

Clark, M. A., & Baumann, P. (1991). Modification of the *Bacillus sphaericus* 51-kilodalton and 42-kilodalton mosquitocidal proteins — effects of internal deletions, duplications, and formation of hybrid proteins. *Appl. Environ. Microbiol., 57*, 267—271.

Claus, H., Jackson, T. A., & Filip, Z. (1995). Characterization of *Serratia entomophila* strains by DNA fingerprints and plasmid profiles. *Microbiol. Res., 150*, 159—166.

Cohen, S., Dym, O., Albeck, S., Ben-Dov, E., Cahan, R., Firer, M., & Zaritsky, A. (2008). High-resolution crystal structure of activated Cyt2Ba monomer from *Bacillus thuringiensis* subsp. *israelensis. J. Mol. Biol., 380*, 820—827.

Collier, R. H., Finch, S., & Davies, G. (2001). Pest insect control in organically-produced crops of field vegetables. *Meded. Rijksuniv. Gent Fak. Landbouwkd. Toegep. Biol. Wet., 66*, 259—267.

Cooper, M. A., Carroll, J., Travis, E. R., Williams, D. H., & Ellar, D. J. (1998). *Bacillus thuringiensis* Cry1Ac toxin interaction with *Manduca sexta* aminopeptidase N in a model membrane environment. *Biochem. J., 333*, 677—683.

Copping, L. G., & Menn, J. J. (2000). Biopesticides: a review of their action, applications and efficacy. *Pest Manag. Sci., 56*, 651—676.

Cordaux, R., Paces-Fessy, M., Raimond, M., Michel-Salzat, A., Zimmer, M., & Bouchon, D. (2007). Molecular characterization and evolution of arthropod-pathogenic *Rickettsiella* bacteria. *Appl. Environ. Microbiol., 73*, 5045—5047.

Correa, M. M., & Yousten, A. A. (2001). Pulsed-field gel electrophoresis for the identification of bacteria causing milky disease in scarab larvae. *J. Invertebr. Pathol., 78*, 278—279.

Crickmore, N., & Ellar, D. J. (1992). Involvement of a possible chaperonin in the efficient expression of a cloned CryIIA delta-endotoxin gene in *Bacillus thuringiensis*. *Mol. Microbiol., 6*, 1533−1537.

Crickmore, N., Bone, E. J., Williams, J. A., & Ellar, D. J. (1995). Contribution of the individual components of the δ-endotoxin crystal to the mosquitocidal activity of *Bacillus thuringiensis* subsp. *israelensis*. *FEMS Microbiol. Lett., 131*, 249−254.

Crickmore, N., Zeigler, D. R., Feitelson, J., Schnepf, E., Van Rie, J., Lereclus, D., Baum, J., & Dean, D. H. (1998). Revision of the nomenclature for the *Bacillus thuringiensis* pesticidal crystal proteins. *Microbiol. Mol. Biol. Rev., 62*, 807−813.

Crickmore, N., Zeigler, D. R., Schnepf, E., Van Rie, J., Lereclus, D., Baum, J., Bravo, A., & Dean, D. H. (2011). Bacillus thuringiensis *toxin nomenclature.* http://www.lifesci.sussex.ac.uk/Home/Neil_Crickmore/Bt/.

Dacheux, D., Goure, J., Chabert, J., Usson, Y., & Attree, I. (2001). Pore-forming activity of type III system-secreted proteins leads to oncosis of *Pseudomonas aeruginosa*-infected macrophages. *Mol. Microbiol., 40*, 76−85.

Daffonchio, D., Cherif, A., & Borin, S. (2000). Homoduplex and heteroduplex polymorphisms of the amplified ribosomal 16S−23S internal transcribed spacers describe genetic relationships in the *Bacillus cereus* group. *Appl. Environ. Microbiol., 66*, 5460−5468.

Dai, J. Y., Yu, L., Wang, B., Luo, X. X., Yu, Z. N., & Lecadet, M. M. (1996). *Bacillus thuringiensis* subspecies *huazhongensis*, serotype H40, isolated from soils in the People's Republic of China. *Lett. Appl. Microbiol., 22*, 42−45.

Dalhammar, G., & Steiner, H. (1984). Characterization of inhibitor A, a protease from *Bacillus thuringiensis* which degrades attacins and cecropins, two classes of antibacterial proteins in insects. *Eur. J. Biochem., 139*, 247−252.

Damgaard, P. H., Larsen, H. D., Hansen, B. M., Bresciani, J., & Jorgensen, K. (1996). Enterotoxin-producing strains of *Bacillus thuringiensis* isolated from food. *Lett. Appl. Microbiol., 23*, 146−150.

Damgaard, P. H., Hansen, B. M., Pedersen, J. C., & Eilenberg, J. (1997). Natural occurrence of *Bacillus thuringiensis* on cabbage foliage and in insects associated with cabbage crops. *J. Appl. Microbiol., 82*, 253−258.

Dang Vu, K., Tyagi, R. D., Brar, S. K., Valero, J. R., & Surampalli, R. Y. (2009). Starch industry wastewater for production of biopesticides − ramifications of solids concentrations. *Environ. Technol., 30*, 393−405.

Daniell, H. (2002). Molecular strategies for gene containment in transgenic crops. *Nat. Biotechnol., 20*, 581−586.

Davidson, E. W. (1979). Ultrastructure of midgut events in the pathogenesis of *Bacillus sphaericus* strain SSII-1 infections of *Culex pipiens quinquefasciatus* larvae. *Can. J. Microbiol., 25*, 178−184.

Davidson, E. W. (1988). Binding of *Bacillus sphaericus* (Eubacteriales: Bacillacae) toxin to midgut cells of mosquito (Diptera: Culicidae) larvae: relationship to host range. *J. Med. Entomol., 25*, 151−157.

Davidson, E. W. (1989). Variation in binding of *Bacillus sphaericus* toxin and wheat germ agglutinin to larval midgut cells of six species of mosquitoes. *J. Invertebr. Pathol., 53*, 251−259.

Davidson, E. W., & Yamamoto, T. (1984). Isolation and assay of the toxic component from the crystals of *Bacillus thuringiensis* var. *israelensis*. *Curr. Microbiol., 11*, 171−174.

Davidson, E. W., Singer, S., & Briggs, J. D. (1975). Pathogenesis of *Bacillus sphaericus* strain SSII-1 infections in *Culex pipiens quinquefasciatus* (= *C. pipiens fatigans*) larvae. *J. Invertebr. Pathol., 25*, 179−184.

Davidson, E. W., Urbina, M., Payne, J., Mulla, M. S., Darwazeh, H., Dulmage, H. T., & Correa, J. A. (1984). Fate of *Bacillus sphaericus* 1593 and 2362 spores used as larvicides in the aquatic environment. *Appl. Environ. Microbiol., 47*, 125−129.

Dean, D. H., Rajamohan, F., Lee, M. K., Wu, S. J., Chen, X. J., Alcantara, E., & Hussain, S. R. (1996). Probing the mechanism of action of *Bacillus thuringiensis* insecticidal proteins by site-directed mutagenesis. A minireview. *Gene., 179*, 111−117.

Dechklar, M., Tiewsiri, K., Angsuthanasombat, C., & Pootanakit, K. (2011). Functional expression in insect cells of glycosylphosphatidylinositol-linked alkaline phosphatase from *Aedes aegypti* larval midgut: a *Bacillus thuringiensis* Cry4Ba toxin receptor. *Insect Biochem. Mol. Biol., 41*, 159−166.

Delattre, D., Rang, C., Lecointe, N., Royer, M., Delécluse, A., Moar, W. J., & Frutos, R. (1999). Expression of orf1 from the *Bacillus thuringiensis* NRD-12 *cry2Aa1* operon. *Curr. Microbiol., 39*, 9−13.

Delello, E., Hanton, W., Bishoff, S., & Mish, D. (1984). Histopathological effects of *Bacillus thuringiensis* on the midgut of tobacco hornworm larvae (*Manduca sexta*): low doses compared with fasting. *J. Invertebr. Pathol., 43*, 169−181.

Delmas, F., & Timon-David, P. (1985). Effect of invertebrate rickettsiae on vertebrates: experimental infection of mice by *Rickettsiella grylli*. *C.R. Acad. Sci. III, 300*, 115−117.

DeLucca, A. J., II, Simonson, J. L., & Larson, A. D. (1979). Two new serovars of *Bacillus thuringiensis*: serovars *dakota* and *indiana* (serovars 15 and 16). *J. Invertebr. Pathol., 34*, 323−324.

DeLucca, A. J., II, Palmgren, M. S., & de Barjac, H. (1984). A new serovar of *Bacillus thuringiensis* from grain dusts: *Bacillus thuringiensis* var. *colmeri* (serovar H-21). *J. Invertebr. Pathol., 43*, 437−438.

Delvecchio, V. G., Connolly, J. P., Alefantis, T. G., Walz, A., Quan, M. A., Patra, G., Ashton, J. M., Whittington, J. T., Chafin, R. D., Liang, X., Grewal, P., Khan, A. S., & Mujer, C. V. (2006). Proteomic profiling and identification of immunodominant spore antigens of *Bacillus anthracis*, *Bacillus cereus*, and *Bacillus thuringiensis*. *Appl. Environ. Microbiol., 72*, 6355−6363.

Demuro, M. A., Mitchell, W. J., & Priest, F. G. (1992). Differentiation of mosquito-pathogenic strains of *Bacillus sphaericus* from nontoxic varieties by ribosomal-RNA gene restriction patterns. *J. Gen. Microbiol., 138*, 1159−1166.

Dennis, R. D., Wiegandt, H., Haustein, D., Knowles, B., & Ellar, D. J. (1986). Thin layer chromatography overlay technique in the analysis of the binding of the solubilized protoxin of *Bacillus thuringiensis* var. *kurstaki* to an insect glycosphingolipid of known structure. *Biomed. Chromatogr., 1*, 31−37.

Derbyshire, D. J., Ellar, D. J., & Li, J. (2001). Crystallization of the *Bacillus thuringiensis* toxin Cry1Ac and its complex with the receptor ligand *N*-acetyl-d-galactosamine. *Acta Crystallogr. Sect. D. Biol. Crystallogr., 57*, 1938−1944.

Dervyn, E., Poncet, S., Klier, A., & Rapoport, G. (1995). Transcriptional regulation of the cryIVD gene operon from *Bacillus thuringiensis* subsp. *israelensis*. *J. Bacteriol., 177*, 2283−2291.

Dillon, R., & Charnley, K. (2002). Mutualism between the desert locust *Schistocerca gregaria* and its gut microbiota. *Res. Microbiol., 153*, 503−509.

Ding, X., Luo, Z., Xia, L., Gao, B., Sun, Y., & Zhang, Y. (2008). Improving the insecticidal activity by expression of a recombinant *cry1Ac* gene with chitinase-encoding gene in acrystalliferous *Bacillus thuringiensis*. *Curr. Microbiol., 56*, 442–446.

Dingman, D. W. (2009). DNA fingerprinting of *Paenibacillus popilliae* and *Paenibacillus lentimorbus* using PCR-amplified 16S–23S rDNA intergenic transcribed spacer (ITS) regions. *J. Invertebr. Pathol., 100*, 16–21.

Doane, C. C. (1960). Bacterial pathogens of *Scolytus multistriatus* as related to crowding. *J. Insect Pathol., 2*, 24–29.

Dodd, S. J., Hurst, M. R. H., Glare, T. R., O'Callaghan, M., & Ronson, C. W. (2006). Occurrence of sep insecticidal toxin complex genes in *Serratia* spp. and *Yersinia frederiksenii*. *Appl. Environ. Microbiol., 72*, 6584–6592.

Dong, C., Ruan, L., Sun, M., & Yu, Z. (2007). Screening and characterization of a thuringiensin mutant from *Bacillus thuringiensis*. *Chin. J. Appl. Environ. Biol., 13*, 526–529.

Donnarumma, F., Paffetti, D., Stotzky, G., Giannini, R., & Vettori, C. (2010). Potential gene exchange between *Bacillus thuringiensis* subsp. *kurstaki* and *Bacillus* spp. in soil *in situ*. *Soil Biol. Biochem., 42*, 1329–1337.

Donovan, W. P., Rupar, M. J., Slaney, A. C., Malvar, T., Gawron-Burke, M. C., & Johnson, T. B. (1992). Characterization of two genes encoding *Bacillus thuringiensis* insecticidal crystal proteins toxic to Coleoptera species. *Appl. Environ. Microbiol., 58*, 3921–3927.

Donovan, W. P., Donovan, J. C., & Engleman, J. T. (2001). Gene knockout demonstrates that *vip3A* contributes to the pathogenesis of *Bacillus thuringiensis* toward *Agrotis ipsilon* and *Spodoptera exigua*. *J. Invertebr. Pathol., 78*, 45–51.

Driss, F., Baanannou, A., Rouis, S., Masmoudi, I., Zouari, N., & Jaoua, S. (2007). Effect of the chitin binding domain deletion from *Bacillus thuringiensis* subsp. *kurstaki* chitinase Chi255 on its stability in *Escherichia coli*. *Mol. Biotechnol., 36*, 232–237.

Driss, F., Tounsi, S., & Jaoua, S. (2011). Relationship between plasmid loss and gene expression in *Bacillus thuringiensis*. *Curr. Microbiol., 62*, 1287–1293.

Du, C., & Nickerson, K. W. (1996a). The *Bacillus thuringiensis* insecticidal toxin binds biotin-containing proteins. *Appl. Environ. Microbiol., 62*, 2932–2939.

Du, C., & Nickerson, K. W. (1996b). *Bacillus thuringiensis* HD-73 spores have surface-localized Cry1Ac toxin: physiological and pathogenic consequences. *Appl. Environ. Microbiol., 62*, 3722–3726.

Du, J., Knowles, B. H., Li, J., & Ellar, D. J. (1999). Biochemical characterization of *Bacillus thuringiensis* cytolytic toxins in association with a phospholipid bilayer. *Biochem. J., 338*, 185–193.

Dubois, N. R., & Dean, D. H. (1995). Synergism between CryIA insecticidal crystal proteins and spores of *Bacillus thuringiensis*, other bacterial spores, and vegetative cells against *Lymantria dispar* (Lepidoptera: Lymantriidae) larvae. *Environ. Entomol., 24*, 1741–1747.

Dulmage, H. T., Graham, H. M., & Martinez, E. (1978). Interactions between the tobacco budworm, *Heliothis virescens*, and the δ-endotoxin produced by the HD-1 isolate of *Bacillus thuringiensis* var. *kurstaki*: relationship between length of exposure to the toxin and survival. *J. Invertebr. Pathol., 32*, 40–50.

Dulmage, H. T., de Barjac, H., Krywienczyk, J., del var Petersen, H., Aizawa, K., Fujiyoshi, N., Ohba, M., Beegle, C. C., Needleman, D. S., Burges, H. D., Jarrett, P., Dubois, N. R., Van der

Geest, L. P., Wassink, H. J., Gingrich, R. E., Haufler, M., Allan, N., Hall, I. M., Arakawa, K. Y., Lewis, L. C., McGaughey, W. H., Dicke, E. B., & Thompson, C. G. (1981). Insecticidal activity of isolates of *Bacillus thuringiensis* and their potential for pest control. In H. D. Burges (Ed.), *Microbial Control of Pests and Plant Diseases, 1970–1980* (pp. 193–222). New York: Academic Press.

Dutky, S. R. (1940). Two new spore-forming bacteria causing milky diseases of Japanese beetle larvae. *J. Agric. Res., 61*, 57–68.

Dutky, S. R. (1963). The milky diseases. In E. A. Steinhaus (Ed.), *Insect Pathology, An Advanced Treatise*, (Vol. 2) (pp. 75–115). New York: Academic Press.

Dutky, S. R., & Gooden, E. L. (1952). *Coxiella popilliae*, n. sp., a *Rickettsia* causing blue disease of Japanese beetle larvae. *J. Bacteriol., 63*, 743–750.

Dykhuizen, D. E., & Green, L. (1991). Recombination in *Escherichia coli* and the definition of biological species. *J. Bacteriol., 173*, 7257–7268.

Edlund, T., Sidén, I., & Boman, H. G. (1976). Evidence for two immune inhibitors from *Bacillus thuringiensis* interfering with the humoral defense system of saturniid pupae. *Infect. Immun., 14*, 934–941.

Elcin, Y. M. (1995a). Control of mosquito larvae by encapsulated pathogen *Bacillus thuringiensis* var. *israelensis*. *J. Microencaps., 12*, 515–523.

Elcin, Y. M. (1995b). *Bacillus sphaericus* 2362-calcium alginate microcapsules for mosquito control. *Enzyme Microb. Technol., 17*, 587–591.

Eleftherianos, I., ffrench-Constant, R. H., Clarke, D. J., Dowling, A. J., & Reynolds, S. E. (2010). Dissecting the immune response to the entomopathogen *Photorhabdus*. *Trends Microbiol., 18*, 552–560.

Ellis, R. T., Stockhoff, B. A., Stamp, L., Schnepf, H. E., Schwab, G. E., Knuth, M., Russell, J., Cardineau, G. A., & Narva, K. E. (2002). Novel *Bacillus thuringiensis* binary insecticidal crystal proteins active on western corn rootworm, *Diabrotica virgifera virgifera* LeConte. *Appl. Environ. Microbiol., 68*, 1137–1145.

Endersby, N. M., Morgan, W. C., Stevenson, B. C., & Waters, C. T. (1992). Alternatives to regular insecticide applications for control of lepidopterous pests of *Brassica oleracea* var. *capitata*. *Biol. Agric. Hortic., 8*, 189–203.

English, L., & Readdy, T. L. (1989). Delta endotoxin inhibits a phosphatase in midgut epithelial membranes of *Heliothis virescens*. *Insect Biochem., 19*, 145–152.

Enigl, M., & Schausberger, P. (2007). Incidence of the endosymbionts *Wolbachia*, *Cardinium* and *Spiroplasma* in phytoseiid mites and associated prey. *Exp. Appl. Acarol., 42*, 75–85.

Espinasse, S., Gohar, M., Chaufaux, J., Buisson, C., Perchat, S., & Sanchis, V. (2002). Correspondence of high levels of beta-exotoxin I and the presence of Cry1B in *Bacillus thuringiensis*. *Appl. Environ. Microbiol., 68*, 4182–4186.

Estruch, J. J., Warren, G. W., Mullins, M. A., Nye, G. J., Craig, J. A., & Koziel, M. G. (1996). Vip3A, a novel *Bacillus thuringiensis* vegetative insecticidal protein with a wide spectrum of activities against lepidopteran insects. *Proc. Natl. Acad. Sci. USA, 93*, 5389–5394.

Euzéby, J. P. (1997). List of bacterial names with standing in nomenclature: a folder available on the Internet. *Int. J. Syst. Bacteriol., 47*, 590–592, Available online as List of Prokaryotic Names with Standing in Nomenclature at http://www.bacterio.net.

Fabrick, J. A., & Tabashnik, B. E. (2007). Binding of *Bacillus thuringiensis* toxin Cry1Ac to multiple sites of cadherin in pink bollworm. *Insect Biochem. Mol. Biol., 37*, 97–106.

Fabrick, J., Oppert, C., Lorenzen, M. D., Morris, K., Oppert, B., & Jurat-Fuentes, J. L. (2009). A novel *Tenebrio molitor* cadherin is a functional receptor for *Bacillus thuringiensis* Cry3Aa toxin. *J. Biol. Chem., 284*, 18401–18410.

Farrar, R. R., Jr., Martin, P. A. W., & Ridgway, R. L. (2001). A strain of *Serratia marcescens* (Enterobacteriaceae) with high virulence *per os* to larvae of a laboratory colony of the corn earworm (Lepidoptera: Noctuidae). *J. Entomol. Sci., 36*, 380–390.

Farrar, R. R., Jr., Shapiro, M., & Shepard, B. M. (2004). Activity of the nucleopolyhedrovirus of the fall armyworm (Lepidoptera: Noctuidae) on foliage of transgenic sweet corn expressing a CryIA(b) toxin. *Environ. Entomol., 33*, 982–989.

Fast, P. G., & Angus, T. A. (1965). Effects of parasporal inclusions of *Bacillus thuirngiensis* var. *sotto* Ishiwata on the permeability of the gut wall of *Bombyx mori* (Linnaeus) larvae. *J. Invertebr. Pathol., 7*, 29–32.

Favret, M. E., & Yousten, A. A. (1985). Insecticidal activity of *Bacillus laterosporus*. *J. Invertebr. Pathol., 45*, 195–203.

Federici, B. A. (1980). Reproduction and morphogenesis of *Rickettsiella chironomi*, an unusual intracellular procaryotic parasite of midge larvae. *J. Bacteriol., 143*, 995–1002.

Federici, B. A. (2005). Insecticidal bacteria: an overwhelming success for invertebrate pathology. *J. Invertebr. Pathol., 89*, 30–38.

Federici, B. A., & Bauer, L. S. (1998). Cyt1Aa protein of *Bacillus thuringiensis* is toxic to the cottonwood leaf beetle, *Chrysomela scripta*, and suppresses high levels of resistance to Cry3Aa. *Appl. Environ. Microbiol., 64*, 4368–4371.

Fedhila, S., Gohar, M., Slamti, L., Nel, P., & Lereclus, D. (2003). The *Bacillus thuringiensis* PlcR-regulated gene *inhA2* is necessary, but not sufficient, for virulence. *J. Bacteriol., 185*, 2820–2825.

Fedhila, S., Guillemet, E., Nel, P., & Lereclus, D. (2004). Characterization of two *Bacillus thuringiensis* genes identified by *in vivo* screening of virulence factors. *Appl. Environ. Microbiol., 70*, 4784–4791.

Felber, F., Kozlowski, G., Arrigo, N., & Guadagnuolo, R. (2007). Genetic and ecological consequences of transgene flow to the wild flora. *Adv. Biochem. Eng. Biotechnol., 107*, 173–205.

Fernández, L. E., Pérez, C., Segovia, L., Rodríguez, M. H., Gill, S. S., Bravo, A., & Soberón, M. (2005). Cry11Aa toxin from *Bacillus thuringiensis* binds its receptor in *Aedes aegypti* mosquito larvae through loop α-8 of domain II. *FEBS Lett., 579*, 3508–3514.

Fernandez, L. E., Aimanova, K. G., Gill, S. S., Bravo, A., & Soberón, M. (2006). A GPI-anchored alkaline phosphatase is a functional midgut receptor of Cry11Aa toxin in *Aedes aegypti* larvae. *Biochem. J., 394*, 77–84.

Fernandez, L. E., Martinez-Anaya, C., Lira, E., Chen, J., Evans, A., Hernández-Martínez, S., Lanz-Mendoza, H., Bravo, A., Gill, S. S., & Soberón, M. (2009). Cloning and epitope mapping of Cry11Aa-binding sites in the Cry11Aa-receptor alkaline phosphatase from *Aedes aegypti*. *Biochemistry, 48*, 8899–8907.

Fernandez-Luna, M. T., Lanz-Mendoza, H., Gill, S. S., Bravo, A., Soberon, M., & Miranda-Rios, J. (2010). An α-amylase is a novel receptor for *Bacillus thuringiensis* ssp. *israelensis* Cry4Ba and Cry11Aa toxins in the malaria vector mosquito *Anopheles albimanus* (Diptera: Culicidae). *Environ. Microbiol., 12*, 746–757.

Fernández-Ruvalcaba, M., Peña-Chora, G., Romo-Martínez, A., Hernández-Velázquez, V., de La Parra, A. B., & De La Rosa, D. P. (2010). Evaluation of *Bacillus thuringiensis* pathogenicity for a strain of the tick, *Rhipicephalus microplus*, resistant to chemical pesticides. *J. Insect Sci., 10*, 186.

Ferrandis, M. D., Andrew, R., Porcar, M., Iriarte, J., Cosmao-Dumanoir, V., Lecadet, M. M., Caballero, P., & Ferré, J. (1999). Characterization of *Bacillus thuringiensis* serovar *bolivia* (serotype H63), a novel serovar isolated from the Bolivian high valleys. *Lett. Appl. Microbiol., 28*, 440–444.

Ferreira, L. M., Romão, T. P., de-Melo-Neto, O. P., & Silva-Filha, M. H. (2010). The orthologue to the Cpm1/Cqm1 receptor in *Aedes aegypti* is expressed as a midgut GPI-anchored α-glucosidase, which does not bind to the insecticidal binary toxin. *Insect Biochem. Mol. Biol., 40*, 604–610.

ffrench-Constant, R., & Waterfield, N. (2006). An ABC guide to the bacterial toxin complexes. *Adv. Appl. Microbiol., 58*, 169–183.

Flanders, S. E., & Hall, I. M. (1965). Manipulated bacterial epizootics in *Anagasta* populations. *J. Invertebr. Pathol., 7*, 368–377.

Flannagan, R. D., Yu, C. G., Mathis, J. P., Meyer, T. E., Shi, X., Siqueira, H. A., & Siegfried, B. D. (2005). Identification, cloning and expression of a Cry1Ab cadherin receptor from European corn borer, *Ostrinia nubilalis* (Hubner) (Lepidoptera: Crambidae). *Insect Biochem. Mol. Biol., 35*, 33–40.

Flores, A. E., Garcia, G. P., Badii, M. H., Rodriguez Tovar, M. A., & Fernnadez Salas, I. (2004). Effects of sublethal concentrations of Vectobac® on biological parameters of *Aedes aegypti*. *J. Am. Mosq. Control Assoc., 20*, 412–417.

Flores, H., Soberón, X., Sánchez, J., & Bravo, A. (1997). Isolated domain II and III from the *Bacillus thuringiensis* Cry1Ab delta-endotoxin binds to lepidopteran midgut membranes. *FEBS Lett., 414*, 313–318.

Fluer, F. S., Ivinskene, V. L., & Zaiachkauskas, P. A. (1981). Detection of thermolabile exotoxin in *B. thuringiensis* and its separation from phospholipase C. *Zh. Mikrobiol. Epidemiol. Immunobiol., 8*, 81–85.

Folcher, L., Delos, M., Marengue, E., Jarry, M., Weissenberger, A., Eychenne, N., & Regnault-Roger, C. (2010). Lower mycotoxin levels in Bt maize grain. *Agron. Sustain. Dev., 30*, 711–719.

Fortier, M., Vachon, V., Marceau, L., Schwartz, J. L., & Laprade, R. (2007). Kinetics of pore formation by the *Bacillus thuringiensis* toxin Cry1Ac. *Biochim. Biophys. Acta, 1768*, 1291–1298.

Foster, S. J., & Johnstone, K. (1990). Pulling the trigger: the mechanism of bacterial spore germination. *Mol. Microbiol., 4*, 137–141.

Franken, E., Krieger, L., & Schnetter, W. (1996). Bacillus popilliae: a difficult pathogen. *Bull. OILB/SROP, 19*, 40–45.

van Frankenhuyzen, K. (2009). Insecticidal activity of *Bacillus thuringiensis* crystal proteins. *J. Invertebr. Pathol., 101*, 1–16.

van Frankenhuyzen, K., & Nystrom, C. (2011). *The* Bacillus thuringiensis *toxin specificity database*. http://cfs.nrcan.gc.ca/subsite/glfc-bacillus-thuringiensis/bacillus-thuringiensis.

van Frankenhuyzen, K., Liu, Y., & Tonon, A. (2010). Interactions between *Bacillus thuringiensis* subsp. *kurstaki* HD-1 and midgut bacteria in larvae of gypsy moth and spruce budworm. *J. Invertebr. Pathol., 103*, 124–131.

Frederiksen, K., Rosenquist, H., Jorgensen, K., & Wilcks, A. (2006). Occurrence of natural *Bacillus thuringiensis* contaminants and residues of *Bacillus thuringiensis*-based insecticides on fresh fruits and vegetables. *Appl. Environ. Microbiol., 72*, 3435–3440.

Fridlender, B., Keren-Zur, M., Hofstein, R., Bar, E., Sandler, N., Keynan, A., & Braun, S. (1989). The development of *Bacillus thuringiensis* and *Bacillus sphaericus* as biocontrol agents: from research to industrial production. *Mem. Inst. Oswaldo Cruz., 84*, 123–127.

From, C., Granum, P. E., & Hardy, S. P. (2008). Demonstration of a cholesterol-dependent cytolysin in a noninsecticidal *Bacillus sphaericus* strain and evidence for widespread distribution of the toxin within the species. *FEMS Microbiol. Lett., 286*, 85−92.

Fu, Q. L., Deng, Y. L., Li, H. S., Liu, J., Hu, H. Q., Chen, S. W., & Sa, T. M. (2009). Equilibrium, kinetic and thermodynamic studies on the adsorption of the toxins of *Bacillus thuringiensis* subsp. *kurstaki* by clay minerals. *Appl. Surf. Sci., 255*, 4551−4557.

Fuchs, T. M., Bresolin, G., Marcinowski, L., Schachtner, J., & Scherer, S. (2008). Insecticidal genes of *Yersinia* spp.: taxonomical distribution, contribution to toxicity towards *Manduca sexta* and *Galleria mellonella*, and evolution. *BMC Microbiol., 8*, 214.

Furlaneto, L., Saridakis, H. O., & Arantes, O. M. N. (2000). Survival and conjugal transfer between *Bacillus thuringiensis* strains in aquatic environment. *Braz. J. Microbiol., 31*, 233−238.

Gaertner, F. H., Quick, T. C., & Thompson, M. A. (1993). CellCap; an encapsulation system for insecticidal biotoxin proteins. In L. Kim (Ed.), *Advanced Engineered Pesticides* (pp. 73−83). New York: Marcel Dekker.

Gahan, L. J., Gould, F., & Heckel, D. G. (2001). Identification of a gene associated with Bt resistance in *Heliothis virescens*. *Science, 293*, 857−860.

Gahan, L. J., Pauchet, Y., Vogel, H., & Heckel, D. G. (2010). An ABC transporter mutation is correlated with insect resistance to *Bacillus thuringiensis* Cry1Ac toxin. *PLoS Genet., 6*, e1001248.

Galitsky, N., Cody, V., Wojtczak, A., Ghosh, D., Luft, J. R., Pangborn, W., & English, L. (2001). Structure of the insecticidal bacterial δ-endotoxin Cry3Bb1 of *Bacillus thuringiensis*. *Acta Crystallogr. Sect. D Biol. Crystallogr., 57*, 1101−1109.

Gamel, P. H., & Piot, J. C. (1992). Characterization and properties of a novel plasmid vector for *Bacillus thuringiensis* displaying compatibility with host plasmids. *Gene., 120*, 17−26.

Gao, Y., Jurat-Fuentes, J. L., Oppert, B., Fabrick, J. A., Liu, C., Gao, J., & Lei, Z. (2011). Increased toxicity of *Bacillus thuringiensis* Cry3Aa against *Crioceris quatuordecimpunctata*, *Phaedon brassicae* and *Colaphellus bowringi* by a *Tenebrio molitor* cadherin fragment. *Pest Manag. Sci., 67*, 1076−1081.

Garcia-Robles, I., Sánchez, J., Gruppe, A., Martínez-Ramírez, A. C., Rausell, C., Real, M. D., & Bravo, A. (2001). Mode of action of *Bacillus thuringiensis* PS86Q3 strain in hymenopteran forest pests. *Insect Biochem. Mol. Biol., 31*, 849−856.

Garczynski, S. F., & Adang, M. J. (1995). *Bacillus thuringiensis* CryIA(c) δ-endotoxin Garnier, M., Foissac, X., Gaurivaud, P., Laigret, F., Renaudin, J., Saillard, C., and Bové, J.M. (2001). Mycoplasmas, plants, insect vectors: a matrimonial triangle. *C.R. Acad. Sci. III, 324*, 923−928.

Garczynski, S. F., Crim, J. W., & Adang, M. J. (1991). Identification of putative insect brush border membrane-binding molecules specific to *Bacillus thuringiensis* delta-endotoxin by protein blot analysis. *Appl. Environ. Microbiol., 57*, 2816−2820.

Gatehouse, H. S., Marshall, S. D., Simpson, R. M., Gatehouse, L. N., Jackson, T. A., & Christeller, J. T. (2008). *Serratia entomophila* inoculation causes a defect in exocytosis in *Costelytra zealandica* larvae. *Insect. Mol. Biol., 17*, 375−385.

Gaviria Rivera, A. M., & Priest, F. G. (2003). Pulsed field gel electrophoresis of chromosomal DNA reveals a clonal population structure to *Bacillus thuringiensis* that relates in general to crystal protein gene content. *FEMS Microbiol. Lett., 223*, 61−66.

Gayathri, V., Jeyalakshmi, T., Shanmugasundaram, R., & Murthy, P. B. (2004). Rotational application of bioinsecticide with deltamethrin − an antilarval measure for the control of filarial vector, *Culex quinquefasciatus* (Culicidae: Diptera). *J. Environ. Biol., 25*, 419−421.

Ge, A. Z., Rivers, D., Milne, R., & Dean, D. H. (1991). Functional domains of *Bacillus thuringiensis* insecticidal crystal proteins. Refinement of *Heliothis virescens* and *Trichoplusia ni* specificity domains on CryIA(c). *J. Biol. Chem., 266*, 17954−17958.

Ge, B., Bideshi, D., Moar, W. J., & Federici, B. A. (1998). Differential effects of helper proteins encoded by the cry2A and cry11A operons on the formation of Cry2A inclusions in *Bacillus thuringiensis*. *FEMS Microbiol. Lett., 165*, 35−41.

Gelernter, W. (2004). The rise and fall of *Bacillus thuringiensis tenebrionis*. *Phytoparasitica, 32*, 321−324.

Genersch, E., Forsgren, E., Pentikäinen, J., Ashiralieva, A., Rauch, S., Kilwinski, J., & Fries, I. (2006). Reclassification of *Paenibacillus larvae* subsp. *pulvifaciens* and *Paenibacillus larvae* subsp. *larvae* as *Paenibacillus larvae* without subspecies differentiation. *Int. J. Syst. Evol. Microbiol., 56*, 501−511.

Gerber, D., & Shai, Y. (2000). Insertion and organization within membranes of the δ-endotoxin pore-forming domain, helix 4-loop-helix 5, and inhibition of its activity by a mutant helix 4 peptide. *J. Biol. Chem., 275*, 23602−23607.

Giannakou, I. O., Karpouzas, D. G., & Prophetou-Athanasiadou, D. (2004). A novel non-chemical nematicide for the control of root-knot nematodes. *Appl Soil Ecol., 26*, 69−79.

Gill, S. S., Singh, G. J., & Hornung, J. M. (1987). Cell membrane interaction of *Bacillus thuringiensis* subsp. *israelensis* cytolytic toxins. *Infect. Immun., 55*, 1300−1308.

Gilliam, M. (1985). Microbes from apiarian sources: *Bacillus* spp. in frass of the greater wax moth. *J. Invertebr. Pathol., 45*, 218−224.

Gilliam, M. (1997). Identification and roles of non-pathogenic microflora associated with honey bees. *FEMS Microbiol. Lett., 155*, 1−10.

Gillis, M., & De Ley, J. (2006). The genera *Chromobacterium* and *Janthinobacterium*. In M. Dworkin, S. Falkow, E. Rosenberg, K. H. Schleifer & E. Stackebrandt (Eds.). (3rd ed.). *The Prokaryotes*, (Vol. 5) (pp. 737−746) New York: Springer.

Glancy, B. M., VanderMeer, R. K., Glover, A., Lofgren, C. S., & Vinson, S. B. (1981). Filtration of microparticles from liquids ingested by the red imported fire ant *Solenopsis invicta* Buren. *Insect. Soc., 28*, 395−401.

Glare, T. R., & O'Callaghan, M. (2000). Bacillus thuringiensis*: Biology, Ecology and Safety*. Chichester: John Wiley & Sons.

Glare, T. R., Corbett, G. E., & Sadler, T. J. (1993). Association of a large plasmid with amber disease of the New Zealand grass grub, *Costelytra zealandic a*, caused by *Serratia entomophila* and *Serratia proteamaculans*. *J. Invertebr. Pathol., 62*, 165−170.

Glinsky, Z. F., & Jarosz, J. (1992). *Varroa jacobsoni* as a carrier of bacterial infections to a recipient bee host. *Apidologie, 23*, 25−31.

Gohar, M., Gilois, N., Graveline, R., Garreau, C., Sanchis, V., & Lereclus, D. (2005). A comparative study of *Bacillus cereus, Bacillus thuringiensis* and *Bacillus anthracis* extracellular proteomes. *Proteomics, 5*, 3696−3711.

Gohar, M., Faegri, K., Perchat, S., Ravnum, S., Økstad, O. A., Gominet, M., Kolstø, A. B., & Lereclus, D. (2008). The PlcR virulence regulon of *Bacillus cereus*. *PLoS ONE, 3*, e2793.

Goldberg, L. J., & Margalit, J. (1977). A bacterial spore demonstraitng rapid larvicidal activity against *Anopheles sergentii, Uranotaenia unguicalata, Culex univitattus, Aedes aegyptii* and *Culex pipiens*. *Mosq. News, 37*, 355−358.

Gómez, I., Sánchez, J., Miranda, R., Bravo, A., & Soberón, M. (2002). Cadherin-like receptor binding facilitates proteolytic cleavage of helix alpha-1 in domain I and oligomer pre-pore formation of *Bacillus thuringiensis* Cry1Ab toxin. *FEBS Lett., 513*, 242−246.

Gómez, I., Arenas, I., Benitez, I., Miranda-Ríos, J., Becerril, B., Grande, R., Almagro, J. C., Bravo, A., & Soberón, M. (2006). Specific epitopes of domains II and III of *Bacillus thuringiensis* Cry1Ab toxin involved in the sequential interaction with cadherin and aminopeptidase-N receptors in *Manduca sexta. J. Biol. Chem., 281*, 34032−34039.

González, J. M. J., & Carlton, B. C. (1980). Patterns of plasmid DNA in crystalliferous and acrystalliferous strains of *Bacillus thuringiensis. Plasmid, 3*, 92−98.

González, J. M. J., Dulmage, H. T., & Carlton, B. C. (1981). Correlation between specific plasmids and delta-endotoxin production in *Bacillus thuringiensis. Plasmid, 5*, 351−365.

González, J. M. J., Brown, B. J., & Carlton, B. C. (1982). Transfer of *Bacillus thuringiensis* plasmids coding for delta-endotoxin among strains of *Bacillus thuringiensis* and *Bacillus cereus. Proc. Natl. Acad. Sci. USA., 79*, 6951−6955.

Goryluk, A., Rekosz-Burlaga, H., & Błaszczyk, M. (2009). Isolation and characterization of bacterial endophytes of *Chelidonium majus* L. *Pol. J. Microbiol., 58*, 355−361.

Götz, P. (1972). *Rickettsiella chironomi*: an unusual bacterial pathogen which reproduces by multiple cell division. *J. Invertebr. Pathol., 20*, 22−30.

Granados, R. R., Fu, Y., Corsaro, B., & Hughes, P. R. (2001). Enhancement of *Bacillus thuringiensis* toxicity to lepidopterous species with the enhancin from *Trichoplusia ni* granulovirus. *Biol. Control, 20*, 153−159.

Grandvalet, C., Gominet, M., & Lereclus, D. (2001). Identification of genes involved in the activation of the *Bacillus thuringiensis* inhA metalloprotease gene at the onset of sporulation. *Microbiology, 147*, 1805−1813.

Greany, P. D., Allen, G. E., Webb, J. C., Sharp, J. L., & Chambers, D. L. (1977). Stress-induced septicemia as an impediment to laboratory rearing of the fruit fly parasitoid *Biosteres (opius) longicaudatus* (Hymenoptera: Braconidae) and the Caribbean fruit fly *Anastrepha suspensa* (Diptera: Tephritidae). *J. Invertebr. Pathol., 29*, 153−161.

Griffitts, J. S., Whitacre, J. L., Stevens, D. E., & Aroian, R. V. (2001). Bt toxin resistance from loss of a putative carbohydrate-modifying enzyme. *Science, 293*, 860−864.

Griffitts, J. S., Haslam, S. M., Yang, T., Garczynski, S. F., Mulloy, B., Morris, H., Cremer, P. S., Dell, A., Adang, M. J., & Aroian, R. V. (2005). Glycolipids as receptors for *Bacillus thuringiensis* crystal toxin. *Science, 307*, 922−925.

Grimont, P. A., & Grimont, F. (1978). The genus. *Serratia. Annu. Rev. Microbiol., 32*, 221−248.

Grimont, F., & Grimont, P. A. D. (2006). The genus *Serratia*. In M. Dworkin, S. Falkow, E. Rosenberg, K. H. Schleifer & E. Stackebrandt (Eds.), (3rd ed.). *The Prokaryotes*, (Vol. 6) (pp. 219−244) New York: Springer.

Grimont, P. A. D., Grimont, F., & Lysenko, O. (1979). Species and biotype identification of *Serratia* strains associated with insects. *Curr. Microbiol., 2*, 139−142.

Grimont, P. A. D., Jackson, T. A., Ageron, E., & Noonan, M. J. (1988). *Serratia entomophila* sp. nov. associated with amber disease in the New Zealand grass grub *Costelytra zealandica. Int. J. Syst. Bacteriol., 38*, 1−6.

Grkovic, S., Glare, T. R., Jackson, T. A., & Corbett, G. E. (1995). Genes essential for amber disease in grass grubs are located on the large plasmid found in *Serratia entomophila* and *Serratia proteamaculans. Appl. Environ. Microbiol., 61*, 2218−2223.

Grochulski, P., Masson, L., Borisova, S., Pusztai-Carey, M., Schwartz, J. L., Brousseau, R., & Cygler, M. (1995). *Bacillus thuringiensis* CryIA(a) insecticidal toxin: crystal structure and channel formation. *J. Mol. Biol., 254*, 447−464.

Groulx, N., Juteau, M., & Blunck, R. (2010). Rapid topology probing using fluorescence spectroscopy in planar lipid bilayer: the pore-forming mechanism of the toxin Cry1Aa of *Bacillus thuringiensis. J. Gen. Physiol., 136*, 497−513.

Guerchicoff, A., Ugalde, R. A., & Rubinstein, C. P. (1997). Identification and characterization of a previously undescribed *cyt* gene in *Bacillus thuringiensis* subsp. *israelensis. Appl. Environ. Microbiol., 63*, 2716−2721.

Güereca, L., & Bravo, A. (1999). The oligomeric state of *Bacillus thuringiensis* Cry toxins in solution. *Biochim. Biophys. Acta, 1429*, 342−350.

Guerineau, M., Alexander, B., & Priest, F. G. (1991). Isolation and Identification of *Bacillus sphaericus* strains pathogenic for mosquito larvae. *J. Invertebr. Pathol., 57*, 325−333.

Guillemet, E., Cadot, C., Tran, S. L., Guinebretière, M. H., Lereclus, D., & Ramarao, N. (2010). The InhA metalloproteases of *Bacillus cereus* contribute concomitantly to virulence. *J. Bacteriol., 192*, 286−294.

Guo, S., Ye, S., Liu, Y., Wei, L., Xue, J., Wu, H., Song, F., Zhang, J., Wu, X., Huang, D., & Rao, Z. (2009a). Crystal structure of *Bacillus thuringiensis* Cry8Ea1: an insecticidal toxin toxic to underground pests, the larvae of *Holotrichia parallela. J. Struct. Biol., 168*, 259−266.

Guo, S., Zhang, Y., Song, F., Zhang, J., & Huang, D. (2009b). Protease-resistant core form of *Bacillus thuringiensis* Cry1Ie: monomeric and oligomeric forms in solution. *Biotechnol. Lett., 31*, 1769−1774.

Guo, S., Li, J., Liu, Y., Song, F., & Zhang, J. (2011). The role of DNA binding with the Cry8Ea1 toxin of *Bacillus thuringiensis. FEMS Microbiol. Lett., 317*, 203−210.

Guttmann, D. M., & Ellar, D. J. (2000). Phenotypic and genotypic comparisons of 23 strains from the *Bacillus cereus* complex for a selection of known and putative *B. thuringiensis* virulence factors. *FEMS Microbiol. Lett., 188*, 7−13.

Haider, M. Z., & Ellar, D. J. (1987). Analysis of the molecular basis of insecticidal specificity of *Bacillus thuringiensis* crystal delta-endotoxin. *Biochem. J., 248*, 197−201.

Hallet, B., Rezsèohazy, R., & Delcour, J. (1991). IS231A from *Bacillus thuringiensis* is functional in *Escherichia coli*: transposition and insertion specificity. *J. Bacteriol., 173*, 4526−4529.

Halpern, M., Landsberg, O., Raats, D., & Rosenberg, E. (2007). Culturable and VBNC *Vibrio cholerae*: interactions with chironomid egg masses and their bacterial population. *Microb. Ecol., 53*, 285−293.

Han, S., Craig, J. A., Putnam, C. D., Carozzi, N. B., & Tainer, J. A. (1999). Evolution and mechanism from structures of an ADP-ribosylating toxin and NAD complex. *Nat. Struct. Biol., 6*, 932−936.

Hannay, C. L., & Fitz-James, P. (1955). The protein crystals of *Bacillus thuringiensis* Berliner. *Can. J. Microbiol., 1*, 694−710.

Hansen, B. M., & Salamitou, S. (2000). Virulence of *Bacillus thuringiensis*. In J. F. Charles, A. Delécluse & C. Nielsen-LeRoux (Eds.),

Entomopathogenic Bacteria: from Laboratory to Field Application (pp. 41–64). Dordrecht: Kluwer Academic Publishers.

Hansen, B. M., Damgaard, P. H., Eilenberg, J., & Pedersen, J. C. (1998). Molecular and phenotypic characterization of *Bacillus thuringiensis* isolated from leaves and insects. *J. Invertebr. Pathol., 71,* 106–114.

Harrison, H., Patel, R., & Yousten, A. A. (2000). *Paenibacillus* associated with milky disease in central and South American scarabs. *J. Invertebr. Pathol., 76,* 169–175.

Hartman, E. (1931). A flacherie disease of silkworm caused by *Bacillus bombysepticus. Lignan Sci. J., 10,* 279–289.

Harvie, D. R., & Ellar, D. J. (2005). A ferric dicitrate uptake system is required for the full virulence of *Bacillus cereus. Curr. Microbiol., 50,* 246–250.

Hayakawa, T., Shitomi, Y., Miyamoto, K., & Hori, H. (2004). GalNAc pretreatment inhibits trapping of *Bacillus thuringiensis* Cry1Ac on the peritrophic membrane of *Bombyx mori. FEBS Lett., 576,* 331–335.

Hazes, B., & Read, R. J. (1995). A mosquitocidal toxin with a ricin-like cell-binding domain. *Nat. Struct. Biol., 2,* 358–359.

Heimpel, A. M. (1967). A critical review of *Bacillus thuringiensis* var. *thuringiensis* Berliner and other crystalliferous bacteria. *Annu. Rev. Entomol., 12,* 287–322.

Heimpel, A. M., & Angus, T. A. (1958). The taxonomy of insect pathogens related to *Bacillus cereus* Frankland and Frankland. *Can. J. Microbiol., 4,* 531–541.

Heimpel, A. M., & Angus, T. A. (1959). The site of action of crystalliferous bacteria in Lepidoptera larvae. *J. Insect Pathol., 1,* 152–170.

Held, G. A., Bulla, L. A., Jr., Ferrari, E., Hoch, J., Aronson, A. I., & Minnich, S. A. (1982). Cloning and localization of the lepidopteran protoxin gene of *Bacillus thuringiensis* subsp. *kurstaki. Proc. Natl. Acad. Sci. USA, 79,* 6065–6069.

Helgason, E., Økstad, O. A., Caugant, D. A., Johansen, H. A., Fouet, A., Mock, M., Hegna, I., & Kolstø, A. B. (2000). *Bacillus anthracis, Bacillus cereus,* and *Bacillus thuringiensis* — one species on the basis of genetic evidence. *Appl. Environ. Microbiol., 66,* 2627–2630.

d'Herelle, F. (1911). Sur une épizootie de nature bacterienne sevissant sur les sauterelles au Mexique. *C.R. Seances Acad. Sci. D., 152,* 1413–1415.

d'Herelle, F. (1912). Sur la propagation, dans la Republique Argentine, de l'epizootie des sauterelles du Mexique. *C.R. Seances Acad. Sci. D., 154,* 623–625.

Herman, R. A., Scherer, P. N., Young, D. L., Mihaliak, C. A., Meade, T., Woodsworth, A. T., Stockhoff, B. A., & Narva, K. E. (2002). Binary insecticidal crystal protein from *Bacillus thuringiensis,* strain PS149B1: effects of individual protein components and mixtures in laboratory bioassays. *J. Econ. Entomol., 95,* 635–639.

Hernández, C. S., & Ferré, J. (2005). Common receptor for *Bacillus thuringiensis* toxins Cry1Ac, Cry1Fa, and Cry1Ja in *Helicoverpa armigera, Helicoverpa zea,* and *Spodoptera exigua. Appl. Environ. Microbiol., 71,* 5627–5629.

Hernández, C. S., Ferré, J., & Larget-Thiéry, I. (2001). Update on the detection of β-exotoxin in *Bacillus thuringiensis* strains by HPLC analysis. *J. Appl. Microbiol., 90,* 643–647.

Hernández, C. S., Martínez, C., Porcar, M., Caballero, P., & Ferré, J. (2003). Correlation between serovars of *Bacillus thuringiensis* and type I β-exotoxin production. *J. Invertebr. Pathol., 82,* 57–62.

Hernandez, J. L. L. (1988). Évaluation de la toxicité de *Bacillus thuringiensis* sur *Spodoptera frugiperda. Entomophaga, 33,* 163–171.

Hernández-Martínez, P., Naseri, B., Navarro-Cerrillo, G., Escriche, B., Ferré, J., & Herrero, S. (2010). Increase in midgut microbiota load induces an apparent immune priming and increases tolerance to *Bacillus thuringiensis. Environ. Microbiol., 12,* 2730–2737.

Herrero, S., Gechev, T., Bakker, P. L., Moar, W. J., & de Maagd, R. A. (2005). *Bacillus thuringiensis* Cry1Ca-resistant *Spodoptera exigua* lacks expression of one of four aminopeptidase N genes. *BMC Genomics., 6,* 96.

Heuberger, S., Ellers-Kirk, C., Tabashnik, B. E., & Carrière, Y. (2010). Pollen- and seed-mediated transgene flow in commercial cotton seed production fields. *PLoS ONE, 5,* e14128.

Hodgman, T. C., & Ellar, D. J. (1990). Models for the structure and function of the *Bacillus thuringiensis* δ-endotoxins determined by compilational analysis. *J. DNA Seq. Map., 1,* 97–106.

Hodgman, T. C., Ziniu, Y., Shen, J., & Ellar, D. J. (1993). Identification of a cryptic gene associated with an insertion sequence not previously identified in *Bacillus thuringiensis. FEMS Microbiol. Lett., 114,* 23–30.

Höfte, H., & Whitely, H. R. (1989). Insecticidal crystal proteins of *Bacillus thuringiensis. Microbiol. Rev., 53,* 242–255.

Holt, J. G. (Ed.). (1994). *Bergey's Manual of Determinative Bacteriology* (9th ed.). Baltimore: Williams & Wilkins.

Hossain, D. M., Shitomi, Y., Moriyama, K., Higuchi, M., Hayakawa, T., Mitsui, T., Sato, R., & Hori, H. (2004). Characterization of a novel plasma membrane protein, expressed in the midgut epithelia of *Bombyx mori,* that binds to Cry1A toxins. *Appl. Environ. Microbiol., 70,* 4604–4612.

Howlader, M. T., Kagawa, Y., Sakai, H., & Hayakawa, T. (2009). Biological properties of loop-replaced mutants of *Bacillus thuringiensis* mosquitocidal Cry4Aa. *J. Biosci. Bioeng., 108,* 179–183.

Hu, H. Y., Liu, X. X., Zhao, Z. W., Sun, J. G., Zhang, Q. W., Liu, X. Z., & Yu, Y. (2009). Effects of repeated cultivation of transgenic Bt cotton on functional bacterial populations in rhizosphere soil. *World J. Microbiol. Biotechnol., 25,* 357–366.

Hu, X., Hansen, B. M., Eilenberg, J., Hendriksen, N. B., Smidt, L., Yuan, Z., & Jensen, G. B. (2004). Conjugative transfer, stability and expression of a plasmid encoding a *cry1Ac* gene in *Bacillus cereus* group strains. *FEMS Microbiol. Lett., 231,* 45–52.

Hu, X., Hansen, B. M., Yuan, Z., Johansen, J. E., Eilenberg, J., Hendriksen, N. B., Smidt, L., & Jensen, G. B. (2005). Transfer and expression of the mosquitocidal plasmid pBtoxis in *Bacillus cereus* group strains. *FEMS Microbiol. Lett., 245,* 239–247.

Hu, X., Fan, W., Han, B., Liu, H., Zheng, D., Li, Q., Dong, W., Yan, J., Gao, M., Berry, C., & Yuan, Z. (2008). Complete genome sequence of the mosquitocidal bacterium *Bacillus sphaericus* C3-41 and comparison with those of closely related *Bacillus* species. *J. Bacteriol., 190,* 2892–2902.

Hua, G., Jurat-Fuentes, J. L., & Adang, M. J. (2004). Fluorescent-based assays establish *Manduca sexta* Bt-R_{1a} cadherin as a receptor for multiple *Bacillus thuringiensis* Cry1A toxins in *Drosophila* S2 cells. *Insect Biochem. Mol. Biol., 34,* 193–202.

Hua, G., Zhang, R., Abdullah, M. A., & Adang, M. J. (2008). *Anopheles gambiae* cadherin AgCad1 binds the Cry4Ba toxin of *Bacillus thuringiensis israelensis* and a fragment of AgCad1 synergizes toxicity. *Biochemistry, 47,* 5101–5110.

Hua, G., Zhang, R., Bayyareddy, K., & Adang, M. J. (2009). *Anopheles gambiae* alkaline phosphatase is a functional receptor of *Bacillus thuringiensis jegathesan* Cry11Ba toxin. *Biochemistry, 48,* 9785–9793.

Huang, L., Cheng, T., Xu, P., Cheng, D., Fang, T., & Xia, Q. (2009). A genome-wide survey for host response of silkworm, *Bombyx mori* during pathogen *Bacillus bombysepticus* infection. *PLoS ONE, 4,* e8098.

Hucko, M. (1984). The role of the house fly (*Musca domestica* L.) in the transmission of. *Coxiella burnetii. Folia Parasitol., 31,* 177–181.

Huger, A. M., & Krieg, A. (1967). New aspects of the mode of reproduction of *Rickettsiella* organisms in insects. *J. Invertebr. Pathol., 9,* 442–445.

Hughes, P. A., Stevens, M. M., Park, H. W., Federici, B. A., Dennis, E. S., & Akhurst, R. (2005). Response of larval *Chironomus tepperi* (Diptera: Chironomidae) to individual *Bacillus thuringiensis* var. *israelensis* toxins and toxin mixtures. *J. Invertebr. Pathol., 88,* 34–39.

Humphreys, M. J., & Berry, C. (1998). Variants of the *Bacillus sphaericus* binary toxins: implications for differential toxicity of strains. *J. Invertebr. Pathol., 71,* 184–185.

Hurst, M. R., & Glare, T. R. (2002). Restriction map of the *Serratia entomophila* plasmid pADAP carrying virulence factors for *Costelytra zealandica. Plasmid, 47,* 51–60.

Hurst, M. R., & Jackson, T. A. (2002). Use of the green fluorescent protein to monitor the fate of *Serratia entomophila* causing amber disease in the New Zealand grass grub, *Costelytra zealandica. J. Microbiol. Methods., 50,* 1–8.

Hurst, M. R., Glare, T. R., Jackson, T. A., & Ronson, C. W. (2000). Plasmid-located pathogenicity determinants of *Serratia entomophila,* the causal agent of amber disease of grass grub, show similarity to the insecticidal toxins of *Photorhabdus luminescens. J. Bacteriol., 182,* 5127–5138.

Hurst, M. R., Beard, S. S., Jackson, T. A., & Jones, S. M. (2007a). Isolation and characterization of the *Serratia entomophila* antifeeding prophage. *FEMS Microbiol. Lett., 270,* 42–48.

Hurst, M. R., Jones, S. M., Tan, B., & Jackson, T. A. (2007b). Induced expression of the *Serratia entomophila* Sep proteins shows activity towards the larvae of the New Zealand grass grub *Costelytra zealandica. FEMS Microbiol. Lett., 275,* 160–167.

Hurst, M. R., Becher, S. A., & O'Callaghan, M. (2011a). Nucleotide sequence of the *Serratia entomophila* plasmid pADAP and the *Serratia proteamaculans* pU143 plasmid virulence associated region. *Plasmid, 65,* 32–41.

Hurst, M. R., Becher, S. A., Young, S. D., Nelson, T. L., & Glare, T. R. (2011b). *Yersinia entomophaga* sp. nov., isolated from the New Zealand grass grub *Costelytra zealandica. Int. J. Syst. Evol. Microbiol., 61,* 844–849.

Hurst, M. R., Jones, S. A., Binglin, T., Harper, L. A., Jackson, T. A., & Glare, T. R. (2011c). The main virulence determinant of *Yersinia entomophaga* MH96 is a broad host-range toxin complex active against insects. *J. Bacteriol., 193,* 1966–1980.

Hussain, S.-R. A., Aronson, A. I., & Dean, D. H. (1996). Substitution of residues on the proximal side of Cry1A *Bacillus thuringiensis* δ-endotoxins affects irreversible binding to *Manduca sexta* midgut membrane. *Biochem. Biophys. Res. Commun., 226,* 8–14.

Hutchison, W. D., Burkness, E. C., Mitchell, P. D., Moon, R. D., Leslie, T. W., Fleischer, S. J., Abrahamson, M., Hamilton, K. L., Steffey, K. L., Gray, M. E., Hellmich, R. L., Kaster, L. V., Hunt, T. E., Wright, R. J., Pecinovsky, K., Rabaey, T. L., Flood, B. R., & Raun, E. S. (2010). Areawide suppression of European corn borer with Bt maize reaps savings to non-Bt maize growers. *Science, 330,* 222–225.

Ibargutxi, M. A., Muñoz, D., Escudero, I. R.d., & Caballero, P. (2008). Interactions between Cry1Ac, Cry2Ab, and Cry1Fa *Bacillus thuringiensis* toxins in the cotton pests *Helicoverpa armigera* (Hübner) and *Earias insulana* (Boisduval). *Biol. Control, 47,* 89–96.

Ibrahim, M. A., Griko, N., Junker, M., & Bulla, L. A. (2010). *Bacillus thuringiensis*: a genomics and proteomics perspective. *Bioeng. Bugs., 1,* 31–50.

Icoz, I., Saxena, D., Andow, D. A., Zwahlen, C., & Stotzky, G. (2008). Microbial populations and enzyme activities in soil *in situ* under transgenic corn expressing cry proteins from *Bacillus thuringiensis. J. Environ. Qual., 37,* 647–662.

Ignoffo, C. M., Hostette, D. L., & Pinnell, R. E. (1974). Stability of *Bacillus thuringiensis* and *Baculovirus heliothis* on soybean foliage. *Environ. Entomol., 3,* 117–119.

Ihara, H., Uemura, T., Masuhara, M., Ikawa, S., Sugimoto, K., Wadano, A., & Himeno, M. (1998). Purification and partial amino acid sequences of the binding protein from *Bombyx mori* for CryIAa δ-endotoxin of *Bacillus thuringiensis. Comp. Biochem. Physiol. B, 120,* 197–204.

Ikezawa, H., Kato, T., Ohta, S., Nakabayashi, T., Iwata, Y., & Ono, K. (1985). Physiological actions of phosphatidylinositol-specific phospholipase C from *Bacillus thuringiensis* on KB III cells: alkaline phosphatase release and growth inhibition. *Toxicon, 23,* 145–155.

Ikezawa, H., Hashimoto, A., Taguchi, R., Nakabayashi, T., & Himeno, M. (1989). Release of PI-anchoring enzymes and other effects of phosphatidylinositol-specific phospholipase C from *Bacillus thuringiensis* on TN-368 cells from a moth ovary. *Toxicon, 27,* 637–645.

Iriarte, J., Bel, Y., Ferrandis, M. D., Andrew, R., Murillo, J., Ferré, J., & Caballero, P. (1998). Environmental distribution and diversity of *Bacillus thuringiensis* in Spain. *Syst. Appl. Microbiol., 21,* 97–106.

Iriarte, J., Dumanoir, V. C., Bel, Y., Porcar, M., Ferrandis, M. D., Lecadet, M., Ferré, J., & Caballero, P. (2000). Characterization of *Bacillus thuringiensis* ser. *balearica* (serotype H48) and ser. *navarrensis* (serotype H50): two novel serovars isolated in Spain. *Curr. Microbiol., 40,* 17–22.

Ishikawa, H., Hoshino, Y., Motoki, Y., Kawahara, T., Kitajima, M., Kitami, M., Watanabe, A., Bravo, A., Soberon, M., Honda, A., Yaoi, K., & Sato, R. (2007). A system for the directed evolution of the insecticidal protein from *Bacillus thuringiensis. Mol. Biotechnol., 36,* 90–101.

Ishiwata, S. (1901). On a kind of flacherie (sotto disease). *Dainihon Sanshi Keiho, 114,* 1–5.

Ishiwata, S. (1905). About sottokin, a bacillus of a disease of the silkworm. *Dainihon Sanshi Keiho, 161,* 1–5.

Jackson, R. E., Marcus, M. A., Gould, F., Bradley, J. R., Jr., & Van Duyn, J. W. (2007). Cross-resistance responses of Cry1Ac-selected *Heliothis virescens* (Lepidoptera: Noctuidae) to the *Bacillus thuringiensis* protein vip3A. *J. Econ. Entomol., 100,* 180–186.

Jackson, T. A. (1992). Scarabs-pests of the past or the future. In T. R. Glare & T. A. Jackson (Eds.), *Use of Pathogens in Scarab Pest Management* (pp. 1–10). Andover: Intercept Press.

Jackson, T. A. (1995). Amber disease reduces trypsin activity in midgut of *Costeltyra zealandica* (Coleoptera: Scarabaeidae) larvae. *J. Invertebr. Pathol., 65,* 68–69.

Jackson, T. A. (2003). Environmental safety of inundative application of a naturally occurring biocontrol agent, *Serratia entomophila.* In H. Hokkanken & A. Hajek (Eds.), *Environmental Impacts of*

Microbial Insecticides: Need and Methods for Risk Assessment (pp. 169−176). Dordrecht: Kluwer Academic Publishers.

Jackson, T. A. (2007). A novel bacterium for control of grass grub. In C. Vincent, M. S. Goettel & G. Lazarovits (Eds.), *Biological Control: A Global Perspective* (pp. 160−168). Wallingford: CABI.

Jackson, T. A., & Klein, M. G. (2006). Scarabs as pests: a continuing problem. *Coleopterists Soc. Monogr., 5*, 102−119.

Jackson, T. A., & McNeill, M. R. (1998). Premature death in parasitized *Listronotus bonariensis* adults can be caused by bacteria transmitted by the parasitoid *Microctonus hyperodae*. *Biocontrol Sci. Technol., 8*, 389−396.

Jackson, T. A., & Zimmermann, G. (1996). Is there a role for *Serratia* spp. in the biocontrol of *Melolontha* spp. *Bull. OILB/SROP, 19*, 47−53.

Jackson, T. A., Pearson, J. F., & Stucki, G. (1986). Control of the grass grub, *Costelytra zealandica* (White) (Coleoptera: Scarabaeidae), by application of the bacteria *Serratia* spp. causing honey disease. *Bull. Entomol. Res., 76*, 69−76.

Jackson, T. A., Pearson, J. F., O'Callaghan, M., Mahanty, H. K., & Willcocks, M. J. (1992). Pathogen to product − development of *Serratia entomophila* (Enterobacteriaceae) as a commercial biological agent for the New Zealand grass grub (*Costelytra zealandica*). In T. A. Jackson & T. R. Glare (Eds.), *Use of Pathogens in Scarab Pest Management* (pp. 191−198). Andover: Intercept Press.

Jackson, T. A., Huger, A. M., & Glare, T. R. (1993). Pathology of amber disease in the New Zealand grass grub *Costelytra zealandica* (Coleoptera: Scarabaeidae). *J. Invertebr. Pathol., 61*, 123−130.

Jackson, T. A., Townsend, R. J., Nelson, T. L., Richards, N. K., & Glare, T. R. (1997). Estimating amber disease in grass grub populations by visual assessment and DNA colony blot analysis. In M. O'Callaghan (Ed.), *Proceedings of the 50th New Zealand Plant Protection Conference* (pp. 165−168). Palmerston North: NZ Plant Protection Society.

Jackson, T. A., Townsend, R. J., & Barlow, N. D. (1999). Predicting grass grub population change in Canterbury. In J. N. Mathiessen (Ed.), *Proceedings of the 7th Australasian Conference on Grassland Invertebrate Ecology* (pp. 21−26). Perth: CSIRO Entomology.

Jackson, T. A., Boucias, D. G., & Thaler, J. O. (2001). Pathobiology of amber disease, caused by *Serratia* spp., in the New Zealand grass grub, *Costelytra zealandica*. *J. Invertebr. Pathol., 78*, 232−243.

Jackson, T. A., Christeller, J. T., McHenry, J. Z., & Laing, W. A. (2004). Quantification and kinetics of the decline in grass grub endopeptidase activity during initiation of amber disease. *J. Invertebr. Pathol., 86*, 72−76.

Jacob, A., Philip, B. M., & Mathew, M. P. (1982). *Bacillus subtilis* as a pathogen on bhindi leaf roller, *Sylepta derogata* (Pyralidae: Lepidoptera). *J. Invertebr. Pathol., 40*, 301−302.

James, C. (2010). *Global Status of Commercialized Biotech/GM Crops: 2010*. Ithaca: International Service for the Acquisition of Agri-Biotech Applications.

James, R. R., & Lighthart, B. (1992). The effect of temperature, diet, and larval instar on the susceptibility of an aphid predator, *Hippodamia convergens* (Coleoptera: Coccinellidae), to the weak bacterial pathogen *Pseudomonas fluorescens*. *J. Invertebr. Pathol., 60*, 215−218.

Jaques, R. P., & Laing, D. R. (1988). Effectiveness of microbial and chemical insecticides in control of the Colorado potato beetle (Coleoptera: Chrysomelidae) on potatoes and tomatoes. *Can. Entomol., 120*, 1123−1131.

Jarosz, J. (1979). Gut flora of *Galleria mellonella* suppressing ingested bacteria. *J. Invertebr. Pathol., 34*, 192−198.

Jarrett, C. O., Deak, E., Isherwood, K. E., Oyston, P. C., Fischer, E. R., Whitney, A. R., Kobayashi, S. D., DeLeo, F. R., & Hinnebusch, B. J. (2004). Transmission of *Yersinia pestis* from an infectious biofilm in the flea vector. *J. Infect. Dis., 190*, 783−792.

Jarrett, P. (1985). Potency factors in the delta-endotoxin of *Bacillus thuringiensis* var. *aizawai* and the significance of plasmids in their control. *J. Appl. Bacteriol., 58*, 437−448.

Jarrett, P., & Stephenson, M. (1990). Plasmid transfer between strains of *Bacillus thuringiensis* infecting *Galleria mellonella* and *Spodoptera litoralis*. *Appl. Environ. Microbiol., 56*, 1608−1614.

Jenkins, J. L., & Dean, D. H. (2001). Binding specificity of *Bacillus thuringiensis* Cry1Aa for purified, native *Bombyx mori* aminopeptidase N and cadherin-like receptors. *BMC Biochem., 2*, 1−8.

Jenkins, J. L., Lee, M. K., Sangadala, S., Adang, M. J., & Dean, D. H. (1999). Binding of *Bacillus thuringiensis* Cry1Ac toxin to *Manduca sexta* aminopeptidase-N receptor is not directly related to toxicity. *FEBS Lett., 462*, 373−376.

Jenkins, J. L., Lee, M. K., Valaitis, A. P., Curtiss, A., & Dean, D. H. (2000). Bivalent sequential binding model of a *Bacillus thuringiensis* toxin to gypsy moth aminopeptidase N receptor. *J. Biol. Chem., 275*, 14423−14431.

Jensen, G. B., Larsen, P., Jacobsen, B. L., Madsen, B., Smidt, L., & Andrup, L. (2002). *Bacillus thuringiensis* in fecal samples from greenhouse workers after exposure to *B. thuringiensis*-based pesticides. *Appl. Environ. Microbiol., 68*, 4900−4905.

Jensen, G. B., Hansen, B. M., Eilenberg, J., & Mahillon, J. (2003). The hidden lifestyles of *Bacillus cereus* and relatives. *Environ. Microbiol., 5*, 631−640.

Jeong, H., Mun, H., Oh, H., Kim, S., Yang, K., Kim, I., & Lee, H. (2010). Evaluation of insecticidal activity of a bacterial strain, *Serratia* sp. EML-SE1 against diamondback moth. *J. Microbiol., 48*, 541−545.

Jiménez-Juárez, N., Muñoz-Garay, C., Gómez, I., Saab-Rincon, G., Damian-Almazo, J. Y., Gill, S. S., Soberón, M., & Bravo, A. (2007). *Bacillus thuringiensis* Cry1Ab mutants affecting oligomer formation are non-toxic to *Manduca sexta* larvae. *J. Biol. Chem., 282*, 21222−21229.

Johnson, D. E., & McGaughey, W. H. (1996). Contribution of *Bacillus thuringiensis* spores to toxicity of purified Cry proteins towards Indianmeal moth larvae. *Curr. Microbiol., 33*, 54−59.

Johnson, D. E., Oppert, B., & McGaughey, W. H. (1998). Spore coat protein synergizes *Bacillus thuringiensis* crystal toxicity for the indianmeal moth (*Plodia interpunctella*). *Curr. Microbiol., 36*, 278−282.

Johnson, V. W., Pearson, J. F., & Jackson, T. A. (2001). Formulation of *Serratia entomophila* for biological control of grass grub. *N.Z. Plant Prot., 54*, 125−127.

Johnston, P. R., & Crickmore, N. (2009). Gut bacteria are not required for the insecticidal activity of *Bacillus thuringiensis* toward the tobacco hornworm, *Manduca sexta*. *Appl. Environ. Microbiol., 75*, 5094−5099.

Jones, G. W., Nielsen-Leroux, C., Yang, Y., Yuan, Z., Dumas, V. F., Monnerat, R. G., & Berry, C. (2007). A new Cry toxin with a unique two-component dependency from *Bacillus sphaericus*. *FASEB J., 21*, 4112−4120.

Jones, G. W., Wirth, M. C., Monnerat, R. G., & Berry, C. (2008). The Cry48Aa-Cry49Aa binary toxin from *Bacillus sphaericus* exhibits highly restricted target specificity. *Environ. Microbiol., 10,* 2418−2424.

Juárez-Pérez, V. M., Jacquemard, P., & Frutos, R. (1994). Characterization of the type strain of *Bacillus thuringiensis* subsp. *cameroun* serotype H32. *FEMS Microbiol. Lett., 122,* 43−48.

Jucovic, M., Walters, F. S., Warren, G. W., Palekar, N. V., & Chen, J. S. (2008). From enzyme to zymogen: engineering Vip2, an ADP-ribosyltransferase from *Bacillus cereus,* for conditional toxicity. *Protein Eng. Des. Sel., 21,* 631−638.

Jung, S., & Kim, Y. (2006). Synergistic effect of entomopathogenic bacteria (*Xenorhabdus* sp. and *Photorhabdus temperata* ssp. *temperata*) on the pathogenicity of *Bacillus thuringiensis* ssp. *aizawai* against *Spodoptera exigua* (Lepidoptera: Noctuidae). *Environ. Entomol., 35,* 1584−1589.

Jurat-Fuentes, J. L., & Adang, M. J. (2001). Importance of Cry1 deltaendotoxin domain II loops for binding specificity in *Heliothis virescens* (L.). *Appl. Environ. Microbiol., 67,* 323−329.

Jurat-Fuentes, J. L., & Adang, M. J. (2004). Characterization of a Cry1Ac-receptor alkaline phosphatase in susceptible and resistant *Heliothis virescens* larvae. *Eur. J. Biochem., 271,* 3127−3135.

Jurat-Fuentes, J. L., & Adang, M. J. (2006). The *Heliothis virescens* cadherin protein expressed in *Drosophila* S2 cells functions as a receptor for *Bacillus thuringiensis* Cry1A but not Cry1Fa toxins. *Biochemistry, 45,* 9688−9695.

Jurat-Fuentes, J. L., Gahan, L. J., Gould, F. L., Heckel, D. G., & Adang, M. J. (2004). The HevCaLP protein mediates binding specificity of the Cry1A class of *Bacillus thuringiensis* toxins in *Heliothis virescens. Biochemistry, 43,* 14299−14305.

Jurat-Fuentes, J. L., Karumbaiah, L., Jakka, S. R. K., Ning, C., Liu, C., Wu, K., Jackson, J., Gould, F., Blanco, C. A., Portilla, M., Perera, O. P., & Adang, M. (2011). Reduced levels of membranebound alkaline phosphatase are common to lepidopteran strains resistant to Cry toxins from *Bacillus thuringiensis. PLoS ONE., 6,* e17606.

Kaaya, G. P., & Darji, N. (1989). Mortality in adult tsetse, *Glossina morsitans morsitans,* caused by entomopathogenic bacteria. *J. Invertebr. Pathol., 54,* 32−38.

Kabaluk, T., & Gazdik, K. (2005). *Directory of microbial pesticides for agricultural crops in OECD countries.* http://www4.agr.gc.ca/resources/prod/doc/pmc/pdf/micro_e.pdf.

Kageyama, D., Anbutsu, H., Shimada, M., & Fukatsu, T. (2007). *Spiroplasma* infection causes either early or late male killing in *Drosophila,* depending on maternal host age. *Naturwissenschaften, 94,* 333−337.

Kalucy, E. C., & Daniel, A. (1972). The reaction of *Anopheles annulipes* larvae to infection by *Aeromonas punctata. J. Invertebr. Pathol., 19,* 189−197.

Kamdar, H., & Jayaraman, K. (1983). Spontaneous loss of a high molecular weight plasmid and the biocide of *Bacillus thuringiensis* var. *israeliensis. Biochem. Biophys. Res. Commun., 110,* 477−482.

Kanbar, G., & Engels, W. (2003). Ultrastructure and bacterial infection of wounds in honey bee (*Apis mellifera*) pupae punctured by *Varroa* mites. *Parasitol. Res., 90,* 349−354.

Kaniewski, W. K., & Thomas, P. E. (2004). The potato story. *AgBioForum, 7,* 41−46.

Katznelson, H. (1950). *Bacillus pulvifaciens* (N. SP.), an organism associated with powdery scale of honeybee larvae. *J. Bacteriol., 59,* 153−155.

Kaur, S. (2000). Molecular approaches towards development of novel *Bacillus thuringiensis* biopesticides. *World J. Microbiol. Biotechnol., 16,* 781−793.

Kawanishi, C. Y., Splittstoesser, C. M., & Tashiro, H. (1978). Infection of the European chafer, *Amphimallon majalis,* by *Bacillus popilliae:* ultrastructure. *J. Invertebr. Pathol., 31,* 91−102.

Kaya, H. K., Klein, M. G., Burlando, T. M., Harrison, R. E., & Lacey, L. A. (1992). Prevalence of 2 *Bacillus popilliae* Dutky morphotypes and blue disease in *Cyclocephala hirta* Leconte (Coleoptera, Scarabaeidae) populations in California. *Pan-Pac. Entomol., 68,* 38−45.

Keeton, T. P., & Bulla, L. A., Jr. (1997). Ligand specificity and affinity of BT-R1, the *Bacillus thuringiensis* Cry1A toxin receptor from *Manduca sexta,* expressed in mammalian and insect cell cultures. *Appl. Environ. Microbiol., 63,* 3419−3425.

Khodyrev, V. P. (1990). *Bacillus thuringiensis* subsp. *toguchini* − a new subspecies of crystal-forming bacteria. *Izv. Akad. Nauk SSSR. Biol.* 789−791.

Khyami-Horani, H., Hajaij, M., & Charles, J. F. (2003). Characterization of *Bacillus thuringiensis* ser. *jordanica* (serotype H71), a novel serovariety isolated in Jordan. *Curr. Microbiol., 47,* 26−31.

Kim, S. H., Park, S. Y., Heo, Y. J., & Cho, Y. H. (2008). *Drosophila melanogaster*-based screening for multihost virulence factors of *Pseudomonas aeruginosa* PA14 and identification of a virulence-attenuating factor, HudA. *Infect. Immun., 76,* 4152−4162.

Kim, S. K., Kim, Y. C., Lee, S., Kim, J. C., Yun, M. Y., & Kim, I. S. (2010). Insecticidal activity of rhamnolipid isolated from *Pseudomonas* sp. EP-3 against green peach aphid (*Myzus persicae. J. Agric. Food Chem., 59,* 934−938.

Klein, M. G. (1981). Advances in the use of *Bacillus popillae* for pest control. In H. D. Burges (Ed.), *Microbial Control of Pests and Plant Diseases 1970−1980* (pp. 183−192). London: Academic Press.

Klein, M. G., & Jackson, T. A. (1992). Bacterial diseases of scarabs. In T. A. Jackson & T. R. Glare (Eds.), *Use of Pathogens in Scarab Pest Management* (pp. 43−61). Andover: Intercept Press.

Klier, A., Lecadet, M., & De Barjac, H. (1974). Inhibition des ARNpolymérasas ADN-dépendantes de *B. thuringiensis* par l'exotoxine de *B. thuringiensis. C.R. Acad. Sci. Hebd. Seances Acad. Sci. D, 277,* 2805−2808.

Klier, A., Bourgouin, C., & Rapaport, G. (1983). Mating between *Bacillus subtilus* and *Bacillus thuringiensis* and transfer of cloned crystal genes. *Mol. Gen. Genet., 191,* 257−262.

Klimowicz, A. K., Benson, T. A., & Handelsman, J. (2010). A quadruple-enterotoxin-deficient mutant of *Bacillus thuringiensis* remains insecticidal. *Microbiology, 156,* 3575−3583.

Knight, P. J., Crickmore, N., & Ellar, D. J. (1994). The receptor for *Bacillus thuringiensis* CrylA(c) delta-endotoxin in the brush border membrane of the lepidopteran *Manduca sexta* is aminopeptidase N. *Mol. Microbiol., 11,* 429−436.

Knowles, B. H. (1994). Mechanism of action of *Bacillus thuringiensis* insecticidal δ-endotoxins. *Adv. Insect Physiol., 24,* 275−308.

Knowles, B. H., & Ellar, D. J. (1987). Colloid-osmotic lysis is a general feature of the mechanism of action of *Bacillus thuringiensis* δ-endotoxins with different insect specificity. *Biochim. Biophys. Acta, 924,* 509−518.

Knowles, B. H., & Farndale, R. W. (1988). Activation of insect cell adenylate cyclase by *Bacillus thuringiensis* δ-endotoxin and melittin. *Biochem. J., 253,* 235−241.

Knowles, B. H., Francis, P. H., & Ellar, D. J. (1986). Structurally related *Bacillus thuringiensis* δ-endotoxins display major differences in insecticidal activity *in vivo* and *in vitro. J. Cell Sci., 84*, 221–236.

Knowles, B. H., Blatt, M. R., Tester, M., Horsnell, J. M., Carroll, J., Menestrina, G., & Ellar, D. J. (1989). A cytolytic delta-endotoxin from *Bacillus thuringiensis* var. *israelensis* forms cation-selective channels in planar lipid bilayers. *FEBS Lett., 244*, 259–262.

Kolstø, A. B., Tourasse, N. J., & Økstad, O. A. (2009). What sets *Bacillus anthracis* apart from other *Bacillus* species? *Annu. Rev. Microbiol., 63*, 451–476.

Koni, P. A., & Ellar, D. J. (1993). Cloning and characterization of a novel *Bacillus thuringiensis* cytolytic delta-endotoxin. *J. Mol. Biol., 229*, 319–327.

Koni, P. A., & Ellar, D. J. (1994). Biochemical characterization of *Bacillus thuringiensis* cytolytic delta-endotoxins. *Microbiology, 140*, 1869–1880.

Koppenhöfer, A. M., Choo, H. Y., Kaya, H. K., Lee, D. W., & Gelernter, W. D. (1999). Increased field and greenhouse efficacy against scarab grubs with a combination of an entomopathogenic nematode and *Bacillus thuringiensis. Biol. Control., 14*, 37–44.

Kramer, K. J., & Muthukrishnan, S. (1997). Insect chitinases: molecular biology and potential use as biopesticides. *Insect Biochem. Mol. Biol., 27*, 887–900.

Krieg, A. (1970). *In vitro* determination of *Bacillus thuringiensis, Bacillus cereus*, and related bacilli. *J. Invertebr. Pathol., 15*, 313–320.

Krieg, A. (1971). Concerning α-exotoxin produced by vegetative cells of *Bacillus thuringiensis* and *Bacillus cereus. J. Invertebr. Pathol., 17*, 134–135.

Krieg, A. (1987). Diseases caused by bacteria and other prokaryotes. In J. R. Fuxa & Y. Tanada (Eds.), *Epizootiology of Insect Diseases* (pp. 323–355). New York: John Wiley and Sons.

Krieg, A., & Langerbruch, G. A. (1981). Susceptibility of arthropod species to *Bacillus thuringiensis*. In H. D. Burges (Ed.), *Microbial Control of Pests and Plant Diseases 1970–1980* (pp. 837–896). London: Academic Press.

Krieg, A., de Barjac, H., & Bonnefoi, A. (1968). A new serotype of *Bacillus thuringiensis* isolated in Germany: *Bacillus thuringiensis* var. *darmstadiensis. J. Invertebr. Pathol., 10*, 428–430.

Krieg, V. A., Huger, A. M., Langenbruch, G. A., & Schnetter, W. (1983). *Bacillus thuringiensis* var. *tenebrionis*: a new pathotype effective against larvae of Coleoptera. *Z. Ang. Entomol., 96*, 500–508.

Krieger, L., Franken, E., & Schnetter, W. (1996). *Bacillus popilliae* var. *melolontha* H1, a pathogen for the May beetles, *Melolontha* spp. In T. A. Jackson & T. R. Glare (Eds.), *Proceedings of the 3rd International Workshop on Microbial Control of Soil Dwelling Pests* (pp. 79–87). Lincoln: AgResearch.

Krishnamoorthy, M., Jurat-Fuentes, J. L., McNall, R. J., Andacht, T., & Adang, M. J. (2007). Identification of novel Cry1Ac binding proteins in midgut membranes from *Heliothis virescens* using proteomic analyses. *Insect Biochem. Mol. Biol., 37*, 189–201.

Kronstad, J. W., & Whiteley, H. R. (1984). Inverted repeat sequences flank a *Bacillus thuringiensis* crystal protein gene. *J. Bacteriol., 160*, 95–102.

Kronstad, J. W., & Whiteley, H. R. (1986). Three classes of homologous *Bacillus thuringiensis* crystal-protein genes. *Gene., 43*, 29–40.

Kronstad, J. W., Schnepf, H. E., & Whiteley, H. R. (1983). Diversity of locations for *Bacillus thuringiensis* crystal protein genes. *J. Bacteriol., 154*, 419–428.

Krych, V. K., Johnson, J. L., & Yousten, A. A. (1980). Deoxyribonucleic acid homologies among strains of *Bacillus sphaericus. Int. J. Syst. Bacteriol., 30*, 476–484.

Kumar, A. S., & Aronson, A. I. (1999). Analysis of mutations in the pore-forming region essential for insecticidal activity of a *Bacillus thuringiensis* δ-endotoxin. *J. Bacteriol., 181*, 6103–6107.

Kumar, N. S., & Venkateswerlu, G. (1998a). Intracellular proteases in sporulated *Bacillus thuringiensis* subsp. *kurstaki* and their role in protoxin activation. *FEMS Microbiol. Lett., 166*, 377–382.

Kumar, N. S., & Venkateswerlu, G. (1998b). Endogenous protease-activated 66-kDa toxin from *Bacillus thuringiensis* subsp. *kurstaki* active against *Spodoptera littoralis. FEMS Microbiol. Lett., 159*, 113–120.

Kumar, P. A., Sharma, R. P., & Malik, V. S. (1996). The insecticidal proteins of *Bacillus thuringiensis. Adv. Appl. Microbiol., 42*, 1–43.

Kumar, S., Chandra, A., & Pandey, K. C. (2008). *Bacillus thuringiensis* (Bt) transgenic crop: an environment friendly insect-pest management strategy. *J. Environ. Biol., 29*, 641–653.

Kumari, S. M., & Neelgund, Y. F. (1985). Preliminary infectivity tests using six bacterial formulations against the red flour beetle, *Tribolium castaneum. J. Invertebr. Pathol., 46*, 198–199.

Kurstak, E. (1964). Le processus de l'infection par *Bacillus thuringiensis* Berl. d' *Ephestia kühniella* Zell. déclenché par le parasitisme de *Nemeritis canescens* Grav. (Ichneumonidae). *C.R. Hebd. Acad. Sci., 259*, 211–212.

Kwon, M. O., Wayadande, A. C., & Fletcher, J. (1999). *Spiroplasma citri* movement into the intestines and salivary glands of its leafhopper vector, *Circulifer tenellus. Phytopathology, 89*, 1144–1151.

Kyei-Poku, G., Gauthier, D., Pang, A., & van Frankenhuyzen, K. (2007). Detection of *Bacillus cereus* virulence factors in commercial products of *Bacillus thuringiensis* and expression of diarrheal enterotoxins in a target insect. *Can. J. Microbiol., 53*, 1283–1290.

Lacey, L. A. (1990). Persistence and formulation of *Bacillus sphaericus*. In H. de Barjac & D. J. Sutherland (Eds.), *Bacterial Control of Mosquitoes & Black Flies: Biochemistry, Genetics, & Applications of* Bacillus thuringiensis israelensis & Bacillus sphaericus (pp. 285–294). New Brunswick: Rutgers University Press.

Lacey, L. A. (2007). *Bacillus thuringiensis* serovariety *israelensis* and *Bacillus sphaericus* for mosquito control. *J. Am. Mosq. Control Assoc., 23*, 133–163.

Lacey, L. A., & Federici, B. A. (1979). Pathogenesis and midgut histopathology of *Bacillus thuringiensis* in *Simulium vittatum* (Diptera: Simuliidae). *J. Invertebr. Pathol., 33*, 171–182.

Lacey, L. A., & Mulla, M. S. (1990). Safety of *Bacillus thuringiensis* (H-14) and *Bacillus sphaericus* to non-target organisms in the aquatic environment. In M. Laird, L. A. Lacey & E. W. Davidson (Eds.), *Safety of Microbial Insecticides* (pp. 169–188). Boca Raton: CRC Press.

Lacey, L. A., & Shapiro-Ilan, D. I. (2008). Microbial control of insect pests in temperate orchard systems: potential for incorporation into IPM. *Annu. Rev. Entomol., 53*, 121–144.

Lacey, L. A., & Undeen, A. H. (1986). Microbial control of black flies and mosquitoes. *Annu. Rev. Entomol., 31*, 265–296.

Lacey, L. A., Day, J., & Heitzman, C. M. (1987). Long-term effects of *Bacillus sphaericus* on *Culex quinquefasciatus. J. Invertebr. Pathol., 49*, 116–123.

Lan, R., & Reeves, P. R. (2000). Intraspecies variation in bacterial genomes: the need for a species genome concept. *Trends Microbiol., 8*, 396–401.

Lan, R., & Reeves, P. R. (2001). When does a clone deserve a name? A perspective on bacterial species based on population genetics. *Trends Microbiol., 9*, 419–424.

Lecadet, M. M., Frachon, E., Dumanoir, V. C., Ripouteau, H., Hamon, S., Laurent, P., & Thiery, I. (1999). Updating the H-antigen classification of *Bacillus thuringiensis. J. Appl. Microbiol., 86*, 660–672.

Leclerque, A. (2008). Whole genome-based assessment of the taxonomic position of the arthropod pathogenic bacterium *Rickettsiella grylli. FEMS Microbiol. Lett., 283*, 117–127.

Leclerque, A., & Kleespies, R. G. (2008). Genetic and electron-microscopic characterization of *Rickettsiella tipulae*, an intracellular bacterial pathogen of the crane fly, *Tipula paludosa. J. Invertebr. Pathol., 98*, 329–334.

Lee, D. H., Machii, J., & Ohba, M. (2002). High frequency of *Bacillus thuringiensis* in feces of herbivorous animals maintained in a zoological garden in Japan. *Appl. Entomol. Zool., 37*, 509–516.

Lee, H. H., Lee, J.-A., Lee, K.-Y., Chung, J.-D., de Barjac, H., Charles, J. F., Dumanoir, V. C., & Frachon, E. (1994). New serovars of *Bacillus thuringiensis: B. thuringiensis* ser. *coreanensis* (serotype H25), *B. thuringiensis* ser. *leesis* (serotype H33), and *B. thuringiensis* ser. *konkukian* (serotype H34). *J. Invertebr. Pathol., 63*, 217–219.

Lee, H. H., Jung, J. D., Yoon, M. S., Lee, K. K., Lecadet, M. M., Charles, J. F., Cosmao Dumanoir, V., Frachon, E., & Shim, J. C. (1995). Distribution of *Bacillus thuringiensis* in Korea. In T. Y. Feng (Ed.), Bacillus thuringiensis *Biotechnology and Environmental Benefits* (pp. 201–215). Taiwan: Hua Shiang Yuan Publishing Company.

Lee, M. K., Young, B. A., & Dean, D. H. (1995). Domain III exchanges of *Bacillus thuringiensis* Cry1A toxins affect binding to different gypsy moth midgut receptors. *Biochem. Biophys. Res. Commun., 216*, 306–312.

Lee, M. K., Curtiss, A., Alcantara, E., & Dean, D. H. (1996). Synergistic effect of the *Bacillus thuringiensis* toxins CryIAA and CryIAc on the gypsy moth, *Lymantria dispar. Appl. Environ. Microbiol., 62*, 583–586.

Lee, M. K., Rajamohan, F., Jenkins, J. L., Curtiss, A. S., & Dean, D. H. (2000). Role of two arginine residues in domain II, loop 2 of Cry1Ab and Cry1Ac *Bacillus thuringiensis* delta-endotoxin in toxicity and binding to *Manduca sexta* and *Lymantria dispar* aminopeptidase N. *Mol. Microbiol., 38*, 289–298.

Lee, M. K., Walters, F. S., Hart, H., Palekar, N., & Chen, J. S. (2003). The mode of action of the *Bacillus thuringiensis* vegetative insecticidal protein Vip3A differs from that of Cry1Ab δ-endotoxin. *Appl. Environ. Microbiol., 69*, 4648–4657.

Lee, M. K., Miles, P., & Chen, J. S. (2006). Brush border membrane binding properties of *Bacillus thuringiensis* Vip3A toxin to *Heliothis virescens* and *Helicoverpa zea* midguts. *Biochem. Biophys. Res. Commun., 339*, 1043–1047.

Lee, X., Johnston, R. A., Rose, D. R., & Young, N. M. (1989). Crystallization and preliminary X-ray diffraction studies of the complex of *Maclura pomifera* agglutinin with the disaccharide Gal beta 1-3GalNAc. *J. Mol. Biol., 210*, 685–686.

Lereclus, D., Mahillon, J., Menou, G., & Lecadet, M. M. (1986). Identification of Tn4430, a transposon of *Bacillus thuringiensis* functional in *Escherichia coli. Mol. Gen. Genet., 204*, 52–57.

Lereclus, D., Vallade, M., Chaufaux, J., Arantes, O., & Rambaud, S. (1992). Expansion of insecticidal host range of *Bacillus thuringiensis* by *in vivo* genetic recombination. *Bio/Technology, 10*, 418–421.

Lereclus, D., Agaisse, H., Gominet, M., Salamitou, S., & Sanchis, V. (1996). Identification of a *Bacillus thuringiensis* gene that positively regulates transcription of the phosphatidylinositol-specific phospholipase C gene at the onset of the stationary phase. *J. Bacteriol., 178*, 2749–2756.

Lertcanawanichakul, M., Wiwat, C., Bhumiratana, A., & Dean, D. H. (2004). Expression of chitinase-encoding genes in *Bacillus thuringiensis* and toxicity of engineered *B. thuringiensis* subsp. *aizawai* toward *Lymantria dispar* larvae. *Curr. Microbiol., 48*, 175–181.

Leuber, M., Orlik, F., Schiffler, B., Sickmann, A., & Benz, R. (2006). Vegetative insecticidal protein (Vip1Ac) of *Bacillus thuringiensis* HD201: evidence for oligomer and channel formation. *Biochemistry, 45*, 283–288.

Levinson, B. L., Kasyan, K. J., Chiu, S. S., Currier, T. C., & González, J. M., Jr. (1990). Identification of beta-exotoxin production, plasmids encoding beta-exotoxin, and a new exotoxin in *Bacillus thuringiensis* by using high performance liquid chromatography. *J. Bacteriol., 172*, 3172–3179.

Li, J., Koni, P. A., & Ellar, D. J. (1996). Structure of the mosquitocidal delta-endotoxin CytB from *Bacillus thuringiensis* sp. *kyushuensis* and implications for membrane pore formation. *J. Mol. Biol., 257*, 129–152.

Li, J., Derbyshire, D. J., Promdonkoy, B., & Ellar, D. J. (2001). Structural implications for the transformation of the *Bacillus thuringiensis* delta-endotoxins from water-soluble to membrane-inserted forms. *Biochem. Soc. Trans., 29*, 571–577.

Li, J. D., Carroll, J., & Ellar, D. J. (1991). Crystal structure of insecticidal delta-endotoxin from *Bacillus thuringiensis* at 2.5 A resolution. *Nature, 353*, 815–821.

Li, R., Gao, M., Dai, S., & Li, X. (1999). The identification of 5 new serotypes of *Bacillus thuringiensis* from soil in China. *Acta Microbiol. Sin., 39*, 154–159.

Li, R., Dai, S., & Gao, M. (2000). A new serovar of *Bacillus thuringiensis: Bacillus thuringiensis* subsp. *sinensis. Virol. Sin., 15*, 224–225.

Liang, Y., Patel, S. S., & Dean, D. H. (1995). Irreversible binding kinetics of *Bacillus thuringiensis* CryIA delta-endotoxins to gypsy moth brush border membrane vesicles is directly correlated to toxicity. *J. Biol. Chem., 270*, 24719–24724.

Likitvivatanavong, S., Chen, J., Bravo, A., Soberón, M., & Gill, S. S. (2011). Cadherin, alkaline phosphatase, and aminopeptidase N as receptors of Cry11Ba toxin from *Bacillus thuringiensis* subsp. *jegathesan* in *Aedes aegypti. Appl. Environ. Microbiol., 77*, 24–31.

Lima, G. M. S., Aguiar, R. W. S., Corrêa, R. F. T., Martins, E. S., Gomes, A. C. M., Nagata, T., De-Souza, M. T., Monnerat, R. G., & Ribeiro, B. M. (2008). Cry2A toxins from *Bacillus thuringiensis* expressed in insect cells are toxic to two lepidopteran insects. *World J. Microbiol. Biotechnol., 24*, 2941–2948.

Lima, M. M., dos Santos, L. M., da Silva, M. H., & Rabinovitch, L. (1994). Effects of the spore-endotoxin complex of a strain of *Bacillus thuringiensis* serovar *morrisoni* upon *Triatoma vitticeps* (Hemiptera: Reduviidae) under laboratory conditions. *Mem. Inst. Oswaldo Cruz., 89*, 403–405.

Limpanawat, S., Promdonkoy, B., & Boonserm, P. (2009). The C-terminal domain of BinA is responsible for *Bacillus sphaericus* binary toxin BinA-BinB interaction. *Curr. Microbiol., 59*, 509–513.

Liu, B., Wang, L., Zeng, Q., Meng, J., Hu, W. J., Li, X. G., Zhou, K. X., Xue, K., Liu, D. D., & Zheng, Y. P. (2009). Assessing effects of transgenic Cry1Ac cotton on the earthworm *Eisenia fetida. Soil Biol. Biochem., 41*, 1841–1846.

Liu, D., Cai, J., Xie, C.c., Liu, C., & Chen, Y.h. (2010). Purification and partial characterization of a 36-kDa chitinase from *Bacillus thuringiensis* subsp. *colmeri*, and its biocontrol potential. *Enzyme Microb. Technol., 46*, 252–256.

Liu, J. W., Porter, A. G., Wee, B. Y., & Thanabalu, T. (1996a). New gene from nine *Bacillus sphaericus* strains encoding highly conserved 35.8-kilodalton mosquitocidal toxins. *Appl. Environ. Microbiol., 62*, 2174–2176.

Liu, J. W., Yap, W. H., Thanabalu, T., & Porter, A. G. (1996b). Efficient synthesis of mosquitocidal toxins in *Asticcacaulis excentricus* demonstrates potential of gram-negative bacteria in mosquito control. *Nat. Biotechnol., 14*, 343–347.

Liu, M., Cai, Q. X., Liu, H. Z., Zhang, B. H., Yan, J. P., & Yuan, Z. M. (2002). Chitinolytic activities in *Bacillus thuringiensis* and their synergistic effects on larvicidal activity. *J. Appl. Microbiol., 93*, 374–379.

Liu, X. Y., Ruan, L. F., Hu, Z. F., Peng, D. H., Cao, S. Y., Yu, Z. N., Liu, Y., Zheng, J. S., & Sun, M. (2010). Genome-wide screening reveals the genetic determinants of an antibiotic insecticide in *Bacillus thuringiensis. J. Biol. Chem., 285*, 39191–39200.

Lonc, E., Lecadet, M. M., Lachowicz, T. M., & Panek, E. (1997). Description of *Bacillus thuringiensis wratislaviensis* (H-47), a new serotype originating from Wrocław (Poland), and other soil isolates from the same area. *Lett. Appl. Microbiol., 24*, 467–473.

Louis, C., Jourdan, M., & Cabanac, M. (1986). Behavioral fever and therapy in a rickettsia-infected Orthoptera. *Am. J. Physiol., 250*, R991–R995.

Lovgren, A., Zhang, M., Engstrom, A., Dalhammar, G., & Landen, R. (1990). Molecular characterization of immune inhibitor A, a secreted virulence protease from *Bacillus thuringiensis. Mol. Microbiol., 4*, 2137–2146.

Lu, Y., Wu, K., Jiang, Y., Xia, B., Li, P., Feng, H., Wyckhuys, K. A., & Guo, Y. (2010). Mirid bug outbreaks in multiple crops correlated with wide-scale adoption of Bt cotton in China. *Science, 328*, 1151–1154.

Ludwig, W., Euzéby, J., & Whitman, W. B. (2006). *Taxonomic outlines of the phyla* Bacteroidetes, Spirochaetes, Tenericutes *(Mollicutes)*, Acidobacteria, Fibrobacteres, Fusobacteria, Dictyoglomi, Gemmatimonadetes, Lentisphaerae, Verrucomicrobia, Chlamydiae, and Planctomycetes. http://www.bergeys.org/outlines/bergeys_vol_4_roadmap_outline.pdf.

Ludwig, W., Euzéby, J., & Whitman, W. B. (2008a). *Taxonomic outline of the phylum* Actinobacteria. http://www.bergeys.org/outlines/bergeys_vol_5_roadmap_outline.pdf.

Ludwig, W., Schleifer, K.-H., & Whitman, W. B. (2008b). *Taxonomic outline of the phylum* Firmicutes. http://www.bergeys.org/outlines/bergeys_vol_3_roadmap_outline.pdf.

Luo, K., Sangadala, S., Masson, L., Mazza, A., Brousseau, R., & Adang, M. J. (1997). The *Heliothis virescens* 170 kDa aminopeptidase functions as "receptor A" by mediating specific *Bacillus thuringiensis* Cry1A delta-endotoxin binding and pore formation. *Insect Biochem. Mol. Biol., 27*, 735–743.

Luttrell, R. G., Ali, A., Young, S. Y., & Knighten, K. (1998). Relative activity of commercial formulations of *Bacillus thuringiensis* against selected noctuid larvae (Lepidoptera: Noctuidae). *J. Entomol. Sci., 33*, 365–377.

Luxananil, P., Atomi, H., Panyim, S., & Imanaka, T. (2001). Isolation of bacterial strains colonizable in mosquito larval guts as novel host cells for mosquito control. *J. Biosci. Bioeng., 92*, 342–345.

Lynch, R. E., Lewis, L. C., & Brindley, T. A. (1976a). Bacteria associated with eggs and first-instar larvae of European corn borer: isolation techniques and pathogenicity. *J. Invertebr. Pathol., 27*, 325–331.

Lynch, R. E., Lewis, L. C., & Brindley, T. A. (1976b). Bacteria associated with eggs and first-instar larvae of the European corn borer: identification and frequency of occurrence. *J. Invertebr. Pathol., 27*, 229–237.

Lysenko, O. (1963). The taxonomy of entomogenous bacteria. In E. A. Steinhaus (Ed.), *Insect Pathology, An Advanced Treatise*, (Vol. 2) (pp. 1–20). New York: Academic Press.

Lysenko, O. (1972). Pathogenicity of *Bacillus cereus* for insects II. Toxicity of phospholipase-C for *Galleria mellonella. Folia Microbiol., 17*, 228–231.

Lysenko, O. (1974). Bacterial exoenzymes toxic for insects; proteinase and lecithinase. *J. Hyg. Epid. Microb. Immmunol., 18*, 347–352.

Macdonald, R., & Kalmakoff, J. (1995). Comparison of pulsed-field gel electrophoresis DNA fingerprints of field isolates of the entomopathogen *Bacillus popilliae. Appl. Environ. Microbiol., 61*, 2446–2449.

MacIntosh, S. C., Kishore, G. M., Perlak, F. J., Marrone, P. G., Stone, T. B., Sims, S. R., & Fuch, R. L. (1990). Potentiation of *Bacillus thuringiensis* insecticidal activity by serine protease inhibitors. *J. Agric. Food Chem., 38*, 1145–1152.

Maduell, P., Armengol, G., Llagostera, M., Orduz, S., & Lindow, S. (2008). *B. thuringiensis* is a poor colonist of leaf surfaces. *Microb. Ecol., 55*, 212–219.

de Maagd, R. A., Bakker, P. L., Masson, L., Adang, M. J., Sangadala, S., Stiekema, W., & Bosch, D. (1999a). Domain III of the *Bacillus thuringiensis* delta-endotoxin Cry1Ac is involved in binding to *Manduca sexta* brush border membranes and to its purified aminopeptidase N. *Mol. Microbiol., 31*, 463–471.

de Maagd, R. A., Bakker, P., Staykov, N., Dukiandjiev, S., Stiekema, W., & Bosch, D. (1999b). Identification of *Bacillus thuringiensis* delta-endotoxin Cry1C domain III amino acid residues involved in insect specificity. *Appl. Environ. Microbiol., 65*, 4369–4374.

de Maagd, R. A., Bravo, A., Berry, C., Crickmore, N., & Schnepf, H. E. (2003). Structure, diversity, and evolution of protein toxins from spore-forming entomopathogenic bacteria. *Annu. Rev. Genet., 37*, 409–433.

Maheswaran, S., Sreeramanan, S., Reena Josephine, C. M., Marimuthu, K., & Xavier, R. (2010). Occurrence of *Bacillus thuringiensis* in faeces of herbivorous farm animals. *Afr. J. Biotechnol., 9*, 8013–8019.

Mahillon, J., Seurinck, J., van Rompuy, L., Delcour, J., & Zabeau, M. (1985). Nucleotide sequence and structural organization of an insertion sequence element (IS231) from *Bacillus thuringiensis* strain *berliner* 1715. *EMBO J., 4*, 3895–3899.

Malvar, T., Gawron-Burke, C., & Baum, J. A. (1994). Overexpression of *Bacillus thuringiensis* HknA, a histidine protein kinase homology, bypasses early Spo mutations that result in CryIIIA overproduction. *J. Bacteriol., 176*, 4742–4749.

Manasherob, R., Ben-Dov, E., Zaritsky, A., & Barak, Z. (1994). Protozoan-enhanced toxicity of *Bacillus thuringiensis* var. *israelensis* delta-endotoxin against *Aedes aegypti* larvae. *J. Invertebr. Pathol., 63*, 244–248.

Manasherob, R., Ben-Dov, E., Margalit, J., Zaritsky, A., & Barak, Z. (1996). Raising activity of *Bacillus thuringiensis* var. *israelensis* against *Anopheles stephensi* larvae by encapsulation in *Tetrahymena pyriformis* (Hymenostomatida: Tetrahymenidae). *J. Am. Mosq. Control Assoc., 12*, 627–631.

Manasherob, R., Otieno-Ayayo, Z. N., Ben-Dov, E., Miaskovsky, R., Boussiba, S., & Zaritsky, A. (2003). Enduring toxicity of transgenic *Anabaena* PCC 7120 expressing mosquito larvicidal genes from *Bacillus thuringiensis* ssp. *israelensis*. *Environ. Microbiol., 5*, 997–1001.

Manceva, S. D., Pusztai-Carey, M., Russo, P. S., & Butko, P. (2005). A detergent-like mechanism of action of the cytolytic toxin Cyt1A from *Bacillus thuringiensis* var. *israelensis*. *Biochemistry, 44*, 589–597.

Manfredini, F., Beani, L., Taormina, M., & Vannini, L. (2010). Parasitic infection protects wasp larvae against a bacterial challenge. *Microbes. Infect., 12*, 727–735.

Margalit, J., & Dean, D. (1985). The story of *Bacillus thuringiensis* var. *israelensis* (B.t.i.). *J. Am. Mosq. Control Assoc., 1*, 1–7.

Marshall, S. D., Gatehouse, L. N., Becher, S. A., Christeller, J. T., Gatehouse, H. S., Hurst, M. R., Boucias, D. G., & Jackson, T. A. (2008). Serine proteases identified from a *Costelytra zealandica* (White) (Coleoptera: Scarabaeidae) midgut EST library and their expression through insect development. *Insect Mol. Biol., 17*, 247–259.

Martin, P. A., & Travers, R. S. (1989). Worldwide abundance and distribution of *Bacillus thuringiensis* isolates. *Appl. Environ. Microbiol., 55*, 2437–2442.

Martin, P. A., Gundersen-Rindal, D., Blackburn, M., & Buyer, J. (2007a). *Chromobacterium subtsugae* sp. nov., a betaproteobacterium toxic to Colorado potato beetle and other insect pests. *Int. J. Syst. Evol. Microbiol., 57*, 993–999.

Martin, P. A., Hirose, E., & Aldrich, J. R. (2007b). Toxicity of *Chromobacterium subtsugae* to southern green stink bug (Heteroptera: Pentatomidae) and corn rootworm (Coleoptera: Chrysomelidae). *J. Econ. Entomol., 100*, 680–684.

Martin, P. A., Farrar, R. R., Jr., & Blackburn, M. B. (2009). Survival of diverse *Bacillus thuringiensis* strains in gypsy moth (Lepidoptera: Lymantriidae) is correlated with urease production. *Biol. Control, 51*, 147–151.

Martin, P. A., Gundersen-Rindal, D. E., & Blackburn, M. B. (2010). Distribution of phenotypes among *Bacillus thuringiensis* strains. *Syst. Appl. Microbiol., 33*, 204–208.

Martins, E. S., Monnerat, R. G., Queiroz, P. R., Dumas, V. F., Braz, S. V., de Souza Aguiar, R. W., Gomes, A. C., Sánchez, J., Bravo, A., & Ribeiro, B. M. (2010). Midgut GPI-anchored proteins with alkaline phosphatase activity from the cotton boll weevil (*Anthonomus grandis*) are putative receptors for the Cry1B protein of *Bacillus thuringiensis*. *Insect Biochem. Mol. Biol., 40*, 138–145.

Marzban, R., He, Q., Liu, X., & Zhang, Q. (2009). Effects of *Bacillus thuringiensis* toxin Cry1Ac and cytoplasmic polyhedrosis virus of *Helicoverpa armigera* (Hubner) (HaCPV) on cotton bollworm (Lepidoptera: Noctuidae). *J. Invertebr. Pathol., 101*, 71–76.

Masson, L., Lu, Y.-j., Mazza, A., Brosseau, R., & Adang, M. J. (1995). The CryIA(c) receptor purified from *Manduca sexta* displays multiple specificities. *J. Biol. Chem., 270*, 20309–20315.

Masson, L., Erlandson, M., Puzstai-Carey, M., Brousseau, R., Juárez-Pérez, V., & Frutos, R. (1998). A holistic approach for determining the entomopathogenic potential of *Bacillus thuringiensis* strains. *Appl. Environ. Microbiol., 64*, 4782–4788.

Masson, L., Tabashnik, B. E., Liu, Y. B., & Schwartz, J. L. (1999). Helix 4 of the *Bacillus thuringiensis* Cry1Aa toxin lines the lumen of the ion channel. *J. Biol. Chem., 274*. 3196–2000.

Masson, L., Schwab, G., Mazza, A., Brousseau, R., Potvin, L., & Schwartz, J. L. (2004). A novel *Bacillus thuringiensis* (PS149B1) containing a Cry34Ab1/Cry35Ab1 binary toxin specific for the western corn rootworm *Diabrotica virgifera virgifera* LeConte forms ion channels in lipid membranes. *Biochemistry, 43*, 12349–12357.

Mathy, N., Bénard, L., Pellegrini, O., Daou, R., Wen, T., & Condon, C. (2007). 5′-to-3′ exoribonuclease activity in bacteria: role of RNase J1 in rRNA maturation and 5′ stability of mRNA. *Cell, 129*, 681–692.

McClintock, J. T., Schaffer, C. R., & Sjoblad, R. D. (1995). A comparative review of the mammalian toxicity of *Bacillus thuringiensis*-based pesticides. *Pestic. Sci., 45*, 95–105.

McConnell, E., & Richards, A. G. (1959). The production by *Bacillus thuringiensis* Berliner of a heat-stable substance toxic for insects. *Can. J. Microbiol., 5*, 161–168.

McLean, K. M., & Whiteley, H. R. (1987). Expression in *Escherichia coli* of a cloned crystal protein gene of *Bacillus thuringiensis* subsp. *israelensis*. *J. Bacteriol., 169*, 1017–1023.

McNall, R. J., & Adang, M. J. (2003). Identification of novel *Bacillus thuringiensis* Cry1Ac binding proteins in *Manduca sexta* midgut through proteomic analysis. *Insect Biochem. Mol. Biol., 33*, 999–1010.

McNeill, M. R. (2000). Effect of container type on suitability of the pathogen *Serratia marcescens* − *Microctonus hyperodae* (Hym., Braconidae) association to indicate parasitoid oviposition attempts. *J. Appl. Entomol., 124*, 93–98.

McNeill, M. R., & Hurst, M. R. H. (2008). *Yersinia* sp. (MH96) − a potential biopesticide of migratory locust *Locusta migratoria* L. *N.Z. Plant Prot., 61*, 236–242.

McVay, J. R., Gudauskas, R. T., & Harper, J. D. (1977). Effects of *Bacillus thuringiensis* nuclear-polyhedrosis virus mixtures on *Trichoplusia ni* larvae. *J. Invertebr. Pathol., 29*, 367–372.

Mehlo, L., Gahakwa, D., Nghia, P. T., Loc, N. T., Capell, T., Gatehouse, J. A., Gatehouse, A. M., & Christou, P. (2005). An alternative strategy for sustainable pest resistance in genetically enhanced crops. *Proc. Natl. Acad. Sci. USA, 102*, 7812–7816.

de Melo, J. V., Jones, G. W., Berry, C., Vasconcelos, R. H., de Oliveira, C. M., Furtado, A. F., Peixoto, C. A., & Silva-Filha, M. H. (2009). Cytopathological effects of *Bacillus sphaericus* Cry48Aa/Cry49Aa toxin on binary toxin-susceptible and -resistant *Culex quinquefasciatus* larvae. *Appl. Environ. Microbiol., 75*, 4782–4789.

Metting, F. B. (1993). Structure and physiological ecology of soil microbial communities. In F. B. Metting (Ed.), *Soil Microbial Ecology: Applications in Agricultural and Environmental Management* (pp. 3–25). New York: Marcel Dekker.

Meunier, L., Préfontaine, G., Van Munster, M., Brousseau, R., & Masson, L. (2006). Transcriptional response of *Choristoneura fumiferana* to sublethal exposure of Cry1Ab protoxin from *Bacillus thuringiensis*. *Insect Mol. Biol., 15*, 475–483.

Midboe, E. G., Candas, M., & Bulla, L. A., Jr. (2003). Expression of a midgut-specific cadherin BT-R1 during the development of *Manduca sexta* larva. *Comp. Biochem. Physiol. B., 135*, 125–137.

Milner, R. J. (1981a). Identification of the *Bacillus popilliae* group of insect pathogens. In H. D. Burges (Ed.), *Microbial control of pests and plant diseases 1970–1980* (pp. 45–59). London: Academic Press.

Milner, R. J. (1981b). A novel milky disease organism from Australian scarabaeids: field occurrence, isolation, and infectivity. *J. Invertebr. Pathol., 37,* 304–309.

Mittal, P. K. (2003). Biolarvicides in vector control: challenges and prospects. *J. Vector Borne Dis., 40,* 20–32.

Miyasono, M., Inagaki, S., Yamamoto, M., Ohba, K., Ishiguro, T., Takeda, R., & Hayashi, Y. (1994). Enhancement of δ-endotoxin activity by toxin-free spore of *Bacillus thuringiensis* against the diamondback moth, *Plutella xylostella. J. Invertebr. Pathol., 63,* 111–112.

Moar, W., Osbrink, W. L. A., & Trumble, J. T. (1986). Potentiation of *Bacillus thuringiensis* var. *kurstaki* with thuringiensin on beet armyworm (Lepidoptera: Noctuidae). *J. Econ. Entomol., 79,* 1443–1446.

Moar, W. J., Trumble, J. T., Hice, R. H., & Backman, P. A. (1994). Insecticidal activity of the CryIIA protein from the NRD-12 isolate of *Bacillus thuringiensis* subsp. *kurstaki* expressed in *Escherichia coli* and *Bacillus thuringiensis* and in a leaf-colonizing strain of *Bacillus cereus. Appl. Environ. Microbiol., 60,* 896–902.

Mohd-Salleh, M. B., & Lewis, L. C. (1982). Feeding deterrent response of corn insects to β-exotoxin of *Bacillus thuringiensis. J. Invertebr. Pathol., 39,* 323–328.

Monk, J., Young., S. D., Vink, C. J., Winder, L. M., & Hurst, M. R. H. (2010). Q-PCR and high-resolution DNA melting analysis for simple and efficient detection of biocontrol agents. In H. J. Ridgway, T. R. Glare, S. A. Wakelin & M. O'Callaghan (Eds.), *Paddock to PCR: Demystifying Molecular Technologies for Practical Plant Protection* (pp. 117–124). Lincoln: New Zealand Plant Protection Society.

Monnerat, R. G., Soares, C. M., Capdeville, G., Jones, G., Martins, E. S., Praça, L., Cordeiro, B. A., Braz, S. V., Dos Santos, R. C., & Berry, C. (2009). Translocation and insecticidal activity of *Bacillus thuringiensis* living inside of plants. *Microb. Biotechnol., 2,* 512–520.

Moonsom, S., Chaisri, U., Kasinrerk, W., & Angsuthanasombat, C. (2007). Binding characteristics to mosquito-larval midgut proteins of the cloned domain II–III fragment from the *Bacillus thuringiensis* Cry4Ba toxin. *J. Biochem. Mol. Biol., 40,* 783–790.

Moran, N. A., Russell, J. A., Koga, R., & Fukatsu, T. (2005). Evolutionary relationships of three new species of Enterobacteriaceae living as symbionts of aphids and other insects. *Appl. Environ. Microbiol., 71,* 3302–3310.

Moreau, G., & Bauce, E. (2003). Lethal and sublethal effects of single and double applications of *Bacillus thuringiensis* variety *kurstaki* on spruce budworm (Lepidoptera: Tortricidae) larvae. *J. Econ. Entomol., 96,* 280–286.

Morin, S., Biggs, R. W., Sisterson, M. S., Shriver, L., Ellers-Kirk, C., Higginson, D., Holley, D., Gahan, L. J., Heckel, D. G., Carrière, Y., Dennehy, T. J., Brown, J. K., & Tabashnik, B. E. (2003). Three cadherin alleles associated with resistance to *Bacillus thuringiensis* in pink bollworm. *Proc. Natl. Acad. Sci. USA, 100,* 5004–5009.

Morón Rios, M. A. (2004). *Escarabajos: 200 Millones de Años de Evolución* (2nd ed.). Zaragoza: Instituto de Ecología, A.C., y Sociedad Entomológica Aragonesa.

Morse, R. J., Yamamoto, T., & Stroud, R. M. (2001). Structure of Cry2Aa suggests an unexpected receptor binding epitope. *Structure, 9,* 409–417.

Mulla, M. S., Darwazeh, H., Davidson, E. W., & Dulmage, H. T. (1984). Efficacy and persistence of the microbial agent *Bacillus sphaericus* for the control of mosquito larvae in organically enriched habitats. *Mosq. News., 44,* 166–173.

Mulligan, F. S., Schaefer, C. H., & Wilder, W. H. (1980). Efficacy and persistence of *Bacillus sphaericus* and *B. thuringiensis* H. 14 against mosquitoes under laboratory and field conditions. *J. Econ. Entomol., 73,* 684–688.

Munkvold, G. P., Hellmich, R. L., & Showers, W. B. (1997). Reduced *Fusarium* ear rot and symptomless infection in kernels of maize genetically engineered for European corn borer resistance. *Phytopathology, 87,* 1071–1077.

Muñoz-Garay, C., Sánchez, J., Darszon, A., de Maagd, R. A., Bakker, P., Soberón, M., & Bravo, A. (2006). Permeability changes of *Manduca sexta* midgut brush border membranes induced by oligomeric structures of different cry toxins. *J. Membr. Biol., 212,* 61–68.

Musser, R. O., Kwon, H. S., Williams, S. A., White, C. J., Romano, M. A., Holt, S. M., Bradbury, S., Brown, J. K., & Felton, G. W. (2005). Evidence that caterpillar labial saliva suppresses infectivity of potential bacterial pathogens. *Arch. Insect Biochem. Physiol., 58,* 138–144.

Myasnik, M., Manasherob, R., Ben-Dov, E., Zaritsky, A., Margalith, Y., & Barak, Z. (2001). Comparative sensitivity to UV-B radiation of two *Bacillus thuringiensis* subspecies and other *Bacillus* sp. *Curr. Microbiol., 43,* 140–143.

Nagamatsu, Y., Koike, T., Sasaki, K., Yoshimoto, A., & Furukawa, Y. (1999). The cadherin-like protein is essential to specificity determination and cytotoxic action of the *Bacillus thuringiensis* insecticidal CryIAa toxin. *FEBS Lett., 460,* 385–390.

Naimov, S., Weemen-Hendriks, M., Dukiandjiev, S., & de Maagd, R. A. (2001). *Bacillus thuringiensis* delta-endotoxin Cry1 hybrid proteins with increased activity against the Colorado potato beetle. *Appl. Environ. Microbiol., 67,* 5328–5330.

Nair, M. S., & Dean, D. H. (2008). All domains of Cry1A toxins insert into insect brush border membranes. *J. Biol. Chem., 283,* 26324–26331.

Nakamura, L. K. (2000). Phylogeny of *Bacillus sphaericus*-like organisms. *Int. J. Syst. Evol. Microbiol., 50* (Pt 5), 1715–1722.

Nambiar, P. T. C., Ma, S. W., & Iyer, V. N. (1990). Limiting an insect infestation of nitrogen-fixing root nodules of the pigeon pea (*Cajanus cajan*) by engineering the expression of an entomocidal gene in its root nodules. *Appl. Environ. Microbiol., 56,* 2866–2869.

Naranjo, S. E. (2009). Impacts of Bt crops on non-target invertebrates and insecticide use patterns. *CAB Reviews: Perspectives in Agriculture, Veterinary Science, Nutrition and Natural Resources, 4,* 1–11.

Nester, E. W., Thomashow, L. S., Metz, M., & Gordon, M. (2002). *100 Years of* Bacillus thuringiensis*: A Critical Scientific Assessment.* Washington, DC: American Academy of Microbiology.

Nicolas, L., Nielsen-Leroux, C., Charles, J. F., & Delécluse, A. (1993). Respective role of the 42- and 51-kDa components of the *Bacillus sphaericus* toxin overexpressed in *Bacillus thuringiensis. FEMS Microbiol. Lett., 106,* 275–279.

Ning, C., Wu, K., Liu, C., Gao, Y., Jurat-Fuentes, J. L., & Gao, X. (2010). Characterization of a Cry1Ac toxin-binding alkaline phosphatase in the midgut from *Helicoverpa armigera* (Hubner) larvae. *J. Insect Physiol., 56,* 666–672.

Nishiitsutsuji-Uwo, J., & Endo, Y. (1980). Mode of action of *Bacillus thuringiensis* δ-endotoxin: general characteristics of intoxicated *Bombyx* larvae. *J. Invertebr. Pathol., 35,* 219–228.

Nishimoto, T., Yoshisue, H., Ihara, K., Sakai, H., & Komano, T. (1994). Functional analysis of block 5, one of the highly conserved amino acid sequences in the 130-kDa CryIVA protein produced by *Bacillus thuringiensis* subsp. *israelensis. FEBS Lett., 348,* 249−254.

Nishiwaki, H., Nakashima, K., Ishida, C., Kawamura, T., & Matsuda, K. (2007). Cloning, functional characterization, and mode of action of a novel insecticidal pore-forming toxin, sphaericolysin, produced by *Bacillus sphaericus. Appl. Environ. Microbiol., 73,* 3404−3411.

Nisnevitch, M., Cohen, S., Ben-Dov, E., Zaritsky, A., Sofer, Y., & Cahan, R. (2006). Cyt2Ba of *Bacillus thuringiensis israelensis*: activation by putative endogenous protease. *Biochem. Biophys. Res. Commun., 344,* 99−105.

Noguera, P. A., & Ibarra, J. E. (2010). Detection of new *cry* genes of *Bacillus thuringiensis* by use of a novel PCR primer system. *Appl. Environ. Microbiol., 76,* 6150−6155.

Norris, J. R. (1964). The classification of *Bacillus thuringiensis. J. Appl. Bacteriol., 27,* 439−447.

Northrup, Z. (1914). A bacterial disease of the larvae of the June beetle, *Lachnosterna* spp. *Mich. Agric. Exp. Stn. Tech. Bull., 18.*

Nuñez-Valdez, M. E., Calderón, M. A., Aranda, E., Hernández, L., Ramírez-Gama, R. M., Lina, L., Rodríguez-Segura, Z., Gutiérrez Mdel, C., & Villalobos, F. J. (2008). Identification of a putative Mexican strain of *Serratia entomophila* pathogenic against root-damaging larvae of Scarabaeidae (Coleoptera). *Appl. Environ. Microbiol., 74,* 802−810.

Nyouki, F. F. R., Fuxa, J. R., & Richter, A. R. (1996). Spore−toxin interactions and sublethal effects of *Bacillus thuringiensis* in *Spodoptera frugiperda* and *Pseudoplusia includens* (Lepidoptera: Noctuidae). *J. Entomol. Sci., 31,* 52−62.

Obukowicz, M. G., Perlak, F. J., & Kusano-Kretzmer, K. (1986). Integration of the delta-endotoxin gene of *Bacillus thuringiensis* into the chromosome of root-colonizing strains of pseudomonads using Tn5. *Gene., 45,* 327−331.

O'Callaghan, M., & Jackson, T. A. (1993). Isolation and enumeration of *Serratia entomophila* − a bacterial pathogen of the New Zealand grass grub, *Costelytra zealandica. J. Appl. Bacteriol., 75,* 307−314.

O'Callaghan, M., Garnham, M. L., Nelson, T. L., Baird, D., & Jackson, T. A. (1996). The pathogenicity of *Serratia* strains to *Lucilia sericata* (Diptera: Calliphoridae). *J. Invertebr. Pathol., 68,* 22−27.

O'Callaghan, M., Jackson, T. A., & Glare, T. R. (1997). *Serratia entomophila* bacteriophages: host range determination and preliminary characterization. *Can. J. Microbiol., 43,* 1069−1073.

O'Callaghan, M., Young, S. D., Barlow, N. D., & Jackson, T. A. (1999). The ecology of grass grub pathogenic *Serratia* spp. in New Zealand pastures. In J. N. Mathiessen (Ed.), *Proceedings of the 7th Australasian Conference on Grassland Invertebrate Ecology* (pp. 85−91). Perth: CSIRO Entomology.

O'Callaghan, M., Glare, T. R., Burgess, E. P., & Malone, L. A. (2005). Effects of plants genetically modified for insect resistance on nontarget organisms. *Annu. Rev. Entomol., 50,* 271−292.

Ochoa-Campuzano, C., Real, M. D., Martínez-Ramírez, A. C., Bravo, A., & Rausell, C. (2007). An ADAM metalloprotease is a Cry3Aa *Bacillus thuringiensis* toxin receptor. *Biochem. Biophys. Res. Commun., 362,* 437−442.

Oehler, D. D., Gingrich, R. E., & Haufler, M. (1982). High-performance liquid chromatographic determination of β-exotoxin produced by the bacterium *Bacillus thuringiensis. J. Agric. Food Chem., 30,* 407−408.

Oei, C., Hindley, J., & Berry, C. (1992). Binding of purified *Bacillus sphaericus* binary toxin and its deletion derivatives to *Culex*

quinquefasciatus gut: elucidation of functional binding domains. *J. Gen. Microbiol., 138,* 1515−1526.

Ohba, M. (1996). *Bacillus thuringiensis* populations naturally occurring on mulberry leaves: a possible source of the populations associated with silkworm-rearing insectaries. *J. Appl. Bacteriol., 80,* 56−64.

Ohba, M., & Aizawa, K. (1979). New subspecies of *Bacillus thuringiensis* possessing 11a:11c flagellar antigenic structure: *Bacillus thuringiensis* subsp. *kyushuensis. J. Invertebr. Pathol., 33,* 387−388.

Ohba, M., & Aizawa, K. (1986). *Bacillus thuringiensis* subsp. *japonensis* (flagellar serotype 23): a new subspecies of *Bacillus thuringiensis* with a novel flagellar antigen. *J. Invertebr. Pathol., 48,* 129−130.

Ohba, M., & Aizawa, K. (1989). New flagellar (H) antigenic subfactors in *Bacillus thuringiensis* H-serotype 3 with description of 2 new subspecies, *Bacillus thuringiensis* subsp. *sumiyoshiensis* (H-serotype 3a:3d) and *Bacillus thuringiensis* subsp. *fukuokaensis* (H-serotype 3a:3d:3e). *J. Invertebr. Pathol., 54,* 208−212.

Ohba, M., Aizawa, K., & Shimizu, S. (1981a). A new subspecies of *Bacillus thuringiensis* isolated in Japan: *Bacillus thuringiensis tohokuensis* (serotype H-17). *J. Invertebr. Pathol., 38,* 307−309.

Ohba, M., Ono, K., Aizawa, K., & Iwana, M. I. (1981b). Two new subspecies of *Bacillus thuringiensis* isolated in Japan; B. *thuringiensis* subsp. *kumamotoensis* (serotype H-18) and *B. thuringiensis* subsp. *tochigiensis* (serotype H-19). *J. Invertebr. Pathol., 38,* 184−190.

Ohba, M., Tantichodok, A., & Aizawa, K. (1981c). Production of heat-stable exotoxin by *Bacillus thuringiensis* and related bacteria. *J. Invertebr. Pathol., 38,* 26−32.

Ohba, M., Saitoh, H., Miyamoto, K., Higuchi, K., & Mizuki, E. (1995). *Bacillus thuringiensis* serovar higo (flagellar serotype 44), a new serogroup with a larvicidal activity preferential for the anopheline mosquito. *Lett. Appl. Microbiol., 21,* 316−318.

Oliveira, C. D., Tadei, W. P., & Abdalla, F. C. (2009). Occurrence of apocrine secretion in the larval gut epithelial cells of *Aedes aegypti* L., *Anopheles albitarsis* Lynch-Arribalzaga and *Culex quinquefasciatus* Say (Diptera: Culicidae): a defense strategy against infection by *Bacillus sphaericus* Neide? *Neotrop. Entomol., 38,* 624−631.

de Oliveira, E. J., Rabinovitch, L., Monnerat, R. G., Passos, L. K. J., & Zahner, V. (2004). Molecular characterization of *Brevibacillus laterosporus* and its potential use in biological control. *Appl. Environ. Microbiol., 70,* 6657−6664.

Opota, O., Charles, J. F., Warot, S., Pauron, D., & Darboux, I. (2008). Identification and characterization of the receptor for the *Bacillus sphaericus* binary toxin in the malaria vector mosquito, *Anopheles gambiae. Comp. Biochem. Physiol. B., 149,* 419−427.

Opota, O., Gauthier, N. C., Doye, A., Berry, C., Gounon, P., Lemichez, E., & Pauron, D. (2011). *Bacillus sphaericus* binary toxin elicits host cell autophagy as a response to intoxication. *PLoS ONE, 6,* e14682.

Oppert, B. (1999). Protease interactions with *Bacillus thuringiensis* insecticidal toxins. *Arch. Insect Biochem. Physiol., 42,* 1−12.

Orduz, S., Rojas, W., Correa, M. M., Montoya, A. E., & de Barjac, H. (1992). A new serotype of *Bacillus thuringiensis* from Colombia toxic to mosquito larvae. *J. Invertebr. Pathol., 59,* 99−103.

Orlova, M. V., Smirnova, T. A., Ganushkina, L. A., Yacubovich, V. Y., & Azizbekyan, R. R. (1998). Insecticidal activity of *Bacillus laterosporus. Appl. Environ. Microbiol., 64,* 2723−2725.

Ounjai, P., Unger, V. M., Sigworth, F. J., & Angsuthanasombat, C. (2007). Two conformational states of the membrane-associated *Bacillus thuringiensis* Cry4Ba δ-endotoxin complex revealed by electron crystallography: implications for toxin-pore formation. *Biochem. Biophys. Res. Commun., 361,* 890−895.

Ozawa, K., & Iwahana, H. (1986). Involvement of a transmissible plasmid in heat-stable exotoxin and delta-endotoxin production in *Bacillus thuringiensis* subspecies *darmstadiensis. Curr. Microbiol., 13,* 337−340.

Ozbek, E., Miller, S. A., Meulia, T., & Hogenhout, S. A. (2003). Infection and replication sites of *Spiroplasma kunkelii* (Class: Mollicutes) in midgut and Malpighian tubules of the leafhopper *Dalbulus maidis. J. Invertebr. Pathol., 82,* 167−175.

Pacheco, S., Gómez, I., Arenas, I., Saab-Rincon, G., Rodríguez-Almazán, C., Gill, S. S., Bravo, A., & Soberón, M. (2009a). Domain II loop 3 of *Bacillus thuringiensis* Cry1Ab toxin is involved in a "ping pong" binding mechanism with *Manduca sexta* aminopeptidase-N and cadherin receptors. *J. Biol. Chem., 284,* 32750−32757.

Pacheco, S., Gómez, I., Gill, S. S., Bravo, A., & Soberón, M. (2009b). Enhancement of insecticidal activity of *Bacillus thuringiensis* Cry1A toxins by fragments of a toxin-binding cadherin correlates with oligomer formation. *Peptides, 30,* 583−588.

Padbidri, V. S., Mourya, D. T., & Dhanda, V. (1982). Multiplication of *Coxiella burnetii*in *Aedes aegypti. Indian J. Med. Res., 76,* 185−189.

Pandian, G. N., Ishikawa, T., Togashi, M., Shitomi, Y., Haginoya, K., Yamamoto, S., Nishiumi, T., & Hori, H. (2008). *Bombyx mori* midgut membrane protein P252, which binds to *Bacillus thuringiensis* Cry1A, is a chlorophyllide-binding protein, and the resulting complex has antimicrobial activity. *Appl. Environ. Microbiol., 74,* 1324−1331.

Pandian, G. N., Ishikawa, T., Vaijayanthi, T., Hossain, D. M., Yamamoto, S., Nishiumi, T., Angsuthanasombat, C., Haginoya, K., Mitsui, T., & Hori, H. (2010). Formation of macromolecule complex with *Bacillus thuringiensis* Cry1A toxins and chlorophyllide binding 252-kDa lipocalin-like protein locating on *Bombyx mori* midgut membrane. *J. Membr. Biol., 237,* 125−136.

Pannucci, J., Okinaka, R. T., Sabin, R., & Kuske, C. R. (2002a). *Bacillus anthracis* pXO1 plasmid sequence conservation among closely related bacterial species. *J. Bacteriol., 184,* 134−141.

Pannucci, J., Okinaka, R. T., Williams, E., Sabin, R., Ticknor, L. O., & Kuske, C. R. (2002b). DNA sequence conservation between the *Bacillus anthracis* pXO2 plasmid and genomic sequence from closely related bacteria. *BMC Genomics, 3,* 34.

Pardo-López, L., Gómez, I., Rausell, C., Sanchez, J., Soberón, M., & Bravo, A. (2006). Structural changes of the Cry1Ac oligomeric pre-pore from *Bacillus thuringiensis* induced by *N*-acetylgalactosamine facilitates toxin membrane insertion. *Biochemistry, 45,* 10329−10336.

Pardo-López, L., Muñoz-Garay, C., Porta, H., Rodríguez-Almazán, C., Soberón, M., & Bravo, A. (2009). Strategies to improve the insecticidal activity of Cry toxins from *Bacillus thuringiensis. Peptides, 30,* 589−595.

Park, H. W., & Federici, B. A. (2004). Effect of specific mutations in helix alpha7 of domain I on the stability and crystallization of Cry3A in *Bacillus thuringiensis. Mol. Biotechnol., 27,* 89−100.

Park, H. W., Bideshi, D. K., Johnson, J. J., & Federici, B. A. (1999). Differential enhancement of Cry2A versus Cry11A yields in *Bacillus thuringiensis* by use of the *cry3A* STAB mRNA sequence. *FEMS Microbiol. Lett., 181,* 319−327.

Park, H. W., Bideshi, D. K., & Federici, B. A. (2000). Molecular genetic manipulation of truncated Cry1C protein synthesis in *Bacillus*

thuringiensis to improve stability and yield. *Appl. Environ. Microbiol., 66,* 4449−4455.

Park, H. W., Delécluse, A., & Federici, B. A. (2001). Construction and characterization of a recombinant *Bacillus thuringiensis* subsp. *israelensis* strain that produces Cry11B. *J. Invertebr. Pathol., 78,* 37−44.

Park, H. W., Bideshi, D. K., & Federici, B. A. (2003). Recombinant strain of *Bacillus thuringiensis* producing Cyt1A, Cry11B, and the *Bacillus sphaericus* binary toxin. *Appl. Environ. Microbiol., 69,* 1331−1334.

Park, H. W., Bideshi, D. K., & Federici, B. A. (2007a). The 20-kDa protein of *Bacillus thuringiensis* subsp. *israelensis* enhances *Bacillus sphaericus* 2362 bin toxin synthesis. *Curr. Microbiol., 55,* 119−124.

Park, H. W., Mangum, C. M., Zhong, H. E., & Hayes, S. R. (2007b). Isolation of *Bacillus sphaericus* with improved efficacy against *Culex quinquefasciatus. J. Am. Mosq. Control Assoc., 23,* 478−480.

Park, H. W., Bideshi, D. K., & Federici, B. A. (2010). Properties and applied use of the mosquitocidal bacterium, *Bacillus sphaericus. J. Asia-Pacif. Entomol., 13,* 159−168.

Park, Y., Abdullah, M. A., Taylor, M. D., Rahman, K., & Adang, M. J. (2009). Enhancement of *Bacillus thuringiensis* Cry3Aa and Cry3Bb toxicities to coleopteran larvae by a toxin-binding fragment of an insect cadherin. *Appl. Environ. Microbiol., 75,* 3086−3092.

Peacock, J. W., Schweitzer, D. F., Carter, J. L., & Dubois, N. R. (1998). Laboratory assessment of the effects of *Bacillus thuringiensis* on native Lepidoptera. *Environ. Entomol., 27,* 450−457.

Pedersen, A., Dedes, J., Gauthier, D., & van Frankenhuyzen, K. (1997). Sublethal effects of *Bacillus thuringiensis* on the spruce budworm, *Choristoneura fumiferana. Entomol. Exp. Appl., 83,* 253−262.

Pedersen, J. C., Damgaard, P. H., Eilenberg, J., & Hansen, B. M. (1995). Dispersal of *Bacillus thuringiensis* var. *kurstaki* in an experimental cabbage field. *Can. J. Microbiol., 41,* 118−125.

Peferoen, M. (1997). Progress and prospects for field use of Bt genes in crops. *Trends Biotechnol., 15,* 173−177.

Peña, G., Miranda-Rios, J., de la Riva, G., Pardo-López, L., Soberón, M., & Bravo, A. (2006). A *Bacillus thuringiensis* S-layer protein involved in toxicity against *Epilachna varivestis* (Coleoptera: Coccinellidae). *Appl. Environ. Microbiol., 72,* 353−360.

Pendleton, I. R. (1970). Sodium and potassium fluxes in *Philosamia ricini* during *Bacillus thuringiensis* protein crystal intoxication. *J. Invertebr. Pathol., 16,* 313−314.

Peng, D., Xu, X., Ruan, L., Yu, Z., & Sun, M. (2010a). Enhancing Cry1Ac toxicity by expression of the *Helicoverpa armigera* cadherin fragment in *Bacillus thuringiensis. Res. Microbiol., 161,* 383−389.

Peng, D., Xu, X., Ye, W., Yu, Z., & Sun, M. (2010b). *Helicoverpa armigera* cadherin fragment enhances Cry1Ac insecticidal activity by facilitating toxin-oligomer formation. *Appl. Microbiol. Biotechnol., 85,* 1033−1040.

Perani, M., & Bishop, A. H. (2000). Effects of media composition on δ-endotoxin production and morphology of *Bacillus thuringiensis* in wild types and spontaneously mutated strains. *Microbios., 101,* 47−66.

Perchat, S., Buisson, C., Chaufaux, J., Sanchis, V., Lereclus, D., & Gohar, M. (2005). *Bacillus cereus* produces several nonproteinaceous insecticidal exotoxins. *J. Invertebr. Pathol., 90,* 131−133.

Pérez, C., Fernandez, L. E., Sun, J., Folch, J. L., Gill, S. S., Soberón, M., & Bravo, A. (2005). *Bacillus thuringiensis* subsp. *israelensis* Cyt1Aa synergizes Cry11Aa toxin by functioning as a membrane-bound receptor. *Proc. Natl. Acad. Sci. USA, 102,* 18303−18308.

Pérez, C., Muñoz-Garay, C., Portugal, L. C., Sánchez, J., Gill, S. S., Soberón, M., & Bravo, A. (2007). *Bacillus thuringiensis* ssp. *israelensis* Cyt1Aa enhances activity of Cry11Aa toxin by facilitating the formation of a pre-pore oligomeric structure. *Cell Microbiol., 9*, 2931–2937.

Pérez-García, G., Basurto-Ríos, R., & Ibarra, J. E. (2010). Potential effect of a putative σ^H-driven promoter on the over expression of the Cry1Ac toxin of *Bacillus thuringiensis. J. Invertebr. Pathol., 104*, 140–146.

Perlak, F. J., Fuchs, R. L., Dean, D. A., McPherson, S. L., & Fischhoff, D. A. (1991). Modification of the coding sequence enhances plant expression of insect control protein genes. *Proc. Natl. Acad. Sci. USA., 88*, 3324–3328.

Pettersson, B., Rippere, K. E., Yousten, A. A., & Priest, F. G. (1999). Transfer of *Bacillus lentimorbus* and *Bacillus popilliae* to the genus *Paenibacillus* with emended descriptions of *Paenibacillus lentimorbus* comb. nov. and *Paenibacillus popilliae* comb. nov. *Int. J. Syst. Bacteriol., 49*, 531–540.

Peyronnet, O., Vachon, V., Schwartz, J. L., & Laprade, R. (2001). Ion channels induced in planar lipid bilayers by the *Bacillus thuringiensis* toxin Cry1Aa in the presence of gypsy moth (*Lymantria dispar*) brush border membrane. *J. Membr. Biol., 184*, 45–54.

Peyronnet, O., Nieman, B., Généreux, F., Vachon, V., Laprade, R., & Schwartz, J. L. (2002). Estimation of the radius of the pores formed by the *Bacillus thuringiensis* Cry1C δ-endotoxin in planar lipid bilayers. *Biochim. Biophys. Acta, 1567*, 113–122.

Phannachet, K., Raksat, P., Limvuttegrijeerat, T., & Promdonkoy, B. (2010). Production and characterization of N- and C-terminally truncated Mtx2: a mosquitocidal toxin from *Bacillus sphaericus. Curr. Microbiol., 61*, 549–553.

Pigott, C. R., & Ellar, D. J. (2007). Role of receptors in *Bacillus thuringiensis* crystal toxin activity. *Microbiol. Mol. Biol. Rev., 71*, 255–281.

Pigott, C. R., King, M. S., & Ellar, D. J. (2008). Investigating the properties of *Bacillus thuringiensis* Cry proteins with novel loop replacements created using combinatorial molecular biology. *Appl. Environ. Microbiol., 74*, 3497–3511.

Pingel, R. L., & Lewis, L. C. (1999). Effect of *Bacillus thuringiensis, Anagrapha falcifera* multiple nucleopolyhedrovirus, and their mixture on three lepidopteran corn ear pests. *J. Econ. Entomol., 92*, 91–96.

Pinnock, D. E., Brand, R. J., Jackson, K. L., & Milstead, J. E. (1974). The field persistence of *Bacillus thuringiensis* spores on *Cercis occidentalis* leaves. *J. Invertebr. Pathol., 23*, 341–346.

Pinnock, D. E., Brand, R. J., Milstead, J. E., & Jackson, K. L. (1975). Effect of tree species on the coverage and field persistence of *Bacillus thuringiensis* spores. *J. Invertebr. Pathol., 25*, 209–214.

Pinyon, R. A., Linedale, E. C., Webster, M. A., & Thomas, C. J. (1996). Tn5-induced *Xenorhabdus bovienii* lecithinase mutants demonstrate reduced virulence for *Galleria mellonella* larvae. *J. Appl. Bacteriol., 80*, 411–417.

Podgwaite, J. D., & Cosenza, B. J. (1976). A strain of *Serratia marcescens* pathogenic for larvae of *Lymantria dispar*: infectivity and mechanisms of pathogenicity. *J. Invertebr. Pathol., 27*, 199–208.

Poinar, G. O., Jr., Wassink, H. J. M., Leegwater-van der Linden, M. E., & van der Geest, L. P. S. (1979). *Serratia marcescens* as a pathogen of tsetse flies. *Acta Trop., 36*, 223–227.

Poncet, S., Dervyn, E., Klier, A., & Rapoport, G. (1997). Spo0A represses transcription of the cry toxin genes in *Bacillus thuringiensis. Microbiology, 143*, 2743–2751.

Poopathi, S., & Abidha. (2008). New bacterial culture medium for production of mosquito pathogenic bacilli using agro-poultry industrial wastes. *Biocontrol Sci. Technol., 18*, 535–540.

Poprawski, T. J., & Yule, W. N. (1990). Bacterial pathogens of *Phyllophaga* spp. (Col. Scarabaeidae) in southern Quebec, Canada. *J. Appl. Entomol., 109*, 414–422.

Porcar, M., & Caballero, P. (2000). Molecular and insecticidal characterization of a *Bacillus thuringiensis* strain isolated during a natural epizootic. *J. Appl. Microbiol., 89*, 309–316.

Porcar, M., Iriarte, J., Cosmao Dumanoir, V., Ferrandis, M. D., Lecadet, M., Ferré, J., & Caballero, P. (1999). Identification and characterization of the new *Bacillus thuringiensis* serovars *pirenaica* (serotype H57) and *iberica* (serotype H59). *J. Appl. Microbiol., 87*, 640–648.

Porter, A. G., Davidson, E. W., & Liu, J. W. (1993). Mosquitocidal toxins of bacilli and their genetic manipulation for effective biological control of mosquitoes. *Microbiol. Rev., 57*, 838–861.

Poulin, B., Lefebvre, G., & Paz, L. (2010). Red flag for green spray: adverse trophic effects of *Bti* on breeding birds. *J. Appl. Ecol., 47*, 884–889.

Pozsgay, M., Fast, P., & Kaplan, H. (1987). The effect of sunlight on the protein crystals from *Bacillus thuringiensis* var. *kurstaki* HD1 and NRD12: a raman spectroscopic study. *J. Invertebr. Pathol., 50*, 246–253.

Prabakaran, G., Hoti, S. L., Manonmani, A. M., & Balaraman, K. (2008). Coconut water as a cheap source for the production of δ endotoxin of *Bacillus thuringiensis* var. *israelensis*, a mosquito control agent. *Acta Trop., 105*, 35–38.

Prasertphon, S., Areekul, P., & Tanada, Y. (1973). Sporulation of *Bacillus thuringiensis* in host cadavers. *J. Invertebr. Pathol., 21*, 205–207.

Priest, F. G., & Dewar, S. J. (2000). Bacteria and insects. In F. G. Priest & M. Goodfellow (Eds.), *Applied Microbial Systematics* (pp. 165–202). Dordrecht: Kluwer Academic Publishers.

Priest, F. G., Kaji, D. A., Rosato, Y. B., & Canhos, V. P. (1994). Characterization of *Bacillus thuringiensis* and related bacteria by ribosomal RNA gene restriction fragment length polymorphisms. *Microbiology, 140*, 1015–1022.

Priest, F. G., Ebdrup, L., Zahner, V., & Carter, P. E. (1997). Distribution and characterization of mosquitocidal toxin genes in some strains of *Bacillus sphaericus. Appl. Environ. Microbiol., 63*, 1195–1198.

Promdonkoy, B., & Ellar, D. J. (2005). Structure–function relationships of a membrane pore forming toxin revealed by reversion mutagenesis. *Mol. Membr. Biol., 22*, 327–337.

Promdonkoy, B., Promdonkoy, P., & Panyim, S. (2005). Co-expression of *Bacillus thuringiensis* Cry4Ba and Cyt2Aa2 in *Escherichia coli* revealed high synergism against *Aedes aegypti* and *Culex quinquefasciatus* larvae. *FEMS Microbiol. Lett., 252*, 121–126.

Promdonkoy, B., Promdonkoy, P., Wongtawan, B., Boonserm, P., & Panyim, S. (2008). Cys31, Cys47, and Cys195 in BinA are essential for toxicity of a binary toxin from *Bacillus sphaericus. Curr. Microbiol., 56*, 334–338.

Puntheeranurak, T., Stroh, C., Zhu, R., Angsuthanasombat, C., & Hinterdorfer, P. (2005). Structure and distribution of the *Bacillus thuringiensis* Cry4Ba toxin in lipid membranes. *Ultramicroscopy, 105*, 115–124.

Qin, Y., Ying, S. H., Chen, Y., Shen, Z. C., & Feng, M. G. (2010). Integration of insecticidal protein Vip3Aa1 into *Beauveria bassiana* enhances fungal virulence to *Spodoptera litura* larvae by cuticle and *per os* infection. *Appl. Environ. Microbiol., 76*, 4611–4618.

Quesada-Moraga, E., García-Tóvar, E., Valverde-García, P., & Santiago-Álvarez, C. (2004). Isolation, geographical diversity and insecticidal activity of *Bacillus thuringiensis* from soils in Spain. *Microbiol. Res., 159*, 59–71.

Rabinovitch, L., de Jesus, F. F., Cavados, C. F., Zahner, V., Momen, H., da Silva, M. H., Dumanoir, V. C., Frachon, E., & Lecadet, M. M. (1995). *Bacillus thuringiensis* subsp. *oswaldocruzi* and *Bacillus thuringiensis* subsp. *brasiliensis*, two novel Brazilian strains which determine new serotype H38 and H39, respectively. *Mem. Inst. Oswaldo Cruz., 90*, 41–42.

Radek, R. (2000). Light and electron microscopic study of a *Rickettsiella* species from the cockroach *Blatta orientalis*. *J. Invertebr. Pathol., 76*, 249–256.

Radnedge, L., Agron, P. G., Hill, K. K., Jackson, P. J., Ticknor, L. O., Keim, P., & Andersen, G. L. (2003). Genome differences that distinguish *Bacillus anthracis* from *Bacillus cereus* and *Bacillus thuringiensis*. *Appl. Environ. Microbiol., 69*, 2755–2764.

Rahmet-Alla, M., & Rowley, A. F. (1989). Studies on the pathogenicity of different strains of *Bacillus cereus* for the cockroach, *Leucophaea maderae*. *J. Invertebr. Pathol., 53*, 190–196.

Rajagopal, R., Sivakumar, S., Agrawal, N., Malhotra, P., & Bhatnagar, R. K. (2002). Silencing of midgut aminopeptidase N of *Spodoptera litura* by double-stranded RNA establishes its role as *Bacillus thuringiensis* toxin receptor. *J. Biol. Chem., 277*, 46849–46851.

Rajamohan, F., Alcantara, E., Lee, M. K., Chen, X. J., Curtiss, A., & Dean, D. H. (1995). Single amino acid changes in domain II of *Bacillus thuringiensis* CryIAb δ-endotoxin affect irreversible binding to *Manduca sexta* midgut membrane vesicles. *J. Bacteriol., 177*, 2276–2282.

Rajamohan, F., Alzate, O., Cotrill, J. A., Curtiss, A., & Dean, D. H. (1996a). Protein engineering of *Bacillus thuringiensis* delta-endotoxin: Mutations at domain II of CryIAb enhance receptor affinity and toxicity toward gypsy moth larvae. *Proc. Natl. Acad. Sci. USA, 93*, 14338–14343.

Rajamohan, F., Cotrill, J. A., Gould, F., & Dean, D. H. (1996b). Role of domain II, loop 2 residues of *Bacillus thuringiensis* CryIAb delta-endotoxin in reversible and irreversible binding to *Manduca sexta* and *Heliothis virescens*. *J. Biol. Chem., 271*, 2390–2396.

Rajamohan, F., Hussain, S. R., Cotrill, J. A., Gould, F., & Dean, D. H. (1996c). Mutations at domain II, loop 3, of *Bacillus thuringiensis* CryIAa and CryIAb delta-endotoxins suggest loop 3 is involved in initial binding to lepidopteran midguts. *J. Biol. Chem., 271*, 25220–25226.

Ramírez-Prado, J. H., Martínez-Márquez, E. I., & Olmedo-Alvarez, G. (2006). cry1Aa lacks stability elements at its 5-UTR but integrity of its transcription terminator is critical to prevent decay of its transcript. *Curr. Microbiol., 53*, 23–29.

Ramoska, W. A., & Hopkins, T. L. (1981). Effects of mosquito larval feeding behavior on *Bacillus sphaericus* efficacy. *J. Invertebr. Pathol., 37*, 269–272.

Randhawa, G. J., Singh, M., & Grover, M. (2011). Bioinformatic analysis for allergenicity assessment of *Bacillus thuringiensis* Cry proteins expressed in insect-resistant food crops. *Food Chem. Toxicol., 49*, 356–362.

Rang, C., Bes, M., Lullien-Pellerin, V., Wu, D., Federici, B. A., & Frutos, R. (1996). Influence of the 20-kDa protein from *Bacillus thuringiensis* ssp. *israelensis* on the rate of production of truncated Cry1C proteins. *FEMS Microbiol. Lett., 141*, 261–264.

Rasko, D. A., Altherr, M. R., Han, C. S., & Ravel, J. (2005). Genomics of the *Bacillus cereus* group of organisms. *FEMS Microbiol. Rev., 29*, 303–329.

Rausell, C., García-Robles, I., Sánchez, J., Muñoz-Garay, C., Martínez-Ramírez, A. C., Real, M. D., & Bravo, A. (2004). Role of toxin activation on binding and pore formation activity of the *Bacillus thuringiensis* Cry3 toxins in membranes of *Leptinotarsa decemlineata* (Say). *Biochim. Biophys. Acta, 1660*, 99–105.

Rausell, C., Ochoa-Campuzano, C., Martínez-Ramírez, A. C., Bravo, A., & Real, M. D. (2007). A membrane associated metalloprotease cleaves Cry3Aa *Bacillus thuringiensis* toxin reducing pore formation in Colorado potato beetle brush border membrane vesicles. *Biochim. Biophys. Acta, 1768*, 2293–2299.

Ravoahangimalala, O., Charles, J. F., & Schoeller-Raccaud, J. (1993). Immunological localization of *Bacillus thuringiensis* serovar *israelensis* toxins in midgut cells of intoxicated *Anopheles gambiae* larvae (Diptera: Culicidae). *Res. Microbiol., 144*, 271–278.

Raymond, B., Elliot, S. L., & Ellis, R. J. (2008a). Quantifying the reproduction of *Bacillus thuringiensis* HD1 in cadavers and live larvae of *Plutella xylostella*. *J. Invertebr. Pathol., 98*, 307–313.

Raymond, B., Lijek, R. S., Griffiths, R. I., & Bonsall, M. B. (2008b). Ecological consequences of ingestion of *Bacillus cereus* on *Bacillus thuringiensis* infections and on the gut flora of a lepidopteran host. *J. Invertebr. Pathol., 99*, 103–111.

Raymond, B., Johnston, P. R., Wright, D. J., Ellis, R. J., Crickmore, N., & Bonsall, M. B. (2009). A mid-gut microbiota is not required for the pathogenicity of *Bacillus thuringiensis* to diamondback moth larvae. *Environ. Microbiol., 11*, 2556–2563.

Raymond, B., Wyres, K. L., Sheppard, S. K., Ellis, R. J., & Bonsall, M. B. (2010). Environmental factors determining the epidemiology and population genetic structure of the *Bacillus cereus* group in the field. *PLoS Pathog., 6*, e1000905.

Reddy, A., Battisti, L., & Thorne, C. B. (1987). Identification of self-transmissable plasmids in four *Bacillus thuringiensis* subspecies. *J. Bacteriol., 169*, 5263–5270.

Reddy, G. V. P. (2011). Comparative effect of integrated pest management and farmers' standard pest control practice for managing insect pests on cabbage (*Brassica* spp.). *Pest Manag. Sci., 67*, 980–985.

Reddy, S. T., Kumar, N. S., & Venkateswerlu, G. (1998). Comparative analysis of intracellular proteases in sporulated *Bacillus thuringiensis* strains. *Biotechnol. Lett., 20*, 279–281.

Rees, J. S., Jarrett, P., & Ellar, D. J. (2009). Peritrophic membrane contribution to Bt Cry delta-endotoxin susceptibility in Lepidoptera and the effect of Calcofluor. *J. Invertebr. Pathol., 100*, 139–146.

Reeves, W. K., & Nayduch, D. (2002). Pathogenic *Bacillus* from a larva of the *Simulium tuberosum* species complex (Diptera: Simuliidae). *J. Invertebr. Pathol., 79*, 126–128.

Regev, A., Keller, M., Strizhov, N., Sneh, B., Prudovsky, E., Chet, I., Ginzberg, I., Koncz-Kalman, Z., Koncz, C., Schell, J., & Zilberstein, A. (1996). Synergistic activity of a *Bacillus thuringiensis* delta-endotoxin and a bacterial endochitinase against *Spodoptera littoralis* larvae. *Appl. Environ. Microbiol., 62*, 3581–3586.

Rehacek, J. (1979). Persistence of *Coxiella burnetii* in beetles. *Dermestes maculatus* (Dermestidae). *Folia Parasitol., 26*, 39–43.

Ren, G. X., Li, K. T., Ying, M. H., & Yi, X. M. (1975). The classification of the strains of *Bacillus thuringiensis* group. *Acta Microbiol. Sin., 15*, 291–305.

Reyes-Ramírez, A., & Ibarra, J. E. (2005). Fingerprinting of *Bacillus thuringiensis* type strains and isolates by using *Bacillus cereus* group-specific repetitive extragenic palindromic sequence-based PCR analysis. *Appl. Environ. Microbiol., 71*, 1346–1355.

Reyes-Ramírez, A., & Ibarra, J. E. (2008). Plasmid patterns of *Bacillus thuringiensis* type strains. *Appl. Environ. Microbiol., 74*, 125–129.

Ribier, J., & Lecadet, M. M. (1973). Etude ultrastructurale et cinâetique de la sporulation de *Bacillus thuringiensis* var. Berliner 1715. Remarques sur la formation de l'inclusion parasporale. *Ann. Microbiol., 124*, 311–344.

Richards, G. R., & Goodrich-Blair, H. (2010). Examination of *Xenorhabdus nematophila* lipases in pathogenic and mutualistic host interactions reveals a role for xlpA in nematode progeny production. *Appl. Environ. Microbiol., 76*, 221–229.

Rigden, D. J. (2009). Does distant homology with Evf reveal a lipid binding site in *Bacillus thuringiensis* cytolytic toxins? *FEBS Lett., 583*, 1555–1560.

del Rincón-Castro, M. C., Barajas-Huerta, J., & Ibarra, J. E. (1999). Antagonism between Cry1Ac1 and Cyt1A1 toxins of *Bacillus thuringiensis*. *Appl. Environ. Microbiol., 65*, 2049–2053.

Rippere, K. E., Tran, M. T., Yousten, A. A., Hilu, K. H., & Klein, M. G. (1998). *Bacillus popilliae* and *Bacillus lentimorbus*, bacteria causing milky disease in Japanese beetles and related scarab larvae. *Int. J. Syst. Bacteriol., 48*, 395–402.

Rivers, D. B., Vann, C. N., Zimmack, H. L., & Dean, D. H. (1991). Mosquitocidal activity of *Bacillus laterosporus*. *J. Invertebr. Pathol., 58*, 444–447.

Rizwan-Ul-Haq, M., Hu, Q. B., Hu, M. Y., Zhong, G., & Weng, Q. (2009). Study of destruxin B and tea saponin, their interaction and synergism activities with *Bacillus thuringiensis kurstaki* against *Spodoptera exigua* (Hübner) (Lepidoptera: Noctuidae). *Appl. Entomol. Zool., 44*, 419–428.

Rodriguez-Almazan, C., Ruiz de Escudero, I., Cantón, P. E., Muñoz-Garay, C., Pérez, C., Gill, S. S., Soberón, M., & Bravo, A. (2011). The amino- and carboxyl-terminal fragments of the *Bacillus thuringensis* Cyt1Aa toxin have differential roles in toxin oligomerization and pore formation. *Biochemistry, 50*, 388–396.

Rodriguez-Padilla, C., Galan-Wong, L., de Barjac, H., Roman-Calderon, E., Tamez-Guerra, R., & Dulmage, H. (1990). *Bacillus thuringiensis* subspecies *neoleonensis* serotype H-24, a new subspecies which produces a triangular crystal. *J. Invertebr. Pathol., 56*, 280–282.

Roh, J. Y., Liu, Q., Lee, D. W., Tao, X., Wang, Y., Shim, H. J., Choi, J. Y., Seo, J. B., Ohba, M., Mizuki, E., & Je, Y. H. (2009a). *Bacillus thuringiensis* serovar *mogi* (flagellar serotype 3a3b3d), a novel serogroup with a mosquitocidal activity. *J. Invertebr. Pathol., 102*, 266–268.

Roh, J. Y., Nair, M. S., Liu, X. S., & Dean, D. H. (2009b). Mutagenic analysis of putative domain II and surface residues in mosquitocidal *Bacillus thuringiensis* Cry19Aa toxin. *FEMS Microbiol. Lett., 295*, 156–163.

Rojas-Avelizapa, L. I., Cruz-Camarillo, R., Guerrero, M. I., Rodríguez-Vázquez, R., & Ibarra, J. E. (1999). Selection and characterization of a proteo-chitinolytic strain of *Bacillus thuringiensis*, able to grow in shrimp waste media. *World J. Microbiol. Biotechnol., 15*, 261–268.

Rolff, J., & Reynolds, S. W. (2009). *Insect Infection and Immunity: Evolution, Ecology, and Mechanisms*. New York: Oxford University Press.

Romeis, J., Dutton, A., & Bigler, F. (2004). *Bacillus thuringiensis* toxin (Cry1Ab) has no direct effect on larvae of the green lacewing *Chrysoperla carnea* (Stephens) (Neuroptera: Chrysopidae). *J. Insect Physiol., 50*, 175–183.

Romeis, J., Meissle, M., & Bigler, F. (2006). Transgenic crops expressing *Bacillus thuringiensis* toxins and biological control. *Nat. Biotechnol., 24*, 63–71.

Rosa, J. C., De Oliveira, P. S., Garratt, R., Beltramini, L., Resing, K., Roque-Barreira, M.-C., & Green, L. J. (1999). KM+, a mannose-binding lectin from *Artocarpus integrifolia*: amino acid sequence, predicted tertiary structure, carbohydrate recognition, and analysis of the β-prism fold. *Protein Sci., 8*, 13–24.

Rosas-García, N. M. (2006). Laboratory and field tests of spray-dried and granular formulations of a *Bacillus thuringiensis* strain with insecticidal activity against the sugarcane borer. *Pest Manag. Sci., 62*, 855–861.

Roux, V., Bergoin, M., Lamaze, N., & Raoult, D. (1997). Reassessment of the taxonomic position of *Rickettsiella grylli*. *Int. J. Syst. Bacteriol., 47*, 1255–1257.

Roy, B. P., Selinger, L. B., & Khachatourians, G. G. (1987). Plasmid stability of *Bacillus thuringiensis* var. *kurstaki* (HD-1) during continuous phased cultivation. *Biotechnol. Lett., 9*, 483–488.

Ruhfel, R. E., Robillard, N. J., & Thorne, C. B. (1984). Interspecies transduction of plasmids among *Bacillus anthracis, B. cereus,* and *B. thuringiensis*. *J. Bacteriol., 157*, 708–711.

Ruiu, L., Floris, I., Satta, A., & Ellar, D. J. (2007a). Toxicity of a *Brevibacillus laterosporus* strain lacking parasporal crystals against *Musca domestica* and *Aedes aegypti*. *Biol. Control, 43*, 136–143.

Ruiu, L., Satta, A., & Floris, I. (2007b). Susceptibility of the house fly pupal parasitoid *Muscidifurax raptor* (Hymenoptera: Pteromalidae) to the entomopathogenic bacteria *Bacillus thuringiensis* and *Brevibacillus laterosporus*. *Biol. Control, 43*, 188–194.

Ruiu, L., Satta, A., & Floris, I. (2008). Immature house fly (*Musca domestica*) control in breeding sites with a new *Brevibacillus laterosporus* formulation. *Environ. Entomol., 37*, 505–509.

Ruiz, L. M., Segura, C., Trujillo, J., & Orduz, S. (2004). *In vivo* binding of the Cry11Bb toxin of *Bacillus thuringiensis* subsp. *medellin* to the midgut of mosquito larvae (Diptera: Culicidae). *Mem. Inst. Oswaldo Cruz., 99*, 73–79.

Salamitou, S., Ramisse, F., Brehélin, M., Bourguet, D., Gilois, N., Gominet, M., Hernandez, E., & Lereclus, D. (2000). The plcR regulon is involved in the opportunistic properties of *Bacillus thuringiensis* and *Bacillus cereus* in mice and insects. *Microbiology, 146*, 2825–2832.

Sampson, M. N., & Gooday, G. W. (1998). Involvement of chitinases of *Bacillus thuringiensis* during pathogenesis in insects. *Microbiology, 144*, 2189–2194.

Sanchis, V. (2011). From microbial sprays to insect-resistant transgenic plants: history of the biopesticide *Bacillus thuringiensis*. A review. *Agron. Sustain. Dev., 31*, 217–231.

Sanchis, V., Agaisse, H., Chaufaux, J., & Lereclus, D. (1996). Construction of new insecticidal *Bacillus thuringiensis* recombinant strains by using the sporulation non-dependent expression system of *cryIIIA* and a site specific recombination vector. *J. Biotechnol., 48*, 81–96.

Sanchis, V., Gohar, M., Chaufaux, J., Arantes, O., Meier, A., Agaisse, H., Cayley, J., & Lereclus, D. (1999). Development and field performance of a broad-spectrum nonviable asporogenic recombinant strain of *Bacillus thuringiensis* with greater potency and UV resistance. *Appl. Environ. Microbiol., 65*, 4032−4039.

Sangadala, S., Walters, F. S., English, L. H., & Adang, M. J. (1994). A mixture of *Manduca sexta* aminopeptidase and phosphatase enhances *Bacillus thuringiensis* insecticidal CryIA(c) toxin binding and $^{86}Rb^{(+)}$-K^{+} efflux *in vitro*. *J. Biol. Chem., 269*, 10088−10092.

Sanitt, P., Promdonkoy, B., & Boonserm, P. (2008). Targeted mutagenesis at charged residues in *Bacillus sphaericus* BinA toxin affects mosquito-larvicidal activity. *Curr. Microbiol., 57*, 230−234.

Sankaranarayanan, R., Sekar, K., Banerjee, R., Sharma, V., Surolia, A., & Vijayan, M. (1996). A novel mode of carbohydrate recognition in jacalin, a *Moraceae* plant lectin with a beta-prism fold. *Nat. Struct. Biol., 3*, 596−603.

Sarrafzadeh, M. H., Bigey, F., Capariccio, B., Mehrnia, M. R., Guiraud, J. P., & Navarro, J. M. (2007). Simple indicators of plasmid loss during fermentation of *Bacillus thuringiensis*. *Enzyme Microb. Technol., 40*, 1052−1058.

Sasaki, K., Jiaviriyaboonya, S., & Rogers, P. L. (1998). Enhancement of sporulation and crystal toxin production by cornsteep liquor feeding during intermittent fed-batch culture of *Bacillus sphaericus* 2362. *Biotechnol. Lett., 20*, 165−168.

Saxena, D., Ben-Dov, E., Manasherob, R., Barak, Z., Boussiba, S., & Zaritsky, A. (2002). A UV tolerant mutant of *Bacillus thuringiensis* subsp. *kurstaki* producing melanin. *Curr. Microbiol., 44*, 25−30.

Sayed, A., Nekl, E. R., Siqueira, H. A., Wang, H. C., ffrench-Constant, R. H., Bagley, M., & Siegfried, B. D. (2007). A novel cadherin-like gene from western corn rootworm, *Diabrotica virgifera virgifera* (Coleoptera: Chrysomelidae), larval midgut tissue. *Insect Mol. Biol., 16*, 591−600.

Sayyed, A. H., Crickmore, N., & Wright, D. J. (2001). Cyt1Aa from *Bacillus thuringiensis* subsp. *israelensis* is toxic to the diamondback moth, *Plutella xylostella*, and synergizes the activity of Cry1Ac towards a resistant strain. *Appl. Environ. Microbiol., 67*, 5859−5861.

Schachter, J., Pérez, M. M., & Quesada-Allué, L. A. (2007). The role of *N*-[beta]-alanyldopamine synthase in the innate immune response of two insects. *J. Insect Physiol., 53*, 1188−1197.

Scherrer, P. S., & Somerville, H. J. (1977). Membrane fractions from the outer layers of spores of *Bacillus thuringiensis* with toxicity to lepidopterous larvae. *Eur. J. Biochem., 72*, 479−490.

Schirmer, J., Just, I., & Aktories, K. (2002). The ADP-ribosylating mosquitocidal toxin from *Bacillus sphaericus*: proteolytic activation, enzyme activity, and cytotoxic effects. *J. Biol. Chem., 277*, 11941−11948.

Schnepf, E., Crickmore, N., Van Rie, J., Lereclus, D., Baum, J., Feitelson, J., Zeigler, D. R., & Dean, D. H. (1998). *Bacillus thuringiensis* and its pesticidal crystal proteins. *Microbiol. Mol. Biol. Rev., 62*, 775−806.

Schnepf, H. E., & Whiteley, H. R. (1981). Cloning and expression of the *Bacillus thuringiensis* crystal protein gene in *Escherichia coli*. *Proc. Natl. Acad. Sci. USA, 78*, 2893−2897.

Schnepf, H. E., Lee, S., Dojillo, J., Burmeister, P., Fencil, K., Morera, L., Nygaard, L., Narva, K. E., & Wolt, J. D. (2005). Characterization of Cry34/Cry35 binary insecticidal proteins from diverse *Bacillus thuringiensis* strain collections. *Appl. Environ. Microbiol., 71*, 1765−1774.

Schroeder, J. M., Chamberlain, C., & Davidson, E. W. (1989). Resistance to the *Bacillus sphaericus* toxin in cultured mosquito cells. *In Vitro Cell Dev. Biol., 25*, 887−891.

Schwartz, J. L., Juteau, M., Grochulski, P., Cygler, M., Préfontaine, G., Brousseau, R., & Masson, L. (1997a). Restriction of intramolecular movements within the Cry1Aa toxin molecule of *Bacillus thuringiensis* through disulfide bond engineering. *FEBS Lett., 410*, 397−402.

Schwartz, J. L., Lu, Y. J., Soehnlein, P., Brousseau, R., Masson, L., Laprade, R., & Adang, M. J. (1997b). Ion channels formed in planar lipid bilayers by *Bacillus thuringiensis* toxins in the presence of *Manduca sexta* midgut receptors. *FEBS Lett., 412*, 270−276.

Seale, J. W. (2005). The role of a conserved histidine−tyrosine inter-helical interaction in the ion channel domain of delta-endotoxins from *Bacillus thuringiensis*. *Proteins: Struct. Funct. Bioinform., 63*, 385−390.

Sebesta, K., & Horská, K. (1968). Inhibition of DNA-dependent RNA polymerase by the exotoxin of *Bacillus thuringiensis* var. *gelechiae*. *Biochim. Biophys. Acta, 169*, 281−282.

Sebesta, K., & Horská, K. (1970). Mechanism of inhibition of DNA-dependent RNA polymerase by exotoxin of *Bacillus thuringiensis*. *Biochim. Biophys. Acta, 209*.

Sebesta, K., Farkas, J., Horská, K., & Vankova, J. (1981). Thuringiensin, the beta-exotoxin of *Bacillus thuringiensis*. In H. D. Burges (Ed.), *Microbial Control of Pests and Plant Diseases 1970−1980* (pp. 249−281). London: Academic Press.

Sekar, V. (1988). The insecticidal crystal protein gene is expressed in vegetative cells of *Bacillus thuringiensis* var. *tenebrionis*. *Curr. Microbiol., 17*, 347−349.

Selander, R. K., Musser, J. M., Caugant, D. A., Gilmour, M. N., & Whittam, T. S. (1987). Population genetics of pathogenic bacteria. *Microb. Pathog., 3*, 1−7.

Seleena, P., Lee, H. L., & Lecadet, M. M. (1995). A new serovar of *Bacillus thuringiensis* possessing 28a28c flagellar antigenic structure: *Bacillus thuringiensis* servoar *jegathesan*, selectively toxic against mosquito larvae. *J. Am. Mosq. Control Assoc., 11*, 471−473.

Selvakumar, G., Mohan, M., Sushil, S. N., Kundu, S., Bhatt, J. C., & Gupta, H. S. (2007). Characterization and phylogenetic analysis of an entomopathogenic *Bacillus cereus* strain WGPSB-2 (MTCC 7182) isolated from white grub, *Anomala dimidiata* (Coleoptera: Scarabaeidae). *Biocontrol Sci. Technol., 17*, 525−534.

Sen, A., Rybakova, D., Hurst, M. R. H., & Mitra, A. K. (2010). Structural study of the *Serratia entomophila* antifeeding prophage: three-dimensional structure of the helical sheath. *J. Bacteriol., 192*, 4522−4525.

Sena, J. A., Hernández-Rodríguez, C. S., & Ferré, J. (2009). Interaction of *Bacillus thuringiensis* Cry1 and Vip3A proteins with *Spodoptera frugiperda* midgut binding sites. *Appl. Environ. Microbiol., 75*, 2236−2237.

Sharma, H. C., Sharma, K. K., Seetharama, N., & Crouch, J. H. (2003). The utility and management of transgenic plants with *Bacillus thuringiensis* genes for protection from pests. *J. New Seeds, 5*, 53−76.

Sharma, P., Nain, V., Lakhanpaul, S., & Kumar, P. A. (2010). Synergistic activity between *Bacillus thuringiensis* Cry1Ab and Cry1Ac toxins against maize stem borer (*Chilo partellus* Swinhoe). *Lett. Appl. Microbiol., 51*, 42−47.

Sharpe, E. S., & Detroy, R. W. (1979). Fat body depletion, a debilitating result of milky disease in Japanese beetle larvae. *J. Invertebr. Pathol., 34*, 92−94.

Shi, Y., Xu, W., Yuan, M., Tang, M., Chen, J., & Pang, Y. (2004). Expression of *vip1/vip2* genes in *Escherichia coli* and *Bacillus thuringiensis* and the analysis of their signal peptides. *J. Appl. Microbiol., 97*, 757–765.

Shi, Y. X., Yuan, M. J., Chen, J. W., Sun, F., & Pang, Y. (2006). Effects of helper protein P20 from *Bacillus thuringiensis* on Vip3A expression. *Acta Microbiol. Sin., 46*, 85–89.

Shida, O., Takagi, H., Kadowaki, K., & Komagata, K. (1996). Proposal for two new genera, *Brevibacillus* gen. nov. and *Aneurinibacillus* gen. nov. *Int. J. Syst. Bacteriol., 46*, 939–946.

Shimada, N., Murata, H., & Miyazaki, S. (2008). Safety evaluation of Bt plants. *Jpn. Agric. Res. Q., 42*, 251–259.

Shimizu, T., Vassylyev, D. G., Kido, S., Doi, Y., & Morikawa, K. (1994). Crystal structure of vitelline membrane outer layer protein I (VMO-I): a folding motif with homologous Greek key structures related by an internal three-fold symmetry. *EMBO J., 13*, 1003–1010.

Shisa, N., Wasano, N., & Ohba, M. (2002). Discrepancy between cry gene-predicted and bioassay-determined insecticidal activities in *Bacillus thuringiensis* natural isolates. *J. Invertebr. Pathol., 81*, 59–61.

Shivakumar, A. G., Vanags, R. I., Wilcox, D. R., Katz, L., Vary, P. S., & Fox, J. L. (1989). Gene dosage effect on the expression of the delta-endotoxin genes of *Bacillus thuringiensis* subsp. *kurstaki* in *Bacillus subtilis* and *Bacillus megaterium. Gene., 79*, 21–31.

Sidén, I., Dalhammar, G., Telander, B., Boman, H. G., & Somerville, H. (1979). Virulence factors in *Bacillus thuringiensis*: purification and properties of a protein inhibitor of immunity in insects. *J. Gen. Microbiol., 114*, 45–52.

Siegel, J. P. (2001). The mammalian safety of *Bacillus thuringiensis*-based insecticides. *J. Invertebr. Pathol., 77*, 13–21.

Siegel, J. P., & Shadduck, J. A. (1990). Mammalian safety of *Bacillus sphaericus*. In H. de Barjac & D. J. Sutherland (Eds.), *Bacterial Control of Mosquitoes & Black Flies: Biochemistry, Genetics, & Applications of* Bacillus thuringiensis israelensis *&* Bacillus sphaericus (pp. 321–331). New Brunswick: Rutgers University Press.

Sikorowski, P. P., & Lawrence, A. M. (1994). Microbial contamination and insect rearing. *Am. Entomol., 40*, 240–253.

Silapanuntakul, S., Pantuwatana, S., Bhumiratana, A., & Charoensiri, K. (1983). The comparative persistence and toxicity of *Bacillus sphaericus* strain 1593 and *Bacillus thuringiensis* serotype H-14 against mosquito larvae in different kinds of environments. *J. Invertebr. Pathol., 42*, 387–392.

Silva-Filha, M. H., Nielsen-Leroux, C., & Charles, J. F. (1999). Identification of the receptor for *Bacillus sphaericus* crystal toxin in the brush border membrane of the mosquito *Culex pipiens* (Diptera: Culicidae). *Insect Biochem. Mol. Biol., 29*, 711–721.

Singer, S. (1981). Potential of *Bacillus sphaericus* and related spore-forming bacteria for pest control. In H. D. Burges (Ed.), *Microbial Control of Pests and Plant Diseases 1970–1980* (pp. 283–298). London: Academic Press.

Singer, S. (1996). The utility of strains of morphological group II *Bacillus*. In L. N. Saul & I. L. Allen (Eds.), *Advances in Applied Microbiology*, (Vol. 42) (pp. 219–261). San Diego: Academic Press.

Singh, G., Rup, P. J., & Koul, O. (2007). Acute, sublethal and combination effects of azadirachtin and *Bacillus thuringiensis* toxins on *Helicoverpa armigera* (Lepidoptera: Noctuidae) larvae. *Bull. Entomol. Res., 97*, 351–357.

Singh, G., Sachdev, B., Sharma, N., Seth, R., & Bhatnagar, R. K. (2010). Interaction of *Bacillus thuringiensis* vegetative insecticidal protein with ribosomal S2 protein triggers larvicidal activity in *Spodoptera frugiperda. Appl. Environ. Microbiol., 76*, 7202–7209.

Singh, G. J. P., & Gill, S. S. (1988). An electron microscope study of the toxic action of *Bacillus sphaericus* in *Culex quinquefasciatus* larvae. *J. Invertebr. Pathol., 52*, 237–247.

Singkhamanan, K., Promdonkoy, B., Chaisri, U., & Boonserm, P. (2010). Identification of amino acids required for receptor binding and toxicity of the *Bacillus sphaericus* binary toxin. *FEMS Microbiol. Lett., 303*, 84–91.

Sirichotpakorn, N., Rongnoparut, P., Choosang, K., & Panbangred, W. (2001). Coexpression of chitinase and the *cry11Aa1* toxin genes in *Bacillus thuringiensis* serovar *israelensis. J. Invertebr. Pathol., 78*, 160–169.

Skot, L., Harrison, S. P., Nath, A., Mytton, L. R., & Clifford, B. C. (1990). Expression of insecticidal activity in *Rhizobium* containing the delta-endotoxin gene cloned from *Bacillus thuringiensis* subsp. *tenebrionis. Plant Soil, 127*, 285–295.

Skovmand, O., & Guillet, P. (2000). Sedimentation of *Bacillus sphaericus* in tap water and sewage water. *J. Invertebr. Pathol., 75*, 243–250.

Skovmand, O., Ouedraogo, T. D. A., Sanogo, E., Samuelsen, H., Toe, L. P., & Baldet, T. (2009). Impact of slow-release *Bacillus sphaericus* granules on mosquito populations followed in a tropical urban environment. *J. Med. Entomol., 46*, 67–76.

Skyler, P. J., Roberts, J. W., & Barry, J. W. (1990). *Aerial Insecticide Projects for Suppression of Western Defoliators: 1970–1989. An Annotated Bibliography.* US Forest Service, Fire Project Management 90-11.

Slatten, B. H., & Larson, A. D. (1967). Mechanism of pathogenicity of *Serratia marcescens*. I. Virulence for the adult boll weevil. *J. Invertebr. Pathol., 9*, 78–81.

Smedley, D. P., & Ellar, D. J. (1996). Mutagenesis of three surface-exposed loops of a *Bacillus thuringiensis* insecticidal toxin reveals residues important for toxicity, receptor recognition and possibly membrane insertion. *Microbiology, 142*, 1617–1624.

Smirnoff, W. A. (1971). Effect of chitinase on the action of *Bacillus thuringiensis. Can. Entomol., 103*, 1829–1831.

Smirnoff, W. A. (1973). Results of tests with *Bacillus thuringiensis* and chitinase on larvae of the spruce budworm. *J. Invertebr. Pathol., 21*, 116–118.

Smirnova, T. A., Minenkova, I. B., Orlova, M. V., Lecadet, M. M., & Azizbekyan, R. R. (1996). The crystal-forming strains of *Bacillus laterosporus. Res. Microbiol., 147*, 343–350.

Smith, A. W., Cámara-Artigas, A., Brune, D. C., & Allen, J. P. (2005). Implications of high-molecular-weight oligomers of the binary toxin from *Bacillus sphaericus. J. Invertebr. Pathol., 88*, 27–33.

Smith, G. P., Merrick, J. D., Bone, E. J., & Ellar, D. J. (1996). Mosquitocidal activity of the CryIC delta-endotoxin from *Bacillus thuringiensis* subsp. *aizawai. Appl. Environ. Microbiol., 62*, 680–684.

Smith, R. A., & Barry, J. W. (1998). Environmental persistence of *Bacillus thuringiensis* spores following aerial application. *J. Invertebr. Pathol., 71*, 263–267.

Smith, R. A., & Couche, G. A. (1991). The phylloplane as a source of *Bacillus thuringiensis* variants. *Appl. Environ. Microbiol., 57*, 311–315.

Soberón, M., Pardo-López, L., López, I., Gómez, I., Tabashnik, B. E., & Bravo, A. (2007). Engineering modified Bt toxins to counter insect resistance. *Science, 318*, 1640–1642.

Soberón, M., Gill, S. S., & Bravo, A. (2009). Signaling versus punching hole: how do *Bacillus thuringiensis* toxins kill insect midgut cells? *Cell. Mol. Life Sci., 66*, 1337–1349.

Soberón, M., Pardo, L., Muñóz-Garay, C., Sánchez, J., Gómez, I., Porta, H., & Bravo, A. (2010). Pore formation by Cry toxins. *Adv. Exp. Med. Biol., 677,* 127–142.

Somerville, H. J., & Pockett, H. V. (1975). An insect toxin from spores of *Bacillus thuringiensis* and *Bacillus cereus. J. Gen. Microbiol., 87,* 359–369.

Song, L., Gao, M., Dai, S., Wu, Y., Yi, D., & Li, R. (2008). Specific activity of a *Bacillus thuringiensis* strain against *Locusta migratoria manilensis. J. Invertebr. Pathol., 98,* 169–176.

Song, R., Peng, D., Yu, Z., & Sun, M. (2008). Carboxy-terminal half of Cry1C can help vegetative insecticidal protein to form inclusion bodies in the mother cell of *Bacillus thuringiensis. Appl. Microbiol. Biotechnol., 80,* 647–654.

Soufiane, B., & Côté, J.-C. (2010). *Bacillus thuringiensis* serovars *bolivia, vazensis* and *navarrensis* meet the description of *Bacillus weihenstephanensis. Curr. Microbiol., 60,* 343–349.

de Souza, M. T., Lecadet, M. M., & Lereclus, D. (1993). Full expression of the cryIIIA toxin gene of *Bacillus thuringiensis* requires a distant upstream DNA sequence affecting transcription. *J. Bacteriol., 175,* 2952–2960.

Spies, A. G., & Spence, K. D. (1985). Effect of sublethal *Bacillus thuringiensis* crystal endotoxin treatment on the larval midgut of a moth, *Manduca sexta. Tissue Cell, 17,* 379–394.

Splittstoesser, C. M., Tashiro, H., Lin, S. L., Steinkraus, K. H., & Fiori, B. J. (1973). Histopathology of the European chafer, *Amphimallon majalis,* infected with *Bacillus popilliae. J. Invertebr. Pathol., 22,* 161–167.

St. Julian, G., Jr., Pridham, T. G., & Hall, H. H. (1963). Effect of diluents on viability of *Popillia japonica* Newman larvae, *Bacillus popilliae* Dutky, and *Bacillus lentimorbus* Dutky. *J. Insect Pathol., 5,* 440–450.

Stackebrandt, E., Frederiksen, W., Garrity, G. M., Grimont, P. A., Kämpfe, P., Maiden, M. C., Nesme, X., Rosselló-Mora, R., Swings, J., Trüper, H. G., Vauterin, L., Ward, A. C., & Whitman, W. B. (2002). Report of the ad hoc committee for the re-evaluation of the species definition in bacteriology. *Int. J. Syst. Evol. Microbiol., 52,* 1043–1047.

Stahly, D. P., & Klein, M. G. (1992). Problems with *in vitro* production of spores of *Bacillus popilliae* for use in biological control of the Japanese beetle. *J. Invertebr. Pathol., 60,* 283–291.

Staples, N., Ellar, D., & Crickmore, N. (2001). Cellular localization and characterization of the *Bacillus thuringiensis* Orf2 crystallization factor. *Curr. Microbiol., 42,* 388–392.

Starr, M. P., Grimont, P. A., Grimont, F., & Starr, P. B. (1976). Caprylate-thallous agar medium for selectively isolating *Serratia* and its utility in the clinical laboratory. *J. Clin. Microbiol., 4,* 270–276.

Starzak, M., & Bajpai, R. K. (1991). A structured model for vegetative growth and sporulation in *Bacillus thuringiensis. Appl. Biochem. Biotechnol., 28–29,* 699–718.

Steinhaus, E. A. (1957). Microbial diseases of insects. *Annu. Rev. Microbiol., 11,* 165–182.

Steinhaus, E. A. (Ed.). (1963), *Insect Pathology: An Advanced Treatise,* (Vols. 1–2). New York: Academic Press.

Stewart, C. N., Jr., Halfhill, M. D., & Warwick, S. I. (2003). Transgene introgression from genetically modified crops to their wild relatives. *Nat. Rev. Genet., 4,* 806–817.

Stragier, P., & Losick, R. (1996). Molecular genetics of sporulation in. *Bacillus subtilis. Annu. Rev. Genet., 30.* 297–241.

Strongman, D. B., Eveleigh, E. S., van Frankenhuyzen, K., & Royama, T. (1997). The occurrence of two types of entomopathogenic bacilli in natural populations of the spruce budworm, *Choristoneura fumiferana. Can. J. For. Res., 27,* 1922–1927.

Stucki, G., Jackson, T. A., & Noonan, M. J. (1984). Isolation and characterisation of *Serratia* strains pathogenic for larvae of the New Zealand grass grub *Costelytra zealandica. N.Z.J. Sci., 27,* 255–260.

Sturtevant, A. P., & Revell, I. L. (1953). Reduction of *Bacillus larvae* spores in liquid food of honey bees by action of the honey stopper, and its relation to the development of American foulbrood. *J. Econ. Entomol., 46,* 855–860.

Su, T. S. (2008). Evaluation of water-soluble pouches of *Bacillus sphaericus* applied as prehatch treatment against *Culex* mosquitoes in simulated catch basins. *J. Am. Mosq. Control Assoc., 24,* 54–60.

Suzuki, M. T., Lereclus, D., & Arantes, O. M. (2004). Fate of *Bacillus thuringiensis* strains in different insect larvae. *Can. J. Microbiol., 50,* 973–975.

Swiecicka, I. (2008). Natural occurrence of *Bacillus thuringiensis* and *Bacillus cereus* in eukaryotic organisms: a case for symbiosis. *Biocontrol Sci. Technol., 18,* 221–239.

Tabashnik, B. E. (1992). Evaluation of synergism among *Bacillus thuringiensis* toxins. *Appl. Environ. Microbiol., 58,* 3343–3346.

Tabashnik, B. E., Malvar, T., Liu, Y. B., Finson, N., Borthakur, D., Shin, B. S., Park, S. H., Masson, L., de Maagd, R. A., & Bosch, D. (1996). Cross-resistance of the diamondback moth indicates altered interactions with domain II of *Bacillus thuringiensis* toxins. *Appl. Environ. Microbiol., 62,* 2839–2844.

Takatsuka, J., & Kunimi, Y. (1998). Replication of *Bacillus thuringiensis* in larvae of the Mediterranean flour moth, *Ephestia kuehniella* (Lepidoptera: Pyralidae): Growth, sporulation and insecticidal activity of parasporal crystals. *Appl. Entomol. Zool., 33,* 479–486.

Takatsuka, J., & Kunimi, Y. (2000). Intestinal bacteria affect growth of *Bacillus thuringiensis* in larvae of the oriental tea tortrix, *Homona magnanima* diakonoff (Lepidoptera: tortricidae). *J. Invertebr. Pathol., 76,* 222–226.

Talalaev, E. V. (1956). Septicemia of the caterpillars of the Siberian silkworm. *Mikrobiologiya, 25,* 99–102.

Tan, B., Jackson, T. A., & Hurst, M. R. (2006). Virulence of *Serratia* strains against *Costelytra zealandica. Appl. Environ. Microbiol., 72,* 6417–6418.

Tan, Y., & Donovan, W. P. (2000). Deletion of *aprA* and *nprA* genes for alkaline protease A and neutral protease A from *Bacillus thuringiensis*: effect on insecticidal crystal proteins. *J. Biotechnol., 84,* 67–72.

Tanada, Y., & Kaya, H. K. (1993). *Insect Pathology.* San Diego: Academic Press.

Tanapongpipat, S., Nantapong, N., Cole, J., & Panyim, S. (2003). Stable integration and expression of mosquito-larvicidal genes from *Bacillus thuringiensis* subsp. *israelensis* and *Bacillus sphaericus* into the chromosome of *Enterobacter amnigenus*: a potential breakthrough in mosquito biocontrol. *FEMS Microbiol. Lett., 221,* 243–248.

Tantimavanich, S., Pantuwatana, S., Bhumiratana, A., & Panbangred, W. (1997). Cloning of a chitinase gene into *Bacillus thuringiensis* subsp. *aizawai* for enhanced insecticidal activity. *J. Gen. Appl. Microbiol., 43,* 341–347.

Taveecharoenkool, T., Angsuthanasombat, C., & Kanchanawarin, C. (2010). Combined molecular dynamics and continuum solvent studies of the pre-pore Cry4Aa trimer suggest its stability in solution and how it may form a pore. *PMC Biophys., 3,* 10.

Tchicaya, E. S., Koudou, B. G., Keiser, J., Adja, A. M., Ciss, G., Tanner, M., Tano, Y., & Utzinger, J. (2009). Effect of repeated application of microbial larvicides on malaria transmission in central Côte d'Ivoire. *J. Am. Mosq. Control Assoc., 25,* 382–385.

Terra, W. R., & Ferreira, C. (1994). Insect digestive enzymes: properties, compartmentalization and function. *Comp. Biochem. Physiol. B, 109,* 1–62.

Thaler, J. O., Duvic, B., Givaudan, A., & Boemare, N. (1998). Isolation and entomotoxic properties of the *Xenorhabdus nematophilus* F1 lecithinase. *Appl. Environ. Microbiol., 64,* 2367–2373.

Thamthiankul, S., Suan-Ngay, S., Tantimavanich, S., & Panbangred, W. (2001). Chitinase from *Bacillus thuringiensis* subsp. *pakistani. Appl. Microbiol. Biotechnol., 56,* 395–401.

Thamthiankul, S., Moar, W. J., Miller, M. E., & Panbangred, W. (2004). Improving the insecticidal activity of *Bacillus thuringiensis* subsp. *aizawai* against *Spodoptera exigua* by chromosomal expression of a chitinase gene. *Appl. Microbiol. Biotechnol., 65,* 183–192.

Thanabalu, T., & Porter, A. G. (1995). Efficient expression of a 100-kilodalton mosquitocidal toxin in protease-deficient recombinant *Bacillus sphaericus. Appl. Environ. Microbiol., 61,* 4031–4036.

Thanabalu, T., & Porter, A. G. (1996). A *Bacillus sphaericus* gene encoding a novel type of mosquitocidal toxin of 31.8 kDa. *Gene., 170,* 85–89.

Thanabalu, T., Hindley, J., Jackson-Yap, J., & Berry, C. (1991). Cloning, sequencing, and expression of a gene encoding a 100-kilodalton mosquitocidal toxin from *Bacillus sphaericus* SSII-1. *J. Bacteriol., 173,* 2776–2785.

Thanabalu, T., Hindley, J., & Berry, C. (1992a). Proteolytic processing of the mosquitocidal toxin from *Bacillus sphaericus* SSII-1. *J. Bacteriol., 174,* 5051–5056.

Thanabalu, T., Hindley, J., Brenner, S., Oei, C., & Berry, C. (1992b). Expression of the mosquitocidal toxins of *Bacillus sphaericus* and *Bacillus thuringiensis* subsp. *israelensis* by recombinant *Caulobacter crescentus,* a vehicle for biological control of aquatic insect larvae. *Appl. Environ. Microbiol., 58,* 905–910.

Thanabalu, T., Berry, C., & Hindley, J. (1993). Cytotoxicity and ADP-ribosylating activity of the mosquitocidal toxin from *Bacillus sphaericus* SSII-1: possible roles of the 27- and 70-kilodalton peptides. *J. Bacteriol., 175,* 2314–2320.

Then, C. (2009). Risk assessment of toxins derived from *Bacillus thuringiensis* – synergism, efficacy, and selectivity. *Environ. Sci. Pollut. Res., 17,* 791–797.

Thiery, I., & Hamon, S. (1998). Bacterial control of mosquito larvae: investigation of stability of *Bacillus thuringiensis* var. *israelensis* and *Bacillus sphaericus* standard powders. *J. Am. Mosq. Control Assoc., 14,* 472–476.

Thiéry, I., Hamon, S., Delécluse, A., & Orduz, S. (1998). The introduction into *Bacillus sphaericus* of the *Bacillus thuringiensis* subsp. *medellin cyt1Ab1* gene results in higher susceptibility of resistant mosquito larva populations to *B. sphaericus. Appl. Environ. Microbiol., 64,* 3910–3916.

Thomas, D. J. I., Morgan, J. A. W., Whipps, J. M., & Saunders, J. R. (2002). Transfer of plasmid pBC16 between *Bacillus thuringiensis* strains in non-susceptible larvae. *FEMS Microbiol. Ecol., 40,* 181–190.

Thomas, W. E., & Ellar, D. J. (1983a). Bacillus thuringiensis var israelensis crystal δ-endotoxin: effects on insect and mammalian cells in vitro and in vivo. J. Cell Sci., 60, 181–197.

Thomas, W. E., & Ellar, D. J. (1983b). Mechanism of action of *Bacillus thuringiensis* var *israelensis* insecticidal δ-endotoxin. *FEBS Lett., 154,* 362–368.

Tilquin, M., Paris, M., Reynaud, S., Despres, L., Ravanel, P., Geremia, R. A., & Gury, J. (2008). Long lasting persistence of *Bacillus thuringiensis* subsp. *israelensis (Bti)* in mosquito natural habitats. *PLoS ONE., 3,* e3432.

Tomasino, S. F., Leister, R. T., Dimock, M. B., Beach, R. M., & Kelly, J. L. (1995). Field performance of *Clavibacter xyli* subsp. *cynodontis* expressing the insecticidal protein gene *cry1Ac* of *Bacillus thuringiensis* against European corn borer in field corn. *Biol. Control, 5,* 442–448.

Torres, J., Lin, X., & Boonserm, P. (2008). A trimeric building block model for Cry toxins *in vitro* ion channel formation. *Biochim. Biophys. Acta, 1778,* 392–397.

Tourasse, N. J., Helgason, E., Klevan, A., Sylvestre, P., Moya, M., Haustant, M., Økstad, O. A., Fouet, A., Mock, M., & Kolstø, A.-B. (2011). Extended and global phylogenetic view of the *Bacillus cereus* group population by combination of MLST, AFLP, and MLEE genotyping data. *Food Microbiol., 28,* 236–244.

Townsend, R. J., Ferguson, C. M., Proffitt, J. R., Slay, M. W. A., Swaminathan, J., Day, S., Gerard, E., O'Callaghan, M., Johnson, V. W., & Jackson, T. A. (2004). Establishment of *Serratia entomophila* after application of a new formulation for grass grub control. *N.Z. Plant Prot., 57,* 10–12.

Trisyono, A., & Whalon, M. E. (1999). Toxicity of neem applied alone and in combinations with *Bacillus thuringiensis* to Colorado potato beetle (Coleoptera: Chrysomelidae). *J. Econ. Entomol., 92,* 1281–1288.

Trought, T. E. T., Jackson, T. A., & French, R. A. (1982). Incidence and transmission of a disease of grass grub (*Costelytra zealandica*) in Canterbury. *N.Z.J. Exp. Agric., 10,* 79–82.

Tyrell, D. J., Bulla, L. A., Jr., Andrews, R. E., Jr., Kramer, K. J., Davidson, L. I., & Nordin, P. (1981). Comparative biochemistry of entomocidal parasporal crystals of selected *Bacillus thuringiensis* strains. *J. Bacteriol., 145,* 1052–1062.

Vachon, V., Schwartz, J. L., & Laprade, R. (2006). Influence of the biophysical and biochemical environment on the kinetics of pore formation by Cry toxins. *J. Invertebr. Pathol., 92,* 160–165.

Vadlamudi, R. K., Ji, T. H., & Bulla, L. A., Jr. (1993). A specific binding protein from *Manduca sexta* for the insecticidal toxin of *Bacillus thuringiensis* subsp. *berliner. J. Biol. Chem., 268,* 12334–12340.

Vadlamudi, R. K., Weber, E., Ji, I., Ji, T. H., & Bulla, L. A., Jr. (1995). Cloning and expression of a receptor for an insecticidal toxin of *Bacillus thuringiensis. J. Biol. Chem., 270,* 5490–5494.

Valaitis, A. P., Mazza, A., Brousseau, R., & Masson, L. (1997). Interaction analyses of *Bacillus thuringiensis* Cry1A toxins with two aminopeptidases from gypsy moth midgut brush border membranes. *Insect Biochem. Mol. Biol., 27,* 529–539.

Valaitis, A. P., Jenkins, J. L., Lee, M. K., Dean, D. H., & Garner, K. J. (2001). Isolation and partial characterization of gypsy moth BTR-270, an anionic brush border membrane glycoconjugate that binds *Bacillus thuringiensis* Cry1A toxins with high affinity. *Arch. Insect Biochem. Physiol., 46,* 186–200.

Vallet-Gely, I., Novikov, A., Augusto, L., Liehl, P., Bolbach, G., Pechy-Tarr, M., Cosson, P., Keel, C., Caroff, M., & Lemaitre, B. (2010). Association of hemolytic activity of *Pseudomonas entomophila,* a versatile soil bacterium, with cyclic lipopeptide production. *Appl. Environ. Microbiol., 76,* 910–921.

Vandamme, P., Pot, B., Gillis, M., de Vos, P., Kersters, K., & Swings, J. (1996). Polyphasic taxonomy, a consensus approach to bacterial systematics. *Microbiol. Rev., 60,* 407–438.

Vanková, J., & Purrini, K. (1979). Natural epizooties caused by bacilli of the species *Bacillus thuringiensis* and *Bacillus cereus*. *Z. Angew. Entomol., 88*, 216–221.

Varma, R. V., & Mohamed Ali, M. I. (1986). *Bacillus firmus*as a new insect pathogen on a Lepidopteran pest of *Ailanthus triphysa*. *J. Invertebr. Pathol., 47*, 379–380.

Vié, V., Van Mau, N., Pomarède, P., Dance, C., Schwartz, J.-L., Laprade, R., Frutos, R., Rang, C., Masson, L., Heitz, F., & Le Grimellec, C. (2001). Lipid-induced pore formation of the *Bacillus thuringiensis* Cry1Aa insecticidal toxin. *J. Membr. Biol., 180*, 195–203.

Vilas-Bôas, G. F. L. T., Vilas-Bôas, L. A., Lereclus, D., & Arantes, O. M. N. (1998). *Bacillus thuringiensis* conjugation under environmental conditions. *FEMS Microbiol. Ecol., 25*, 369–374.

Vilas-Boas, G., Sanchis, V., Lereclus, D., Lemos, M. V. F., & Bourguet, D. (2002). Genetic differentiation between sympatric populations of *Bacillus cereus* and *Bacillus thuringiensis*. *Appl. Environ. Microbiol., 68*, 1414–1424.

Vilas-Bôas, G. T., Peruca, A. P. S., & Arantes, O. M. N. (2007). Biology and taxonomy of *Bacillus cereus, Bacillus anthracis*, and *Bacillus thuringiensis*. *Can. J. Microbiol., 53*, 673–687.

Vilas-Bôas, L. A., Vilas-Bôas, G. F. L. T., Saridakis, H. O., Lemos, M. V. F., Lereclus, D., & Arantes, O. M. N. (2000). Survival and conjugation of *Bacillus thuringiensis* in a soil microcosm. *FEMS Microbiol. Ecol., 31*, 255–259.

Villalon, M., Vachon, V., Brousseau, R., Schwartz, J. L., & Laprade, R. (1998). Video imaging analysis of the plasma membrane permeabilizing effects of *Bacillus thuringiensis* insecticidal toxins in Sf9 cells. *Biochim. Biophys. Acta, 1368*, 27–34.

Vilmos, P., & Kurucz, E. (1998). Insect immunity: evolutionary roots of the mammalian innate immune system. *Immunol. Lett., 62*, 59–66.

Visick, J. E., & Whiteley, H. R. (1991). Effect of a 20-kilodalton protein from *Bacillus thuringiensis* subsp. *israelensis* on production of the CytA protein by *Escherichia coli*. *J. Bacteriol., 173*, 1748–1756.

Visnovsky, G. A., Smalley, D. J., O'Callaghan, M., & Jackson, T. A. (2008). Influence of culture medium composition, dissolved oxygen concentration and harvesting time on the production of *Serratia entomophila*, a microbial control agent of the New Zealand grass grub. *Biocontrol Sci. Technol., 18*, 87–100.

Vodovar, N., Vinals, M., Liehl, P., Basset, A., Degrouard, J., Spellman, P., Boccard, F., & Lemaitre, B. (2005). *Drosophila* host defense after oral infection by an entomopathogenic *Pseudomonas* species. *Proc. Natl. Acad. Sci. USA, 102*, 11414–11419.

Vodovar, N., Vallenet, D., Cruveiller, S., Rouy, Z., Barbe, V., Acosta, C., Cattolico, L., Jubin, C., Lajus, A., Segurens, B., Vacherie, B., Wincker, P., Weissenbach, J., Lemaitre, B., Médigue, C., & Boccard, F. (2006). Complete genome sequence of the entomopathogenic and metabolically versatile soil bacterium *Pseudomonas entomophila*. *Nat. Biotechnol., 24*, 673–679.

Vu, K. D., Yan, S., Tyagi, R. D., Valero, J. R., & Surampalli, R. Y. (2009). Induced production of chitinase to enhance entomotoxicity of *Bacillus thuringiensis* employing starch industry wastewater as a substrate. *Bioresour. Technol., 100*, 5260–5269.

Waalwijck, C., Dullemans, A. M., van Workum, M. E. S., & Visser, B. (1985). Molecular cloning and the nucleotide sequence of the Mr 28,000 crystal protein gene of *Bacillus thuringiensis* subsp. *israelensis*. *Nucleic Acids Res., 13*, 8206–8217.

Wai Nam, T., Gudauskas, R. T., & Harper, J. D. (1975). Pathogenicity of *Bacillus cereus* isolated from *Trichoplusia ni* larvae. *J. Invertebr. Pathol., 26*, 135–136.

Walker, K., Mendelsohn, M., Matten, S., Alphin, M., & Ave, D. (2003). The role of microbial Bt products in US crop protection. *J. New Seeds, 5*, 31–51.

Wallner, W. E. (1971). Suppression of four hardwood defoliators by helicopter application of concentrate and dilute chemical and biological sprays. *J. Econ. Entomol., 64*, 1487–1490.

Walters, F. S., & English, L. H. (1995). Toxicity of *Bacillus thuringiensis* delta-endotoxins toward the potato aphid in an artificial diet bioassay. *Entomol. Exp. Appl., 77*, 211–216.

Walters, F. S., Kulesza, C. A., Phillips, A. T., & English, L. H. (1994). A stable oligomer of *Bacillus thuringiensis* delta-endotoxin, CryIIIA. *Insect Biochem. Mol. Biol., 24*, 963–968.

Walters, F. S., Stacy, C. M., Lee, M. K., Palekar, N., & Chen, J. S. (2008). An engineered chymotrypsin/cathepsin G site in domain I renders *Bacillus thuringiensis* Cry3A active against Western corn rootworm larvae. *Appl. Environ. Microbiol., 74*, 367–374.

Walters, F. S., deFontes, C. M., Hart, H., Warren, G. W., & Chen, J. S. (2010). Lepidopteran-active variable-region sequence imparts coleopteran activity in eCry3.1Ab, an engineered *Bacillus thuringiensis* hybrid insecticidal protein. *Appl. Environ. Microbiol., 76*, 3082–3088.

Ward, E. S., & Ellar, D. J. (1983). Assignment of the delta-endotoxin gene of *Bacillus thuringiensis* var *Israelensis* to a specific plasmid by curing analysis. *FEBS Lett., 158*, 45–49.

Warren, G. W. (1997). Vegetative insecticidal proteins: novel proteins for control of corn pests. In N. Carozzi & M. Koziel (Eds.), *Advances in Insect Control: the Role of Transgenic Plants* (pp. 109–121). Bristol: Taylor & Francis.

Warren, G. W., Koziel, M. G., Mullins, M. A., Nye, G. J., Carr, B., Desai, N. M., Kostichka, K., Duck, N. B., & Estruch, J. J. (1996). *Novel pesticidal proteins and strains*. WIP Organization. WO 96/10083.

Wasano, N., Imura, S., & Ohba, M. (1999). Failure to recover *Bacillus thuringiensis* from the Lutzow-Holm Bay region of Antarctica. *Lett. Appl. Microbiol., 28*, 49–51.

Waterfield, N. R., Daborn, P. J., & ffrench-Constant, R. H. (2002). Genomic islands in *Photorhabdus*. *Trends Microbiol., 10*, 541–545.

Wayne, L. G., Brenner, D. J., Colwell, R. R., Grimont, P. A. D., Kandler, O., Krichevsky, M. I., Moore, L. H., Moore, W. E. C., Murray, R. G. E., Stackebrandt, E., Starr, M. P., & Truper, H. G. (1987). Report of the ad-hoc-committee on reconciliation of approaches to bacterial systematics. *Int. J. Syst. Bacteriol., 37*, 463–464.

Wei, S., Cai, Q., & Yuan, Z. (2006). Mosquitocidal toxin from *Bacillus sphaericus* induces stronger delayed effects than binary toxin on *Culex quinquefasciatus* (Diptera: Culicidae). *J. Med. Entomol., 43*, 726–730.

Weiser, J. (1984). A mosquito-virulent *Bacillus sphaericus* in adult *Simulium damnosum* from northern Nigeria. *Zentralbl. Mikrobiol., 139*, 57–60.

Weiser, J., & Prasertphon, S. (1984). Entomopathogenic spore-formers from soil samples of mosquito habitats in northern Nigeria. *Zentralbl. Mikrobiol., 139*, 49–55.

West, A. W., Burges, H. D., White, R. J., & Wyborn, C. H. (1984). Persistence of *Bacillus thuringiensis* parasporal crystal insecticidal activity in soil. *J. Invertebr. Pathol., 44*, 128–133.

Whalon, M. E., & Wingerd, B. A. (2003). Bt: Mode of action and use. *Arch. Insect Biochem. Physiol., 54*, 200−211.

White, G. F. (1907). *The cause of American foulbrood.* US Department of Agriculture. Bureau of Entomology, Circular 94.

Wickremesinghe, R. S. B., & Mendis, C. L. (1980). *Bacillus sphaericus* from Sri Lanka demonstrating rapid larvicidal activity on *Culex quinquefasciatus. Mosquito News, 40,* 387−389.

Widner, W. R., & Whiteley, H. R. (1989). Two highly related insecticidal crystal proteins of *Bacillus thuringiensis* subsp. *kurstaki* possess different host range specificities. *J. Bacteriol., 171,* 965−974.

Widner, W. R., & Whiteley, H. R. (1990). Location of the dipteran specificity region in a lepidopteran-dipteran crystal protein from *Bacillus thuringiensis. J. Bacteriol., 172,* 2826−2832.

Wilcks, A., Smidt, L., Okstad, O. A., Kolsto, A. B., Mahillon, J., & Andrup, L. (1999). Replication mechanism and sequence analysis of the replicon of pAW63, a conjugative plasmid from *Bacillus thuringiensis. J. Bacteriol., 181,* 3193−3200.

Wilson, C. J., Mahanty, H. K., & Jackson, T. A. (1992). Adhesion of bacteria (*Serratia* spp.) to the foregut of grass grub (*Costelytra zealandica* White) larvae and its relationship to the development of amber disease. *Biocontrol Sci. Technol., 2,* 59−64.

Wilson, K., Thomas, M. B., Blanford, S., Doggett, M., Simpson, S. J., & Moore, S. L. (2002). Coping with crowds: density-dependent disease resistance in desert locusts. *Proc. Natl. Acad. Sci. USA, 99,* 5471−5475.

Wirth, M. C., Georghiou, G. P., & Federici, B. A. (1997). CytA enables CryIV endotoxins of *Bacillus thuringiensis* to overcome high levels of CryIV resistance in the mosquito, *Culex quinquefasciatus. Proc. Natl. Acad. Sci. USA, 94,* 10536−10540.

Wirth, M. C., Delécluse, A., & Walton, W. E. (2001). Cyt1Ab1 and Cyt2Ba1 from *Bacillus thuringiensis* subsp. *medellin* and *B. thuringiensis* subsp. *israelensis* synergize *Bacillus sphaericus* against *Aedes aegypti* and resistant *Culex quinquefasciatus* (Diptera: Culicidae). *Appl. Environ. Microbiol., 67,* 3280−3284.

Wirth, M. C., Jiannino, J. A., Federici, B. A., & Walton, W. E. (2004). Synergy between toxins of *Bacillus thuringiensis* subsp. *israelensis* and *Bacillus sphaericus. J. Med. Entomol., 41,* 935−941.

Wirth, M. C., Jiannino, J. A., Federici, B. A., & Walton, W. E. (2005a). Evolution of resistance toward *Bacillus sphaericus* or a mixture of *B. sphaericus* + Cyt1A from *Bacillus thuringiensis,* in the mosquito, *Culex quinquefasciatus* (Diptera: Culicidae). *J. Invertebr. Pathol., 88,* 154−162.

Wirth, M. C., Park, H. W., Walton, W. E., & Federici, B. A. (2005b). Cyt1A of *Bacillus thuringiensis* delays evolution of resistance to Cry11A in the mosquito *Culex quinquefasciatus. Appl. Environ. Microbiol., 71,* 185−189.

Wirth, M. C., Yang, Y., Walton, W. E., Federici, B. A., & Berry, C. (2007). Mtx toxins synergize *Bacillus sphaericus* and Cry11Aa against susceptible and insecticide-resistant *Culex quinquefasciatus* larvae. *Appl. Environ. Microbiol., 73,* 6066−6071.

Wiwat, C., Panbangred, W., & Bhumiratana, A. (1990). Transfer of plasmids and chromosomal genes amongst subspecies of *Bacillus thuringiensis. J. Ind. Microbiol., 6,* 19−27.

Wiwat, C., Panbangred, W., Mongkolsuk, S., Pantuwatana, S., & Bhumiratana, A. (1995). Inhibition of a conjugation-like gene transfer process in *Bacillus thuringiensis* subsp. *israelensis* by the anti-S-layer protein antibody. *Curr. Microbiol., 30,* 69−75.

Wiwat, C., Thaithanun, S., Pantuwatana, S., & Bhumiratana, A. (2000). Toxicity of chitinase-producing *Bacillus thuringiensis* ssp. *kurstaki*

HD-1 (G) toward *Plutella xylostella. J. Invertebr. Pathol., 76,* 270−277.

Wolfersberger, M. G., Chen, X. J., & Dean, D. H. (1996). Site-directed mutations in the third domain of *Bacillus thuringiensis* delta-endotoxin CryIAa affect its ability to increase the permeability of *Bombyx mori* midgut brush border membrane vesicles. *Appl. Environ. Microbiol., 62,* 279−282.

Wong, H. C., & Chang, S. (1986). Identification of a positive retro-regulator that stabilizes mRNAs in bacteria. *Proc. Natl. Acad. Sci. USA, 83,* 3233−3237.

Wong, H. C., Schnepf, H. E., & Whiteley, H. R. (1983). Transcriptional and translational start sites for the *Bacillus thuringiensis* crystal protein gene. *J. Biol. Chem., 258,* 1960−1967.

Woodburn, M. A., Yousten, A. A., & Hilu, K. H. (1995). Random amplified polymorphic DNA-fingerprinting of mosquito-pathogenic and nonpathogenic strains of *Bacillus sphaericus. Int. J. Syst. Bacteriol., 45,* 212−217.

World Health Organization. (1999). *Guidelines Specification for Bacterial Larvicides for Public Health Use. Report of the WHO Informal Consultation.* WHO/CDS/CPC/WHOPES/99.92. Geneva: World Health Organization.

Wraight, S. P., & Ramos, M. E. (2005). Synergistic interaction between *Beauveria bassiana-* and *Bacillus thuringiensis tenebrionis-*based biopesticides applied against field populations of Colorado potato beetle larvae. *J. Invertebr. Pathol., 90,* 139−150.

Wraight, S. P., Molloy, D. P., & Singer, S. (1987). Studies on the culicine mosquito host range of *Bacillus sphaericus* and *Bacillus thuringiensis* var. *israelensis* with notes on the effects of temperature and instar on bacterial efficacy. *J. Invertebr. Pathol., 49,* 291−302.

Wu, D., & Aronson, A. I. (1992). Localized mutagenesis defines regions of the *Bacillus thuringiensis* delta-endotoxin involved in toxicity and specificity. *J. Biol. Chem., 267,* 2311−2317.

Wu, D., & Chang, F. N. (1985). Synergism in mosquitocidal activity of 26 and 65 kDa proteins from *Bacillus thuringiensis* subsp. *israelensis* crystal. *FEBS Lett., 190,* 232−236.

Wu, D., & Federici, B. A. (1995). Improved production of the insecticidal CryIVD protein in *Bacillus thuringiensis* using cryIA(c) promoters to express the gene for an associated 20-kDa protein. *Appl. Microbiol. Biotechnol., 42,* 697−702.

Wu, D., Johnson, J. J., & Federici, B. A. (1994). Synergism of mosquitocidal toxicity between CytA and CryIVD proteins using inclusions produced from cloned genes of *Bacillus thuringiensis. Mol. Microbiol., 13,* 965−972.

Wu, J., Zhao, F., Bai, J., Deng, G., Qin, S., & Bao, Q. (2007a). Evidence for positive Darwinian selection of Vip gene in *Bacillus thuringiensis. J. Genet. Genomics, 34,* 649−660.

Wu, J. Y., Zhao, F. Q., Bai, J., Deng, G., Qin, S., & Bao, Q. Y. (2007b). Adaptive evolution of *cry* genes in *Bacillus thuringiensis:* implications for their specificity determination. *Genomics Proteomics Bioinformatics, 5,* 102−110.

Wu, K. M., Lu, Y. H., Feng, H. Q., Jiang, Y. Y., & Zhao, J. Z. (2008). Suppression of cotton bollworm in multiple crops in China in areas with Bt toxin-containing cotton. *Science, 321,* 1676−1678.

Xia, L., Sun, Y., Ding, X., Fu, Z., Mo, X., Zhang, H., & Yuan, Z. (2005). Identification of *cry*-type genes on 20-kb DNA associated with Cry1 crystal proteins from *Bacillus thuringiensis. Curr. Microbiol., 51,* 53−58.

Xia, L., Zhao, X. M., Ding, X. Z., Wang, F. X., & Sun, Y. J. (2008). The theoretical 3D structure of *Bacillus thuringiensis* Cry5Ba. *J. Mol. Model, 14,* 843–848.

Xia, L., Long, X., Ding, X., & Zhang, Y. (2009a). Increase in insecticidal toxicity by fusion of the cry1Ac gene from *Bacillus thuringiensis* with the neurotoxin gene hwtx-I. *Curr. Microbiol., 58,* 52–57.

Xia, L., Zeng, Z., Ding, X., & Huang, F. (2009b). The expression of a recombinant *cry1Ac* gene with subtilisin-like protease CDEP2 gene in acrystalliferous *Bacillus thuringiensis* by Red/ET homologous recombination. *Curr. Microbiol., 59,* 386–392.

Xie, R., Zhuang, M., Ross, L. S., Gomez, I., Oltean, D. I., Bravo, A., Soberon, M., & Gill, S. S. (2005). Single amino acid mutations in the cadherin receptor from *Heliothis virescens* affect its toxin binding ability to Cry1A toxins. *J. Biol. Chem., 280,* 8416–8425.

Xu, X., Yu, L., & Wu, Y. (2005). Disruption of a cadherin gene associated with resistance to Cry1Ac δ-endotoxin of *Bacillus thuringiensis* in *Helicoverpa armigera. Appl. Environ. Microbiol., 71,* 948–954.

Yamagiwa, M., Esaki, M., Otake, K., Inagaki, M., Komano, T., Amachi, T., & Sakai, H. (1999). Activation process of dipteran-specific insecticidal protein produced by *Bacillus thuringiensis* subsp. *israelensis. Appl. Environ. Microbiol., 65,* 3464–3469.

Yamaguchi, T., Sahara, K., Bando, H., & Asano, S.-I. (2010). Intramolecular proteolytic nicking of *Bacillus thuringiensis* Cry8Da toxin in BBMVs of Japanese beetle. *J. Invertebr. Pathol., 105,* 243–247.

Yamamoto, T., Iizuka, T., & Aronson, J. N. (1983). Mosquitocidal protein of *Bacillus thuringiensis* subsp. *israelensis:* identification and partial isolation of the protein. *Curr. Microbiol., 9,* 279–284.

Yang, Y., Zhu, Y. C., Ottea, J., Husseneder, C., Rogers Leonard, B., Abel, C., & Huang, F. (2010). Molecular characterization and RNA interference of three midgut aminopeptidase N isozymes from *Bacillus thuringiensis*-susceptible and -resistant strains of sugarcane borer, *Diatraea saccharalis. Insect Biochem. Mol. Biol., 40,* 592–603.

Yap, H. H. (1990). Field trials of *Bacillus sphaericus* for mosquito control. In H. de Barjac & D. J. Sutherland (Eds.), *Bacterial Control of Mosquitoes & Black Flies: Biochemistry, Genetics, & Applications of* Bacillus thuringiensis israelensis & Bacillus sphaericus (pp. 307–320). New Brunswick: Rutgers University Press.

Yara, K., Kunimi, Y., & Iwahana, H. (1997). Comparative studies of growth characteristic and competitive ability in *Bacillus thuringiensis* and *Bacillus cereus* in soil. *Appl. Entomol. Zool., 32,* 625–634.

Ying, W., Jie, W., & Xichang, F. (1986). A new serovar of *Bacillus thuringiensis. Acta Microbiol. Sin., 26,* 1–6.

Yokoyama, T., Tanaka, M., & Hasegawa, M. (2004). Novel cry gene from *Paenibacillus lentimorbus* strain Semadara inhibits ingestion and promotes insecticidal activity in *Anomala cuprea* larvae. *J. Invertebr. Pathol., 85,* 25–32.

Yoshisue, H., Fukada, T., Yoshida, K., Sen, K., Kurosawa, S., Sakai, H., & Komano, T. (1993). Transcriptional regulation of *Bacillus thuringiensis* subsp. *israelensis* mosquito larvicidal crystal protein gene cryIVA. *J. Bacteriol., 175,* 2750–2753.

Yoshiyama, M., & Kimura, K. (2009). Bacteria in the gut of Japanese honeybee, *Apis cerana japonica,* and their antagonistic effect against *Paenibacillus larvae,* the causal agent of American foulbrood. *J. Invertebr. Pathol., 102,* 91–96.

Youngjin, P., Kim, K., & Kim, Y. (2002). A pathogenic bacterium, *Enterococcus faecalis,* to the beet armyworm, *Spodoptera exigua. J. Asia-Pacific Entomol., 5,* 221–225.

Yousten, A. A. (1984). *Bacillus sphaericus:* microbiological factors related to its potential as a mosquito larvicide. *Adv. Biotechnol. Processes, 3,* 315–343.

Yousten, A. A., Genthner, F. J., & Benfield, E. F. (1992). Fate of *Bacillus sphaericus* and *Bacillus thuringiensis* serovar *israelensis* in the aquatic environment. *J. Am. Mosq. Control Assoc., 8,* 143–148.

Yu, C. G., Mullins, M. A., Warren, G. W., Koziel, M. G., & Estruch, J. J. (1997). The *Bacillus thuringiensis* vegetative insecticidal protein Vip3A lyses midgut epithelium cells of susceptible insects. *Appl. Environ. Microbiol., 63,* 532–536.

Yu, C. Z., Adamczyk, J. J., Jr., & West, S. (2005). Avidin, a potential biopesticide and synergist to *Bacillus thuringiensis* toxins against field crop insects. *J. Econ. Entomol., 98,* 1566–1571.

Yu, J., Xie, R., Tan, L., Xu, W., Zeng, S., Chen, J., Tang, M., & Pang, Y. (2002). Expression of the full-length and 3'-spliced *cry1Ab* gene in the 135-kDa crystal protein minus derivative of *Bacillus thuringiensis* subsp. *kyushuensis. Curr. Microbiol., 45,* 133–138.

Yu, Z., Dai, J., Zhou, H., Dong, Z., & Wang, W. (1984). A new serotype of *Bacillus thruringiensis. Acta Microbiol. Sin., 24,* 117–121.

Yuan, Y. M., Hu, X. M., Liu, H. Z., Hansen, B. M., Yan, J. P., & Yuan, Z. M. (2007). Kinetics of plasmid transfer among *Bacillus cereus* group strains within lepidopteran larvae. *Arch. Microbiol., 187,* 425–431.

Yuan, Z., Rang, C., Maroun, R. C., Juarez-Perez, V., Frutos, R., Pasteur, N., Vendrely, C., Charles, J. F., & Nielsen-Leroux, C. (2001). Identification and molecular structural prediction analysis of a toxicity determinant in the *Bacillus sphaericus* crystal larvicidal toxin. *Eur. J. Biochem., 268,* 2751–2760.

Yue, C., Sun, M., & Yu, Z. (2005a). Improved production of insecticidal proteins in *Bacillus thuringiensis* strains carrying an additional *cry1C* gene in its chromosome. *Biotechnol. Bioeng., 92,* 1–7.

Yue, C., Sun, M., & Yu, Z. (2005b). Broadening the insecticidal spectrum of Lepidoptera-specific *Bacillus thuringiensis* strains by chromosomal integration of *cry3A. Biotechnol. Bioeng., 91,* 296–303.

Zeilinger, A. R., Andow, D. A., Zwahlen, C., & Stotzky, G. (2010). Earthworm populations in a northern US Cornbelt soil are not affected by long-term cultivation of Bt maize expressing Cry1Ab and Cry3Bb1 proteins. *Soil Biol. Biochem., 42,* 1284–1292.

Zghal, R. Z., Tounsi, S., & Jaoua, S. (2005). Characterization of a *cry4Ba*-type gene of *Bacillus thuringiensis israelensis* and evidence of the synergistic larvicidal activity of its encoded protein with Cry2A delta-endotoxin of *B. thuringiensis kurstaki* on *Culex pipiens. Biotechnol. Appl. Biochem., 44,* 19–25.

Zhang, B., Liu, M., Yang, Y., & Yuan, Z. (2006). Cytolytic toxin Cyt1Aa of *Bacillus thuringiensis* synergizes the mosquitocidal toxin Mtx1 of *Bacillus sphaericus. Biosci. Biotechnol. Biochem., 70,* 2199–2204.

Zhang, H., & Jackson, T. A. (2008). Autochthonous bacterial flora indicated by PCR-DGGE of 16S rRNA gene fragments from the alimentary tract of *Costelytra zealandica* (Coleoptera: Scarabaeidae). *J. Appl. Microbiol., 105,* 1277–1285.

Zhang, J., Hodgman, T. C., Krieger, L., Schnetter, W., & Schairer, H. U. (1997). Cloning and analysis of the first cry gene from *Bacillus popilliae. J. Bacteriol., 179,* 4336–4341.

Zhang, J. T., Yan, J. P., Zheng, D. S., Sun, Y. J., & Yuan, Z. M. (2008). Expression of mel gene improves the UV resistance of *Bacillus thuringiensis. J. Appl. Microbiol., 105,* 151–157.

Zhang, M. Y., Lovgren, A., Low, M. G., & Landen, R. (1993). Characterization of an avirulent pleiotropic mutant of the insect pathogen

Bacillus thuringiensis: reduced expression of flagellin and phospholipases. *Infect. Immun., 61,* 4947–4954.

Zhang, R., Hua, G., Andacht, T. M., & Adang, M. J. (2008). A 106-kDa aminopeptidase is a putative receptor for *Bacillus thuringiensis* Cry11Ba toxin in the mosquito *Anopheles gambiae. Biochemistry, 47,* 11263–11272.

Zhang, X., Candas, M., Griko, N. B., Rose-Young, L., & Bulla, L. A. (2005). Cytotoxicity of *Bacillus thuringiensis* Cry1Ab toxin depends on specific binding of the toxin to the cadherin receptor BT-R$_1$ expressed in insect cells. *Cell Death Differ., 12,* 1407–1416.

Zhang, X., Candas, M., Griko, N. B., Taussig, R., & Bulla, L. A., Jr. (2006). A mechanism of cell death involving an adenylyl cyclase/PKA signaling pathway is induced by the Cry1Ab toxin of *Bacillus thuringiensis. Proc. Natl. Acad. Sci. USA, 103,* 9897–9902.

Zhao, J., Jin, L., Yang, Y., & Wu, Y. (2010). Diverse cadherin mutations conferring resistance to *Bacillus thuringiensis* toxin Cry1Ac in *Helicoverpa armigera. Insect Biochem. Mol. Biol., 40,* 113–118.

Zhao, J. Z., Cao, J., Li, Y., Collins, H. L., Roush, R. T., Earle, E. D., & Shelton, A. M. (2003). Transgenic plants expressing two *Bacillus*

thuringiensis toxins delay insect resistance evolution. *Nat. Biotechnol., 21,* 1493–1497.

Zhong, C., Peng, D., Ye, W., Chai, L., Qi, J., Yu, Z., Ruan, L., & Sun, M. (2011). Determination of plasmid copy number reveals the total plasmid DNA amount is greater than the chromosomal DNA amount in *Bacillus thuringiensis* YBT-1520. *PLoS ONE, 6,* e16025.

Zhou, X., Konkel, M. E., & Call, D. R. (2009). Type III secretion system 1 of *Vibrio parahaemolyticus* induces oncosis in both epithelial and monocytic cell lines. *Microbiology, 155,* 837–851.

Zhuang, M., Oltean, D. I., Gómez, I., Pullikuth, A. K., Soberón, M., Bravo, A., & Gill, S. S. (2002). *Heliothis virescens* and *Manduca sexta* lipid rafts are involved in Cry1A toxin binding to the midgut epithelium and subsequent pore formation. *J. Biol. Chem., 277,* 13863–13872.

Zubasheva, M. V., Ganushkina, L. A., Smirnova, T. A., & Azizbekyan, R. R. (2010). Larvicidal activity of crystal-forming strains of *Brevibacillus laterosporus. Appl. Biochem. Microbiol., 46,* 755–762.

Chapter 9

Wolbachia Infections in Arthropod Hosts

Grant L. Hughes and Jason L. Rasgon

College of Agricultural Sciences, Department of Entomology, Pennsylvania State University, USA

Chapter Outline

Summary

The α-proteobacteria *Wolbachia* are obligate endosymbionts that infect a diverse range of invertebrate taxa and are possibly the most common endosymbiotic bacteria on Earth. In their arthropod hosts, *Wolbachia* induce a variety of reproductive manipulations that enhance the fitness of infected females compared to their uninfected counterparts. Adding to their phenotypic repertoire, *Wolbachia* have recently been shown to interfere with pathogen infection and transmission in both naturally infected and artificially transinfected insects. These properties make *Wolbachia* systems interesting to study on numerous levels, including evolution and speciation, microbe–host interactions, and applied strategies to minimize the impact of arthropod-borne diseases and insect pests. This chapter will review the current state of knowledge of *Wolbachia* reproductive phenotypes and *Wolbachia*-induced pathogen interference, and discuss these in the context of applied use of *Wolbachia* for arthropod-borne disease and pest control.

9.1. INTRODUCTION

The maternally inherited α-proteobacteria *Wolbachia* is possibly the most common endosymbiotic bacterial species on the planet (Hilgenboecker *et al.*, 2008), infecting a diverse range of arthropods and filarial nematodes. The intracellular lifestyle of the bacterium ensures that it has an obligate relationship with its host. This dependency on the host has not hampered the proliferation of *Wolbachia* into arthropods, with an estimated 66% of insect species infected by the bacterium (Hilgenboecker *et al.*, 2008). The widespread nature of *Wolbachia* infection in arthropods has been attributed to the bacteria's ability to manipulate the reproduction of the host; however, it is becoming increasingly more evident that other phenotypes can arise from infection. These include provision of nutrients to the host and protection from pathogens (Hedges *et al.*, 2008; Hosokawa *et al.*, 2010), although it is yet to be empirically determined whether these fitness benefits assist *Wolbachia* dissemination in insect populations. In addition, cryptic infections, which occur at low frequency in the population and at low density within individuals, have been reported (Arthofer *et al.*, 2009; Hughes *et al.*, 2011a). Even though the biology of these infections is unknown, their occurrence suggests that *Wolbachia* infections within insect species may be more prominent than previously assumed.

9.2. CLASSIFICATION AND PHYLOGENY

Wolbachia pipientis is the sole member of the genus, which is included in the family Anaplasmataceae of the order Rickettsiales. Two other species, *W. persica* and *W. melophagi*, have historically been included in the genus but are now recognized to be unrelated (Dumler *et al.*, 2001). *Ehrlichia*, *Anaplasma*, and *Neorickettsia* are closely related to *Wolbachia* (Dumler *et al.*, 2001). Within the genus *Wolbachia*, eight phylogenetic "supergroups" (A—H) are currently recognized based on *Wsp*, *FtsZ*, and *16S* gene sequences (Lo *et al.*, 2002); however, more robust multi-locus sequence typing is now being used for classification (Baldo *et al.*, 2006). The majority of insect infections fall into supergroups A and B, while supergroups E, G, and H are more specialized. Supergroups C and D infect filarial nematodes and supergroup F is unique in that members of this clade infect insects, arachnids, and filarial nematodes (Hughes *et al.*, 2011a; Zhang *et al.*, 2011).

9.3. PATHOGEN VERSUS MUTUALIST

When de Bary (1879) coined the term symbiosis as the living together of unlike organisms, there was considerable room on the spectra between parasitism and mutualism. Indeed, *Wolbachia* lies on both ends of this continuum. In nematodes, *Wolbachia* are obligate mutualists, essential for the survival and reproduction of the roundworm, while in some insects, pathogenic strains occur that decrease host life expectancy (Min and Benzer, 1997; Braquart-Varnier *et al.*, 2008). Mutualist *Wolbachia* in nematodes are the causal agent of pathology in such diseases as river blindness in humans (Pearlman and Gilliette-Ferguson, 2007). However, if we consider strains that only infect arthropods, these bacteria still have a great range of interactions with their host. *Wolbachia* are well known as a reproductive parasite, inducing phenotypes such as cytoplasmic incompatibility, sex-ratio distorters (e.g., male-killing and feminization), and parthenogenesis (Werren *et al.*, 2008), but they can also enhance host reproductive fitness (Dobson *et al.*, 2002, 2004). In *Asobara* wasps, *Wolbachia* is an obligate mutualist that is essential for oogenesis and mating success, with loss of symbionts resulting in unsuccessful reproduction (Dedeine *et al.*, 2001; Pannebakker *et al.*, 2007; Kremer *et al.*, 2009). In the neotropical *Drosophila paulistorum* species complex, *Wolbachia* have co-evolved towards obligate mutualism with their respective native hosts (Miller *et al.*, 2010). In semi-species hybrids, mutualistic *Wolbachia* become pathogenic against the host, triggering embryonic lethality and male sterility via over-replication (Miller *et al.*, 2010). Transinfection, the transfer of *Wolbachia* between insects, results in unpredictable phenotypic shifts. For example, the transfer of *Wolbachia* from *Drosophila melanogaster* to *Aedes albopictus* resulted in severe pathogenicity (Suh *et al.*, 2009). Furthermore, infection of the isopod *Porcellio dilatatus* males with the wVul strain from the isopod *Armadillidium vulgare* was lethal (Bouchon *et al.*, 1998). Maladaptation between the bacteria and the host has been postulated as a cause of these drastic phenotype changes. Alternatively, transinfection

can lead to changes in the reproductive phenotype induced by the bacteria in the new host (i.e., feminization to male killing), which underscores the influence of the host on the overall phenotype (Fujii *et al.*, 2001).

In addition to reproductive manipulations in arthropods, some *Wolbachia* strains have been shown to be nutritional mutualists. The most striking example of this is the F supergroup *Wolbachia* found in the bedbug *Cimex lectularius*, where the symbionts reside in specialized bacteriome cells and provide B vitamins to the insect host (Hosokawa *et al.*, 2010). *Wolbachia* infection can confer a positive fecundity benefit for *D. melanogaster* when iron is either limited or abundant (Brownlie *et al.*, 2009) and can increase fly longevity (Toivonen *et al.*, 2007). Under different growth regimes, *Wolbachia* can shift from mutualist to parasite. When reared under high density, *Wolbachia*-infected *Ae. albopictus* were less competitive than their uninfected counterparts, yet at low competitive pressures infected females experience higher survivorship (Gavotte *et al.*, 2010). Moreover, *Wolbachia* have recently been shown to protect the host against pathogens (pathogen interference), which has important application for arthropod-borne disease control (Brownlie and Johnson, 2009; Cook and McGraw, 2010). These examples show not only how *Wolbachia* can cause a diverse range of phenotypes, but that these phenotypes exist on a sliding continuum between mutualist and pathogen dictated by the *Wolbachia*—host association and environment.

An intriguing report from Kaiser *et al.* (2010) suggests that *Wolbachia* can affect plant physiology. When the *Wolbachia*-infected phytophagous leaf-mining moth *Phyllonorycter blancardella* feeds on senescent leaves, the area proximal to the feeding site becomes a "green island" of photosynthetically active leaf tissue (Kaiser *et al.*, 2010). The phenotype in the plant has been attributed to localized elevation of cytokinins (Giron *et al.*, 2007). Insects cured of *Wolbachia* by antibiotic treatment had lower cytokinin levels and could not induce the "green island" effect in the plant (Kaiser *et al.*, 2010). Thus, *Wolbachia* provides an indirect fitness benefit to the host by manipulating the quality of the host's food source.

9.4. HISTORICAL OVERVIEW OF *WOLBACHIA* AND VECTOR CONTROL

After initial reports of microbes present in *Culex pipiens* reproductive organs (Hertig and Wolbach, 1924), Hertig (1936) named these bacteria *Wolbachia pipientis* after his colleague, S. Burt Wolbach. The term cytoplasmic incompatibility (CI) was developed to describe the cytoplasmically inherited infertility resulting from crosses between different strains of *Culex* mosquitoes (Laven, 1951). Laven (1967a) proposed that this incompatibility may be a driving force for speciation, while successful population suppression field trials were conducted whereby incompatible males were released which caused CI with females from local mosquito populations (Laven, 1967b). After curing mosquitoes with antibiotics, Yen and Barr (1971) determined that CI was a result of *Wolbachia* infection. Further laboratory and field trials were completed to evaluate the use of CI for population replacement in *Culex fatigans* populations (Curtis and Adak, 1974; Curtis, 1976). These early field trials set the stage for the use of *Wolbachia* for arthropod-borne disease vector control.

Today, similar approaches are being evaluated and implemented. The major vector of dengue virus, *Aedes aegypti*, is naturally uninfected with *Wolbachia*. Stable *Wolbachia* infected lines were developed by embryonic microinjection (McMeniman *et al.*, 2009), and intriguingly, these mosquitoes had *Wolbachia*-induced characteristics which made them inefficient vectors, including refractoriness to the virus itself (Moreira *et al.*, 2009a, b; Turley *et al.*, 2009). Field trials are currently underway to ascertain whether these mosquitoes can have an impact on pathogen transmission in far north Queensland, Australia (http://eliminatedengue.org). Another *Wolbachia*-based vector control strategy is being investigated to suppress lymphatic filariasis in the South Pacific region. Introgression of *Wolbachia* from *Aedes riversi* into *Aedes polynesiensis* resulted in bidirectional incompatibility with naturally infected mosquitoes, resulting in female sterility (Brelsfoard *et al.*, 2008). The release of *Wolbachia*-infected males is an approach that may suppress natural mosquito populations. Moreover, strategies that use *Wolbachia* to directly inhibit transmission of the nematode which causes lymphatic filariasis are under investigation (Kambris *et al.*, 2009). Further work is underway on other mosquito vectors including approaches using *Wolbachia* to control malaria (Jin *et al.*, 2009; Kambris *et al.*, 2010; Hughes *et al.*, 2011b).

9.5. REPRODUCTIVE MANIPULATIONS

Wolbachia can manipulate the reproduction of its host in a variety of ways. The commonality among all these methods is that they all increase the number of female offspring generated by infected females relative to their uninfected counterparts. As *Wolbachia* is maternally inherited, this increases the proportion of infected individuals in the population and ensures the spread of the bacteria. This selective advantage allows *Wolbachia* to sweep rapidly into insect populations. The large number of infected arthropod species is testament to the evolutionary success of this approach.

9.5.1. Cytoplasmic Incompatibility

The most well-characterized type of reproductive manipulation is CI. In its simplest form, CI occurs when an infected male mates with an uninfected female. Reciprocal crosses between uninfected males and infected females are fertile (Fig. 9.1). As infected females can mate with either infected or uninfected males, these females have a fitness advantage over uninfected females, who can only reproduce by mating with uninfected males. This fitness advantage enables *Wolbachia* to spread through insect populations (Turelli and Hoffmann, 1991). CI can be further complicated when multiple *Wolbachia* strains are present in the population, resulting in complex crossing patterns (Fig. 9.1).

The molecular mechanism of CI is unknown. However, it is thought that the sperm of *Wolbachia*-infected males is modified, and for successful fertilization to occur, females must be infected to "unlock" this modification. After cytological examination of an incompatible cross, the events in early mitosis are affected. Asynchronous development of the pronuclei occurs, resulting in female-derived chromatid development but incomplete condensation of male chromatids (Lassy and Karr, 1996). Specifically, there is a delay in the nuclear envelope breakdown and paternal chromosome condensation (Tram and Sullivan, 2002). In diploid insects this results in impaired zygote development, leading to embryonic mortality; however, in haplodiploids, haploid offspring can develop into males, although death of the embryo is also common (Reed and Werren, 1995).

Some *Wolbachia* strains differ in either inducing or rescuing CI, which has led to the modification—rescue (mod-resc) hypothesis (Werren, 1997). "Modification"

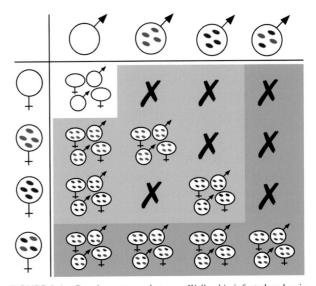

FIGURE 9.1 Crossing patterns between *Wolbachia*-infected and uninfected individuals. Different *Wolbachia* strains are depicted by red and blue dots. Crosses that induce cytoplasmic incompatibility (CI) do not produce viable offspring and are depicted with a cross (X). *Wolbachia* is inherited maternally in fertile crosses that produce viable offspring. Nonviable crosses in the areas shaded light and dark green are examples of unidirectional CI and bidirectional CI, respectively.

(mod) refers to the ability of *Wolbachia* to modify the male's sperm in some way to induce CI, while "rescue" (resc) refers to the ability of the *Wolbachia* in the female to counteract or complement the sperm modification and allow for successful fertilization to take place. *Wolbachia* strains can be assigned to four possible categories (mod+ resc+, mod+ resc−, mod− resc+, and mod− resc−) based on the strain's propensity for modification and rescue abilities (Werren and O'Neill, 2004). The mod+ resc− or "suicide strain" has never been found in nature; however, when wTei (*Wolbachia* from *Drosophila teissieri*) was transinfected into *Drosophila simulans*, this strain appeared to have a mod function but no rescue ability (Zabalou *et al.*, 2008). It is easy to reconcile that if such transfers occurred in nature, the selection pressure against this association would be too great for maintenance in the population.

Three models have been postulated to explain CI. The "lock and key" model (Breeuwer and Werren, 1990; L. D. Hurst, 1991; Werren, 1997) suggests that the mod function locks the sperm and *Wolbachia* is required within the female to unlock (rescue) it. Removal of essential proteins required for development in sperm by *Wolbachia* and supply within the egg in infected females is the theory behind the "titration−restitution" model (Kose and Karr, 1995; Werren, 1997), while the "slow-motion" or "timing" model (Callaini *et al.*, 1997; Tram and Sullivan, 2002) suggests that the paternal chromosomes are merely delayed, and rescue would involve a similar delay on the maternal side. Recently, Bossan *et al.* (2011) proposed the "goalkeeper" model, which is an expansion of the "timing" model and relies on two traits that vary quantitatively. Poinsot *et al.* (2003) compared the evidence from the literature to the models and concluded that the "lock and key" model was the most parsimonious explanation for CI, although the other hypotheses still fit certain data. It is likely that multiple mechanisms may occur in different strains of *Wolbachia* and host insects.

Case Study: Drosophila

The fruit fly has been, and continues to be, a model organism for many systems, and *Wolbachia* and CI is no exception. To date, five *Wolbachia* strains have been detected in *D. simulans* (wRi, wHa, wMa, wAu and wNo) (Merçot and Charlat, 2004), while only one strain, wMel is found in wild *D. melanogaster* populations. However, molecular analysis of wMel has identified five variants of this strain, one of which (wMelCS) has swept into fly populations and replaced the other variants over time (Riegler *et al.*, 2005). In *D. melanogaster*, CI levels vary from 10 to 77% (Solignac *et al.*, 1994). This variation has previously been attributed to differences in host factors, *Wolbachia* density, environmental conditions, or experimental conditions. Recent work

by Yamada *et al.* (2007) elegantly describes male developmental time as a key component in CI. Males that develop faster induce greater CI, whereas their "younger brothers" (i.e., males that have longer developmental times as larvae) lose their ability to express CI over time (Yamada *et al.*, 2007). Additional investigation into the causality of this effect may help to explain the variation in reports of CI in *D. melanogaster* and allow this fly species (for many molecular tools exist) to be used to further dissect the mechanism behind CI.

Cytoplasmic incompatibility in *D. simulans* has been extensively studied compared to other host species. Like *D. melanogaster*, *D. simulans* also exhibits variability in CI, but this variation is mostly attributed to specific *Wolbachia* strains (Merçot and Charlat, 2004). Three strains, wRi, wHa, and wNo, exhibit the mod+ resc+ phenotype. Of these, wRi displays complete CI, whereas subtler CI is induced by wHa, followed by wNo. Introgression and transinfection experiments show that these three strains are also bidirectionally incompatible, and that CI is caused by the bacteria and is not influenced by host genotype (Merçot and Charlat, 2004; Zabalou *et al.*, 2008). The other two strains, wAu and wMa, do not appear to induce CI (mod−); however, wMa was seen to rescue wNo, while it appears that wAu has no rescue ability (James and Ballard, 2000; Charlat *et al.*, 2003b). In theory, the mod− resc− phenotype should be lost from a population due to imperfect maternal transmission and fitness costs in infected insects. The ability of the wAu strain to persist in *Drosophila* populations has perplexed scientists, as no fitness benefit had been identified (Harcombe and Hoffmann, 2004). However, the recent finding that wAu-infected *D. simulans* have a greater lifespan compared to either uninfected flies or flies infected with wRi, wHa, or wNo when challenged with viral pathogens may explain this strain's existence in nature (Osborne *et al.*, 2009).

Case Studies: Aedes and Culex

Wolbachia was first described in *Cx. pipiens* mosquitoes by Hertig and Wolbach (1924). Since then, *Wolbachia* has been identified in other medically important mosquito species (Kittayapong *et al.*, 2000; Ricci *et al.*, 2002; Rasgon and Scott, 2004b). The major vectors of dengue and malaria are not naturally infected with *Wolbachia*, and as such represent an open niche for *Wolbachia*-based vector control strategies (Sinkins *et al.*, 1997). Artificial transinfection with multiple *Wolbachia* strains has been successful in the *Aedes* genus. Cultured *Anopheles* cells have been infected with *Wolbachia* (Fig. 9.2) (Rasgon *et al.*, 2006) and somatic infections can be established in live mosquitoes (Fig. 9.3) (Jin *et al.*, 2009), but in spite of numerous attempts, *Anopheles* mosquitoes have never been stably transinfected.

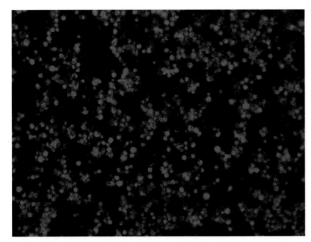

FIGURE 9.2 Fluorescence *in situ* hybridization (FISH) visualization of *Wolbachia* strain wAlbB in *Anopheles gambiae* Sua5B cells. *Wolbachia* are stained red; mosquito cell nuclei are stained blue.

The Asian tiger mosquito, *Ae. albopictus*, is infected with two strains of *Wolbachia*, wAlbA and wAlbB, with these strains residing in the A and B supergroups, respectively (Sinkins *et al.*, 1995b). This mosquito is an important vector of chikungunya virus, a secondary vector of dengue virus, and it also can transmit filarial nematodes (Kow *et al.*, 2001; Cancrini *et al.*, 2003; Reiter *et al.*, 2006). These *Wolbachia*-superinfected mosquitoes displayed bidirectional incompatibility when mated with mosquitoes only infected with wAlbA, or an artificially created wAlbB line and unidirectional CI when mated with uninfected females (Kambhampai *et al.*, 1993; Sinkins *et al.*, 1995b; Xi *et al.*, 2005a). The occurrence of CI in crosses between superinfected males and females with only one strain suggests these two strains employ different CI-inducing mechanisms. High unidirectional CI and maternal transmission rates were observed in *Ae. albopictus* in the laboratory (Sinkins *et al.*, 1995a; Kittayapong *et al.*, 2002a). In other insects, the CI phenotype can diminish with age, yet in field-caught *Ae. albopictus*, high rates of CI occurred when males were mated with uninfected females or females with only the wAlbA strain (Kittayapong *et al.*, 2002b). However, a singly infected wAlbA line did display reduced CI with age (Kittayapong *et al.*, 2002b). In males, wAlbA density has been shown to decrease with age, whereas the density of wAlbB increases (Tortosa *et al.*, 2010),

suggesting that CI results could be explained by *Wolbachia* density changing over time. Recently, Mousson *et al.* (2010) found that chikungunya virus infection decreases the density of both strains of *Wolbachia* in the mosquito. The effect of this viral-induced density shift on CI in the mosquito would be interesting to determine, particularly if mosquito age was factored into the experiment.

Embryonic microinjection has been used to create novel *Wolbachia*−*Aedes* interactions which induce CI, such as *Ae. albopictus* with wRi infection (Xi *et al.*, 2006), *Ae. aegypti* with wAlbB (Xi *et al.*, 2005b), *Ae. aegypti* with wMelPop (McMeniman *et al.*, 2009), and wAlbA and wAlbB in *Ae. aegypti* (Ruang-Areerate and Kittayapong, 2006). In the latter example, the density of *Wolbachia* in the host was seen to influence intensity of CI (Ruang-Areerate and Kittayapong, 2006). In naturally infected *Ae. albopictus*, density is also a critical component of CI, with strains that induced more intense CI possessing higher bacterial loads (Sinkins *et al.*, 1995a). In other systems, *Wolbachia* density and CI expression do not necessarily correlate (Yamada *et al.*, 2007).

The *Cx. pipiens* species complex has become a model system to examine CI dynamics because these species display the greatest variation in CI patterns of any species (Laven, 1951). Many studies have identified both unidirectional and bidirectional incompatibility between crosses of *Cx. pipiens* mosquitoes (Barr, 1980; Curtis and Suya, 1981; O'Neill and Paterson, 1992; Guillemaud *et al.*, 1997). Nearly all *Cx. pipiens* individuals (more than 99%) are infected with the strain wPip (Rasgon and Scott, 2003; Duron and Weill, 2006). The high prevalence of infection can be attributed to the induction of strong CI and a high maternal transmission rate (Rasgon and Scott, 2003; Duron and Weill, 2006).

Although there is incompatibility among *Cx. pipiens* mosquitoes from different geographical locations, minimal molecular variation was found between incompatible wPip isolates in loci commonly used to discriminate *Wolbachia* strains (Guillemaud *et al.*, 1997; Baldo *et al.*, 2006), necessitating the development of higher resolution markers. Prophage inserts and transposable elements in different *Wolbachia* strains may provide this resolution, with approximately 70 molecular variants circulating in natural populations (Duron *et al.*, 2006; Sanogo and Dobson, 2006). The advent of the first sequenced *Wolbachia* genome revealed an unusually high number of ankryin domain-containing genes (Wu *et al.*, 2004). Sinkins *et al.* (2005) observed variation in some of these genes that correlated with CI crossing type between two *Culex quinquefasciatus* laboratory strains, although Duron *et al.* (2007) showed that ankryin variation was probably not responsible for CI variation in nature. Crosses between other *Cx. pipiens* populations in La Réunion (an island in the Indian Ocean near Madagascar) indicate that there are

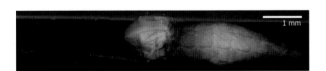

FIGURE 9.3 Whole mount confocal fluorescence *in situ* hybridization (FISH) visualization of *Wolbachia* strain wMelPop somatically infecting an *Anopheles gambiae* adult female. *Wolbachia* are stained orange; mosquito DNA is stained green.

at least five distinct incompatibility groups controlled by several *Wolbachia* mod/resc factors (Atyame *et al.*, 2010).

9.5.2. Sex Ratio Distorters

Owing to the strict maternal inheritance of *Wolbachia*, males are dead-end hosts. As horizontal transmission rates are low, male infections are essentially superfluous. Therefore, strategies that increase the proportion of infected females in a population are advantageous for the bacteria. As such, *Wolbachia* with sex-ratio distorting phenotypes are prominent in multiple insect orders.

Male Killing

Male killing (MK) is a trait not only restricted to *Wolbachia* but which can also be caused by other symbiotic bacteria such as *Spiroplasma* and *Arsenophonus*. The phenotype is typified by death of male offspring early in development, resulting in a female-biased sex ratio (Hurst *et al.*, 1999; Ferree *et al.*, 2008). The widespread nature of MK in several bacteria has led to the view that the MK phenotype is easily evolved (L. Hurst *et al.*, 1997). *Wolbachia* is able to induce MK in numerous arthropod orders including Diptera (Dyer and Jaenike, 2004), Lepidoptera (Jiggins *et al.*, 2001), Coleoptera (Fialho and Stevens, 2000), Araneae (Vanthournout *et al.*, 2011), and Pseudoscorpionida (Zeh *et al.*, 2005). It has been hypothesized that *Wolbachia* kills males to confer a fitness benefit to infected female hosts relative to uninfected females in the population. These benefits include reductions in sibling competition, inbreeding avoidance, egg cannibalism, and resource reallocation (Hurst, 1991; Jaenike *et al.*, 2003).

The molecular basis behind MK is poorly understood, although like CI, the interaction can occur during embryogenesis. However, it seems that death can also arise later in the developmental process. In *Drosophila willistoni* infected with an MK *Spiroplasma*, death can arise both before and after gastrulation (Counce and Poulson, 1962). Results by Charlat *et al.* (2007a) suggest that MK *Wolbachia* can vary in timing, acting both at embryogenesis and at later larval instars. Treatments that cleared *Wolbachia* from the host resulted in a loss of the MK phenotype and a reduction in the female-biased broods, suggesting a density threshold required for induction of the MK phenotype (Hurst *et al.*, 2000). Furthermore, mutation of *Drosophila* genes encoding the dosage compensation complex rendered the fly resistant to MK *Spiroplasma* (Koop *et al.*, 2009). Regardless of the specific mechanism employed by *Wolbachia*, the bacteria must either detect host sex before the onset of embryogenesis and then act to kill males, or interfere directly with sex determination to produce male-specific death (Charlat *et al.*, 2003a). The strategy is very effective, with surveys estimating that MK *Wolbachia* can spread into a population very quickly

(Charlat *et al.*, 2009; Duplouy *et al.*, 2010). Still puzzling, however, is the method by which MK *Wolbachia* initially disseminates into the insect population. It has been postulated that *Wolbachia* can provide direct and substantial fitness benefits to their hosts (McGraw and O'Neill, 2007). In light of recent findings that describe *Wolbachia*-mediated host protection from pathogens, it would be interesting to determine whether MK *Wolbachia* provides a similar protective benefit to their female hosts. Recent empirical evidence in the pseudoscorpion *Cordylochernes scorpioides* suggests that females infected with an MK *Wolbachia* produced more and larger daughters than a cured line (Koop *et al.*, 2009).

Insects infected with MK *Wolbachia* can produce broods with a female-biased sex ratio of around 90−100%. This causes a strong selection pressure for resistance in the host as it kills a large proportion of offspring (Hurst *et al.*, 2002). In populations where the sex ratio is biased towards females, the cost of MK on the host will be reduced. In the butterfly *Hypolimnas bolina*, some populations infected with the *Wolbachia* strain wBol1 have a sex ratio of 1:1, whereas other populations infected with the same strain exhibit MK, suggesting evolution of a suppressor of the MK phenotype (Hornett *et al.*, 2006). Modeling and empirical evidence show a rapid spread of a suppressor gene through butterfly populations (Hornett *et al.*, 2006; Charlat *et al.*, 2007b). Recently, it was identified that a single dominant allele inherited in a Mendelian fashion suppresses the MK phenotype of a γ-proteobacterium in the ladybird beetle, *Cheilomenes sexmaculata* (Majerus and Majerus, 2010), demonstrating evolution of a suppressor gene in other associations.

Shifts in the form of reproductive manipulation induced by *Wolbachia* on the host exemplify the phenotypic plasticity in these associations. Male *H. bolina* butterflies surviving the MK wBol infection through suppression of the MK phenotype have been shown to be reproductively incompatible with uninfected females, demonstrating that an MK strain of *Wolbachia* can also cause CI (Hornett *et al.*, 2008). Similarly, after experimental manipulation of the host which nullified the MK phenotype, *Wolbachia* induced weak CI in *Drosophila bifasciata* (Hurst *et al.*, 2000). Transinfection experiments have also revealed changes in reproductive phenotypes, where a strain of *Wolbachia* that induces CI in one host causes MK in another, suggesting that the interaction between host and bacteria is influential in determining the reproductive phenotype (Sasaki *et al.*, 2002). In a similar vein, when *Wolbachia* from *Drosophila recens* was introgressed into the *Drosophila subquinaria* background, the phenotype changed from CI in the natural host to MK in the novel host (Jaenike, 2007). The ability of *Wolbachia* to induce both MK and CI suggests a mechanistic link between these phenotypes. Moreover, it may be plausible that MK was once more prevalent in insect populations but the spread of suppressor gene(s) in host insects has unmasked CI as a more common phenotype. If and when an MK suppressor gene(s) is identified, it would be interesting to determine whether other insect hosts, which possess CI-inducing *Wolbachia*, are masking an MK effect. If this is the case, knockdown or abolition of this host-derived suppressor may result in a reversion to MK, suggesting that MK has been a prominent evolutionary force in many insects. An alternative scenario may be that CI is indeed the more common phenotype, and where host biology allows, MK evolves from it. In this case, suppression of MK action would revert the phenotype back to CI. Another intriguing aspect of some MK *Wolbachia*–host interactions is evident when the bacteria are cleared from the host. When the MK wSca strain is removed from the host, the adzuki bean borer *Ostrinia scapulalis* during the larval stage, the sex ratio of the brood changes from all female to all male (Kageyama *et al.*, 2003). Moreover, if heat or antibiotic treatment is applied at the adult stage, the resulting progeny are genotypically male but are a sexual mosaic, possessing both male and female phenotypic traits (Kageyama *et al.*, 2003; Sakamoto *et al.*, 2008). The sexual mosaic phenotype has been attributed to the *Wolbachia* modulating the expression of the sex-determining gene *doublesex* (Sugimoto *et al.*, 2010). These studies demonstrate that *Wolbachia* can profoundly influence the host sex determination pathway.

Feminization

Feminizing *Wolbachia* alter the normal sex determination pathway of their host, leading to genotypic males that become fully functional females. This phenotype has obvious advantages for a maternally transmitted symbiont. Examples of hosts affected by *Wolbachia*-induced feminization include the Asian corn borer *Ostrinia fulnacalis* (Kageyama *et al.*, 1998), the pierid *Eurema hecabe* (Hiroki *et al.*, 2002), the leafhopper *Zyginidia pullula* (Negri *et al.*, 2006), and the terrestrial isopods *Oniscus asellus* (Rigaud *et al.*, 1999) and *A. vulgare* (Rigaud *et al.*, 1991). Like other *Wolbachia*-induced phenotypes, the mechanistic basis for feminization is unclear, although in isopods, feminizing *Wolbachia* have been shown to proliferate within the androgenic gland, causing gland hypertrophy and inhibited function (Vandekerckhove *et al.*, 2003), leading genetic males to develop as females. In the butterfly *E. hecabe*, the removal of *Wolbachia* can lead to intersex phenotypes (Narita *et al.*, 2007) or all-male progeny (Hiroki *et al.*, 2002). In the leafhopper *Z. pullula*, *Wolbachia* was found to induce host methylation patterns influencing the expression of genes involved in sex differentiation and development in a density-dependent manner (Negri *et al.*, 2009).

Perhaps the greatest understanding of the feminizing process comes from studies on *A. vulgare*. Two feminizing

strains of *Wolbachia* have been detected in *A. vulgare* (wVulC and wVulM), with the former possessing higher vertical transmission rates (Cordaux *et al.*, 2004). The sex determination system in this isopod is complex, with both host and bacterial factors influencing the overall sexual phenotype. Feminization can be caused not only by *Wolbachia* in *A. vulgare*, but also by a region of DNA of bacterial origin (possibly derived from *Wolbachia*) inserted unstably into the host genome (Juchault *et al.*, 1992, 1993). Either the presence of *Wolbachia* or the insert can lead to homogametic males (ZZ) with female characteristics. The feminization process occurs in the initial embryonic state with inhibition of the growth of the androgenous gland, leading to impaired development of the male gonads, resulting in "neo-females" (Suzuki and Yamasaki, 1991). Rigaud and Juchault (1993) suggest that the host possesses a dominant autosomal masculinizing allele that can override the feminizing affects of the insert, but which cannot outcompete the affects of *Wolbachia* (Rigaud and Juchault, 1993). However, Martin *et al.* (2010) have proposed that *Wolbachia*-derived DNA inserted into the host genome is responsible for impairing the expression of genes responsible for male differentiation. *Wolbachia*-infected males have been observed, suggesting that the host may possess a mechanism to counteract the influence of *Wolbachia* (Juchault *et al.*, 1994). To add further complexity to the system, the host also has a transmission suppressor system that limits transovarial transmission of *Wolbachia* to offspring (Rigaud and Juchault, 1992).

9.5.3. Parthenogenesis

Wolbachia-induced parthenogenesis (thelytoky) in haplodiploid systems enables infected females to produce daughters from unfertilized eggs, resulting in populations without the males that are dead-end hosts for *Wolbachia*. In thelytokous parthenogenesis, all eggs are unfertilized and give rise to females (Stouthamer and Kazmer, 1994). *Wolbachia* was first associated with thelytoky when after curing *Trichogramma* wasps with antibiotics, male progeny were observed (Stouthamer *et al.*, 1990, 1993; Stouthamer and Werren, 1993). *Wolbachia* from both A and B supergroups induce parthenogenesis in mites and wasps, with well-characterized examples including *Trichogramma* (Huigens *et al.*, 2004) and *Drosophila* parasitoids (Vavre *et al.*, 2009).

Variation in the chromosomal mechanism for parthenogenesis induction has been observed between different *Wolbachia* strains and hosts. In some species (such as *Leptopilina clavipes* and *Trichogramma* spp.), *Wolbachia* interferes with anaphase during the first mitotic event, leading to gamete duplication and diploid eggs which develop into females (Stouthamer and Kazmer, 1994; Pannebakker *et al.*, 2004), In other hymenopteran species (e.g., *Diplolepis rosae* and *Muscidifurax uniraptor*), gamete duplication occurs via the fusion of two cell nuclei (Stille and Davring, 1980; Gottlieb *et al.*, 2002), while in the mite *Bryobia praetiosa*, meiosis is altered, resulting in diploid gametes (Weeks and Breeuwer, 2001). Multiple mechanisms for inducing parthenogenesis have led to speculation that this phenotype can evolve readily (Werren, 1997).

Removal of males is an elegant strategy for enhanced transmission of *Wolbachia*, but comes with significant evolutionary ramifications for the host. Theory predicts that parthenogenesis-inducing *Wolbachia* can lead to irreversible loss of sexual reproduction (Stouthamer *et al.*, 2010). Thelytokous parthenogenesis can decrease selection on host genes leading to accumulation of mutations, particularly in genes involved in sexual reproduction. This has been confirmed empirically in some species, such as *Asobara* parasitoid wasps where *Wolbachia* is now an obligate symbiont essential for oocyte maturation and mating success (Pannebakker *et al.*, 2007; Kremer *et al.*, 2009).

9.6. PATHOGEN INTERFERENCE AND PATHOGEN PROTECTION

Wolbachia has the ability to interfere with pathogen development in the host, providing the insect with a fitness benefit (pathogen protection) or altering its ability to become infected with and transmit vertebrate pathogens (pathogen interference). In some instances, both pathogen protection and pathogen interference may occur if the transmitted pathogen itself reduces fitness of the insect host. First identified in *D. melanogaster*, *Wolbachia*-infected females were more resistant to the fungal entomopathogen *Beauveria bassiana* than uninfected females (Panteleev *et al.*, 2007). Infected *D. simulans* had a greater lifespan when challenged with *Drosophila C virus* (DCV), *Cricket paralysis virus* (CrPV), or *Flock house virus* (FHV) compared to their non-infected conspecifics (Hedges *et al.*, 2008). Similar results were recorded by Teixeira *et al.* (2008), who demonstrated that *Wolbachia*-infected flies had increased resistance to the RNA viruses, DCV, FHV and *Nora virus*, but the bacteria did not convey resistance against the DNA virus, *Insect Iridescent Virus 6*. However, not all strains of *Wolbachia* protect the host. In *D. simulans*, both wAu and wRi significantly increase longevity of flies challenged with DCV and FHV, whereas the strain wHa provided limited protection against DCV, while no protective phenotype was observed for flies infected with the wNo strain (Osborne *et al.*, 2009). In flies whose bacteria provided protection, a higher density of *Wolbachia* was recorded (Osborne *et al.*, 2009), suggesting protection may occur in a density-dependent manner.

This protective phenotype in not limited to hosts naturally infected with *Wolbachia*. *Aedes aegypti* artificially transinfected with wMelPop are less susceptible to dengue virus, chikungunya virus, *Plasmodium gallinaceum*, and filarial nematodes (Moreira *et al.*, 2009a; Kambris *et al.*, 2010). In addition, the wAlbB strain from *Ae. albopictus* confers resistance to dengue virus in *Ae. aegypti* (Bian *et al.*, 2010). *Plasmodium berghei* and *P. falciparum* are both susceptible to pathogen interference in *Anopheles gambiae* mosquitoes somatically infected with wMelPop (Kambris *et al.*, 2010; Hughes *et al.*, 2011b). The wAlbB strain also inhibits *P. falciparum* (Hughes *et al.*, 2011b). Other arboviruses appear susceptible to *Wolbachia* protection. *Drosophila melanogaster* infected with wMel had lower *West Nile virus* (WNV) titers compared to uninfected flies (Glaser and Meola, 2010), although *D. melanogaster* is not the natural host of WNV. When the natural host *Cx. quinquefasciatus* was cleared of *Wolbachia* infection and given a WNV-infected blood meal, a higher virus titer was seen in *Wolbachia*-uninfected mosquitoes than in their infected counterparts (Glaser and Meola, 2010).

Although the mechanisms that cause pathogen interference in the host are poorly understood, several nonexclusive hypotheses have been postulated. One such hypothesis proposes that pathogen interference may be the result of an increase in host basal immunity. For example, many host immune genes are upregulated in *Ae. aegypti* transinfected with *Wolbachia* (Kambris *et al.*, 2009; Moreira *et al.*, 2009a; Bian *et al.*, 2010). It is well known that priming of the immune response before exposure to pathogens decreases the susceptibility of the host (Brown *et al.*, 2003; Eleftherianos *et al.*, 2006; Sadd and Schmid-Hempel, 2006; Pham *et al.*, 2007; Rodrigues *et al.*, 2010). In *An. gambaie*, there is some evidence that the complement-like peptide TEP1 mediates the interference phenotype against *P. berghei*, with knockdown of TEP1 in wMelPop-infected mosquitoes reverting oocyst numbers to similar levels to those in uninfected mosquitoes (Kambris *et al.*, 2010). However, further work is still required to determine whether TEP1 is involved in the pathogen interference pathway. The elevation of *Plasmodium* oocysts after TEP1 knockdown in *Wolbachia*-infected mosquitoes may simply be due to the removal of a critical immune gene, rather than a reversion of the *Wolbachia*-mediated phenotype.

The effect of *Wolbachia* on host immunity may be variable between insect species. In native associations, *Wolbachia* is thought to evade immune detection and therefore should have limited effect on immunity. Bourtzis *et al.* (2000) found no change in levels of defensin, an antimicrobial effector molecule, in *Ae. albopictus* infected by wAlbA and wAlbB compared to uninfected mosquitoes. Furthermore, expression of cecropin or diptericin in *D. melanogaster* infected with wMel was similar to that in uninfected flies (Bourtzis *et al.*, 2000); yet, this strain of *Wolbachia* provides flies protection from viruses (Teixeira *et al.*, 2008). Evidence from cell culture experiments suggests that native associations also evade host immunity even after reinfection into cells. Microarray analysis of *Drosophila* S2 cells revealed little change in gene expression between cells recently infected with wRi compared to uninfected cells (Xi *et al.*, 2008). This was in stark contrast to *Wolbachia* infection in cells of a novel host, *An. gambiae* (Sua5B cells), where infection profoundly altered gene expression (Hughes *et al.*, 2011c). However, the strains wRi and wAlbB were seen to suppress gene expression in Sua5B cells (Hughes *et al.*, 2011c).

When immune expression of *Anopheles* mosquitoes somatically infected with *Wolbachia* was investigated, the effect was more complex, with temporal modulation of immunity (Hughes *et al.*, 2011b). After injection into adults, *Wolbachia* initially had limited effect on immunity, while slight induction and then suppression of immune transcripts were observed at later time-points (Hughes *et al.*, 2011b). Mosquitoes with modulating immunity were shown to have suppressed *Plasmodium* infection (Hughes *et al.*, 2011b). The seemingly different effect of *Wolbachia* on immunity in *Anopheles* and *Aedes* may be a feature of the host, or due to the type of infection, with *Aedes* stably infected and *Anopheles* somatically infected by adult microinjection. If induction of immunity is the mechanism behind pathogen interference in *Anopheles*, exposing *Plasmodium* to the mosquito at different times after somatic infection may result in varying levels of protection. However, to sustain the immune activation hypothesis, it would seem imperative to have an understanding of the influence of *Wolbachia* infection on genome-wide transcription profiles of a wide range of insects at varying ages.

A second hypothesis suggests that metabolic competition between *Wolbachia* and the pathogen may cause pathogen interference. If the bacteria compete for essential nutrients required for pathogen survival, the host environment may no longer be suitable for pathogen infection. Evidence for this is provided by the exclusion of dengue virus in cells infected by *Wolbachia* in *Ae. aegypti* (Moreira *et al.*, 2009a). Cell lines may provide a tractable system to elucidate critical cellular requirements for both *Wolbachia* and pathogen. Frentiu *et al.* (2010) found that cells with higher *Wolbachia* infection had lower DENV-2 viral titer. This was similar to the response seen in fruit flies, whereby strains with higher density in the host provided greater protection (Osborne *et al.*, 2009). This density effect would be consistent with the metabolic competition hypothesis, as it would be expected that increases in *Wolbachia* density would result in a depletion of the essential nutrient(s) and thus lead to a lower pathogen titer. Alternatively, Cook and McGraw (2010) suggest that the mechanism for competitive exclusion may be

induced by RNAi or *Wolbachia*-mediated membrane modifications which prevent pathogen entry into host cells.

Another possibility is that *Wolbachia*-mediated changes in reactive oxygen species (ROS) and melanization cause pathogen interference. Pathogen melanization is an essential part of insect innate immunity (Cerenius *et al.*, 2008), and ROS influences pathogen levels, particularly *Plasmodium*, in mosquitoes (Peterson *et al.*, 2007; Molina-Cruz *et al.*, 2008). Furthermore, ROS levels have been implicated as a factor contributing to the melanotic encapsulation of *Plasmodium* in an *Anopheles* refractory strain (Kumar *et al.*, 2003). *Wolbachia* infection induces melanization in *Aedes* and *Drosophila* (Thomas *et al.*, 2010), while wMelPop increases levels of many prophenoloxidase genes (part of the melanization pathway) in *Ae. aegypti* (Kambris *et al.*, 2009; Moreira *et al.*, 2011). Modulation of ROS was hypothesized to involved in the blood meal-induced mortality in *Anopheles* somatically infected with wMelPop (Hughes *et al.*, 2011b). It is evident from biochemical and genetic analysis that *Wolbachia* influences melanization and potentially mediates ROS within insects; however, further investigations are needed to determine whether this is related to the pathogen interference phenotype.

Although there are multiple hypotheses for the mechanism of *Wolbachia*-mediated pathogen interference, with the current data at hand, it seems that the most parsimonious explanation is induction of host immunity. However, this hypothesis may not fit all associations and pathogens. It is more likely that various mechanisms occur within different hosts and act upon divergent pathogens. Furthermore, multiple mechanisms could occur within an insect species, leading to a cumulative inhibitive effect. Regardless of the mechanism, the ability of *Wolbachia* to inhibit the development of a diverse array of pathogens underscores the value of these bacteria for disease control.

9.7. APPLIED USE OF *WOLBACHIA*

Wolbachia has great potential to be used as an agent for controlling arthropod-borne diseases. Applied approaches have predominantly focused on exploiting the reproductive manipulations that *Wolbachia* employ on their host. These manipulations of host reproduction may allow the use of *Wolbachia* for population replacement, to drive desirable genes into an insect population, and the development of incompatible insects for population suppression (Dobson, 2003; Sinkins and Gould, 2006; Bourtzis, 2008). The use of strains that confer pathogen interference and virulent strains that shorten insect lifespan are alternative strategies for pathogen control (Cook *et al.*, 2008; Cook and McGraw, 2010).

The incompatible insect technique (IIT) uses *Wolbachia*-mediated CI to suppress insect populations, in a similar fashion to the sterile insect technique (SIT).

Matings between infected males released into the population and wild-type females are incompatible and do not produce progeny. For insects that harbor a natural *Wolbachia* infection, males with an alternative strain or super-infected males can be used for suppression. Such strategies are being devised for the medfly, *Ceratitis capitata* (Zabalou *et al.*, 2004), and for *Ae. albopictus* (Calvitti *et al.*, 2010) and *Ae. polynesiensis* (Brelsfoard *et al.*, 2008).

Advances in genetic transformation technology have allowed several vector species to be genetically engineered to be refractory to pathogens (Olson *et al.*, 1996; de Lara Capurro *et al.*, 2000; Ito *et al.*, 2002; Franz *et al.*, 2006). However, it is impractical to release these insects on a scale that would have an impact on pathogen transmission. Such release strategies would need to encompass a gene drive mechanism that would enable the spread of insects with the desired genotype into the population. A *Wolbachia*-mediated CI sweep may be harnessed to drive a transgene into an insect population. If a transgene of interest were placed into the *Wolbachia* genome, the transgene would be expected to increase in frequency despite potential fitness costs as *Wolbachia* spread, driving the desired genotype through the population (Rasgon and Scott, 2004a). This hypothesis is also applicable for population replacement, whereby a strain of insect with a desirable phenotype, such as impaired vector competence (Blandin *et al.*, 2004; Osta *et al.*, 2004; Rottschaefer *et al.*, 2011), is spread into wild-type populations.

Owing to the extrinsic incubation period of many pathogens, age is a critical component of vector competence. Therefore, manipulation of vector survival is theoretically a highly effective control strategy (Cook *et al.*, 2008). The *Wolbachia* strain wMelPop causes life shortening in *D. melanogaster* (Min and Benzer, 1997). Modeling suggests that *Wolbachia*-induced mortality can have a significant effect on reducing pathogen transmission (Rasgon *et al.*, 2003). The transfer of wMelPop into insect species of medical importance, and the induction of life shortening in these new hosts, demonstrate the validity of this concept (McMeniman *et al.*, 2009). Theoretically, this approach would be applicable for any pathogen with an extended extrinsic incubation period.

9.8. FUTURE RESEARCH DIRECTIONS

Recent research has exposed many aspects of *Wolbachia* biology and revealed the multitude of effects that these bacteria can have on their hosts. Yet, much still remains to be elucidated. An important research priority is the development of transgenic technology for *Wolbachia*. This will not only allow analysis of genes that are critical for the interaction of *Wolbachia* with the host, but also open up the prospect for *Wolbachia* to be used in a paratransgenic capacity. Most of the attention afforded to *Wolbachia* for control of arthropod-borne diseases centers on the ability of *Wolbachia* to manipulate host reproduction and, recently, to interfere with pathogens of the host. However, the mechanistic basis of such effects remains largely unknown and awaits further investigation. In addition, many vectors of arthropod-borne diseases are uninfected by these bacteria, and while successful transfer of *Wolbachia* has been accomplished in some species, others remain seemingly recalcitrant to infection, particularly species in the genus *Anopheles*. Greater understanding is also required of the behavioral and fitness effects that *Wolbachia* has upon on its host. Such knowledge may help to shed light on the evolutionary forces that *Wolbachia* has imparted on its host. If these challenges can be overcome, *Wolbachia* holds great promise to be used to control many arthropod-borne diseases.

REFERENCES

Arthofer, W., Riegler, M., Schneider, D., Krammer, M., Miller, W. J., & Stauffer, C. (2009). Hidden *Wolbachia* diversity in field populations of the European cherry fruit fly, *Rhagoletis cerasi* (Diptera, Tephritidae). *Mol. Ecol., 18*, 3816−3830.

Atyame, C. M., Duron, O., Tortosa, P., Pasteur, N., Fort, P., & Weill, M. (2010). Multiple *Wolbachia* determinants control the evolution of cytoplasmic incompatibilities in *Culex pipiens* mosquito populations. *Mol. Ecol., 20*, 286−298.

Baldo, L., Hotopp, J. C. D., Jolley, K. A., Bordenstein, S. R., Biber, S. A., Choudhury, R. R., Hayashi, C., Maiden, M. C. J., Tettelin, H., & Werren, J. H. (2006). Multilocus sequence typing system for the endosymbiont *Wolbachia pipientis*. *Appl. Environ. Microbiol., 72*, 7098−7110.

Barr, A. R. (1980). Cytoplasmic incompatibility in natural populations of a mosquito, *Culex pipiens* L. *Nature, 283*, 71−72.

de Bary, H. A. (1879). Die Erscheinung der Symbiose. *Verlag von Karl J. Strasburg.*

Bian, G., Xu, Y., Lu, P., Xie, Y., & Xi, Z. (2010). The endosymbiotic bacterium *Wolbachia* induces resistance to dengue virus in *Aedes aegypti. PLoS Pathog, 6.* e1000833.

Blandin, S. A., Shiao, S.-H., Moita, L. F., Janse, C. J., Waters, A. P., Kafatos, F. C., & Levashina, E. A. (2004). Complement-like protein TEP1 is a determinant of vectorial capacity in the malaria vector Anopheles gambiae. *Cell, 116*, 661−670.

Bossan, B., Koehncke, A., & Hammerstein, P. (2011). A new model and method for understanding *Wolbachia*-induced cytoplasmic incompatibility. *PLoS ONE, 6.* e19757.

Bouchon, D., Rigaud, T., & Juchault, P. (1998). Evidence for widespread *Wolbachia* infection in isopod crustaceans: molecular identification and host feminization. *Proc. R. Soc. Lond. B, 265*, 1081−1090.

Bourtzis, K. (2008). *Wolbachia*-based technologies for insect pest population control. *Adv. Exp. Med. Biol., 627*, 104−113.

Bourtzis, K., Pettigrew, M. M., & O'Neill, S. L. (2000). *Wolbachia* neither induces nor suppresses transcripts encoding antimicrobial peptides. *Insect Mol. Biol., 9*, 635−639.

Braquart-Varnier, C., Lachat, M., Herbinière, J., Johnson, M., Caubet, Y., Bouchon, D., & Sicard, M. (2008). *Wolbachia* mediate variation of host immunocompetence. *PLoS ONE, 3.* e3286.

Breeuwer, J. A. J., & Werren, J. H. (1990). Microorganisms associated with chromosome destruction and reproductive isolation between two insect species. *Nature, 346*, 558−560.

Brelsfoard, C. L., Séchan, Y., & Dobson, S. L. (2008). Interspecific hybridization yields strategy for South Pacific filariasis vector elimination. *PLoS Negl. Trop. Dis, 2*. e129.

Brown, M. J. F., Moret, Y., & Schmid-Hempel, P. (2003). Activation of host constitutive immune defence by an intestinal trypanosome parasite of bumble bees. *Parasitology, 126*, 253−260.

Brownlie, J. C., & Johnson, K. N. (2009). Symbiont-mediated protection in insect hosts. *Trends Microbiol., 17*, 348−354.

Brownlie, J. C., Cass, B. N., Riegler, M., Witsenburg, J. J., Iturbe-Ormaetxe, I., McGraw, E. A., & O'Neill, S. L. (2009). Evidence for metabolic provisioning by a common invertebrate endosymbiont, *Wolbachia pipientis*, during periods of nutritional stress. *PLoS Pathog, 5*. e1000368.

Callaini, G., Dallai, R., & Ripardelli, M. G. (1997). *Wolbachia* induced delay of paternal chromatin condensation does not prevent maternal chromosomes from entering anaphase in incompatible crosses in Drosophila simulans. *J. Cell Sci., 110*, 271−280.

Calvitti, M., Moretti, R., Lampazzi, E., Bellini, R., & Dobson, S. L. (2010). Characterization of a new *Aedes albopictus* (Diptera: Culicidae)−*Wolbachia pipientis* (Rickettsiales: Rickettsiaceae) symbiotic association generated by artificial transfer of the wPip strain from *Culex pipiens* (Diptera: Culicidae). *J. Med. Entomol., 47*, 179−187.

Cancrini, G., Frangipane di Regalbono, A., Ricci, I., Tessarin, C., Gabrielli, S., & Pietrobelli, M. (2003). *Aedes albopictus* is a natural vector of *Dirofilaria immitis* in Italy. *Vet. Parasitol., 118*, 195−202.

Cerenius, L., Lee, B. L., & Söderhäll, K. (2008). The proPO-system: pros and cons for its role in invertebrate immunity. *Trends Immunol., 29*, 263−271.

Charlat, S., Hurst, G. D. D., & Merçot, H. (2003a). Evolutionary consequences of *Wolbachia* infections. *Trends Genet., 19*, 217−223.

Charlat, S., Le Chat, L., & Merçot, H. (2003b). Characterization of non-cytoplasmic incompatibility inducing *Wolbachia* in two continental African populations of *Drosophila simulans*. *Heredity, 90*, 49−55.

Charlat, S., Davies, N., Roderick, G. K., & Hurst, G. D. D. (2007a). Disrupting the timing of *Wolbachia*-induced male-killing. *Biol. Lett., 3*, 154−156.

Charlat, S., Hornett, E. A., Fullard, J. H., Davies, N., Roderick, G. K., Wedell, N., & Hurst, G. D. D. (2007b). Extraordinary flux in sex ratio. *Science, 317*, 214.

Charlat, S., Duplouy, A., Hornett, E. A., Dyson, E. A., Davies, N., Roderick, G. K., Wedell, N., & Hurst, G. D. D. (2009). The joint evolutionary histories of *Wolbachia* and mitochondria in *Hypolimnas bolina*. *BMC Evol. Biol., 9*, 64.

Cook, P. E., & McGraw, E. A. (2010). *Wolbachia pipientis*: an expanding bag of tricks to explore for disease control. *Trends Parasitol., 26*, 373−375.

Cook, P., McMeniman, C. J., & O'Neill, S. L. (2008). Modifying insect population age structure to control vector-borne disease. *Adv. Exp. Med. Biol., 627*, 126−140.

Cordaux, R., Michel-Salzat, A., Frelon-Raimond, M., Rigaud, T., & Bouchon, D. (2004). Evidence for a new feminizing *Wolbachia* strain in the isopod *Armadillidium vulgare*: evolutionary implications. *Heredity, 93*, 78−84.

Counce, S., & Poulson, D. (1962). Developmental effects of sex-ratio agent in embryos of *Drosophila willistoni*. *J. Exp. Zool., 151*, 17−31.

Curtis, C. F. (1976). Population replacement in *Culex fatigans* by means of cytoplasmic incompatibility. 2. Field cage experiments with overlapping generations. *Bull. World Health Organ., 53*, 107−119.

Curtis, C. F., & Adak, T. (1974). Population replacement in *Culex fatigans* by means of cytoplasmic incompatibility. Laboratory experiments with non-overlapping generations. *Bull. World Health Organ., 51*, 249−255.

Curtis, C. F., & Suya, T. B. (1981). Variations in cytoplasmic incompatibility properties in Tanzanian populations of *Culex quinquefasciatus*. *Ann. Trop. Med. Parasitol., 75*, 101−106.

Dedeine, F., Vavre, F., Fleury, F., Loppin, B., Hochberg, M. E., & Bouletreau, M. (2001). Removing symbiotic *Wolbachia* bacteria specifically inhibits oogenesis in a parasitic wasp. *Proc. Natl. Acad. Sci. USA, 98*, 6247−6252.

Dobson, S. L. (2003). Reversing *Wolbachia*-based population replacement. *Trends Parasitol., 19*, 128−133.

Dobson, S. L., Marsland, E., & Rattanadechakul, W. (2002). Mutualistic *Wolbachia* infection in *Aedes albopictus*: accelerating cytoplasmic drive. *Genetics, 160*, 1087−1094.

Dobson, S. L., Rattanadechakul, W., & Marsland, E. J. (2004). Fitness advantage and cytoplasmic incompatibility in *Wolbachia* single- and superinfected *Aedes albopictus*. *Heredity, 93*, 135−142.

Dumler, J. S., Barbet, A. F., Bekker, C. P., Dasch, G. A., Palmer, G. H., Ray, S. C., Rikihisa, Y., & Rurangirwa, F. R. (2001). Reorganization of genera in the families Rickettsiaceae and Anaplasmataceae in the order Rickettsiales: unification of some species of *Ehrlichia* with *Anaplasma*, *Cowdria* with *Ehrlichia* and *Ehrlichia* with *Neorickettsia*, descriptions of six new species combinations and designation of *Ehrlichia equi* and "*HGE agent*" *as subjective synonyms of Ehrlichia phagocytophila*. *Int. J. Syst. Evol. Microbiol., 51*, 2145−2165.

Duplouy, A., Hurst, G. D. D., O'Neill, S. L., & Charlat, S. (2010). Rapid spread of male-killing *Wolbachia* in the butterfly *Hypolimnas bolina*. *J. Evol. Biol., 23*, 231−235.

Duron, O., & Weill, M. (2006). *Wolbachia* infection influences the development of *Culex pipiens* embryo in incompatible crosses. *Heredity, 96*, 493−500.

Duron, O., Fort, P., & Weill, M. (2006). Hypervariable prophage WO sequences describe an unexpected high number of *Wolbachia* variants in the mosquito *Culex pipiens*. *Proc. R. Soc. B, 273*, 495−502.

Duron, O., Boureux, A., Echaubard, P., Berthomieu, A., Berticat, C., Fort, P., & Weill, M. (2007). Variability and expression of ankyrin domain genes in *Wolbachia* variants infecting the mosquito *Culex pipiens*. *J. Bacteriol, 189*, 4442−4448.

Dyer, K. A., & Jaenike, J. (2004). Evolutionarily stable infection by a male-killing endosymbiont in *Drosophila innubila*: molecular evidence from the host and parasite genomes. *Genetics, 168*, 1443−1455.

Eleftherianos, I., Marokhazi, J., Millichap, P. J., Hodgkinson, A. J., Sriboonlert, A., Ffrench-Constant, R. H., & Reynolds, S. E. (2006). Prior infection of *Manduca sexta* with non-pathogenic *Escherichia coli* elicits immunity to pathogenic *Photorhabdus luminescens*: roles of immune-related proteins shown by RNA interference. *Insect Biochem. Mol. Biol., 36*, 517−525.

Ferree, P. M., Avery, A., Azpurua, J., Wilkes, T., & Werren, J. H. (2008). A bacterium targets maternally inherited centrosomes to kill males in. *Nasonia. Curr. Biol., 18*, 1409−1414.

Fialho, R., & Stevens, L. (2000). Male-killing *Wolbachia* in a flour beetle. *Proc. R. Soc. Lond. B, 267*, 1469−1473.

Franz, A. W. E., Sanchez-Vargas, I., Adelman, Z. N., Blair, C. D., Beaty, B. J., James, A. A., & Olson, K. E. (2006). Engineering RNA interference-based resistance to dengue virus type 2 in genetically modified *Aedes aegypti*. *Proc. Natl. Acad. Sci. USA, 103,* 4198−4203.

Frentiu, F. D., Robinson, J., Young, P. R., McGraw, E. A., & O'Neill, S. L. (2010). *Wolbachia*-mediated resistance to dengue virus infection and death at the cellular level. *PLoS ONE 5.* e13398.

Fujii, Y., Kageyama, D., Hoshizaki, S., Ishikawa, H., & Sasaki, T. (2001). Transfection of *Wolbachia* in Lepidoptera: the feminizer of the adzuki bean borer *Ostrinia scapulalis* causes male killing in the Mediterranean flour moth *Ephestia kuehniella*. *Proc. R. Soc. Lond. B, 268,* 855−859.

Gavotte, L., Mercer, D. R., Stoeckle, J. J., & Dobson, S. L. (2010). Costs and benefits of *Wolbachia* infection in immature *Aedes albopictus* depend upon sex and competition level. *J. Invertebr. Pathol., 105,* 341−346.

Giron, D., Kaiser, W., Imbault, N., & Casas, J. (2007). Cytokinin-mediated leaf manipulation by a leafminer caterpillar. *Biol. Lett., 3,* 340−343.

Glaser, R. L., & Meola, M. A. (2010). The native *Wolbachia* endosymbionts of *Drosophila melanogaster* and *Culex quinquefasciatus* increase host resistance to West Nile virus infection. *PLoS ONE, 5.* e11977.

Gottlieb, Y., Zchori-Fein, E., Werren, J. H., & Karr, T. L. (2002). Diploidy restoration in *Wolbachia*-infected *Muscidifurax uniraptor* (Hymenoptera: Pteromalidae). *J. Invertebr. Pathol., 81,* 166−174.

Guillemaud, T., Pasteur, N., & Rousset, F. (1997). Contrasting levels of variability between cytoplasmic genomes and incompatibility types in the mosquito *Culex pipiens*. *Proc. R. Soc. Lond. B, 264,* 245−251.

Harcombe, W. R., & Hoffmann, A. A. (2004). *Wolbachia* effects in *Drosophila melanogaster*: in search of fitness benefits. *J. Invertebr. Pathol., 87,* 45−50.

Hedges, L. M., Brownlie, J. C., O'Neill, S. L., & Johnson, K. N. (2008). *Wolbachia* and virus protection in insects. *Science, 322,* 702.

Hertig, M. (1936). The rickettsia, *Wolbachia pipientis* (gen. et sp. n.) and associated inclusions of the mosquito *Culex pipiens*. *Parasitology, 28,* 453−458.

Hertig, M., & Wolbach, S. B. (1924). Studies on rickettsia-like microorganisms in insects. *J. Med. Res., 44,* 329−374.

Hilgenboecker, K., Hammerstein, P., Schlattmann, P., Telschow, A., & Werren, J. H. (2008). How many species are infected with *Wolbachia*? − A statistical analysis of current data. *FEMS Microbiol. Lett., 281,* 215−220.

Hiroki, M., Kato, Y., Kamito, T., & Miura, K. (2002). Feminization of genetic males by a symbiotic bacterium in a butterfly, *Eurema hecabe* (Lepidoptera: Pieridae). *Naturwissenschaften, 89,* 167−170.

Hornett, E. A., Charlat, S., Duplouy, A. M. R., Davies, N., Roderick, G. K., Wedell, N., & Hurst, G. D. D. (2006). Evolution of male-killer suppression in a natural population. *PLoS Biol, 4.* e283.

Hornett, E. A., Duplouy, A. M. R., Davies, N., Roderick, G. K., Wedell, N., Hurst, G. D. D., & Charlat, S. (2008). You can't keep a good parasite down: evolution of a male-killer suppressor uncovers cytoplasmic incompatibility. *Evolution, 62,* 1258−1263.

Hosokawa, T., Koga, R., Kikuchi, Y., Meng, X.-Y., & Fukatsu, T. (2010). *Wolbachia* as a bacteriocyte-associated nutritional mutualist. *Proc. Natl. Acad. Sci. USA, 107,* 769−774.

Hughes, G. L., Allsopp, P. G., Brumbley, S. M., Woolfit, M., McGraw, E. A., & O'Neill, S. L. (2011a). Variable infection frequency and high diversity of multiple strains of *Wolbachia* in *Perkinsiella* planthoppers. *Appl. Environ. Microbiol., 77,* 2165−2168.

Hughes, G. L., Koga, R., Xue, P., Fukastu, T., & Rasgon, J. L. (2011b). *Wolbachia* infections are virulent and inhibit the human malaria parasite *Plasmodium falciparum* in *Anopheles gambiae*. *PLoS Pathog, 7.* e1002043.

Hughes, G. L., Ren, X., Ramirez, J. L., Sakamoto, J. M., Bailey, J. A., Jedlicka, A. E., & Rasgon, J. L. (2011c). *Wolbachia* infections in *Anopheles gambiae* cells: transcriptomic characterization of a novel host−symbiont interaction. *PLoS Pathog, 7.* e1001296.

Huigens, M. E., de Almeida, R. P., Boons, P. A. H., Luck, R. F., & Stouthamer, R. (2004). Natural interspecific and intraspecific horizontal transfer of parthenogenesis-inducing *Wolbachia* in *Trichogramma* wasps. *Proc. R. Soc. Lond. B, 271,* 509−515.

Hurst, G. D. D., Hurst, L. D., & Majerus, M. E. N. (1997). Cytoplasmic sex ratio distorters. In S. L. O'Neill, A. A. Hoffmann & J. H. Werren (Eds.), *Influential Passengers: Inherited Microorganisms and Arthropod Reproduction* (pp. 125−154). Oxford: Oxford University Press.

Hurst, G. D., von der Schulenburg, J., Majerus, T., Bertrand, D., Zakharov, I., Baungaard, J., Volkl, W., Stouthamer, R., & Majerus, M. (1999). Invasion of one insect species, *Adalia bipunctata*, by two different male-killing bacteria. *Insect Mol. Biol., 8,* 133−139.

Hurst, G. D. D., Johnson, A. P., Schulenburg, J. H. G., & Fuyama, Y. (2000). Male-killing *Wolbachia* in *Drosophila*: a temperature-sensitive trait with a threshold bacterial density. *Genetics, 156,* 699−709.

Hurst, G. D. D., Jiggins, F. M., & Pomiankowski, A. (2002). Which way to manipulate host reproduction? *Wolbachia* that cause cytoplasmic incompatibility are easily invaded by sex ratio distorting mutants. *Am. Nat., 160,* 360−373.

Hurst, L. (1991). The incidences and evolution of cytoplasmic male killers. *Proc. R. Soc. Lond. B, 244,* 91−99.

Hurst, L. D. (1991). The evolution of cytoplasmic incompatibility or when spite can be successful. *J. Theor. Biol., 148,* 269−277.

Ito, J., Ghosh, A. K., Moreira, L. A., Wimmer, E. A., & Jacobs-Lorena, M. (2002). Transgenic anopheline mosquitoes impaired in transmission of a malaria parasite. *Nature, 417,* 452−455.

Jaenike, J. (2007). Spontaneous emergence of a new *Wolbachia* phenotype. *Evolution, 61,* 2244−2252.

Jaenike, J., Dyer, K., & Reed, L. (2003). Within-population structure of competition and the dynamics of male-killing. *Wolbachia. Evol. Ecol. Res., 5,* 1023−1036.

James, A. C., & Ballard, J. W. (2000). Expression of cytoplasmic incompatibility in *Drosophila simulans* and its impact on infection frequencies and distribution of. *Wolbachia pipientis. Evolution, 54,* 1661−1672.

Jiggins, F. M., Hurst, G. D., Schulenburg, J. H., & Majerus, M. E. (2001). Two male-killing *Wolbachia* strains coexist within a population of the butterfly *Acraea encedon*. *Heredity, 86,* 161−166.

Jin, C., Ren, X., & Rasgon, J. L. (2009). The virulent *Wolbachia* strain wMelPop efficiently establishes somatic infections in the malaria vector *Anopheles gambiae*. *Appl. Environ. Microbiol., 75,* 3373−3376.

Juchault, P., Rigaud, T., & Mocquard, J.-P. (1992). Evolution of sex-determining mechanisms in a wild population of *Armadillidium vulgare* Latr. (Crustacea, Isopoda): competition between two feminizing parasitic sex factors. *Heredity, 69,* 382−390.

Juchault, P., Rigaud, T., & Mocquard, J.-P. (1993). Evolution of sex determination and sex ratio variability in wild populations of

Armadillidium vulgare (Latr.) (Crustacea, Isopoda): a case study in conflict resolution. *Acta Oecol., 14,* 547–562.

Juchault, P., Frelon, M., Bouchon, D., & Rigaud, T. (1994). New evidence for feminizing bacteria in terrestrial isopods – evolutionary implications. *C.R. Acad. Sci. III, 317,* 225–230.

Kageyama, D., Hoshizaki, S., & Ishikawa, Y. (1998). Female-biased sex ratio in the Asian corn borer, *Ostrinia furnacalis*: evidence for the occurrence of feminizing bacteria in an insect. *Heredity, 81,* 311–316.

Kageyama, D., Ohno, S., Hoshizaki, S., & Ishikawa, Y. (2003). Sexual mosaics induced by tetracycline treatment in the *Wolbachia*-infected adzuki bean borer, *Ostrinia scapulalis. Genome, 46,* 983–989.

Kaiser, W., Huguet, E., Casas, J., Commin, C., & Giron, D. (2010). Plant green-island phenotype induced by leaf-miners is mediated by bacterial symbionts. *Proc. R. Soc. B, 277,* 2311–2319.

Kambhampai, S., Rai, K., & Burgun, S. (1993). Unidirectional cytoplasmic incompatibility in the mosquito, *Aedes albopictus. Evolution, 47,* 673–677.

Kambris, Z., Cook, P. E., Phuc, H. K., & Sinkins, S. P. (2009). Immune activation by life-shortening *Wolbachia* and reduced filarial competence in mosquitoes. *Science, 326,* 134–136.

Kambris, Z., Blagborough, A. M., Pinto, S. B., Blagrove, M. S. C., Godfray, H. C. J., Sinden, R. E., & Sinkins, S. P. (2010). *Wolbachia* stimulates immune gene expression and inhibits *Plasmodium* development in *Anopheles gambiae. PLoS Pathog, 6.* e1001143.

Kittayapong, P., Baisley, K. J., Baimai, V., & O'Neill, S. L. (2000). Distribution and diversity of *Wolbachia* infections in southeast Asian mosquitoes (Diptera: Culicidea). *J. Med. Entomol., 37,* 340–345.

Kittayapong, P., Baisley, K., Sharpe, R., Baimai, V., & O'Neill, S. L. (2002a). Maternal transmission efficiency of *Wolbachia* superinfections in *Aedes albopictus* populations in Thailand. *Am. J. Trop. Med. Hyg., 66,* 103–107.

Kittayapong, P., Mongkalangoon, P., Baimai, V., & O'Neill, S. L. (2002b). Host age effect and expression of cytoplasmic incompatibility in field populations of *Wolbachia*-superinfected *Aedes albopictus. Heredity, 88,* 270–274.

Koop, J. L., Zeh, D. W., Bonilla, M. M., & Zeh, J. A. (2009). Reproductive compensation favours male-killing *Wolbachia* in a live-bearing host. *Proc. R. Soc. B, 276,* 4021–4028.

Kose, H., & Karr, T. L. (1995). Organization of *Wolbachia pipientis* in the *Drosophila* fertilized egg and embryo revealed by anti- *Wolbachia* monoclonal antibody. *Mech. Dev., 5,* 275–288.

Kow, C. Y., Koon, L. L., & Yin, P. F. (2001). Detection of dengue viruses in field caught male *Aedes aegypti* and *Aedes albopictus* (Diptera: Culicidae) in Singapore by type-specific PCR. *J. Med. Entomol., 38,* 475–479.

Kremer, N., Charif, D., Henri, H., Bataille, M., Prévost, G., Kraaijeveld, K., & Vavre, F. (2009). A new case of *Wolbachia* dependence in the genus *Asobara*: evidence for parthenogenesis induction in *Asobara japonica. Heredity, 103,* 248–256.

Kumar, S., Christophides, G. K., Cantera, R., Charles, B., Han, Y. S., Meister, S., Dimopoulos, G., Kafatos, F. C., & Barillas-Mury, C. (2003). The role of reactive oxygen species on *Plasmodium* melanotic encapsulation in *Anopheles gambiae. Proc. Natl. Acad. Sci. USA, 100,* 14139–14144.

de Lara Capurro, M., Coleman, J., Beerntsen, B. T., Myles, K. M., Olson, K. E., Rocha, E., Krettli, A. U., & James, A. A. (2000). Virus-expressed, recombinant single-chain antibody blocks sporozoite infection of salivary glands in *Plasmodium gallinaceum*-infected *Aedes aegypti. Am. J. Trop. Med. Hyg., 62,* 427–433.

Lassy, C. W., & Karr, T. L. (1996). Cytological analysis of fertilization and early embryonic development in incompatible crosses of *Drosophila simulans. Mech. Dev., 57,* 47–58.

Laven, H. (1951). Crossing experiments with *Culex* strains. *Evolution, 5,* 370–375.

Laven, H. (1967a). A possible model for speciation by cytoplasmic isolation in the *Culex pipiens* complex. *Bull. World Health Organ., 37,* 263–266.

Laven, H. (1967b). Eradication of *Culex pipiens* fatigans through cytoplasmic incompatibility. *Nature, 216,* 383–384.

Lo, N., Casiraghi, M., Salati, E., Bazzocchi, C., & Bandi, C. (2002). How many *Wolbachia* supergroups exist? *Mol. Biol. Evol., 19,* 341–346.

Majerus, T. M. O., & Majerus, M. E. N. (2010). Intergenomic arms races: detection of a nuclear rescue gene of male-killing in a ladybird. *PLoS Pathog, 6.* e1000987.

Martin, G., Delaunay, C., Braquart-Varnier, C., & Azzouna, A. (2010). Prophage elements from the endosymbiont, *Wolbachia* Hertig, 1936 transferred to the host genome of the woodlouse, *Armadillidium vulgare* Latreille, 1804 (Peracarida, Isopoda). *Crustaceana, 83,* 539–548.

McGraw, E. A., & O'Neill, S. L. (2007). *Wolbachia*: invasion biology in South Pacific butterflies. *Curr. Biol., 17,* R220–R221.

McMeniman, C. J., Lane, A. M., Cass, B. N., Fong, A. W. C., Sidhu, M., Wang, Y. F., & O'Neill, S. L. (2009). Stable introduction of a life-shortening *Wolbachia* infection into the mosquito *Aedes aegypti. Science, 323,* 141–144.

Merçot, H., & Charlat, S. (2004). *Wolbachia* infections in *Drosophila melanogaster* and *D. simulans*: polymorphism and levels of cytoplasmic incompatibility. *Genetica, 120,* 51–59.

Miller, W. J., Ehrman, L., & Schneider, D. (2010). Infectious speciation revisited: impact of symbiont-depletion on female fitness and mating behavior of *Drosophila paulistorum. PLoS Pathog, 6.* e1001214.

Min, K. T., & Benzer, S. (1997). *Wolbachia*, normally a symbiont of *Drosophila*, can be virulent, causing degeneration and early death. *Proc. Natl. Acad. Sci. USA, 94,* 10792–10796.

Molina-Cruz, A., DeJong, R. J., Charles, B., Gupta, L., Kumar, S., Jaramillo-Gutierrez, G., & Barillas-Mury, C. (2008). Reactive oxygen species modulate *Anopheles gambiae* immunity against bacteria and *Plasmodium. J. Biol. Chem., 283,* 3217–3223.

Moreira, L. A., Iturbe-Ormaetxe, I., Jeffery, J. A., Lu, G., Pyke, A. T., Hedges, L. M., Rocha, B. C., Hall-Mendelin, S., Day, A., Riegler, M., Hugo, L. E., Johnson, K. N., Kay, B. H., McGraw, E. A., van den Hurk, A. F., Ryan, P. A., & O'Neill, S. L. (2009a). A *Wolbachia* symbiont in *Aedes aegypti* limits infection with dengue, chikungunya, and *Plasmodium. Cell, 139,* 1268–1278.

Moreira, L. A., Saig, E., Turley, A. P., Ribeiro, J. M. C., O'Neill, S. L., & McGraw, E. A. (2009b). Human probing behavior of *Aedes aegypti* when infected with a life-shortening strain of *Wolbachia. PLoS Negl. Trop. Dis, 3.* e568.

Moreira, L. A., Ye, Y. H., Turner, K., Eyles, D. W., McGraw, E. A., & O'Neill, S. L. (2011). The wMelPop strain of *Wolbachia* interferes with dopamine levels in *Aedes aegypti. Parasit. Vectors, 4,* 28.

Mousson, L., Martin, E., Zouache, K., Madec, Y., Mavingui, P., & Failloux, A. B. (2010). *Wolbachia*modulates Chikungunya replication in *Aedes albopictus. Mol. Ecol., 19,* 1953–1964.

Narita, S., Kageyama, D., Nomura, M., & Fukatsu, T. (2007). Unexpected mechanism of symbiont-induced reversal of insect sex: feminizing

Wolbachia continuously acts on the butterfly *Eurema hecabe* during larval development. *Appl. Environ. Microbiol., 73,* 4332–4341.

Negri, I., Pellecchia, M., Mazzoglio, P. J., Patetta, A., & Alma, A. (2006). Feminizing *Wolbachia* in *Zyginidia pullula* (Insecta, Hemiptera), a leafhopper with an XX/X0 sex-determination system. *Proc. R. Soc. B, 273,* 2409–2416.

Negri, I., Franchini, A., Gonella, E., Daffonchio, D., Mazzoglio, P. J., Mandrioli, M., & Alma, A. (2009). Unravelling the *Wolbachia* evolutionary role: the reprogramming of the host genomic imprinting. *Proc. R. Soc. B, 276,* 2485–2491.

Olson, K. E., Higgs, S., Gaines, P. J., Powers, A. M., Davis, B. S., Kamrud, K. I., Carlson, J. O., Blair, C. D., & Beaty, B. J. (1996). Genetically engineered resistance to dengue-2 virus transmission in mosquitoes. *Science, 272,* 884–886.

O'Neill, S. L., & Paterson, H. E. (1992). Crossing type variability associated with cytoplasmic incompatibility in Australian populations of the mosquito *Culex quinquefasciatus* Say. *Med. Vet. Entomol., 6,* 209–216.

Osborne, S. E., Leong, Y. S., O'Neill, S. L., & Johnson, K. N. (2009). Variation in antiviral protection mediated by different *Wolbachia* strains in *Drosophila simulans*. *PLoS Pathog, 5.* e1000656.

Osta, M. A., Christophides, G. K., & Kafatos, F. C. (2004). Effects of mosquito genes on *Plasmodium* development. *Science, 303,* 2030–2032.

Pannebakker, B. A., Pijnacker, L. P., Zwaan, B. J., & Beukeboom, L. W. (2004). Cytology of *Wolbachia*-induced parthenogenesis in *Leptopilina clavipes* (Hymenoptera: Figitidae). *Genome, 47,* 299–303.

Pannebakker, B. A., Loppin, B., Elemans, C. P. H., Humblot, L., & Vavre, F. (2007). Parasitic inhibition of cell death facilitates symbiosis. *Proc. Natl. Acad. Sci. USA, 104,* 213–215.

Panteleev, D. I., Goriacheva, I. I., Andrianov, B. V., Reznik, N. L., Lazebnyĭ, O. E., & Kulikov, A. M. (2007). The endosymbiotic bacterium *Wolbachia* enhances the nonspecific resistance to insect pathogens and alters behavior of *Drosophila melanogaster*. *Genetika, 43,* 1277–1280.

Pearlman, E., & Gillette-Ferguson, I. (2007). *Onchocerca volvulus, Wolbachia* and river blindness. *Chem. Immunol. Allergy, 92,* 254–265.

Peterson, T. M. L., Gow, A. J., & Luckhart, S. (2007). Nitric oxide metabolites induced in *Anopheles stephensi* control malaria parasite infection. *Free Radic. Biol. Med., 42,* 132–142.

Pham, L. N., Dionne, M. S., Shirasu-Hiza, M., & Schneider, D. S. (2007). A specific primed immune response in *Drosophila* is dependent on phagocytes. *PLoS Pathog, 3.* e26.

Poinsot, D., Charlat, S., & Merçot, H. (2003). On the mechanism of *Wolbachia*-induced cytoplasmic incompatibility: confronting the models with the facts. *BioEssays, 25,* 259–265.

Rasgon, J. L., & Scott, T. W. (2003). *Wolbachia* and cytoplasmic incompatibility in the California *Culex pipiens* mosquito species complex: parameter estimates and infection dynamics in natural populations. *Genetics, 165,* 2029–2038.

Rasgon, J. L., & Scott, T. W. (2004a). Impact of population age structure on *Wolbachia* transgene driver efficacy: ecologically complex factors and release of genetically modified mosquitoes. *Insect Biochem. Mol. Biol., 34,* 707–713.

Rasgon, J. L., & Scott, T. W. (2004b). An initial survey for *Wolbachia* (Rickettsiales: Rickettsiaceae) infections in selected California mosquitoes (Diptera: Culicidae). *J. Med. Entomol., 41,* 255–257.

Rasgon, J. L., Styer, L. M., & Scott, T. W. (2003). *Wolbachia*-induced mortality as a mechanism to modulate pathogen transmission by vector arthropods. *J. Med. Entomol., 40,* 125–132.

Rasgon, J. L., Ren, X., & Petridis, M. (2006). Can *Anopheles gambiae* be infected with *Wolbachia pipientis*? Insights from an *in vitro* system. *Appl. Environ. Microbiol., 72,* 7718–7722.

Reed, K. M., & Werren, J. H. (1995). Induction of paternal genome loss by the paternal-sex-ratio chromosome and cytoplasmic incompatibility bacteria (*Wolbachia*): a comparative study of early embryonic events. *Mol. Reprod. Dev., 40,* 408–418.

Reiter, P., Fontenille, D., & Paupy, C. (2006). *Aedes albopictus* as an epidemic vector of chikungunya virus: another emerging problem? *Lancet Infect. Dis., 6,* 463–464.

Ricci, I., Cancrini, G., Gabrielli, S., D'Amelio, S., & Favi, G. (2002). Searching for *Wolbachia* (Rickettsiales: Rickettsiaceae) in mosquitoes (Diptera: Culicidae): large polymerase chain reaction survey and new identifications. *J. Med. Entomol., 39,* 562–567.

Riegler, M., Sidhu, M., Miller, W., & O'Neill, S. L. (2005). Evidence for a global *Wolbachia* replacement in *Drosophila melanogaster*. *Curr. Biol., 15,* 1428–1433.

Rigaud, T., & Juchault, P. (1992). Genetic control of the vertical transmission of a cytoplasmic sex factor in *Armadillidium vulgare* Latr. (Crustacea, Oniscidea). *Heredity, 68,* 47–52.

Rigaud, T., & Juchault, P. (1993). Conflict between feminizing sex ratio distorters and an autosomal masculinizing gene in the terrestrial isopod *Armadillidium vulgare* Latr. *Genetics, 133,* 247–252.

Rigaud, T., Souty-Gosset, C., Raimond, R., Mocquard, J.-P., & Juchault, P. (1991). Feminizing endocytobiosis in the terrestrial crustacean *Armadillidium vulgare* Latr. (Isopoda): recent acquisitions. *Endocyt. Cell Res., 7,* 259–273.

Rigaud, T., Moreau, J., & Juchault, P. (1999). *Wolbachia* infection in the terrestrial isopod *Oniscus asellus*: sex ratio distortion and effect on fecundity. *Heredity, 83,* 469–475.

Rodrigues, J., Brayner, F. A., Alves, L. C., Dixit, R., & Barillas-Mury, C. (2010). Hemocyte differentiation mediates innate immune memory in *Anopheles gambiae* mosquitoes. *Science, 329,* 1353–1355.

Rottschaefer, S. M., Riehle, M. M., Coulibaly, B., Sacko, M., Niaré, O., Morlais, I., Traoré, S. F., Vernick, K. D., & Lazzaro, B. P. (2011). Exceptional diversity, maintenance of polymorphism, and recent directional selection on the APL1 malaria resistance genes of *Anopheles gambiae*. *PLoS Biol, 9.* e1000600.

Ruang-Areerate, T., & Kittayapong, P. (2006). *Wolbachia* transinfection in *Aedes aegypti*: a potential gene driver of dengue vectors. *Proc. Natl. Acad. Sci. USA, 103,* 12534–12539.

Sadd, B. M., & Schmid-Hempel, P. (2006). Insect immunity shows specificity in protection upon secondary pathogen exposure. *Curr. Biol., 16,* 1206–1210.

Sakamoto, H., Kageyama, D., Hoshizaki, S., & Ishikawa, Y. (2008). Heat treatment of the Adzuki bean borer, *Ostrinia scapulalis* infected with *Wolbachia* gives rise to sexually mosaic offspring. *J. Insect Sci., 8,* 67.

Sanogo, Y., & Dobson, S. L. (2006). WO bacteriophage transcription in *Wolbachia*-infected *Culex pipiens*. *Insect Biochem. Mol. Biol., 36,* 80–85.

Sasaki, T., Kubo, T., & Ishikawa, H. (2002). Interspecific transfer of *Wolbachia* between two lepidopteran insects expressing cytoplasmic incompatibility: a *Wolbachia* variant naturally infecting *Cadra cautella* causes male killing in Ephestia kuehniella. *Genetics, 162,* 1313–1319.

Sinkins, S. P., & Gould, F. (2006). Gene drive systems for insect disease vectors. *Nat. Rev. Genet., 7*, 427–435.

Sinkins, S. P., Braig, H. R., & O'Neill, S. L. (1995a). *Wolbachia pipientis*: bacterial density and unidirectional cytoplasmic incompatibility between infected populations of *Aedes albopictus*. *Exp. Parasitol., 81*, 284–291.

Sinkins, S. P., Braig, H. R., & O'Neill, S. L. (1995b). *Wolbachia* superinfections and the expression of cytoplasmic incompatibility. *Proc. R. Soc. Lond. B, 261*, 325–330.

Sinkins, S. P., Curtis, C. F., & O'Neill, S. L. (1997). The potential application of inherited symbiont systems to pest control. In S. L. O'Neill, A. A. Hoffmann & J. H. Werren (Eds.), *Influential Passengers: Inherited Microorganisms and Arthropod Reproduction* (pp. 155–175). Oxford: Oxford University Press.

Sinkins, S. P., Walker, T., Lynd, A., Steven, A., Makepeace, B., Godfray, H. C. J., & Parkhill, J. (2005). *Wolbachia* variability and host effects on crossing type in *Culex* mosquitoes. *Nature, 436*, 257–260.

Solignac, M., Vautrin, D., & Rousset, F. (1994). Widespread occurrence of the proteobacteria *Wolbachia* and partial cytoplasmic incompatibility in *Drosophila melanogaster*. *C.R. Acad. Sci, III*(317), 461–470.

Stille, B., & Davring, L. (1980). Meiosis and reproductive strategy in the parthenogenetic gall wasp *Diplolepis rosae* (L.) (Hymenoptera, Cynipidae). *Hereditas, 92*, 353–362.

Stouthamer, R., & Kazmer, D. (1994). Cytogenetics of microbe-associated parthenogenesis and its consequences for gene flow in *Trichogramma* wasps. *Heredity, 73*, 317–327.

Stouthamer, R., & Werren, J. (1993). Microbes associated with parthenogenesis in wasps of the genus *Trichogramma*. *J. Invertebr. Pathol., 61*, 6–9.

Stouthamer, R., Luck, R. F., & Hamilton, W. D. (1990). Antibiotics cause parthenogenetic *Trichogramma* (Hymenoptera/Trichogrammatidae) to revert to sex. *Proc. Natl. Acad. Sci. USA, 87*, 2424–2427.

Stouthamer, R., Breeuwer, J., Luck, R., & Werren, J. (1993). Molecular identification of microorganisms associated with parthenogenesis. *Nature, 361*, 66–68.

Stouthamer, R., Russell, J. E., Vavre, F., & Nunney, L. (2010). Intragenomic conflict in populations infected by parthenogenesis inducing *Wolbachia* ends with irreversible loss of sexual reproduction. *BMC Evol. Biol., 10*, 229.

Sugimoto, T. N., Fujii, T., Kayukawa, T., Sakamoto, H., & Ishikawa, Y. (2010). Expression of a doublesex homologue is altered in sexual mosaics of *Ostrinia scapulalis* moths infected with *Wolbachia*. *Insect Biochem. Mol. Biol., 40*, 847–854.

Suh, E., Mercer, D., Fu, Y., & Dobson, S. (2009). Pathogenicity of life-shortening *Wolbachia* in *Aedes albopictus* after transfer from *Drosophila melanogaster*. *Appl. Environ. Microbiol., 75*, 7783–7788.

Suzuki, S., & Yamasaki, K. (1991). Sex-reversal of male *Armadillidium vulgare* (Isopoda, Malacostraca, Crustacea) following andrectomy and partial gonadectomy. *Gen. Comp. Endocrinol., 83*, 375–378.

Teixeira, L., Ferreira, A., & Ashburner, M. (2008). The bacterial symbiont *Wolbachia* induces resistance to RNA viral infections in *Drosophila melanogaster*. *PloS Biol, 6*. e1000002.

Thomas, P., Kenny, N., Eyles, D., Moreira, L. A., O'Neill, S. L., & Asgari, S. (2010). Infection with the wMel and wMelPop strains of *Wolbachia* leads to higher levels of melanization in the hemolymph of *Drosophila melanogaster, Drosophila simulans* and *Aedes aegypti*. *Dev. Comp. Immunol, 52*, 360–365.

Toivonen, J. M., Walker, G. A., Martinez-Diaz, P., Bjedov, I., Driege, Y., Jacobs, H. T., Gems, D., & Partridge, L. (2007). No influence of Indy on lifespan in *Drosophila* after correction for genetic and cytoplasmic background effects. *PLoS Genet, 3*. e95.

Tortosa, P., Charlat, S., Labbé, P., Dehecq, J.-S., Barré, H., & Weill, M. (2010). *Wolbachia* age-sex-specific density in *Aedes albopictus*: a host evolutionary response to cytoplasmic incompatibility? *PLoS ONE, 5*. e9700.

Tram, U., & Sullivan, W. (2002). Role of delayed nuclear envelope breakdown and mitosis in *Wolbachia*-induced cytoplasmic incompatibility. *Science, 296*, 1124–1126.

Turelli, M., & Hoffmann, A. A. (1991). Rapid spread of an inherited incompatibility factor in California *Drosophila*. *Nature, 353*, 440–442.

Turley, A. P., Moreira, L. A., O'Neill, S. L., & McGraw, E. A. (2009). *Wolbachia* infection reduces blood-feeding success in the dengue fever mosquito. *Aedes aegypti. PLoS Negl. Trop. Dis., 3*. e516.

Vandekerckhove, T. T. M., Watteyne, S., Bonne, W., Vanacker, D., Devaere, S., Rumes, B., Maelfait, J.-P., Gillis, M., Swings, J. G., Braig, H. R., & Mertens, J. (2003). Evolutionary trends in feminization and intersexuality in woodlice (Crustacea, Isopoda) infected with *Wolbachia pipientis* (α-Proteobacteria). *Belg. J. Zool., 133*, 61–69.

Vanthournout, B., Swaegers, J., & Hendrickx, F. (2011). Spiders do not escape reproductive manipulations by *Wolbachia*. *BMC Evol. Biol., 11*, 15.

Vavre, F., Mouton, L., & Pannebakker, B. A. (2009). *Drosophila*-parasitoid communities as model systems for host– *Wolbachia* interactions. *Adv. Parasitol., 70*, 299–331.

Weeks, A. R., & Breeuwer, J. A. (2001). *Wolbachia*-induced parthenogenesis in a genus of phytophagous mites. *Proc. R. Soc. Lond. B, 268*, 2245–2251.

Werren, J. H. (1997). Biology of Wolbachia. *Annu. Rev. Entomol., 42*, 587–609.

Werren, J. H., & O'Neill, S. L. (2004). Influential Passengers: Inherited Microorganisms and Arthropod Reproduction. In S. L. O'Neill, A. A. Hoffmann & J. H. Werren (Eds.), *The evolution of heritable symbionts* (pp. 1–41). Oxford: Oxford University Press.

Werren, J. H., Baldo, L., & Clark, M. E. (2008). Wolbachia: master manipulators of invertebrate biology. *Nat. Rev. Microbiol., 6*, 741–751.

Wu, M., Sun, L. V., Vamathevan, J., Riegler, M., Deboy, R., Brownlie, J. C., McGraw, E. A., Martin, W., Esser, C., Ahmadinejad, N., Wiegand, C., Madupu, R., Beanan, M. J., Brinkac, L. M., Daugherty, S. C., Durkin, A. S., Kolonay, J. F., Nelson, W. C., Mohamoud, Y., Lee, P., Berry, K., Young, M. B., Utterback, T., Weidman, J., Nierman, W. C., Paulsen, I. T., Nelson, K. E., Tettelin, H., O'Neill, S. L., & Eisen, J. A. (2004). Phylogenomics of the reproductive parasite *Wolbachia pipientis* wMel: a streamlined genome overrun by mobile genetic elements. *PLoS Biol., 2*. e69.

Xi, Z., Dean, J., Khoo, C., & Dobson, S. L. (2005a). Generation of a novel *Wolbachia* infection in *Aedes albopictus* (Asian tiger mosquito) via embryonic microinjection. *Insect Biochem. Mol. Biol., 35*, 903–910.

Xi, Z., Khoo, C., & Dobson, S. (2005b). *Wolbachia* establishment and invasion in an *Aedes aegypti* laboratory population. *Science, 310*, 326–328.

Xi, Z., Khoo, C., & Dobson, S. L. (2006). Interspecific transfer of *Wolbachia* into the mosquito disease vector *Aedes albopictus*. *Proc. R. Soc. Lond. B, 273*, 1317–1322.

Xi, Z., Gavotte, L., Xie, Y., & Dobson, S. L. (2008). Genome-wide analysis of the interaction between the endosymbiotic bacterium *Wolbachia* and its *Drosophila* host. *BMC Genomics, 9*, 1.

Yamada, R., Floate, K. D., Riegler, M., & O'Neill, S. L. (2007). Male development time influences the strength of *Wolbachia*-induced cytoplasmic incompatibility expression in *Drosophila melanogaster*. *Genetics, 177*, 801–808.

Yen, J., & Barr, A. (1971). New hypothesis of the cause of cytoplasmic incompatibility in *Culex pipiens*. *Nature, 232*, 657–658.

Zabalou, S., Riegler, M., Theodorakopoulou, M., Stauffer, C., Savakis, C., & Bourtzis, K. (2004). *Wolbachia*-induced cytoplasmic incompatibility as a means for insect pest population control. *Proc. Natl. Acad. Sci. USA, 101*, 15042–15045.

Zabalou, S., Apostolaki, A., Pattas, S., Veneti, Z., Paraskevopoulos, C., Livadaras, I., Markakis, G., Brissac, T., Merçot, H., & Bourtzis, K. (2008). Multiple rescue factors within a *Wolbachia* strain. *Genetics, 178*, 2145–2160.

Zeh, D. W., Zeh, J. A., & Bonilla, M. M. (2005). *Wolbachia*, sex ratio bias and apparent male killing in the harlequin beetle riding pseudoscorpion. *Heredity, 95*, 41–49.

Zhang, X., Norris, D. E., & Rasgon, J. L. (2011). Distribution and molecular characterization of *Wolbachia* endosymbionts and filarial nematodes in Maryland populations of the lone star tick (*Amblyomma americanum*). *FEMS Microbiol. Ecol., 77*, 50–56.

Chapter 10

Protistan Entomopathogens

Carlos E. Lange* and Jeffrey C. Lord[†]

*Comisión de Investigaciones Científicas (CIC) de la provincia de Buenos Aires, CCT La Plata, CEPAVE-CONICET-UNLP Argentina, [†]United States Department of Agriculture, Agricultural Research Service, Manhattan, Kansas, USA

SUMMARY

Protists, eukaryotes of mainly unicellular organization, are among the most diverse and numerous of insect pathogens. As a group, protists exhibit the full range of symbiotic associations with insects, from mutualism and commensalism to parasitism. However, most protistan etiological agents of insects cause chronic rather than acute diseases that tend to be unapparent and may cause population effects that are poorly researched and understood. The current knowledge on entomopathogenic protists in the taxa Amoebozoa, Apicomplexa, Ciliophora, Euglenozoa, and Helicosporidia is reviewed. Topics include morphology, development, transmission, host range, host—pathogen associations, and potential use in biological pest control. Complex life cycles are described for each group. Although protistan entomopathogens are often prevalent and persistent in nature, they are generally not regarded as likely microbial insecticides owing to high host specificity and difficulties associated with mass production. Inoculative release and conservation are more feasible approaches. The research needs, problems of taxonomy, and population dynamics are discussed.

10.1. INTRODUCTION

Protists are arguably among the most diverse and numerous of insect pathogens and exhibit many variations in form and function. They are essentially aquatic or semi-aquatic microorganisms that are suited for an endosymbiotic lifestyle. All have at least one motile stage driven with flagella, cilia, pseudopodia, or flexing. The vegetative stages of most protists lack a rigid wall, but many are covered by a thin pellicle or glycocalyx that functions in host recognition and attachment, and further serves as a chemical barrier. Many symbiotic protists alternate between proliferative stages, such as trophozoites, and resistant cysts that facilitate transmission and enable survival outside the host during unfavorable conditions.

Protists reproduce asexually by binary fission or multiple division by either merogony (schizogony) or plasmotomy. In merogony, the nucleus and other organelles replicate repeatedly as the cytoplasm expands within the original cell membrane prior to cytokinesis into multiple identical uninucleate daughter cells. In plasmotomy, the daughter cells have more than one nucleus. Protists may also undergo sexual reproduction through gametogony that involves meiosis and subsequent fusion of gametes to form a zygote that then undergoes sporogony.

The terminology that has been used to describe protists in the literature has been notably inconsistent. For the purpose of consistency, the present text adopts the terminology of Levine (1971) for the Apicomplexa.

Insect Pathology. DOI: 10.1016/B978-0-12-384984-7.00010-5

10.2. CLASSIFICATION AND PHYLOGENY

As a taxonomic group, protists are a heterogeneous assemblage of diverse taxa that have been reclassified several times, always with controversy (Cavalier-Smith, 2003, 2010; Adl *et al.*, 2005). They are polyphyletic and represented in all five of the eukaryotic kingdoms of Cavalier-Smith (1998). The entomopathogenic protistan taxa described in this chapter fall into four of five supergroups of the phylogenetic tree of eukaryotes presented by Keeling *et al.* (2005). The term protist was coined by Ernst Haeckel in his 1866 volume "Generelle Morphologie der Organismen", which included many microbes that have since been transferred to other assemblages. The complexity of microbial diversity precludes a single natural taxonomic unit that encompasses all of the organisms designated as protists or even protozoa under current common usage (Huisman and Saunders, 2007). Indeed, Cavalier-Smith (2010) considers the Euglenozoa to be distinct from all other eukaryotes. Nevertheless, it is the broadly encompassing usage of Protista that includes algae as well as protozoa that is applied in this chapter. This grouping comprises all eukaryotic, mainly unicellular organisms that do not fit into the other taxa. Although each of the individual groups treated here is monophyletic, arrangements into taxa above the generic level are fluid, and this chapter will follow Adl *et al.* (2005) in referring to such groupings simply as taxa. There is also a great deal of uncertainty at the generic and species levels, which will be addressed in discussions of individual groups.

10.3. ASSOCIATIONS, SIGNS, AND SYMPTOMS

Protists exhibit the full range of symbiotic associations, from mutualism and commensalism to parasitism. Mutualistic associations, such as those of nutrient-providing protists in insect alimentary tracts, have evolved in some cases to a level of interdependency where the host has developed structures and behaviors that facilitate transmission and maintenance of specific groups of symbionts. For some associations, there is no discernible positive or negative effect on the host, whereas others are fatal. Likewise, the degree of intimacy between symbiont and host is not closely linked with pathogenicity. Among the Apicomplexa, for example, which exhibit intracellular development and are among the most intimately associated symbionts, most eugregarines are commensalistic while others such as neogregarines and coccidians are highly pathogenic. Host specificity among protistan parasites is highly variable. Most pathogens are host specific, infecting a few closely related species, but a few exhibit a wide host range. The algal Helicosporidia, which infect insects of

several orders, arthropods of other classes, and even trematodes, are a good example of the latter. Only members of the Helicosporidia and some kinoplastid flagellates are known to be culturable in cell-free media.

With a few notable exceptions, horizontal transmission via oral ingestion (*per os*) of infectious stages is the most commonly recognized mode of transmission for most insect pathogenic protists. One exception is the ciliate *Lambornella clarki*, which encysts on and penetrates the cuticle of larval mosquitoes (Clark and Brandl, 1976). Other ciliates are also thought to invade though compromised cuticle. There are no recognized instances of transovarial transmission among insect protists. Vertical transmission has been demonstrated for *Helicosporidium* sp. infecting beet armyworm, *Spodoptera exigua*, and *Ophryocystis electroscirrha* in monarch butterflies, *Danaus plexippus*; however, the mechanism(s) are unclear as no infection could be detected in the eggs (Leong *et al.*, 1992; Bläske and Boucias, 2004).

There are few obvious external signs or symptoms associated with protistan infections, which are typically chronic and occult with no overt color change or extracorporeal growth. Insect larvae that are heavily infected with neogregarines or coccidians appear swollen and whitish and are sluggish in their movements. These signs and behavior can be easily confused with pupation in many hosts and require a trained eye to distinguish. A common diagnostic sign of infection with *Mattesia* spp. is fluorescence when examined under ultraviolet illumination (Marzke and Dicke, 1958), but this does not occur with the related *Farinocystis tribolii*. Since parasite-induced host fluorescence is a common feature associated with infection by other insect pathogens, such as Microsporidia, its usefulness as a diagnostic feature has obvious limitations. The large ciliates that infect aquatic Diptera larvae with semi-transparent integuments are readily seen with the aid of a stereo microscope.

Postmortem diagnosis of protistan infections is less difficult than for many diseases because of their relatively large size. All are readily visible with light microscopy, and most have characteristic shapes and features. Cadavers are usually pliable until very dry and, when pricked, may release a milky fluid consisting primarily of the disease-causing organisms. The diagnosis is simplified when they are restricted to specific tissues, such as gut epithelium or Malpighian tubules.

10.4. AMOEBOZOA

Amoebae (amebas) are unicellular eukaryotes that have been historically included in various polyphyletic higher ranks including Sarcodina, Rhizopoda, or Sarcomastighophora (Hausmann *et al.*, 2003). They are now placed in the Amoebozoa, a supergroup clustered by molecular phylogenies (Adl *et al.*, 2005). Except when cysts or shells are present, there is no fixed shape but rather a flowing, changeable body form that has been referred to as amoeboid. The presence of pseudopodia ("false feet"), transient cytoplasmic projections for locomotion and/or feeding (phagotrophy), is a salient feature of the group (Schuster, 1990; Rogerson and Patterson, 2000). There is a great range in size variation, from a few micrometers to a few millimeters. Amoebae have a relatively simple structure. Trophozoites, the trophic forms (i.e., active, feeding stages), are typically uninucleate but may also exist as binucleate or multinucleate cells. They contain vacuolated cytoplasm and normally divide into a granular endoplasm having inclusions and a hyaline ectoplasm. Although many species have either shells (tests, thecae, loricas) or flagella, the entomopathogenic species are naked cells devoid of flagella.

Reproduction is typically asexual by binary fission, schizogony, or plasmotomy. Typical centrioles are absent. Mitochondria, when not secondarily lost, have tubular cristae that are frequently branched (ramicristate). Microtubules appear to occur only in connection with the mitotic spindle apparatus. Life cycles are also rather simple compared to other protists. Meiosis has been reported in the testate species *Arcella vulgaris* (Mignot and Raikov, 1992), and the genus *Hartmannina* is believed to exhibit sexuality (Patterson *et al.*, 2000).

The formation of resistant, walled stages called cysts is common, and Amoebae are extremely ubiquitous. As free-living organisms they have exploited all major habitats: terrestrial, marine, and freshwater. Terrestrial forms abound wherever relatively high moisture is present but are not entirely restricted to its presence. In many species, their ubiquity is enhanced by the ability to form resistant cysts that facilitate persistence and dispersal. As symbionts of animals, species of amoebae range in associations from commensalistic to pathogenic. There are species, such as *Acanthamoeba*, that are amphizoic (i.e., usually free living but can become parasitic). Comparatively few species of amoebae are pathogenic (either obligate or facultative) (Rogerson and Patterson, 2000) to insects. Most species of amoebae associated with insects belong to genera *Entamoeba*, *Endamoeba*, *Endolimax*, and *Dobellina*, and are considered commensals found in the digestive tracts of cockroaches, termites, and crane flies (Purrini and Žižka, 1983; Patterson *et al.*, 2000).

As insect pathogens, amoebae have been most recently reviewed by Tanada and Kaya (1993) and Boucias and Pendland (1998). Earlier, a detailed review of the most well-known species, *Malameba locustae*, was published by Brooks (1988). Since then, limited research has been conducted and old nomenclatural issues and deeper classification themes remain unresolved.

To the authors' knowledge, there are currently only six reported species of insect-pathogenic amoebae. Two species, *Malameba locustae* and *Malamoeba indica*, have been described from grasshoppers, and one each from bark beetles (*Malamoeba scolyti*), honey bees (*Malpighamoeba mellificae*), fleas (*Malpighiella refringens*), and a bristletail (*Vahlkampfia* sp.) (Table 10.1). All these species are known to form cysts and infect host Malpighian tubules and midgut, except for *Vahlkampfia*, which appears not to extend to the excretory tubules (Larsson *et al.*, 1992). Conditions or factors leading to encystation are not known in these amoebae, and it is conceivable that they might spend extended periods as trophozoites only. In general, amoebae form cysts in response to unfavorable conditions,

---

TABLE 10.1 Known Entomopathogenic Amoebae

Amoebae Species	Host(s)	References
Malameba locustae	Grasshoppers, locusts, crickets (Orthoptera: Acrididae, Gryllidae); *Lepisma sacharina* (accidental) (Thysanura: Lepismatidae)	King and Taylor (1936), Taylor and King (1937), Harry and Finlayson (1976), Larsson (1976), Papillion and Cassier (1978), Braun *et al.* (1988), Lange (2002)
Malamoeba indica	*Poicilocera picta* (Orthoptera: Acrididae)	Narasimhamurti and Ahamed (1980)
Malamoeba scolyti	*Dryocoetes autographus* and other bark beetles (Coleoptera: Curculionidae: Scolytinae)	Purrini (1980), Purrini and Žižka (1983), Kirchhoff and Führer (1990), Kirchhoff *et al.* (2005)
Malpighamoeba mellificae	*Apis mellifera* (Hymenoptera: Apidae)	Massen (1916), Prell (1926), Schwantes and Eichelberg (1984), Liu (1985a, b)
Malpighiella refringens	*Ceratophyllus fasciatus* (Siphonaptera: Ceratophyllidae)	Minchin (1910)
Vahlkampfia sp.	*Promesomachilis hispanica* (Microcoryphia: Machilidae)	Larsson *et al.* (1992)

during periods of desiccation or limited food supply (Schuster, 1990). Detection of amoebae in insects has historically been by microscopic observation of the conspicuous and characteristic cysts (Fig. 10.1). In the absence of cysts, amoebae may pass unnoticed, particularly if trophozoite burden is low. The development and use of molecular diagnostic techniques would provide a much needed tool for a more accurate assessment of the diversity of insect-pathogenic amoebae.

Of the six recognized entomopathogenic amoebae, *M. locustae* has been the most studied species and serves as a model among species affecting the Malpighian tubules of insects. The first record and description of *M. locustae* was by King and Taylor (1936) from three Nearctic species of *Melanoplus* grasshoppers maintained in rearing facilities. It was originally described as *Malpighamoeba locustae*, but later transferred by the same authors (Taylor and King, 1937) to the genus *Malameba* because it differed from a species infecting the honey bee,

Malpighamoeba mellificae, described earlier by Massen (1916) and Prell (1926). *Malameba locustae* is largely cosmopolitan and has been recorded in grasshopper populations in North and South America, Africa, and Australia (Venter, 1966; Henry, 1969; Ernst and Baker, 1982; Henry *et al.*, 1985; Lange, 2002, 2004). However, infections of grasshoppers and locusts in rearing facilities worldwide have been reported more frequently (Davies, 1973; Larsson, 1976; Lipa, 1982; Henry, 1985; Hinks and Erlandson, 1994; C. Lange, unpubl.). It is common for insect disease prevalence to be greater in rearing facilities than in the field for several reasons, but that difference may be exaggerated for *M. locustae* because detection is almost exclusively based on microscopic observation of the highly refractive ellipsoid cysts. Conditions in rearing facilities may favor encystment, while cyst formation may be less common or delayed under natural conditions. Trophozoite-only infections are difficult to detect and diagnose, particularly if parasite burden is low. Adverse

FIGURE 10.1 Entomopathogenic amoebae. (A—E) *Malameba locustae* in grasshoppers; (F) *Malpighamoeba mellificae* in honey bees. (A) Several trophozoites showing a single large pseudopodium each and two subspherical precysts; (B) trophozoite showing numerous smaller pseudopodia; (C) trophozoite undergoing binary fission; (D) mature cysts; (E) cysts accumulated in the lumen of a grasshopper's Malpighian tubule; (F) part of a honey bee Malpighian tubule containing numerous cysts. Photographs taken using phase-contrast microscopy.

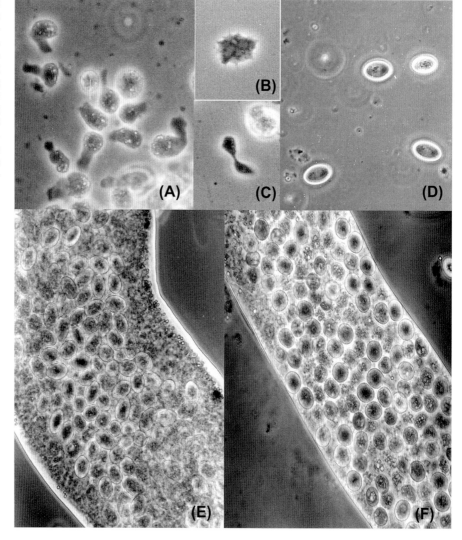

conditions are known to cause encystment in other amoebae (Schuster, 1990).

Malameba locustae has a wide host range that includes juveniles and adults of at least 55 naturally or experimentally susceptible species of grasshoppers and locusts, one cricket, and a silverfish (Larsson, 1976; Brooks, 1988; Lange, 2002). Diagnosis of infection has been based solely on parasite morphology and, given the wide geographical and host range of *M. locustae*, the existence of cryptic species cannot be ruled out. *Malamoeba indica* from the Indian grasshopper *Poecilocera picta* might well be another isolate of *M. locustae* because its differentiation was based on morphological characters likely to vary between isolates (Narasimhamurti and Ahamed, 1980). *Malamoeba scolyti* also has a wide host range, infecting naturally or experimentally juveniles and adults of at least 12 species of bark beetle (Kirchhoff *et al.*, 2005). Conversely, *M. mellificae* appears to infect only adult honey bees (Bailey, 1968).

The life cycle of *M. locustae* has been further studied by Hanrahan (1975), Harry and Finlayson (1976), Žižka (1987), and Braun *et al.* (1988) (Fig. 10.1). Trophozoites have one or two nuclei, cannot survive outside the host, and exist in the lumen and epithelia of midgut and Malpighian tubules. The cysts are uninucleate, pack Malpighian tubules in heavy infections, and are released in feces. It is still unclear how the excystation process proceeds. Hanrahan (1979) suggested that the cyst wall is digested in the gut of the host, while Prinsloo (1960) believed that the trophozoite emerges through a lateral break in the cyst wall. Braun *et al.* (1988) indicated that trophozoites do not reproduce in the intestine, and invade the Malpighian tubules by way of the digestive tract, not through the hemocoel after disrupting midgut epithelium cells, as proposed earlier by Evans and Elias (1970). Once trophozoites reach the Malpighian tubules, they reproduce by binary fission. Trophozoites are observed in the Malpighian tubules five and eight days after inoculation of cysts in *Melanoplus sanguinipes* and *Dichroplus schulzi*, respectively (Braun *et al.*, 1988; Lange, 2002), while cysts begin to appear in feces 14 days after inoculation in *D. schulzi* (Lange, 2002).

Like most entomopathogenic protists, *M. locustae* is not highly virulent. It produces a largely chronic disease, characterized by general debilitation and reduction of host vigor. Hosts with early or light infections do not exhibit external signs or symptoms. However, when infections are heavy, many abnormalities have been noted including ventral and lateral melanic spots in the thorax and abdomen, lethargy, hyperactivity, loss of appetite, premature death, inability to remain in an upright position, and tetanic twitches of the hind legs (King and Taylor, 1936; Henry, 1968; Harry and Finlayson, 1976; Hinks and Ewen, 1986).

Internally, with gradual invasion of the Malpighian tubules by trophozoites and cysts, pathological changes

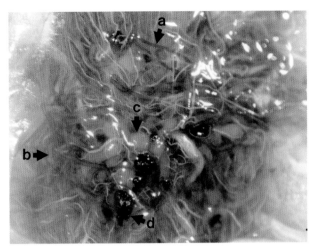

FIGURE 10.2 Gross pathology caused by *Malameba locustae* in the Malpighian tubules of a grasshopper. Malpighian tubules showing infections at relatively early stage (a); intermediate stage with loss of color and slight swelling (b); late stage with gross swelling (c); and encapsulation and melanization (d).

begin to become obvious (Fig. 10.2). The Malpighian tubules lose their normal appearance as thin, usually dark, highly movable threads, and become increasingly hypertrophied, light in color (typically from brown to milky white), and immobile (contortion and peristaltic movements are blocked) (Martoja, 1969; Harry and Finlayson, 1976; Papillion and Cassier, 1978). Swollen Malpighian tubules displace other organs and interfere with egg development in females. In very heavy or terminal infections, the pressure exerted by the huge quantity of cysts produces breaks in the tubules' basement membrane that allow cysts to invade the hemocoel, where they are phagocytosed and encapsulated (Hanrahan, 1980) (Fig. 10.2). In severe infections, the excretory function is altered, resulting in an accumulation of toxic products (Henry, 1968, 1969). Proux (1991) reported that *M. locustae* negatively affects the serotic membrane of the tubules, inhibiting its response to the diuretic hormone. According to Jackson *et al.* (1968), the reproductive potential of the host is diminished not only as a result of reduced longevity but also by an alteration of lipids in the eggs that results in less viability.

Venter (1966) reported that an epizootic by *M. locustae* in the South African locust *Locustana pardalina* prevented the occurrence of a second generation by altering reproduction of first generation adults. Taylor and King (1937) applied grasshopper feces with cysts mixed with bran and molasses for control of grasshoppers in the USA. Although effective control was not achieved, some infections were recovered several weeks later. Similar results were obtained by Lange (2002) after a small application of cysts on wheat bran against grasshoppers in the Pampas of Argentina. Although several authors have continued to acknowledge the potential use of *M. locustae* as a biocontrol agent (Brooks, 1988; Odindo, 1991; Raina, 1992; Lomer *et al.*, 1999), interest has waned and further studies have not been conducted. Clearly, molecular diagnostic techniques and phylogenetic studies are needed with *M. locustae* to clarify and increase knowledge on natural prevalence and distribution, taxonomic status of isolates from different hosts and geographical regions, and phylogenetic affinities with other amoebae.

As previously noted, *M. locustae* possesses several attributes that may justify revitalizing research towards its development as a biological alternative for grasshopper and locust long-term control, possibly as seasonal or one-time inoculative releases as defined by Roberts *et al.* (1991): (1) transmission is easily induced; (2) *M. locustae* is widely distributed and occurs naturally in a number of grasshopper and locust species (a feature that may bridge legal constraints associated with the introduction of non-native species); (3) it produces chronic infections that favor long-term persistence; and (4) it produces resistant cysts that can be stored for prolonged intervals. However, a major disadvantage and limiting factor includes the inability to economically produce sufficient quantities of cysts for sustainable use (Streett and Henry, 1990). *In vivo* production is laborious and inefficient. However, since other amoebae are amenable to *in vitro* cultivation, including encystation and excystation processes (Byers *et al.*, 1980; Vázquezdelara-Cisneros and Arroyo-Begovich, 1984; Avron *et al.*, 1986; Schuster, 1990; Chávez-Mungía, 2007), efforts to produce *M. locustae in vitro* should be pursued.

10.5. APICOMPLEXA

10.5.1. Eugregarinorida

Eugregarines are large, single-celled obligate parasites of invertebrates, particularly insects and annelids. They develop in cavities of their hosts, such as the digestive tract, the body cavity, and the reproductive vesicles. In insects, enteric eugregarines predominate. Because eugregarines produce spores, they were traditionally included in the old taxon Sporozoa (Perkins, 1991; Hausmann *et al.*, 2003). However, owing to the presence of stages (sporozoites) with a unique structure called apical complex, as originally proposed by Levine (1970, 1988a), they are currently classified (along with neogregarines, archigregarines, coccidians, haemosporidians, and piroplasmids) within the taxon Apicomplexa in the higher rank Alveolata (Adl *et al.*, 2005). Eugregarines are considered a monophyletic group within the monophyletic Apicomplexa (Clopton, 2009). Since insects constitute a hyperdiverse group, and a single insect species may be host of several eugregarine species (Clopton and Janovy, 1993), eugregarines may be some of the most ubiquitous and species-rich parasites. A survey of

almost any insect species is likely to yield at least one eugregarine species. In the last taxonomic review, Clopton (2000) mentioned that there are approximately 1656 named species of eugregarines distributed in 244 genera, the majority of which were reported from insects.

Most eugregarines are monoxenous parasites (only the Porosporicae complete their development in two host species, a crustacean and a mollusk) and there is no asexual reproduction. The life cycle of eugregarines is straightforward and homogeneous within the group (Vivier and Desportes, 1990; Perkins *et al.*, 2000). After being ingested by a susceptible host, the eugregarine propagules (spores termed oocysts) release the sporozoites, the infective forms. Sporozoites are haploid and typically number eight per oocyst but sometimes four or 16. Infection begins when sporozoites either penetrate temporarily or attach to intestinal host cells, depending on whether or not there is early intracellular development. Through morphogenesis and a remarkable increase in size, sporozoites develop into trophozoites (sometimes also named trophonts or cephalonts) that in most species remain attached by means of special anchoring structures (either mucron or epimerite) at the anterior end of the cell. Release, breakdown or retraction of mucron or epimerite leads to detachment, and trophozoites become gamonts (often also called sporadins). Gamonts associate with each other intimately (normally in pairs but also in longer tandems of more individuals) relatively soon or remain unassociated in the digestive tract lumen while growth continues. Usually, paired cells are not equal in size, the smaller one being the satellite, while the larger one is termed the primite. Gamogony begins when associated gamonts initiate development of a membrane around them (encystment) and rotational movements start, normally coupled processes unified under the term syzygy. Within the forming spherical gametocyst, each gamont produces either isogametes or anisogametes (gametes of similar or dissimilar size and structure, respectively). Union of gametes initiates sporogony and each zygote, the only diploid stage in the eugregarine life cycle, develops a wall to become an oocyst with the sporozoites inside. After leaving the host in the feces, the oocysts are freed from the gametocysts by simple rupture or dehiscence through sporoducts.

Morphologically, structurally, and behaviorally, eugregarines are among the most characteristic and unmistakable protists (Vivier and Desportes, 1990; Perkins, 1991; Schrével and Philippe, 1993) (Figs. 10.3 and 10.4). Normally, mature gamonts and gametocysts are the stages that are initially detected in infected insects. Gamonts are usually the largest and most conspicuous developmental stages. According to species and state of development, they may range in size from several micrometers to cells readily visible under the dissecting microscope. Shape varies considerably too, but within either of two body plans. In

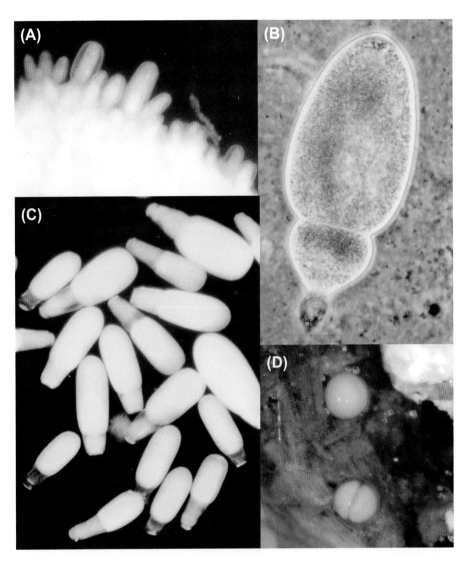

FIGURE 10.3 Eugregarines from grasshoppers. (A, B, D) *Gregarina ronderosi*; (C) *Leidyana ampulla*. (A) Trophozoites attached to host intestinal epithelium; (B) trophozoite; (C) gamonts; (D) mature (top) and immature (bottom) gametocysts in feces of the host. Photographs taken using phase-contrast microscopy except for (D), which is under a dissecting microscope.

aseptate or monocystid eugregarines, also termed acephalines, the body of the gamont (and preceding trophozoite) is not segmented. In septate or polycystid eugregarines, also called cephalines, the body is segmented. Gamonts of cephalines have an anterior (protomerite) and a larger, posterior (deutomerite) segment. The deutomerite contains the nucleus. An additional segment, the epimerite, is present at the anterior end of the protomerite in trophozoites of cephalines. Trophozoites and gamonts of most eugregarines are motile. Deep longitudinal folds supported by cross-linked actin filaments on the surface of trophozoites and gamonts produce a characteristic gliding motion that moves the cell forward over a layer of secreted viscous material. The linear locomotion movement is complemented by a capacity to bend the body. Gametocysts are also conspicuous. They are clear-colored, usually white or yellowish spheres surrounded by a thick, translucent hyaline coat (ectocyst). Gametocysts can sometimes be visible even to the naked eye.

Most species descriptions, taxonomy, and classification of eugregarines have largely relied on morphological characters of trophozoites, gamonts, and oocysts under light microscopic observations. Since unique characters were seldom clearly recognized, unique combinations of non-unique characters and host range (different host, different parasite concept) were historically employed. Descriptions based on whole life cycle observations were less common, and only exceptionally were cross-infection experiments conducted to check for specificity (Lange and Wittenstein, 2002; Lantova *et al.*, 2010). Studies of insect eugregarine species at the molecular level lagged behind those of other protists and are just beginning to emerge (Clopton, 2009; Lantova *et al.*, 2010).

The absence of asexual reproduction in the life cycle of eugregarines accounts for the fact that infection intensity is directly proportional to the amount of viable oocysts ingested by the susceptible host. The number of gamonts in an infected host can never be higher than the amount of

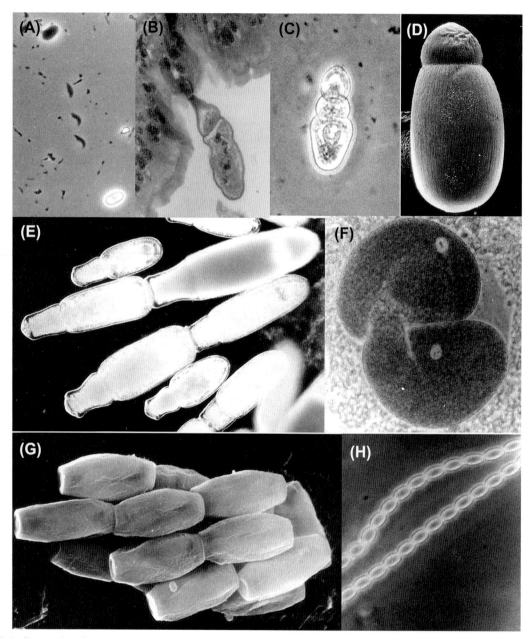

FIGURE 10.4 Eugregarines from grasshoppers. (A—D, F, H) *Gregarina ronderosi*; (E, G) *Leidyana ampulla*. (A) Mature oocyst (lower right), empty oocyst (upper left), and just emerged sporozoites; (B) trophozoite attached to intestinal epithelium; (C) young, unattached trophozoite; (D) gamont; (E) associated and unassociated gamonts; (F) rotation during caudofrontal syzygy; (G) oocysts; (H) part of an oocyst chain just after gametocyst dehiscence. Photographs taken using phase-contrast microscopy except for (D) and (G), which are taken using scanning electron microscopy.

sporozoites released from the ingested oocysts. This characteristic raises the question of whether eugregarines should actually be considered true pathogens or classical parasites. One of the defining features of pathogens (sometimes also called microparasites) is that they have extremely high rates of reproduction within the host (Anderson and May, 1981; Boucias and Pendland, 1998). As a result of gamogony, eugregarines produce a relative large amount of progeny, but gamogony does not normally compensate for the absence of asexual proliferation and eugregarines do not invade tissues or organs of the host but remain confined in gametocysts within cavities until egress from the host. Brooks and Jackson (1990) mentioned that adults of the chrysomelid *Diabrotica virgifera* lose an infection with an undescribed species of *Gregarina* within two weeks unless additional oocysts are ingested. This scenario seems, at least in part, more reminiscent of a classical parasite than of a true pathogen.

Several excellent reviews (Maddox, 1987; Brooks, 1988; Brooks and Jackson, 1990; Tanada and Kaya, 1993; Boucias and Pendland, 1998) have summarized the consensus among most authors that eugregarines are symbionts that fluctuate along the mutualism–commensalism–parasitism continuum depending on many factors. Perhaps clearer than in any other group of insect pathogenic protists, the key factor governing the onset of pathogenicity for eugregarines is parasite burden. Although eugregarines do divert host nutrients to their own use, damage host cell integrity, and occupy space, such actions are generally of low impact in light infections, as under most reported field situations. Since parasite burden depends directly on oocyst acquisition, any situation that favors a sustained ingestion of oocysts, such as heavy oocyst release and cage confinement in overcrowded rearing facilities (Henry, 1985; Brooks and Jackson, 1990), would lead to high parasite loads. Pathologies and effects that might be negligible can then become obvious when infections are heavy. In such cases, enteric eugregarines may cause intestinal blockages and considerable epithelium disruption leading to mortality and reduced vigor, longevity, and fecundity. As summarized in Table 10.2, the effects on hosts that have been recently mentioned for various eugregarine species are subtle rather than overt.

Among eugregarines, only the aseptate members of the genus *Ascogregarina* have been considered potential biological control agents (Tseng, 2007). They infect mainly container-inhabiting mosquitoes (Chen, 1999). *Ascogregarina taiwanensis* and *A. culicis* have received the most attention because they infect the important disease vectors *Aedes albopictus* and *Ae. aegypti*, respectively (Reyes-Villanueva *et al.*, 2003; Passos and Tadei, 2008). Their life cycles are similar to other species in the genus. Oocysts ingested by early mosquito instars release sporozoites, which then enter gut epithelial cells and develop into trophozoites. Their development is synchronized with that of the host (Chen and Yang, 1996), and before mosquito pupation, trophozoites migrate from the midgut into the Malpighian tubules where they transform into microgametes or macrogametes. During the pupal stage, gametes fuse to form a gametocyst, within which many oocysts are formed, each containing eight sporozoites (Chen, 1999). Oocysts are shed into water by adults emerging from pupae as well as by dying adults. Oocysts are also expelled by females through the rectum during oviposition in larval habitats, or when infected adults of either sex defecate.

Several field surveys have documented the prevalence rates, seasonality, and distribution in different regions of *A. taiwanensis* and *A. culicis* (García *et al.*, 1994; Passos and Tadei, 2008; Albicocco and Vezzani, 2009). There is reported geographical variation in pathogenicities. Some Asian strains of *A. culicis* are pathogenic to *Ae. aegypti* (Sulaiman, 1992), while the US strains of *A. culicis*

TABLE 10.2 A Selection of Recent Studies on Effects of Eugregarines on Hosts

Eugregarine Species	Host Species	Effects	References
Ascogregarina spp.	*Aedes albopictus, Aedes aegypti, Ochlerotatus triseriatus* (Diptera: Culicidae)	Alteration of interspecific competitive interactions; higher mortality under low nutrient conditions; reduced male size at pupal emergence; increased mortality	Aliabadi and Juliano (2002), Comiskey *et al.* (1999a, b), Tseng (2007), Siegel *et al.* (1992)
Diplocystis tipulae	*Tipula paludosa* (Diptera: Tipulidae)	Reduced larval size	Er and Gökçe (2005)
Gregarina niphandrodes	*Tenebrio molitor* (Coleoptera: Tenebrionidae)	Reduced longevity	Rodríguez *et al.* (2007)
Gregarina sp.	*Blatella germanica* (Blattaria: Blattellidae)	Swollen abdomens, slower movement, shorter antennae, more susceptible to other diseases, factors	Lopes and Alves (2005)
Haplorhynchus polyhamatus	*Mnais costalis* (Odonata: Calopterygidae)	Reduced longevity depending on food intake	Tsubaki and Hooper (2004)
Haplorhynchus sp.	*Enallagma boreale* (Odonata: Coenagrionidae)	Possible sex biases and reduced longevity; subtle and likely to be masked by other factors	Hecker *et al.* (2002)
Unspecified	*Requena verticalis* (Orthoptera: Tettigoniidae)	Alteration of mating in relation to rich/poor diet	Simmons (1993)

(Barrett, 1968) or *A. taiwanensis* (Fukuda *et al.*, 1997) are not considered pathogenic to their mosquito hosts. However, Comiskey *et al.* (1999a, b) have demonstrated that *A. taiwanensis* can exert a detrimental impact upon the fitness (mortality, adult size) of *Ae. albopictus* adults and enhance its vector competence for the dog heartworm *Dirofilaria immitis*. Mortality of all stages of *Ae. albopictus* is increased by *A. taiwanensis* only when they are under nutritional stress, when it may increase by as much as seven-fold (Comiskey *et al.*, 1999a, b). Although gregarine infection can reduce the fitness of both natural and non-natural hosts, the fact that gregarines require their host to live to adulthood in order for parasites to be transmitted reduces their efficacy as sustainable biocontrol agents.

Recent phylogenetic analyses have demonstrated paraphyly of the genus *Ascogregarina* and revealed disparate phylogenetic positions of gregarines parasitizing mosquitoes and gregarines retrieved from sand flies (Votýpka *et al.*, 2009). The new genus *Psychodiella* was created to accommodate gregarines from sand flies. Furthermore, those two genera and some others of aseptate eugregarines were grouped with neogregarines rather than septate eugregarines based on small-subunit rDNA sequences.

10.5.2. Neogregarinorida

Neogregarines, previously known as schizogregarines, are found almost exclusively in insects. They are defined by having septate gamonts and by having merogony (schizogony) present in their life cycles in addition to gametogony and sporogony, which differentiates them from eugregarines (Levine, 1988b). There is a great deal of uncertainty in the taxonomy of the neogregarines owing to host range overlap, fragmentary data, and the difficulty of observing all of the salient developmental stages. Few of the species and genera have been described adequately. Consequently, there is no doubt that many of the host records and generic assignments are invalid. Gene sequencing has only recently been applied to the neogregarines (Valles and Pereira, 2003; Votýpka *et al.*, 2009). In addition, there is no existing type material for most taxa, and even the type localities are often unknown.

Neogregarine life cycles are among the most complex of insect pathogens. They produce pellicle-covered, motile sporozoite and merozoite stages that bear apical complexes that are typical of the phylum and lack a mucron. The oocysts of neogregarines, which are formed within gametocysts, are typically ovoid or lemon shaped with thick, layered walls and distinct plugs at both poles (Fig. 10.5). *Syncystis* spp. are exceptions that lack polar plugs and are ornamented with four filamentous projections at each pole (Larsson, 1991). Vermiform sporozoites are released by digestion of the polar plugs or by breakage of oocyst walls. The sporozoites penetrate the gut to susceptible tissues

where they divide into plasmodia in the initiation of merogony, which may be intracellular or extracellular. In some species there is a single merogony designated macronuclear merogony, in which the meronts have notably large nuclei. When there are two merogonial sequences, a micronuclear merogony precedes the macronuclear merogony. In micronuclear merogony, plasmodia are produced with up to 200 nuclei. Micronuclei migrate to the periphery of the plasmodium for division into micronuclear merozoites often by budding, leaving from a large residuum (Canning and Sinden, 1975). Micronuclear merogony may be repeated, or merozoites may become transformed with enlarged nuclei, giving rise to macronuclear merogony that generally produces merozoites with large nuclei and vacuoles that are usually broadly pear shaped and less motile than the micronuclear merozoites. These macronuclear merozoites develop into gametocytes. Uninucleate and multinucleate gametocytes associate in pairs to form gametocysts that develop in the manner of eugregarines. Within the gametocytes, the gametes unite to form zygotes that undergo sporogony to form oocysts with eight sporozoites that are usually aligned in alternating directions. The number of oocysts per gametocyst, which is used as a taxonomic character for determination of genera, may be one, two, four or a subsequent multiple. Oocysts in massive quantities are usually the only stage present in late-stage infections.

There are reports of neogregarine infections in a millipede and mite species, but all other records are from insects. Host insects are from 11 orders, but the predominance of cases is in Lepidoptera, Diptera, and Coleoptera (Levine, 1988b). Most are relatively host specific, but some of the better studied species are very closely aligned and have overlapping physiological host ranges. Indeed, the host ranges may extend across insect orders. Hymenopteran parasitoids have been infected with *Mattesia dispora*, *M. grandis, and M. oryzaephili* as a result of attacking infected moth and beetle larvae (McLaughlin and Adams, 1966; Leibenguth, 1972; Lord, 2006). *Mattesia dispora* and *M. oryzaephili* share many of the same lepidopteran and coleopteran hosts but with differing infectivity (Lord, 2003). Among species of the commonly encountered genera *Mattesia* and *Farinocystis*, there seems to be a disproportionate affinity for hosts that frequent stored products (Purrini, 1976a; Lord, 2003). Common species are *M. dispora, M. oryzaephili*, and *M. trogodermae*, and *Farinocystis tribolii*, all of which infect pests of stored grains, fruits, and nuts. In general, congeners tend to infect related hosts. For example, all species of *Caulleryella* infect the Malpighian tubules of lower Diptera, mainly mosquitoes (Levine, 1988b).

Like eugregarines, neogregarines are transmitted by way of contaminated food. Unlike eugregarines, in which the oocysts are continuously released into the environment

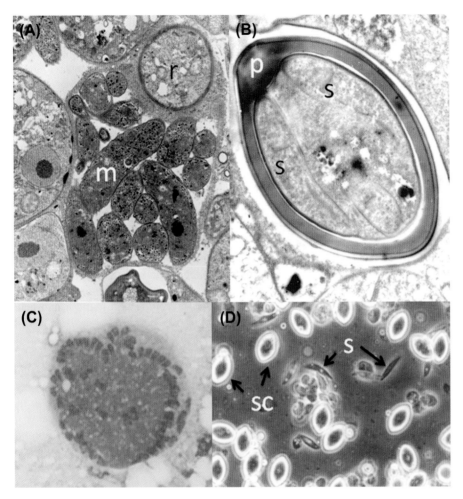

FIGURE 10.5 *Mattesia oryzaephili.* (A) Transmission electron micrograph (TEM) of merozoites (m) still grouped with residual body (r) at the completion of mergony; (B) TEM of a sporocyst showing sporozoites (s) and polar plug (p); (C) light micrograph of Giemsa-stained micronuclear meront with peripheral nuclei in the early stage of merozoites formation; (D) fresh sporocysts (sc) and sporozoites (s) under phase-contrast microscopy.

with host feces at completion of their development, neogregarines are released by predation, scavenging, and the breakdown of cadavers. However, after being eaten, many oocysts pass ungerminated into feces and are thus disseminated by host and non-host species (Lord, 2003). The merogony of neogregarines provides massive proliferation capacity and results in neogregarines being more pathogenic than eugregarines. In fat body, the tissue in which most species proliferate, host cells become vacuolated and degenerate, releasing oocysts into the hemolymph (Valigurová and Koudela, 2006). Furthermore, the oocysts can germinate within host tissues, reinitiating the life cycle in a process that was called autoinfection by Žižka (1972). Thus, they kill their hosts with overwhelming numbers, destroying tissues and organs, particularly the fat body, Malpighian tubules, and alimentary canal. Nevertheless, the diseases that they cause are chronic, and hosts can survive for extended periods with infections. Often, infected individuals remain in the larval stage long after healthy individuals of the same age have pupated. This effect appears to be due to more than general debilitation. Rabindra *et al.* (1981) reported that *F. tribolii* has juvenile

hormone effects on *Tribolium castaneum* and found that ether extracts of oocysts applied to *T. castaneum* pupae or last instar nymphs of the pyrrhocorid *Dysdercus cingulatus* produced immature-adult intermediates.

Like most other protistan pathogens, neogregarines appear to evade host recognition and do not trigger melanization or phagocytosis. Consequently, there has been little study of host immune responses to them. Lindsey and Altizer (2009) reported that male but not female larvae of *D. plexippus* had increased concentrations of hemocytes in the presence of *Ophryocystis elektroscirrha* infection, while midgut phenoloxidase activity was not affected. Clearly, gregarines are well adapted to their hosts.

Neogregarines of the genus *Mattesia* have been described from Coleoptera, Lepidoptera, Siphonaptera, and Hymenoptera. As with many of the protists that have sparse descriptions and little ensuing study, some of the species are probably not legitimately congeneric, and some renaming is likely when more genomic data become available. *Mattesia* species have a convenient characteristic for diagnosis in that heavily infected hosts fluoresce under ultraviolet light (Burkholder and Dicke, 1964).

The genus includes several important pathogens of major pests. *Mattesia grandis* is a well-studied pathogen of the cotton boll weevil, *Anthonomus grandis*. An interesting feature of its biology is that its development is arrested when the host's adipose tissue is scant as a result of poor nutrition. This is apparently a consequence of lack of polyunsaturated fatty acids that the pathogen must acquire from the beetle (Thompson and McLaughlin, 1977). *Mattesia* was researched as a control agent for cotton boll weevil for several years and deemed of such potential that a mass production system was developed (McLaughlin and Bell, 1970). However, when it was field tested in bait form, the results were not promising (McLaughlin *et al.*, 1969) and interest waned.

Mattesia trogodermae was considered to have great potential as a suppressive agent for dermestid beetles that are pests of stored products, including the destructive khapra beetle, *Trogoderma granarium*. It is a cosmopolitan pathogen of *Trogoderma* species and appears to have a host range that does not extend beyond that genus (Shapas *et al.*, 1977). It has been shown to be capable of suppressing *Trogoderma glabrum* populations under simulated warehouse conditions when used with a pheromone lure for attraction of male beetles and dissemination of oocysts (Shapas *et al.*, 1977). Its spread is assisted by transfer from oocyst-contaminated males to females during mating (Schwalbe *et al.*, 1974). The disease was observed in stored wheat in Kansas that was infested with *Trogoderma* spp. and may have contributed to periodic population collapses there (Marzke and Dicke, 1958).

Mattesia oryzaephili and *M. dispora* are morphologically similar and have considerable host range overlap. They can be easily confused by scientists not familiar with them; most of the *M. dispora* reported from beetles are probably *M. oryzaephili* that were misidentified (Lord, 2003). They are commonly encountered in grain stores and mills infecting beetles and larvae of the Indian meal moth, *Plodia interpunctella*, and the Mediterranean flour moth, *Ephestia kuehniella* (Purrini, 1976a; Moore *et al.*, 2000). Because both of these species infect wasp parasitoids of their pest hosts (Leibenguth, 1972; Lord, 2006), they have an alternate long-distance dispersal mechanism.

There are at least two *Mattesia* species that infect ants. *Mattesia geminata* is common in the red imported fire ant, *Solenopsis invicta* (Jouvenaz and Anthony, 1979), and has been found in *Leptothorax* ants (Kleepsies *et al.*, 1997). A distinct undescribed *Mattesia* species designated yellowhead disease has also been reported in *S. invicta* (Pereira *et al.*, 2002).

Like *Mattesia*, the genus *Farinocystis* is a member of the family Lipotrophidae. The oocysts closely resemble those of *Mattesia* spp., and it is most easily distinguished by having six nuclear divisions in gametogony resulting in up to 16 oocysts per gametocyst, compared to only two in

Mattesia (Weiser, 1955). The usual number of oocysts is eight or nine, since some gametes fail to associate and some zygotes do not develop (Ashford, 1968). Another distinguishing characteristic demonstrated by Žižka (1978) is a deep transverse invagination in the merozoites that he interpreted as a septum. The merozoites of *Mattesia* spp. have only a shallow invagination or none (Vavra and McLaughlin, 1970; Valigurová and Koudela, 2006). *Farinocystis tribolii* is the only species in the genus with a supportable description. It is a common pathogen of stored grain beetles, especially *Tribolium castaneum* and *T. confusum*, from which it was described by Weiser (1953). It has been found in other Tenebrionidae as well as in Curculionidae and Cerambycidae (Purrini, 1976a; Steinkraus *et al.*, 1992). An undetermined species of *Farinocystis* that may be *F. tribolii* reduced the longevity and fecundity of the weevil *Euscepes postfasciatus* (Kumano *et al.*, 2010).

Farinocystis tribolii has been found at high prevalence rates in nature. Steinkraus *et al.* (1992) found it in a field population of the poultry production pest *Alphitobius diaperinus*, with a prevalence of 44%. The high prevalence is due, in part, to the chronic nature of the disease. Its chronicity was shown by Rabindra *et al.* (1983), who reported that fifth instar *T. castaneum* larvae that were fed a very high dose of oocysts had a 50% survival time of 24 days. The chronic effects appear to reduce fitness in a variety of ways. According to Rabindra *et al.* (1985), *F. tribolii* infection causes an increased rate of respiration in *T. castaneum*. As is frequently the case with chronic diseases, insects infected with *F. tribolii* also exhibit increased susceptibility to chemical insecticides (Rabindra *et al.*, 1988).

Ophryocystis species are predominantly found in the Malpighian tubules of beetles. However, the most studied species, *O. elektroscirrha*, infects monarch and queen butterflies in the New World and Australia with prevalence rates that can exceed 70% (Altizer *et al.*, 2000). The main transmission route is from females to their progeny when oocysts are scattered on eggs and host plants during oviposition. An alternative route is horizontal transmission during mating. Once ingested, the oocysts germinate, releasing sporozoites that penetrate to the midgut epithelium, where merogony takes place in two stages. The infection is restricted to epithelial cells (McLaughlin and Myers, 1970) and progression is coordinated with host development. In pupae, sexual reproduction produces haploid sporozoites in oocysts in the epidermis around developing scales before adult eclosure. Adults emerge with oocysts on their integuments, facilitating horizontal transmission (Leong *et al.*, 1992). The negative effects include mortality of immatures and reduced size, longevity, and flight capacity in adults (Altizer and Oberhauser, 1999; Bradley and Altizer, 2005; de Roode *et al.*, 2007). In spite of its high prevalence, under natural conditions, the disease does not appear to be a significant mortality factor (Leong *et al.*, 1992).

10.5.3. Coccidia

Coccidia comprise the subgroup Coccidiasina of the Conoidasida. They are closely related to gregarines (Kopecna *et al.*, 2006; Templeton *et al.*, 2010) and resemble neogregarines in the size of vegetative stages, having merogony, and lacking a mucron or epimerite associated with the apical complex. They differ from gregarines in lacking syzygy and usually having sporocysts that contain the sporozoites within the oocysts (Fig. 10.6). The ultrastructure of the sporozoites and merozoites of the group is typical of the Apicomplexa. The gametes are anisogamous, and all development is intracellular. While there are no obvious external signs of disease in insect hosts, the large ovoid or spherical oocysts are easily seen with a compound microscope at low power.

Coccidia are mainly parasites of vertebrates, with less than 1% of the known species infecting insects and some of those being only transmitted by insect vectors rather than having insects as definitive hosts. There are currently eight genera with species that have been reported as having insects as sole or primary hosts. All of those are rare and are known only from sparse species descriptions except for *Adelina*, which includes several species that are commonly encountered. The genus *Barrouxia*, which has species that infect Nemertea, centipedes and snails in addition to Hemiptera, is characterized by oocysts with numerous sporocysts, each sporocyst with a bivalved wall and containing a single sporozoite (Levine, 1988b). *Chagasella* species infect the gut and salivary glands of Hemiptera or

termites and do not spread to the fat body. There have been no records of any of its four species since 1944. *Legerella*, a genus in which no sporocysts are formed and sporozoites are free within the oocysts, has three species that infect the Malpighian tubules of fleas and beetles, as well as species that infect millipedes and nematodes. *Ithania* and *Rasajeyna* are monotypic genera, both of which have species that infect the midgut cells of crane flies (Ludwig, 1947; Beesley 1977). The genus *Ganapatiella* also has only one species, *G. odontotermi*, which infects the fat body of the termite *Odontotermes obesus* (Kalavati, 1977).

Most, if not all, transmission is by ingestion of contaminated food. In some species that inhabit the gut, oocysts are continuously disseminated with feces, while those that inhabit the fat body are transmitted and disseminated when predators, cannibals, and scavengers consume the hosts' tissues. Even the plant-feeding Hemiptera that are known hosts of coccidia practice cannibalism, which can be a means of transmission. Host susceptibility decreases with age, but in an apparent anomaly reported by Weiser and Beard (1959), there was a high prevalence of *Adelina sericesthis* in scarab larvae at all collection sites, but laboratory transmission frequency was low. The effect may have been due to unavailability of very young and susceptible larvae.

The only insect-infecting coccidian genus that is well studied is *Adelina*. It was first found in oligochaetes and has five species that infect centipedes, but most of the known species are pathogens of insects. *Adelina* species infect Coleoptera, Lepidoptera, Orthoptera, Diptera, Embioptera, and Blattaria. They are cosmopolitan and are commonly encountered in surveys of insects in grain facilities (Steinhaus, 1947; Loschiavo, 1969; Purrini, 1976b; Ghosh *et al.*, 2000) and turf-infesting scarabs (Weiser and Beard, 1959; Hanula and Andreadis, 1988; Malone and Dhana, 1988). Larvae of the Egyptian alfalfa weevil, *Hypera brunneipennis*, collected from Davis, California, USA, and larvae of alfalfa weevil, *Hypera postica*, collected from Abu-Ghraib, Iraq, were found to be infected with a species of *Adelina* (Merritt *et al.*, 1975). An isolate was found in field crickets in Argentina (Lange and Wittenstein, 2001), and it may be closely related to *Adelina grylli*, which was isolated from crickets in central Asia (Sokolova *et al.*, 1999). Most species, including *Adelina tenebrionis* and *A. sericesthis*, are thought to have broad host ranges (Weiser and Beard, 1959; Malone and Dhana, 1988), while others appear to be more host specific (Yarwood, 1937), but the application of molecular phylogenetics may change this. Host specificities are subject to modification because, as with many protists, it is likely that some of the species delineations are invalid. Because of a paucity of studies of ecological and physiological host ranges, as well as gene sequences, there may be synonymies among the described species. Indeed, it is not clear that there is no synonymy

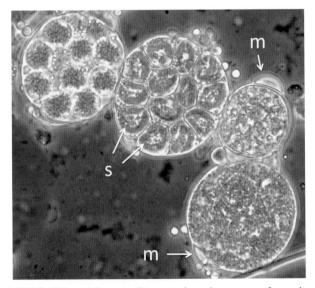

FIGURE 10.6 *Adelina mesnili* oocysts in various stages of maturity. Note sporozoites (s) in sporocysts of the mature oocyst and residues of microgametocytes (m) attached to the oocysts. Phase-contrast microscopy.

among three of the most reported species, *Adelina mesnili*, *A. tenebrionis*, and *A. tribolii*.

There has been only one published study of host population effects of coccidia. It was carried out with *A. tribolii*, which infects both *T. castaneum* and *T. confusum* (Park and Frank, 1950). Density and growth patterns were similar for non-parasitized and parasitized populations, except that the former had approximately 20% more adults and 20% fewer immature than the latter. When both species were infected, *T. confusum* outcompeted *T. castaneum* and became the dominant species, while the reverse was true in uninfected populations.

The life cycles of *Adelina* spp. are very similar to those of neogregarines (Fig. 10.7). The infection process begins with oocyst and sporocyst ingestion and digestion by a suitable host. Each oocyst releases three to 20 spherical to ellipsoid sporocysts. Each sporocyst then releases its two vermiform sporozoites that penetrate the gut wall. The sporozoites undergo some morphological changes after entering the hemocoel, becoming vacuolated, more intensely staining with Giemsa, and indistinguishable from merozoites. They are typical of coccidian motile stages with an apical complex (Malone and Dhana, 1988;

Sokolova *et al.*, 1999). They are tapered at both ends, have a centrally located nucleus, and are surrounded by a pellicle with subpellicular microtubules (Sokolova *et al.*, 1999). There is a conoidal complex surrounded by a three-layered cytoplasmic membrane, eight to 10 rhoptries, which are specialized enzyme-secreting organelles that extend from the apical end, and abundant micronemes, which are secretory organelles that function in motility and host cell invasion (Sokolova *et al.*, 1999). Weiser (1963) stated that the stages that enter the hemocoel are schizonts (merozoites), but merogony has been detected only within cells of the fat body and, to a lesser extent, subcuticular connective tissue. The first spherical meronts of *A. tenebrionis* were detected 19 days after dosing by Malone and Dhana (1988). They are enveloped by a parasitophorus vacuole that may have host-derived inclusions (Sokolova *et al.*, 1999). After nuclear proliferation, an oval membrane develops around each nucleus forming cells that elongate and differentiate into merozoites that spread to surrounding healthy tissue. The plasmodial meronts divide into eight to 25 merozoites for most species. Although authors working with other species do not mention differences between the first merogony and subsequent merogonies, Weiser and Beard

FIGURE 10.7 Life cycle of *Adelina cryptocerci* adapted from Yarwood (1937) showing sexual and asexual development and multiplication in the host *Cryptocercus punctulatus*.

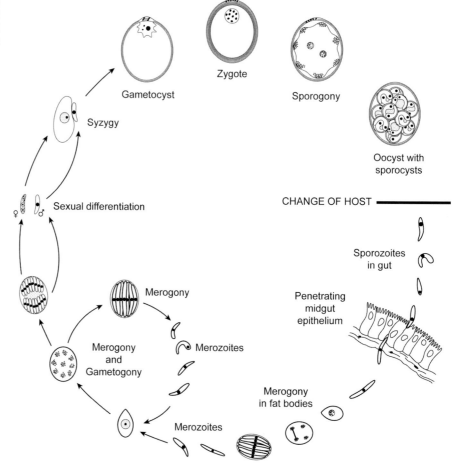

(1959) state that the first merogony of *A. sericesthis* produces 30–60 merozoites that are 16–18 μm long and 3–4 μm wide, while the second produces eight to 32 merozoites that are 7–8 μm long and 5–6 μm wide. The merozoites are capable of coiling and uncoiling, spiral, and gliding movements (Steinhaus, 1947).

After the first or a subsequent merogony, the merozoites transform and initiate gametogony. The macrogametocyte, often referred to as female, secretes a covering layer, becomes oval, and grows to 25–65 μm. The cytoplasm has many large vacuoles and abundant granules that are thought to contain a polysaccharide, as was confirmed in the coccidian *Barrouxia schneideri* by the periodic acid–Schiff test (Canning, 1962). The carbohydrate has been referred to as amylopectin (Sokolova *et al.*, 1999), but that is not likely in the absence of photosynthesis. Microgametocytes are much smaller and become lenticular or cupped when they attach to macrogametocytes. The gametocyst wall forms around the gametocytes. The nucleus of the microgametocyte undergoes two divisions to form four microgametes, one of which fertilizes the macrogamete. After fertilization, the gametocyst becomes an oocyst, and the zygote begins dividing into sporoblasts with a cyst wall around each to form spherical sporocysts. As is the case with other coccidians, the first division of the zygote is meiosis. Therefore, the organism is haploid (Canning, 1962). The number of sporocysts within an oocyst varies from three to 30 and has been used as a character for describing species, although Weiser (1963) states that the number of sporocysts in one oocyst is variable even within a host species from one larva to another. Within the sporocysts, there are two sporozoites that lie along the wall with their concave sides facing each other. *Adelina* species are slow developing and some species may require as long as 46 days until the first appearance of oocysts (Malone and Dhana, 1988), while in other species they may appear in as little as 10 days at the proper temperature (Weiser, 1963).

10.6. CILIOPHORA

The ciliates are common freshwater Alveolata that are characterized by the presence of cilia that are structurally identical to flagella but typically shorter and much more profuse. All have a compound subpellicular infraciliature that is composed of kinetosomes, or basal granules, interconnected by longitudinal fibrils. They have a heterokaryotic condition, with each cell usually having one or more polyploid macronuclei that contain multiple copies of gene-sized DNA molecules and one or more diploid micronuclei (Corliss, 1975). Asexual reproduction is by binary fission, with micronuclei undergoing mitosis and macronuclei elongating and dividing by amitosis. Sexual reproduction is by conjugation of complementary mating types followed by division of the paired cells into four new

cells. Macronuclei are destroyed during sexual reproduction, and new macronuclei are formed from mitotic copies of the polytene chromosomes of the micronucleus. In free-living forms, there is often a ciliated oral groove, or cytostome, with a cytopharynx through which food consisting of microbes and detritus is swept. Most ciliates are free living, but many are commensals of vertebrates and invertebrates. The ciliates associated with insects are primarily commensal, with only a few species that are pathogenic and are able to penetrate the host cuticle and invade the underlying tissues (Corliss and Coats, 1976). Commensals, such as *Balantidium* spp., inhabit the digestive tracts of insects, especially cockroaches and termites. Others ciliates, such as *Rhabdostyla* spp., are epibionts on a variety of invertebrates including insects (Regali-Seleghim and Godinho, 2004) and are not pathogenic. All aspects of the Ciliophora are discussed in detail by Lynn (2008).

With the exception of reports of *Chilodonella uncinata* from the larvae of culicine and anopheline mosquitoes in India (Das, 2003) and what was thought to be *Colpoda* sp. in a tumor on *Nepa cinerea* (Mercier and Poisson, 1923), the ciliates that are known pathogens of insects belong to the taxon Hymenostomatida. A few species of *Ophryoglena* (Ophryoglenidae) infect mayfly nymphs, destroying several tissues and sterilizing females (Gaino and Rebora, 2000). All other known ciliates that are pathogenic for insects belong to Tetrahymenidae and have been assigned to either *Tetrahymena* or *Lambornella*. Records include ciliatosis (ciliate infection) in mosquitoes, chironomids, and black flies (Corliss *et al.*, 1979). They are characterized by uniform body ciliature and a well-defined buccal cavity with ciliature consisting of an undulating membrane and three sheets of fused cilia called adoral membranelles. Free-living stages are pyriform to ovoid, and most range from 40 to 80 μm in length. The most important differentiating characteristic for separating the two genera is the cuticular "invasion" cyst formed by *Lambornella* when it attaches to the cuticle of a host (Keilin, 1921a; Corliss and Coats, 1976). However, recent comparisons of both long- and short-segment ribosomal DNA sequences demonstrated that the genus *Lambornella* arises within the *Tetrahymena* clade, and it is therefore not a defensible one (Nanney *et al.*, 1998; Strüder-Kypke *et al.*, 2001). Nevertheless, it will be used here since there has been no formal revision of the genus.

Ciliatosis in container-breeding mosquitoes appears to be widespread. The tetrahymenid ciliates that infect container-inhabiting mosquitoes (and perhaps those of ground pools) have a unique relationship with insect larvae in that they are both prey and pathogens. Free-living ciliates are often very abundant in the habitats of container-inhabiting mosquitoes and midges, and can be an important component of the insects' larval diet. There are numerous reports of parasitic ciliates in a variety of host species, but apparently all belong to one of two *Lambornella* species. The best studied of these is *L. clarki*, which is important as both prey and pathogen of *Ochlerotatus sierrensis*, a tree-hole mosquito in forested habitats of western North America. Mosquito eggs and drought-resistant cysts of *L. clarki* both hatch when tree holes are filled at the onset of the rainy season. The ciliates are only found in tree-hole water with its pH within one or two points of neutral (Egerter and Anderson, 1985; Washburn and Anderson, 1986). In tree holes that lack mosquitoes, *L. clarki* is present as free-living trophonts. When larvae of *O. sierrensis* are present, they release an unidentified water-borne factor that induces *L. clarki* to undergo a synchronous response in which cells divide and transform into parasitic cells called theronts (Washburn *et al.*, 1988b). The transformation can also be induced by tannic acid, which is a major solute in tree-hole water that also affects mosquito development (Mercer and Anderson, 1994). The transformation occurs within 16 h of exposure to host larvae. The theronts are spherical cells lacking an oral opening. They attach to the larval cuticle and form hemispherical invasion cysts. The parasitic ciliates penetrate the cuticle and enter the larva's hemocoel. The ciliatosis is chronic and fatal, with the predator-host usually dying in three to four weeks, but the cuticular penetration can provide an entry route for opportunistic microbes and result in rapid death (Washburn *et al.*, 1988a). Melanization and death of the ciliates can occur between cuticle and epidermis, but not after successful penetration to the hemocoel (Clark and Brandl, 1976). Transmission occurs when trophonts that escape from dying hosts transform into the infective theronts and attack other larvae. In nature, this trophic shift can lead to 100% infection and local extinction of mosquito larvae and dramatic changes in microbial populations (Washburn and Anderson, 1986). Facultative parasitism by this polymorphic ciliate is thought to have evolved as an antipredator strategy (Washburn *et al.*, 1988a; Mercer and Anderson, 1994).

Infections of mosquito larvae had been assumed to be fatal in all cases because of the thousands of ciliates observed in the body, but scattered reports in the literature of the appearance of ciliates in adult mosquitoes raised the question of how they got there (Corliss and Coats, 1976). The few infected *O. sierrensis* that survive to the adult stage are a dispersal mechanism for *L. clarki*. The ciliate invades the ovaries and castrates females, which then deposit the ciliates as they would their eggs. The infection reduces flight time in females, but not in males (Yee and Anderson, 1995), and host-seeking and blood-feeding are inhibited by the infection (Egerter and Anderson, 1989). Oviposition behavior of infected females mimics that of normal gravid females in their first gonotrophic cycle except that some aspects are prolonged (Egerter *et al.*, 1986; Yee, 1995).

Adults of both sexes also passively disperse ciliates by dying on water surfaces, and infected adults are more likely to die on water than uninfected adults. Ciliates dispersed by infected adults can infect larvae or form desiccation-resistant cysts. There is evidence that the hosts have greater dispersal ability than the ciliates do (Ganz and Washburn, 2006). Hosts colonized artificial tree holes in the field at a much higher rate than the parasite, and field releases of the pathogen demonstrated that it colonizes and persists in natural tree holes where it was previously absent, suggesting that its distribution is limited by its dispersal ability.

The impact of *L. clarki* infections on *O. sierrensis* populations varied with resource availability (Washburn *et al.*, 1991). When host populations developed with sufficient food, mortality from parasites was additive and reduced the number of emerging mosquitoes. For food-limited populations, pathogen-induced mortality resulted in as many or more emerging adults with higher average fitness than those from uninfected populations. When nutrients were scarce, parasitic infections reduced larval competition and increased food per individual owing to reduced host abundance. Food limitation altered larval feeding behavior, reducing horizontal transmission and subsequent mortality from parasitism.

Lambornella stegomyiae is a poorly characterized species that infects the important arbovirus vectors, *Aedes scutellaris*, *Ae. albopictus*, and *Ae. aegypti*. Vythilingam *et al.* (1996) found low prevalence in *Amigeres* sp. and in *Anopheles subpictus*, as well as *Ae. albopictus*. In contrast to Arshad and Sulaiman (1995), who reported high infectivity and virulence for *Ae. albopictus*, and *Ae. aegypti*, Vythilingam *et al.* (1996) found only weak infectivity under laboratory conditions for these species, suggesting a strain or species difference.

The genus *Tetrahymena* includes species that associate facultatively with mosquitoes and other Nematocera in a variety of habitats (Corliss and Coats, 1976). Hill (1972) made the observation that *Tetrahymena* species must be highly derived based on their loss of biosynthetic abilities. They require 10 amino acids, six vitamins, guanine, and uracil. They have no urea cycle enzymes, and they probably make neither sterols nor glutathione. There is a large group of morphologically indistinguishable species referred to as the *T. pyriformis* complex that includes species that are reproductively isolated by mating type as well as asexual species characterized by the absence of a micronucleus. An example of the cryptic species in the complex is a poorly characterized ciliate that was found in 0.25% of adults of an unidentified *Aedes* species in Canada and was described as a new species, *Tetrahymena empidokyrea*, in the *T. pyriformis* complex based on morphology and small-subunit rDNA (Jerome *et al.*, 1996). No doubt it is but one of many such cryptic species within the complex that are likely to have different ecological roles (Simon *et al.*, 2008).

Tetrahymena pyriformis is the best studied member of the genus. It has cosmopolitan distribution and has been the subject of a great deal of study (Elliot, 1972; Corliss and Coats, 1976). There have been reported infections of mosquitoes and black flies by ciliates identified as *T. pyriformis*. In these latter cases, however, the ciliates may not have been correctly identified (Jerome *et al.*, 1996). Identification is difficult because species in the *T. pyriformis* complex have trophonts that do not transform into other life-cycle stages (Corliss, 1972). Invasion by *Tetrahymena* spp. is thought to be through abrasions of the integument or incompletely sclerotized cuticle after a molt rather than by invasion cysts (Corliss, 1960). The host range of *Tetrahymena pyriformis* is broader than that of *Lambornella* spp., which is consistent with the opportunistic host invasion. The physiological host range appears to be broad, as demonstrated by Thompson (1958), who established infections in insects of five orders by inoculation with *T. pyriformis* or any of four other species, but it may not be reflective of the ecological host range for which abundance is critical. Under laboratory conditions, *T. pyriformis* is much more infective for *Culex tarsalis*, whose larvae inhabit ground pools, than for the container species *Ae. aegypti* (Grassmick and Rowley, 1973). However, it is generally more abundant in containers than in ground pools, and since abundance in the same ecological niche as the potential host is necessary for free-living ciliates to colonize the host (Corliss, 1960), it is not found in *Cx. tarsalis* in nature.

Tetrahymena pyriformis is a bacteria-feeding species and has been proposed as a bioencapsulation delivery vehicle for the microbial insecticide *Bacillus thuringiensis* subsp. *israelensis* (Bti) to mosquito larvae (Zaritsky *et al.*, 1991; Manasherob *et al.*, 1998) in a control approach that depends only on its status as prey for mosquitoes, independent of pathogenicity. The Bti δ-endotoxin remains stable after encapsulation in *T. pyriformis* food vacuoles, and its impact on larvae of *Ae. aegypti* and *Anopheles stephensi* is increased by three- and eight-fold, respectively (Manasherob *et al.*, 1994, 1996) because each *T. pyriformis* cell concentrates between 180 and 240 spores in food vacuoles. Larvae of mosquitoes fed on Bti-loaded *T. pyriformis* ingest large, lethal quantities of toxin and rapidly die. It was found that the interaction between ingested spores and *T. pyriformis* creates natural sites for recycling of Bti. Spores germinate, grow, and sporulate in excreted food vacuoles of *T. pyriformis*, forming new active insecticidal crystal protein during the cycle, thereby providing a mechanism for residual mosquito control (Manasherob *et al.*, 1998). Because *Tetrahymena* spp. feed in both the sediment and water column, they consume and accumulate the Bti spores and toxin crystals from the sediment and move them into the water column, where they are more available to mosquitoes, especially surface-feeding

anophelines. The ciliates themselves are unaffected by the Bti protein crystals. How effective the bioencapsulation may be in increasing efficacy of Bti under field conditions remains to be tested.

Tetrahymena chironomi is a small, micronucleus-bearing facultative pathogen of the midge, *Chironomus plumosus*, that is easily cultured on synthetic media. It is an uncommon species and is known only from the detailed species description of Corliss (1960). It is not clear that it was among the unidentified *Tetrahymena* spp. that have been found in several chironomid species in the Laurentian Great Lakes (Golini and Corliss, 1981). In three months of observation at sites near Paris and Berlin, the prevalence rate was limited to 9%. As with other insect-pathogenic *Tetrahymena* species, the mode of transmission is unknown, but Corliss (1960) suspected entry via wounds, which are common on chironomid larvae. He considered the species to be monomorphic, but stated that it was frequently found in the same specimens with an amicronucleate *T. pyriformis*. *Tetrahymena chironomi* exhibits a curious phenomenon that indicates a poorly developed association with the host. The ciliates proliferate to fill the host's body cavity, and then enter into sexual conjugation that is consistently suicidal. All of the exconjugants die shortly before or after the death of the host. There is no known explanation for the phenomenon.

A black fly pathogen, *Tetrahymena rotunda*, is a large, globose species that is a pathogen of *Simulium* spp. in upper New York State. In a five-year period (1976–1980), the infection rate in *Simulium* spp. was usually less than 2%. Ciliates were found in *Simulium tuberosum* larvae and in the larval, pupal, and adult stages of *Simulium venustumn*. Hosts typically harbored around 10 ciliates, a number far less than generally reported for ciliate-infected dipterans (Lynn *et al.*, 1981).

Tetrahymena dimorpha (Batson, 1983) and *T. sialidos* (Batson, 1985) resemble *Lambornella* in being dimorphic with free-living and parasitic forms that were whimsically designated Jekyll and Hyde. *Tetrahymena dimorpha* infects all stages of the black flies *Simulium equinum* and *S. ornatum*. The Hyde form, which is large and broadly oval and has many abnormal kineties, is parasitic in larvae. The Jekyll form is either parasitic in adults or free living. Within larval hosts, the total number of parasites never exceeds about 240 and the infection is benign. When the host matures to the adult stage there is a form change and dramatic increase in the number of ciliates (up to 19,000 per host), which soon causes host death. In adult hosts, the Jekyll form ciliates are small and pyriform with somatic kineties reduced in number and in a more ordered arrangement, typical of the *T. pyriformis* complex. Both forms are culturable and the form shift can be induced *in vitro* by manipulation of medium components or by the presence of bacteria (Batson, 1983). *Tetrahymena sialidos* infects larval *Sialis lutaria*, an alderfly. In the parasitic phase, *T. sialidos* is broadly oval with dense granular cytoplasm. Infection is synchronized with host life cycle and lethal to the host after 11—12 months, with the number of ciliates per larva reaching 55,000. The free-living phase that develops after the ciliates emerge from moribund or dead larvae is elongate and pyriform with sparsely granular cytoplasm. It undergoes characteristic synchronous conjugation. Host death and emergence of ciliates coincide with the appearance of the next generation of host larvae. Prevalence rates in nature of 40—70% were reported by Batson (1985).

10.7. EUGLENOZOA

Protists in the higher rank Euglenozoa are flagellates (mastigotes) having one or two (rarely more) flagella arising from an apical or subapical pocket-like invagination of the cell called the reservoir (Leedale and Vickerman, 2000; Adl *et al.*, 2005). The axoneme of flagella is frequently accompanied by paraxonemal rods, mitochondria have mostly discoid cristae, and there is usually a tubular feeding apparatus associated with the flagellar apparatus. Euglenozoa usually reproduce asexually by longitudinal binary fission. Among the Euglenozoa, the family Trypanosomatidae in the class Kinetoplastea contains the majority of entomogenous, pathogenic flagellates (Tanada and Kaya, 1993). Other flagellate protists, such as taxa Retortamonadida, Diplomonadida, Parabasalia, and Pyrsonympha, occur as mutualistic or commensalistic symbionts in the digestive tract of various species of Blattaria, Isoptera, and other insects.

Members of Kinetoplastea are mainly characterized by the presence of the kinetoplast, a large mass of fibrillar DNA in the mitochondrion, normally in close association with the flagellar area of attachment. Trypanosomatids are small (usually 4—15 μm) heteroxenous or monoxenous parasitic osmotrophs or phagotrophs with one flagellum without mastigonemes and emerging anteriorly or laterally from the reservoir. In addition to the basal attachment, the flagellum may be attached to several other spots along the cell surface, thus constituting an undulating membrane. Trypanosomatids are also characterized by the occurrence of the unique glycosomes, small (0.2—0.3 μm), spherical or ovoid organelles containing glycolytic enzymes (Hausmann *et al.*, 2003).

Heteroxenous species are transmitted by hematophagous or phytophagous insects (Diptera, Hemiptera) and develop in vertebrates (trypanosomatid genera *Leishmania*, *Sauroleishmania*, *Trypanosoma*, *Endotrypanum*) or plants (*Phytomonas*), respectively. Some major human and domestic animal infectious diseases have trypanosomatids as etiological agents, notably Chagas disease (*T. cruzi*) in areas of Latin America, African sleeping sickness (*T. brucei*), and Old and New World leishmaniasis. Similarly, several species of *Phytomonas* are economically important pathogens of plants and develop in phytophagous Hemiptera (Camargo and Wallace, 1994). In most cases, as opposed to the relatively strong effects on vertebrate or plant hosts, most heteroxenous trypanosomatids do not seem to seriously affect their vectors under normal conditions.

Most monoxenous (i.e., entomophilic) trypanosomatids occur in the alimentary canal, Malpighian tubules, or salivary glands, and some invade the hemocoel. Many species tend to attach to the intestinal epithelium in a carpet-like layer while others remain free in the lumen. Life cycles are less complex than in heteroxenous trypanosomatids but still they mostly occur in alternating morphotypes. Transmission is usually through ingestion of stages liberated from feces of infected insects. Host insects include species of Hemiptera, Diptera, Hymenoptera, Blattaria, Lepidoptera, Siphonaptera, and Anoplura. Under laboratory conditions, most insect trypanosomatids can be transmitted to hosts other than the ones from which they have been isolated in nature (natural hosts as defined by Onstad *et al.*, 2006). Monoxenous trypanosomatid species, commonly referred to as "lower trypanosomatids", are in the genera *Herpetomonas*, *Crithidia*, *Blastocrithidia*, *Leptomonas*, *Wallaceina*, *Sergeia*, and *Rhynchoidomonas*. There are approximately 1000 descriptions of monoxenous trypanosomatids, of which about 300 would be valid species (Svobodová *et al.*, 2007).

Trypanosomatids are polymorphic, having two or more developmental forms (phenotypes, morphotypes, or morphs) in their life cycles. Based on cell shape and the relative position of the flagellum base/kinetoplast, the nucleus, and the arrangement and emergence of the flagellum, Hoare and Wallace (1966) provided a terminology scheme for referring to the different forms. Morphotypes are generally either elongate or somewhat fusiform in shape (promastigote, opisthomastigote, epimastigote, trypanomastigote) or somewhat spherical or rounded (amastigote, choanomastigote, nectomonad, haptomonad) (see Leedale and Vickerman, 2000 for diagrams). Cyst-like stages are produced in some species. Some insect trypanosomatids contain a bacterial endosymbiont in their cytoplasm that divides synchronously with the trypanosomatid. Although the different morphotypes are not equally amenable to culturing, most trypanosomatids can be maintained in culture, usually on blood-containing agar or insect tissue culture media (Vickerman, 2000; Podlipaev and Naumov, 2000). Traditionally, genus assignment was largely based on the presence or absence of certain combinations of morphotypes in life cycles, while species description was mostly dependent on the "new host—new species" approach. Such criteria are no longer considered tenable (Yurchenko *et al.*, 2008). Most trypanosomatids are morphologically more variable than previously thought and a single species may infect more than one host species and vice versa (Zidkova et al., 2010). The advent of molecular sequencing techniques is drastically reshaping the taxonomy and phylogeny of trypanosomatids (Svobodová et al., 2007; Schmid-Hempel and Tognazzo, 2010).

The pathogenicity of trypanosomatids for insects was reviewed by Schaub (1994), who more recently revisited the subject with an emphasis on trypanosomatids associated with triatomines (Hemiptera: Triatominae), one of the most common host groups (Schaub, 2009). Most trypanosomatids were considered apathogenic (would not adversely affect the host) or subpathogenic (would affect the host only under adverse conditions). Insect pathogenic trypanosomatids are relatively few compared to the estimated diversity of the group. As with most other entomopathogenic protists, infections by most trypanosomatids on insects are typically debilitative and chronic in nature rather than acute. However, trypanosomatids that reach the hemocoel of the infected insect tend to be more pathogenic. Numerous effects of trypanosomatids on insect hosts are known, most being in direct relationship to parasite burden. They range from alterations of behavior and food intake and utilization to disruption of organs, delayed development, and reductions in lifespan and reproductive rate. Such effects are usually enhanced by biotic or abiotic stressors.

Historically, trypanosomatids in genus *Herpetomonas* were characterized by having promastigotes and opisthomastigotes in their life cycles. However, recent research at the molecular level showed that taxonomy based on morphology only is not valid, and the genus, as the other monoxenous trypanosomatid genera, does not constitute a monophyletic assemblage of species (Lukes, 2009). Over 30 species have been described, most of which parasitize the digestive tract of Diptera and Hymenoptera, although they have also been isolated from plants and even mammals (Marín et al., 2007; Zídkova et al., 2010). Tanada and Kaya (1993) reviewed the two most widely known species from a pathology perspective, *H. muscarum* and *H. swainei*. *Herpetomonas muscarum*, the type species for the genus, was reported from more than 200 species of flies. It is usually transmitted through contaminated feces, and when it invades and develops in the hemocoel it usually kills the host. *Herpetomonas swainei* infects the digestive tract, Malpighian tubules, and hemolymph of *Neodiprion swainei* and other sawfly species, causing 15–20% mortality.

The genus *Crithidia* is classically defined as those trypanosomatids with choanomastigote morphotypes in the lumen of the digestive tracts of insects in the orders Diptera, Hemiptera, Trichoptera, and Hymenoptera. The type species is *C. fasciculata* from the mosquito *Anopheles maculipennis*. Over 30 species have been described, but the current assignments at the genus level need to be revised integrating molecular data with morphology-based descriptions (Yurchenko et al., 2008). Although *Crithidia* species have been typically considered as non-pathogenic, *Crithidia bombi*, a parasite of *Bombus terrestris* and other European bumble bees, reduces the fitness of infected queens (Brown et al., 2003). Since host specificity seems low in *C. bombi* and *B. terrestris* is commercialized worldwide for pollination services, there is fear that *C. bombi* might spread to previously unexposed populations of other bumble bee species (Colla et al., 2006; Plischuck and Lange, 2009). *Crithidia expoeki*, a second species of *Crithidia* associated with bumble bees, was recently described by Schmid-Hempel and Tognazzo (2010) from *Bombus lucorum* in Switzerland.

According to the traditional morphology-based taxonomy, the genus *Blastocrithidia* is characterized by the presence of epimastigotes and cyst-like stages called straphangers. About 40 species have been described, mostly from the digestive tract of insects in the orders Diptera, Hemiptera, Hymenoptera, and Siphonaptera. The type species is *B. gerridis* from the water strider, *Gerris fossarum*, and other Hemiptera in the families Gerridae and Veliidae. The most intensively studied species is *B. triatomae* from the intestine of *Triatoma infestans* and other triatomine vectors of Chagas disease. Unlike most other monoxenous trypanosomatids, *B. triatomae* is pathogenic to its hosts, causing alterations in food ingestion, digestion, excretion, molting, immunity response, starvation capacity, development, and longevity (Schaub, 1994, 2009).

As for other monoxenous trypanosomatids, molecular studies have shown that the genus *Leptomonas* is not monophyletic (Merzlyak et al., 2001; Yurchenko et al., 2008). Morphologically, promastigotes dominate but amastigotes and cyst-like stages also occur. Some 40 species of *Leptomonas* have been described parasitizing the digestive tract of Hemiptera, Diptera, Hymenoptera, Blattaria, Lepidoptera, Siphonaptera, and Anoplura. However, the type species, *L. buetschlii*, is from a nematode, and species are known from ciliate protists. Historically, the genus has been used for assignment of some species that did not have a clear affinity with other existing genera. Some relatively well-known species are *L. ctenocephalidae* and *L. pulexsimulantis* from fleas.

10.8. HELICOSPORIDIA

Helicosporidia (Chlorophyta: Trebouxiophyceae) are the first known invertebrate-pathogenic green algae (Tartar et al., 2002). *Helicosporidium parasiticum*, originally found in the ceratopogonid midge *Dasyhelea obscura* (Keilin, 1921b) is the only described species, although it has been isolated from or transmitted to many invertebrates including Coleoptera, Collembola, Diptera, Lepidoptera, mites, *Daphnia*, and a trematode (Weiser,

1970; Kellen and Lindegren, 1973; Purrini, 1984; Sayre and Clark, 1978; Pekkarinen, 1993; Seif and Rifaat, 2001; Yaman and Radek, 2005). Other species of *Helicosporidium* have not been identified from the available morphological data, and molecular data suggest that those found in the various hosts represent different isolates or strains of the originally described species, *H. parasiticum* (Tartar *et al.*, 2003).

The life cycle of *H. parasiticum* was described by Boucias *et al.* (2001) and Bläske-Lietze *et al.* (2006) and is characterized by an infectious pellicle-covered cyst that contains a coiled filamentous cell and three ovoid cells (Fig. 10.8). Following ingestion by the host, digestive fluids cause the cyst to dehisce releasing the ovoid cells, which deteriorate, and the invasive filamentous cell, which penetrates the columnar epithelium and enters the hemocoel. The filamentous cells have surface barbs that may aid in the passage through the midgut epithelial layer (Boucias *et al.*, 2001). Within the nutrient-rich hemolymph, Helicosporidia undergo multiple cycles of vegetative replication. Like the cyst stage, vegetative cells are non-motile and enclosed within a pellicle that can contain two, four, or eight daughter cells (Lindegren and Hoffmann, 1976; Boucias *et al.*, 2001). Eventually, the host hemocoel is colonized by massive numbers of vegetative cells, some of which

differentiate into mature cysts. The resulting *in vivo* cells can be harvested and cultured *in vitro* (Boucias *et al.*, 2001). Purified *in vivo*-produced cysts can be experimentally dehisced with larval digestive fluids from different lepidopteran species (Boucias *et al.*, 2001).

While they lack chlorophyll, Helicosporidia are reported to have modified but functional plastids. Although none has been detected by microscopic methods, their existence has been confirmed by molecular evidence (Tartar and Boucias, 2004; de Konig and Keeling, 2004), and the plastid genome is reduced to minimal size and complexity (de Konig and Keeling, 2006).

Where pathological effects of Helicosporidia have been observed, they comprise reduced robustness and moderate mortality. Bläske and Boucias (2004) treated early instar noctuids with a weevil isolate and found that while most pupated, only 20−30% of the pupae survived to adulthood, and most of those had malformed wings and reduced longevity. In experimentally infected *S. exigua*, Bläske and Boucias (2004) observed reduced fecundity and a low level of transmission to progeny but no evidence of invasion of the reproductive tissue. Conklin *et al.* (2005) found a virulence difference between two isolates, but both were pathogenic to mosquitoes and citrus root weevils, which were markedly reduced in size when infected.

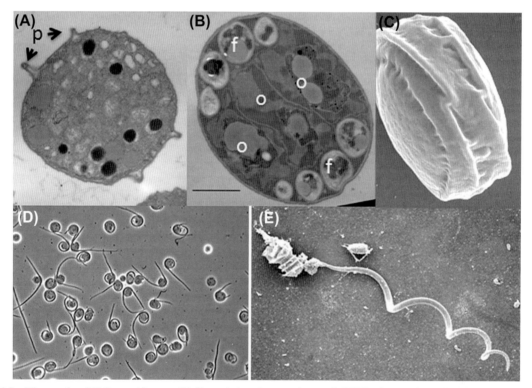

FIGURE 10.8   Micrographs of *Helicosporidium* sp. (A) Transmission electron micrograph (TEM) of vegetative cell with a developing pellicle (p); (B) TEM of mature cyst demonstrating the characteristic three ovoid cells (o) surrounded by the filamentous cell (f); scale bar = 1 μm; (C) scanning electron micrograph (SEM) of mature cyst; (D) light micrograph dehisced cysts releasing filamentous and ovoid cells; (E) SEM of dehisced cyst with extending uncoiled filamentous cell. [*Photo credits: (A, B) J. J Becnel; (C−E) D. G. Boucias.*]

10.9. FUTURE RESEARCH DIRECTIONS

The research needs that relate to pest control with protistan pathogens are markedly different from those of pathogens that are readily mass produced and can be applied for short-term or seasonal pest control. Because of host specificities that limit their commercial viability and the difficulties in mass production, pathogenic protists are generally not regarded as likely microbial insecticides. If they are to have a place in insect control, inoculative release and perhaps conservation are more feasible approaches. Before those approaches will be justified, a great deal more information relating to their effects on host populations must be obtained.

Protistan pathogens are often prevalent and persistent. Accordingly, they afford ample opportunity for epizootiological studies. Unfortunately, there is a dearth of long-term epizootiological studies conducted with pathogenic protists in insects. Accounts of protistan prevalence in natural insect populations, whether or not they are accompanied by host density data, usually consist of single observations in a single or multiple locations. A rare exception is the long-term study of *O. elektroscirrha* prevalence in monarch butterflies by Altizer *et al.* (2000), which tracked prevalence but not host population trends. The need for such work is apparent because the population effects of protistan pathogens, although they may be consequential, are usually unapparent. New life tables and population models could be constructed, and existing ones could be made more accurate and useful. Unfortunately, it is extremely difficult to obtain long-term epizootiological data in most production agriculture systems, from which the data are most needed, because practices such as the use of broad-spectrum pesticides and crop rotation disrupt natural population changes. Nevertheless, it would be useful to have an understanding of how these disruptions affect pathogenic protists as well as their hosts.

There are many important questions to be addressed by long-term epizootiology studies. What are host population density thresholds for sustained transmission and spread? How are pathogens maintained at low host density? How does pathogen–host epizootiology differ in natural and artificial habitats? Insights could also be obtained on evolutionary strategies. Other topics that are in need of data and are relevant to epizootiology are transmission methods, virulence, persistence capacity, generation time, and effects of pathogens on host behavior. There are very sparse data related to sublethal effects and their impact when various stresses including control measures are applied.

Perhaps the most obvious need for research is that for untangling the confusion in systematics and taxonomy. The problem is exemplified by the 198 protistan genera of uncertain affiliation that were listed in a recent reclassification by Adl *et al.* (2005). The advent of ribosomal DNA and whole genome sequencing has made it possible to correctly assign the taxa that are currently available, but what is to be done about those with inadequate descriptions and lack of type material?

REFERENCES

Adl, S. M., Simpson, A. G. B., Farmer, M. A., Andersen, R. A., Anderson, O. R., Barta, J. R., Bowser, S. S., Brugerolle, G., Fensome, R. A., Fredericq, S., James, T. Y., Karpov, S., Kugrens, P., Krug, J., Lane, C. E., Lewis, L. A., Lodge, J., Lynn, D. H., Mann, D. G., McCourt, R. M., Mendoza, L., Moestrup, O., Mozley-Standridge, S. E., Nerad, T. A., Shearer, C. A., Smirnov, A. V., Spiegel, F. W., & Taylor, M. F. J. R. (2005). The new higher level classification of eukaryotes with emphasis on the taxonomy of protists. *J. Eukaryot. Microbiol., 52,* 399–451.

Albicocco, A. P., & Vezzani, D. (2009). Further study on *Ascogregarina culicis* in temperate Argentina: prevalence and intensity in *Aedes aegypti* larvae and pupae. *J. Invertebr. Pathol., 101,* 210–214.

Aliabadi, B. W., & Juliano, S. A. (2002). Escape from gregarine parasites affects the competitive interactions of an invasive mosquito. *Biol. Inv., 4,* 283–297.

Altizer, S. M., & Oberhauser, K. S. (1999). Effects of the protozoan parasite *Ophryocystis elektroscirrha* on the fitness of monarch butterflies (*Danaus plexippus*). *J. Invertebr. Pathol., 74,* 76–88.

Altizer, S. M., Oberhauser, K., & Brower, L. P. (2000). Associations between host migration and the prevalence of a protozoan parasite in natural populations of adult monarch butterflies. *Ecol. Entomol., 25,* 125–139.

Anderson, R. M., & May, R. M. (1981). The population dynamics of microparasites and their invertebrate hosts. *Philos. Trans. R. Soc. Lond. B Biol. Sci., 291,* 451–524.

Arshad, H. H., & Sulaiman, I. (1995). Infection of *Aedes albopictus* (Diptera: Culicidae) and *Ae. aegypti* with *Lambornella stegomyiae* (Ciliophora: Tetrahymenidae). *J. Invertebr. Pathol., 66,* 303–306.

Ashford, R. W. (1968). Sporozoan parasites of *Tribolium castaneum* (Herbst): *Triboliocystis garnhami* Dissanaike a synonym of *Farinocystis tribolii* Weiser. *J. Protozool., 15,* 466–470.

Avron, B., Stolarsky, T., Chayen, A., & Mirelman, D. (1986). Encystation of *Entamoeba invadens* IP-1 is induced by lowering the osmotic pressure and depletion of nutrients from the medium. *J. Protozool. 33,* 522–525.

Bailey, L. (1968). The measurement and interrelationships of infections with *Nosema apis* and *Malpighamoeba mellificae* of honey-bee populations. *J. Invertebr. Pathol., 12,* 175–179.

Barrett, W. L. (1968). Damage caused by *Lankesteria culicis* (Ross) to *Aedes aegypti* (L.). *Mosq. News, 28,* 441–444.

Batson, B. S. (1983). *Tetrahymena dimorpha* sp. nov. (Hymenostomatida: Tetrahymenidae), a new ciliate parasite of Simuliidae (Diptera) with potential as a model for the study of ciliate morphogenesis. *Philos. Trans. R. Soc. Lond. B Biol. Sci., 301,* 345–363.

Batson, B. S. (1985). A paradigm for the study of insect–ciliate relationships: *Tetrahymena sialidos* sp. nov. (Hymenostomatida: Tetrahymenidae), parasite of larval *Sialis lutaria* (Linn.) (Megaloptera: Sialidae). *Philos. Trans. R. Soc. Lond. B Biol. Sci., 310,* 123–144.

Beesley, J. E. (1977). Life-cycle of *Rasajeyna nannyla* n.gen., n.sp., a coccidian pathogen of *Tipula paludosa* Meigen. *Parasitology, 74,* 273–283.

Bläske, V. U., & Boucias, D. G. (2004). Influence of *Helicosporidium* spp. (Chlorophyta: Trebouxiophyceae) infection on the development and survival of three noctuid species. *Environ. Entomol., 33*, 54−61.

Bläske-Lietze, V.-U., Shapiro, A., Denton, J. S., Botts, M., Becnel, J. J., & Boucias, D. G. (2006). Development of the insect pathogenic alga *Helicosporidium. J. Eukaryot. Microbiol., 53*, 165−176.

Boucias, D. J., & Pendland, J. C. (1998). *Principles of Insect Pathology.* Boston: Kluwer.

Boucias, D. G., Becnel, J. J., White, S. E., & Bott, M. (2001). *In vivo* and *in vitro* development of the protist *Helicosporidium* sp. *J. Eukaryot. Microbiol., 48*, 460−470.

Bradley, C. A., & Altizer, S. (2005). Parasites hinder monarch butterfly flight: implications for disease spread in migratory hosts. *Ecol. Lett., 8*, 290−300.

Braun, L., Ewen, A. B., & Gillot, C. (1988). The life cycle and ultrastructure of *Malameba locustae* (K. & T.) (Amoebidae) in the migratory grasshopper *Melanoplus sanguinipes. Can. Entomol., 120*, 759−772.

Brooks, W. M. (1988). Entomogenous protozoa. In C. M. Ignoffo (Ed.), *CRC Handbook of Natural Pesticides: Vol. 5, Microbial Insecticides, Part A: Entomogenous Protozoa and Fungi*, Vol. 5 (pp. 1−149). Boca Raton: CRC Press.

Brooks, W.M., and Jackson, J. J. (1990). Eugregarines: current status as pathogens, illustrated in corn rootworms. In: *Proceedings of the Vth International Colloquium on Invertebrate Pathology and Microbial Control*, (pp. 512−515) Adelaide.

Brown, M. J. F., Schmid-Hempel, R., & Schmid-Hempel, P. (2003). Strong context-dependent virulence in a host-parasite system: reconciling genetic evidence with theory. *J. Anim. Ecol., 72*, 994−1002.

Burkholder, W. E., & Dicke, R. J. (1964). Detection by ultraviolet light of stored-product insects infected with *Mattesia dispora. J. Econ. Entomol., 57*, 818−819.

Byers, T. J., Akins, R. A., Maynard, B. J., Lefken, R. A., & Martin, S. M. (1980). Rapid growth of *Acanthamoeba* in defined media: induction of encystment by glucose-acetate starvation. *J. Protozool, 27*, 216−219.

Camargo, E. P., & Wallace, F. G. (1994). Vectors of plant parasites of the genus *Phytomonas* (Protozoa, Zoomastigophorea, Kinetoplastida). *Adv. Dis. Vector Res., 10*, 333−359.

Canning, E. U. (1962). Sexual differentiation of merozoites of *Barrouxia schneideri* (Butschli). *Nature, 195*, 720−721.

Canning, E. U., & Sinden, R. E. (1975). Development of *Farinocystis tribolii* Weiser (Sporozoa, Neogregarinida) in *Palembus ocularis* Casey (Coleoptera, Tenebrionidae) and some observations on its fine structure. *Protistologica, 9*, 221−231.

Cavalier-Smith, T. (1998). A revised six-kingdom system of life. *Biol. Rev. Camb. Philos. Soc., 73*, 203−266.

Cavalier-Smith, T. (2003). Protist phylogeny and the high-level classification of Protozoa. *Eur. J. Protistol., 39*, 338−348.

Cavalier-Smith, T. (2010). Kingdoms Protozoa and Chromista and the eozoan root of the eukaryotic tree. *Biol. Lett., 6*, 342−345.

Chávez-Munguía, B., Omana-Molina, M., González-Lázaro, M., González Robles, A., Bonilla, P., & Martínez-Palomo, A. (2007). Ultrastructure of cyst differentiation in parasitic protozoa. *Parasitol. Res., 100*, 1169−1175.

Chen, W. J. (1999). The life cycle of *Ascogregarina taiwanensis* (Apicomplexa: Lecudinidae). *Parasitol. Today, 15*, 153−156.

Chen, W. J., & Yang, C. H. (1996). Developmental synchrony of *Ascogregarina taiwanensis* (Apicomplexa: Lecudinidae) within *Aedes albopictus* (Diptera: Culicidae). *J. Med. Entomol., 33*, 212−215.

Clark, T. B., & Brandl, D. G. (1976). Observations on the infection of *Aedes sierrensis* by a tetrahymenine ciliate. *J. Invertebr. Pathol., 23*, 341−349.

Clopton, R. E. (2000). In J. J. Lee, G. F. Leedale & P. Bradbury (Eds.), *The Illustrated Guide to the Protozoa. Order Eugregarinorida, Vol. 1* (pp. 205−288). Lawrence: Society of Protozoologists.

Clopton, R. E. (2009). Phylogenetic relationships, evolution, and systematic revision of the septate gregarines (Apicomplexa: Eugregarinorida: Septatorina). *Comp. Parasitol., 76*, 167−190.

Clopton, R. E., & Janovy, J. (1993). Developmental niche structure in the gregarine assemblage parasitizing *Tenebrio molitor. J. Parasitol., 79*, 701−709.

Colla, S. R., Otterstatter, M. C., Gregear, R. J., & Thomson, J. D. (2006). Plight of the bumble bee: pathogen spillover from commercial to wild populations. *Biol. Conserv., 129*, 461−467.

Comiskey, N. M., Lowrie, R. C., & Wesson, D. M. (1999a). Effect of nutrient levels and *Ascogregarina taiwanensis* (Apicomplexa: Lecudinidae) infections on the vector competence of *Aedes albopictus* (Diptera: Culicidae) for *Dirofilaria immitis* (Filarioidea: Onchocercidae). *J. Med. Entomol., 36*, 55−61.

Comiskey, N. M., Lowrie, R. C., & Wesson, D. M. (1999b). Role of habitat components on the dynamics of *Aedes albopictus* (Diptera: Culicidae) from New Orleans. *J. Med. Entomol., 36*, 313−320.

Conklin, T., Bläske-Lietze, V.-U., Becnel, J. J., & Boucias, D. G. (2005). Infectivity of two isolates of *Helicosporidium* spp. (Chlorophyta: Trebouxiphyceae) in heterologous host insects. *Fla. Entomol., 88*, 431−439.

Corliss, J. P. (1960). *Tetrahymena chironomi* sp. nov., a ciliate from midge larvae, and the current status of facultative parasitism in the genus *Tetrahymena. Parasitology, 50*, 111−153.

Corliss, J. O. (1972). History, taxonomy, ecology, and evolution of species of *Tetrahymena*. In A. M. Elliot (Ed.), *Biology of Tetrahymena* (pp. 1−55). Stroudsburg: Dowden, Hutchinson and Ross.

Corliss, J. O. (1975). Nuclear characteristics and physiology in the protistan phylum Ciliophora. *Biosystems, 7*, 338−349.

Corliss, J. O., & Coats, D. W. (1976). A new cuticular cyst-producing tetrahymenid ciliate, *Lambornella clarki* n. sp., and the current status of ciliatosis in culicine mosquitoes. *Trans. Am. Microsc. Soc., 95*, 725−739.

Corliss, J. O., Berl, D., & Laird, M. (1979). A note on the occurrence of the ciliate *Tetrahymena*, potential biocontrol agent, in the blackfly vector of onchocerciasis from Ivory Coast. *Trans. Am. Microsc. Soc., 98*, 587−591.

Das, B. P. (2003). *Chilodonella uncinata* − a protozoa pathogenic to mosquito larvae. *Curr. Sci., 85*, 483−489.

Davies, K. A. (1973). Observations on *Malameba locustae* from *Chortoicetes terminifera* cultures in Australia. *J. Invertebr. Pathol., 22*, 475.

Egerter, D. E., & Anderson, J. R. (1985). Infection of the western treehole mosquito, *Aedes sierrensis* (Diptera: Culicidae), with *Lambornella clarki* (Ciliophora: Tetrahymenidae). *J. Invertebr. Pathol., 46*, 296−304.

Egerter, D. E., & Anderson, J. R. (1989). Blood-feeding drive inhibition of *Aedes sierrensis* (Diptera: Culicidae) induced by the parasite

Lambornella clarki (Ciliophora: Tetrahymenidae). *J. Med. Entomol.,* *26,* 46–54.

Egerter, D. E., Anderson, J. R., & Washburn, J. O. (1986). Dispersal of the parasitic ciliate *Lambornella clarki*: Implications for ciliates in the biological control of mosquitoes. *Proc. Natl. Acad. Sci. USA, 83,* 7335–7339.

Elliot, A. M. (1972). *Biology of* Tetrahymena. Stroudsburg: Dowden, Hutchinson and Ross.

Er, M. K., & Gökçe, A. (2005). Effect of *Diplocystis tipulae* (Eugregarinida: Apicomplexa), a coelomic gregarine pathogen of tipulids, on the larval size of *Tipula paludosa* Meigen (Tipulidae: Diptera). *J. Invertebr. Pathol., 89,* 112–115.

Ernst, H. P., & Baker, G. L. (1982). *Malameba locustae* (K. & T.) (Protozoa: Amoebidae) in field populations of Orthoptera in Australia. *J. Aust. Entomol. Soc., 21,* 295–296.

Evans, W. A., & Elias, R. G. (1970). The life cycle of *Malameba locustae* (King and Taylor) in *Locusta migratoria migratorioides* (R. & F.). *Acta Protozool., 7,* 229–241.

Fukuda, T., Willis, O. R., & Barnard, D. R. (1997). Parasites of the Asian tiger mosquito and other container-inhabiting mosquitoes (Diptera: Culicidae) in northcentral Florida. *J. Med. Entomol., 34,* 226–233.

Gaino, E., & Rebora, M. (2000). *Ophryoglena* sp. (Ciliata: Oligohymenophora) in *Caenis luctuosa* (Ephemeroptera: Caenidae). *Acta Protozool., 39,* 225–231.

Ganz, H. H., & Washburn, J. O. (2006). Relative migration rates and local adaptation in a mosquito–protozoan interaction. *J. Evol. Biol., 19,* 816–824.

García, J. J., Fukuda, T., & Becnel, J. J. (1994). Seasonality, prevalence and pathogenicity of the gregarine *Ascogregarina taiwanensis* (Apicomplexa: Lecudinidae) in mosquitoes from Florida. *J. Am. Mosq. Control Assoc., 10,* 413–418.

Ghosh, C., Choudhrury, A., & Misa, K. K. (2000). Life histories of three new coccidian parasites from three coleopteran stored-product pests of India. *Acta Protozool., 39,* 233–240.

Golini, V. I., & Corliss, J. O. (1981). A note on the occurrence of the hymenostome ciliate *Tetrahymena* in chironomid larvae (Diptera: Chironomidae) from the Laurentian Great Lakes. *Trans. Am. Microsc. Soc., 100,* 89–93.

Grassnick, R. A., & Rowley, W. A. (1973). Larval mortality of *Culex tarsalis* and *Aedes aegypti* when reared with different concentrations of *Tetrahymena pyriformis*. *J. Invertebr. Pathol., 22,* 86–93.

Haeckel, E. (1866). *Generelle Morphologie der Organismen.* Berlin: Georg Reimer.

Hanrahan, S. A. (1975). Ultrastructure of *Malameba locustae* (K. & T.), a protozoan parasite of locusts. *Acrida, 4,* 235–249.

Hanrahan, S. A. (1979). Malameba locustae*: its ultrastructure and its influence on the locust host.* PhD thesis. University of the Witwatersrand.

Hanrahan, S. A. (1980). Locust haemocyte reaction to a parasitic infection. *Proc. Electron Microsc. Soc. S. Afr., 10,* 107–108.

Hanula, J. L., & Andreadis, T. G. (1988). Parasitic microorganisms of Japanese beetle (Coleoptera: Scarabaeidae) and associated scarab larvae in Connecticut soils. *Environ. Entomol., 17,* 709–714.

Harry, O. G., & Finlayson, L. H. (1976). The life-cycle, ultrastructure, and mode of feeding of the locust amoeba *Malpighamoeba locustae*. *Parasitology, 72,* 127–135.

Hausman, K., Hülsmann, N., & Radek, R. (2003). *Protistology* (3rd ed.). Berlin: E. Schweizerbart'sche Verlagbuchhandlung.

Hecker, K. R., Forbes, M. R., & Léonard, N. J. (2002). Parasitism of damselflies (*Enallagma boreale*) by gregarines: sex biases and relations to adult survivorship. *Can. J. Zool., 80,* 162–168.

Henry, J. E. (1968). *Malameba locustae* and its antibiotic control in grasshopper cultures. *J. Invertebr. Pathol., 11,* 224–233.

Henry, J. E. (1969). *Protozoan and viral pathogens of grasshoppers.* PhD thesis. Bozeman: Montana State University.

Henry, J. E. (1985). *Melanoplus* spp. In P. Singh & R. F. Moore (Eds.), *Handbook of Insect Rearing, Vol. 1* (pp. 451–464). Amsterdam: Elsevier.

Henry, J. E., Wilson, M. C., Oma, E. A., & Fowler, J. L. (1985). Pathogenic microorganisms isolated from West African grasshoppers (Orthoptera: Acrididae). *Trop. Pest Manag, 31,* 192–195.

Hill, D. L. (1972). *The Biochemistry and Physiology of* Tetrahymena. New York: Academic Press.

Hinks, C. F., & Erlandson, M. A. (1994). Rearing grasshoppers and locusts: review, rationale and update. *J. Orth. Res., 3,* 1–10.

Hinks, C. F., & Ewen, A. B. (1986). Pathological effects of the parasite *Malameba locustae* in males of the migratory grasshopper *Melanoplus sanguinipes* and its interaction with the insecticide cypermethrin. *Entomol. Exp. Appl., 42,* 39–44.

Hoare, C. A., & Wallace, F. G. (1966). Developmental stages of trypanosomatid flagellates: a new terminology. *Nature, 212,* 1385–1386.

Huisman, J. M., & Saunders, G. W. (2007). Phylogeny and classification of the algae. In P. M. McCarthy & A. E. Orchard (Eds.), *Algae of Australia: Introduction* (pp. 66–103). Melbourne: CSIRO Publishing.

Jackson, L. L., Baker, G. L., & Henry, J. E. (1968). Effect of *Malamoeba locustae* infection on the egg lipids of the grasshopper *Melanoplus bivittatus*. *J. Insect Physiol., 14,* 1773–1778.

Jerome, C. A., Simon, E. M., & Lynn, D. H. (1996). Description of *Tetrahymena empidokyrea* n.sp., a new species in the *Tetrahymena pyriformis* sibling species complex (Ciliophora, Oligohymenophorea), and an assessment of its phylogenetic position using small-subunit rRNA sequences. *Can. J. Zool., 74,* 1898–1906.

Jouvenaz, D. P., & Anthony, D. W. (1979). *Mattesia geminata* sp. n. (Neogregarinida: Ophrocystidae) a parasite of the tropical fire ant, *Solenopsis geminata* (Fabricius). *J. Protozool., 26,* 354–356.

Kalavati, C. (1977). Morphology and life-cycle of a new adeleid coccidian, *Ganapatiella odontotermi* n.gen. n.sp. from the adipose tissue of *Odontotermes obesus*. *Arch. Protistenkd., 119,* 217–223.

Keeling, P. J., Burger, G., Durnford, D. G., Lang, B. F., Lee, R. W., Pearlman, R. E., Roger, A. J., & Gray, M. W. (2005). The tree of eukaryotes. *Trends Ecol. Evol., 20,* 670–676.

Keilin, D. (1921a). On a new ciliate: *Lambornella stegomyiae* n. g., n. sp., parasitic in the body-cavity of the larvae of *Stegomyia scutellaris* Walker (Diptera, Nematocera, Culicidae). *Parasitology, 13,* 216–224.

Keilin, D. (1921b). On the life-history of *Helicosporidium parasiticum*, n,g., n.sp., a new type of protist parasitic in the larva of *Dasyhelea obscura* Winn. (Diptera, Ceratopogonidae) and in some other arthropods. *Parasitology, 13,* 97–113.

Kellen, W. R., & Lindegren, J. E. (1973). New host records for *Helicosporidium parasiticum* Keilin. *J. Invertebr. Pathol., 22,* 296–297.

King, R. L., & Taylor, A. B. (1936). *Malpighamoeba locustae* n. sp. (Amoebidae), a protozoan parasitic in the Malpighian tubes of grasshoppers. *Trans. Am. Microsc. Soc., 55,* 6–10.

Kirchhoff, J.-F., & Führer, E. (1990). Experimental analysis of infection and developmental cycle of *Malamoeba scolyti* (Purrini) in

Dryocoetes autographus (Ratz.) [Coleoptera: Scolytidae]. *Entomophaga, 35,* 537–544.

Kirchhoff, J.-F., Wegensteiner, R., Weiser, J., & Führer, E. (2005). Laboratory evaluation of *Malamoeba scolyti* (Rhizopoda, Amoebidae) in different bark beetle hosts (Coleoptera, Scolytidae). *IOBC/ wprs Bull., 28,* 159–164.

Kleespies, R. G., Huger, A. M., Buschinger, A., Nähring, S., & Schumann, R. D. (1997). Studies on the life history of a neogregarine parasite found in *Leptothorax* ants from North America. *Biocontrol Sci. Technol., 7,* 117–129.

de Koning, A. P., & Keeling, P. J. (2004). Nucleus-encoded genes for plastid targeted proteins in *Helicosporidium*: functional diversity of a cryptic plastid in a parasitic alga. *Eukaryot. Cell, 3,* 1198–1205.

de Koning, A. P., & Keeling, P. J. (2006). The complete plastid genome sequence of the parasitic green alga *Helicosporidium* sp. is highly reduced and structured. *BMC Biol., 4,* 12.

Kopecna, J., Jirku, M., Obornik, M., Tokarev, Y. S., Lukes, J., & Modry, D. (2006). Phylogenetic analysis of coccidian parasites from invertebrates: search for missing links. *Protist, 157,* 173–183.

Kumano, N., Iwata, N., Kuriwada, T., Shiromoto, K., Haraguchi, D., Yasunaga-Aoki, C., & Kohama, T. (2010). The neogregarine protozoan *Farinocystis* sp. reduces longevity and fecundity in the West Indian sweet potato weevil, *Euscepes postfasciatus* (Fairmaire). *J. Invertebr. Pathol., 105,* 298–304.

Lange, C. E. (2002). La amebiasis debilitativa de los ortópteros y su potencial para el control biológico de acridios (Orthoptera: Acridoidea) en la Argentina. *Rev. Investig. Agrop., 31,* 25–38.

Lange, C. E. (2004). Presencia de *Malameba locustae* (Protozoa: Rhizopoda) en acridios (Orhtoptera: Acrididae) de la provincia de Misiones, Argentina. *Rev. Soc. Entomol. Arg., 63,* 55–57.

Lange, C. E., & Wittenstein, E. (2001). An *Adelina* sp. (Apicomplexa: Coccidia) found in *Anurogryllus muticus* (De Geer) (Orthoptera: Gryllidae). *J. Invertebr. Pathol., 77,* 83–84.

Lange, C. E., & Wittenstein, E. (2002). The life cycle of *Gregarina ronderosi* n. sp. (Apicomplexa: Gregarinidae) in the argentine grasshopper *Dichroplus elongatus* (Orthoptera: Acrididae). *J. Invertebr. Pathol., 79,* 27–36.

Lantova, L., Ghosh, K., Svobodova, M., Braig, H. R., Rowton, E., Weina, P., Volf, P., & Votýpka, J. (2010). The life cycle and host specificity of *Psychodiella sergenti* n. sp. and *Ps. tobbi* n. sp. (Protozoa: Apicomplexa) in sand flies *Phlebotomus sergenti* and *Ph. tobbi* (Diptera: Psychodidae). *J. Invertebr. Pathol., 105,* 182–189.

Larsson, J. I. R. (1991). On the cytology and fine structure of the neogregarine *Syncystis aeshnae* Tuzet and Manier, 1953 (Apicomplexa, Syncystidae). *J. Protozool., 38,* 383–392.

Larsson, J. I. R., Bach de Roca, C., & Gaju-Ricart, M. (1992). Fine structure of an amoeba of the genus *Vahlkampfia* (Rhizopoda, Vahlkampfiidae), a parasite of the gut epithelium of the bristletail *Promesomachilis hispanica* (Microcoryphia, Machilidae). *J. Invertebr. Pathol., 59,* 81–89.

Larsson, R. (1976). Insect pathological investigations on Swedish Thysanura I. Observations on *Malamoeba locustae* (Protozoa, Amoebidae) from *Lepisma saccharina* (Thysanura, Lepismatidae). *J. Invertebr. Pathol., 28,* 43–46.

Leedale, G. F., & Vickerman, K. (2000). Phylum Euglenozoa. In J. J. Lee, G. F. Leedale & P. Bradbury (Eds.), (2nd ed.). *The Illustrated Guide to the Protozoa, Vol. II* (pp. 1135–1136) Lawrence: Society of Protozoologists.

Leibenguth, F. (1972). Development of *Mattesia dispora* in Habrobracon juglandis. *Z. Parasitenk., 38,* 162–173.

Leong, K. L. H., Kaya, H. K., Yoshimura, M. A., & Frey, D. F. (1992). The occurrence and effect of a protozoan parasite, *Ophryocystis elektroscirrha* (Neogregarinida, Ophryocystidae) on overwintering monarch butterflies, *Danaus plexippus* (Lepidoptera, Danaidae) from two California winter sites. *Ecol. Entomol., 17,* 338–342.

Levine, N. D. (1970). Taxonomy of the Sporozoa. *J. Parasitol., 56,* 208–209.

Levine, N. D. (1971). Uniform terminology for the protozoan phylum Apicomplexa. *J. Protozool., 18,* 352–355.

Levine, N. D. (1988a). Progress in taxonomy of the Apicomplexan Protozoa. *J. Protozool., 35,* 518–520.

Levine, N. D. (1988b). *The Protozoan Phylum Apicomplexa.* Boca Raton: CRC Press.

Lindegren, J. E., & Hoffman, D. F. (1976). Ultrastructure of some developmental stages of *Helicosporidium* sp. in the navel orangeworm *Paramyelois transitella. J. Invertebr. Pathol., 27,* 105–113.

Lindsey, E., & Altizer, S. (2009). Sex differences in immune defenses and response to parasitism in monarch butterflies. *Evol. Ecol., 23,* 607–620.

Lipa, J. J. (1982). An amoeba (*Malamoeba locustae* King et Taylor) and an eugregarine (*Gregarina garnhami* Canning) parasitizing in the desert locust (*Schistocerca gregaria* L.) in the Poznan zoological garden. *Wiad. Parazytol., 3/4,* 489–492.

Liu, T. P. (1985a). Scanning electron microscopy of developmental stages of *Malpighamoeba mellificae* Prell in the honey bee. *J. Protozool., 32,* 139–144.

Liu, T. P. (1985b). Scanning electron microscope observations on the pathological changes of Malpighian tubules in the worker honeybee, *Apis mellifera*, infected by *Malpighamoeba mellificae. J. Invertebr. Pathol., 46,* 125–132.

Lomer, C. J., Bateman, R. P., De Groote, H., Douro-Kpindou, O. K., Kooyman, C., Langewald, J., Ouambama, Z., Peveling, R., & Thomas, M. (1999). Development of strategies for the incorporation of biological pesticides into the integrated management of locusts and grasshoppers. *Agric. For. Entomol., 1,* 71–88.

Lopes, R. B., & Alves, S. B. (2005). Effect of *Gregarina* sp. parasitism on the susceptibility of *Blatella germanica* to some control agents. *J. Invertebr. Pathol., 88,* 261–264.

Lord, J. C. (2003). *Mattesia oryzaephili* (Neogregarinorida: Lipotrophidae), a pathogen of stored-grain insects: virulence, host range and comparison with *Mattesia dispora. Biocontrol Sci. Technol., 13,* 589–598.

Lord, J. C. (2006). Interaction of *Mattesia oryzaephili* (Neogregarinorida: Lipotrophidae) with *Cephalonomia* spp. (Hymenoptera: Bethylidae) and their hosts *Cryptolestes ferrugineus* (Coleoptera: Laemophloeidae) and *Oryzaephilus surinamensis* (Coleoptera: Silvanidae). *Biol. Control, 37,* 167–172.

Loschiavo, S. R. (1969). A coccidian pathogen of the dermestid *Trogoderma parabile. J. Invertebr. Pathol., 14,* 89–92.

Ludwig, F. W. (1947). Studies on the protozoan fauna of the larvae of the crane-fly, *Tipula abdominalis*. II. The life history of *Ithania wenrichi* n. gen., n. sp., a coccidian, found in the caeca and mid-gut, and a diagnosis of Ithaniinae, n. subfamily. *Trans Am. Microsc. Soc., 66,* 22–33.

Lukes, J. (2009). *Herpetomonas* Kent 1880. In *The Tree of Life Web Project*. http://tolweb.org

Lynn, D. H. (2008). *The Ciliated Protozoa: Characterization, Classification, and Guide to the Literature* (3rd ed.). New York: Springer.

Lynn, D. H., Molloy, D., & LeBrun, R. (1981). *Tetrahymena rotunda* n.sp. (Hymenostomatida: Tetrahymenidae), a ciliate parasite of the hemolymph of *Simulium* (Diptera: Simuliidae). *Trans. Am. Microsc. Soc., 100*, 134–141.

Maddox, J. V. (1987). Protozoan diseases. In J. R. Fuxa & Y. Tanada (Eds.), *Epizootiology of Insect Diseases* (pp. 417–452). New York: Wiley-Interscience.

Malone, L. A., & Dhana, S. (1988). Life cycle and ultrastructure of *Adelina tenebrionis* (Sporozoea: Adeleidae) from *Heteronychus arator* (Coleoptera: Scarabaeidae). *Parasitol. Res., 74*, 201–207.

Manasherob, R., Ben-Dov, E., Zaritsky, A., & Barak, Z. (1994). Protozoan-enhanced toxicity of *Bacillus thuringiensis* var. *israelensis* δ-endotoxin against *Aedes aegypti* larvae. *J. Invertebr. Pathol., 63*, 244–248.

Manasherob, R., Ben-Dov, E., Margalit, J., Zaritsky, A., & Barak, Z. (1996). Raising activity of *Bacillus thuringiensis* var. *israelensis* against *Anopheles stephensi* larvae by encapsulation in *Tetrahymena pyriformis* (Hymenostomatida: Tetrahymenidae). *J. Am. Mosq. Control Assoc., 12*, 627–631.

Manasherob, R., Ben-Dov, E., Zaritsky, A., & Barak, Z. (1998). Germination, growth, and sporulation of *Bacillus thuringiensis* subsp. *israelensis* in excreted food vacuoles of the protozoan *Tetrahymena pyriformis*. *Appl. Environ. Microbiol., 64*, 1750–1758.

Marín, C., Fabre, S., Sánchez-Moreno, M., & Dollet, M. (2007). *Herpetomonas* spp. isolated from tomato fruits (*Lycopersicon esculentum*) in southern Spain. *Exp. Parasitol., 116*, 88–90.

Martoja, M. R. (1969). Données histopathologiques sur une amibiase d'Acridiens. *C.R. Acad. Sci. Paris, 268*, 2442–2445.

Marzke, F. O., & Dicke, R. J. (1958). Disease-producing protozoa in species of *Trogoderma*. *J. Econ. Entomol., 51*, 916–917.

Massen, M. H. (1916). Uber Bienenkrankheiten. *Mitt. Kas. Biol. Land Forstwirtschaft, 16*, 51–58.

McLaughlin, R. E., & Adams, C. H. (1966). Infection of *Bracon mellitor* (Hymenoptera: Braconidae) by *Mattesia grandis* (Protozoa: Neogregarinida). *Ann. Entomol. Soc. Am., 59*, 800–802.

McLaughlin, R. E., & Bell, M. R. (1970). Mass production *in vivo* of two protozoan pathogens, *Mattesia grandis* and *Glugea gasti*, of the boll weevil, *Anthonomus grandis*. *J. Invertebr. Pathol., 16*, 84–88.

McLaughlin, R. E., & Myers, J. (1970). *Ophryocystis elektroscirrha* sp. n., a neogregarine pathogen of the monarch butterfly *Danaus plexippus* (L.) and the Florida queen butterfly *D. gilippus berenice* Cramer. *J. Protozool., 17*, 300–305.

McLaughlin, R. E., Cleveland, T. C., Daum, R. J., & Bell, M. R. (1969). Development of the bait principle for boll weevil control. IV. Field tests with a bait containing a feeding stimulant and the sporozoans *Glugea gasti* and *Mattesia grandis*. *J. Invertebr. Pathol., 13*, 429–441.

Mercer, D. R., & Anderson, J. R. (1994). Tannins in treehole habitats and their effects on *Aedes sierrensis* (Diptera: Culicidae) production and parasitism by *Lambornella clarki* (Ciliophora: Tetrahymenidae). *J. Med. Entomol., 31*, 159–167.

Mercier, L., & Poisson, R. (1923). Un cas de parasitisme accidental d'une Nepe par un Infusoire. *C.R. hebd. Séanc. Acad. Sci. Paris, 176*, 1838–1841.

Merritt, C. M., Thomas, G. M., & Christensen, J. (1975). A natural epizootic of a coccidian in a population of the Egyptian alfalfa weevil, *Hypera brunneipennis* and the alfalfa weevil,. *H. postica*. *J. Invertebr. Pathol., 26*, 413–414.

Merzlyak, E., Yurchenko, V., Kolesnikov, A. A., Alexandrov, K., Podlipaev, S. A., & Maslov, D. A. (2001). Diversity and phylogeny of insect trypanosomatids based on small subunit rRNA genes: polyphyly of *Leptomonas* and *Blastocrithidia*. *J. Eukaryot. Microbiol., 48*, 161–169.

Mignot, J. P., & Raikov, I. B. (1992). Evidence for meiosis in the testate amoeba. *Arvella*. *J. Protozool., 39*, 287–289.

Minchin, E. A. (1910). On some parasites observed in the rat flea (*Ceratophyllus fasciatus*). *Festschr. 60 Geburstag R. Hertwigs, 1*, 289–302.

Moore, D., Lord, J. C., & Smith, S. (2000). Pathogens. In B. Subramanyan & D. W. Hagstrum (Eds.), *Alternatives to Pesticides in Stored-Product IPM* (pp. 193–227). Dordrecht: Kluwer Academic Publishers.

Nanney, D. L., Park, C., Preparata, R., & Simon, E. M. (1998). Comparison of sequence differences in a variable 23S rRNA domain among sets of cryptic species of ciliated protozoa. *J. Eukaryot. Microbiol., 45*, 91–100.

Narasimhamurti, C. C., & Ahamed, S. N. (1980). *Malamoeba indica* n. sp. from the Malpighian tubules of *Poecilocera picta*. *Proc. Indian Acad. Sci., 89*, 141–145.

Odindo, M. O. (1991). Potential of microorganisms for locust and grasshopper control. *Insect Sci. Appl., 12*, 717–722.

Onstad, D. W., Fuxa, J. R., Humber, R. A., Oestergaard, J., Shapiro-Ilan, D. I., Gouli, V. V., Anderson, R. S., Andreadis, T. G., & Lacey, L. A. An Abridged Glossary of Terms Used in Invertebrate Pathology, (3rd ed.) Society for Invertebrate Pathology. http://sipweb.org/glossary.

Papillion, M., & Cassier, P. (1978). Morphological and physiological disturbances caused by the presence of the protozoan parasite *Malameba locustae* (K. & T.) in *Schistocerca gregaria* (Forsk.). *Acrida, 7*, 101–114.

Park, T., & Frank, M. B. (1950). The population history of *Tribolium* free of sporozoan infection. *J. Anim. Ecol., 19*, 95–105.

Passos, R. A., & Tadei, W. P. (2008). Parasitism of *Ascogregarina taiwanensis* and *Ascogregarina culicis* (Apicomplexa: Lecudinidae) in larvae of *Aedes albopictus* and *Aedes aegypti* (Diptera: Culicidae) from Manaus, Amazon region. *Brazil. J. Invertebr. Pathol., 97*, 230–236.

Patterson, D. J., Simpson, A. G. B., & Rogerson, A. (2000). Amoeba of uncertain affinities. In J. J. Lee, G. F. Leedale & P. Bradbury (Eds.), (2nd ed.). *The Illustrated Guide to the Protozoa, Vol. II* (pp. 804–827) Lawrence: Society of Protozoologists.

Pekkarinen, M. (1993). Bucephalid trematode sporocysts in brackish-water *Mytilus edulis*, new host of a *Helicosporidium* sp. (Protozoa: Helicosporida). *J. Invertebr. Pathol., 61*, 214–216.

Pereira, R. M., Williams, D. F., Becnel, J. J., & Oi, D. H. (2002). Yellowhead disease caused by a newly discovered *Mattesia* sp. in populations of the red imported fire ant, *Solenopsis invicta*. *J. Invertebr. Pathol., 81*, 45–48.

Perkins, F. O. (1991). "Sporozoa": Apicomplexa, Microsporidia, Haplosporidia, Paramyxea, Myxosporidia, and Actinosporidia. In F. W. Harrison & J. O. Corliss (Eds.), *Microscopic Anatomy of Invertebrates, Vol. 1* (pp. 261–331). New York: Wiley-Liss, Protozoa.

Perkins, F. O., Barta, J. R., Clopton, R. E., Peirce, M. A., & Upton, S. J. (2000). Phylum Apicomplexa. In J. J. Lee, G. F. Leedale &

P. Bradbury (Eds.), (2nd ed.). *The Illustrated Guide to the Protozoa*, *Vol. 1* (pp. 190–369). Lawrence: Society of Protozoologists.

Plischuk, S., & Lange, C. E. (2009). Invasive *Bombus terrestris* (Hymenoptera: Apidae) parasitized by a flagellate (Euglenozoa: Kinetoplastea) and a neogregarine (Apicomplexa: Neogregarinorida). *J. Invertebr. Pathol., 102*, 263–265.

Podlipaev, S. A., & Naumov, A. D. (2000). Colonies of trypanosomatids on agar plates: the tool for differentiation of species and isolates. *Protistology, 1*, 113–119.

Prell, H. (1926). Beitrage sur kenntnis der Amoebenseuche der erwachsenen Honigbieue. *Archiv. Bienenkd, 7*, 113–121.

Prinsloo, H. E. (1960). Parasitic microorganisms of the brown locust, *Locustana pardalina* (Walk.). *S. Afr. J. Agric. Sci., 3*, 551–560.

Proux, J. (1991). Lack of responsiveness of Malpighian tubules to the AVP-like insect diuretic hormone on migratory locusts infected with the protozoan *Malameba locustae. J. Invertebr. Pathol., 58*, 353–361.

Purrini, K. (1976a). Zwei Schizogregarinen-Arten (Protozoa, Sporozoa) bei vorratsschadlichen Insekten in Jugoslawishen Muhlen. *Anz. Schadlingskd. Pfl., 49*, 83–85.

Purrini, K. (1976b). *Adelina tribolii* Bhatia und *A. mesnili* Pérez (Sporozoa, Coccidia) als Krankheitserreger bei vorratsschädlichen Insekten im Gebiet von Kosova. *Jugoslawien. Anz. Schadlingskd. Pflanzenschutz Umweltschutz, 49*, 83–85.

Purrini, K. (1980). *Malamoeba scolyti* n. sp. (Amoebidae, Rhizopoda, Protozoa) parasitizing the bark beetles *Dryocoetes autographus* Ratz., and *Hyturgops palliates* Gyll. (Scolytidae, Coleoptera). *Arch. Protistenk., 123*, 358–366.

Purrini, K. (1984). Light and electron microscope studies on *Helicosporidium* sp. parasitizing oribatid mites (Oribatei, Acarina) and Collembola (Apterygota, Insecta) in forest soils. *J. Invertebr. Pathol., 44*, 18–27.

Purrini, K., & Žižka, Z. (1983). More on the life cycle of *Malamoeba scolyti* (Amoebidae: Sarcomastigophora) parasitizing the bark beetle *Dryocoetes autographus* (Scolytidae, Coleoptera). *J. Invertebr. Pathol., 42*, 96–105.

Rabindra, R. J., Balasubramanian, M., & Jayaraj, S. (1981). The effects of *Farinocystis tribolii* on the growth and development of the flour beetle *Tribolium castaneum. J. Invertebr. Pathol., 38*, 345–351.

Rabindra, R. J., Balasubramanian, M., & Jayaraj, S. (1983). The susceptibility of *Tribolium castaneum* (Col.: Tenebrionidae) to *Farinocystis tribolii* (Protozoa: Schizogregarinida). *Entomophaga, 28*, 73–81.

Rabindra, R. J., Balasubramanian, M., & Jayaraj, S. (1985). Effect of *Farinocystis tribolii* on the respiration of *Tribolium castaneum* larvae. *J. Invertebr. Pathol., 45*, 115–116.

Rabindra, R. J., Jayaraj, S., & Balasubramanian, M. (1988). *Farinocystis tribolii*-induced susceptibility to some insecticides in *Tribolium castaneum* larvae. *J. Invertebr. Pathol., 52*, 389–392.

Raina, S. K. (1992). ICIPE, development of a biocontrol strategy for the management of the desert locust, *Schistocerca gregaria*. In C. J. Lomer & C. Prior (Eds.), *Biological Control of Locusts and Grasshoppers* (pp. 54–56). Wallingford: CABI.

Regali-Seleghim, M. H., & Godinho, M. J. L. (2004). Peritrich epibiont protozoans in the zooplankton of a subtropical shallow aquatic ecosystem (Monjolinho Reservoir, São Carlos, Brazil). *J. Plankton Res., 26*, 501–508.

Reyes-Villanueva, F., Becnel, J. J., & Butler, J. F. (2003). Susceptibility of *Aedes aegypti* and *Aedes albopictus* larvae to *Ascogregarina culicis*

and *Ascogregarina taiwanensis* (Apicomplexa: Lecudinidae) from Florida. *J. Invertebr. Pathol., 84*, 47–53.

Roberts, D. W., Fuxa, J. R., Gaugler, R., Goettel, M., Jaques, R., & Maddox, J. V. (1991). Use of pathogens in insect control. In D. Pimentel (Ed.), (2nd ed). *CRC Handbook of Pest Management in Agriculture, Vol. 2* (pp. 243–278). Boca Raton: CRC Press.

Rodríguez, Y., Omoto, C. K., & Gomulkiewicz, R. (2007). Individual and population effects of eugregarine, *Gregarina niphandrodes* (Eugregarinida: Gregarinidae), on *Tenebrio molitor* (Coleoptera: Tenebrionidae). *Environ. Entomol., 36*, 689–693.

Rogerson, A., & Patterson, J. J. (2000). The naked ramicristate amoebae (Gymnamoebae). In J. J. Lee, G. F. Leedale & P. Bradbury (Eds.), (2nd ed.). *The Illustrated Guide to the Protozoa, Vol. II* (pp. 1023–1053). Lawrence: Society of Protozoologists.

de Roode, J. C., Gold, L. R., & Altizer, S. (2007). Virulence determinants in a natural butterfly–parasite system. *Parasitology, 134*, 657–668.

Sayre, R. M., & Clark, T. B. (1978). *Daphnia magna* (Cladocera: Chydoroidea), a new host of a *Helicosporidium* sp. (Protozoa: Helicosporida). *J. Invertebr. Pathol., 31*, 260–261.

Schaub, G. A. (1994). Pathogenicity of trypanosomatids on insects. *Parasitol. Today, 10*, 463–468.

Schaub, G. A. (2009). Interactions of trypanosomatids and triatomines. *Adv. Insect. Physiol., 37*, 177–242.

Schmid-Hempel, R., & Tognazzo, M. (2010). Molecular divergence defines two distinct lineages of *Crithidia bombi* (Trypanosomatidae), parasites of bumblebees. *J. Eukaryot. Microbiol., 57*, 337–345.

Schrevel, J., & Phillipe, M. (1993). The gregarines. In J. P. Kreir (Ed.), *Parasitic Protozoa, Vol. 4* (pp. 133–245). San Diego: Academic Press.

Schuster, F. L. (1990). Phylum Rhizopoda. In L. Margulis, J. O. Corliss, M. Melkonian & D. J. Chapman (Eds.), *Handbook of Protoctista* (pp. 3–18). Boston: Jones and Bartlett.

Schwalbe, C., Burkholder, W., & Boush, G. (1974). *Mattesia trogodermae* infection rates as influenced by mode of transmission, dosage and host species. *J. Stored Prod. Res., 10*, 161–166.

Schwantes, U., & Eichelberg, D. (1984). Elektronenmikroskopische Untersuchungen zum Parasitenbefall der Malpighischen Gefasse von *Apis mellifera* durch *Malpigamoeba mellificae. Apidologie, 15*, 435–450.

Seif, A. I., & Rifaat, M. M. (2001). Laboratory evaluation of a *Helicosporidium* sp. (Protozoa: Helicosporida) as an agent for the microbial control of mosquitoes. *J. Egypt Soc. Parasitol., 31*, 21–35.

Shapas, T. J., Burkholder, W. E., & Boush, G. M. (1977). Population suppression of *Trogoderma glabrum* by using pheromone luring for protozoan pathogen dissemination. *J. Econ. Entomol., 70*, 469–474.

Siegel, J. P., Novak, R. J., & Maddox, J. V. (1992). Effects of *Ascogregarina barretti* (Eugregarinida, Lecudinidae) infection on *Aedes triseriatus* (Diptera, Culicidae) in Illinois. *J. Med. Entomol., 29*, 968–973.

Simmons, L. W. (1993). Some constraints on reproduction for male bushcrickets, *Requena verticalis* (Orthoptera: Tettigonidae): diet, size and parasite load. *Behav. Ecol. Sociobiol., 32*, 135–139.

Simon, E. M., Nanney, D. L., & Doerder, F. P. (2008). The *Tetrahymena pyriformis* complex of cryptic species. *Biodivers. Conserv., 17*, 365–380.

Sokolova, Y. Y., Butaeva, F. B., & Dolgikh, V. V. (1999). Light and electron microscopic observations on life cycle stages of *Adelina grylli* Butaeva 1996 (Sporozoa, Adeleidae) from fat body of the cricket *Gryllus bimaculatus. Protistology, 1*, 4–42.

Steinhaus, E. A. (1947). A coccidian parasite of *Ephestia kühniella* Keller and *Plodia interpunctella* (Hbn.) (Lepidoptera, Phycitidae). *J. Parasitol., 33*, 29–32.

Steinkraus, D. C., Brooks, W. M., & Geden, C. G. (1992). Discovery of the neogregarine *Farinocystis tribolii* and an eugregarine in the lesser mealworm, *Alphitobius diaperinus. J. Invertebr. Pathol., 59*, 203–205.

Streett, D. A., & Henry, J. E. (1990). Microbial control of locusts and grasshoppers in the semi-arid tropics. *Bol. San. Veg. Plagas, 20*, 21–27.

Strüder-Kypke, M. C., Wright, A.-D. G., Jerome, C. A., & Lynn, D. H. (2001). Parallel evolution of histophagy in ciliates of the genus *Tetrahymena. BMC Evol. Biol., 1*, 5.

Sulaiman, I. (1992). Infectivity and pathogenicity of *Ascogregarina culicis* (Eugregarinida, Lecudinidae) to *Aedes aegypti* (Diptera, Culicidae). *J. Med. Entomol., 29*, 1–4.

Svobodová, M., Zídková, L., Čepička, I., Oborník, M., Lukeš, J., & Votýpka, J. (2007). *Sergeia podlipaevi* gen. nov., sp. nov. (Trypanosomatidae, Kinetoplastida), a parasite of biting midges (Ceratopogonidae, Diptera). *Int. J. Syst. Evol. Microbiol., 57*, 423–432.

Tanada, Y., & Kaya, H. K. (1993). *Insect Pathology.* San Diego: Academic Press.

Tartar, A., & Boucias, D. G. (2004). The non-photosynthetic, pathogenic green alga *Helicosporidium* sp. has retained a modified, functional plastid genome. *FEMS Microbiol. Lett., 233*, 153–157.

Tartar, A., Boucias, D. G., Adams, B. J., & Becnel, J. J. (2002). Phylogenetic analysis identifies the invertebrate pathogen *Helicosporidium* sp. as a green algae (Chlorophyta). *Int. J. Syst. Evol. Microbiol., 52*, 273–279.

Tartar, A., Boucias, D. G., Becnel, J. J., & Adams, B. J. (2003). Comparison of plastid 16S rDNA genes from *Helicosporidium* spp.: evidence supporting the reclassification of Helicosporidia as green algae (Chlorophyta). *Int. J. Syst. Evol. Microbiol., 53*, 1719–1723.

Taylor, A. B., & King, R. L. (1937). Further studies on the parasitic amebae found in grasshoppers. *Trans. Am. Microsc. Soc., 56*, 172–176.

Templeton, T. J., Enomoto, S., Chen, W.-J., Huang, C.-G., Lancto, C. A., Abrahamsen, M. S., & Zhu, G. (2010). A genome-sequence survey for *Ascogregarina taiwanensis* supports evolutionary affiliation but metabolic diversity between a gregarine and *Cryptosporidium. Mol. Biol. Evol., 27*, 235–248.

Thompson, J. C. (1958). Experimental infections of various animals with strains of the genus *Tetrahymena. J. Protozool., 5*, 203–205.

Thompson, A. C., & McLaughlin, R. E. (1977). Comparison of the lipids and fatty acids of *Mattesia grandis* and the fat body of the host, *Anthonomus grandis. J. Invertebr. Pathol., 30*, 108–109.

Tseng, M. (2007). Ascogregarine parasites as possible biocontrol agents of mosquitoes. *J. Am. Mosq. Control Assoc., 23*, 30–34.

Tsubaki, Y., & Hooper, R. E. (2004). Effects of eugregarine parasites on adult longevity in the polymorphic damselfly *Mnais costalis* Selys. *Ecol. Entomol., 29*, 361–366.

Valigurová, A., & Koudela, B. (2006). Ultrastructural study of developmental stages of *Mattesia dispora* (Neogregarinorida: Lipotrophidae), a parasite of the flour moth *Ephestia kuehniella* (Lepidoptera). *Eur. J. Protistol., 42*, 313–323.

Valles, S. M., & Pereira, R. M. (2003). Use of ribosomal DNA sequence data to characterize and detect a neogregarine pathogen of *Solenopsis*

invicta (Hymenoptera: Formicidae). *J. Invertebr. Pathol., 84*, 114–118.

Vavra, J., & McLaughlin, R. E. (1970). The fine structure of some developmental stages of *Mattesia grandis* McLaughlin (Sporozoa, Neogregarinida), a parasite of the boll weevil *Anthonomus grandis* Boheman. *J. Protozool., 17*, 483–496.

Vázquezdelara-Cisneros, L. G., & Arroyo-Begovich, A. (1984). Induction of encystations of *Entamoeba invadens* by removal of glucose from the culture medium. *J. Parasitol, 70*, 629–633.

Venter, I. G. (1966). Egg development in the brown locust, *Locustana pardalina* (Walker), with special reference to the effect of infestation by *Malameba locustae. S. Afr. J. Agric. Sci., 9*, 429–433.

Vickerman, K. (2000). Order Kinetoplastea. In J. J. Lee, G. F. Leedale & P. Bradbury (Eds.), (2nd ed.). *The Illustrated Guide to the Protozoa, Vol. II* (pp. 1159–1185). Lawrence: Society of Protozoologists.

Vivier, E., & Desportes, I. (1990). Phylum Apicomplexa. In L. Margulis, J. O. Corliss, M. Melkonian & D. J. Chapman (Eds.), *Handbook of Protoctista* (pp. 549–573). Boston: Jones and Bartlett.

Votýpka, J., Lantová, L., Ghosh, K., Braig, H., & Volf, P. (2009). Molecular characterization of gregarines from sand flies (Diptera: Psychodidae) and description of *Psychodiella* n. g. (Apicomplexa: Gregarinida). *J. Eukaryot. Microbiol., 56*, 583–588.

Vythilingam, I., Mahadevan, S., Ong, K. K., Abdullah, G., & Ong, Y. F. (1996). Distribution of *Lambornella stegomyiae* in Malaysia and its potential for the control of mosquitoes of public health importance. *J. Vector Ecol., 21*, 89–93.

Washburn, J. O., & Anderson, J. R. (1986). Distribution of *Lambornella clarki* (Ciliophora: Tetrahymenidae) and other mosquito parasites in California treeholes. *J. Invertebr. Pathol., 48*, 296–309.

Washburn, J. O., Egerter, D. E., Anderson, J. R., & Saunders, G. A. (1988a). Density reduction in larval mosquito (Diptera: Culicidae) populations by interactions between a parasitic ciliate (Ciliophora: Tetrahymenidae) and an opportunistic fungal (Oomycetes: Pythiaceae) parasite. *J. Med. Entomol., 25*, 307–314.

Washburn, J. O., Gross, M. E., Mercer, D. R., & Anderson, J. R. (1988b). Predator-induced trophic shift of a free-living ciliate: parasitism of mosquito larvae by their prey. *Science, 240*, 1193–1195.

Washburn, J. O., Mercer, D. R., & Anderson, J. R. (1991). Regulatory role of parasites: impact on host population shifts with resource availability. *Science, 253*, 185–188.

Weiser, J. (1953). Schizogregariny z hmyzu škodícího zásobám mouky. *Vest. Cesk. Spol. Zool., 17*, 199–212.

Weiser, J. (1955). A new classification of the Schizogregarina. *J. Protozool., 2*, 6–12.

Weiser, J. (1963). Sporozoan infections. In E. A. Steinhaus (Ed.), *Insect Pathology: An Advanced Treatise, Vol. 2* (pp. 291–334). New York: Academic Press.

Weiser, J. (1970). *Helicosporidium parasiticum* Keilin infection in the caterpillar of a hepialid moth in Argentina. *J. Protozool., 17*, 440–445.

Weiser, J., & Beard, R. L. (1959). *Adelina sericesthis* n. sp., a new coccidian parasite of scarabaeid larvae. *J. Insect. Pathol., 1*, 99–106.

Yaman, M., & Radek, R. (2005). *Helicosporidium* infection of the great European spruce bark beetle, *Dendroctonus micans* (Coleoptera: Scolytidae). *Eur. J. Protistol., 41*, 203–207.

Yarwood, E. A. (1937). The life cycle of *Adelina cryptocerci* sp. nov., a coccidian parasite of the roach *Cryptocercus punctulatus. Parasitology, 29*, 370–390.

Yee, W. L. (1995). Behaviors associated with egg and parasite deposition by gravid and *Lambornella clarki*-infected *Aedes sierrensis*. *J. Parasitol., 81*, 694—697.

Yee, W. L., & Anderson, J. R. (1995). Free flight of *Lambornella clarki*-infected, blood-fed, and gravid *Aedes sierrensis* (Diptera: Culicidae). *J. Med. Entomol., 32*, 407—412.

Yurchenko, V. Y., Lukes, J., Tesarová, M., Jirku, M., & Maslov, D. A. (2008). Morphological discordance of the new trypanosomatid species phylogenetically associated with genus *Crithidia. Protist., 159*, 99—114.

Zaritsky, A., Zalkinder, V., Ben-Dov, E., & Barak, Z. (1991). Bioencapsulation and delivery to mosquito larvae of *Bacillus thuringiensis* H-14 toxicity by *Tetrahymena pyriformis. J. Invertebr. Pathol., 58*, 455—457.

Zídkova, L., Cepicka, I., Votypka, J., & Svobodová, M. (2010). *Herpetomonas trimorpha* sp. nov. (Trypanosomatidae, Kinetoplastida), a parasite of the biting midge *Culicoides truncorum* (Ceratopogonidae, Diptera). *Int. J. Syst. Evol. Microbiol., 60*, 2236—2246.

Žižka, Z. (1972). An electron microscope study of autoinfection in neogregarines (Sporozoa, Neogregarinida). *J. Protozool., 19*, 275—280.

Žižka, Z. (1978). Fine structure of the neogregarine *Farinocystis tribolii* Weiser, 1953. Free gametocytes. *Acta Protozool., 17*, 255—259.

Žižka, Z. (1987). Ultrastructure of trophozoites and cysts of the amoeba *Malamoeba locustae* K. & T., 1936, parasitizing the locust *Locusta migratoria* R. et F. *Acta Protozool., 26*, 285—290.

Chapter 11

Nematode Parasites and Entomopathogens

Edwin E. Lewis* and David J. Clarke†

*University of California, Davis, California, USA, † University College Cork, Cork, Ireland

Chapter Outline

SUMMARY

This chapter covers recent advances in the study of nematodes that are parasites of insects and focuses primarily on those that are entomopathogenic in the genera *Heterorhabditis* and *Steinernema*, characterized by their symbiotic association with the bacteria *Photorhabdus* and *Xenorhabdus*, respectively. These entomopathogenic nematodes are extremely virulent insect pathogens that have been successfully marketed as biocontrol agents for the protection of certain high-value crops from a range of insect pests. Entomopathogenic nematodes are useful models for studying the ecology of soil food webs. The obligate interaction between the bacterium and the nematode has also established entomopathogenic nematodes as a very useful model for the study of the molecular mechanisms underpinning bacteria–host interactions. Although they appear to have very similar lifestyles, the two nematode genera are not phylogenetically close, and recent developments have shown that the *Heterorhabditis*–*Photorhabdus* and *Steinernema*–*Xenorhabdus* associations are very different at the molecular level.

11.1. INTRODUCTION

Nematodes, also known as roundworms, eelworms, or threadworms, comprise a tremendously diverse phylum whose members exploit habitats more varied than any

Insect Pathology. DOI: 10.1016/B978-0-12-384984-7.00011-7

other group of animals except arthropods (Tanada and Kaya, 1993). They are aquatic organisms requiring water for survival for at least part of their life cycle. Although some nematode species can survive anhydrobiotically (i.e., ametabolic life in the absence of water), all nematodes require moisture for reproduction. They occur as free-living animals or as facultative or obligate parasites of plants and animals. Inasmuch as nematodes and insects occur in many of the same habitats, it is not surprising that these animals are often associated with each other.

Nematodes are associated with insects in diverse ways. They occur as direct, or monoxenic, obligate parasites (e.g., Mermithidae, Iotonchiidae), direct pathogens that are symbiotically associated with bacteria (e.g., Steinernematidae, Heterorhabditidae, some Rhabditidae), facultative parasites of insects (Phaenopsitylenchidae), obligate parasites vectored by insects to vertebrate hosts (e.g., Onchocercidae), facultative parasites vectored by insects to plant hosts (e.g., Aphelenchoididae), and as commensals or harmless phoretic associates (e.g., Rhabditidae, Diplogasteridae, Cephalobidae, etc.) (Poinar, 1975). This functional and phylogenetic diversity presents a challenge when preparing a chapter such as this, in that the focus must be narrowed to include only a few examples of these types of relationships.

The biology and impact of three representative species of monoxenous parasites (parasites that complete their life cycle within a single host) of insects are discussed, but this chapter focuses mainly on a group of nematodes that are symbiotically associated with bacteria, known as entomopathogenic nematodes (EPNs). The focus is on this group for two reasons. First, the overwhelming majority of nematodes used as biological control agents of insects are confined to the Steinernematidae and Heterorhabditidae. Second, their biology currently receives a great deal of attention owing to their value as model systems for studying basic processes ranging from the genetics of symbiotic relationships to the spatial ecology of soil ecosystems.

11.2. CLASSIFICATION AND PHYLOGENY

Nematodes belong to the clade of metazoans known as the Ecdysozoa, or animals that molt (Aguinaldo et al., 1997). This group includes nematodes, arthropods, kinorhynchs, nematomorphs, onychophorans, priapulids, and tardigrades. The nematodes associated with insects are diverse phylogenetically, belonging to 13 different suborders of the Nematoda (Blaxter, 2011). Nematode species belonging to five families serve as examples in this chapter.

11.2.1. Insect-parasitic Nematodes

Three families of non-EPNs are included in this chapter: Phaenopsitylenchidae, Mermithidae, and Iotonchiidae. The Mermithidae are phylogenetically located within their own order in the Dorylamia, whereas the other two families are both within the Tylenchida. The order Tylenchida includes an enormously diverse group of nematodes, some of which are parasitic to insects, but most species are plant parasites.

11.2.2. Entomopathogenic Nematodes and Symbiotic Bacteria

The other two nematode families covered in the chapter, which comprise the EPNs (Heterorhabditidae and Steinernematidae), are the focus of this chapter. Blaxter et al. (1998) provide molecular data as to the placement of these families within the Nematoda and this information is summarized by Adams et al. (2006). The Heterorhabditidae are thought to be most closely aligned with a group of parasites of vertebrates, the Strongylida, which as a group shares a common ancestor with a group of free-living, bacterial feeding nematodes. The Steinernematidae are hypothesized to be most closely aligned with the Panagrolaimoidea, of which some are free living and others are associated with insects, and they may also be allied with the Strongyloididae, which are parasites of vertebrates. The Steinernematidae share a clade with nematodes that are fungal feeding and plant parasites as well. Thus, while the Heterorhabditidae are thought to have arisen from a free-living bacterial-feeding ancestor, the phylogenetic background of the Steinernematidae is less easily reconstructed.

The classification of the bacteria and the nematodes is considered together because they are paired in nature. Nematodes in the genus *Steinernema* are associated with bacteria in the genus *Xenorhabdus*, whereas nematodes in the genus *Heterorhabditis* spp. are associated with *Photorhabdus* spp. bacteria. Tables 11.1 and 11.2 list more described species of nematodes than there are of bacteria, a situation that arises for two reasons. With one exception, each species of nematode is associated exclusively with a single species of bacteria, e.g., *Steinernema carpocapsae* is associated with *Xenorhabdus nematophila* (Table 11.1). The exception occurs because *Steinernema anomoli* was synonymized with *S. arenarium*; each former species was associated with a different bacterial species before reclassification. On the other hand, there are several species of bacteria that are associated with more than a single nematode species, e.g., *Xenorhabdus kozodoii* is found in *Steinernema arenarium*, *S. apuliae*, and an as yet undescribed *Steinernema* species (Table 11.1) (Adams et al., 2006). Second, there are several new species of nematode for which the bacteria have not been yet described. Species

TABLE 11.1 List of Described Species of Nematodes in the Family Steinernematidae and their Respective *Xenorhabdus* Bacterial Symbionts

Nematode Species	Bacteria Species
Steinernema abbasi	Unknown
S. aciari	Unknown
S. affine	Unknown
S. anatoliense	Unknown
S. arenarium	*X. hominickii, X. kozodoii*
S. ashiuense [a]	Unknown
S. asiaticum	Unknown
S. apuliae	*X. kozodoii*
S. australe [a]	Unknown
S. backanense [a]	Unknown
S. boemarei [a]	Unknown
S. brazilense [a]	Unknown
S. bicornutum	*X. budapestensis*
S. carpocapsae	*X. nematophila*
S. caudatum	Unknown
S. ceratophorum	Unknown
S. cholashanense [a]	Unknown
S. costaricense [a]	Unknown
S. colombiense [a]	Unknown
S. cubanum	*X. poinari*
S. diaprepesi	*X. doucetiae*
S. eapokense [a]	Unknown
S. everestense [a]	Unknown
S. feltiae	*X. bovienii*
S. glaseri	*X. poinari*
S. guangdongense	Unknown
S. hebeiense [a]	Unknown
S. hermaphroditum	*X. griffiniae*
S. ichnusae [a]	Unknown
S. intermedium	Unknown
S. jollieti	Unknown
S. karii	*X. hominickii*
S. khoisanae [a]	Unknown
S. kraussei	Unknown
S. kushidai	*X. japonica*
S. loci	Unknown
S. leizhouense [a]	Unknown
S. longicaudum	Unknown

TABLE 11.1 List of Described Species of Nematodes in the Family Steinernematidae and their Respective *Xenorhabdus* Bacterial Symbionts—cont'd

Nematode Species	Bacteria Species
S. masoodi [a]	Unknown
S. minutum [a]	Unknown
S. monticolum	*X. hominickii*
S. neocurtillae	Unknown
S. oregonense	Unknown
S. pakistanense	Unknown
S. puertoricense	*X. romanii*
S. pui [a]	Unknown
S. puntauvense [a]	Unknown
S. rarum	*X. szentirmaii*
S. riobrave	*X. cabanillasii*
S. ritteri	Unknown
S. robustispiculum	Unknown
S. sangi	*X. vietnamensis*
S. sasonense [a]	Unknown
S. scapterisci	*X. innexi*
S. scarabaei	*X. koppenhoeferi*
S. seemae [a]	Unknown
S. serratum	*X. ehlersii*
S. siamkayai	*X. stockiae*
S. sichuanense [a]	Unknown
S. silvaticum [a]	Unknown
S. tami	Unknown
S. thanhi [a]	Unknown
S. thermophilum	*X. indica*
S. unicornum [a]	Unknown
S. websteri	Unknown
S. weiseri	Unknown
S. xueshanense	Unknown
S. yirgalemense	Unknown
Steinernema sp.	*X. miraniensis*
Steinernema sp.	*X. mauleonii*
Neosteinernema longicurvicauda	Unknown

Where the nematode has been described, but the bacteria have not been identified, the bacterial species is listed as "Unknown." Where the bacterium has been described, but the nematode remains undescribed, the nematode species is listed as "*Steinernema* sp."
[a]Indicates a nematode species that was not included in Adams et al. (2006).

TABLE 11.2 List of Described Species of Nematodes in the Family Heterorhabditidae and their Respective *Photorhabdus* Bacterial Symbionts

Nematode Species	Bacteria Species/subspecies
Heterorhabditis amazonensis [a]	Unknown
H. bacteriophora	*P. luminescens* subsp. *kayaii*, *P. luminescens* subsp. *laumondii*, *P. luminescens* subsp. caribbeanensis, *P. temperata* subsp. *khanii*, *P. temperata* subsp. *thracensis*
H. baujardi	Unknown
H. brevicaudis	Unknown
H. downesi	*Photorhabdus temperata* subsp. *cinerea*
H. floridensis [a]	Unknown
H. georgiana [a]	*Photorhabdus luminescens* subsp. *akhurstii*
H. gerrardi [a]	*P. asymbiotica kingscliffe*
H. indica	*P. luminescens* subsp. *akhurstii*
H. marelatus	Unknown
H. megidis	*P. temperata* subsp. *cinerea*
H. mexicana	Unknown
H. poinari	Unknown
H. safricana [a]	Unknown
H. sonorensis [a]	Unknown
H. taysearae	Unknown
H. zealandica	*P. temperata* subsp. *tasmaniensis*
None known	*P. asymbiotica*
None known	*P. asymbiotica* subsp. *australis*

Where the nematode has been described, but the bacteria have not been identified, the bacterial species is listed as "Unknown." Where the bacterium has been described, but has never been associated with a *Heterorhabditis* host, "None known" is the designation used for the nematode.
[a]Indicates a nematode species that was not included in Adams et al. (2006).

descriptions of symbiotic bacteria have tended to be slower than the descriptions of nematode species.

Systematic activity in this group is intense, most likely because of the economic importance of the nematodes to agriculture, despite the limited number of systematists with the necessary expertise. A search of the current literature shows that from 2005 to 2010, more than 30 surveys have been conducted for EPNs. It should be remembered that merely isolating an EPN strain will not, in most cases, lead to a publication of the account unless there is more

information included, such as the description of a new species, some special characteristic of the isolate found through experimentation, or something unusual about the habitat from which it was isolated. So, the actual number of new isolates of nematodes that have been collected and are still awaiting identification, or description if they are new species, probably far outnumbers the list of published accounts of successful surveys for the nematodes.

11.3. INSECT-PARASITIC NEMATODES

Tanada and Kaya (1993) provide a broader treatment of nematodes, nematomorphs, and platyhelminths associated with insects. The focus here is on three types of parasitic relationships between nematodes and insects: facultative parasites, monoxenous parasites, and EPNs. For clarification of the terminology, the general life cycle of nematodes includes six stages: the egg, four juvenile stages, and the adult. The transmission stage of the parasite may be any of these stages. When one of the juvenile stages is responsible for transmission, it is usually referred to as the "infective stage juvenile", but sometimes is named the "pre-parasite". Adults can also serve as the infective stage. This terminology is used throughout the chapter.

11.3.1. Facultative Parasite: *Beddingia siricidicola*

A facultative parasite is an organism that can have a parasitic relationship with a host, but can also live and reproduce independently when the host is absent. Nematodes in the genus *Beddingia* (*Deladenus*) (Phaenopsitylenchidae) have a complicated life cycle that includes both parasitic and free-living generations. There are seven species in the genus *Beddingia* (Bedding, 1967; Bedding and Akhurst 1974), although the taxonomic status of the genus is unsettled (Kerry and Hominick, 2002). The life history and biology of *B. siricidicola*, which has been used successfully as a classical biological control agent of the Sirex woodwasp, *Sirex noctilio* (Bedding, 1984), are treated here. The insect, native to Europe, is a devastating pest of several North American species of pine. Large plantations of susceptible North American Monterey pine (*Pinus radiata*) were established in New Zealand, Australia, and elsewhere, and in the early 1900s and 1950s, the woodwasp was inadvertently released in New Zealand and Australia, respectively. A more complete summary of the implementation of this nematode as a biological control agent is detailed in Bedding and Iede (2005) and in Chapter 3.

When the woodwasp is not nearby (not perceived by the nematode), the nematodes have a free-living life cycle with the normal six life stages; the four juvenile stages and the adults feed on the fungus *Amylostereum chailletti*, which occurs in the tree trunk and is the only fungus the nematode consumes (Bedding and Akhurst, 1978). This free-living life cycle can persist for long periods within pines that have been inoculated with the nematode and the fungus by the woodwasp host. The juvenile nematodes develop into pre-parasitic adults when stimulated by elevated carbon dioxide levels and low pH, conditions that occur near woodwasp larvae (Bedding, 1993).

The parasitic life cycle of *B. siricidicola* is in some ways similar to that of *Paraiotonchium autumnale* described below (see Section 11.3.2), especially in that both render the host sterile and that the parasitized host transports the nematodes. Larval woodwasps are invaded by mated, pre-parasitic female nematodes, which penetrate through the insect's cuticle using a stylet. The female increases in size over a few weeks (sometimes by a factor of 1000, depending on the size of the host) by taking in nourishment through its cuticle, and can remain in this condition for months. At about the time the adult wasp emerges from the tree, the female nematode produces thousands of juvenile nematodes that invade the woodwasp ovaries, enter the eggs, and are nemaposited (that is, fourth stage juvenile nematodes from the insect's ovaries are deposited instead of insect eggs; sometimes the nematodes may occur within the egg) into a new host tree by the wasp, thus inoculating a new individual tree with nematodes. If the juvenile nematodes are deposited into a tree not infested with woodwasp larvae the free-living life cycle ensues.

The ability to switch from parasitic to free living is a key to the success of *B. siricidicola* as a biological control agent, in contrast to the life cycle of *P. autumnale*, which does not have a non-parasitic life cycle. If *P. autumnale* juvenile nematodes are deposited into a dung pat without hosts, they will perish. *Beddingia siricidicola* can persist and multiply on a symbiotic fungus they inoculate into the tree while waiting for a host. Another reason for their success is that these nematodes can be cultured *in vivo* by growing the fungus on which they feed, and inoculating it with the nematodes into dying or felled pine trees (Bedding and Akhurst, 1974). However, *B. siricidicola* must pass through a parasitic phase periodically, or they can lose the capability of infecting their host (Bedding and Iede, 2005).

11.3.2. Monoxenous Obligate Parasites: No Symbionts

Monoxenous parasites, as a large and diverse group of nematodes, have equally diverse relationships with their hosts. They are characterized by completing their life cycle within a single host. The life cycles of two nematode species, *Romanomermis culicivorax* and *P. autumnale*, provide examples. These two species were selected as representatives because (1) they are both reasonably well-studied species (*R. culicivorax* is a biological control agent for mosquitoes and *P. autumnale* is a biological control

agent for face flies); and (2) they can serve as models of a very simple life cycle and outcome to the host in the case of *R. culicivorax* and a very complicated life cycle and impact on hosts in the case of *P. autumnale*.

Romanomermis culicivorax: *Obligate, Lethal Parasite*

Mermithids are all obligate, lethal parasites of arthropods (Platzer *et al.*, 2005). *Romanomermis culicivorax* parasitizes aquatic larval mosquitoes and has a relatively simple life cycle (Fig. 11.1). Pre-parasites (second stage juveniles) hatch from eggs that were deposited in the substrate of an aquatic habitat by mated female nematodes and swim to the top of the pool actively searching for larval mosquitoes. The pre-parasites are fairly short lived (less than a day), so they must find and infect a host soon after hatching. They enter the host through the cuticle and begin development within the host hemocoel. After seven to 10 days of taking nourishment from the host through their own cuticle, the host dies and the postparasite nematodes leave the host and continue to develop into adults in the sediment. There, they mate and females oviposit. The eggs hatch in about three weeks at 27°C (Platzer, 1981) and the cycle starts anew.

Romanomermis culicivorax as a parasite has an unusually broad host range, being able to infect more than 90 species of mosquitoes (Peng *et al.*, 1992). This nematode and many of its congeners have been used successfully in inoculative biological control programs against several different groups of mosquito (Platzer *et al.*, 2005). The parasites are reared in laboratories and released into a mosquito habitat with the intent of establishing populations of parasites that suppress mosquito populations over the long term. As they must be produced *in vivo*, their use is very costly because a colony of live mosquitoes must be maintained to produce the nematodes. Thus, they are not economically competitive with products based on *Bacillus*

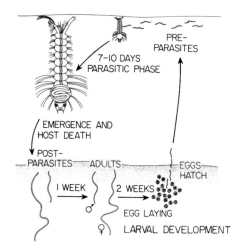

FIGURE 11.1 Life cycle of *Romanomermis culicivorax*. (From Tanada and Kaya, 1993.)

thuringiensis subsp. *israelensis*, which is an effective and cheaply produced pathogen of mosquitoes (Kerry and Hominick, 2002).

Paraiotonchium autumnale: Obligate, Non-lethal Parasite

The family Iotonchiidae includes species, such as *P. autumnale*, that have quite complex life cycles and they often render their insect hosts sterile rather than killing them outright. Thus, they are obligate parasites, but are not immediately lethal to their hosts. *Paraiotonchium autumnale* has a very narrow host range that consists only of the face fly, *Musca autumnalis*, but there are closely related species that infect other flies in the genus *Musca*.

The nematode's life cycle was originally described by Nickle (1967), and is closely intertwined with the host life cycle. Female flies mate and lay eggs in dung pats shortly after they are defecated by cattle. The eggs hatch and the larvae progress through three instars. The third instars leave the dung and pupate in the soil beneath the dung pat, after which they eclose as adults. The complete cycle takes a few days during summer months (Krafsur and Moon, 1997). Nematodes deposited in the dung as fourth stage juveniles molt to the adult stage and are the sexually reproducing generation. Male and female nematodes mate in the dung, and the mated females then enter face fly larvae by puncturing the host cuticle with their stylet. Within the host hemocoel, the females produce a new generation, which is parthenogenic. This generation of nematodes gives rise to sexually reproducing females and males that invade the female fly's ovaries (Fig. 11.2). The cycle is completed when the female face fly nemaposits new generations of males and females into fresh cattle dung. In a behavioral change caused by the nematode, the female flies with advanced parasitism are termed permanent dung seekers (Kaya and Moon, 1980) because they go from dung pat to dung pat and do not return to cattle, whereas unparasitized flies feed on lachrymal secretions of cattle before each normal oviposition event. Although *P. autumnale* parasitizes both male and female flies, male flies are "dead-end hosts" because males do not oviposit, and therefore cannot nemaposit (Kaya and Moon, 1980).

The face fly is a serious pest of cattle throughout the USA and negatively affects cattle in two ways. Female flies feed on lachrymal secretions from cattle because they need the protein to produce vitellogenin. The first effect of the flies on the cattle is the disturbance, which can result in retarded growth or reduced milk production. The second effect is that these flies can mechanically transmit the causative agent of pinkeye, the bacterium *Moraxella bovis*. Consequently, there is a need to control face flies.

Like *R. culicivorax* and unlike *B. siricidicola*, *P. autumnale* must be reared *in vivo*, which makes them

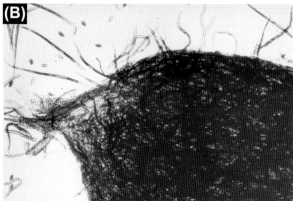

FIGURE 11.2 Ovaries of the face fly, *Musca autumnalis*. (A) Ovaries on the right contain mature eggs of the face fly and those on the left are infested with the juveniles of *Paraiotonchium autumnale*. (B) Close-up of an ovary infested with the nematodes. (From Tanada and Kaya, 1993.)

quite difficult to produce and use, especially compared to feed-through insecticides which are commonly used to control veterinary insect pests in cattle. The level of control provided by these flies is difficult to assess because of the cryptic nature of their life cycle (i.e., the rate of mortality of parasitized larvae is not incorporated into population estimates for face flies) and the fact that the flies can travel significant distances from their natal site. A mass-release study of face flies parasitized by *P. autumnale* has shown that in an area where laboratory-reared, parasitized face flies were released, a significant increase was measured in the prevalence of parasitism over time compared to control areas without augmentation (Chirico, 1996).

11.4. ENTOMOPATHOGENIC NEMATODES: MONOXENOUS, LETHAL PARASITES WITH SYMBIONTS

EPNs cause disease to insects by vectoring pathogenic bacteria into the insect hemocoel. Some aspects of their biology, for example, those that relate directly to their economic value via their use in agriculture, are very well known because of their importance as biological control agents of insect pests. Other characteristics such as their phylogenetic relationships or their host ranges are less known.

11.4.1. Life Cycle

A brief overview of the life cycle of EPNs will help in understanding the more detailed information below. The infective stage juvenile (IJ) is the only stage of the life cycle that exists outside the host and serves as the transmission stage (Fig. 11.3). The IJ is also the life stage that is applied as a biological control agent in most commercial products, although there are a few formulations based on infected insect hosts. As with all nematodes, there are six life stages: the egg, four juvenile stages, and the adult. The IJ is a specialized third stage juvenile that is analogous to the dauer stage of the well-studied nematode, *Caenorhabditis elegans*. "Dauer" is a term meaning enduring and, hence, the dauer stage is able to survive without nourishment for prolonged periods. The ultimate function of the IJ is to infect a host. They are sealed within the cuticle of the second stage juvenile, which is referred to as the sheath. The sheath is thought to afford extra tolerance to environmental extremes (Rickert-Campbell and Gaugler, 1991). The IJ penetrates a host through natural openings like the mouth, anus, or spiracles. In some cases with heterorhabditids, they can make a hole in thin cuticle with an apical tooth and some steinernematids also enter via the cuticle. Once inside the host hemocoel, the IJ molts and releases symbiotic bacteria within a few hours. The bacteria kill the host by a combination of toxins and septicemia, a morbid condition caused by the multiplication of microorganisms in the blood (Steinhaus and Martignoni, 1970) (details of the pathogenesis of the bacteria can be found below). At this point, heterorhabditid IJs develop into adult hermaphrodites and produce a second generation. This second generation has three sexes: hermaphrodites, true females, and males. Depending on the size of host, there may be one or two more

generations of nematodes before IJs form, in response to cues associated with food and declining nutritive value inside the host (Johnigk and Ehlers, 1999). For *Steinernema*, all species were thought to be amphimictic (i.e., reproduce sexually) until Griffin *et al.* (2000) described *Steinernema hermaphroditum*. Hermaphroditic species of *Steinernema* have a life cycle very close to *Heterorhabditis* spp. For amphimictic species of *Steinernema*, each IJ develops into either a male or a female, and they need to mate to reproduce. So a host must be invaded by at least one male IJ and one female IJ to enable reproduction. The time elapsed from initial infection to the first emergence of IJs ranges from fewer than 10 days to as many as 30 days, depending on nematode species, host species, host size, and temperature.

Entomopathogenic nematodes are considered bacterial feeders, and examination of their biology and life cycle shows that they are remarkably similar to the free-living nematode *C. elegans*. Once inside the insect hemocoel, they release their bacteria by regurgitation or defecation, after which the bacteria use the insect tissues as substrate. The nematodes develop by feeding on the bacteria. The nematodes are not efficient at killing the insects by themselves, and the bacteria lack an efficient transmission mechanism, thus the need for symbiosis.

While most of the species of nematode that are considered as entomopathogenic are classified into two families, the Steinernematidae and Heterorhabditidae, more recently, some members of the Rhabditidae have been suggested to be entomopathogenic (Zhang *et al.*, 2008) (see Section 11.5.1). There are also a few descriptions of nematodes that are symbiotically linked with pathogenic bacteria and have life cycles very similar to EPNs, but do not infect insects. The best known is the nematode *Phasmarhabditis hermaphrodita* (Rhabditidae), a parasite of molluscs that is used commercially throughout Europe for slug and snail control (Wilson *et al.*, 1993; Wilson and Grewal, 2005). Although this nematode species does not have a single symbiotic partner in nature and the role of

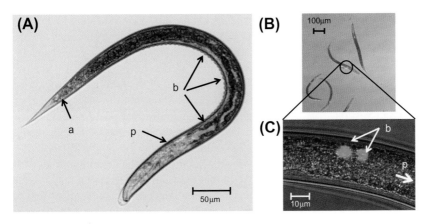

FIGURE 11.3 Colonization of infective juveniles (IJs) by green fluorescent protein-labeled symbiotic bacteria. (A) *Photorhabdus* bacteria (b) colonize the IJ gut starting at a position proximal to the pharynx (p) and extending through the gut towards the anus (a). (B, C) *Xenorhabdus* bacteria (b) are also found proximal to the pharynx (p) but, in contrast to *Photorhabdus*, *Xenorhabdus* colonize a specific vesicle in the gut. [*Photo credits: (A) C. Easom; (B, C) Heidi Goodrich-Blair.*]

bacteria in the life cycle seems to be one of nutrition alone. When, *P. hermaphrodita* is cultured *in vitro*, it is grown with the bacterium *Moraxella osloensis*.

11.4.2. History

The history of EPNs as biological control agents, and hence their extensive coverage in the scientific literature, began when Glaser and Fox (1930) discovered larvae of the Japanese beetle, *Popillia japonica*, on a New Jersey golf course infected with parasitic nematodes. The parasites were identified for the authors of the study by Steiner as *Neoaplectana* (now *Steinernema*) *glaseri* (Glaser and Fox, 1930). The identity and importance of the bacteria to the nematode life cycle were elucidated later, and the pathway by which the symbiosis became known is more completely described by Gaugler *et al.* (1992). In brief, Glaser *et al.* (1942) reported that when trying to develop an axenic (i.e., without any other organism) culture of *S. glaseri* for biological control, there was a bacterial contaminant in the culture that was tolerated by the nematodes. This contaminant was probably *Xenorhabdus poinari*, the bacterial symbiont of *S. glaseri*. Glaser's group was able to culture the nematodes axenically, but without the bacteria, biological control of Japanese beetle by this nematode met with limited success (Gaugler *et al.*, 1992). This attempt did serve to stimulate research activity toward developing the nematodes as biological control agents.

11.4.3. Symbiotic Relationship

Dutky (1937) was the first to identify the bacteria as symbiotic partners of the nematodes, but his research was conducted with *Steinernema carpocapsae* (referred to as DD-136). This phase of the history of EPNs illustrates the central role played by the symbiosis itself. The symbiosis is so central, in fact, that when the term "entomopathogenic nematode" is used in the scientific literature, the terminology is referring to the nematode—bacterium complex that formulates the functional pathogen. Unless indicated otherwise, the term "nematode" actually refers to the nematode—bacterium complex throughout this chapter.

That the functional pathogen must consist of both symbiotic partners is of great importance to implementing the nematodes to manage insect pests, especially when working to determine whether or not a particular nematode species might be effective against a given pest. Several studies have shown that the efficacy of EPNs depends on both the nematodes' ability to locate, recognize, and invade a host, and the virulence of the bacteria to the host (Griffin *et al.*, 2005). Furthermore, it is essential that when grown commercially, the nematodes retain bacteria as IJs, and that the bacteria retain their virulence to insects.

11.4.4. Sampling

EPNs are exceedingly common soil organisms and have been isolated from every continent, excluding Antarctica (Hominick, 2002). Although negative results are seldom reported in the literature, it seems that wherever a concerted effort has been made to isolate EPNs from most soil systems, researchers have met with at least some success, from jungles to deserts and tropics to tundra. Part of the success in their isolation has to do with how simple it is. As described by Kaya and Stock (1997), the "*Galleria*-bait method" is used for isolating EPNs around the world. In brief, soil samples are collected and returned to the laboratory. If necessary, they are moistened to 6—10% moisture by weight, after which several last instar *Galleria mellonella* are exposed to the soil. When the insects die, they are placed on a White trap (White, 1927), which consists of the lid of a 60 mm Petri dish lined with filter paper which is floating in water that is contained in a 90 mm Petri dish. The purportedly infected host is placed on the filter-paper lined dish and IJs emerging from the host are entrapped in water in the 90 mm Petri dish. After the supposed IJs have been isolated, they are subjected to Koch's postulates to determine whether or not an organism is a pathogen (see Chapter 1). EPNs are easy to find and isolate, but identifying them and understanding their biology remain very challenging.

11.5. BIODIVERSITY

11.5.1. Entomopathogenic Nematodes

The most current review of the phylogeny and classification of both the nematodes and the bacteria (Adams *et al.*, 2006) provides an excellent overview of the bases of classification of both groups. There are two genera in the Steinernematidae, *Neosteinernema* and *Steinernema*. *Neosteinernema* contains a single species, *N. longicurvicauda*, which unfortunately no longer exists in culture and has not been reisolated from nature. *Steinernema* contained 42 recognized species as of 2006. Twenty-five new species have been described since this last review (Table 11.1). The genus *Heterorhabditis*, which is the sole genus of the family Heterorhabditidae, contained 11 recognized species as of 2006, and six new species have been described since then (Table 11.2). Thus, each of the families has increased their number of species by about 50% in only six years.

Another genus of nematodes that could potentially be considered to be entomopathogenic has also been recognized since 2006, and belongs to the family Rhabditidae. These nematodes may conform to the definition of an EPN provided above in that they carry bacteria that may be true symbionts on which they depend for nourishment and to kill their host. The genus *Heterorhabditoides*, with the type and only species, *H. chongmingensis*, was isolated and described from soil collected from the Chongming Islands in the south-eastern area of Shanghai, China (Zhang *et al.*, 2008). However, based on a number of morphological characters, this genus was combined into the genus *Oscheius* by Ye *et al.* (2010). Zhang *et al.* (2009) proposed the bacterial symbiont of this nematode to be the recently described *Serratia nematodiphila*. The other new potential EPN is *Oscheius carolinensis*, described by Ye *et al.* (2010). According to Torres-Barragan *et al.* (2011), the relationship between *O. carolinensis* and potential bacterial symbionts is less clear, in that there were at least four different species of bacteria associated with the nematode in some way — one of which was *Serratia marcescens*. This could be a similar situation to that of *P. hermaphrodita*, which is described above. Both of these species are within the *Insectivorus* group of the genus *Oscheius*. Should further study uncover a true symbiotic relationship between these nematodes and their bacteria, their occurrence and biology will expand the taxonomic base of what are now considered "entomopathogenic nematodes."

11.5.2. Symbiotic Bacteria

The genera *Photorhabdus* and *Xenorhabdus* are phylogenetic sister groups within the family Enterobacteriaceae, a well-studied family of Gram-negative bacteria that includes important mammalian pathogens such as *Escherichia coli*, *Salmonella* and *Yersinia* spp. (Fischer-Le Saux *et al.*, 1999; Tailliez *et al.*, 2010). This close relationship highlights the relevance of *Photorhabdus* and *Xenorhabdus* as models for the study of bacteria—host interactions. Indeed, the closest phylogenetic neighbor of *Photorhabdus* and *Xenorhabdus* is *Proteus*, a genus that is the primary causative agent of hospital-acquired infections in catheterized patients (Tailliez *et al.*, 2010).

A recent phylogenetic study of the *Photorhabdus* and *Xenorhabdus* genera, using the concatenated partial sequences of four protein-coding genes (*recA*, *gyrB*, *dnaN*, and *gltX*), described three species of *Photorhabdus*. These were *P. luminescens*, *P. temperata*, and *P. asymbiotica*, and more than 20 species of *Xenorhabdus*, including the best studied species, *X. nematophila* (Tailliez *et al.*, 2010). Complete genome sequences of *P. luminescens* TT01, *P. asymbiotica* ATCC43949, *X. nematophila* ATCC19061, and *X. bovienii* SS-2004 are publicly available at the National Center for Biotechnology Information (NCBI) (http://www.ncbi.nlm.nih.gov) (Duchaud *et al.*, 2003; Wilkinson *et al.*, 2009). Comparative analyses of these genome sequences will undoubtedly provide important insights into the evolution of these bacteria and, perhaps more importantly, identify genes that are important for initiating and maintaining the bacteria—insect and bacteria—nematode associations.

11.6. INFECTIVE STAGE JUVENILE AND BACTERIAL STORAGE

The life cycle of *Steinernema* and *Heterorhabditis* can be considered to begin and end with the IJ (Fig. 11.3). This is the only stage of the nematode that can survive outside the insect and the role of the IJ is to disperse in the soil and infect new insect hosts. In both EPN genera, the IJ is non-feeding and, consequently, the pharynx and gut are collapsed. In *Heterorhabditis*, IJ viability is maintained over extended periods (in the order of several months to years) by metabolizing lipid stores that were deposited during development. Dauer formation in *C. elegans* is a developmental program regulated by a complex network that involves parallel signaling pathways that are controlled by environmental signals such as food availability (Hu, 2007). Although the molecular mechanisms of IJ formation have not been studied in either *Heterorhabditis* or *Steinernema*, it is anticipated that there will be significant overlap with *C. elegans* (Ciche, 2007). Indeed, analysis of the available *Heterorhabditis* genome dataset does identify homologues of many of the genes already identified in *C. elegans* as playing a role in dauer formation (Ciche, 2007; Bai *et al.*, 2009; Clarke, unpubl. data).

Steinernema and *Heterorhabditis* IJs carry a population of around 100—300 colony-forming units (cfu) of either *Xenorhabdus* or *Photorhabdus*, respectively (Ciche and Ensign, 2003; Martens *et al.*, 2003). Typically, over 90% of IJs in a population are colonized by their bacterial symbiont. *Xenorhabdus* spp. appear to colonize a specialized caecum-like vesicle in the anterior region of the IJ gut (Fig. 11.3B, C) (Martens and Goodrich-Blair, 2005). This vesicle contains an anucleate body called the intravesicular structure, which appears to play a role in the colonization of the vesicle by *Xenorhabdus* (Martens and Goodrich-Blair, 2005). Several *nil* mutants have been identified in *Xenorhabdus*; *nil* indicates that they are unable to colonize this vesicle (Heungens *et al.*, 2002; Cowles and Goodrich-Blair, 2004, 2006).

Unlike *Steinernema*, *Heterorhabditis* does not contain a specific vesicle, but rather *Photorhabdus* appears to attach to the preintestinal valve cell of the IJ, situated just below the pharynx (Fig. 11.3A) (Ciche *et al.*, 2008). In order to colonize the IJ, *Photorhabdus* must first bind to, and invade, rectal gland cells situated at the distal end of the gut of the adult hermaphrodite nematode. The bacterial cells replicate inside vacuoles within the rectal gland cells before rupturing the vacuoles and emerging into the body cavity of the hermaphrodite. At the same time, the hermaphrodite ceases laying eggs and the remaining eggs hatch internally (in a process called *endotokia matricida*) (Johnigk and Ehlers, 1999). These larvae develop into IJs within the mother's body cavity, where they encounter *Photorhabdus*. At the appropriate time, the IJs become permissive to colonization by *Photorhabdus* and transmission is then complete (Ciche *et al.*, 2008). The relatively short duration of the permissive phase of IJ development suggests that there are specific, regulated factors involved in this part of the nematode—bacterium interaction (Ciche *et al.*, 2008). Indeed, recent genetic studies have shown that IJ colonization by *Photorhabdus* requires surface structures such as lipopolysaccharide (LPS) and a type I fimbriae encoded by the *mad* locus (Bennett and Clarke, 2005; Easom *et al.*, 2010; Somvanshi *et al.*, 2010). The type I Mad fimbriae are required for attachment to the rectal gland cells, while the role of LPS appears to be at a later stage (Easom *et al.*, 2010; Somvanshi *et al.*, 2010). The production of the Mad fimbriae is phase variable and, therefore, only a subset of *Photorhabdus* cells within a population produces these surface appendages, giving these cells a clear advantage during this stage of the life cycle (Somvanshi *et al.*, 2010). Thus, production of the Mad fimbriae is associated with the formation of small colony variants on agar plates, indicating that expression of the *mad* locus has other physiological consequences (T. A. Ciche, pers. comm.). Type I fimbriae (encoded by the *mrxACDGH* operon) are also required for colonization of the *Steinernema* IJ, although the production of these fimbriae does not appear to be phase variable (Chandra *et al.*, 2008).

In both systems, the IJ is initially colonized by only one or two bacterial cells and these cells replicate to achieve the final observed level of symbiont (Martens *et al.*, 2003; Ciche *et al.*, 2008). Therefore, the IJ must provide nutrients to the growing bacterial cells. Clearly, any cost associated with nutrient provision must be outweighed by the benefits associated with the bacterial contribution to insect death and bioconversion. Studies with *Xenorhabdus* have shown that the vesicle in the *Steinernema* IJ appears to be replete with a number of important nutrients, including amino acids such as serine, histidine, and isoleucine (Martens *et al.*, 2005). Other nutrients, such as para-aminobenzoic acid, appear to be limiting or absent from the vesicle, and although mutants unable to make these compounds do colonize the vesicle, these mutants fail to replicate *in vivo* (Martens *et al.*, 2005). Less is known about the nutritional conditions found within the gut of the *Heterorhabditis* IJ. However, there is some evidence that this environment is limited for Mn^{2+}, suggesting that *Photorhabdus* must have adaptive features to permit colonization of this niche (Watson *et al.*, 2010).

11.7. INFECTION

11.7.1. Infective Stage Juvenile Host Finding, Recognition, and Penetration

The process of infection by EPNs is described chronologically, starting with how IJs find and invade a host. The process can be broken down into four categories: (1) host habitat finding; (2) host finding; (3) host recognition; and (4) host acceptance (Campbell and Lewis, 2002). Where the IJ searches is related to its behavioral repertoire concerning host finding. Host-finding behavior for EPN IJs varies among species along a continuum of foraging strategies that ranges in behavior from ambushing, where a nematode would move relatively little and await a passing host (usually at or near the surface), to cruising, where a nematode would move throughout the soil column in search of more sedentary hosts (Lewis *et al.*, 1992, 2006; Grewal *et al.*, 1994). Foraging ecology was found to be a useful and practical area of study because it was invoked as a plausible explanation of why some species of nematodes that were able to kill particular insect species in the laboratory proved to be less effective than predicted in the field (Georgis and Gaugler, 1991; Gaugler *et al.*, 1998). Much of the work on EPN behavior and ecology was first engaged as a strategy to improve the performance of the nematodes as biological control agents.

Description of the nematodes' foraging strategies was first based on the type of activity demonstrated by the IJ. Because the IJ does not feed or mate during this life stage, all activity should be directed toward finding and infecting a new host, and therefore, anything the IJ does should increase the probability of the IJ encountering a suitable host. The argument was made that the mode of foraging of an IJ predicted, at least to some extent, what insect species it was likely to encounter (Lewis *et al.*, 2006). For instance, a nematode species that foraged by ambushing on the surface of the soil was unlikely to find an insect that was sedentary and lived deep in the soil. This behavior obviously affects how EPNs would be used in biological control and helps to explain, for example, why applications of *S. carpocapsae*, an ambushing surface-dwelling nematode, were not at all effective against the subterranean larvae of scarabs (Gaugler *et al.*, 1998).

Foraging strategy was intended to serve as a theoretical framework, not a way to pigeon-hole species of EPNs into particular categories. Campbell *et al.* (2003) suggested that foraging could be categorized as "ambushing", "intermediate", or "cruising", based on a number of behavioral tests that measured specific behaviors of IJs. Campbell *et al.* (2003) mapped foraging behavior on to the phylogeny of the *Steinernema*. For instance, *S. carpocapsae* IJs spend long periods performing a behavior called "standing" (originally described as "nictating", but this term is no longer in use), as described by Campbell and Gaugler (1993). This behavior is typified by the IJ lifting 99% of its body length from the substrate and remaining more or less motionless, sometimes for hours at a time. Functionally, this behavior would facilitate an IJ detaching from the substrate to attach to a passing potential host. Campbell *et al.* (2003) showed that three EPN species,

S. carpocapsae, *S. scapterisci*, and *S. siamkayai*, demonstrate standing behavior for prolonged periods. This behavior was hypothesized to have arisen in one clade of the Steinernematidae that includes those three species. Some species that were classified as intermediate, such as *Steinernema bicornutum*, *S. riobrave*, and a few others, stand for only a few seconds. None of the species designated as cruisers is able to stand.

A group of behaviors related to standing have been shown to increase an ambushing IJ's probability of locating a host and have been termed "waving" and "jumping". Both of these behaviors are demonstrated by standing IJs when they are stimulated by various environmental conditions (e.g., air movement) or volatile host cues (Campbell and Kaya, 2000). Waving by IJ nematodes is characterized by the standing IJ remaining in the same place and waving back and forth while maintaining a straight posture. Campbell and Kaya (2000) suggested that this behavior could increase the effective search area of a stationary nematode and thereby increase the probability of encountering a host.

Jumping is a behavior associated with ambushing species of the Steinernematidae, first described by Reed and Wallace (1965) for *S. carpocapsae* IJs. Campbell and Gaugler (1993) showed that several species of *Steinernema* are able to jump, but the role of this behavior in host finding was unknown at the time. Campbell and Kaya (1999, 2000) described the mechanics and function of jumping behavior, as it pertains to host location. The following description of jumping is from Campbell and Kaya (1999):

Jumping … is initiated when a standing nematode quickly bends the anterior half of its body until its head region makes contact with the ventral side of its body. The two body regions appear to be held together by the film of water covering the nematode's body. The nematode now has resistance to the bending of its body, so it can use its normal sinusoidal crawling behavior to slide its body in a posterior direction, causing the loop to become progressively smaller and the bend in its body to become more acute. Eventually, the body becomes so contorted that the cuticle on the dorsal side becomes extremely stretched and the cuticle on the ventral side kinks, generating sufficient force to break the surface tension forces holding the two body parts together. As its body straightens out, enough force is applied to break the surface tension forces holding the nematode to the substrate and propel it through the air. The forces generated by this jumping mechanism are sufficient to propel nematodes an average distance of 4.8 ± 0.8 mm (nine times the nematode's body length) and an average height of 3.9 ± 0.1 mm (seven times the nematode's body length).

Furthermore, Campbell and Kaya (1999) state that jumping by *S. carpocapsae* IJs is directional; they tend to jump towards the source of volatile host cues. Thus, jumping is a component of the suite of host-finding behaviors of *S. carpocapsae*.

Based on the suite of behaviors of IJs associated with host finding, one can develop hypotheses about other aspects of the nematodes' life history. These suites of behaviors are sometimes referred to as adaptive syndromes. For example, Lewis *et al.* (1992, 1993) tested the hypothesis that if ambushing nematode species, like *S. carpocapsae*, encountered hosts by waiting for contact with them, they should be less responsive to distant host cues than would be a cruising forager, like *S. glaseri*, which indeed is the case. The work on foraging strategies of EPNs led to a series of hypotheses about metabolic rates, where ambushers should have lower rates than cruisers (Lewis *et al.*, 1995b), lifespan, where ambushers should live longer than cruisers (Lewis *et al.*, 1995a), host affiliations, where cruisers should have more sedentary hosts than ambushers (Grewal *et al.*, 1994), horizontal and vertical distribution within the soil habitat (Campbell *et al.*, 1995), etc. Some of these hypotheses were supported and others were refuted in these tests, but the foraging strategy served as a useful theoretical framework, whether or not any specific hypothesis was supported.

In addition to variation in behaviors among species and isolates of EPNs, there is variation among individuals within the same species, and even among individuals that emerge from the same host cadaver. In some EPN species, e.g., *S. glaseri*, *S. longicaudum*, and *S. kraussei*, males tend to emerge from the host before females do (Lewis and Gaugler, 1994; Alsaiyah *et al.*, 2009), whereas for *Steinernema feltiae* and *S. carpocapsae*, females seem to emerge first (Bohan and Hominick, 1997; Renn, 1998). Further, Grewal *et al.* (1993) proposed that male EPNs are usually the sex that initiates infection based on their greater sensitivity to host cues, but subsequent studies did not demonstrate that males actually infect hosts before females (Stuart *et al.*, 1998; Alsaiyah *et al.*, 2009). Sensitivity to host cues varies with species and among individuals, and determining which IJs actually initiate the infection when there are tens to hundreds of individuals ultimately infecting a single host is just beginning to be understood.

When considering the dynamics of infection by EPNs, it is important to remember that most studies of how IJs respond to hosts are based on recording their responses to uninfected hosts. However, in a system where tens to hundreds of IJs invade a single insect, only the first IJ makes the decision to infect based on evaluation of the uninfected host. The remainder must assess the quality of an infected host that changes rapidly with time (Lewis *et al.*, 2006). Thus, host recognition can be broken into two categories: recognition of a host species and recognition of a host of the best quality within a species. The following describes host species recognition.

Considering how many studies have been published on EPN efficacy against various insect pests in agriculture, surprisingly little is known of natural host affiliations of most species. Lewis *et al.* (2006) suggest that this paucity of information can be attributed to the ease of their isolation from soil using a baiting technique, and the relative fragility of a nematode-killed insect (i.e., a host cadaver containing nematodes). Only a few accounts of discovering new nematode isolates from nematode-killed hosts in the field have been published (Lewis *et al.*, 2006). The over-whelming majority of nematode species and isolates have been recovered from soil samples by exposing a susceptible host (usually *G. mellonella*) to soil and waiting to find nematode-killed insects after a period of a few days (as described in Section 11.4.4). All that can be certain about host range in this case is that the nematodes isolated can infect the species of host that was used as bait.

Once the IJ finds and comes into contact with a potential host, it must evaluate the host before invasion. Because the decision to invade, or infect, a host is irreversible (i.e., an IJ cannot leave the host after its bacteria have been released and development has resumed), strong selection pressure against individuals who enter an unsuitable host should shape host recognition capabilities. Studies of host recognition behavior have shed some light on host range for a few nematode species. Lewis *et al.* (1995b) developed a two-step assay to measure the behavioral response of *S. carpocapsae* to a variety of suspected host and non-host species, hypothesizing that the nematodes should respond most strongly behaviorally to those species in which reproduction was most efficient. In brief, Lewis *et al.* (1993) showed that *S. carpocapsae* IJs were not attracted to volatile cues produced by hosts, as were the cruising *S. glaseri* IJs. However, after *S. carpocapsae* IJs had been exposed to host cuticle, they responded strongly to volatiles, suggesting that this was due to the route of entry into hosts by *S. carpocapsae* being spiracles (Lewis *et al.*, 1995b). Lewis *et al.* (1996) hypothesized that different species of potential hosts should elicit differential behavioral responses from *S. carpocapsae*, based on the quality of the host species. The final result was that the behavioral responses of *S. carpocapsae* IJs to various host species correlated with the number of IJs produced per milligram of host tissue. This assay has been used to add some detail to studies seeking to show a relationship between a nematode species and a particular host. For example, Hodson *et al.* (2011) used this assay in a study of the relationship between *S. carpocapsae* and the European earwig, *Forficula auricularia*, in pistachio production in California to elucidate interactions that occurred after the nematodes had been applied to orchards and a decrease in earwig population densities was recorded. Hodson *et al.* (2011) originally discovered that applications of *S. carpocapsae* to pistachio orchards resulted in significant decreases in earwig populations. Upon further testing in the laboratory, they found that *S. carpocapsae* IJs engaged in behaviors that indicated that they recognized European earwig as a host.

Nevertheless, the species of host is not the only factor that affects host quality for EPNs; the infection status of a host is also important. Fushing *et al.* (2008) created a mathematical model that described the dynamics of EPN infection in terms of mass-event history analysis. This approach is valuable in that it considers how the decisions made by individuals within a group (i.e., IJs in the vicinity of a host) are driven, in part, by the decisions of others. The premise is that an uninfected host is not an ideal resource; there are risks associated with being the first individual to invade a host, including attack by the host immune system, and the probability of entering a host and not finding a mate within (for most *Steinernema* spp.). Once the host is infected by a conspecific individual IJ, its value as a resource increases to IJs outside the host because the immune response will soon be diminished and there is a readily available food supply since the bacteria are released within a few hours of an IJ entering a host. This contention is supported by other work showing that volatile cues associated with infected hosts are more attractive than those produced by uninfected ones (Grewal *et al.*, 1996). Thus, a typical temporal pattern of invasion of a host by a mass of IJs could be characterized by the host remaining uninfected for substantial periods, until it is infected by an individual, after which following nematodes would be expected to enter as quickly as possible. The conditions that stimulate the first IJ to invade are unclear but are under investigation. This situation has implications for both temporal and spatial distributions of EPNs in the field in that it reinforces the patchy nature of their distributions.

Regardless of their strategy, all IJs must be able to determine when an insect is close by; i.e., they need to be able to respond to chemical cues produced by their hosts. For many parasitic nematodes including EPNs, these cues include carbon dioxide (CO_2) (as a byproduct of host respiration) and volatile compounds produced by the host. *Heterorhabditis* IJs are strongly attracted to CO_2 and this is dependent on the BAG neurons, which are sensory neurons in the head that have been shown to regulate parasitic and free-living nematodes' responses to CO_2 (Hallem *et al.*, 2011). Importantly, the BAG neurons were also shown to be important for IJ attraction to *G. mellonella* larvae, providing strong evidence that CO_2 is a major chemical cue used for host finding by *Heterorhabditis* nematodes (Hallem *et al.*, 2011). Another source of environmental cues is volatile compounds that may be released by potential hosts or as the result of insect damage to plants. Maize roots release a sesquiterpene compound (E)-b-caryophyllene, when they suffer feeding damage by larvae of the western corn rootworm, *Diabrotica virgifera virgifera*. This compound was shown to act as a below-ground chemoattractant for *Heterorhabditis* IJs with a range, measured by diffusion experiments, exceeding 250 times the body length of the nematode (Rasmann *et al.*, 2005). By attracting EPNs to an injured plant, this compound provides indirect protection to the maize. Most North American lines of maize have lost the ability to produce the sesquiterpene, perhaps explaining the relatively poor protection provided to these lines by EPN application.

Once a potential host has been found and the decision to infect has been made, the IJs generally enter the host through natural openings in the insect cuticle such as the spiracles, the mouth, or the anus and penetrate into the hemocoel, where the bacterial symbionts are released (Ciche and Ensign, 2003). *Steinernema* IJs secrete hydrolytic enzymes (i.e., serine proteases) that may facilitate transmission from the insect gut into the hemolymph (Peters and Ehlers, 1994; AbuHatab *et al.*, 1995; Toubarro *et al.*, 2010). In contrast, *Heterorhabditis* IJs do not produce hydrolytic enzymes but, rather, they have a dorsal tooth-like appendage that may also be involved in tearing the insect cuticle and/or gut to provide access to the hemocoel (Bedding and Molyneux, 1982). Indeed, for *Heterorhabditis bacteriophora* IJs, there is some evidence that infectivity (i.e., the number of IJs per insect) may be correlated with the percentage time IJs spend thrusting their heads in a motion that resembles penetration (Dempsey and Griffin, 2002), and that their primary route of entry may be through the thin intersegmental membranes on the exterior of the insect (Wang and Gaugler, 1998).

11.7.2. Release of Bacteria

Upon entering the hemocoel of the insect, the IJ is stimulated to release the bacterial symbionts into the insect hemolymph (Ciche and Ensign, 2003). Bacterial release is the first step in a process called recovery, whereby the IJ exits diapause (the condition of the IJ, since no growth, development, or reproduction occurs during this stage) and develops into an adult hermaphrodite nematode (in the case of *Heterorhabditis*) or either an adult male or female nematode in the case of *Steinernema*, with the exception of *S. hermaphroditum* (Stock *et al.*, 2004). The signals present within the insect hemolymph responsible for initiating IJ recovery have not been identified, but in the case of *Heterorhabditis* at least, the signal(s) has been shown to be a small [< 3 kilodalton (kDa)], heat-stable molecule(s) (Ciche and Ensign, 2003). *Xenorhabdus* are released by defecation through the anus, whereas *Photorhabdus* are regurgitated, over a period of several hours (Ciche and Ensign, 2003; Snyder *et al.*, 2007). In both cases, it appears that release is independent of bacterial motility (Ciche and Ensign, 2003; Snyder *et al.*, 2007; Easom and Clarke, 2008).

Although the dominant signal for IJ recovery in the environment is likely to be the uncharacterized molecule(s) present in the insect hemolymph, there is strong evidence to suggest that *Photorhabdus* produces signals that control nematode development. It has been appreciated for some time that during *in vitro* culturing of the nematode−bacterium complex, IJ recovery is dependent on a "food signal" produced by *Photorhabdus* (Strauch and Ehlers, 1998). Recently, the ASJ neuron, a sensory neuron in the head, has been shown to be required for the detection of this food signal by *Heterorhabditis*, and this is the same neuron that is required for dauer recovery in *C. elegans* and the human parasitic nematode *Strongyloides stercoralis* (Hallem *et al.*, 2007). A major component of this food signal is a molecule called 3′,5′-dihydroxy-4-isopropylstilbene (ST or IPS) (Joyce *et al.*, 2008). All *Photorhabdus* strains tested have been shown to produce ST in relatively large quantities during the post-exponential phase of bacterial growth, and recent genetic and metabolic studies have shown that this compound is derived from phenylalanine metabolism and branched chain fatty acid (BCFA) biosynthesis (Hu and Webster, 2000; Williams *et al.*, 2005; Joyce *et al.*, 2008). ST is a member of the stilbene family of compounds, an important group of pharmacologically active chemicals that includes resveratrol, a compound found in red wine that has been associated with a number of beneficial activities including increased longevity in mammals (Pervaiz, 2003; Baur *et al.*, 2006; Valenzano *et al.*, 2006). Stilbene production is normally associated with plants, and in fact, ST is the only reported non-plant-derived stilbene. Although ST production is not required for IJ recovery in the insect, the absence of this molecule in the insect significantly affects nematode growth and development *in vivo*, confirming the role of ST as an inter-kingdom signaling molecule (Joyce *et al.*, 2008). Moreover, ST appears to be a multipotent compound that has, in addition to its role as a signaling molecule, antimicrobial (particularly against Gram-positive bacteria) and nematicidal activity, as well as being involved in insect virulence (see Section 11.7.3) (Li *et al.*, 1995; Hu *et al.*, 1999; Eleftherianos *et al.*, 2007; Boina *et al.*, 2008).

ST is one of a number of small, bioactive compounds produced by the secondary metabolism of *Photorhabdus*. Other secondary metabolites, such as the anthraquinone pigment and a carbapenem antibiotic, have also been characterized (Hu *et al.*, 1998; Derzelle *et al.*, 2002; Brachmann *et al.*, 2007). *Xenorhabdus* also have the potential to produce a number of bioactive compounds, some of which have been at least partially characterized and been shown to have antimicrobial or immunosuppressive activities (Brachmann *et al.*, 2006; Gualtieri *et al.*, 2009; Bode, 2009; Song *et al.*, 2011). Perhaps the best characterized metabolite is the major antimicrobial compound, xenocoumacin, produced by *X. nematophila* (Park *et al.*, 2009). However, none of the metabolites produced by *Xenorhabdus* appears to have a role in the interaction with the *Steinernema* nematode.

Several recent sequencing projects have revealed that the genomes of both *Photorhabdus* and *Xenorhabdus* contain many uncharacterized genetic loci that are predicted to be involved in the production of small, bioactive compounds (Bode, 2009). The vast majority of these loci are cryptic (i.e., not expressed under laboratory conditions), and therefore the compound(s) produced by these loci remains unidentified. Nonetheless, it is possible that these loci are expressed during the different interactions between the bacteria and their nematode and/or insect hosts, and the characterization of the compounds produced by these loci will be an important goal for future research.

11.7.3. Overcoming the Insect Immune System

Insects do not have antibody-based adaptive immunity but these organisms do have an advanced innate immune system with striking parallels to the innate immune system observed in mammals (Uvell and Engström, 2007). *Xenorhabdus* and *Photorhabdus* have been shown to be sensitive to the insect's immune system and bacterial virulence is greatly reduced if the immune system has been previously activated by the injection of non-pathogenic bacteria (Eleftherianos *et al.*, 2006a, b; Aymeric *et al.*, 2010). Therefore, in order to establish an infection, the nematode–bacterium complex must overcome, avoid, or subdue the insect immune system.

Nematode Contribution

Axenic *Steinernema* IJ nematodes are able to kill insects. Virulence is likely to be due, at least in part, to the secretion of a number of proteases during IJ recovery (Simões *et al.*, 2000; Toubarro *et al.*, 2009; Jing *et al.*, 2010). Several of these proteases have been purified and their contribution to virulence appears to be through immune suppression, e.g., inhibiting phenoloxidase activity and/or affecting hemocyte function (Balasubramanian *et al.*, 2009; Jing *et al.*, 2010). *Heterorhabditis* do not secrete any proteases and are not virulent in the absence of their bacterial symbionts. Axenic *Heterorhabditis* IJs elicit an immune response to a lesser extent than either colonized nematodes or *Photorhabdus* alone, suggesting that they are not easily recognized by the insect immune system (Eleftherianos *et al.*, 2010). Indeed, *Heterorhabditis* appears to actively modulate the insect immune system by reducing the phagocytic capability of the circulating hemocytes in *Manduca sexta* (Eleftherianos *et al.*, 2010). The mechanisms underlying this immune system modulation by *Heterorhabditis* are not understood. What is known about how EPNs interact with insect immune systems is based on studies with only a few host–parasite combinations, so it is also possible that the immune response against the EPN

(*Steinernema* or *Heterorhabditis*) may differ depending on the insect host.

Bacterial Contribution

Innate immunity in the model dipteran host, *Drosophila melanogaster*, is well studied and has been shown to be controlled by two parallel signaling pathways, the Toll and immune deficiency (Imd) pathways (Lemaitre and Hoffmann, 2007). The Toll pathway is involved in the recognition of fungi and Gram-positive bacteria, while the Imd pathway recognizes Gram-negative bacteria (Lemaitre *et al.*, 1995; de Gregorio *et al.*, 2002; Leulier *et al.*, 2003). As expected, recent work using the *Drosophila* model has indicated that the Imd pathway is activated following infection with either *Xenorhabdus* or *Photorhabdus* (Hallem *et al.*, 2007; Aymeric *et al.*, 2010). The Imd pathway controls the production of antimicrobial peptides, suggesting an important role for this component of the innate immune system in protection against these bacteria. Many bacteria respond to the presence of antimicrobial peptides by altering their surface, specifically by changing the composition of their LPS, an important surface molecule found exclusively on Gram-negative bacteria (Raetz *et al.*, 2007). Antimicrobial peptides are generally positively charged molecules and are normally attracted to the negatively charged LPS molecules on the surface of the bacteria. In *Salmonella*, these LPS modifications include the addition of L-aminoarabinose to the lipid A moiety of the LPS (Raetz *et al.*, 2007). This has the effect of reducing the overall negative charge of the LPS and therefore reducing the electrostatic interaction between the bacterial surface and the antimicrobial peptides. The biosynthesis of L-aminoarabinose is dependent on the *pbgPE* operon, the expression of which is controlled by a two-component signaling pathway, PhoPQ (Guo *et al.*, 1997; Gunn *et al.*, 1998). In *Photorhabdus*, mutations in either *phoPQ* or *pbgPE* have been shown to be hypersensitive to antimicrobial peptides and greatly reduced in virulence against insects, highlighting the similarities between this insect pathogen and important mammalian pathogens such as *Salmonella* (Derzelle *et al.*, 2004; Bennett and Clarke, 2005; Easom *et al.*, 2010).

While *Photorhabdus* is recognized by the insect immune system, it appears that *Xenorhabdus* actively suppresses the insect immune system. One strategy involves the production of a soluble inhibitor of insect phospholipase A_2, an enzyme required for nodule maturation (Kim *et al.*, 2005). There is also evidence that *Xenorhabdus* can suppress the production of antimicrobial peptides, although the factors responsible for this activity have not been characterized. Mutations in either *lrp* (encoding an important global regulator) or *cpxR* (encoding the response regulator component of the

CpxAR two-component pathway) produce strains that, when injected into insects, result in a significant increase in the production of the antimicrobial peptide cecropin (Cowles *et al.*, 2007; Herbert Tran and Goodrich-Blair, 2009). These mutants are also attenuated in virulence, suggesting that *Xenorhabdus* may not be able to adapt to the presence of the increased level of antimicrobial peptides. Therefore, *Photorhabdus* and *Xenorhabdus* appear to have fundamentally different strategies when interacting with the insect immune system (Goodrich-Blair and Clarke, 2007).

Photorhabdus spp. produce a molecule, ST, during the postexponential phase of growth. Mutants unable to produce ST were attenuated for virulence when injected into the model insect host, *M. sexta* (ST-minus *Photorhabdus* were unaffected for virulence in another insect, *G. mellonella*, suggesting that some host-dependent factors are important) (Williams *et al.*, 2005; Eleftherianos *et al.*, 2007). Further work established that the role of ST in virulence was through the inhibition of the insect phenoloxidase enzyme; when phenoloxidase is inhibited, decreased levels of melanization and a reduction in the number of nodules produced by the insect are seen (Eleftherianos *et al.*, 2007).

Both *Photorhabdus* and *Xenorhabdus* produce a number of toxins and extracellular enzymes that are expected to affect the cellular innate response, e.g., binary toxin systems such as PirA/PirB in *P. luminescens* and XaxA/XaxB in *X. nematophila* (Vigneux *et al.*, 2007; Ahantarig *et al.*, 2009). These binary toxins are lethal when injected into a number of different insect hosts and have potent activity against a number of insect and mammalian cell lines. Another toxin produced by *Photorhabdus*, Mcf (makes caterpillars floppy), has been shown to induce apoptosis in cultured cells (Daborn *et al.*, 2002; Waterfield *et al.*, 2003; Dowling *et al.*, 2004, 2007). Perhaps the best characterized of all the toxins produced by either *Photorhabdus* or *Xenorhabdus* are the Tc toxins that were first described in the supernatants of cultures of *P. luminescens* W14 (Bowen *et al.*, 1998; Blackburn *et al.*, 1998; Waterfield *et al.*, 2001). When fed to insect larvae, the Tc toxins had a striking phenotype whereby the insects stopped feeding and quickly died. Analysis of the Tc toxins has shown that they are genetically complex and it is now apparent that the genes encoding the Tc complex are not restricted to *Photorhabdus* but, rather, can be identified in many other genera such as *Xenorhabdus* (Sheets *et al.*, 2011), *Yersinia* (Hares *et al.*, 2008; Waterfield *et al.*, 2007), *Serratia* (Hurst *et al.*, 2004), and *B. thuringiensis* (Blackburn *et al.*, 2011). Two of the *Photorhabdus* Tc toxins, TccC3 and TccC5, have recently been shown to have ADP-ribosyltransferase activity that targets host cell actin and Rho GTPases, respectively, resulting in the inhibition of phagocytosis (Lang *et al.*, 2010).

Many Gram-negative pathogenic bacteria have type III secretion systems (T3SS) that are required to deliver proteins called effectors directly into the cytosol of host cells (Cornelis, 2006). These effectors interact with target proteins in the host and generally play an important role in the virulence of the bacteria that carry them. All *Photorhabdus* species studied have at least one T3SS and recent genomic studies have shown that the sequenced genome of *P. asymbiotica* ATCC43949 has two T3SS (Wilkinson *et al.*, 2009). However, the number and type of effectors secreted by these systems appear to be species, and even strain, dependent (Brugirard-Ricaud *et al.*, 2004, 2005). This diversity in effectors is probably due to horizontal gene transfer, as many effectors are encoded in loci alongside genes with predicted roles in DNA mobility. In the case of *P. luminescens* TT01, the T3SS has been shown to be involved in the delivery of at least one effector, LopT, into target cells (Brugirard-Ricaud *et al.*, 2005). LopT has striking homology to the YopT effector produced by *Yersinia pestis*, the causative agent of plague. The role of YopT is to inhibit phagocytosis, and there is some evidence that this may also be the function of LopT in *P. luminescens* (Brugirard-Ricaud *et al.*, 2005). Another potential effector identified in *P. luminescens* is a protein called Cif. This protein appears to be a hydrolytic enzyme that can arrest the host cell cycle at G2/M, and it is also found in enteropathogenic *E. coli*, *Yersinia*, and *Burkholderia* (Yao *et al.*, 2009). This highlights potential similarities between effector proteins targeting insect and mammalian cells. The recent completion of the *X. nematophila* genome sequence has revealed that this bacterium does not encode a T3SS, further highlighting the differences in virulence strategies employed by these bacteria (H. Goodrich-Blair, pers. comm.).

11.8. NUTRITION WITHIN THE INSECT

Arguably the most significant selection pressure in the formation and maintenance of a mutualistic interaction is the provision of nutrients. In the case of the EPNs, it is clear that one of the primary roles of the bacterial symbiont is to convert tissues and organs of the insect host into a nutrient soup that supports nematode growth and development. Therefore, by carrying specific bacteria for the purpose of seeding, propagation and harvesting food in their chosen host, it might be considered that the nematodes are carrying out a primitive form of agriculture. Similar farming symbioses are relatively rare in nature (e.g., fungus-harvesting ants) and this feature distinguishes the EPN association from other symbiosis models. Thus, the *Heterorhabditis* nematode has an obligate requirement for *Photorhabdus* for growth and development and this association has evolved to such a degree that the nematode may only grow on its cognate bacterial partner or a very close relative (Watson *et al.*, 2010). The nutritional relationship between *Steinernema* and *Xenorhabdus* appears to be somewhat more relaxed and there are (anecdotal) reports that *Steinernema* will undergo limited growth and development on bacteria such as *E. coli*.

The obligate relationship between *Heterorhabditis* and *Photorhabdus* has facilitated some studies into the nutritional basis of this symbiosis. Activities produced during bacterial growth, in particular the postexponential phase of growth, are essential for nematode growth and development. These activities have been termed symbiosis factors and they include, but are not limited to, proteases, lipases, bioluminescence, and ST. The production of these symbiosis factors is coordinately regulated by a transcriptional regulator called HexA and mutations in *hexA* result in a strain that constitutively expresses the symbiosis factors but, as a consequence, is attenuated for virulence (Joyce and Clarke, 2003; Kontnik *et al.*, 2010). In contrast, the HexA homologue in *X. nematophila*, LrhA, is required for virulence but has no role in the symbiosis with *Steinernema* (Richards *et al.*, 2008). In *P. luminescens*, the production of ST and bioluminescence has been shown to require a functional TCA cycle, leading to the suggestion that a metabolic switch is involved in the production of the symbiosis factors (Lango and Clarke, 2010). The presence of such a switch would imply that the availability of particular nutrients within the insect cadaver might, in some way, control the symbiosis between *Photorhabdus* and *Heterorhabditis*. It has been suggested that the assimilation of the amino acid proline, which is present at high levels in the insect hemolymph, is perceived as a signal by *Photorhabdus* that they are in an insect host (Crawford *et al.*, 2010). Moreover, the Lrp protein, a global regulator that responds to the availability of nutrients in *Xenorhabdus*, has been reported to have an important role in the regulation of pathogenicity and mutualism in this bacterium (Cowles *et al.*, 2007). Therefore, nutritional signaling may be an important regulator of pathogenicity and mutualism in both *Photorhabdus* and *Xenorhabdus*.

In addition to the identification of metabolic regulators, specific genes in *Photorhabdus* that are important for nematode growth and development have been identified. The *cipA* and *cipB* genes encode for proteins that form intracellular inclusion bodies during the postexponential phase of bacterial growth (Bintrim and Ensign, 1998). The composition of the CipA and CipB proteins is striking as these proteins have a high proportion of essential amino acids, indicating a possible role for these inclusions in nematode nutrition (Bintrim and Ensign, 1998; You *et al.*, 2006). In another study, it was shown that the level of iron in the bacteria had implications for nematode nutrition (Watson *et al.*, 2010). Therefore, a mutant of *P. temperata* K122 that was defective in its ability to scavenge iron from the environment (and therefore had lower iron levels than the wild-type bacteria) was unable to support nematode growth and development (Watson *et al.*, 2005). This defect was remedied by the addition of exogenous iron to the medium, suggesting that the bacteria were required to deliver iron to their nematode partner. In a follow-up study, it was shown that the requirement for iron was nematode dependent and the *P. temperata* K122 partner, *Heterorhabditis downesi*, was much more sensitive to perturbations in the levels of iron in their bacterial symbiont than was the *P. luminescens* TTO1 partner, *H. bacteriophora* (Watson *et al.*, 2010).

As previously mentioned, *Photorhabdus* produces a food signal that is perceived by the nematode as an indication that the insect cadaver is suitable for nematode growth and development. A component of this food signal is the ST molecule and the biosynthesis of ST is closely integrated with BCFA biosynthesis. In fact, BCFAs are the dominant fatty acid group found in the membranes of *P. luminescens* TTO1, and this property appears to be unique among the Enterobacteriaceae (Joyce *et al.*, 2008; Kazakov *et al.*, 2009). Mutations in the *bkdABC* operon, encoding a putative branched chain α-keto acid dehydrogenase required for BCFA production, result in cells that are significantly affected in their ability to support nematode growth and development (Joyce *et al.*, 2008). This result might be expected as these mutants are also unable to produce ST but, importantly, this defect in growth and development could only partially be rescued by the addition of exogenous ST (Clarke, unpubl. data). Consequently, BCFAs appear to have an ST-independent role in nematode growth and development. Intriguingly, BCFAs have been shown to be essential for normal growth and development in *C. elegans*, where the nematode has the necessary genes to produce its own supply of these important fatty acids (Kniazeva *et al.*, 2004, 2008). It is possible that *Photorhabdus* have taken on the responsibility for the production of BCFAs for their nematode partner. Such a commitment would be readily explainable from an evolutionary perspective, as the nematodes would now have an obligate requirement to maintain their association with the bacteria. As the *Photorhabdus* genome is likely to reflect the environment in which it has evolved, it is likely that further genetic studies will identify a more diverse range of nutritional commitments that will fully explain why the nematode is locked into its relationship with *Photorhabdus*. More challenging perhaps is a satisfactory explanation of what the nematode offers the bacteria in return for all of this food.

11.9. NATURAL POPULATIONS AND HOST ASSOCIATIONS

11.9.1. Natural Host Affiliations

As mentioned above, surprisingly little is known about the natural host ranges of EPNs currently in culture. Peters

(1996) summarized the known associations, based on isolates of EPNs that were isolated from natural infections found in the field (Table 11.3). This list includes most of the recorded natural associations between EPNs and hosts that have been discovered in field conditions. One obvious point from this listing is that host ranges vary in breadth among species. For example, *S. glaseri* has been found infecting only four species from two families of Coleoptera, whereas *S. carpocapsae* has been found infecting 14 species in 10 families and four orders of insects. Of course, there are some species that have been recorded from only one species of insect, like *Steinernema diaprepesi*, but this very limited host range could be an artifact caused by the relatively recent discovery of this species, whereas *S. carpocapsae* and *S. glaseri* have both been studied for decades. The main limitation of the information available about natural host associations is the narrowness of the taxonomic representation of the existing nematode species; of more than 70 species of *Steinernema*, there are only 14 for which there are natural hosts recorded. In terms of biological control, these associations are of great consequence, since EPN species are most likely to reduce the populations of insect species that they naturally infect. In addition, by learning the natural host affiliations of EPN species, at least reasonable hypotheses can be made about natural host ranges based on host phylogenies.

Probably the best characterized example of natural populations of EPNs is the system found in populations of bush lupine (lupin), *Lupinus arboreus*, that are fed upon by caterpillars of the ghost moth, *Hepialus californicus*, which in turn are infected by the EPN *Heterorhabditis marelatus*. This system was described by Strong *et al.* (1995) as an example of a terrestrial trophic cascade and has been studied extensively since that time. In brief, Strong and colleagues noticed large die-offs of bush lupines in the Bodega Bay Marine Laboratory, located on the coast of northern California just north of San Francisco, USA. In these die-offs, thousands of bush lupines would succumb to herbivory by ghost moth caterpillars feeding on their roots and tunneling inside the stems. Careful sampling of surrounding soil revealed that some ghost moth caterpillars were infected by *H. marelatus* (noted as an undescribed species of *Heterorhabditis* at that time), and a simple garden experiment showed that when bush lupines were planted in soil where *H. marelatus* existed, death due to herbivory was reduced (Strong *et al.*, 1999). This series of studies continues today and has elucidated much about the impact of EPNs on natural host populations, nematode survivorship (Polis and Strong, 1996), population structure (Strong *et al.*, 1999; Preisser, 2003), and dispersal capabilities and propensity (Ram *et al.*, 2008).

Of course, the system is more complicated than what would be indicated by the three players mentioned

TABLE 11.3 Natural Host Breadths of Various Entomopathogenic Nematode Species Based on Reports of Isolations from Hosts in the Field

Nematode Species	Order	Families	No. of Species
Steinernema affinis	Diptera	2	2
S. anomaly (now *S. arenarium*)	Coleoptera	1	1
S. carpocapsae	Coleoptera	3	6
	Hymenoptera	2	2
	Diptera	1	1
	Lepidoptera	4	5
S. diaprepesi	Coleoptera	1	1
S. feltiae	Coleoptera	6	14
	Diptera	1	1
	Lepidoptera	1	5
S. glaseri	Coleoptera	2	4
S. kraussei	Hymenoptera	1	2
S. kushidai	Coleoptera	1	1
S. rarum	Lepidoptera	1	1
S. riobrave	Lepidoptera	1	2
S. scapterisci	Orthoptera	1	4
S. scarabaei [a]	Coleoptera	1	2
S. neocurtillae	Orthoptera	1	1
Steinernema sp.	Coleoptera	2	8
	Lepidoptera	1	3
Neosteinernema longicurvicaudum	Isoptera	1	1
Heterorhabditis bacteriophora	Coleoptera	3	6
	Lepidoptera	2	3
H. megidis	Coleoptera	2	4
H. zealandica	Coleoptera	1	1
H. marelatus [a]	Lepidoptera	1	1
Heterorhabditis sp.	Coleoptera	3	9

[a]Indicates entomopathogenic nematode species isolated naturally after Peters (1996) was published.
Adapted from Peters (1996).

above; a plant, and herbivore, and the herbivore's natural enemy never exist in a vacuum. For example, subsequent studies have shown that *H. marelatus* also has natural enemies, at this site and others, in the form of nematode-trapping fungi that reduce their populations, which also reduces their ability to protect the bush lupines from herbivory (Jaffee *et al.*, 1996). Indeed, the impact of soil biota on the survival of EPNs has long been implicated in the poor persistence of their applications as biological control agents (Kaya and Koppenhöfer, 1996). More recent work by Hodson *et al.* (2011) suggests that populations of soil mites can reduce applied nematodes in pistachio orchards. Thus, even in agricultural systems, which are considered simplified compared with natural ones, habitat complexity renders it difficult to predict EPN survival.

Steinernema scapterisci is one of the few nematodes isolated from its original host that has been employed commercially as a biological control agent. This nematode species was first isolated from infected mole crickets in Uruguay (Nyguen and Smart, 1990). The reason behind the search for this nematode was to provide classical biological control in Florida for the accidentally introduced tawny mole cricket, *Scapteriscus vicinus*. Parkman and Smart (1996) provide a detailed history of the first few years after the initial release of *S. scapterisci* for tawny mole cricket control and later for controlling the southern mole cricket, *Scapteriscus borellia*. To summarize, after the nematodes were isolated and delivered to Florida, they were released into three pastures that were infested with the mole crickets. After three years, the mole cricket populations were reduced by up to 98% with no additional releases of the nematodes. Furthermore, the nematodes were confirmed to have been transported by infected insects to areas where they were not released. The continued success of this nematode species in reducing a target pest indicates the importance of learning more about natural host ranges of EPNs.

11.9.2. Population Structure in Nature

Methods of predicting species distribution of soil fauna has been impeded by the cryptic nature of this habitat (Tiedje *et al.*, 1999). For years, species distribution has been studied in above-ground systems using many well-characterized models, and these systems serve as the bases for almost all ecological theory on the subject of species distribution and population structure. EPNs have several characteristics that make them an ideal model for studying soil fauna populations, especially in agroecosystems. The EPNs, as a group, are common in most soils (Hominick, 2002), are simple and easy to isolate (Kaya and Stock, 1997), and have significant impact on soil food webs (Kaya, 1990; Lewis, 2002), and most aspects of their biology are quite well known (Gaugler and Kaya, 1990; Gaugler, 2002; Grewal *et al.*, 2005). The population structure and distribution of EPNs currently receive attention primarily to improve their success as biological control

agents in agriculture, but they may also prove to be essential to the study of soil food webs.

From the standpoint of exploiting EPNs for biological control of insect pests, unpredictable persistence and distribution are impediments. For example, even when a target pest is known to be susceptible to infection, field efficacy can vary significantly in space and time (Shapiro-Ilan *et al.*, 2002). Their success depends on their ability to move, survive, and remain virulent long enough to reach and infect a host (Shapiro-Ilan *et al.*, 2006). For an application of EPNs to reduce pests, the nematodes must be able to move to reach the target, survive long enough to reach it, and remain able to infect the target once it is reached. The reasons that EPNs persist longer after some applications than others are not at all clear, but what is clear is that they are intimately tied to the nematodes' population dynamics (Stuart *et al.*, 2006). Thus, studying natural populations will provide clues as to why some populations persist and others do not. This approach should lead to hypotheses about why some applications persist and some do not.

Stuart *et al.* (2006) provide an extensive review of the spatial and temporal structure of EPNs, both in natural systems and in artificial settings. What drives the population structure of soil organisms is a challenging topic, mainly because of the difficulties associated with working in soil ecosystems; soil can be sampled to locate EPN IJs by baiting with susceptible hosts such as *G. mellonella*, or by isolating them directly from soil using various techniques such as the Baermann funnel or flotation and decanting (Barker, 1985), but this only provides snapshots of what is happening to the populations themselves. Simply stated, working with transparent organisms, most of which are less than a millimeter long, in the soil, is difficult. Nevertheless, some aspects of EPN population biology are quite well known, especially in the context of attempting to better exploit their capabilities to reduce herbivore populations.

Most EPN surveys have been conducted without considering how the populations are distributed within the sampled area. This is because most of these surveys have the goal of discovering new species and strains of nematodes to further screen for their utility as biological control agents. Thus, the information they provide is just presence or absence in a particular area. When studies do focus on the distribution of nematodes, they are found to be patchy in nature (Stuart *et al.*, 2006). However, not all species are distributed in the same way; some are more patchy than others. For instance, Campbell *et al.* (1995, 1996) studied the distribution of two EPN species, *H. bacteriophora* and *S. carpocapsae*, in a turfgrass ecosystem in New Jersey, USA. They found that *S. carpocapsae* populations were more evenly distributed than were *H. bacteriophora* and attributed this difference at least partially to their vastly different foraging behavior; *S. carpocapsae* is an ambusher and *H. bacteriophora* is a cruiser. All studies to date agree that EPNs are distributed in a patchy manner, but several different drivers for this distribution have been proposed which are surely not mutually exclusive. One possibility is that the large production of IJs from a single infected host combined with the limited dispersal capabilities of the IJs causes this patchy distribution (Efron *et al.*, 2001), and so to some extent, the distribution of the nematodes must reflect the distribution of their hosts. In addition, local extinctions of populations could be caused by changes in moisture levels of soil on a small level, or by patchily distributed natural enemies, such as mites or fungi. Finally, nematodes can be transported passively by running water, or potentially by hosts that are not yet infected (Campbell *et al.*, 1998), or phoretically by mites (Epsky *et al.*, 1988), or earthworms (Shapiro *et al.*, 1993). Drivers of this distribution pattern must be very common and strong, since within a few days or, at the most, weeks, homogeneous applications of *H. bacteriophora* return to the typical patchy distribution found so commonly in nature (Campbell *et al.*, 1998; Wilson *et al.*, 2003). Although manipulating these distributions would be of great benefit to biological control, the methods to do so remain enigmatic.

11.10. AGING AND LIFESPAN

There exists a vast literature concerned with the aging and lifespan of nematodes, primarily stemming from the use of free-living nematodes such as *Caenorhabditis briggsae*, *C. elegans*, *Turbatrix aceti*, and *Panagrellus redivivus* as models for studies of human aging (Lewis and Perez, 2004). When considering the aging and lifespan of EPNs in particular, however, only a fraction of those studies is directly relevant, owing to the life stage that is the target of study. "Aging" is generally defined in the gerontology literature as "the time-independent series of cumulative, progressive, intrinsic and deleterious functional and structural changes that usually begin to manifest themselves at reproductive maturity and eventually culminate in death" (Arking, 1999). Thus, most of the studies with the models mentioned above have the adult worms as the subject. Studies of EPNs are primarily concerned not with aging, as strictly defined by Arking (1999), but with the persistence of the IJ, or dauer larva, which is not reproductively mature. One important distinction between studying IJs versus adults is that the changes that take place in adults are irreversible, whereas once the IJ infects a host and resumes development, there are no lingering effects of a prolonged IJ stage. Arking (1999) goes on to suggest that to determine whether or not a condition is related to age, it must be "cumulative, progressive, intrinsic and deleterious". Studies of IJ persistence are commonly called aging studies, or lifespan studies, so for convention the term will

be used here. Although it is a semantic difference, it is important to highlight the difference between these two types of aging study — those that focus on adults versus those that focus on the IJ.

A key to the above definition is the idea of "time independence". Aging is generally thought of in terms of a calendar. That is, the objective of the study is to determine how many months an IJ lives. For many questions about EPNs, this is a reasonable approach, since shelf-life, field persistence, etc., can be measured in no other way. But for questions about how nematodes change with age, non-chronological time markers are superior because when comparing individuals, time to death is a more useful number than how long an organism has been alive. One way to think of this is that a biological marker instead of time would serve as the independent variable in studies of aging. Several different biomarkers have been proposed to serve as the independent variable for nematodes, including lipid content, muscle tissue degradation, acetylcholinesterase activity, esterase and phosphatase activity, and specific gravity.

Attempts at becoming less reliant on chronology to study aging have been made for EPNs. Selvan *et al.* (1993) measured biochemical energy reserves of six EPN species in an effort to understand the links between energy reserves, lifespan, and activity levels. Lipids, glycogen, and protein content were measured by extracting them from large batches (several thousand individuals) of IJs. A subsequent study by Lewis *et al.* (1995a) showed that as lipid levels dropped for *S. glaseri* IJs, their rate of establishment in hosts dropped as well. Fitters *et al.* (1997) first used image analysis densitometry to estimate the amount of neutral lipid remaining in IJs. Essentially, the technique entails obtaining a digital micrograph of an IJ and measuring the density in each pixel of the photo. The advantage to this technique is that one can obtain estimates of remaining energy reserves from individual nematodes. Fitters and Griffin (2006) studied the differences in lifespan among eight isolates of *Heterorhabditis megidis* by photographing heat-killed IJs; using this technique, they found that 50% energy depletion, as estimated using optical density, was very tightly correlated with 50% survival time across all eight isolates. This work strongly suggests that IJs actually die from starvation, as their energy reserves wane. This technique is a very useful tool for studying changes with age, especially when a single IJ can be photographed alive multiple times (Fig. 11.4).

As stated above, however, measuring changes chronologically is useful for many purposes. There is great variation in lifespan among EPN species. Lewis *et al.* (1995a) measured differences in IJ behavior related to age and found that rate of movement, number of bacteria carried per IJ, amount of lipid, and infectivity all decreased with time for *S. carpocapsae*, *S. glaseri*, and *H. bacteriophora*, albeit at different rates. Comparing maximum lifespans while

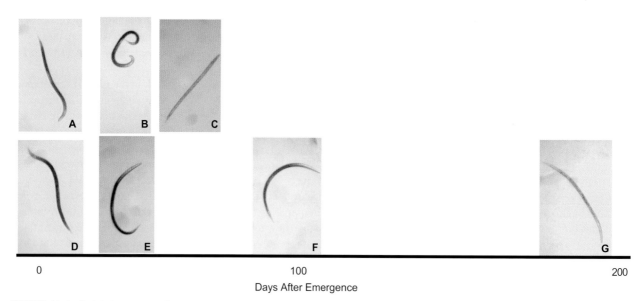

0   100   200

Days After Emergence

FIGURE 11.4 Serial photographs of two *Steinernema feltiae* infective juveniles over approximately 26 weeks. Both infective juveniles came from the same host cadaver and emerged at about the same time. (A–C) A short-lived individual; (D–G) a longer lived individual. (B) Median point of the short-lived nematode's lifespan; (F) median point of the others. (C, G) Nematodes on the day of death; notice the clear appearance indicating the loss of lipid energy reserves.

stored in deionized water at 25°C, *S. glaseri* was the longest lived, surviving for 36 weeks, *S. carpocapsae* was intermediate at 16 weeks, and *H. bacteriophora* had the shortest maximum lifespan at only seven weeks. These differences in lifespan were correlated somewhat with foraging strategy, in that *H. bacteriophora* was most active and a cruise forager and *S. carpocapsae* was least active and an ambush forager. However, the longest lived nematode species was *S. glaseri*, which is a cruise forager. Lewis *et al.* (1995a) suggested that the size of *S. glaseri* was a contributing factor, since they held so much more lipid than the other two species. Patel *et al.* (1997) support this contention, as they found that survival of *Steinernema* species IJs was a function of body size; *S. feltiae* [IJ length = 849 (736–950) μm] and *S. glaseri* [IJ length = 1130 (864–1448) μm] IJs persisted longer in storage than *S. carpocapsae* [IJ length = 558 (438–650) μm] and *S. riobrave* [IJ length = 622 (561–701) μm]. But lifespan also varies among strains of the same species. Fitters and Griffin (2006) reported that median lifespan varied from 5.6 to 12.5 weeks among eight isolates of *H. megidis*, when they were stored in deionized water at 20°C. No studies to date have been published on individual survival, but even among individual *S. feltiae* that emerge from the same cadaver, there seems to be significant variation in lifespan (Fig. 11.4) (E. E. Lewis, pers. obs.). Lifespan measured in this way is of great importance when evaluating a commercial product based on formulated EPNs, since companies strive to avoid selling dead IJs. These measures are also integral to developing models of population biology and persistence for EPNs.

IJ behaviors change with age, but not always in terms of slowing, becoming weaker, and eventually stopping. *Heterorhabditis* spp. IJs have been shown to have "phased" behavioral repertoires. That is, instead of a steady decline in movement or infectivity, there seem to be fluctuating patterns of peaks in these activities. Fitters and Griffin (2004) monitored spontaneous activity, stimulated activity and infectivity for three strains of *H. megidis* over a period of 10 weeks. Spontaneous activity generally declined with time, indicating that this is an energy-saving measure that would increase the lifespan of the IJs (Fitters and Griffin, 2004). However, activity that resulted from mechanical stimulation showed a different pattern with time, as IJs that were eight to 10 weeks old were stimulated to move more quickly after stimulation than they were at four to six weeks of age (Ryder and Griffin, 2003). Fitters and Griffin (2004) suggest here that as the IJs approach the point at which they will starve to death, they become more active and perhaps less risk averse. Because of the level of risk associated with initiating an infection, as pointed out by Fushing *et al.* (2008), this seems a plausible explanation of this pattern.

The temporal pattern in infectivity has been recorded multiple times for species in the *Heterorhabditis* and has been described as "phased infectivity." In this scenario, the population-level peak of infectivity is delayed by some time after a cohort of nematodes emerges from their parent cadaver. Hominick and Reid (1990) suggested that only a subset of IJs that emerge from a cadaver should be infective immediately and that phasing infectivity would be adaptive in that it would reduce the amount of competition among individual IJs that emerge from the same host cadaver. However, Dempsey and Griffin (2002) suggested an alternative explanation, based on the idea that IJs go through three phases of behavior. The first phase is one of dispersal, where the IJs have relatively low levels of infectivity; the second phase is one of high levels of infectivity and low levels of dispersal; and in the third phase, both of these behaviors decline. The temporal patterns of dispersal and infectivity in IJs turn out to be extremely complex and are based not only on IJ age, but also on the time at which they emerge from their host cadaver and how crowded the conditions within the cadaver were (Ryder and Griffin, 2002).

Studies of aging of EPN IJs remain primarily focused on how long they can persist, either as a product or after they have been applied to protect a crop. A great deal of effort is put forth to extend IJ persistence and will be addressed in the next section. However, it is important to understand the variability in longevity among species, strains, and even individuals. Average or median lifespan is only a small part of the information provided by studies of aging. It is often the case that the variation among individuals is where valuable information can be found.

11.11. SURVIVAL MECHANISMS

The nematode–bacterium complex is subjected to conditions that are detrimental and lethal on a regular basis, either in the field or as the active ingredients of commercial products. Some species of nematodes survive unfavorable conditions in a dormant state which prolongs lifespan, known as either diapause or quiescence. EPNs can be brought to a partial dormancy (Glazer, 2002) via exposure to desiccating conditions, but for the most part, they are unable to shut down metabolic functions in response to adverse conditions. For example, they do not exhibit anhydrobiosis in response to prolonged exposure to desiccating conditions. The mechanisms of how they cope with adverse environmental conditions have been studied and can serve as a basis for efforts to enhance their survival both as commercial products and after application to areas with less than optimal conditions.

Several categories of conditions have been the focus of research about EPN survival: desiccation, osmotic stress, temperature extremes, and ultraviolet light exposure. Most of these studies focus on the IJ stage, since this is the life stage that is adapted to life outside the host cadaver, and in most cases, this is also the life stage that is the active ingredient in commercial products. However, there are some examples of how the cadaver itself may be instrumental in surviving extreme environmental conditions and may also be of commercial interest.

11.11.1. Desiccation

When considering the effects of desiccation on nematodes, it must be remembered that nematodes are aquatic organisms and need a film of water surrounding them to move (Norton, 1978). Thus, very specialized responses to dry conditions exist for many nematode species. EPNs do not tolerate fast dehydration, and so dehydration must be done by exposure to conditions of high relative humidity (RH), but no free water. This process allows biochemical changes and behavioral responses to occur before a tolerable level of desiccation is achieved. Most studies of EPN desiccation tolerance have been undertaken with the goal of improving the storage stability and longevity of formulated commercial products based on the IJ stage, because the limited shelf-life of these products has long been recognized as a primary impediment to their wider use. The first step to extending life using desiccation would be to understand how desiccation affects their biology.

One way to find EPN isolates that can withstand desiccation is to search for them in areas that are dry. Solomon *et al.* (1999) found that two strains of *H. bacteriophora* that were isolated from desert regions of Israel, designated IS-5 and IS-6, were both more tolerant to desiccating conditions than was an isolate of the same species from Germany. Surviving nematodes showed typical aggregation and coiling behaviors in response to drying.

The behavioral strategies of nematodes that are stimulated by exposure to adverse conditions were reviewed by Wharton (2004). Two behaviors have been associated with many species of nematodes, including EPNs, when they are exposed to desiccating conditions; usually, these experimental conditions consist of exposure to reasonably high RH (above 80%) while the nematodes are on a non-absorbent surface (often a microscope slide or similar). The experimental conditions do not resemble the conditions that the nematodes would experience in nature, but it is thought that the rate of desiccation is much more important than the physical appearance of the experimental area. Both steinernematid and heterorhabditid IJs are known to aggregate into clumps of nematodes in dry conditions (Glazer, 2002). Aggregating is thought to offer some level of protection from desiccation, at least to those individuals at the center of the aggregation. The other behavioral response to desiccation is coiling. Nematodes form a coil with their bodies, and this may afford nematodes a way to slow their desiccation by reducing their surface area to volume ratio (Glazer, 2002). Whereas most species of *Steinernema* tested coil in response to desiccation, none of the *Heterorhabditis* species tested so far does. As the nematodes desiccate, they also shrink, presumably because of the removal of water from their bodies.

The primary biochemical change that occurs in nematodes, and many other metazoans, in desiccating conditions is an increase in trehalose and/or glycogen concentrations. Trehalose is found in almost every animal that undergoes anhydrobiosis. Solomon *et al.* (1999) measured an increase in trehalose levels in *S. feltiae* that had been exposed to low-humidity conditions (97% RH for three days). The function of trehalose is to stabilize membranes during desiccation, and its concentration increases in many animals when they are exposed to an array of environmental stressors, including temperature extremes, along with desiccation.

A possible mechanism for surviving desiccation has been suggested by Spence *et al.* (2010). When *S. carpocapsae*-killed *G. mellonella* are exposed to 0% humidity for up to several days, there is significant survival of the nematodes therein, even when up to 50% of the cadaver's mass has been lost owing to desiccation, for *S. riobrave*, *S. carpocapsae*, and *H. bacteriophora*. This finding suggests a strategy for survival that depends on the host cadaver for some shelter from dehydration and may represent a method of application or storage for EPN products (Koppenhöfer *et al.*, 1997; Spence *et al.*, 2010).

11.11.2. Osmotic Stress

Osmotic stress differs from desiccation because water is not removed from the surroundings of the nematode, but the level of solutes within the water imposes some stress on the nematode. High salinity levels inhibit nematode movement, but do not generally seem to be lethal (Griffin *et al.*, 1994; Nielsen *et al.*, 2011). The most common way for EPNs to be exposed to osmotic stress is via high-salinity soils. When IJs are applied to agricultural areas, especially where irrigation is common, they are often found to be ineffective. Kaspi *et al.* (2010) found that there were two main characteristics of soil that were detrimental to EPN efficacy: small particle size and high levels of salinity.

11.11.3. Temperature Extremes

Nematodes have two strategies for surviving freezing conditions: freeze tolerance, where the nematodes can actually survive when their tissues experience ice crystallization; and freeze avoidance, where the nematodes are supercooled, but their fluids do not actually freeze (Sayre, 1964). EPN differ in their response to freezing. Most EPN species studied to date, with the exception of *Heterorhabditis zealandica*, are freeze tolerant (Brown and Gaugler, 1998). Thus, *S. riobrave*, *S. carpocapsae*, and *S. glaseri* were all capable of surviving exposure to −4°C and −20°C, and *S. carpocapsae* had the best survival and *S. glaseri* the worst. Brown and Gaugler (1998) also showed that when IJs were exposed to these cold temperatures, their infectivity was reduced.

Cryopreservation is a common way of storing and maintaining cultures of nematodes. In general, the nematodes are pretreated in a solution of about 15% (v/v) glycerol for 48 h (Curran *et al.*, 1992). Exposure to glycerol elicits a similar response by the nematodes to exposure to desiccation, and indeed desiccation is sometimes used as a pretreatment for animals that are destined for cryopreservation. After the 48 h incubation period, the glycerol is removed by washing and the nematodes are placed into a cryovial and plunged into liquid nitrogen. This technique, sometimes with minor variation, provides high rates of survival for years and is commonly used.

Lewis and Shapiro-Ilan (2002) found that freezing EPNs while inside their cadavers allowed survival, and that IJs emerged from the previously frozen cadavers. As stated above, IJs can survive freezing conditions for limited durations when they are not pretreated in some way. Lewis and Shapiro-Ilan (2002) did not pretreat the insect cadavers, but just placed them into freezing conditions in the laboratory. While there may be limited practical use for this information in terms of storage, the data suggested that this could be a possible survival mechanism for over-wintering nematodes.

High temperatures are often fatal to nematodes. However, the degree to which nematodes can tolerate heat varies with species and strains, and depends in particular on the environmental conditions that occur where the strain is isolated. For example, the IS-5 isolate of *H. bacteriophora* mentioned above for its tolerance to desiccation was also found to be tolerant of very high temperatures; they can survive short-term exposure to 37°C, whereas another strain of *H. bacteriophora* had very low survival at this temperature (Glazer *et al.*, 1996). The IS-5 strain was also able to infect a host and reproduce successfully at temperatures up to 30°C, whereas the success of the HP88 strain of *H. bacteriophora* was much reduced at this temperature.

It is logical that nematodes isolated from warm climates should be able to tolerate warm temperatures, but they can also respond to high temperatures physiologically, by producing heat shock proteins (*hsp*) (Selvan *et al.*, 1996). Heat shock proteins are produced by virtually all animals in response to shock (high or low temperatures, toxins, or other stressors). Their function is to protect other proteins from the degradation that is caused by high temperatures and other adverse conditions. Their positive impact on EPN has been exploited by developing a transgenic heterorhabditid nematode that contained extra copies of the *hsp*70a gene from *C. elegans* (Hashmi *et al.*, 1998). Overexpression of heat shock proteins may afford increased protection of transgenic nematodes from exposure to high temperatures. In this series of experiments, Hashmi *et al.* (1998) preconditioned IJs to warm temperatures by exposure to 35°C for 2 h to induce heat shock protein production. When the preconditioned nematodes were challenged with 1 h exposure at 40°C, wild-type *H. bacteriophora* suffered almost 100% mortality, whereas the transgenic nematodes suffered less than 10% mortality. Genetic transformation did not affect the nematodes' persistence in the field (Gaugler *et al.*, 1997), virulence to different host species of insect, or reproduction (Wilson *et al.*, 1999). Transgenic nematodes were never commercialized, but this work opened the door to using transgenesis as a means of improving particular EPN characteristics.

11.12. FUTURE RESEARCH DIRECTIONS

Several different areas of research are promising for EPNs in the near future. One difficult, or perhaps deficient, aspect of the current body of research is the taxonomic narrowness of the discipline in general. Most of the work on EPNs, with the exception of the identification of new isolates, is conducted using those species about which the most is already known. For example, Lewis (2002) stated that 90% of all citations about EPN behavior were about only four species, and that only half of the species at the time had any citations about them other than their species descriptions. Sadly, this condition remains for the most part concerning all aspects of EPN biology. However, interest in new EPN species is only spurred when they have particular attributes that are desirable.

Most studies of the behavior, ecology, and physiology of EPNs are based on groups of individuals, but ignore interindividual variability. Indeed, much emphasis is placed on reducing experimental error rates; not just with EPNs, but with these types of studies in general. However, the variability among individual IJs, even from the same cadaver, can be significant and important. With the development of more sensitive tools, better imaging systems, and higher resolution measurement systems, it is now becoming possible to collect data from individual nematodes. Future studies should look at variability among individuals not as annoying experimental errors, but as a biological phenomenon deserving consideration.

Along the lines of taxonomic breadth is the question of which genera should be included in the "entomopathogenic nematodes". Traditionally, the Heterorhabditidae and the Steinernematidae have been included, but the description of two species in the genus *Oscheius* that infect insects and are associated with bacterial symbionts may challenge this classification. There are likely to be more species in the Rhabditidae that fall into this classification, and these species may turn out to be valuable biological control agents as well.

The level of understanding of nematode nutrition can be improved, and provide some very practical advances. For example, analysis of the nutritional contributions of *Photorhabdus* and *Xenorhabdus* to the symbiotic relationship with their nematode hosts could lead to increasing nematode yields for *in vitro* production. Production could be made more efficient by using a rationale-based approach; for example, by increasing the production of food signal to increase the level of IJ recovery or increasing the production of nutrients required by the nematode (e.g., vitamin B_{12}, BCFA).

Genetic and functional analysis of the wide range of bioactive molecules produced as a result of the extensive secondary metabolisms present in both *Photorhabdus* and *Xenorhabdus* can also be further explored. The new compounds could be used as lead chemicals for new insecticides or nematicides or may be developed as antibiotics.

Ecology beneath the soil is becoming a more common topic of research, but still lags far behind what is understood about above-ground systems. Here, the EPN community may play a role in providing a new "model system" for this study. With a large literature and a complete understanding of many components of EPN biology, these nematodes can be easily used to understand the workings of underground biological systems.

REFERENCES

AbuHatab, M., Selvan, S., & Gaugler, R. (1995). Role of proteases in penetration of insect gut by the entomopathogenic nematode *Steinernema glaseri* (Nematoda: Steinernematidae). *J. Invertebr. Pathol., 66*, 125–130.

Adams, B. J., Fodor, A., Koppenhöfer, H. S., Stackebrandt, E., Stock, S. P., & Klein, M. G. (2006). Biodiversity and systematics of nematode–bacterium entomopathogens. *Biol. Control, 37*, 32–49.

Aguinaldo, A. M. A., Turbeville, J. M., Linford, L. S., Rivera, M. C., Garey, J. R., Raff, R. A., & Lake, J. A. (1997). Evidence for a clade of nematodes, arthropods and other moulting animals. *Nature, 387*, 489–493.

Ahantarig, A., Chantawat, N., Waterfield, N. R., ffrench-Constant, R., & Kittayapong, P. (2009). PirAB toxin from *Photorhabdus asymbiotica* as a larvicide against dengue vectors. *Appl. Environ. Microbiol., 75*, 4627–4629.

Alsaiyaha, M. A. M., Ebssa, L., Zennera, A., O'Callaghana, K. M., & Griffin, C. T. (2009). Sex ratios and sex-biased infection behaviour in the entomopathogenic nematode genus *Steinernema. Int. J. Parasitol., 39*, 725–734.

Arking, R. (1999). *Biology of Aging* (2nd ed.). Sunderland: Sinauer Associates.

Aymeric, J. L., Givaudan, A., & Duvic, B. (2010). Imd pathway is involved in the interaction of *Drosophila melanogaster* with the entomopathogenic bacteria, *Xenorhabdus nematophila* and *Photorhabdus luminescens. Mol. Immunol., 47*, 2342–2348.

Bai, X., Adams, B. J., Ciche, T. A., Clifton, S., Gaugler, R., Hogenhout, S. A., Spieth, J., Sternberg, P. W., Wilson, R. K., & Grewal, P. S. (2009). Transcriptomic analysis of the entomopathogenic nematode *Heterorhabditis bacteriophora* TTO1. *BMC Genomics, 10*, 205.

Balasubramanian, N., Hao, Y. J., Toubarro, D., Nascimento, G., & Simões, N. (2009). Purification, biochemical and molecular analysis of a chymotrypsin protease with prophenoloxidase suppression activity from the entomopathogenic nematode *Steinernema carpocapsae*. *Int. J. Parasitol., 39*, 975−984.

Barker, K. R. (1985). Nematode extraction and bioassays. In K. R. Barker, C. C. Carter & J. N. Sasser (Eds.), *An Advanced Treatise on Meloidogyne, Vol. 2. Methodology* (pp. 19−35). Raleigh: North Carolina State University Graphics.

Baur, J. A., Pearson, K. J., Price, N. L., Jamieson, H. A., Lerin, C., Kalra, A., Prabhu, V. V., Allard, J. S., Lopez-Lluch, G., Lewis, K., Pistell, P. J., Poosala, S., Becker, K. G., Boss, O., Gwinn, D., Wang, M. Y., Ramaswamy, S., Fishbein, K. W., Spencer, R. G., Lakatta, E. G., Le Couteur, D., Shaw, R. J., Navas, P., Puigserver, P., Ingram, D. K., de Cabo, R., & Sinclair, D. A. (2006). Resveratrol improves health and survival of mice on a high-calorie diet. *Nature, 444*, 337−342.

Bedding, R. A. (1967). Parasitic and free-living cycles in the entomogenous nematodes of the genus *Deladenus*. *Nature, 214*, 174−175.

Bedding, R. A. (1984). Large scale production, storage and transport of the insect-parasitic nematodes *Neoaplectana* spp. and *Heterorhabditis* spp. *Ann. Appl. Biol., 104*, 117−120.

Bedding, R. A. (1993). Biological control of *Sirex noctilio* using the nematode *Deladenus siridicola*. In R. Bedding, R. Akhurst & H. Kaya (Eds.), *Nematodes and the Biological Control of Insect Pests* (pp. 11−20). East Melbourne: CSIRO Publications.

Bedding, R. A., & Akhurst, R. J. (1974). Use of the nematode *Deladenus siridicola* in the biological control of *Sirex noctilio* in Australia. *Aust. J. Entomol., 13*, 129−135.

Bedding, R. A., & Akhurst, R. J. (1978). Geographic distribution and host preference of *Deladenus* species (Nematoda: Neotylenchidae) parasitic in siricid woodwasps associated hymenopterous parasitoids. *Nematologica, 24*, 286−294.

Bedding, R. A., & Iede, E. T. (2005). Application of *Beddingia siricidicola* for *Sirex* woodwasp control. In P. Grewal, R.-U. Ehlers & D. I. Shapiro-Ilan (Eds.), *Nematodes as Biocontrol Agents* (pp. 411−418). Wallingford: CABI.

Bedding, R. A., & Molyneux, A. S. (1982). Penetration of insect cuticle by infective juveniles of *Heterorhabditis* spp. (Heterorhabditidae: Nematoda). *Nematologica, 28*, 354−359.

Bennett, H. P., & Clarke, D. J. (2005). The *pbgPE* operon in *Photorhabdus luminescens* is required for pathogenicity and symbiosis. *J. Bacteriol., 187*, 77−84.

Bintrim, S. B., & Ensign, J. C. (1998). Insertional inactivation of genes encoding the crystalline inclusion proteins of *Photorhabdus luminescens* results in mutants with pleiotropic phenotypes. *J. Bacteriol., 180*, 1261−1269.

Blackburn, M., Golubeva, E., Bowen, D., & ffrench-Constant, R. H. (1998). A novel insecticidal toxin from *Photorhabdus luminescens*, toxin complex a (Tca), and its histopathological effects on the midgut of *Manduca sexta*. *Appl. Environ. Microbiol., 64*, 3036−3041.

Blackburn, M. B., Martin, P. A., Kuhar, D., Farrar, R. R., Jr., & Gundersen-Rindal, D. E. (2011). The occurrence of *Photorhabdus*-like toxin complexes in *Bacillus thuringiensis*. *PLoS ONE, 6*. e18122.

Blaxter, M. (2011). Nematodes: the worm and its relatives. *PLoS Biol, 9*. e1001050.

Blaxter, M. L., Ley, P. D., Garey, J. R., Liu, L. X., Scheldeman, P., Vierstraete, A., Vanfleteren, J. R., Mackey, L. Y., Dorris, M.,

Frisse, L. M., Vida, J. T., & Thomas, W. K. (1998). A molecular evolutionary framework for the phylum nematoda. *Nature, 392*, 71−75.

Bode, H. B. (2009). Entomopathogenic bacteria as a source of secondary metabolites. *Curr. Opin. Chem. Biol., 13*, 224−230.

Bohan, D. A., & Hominick, W. M. (1997). Long-term dynamics of infectiousness within the infective-stage pool of the entomopathogenic nematode *Steinernema feltiae* (Site 76 strain) *Filipjev*. *Parasitology, 114*, 301−308.

Boina, D. R., Lewis, E. E., & Bloomquist, J. R. (2008). Nematicidal activity of anion transport blockers against *Meloidogyne incognita*, *Caenorhabditis elegans* and *Heterorhabditis bacteriophora*. *Pest Manag. Sci., 64*, 646−653.

Bowen, D., Rocheleau, T. A., Blackburn, M., Andreev, O., Golubeva, E., Bhartia, R., & ffrench-Constant, R. H. (1998). Insecticidal toxins from the bacterium *Photorhabdus luminescens*. *Science, 280*, 2129−2132.

Brachmann, A. O., Forst, S., Furgani, G. M., Fodor, A., & Bode, H. B. (2006). Xenofuranones A and B: phenylpyruvate dimers from *Xenorhabdus szentirmaii*. *J. Nat. Prod., 69*, 1830−1832.

Brachmann, A. O., Joyce, S. A., Jenke-Kodama, H., Schwar, G., Clarke, D. J., & Bode, H. B. (2007). A type II polyketide synthase is responsible for anthraquinone biosynthesis in *Photorhabdus luminescens*. *Chembiochem, 8*, 1721−1728.

Brown, I. M., & Gaugler, R. (1998). Survival of steinernematid nematodes exposed to freezing. *J. Therm. Biol., 23*, 75−80.

Brugirard-Ricaud, K., Givaudan, A., Parkhill, J., Boemare, N., Kunst, F., Zumbihl, R., & Duchaud, E. (2004). Variation in the effectors of the type III secretion system among *Photorhabdus* species as revealed by genomic analysis. *J. Bacteriol., 186*, 4376−4381.

Brugirard-Ricaud, K., Duchaud, E., Givaudan, A., Girard, P. A., Kunst, F., Boemare, N., Brehelin, M., & Zumbihl, R. (2005). Site-specific antiphagocytic function of the *Photorhabdus luminescens* type III secretion system during insect colonization. *Cell. Microbiol., 7*, 363−371.

Campbell, J. F., & Gaugler, R. (1993). Nictation behavior and its ecological implications in the host search strategies of entomopathogenic nematodes (Heterorhabditidae and Steinernematidae). *Behaviour, 126*, 155−170.

Campbell, J. F., & Kaya, H. K. (1999). How and why a parasitic nematode jumps. *Nature, 397*, 485−486.

Campbell, J. F., & Kaya, H. K. (2000). Influence of insect associated cues on the jumping behavior of entomopathogenic nematodes (*Steinernema* spp.). *Behaviour, 137*, 591−609.

Campbell, J. F., & Lewis, E. E. (2002). Entomopathogenic nematode search strategies. In E. E. Lewis, J. F. Campbell & M. V. K. Suhkdeho (Eds.), *Behavioral Ecology of Parasites* (pp. 13−38). Wallingford: CABI.

Campbell, J. F., Lewis, E. E., Yoder, F., & Gaugler, R. (1995). Entomopathogenic nematode (Heterorhabditidae and Steinernematidae) seasonal population dynamics and impact on insect populations in turfgrass. *Biol. Control, 5*, 598−606.

Campbell, J. F., Lewis, E. E., Yoder, F., & Gaugler, R. (1996). Spatial and temporal distribution of entomopathogenic nematodes in turf. *Parasitology, 113*, 473−482.

Campbell, J. F., Orza, G., Yoder, R., Lewis, E., & Gaugler, R. (1998). Spatial and temporal distribution of endemic and released entomopathogenic nematode populations in turfgrass. *Entomol. Exp. Appl., 86*, 1−11.

Campbell, J. F., Lewis, E. E., Stock, S. P., Nadler, S., & Kaya, H. K. (2003). Evolution of host search strategies in entomopathogenic nematodes. *J. Nematol., 35,* 142–145.

Chandra, H., Khandelwal, P., Khattri, A., & Banerjee, N. (2008). Type 1 fimbriae of insecticidal bacterium *Xenorhabdus nematophila* is necessary for growth and colonization of its symbiotic host nematode *Steinernema carpocapsiae. Environ. Microbiol., 10,* 1285–1295.

Chirico, J. (1996). Transmission of *Heterotylenchus autumnalis* nematodes into field populations of *Musca autumnalis* by release of laboratory-reared specimens. *Med. Vet. Entomol., 10,* 187–189.

Ciche, T. (2007). The biology and genome of *Heterorhabditis bacteriophora.* In The C. elegans Research Community. (Ed.), *WormBook* (pp. 1–9), doi/10.1895/wormbook.1.7.1. http://www.wormbook.org.

Ciche, T. A., & Ensign, J. C. (2003). For the insect pathogen *Photorhabdus luminescens,* which end of a nematode is out? *Appl. Environ. Microbiol., 69,* 1890–1897.

Ciche, T. A., Kim, K., Kaufmann-Daszczuk, B., Nguyen, K. C. Q., & Hall, D. H. (2008). Cell invasion and matricide during *Photorhabdus luminescens* transmission by *Heterorhabditis bacteriophora* nematodes. *Appl. Environ. Microbiol., 74,* 2275–2287.

Cornelis, G. R. (2006). The type III secretion injectisome. *Nat. Rev. Microbiol., 4,* 811–825.

Cowles, C. E., & Goodrich-Blair, H. (2004). Characterization of a lipoprotein, NilC, required by *Xenorhabdus nematophila* for mutualism with its nematode host. *Mol. Microbiol., 54,* 464–477.

Cowles, C. E., & Goodrich-Blair, H. (2006). nilR is necessary for coordinate repression of *Xenorhabdus nematophila* mutualism genes. *Mol. Microbiol., 62,* 760–771.

Cowles, K. N., Cowles, C. E., Richards, G. R., Martens, E. C., & Goodrich-Blair, H. (2007). The global regulator Lrp contributes to mutualism, pathogenesis and phenotypic variation in the bacterium *Xenorhabdus nematophila. Cell. Microbiol., 9,* 1311–1323.

Crawford, J. M., Kontnik, R., & Clardy, J. (2010). Regulating alternative lifestyles in entomopathogenic bacteria. *Curr. Biol., 20,* 69–74.

Curran, J., Gilbert, C., & Butler, K. (1992). Routine cryopreservation of isolates of *Steinernema* and *Heterorhabditis* spp. *J. Nematol., 24,* 269–270.

Daborn, P. J., Waterfield, N., Silva, C. P., Au, C. P., Sharma, S., & ffrench-Constant, R. H. (2002). A single *Photorhabdus* gene, makes caterpillars floppy (mcf), allows *Escherichia coli* to persist within and kill insects. *Proc. Natl. Acad. Sci. USA, 99,* 10742–10747.

Dempsey, C. M., & Griffin, C. T. (2002). Phased activity in *Heterorhabditis megidis* infective juveniles. *Parasitology, 124,* 605–613.

Derzelle, S., Duchaud, E., Kunst, F., Danchin, A., & Bertin, P. (2002). Identification, characterization, and regulation of a cluster of genes involved in carbapenem biosynthesis in *Photorhabdus luminescens. Appl. Environ. Microbiol., 68,* 3780–3789.

Derzelle, S., Turlin, E., Duchaud, E., Pages, S., Kunst, F., Givaudan, A., & Danchin, A. (2004). The PhoP–PhoQ two-component regulatory system of *Photorhabdus luminescens* is essential for virulence in insects. *J. Bacteriol., 186,* 1270–1279.

Dowling, A. J., Daborn, P. J., Waterfield, N. R., Wang, P., Streuli, C. H., & ffrench-Constant, R. H. (2004). The insecticidal toxin Makes caterpillars floppy (Mcf) promotes apoptosis in mammalian cells. *Cell Microbiol., 6,* 345–353.

Dowling, A. J., Waterfield, N. R., Hares, M. C., Le Goff, G., Streuli, C. H., & ffrench-Constant, R. H. (2007). The Mcf1 toxin induces apoptosis via the mitochondrial pathway and apoptosis is attenuated by mutation of the BH3-like domain. *Cell Microbiol., 9,* 2470–2484.

Duchaud, E., Rusniok, C., Frangeul, L., Buchrieser, C., Givaudan, A., Taourit, S., Bocs, S., Boursaux-Eude, C., Chandler, M., Charles, J. F., Dassa, E., Derose, R., Derzelle, S., Freyssinet, G., Gaudriault, S., Medigue, G., Lanois, A., Powell, K., Siguier, P., Vincent, R., Wingate, V., Zouine, M., Glaser, P., Boemare, N., Danchin, A., & Kunst, F. (2003). The genome sequence of the entomopathogenic bacterium *Photorhabdus luminescens. Nat. Biotechnol., 21,* 1307–1313.

Dutky, S. R. (1937). *Investigation of disease of the immature stages of the Japanese beetle.* PhD thesis. New Brunswick: Rutgers University.

Easom, C. A., & Clarke, D. J. (2008). Motility is required for the competitive fitness of entomopathogenic *Photorhabdus luminescens* during insect infection. *BMC Microbiol., 8,* 168.

Easom, C. A., Joyce, S. A., & Clarke, D. J. (2010). Identification of genes involved in the mutualistic colonization of the nematode *Heterorhabditis bacteriophora* by the bacterium *Photorhabdus luminescens. BMC Microbiol., 10,* 45.

Efron, D., Nestel, D., & Glazer, I. (2001). Spatial analysis of entomopathogenic nematodes and insect hosts in a citrus grove in a semi-arid region in Israel. *Environ. Entomol., 30,* 254–261.

Eleftherianos, I., Marokhazi, J., Millichap, P. J., Hodgkinson, A. J., Sriboonlert, A., ffrench-Constant, R. H., & Reynolds, S. E. (2006a). Prior infection of *Manduca sexta* with non-pathogenic *Escherichia coli* elicits immunity to pathogenic *Photorhabdus luminescens*: roles of immune-related proteins shown by RNA interference. *Insect Biochem. Mol. Biol., 36,* 517–525.

Eleftherianos, I., Millichap, P. J., ffrench-Constant, R. H., & Reynolds, S. E. (2006b). RNAi suppression of recognition protein mediated immune responses in the tobacco hornworm *Manduca sexta* causes increased susceptibility to the insect pathogen *Photorhabdus. Dev. Comp. Immunol., 30,* 1099–1107.

Eleftherianos, I., Boundy, S., Joyce, S. A., Aslam, S., Marshall, J. W., Cox, R. J., Simpson, T. J., Clarke, D. J., ffrench-Constant, R. H., & Reynolds, S. E. (2007). An antibiotic produced by an insect-pathogenic bacterium suppresses host defenses through phenoloxidase inhibition. *Proc. Natl. Acad. Sci. USA, 104,* 2419–2424.

Eleftherianos, I., Joyce, S., ffrench-Constant, R. H., Clarke, D. J., & Reynolds, S. E. (2010). Probing the tri-trophic interaction between insects, nematodes and *Photorhabdus. Parasitology, 137,* 1695–1706.

Epsky, N. D., Walter, D. E., & Capinera, J. L. (1988). Potential role of nematophagous microarthropods as biotic mortality factors of entomopathogenic nematodes (Rhabditida: Steinernematidae, Heterorhabditidae). *J. Econ. Entomol., 81,* 821–825.

Fischer-Le Saux, M., Viallard, V., Brunel, B., Normand, P., & Boemare, N. E. (1999). Polyphasic classification of the genus *Photorhabdus* and proposal of new taxa: *P. luminescens* subsp. *luminescens* subsp. nov., *P. luminescens* subsp. *akhurstii* subsp. nov., *P. luminescens* subsp. *laumondii* subsp. nov., *P. temperata* sp. nov., *P. temperata* subsp. *temperata* subsp. nov. and *P. asymbiotica* sp. nov. *Int. J. Syst. Bacteriol., 49,* 1645–1656.

Fitters, P. F. L., & Griffin, C. T. (2004). Spontaneous and induced activity of *Heterorhabditis megidis* infective juveniles during storage. *Nematology, 6,* 911–917.

Fitters, P. F. L., & Griffin, C. T. (2006). Survival, starvation, and activity in *Heterorhabditis megidis* (Nematoda: Heterorhabditidae). *Biol. Control, 37,* 82–88.

Fitters, P. F. L., Wright, D. J., Miejer, E. M. J., & Griffin, C. T. (1997). Estimation of lipid reserves in unstained living and dead nematodes by image analysis. *J. Nematol., 29*, 160–167.

Fushing, H., Zhu, L., Shapiro-Ilan, D. I., Campbell, J. F., & Lewis, E. E. (2008). State-space based mass event-history model I: many decision-making agents with one target. *Ann. Appl. Stat., 2*, 1503–1522.

Gaugler, R. (Ed.). (2002). *Entomopathogenic Nematology.* Wallingford: CABI.

Gaugler, R. & Kaya, H. K. (Eds.). (1990). *Entomopathogenic Nematodes in Biological Control.* Boca Raton: CRC Press.

Gaugler, R., Campbell, J. F., Selvan, S., & Lewis, E. E. (1992). Large-scale inoculative release of the entomopathogenic nematode *Steinernema glaseri*: assessment 50 years later. *Biol. Control, 2*, 181–187.

Gaugler, R., Wilson, M., & Shearer, P. (1997). Field release and environmental fate of a transgenic entomopathogenic nematode. *Biol. Control, 9*, 75–80.

Gaugler, R., Lewis, E. E., & Stuart, R. J. (1998). Ecology in the service of biological control: the case of entomopathogenic nematodes. *Oecologia, 109*, 483–489.

Georgis, R., & Gaugler, R. (1991). Predictability in biological control using entomopathogenic nematodes. *J. Econ. Entomol., 84*, 713–720.

Glaser, R. W., & Fox, H. (1930). A nematode parasite of the Japanese beetle (*Popillia japonica* Newm.). *Science, 71*, 16–17.

Glaser, R. W., McCoy, E. E., & Girth, H. B. (1942). The biology and culture of *Neoaplectana chresima*, a new nematode parasitic in insects. *J. Parasitol., 28*, 123–126.

Glazer, I. (2002). Survival biology. In R. Gaugler (Ed.), *Entomopathogenic Nematology* (pp. 169–187). Wallingford: CABI.

Glazer, I., Kozodoi, E., Hashmi, G., & Gaugler, R. (1996). Biological characteristics of the entomopathogenic nematode *Heterorhabditis* sp. Is-5: a heat tolerant isolate from Israel. *Nematologica, 42*, 481–492.

Goodrich-Blair, H., & Clarke, D. J. (2007). Mutualism and pathogenesis in *Xenorhabdus* and *Photorhabdus*: two roads to the same destination. *Mol. Microbiol., 64*, 260–268.

de Gregorio, E., Spellman, P. T., Tzou, P., Rubin, G. M., & Lemaitre, B. (2002). The Toll and Imd pathways are the major regulators of the immune response in. *Drosophila. EMBO J., 21*, 2568–2579.

Grewal, P. S., Selvan, S., Lewis, E. E., & Gaugler, R. (1993). Males as the colonizing sex in insect parasitic nematodes. *Experientia, 49*, 605–608.

Grewal, P. S., Lewis, E. E., Campbell, J. F., & Gaugler, R. (1994). Searching behavior as a predictor of foraging strategy for entomopathogenic nematodes. *Parasitology, 108*, 207–215.

Grewal, P., Lewis, E. E., & Gaugler, R. (1996). Response of infective stage parasites (Rhabditida: Steinernematidae) to volatile cues from infected hosts. *J. Chem. Ecol., 23*, 503–515.

Grewal, P. S., Ehlers, R.-U. & Shapiro-Ilan, D. I. (Eds.). (2005). *Nematodes as Biocontrol Agents.* Wallingford: CABI.

Griffin, C. T., Finnegan, M. M., & Downes, M. J. (1994). Environmental tolerances and the dispersal of *Heterorhabditis*: survival and infectivity of European *Heterorhabditis* following prolonged immersion in seawater. *Fund. Appl. Nematol., 17*, 415–421.

Griffin, C. T., Chaerani, R., Fallon, D., Reid, A. P., & Downes, M. J. (2000). Occurrence and distribution of the entomopathogenic nematodes *Steinernema* spp. and *Heterorhabditis indica* in Indonesia. *J. Helminth., 74*, 143–150.

Griffin, C. T., Boemare, N. E., & Lewis, E. E. (2005). Biology and behaviour. In P. Grewal, R.-U. Ehlers & D. I. Shapiro-Ilan (Eds.), *Nematodes as Biocontrol Agents* (pp. 47–64). Wallingford: CABI.

Gualtieri, M., Aumelas, A., & Thaler, J. O. (2009). Identification of a new antimicrobial lysine-rich cyclolipopeptide family from *Xenorhabdus nematophila. J. Antibiot., 62*, 295–302.

Gunn, J. S., Lim, K. B., Krueger, J., Kim, K., Guo, L., Hackett, M., & Miller, S. I. (1998). PmrA–PmrB-regulated genes necessary for 4-aminoarabinose lipid A modification and polymyxin resistance. *Mol. Microbiol., 27*, 1171–1182.

Guo, L., Lim, K. B., Gunn, J. S., Bainbridge, B., Darveau, R. P., Hackett, M., & Miller, S. I. (1997). Regulation of lipid A modifications by *Salmonella typhimurium* virulence genes phoP–phoQ. *Science, 276*, 250–253.

Hallem, E. A., Rengarajan, M., Ciche, T. A., & Sternberg, P. W. (2007). Nematodes, bacteria, and flies: a tripartite model for nematode parasitism. *Curr. Biol., 17*, 898–904.

Hallem, E. A., Dillman, A. R., Hong, A. V., Zhang, Y., Yano, J. M., DeMarco, S. F., & Sternberg, P. W. (2011). A sensory code for host seeking in parasitic nematodes. *Curr. Biol., 21*, 377–383.

Hashmi, W., Hashmi, G., Glazer, I., & Gaugler, R. (1998). Thermal response of *Heterorhabditis bacteriophora* transformed with the *Caenorhabditis elegans hsp*70 encoding gene. *J. Exp. Zool., 281*, 164–170.

Hares, M. C., Hinchliffe, S. J., Strong, P. C., Eleftherianos, I., Dowling, A. J., ffrench-Constant, R. H., & Waterfield, N. (2008). The *Yersinia pseudotuberculosis* and *Yersinia pestis* toxin complex is active against cultured mammalian cells. *Microbiology, 154*, 3503–3517.

Herbert Tran, E. E., & Goodrich-Blair, H. (2009). CpxRA contributes to *Xenorhabdus nematophila* virulence through regulation of *lrhA* and modulation of insect immunity. *Appl. Environ. Microbiol., 75*, 3998–4006.

Heungens, K., Cowles, C. E., & Goodrich-Blair, H. (2002). Identification of *Xenorhabdus nematophila* genes required for mutualistic colonization of *Steinernema carpocapsae* nematodes. *Mol. Microbiol., 45*, 1337–1353.

Hodson, A. K., Friedman, M. L., Wu, L. N., & Lewis, E. E. (2011). European earwig (*Forficula auricularia*) as a novel host for the entomopathogenic nematode *Steinernema carpocapsae. J. Invertebr. Pathol., 107*, 60–64.

Hominick, W. M. (2002). Biogeography. In R. Gaugler (Ed.). *Entomopathogenic Nematology* (pp. 115–144). Wallingford: CABI.

Hominick, W. M., & Reid, A. P. (1990). Behaviour of infective juveniles. In R. Gaugler & H. K. Kaya (Eds.). *Entomopathogenic Nematodes in Biological Control* (pp. 327–345). Boca Raton: CRC Press.

Hu, K., & Webster, J. M. (2000). Antibiotic production in relation to bacterial growth and nematode development in *Photorhabdus–Heterorhabditis* infected *Galleria mellonella* larvae. *FEMS Microbiol. Lett., 189*, 219–223.

Hu, K. J., Li, J. X., Wang, W. J., Wu, H. M., Lin, H., & Webster, J. M. (1998). Comparison of metabolites produced *in vitro* and *in vivo* by *Photorhabdus luminescens*, a bacterial symbiont of the entomopathogenic nematode *Heterorhabditis megidis. Can. J. Microbiol., 44*, 1072–1077.

Hu, K. J., Li, J. X., & Webster, J. M. (1999). Nematicidal metabolites produced by *Photorhabdus luminescens* (Enterobacteriaceae), bacterial symbiont of entomopathogenic nematodes. *Nematology, 1*, 457–469.

Hu, P. J. (2007). Dauer. In The C. elegans Research Community. (Ed.), *WormBook* (pp. 1–19), doi/10.1895/wormbook.1.7.1. http://www.wormbook.org.

Hurst, M. R., Glare, T. R., & Jackson, T. A. (2004). Cloning *Serratia entomophila* antifeeding genes – a putative defective prophage active against the grass grub *Costelytra zealandica*. *J. Bacteriol., 186*, 5116–5128.

Jaffee, B. A., Strong, D. R., & Muldoon, A. E. (1996). Nematode-trapping fungi of a natural shrubland: tests for food chain involvement. *Mycologia, 88*, 554–564.

Jing, Y., Toubarro, D., Hao, Y., & Simões, N. (2010). Cloning, characterisation and heterologous expression of an astacin metalloprotease, *Sc*-AST, from the entomoparasitic nematode *Steinernema carpocapsae*. *Mol. Biochem. Parasitol., 174*, 101–108.

Johnigk, S.-A., & Ehlers, R.-U. (1999). *Endotokia matricida* in hermaphrodites of *Heterorhabditis* spp. and the effect of the food supply. *Nematology, 1*, 717–726.

Joyce, S. A., & Clarke, D. J. (2003). A *hexA* homologue from *Photorhabdus* regulates pathogenicity, symbiosis and phenotypic variation. *Mol. Microbiol., 47*, 1445–1457.

Joyce, S. A., Brachmann, A. O., Glazer, I., Lango, L., Schwar, G., Clarke, D. J., & Bode, H. B. (2008). Bacterial biosynthesis of a multipotent stilbene. *Angewantde Chemie, 47*, 1942–1945.

Kaspi, R., Ross, A., Hodson, A., Stevens, G., Kaya, H., & Lewis, E. E. (2010). Foraging efficacy of the entomopathogenic nematode *Steinernema riobrave* in different soil types from California citrus groves. *Appl. Soil Ecol., 45*, 243–253.

Kaya, H. K. (1990). Soil ecology. In R. Gaugler & H. K. Kaya (Eds.), *Entomopathogenic Nematodes in Biological Control* (pp. 93–116). Boca Raton: CRC Press.

Kaya, H. K., & Koppenhöfer, A. M. (1996). Effects of microbial and other antagonistic organism and competition on entomopathogenic nematodes. *Biocontrol Sci. Technol., 6*, 357–371.

Kaya, H. K., & Moon, R. D. (1980). Influence of protein in the diet of face fly on the development of its nematode parasite, *Heterotylenchus autumnalis. Ann. Entomol. Soc. Am., 73*, 547–552.

Kaya, H. K., & Stock, S. P. (1997). Techniques in insect nematology. In L. A. Lacey (Ed.), *Manual of Techniques in Insect Pathology* (pp. 281–324). New York: Academic Press.

Kazakov, A. E., Rodionov, D. A., Alm, E., Arkin, A. P., Dubchak, I., & Gelfand, M. S. (2009). Comparative genomics of regulation of fatty acid and branched-chain amino acid utilization in proteobacteria. *J. Bacteriol., 191*, 52–64.

Kerry, B. R., & Hominick, W. M. (2002). Biological control. In D. L. Lee (Ed.), *The Biology of Nematodes* (pp. 483–509). London: Taylor and Francis.

Kim, Y., Ji, D., Cho, S., & Park, Y. (2005). Two groups of entomopathogenic bacteria, *Photorhabdus* and *Xenorhabdus*, share an inhibitory action against phospholipase A2 to induce host immunodepression. *J. Invertebr. Pathol., 89*, 258–264.

Kniazeva, M., Crawford, Q. T., Seiber, M., Wang, C. Y., & Han, M. (2004). Monomethyl branched-chain fatty acids play an essential role in *Caenorhabditis elegans* development. *PLoS Biol., 2*, E257.

Kniazeva, M., Euler, T., & Han, M. (2008). A branched-chain fatty acid is involved in post-embryonic growth control in parallel to the insulin receptor pathway and its biosynthesis is feedback-regulated in C. elegans. *Genes Dev., 22*, 2102–2110.

Kontnik, R., Crawford, J. M., & Clardy, J. (2010). Exploiting a global regulator for small molecule discovery in *Photorhabdus luminescens. ACS Chem. Biol., 5*, 659–665.

Koppenhöfer, A. M., Baur, M. E., Stock, S. P., Choo, H. Y., Chinnasri, B., & Kaya, H. K. (1997). Survival of entomopathogenic nematodes within host cadavers in dry soil. *Appl. Soil Ecol., 6*, 231–240.

Krafsur, E. S., & Moon, R. D. (1997). Bionomics of the face fly, *Musca autumnalis. Annu. Rev. Entomol., 42*, 503–523.

Lang, A. E., Schmidt, G., Schlosser, A., Hey, T. D., Larrinua, I. M., Sheets, J. J., Mannherz, H. G., & Aktories, K. (2010). *Photorhabdus luminescens* toxins ADP-ribosylate actin and RhoA to force actin clustering. *Science, 327*, 1139–1142.

Lango, L., & Clarke, D. J. (2010). A metabolic switch is involved in lifestyle decisions in *Photorhabdus luminescens. Mol. Microbiol., 77*, 1394–1405.

Lemaitre, B., & Hoffmann, J. (2007). The host defense of *Drosophila melanogaster. Annu. Rev. Immunol., 25*, 697–743.

Lemaitre, B., Kromer-Metzger, E., Michaut, L., Nicolas, E., Meister, M., Georgel, P., Reichart, J.-M., & Hoffmann, J. A. (1995). A recessive mutation, immune deficiency (*imd*), defines two distinct control pathways in the *Drosophila* host defense. *Proc. Natl. Acad. Sci. USA, 92*, 9465–9469.

Leulier, F., Parquet, C., Pili-Floury, S., Ryu, J. H., Caroff, M., Lee, W. J., Mengin-Lecreulx, D., & Lemaitre, B. (2003). The *Drosophila* immune system detects bacteria through specific peptidoglycan recognition. *Nat. Immunol., 4*, 478–484.

Lewis, E. E. (2002). Behavioral ecology. In R. Gaugler (Ed.), *Entomopathogenic Nematology* (pp. 205–224). Wallingford: CABI.

Lewis, E. E., & Gaugler, R. (1994). Entomopathogenic nematode sex ratio relates to foraging strategy. *J. Invertebr. Pathol., 64*, 238–242.

Lewis, E. E., & Perez, E. E. (2004). Aging and developmental behavior. In R. Gaugler & A. Bilgrami (Eds.), *Nematode Behaviour* (pp. 151–176). Wallingford: CABI.

Lewis, E. E., & Shapiro-Ilan, D. I. (2002). Host cadavers protect entomopathogenic nematodes during freezing. *J. Invertebr. Pathol., 81*, 25–32.

Lewis, E. E., Gaugler, R., & Harrison, R. (1992). Entomopathogenic nematode host finding: response to contact cues by cruise and ambush foragers. *Parasitology, 105*, 309–315.

Lewis, E. E., Gaugler, R., & Harrison, R. (1993). Response of cruiser and ambusher entomopathogenic nematodes (Steinernematidae) to host volatile cues. *Can. J. Zool., 71*, 765–769.

Lewis, E. E., Selvan, S., Campbell, J. F., & Gaugler, R. (1995a). Changes in foraging behaviour during the infective stage of entomopathogenic nematodes. *Parasitology, 110*, 583–590.

Lewis, E. E., Grewal, P. S., & Gaugler, R. (1995b). Hierarchical order of host cues in parasite foraging strategies. *Parasitology, 110*, 207–213.

Lewis, E. E., Ricci, M., & Gaugler, R. (1996). Host recognition behavior reflects host suitability for the entomopathogenic nematode, *Steinernema carpocapsae. Parasitology, 113*, 573–579.

Lewis, E. E., Campbell, J., Griffin, C., Kaya, H., & Peters, A. (2006). Behavioral ecology of entomopathogenic nematodes. *Biol. Control, 38*, 66–79.

Li, J., Chen, G., Wu, H., & Webster, J. M. (1995). Identification of two pigments and a hydroxystilbene antibiotic from *Photorhabdus luminescens. Appl. Environ. Microbiol., 61*, 4329–4333.

Martens, E. C., & Goodrich-Blair, H. (2005). The *Steinernema carpocapsae* intestinal vesicle contains a subcellular structure with which

Xenorhabdus nematophila associates during colonization initiation. *Cell. Microbiol., 7,* 1723−1735.

Martens, E. C., Heungens, K., & Goodrich-Blair, H. (2003). Early colonization events in the mutualistic association between *Steinernema carpocapsae* nematodes and *Xenorhabdus nematophila* bacteria. *J. Bacteriol., 185,* 3147−3154.

Martens, E. C., Russell, F. M., & Goodrich-Blair, H. (2005). Analysis of *Xenorhabdus nematophila* metabolic mutants yields insight into stages of *Steinernema carpocapsae* nematode intestinal colonization. *Mol. Microbiol., 58,* 28−45.

Nguyen, K. B., & Smart, G. C. (1990). *Steinernema scapterisci* n. sp. (Rhabditida: Steinernematidae). *J. Nematol., 22,* 187−199.

Nickle, W. R. (1967). *Heterotylenchus autumnalis* sp. n. (Nematoda: Sphaerulariidae), a parasite of the face fly, *Musca autumnalis* De Geer. *J. Parasitol., 53,* 398−401.

Nielsen, A. L., Spence, K. O., Nakatani, J., & Lewis, E. E. (2011). Effect of soil salinity on entomopathogenic nematode survivorship and behavior. *Nematology, 13.* (in press).

Norton, D. C. (1978). *Ecology of Plant-Parasitic Nematodes.* New York: John Wiley & Sons.

Park, D., Ciezki, K., van der Hoeven, R., Singh, S., Reimer, D., Bode, H. B., & Forst, S. (2009). Genetic analysis of xenocoumacin antibiotic production in the mutualistic bacterium *Xenorhabdus nematophila. Mol. Microbiol., 73,* 938−949.

Parkman, J. P., & Smart, G. C. (1996). Entomopathogenic nematodes, a case study: introduction of *Steinernema scapterisci* in Florida, USA. *Biocontrol Sci. Technol., 6,* 413−419.

Patel, M. N., Stolinski, M., & Wright, D. J. (1997). Neutral lipids and the assessment of infectivity in entomopathogenic nematodes: observations on four *Steinernema* species. *Parasitology, 114,* 489−496.

Peng, Y., Platzer, E. G., Song, J., & Peloquin, J. (1992). Susceptibility of thirty-one species of mosquitoes to *Romanomermis yunanensis. J. West China Univ. Med. Sci., 23,* 412−415.

Pervaiz, S. (2003). Resveratrol: from grapevines to mammalian biology. *FASEB J., 17,* 1975−1985.

Peters, A. (1996). The natural host range of *Steinernema* and *Heterorhabditis* spp. and their impact on insect populations. *Biocontrol Sci. Technol., 6,* 389−402.

Peters, A., & Ehlers, R.-U. (1994). Susceptibility of leatherjackets (*Tipula paludosa* and *Tipula oleracea*; Tipulidae; Nematocera) to the entomopathogenic nematode *Steinernema feltiae. J. Invertebr. Pathol., 63,* 163−171.

Platzer, E. G. (1981). Biological control of mosquitoes with mermithids. *J. Nematol., 13,* 257−262.

Platzer, E. G., Mullens, B. A., & Shamseldean, M. M. (2005). Mermithid nematodes. In P. Grewal, R.-U. Ehlers & D. I. Shapiro-Ilan (Eds.), *Nematodes as Biocontrol Agents* (pp. 411−418). Wallingford: CABI.

Poinar, G. O. (1975). Description and biology of a new insect parasitic rhabditoid, *Heterorhabditis bacteriophora* n. gen. n. sp. (Rhabditida: Heterorhabditidae n. fam.). *Nematologica, 21,* 463−470.

Poinar, G. O., Jr. (1990). Taxonomy and biology of Steinernematidae and Heterorhabditidae. In R. Gaugler & H. K. Kaya (Eds.), *Entomopathogenic Nematodes in Biological Control* (pp. 23−61). Boca Raton: CRC Press.

Polis, G. A., & Strong, D. R. (1996). Food web complexity and community dynamics. *Am. Nat., 147,* 813−846.

Preisser, E. L. (2003). Field evidence for a rapidly cascading underground food web. *Ecology, 84,* 869−874.

Raetz, C. R. H., Reynolds, C. M., Trent, M. S., & Bishop, R. E. (2007). Lipid A modification systems in gram-negative bacteria. *Annu. Rev. Biochem., 76,* 295−329.

Ram, K., Gruner, D. S., McLaughlin, J. P., Evan, L., Preisser, E. L., & Strong, D. R. (2008). Dynamics of a subterranean trophic cascade in space and time. *J. Nematol., 40,* 85−92.

Rasmann, S., Kollner, T. G., Degenhardt, J., Hiltpold, I., Toepfer, S., Kuhlmann, U., Gershenzon, J., & Turlings, T. C. J. (2005). Recruitment of entomopathogenic nematodes by insect-damaged maize roots. *Nature, 434,* 732−737.

Reed, E. E., & Wallace, H. R. (1965). Leaping locomotion by an insect-parasitic nematode. *Nature, 206,* 210−211.

Renn, N. (1998). Routes of penetration of the entomopathogenic nematode *Steinernema feltiae* attacking larval and adult houseflies (*Musca domestica*). *J. Invertebr. Pathol., 72,* 281−287.

Richards, G. R., Herbert, E. E., Park, Y., & Goodrich-Blair, H. (2008). *Xenorhabdus nematophila* lrhA is necessary for motility, lipase activity, toxin expression, and virulence in *Manduca sexta* insects. *J. Bacteriol., 190,* 4870−4879.

Rickert-Campbell, L., & Gaugler, R. (1991). Role of the sheath in desiccation tolerance of two entomopathogenic nematodes. *Nematologica, 37,* 324−332.

Ryder, J. J., & Griffin, C. T. (2002). Density-dependent fecundity and infective juvenile production in the entomopathogenic nematode, *Heterorhabditis megidis. Parasitology, 125,* 83−92.

Ryder, J. J., & Griffin, C. T. (2003). Phased infectivity in *Heterorhabditis megidis*: the effects of infection density in the parental host and filial generation. *Int. J. Parasitol., 33,* 1013−1018.

Sayre, R. M. (1964). Cold-hardiness of nematodes I. Effects of rapid freezing on the eggs and larvae of *Meloidogyne incognita* and M. *hapla. Nematologica, 10,* 168−179.

Selvan, S., Gaugler, R., & Lewis, E. E. (1993). Biochemical energy reserves of entomopathogenic nematodes. *J. Parasitol., 79,* 167−172.

Selvan, S., Grewal, P. S., Leustek, T., & Gaugler, R. (1996). Heat shock enhances thermotolerance of infective juvenile insect-parasitic nematodes *Heterorhabditis bacteriophora* (Rhabditida: Heterorhabditidae). *Experientia, 52,* 727−730.

Shapiro, D. I., Berry, E. C., & Lewis, L. C. (1993). Interactions between nematodes and earthworms: enhanced dispersal of *Steinernema carpocapsae. J. Nematol., 25,* 189−192.

Shapiro-Ilan, D. I., Gouge, D. H., & Köppenhöfer, A. M. (2002). Factors affecting commercial success: case studies in cotton, turf, and citrus. In R. Gaugler (Ed.), *Entomopathogenic Nematology* (pp. 333−355). Wallingford: CABI.

Shapiro-Ilan, D. I., Stuart, R. J., & McCoy, C. W. (2006). A comparison of entomopathogenic nematode longevity in soil under laboratory conditions. *J. Nematol, 38,* 119−129.

Sheets, J. J., Hey, T. D., Fencil, K. J., Burton, S. L., Ni, W., Lang, A. E., Benz, R., & Aktories, K. (2011). Insecticidal toxin complex proteins from *Xenorhabdus nematophilus*: structure and pore formation. *J. Biol. Chem., 286,* 22742−22749.

Simões, N., Caldas, C., Rosa, J. S., Bonifassi, E., & Laumond, C. (2000). Pathogenicity caused by high virulent and low virulent strains of *Steinernema carpocapsae* to *Galleria mellonella. J. Invertebr. Pathol., 75,* 47−54.

Snyder, H., Stock, S. P., Kim, S. K., Flores-Lara, Y., & Forst, S. (2007). New insights into the colonization and release processes of *Xenorhabdus nematophila* and the morphology and ultrastructure of the bacterial receptacle of its nematode host, *Steinernema carpocapsae*. *Appl. Environ. Microbiol., 73*, 5338–5346.

Solomon, A., Paperna, I., & Glazer, I. (1999). Desiccation survival of the entomopathogenic nematode *Steinernema feltiae*: induction of anhydrobiosis. *Nematology, 1*, 61–68.

Somvanshi, V. S., Kaufmann-Daszczuk, B., Kim, K. S., Mallon, S., & Ciche, T. A. (2010). *Photorhabdus* phase variants express a novel fimbrial locus, *mad*, essential for symbiosis. *Mol. Microbiol., 77*, 1021–1038.

Song, C., Seo, S., Shrestha, S., & Kim, Y. (2011). Bacterial metabolites of an entomopathogenic bacterium, *Xenorhabdus nematophila*, inhibit a catalytic activity of phenoloxidase of the diamondback moth, *Plutella xylostella*. *J. Microbiol. Biotechnol., 21*, 317–322.

Spence, K. O., Stevens, G. N., Arimoto, H., Ruiz-Vega, J., Kaya, H. K., & Lewis, E. E. (2010). Effect of insect cadaver desiccation and soil water potential during rehydration on entomopathogenic nematode (Rhabditida: Steinernematidae and Heterorhabditidae) production and virulence. *J. Invertebr. Pathol., 106*, 268–273.

Steinhaus, E. A., & Martignoni, M. E. (1970). An Abridged Glossary of Terms Used. In *Invertebrate Pathology, 2nd ed. Pacific Northwest Forest and Range Experimental Station*. Portland: USDA Forest Service.

Stock, S. P., Griffin, C. T., & Chaerani, R. (2004). Morphological and molecular characterisation of *Steinernema hermaphroditum* n. sp. (Nematoda: Steinernematidae), an entomopathogenic nematode from Indonesia, and its phylogenetic relationships with other members of the genus. *Nematology, 6*, 401–412.

Strauch, O., & Ehlers, R.-U. (1998). Food signal production of *Photorhabdus luminescens* inducing the recovery of entomopathogenic nematodes *Heterorhabditis* spp. in liquid culture. *Appl. Microbiol. Biotechnol., 50*, 369–374.

Strong, D. R., Maron, J. L., Connors, P. G., Whipple, A., Harrison, S., & Jeffries, R. L. (1995). High mortality, fluctuation in numbers, and heavy subterranean insect herbivory in bush lupine, *Lupinus arboreus*. *Oecologia, 104*, 85–92.

Strong, D. R., Whipple, A. V., Child, A. L., & Dennis, B. (1999). Model selection for a subterranean trophic cascade: root-feeding caterpillars and entomopathogenic nematodes. *Ecology, 80*, 2750–2761.

Stuart, R. J., Abu Hatab, M., & Gaugler, R. (1998). Sex ratio and the infection process in entomopathogenic nematodes: are males the colonizing sex? *J. Invertebr. Pathol., 72*, 288–295.

Stuart, R. J., Barbercheck, M. E., Grewal, P. S., Taylor, R. A. J., & Hoy, C. W. (2006). Population biology of entomopathogenic nematodes: concepts, issues, and models. *Biol. Control, 38*, 80–102.

Tailliez, P., Laroui, C., Ginibre, N., Paule, A., Pages, S., & Boemare, N. (2010). Phylogeny of *Photorhabdus* and *Xenorhabdus* based on universally conserved protein-coding sequences and implications for the taxonomy of these two genera. Proposal of new taxa: *X. vietnamensis* sp. nov., *P. luminescens* subsp. *caribbeanensis* subsp. nov., *P. luminescens* subsp. *hainanensis* subsp. nov., *P. temperata* subsp. *khanii* subsp. nov., *P. temperata* subsp. *tasmaniensis* subsp. nov., and the reclassification of *P. luminescens* subsp. *thracensis* as *P. temperata* subsp. *thracensis* comb. nov. *Int. J. Syst. Evol. Microbiol., 60*, 1921–1937.

Tanada, Y., & Kaya, H. K. (1993). *Insect Pathology*. San Diego: Academic Press.

Tiedje, J. M., Asuming-Brempong, S., Nüsslein, K., Marsh, T. L., & Flynn, S. J. (1999). Opening the black box of soil microbial diversity. *Appl. Soil. Ecol., 13*, 109–122.

Torres-Barragana, A., Suazob, A., Buhlerc, W. G., & Cardoza, Y. J. (2011). Studies on the entomopathogenicity and bacterial associates of the nematode *Oscheius carolinensis*. *Biol. Control*. (in press).

Toubarro, D., Lucena-Robles, M., Nascimento, G., Costa, G., Montiel, R., Coelho, A. V., & Simões, N. (2009). An apoptosis-inducing serine protease secreted by the entomopathogenic nematode *Steinernema carpocapsae*. *Int. J. Parasitol., 39*, 1319–1330.

Toubarro, D., Lucena-Robles, M., Nascimento, G., Santos, R., Montiel, R., Verissimo, P., Pires, E., Faro, C., Coelho, A. V., & Simões, N. (2010). Serine protease-mediated host invasion by the parasitic nematode *Steinernema carpocapsae*. *J. Biol. Chem., 285*, 30666–30675.

Uvell, H., & Engström, Y. (2007). A multilayered defense against infection: combinatorial control of insect immune genes. *Trends Genet., 23*, 342–349.

Valenzano, D. R., Terzibasi, E., Genade, T., Cattaneo, A., Domenici, L., & Cellerino, A. (2006). Resveratrol prolongs lifespan and retards the onset of age-related markers in a short-lived vertebrate. *Curr. Biol., 16*, 296–300.

Vigneux, F., Zumbihl, R., Jubelin, G., Ribeiro, C., Poncet, J., Baghdiguian, S., Givaudan, A., & Brehelin, M. (2007). The *xaxAB* genes encoding a new apoptotic toxin from the insect pathogen *Xenorhabdus nematophila* are present in plant and human pathogens. *J. Biol. Chem., 282*, 9571–9580.

Wang, Y., & Gaugler, R. (1998). Host and penetration site location by entomopathogenic nematodes against Japanese beetle larvae. *J. Invertebr. Pathol., 72*, 313–318.

Waterfield, N. R., Bowen, D. J., Fetherston, J. D., Perry, R. D., & ffrench-Constant, R. H. (2001). The tc genes of *Photorhabdus*: a growing family. *Trends Microbiol., 9*, 185–191.

Waterfield, N. R., Daborn, P. J., Dowling, A. J., Yang, G., Hares, M., & ffrench-Constant, R. H. (2003). The insecticidal toxin makes caterpillars floppy 2 (Mcf2) shows similarity to HrmA, an avirulence protein from a plant pathogen. *FEMS Microbiol. Lett., 229*, 265–270.

Waterfield, N., Hares, M., Hinchliffe, S., Wren, B., & ffrench-Constant, R. (2007). The insect toxin complex of *Yersinia*. *Adv. Exp. Med. Biol., 603*, 247–257.

Watson, R. J., Joyce, S. A., Spencer, G. V., & Clarke, D. J. (2005). The *exbD* gene of *Photorhabdus temperata* is required for full virulence in insects and symbiosis with the nematode *Heterorhabditis*. *Mol. Microbiol., 56*, 763–773.

Watson, R. J., Millichap, P., Joyce, S. A., Reynolds, S., & Clarke, D. J. (2010). The role of iron uptake in pathogenicity and symbiosis in *Photorhabdus luminescens* TT01. *BMC Microbiol., 10*, 177.

Wharton, D. (2004). Survival strategies. In R. Gaugler & A. Bilgrami (Eds.), *Nematode Behaviour* (pp. 371–399). Wallingford: CABI.

White, G. F. (1927). A method for obtaining infective nematode larvae from cultures. *Science, 30*, 302–303.

Wilkinson, P., Waterfield, N. R., Crossman, L., Corton, C., Sanchez-Contreras, M., Vlisidou, I., Barron, A., Bignell, A., Clark, L., Ormond, D., Mayho, M., Bason, N., Smith, F., Simmonds, M., Churcher, C., Harris, D., Thompson, N. R., Quail, M., Parkhill, J., & ffrench-Constant, R. H. (2009). Comparative genomics of the emerging human pathogen *Photorhabdus asymbiotica* with the insect pathogen *Photorhabdus luminescens*. *BMC Genomics, 10*, 302.

Williams, J. S., Thomas, M., & Clarke, D. J. (2005). The gene stlA encodes a phenylalanine ammonia-lyase that is involved in the production of a stilbene antibiotic in *Photorhabdus luminescens* TT01. *Microbiology, 151,* 2543–2550.

Wilson, M. J., & Grewal, P. S. (2005). Biology, production and formulation of slug-parasitic nematodes. In P. S. Grewal, R.-U. Ehlers & D. I. Shapiro-Ilan (Eds.), *Nematodes as Biocontrol Agents* (pp. 421–429). Wallingford: CABI.

Wilson, M. J., Glen, D. M., & George, S. K. (1993). The rhabditid nematode *Phasmarhabditis hermaphrodita* as a biological control agent for slugs. *Biocontrol Sci. Technol., 3,* 503–511.

Wilson, M., Xin, W., Hashmi, S., & Gaugler, R. (1999). Risk assessment and fitness of a transgenic entomopathogenic nematode. *Biol. Control, 15,* 81–87.

Wilson, M. J., Lewis, E. E., Yoder, F., & Gaugler, R. (2003). Application pattern and persistence of the entomopathogenic nematode *Heterorhabditis bacteriophora. Biol. Control, 26,* 180–188.

Yao, Q., Cui, J. X., Zhu, Y. Q., Wang, G. L., Hu, L. Y., Long, C. Z., Cao, R., Liu, X. Q., Huang, N., Chen, S., Liu, L. P., & Shao, F. (2009). A bacterial type III effector family uses the papain-like hydrolytic activity to arrest the host cell cycle. *Proc. Natl. Acad. Sci. USA, 106,* 3716–3721.

Ye, W., Torres-Barragan, A., & Cardoza, Y. J. (2010). *Oscheius carolinensis* n. sp. (Nematoda: Rhabditidae), a potential entomopathogenic nematode from vermicompost. *Nematology, 12,* 121–135.

You, J., Liang, S., Cao, L., Liu, X., & Han, R. (2006). Nutritive significance of crystalline inclusion proteins of *Photorhabdus luminescens* in *Steinernema* nematodes. *FEMS Microbiol. Ecol., 55,* 178–185.

Zhang, C., Liu, J., Xu, M., Sun, J., Yang, S., An, X., Gao, G., Lin, M., Lai, R., He, Z., Wu, Y., & Zhang, K. (2008). *Heterorhabditidoides chongmingensis* gen. nov., sp. nov. (Rhabditida: Rhabditidae), a novel member of the entomopathogenic nematodes. *J. Parasitol., 98,* 153–168.

Zhang, C. X., Yang, S. Y., Xu, M. U., Sun, J., Liu, H., Liu, J. R., Liu, H., Kan, F., Sun, J., Lai, R., & Zhang, K. Y. (2009). *Serratia nematodiphilia* sp. nov., associated symbiotically with the entomopathogenic nematode *Heterorhabditidoides chongmingensis* (Rhabditida: Rhabditidae). *Int. J. Syst. Evol. Microbiol., 59,* 1603–1608.

Chapter 12

From Silkworms to Bees: Diseases of Beneficial Insects

Rosalind R. James* and Zengzhi Li[†]

*United States Department of Agriculture, Agricultural Research Service, Logan, Utah, USA, [†]Anhui Agricultural University, Hefei, Anjui, China

Chapter Outline

Summary

Diseases of the silkworm (*Bombyx mori*) and managed bees, including the honey bee (*Apis mellifera*), bumble bee (*Bombus* spp.), alfalfa leafcutting bee (*Megachile rotundata*), and mason bee (*Osmia* spp.), are reviewed, with diagnostic descriptions and a summary of control methods for production systems. Silkworms and honey bees have been managed by humans for a few thousand years. This close association has provided insights into the nature of infectious diseases, influencing microbiologists such as Agostino Bassi and Louis Pasteur. Silkworms and bees are susceptible to a wide range of pathogens, including viruses, bacteria, fungi, and protists. However, pathogenic nematodes are conspicuously absent. These insects are also susceptible to parasitoids, mite parasites, pesticides, and environmental toxins. Control measures for silkworm diseases center around sanitation, as chemical treatment options are extremely limited. During the past century, honey beekeepers have developed a reliance on chemical treatment options, but this approach has produced resistant pathogens and parasites, and beekeepers are vulnerable to the introduction of new diseases that do not yet have control methods. An integrated approach, with an emphasis on prevention and sanitation measures, would help to reduce these problems.

12.1. INTRODUCTION

Despite the predominance of attention to insects as pests, many insects are directly beneficial to humans, and as a result, mass-production methods have been developed. The most notable examples are silkworms and honey bees, although many other insects are mass produced for various purposes. For example, screw worms and fruit flies are mass produced for the sole purpose of sterilizing the males and releasing them to mate with wild females in what is known as the sterile insect technique. Lacewings, ladybird beetles, and some parasitoids are mass produced and sold as biological control agents against various pests. Bumble bees and alfalfa leafcutting bees are mass produced as pollinators for greenhouse crops and alfalfa seed production, respectively. A major concern in any kind of mass rearing facility is the risk for disease outbreaks. If diseases are not controlled, a sudden collapse of the insect colony can occur and the quality and quantity of their product may be reduced.

In contrast to other chapters, where the main focus is the use of pathogens to kill insects, what is special about

Insect Pathology. DOI: 10.1016/B978-0-12-384984-7.00012-9

beneficial insects is our desire to protect them from pathogens. Diseases occur in beneficial insects in natural and agricultural settings, and undoubtedly have an impact on the population dynamics of these insects, just as they do on herbivorous insects. However, insect production systems lead to very large, crowded populations, conditions that provide ideal opportunities for disease development and spread. In addition, insect production systems provide a unique opportunity to observe pathogens that may be more difficult to find in nature because the insects are confined and reared in high numbers, and are under careful observation.

This chapter presents a discussion of diseases that occur in two insect groups, silkworms and bees, and methods for controlling these in productions systems. Many of the most well-studied insect pathogens were first found in silkworms or honey bees, as these insects have been managed by humans for millennia. Silkworms have provided scientists with key insights into the very existence of pathogens (see Chapter 2). Likewise, important discoveries are certain to be made as researchers seek to identify the causes of current problems in honey bee production.

12.2. DISEASES OF SILKWORMS

Several silkworm species that are indigenous to Asia and Africa have been domesticated and are raised for silk production throughout most of the temperate zone. These species vary in the quality of the silk they produce, the host plant on which they feed, and the number of generations produced per year. The most widely raised species and the producer of the finest silk is *Bombyx mori*, which is of Chinese origin and feeds on mulberry (*Morus* spp.) leaves. Silkworm rearing and silk weaving can be traced back to the ancient Chinese, as early as at least 6000 years ago. Some classical Chinese books such as "The Book of Songs" (sixth to eleventh century before the common era, BCE) and "The Chronicle of Zuo" (453 BCE) described thriving sericulture and silk textile industries in northern and southern China 2500−3000 years ago. Silk became one of the main cloth materials of the ancient Chinese, and as a result, sericulture became an important component of Chinese agriculture. The westward exportation of silk started at least 3000 years ago, based on Chinese silk found on Egyptian mummies. Starting with the Han Dynasty (202 BCE−220 common era, CE), large quantities of silk and silk products were exported along the famous Silk Road, westwards to Central and Western Asia, the Mediterranean, and Europe, and eastwards to Korea and Japan. Silkworm eggs and sericultural techniques were introduced into Japan via Korea in the second century, and then into Turkey, Egypt, Arabia, the Mediterranean, and Rome in the sixth century. Since then, sericulture has spread throughout much of the world.

The history of sericulture is also a history of the war against silkworm diseases (see Chapter 2). The earliest record of a silkworm disease was the announcement by Guan Zhong (725−645 BCE) that any man who could produce silkworms free of diseases in the spring and autumn (fall) would be awarded 0.5 kg of gold. In the earliest Chinese pharmaceutical work, "Classic of Herbal Medicine" (221−220 BCE), *Beauveria bassiana*-infected silkworms were used as a folk medicine for childhood convulsions. Chen Fu's "Book on Agriculture", published in 1149, describes black, white, and red muscardines (diseases caused by fungi) and some typical symptoms of polyhedrosis-like infections in silkworms. The relationship between the occurrence of the muscardines and weather conditions was analyzed in the "Handbook of Agriculture and Sericulture", published in 1273. This book also describes the symptoms of silkworm adults suffering from pébrine (a disease caused by a microsporidium). In the famous ancient Chinese encyclopedia written by Yingxing Song in 1637 and entitled "Tiangong Kaiwu" ("The Exploration of the Works of Nature"), typical symptoms of grasserie (viral disease) and flacherie (either viral or bacterial disease) are described, including methods of control that involved segregating and eliminating diseased larvae, suggesting that the contagious nature of these diseases had been recognized.

The first scientific study on silkworm diseases was conducted by Agostino Bassi in 1807 on the white muscardine (known as "*mal de segno*" and "*calcinaccio*" in Italian) (see Chapter 2). This disease initially appeared in Italy around 1805, and in France by 1841. Bassi (1835) determined that muscardine was caused by a living entity (a fungus) that formed the powdery appearance on the dead, diseased silkworms, and that this condition was contagious. He is credited with rescuing the economically important silk industry with his recommendations for the use of disinfectants, separating the rows of feeding caterpillars, isolating and destroying infected caterpillars, and keeping the farms clean. His paper was translated into French and distributed throughout Europe and greatly influenced Louis Pasteur (Porter, 1973).

In 1865, epizootics were devastating the silk industry in France. Louis Pasteur was asked by the French Government to investigate, and after several years, he discovered that the silkworms had been suffering from two diseases, called pébrine and flacherie, and proved that the spread of these diseases could be prevented by carefully segregating healthy and diseased larvae from each other (Debré, 1998). He published a two-volume treatise on silkworm diseases to describe his findings (Pasteur, 1870). From the 1870s onwards, advanced sericultural techniques were further developed in Japan. Japanese studies on silkworm viruses also improved the understanding of other silkworm diseases. In the second half of the twentieth century, silkworm pathology entered the age of biochemical and molecular biology. Currently used mainly for disease diagnosis, these techniques could lead to molecular marker-assisted selection, transgenic breeding of resistant varieties, and gene therapy.

To understand the biology of silkworm diseases today, it is important to become familiar with some aspects of modern silkworm production. Silkworm rearing can be "batched" or "multibatched", depending on the duration of the frost-free period and availability of mulberry trees in the region. In batched rearing, larvae are reared five or fewer times in a year, while 12 or more cycles are produced per year in multibatched rearing. The batches overlap in time with multibatched rearing but are consecutive in batched rearing. Accordingly, there is time to disinfect rearing facilities between batches for batched rearing but not for multibatched rearing. Thus, multibatched rearing systems are more prone to disease outbreaks.

In a typical production system, silkworm eggs and young instars (first to third) are produced initially in large professional rearing facilities or by experienced and well-equipped farmers. The fourth instars are then distributed to smaller facilities at farmers' homes. In China, the facilities used to produce the eggs and early instars are mostly state owned. These facilities have specially designed rearing rooms that are easy to clean and disinfect (commonly, rooms with steel walls and ceilings and cement floors). The rearing rooms are also well ventilated and somewhat heat insulated, often with heating and/or cooling systems. Young silkworms are usually reared on rearing trays made of either bamboo or wood and fed the tender new leaves of mulberry. The trays are stacked in racks, seven to 10 tiers high, and the racks are usually either wood or metal. The facilities for the late instars are simpler, with the silkworms reared directly on the floor of a small room (often on a wooden or dirt floor), and heating and cooling systems may be lacking. These are called "rearing beds", The late instars can be fed mature, fresh mulberry leaves. Thus, the production system is not entirely self-contained, as insects are shipped from one facility to another and field-collected fresh mulberry leaves are brought in daily.

Disease control is an important part of sericulture, even in the best facilities. The infectious diseases of silkworms are caused by viruses, bacteria, fungi, and protists (Table 12.1). The most serious diseases are the blood-type grasserie, white muscardine, and pébrine. Most of the diseases can be readily diagnosed based on signs and symptoms in the larvae or cadavers and by microscopic examination (Table 12.1).

12.2.1. Viruses

In silkworms, nucleopolyhedrosis virus and viral flacherie are common and can cause very serious larval losses. The incubation period for viral diseases (the time from infection until symptoms appear) is usually six to 10 days. Most silkworms become infected during the fourth to early fifth instars, whereas the disease outbreaks usually occur in late fifth instars.

Bombyx mori nucleopolyhedrovirus (BmNPV) causes a disease referred to as nuclear polyhedrosis, jaundice, or blood-type grasserie (from the French word "grasse", which means thick and sticky and describes the exudates seen in dead silkworms). BmNPV is a double-stranded circular DNA (dsDNA) virus that has a genome consisting of 128,413 nucleotides, a GC content of 40% (GenBank Accession No. L33180), and 143 genes (Kamita and Maeda, 1997; Gomi *et al.*, 1999) (see Chapter 4). The virions are occluded in an occlusion body. Healthy (i.e., uninfected) larvae are slightly green and appear somewhat transparent white, whereas infected larvae become ivory with swollen intersegmental membranes. Diseased larvae behave differently, becoming easily agitated when disturbed, and are frequently found lying at the edge of the rearing beds before death. In addition, the integument becomes very fragile, and the hemolymph turns milky and turbid.

Bombyx mori cypovirus-1 (BmCPV-1) (also called cytoplasmic polyhedrosis or cypovirus) is a circular, double-stranded RNA (dsRNA) virus. Its genome consists of 10 dsRNA segments that range in size from 0.3–0.8 to 2.2–2.6 kilobases (kb), each possessing a single, complete open reading frame. Segments I–IV, VI, and VII encode structural capsid and occlusion body proteins, segments VIII–X encode non-structural proteins, and the function of segment V is still uncertain (Ikeda *et al.*, 2001; Hagiwara *et al.*, 2002; Lin *et al.*, 2002) (see Chapter 4). This virus forms occlusion bodies that are mostly hexagonal, icosahedral, tetragonal, or occasionally triangular, ranging in size from 0.5 to 10 μm and averaging 2.6 μm (Jin, 2001). Larval development in infected insects becomes asynchronous, with prolonged instar duration. Infected individuals are thin and small, show a loss of appetite, and respond slowly to stimuli. The thorax becomes translucent, the abdomen appears shrunken, and grayish white feces are produced when the larvae are squeezed. The midgut and hindgut become milky, striated in appearance, and shrunken.

In sericultural practice, the French term "flacherie" has been used for two forms of dysentery, non-infectious and infectious (viral), where the silkworm larvae become flaccid. Non-infectious flacherie, also known as "touffee flacherie", is caused by exposure of the larvae to excessively high temperatures. Viral flacherie can be caused by *B. mori* infectious flacherie virus (BmIFV), *B. mori* densovirus (BmDNV) (= *B. mori* densonucleosis virus), or *B. mori* cypovirus-1. These viral infections, either alone or in combination with bacterial infections, destroy the gut tissue, resulting in dysentery and larval flaccidity.

TABLE 12.1 Some Useful Characteristics for Preliminary Diagnosis of Infectious Silkworm Diseases

Pathogen	Disease	General Signs and Symptoms	Hemolymph	Some Distinct Characteristics	
				Midgut	Integument
Viruses [a]					
Bombyx mori nucleopolyhedrovirus (BmNPV)	Nuclear polyhedrosis, jaundice, or blood-type grasserie	Larvae display defensive postures readily; molting is delayed or lacking; posterior abdominal segments swollen; cadavers grayish black and flaccid	Turbid and white, occlusion bodies are present	Not distinct	Shiny just before molting. Fragile, exuding fluids. In late instars, spiracles black, and cuticle with black spots
B. mori cypovirus-1 (BmCPV-1)	Cytoplasmic polyhedrosis or cypovirus	Larvae not easily agitated, pale, thin and small; growth slow and asynchronous; feces remains attached to the anus; cadavers flaccid	Clear, no occlusion bodies present	Posterior midgut ivory in color, shrunken, and striated. Occlusion bodies are present	Not distinct
B. mori infectious flacherie virus (BmIFV)	Viral flacherie	Larvae vomit; head raised; growth and molting asynchronous; head and thorax discolored, flaccid; larvae translucent; feces watery; cadavers have a flattened shape	Clear, without occlusion bodies	Contains yellow–brown fluids. Pink viral spherical bodies present. Stains with pyronin-methyl green	Not fragile
B. mori densovirus (BmDNV) (= *B. mori* densonucleosis virus)	Densoviral flacherie	Like flacherie or bacterial flacherie; anorexia; late molting; bodies shrunk; integument not fragile; thorax mostly translucent	Clear, without occlusion bodies	Contains yellowish green fluids. Cocci and diplococci very abundant but no occlusion bodies	Not distinct
Bacteria					
Bacillus spp.	Fuliginosa septicemia	Large thorax and small abdomen; anterior segments of abdomen with greenish black spots; vomiting and diarrhea; cadavers turn black, decayed and fetid	Brownish black. Bacteria abundant, large bacilliform rods	Not distinct	Not distinct
Serratia marcescens	*Serratia*-type septicemia	Larvae with light specks; cadavers appear shortened, become flaccid, and turn pink or dark pink	Reddish brown. Bacteria abundant, short bacilliform rods	Not distinct	Not distinct
Aeromonas sp.	Green thorax septicemia	Head and thorax of cadavers translucent green, and turned down ventrally; cadavers are flaccid and become putrid	Turbid. Bacteria present, small bacilliform rods	Not distinct	Not distinct
Bacillus thuringiensis subsp. *sotto* (Bt *sotto*)	Sotto disease or bacillary paralysis	Larvae stop feeding, turn dark with internal nodules; posterior is translucent; cadavers turn flaccid	Not distinct	Midgut cells disassociating. Bacteria present, long bacilliform rods.	Not distinct

Enterococcus spp. and other intestinal bacterial species	Bacterial flacherie; thoracic or wrinkling disease	Larvae not uniform in development, become thin and small. Diarrhea	Not distinct	Empty except for yellow–green mucus. Bacteria present, large, bacilliform rods	Translucent on the head and thorax.
Fungi: Filamentous					
Beauveria bassiana	White muscardine	Larvae sluggish, with oil-colored specks; cadavers soft at first and then stiffen, and finally covered with white conidia	Turbid. Becomes full of cylindrical hyphal bodies or hyphae	Not distinct	Larvae with powdery white appearance due to presence of large masses of globose conidia
Nomuraea rileyi	Green muscardine	Larvae sluggish, with oil-colored specks; cadavers soft at first and then stiffen, and finally covered with bright green conidia	Turbid. Gradually full of beaded hyphal bodies or hyphae	Not distinct	Presence of large masses of ovoid conidia results in bright green appearance
Aspergillus spp., e.g., *A. flavus*, *A. ochraceus*, *A. oryzae*, *A. parasiticus*, *A. tamari*	Aspergillosis	Newly hatched larvae stop feeding, stop moving, frequent vomiting, and die in 2–3 days; late instars develop a lesion, thorax protrudes, frequent vomiting, die in 4–5 days; cadavers stiff with hyphae penetrating	Turbid. Hyphal bodies not present	Not distinct	Sporulation occurs in localized spots, color varies, often brown, usually dark. Hyphae visible around spots
Fungi: Microsporidia					
Nosema bombycis	Pébrine	Growth slow and typically asynchronous, molting may be incomplete; no cocoon formation; spores present in the hemolymph, midgut, silk gland, feces and eggs; ovoid 2.9–4.1 × 1.5–2.1 μm	Turbid, spores present	Swollen in appearance, ivory color with black specks	Sometimes peppered with black specks and thin, irregular setae
Protists					
Entamoeba spp.	Amoebiasis	Late instar larvae, growth slowed, the caudal end of the larvae collapses, posterior end shrunken; feces green; cadavers black, mummified	Not distinct	Cells swell due to presence of amoebic cysts, pathogen released into the gut and feces. Cysts tetranucleate	Not distinct
Herpetomonas bombycis	Trypanosome infection	Infected larvae are slightly grayish white, with decreased appetite, turbid hemolymph, pulsating dorsal vessel, and die gradually; cadavers turn blackish brown	Not distinct	Not distinct	Not distinct
Undescribed coccid	Coccid disease	Larvae develop poorly, shrunken, diarrhea, brown mucus forms around the anus; cysts detected in the feces	Not distinct	Infection occurs though midgut epithelial cells	Not distinct

In viral nomenclature, only the type species is in italics.

BmIFV is the most common cause of viral flacherie in silkworms (see Chapter 5). It is a single-stranded, positive-strand RNA virus with no occlusion bodies. The virions are equilateral icosahedron with a diameter of 26 ± 2 nm. The viral RNA sequence has been completed for a Japanese and two Chinese BmIFV strains (GenBank Accession No. EF422865, HM245295, and AB000906, respectively), and range from 9650 to 9675 base pairs (bp), with 99% homology among these three strains (Wang *et al.* 2006a, b). BmIFV-infected larvae are stunted, uneven in growth, show a gradual loss of appetite, and molt asynchronously. As the disease develops, the larvae become discolored, shrunken, and flaccid. The prolegs lose their ability to clasp the surface, the larvae become translucent, and vomiting and diarrhea commonly occur in the fifth instar. The symptoms are similar to BmCPV-1, but the feces are thin, form in a chain, and appear yellowish brown. The cadavers have a foul odor and, upon dissection, the midgut is not milky.

Six strains of BmDNV have been described and categorized into two types: BmDNV-I and BmDNV-II. These densoviruses are single-stranded DNA (ssDNA) with no occlusion body. The virions are globose, with a diameter of 22 nm for BmDNV-I and 24 nm for BmDNV-II. BmDNV-II has two sets of non-homogeneous linear ssDNAs (VD1 and VD2) that are separately encapsulated. The genome of two Chinese strains, VD1 (GenBank DQ017268) and VD2 (GenBank DQ017269), consists of 6543 and 6022 bp, respectively. Symptoms of diarrhea and vomiting are very similar to those caused by BmIFV, but the occurrence of a translucent thorax in a BmDNV-infected larva is more prominent. In addition, the midgut is full of yellowish brown fluid, but without any remains of mulberry leaves because the larvae usually have ceased feeding for an entire week. Large numbers of cocci and diplococci bacteria are typically present in the cadaver and can be detected microscopically (Jin, 2001), but these are not the cause of the disease.

The silkworm viruses can be transmitted both vertically and horizontally, making it difficult to control these pathogens (see for example, BmNPV, Fig. 12.1). Horizontal transmission occurs when the viruses contaminate

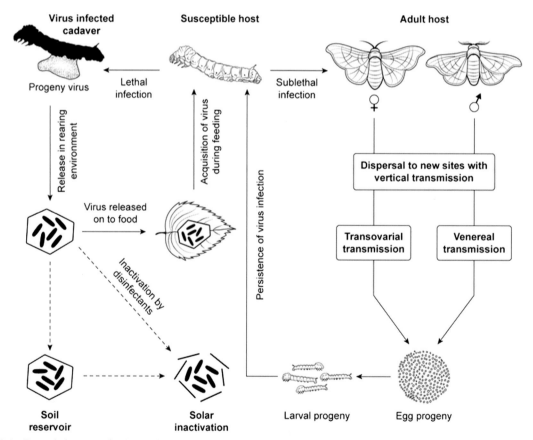

FIGURE 12.1 Transmission routes for the *Bombyx mori nucleopolyhedrovirus* (BmNPV). Cadavers of virus-infected silkworm larvae become a source of inoculum for *per os* infections in other larvae by contaminating the mulberry leaves with occlusion bodies. Occlusion bodies can also be released though the feces of infected larvae. Infected larvae may acquire the virus and then die (repeating the cycle), or they may survive infections and serve as a source of infection through vertical transmission to their progeny. The dispersal of adults to new sites occurs primarily via humans when silkworm cocoons are harvested and the new adults are transferred to egg production facilities. *(From Khurad et al., 2004, with permission from the publisher.)*

mulberry leaves and rearing trays, and infection of the larvae occurs *per os*. The viruses can persist in the cadavers, feces, and exuviae, leading to contamination of the rearing equipment, rearing beds, and leaves (Fig. 12.1). Mulberry leaves can also become contaminated with virus particles in the field from other lepidopteran insects that serve as carriers for the pathogen. Such carriers include the wild silkmoth, *Bombyx mandarina*, the mulberry pyralid, *Glyphodes pyloalis*, the mulberry geometrid, *Phthonandria atrilineata*, the mulberry white caterpillar, *Rondotia menciana*, and the mulberry yellow-tail moth, *Porthesia xanthocampa* (Watanabe *et al.*, 1988). However, most viral transmission occurs within rearing beds. The cadavers serve as the main reservoir for infection, but infected larvae can also contaminate the mulberry leaves and rearing beds. Workers can then further spread the virus particles with their hands, and flies in the rearing rooms can act as carriers. Vertical transmission occurs with BmNPV and BmCPV-1, which can persist in a batch of larvae as sublethal infections. The moths derived from such larvae can successfully spread the viruses venereally to each other during mating and to their offspring by transovarial transmission (Fig. 12.1) (Aruga, 1971; Khurad *et al.*, 2004).

12.2.2. Bacteria

Facultative bacterial pathogens occur widely in nature, have broad host ranges, and cause mostly acute diseases in silkworms. The signs and symptoms of diseased silkworms are often generic and it may be difficult to identify a specific bacterium as the causal agent without performing Koch's postulates. However, diagnosis according to the characteristics of moribund larvae and cadavers, including tissue discoloration, may provide sufficient information as to the probable causal agent (Table 12.1). Bacterial diseases typically occur when larvae are reared in overcrowded conditions.

Bacteria that cause septicemia (a morbid condition caused by the multiplication of microorganisms in the blood) (Steinhaus and Martignoni, 1970) in silkworms belong to many taxa. The most common bacteria associated with septicemia are *Bacillus bombysepticus* (Huang *et al.*, 2009), *Serratia marcescens*, and *Aeromonas mundii* (Cappelloza *et al.*, 2011). *Streptococcus* spp. and *Staphylococcus* spp. also cause septicemia in larvae. The common symptoms of septicemia are larvae becoming dull and motionless with reduced feeding rates. Moribund larvae can be found vomiting and exhibiting spasms while lying on their sides. Signs of a bacterial infection include a swollen thorax and shrunk abdominal segments. Cadavers are often discolored and are characteristically flaccid.

Fuliginosa septicemia is caused by *Bacillus* spp., most commonly *B. bombysepticus* (Li *et al.*, 2011). The first signs are typically a stiffening of the thorax and the first three abdominal segments. The stiffening then extends anteriorly and posteriorly, and the segments darken rapidly such that the entire cadaver turns black and exudes a dark brown fluid. With *Serratia*-type septicemia, cadavers become slightly shortened and develop small brown specks. Cadavers of larvae become flaccid and pink and those of pupae become black or pink, whereas cadavers of the adult abdomen become partially black or pink and exude a pink fluid. Septicemia caused by *Aeromonas* is also called green thorax septicemia, with mortality occurring in late instar larvae. Translucent green or greenish specks appear on the dorsal part of the thorax, and greenish or brownish bubbles form under the integument. The hemolymph from cadavers is turbid and the cadavers emit a putrid odor (Jin, 2001).

Bacteria that induce septicemia are transmitted mainly through wounds; therefore, occurrence of bacterial infections is most closely related to silkworm age, handling technique, and rearing conditions. Late instar larvae have sharp, well-developed tarsal claws and often injure each other under crowded conditions. In fact, up to 80% of the larvae can become injured (Jin, 2001). The large late instar larvae also have an enormous feeding rate, resulting in a copious production of fecal material. When large amounts of feces are combined with high humidity, excessive bacterial growth can result in the rearing beds. Densely populated rearing beds and excessive handling can exacerbate the problem, resulting in higher frequency of larval wounding and bacterial infections due to fecal contamination. Accordingly, septicemia rarely occurs in neonates and is most commonly a problem in mature larvae. Newly hatched larvae have undeveloped tarsal claws and their bodies are protected to some extent by setae.

Another common bacterial disease is Sotto disease, also known as bacillary paralysis, Sottokin, or "sudden death bacillus" (see Chapters 2 and 8). This disease is caused by *Bacillus thuringiensis* subsp. *sotto* (Bt *sotto*). During sporulation, cells of Bt *sotto* produce endospores with parasporal bodies containing the δ-endotoxins, and secrete the α-, β-, and γ-exotoxins into the medium (i.e., hemolymph or nutrient-based agar). Silkworms are very susceptible to these toxins. The disease occurs primarily in late instar larvae. Both acute and chronic signs and symptoms can occur. Acute symptoms include a sudden cessation of larval feeding; the larvae raise their body anteriorly and have convulsive tremors and vomiting, and their thorax becomes slightly swollen, followed by sudden death with their whole body paralyzed. Infected larvae may die 10 min to several hours after consuming the Bt *sotto* toxins. The chronic signs and symptoms associated with Sotto disease include reduced larval feeding, darkening of the body, and the posterior midgut being empty; the feces may be reddish brown in color or be a foul-smelling liquid discharge. Infected larvae become flaccid on the mulberry leaves, and after two to three days, die hanging upside down or lying on their sides. Brownish specks appear on the cadavers. The cadavers become discolored and then flaccid, starting at the dorsal anterior part of the abdomen (Jin, 2001).

A third bacterial disease, bacterial flacherie, also known as thoracic or wrinkling disease, has symptoms that are less distinct than the first two bacterial diseases. Bacterial flacherie occurs sporadically and does not cause heavy losses. It is caused by bacteria that normally occur in the silkworm's digestive tract, including *Enterococcus* spp. (mainly *E. faecalis* and *E. faecium*) and some species of *Staphylococcus*, *Bacillus*, *Klebsiella*, *Alcaligenes*, and *Pseudomonas* (Nagae, 1974, 1982). Normally, silkworms are not susceptible to these bacterial species unless the larvae are under stress owing to starvation, low-quality mulberry leaves, or unsuitable rearing conditions (e.g., high temperature and humidity). Common signs and symptoms of infected larvae include sluggishness, cessation of feeding, vomiting, diarrhea, reduced growth, asynchronous development, shrunken bodies, and translucent thoraxes. The larvae usually die during the molting process. When death occurs between molts, both ends of the larvae become swollen and the cadavers become flaccid (Jin, 2001).

Rearing conditions can increase the prevalence of septicemia, Sotto disease, and bacterial flacherie. As described above, the bacteria that cause septicemia can persist in cadavers and feces of silkworms and contaminate mulberry leaves, rearing beds and equipment, and the floors in the rearing facility. Both Bt *sotto* and *Enterococcus* spp. commonly occur in the environment at non-epizootic levels, and usually the prevalence of these diseases in the silkworm colonies is low or not detectable (Ohba, 1996). However, unfavorable rearing conditions (e.g., high temperatures and humidity) can cause a bacterial disease outbreak. High humidity and a continual feeding of wet mulberry leaves will result in the rearing beds becoming humid and hot, and this situation commonly occurs during the rainy season. Such conditions not only lead to the proliferation of the bacteria on the leaves but are also harmful to silkworm development, probably increasing their susceptibility to bacterial toxins and infections.

12.2.3. Fungi: Filamentous

The muscardines are fungal diseases and are the second most important silkworm diseases in China and Japan (following blood-type grasserie in China and viral flacherie by BmIFV and BmDNV in Japan). The muscardine fungi produce asexual spores (conidia) on the cuticle of cadavers, and these conidia are the infective propagules. Depending on the pathogen species, these conidia are white, green, yellow, black, gray, or red, and the muscardine diseases are named based on these colors. The most common muscardines are white and green.

White muscardine is caused by *B. bassiana*. Infected larvae lose their appetite, become sluggish, and eventually cease to move. Small irregular specks appear on the body, or two or three large, oil-colored specks appear. Immediately after death, the cadavers are soft, but they soon harden and appear mummified, and then white mycelia emerge from the intersegmental membranes and gradually cover the whole body (if conditions are humid enough) and then form a white layer of conidia, making the cadavers look chalky white.

To insect mycologists and microbial control specialists, the term "green muscardine" indicates a mycosis caused by *Metarhizium* spp., mainly *M. anisopliae*. However, in sericulture, this term is used for a common mycosis caused by *Nomuraea rileyi*. Mycoses caused by *M. anisopliae* are rare and are called black muscardine. This nomenclature arose because the silkworm cadavers infected with *N. rileyi* become bright or grass green, while those killed by *M. anisopliae* infections turn dark green. Both fungi infect larvae and pupae. Early stage signs and symptoms are not obvious. In late stages of fungal infection, the larvae lose their appetite, become sluggish, and develop large, dark brown specks on the lateral or dorsal sides of the abdomen. Immediately after death, the soft cadavers appear ivory, with the thorax overextended. These soft cadavers harden gradually, white mycelia emerge from the cuticle, and as sporulation occurs, a green, powdery layer of conidia covers the cadavers. The early signs and symptoms of green muscardine can overlap with those of other diseases. For example, larvae that die in the third and fourth instars from NPV infections also look ivory and shiny. Moribund larvae are easily agitated and their integument ruptures easily, releasing the hemolymph contents, differentiating such infections from the muscardines, where this condition does not occur.

Diseases caused by *Aspergillus* spp. are called aspergillosis. As with the muscardine infections, silkworm cadavers with aspergillosis become stiff and mycelia emerge from the integument. However, with aspergillosis, sporulation occurs only in localized spots, and the color of the conidial layer greatly varies but is most commonly brown. That is why this disease was once called brown muscardine. The fungi causing this disease in silkworms include more than 10 *Aspergillus* species, mainly *A. flavus*, *A. ochraceus*, *A. oryzae*, *A. parasiticus*, and *A. tamarii*. The fungus kills the early instars in two to three days and the later instars in four to six days. The first three instars are readily infected, but the infection is infrequent in mature larvae. When neonates become infected, they stop feeding, vomit, become sluggish, and do not move around in the rearing beds. Mycelia emerge throughout the body and produce flocculent-like (i.e., woolly) conidiogenous structures after one to two days. The cadavers do not decay. When grown larvae are infected, brown specks appear which increase in size to become lesions, usually on an intersegmental membrane or the anal area. The thorax protrudes, and the larvae vomit. After death, the cadavers harden around the lesion, and the rest of the body becomes black. Mycelia form and conidia are produced one to two days postmortem. Infected pupae become dark brown. Eggs can also become infected. The surface of the eggs becomes moldy and depressed, and soon shrivels.

Other mycoses from filamentous fungi are rare, but include gray muscardine caused by *Isaria javanica* (formerly *Paecilomyces javanicus*) (Huang *et al.*, 2008), yellow muscardine caused by *I. farinosa* (formerly *P. farinosus*), and red muscardine caused by *I. fumosoroseus* (formerly *P. fumosoroseus*). Grassy muscardine is caused by *Hirsutella necatrix*, and fusariosis is caused by *Fusarium* species (Jin, 2001). These fungal pathogens usually do not pose a threat to sericulture. The cadavers exhibiting gray, yellow, or red muscardine infection become grayish white, yellowish white, dark green to black, or pink, respectively. Those that succumb to grassy muscardine look reddish brown or have synnemata emerging from the cadaver. After infection is initiated, the larvae develop blackish brown specks of various sizes. These appear on almost every segment before death. Fusariosis is characterized by a fecal mass on the anus premortem and postmortem.

For all of the fungi discussed so far, the infective propagules are conidia, which infect through the insect cuticle (see Chapter 6), although *per os* infections can occur (Yanagita and Iwashita, 1987). Conidia can be transported very easily in air and water, thus having the potential to contaminate rearing rooms and all the rearing equipment. As long as the conidia are not exposed to solar radiation or disinfectants, they can persist for long periods, especially in soil, although persistence is strongly affected by several biotic and abiotic factors (see Chapter 6) (Keller and Zimmermann, 1989; Zimmermann 2007a, b). In rearing rooms conidia can persist for six to 12 months on cadavers (Jin, 2001).

The effect of environmental factors on fungal entomopathogens has been widely studied, especially for *B. bassiana* and *M. anisopliae* (Zimmermann, 2007a, b). The most important abiotic environmental constraints for fungi are temperature, humidity, and solar radiation (see Chapter 6). Silkworm rearing rooms, however, are a special artificial ecosystem, where temperature extremes do not commonly occur, moisture levels can be controlled, and solar radiation is limited. Yet, high humidity can occur on rainy days or when mulberry leaves are brought in wet. Under these conditions, microhabitats with high humidity and temperature can easily form on rearing beds, especially in poorly ventilated rooms without sufficient

windows, creating conditions favorable for fungal infections.

Silkworm resistance to fungi can be high in late larval instars. For example, Mu *et al.* (1999) reported that compared to newly hatched larvae, newly molted fifth instars were 86 times more resistant to white muscardine, 79 times more resistant to green muscardine, and 95 times more resistant to aspergillosis. Larvae are also more susceptible to fungal infection during molting than just before or after molting (Mu *et al.*, 1999; Chandrasekharan and Nataraju, 2011). Thus, both environmental conditions and the age of the larvae can play an important role in fungal outbreaks.

Outbreaks of white muscardine on silkworms are sometimes attributed to the application of *B. bassiana* to control the Masson's pine caterpillar, *Dendrolimus punctatus* (Zhu, 2008). Although *B. bassiana* has a very broad host range, the pathogenicity and virulence of each strain tend to be greatest for the hosts from which it was isolated (Li, 1988). The LC_{25}, LD_{25}, and LT_{25} for silkworms caused by the strain used for control of pine caterpillars were 1327, 1378, and 1.5 times that of the LC_{75}, LD_{75}, and LT_{75} caused by a typical pathogenic strain isolated from silkworms. Silkworm white muscardine is characterized by an enzootic nature, being commonly present at a low level (usually below 5%) (Li *et al.*, 2010, 2011), and the application of *B. bassiana* against pine caterpillars probably has little impact on silkworm white muscardine levels.

12.2.4. Fungi: Microsporidia

Pébrine is caused by various microsporidia, predominantly *Nosema bombycis* and, to a lesser extent, *Vairimorpha*, *Pleistophora*, and *Thelohania* species. Although microsporidia belong to the kingdom Fungi (see Chapter 7), the diseases they cause have symptoms (Table 12.1) that are quite different from other mycoses, mainly in that the cadavers are soft and the microsporidia do not produce fruiting bodies. For a long time, microsporidia were thought to be protists and are often listed among this taxon in silkworm pathology texts and other older literature.

Pébrine was first recorded in France in 1845 and then spread to Italy, Spain, Syria, and Romania. It was responsible for the collapse of the French and Italian silkworm industries in 1865 (see Chapter 2). Pasteur discovered that the microsporidium was transmitted both on contaminated mulberry leaves and transovarially. This discovery led to disease control methods that involved sampling and examining the egg-laying moths, culling the infected eggs, and supplying disease-free superior stock. Despite the success of this method, *N. bombycis* is still the most serious pathogen of silkworms throughout the world.

Nosema bombycis attacks all insect tissues and developmental stages, and the signs and symptoms are observed in all life stages from egg to adult. Infected eggs are irregularly shaped, hatch asynchronously, and do not adhere to the substrate. Some may be dead or infertile. Larvae infected by pébrine exhibit a loss of appetite, have reduced growth rate, vary in size, and display incomplete molts. The majority of mortality occurs right after the second molt for infections in larvae that acquired the infection transovarially from their mother (i.e., infected eggs). Those that survive past the third instar are pale and flaccid, small, and vomit frequently. Eventually, these larvae become covered with dark brown spots and either are unable to spin silk or spin the cocoons loosely. However, the spots do not always occur, and are not common on most of the Chinese silkworm varieties. Larval cadavers remain rubbery and do not become flaccid after death. Infected adult moths have deformed wings and distorted antennae, and the scales on the wings and abdomen fall off easily. Adults with pébrine mate poorly and have poor egg production.

The spores of *N. bombycis* are ovoid. A micropyle is located at the anterior end, and in the center is a belt-like sporoplasm, containing four nuclei. A vacuole occurs at each end. A polar capsule, polar capsule nucleus, and polar filament are present. The polar filament is tubular and filamentous in shape, and wound into a coil, as is typical for *Nosema* (see Chapter 7). The spores are highly retractile, appearing light green under the microscope, with a smooth outline. Spore dimorphism was detected during spore development in the epithelium of silkworm larvae (Iwano, 1991).

Nosema bombycis cannot be differentiated from other *Nosema* species using morphological methods; therefore, various serological, protein, and molecular approaches have been developed. For example, Qiu *et al.* (2002) developed a nucleic acid hybridization method that can precisely distinguish 10 *Nosema* species. Research conducted to characterize the genome of *N. bombycis* should improve molecular detection methods (Sironmani, 2000; Xiang, 2010).

The prevalence of pébrine varies with the variety of silkworm, the developmental stage, and the rearing environment (Lu, 1991). The Chinese silkworm varieties tend to be the most pébrine resistant, the Japanese varieties less so, and the European varieties are the most susceptible. This resistance may be related to voltinism. The multivoltine varieties are relatively pébrine resistant, followed by the bivoltine, and then univoltine varieties. Also, young larvae, newly molted larvae, and starving larvae are highly susceptible and show high mortality rates.

Several species of Lepidoptera are susceptible to other *Nosema* species that are also pathogenic to silkworms. These other lepidopteran hosts include *Antheraea yamamai*, *Chilo suppressalis*, *Diaphania pyralis*, *Hyphantria cunea*, *Philosamia cynthia*, *Phthonandria atrilineata*,

Pieris rapae, *Porthesia xanthocampa*, *Spilarctia subcarnea*, and *Spilosoma menthastri* (Hirose, 1979; Wan *et al.*, 1991; Shen *et al.*, 1996). Some of these moths are also susceptible to *N. bombycis*. However, the host–pathogen relationship between *B. mori* and *N. bombycis* still appears to be fairly specialized. For instance, *N. bombycis* invades *B. mori* both *per os* and transovarially, but transovarial transmission does not occur with any other microsporidia that affect silkworms.

Other silkworm pathogenic microsporidia include *Endoreticulatus*, *Vairimorpha*, *Pleistophora*, and *Thelohania*. The taxonomic position of these silkworm pathogenic genera remains uncertain (see Chapter 7). One species of *Endoreticulatus* was reported in China (Wan *et al.*, 1995) and has oval spores with polar filaments arranged in seven to nine rings, and develops inside a parasitic sack formed by the membrane of the endoplasmic reticulum of host cells. This pathogen has high oral infectivity. Several isolates of *Vairimorpha* spp. have been reported from silkworm cultures in Japan (Hatakeyama *et al.*, 1997), China (Zheng *et al.*, 1999; Yang *et al.*, 2001) and India (Rao *et al.*, 2004). Spores are ovoid and variable in size. *Vairimorpha* spp. have a moderate level of oral infectivity. An unnamed species of *Pleistophora* (Fujiwara, 1984a) and *Thelohania* (Fujiwara, 1984b) has been reported in Japan.

Epizootiological studies of microsporidia in silkworms have focused on pébrine caused by *N. bombycis*, and very little is known regarding the epizootiology of the other microsporidia. The extent of horizontal transmission on a rearing tray can be extremely high. If only a few diseased larvae are mixed with healthy ones at an early date, transmission will increase infection levels quickly. For example, it has been demonstrated that if 3% of the first instar larvae are infected, 50–60% of the silkworms will die from pébrine by the time they reach the adult stage, and all the emerging moths will be infected (Jin, 2001).

Spores of *N. bombycis* in the dormant stage are also very persistent, remaining infective for at least three years in the dry body of female moths and after being submerged in water for five months. *Nosema bombycis* spores are inactivated by the following treatments: (1) direct sunlight (39–40°C) for 6–7 h; (2) boiling at 100°C for 5 min; (3) moist heat at 100°C for 10 min; (4) 2% formalin for 40 min; (5) 4% formalin for 5 min; (6) 1% chlorine for 30 min; or (7) 3% chlorine for 10 min (Lu *et al.*, 2000). The spores are very resistant in the environment, surviving not only in various infected stages of the silkworms and their cadavers, but also in silkworm frass, the egg chorion of hatched eggs, exuviae, cocoons, moth scales, and on mulberry leaves, rearing room floors and walls, and rearing instruments. All these survival places should be considered when control measures are undertaken.

Temperature plays a critical role in microsporidian infection (Maddox, 1973). Pébrine mortality can be reduced by keeping the incubation temperatures for the larvae and pupae high and by soaking eggs in hot water (53–55°C) (Lu *et al.*, 2000). A 40-year study on *Nosema* prevalence in stock female moths at the Guangdong Institute of Silkworm Egg Production (China) was conducted between 1957 and 1997, and found that ambient temperatures in one year affected disease prevalence the next year, with high temperatures leading to disease outbreaks and low temperatures leading to low disease prevalence (Liu *et al.*, 2001). Humidity also affects *Nosema* prevalence. High humidity increases the attachment of spores to mulberry leaves; thus, rearing on moist trays increases the chance of silkworms ingesting contaminated mulberry leaves (Lu *et al.*, 2000).

12.2.5. Protists

Various protists (see Chapter 10) can infect silkworms, including amoebae, trypanosomes and coccids, and infections have been reported mostly in Japan and Korea. Amoebae and trypanosomes are transmitted *per os* or via wounds. No transovarial transmission occurs, but little is known about their epizootiology in silkworms.

Infections ensue after amoebic cysts are ingested by silkworms and become active in the digestive tract. The amoebae emerge from the cysts and infect silkworm larvae along the epithelial cells of the midgut, causing them to swell and collapse, releasing more pathogenic cells into the gut. When cysts are produced, they are released from the host into the feces. The species identification of the etiological agent is still uncertain, although it has been referred to *Entamoeba* spp. (Jin, 2001). The cells are globose or subglobose, with a strongly refractive central nucleus and one to three vacuoles in the cytoplasm, and reproduce by binary fission. The cells can form cysts, which are predominantly tetranucleate. The disease occurs most frequently in third instars or later, and larval development becomes retarded. When heavily infected, the caudal end of the larvae collapses and the posterior end shrinks. The feces are green, and the cadavers turn black and mummified.

Trypanosomes were discovered in Italy in the early part of the twentieth century. The etiological agent in silkworms is *Herpetomonas bombycis*, which has 6–12 μm long conical cells with a 10–20 μm long anterior flagellum (Levaditi, 1905). Within the cell, a kinetoplast occurs between the centralized nucleus and the flagellum. The cells can shrink to form a cyst. Reproduction is by binary fission. Infected larvae are slightly grayish white, with decreased appetite, turbid hemolymph, and pulsating dorsal vessel, and die gradually. Cadavers turn blackish brown (Table 12.1). When healthy larvae are injected with body fluid from an infected larva, *H. bombycis* can be detected in 24 h, and the injected larvae die in two to three days. *Leptomonas*, another genus of Trypanosomatidae, was discovered in Japan (Aratake and Kayamura, 1977). Cells are 15 μm long, with a flagellum as long as the cell. Little is known about this *Leptomonas* in silkworms.

A coccid disease of silkworms has been reported in Japan (Jin, 2001) (Table 12.1), but the taxonomic position of the etiological agent has not been determined. The pathogen cells are globose with a large nucleus. This pathogen can develop to form large, globose cysts. Cells inside the cyst can grow to about 100 μm (visible to the naked eye), and then several spores form as the nucleus divides and the protoplasm splits. After the thick membrane of the cysts breaks, the spores are released and can invade other tissues. The coccids infect silkworms through the epithelial cells of the midgut. Diseased larvae develop poorly, become shrunken, have diarrhea, and a brown mucus forms around the anus. Cysts can be detected in the feces.

12.2.6. Non-infectious Biotic Agents

Silkworms can suffer from injuries caused by non-infectious biotic agents such as parasitoids, parasitic mites, and urticating hairs of other Lepidoptera.

Silkworm larvae are attacked by tachinid parasitoids (called uji flies), mainly in the genus *Exorista* spp. (*E. sorbillans* and *E. bombycis*), but also including *Crossococmia sericariae*, *Ctenophorocera pavida*, and *Blepharipa zebina*. These uji flies can threaten the sericultural industry and are most serious during the summer. Losses due to tachinids can reach 10–15% in China, and exceed 30% in areas of southern China that are hot and humid (Jin, 2001). *Exorista sorbillans* typically produces four to five generations in northern and north-eastern China, six to seven generations in eastern China, and 10–14 generations in southern China. Both males and females mate multiple times before females start laying eggs, usually on the second day after sufficient mating events have occurred. *Exorista sorbillans* uses odor to locate hosts, laying eggs on fifth instar silkworm larvae. Larvae require only four to five days to mature, with the pupal stage lasting 10–12 days or for several months if the pupae go into winter diapause. *Exorista sorbillans* is a generalist parasitoid with a host range that includes more than 10 other species of Lepidoptera (Jin, 2001).

The first sign of attack is a small brown lesion, where the parasitoid larva penetrated through the cuticle into the hemocoel. The eggshell of the parasitoid can sometimes be seen on the lesion, but normally it drops off the host's cuticle. The lesion enlarges over time, becoming a black–brown, horn-shaped sheath as the silkworm larva grows. The yellowish parasitoid larva can be found if the lesion is dissected. The silkworms sometimes take on a purplish color, due to oxidized hemolymph, and larvae often cannot complete development to the adult stage and often do not even pupate. Parasitoids pupate outside the host.

More than 10 species of mite in various families are known to parasitize silkworm larvae, pupae, and adults. Among them, the straw itch mite *Pyemotes ventricosus* is the most common. It reproduces ovoviviparously (i.e., eggs, larvae, and nymphs develop inside the maternal body), leading to the emergence of adult mites. The mites are yellow with a large, globose abdomen, and visible without magnification. Newly emerged adult females mate immediately and actively search for hosts, which they pierce with their needle-shaped chelicerae, injecting a toxin that paralyzes the silkworm. The mites continue to feed on the hemolymph until the host dies. *Pyemotes ventricosus* produces several generations per year, and as many as 17–18 generations can occur in eastern China. Winter diapause occurs in the mated females. Another host of *P. ventricosus* is the pink bollworm, *Pectinophora gossypiella*, and therefore, silkworm infestations can be more serious in cotton-producing areas where the rearing rooms are also used for storing cotton.

When early instar silkworms are parasitized, the larvae stop feeding and often experience spasms and vomiting. When the larvae die, the body is typically curved, with the head protruded and the thorax swollen. First and second instar silkworms are usually killed in less than 10 h. Third instar silkworms develop asynchronously and become grayish yellow and shrunken. The larvae then become slightly reddish and swollen, especially the posterior segments, and a reddish brown effluent is defecated. The larvae die hanging upside-down.

The mature larvae are rarely infested, but when attacked, they shrink and exhibit a prolapsed anus. The posterior segments produce a black–brown or red–brown effluent, and spots appear on the thorax and abdomen. These infested larvae usually die just before molting, or during molting, with symptoms similar to those in the younger larvae. Pupae are usually parasitized in the intersegmental membranes on the dorsal side of the abdomen. Infested pupae also develop a large number of black spots and die before emergence. The cadavers are black–brown and show no signs of decay. Adult moths do not show any obvious symptoms when infested. The female mite usually parasitizes adult silkworms on the intersegmental membranes of the abdomen, and the moths become less flexible. Infested males do not behave normally, and the infested females lay fewer eggs, many of which are not viable (Jin, 2001).

The mulberry tussock moth, *Euproctis similis*, has urticating hairs that can poison silkworm larvae. The moth is a common pest on mulberry plantations, producing up to three generations, and overwinters as larvae. Larvae have needle-shaped, aggregate setae that secrete formic acid and toxic proteins. Stings from these setae can occur on silkworms throughout the year. Young silkworm larvae and newly molted larvae are affected the most, exhibiting tissue necrosis from the poisonous setae. Brown or black–brown spots appear first on the abdomen, extend gradually to the prolegs, and finally aggregate in large numbers along the intersegmental membranes, resulting in an irregular belt of black spots or black feet if the spots aggregate on the prolegs. The toxins can also enter the hemocoel, affecting other tissues and organs. Physiological effects include retarded development and delayed molts, and in some cases, death of small larvae. Some poorly developed larvae can spin cocoons and pupate, but the quality of the silk is low.

Consumption of mulberry leaves contaminated with these poisonous setae results in damage to the epithelial cells of the foregut, extending to the posterior part of the midgut. As a result, the silkworm larvae reduce their feeding rate and develop slowly into small-sized larvae. Larvae that have consumed a large number of setae stop feeding entirely, then swing their head from side to side, vomiting, and eventually die. If first instars feed on mulberry leaves contaminated with setae, they die very quickly.

Various slug caterpillars (Limacodidae), such as the mulberry slug caterpillar, *Thosea postornata*, are pests of mulberries. The slug caterpillar has a broad food preference and usually produces two generations per year, with mature larvae overwintering inside their cocoons. The larvae pupate the following spring. The moths emerge and lay eggs that hatch in early July. The larvae have tufted stinging hairs that are long and stiff and secrete acidic toxins. The stinging hairs damage the integument of silkworm larvae, but the toxins can also enter the hemolymph. Damaged silkworm integument will bleed small amounts of hemolymph, which coagulates into small black–brown round spots. These spots are larger than those caused by mulberry tussock moths. When exposure to the urticating hairs is light, the silkworm larvae stop feeding for 30–60 min before gradually resuming their feeding, but larval development will be retarded. Heavy exposure will cause the silkworms to die within a few hours to a couple of days. The cadavers become flaccid and black (Jin, 2001).

12.2.7. Abiotic Agents

Nosotoxicosis is a disease caused by or associated with a toxin. In sericulture, nosotoxicosis occurs when silkworms consume or are exposed to insecticide-contaminated mulberry leaves, a common problem during the summer-to-autumn rearing period. This period is when many pests occur in the mulberry fields, and as a result, pesticides are intensively used. Owing to the wide variety of pesticides used and their timing and dosage, the effect on silkworms varies greatly (Zhang, 2010). Poisonings may be acute or sublethal. Sublethal poisonings may cause a reduction in silkworm resistance to viral and other diseases, negatively affecting the cocooning process, as well as negatively affecting the number and viability of eggs. The most common effects are immediate cessation of feeding, loss of mobility, convulsions, and ultimately death within a matter of minutes or hours after exposure to a lethal concentration of the insecticide.

Besides insecticides, industrial wastes can adversely affect silkworm larvae. For example, fluorides may occur as an industrial waste gas (from aluminum smelting or phosphate processing plants) that can contaminate mulberry leaves. Fluorides can damage the leaves, causing them to become necrotic or show signs of chlorosis. Fluorides once affected sericultural production heavily in Japan, and they still can cause losses in various industrial areas in China (Miao *et al.*, 2005). Poisoned larvae develop asynchronously, molt late, or have supernumerary instars, often two or three extra instars. Lesions form on intersegmental membranes, causing them to swell and look like bamboo segments. The lesions are easily ruptured, releasing a yellow fluid. The larvae have difficultly defecating or defecate bead-like feces. The caudal segments (anything posterior to the fifth abdominal segment) become translucent, and the larvae usually die from losing food and liquids through vomiting.

Gases produced by coal-burning stoves (sometimes used to heat rearing rooms) can be toxic to eggs, especially when they are incubated at high temperatures to promote early hatching. Eggs exposed to the stove gases do not hatch or only a portion of them hatches, whereas larvae have a reduced feeding rate, become sluggish, or lie still on rearing trays. Larvae exposed to high concentration of these gases stop feeding, vomit, and die with a swollen thorax and shriveled caudal segments. Larvae may die halfway through the molting process or without molting. The cadavers sometimes assume a W-shaped position, develop massive black spots, turning completely black, and are rigid and extremely fragile (Jin, 2001).

12.2.8. Disease Control Methods for Sericulture

Two key components are central to disease control in any insect rearing facility: prevention and avoidance. Once a pathogen or parasite establishes within an insect colony, it can spread rapidly and be difficult to control. Chemical treatments can reduce parasites and pathogens, but they are rarely completely effective at eliminating them and they may not stop pathogen transmission. In addition, no treatments exist for many of the silkworm diseases, such as those caused by viruses. The following preventive measures should always be taken:

- Carry out strict disinfection methods for the rearing houses, rearing equipment, and surrounding areas.

- Maintain strict sanitation and hygienic conditions during rearing.

- Rear resistant varieties and use disease-free eggs. Moths used for breeding should be sampled for the presence of infections, and if infections are found, the entire batch of adults and any eggs produced from it should be destroyed.

- Practice intensive rearing management to enhance the vigor of silkworms by maintaining appropriate larval densities and proper humidity. Also, provide suitable food, adding fresh uncontaminated mulberry leaves as needed during rearing.

- Dispose of silkworm litter and other waste materials properly by burying, burning, or disinfecting (such as a steam heat treatment).

- Closely monitor, collect, and destroy weak, diseased, and undernourished larvae.

- Properly control pests in the mulberry field so that high-quality leaves can be supplied to the silkworms that are both pest free and pesticide free.

Silkworm cadavers, feces, and all contaminated materials should be properly disposed of, and all ceilings, walls, grounds, rearing beds, trays, racks, and other rearing appliances in and around rearing rooms properly cleaned and disinfected. Balavenkatasubbaiah *et al.* (1989) found that rearing rooms, trays, stands, and the paper linings used in the trays serve as important sources of contamination, and when these were disinfected between uses, disease prevalence was significantly reduced. Disinfection is carried out using approved chemical disinfectants, such as chloride or quaternary ammonium compounds (product registration for disinfectants varies greatly among countries). Sulfur is used as a disinfectant fumigation against muscardine fungi and mites.

The silkworms have a very poor ability for flight, which aids in the control of disease spread. Methods used to ensure that eggs are disease free are an important aspect of pébrine control. Such efforts in France, Japan, India, and China have brought the infection rates down from 20% in the nineteenth century to approximately 1% at present (Jin, 2001). To control egg infection levels in China, female moth inspections are conducted by provincial authorities. Moths used as breeding or foundation stock are routinely sampled for the presence of the microsporidium. The inspected female moths are ground, filtered, centrifuged, and then inspected microscopically at 400× for the presence of microsporidian spores. For foundation stock, all females are inspected after they produce eggs, whereas occasional samples are taken for breeding stock. If a female is found to be infected, all her egg batches are destroyed.

To further evaluate the health of foundation females, their eggs are also inspected for viability. Eggs are placed at 28−30°C and 80% RH to accelerate hatching and subsequently checked under a microscope to determine whether a high percentage of hatch has occurred.

The health of mature larvae can be assessed by placing them at 27–29°C and 80–85% RH. The larvae will complete development early, and moths will emerge two to four days earlier than usual and can then be examined for presence of the pathogen. Larval health should also be regularly inspected in the following manner. Several larvae are selected from those that appear to molt normally, those that molt late, and those that are half molted, and then are placed at 29°C and 90–95% RH. High heat and humidity increase disease levels in infected larvae, where infection levels may otherwise go undetected until an outbreak occurs. Dead larvae are then examined microscopically to determine whether any infectious disease has occurred. Any dead pupae in cocoons are also inspected.

In addition to sanitation, resistance can be an important tool for reducing diseases. Resistance can be affected by various environmental factors. For example, silkworm resistance to viruses is reduced when the larvae are starved or fed on low-quality mulberry leaves, with some seasonal weather changes, or by exposure to certain chemicals, such as those used for disinfection (Jin, 2001). High and low temperatures tend to affect the larvae and not the pupae. Changes in susceptibility to viral infections vary with the rearing season: blood-type grasserie occurs most often in the spring, while BmCPV-1 and densovirus occur most often in the summer and fall. Thus, disease levels can rise when the proper conditions are not maintained.

Resistance also varies among silkworm strains (Sen et al., 1997) and for different ages and genders. For example, resistance to BmNPV increases as larvae mature, such that resistance increases 10-fold with each instar (Wu et al., 1983a, b). Resistance to BmIFV increases 1.5-fold with the first molt, 3-fold with the second molt, 134-fold with the third molt, and 10,000 to 12,000-fold with the last molt (Wu et al., 1983a, b). Conversely, resistance to BmCPV-1 and BmDNV does not change much as the larvae age (Wu et al., 1983a, b). As for gender differences in susceptibility, the male larvae and pupae are twice as resistant as females (Wu et al., 1983a, b).

Theoretically, resistance could also be achieved with breeding programs, especially since resistance appears to vary between strains, suggesting a genetic component. The use of disease-resistant silkworm varieties would greatly assist in disease management (Eguchi et al., 1996, 1998). Resistance to per os infections of BmNPV is usually greater in those silkworm strains that have a multivoltine lineage, and this resistance is controlled polygenically, with a major dominant gene on a euchromosome and some minor genes on the sex chromosome (Qian et al., 2006). The mechanisms for resistance to viral infections are antiviral activity of the gastric secretions of the midgut and resistance in the peritrophic membrane to viral penetration. Resistance to

the densovirus is controlled by a few major genes (Qian et al., 2006). Resistance to BmCPV-1, BmNPV, and BmIFV is controlled by several minor genes that may be unrelated to each other (Qian et al., 2006). Host resistance to BmCPV-1 appears to be correlated with resistance to BmNPV, as well, but not to BmIFV. Thus, resistances to BmCPV-1 and BmNPV are probably controlled by some common genes, but those to BmCPV-1 and BmIFV are not. Resistance to muscardine diseases is also associated with silkworm variety. Silkworm strains from Chinese lineages are more resistant to B. bassiana than varieties from a European lineage, and hybrids of Chinese lineages with Japanese lineages were even more resistant (Mu et al., 1999).

Uzigawa and Aruga (1966) developed a silkworm variety resistant to BmIFV after selecting survivors from virus-fed larvae for five consecutive generations. Similarly, Watanabe (1967) developed a silkworm variety resistant to BmIFV by selecting survivors from virus-fed larvae for eight generations. Dingle et al. (2005) successfully transferred Bt resistance through a conventional hybridization breeding system. However, more efforts are needed to develop silkworm varieties that are resistant to BmNPV, the most serious viral disease of silkworms. Molecular methods may provide some new assistance to breeding efforts. For example, Yao et al. (2005) constructed an isogenic silkworm line using molecular marker-assisted breeding, which resulted in a new silkworm variety resistant to BmNPV. In spite of this progress, however, no varieties with any substantial resistance have been adopted for silk production.

The use of genetic resistance to pathogens is still a long way from commercial use; therefore, sanitation and maintaining larvae in healthy conditions are the primary means for disease control. For many of the diseases, silkworms generally cannot be cured with medications. The fungicides benomyl, thiophanate, and carbendazim are reported to be therapeutically effective against microsporidia. However, treatment methods are generally lacking and, therefore, disease prevention is key to maintaining productive silkworm cultures.

12.3. DISEASES OF BEES

Diseases of bees have been summarized by Morse and Nowogrodzki (1990) and Bailey and Ball (1991). However, these books deal only with honey bees, and primarily *Apis mellifera*. Furthermore, critical new diseases have developed in *A. mellifera* production systems since these books were published, and so a new summary of bee diseases is needed.

Beekeeping is an ancient industry, and as with silkworms, disease control is critical to its success. The earliest records of hive beekeeping date back to approximately

3500 years ago in Mesopotamia (Crane, 1999). Early beekeepers kept *A. mellifera* and *A. cerana* in hollow logs or tree boles, and later in baskets, jars, and pots of various sorts. Modern honey beekeeping is based on wooden hive designs with removable frames for the honey comb, as developed by Langestroth and Dadant in the 1850s (Crane, 1999). The basic structure of a movable frame hive is a wooden box called a "super" that is filled with wooden frames that hang downward (a typical standard is 10 frames per super). A sheet of wax with a hexagonal grid is mounted across each frame, and this special sheet of wax is called a "foundation" because it serves as the foundation for the bees to build their comb. The purpose of the frames and foundation is to entice the bees to build their comb where the beekeeper wants it, and when the bees build their comb in the frames, the beekeeper can easily remove the frame to access the comb. The bees use the comb for raising their brood, storing pollen, and making and storing honey. The physical structure that houses the bees is referred to as the "hive" and the bees that occupy the hive are called a "colony". *Apis cerana* is sometimes kept in Asia, but *A. mellifera*, a bee originally from Africa and Eurasia, has behavioral characteristics that make it especially amenable to human manipulation to improve honey and colony production, and so it is the most commonly kept bee throughout the world.

Honey bees are only a small portion of all the bees in the world (Michener, 2000). The vast majority of bees are actually solitary, and their life cycles differ greatly. In the last half-century or so, people began to recognize the importance of some of these bees for pollination of both crops and native plant communities. Although *A. mellifera* is a broad generalist in its host plant preferences, this species can be a poor pollinator in some situations. The need for better pollinators in certain situations has led to the development of rearing methods for other kinds of bees. Bumble bees (especially *Bombus terrestris* in Europe, and *B. impatiens* in eastern North America) are produced in insectaries and sold as pollinators for greenhouse crops, especially greenhouse tomatoes. The alkali bee (*Nomia melanderi*) is a solitary ground-nesting bee, and alfalfa seed producers have developed methods to create and maintain specialized nesting beds that are attractive for this bee because it is a better pollinator for alfalfa than are honey bees (Pitts-Singer, 2008). The alfalfa leafcutting bee (*Megachile rotundata*) is a solitary cavity-nesting bee that is used extensively for pollination in alfalfa seed crops in North America (Pitts-Singer and Cane, 2011). Production systems have been developed where *M. rotundata* is enticed to nest in polystyrene blocks that contain rows of holes the size this bee prefers for nesting. Various species of mason bees (*Osmia* spp.) are used for pollinating fruit trees and almonds, and they are often provided with either wooden nesting boards or hollow reeds to nest in. All of these bees are very important for agriculture, and their diseases are discussed below.

12.3.1. Viruses

A fairly large number of viruses is known to infect honey bees and they are generally named after some of the signs or symptoms they cause in the host, although these signs or symptoms are not always a reliable method for identification (Table 12.2 and Fig. 12.2A). These viruses have generally been thought of as being highly specific to honey bees because they cannot be grown on other insects or in cell culture. However, with the advent of reliable molecular techniques for identifying viruses, some of the honey bee viruses are now being found infecting a fairly broad host range within the Apoidea (bees) or even the Aculeate (bees and wasps) (Singh *et al.*, 2010) (Table 12.2). Most viruses that infect bees are non-occluded, picorna-like RNA viruses that are isometrically symmetrical, ovoid or spherical, and range in size from 20 to 30 nm (see Chapter 5). Most identifications rely on serological (Ball, 1999) or nucleic acid techniques (Meeus *et al.*, 2010), the latter being the most commonly used today.

It has been known for a long time that bees can transmit plant viruses through pollen; that is, pollinators can serve as vectors for some plant pathogens, especially those that infect fruit via blossoms. Therefore, it is not surprising that flowers and pollen can serve as a reservoir for bee-pathogenic viruses (Bailey, 1975), but the fact that bees can actually become infected orally from this pollen was demonstrated only recently (Singh *et al.*, 2010). Other routes of transmission within a colony have been better studied and include both horizontal and vertical transmission (Chen *et al.*, 2006). In addition, varroa mites vector some viruses, especially *Deformed wing virus* (Table 12.2) and the *Kashmir bee virus* (Table 12.2) (Chen *et al.*, 2006; de Miranda and Genersch, 2009), and this is one likely reason why varroa mite infestations are detrimental to a colony. Other hive pests may also contribute to the spread of viruses. For example, small hive beetles have been shown to be susceptible to *Sacbrood virus* (Table 12.2 and Fig. 12.2A) (Eyer *et al.*, 2009) and may serve to transmit viruses in honey bees (Eyer *et al.*, 2008).

12.3.2. Bacteria

Bacterial pathogens are not very diverse among bees. None are yet known to be pathogenic to bumble bees and solitary bees, although Skou *et al.* (1963) describes the presence of unidentified, Gram-negative bacilliform bacteria isolated from diseased organs in queens of *B. terrestris* in Denmark. In honey bees, three bacterial pathogens are well described: (1) American foulbrood (AFB) caused by *Paenibacillus larvae*; (2) European foulbrood, caused by *Melissococcus*

TABLE 12.2 Examples of Infectious Diseases and Mite Parasites of Bees

Pathogen/mite	Disease	Known Hosts	Affected Host Stages	Signs and Symptoms
Viruses[a]				
Acute bee paralysis virus (ABPV)	Acute bee paralysis	*Apis mellifera, Bombus terrestris*	Adults	Adults may be paralyzed or asymptomatic. Not thought to be transmitted orally, but found in pollen loads from foragers and thoracic salivary glands. May be vectored by *Varroa destructor*
Black queen cell virus (BQCV)	Black queen cells	*Adrena* sp., *Apis mellifera, Bemix* sp., *Bombus* spp., *Megachile rotundata, Nomia melanderi*	Larval queens, adults	In honey bees, the queen cells develop dark brown to black cell walls. Immature queens die as prepupae or pupae, turn flaccid. Larvae do not become diseased when fed virus particles, but infections are common and asymptomatic in adult bees. Infection is enhanced by the presence of *Nosema apis* infections. Signs and symptoms of other bees are undescribed
Apis iridescent virus (IV-24)	Clustering disease	*Apis cerana*	Adults	During summer, flightless bees detach from the colony and cluster or occur crawling on the ground in large numbers. Infected tissues appear bright blue. Occurs in the fat body, alimentary tract, hypopharyngeal glands and ovaries
Chronic paralysis virus (CPV)	Paralysis, chronic paralysis, or hairless black syndrome	*Apis mellifera*	Adults	Sections of the hindgut epithelium have basophilic cytoplasmic bodies. Type I: abnormal trembling, failure to fly, adults crawling on the ground in large groups, adults cluster at the top of the hive, bloated abdomens and dislocated wings. Type II: adults become black and hairless, abdomens distended and shiny, get slightly attacked by older bees in the colony, often excluded by guard bees; appear like black robber bees
Deformed wing virus (DWV)	Deformed wing disease	*Adrena* sp., *Augochlora pura, Apis mellifera, Apis cerana, Apis florea, Bembix* sp., *Bombus* spp., *Ceratina dupla, Megachile rotundata, Nomia melanderi, Xylocopa* sp.	Adults	In honey bees and bumble bees, newly emerging adults have small, deformed wings. Honey bees may also have bloated, shortened and discolored abdomens. *Varroa destructor* is a vector in honey bees and increases disease symptoms
Kashmir bee virus (KBV)		*Apis cerana, Apis mellifera, Bombus terrestris*	Larvae, adults	KBV and IAPV are closely related, and best described from honey bees, where all life stages can become infected. Transmission: horizontal from infected adults via glandular secretions applied to food sources; transovarial; and vector-borne via *Varroa destructor*. Infected brood either dies before the cell is capped or recovers from infections. Infections can persist in adults without symptoms, or cause rapid mortality within a few days. Virions are non-enveloped, icosahedral capsids ~ 30 nm wide

Pathogen	Disease	Host	Stage	Signs and symptoms
Israeli acute paralysis virus (IAPV)		Found in many species of bees	Adults	Similar to KBV. Presence of this virus in colonies has been correlated with colony collapse disorder in the USA
Sacbrood virus (SBV)	Sacbrood	*Adrena* sp., *Apis cerana*, *Apis mellifera*, *Bombus* spp., *Nomia melanderi*	Larvae, adults	In honey bees, mature larvae fail to molt to pupal stage then become flaccid, turn from white, to yellow, to brown in a few days. The dry scale does not stick firmly to the cell as it does for foulbrood. Larvae eventually become dry and flat in capped cells. Signs and symptoms unknown for other bees
Bacteria				
Melissococcus pluton	Foulbrood, European	*Apis mellifera*, *Apis cerana*	Larvae	Larvae typically die when 4–5 days old, and outbreaks usually occur in early summer. Dead larvae turn brown and flaccid. Midguts filled with bacteria in opaque white clumps. Dead larvae may dry to a scale that is rubbery and does not stick to the wax. Colonies typically recover. *Melissococcus pluton* cells ovoid to lanceolate (0.8 μm wide × 1.0 μm long), forms chains. Facultative anaerobe requires CO_2
Paenibacillus larvae	Foulbrood, American	*Apis mellifera*, *Apis cerana*	Prepupae, pupae	Larvae die after spinning the cocoon, then turn putrid and dark brown, with a distinct fishy odor. If a toothpick is stuck into the larva and pulled out slowly, the remains draw out as brown, ropy thread. Discoloration is first seen when larvae reach 10–15 days old. Cell cappings sink in slightly, often with small hole in the center. After one month, the dead larvae dry to a small, flat scale that sticks to the wax. The bacterium is a Gram-positive bacillus, motile, with oval endospores (~ 2.5 × 0.5 μm, but highly variable in size)
Spiroplasma apis	May disease	*Apis mellifera*	Adults	Large numbers of adults found quivering and unable to fly, moribund, or dead. Abdomens swollen and hard. Midgut full of undigested pollen. Other adults from the colony cluster in small groups away from the hive. Bees not dark and shiny, no obvious hair loss. Large numbers of adults may die in 4–5 weeks, but colonies usually recover. Wall-less, helical bacteria found in hemolymph and digestive tract
Fungi: filamentous				
Ascosphaera apis	Chalkbrood	*Apis mellifera*	Larvae	Infected larvae die at a late stage; sometimes after the cell is capped. The dead larvae are hard, chalk-white, but often mottled with black spots (the fungal spores). Typically, adult bees remove infected larvae from the hive, and these dead larvae can be seen in large numbers near the hive entrance. Microscopic examination will reveal spores formed in sacs called ascomata, and within the ascomata are spore balls (ascospores). The ascospores are ovoid, but only slightly so, nearly spherical

(Continued)

TABLE 12.2 Examples of Infectious Diseases and Mite Parasites of Bees—cont'd

Pathogen/mite	Disease	Known Hosts	Affected Host Stages	Signs and Symptoms
Ascosphaera aggregata	Chalkbrood	Megachile rotundata	Larvae	Infected larvae most often die during the last instar, after they have consumed their entire pollen provision and reached full size, but before cocooning. Dead larvae are hard and black. The surface may easily disintegrate to a black powder (the fungal spores). Larvae may be mottled black and white, or be mostly white, but less commonly so than with chalkbrood in honey bee larvae. Ascospores are long ovoid, typically 2 μm wide and 4–5 μm long
Ascosphaera torchioi	Chalkbrood	Osmia lignaria	Larvae	Similar to A. aggregata above. The ascospores are typically 2 μm wide and 4–5 μm long
Aspergillus flavus and Aspergillus fumigatus	Stone brood	Apis mellifera	Larvae	Infected larvae die soon after cells are capped, before pupation, then turn very hard and pale brownish, gray, or yellow–green. The color comes from the spores. Spores are spherical and highly sculptured
Fungi: Microsporidia				
Nosema apis	Dysentery	Apis mellifera	Adults	Disease characterized by the presence of spores in the midgut epithelium (~ 6 × 3 μm), and milky white appearance of the ventriculus. Usually < 30 coils to the polar filaments in the spores. No outward signs of infection, infected adults live about half as long as uninfected bees. Associated with dysentery in winter and early spring, but probably not the cause. Spread via fecal contamination
Nosema ceranae	Dysentery	Apis mellifera, Apis cerana	Adults	Disease characterized by the presence of spores in the midgut epithelium (~ 5 × 2 μm), with few distinct marks. Approx. 20 coils to the polar filaments in the spores. No outward signs of infection, infected adults live about half as long as uninfected bees. Associated with dysentery in winter and early spring, but probably not the cause. Spread via fecal contamination
Nosema bombi	Nosema disease	Bombus terrestris, Bombus occidentalis, Bombus impatiens	Adults	Disease characterized by the presence of spores throughout the body cavity (~ 4.2–5.4 × 2.7–3.5 μm), with few distinct marks. In B. terrestris and B. impatiens, colony growth and the production of reproductives are greatly reduced, but not with B. occidentalis. Transmission is transovarial and oral
Protista				
Apicysti bombi	No name	Bombus affinis, Bombus bimaculatus, Bombus fervidus, Bombus griseocollis, Bombus impatiens, Bombus perplexus, Bombus terricola, Bombus vagans	Adults	Poor colony growth, fat body destroyed in infected workers. Large spores detected in feces

Species	Common name	Host	Stage	Description
Crithidia bombi and *C. expoeki*	No name	*Bombus terrestris*	Adults	Trypanosome that occurs in hindgut of bumble bees. Transmission occurs horizontally within a nest, between nests via flower visitations, and transovarially. Infections affect founding ability of new queens, survival of workers, and colony reproduction
Crithidia mellificae	Flagellate disease	*Apis mellifera*	Adults	Trypanosome that occurs in gut lumen and epithelium of honey bees. Does not cause overt disease
Gregarines (e.g., *Monoica apis, Apigregarina stammeri, Acuta rousseaui, Leidyana apis*)	Gregarine disease	*Apis mellifera*	Adults	Flagellates occurring in midgut of adult bees. Cephalont stage oval, ~ 16 × 44 μm, in two segments. Sporont stage ~ 35 × 86 μm. Infections do not cause overt disease
Malpighamoeba mellificae	Amoeba disease	*Apis mellifera*	Adults	Cysts occur in the hindgut and rectum. Infections occur in the Malpighian tubules. Infected colonies occur most commonly with other diseases
Invertebrates: Acari				
Acarapis woodi	Tracheal mites	*Apis* spp., *Bombus* spp.	Adults	Affected bees may have disjointed wings and difficulty flying, distended abdomens, or may be asymptomatic. Mites occur in the prothoracic tracheae. Females are 143–174 μm long, males 125–136 μm, white, somewhat transparent, smooth cuticle with a few long hairs on the body and legs. The impact on the colony size and productivity is uncertain
Tropilaelaps clareae	Tropilaelaps	*Apis mellifera, Apis dorsata*	Adults, pupae, larvae	External parasitic mite found in Southeast Asia. Parasitize bee adults and brood, mites ~ 1 mm long × 0.6 mm wide. To sample, hit a frame of comb onto a light-colored surface to dislodge mites, observe under magnification
Varroa destructor	Varroa mite	*Apis cerana, Apis mellifera*	Adults, pupae, larvae	External parasitic mite. Extremely serious pest of *A. mellifera*, eventually causing colony death. Mites parasitize both adults and brood, but reproduce only on capped brood. Able to vector some viruses. Female adults oval, flat, red–brown, ~ 1 × 1.5 mm. Males smaller, white, less frequent. Mites can be sampled by dislodging from adult bees using ethyl alcohol, isopropyl alcohol, ether, or powdered sugar placed in a jar with the bees, then shaking the jar

[a] In viral nomenclature, the type species is in italics.

pluton; and (3) *Spiroplasma* spp., which are quite rare (Table 12.2). AFB is perhaps the most well-known bacterial disease of honey bees. *Paenibacillus larvae* infects the brood and is highly contagious, and thus, it is also strictly regulated in many countries. Apiary inspection services in many parts of the world regularly inspect beekeepers' colonies to ascertain the status of this disease and prevent its spread. The bacterium forms spores that are extremely hardy, capable of persisting for many years on hive materials (Lindström, 2006) and able to withstand high temperatures (Forsgren *et al.*, 2008; Genersch, 2008). In addition, *P. larvae* is resistant to many antibiotics (Kochansky *et al.*, 2001). The infection process begins when honey bee larvae are inadvertently fed spores that contaminate food material in the hive (Fig. 12.3). Transmission is *per os*, and infected larvae usually die after they spin a cocoon. The cadavers then turn dark brown and have a distinct, fishy, putrid odor. If a toothpick is stuck into such a cadaver and pulled out slowly, the material will come out as a thick, ropey thread (Fig. 12.2B). Eventually, the dead larva dries down to a hard scale that sticks to the wax along the side of the brood cell in which it died. This scale will be full of *P. larvae* spores, up to 7.5×10^8/cadaver (James, 2011a), which then serve as a source of contamination and spread the disease.

Melissococcus pluton also infects the larvae of honey bees *per os*. Larvae typically die when four to five days old, that is, before a cocoon has been spun. Cadavers can be found in the brood cells that are brown and flaccid, similar to AFB, but generally European foulbrood cadavers do not have the same characteristics as described for AFB. Similarly, when the cadavers dry, they become rubbery and do not stick to the wax. Colonies typically can recover from infections on their own.

Two species of *Spiroplasma* have been reported as pathogens of honey bees: *S. apis* (Mouches *et al.*, 1982, 1983, 1984) and *S. mellifera* (Clark *et al.*, 1985). Even though these are typically rare, a recent molecular survey of parasites in honey bee colonies was positive for the presence of *Spiroplasma* (Runckel *et al.*, 2011), indicating that they may be present in honey bees as non-disease-causing infections, and thus go undetected much of the time. *Spiroplasma* are believed to cause a disease called May disease. This disease is prevalent in the spring, causing large numbers of moribund adults to cluster around the hive, often quivering and with swollen and distended abdomens. Other adults may cluster together away from the hive. Colonies usually recover from these infections.

12.3.3. Fungi: Filamentous

The most common fungal disease of bees is chalkbrood, which occurs in the larvae. Chalkbrood is caused by fungi in the genus *Ascosphaera*, and it affects many different taxa of bees, although it has never been reported in *Bombus*. Currently, 22 species of *Ascosphaera* have been described and all are associated with bees either as saprophytes on the pollen and feces found in nests, or as pathogens, or as both (i.e., as facultative and opportunistic pathogens) (Anderson and Gibson, 1998; Youssef and McManus, 2001). *Ascosphaera* spp. are found associated with bees as diverse as *A. mellifera, Megachile rotundata, M. centuncularis, Osmia lignaria, O. cornifrons, Trigona carbonaria*, and *Chalicodoma* spp. The most common of these pathogens are described in Table 12.2. In honey bees, the disease is caused by *A. apis*, but generally, chalkbrood is not considered a serious disease in honey bees, even though the pathogen is probably present most of the time and in most colonies (Fig. 12.2C, D). Outbreaks are most common during cold, wet weather, especially in the spring, or if hives are kept in standing water or in humid regions when the hives do not have adequate ventilation. In addition, varroa mite infestations can increase chalkbrood prevalence in honey bees (Getchev and Kantchev, 1998).

Conversely, *Ascosphaera aggregata* can be a very serious disease in *M. rotundata*, and easily the most serious disease of this bee when it is managed for alfalfa seed pollination (Fig. 12.2E). This fungal species kills, on average, about 15% of the larvae in commercial systems, but in some alfalfa seed fields, as many as 45% of the larvae may be killed (R. James, unpubl.). *Ascosphaera aggregata* probably originated in Europe, where this bee is native, and migrated into North America with its host. In Europe, this pathogen has been found to infect *M. centuncularis* and *Osmia rufa* (Skou, 1975), but it has never been isolated from any bee species other than *M. rotundata* in North America. Likewise, *A. apis* is not found associated with *M. rotundata*, either. However, not all species of *Ascosphaera* are host specific, and many have broad host ranges.

Chalkbrood in *M. rotundata* spreads when spores contaminate the body of nesting females, which in turn deposit the spores in the pollen they collect to feed their young (James, 2008). When adult bees emerge from managed nesting systems, they become contaminated with spores from the larvae that died in the previous year (James, 2011b). Chalkbrood occurs at much lower levels in Canada than in the USA. It is unusual for the disease to reach 5% in any Canadian population, and it generally affects less than 1% of the larvae. As a result, Canada supplies a significant portion of the *M. rotundata* used in the USA (approximately 3×10^8 bees/year, which is about half of the total used in the USA). Several other *Ascosphaera* infect *M. rotundata* (Table 12.2 and Fig. 12.2E), but none is as common as *A. aggregata*.

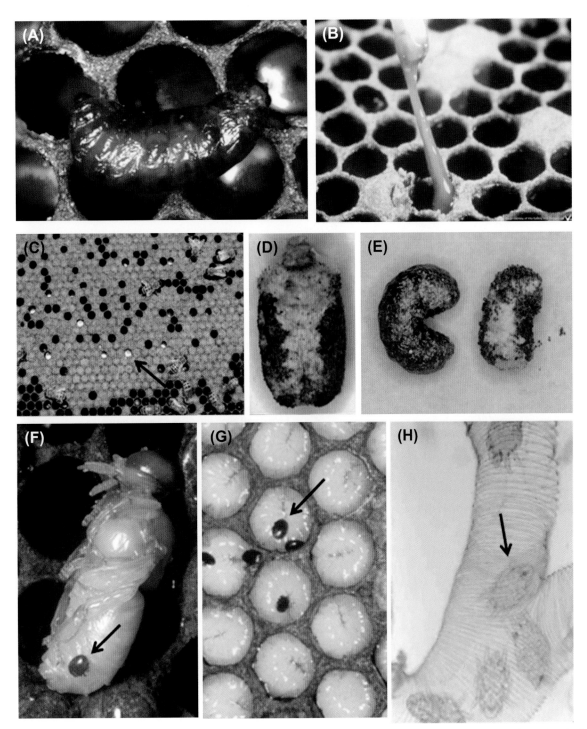

FIGURE 12.2   Diseases of bees. (A) Typical cadaver of *Apis mellifera* larva infected with SBV. (B) Diagnostic characteristic of American foulbrood. *Apis mellifera* larva infected with *Paenibacillus larvae* has decomposed to a soft mass that draws out a "ropey" thread. (C) Brood comb from an *Apis mellifera* colony showing chalkbrood. Larvae infected with *Ascosphaera apis* turn white and chalky (arrow). (D) *Apis mellifera* larva that has died from an *Ascosphaera apis* infection. (E) *Megachile rotundata* larvae that have died from chalkbrood, due to infections with *Ascosphaera aggregata* (left) and *Ascosphaera proliperda* (right). (F) An adult female of the parasitic mite *Varroa destructor* on the posterior end of a pupa of *Apis mellifera* (arrow). (G) *Varroa destructor* females invading the cells of mature larvae of *Apis mellifera* just before the cells will be capped by nurse bees. (H) Tracheal mites, *Acarapis woodi*, inside the trachea of an *Apis mellifera* adult worker. *[Photo credits: (A−D) B. Smith; (E) R. James; (F) S. Bauer; (G) Vita-Bee Ltd; (H) L. I. de Guzman.]*

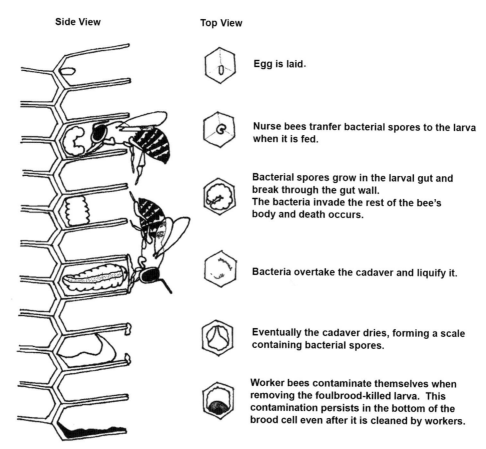

| | Side View | | Top View | |

FIGURE 12.3 Life cycle of American foulbrood (*Paenibacillus larvae*) in *Apis mellifera*, showing side and top view of a brood cell.

12.3.4. Fungi: Microsporidia

Nosema disease is caused by microsporidia in the genus *Nosema*. Two species that infect *A. mellifera* are widespread throughout the world: *N. apis* and *N. ceranae* (Klee *et al.*, 2006). These pathogens infect *per os*, and transmission occurs when bees ingest the spores, probably in contaminated water, pollen, or honey. *Nosema apis* infects the gut and muscle tissue of honey bees and causes both dysentery and crawling bees. Another typical symptom is a milky-white coloration of the gut (Table 12.2).

Nosema ceranae was first identified as a pathogen in the Asian honey bee, *A. cerana* (Fries *et al.*, 1996). It is now widely found in *A. mellifera* in Asia, Europe, and North America (Klee *et al.*, 2007; Williams *et al.*, 2008; Chen *et al.*, 2010). It is not clear whether it has spread rapidly as a new pathogen of *A. mellifera*, or whether it was always present but mistaken for *N. apis* in the past. However, *N. ceranae* does not cause the same symptoms as *N. apis*. *Nosema ceranae* is typically associated with neither dysentery nor crawling bees, but it does affect the feeding behavior of honey bees, making them less inclined to participate in trophallaxis. The main effects of this pathogen are increased bee mortality and decreased colony vigor. *Nosema ceranae* appears to be more sensitive to cold and is thought to be more common in warmer climates (Gisder *et al.*, 2010).

Nosema bombi is a pathogen of bumble bees and occurs widely within the genus *Bombus* (Cameron *et al.*, 2011). Unlike *Nosema* in honey bees, *N. bombi* can be found infecting all tissues throughout the body except the ovaries and proctodeum (Fries *et al.*, 2001). The spores are slightly smaller than *N. apis* and *N. ceranae* (Table 12.2). Although *N. bombi* has been found in many parts of the world, no other *Nosema* has been described from *Bombus* hosts. This pathogen has been investigated as a possible cause for the current decline of some bumble bees, perhaps after being distributed outside Europe through cultivation and release for greenhouse pollination (Colla *et al.*, 2006).

12.3.5. Protists

Flagellate trypanosomes in the genus *Crithidia* are known to infect both honey bees and bumble bees (Table 12.2). Recently, it has been proposed that *Crithidia bombi* is actually composed of two different species, *C. bombi* and *C. expoeki* (Schmid-Hempel and Tognazzo, 2010). Both are gut parasites of bumble bees. *Crithidia mellificae* is found in *A. mellifera*. These parasites infect *per os*, and may be transmitted from one colony to the next through contaminated flowers (Durrer and Schmid-Hempel, 1994).

The gregarines are also gut parasites of *A. mellifera*, but they do not cause overt symptoms in infected colonies (Table 12.2). A neogregarine, *Apicystis bombi* (formerly *Mattesia bombi*), has also been found in several bumble bee species (Liu *et al.*, 1974). This parasite occurs mainly in the fat body of the host. Mobile sporozoites release sporocysts into the midgut, and these invade new fat body cells. Schizonts are produced intracellularly and extracellularly and are multinucleated.

Another protist disease is caused by the amoebozoan *Malpighamoeba mellificae*, and it occurs in managed colonies of *A. mellifera*. Infections with this pathogen are called amoeba disease, and it has been reported in various parts of the world, mainly Europe, but generally it is quite rare. It is not clear whether *M. mellificae* infections actually cause any disease symptoms or mortality in the bees, but infections probably affect colony vigor (Morse and Nowogrodzki, 1990). The pathogen infects the Malpighian tubules of adult bees, and diagnosis of infection is generally based on the presence of cysts in the hindgut of adult bees. The pathogen is probably transmitted *per os* (Bailey and Ball, 1991).

12.3.6. Non-infectious Biotic Agents

Varroa destructor (Fig. 12.2F) is easily the most serious invertebrate parasite of *A. mellifera*. Mated, adult female mites can be found as phoretic passengers on adult worker bees and drones. Typically, adults are found on the intersegmental membrane on the ventral side of the abdomen, particularly between tegula one and two, or two and three. Mites probably feed on adult hemolymph during this time (Bowen-Walker and Gunn, 1998). When a colony is heavily infested, phoretic mites are also found on the dorsal side of the thorax. Phoretic females eventually search for developing bee brood that are within a few hours of being capped (honey bees cap over brood cells with wax when the brood are ready to spin a cocoon) (Fig. 12.2G). The mites crawl inside the cell, wait for it be capped, then they lay a single female egg. Within the next couple of days, the mother mite will lay another egg, usually a male. After that, the mother mite may lay a few more female eggs, but no more than one per day. The mites develop inside the capped brood cell, feeding on the hemolymph of the developing honey bee brood (Fig. 12.2F). The new female mites will mate with the male before they emerge from the host cell. Varroa mites typically emerge from the cell when the host emerges as an adult, but if the host dies, other bees in the hive will uncap the cell to remove the cadaver, and the mites may

exit at that time. The mother mite can live long enough to infest more than one brood cell.

Varroa destructor originated as a parasite on *A. cerana* and adapted to *A. mellifera*, where it became a serious pest worldwide. Most populations of the mite in North America probably originated from Korea and are likely to be the result of several introductions by way of Europe (de Guzman *et al.*, 1997; Anderson and Trueman, 2000). The varroa mite is important not only as a parasite, but also because it serves as a vector for honey bee viruses (see Section 12.3.1). In addition, the feeding activity of these mites compromises the honey bee immune system, leaving parasitized bees more vulnerable to infections (Kanbar and Engles, 2003).

The tracheal mite, *Acarapis woodi*, is a parasite of bees that lives within the trachea of the thorax and head (Fig. 12.2H), where it feeds on the host hemolymph. Adult females invade the host through the spiracles, and then lay nearly one egg a day, and as many as 21 eggs in a lifetime. The only time that the mites live outside the host is when young, new females transfer to new hosts (Pettis and Wilson, 1995). In warm climates, honey bees appear to be able to sustain heavy infestations with little impact on bee health. However, in colder climates, the impact is much greater. Tracheal mites cause adult bees to leave the hive, and affect their ability to fly and defecate, leading to large losses of young adult bees in the winter and spring (Webster and Delaplane, 2001).

It is possible that many of the treatments for varroa mites also control tracheal mites. These mites are microscopic and infested bees can be asymptomatic; thus, the presence of this disease is easy to miss. A typical sign is a weak colony in the early spring, where the returning foragers do not make it all the way back to the hive but must crawl the last few centimeters or so to the hive entrance, resulting in large numbers of bees crawling on the ground in front of the hive.

Bumble bees are also susceptible to *Locustacarus buchneri*, another parasitic mite that infests the trachea. This mite is thought to have originated as a parasite of *B. terrestris* in Europe but has increased its host range as a result of the importation of *B. terrestris* to other parts of the world, such as Japan (Goka *et al.*, 2001)

Conopid flies are dipteran parasitoids in the family Conopidae. Conopid flies have been collected from bumble bees in both Europe and North America, including *Bombus pascuorum*, *B. bifarious*, *B. occidentalis*, *B. impatiens*, *B. griseocollis*, and *B. fervidus* (Gillespie, 2010; Otterstatter, 2011). Several species of conopids probably occur on bumble bees, but the two that are described are *Sicus ferrugineus* and *Physocephala rufipes*. Development time in the host for both parasitoids is about 11 days. The parasitoids can sometimes occur together in a single host, but then must compete with one another, as the host can only support the complete development of one. The first instar larvae have strong pointed mandibles that are probably used to kill other eggs found in the same host (Schmid-Hempel and Schmid-Hempel, 1996).

Another dipteran group of parasitoids is the phorid flies, in the family Phoridae. The phorids that attack bees belong to three genera: *Melaloncha*, *Styleta*, and *Apocephalus*. *Melaloncha* were last revised by Brown and Smith (2010) and currently include 167 species. *Melaloncha* species are neotropical and attack most tropical genera of Apidae (bumble bees, honey bees, and stingless honey bees) (Disney, 1994; Brown and Smith, 2010). *Apocephalus* spp. attack *Bombus* in North America and stingless bees in the neotropical region (Brown, 1996, 1997). Overall, phorid flies do not pose any real threat to beekeeping operations.

Several hymenopteran parasitoids are a more serious problem for beekeeping, including *Leucospis*, *Monotontomeris*, *Mellitobia*, and *Chrysura* species. These parasitoids enter nests through cracks and crevices, locate a developing larva, and then lay their eggs on this host. The parasitoids can be identified by the presence of a large number of small larvae in a cell where there should be only one developing bee larva. These parasitoids are most common in the managed solitary bees, especially *M. rotundata* and *O. lignaria*, becoming a problem when the bees are managed in large, dense populations. Some of these parasitoids also attack Apidae. For example, *Mellitobia acasta* has a long list of known bee hosts, including *Megachile*, *Osmia*, *Chalicodoma*, *Anthophora*, *Anthidium*, *Bombus*, and *Apis* (De Wael *et al.*, 1993, 1995).

12.3.7. Abiotic Agents

Bees are susceptible to many insecticides and, as pollinators, they fly about visiting a large number of flowers to collect pollen and nectar. This activity can expose them to pesticides that are on or in flowering plants. This route of exposure usually occurs when pesticides are applied to agricultural crops or home gardens for pest control. Many countries have environmental regulations restricting the use of insecticides on flowering plants, especially insecticides known to be acutely toxic to bees. However, bees still continue to be exposed to these chemicals, and pesticides can accumulate in the hive (Mullin *et al.*, 2010). A typical sign of pesticide poisoning in honey bees is a large number of dead adult bees found in and around the hives, or large numbers of adult bees found writhing on the ground, often on their dorsum. Typical symptoms of pesticide poisoning in *M. rotundata* include a loss of bee activity at the nesting shelters and large numbers of dead bees around these shelters, especially males (which tend to remain at the shelters to mate). This occurs when the pesticide has acute toxicity, causing rapid bee mortality (Riedl *et al.*, 2006). Little is known about the chronic affects of pesticides when

the bees are exposed to subacute doses, but recent research indicates that it can have serious consequences. Until recently, for example, the insect growth regulator novaluron was considered safe for bees because it lacked acute toxicity to adults and so was approved in the USA for use in alfalfa seed fields to control lygus bugs during the pollination season. However, applications of this pesticide corresponded with very poor reproduction of *M. rotundata*. It was later determined that *M. rotundata* eggs are very sensitive to novaluron. In addition, when reproducing females are fed novaluron, the progeny produced by these bees have poor survivorship (Hodgson *et al.*, 2011).

Another source of pesticide exposure can be the beekeepers themselves. Until very recently, the two most commonly used pesticides to control varroa mite in the USA were coumaphos and *tau*-fluvalinate. Both are relatively persistent acaricides that contaminate the wax, pollen and bees in honey bee hives. Furthermore, varroa mites have developed resistance to both of these, rendering these products less useful to most beekeepers. Some beekeepers have tried to compensate for pesticide resistance by increasing the concentration they put into the hive, especially with regard to *tau*-fluvalinate, which is readily available in liquid formulations for field crops (in both the USA and Europe); however, this practice is not a legal pesticide use. Contamination levels in the wax in honey bee hives are greater for these two insecticides than for any other insecticide, typically by 10-fold or more (Mullin *et al.*, 2010). Although the oral toxicities to honey bees have been quantified for these pesticides, it is not clear what the real exposure levels are for contaminated wax, since the pesticide may be bound to the wax in such a manner that bee exposure is minimal.

12.3.8. Colony Collapse Disorder

In around the year 2005, commercial honey beekeepers in the USA began suffering large losses of bees, and these colony losses recurred annually after that. One beekeeper once described the situation as "The law of thirds: one third dead, one third weak, and one third fine." This high mortality has been attributed to a new disease (not necessarily an infectious disease) named colony collapse disorder (CCD). The disease has been defined by the following symptoms: (1) a rapid loss of adult worker bees, i.e., adult numbers decline over a few weeks, but not with the sudden mortality that is commonly seen with pesticide poisonings; (2) the colony is rapidly dying, but with only a few or no dead bees found in or around the hive; (3) a small cluster of adult bees is present with an egg-laying queen and brood; or (4) the colony is dead, but with brood present and honey and pollen stores intact, but no invasion by hive pests such as wax moths (Cox-Foster *et al.*, 2007). At the time of writing, no single pathogen, parasite, or toxin has yet been found that fully explains the cause of CCD.

In 2006, a metagenomic survey of bees from several states of the USA was undertaken to determine whether any pathogens or parasites could be correlated with CCD prevalence. The only pathogen that occurred in all the CCD samples and none of the non-CCD samples was a virus that had recently been described in Israel, the *Israeli acute paralysis virus* (IAPV, Table 12.2) (Cox-Foster *et al.*, 2007). At the time of that survey, IAPV was not known to be present in the USA, but it has since been shown to be present since at least 2002 (Chen and Evans, 2007). It has not yet been confirmed whether IAPV is responsible for CCD. European beekeepers have also experienced large losses of honey bees, and some scientists have proposed that those losses are due to the invasion of *N. ceranae* (see Section 12.3.4), while others think that multiple factors, including chronic exposure to pesticides, may be to blame.

Indeed, *N. ceranae* is now widespread in the USA (Chen and Huang, 2010), but the presence of this pathogen was not strongly tied to CCD in the metagenomic survey. It has also been proposed that CCD is due to a co-infection of *N. ceranae* and a new undescribed iridovirus (Bromenshenk *et al.*, 2010). An iridovirus has previously been described from *A. cerana*, but not from *A. mellifera* (Bailey *et al.*, 1976; Bailey and Ball, 1978; Verma and Phogat, 1982), and further confirmation on its presence in this host, and in the USA, is still needed.

In 2009, a large epidemiological approach was taken in the USA to identify the cause of CCD, and although more than 200 variables were quantified, no single factor could be identified as the cause of this disorder. However, low coumaphos levels in the hive did correspond with the presence of CCD, although the intensity of varroa mite infestation was not a good predictor (vanEngelsdorp *et al.*, 2009, 2010). Two other pesticides were also found to be predictors for the prevalence of CCD, but the authors did not indicate whether CCD prevalence was related to high or low levels of these pesticides (vanEngelsdorp *et al.*, 2010).

A very thorough and extensive survey of pesticide contamination levels in honey bees and honey bee hives was conducted in the USA in the 2000s (Mullin *et al.*, 2010). Coumaphos and *tau*-fluvalinate were found to occur at high concentrations in nearly every honey bee colony in the USA. As previously discussed (see Section 12.3.7), low levels of pesticides can have unexpected chronic or delayed effects on bees, and this survey revealed that honey bees are exposed to a very long list of insecticides, miticides, and fungicides. Therefore, the role of pesticide exposure, especially exposures that are below the acute toxicity levels, merits further investigation, but it is not clear whether poisonings play a role in CCD.

Large annual surveys of commercial and hobbyist beekeepers in the USA have found that approximately 30%

of all colonies die during the winter every year (vanEngelsdorp *et al.*, 2008, 2010, 2011). However, CCD is not the most prominent cause. Beekeepers most frequently reported queen failure, starvation (this is especially true among hobbyist beekeepers), varroa mites, and weather as the causes for their colony losses during the winter. Thus, although honey bee colony losses may be on the increase worldwide (IBRA, 2010), the cause for this increase is not clearly CCD, but CCD is certainly a factor.

As shown above, CCD is a complex disorder that is not well defined. Although a group of authors has agreed on the symptoms associated with CCD (as described above), the symptoms are vague and could result from many different causes. For this reason, not all bee researchers are in agreement that CCD is actually a specific disease, suggesting instead that it is a collection of symptoms with many possible causes, including *N. ceranae* infections, varroa mite infestation, and abiotic factors such as chronic pesticide exposure (Anderson and East, 2008). From personal experience (R. R. James), commercial beekeepers and apiary inspectors use the term loosely to describe any large losses they experience, even when they are certain of the cause (such as varroa mite infestation or pesticide exposure). For example, one apiary inspector said that CCD in his area was caused mainly by varroa mites, and a commercial beekeeper in the same area expressed that "his" CCD was due to pesticide poisonings (R. R. James, unpubl.).

The loose manner in which CCD is defined by beekeepers makes records and surveys taken from beekeepers difficult to interpret. The situation may be similar to that described by Bailey and Ball (1991) for the Isle of Wight disease in the British Isles: "There are all kinds of possible reasons for the death of bees, apart from infections, and there is little doubt that bees dying of non-infectious diseases were often included in the casualties attributed to the Isle of Wight disease". At that time, the conviction that Isle of Wight disease was indeed an infectious disease led to the determination that it was caused by tracheal mites, a conclusion that was disputed by some scientists (Bailey and Ball, 1991). It can be very difficult to determine the reason why a colony dies, and therefore, many kinds of mortalities are often included as CCD, complicating the discovery of the real cause.

12.3.9. Disease Control Methods for Managed Bees

Disease control for bees is challenging. Bees are susceptible to many diseases, and they are kept outdoors where it is difficult to keep the colonies isolated from other bees that may be carriers of pathogens. Often, chemical treatments can reduce parasites and pathogens, but rarely are they completely effective at eliminating them, and the parasites and pathogens can become resistant. In addition, no treatments exist for many insect diseases, especially viruses, and sanitation can be difficult to accomplish.

Thus, a multifaceted approach is needed for disease control. Disease management can be broken down into six main components: (1) quarantine; (2) sanitation; (3) monitoring; (4) resistance; (5) control disease vectors; and (6) medication (chemical and/or antibiotic) treatment. All of these components are equally important and all can be applied to *A. mellifera*, but they are not always used by beekeepers. Efforts to develop integrated disease and pest control in beekeeping have not been widely adopted by beekeepers. Furthermore, methods for disease management are poorly developed for the other managed bees, and more research is needed to provide beekeepers with effective tools. Below, currently available methods are discussed and areas where more development is needed are identified.

The first two components of disease management, quarantine and sanitation, are employed to avoid disease establishment and prevent its spread. Quarantines are forced isolations or restrictions on movement to prevent the spread of contagious disease, and are usually established by governments. For example, many countries have regulations that restrict the importation of bees and bee pollen. In addition, when a new pathogen or parasite is found, beekeepers may be restricted from moving bees out of the area where it has established. Thus, the state of Wyoming in the USA requires not only *A. mellifera*, but also *M. rotundata* colonies to be inspected for disease. If more than 5% of the overwintering cocoons in a batch of *M. rotundata* show signs of chalkbrood, the producer cannot sell that batch of bees. Wyoming is the only state with a chalkbrood quarantine for *M. rotundata*.

Quarantines can also be implemented by beekeepers themselves, although this is unusual. The main premise to such a quarantine would be to prevent bringing diseased bees into an apiary. One method to do this is to keep the bees in an isolated area for a few weeks and watch for signs and symptoms of disease that may not have been apparent at first. This type of quarantine is critical for bumble bee producers, where the bees are produced in isolation, indoors. Before any new stock is introduced into a rearing facility (such as field-caught queens), the beekeeper should verify that those bees are disease free. With queens, this can be accomplished by allowing the queen to produce 20–30 workers in isolation, then sampling the workers and testing them for the presence of *N. bombi*. This method allows time for the pathogen to build up to levels where infections can be detected. In addition, bumble bee colonies that have been used for field or greenhouse pollination should not be returned to the rearing facility. This self-imposed quarantine method remains the primary means of disease control in commercial bumble bee production.

The second component of disease management is sanitation and involves regular cleaning and disinfection to prevent the spread of disease. Pathogen transmission can occur when nesting materials are reused without sanitation, such as when *A. mellifera* hives are reused for different colonies without first being disinfected, or when tools used by workers (e.g., the hive tools used by beekeepers or grafting tools used in queen production) are not disinfected between uses. Another measure that could be taken to prevent human workers from spreading disease is the decontamination of their outer clothing (e.g., gloves and bee suits).

Apis mellifera beekeepers prefer to reuse old combs because it takes the bees more energy to produce wax than honey; therefore, the comb is quite valuable in terms of the honey production capacity of the colony. However, old combs can also harbor pathogens, such as the causal agents for AFB and chalkbrood, and can also accumulate pesticides. Thus, old comb and hive bodies should be regularly replaced or sanitized. Unfortunately, the methods for sanitation are somewhat limited. Sanitation methods for wax combs include gamma radiation (Studier, 1958; Shimanuki *et al.*, 1984), ethylene oxide (Shimanuki, 1967), heat (Cantwell and Shimanuki, 1970), and high concentrations of ozone (James, 2011a).

Another approach is to replace the comb regularly, perhaps as frequently as every two years. In practice, this could mean replacing half the comb every year, as is done in Denmark (P. Kryger, pers. comm.). Pathogenic organisms have been shown to spread on combs, especially brood combs, where they can contaminate the pollen fed to larvae (see Sections 12.3.2 and 12.3.3). Beekeepers often transfer brood from strong colonies to weaker colonies in an effort to bolster the bee population in the smaller colony; however, this can lead to the direct transfer of contaminated combs. This transmission can also be reduced if measures are taken to keep the brood out of the honey supers. Honey combs are typically exchanged between hives/colonies by beekeepers, and if the brood is not kept out of the honey comb, then the brood comb will also be exchanged between colonies. Brood can be kept out of the honey comb using queen excluders, which are screens that have holes large enough to allow worker bees to pass through but too small to allow queens and drones to pass through. Queen excluders should be placed between the brood chamber and the honey supers to confine the queen to desired parts of the hive, restricting where she can lay eggs. In addition, beekeepers should avoid the practice of transferring material from the brood chamber of one colony to another colony.

Megachile rotundata nesting boards and cocoons are sanitized with formaldehyde gas in Canada to kill *A. aggregata* spores and prevent chalkbrood transmission (Goerzen and Watts, 1991). Formaldehyde is not registered for this use in the USA; therefore, chlorine dips have been recommended (Stephen, 1982; Richards, 1984). However, sanitizing a large number of nesting boards with chlorine is messy and hazardous, and is not often done. Methyl bromide has been used by some farmers (Mayer *et al.*, 1991), but this has poor efficacy against *A. aggregata* spores (James, 2005a). This compound is still legal in the USA for treating wood, but it causes a deterioration of the ozone layer in the Earth's upper atmosphere and a complete ban will eventually come into effect. Heat can be used to kill the spores on wooden nesting boards (Kish, 1983), but melts polystyrene boards. Nesting board sanitation is widely used in *M. rotundata* rearing, but when it is used alone, it is not effective because the nesting boards are not the primary source of pathogen transmission for chalkbrood (James, 2005a, 2011b).

Hygiene is also important in bumble bee rearing. All used bumble bee nesting materials should be sanitized before disposal, or removed to places where bumble bees cannot reach them (e.g., buried or burned). Other sanitation measures include cleaning all equipment (e.g., forceps, nesting boxes, food containers, mating chambers, etc.) with a disinfectant after every use; never transferring food from one colony to another; never transferring food containers between colonies without first sanitizing them; storing food in the refrigerator, and disposing of any that spoils; and maintaining a neat and clean rearing facility.

The third component of disease management is monitoring. Bees should be regularly monitored for disease and general overall health. Many diseases can go undetected in a colony for long periods if they are not watched for carefully and regularly, and then once an outbreak is obvious, losses can be catastrophic. An example of this is *Nosema* in bumble bees. Once *Nosema* becomes established in a facility, it becomes necessary to destroy all the bees, sterilize the equipment and rearing room, and start production all over again. Therefore, it is much more cost-effective to monitor for the disease so that appropriate prevention measures can be taken. Accurate monitoring of bee diseases, especially the viruses and microsporidia, may require laboratory analyses, which are not practical for some beekeepers. However, many countries provide disease-testing laboratories where beekeepers can send samples. In addition, simple, economical testing strips are now available for use to test for the foulbrood diseases.

Many sampling methods have been developed for detecting and quantifying varroa mite infestations. One of the earliest methods was an ether roll, where a sample of approximately 300 bees is placed in a 1-liter (quart) jar with a small amount of ether and shaken vigorously for 1 min. The sampled bees and mites are killed by this method. The ether causes the mites to fall off the bees and stick to the side of the jar, which has become wet from the ether (Delaplane and Hood, 1999). The use of ether in the jar has

recently been replaced by powdered sugar (confectioners' sugar). Powdered sugar disrupts the ability of the mites to stay attached to the bees, but the bees live and can be released after sampling. The mites are then counted in the powdered sugar (Macedo *et al.*, 2002). However, the measurement is more accurate if water is added to the jar after the bees are released, to dissolve the sugar and make the mites more visible. Powdered sugar can also be applied directly to the bee colony, and then the residue collected on a board or tray at the bottom of the hive.

Varroa mites can also be sampled using a sticky board trap at the bottom of a hive. A piece of stiff cardboard is cut to fit inside the bottom of the hive. The top of this board is covered with a thin film of cooking oil, or some other type of adhesive, which is in turn covered with hardware cloth or some other mesh screening (using a 3×3 mm mesh). This apparatus is placed at the bottom of the hive. The mesh keeps the bees from sticking to the board, but mites that fall in the hive are trapped in the adhesive. This sampling method can be used to measure the natural fall rate of the mites or it can be used in conjunction with a miticide (Branco *et al.*, 2006). Evaluating varroa mite infestation levels allows beekeepers to determine when to use treatments for this parasite and disease vector (Delaplane and Hood, 1999; Strange and Sheppard, 2001).

Disease monitoring methods are used in *M. rotundata* beekeeping for chalkbrood. Larvae are sampled during the winter when they are in the cocoon stage. Typically, cocoons are removed from the nesting boards in December and stored at 4°C during the winter and spring. Cocoons are checked for mortality and signs of chalkbrood using X-ray analysis (Stephen and Undurraga, 1976) or by cutting the cells open. Monitoring for chalkbrood is used to evaluate the quality of the bees when they are sold and purchased, and before they are incubated for adult bee release. As stated earlier, it can also be used for quarantine purposes.

Bumble bees in rearing facilities are monitored most often for *N. bombi*. Several methods have been developed for detecting spores in the host and pollen, including immunofluorescent staining and electron microscopy, as well as antigenic, biochemical, and molecular analyses (Weber *et al.*, 1999; Klee *et al.*, 2006). Bumble bee colonies should be sampled regularly during production to verify that they are disease free, and infected colonies should be immediately destroyed. Pollen should also be monitored for the disease to determine that it is safe before it is used as food.

Resistance is the fourth disease management component. Bee resistance to disease can be maintained through both genetics and good rearing practices. It is possible to select for disease resistance in a breeding program, but maintaining genetic resistance also means being careful not to select for susceptibility. Insects are often bred in production facilities to optimize production traits, and breeders must be careful not to inadvertently select for deleterious traits linked to the desired trait. For example, *A. mellifera* are commonly bred for gentleness and honey production, yet little to nothing is known about how these traits are linked to disease resistance or overwintering hardiness. In addition, inbreeding depression can be a challenge in bee breeding programs owing to the ease with which several reproductive females can be produced by a few parents (Rinderer, 1986).

Environmental factors can also play a role in disease resistance. For example, nutrition, temperature, and humidity are all known to affect the rate of infectious diseases in insects. Sometimes environmental conditions may enhance the ability of a pathogen to infect a host (e.g., high humidity might improve fungal spore germination; see Chapter 6), and some environmental conditions enhance insect immunity (e.g., slightly stressful temperatures can increase resistance to infection in bees and lady beetles) (James *et al.*, 1998; James, 2005b). For *A. mellifera*, nutritional supplements can enhance bee health in the spring, when adequate floral resources are scarce (Hoffman *et al.*, 2008). For more information on resistance, see Chapter 13.

The goal is to keep disease resistance high in the bee colonies through a combination of proper care and proper genetics. For instance, population models have shown that an *A. mellifera* colony can withstand varroa mite infestations for many years if the queen lays a sufficient number of eggs. It is during the fall and winter, when brood production declines, that varroa mites often cause the most damage, and similarly, they can quickly devastate a colony with a weak queen (e.g., poor egg layer). Varroa mites also directly inhibit the immune system in *A. mellifera* (Yang and Cox-Foster, 2005, 2007). The following general practices can help to maintain brood health and disease resistance.

- Keep the hives out of standing water; if they flood, move them to higher ground as soon as possible. If the bees are kept in an area where the ground gets saturated, place the hives on a hive stand or concrete pad, above the water level. Use screened bottom boards to increase ventilation in humid or wet environments. Controlling excessive humidity can greatly reduce the incidence of chalkbrood.
- Avoid opening the hives when it is cold outside (below 18°C), especially on cloudy or rainy days, to avoid exposing the brood to the cold. If colonies must be worked during cool weather, minimize the time the hive is open. Again, this is especially important for controlling chalkbrood.
- Use queens from a hygienic breed, such as bees with the VSR (varroa sensitive hygiene, sometimes referred to as SMR, suppression of mite reproduction) trait (Spivak

and Reuter, 2001a, b) or USDA Russian queens (Rinderer *et al.*, 2001). These breeds are especially capable of identifying diseased and dead brood and removing them from the colony. The Russian and VSH breeds are also resistant to varroa mites and may be resistant to tracheal mites.

- When producing queens, avoid inbreeding, use hybrid vigor, mate with many drones, and limit the use of artificial insemination.
- In mid-summer, locate bees to an area with a strong nectar and pollen flow. This will increase both brood and honey production (Rinderer and Rothenbuhler, 1974), and pollen resources reduce viral infections (Hoffman *et al.*, 2010).

Little research has been done regarding the use of resistance for disease control in other bees.

The fifth component in disease management is to control disease vectors. That is, measures need to be taken to control infestations by pests, parasites, and parasitoids that may spread disease within the colony. Controlling such pests is not always easy. The most notable bee vectors are the varroa mites in *A. mellifera*. Varroa mites transmit viruses, especially deformed wing virus (see Section 12.3.1). Varroa mites should be monitored for as discussed above, and treated with medications when needed.

The sixth and last disease management component is medication and chemical treatment. Medications should be reserved as a last resort because most do not completely eliminate the targeted pathogen or parasite and, consequently, when treatments are stopped, the disease may recur. Furthermore, if this cycle is repeated often enough, the pathogen or parasite may develop resistance to the treatments. Resistance to medication has been a serious problem in *A. mellifera*. *Paenibacillus larvae*, the pathogen causing AFB, has developed resistance to the antibiotic oxytetracycline hydrochloride in the USA (Kochansky *et al.*, 2001) and varroa mites have developed resistance to several miticides in the USA and Europe, rendering these treatments useless (Elzen *et al.*, 1999a, b; Elzen and Westervelt, 2002).

Methods for medicating bees other than *A. mellifera* have not been developed. James (2011b) describes a method for treating *M. rotundata* with a fungicide to control chalkbrood, but the results are not consistent. No medications are currently available for viral diseases in *A. mellifera* either, but a new product for treatment is currently being developed for IAPV (Maori *et al.*, 2009) and has been field tested with good results (Hunter *et al.*, 2010). The product is based on gene silencing by RNA interference (RNAi). If this technique works, it will undoubtedly lead to the development of new treatments for other RNA viruses and possibly other pathogens.

Bacterial diseases in *A. mellifera*, such as the foulbroods, can be treated with antibiotics. Currently, two antibiotics are registered in the USA for controlling AFB, oxytetracycline hydrochloride (Terramycin®) and tylosin (Tylan®). In the USA, Terramycin has traditionally been used as a prophylactic. The concept behind its use was to keep the pathogen load low enough that disease never develops. Beekeepers typically treat in the spring or fall (before or after the honey flow), or both. To prolong the presence of the antibiotic in the hive, a method called "extender patties" was developed, where beekeepers would mix the antibiotic with vegetable shortening to slow the rate at which the bees consumed it (Wilson and Elliott, 1971; Wilson *et al.*, 1973). Unfortunately, the prophylactic approach to using Terramycin probably promoted the development of antibiotic resistance (Kochansky *et al.*, 2001). Terramycin was used prophylactically because it was relatively inexpensive and preventive, but more importantly, this antibiotic is not effective in colonies with apparent infections. Terramycin cannot eliminate foulbrood as it only reduces the pathogen load. Tylan is effective after the disease symptoms are present in a colony.

A few chemical-free methods exist for controlling an outbreak of AFB in an *A. mellifera* colony. All these methods are probably best summarized by the "shaking colony method", which is used in Denmark and France. In brief, adult bees (only) are moved to a new hive and forced to build new comb for two weeks; then they are moved to another new hive and forced to build new comb again. The purpose of removing the brood and moving the adult bees to a new hive is to reduce the spore load in the hive and colony, taking advantage of the fact that the pathogen reproduces only in the brood. This is an organic method for foulbrood control, but it is labor intensive, and honey yield will be severely reduced in the year of the treatment. This method was first recommended in 1895 (McEvoy, 1895). A more recent study found this method insufficient to eliminate the pathogen; however, in that study, the bees were transferred to a new hive only once (Del Hoyo *et al.*, 2001).

Fumagillin-B has been the drug of choice for treating *N. apis* infections in *A. mellifera*, but it is still unclear how effective this medication is for treating *N. ceranae* infections. Fumagillin-B is also ineffective in controlling *N. bombi* in bumble bees (Whittington and Winston, 2003) and no other medications are available to control this pathogen.

Several chemical treatments for varroa mites have entered the market in the past several years. The current compounds include formic acid (Mite-Away II®), thymol (Apiguard®), and sucrose octanoate esters (Sucrocide®). *Tau*-fluvalinate and coumaphos can still be purchased. Chemical treatments should be applied in the fall or spring, but not during honey flow. The method of application varies with each product, and some products must be applied multiple times at two-week intervals.

A few different methods are also available for controlling tracheal mites. Vegetable shortening mixed with white granulated sugar is formed into patties and placed on top of the frames in the brood chamber. Vapors from the shortening are thought to disrupt the life cycle of the mites and thus suppress mite populations. Menthol is another tracheal mite control, and it is essentially a fumigation method.

Unfortunately, no effective control methods are available for some bee diseases. The extent of some major bee health issues in managed bees, such as CCD in *A. mellifera*, chalkbrood in *M. rotundata*, and nosema disease in bumble bees, demonstrates that better disease management methods are needed. The integrated approach to disease management described here highlights some of the areas where more research is needed to improve disease control of managed bees.

12.4. FUTURE RESEARCH DIRECTIONS

Although we often think of insects as pests, understanding their pathogens and diseases is important to human survival and well-being. Many insects are beneficial, providing critical ecological services in wild and agricultural ecosystems, and the health of these beneficial insects is important. The importance of diseases in the population dynamics of insects, especially beneficial insects, has received very little attention, and now that bees (both managed bees and wild bees) are experiencing dramatic population declines for no apparent reason, it is clear that this area of research has been neglected for too long. A better understanding of the diseases present in native populations of insects and their ability to move to new hosts is needed, and this information can then be used to improve importation and quarantine regulations, with the protection of beneficial insects in mind.

More emphasis needs to be placed on utilizing an integrated approach to disease management for bees. An emphasis on avoidance and prevention, as has been made in disease control efforts for silkworms, and some of the principles learned could be better applied to bee management. Past research efforts on disease control in *A. mellifera* have overly emphasized the use of medications to treat diseased bees. The problem with this approach is that it cannot be attained for all pathogens and it does not guard against new pathogens and parasites. Every time a new honey bee pest invades a region, it causes serious declines in bees until a new chemical control is developed, as has been experienced with AFB, varroa mites, and CCD. Furthermore, when chemicals are used in a hive, they have the potential to be harmful to the bees themselves and care must be taken to reduce contaminating hive products intended for human use, such as honey, wax, and pollen.

The multifaceted, integrated approach described here adopts a general approach to disease control used with livestock (Mourits and Oude Lanskink, 2006), including principles that reduce the need for chemical treatments, and perhaps more importantly, help to guard against the invasion of new, currently unknown diseases.

For both silkworms and bees, the area needing the most attention with regard to disease control is improving resistance. More research is needed to provide several disease resistance varieties with good production characteristics. For *A. mellifera*, research is also needed to develop effective, but attainable, quarantine measures, especially methods that beekeepers can use to quarantine new bees that they acquire. For example, no information is currently available on how long quarantine periods should be, once isolation has been adequately achieved. Furthermore, knowledge is needed regarding the levels of disease spread that occurs between wild and managed populations, and between different species of bees. Research in these areas would help beekeeping to be more recalcitrant against the accidental introduction of new pathogens and parasites. The development of completely new approaches for treating viruses in beneficial insects, such as gene silencing by RNAi, would greatly benefit both silkworm and bee production.

REFERENCES

Anderson, D., & East, I. J. (2008). The latest buzz about colony collapse disorder. *Science, 319,* 724–725.

Anderson, D., & Gibson, N. (1998). New species and isolates of sporecyst fungi (Plectomycetes: Ascosphaerales) from Australia. *Aust. Syst. Bot., 11,* 53–72.

Anderson, D. L., & Trueman, J. W. H. (2000). *Varroa jacobsoni* (Acari: Varroidae) is more than one species. *Exp. Appl. Acarol., 24,* 165–189.

Aratake, Y., & Kayamura, T. (1977). A new record of a flagellate parasite (Trypanosomatidae: Protomonadina) pathogenic to *Bombyx mori*. *J. Sericult. Sci. Jpn., 46,* 87–88, (In Japanese.).

Aruga, H. (1971). Cytoplasmic polyhedrosis of the silkworm − historical, economical and epizootiological aspects. In H. Aruga & Y. Tanada (Eds.), *The Cytoplasmic-Polyhedrosis Virus of the Silkworm* (pp. 3–21). Tokyo: University of Tokyo Press.

Bailey, L. (1975). Recent research on honey bee viruses. *Bee World, 56,* 55–64.

Bailey, L., & Ball, B. V. (1978). *Apis* iridescent virus and "clustering disease" of *Apis cerana. J. Invertebr. Pathol., 31,* 368–371.

Bailey, L., & Ball, B. V. (1991). *Honey Bee Pathology.* New York: Academic Press.

Bailey, L., Ball, B. V., & Woods, R. D. (1976). An iridovirus from bees. *J. Gen. Virol., 31,* 459–461.

Balavenkatasubbaiah, M., Sharma, S. D., Baig, M., Singh, B. D., Reddy, S. V., & Noamani, M. K. R. (1989). Role of disinfection of rearing appliances and sunlight exposure on the inactivation of a disease causing pathogen of silkworms, *Bombyx mori* L. *Indian J. Sericult., 28,* 200–206.

Ball, B. V. (1999). An introduction to viruses and techniques for their identification and characterization. In M. E. Colin, B. V. Ball & M. Kilani (Eds.), *Bee Disease Diagnosis* (pp. 69–80). Zaragoza: Centre International de Hautes Études Agronomiques Méditerranéennes.

Bassi, A. (1835). *Del Mal del Segno*. Lodi: Tipografia Orcesi.

Bowen-Walker, P. L., & Gunn, A. (1998). Inter-host transfer and survival of *Varroa jacobsoni* under simulated and natural winter conditions. *J. Apicult. Res., 37,* 199–204.

Branco, M. R., Kidd, N., & Pickard, R. S. (2006). A comparative evaluation of sampling methods for *Varroa destructor* (Acari: Varroidae) population estimation. *Apidologie, 37,* 1–10.

Bromenshenk, J. J., Henderson, C. B., Wick, C. H., Stanford, M. F., Zulich, A. W., Jabbour, R. E., Deshpande, S. V., McCubbin, P. E., Seccomb, R. A., Welch, P. M., Williams, T., Firth, D. R., Skowronski, E., Lehmann, M. M., Bilimoria, S. L., Gress, J., Wanner, K. W., & Cramer, R. A., Jr. (2010). Iridovirus and microsporidian linked to honey bee colony decline. *PLoS ONE, 5,* e13181.

Brown, B. V. (1996). Preliminary analysis of a host shift: revision of the Neotropical species of *Apocephalus,* subgenus *Mesophora* (Diptera: Phoridae). *Natural History Museum of Los Angeles County.* Contributions in Science No. 462.

Brown, B. V. (1997). Parasitic phorid flies: a previously unrecognized cost to aggregation behavior of male stingless bees. *Biotropica, 29,* 370–372.

Brown, B. V., & Smith, P. T. (2010). The bee-killing flies, genus *Meloncha* Brues (Diptera: Phoridae): a combined molecular and morphological phylogeny. *Syst. Entomol., 35,* 649–657.

Cameron, S. A., Lozier, J. D., Strange, J. P., Koch, J. B., Cordes, N., Solter, L. F., & Griswold, T. L. (2011). Patterns of widespread decline in North American bumble bees. *Proc. Natl. Acad. Sci. USA, 108,* 662–667.

Cantwell, G. E., & Shimanuki, H. (1970). The use of heat to control *Nosema* and increase production for the commercial beekeeper. *Am. Bee J., 110,* 263.

Cappellozza, S., Saviane, A., Tettamanti, G., Squadrin, M., Vendramin, E., Paolucci, P., Franzetti, E., & Squartini, A. (2011). Identification of *Enterococcus mundtii* as a pathogenic agent involved in the "flacherie" disease in *Bombyx mori* L. larvae reared on artificial diet. *J. Invertebr. Pathol., 106,* 386–393.

Chandrasekharan, K., & Nataraju, B. (2011). *Beauveria bassiana* (Hyphomycetes: Moniliales) infection during ecdysis of silkworm *Bombyx mori* (Lepidoptera: Bombycidae). *Munis Entomol. Zool., 6,* 312–316.

Chen, Y., & Evans, J. D. (2007). Historical presence of Israeli acute paralysis virus in the United States. *Am. Bee J., 147,* 1027–1028.

Chen, Y., & Huang, Z. Y. (2010). *Nosema ceranae,* a newly identified pathogen of *Apis mellifera* in the USA and Asia. *Apidologie, 41,* 364–374.

Chen, Y., Evans, J., & Feldlaufer, M. (2006). Horizontal and vertical transmission of viruses in the honey bee, *Apis mellifera. J. Invertebr. Pathol., 92,* 152–159.

Chen, Y., Evans, J. D., Zhou, L., Boncristiani, H., Kimura, K., Xaio, T., Litkowski, A. M., & Pettis, J. S. (2010). Asymmetrical coexistence of *Nosema ceranae* and *Nosema apis* in honey bees. *J. Invertebr. Pathol., 101,* 204–209.

Clark, T. B., Whitcomb, R. F., Tully, J. G., Mouches, C., Saillard, C., Bové, J. M., Wróblewski, H., Carle, P., Rose, D. L., Henegar, R. B., &

Williamson, D. L. (1985). *Spiroplasma melliferum*, a new species from the honeybee (*Apis mellifera*). *Int. J. Syst. Bacteriol., 35,* 296−308.

Colla, S. R., Otterstatter, M. C., Gagear, R. J., & Thomas, J. D. (2006). Plight of the bumble bee: pathogen spillover from commercial to wild populations. *Biol. Conserv., 129,* 461−467.

Cox-Foster, D. L., Conlan, S., Holmes, E. C., Palacios, G., Evans, J. D., Moran, N. A., Quan, P.-L., Briese, T., Hornig, M., Geiser, D. M., Martinson, V., vanEngelsdorp, D., Kalkstein, A. L., Drysdale, A., Hui, J., Zhai, J., Cui, L., Hutchison, S. K., Simons, J. F., Egholm, M., Pettis, J. S., & Lipkin, W. I. (2007). A metagenomic survey of microbes in honey bee colony collapse disorder. *Science, 318,* 283−287.

Crane, E. (1999). *Beekeeping and Honey Hunting*. New York: Routledge.

De Wael, L., De Greef, M., & Van Laere, O. (1993). *Melittobia acasta* and *Bombacarus buchneri*, dangerous parasites in the *in vitro* rearing of bumblebees. *Apiacta, 28,* 93−101.

De Wael, L., De Greef, M., & Van Laere, O. (1995). Biology and control of *Melittobia acasta*. *Bee World, 76,* 72−76.

Debré, P. (1998). *Louis Pasteur − Translated by E. Forster*. Baltimore: John Hopkins University Press.

Del Hoyo, M. L., Basualdo, M., Lorenzo, A., Palacio, M. A., Rodriguez, E. M., & Bedascarrasbure, E. (2001). Effect of shaking honey bee colonies affected by American foulbrood on *Paenibacillus larvae* larvae spore loads. *J. Apicult. Res., 40,* 65−69.

Delaplane, K. S., & Hood, W. M. (1999). Economic threshold for *Varroa jacobsoni* Oud. in the southeast USA. *Apidologie, 30,* 383−395.

Dingle, J. G., Hassan, E., Gupta, M., George, D., Anota, L., & Begum, H. (2005). *Silk Production in Australia. A report for the Rural Industries Research and Development Corporation*. Kingston: RIRDC Publication No 05/145.

Disney, H. (1994). *Scuttle Flies: The Phoridae*. London: Chapman & Hall.

Durrer, S., & Schmid-Hempel, P. (1994). Shared use of flowers leads to horizontal pathogen transmission. *Proc. R. Soc. Lond. B, 258,* 299−302.

Eguchi, R., Furuta, Y., & Ninagi, O. (1986). Dominant non-susceptibility to densonucleosis virus in the silkworm, *Bombyx mori. J. Sericult. Sci. Jpn., 55,* 177−178.

Eguchi, R., Hara, W., Shimazaki, A., Hirota, K., Ichiba, M., Ninagi, O., & Nagayasu, K.-I. (1998). Breeding of the silkworm race "Taisei" non-susceptible to a densonucleosis virus type-1. *J. Sericult. Sci. Jpn., 67,* 361−366.

Elzen, P. J., & Westervelt, D. (2002). Detection of coumaphos resistance in *Varroa destructor* in Florida. *Am. Bee J., 142,* 291−292.

Elzen, P. J., Baxter, J., Spivak, M., & Wilson, W. T. (1999a). Amitraz resistance in varroa: new discovery in North America. *Am. Bee J., 139,* 362.

Elzen, P. J., Eischen, F., Baxter, J., Elzen, G. W., & Wilson, W. T. (1999b). Detection of resistance in US *Varroa jacobsoni* Oud. (Mesostigmata: Varroidae) to the acaricide fluvalinate. *Apidologie, 30,* 13−17.

vanEngelsdorp, D., Hayes, J., Jr., Underwood, R. M., & Pettis, J. (2008). A survey of honey bee colony losses in the US, fall 2007 to spring 2008. *PLoS ONE, 3,* e4071.

vanEngelsdorp, D., Evans, J. D., Saegerman, C., Mullen, C., Haubruge, E., Nguyen, B. K., Frazier, M., Frazier, J., Cox-Foster, D., Chen, Y., Underwood, R., Tarpy, D. R., & Pettis, J. S. (2009). Colony collapse disorder: a descriptive study. *PLoS ONE, 4,* e6481.

vanEngelsdorp, D., Hayes, J., Jr., Underwood, R. M., & Pettis, J. (2010). A survey of honey bee colony losses in the United States, fall 2008 to spring 2009. *J. Apicult. Res., 49,* 7−14.

vanEngelsdorp, D., Hayes, J., Jr., Underwood, R. M., Caron, D., & Pettis, J. (2011). A survey of managed honey bee colony losses in the USA, fall 2009 to winter 2010. *J. Apicult. Res., 50,* 1−10.

Eyer, M., Chen, Y., Pettis, J. S., & Neumann, P. (2008). Small hive beetle, *Aethina tumida*, is a potential biological vector of honeybee viruses. *Apidologie, 40,* 419−428.

Eyer, M., Chen, Y., Schaefer, M., Pettis, J. S., & Neumann, P. (2009). Honeybee sacbrood virus infects adult small hive beetles, *Aethina tumida* (Coleoptera: Nitidulidae). *J. Apicult. Res., 48,* 296−297.

Forsgren, E., Stevanovic, J., & Fries, I. (2008). Variability in germination and in temperature and storage resistance among *Paenibacillus larvae* genotypes. *Vet. Microbiol., 129,* 342−349.

Fries, I., Feng, F., da Silva, A., Slemenda, S. B., & Pieniazek, N. J. (1996). *Nosema ceranae* n. sp. (Microspora, Nosematidae), morphological and molecular characterization of a microsporidian parasite of the Asian honey bee *Apis cerana* (Hymenoptera, Apidae). *Eur. J. Protistol., 32,* 356−365.

Fries, I., de Ruijter, A. A. D., Paxton, R. J., da Silva, A. J., Slemenda, S. B., & Pieniazek, N. J. (2001). Molecular characterization of *Nosema bombi* (Microsporidia: Nosematidae) and a note on its sites of infection in *Bombus terrestris* (Hymenoptera: Apoidea). *J. Apicult. Res., 40,* 91−96.

Fujiwara, T. (1984a). A *Pleistophora* like microsporidian isolated from the silkworm. *J. Sericult. Sci. Jpn., 53,* 398−402, (In Japanese.).

Fujiwara, T. (1984b). *Thelohania* sp. (Microsporidia: Thelohaniidae) isolated from the silkworm *Bombyx mori. J. Sericult. Sci. Jpn., 53,* 459−460, (In Japanese.).

Genersch, E. (2008). *Paenibacillus larvae* and American foulbrood − long since known and still surprising. *J. Verbr. Lebensm., 3,* 429−434.

Getchev, I., & Kantchev, K. (1998). Influence of the mite control products on the resistance of the honey bee to chalkbrood. *Bull. Vet. Inst. Pulawy, 42,* 139−143.

Gillespie, S. (2010). Factors affecting parasite prevalence among wild bumblebees. *Ecol. Entomol., 35,* 737−747.

Gisder, S., Hedtke, K., Möckel, N., Frielitz, M., Linde, A., & Genersch, E. (2010). Five-year cohort study of *Nosema* spp. in Germany: does climate shape virulence and assertiveness of *Nosema ceranae? Appl. Environ. Microbiol., 76,* 3032−3038.

Goerzen, D. W., & Watts, T. C. (1991). Efficacy of the fumigant paraformaldehyde for control of microflora associated with the alfalfa leafcutting bee, *Megachile rotundata* (Fabricius) (Hymenopotera: Megachilidae). *Bee Sci., 1,* 212−218.

Goka, K., Okabe, K., Yoneda, M., & Niwas, S. (2001). Bumblebee commercialization will cause worldwide migration of parasitic mites. *Mol. Ecol., 10,* 2095−2099.

Gomi, S., Majima, K., & Maeda, S. (1999). Sequence analysis of the genome of *Bombyx mori* nuclear polyhedrosis virus. *J. Gen. Virol., 80,* 1323−1337.

de Guzman, L. I., Rinderer, T. E., & Stelzer, J. A. (1997). DNA evidence of the origin of *Varroa jacobsoni* Oudemans in the Americas. *Biochem. Genet., 35,* 327−335.

Hagiwara, K., Rao, S., Scott, S. W., & Carner, G. R. (2002). Nucleotide sequences of segments 1, 3 and 4 of the genome of *Bombyx mori* cypovirus 1 encoding putative capsid proteins VP1, VP3 and VP4, respectively. *J. Gen. Virol., 83,* 1477−1482.

Hatakeyama, Y., Kawakami, Y., & Iwano, H. (1997). Analysis and taxonomic inferences of small subunit ribosomal RNA sequences of five microsporidia pathogenic to silkworm, *Bombyx mori*. *J. Sericult. Sci. Jpn., 66*, 241–252, (In Japanese.).

Hirose, Y. (1979). On the entomoparasitic microsporidia. *Sericult. Res., 111*, 118–123, (In Japanese.).

Hodgson, E. W., Pitts-Singer, T. L., & Barbour, J. D. (2011). Effects of the insect growth regulator, novaluron on immature alfalfa leafcutting bees, *Megachile rotundata*. *J. Insect Sci., 11*, 43.

Hoffman, G. D., Wardell, G., Ahumada-Secura, F., Rinderer, T. E., Danka, R. G., & Pettis, J. S. (2008). Comparisons of pollen substitute diets for honey bees: consumption rates by colonies and effects on brood and adult populations. *J. Apicult. Res., 47*, 265–270.

Hoffman, G. D., Chen, Y., Huang, E., & Huange, M. H. (2010). The effect of diet on protein concentration, hypopharyngeal gland development and virus load in worker honey bees (*Apis mellifera* L.). *J. Insect Physiol., 56*, 1184–1191.

Huang, B., Wang, S. B., Jin, W., Fan, M. Z., & Li, Z. Z. (2008). Pathogen identification of grey muscardine of the silkworm, *Bombyx mori*. *Sci. Sericult., 34*, 257–261, (In Chinese.).

Huang, L., Cheng, T., Xu, P., Cheng, D., Fang, T., & Xia, Q. (2009). A genome-wide survey for host response of silkworm, *Bombyx mori* during pathogen *Bacillus bombyseptieus* infection. *PLoS ONE, 4*, e8098.

Hunter, W., Ellis, J., vanEngelsdorp, D., Hayes, G. W., Westervelt, D., Glick, E., Williams, M., Sela, I., Maori, E., Pettis, J., Cox-Foster, D., & Paldi, N. (2010). Large-scale field application of RNAi technology reducing Israeli acute paralysis virus disease in honey bees (*Apis mellifera*, Hymenoptera: Apidae). *PLoS Pathol, 6*, e1001160.

IBRA –; International Bee Research Association. (2010). Special Issue: Colony losses. *J. Apicult. Res., 49*, 1–128.

Ikeda, K., Nagaoka, S., Winkler, S., Kotani, K., Yagi, H., Nakanishi, K., Miyajima, S., Kobayashi, J., & Mori, H. (2001). Molecular characterization of *Bombyx mori* cytoplasmic polyhedrosis virus genome segment 4. *J. Virol., 75*, 988–995.

Iwano, H. (1991). Dimorphic development of *Nosema bombycis* spores in gut epithelium of larvae of the silkworm, *Bombyx mori*. *J. Sericult. Sci. Jpn., 60*, 249–256, (In Japanese.).

James, R. R. (2005a). Impact of disinfecting nesting boards on chalkbrood control in the alfalfa leafcutting bee. *J. Econ. Entomol., 98*, 1094–1100.

James, R. R. (2005b). Temperature and chalkbrood development in the alfalfa leafcutting bee. *Apidologie, 36*, 15–23.

James, R. R. (2008). The problem of disease when domesticating bees. In R. R. James & T. L. Pitts-Singer (Eds.), *Bee Pollination in Agricultural Ecosystems* (pp. 124–141). New York: Oxford University Press.

James, R. R. (2011a). Potential of ozone as a fumigant to control pests in honey bee (Hymenoptera: Apidae) hives. *J. Econ. Entomol., 104*, 353–359.

James, R. R. (2011b). Chalkbrood transmission in the alfalfa leafcutting bee: the impact of disinfecting bee cocoons in loose cell management systems. *Environ. Entomol., 40*, 782–787.

James, R. R., Croft, B. A., Shaffer, B. T., & Lighthart, B. (1998). Impact of temperature and humidity on host–pathogen interactions between *Beauveria bassiana* and a coccinellid. *Environ. Entomol., 27*, 1506–1513.

Jin, W. (Ed.). (2001). *Silkworm Pathology.* Beijing: China Agricultural Press. (In Chinese.).

Kamita, S. G., & Maeda, S. (1997). Sequencing of the putative DNA helicase-encoding gene of the *Bombyx mori* nuclear polyhedrosis virus and fine-mapping of a region involved in host range expansion. *Gene, 190*, 173–179.

Kanbar, G., & Engles, W. (2003). Ultrastructure and bacterial infection of wounds in honey bee (*Apis mellifera*) pupae punctured by *Varroa* mites. *Parasitol. Res., 90*, 349–354.

Keller, S., & Zimmermann, G. (1989). Mycopathogens of soil insects. In N. Wilding, M. Collins, P. M. Hammond & J. F. Webber (Eds.), *Insect–Fungus Interactions* (pp. 239–270). London: Academic Press.

Khurad, A. M., Mahulikar, A., Rathod, M. K., Rai, M. M., Kanginakudru, S., & Nagaraju, J. (2004). Vertical transmission of nucleopolyhedrovirus in the silkworm, *Bombyx mori* L. *J. Invertebr. Pathol., 87*, 8–15.

Kish, L. P. (1983). The effect of high temperature on spore germination of *Ascosphaera aggregata*. *J. Invertebr. Pathol., 42*, 244–248.

Klee, J., Tay, W. T., & Paxton, R. J. (2006). Specific and sensitive detection of *Nosema bombi* (Microsporidia: Nosematidae) in bumble bees (*Bombus* spp.; Hymenoptera: Apidae) by PCR of partial rRNA gene sequences. *J. Invertebr. Pathol., 91*, 98–104.

Klee, J., Besana, A. M., Genersch, E., Gisder, S., Nanetti, A., Tam, D. Q., Chinh, T. X., Puerta, F., Ruz, J. M., Kryger, P., Message, D., Hatjina, F., Korpela, S., Fries, I., & Paxton, R. J. (2007). Widespread dispersal of the microsporidian *Nosema ceranae*, an emergent pathogen of the western honey bee, *Apis mellifera*. *J. Invertebr. Pathol., 96*, 1–10.

Kochansky, J., Knox, D. A., Feldlaufer, M., & Pettis, J. S. (2001). Screening alternative antibiotics against oxytetracycline-susceptible and -resistant *Paenibacillus larvae*. *Apidologie, 32*, 215–222.

Levaditi, C. (1905). Sur un nouveau flagellé parasite du *Bombyx mori* (*Herpetomonas bombycis*). *C. R. Hebd. Seances Acad. Sci., 141*, 631–634.

Li, J. L., Cai, Y., Luan, F. G., Wang, B., & Li, Z. Z. (2010). Population genetic structure of *Beauveria bassiana* from south and southwest Anhui sericultural regions: ISSR analysis. *Chin. J. Appl. Ecol., 21*, 3239–3247, (In Chinese.).

Li, J. L., Luan, F. G., & Li, Z. Z. (2011). Tracing of the origin and the spreading track of silkworm white muscardine in southwestern Anhui. *Sci. Agric. Sin., 44*, 143–152, (In Chinese.).

Li, Z. Z. (1988). *List on the insect hosts of* Beauveria bassiana. In: *Study and Application of Entomogenous Fungi in China*, Vol. 1. Beijing: Academic Periodical Press. (In Chinese.)241–255.

Lin, W., Xu, A. L., Yang, W. L., Sun, J. C., Lu, X. Y., & Zhang, J. Q. (2002). Sequencing of genomic dsRNA segment V of *Bombyx mori* cytoplasmic polyhedrosis virus. *Chin. J. Virol., 18*, 375–377, (In Chinese.).

Lindström, A. (2006). *Distribution and transmission of American foulbrood in honey bees.* PhD dissertation. Uppsala: Swedish University of Agricultural Sciences.

Liu, H. J., Macfarlane, R. P., & Pengelly, D. H. (1974). *Mattesia bombi* n. sp. (Neogregarinidae: Ophrocystidae), a parasite of *Bombus* (Hymenoptera: Apidae). *J. Invertebr. Pathol., 23*, 225–231.

Liu, J. P., Zhang, G. Q., & Xu, X. Y. (2001). Analysis of the prevalence of the pebrine disease of the silkworm with the methods of key years. *Sci. Sericult., 27*, 55–58, (In Chinese.).

Lu, X. M., Wu, H. P., & Li, Y. R. (2000). Epizootiological analysis on the occurrence of pebrine in silkworm eggs production. *Sci. Sericult., 26*, 165–171, (In Chinese.).

Lu, Y. L. (1991). *Silkworm Diseases*. Rome: FAO Agricultural Services Bulletin 73/74.

Macedo, P. A., Wu, J., & Ellis, M. D. (2002). Using inert dusts to detect and assess *Varroa* infestations in honey bee colonies. *J. Apicult. Res., 40*, 3−7.

Maddox, J. V. (1973). The persistence of the microsporidia in the environment. *Misc. Publ. Entomol. Soc. Am., 9*, 99−104.

Maori, E., Shafir, S., Kalev, H., Tsur, E., Glick, E., & Sela, I. (2009). IAPV, a bee-affecting virus associated with colony collapse disorder can be silenced by dsRNA ingestion. *Insect Mol. Biol., 18*, 55−60.

Mayer, D. F., Lunden, J. D., Goerzen, D. W., & Simko, B. (1991). Fumigating alfalfa leafcutting bee (*Megachile rotundata*) nesting materials for control of chalkbrood disease). *Bee Sci., 1*, 162−165.

McEvoy, W. (1895). *Foul Brood: Its Cause and Cure*. Trenton: New Jersey State Board of Agriculture.

Meeus, I., Smagghe, G., Siede, R., Jans, K., & de Graaf, D. C. (2010). Multiplex RT-PCR with broad-range primers and an exogenous internal amplification control for the detection of honeybee viruses in bumblebees. *J. Invertebr. Pathol., 105*, 200−203.

Miao, Y. G., Jiang, L. J., & Bharathi, D. (2005). Effects of fluoride on the activities of alkaline phosphatase, adenosine triphosphatase, and phosphorylase in the midgut of silkworm. *Bombyx mori. Fluoride, 38*, 32−37.

Michener, C. D. (2000). *The Bees of the World*. Baltimore: John Hopkins University Press.

de Miranda, J. R., & Genersch, E. (2009). Deformed wing virus. *J. Invertebr. Pathol., 103*, S48−S61.

Morse, R. A., & Nowogrodzki, R. (1990). *Honey Bee Pests, Predators, and Diseases* (2nd ed.). Ithaca: Comstock Publishing.

Mouches, C., Bové, J. M., Albisetti, J., Clark, T. B., & Tully, J. G. (1982). A spiroplasma of serogroup IV causes May-disease-like disorder of honeybees in southwestern France. *Microbiol. Ecol., 8*, 387−399.

Mouches, C., Bové, J. M., Tully, J. G., Rose, D. L., McCoy, R. E., Carle-Junca, P., Garnier, M., & Sailard, C. (1983). *Spiroplasma apis*, a new species from the honey-bee *Apis mellifera. Ann. Microbiol., 134A*, 383−397.

Mouches, C., Bové, J. M., & Albisetti, J. (1984). Pathogenicity of *Spiroplasma apis* and other spiroplasmas for honey bees in southwestern France. *Ann. Microbiol., 135A*, 151−155.

Mourits, M. C. M., & Oude Lansink, A. G. J. M. (2006). Multi-criteria decision making to evaluate quarantine disease control strategies. In A. G. J. M. Oude Lansink (Ed.), *New Approaches to the Economics of Plant Health* (pp. 131−144). Dordrecht: Springer.

Mu, Z. M., Wang, Y. W., Cui, W. Z., & Li, W. G. (1999). Study on the outbreak law of mycosis of silkworm *Bombyx mori. L. J. Shandong Agric. Univ., 30*, 199−207, (In Chinese.).

Mullin, C. A., Frazier, M., Frazier, J. L., Ashcroft, S., Simonds, R., vanEngelsdorp, D., & Pettis, J. S. (2010). High levels of miticides and agrochemicals in North American apiaries: implications for honey bee health. *PLoS ONE, 5*, e9754.

Nagae, T. (1974). The pathogenicity of *Streptococcus* bacteria isolated from the silkworm reared on an artificial diet. 1. Difference in the pathogenesis of the bacteria to silkworm larvae on an artificial diet and to those reared on mulberry leaves. *J. Sericult. Sci. Jpn., 43*, 471−477.

Nagae, T. (1982). The pathogenicity of *Streptococcus* bacteria isolated from the silkworm reared on an artificial diet. 5: Effect of B vitamins in the diet on the pathogenicity of *Streptococcus faecalis* to the silkworm. *J. Sericult. Sci. Jpn., 51*, 40−45.

Ohba, M. (1996). *Bacillus thuringiensis* populations naturally occurring on mulberry leaves: a possible source of the populations associated with silkworm-rearing insectaries. *J. Appl. Microbiol., 80*, 56−64.

Otterstatter, M. C. (2011). Patterns of parasitism among conopid flies parasitizing bumblebees. *Entomol. Exp. Appl., 111*, 133−139.

Pasteur, L. (1870). *Études sur la Maladie des Vers à Soie. Tome I et II*. Paris: Gauthier-Villars.

Pettis, J. S., & Wilson, W. T. (1995). Life history of the honey bee tracheal mite (Acari: Tarsonemidae). *Ann. Entomol. Soc. Am., 89*, 368−374.

Pitts-Singer, T. L. (2008). Past and present management of alfalfa bees. In R. R. James & T. L. Pitts-Singer (Eds.), *Bee Pollination in Agricultural Ecosystems* (pp. 105−123). New York: Oxford University Press.

Pitts-Singer, T. L., & Cane, J. H. (2011). The alfalfa leafcutting bee, *Megachile rotundata*: the world's most intensively managed solitary bee. *Annu. Rev. Entomol., 56*, 221−237.

Porter, J. R. (1973). Agostino Bassi bicentennial (1773−1973). *Bacteriol. Rev., 37*, 284−288.

Qian, H. Y., Xu, A. Y., Lin, C. Q., Zhao, Y. P., Sun, P. J., & Zhang, Y. H. (2006). Studies on the resistance to nuclear polyhedrosis virus (NPV) and its inheritance law in silkworm *Bombyx mori. J. Agric. Univ. Hebei, 29*, 77−79, (In Chinese.).

Qiu, B. L., Xu, X. Y., Mu, Z. M., & Zhou, P. (2002). Identification and detection of *Nosema bombycis* by nucleic acid hybridization. *J. Shandong Agric. Univ., 33*, 14−18, (In Chinese.).

Rao, S. N., Muthulakshmi, M., Kanginakudru, S., & Nagaraju, J. (2004). Phylogenetic relationships of three new microsporidian isolates from the silkworm, *Bombyx mori. J. Invertebr. Pathol., 86*, 87−95.

Richards, K. W. (1984). *Alfalfa Leafcutter Bee Management in Western Canada*. Publication 1495/E. Ottawa: Agriculture Canada. 1−53.

Riedl, H., Johansen, E., Brewer, L., & Barbour, J. (2006). *How to reduce bee poisoning from pesticides*. Corvallis: Pacific Northwest Extension Publication 591, Oregon State University.

Rinderer, T. E. (1986). *Bee Genetics and Breeding*. Orlando: Academic Press.

Rinderer, T. E., & Rothenbuhler, W. C. (1974). The influence of pollen on the susceptibility of honey-bee larvae to *Bacillus larvae. J. Invertebr. Pathol., 23*, 347−350.

Rinderer, T. E., de Guzman, L. I., Delatte, G. T., Stelzer, J. A., Williams, J. L., Beaman, L. D., Kuznetsov, V., Bigalk, M., Bernard, S. J., & Tubbs, H. (2001). Multi-state field trials of ARS Russian honey bees 1. Responses to *Varroa destructor* 1999, 2000. *Am. Bee J., 141*, 658−661.

Runckel, C., Flenniken, M. L., Engel, J. C., Ruby, J. G., Ganem, D., Andino, R., & DeRisi, J. L. (2011). Temporal analysis of the honey bee microbiome reveals four novel viruses and seasonal prevalence of known viruses, *Nosema*, and *Crithidia. PLoS ONE, 6*, e20656.

Schmid-Hempel, R., & Schmid-Hempel, P. (1996). Larval development of two parasitic flies (Conopidae) in the common host *Bombus pascuorum. Ecol. Entomol., 21*, 63−76.

Schmid-Hempel, R., & Tognazzo, M. (2010). Molecular divergence defines two distinct lineages of *Crithidia bombi* (Trypanosomatidae), parasites of bumblebees. *J. Euk. Microbiol., 57*, 337−345.

Sen, R., Patnaik, A. K., Maheswari, M., & Datta, R. K. (1997). Susceptibility status of the silkworm (*Bombyx mori*) germplasm stocks in India to *Bombyx mori* nuclear polyhedrosis virus. *Indian. J. Sericult., 36*, 51−55.

Shen, Z. Y., Xu, L., Wang, H. L., & Huang, K. W. (1996). Pathogenicity of two *Nosema* spp. to the silkworm, *Bombyx mori*. *Sericult. Chin., 3*, 10−11, (In Chinese.).

Shimanuki, H. (1967). Ethylene oxide and control of American foulbrood. *Am. Bee J., 107*, 290−291.

Shimanuki, H., Herbert, H., & Knox, D. A. (1984). High velocity electronic beams for bee disease control. *Am. Bee J., 124*, 865−867.

Singh, R., Levitt, A. L., Rajotte, E. G., Holmes, E. C., Ostiguy, N., vanEnglesdorp, D., Lipkin, W. I., dePamphilis, C. W., Toth, A. L., & Cox-Foster, D. L. (2010). RNA viruses in Hymenopteran pollinators: evidence of inter-taxa virus transmission via pollen and potential impact on non-*Apis* Hymenoptera species. *PLoS ONE, 5*, e14357.

Sironmani, T. A. (2000). Preliminary genomic characterization of microsporidian *Nosema bombycis*. *World J. Microbiol. Biotechnol., 16*, 533−540.

Skou, J. P. (1975). Two new species of *Ascosphaera* and notes on the conidial state of *Bettsia alvei*. *Friesia, 11*, 62−74.

Skou, J. P., Holm, S. N., & Haas, H. (1963). Preliminary investigations on diseases in bumble-bees (*Bombus* Latr.). *Roy. Vet. Agric. Coll. Copenhagen Yearbook, 1963*, 27−41.

Spivak, M., & Reuter, G. S. (2001a). Resistance to American foulbrood disease by honey bee colonies *Apis mellifera* bred for hygienic behavior. *Apidologie, 32*, 555−565.

Spivak, M., & Reuter, G. S. (2001b). *Varroa destructor* infestation in untreated honey bee (Hymenoptera: Apidae) colonies selected for hygienic behavior. *J. Econ. Entomol., 94*, 326−331.

Steinhaus, E. A., & Martignoni, M. E. (1970). An Abridged Glossary of Terms Used. In *Invertebrate Pathology*, (2nd ed.). *Pacific Northwest Forest and Range Experimental Station*. Portland: USDA Forest Service.

Stephen, W. P. (1982). Chalkbrood control in the leafcutting bee. In G. H. Rank (Ed.), *Proceedings of the First International Symposium on Alfalfa Leafcutting Bee Management* (pp. 98−107). Saskatoon: University of Saskatchawan.

Stephen, W. P., & Undurraga, J. M. (1976). X-radiography, an analytical tool in population studies of the leafcutter bee, *Megachile pacifica*. *J. Apicult. Res., 15*, 81−87.

Strange, J. P., & Sheppard, W. S. (2001). Optimum timing of miticide applications for control of *Varroa destructor* (Acari: Varroidae) in *Apis mellifera* (Hymenoptera: Apidae) in Washington State, USA. *J. Econ. Entomol., 94*, 1324−1331.

Studier, H. (1958). The sterilization of American foulbrood by irradiation with gamma rays. *Am. Bee J., 98*, 192.

Uzigawa, K., & Aruga, H. (1966). On the selection of resistance strains of the infectious flacherie virus in the silkworm, *Bombyx mori* L. *J. Sericult. Sci. Jpn., 35*, 23−26.

Verma, S., & Phogat, K. P. S. (1982). Seasonal incidence of *Apis iridescent* virus in *Apis cerana indica* Fab. in Uttar Pradesh, India. *Indian Bee J., 44*, 36−37.

Wan, Y. J., Chen, Z. P., Zhang, L., Du, Y., Ao, M. J., & Yang, B. (1991). A new pathogenic microsporidium (*Nosema* sp.) isolated from the larvae of silkworm (*Bombyx mori* L.). *J. Southwest Agric. Univ., 13*, 621−625, (In Chinese.).

Wan, Y. J., Zhang, L., Chen, Z. P., Pan, M. H., & Du, Y. (1995). Study of a pathogenic microsporidium SCM₇ (*Endoreticullatus* sp.) isolated from the larvae of silkworm. *Bombyx mori. Sci. Sericult., 21*, 168−172, (In Chinese.).

Wang, Y. J., Chen, K. P., Yao, Q., Gao, G. T., & Han, X. (2006a). Complete nucleotide sequence analysis of *Bombyx mori* densonucleosis virus type 3 VD2 (China isolate). *Acta Mycol. Sin., 46*, 363−367, (In Chinese.).

Wang, Y. J., Yao, Q., Chen, K. P., & Han, X. (2006b). Organization and transcription strategy of genome of *Bombyx mori* densonucleosis virus (China isolate) VDl. *Chin. J. Biotechnol., 22*, 707−712, (In Chinese.).

Watanabe, H. (1967). Development of resistance in the silkworm, *Bombyx mori* to preoral infection of a cytoplasmic polyhedrosis virus. *J. Invertebr. Pathol., 9*, 474−479.

Watanabe, H., Kurihara, Y., Wang, Y. X., & Shimizu, T. (1988). Mulberry pyralid, *Glyhodes pyloalis*: habitual host of nonoccluded viruses pathogenic to the silkworm. *Bombyx mori. J. Invertebr. Pathol., 52*, 401−408.

Weber, R., Schwartz, D. A., & Deplazes, P. (1999). Laboratory diagnosis of microsporidiosis. In M. Wittner & L. M. Weiss (Eds.), *The Microsporidia and Microsporidiosis* (pp. 315−362). Washington, DC: American Society of Microbiology.

Webster, T. C., & Delaplane, K. S. (2001). *Mites of the Honey Bee*. Hamilton: Dadant & Sons.

Whittington, R., & Winston, M. L. (2003). Effects of *Nosema bombi* and its treatment fumagillin on bumble bee (*Bombus occidentalis*) colonies. *J. Invertebr. Pathol., 84*, 54−58.

Williams, G. R., Shafer, A. B. A., Rogers, R. E. L., Shutler, D., & Stewart, D. T. (2008). First detection of *Nosema ceranae*, a microsporidian parasite of European honey bees (*Apis mellifera*), in Canada and central USA. *J. Invertebr. Pathol., 97*, 189−192.

Wilson, W. T., & Elliott, J. R. (1971). Prophylactic value of antibiotic extender patties in honey-bee colonies inoculated with *Bacillus larvae*. *Am. Bee J., 111*, 308−309.

Wilson, W. T., Elliot, J. R., & Hitchcock, J. D. (1973). Treatment of American foulbrood with antibiotic extender patties and antibiotic paper packs. *Am. Bee J., 112*, 341−344.

Wu, Y. L. (1983a). Resistance of the silkworm, *Bombyx mori*, to NPV. I. Relation to developmental stage. *Sci. Sericult., 9*, 29−33, (In Chinese.).

Wu, Y. L. (1983b). Resistance of the silkworm, *Bombyx mori*, to NPV, III. Relation to the added proportion of mulberry powder, Vc and water. *Sci. Sericult., 9*, 167−171, (In Chinese.).

Xiang, H. (2010). *Studies on* Nosema bombycis *genomics*. PhD dissertation. China: Southwest University. (In Chinese.).

Yanagita, T., & Iwashita, Y. (1987). Histological observation of larvae of the silkworm, *Bombyx mori*, orally infected with *Beauveria bassiana*. *J. Sericult. Sci. Jpn., 56*, 285−291, (In Japanese.).

Yang, Q., Xu, X. Y., Lu, K. M., Zheng, X. M., Fang, D. J., Liao, S. T., & Yao, X. Z. (2001). Studies on a *Vairimorpha* microsporidium isolated from *Pieris canidia* S. *Sci. Sericult., 27*, 119−123, (In Chinese.).

Yang, X., & Cox-Foster, D. L. (2005). Impact of an ectoparasite on the immunity and pathology of an invertebrate: evidence for host immunosuppression and viral amplification. *Proc. Natl. Acad. Sci. USA, 102*, 7470−7475.

Yang, X., & Cox-Foster, D. L. (2007). Effect of parasitization by *Varroa destructor* on survivorship and physiological traits of *Apis mellifera* in correlation with viral incidence and microbial challenge. *Parasitology, 134*, 405−412.

Yao, Q., Liu, X. Y., Tang, X. D., & Chen, K. P. (2005). Molecular markers-assisted breeding for silkworm resistant variety to BmNPV. *Mol. Plant Breed., 3*, 537−542, (In Chinese.).

Youssef, N. N., & McManus, W. R. (2001). *Ascosphaera torchioi* sp. nov., a pathogen of *Osmia lignaria propinqua* Cresson (Hymenoptera). *Mycotaxon, 127*, 7–13.

Zhang, H. X. (2010). Symptoms of and quick response measures for silkworm poisonings. *Jiangsu Sericult., 1*, 16–17, (In Chinese.).

Zheng, X. M., Yang, Q., Huang, B. H., Liao, S. T., Fang, D. J., Huang, X. G., Yu, A. Q., Xu, X. Y., & Lu, K. M. (1999). Study on a microsporidian *Vairimorpha* MG4 separated from silkworm. *Bombyx mori. Sci. Sericult., 25*, 225–229, (In Chinese.).

Zhu, F. R. (2008). Reasons and controlling strategies for the prevalence of silkworm muscardines in some area of Guangxi. *Guangxi Sericult., 45*, 229–232, Chinese.

Zimmermann, G. (2007a). Review on safety of the entomopathogenic fungi *Beauveria bassiana* and *Beauveria brongniartii*. *Biocontrol Sci. Technol., 17*, 553–596.

Zimmermann, G. (2007b). Review on safety of the entomopathogenic fungus *Metarhizium anisopliae*. *Biocontrol Sci. Technol., 17*, 879–920.

Chapter 13

Physiology and Ecology of Host Defense Against Microbial Invaders

Jonathan G. Lundgren* and Juan Luis Jurat-Fuentes[†]

*United States Department of Agriculture, Agricultural Research Service, Brookings, South Dakota, USA, [†] University of Tennessee, Knoxville, Tennessee, USA

Chapter Outline

Summary

Insects mount a complex hierarchy of defenses that pathogens must overcome before successful infection is achieved. Behavioral avoidance and antiseptic behaviors by host insects reduce the degree of encounters between the insect and pathogens. Any pathogen that contacts or establishes on a potential host faces a series of barriers that restrict entrance into the hemocoel. Pathogens that enter the hemocoel are faced with a multipronged innate immune system. Humoral defenses primarily produce toxic molecules; hemocytes (granulocytes, plasmatocytes, oenocytoids) of the cellular defense have the capacity to phagocytose or encapsulate the target pathogen; and melanization involves both humoral and cellular components to produce several responses that are lethal to the pathogen. Intracellular pathogens must overcome cellular xenophagy and RNA interference defenses. Understanding how host resistance evolves and spreads throughout a population becomes important to preserve entomopathogens as a pest management tool. This phenomenon has been well studied in *Bacillus thuringiensis*, and the development of host resistance to this pathogen, as well as insect resistance management strategies, is discussed.

13.1. INTRODUCTION

This book focuses on the wide variety of entomopathogens and their complex ecology and physiology that

Insect Pathology. DOI: 10.1016/B978-0-12-384984-7.00013-0

make them such a formidable threat to insects and such a useful ally to humans. Given their abundance and the variety of virulence factors that these pathogens produce, it is startling that healthy insects can be found in nature. The insects do not idly wait to be devoured by pathogens; they detect and avoid pathogens, practice hygiene, and present multilayered barriers that block pathogens from establishing an infection. Those pathogens that do breach the behavioral and external defenses of the insect are subjected to the noxious chemicals, deadly cellular responses, and restrictive and toxic melanization responses that are presented by the innate immune systems of the host. The effectiveness of these defenses and how they evolve and spread throughout a population form the basis of resistance management plans that seek to preserve entomopathogen-based control methods for as long as possible. Insect defenses, and how we can predict and delay their effectiveness, are the focus of this chapter.

Pathogens and hosts are in an arms race. The pathogen tries to exploit the host resource to the maximum, but the condition of the host population obviously affects the capacity for a pathogen population to persist. With this in mind, pathogens fall along a continuum of virulence that ranges from causing mere sublethal effects on the host to killing the host quickly. Those pathogens with relatively mild effects on the host result in a more stable host population (i.e., the host population is not rapidly driven to local extinction). Lethal pathogens often have greater persistence in the environment so that they are present when the host population resurges; these virulent pathogens also tend to be more mobile (i.e., they have inherent dispersal means or symbioses with other sources of dispersal), which allows them to locate new host populations (Centofanti, 1995; Ewald, 1995). Vertically (mother to offspring) transmitted pathogens tend to be less virulent than those that are primarily transmitted horizontally (between individuals) (Herre, 1993). Of course, the whole concept of virulence becomes more complicated when one considers that hosts respond against infection. Indeed, virulence of a pathogen is not a consistent characteristic when considering the relative defenses possessed by different potential hosts, or even a single host under different circumstances.

Insect immunity has received substantial attention in several recent reviews, for example Beckage (2007) and Rolff and Reynolds (2009). The intent of this chapter is not to supplant these other sources. Rather, the purpose of this chapter is to introduce the hierarchy of defenses encountered by pathogens and discuss how these defenses help us to understand the similarities and differences among entomopathogens. It is hoped that understanding these defensive systems will clarify the biology and ecology of the pathogens.

13.2. BEHAVIORAL AND PHYSICAL BARRIERS TO INFECTION

Before encountering the immune responses of the host, the pathogen must breach several levels of defenses that reduce both the encounters between the host and pathogen and the likelihood that the pathogen will reach the hemocoel. Reducing pathogen encounters and establishment involves insect behavior in the case of avoidance and antiseptic practices, and it comprises the use of physical barriers that restrict the entrance of the pathogen. It is important to note that these physical barriers may be fortified by antimicrobial compounds, regulated by either the innate immune response or other sources.

13.2.1. Behavioral Avoidance of Pathogens

Avoidance Behavior

The most effective defense against disease is avoidance of contact with the disease-causing agent, which has been observed for some insect species (Alma *et al.*, 2010; Ormond *et al.*, 2011). The omnivorous pirate bug, *Anthocoris nemorum*, avoids foraging on nettle leaves containing spores of *Beauveria bassiana*. Even more striking is that this bug avoids ovipositing on tissue when it contains the fungus, thereby reducing the risk to its offspring (Meyling and Pell, 2006). The gypsy moth, *Lymantria dispar*, detects cadavers and foliage that contain nucleopolyhedrovirus (NPV), and avoids these potential sources of occlusion bodies (Parker *et al.*, 2010). In addition to being able to detect pathogenic organisms, some insects can detect and avoid some of the toxins produced by entomopathogens. For example, larvae of the beet armyworm, *Spodoptera exigua*, avoid consuming diet that contains Cry1C toxin, one of the lethal agents expressed by *Bacillus thuringiensis* (Berdegué *et al.*, 1996). Most insects must come in contact with the pathogen or a cadaver containing the pathogen in order to recognize it as something to be avoided (Meyling and Pell, 2006; Thompson *et al.*, 2007; Alma *et al.*, 2010), but this is not always the case. The termite *Macrotermes michaelseni* not only can detect and avoid sources of *B. bassiana* and *Metarhizium anisopliae*, but also assesses their virulence from a distance and is more strongly repelled by the more virulent strains (Mburu *et al.*, 2009).

Other insects are clearly unable to detect certain pathogens (Boots, 1998; Klinger *et al.*, 2006; Meyling and Pell, 2006). For instance, *Cephalonomia tarsalis*, a parasitoid of the small toothed grain beetle, *Oryxaephilus surinamensis*, is susceptible to *B. bassiana* infection but does not avoid contaminated grain (Lord, 2001). Moreover, this parasitoid oviposits equally in infected and uninfected hosts, thereby dooming the offspring (Lord, 2001). Similarly, of two mole crickets evaluated (*Scapteriscus borellii* and *S. vicinus*), only *S. borellii* could distinguish substrates infested with *B. bassiana* and it adjusted its residence time to reduce exposure (Thompson *et al.*, 2007). Many of these studies that find instances of non-avoidance suggest that the selection pressure on the insect is simply too weak to produce positive selection of the avoidance behavior trait in the specific interactions tested. The majority of studies on this topic have focused on insects' ability to detect potential fungal infections, and it would be interesting to compare how easily some pathogen groups are recognized by insect hosts relative to other groups.

Antiseptic Behavior

Simple grooming behavior is effective in removing many pathogens from the exterior surface of the insect's body before they can cause disease. Although nearly all insects groom themselves in one way or another (many have specialized combs and brushes for just such behavior), grooming and hygienic behaviors of social insects are particularly well studied in this regard. Social insects have limited genetic variability within a colony, live in close quarters with their nestmates, and frequently store food in their nest. These characteristics intuitively should make social insects particularly prone to diseases, but in reality these insects are fairly resistant to non-specialized pathogens. This is in spite of social insects (at least honey bees, *Apis mellifera*) having only one-third of the genes responsible for innate immunity that some other insects possess (Evans *et al.*, 2006). In addition to innate immunity, social insects also practice social immunity, or those behaviors that reduce the exposure of nestmates to potential pathogens (Cremer *et al.*, 2007; Stow and Beattie, 2008). One current argument is that innate immunity incurs a cost to the host (discussed below), but that social insects have been able to replace this physiological immunity with behavioral immunity. Moreover, because behaviors are often pleiotropic and code for other useful behaviors, using these behavioral processes in immune functions may actually be more efficient than having separate genetic machinery for innate immunity (Le Conte *et al.*, 2011).

There is a wide range of examples of this social immunity in insects. Grooming in social insects is divided into autogrooming and allogrooming, depending on whether the insect is grooming itself or its nestmate (Evans and Spivak, 2010). Hygienic behavior is another form of social immunity that is well studied in honey bees. Here, the workers detect larvae that are infected with the fungus *Ascosphaera apis* (chalkbrood) or the bacterium *Paenibacillus larvae* (American foulbrood). They uncap the larval cell, remove the infected larva, and deposit it outside the nest (Evans and Spivak, 2010), and each step in this process is controlled by a suite of genes (Oxley *et al.*, 2010; Le Conte *et al.*, 2011). The chemical cue that bees use to identify infected brood is phenethyl acetate (Swanson *et al.*, 2009). Indeed, even healthy larvae that are marked with this chemical are removed and tossed from the nest by hygienic bees within 24 h. In another social insect example, ants practice necrophagy, which is similar to hygienic behavior, except that the ants sometimes deposit their dead within isolated chambers of the nest (Evans and Spivak, 2010). Both hygienic behavior and necrophagy are carried out by older members of the nest that typically do not care for brood, thereby limiting the exposure to their progeny (Evans and Spivak, 2010). If a nestmate becomes infected with a pathogen, the results are not necessarily catastrophic. Thus, when an ant colony of *Lasius neglectus* was infected with *M. anisopliae*, the workers instantly changed their behaviors to avoid contact with the brood. Even more surprising is that the uninfected ants were more likely to develop resistance to the pathogen after exposure to an infected nestmate (Ugelvig and Cremer, 2007). Hence, there may be a form of transference of immunity among nestmates. Another form of social immunity is when the workers coat the nest (and themselves or each other) in antibiotic substances. A great example of this is propolis, which is a resinous substance that honey bees encase their hives in that is simultaneously waterproof and antimicrobial (Evans and Spivak, 2010). Honey bees will even coat foreign objects that are too large to carry from their nest in propolis to seal them from the colony (Evans and Spivak, 2010). Finally, diseased social insects are frequently relegated to more risk-prone castes that operate outside the nests (e.g., foraging). It is not uncommon for these diseased insects to become disoriented outside the nest or be too weak to return to the nest, thereby limiting exposure to nestmates. It also appears that some social insects commit suicide when they are sick by quickly leaving the nest and not returning (Ruepell *et al.*, 2010). Antiseptic behaviors consisting of suicidal or heat-seeking behaviors have also been described for crickets and grasshoppers to inhibit parasite or pathogen development (Adamo, 1998). Clearly, antiseptic behaviors can have an important effect on insect–pathogen interactions.

13.2.2. Morphological Barriers to Infections

In addition to avoidance and behavioral removal of pathogens, insects possess a variety of morphological characteristics that present barriers to the establishment of pathogens. The precise barriers encountered depend on a pathogen's mode of transmission. The cuticle, tracheal system, and midgut are the major sites of invasion, and each of these fronts has its own set of physical defenses.

Cuticle

The physical and inherent chemical characteristics of the cuticle have a great influence on which pathogens are able to establish on the external surface of a host. The cuticle covers all structures that come in contact with the external environment, including the exterior of the insect, the foregut, hindgut, and tracheal system (Chapman, 1998). The integument offers physical protection from predators, functions as an exoskeleton for muscle attachment, and reduces water loss, which allowed insects to permeate terrestrial ecosystems (Moussain, 2010). Another function of the integument is that it resists pathogen establishment.

Fungi are the main entomopathogens using the cuticular integument as a point of entry and start of infection. The entire external covering of insects is coated in a waxy layer, and the fatty acids that comprise this layer dictate which fungal entomopathogens are able to establish. Some fatty acids enhance the ability of certain fungi to adhere and germinate on the cuticle, while others restrict these processes (Boguś et al., 2010). So far, it appears that fungi respond differently to the various fatty acids encountered on insect cuticles, so the insect needs to possess a wide variety of these fatty acids to have the plasticity necessary to repel the diversity of fungi they encounter. The exact mechanisms for how these fatty acids affect fungal adhesion and germination remain to be fully resolved.

Other factors, namely gland secretions and the microbial community, add antibiotic properties to the cuticle, making it an even more formidable barrier to infection. Some insects coat themselves in gland secretions that have antimicrobial properties. For instance, ants cover themselves in an antibiotic milieu produced by thoracic metapleural glands, which are unique to Formicidae (Poulsen et al., 2003; Stow and Beattie, 2008). In a recent survey, all 26 species of ants examined wiped metapleural gland secretions on their bodies (Fernández-Marín et al., 2006). Ants increase their use of this fungicidal secretion when challenged by a fungal entomopathogen, and this response is activated in as little as 1 h postinfection (Fernández-Marín et al., 2006). When these glands are covered, ants infected with fungus die within a few days. In addition, the beneficial microbial community that resides on insect cuticles is an important source of antibiotics to resist some infections. Two hemipteran herbivores, *Dalbulus maidis* and *Delphacodes kuscheli*, are well known for their ability to resist infections by *B. bassiana*. Examining the bacterial community found on the cuticle revealed that 91 of the 155 bacterial isolates inhibited growth of the fungus, with the most effective isolates belonging to *Bacillus* (Toledo et al., 2011). Only one bacterium, *Bacillus pumilis* Dm-B23,

inhibited germination of the fungus. Clearly, the contributions of the symbiotic microbial community of the cuticle warrant further attention.

Digestive and Tracheal Systems

Many pathogens cannot penetrate the external defenses of the insect host, and circumvent this barrier by attempting to enter the hemocoel through the mouth (*per os*) or tracheal system. These pathogens (many viruses, bacteria, microsporidia, and protists are noteworthy examples) are confronted by a series of formidable defenses that necessitate specialized adaptations by the pathogen before successful infection is achieved. Much of the battle is waged in or near the midgut, where the cuticle is absent.

Microbes that enter the midgut must be adapted to survive and infect under the environmental conditions present in the particular insect, and these conditions vary widely among hosts. The pH, chemical milieu, and resident microbial community all play a role in which microbes are able to penetrate into the hemocoel. For example, *B. thuringiensis* protoxins are only activated under specific pHs and with specific proteases, which in part narrows the host ranges of the various strains and toxins of this pathogen (Haider et al., 1986). Also, the salivary glands of some insects may produce antimicrobial compounds that permeate throughout the gut (Chouvenc et al., 2010). Finally, the resident microbial community within insect guts resists some invaders in trying to reach a homeostatic condition, and even ordinarily pathogenic microbes can subsist as benign or even beneficial components of a healthy gut microbial community (Dillon and Dillon, 2004). The conditions fluctuate widely depending on the physiological status of the host and the environment in which this host lives.

If a microbe finds the environment of the gut suitable, the next barrier to infection is the peritrophic matrix. The peritrophic matrix is a semi-permeable, non-cellular envelope that encases the lumen of the midgut, and offers a barrier against physical abrasion by food and microbial invaders such as bacteria and viruses. This membrane is primarily composed of microfibrils, but also proteoglycans, proteins, and glycoproteins, and acts as a filter for microorganisms seeking access to the epithelial cells of the midgut lining (Lehane, 1997). The thickness and permeability of the peritrophic matrix can vary within an individual as well as among species (Chapman, 1998; Plymale et al., 2008). Most successful pathogens that invade via the midgut (1) are small enough to pass through the peritrophic matrix (e.g., some viruses), (2) send lethal chemicals through the membrane that kill the cells that produce the envelope (e.g., δ-endotoxins of *B. thuringiensis*), or (3) degrade the membrane chemically (Chen et al., 2008). This last approach is well explored in baculoviruses, which produce various enhancins, e.g., virus enhancing protein (vep) and synergistic factor (sf) (Hoover *et al.*, 2010) (see Chapter 4). Another example of this approach comes from entomopoxviruses, which produce spindles (proteinaceous crystalline bodies) containing fusolin that disrupt the peritrophic matrix (Mitsuhashi *et al.*, 2007).

Once the peritrophic matrix is overcome, the midgut epithelial cells and basal lamina are the next barriers that must be breached. The midgut epithelial cells produce various antimicrobial proteins that are effective against Gram-positive and negative bacteria and fungal pathogens, and this process is governed by the immune deficiency (Imd) signaling pathway (discussed in Section 13.3.2) (Brennan and Anderson, 2004; Govind, 2008). In the case of intracellular pathogens, if the midgut cells are infiltrated, then the insect can simply slough them off in order to stop the infection process. At the base of the midgut epithelium (on the hemocoel side of the epithelium) lies a final acellular barrier, called the basal lamina, that must be circumvented before the hemocoel is reached (Passarelli, 2011). The basal lamina is particularly restrictive to viral infections, although viruses may pass through the basal lamina as it is recycled. In this case, the viruses express a gene that forces the cell to recycle layers of the lamina, and as this is happening, viral particles are encased in the layers of the lamina and invade the hemocoel as the lamina degrades (Passarelli, 2011). Viruses also can produce molecules that degrade the basal lamina, such as caspase and viral fibroblast growth factor (Passarelli, 2011). Also, there are occasionally small holes in the basal lamina that can be used for viral escape. Finally, some pathogens bypass the basal lamina of the midgut by infecting the tracheal system (Passarelli, 2011) (see Chapter 4).

The tracheal system is coated in cuticle and is composed of tracheae and much finer tracheoles (or tracheoblasts) (Chapman, 1998). The cuticle of the tracheoles is retained during molts, and these cells sometimes permeate other cells to oxygenate them. Tracheoles oxygenate the midgut epithelium and may be an additional conduit by which viruses infiltrate the hemocoel. But the tracheal system also possesses a basal lamina that represents a similar barrier as in the midgut epithelium (Passarelli, 2011).

13.3. PHYSIOLOGICAL RESPONSE TO INFECTIONS

If a pathogen successfully navigates the external defenses meant to keep it from becoming established, it could be faced with the defensive responses found within the insect hemocoel. Traditionally, the physiological responses of insects have been categorized as humoral or cellular in nature, although in reality the two systems are intertwined

in many aspects. This notwithstanding, these categories have some function, as will be discussed below. It is noteworthy that much of the recent work in this area has focused on *Drosophila melanogaster* and *Anopheles gambiae* as models. This polarized interest is due to the availability of genomic tools and the sequenced genomes for both insects, which have provided unprecedented opportunities to explore the genetic basis of immunity. However, it is likely that important differences exist in the innate immunity expressed among the approximately 10 million species of insects (a conservative estimate) and care should be taken not to overgeneralize the important findings from these two model organisms.

Innate physiological immunity follows a series of pathways to kill and eliminate pathogens from the insect's hemocoel. The first step in the process is that the insect must recognize potentially dangerous pathogens and alert the innate immunity systems. Once this recognition is achieved, one or more of several signaling pathways are initiated. The humoral response involves the production of antibiotics that kill the pathogen. Simultaneously, a cellular response could be initiated, whereby hemocytes aggregate around the pathogen, immobilizing and killing it. Finally, these two defense systems often work together to produce melanin, a third deadly weapon targeted against the extracellular pathogens in the hemocoel. Defenses against intracellular pathogens are less well studied, although it appears that similar signaling systems are involved in at least triggering the intracellular responses (which involve antimicrobial molecules and RNA interference).

13.3.1. Distinguishing Self from Non-self from Altered Self

The first step in the innate physiological responses to pathogens is distinguishing the pathogen or pathogen-affected cells from the rest of milieu in the hemocoel. Essentially, the immune system needs to distinguish self from non-self (e.g., the pathogen or abiotic material) and altered self (e.g., dead or infected cells) (Kanost and Nardi, 2010). To do this, the immune system takes advantage of unique molecules either on or associated with a specific invader. These pathogen-associated molecular pattern (PAMP) molecules include lipopolysaccharides (LPS), lipoteichoic acid (LTA), peptidoglycans (PGN), and β-1,3-glucan (Hoffmann, 1995; Lavine and Strand, 2002; Royet, 2004a; Strand, 2008; Kanost and Nardi, 2010). While LPS, LTA, and PGN are bacterial elicitors, β-1,3-glucan triggers an antifungal response (Royet, 2004b). Although there is little research on the topic of antiviral PAMPs, viral glycoprotein VSV-G is one example that indicates that these molecules play a role in viral immunity (Sabin *et al.*, 2010).

Occasionally, endogenous signals associated with the pathogen are not needed to initiate the immune response. Wounding or stress can elicit such a response without foreign invaders (Brennan and Anderson, 2004; Kanost and Nardi, 2010). Extracellular nucleic acids can also trigger an immune response, presumably because these are a sign of cell damage that sometimes indicates infection (Kanost and Nardi, 2010; Vilcinskas, 2010). In fact, oenocytoids (a type of hemocyte) of some insects rupture in response to pathogen invasions, and their contents thereby become the signal that triggers downstream immune reactions (Vilcinskas, 2010).

Abiotic materials also can elicit immune responses, although the underlying mechanisms that drive this recognition are poorly understood. For example, Lavine and Strand (2001) injected 19 different types of chromatography beads into the lepidopteran *Pseudoplusia includens*, and while the immune system frequently encapsulated the foreign materials, there was little consensus in what signaling pathway or even hemocytes were involved in this process. Clearly, there are many more classes of recognition site than simply the PAMPs often discussed in the literature.

The molecules produced by the insect that recognize, bind to, and mark these PAMPs are called pathogen recognition receptors (PRRs) or pathogen recognition proteins (PRPs). Dozens of these receptors have been identified to date, and they are categorized by Strand (2008) as often being extracellular, cell surface, or transmembrane in nature. Gram-negative binding protein (GNBP), LPS-binding protein C-type lectins, β-1,3-glucan recognizing proteins, immunolectins, tep-proteins, Down's syndrome cell adhesion molecule (Dscam), eicosanoids, and hemolin are examples of PRRs that act against various pathogens (Christensen *et al.*, 2005; Govind, 2008; Jiang, 2008; Strand, 2008; Kanost and Nardi, 2010). Some of these PRRs are inducible, only becoming active when the host is immunochallenged. Other PRRs are produced constitutively, and thus are always probing for potential invaders.

The specificity of these PRRs allows the insect host to mount targeted defenses against the diverse army of invaders it experiences. For example, a commonly studied group of PRRs that displays specificity is the peptidoglycan receptor protein (PGRP) group (Kanost and Nardi, 2010). This group shares a 160 unit peptidoglycan recognition domain (Brennan and Anderson, 2004). Within this group of constitutively produced recognition proteins (Takehana *et al.*, 2004), the PRRs are categorized as PGRP-L and PGRP-S, based on size (Royet, 2004a). PGRP-S have fewer than 200 amino acids, and PGRP-L are proteins that are larger than 200 amino acids (Royet, 2004a). In *Drosophila*, the protein PGRP-LC protein is anchored to the hemocyte membrane, and functions against Gram-negative bacteria (Brennan and Anderson, 2004). In contrast, PGRP-S is a secreted protein in *Drosophila* acting against Gram-positive bacteria (Brennan and Anderson, 2004). Other PGRP-S proteins can act against Gram-negative infections as well (Royet, 2004b). At least two forms of PGRP-L (-LE and -LC) are present on the insect's epithelium, thus representing a first line of defense against infections. Takehana *et al.* (2004) studied these interactions on the tracheal epithelium. C-type lectins do not recognize proteins; rather, they bind to microbial-based polysaccharides (Jiang, 2008). Not all PRRs display specificity for certain pathogens. Hemolin, a hemolymph protein induced by bacteria and which is only present in Lepidoptera, is fairly non-specific to particular pathogen groups and even binds to hemocytes (Eleftherianos *et al.*, 2006; Jiang, 2008). Finally, some genes that encode for PRRs have been identified so far; *semmelweis* and *osiris* are two examples that encode specific PGRP and GNBP PRRs (Hoffmann, 2003).

Even hemocytes can function as PRRs when they adhere to a foreign target (Lavine and Strand, 2002; Carton *et al.*, 2008). In some cases, hemocytes bind to PRRs just like the humoral system does. Some hemocytes, for example, bind to immunolectins that have attached to foreign invaders (Brennan and Anderson, 2004; Strand, 2008), whereas other hemocytes possess transmembrane PRRs in their cell membranes that recognize conserved foreign molecules associated with pathogens. Within Lepidoptera, these poorly understood cell adhesion molecules include integrins, tetraspanins, and neuroglian (Lavine and Strand, 2002; Kanost and Nardi, 2010). Finally, granulocytes can sometimes produce PRRs; lacunin and hemocytin are two examples of these (Kanost and Nardi, 2010).

13.3.2. Humoral Response System

Antimicrobial Peptides

Soluble peptides constitute the main weapon that insects use during humoral responses to pathogens. Dozens of antimicrobial peptides (AMPs) (at least 50) have been identified to date, and these peptides can be very specific for specific pathogens, or classes thereof (Hoffmann, 2003; Govind, 2008; Kanost and Nardi, 2010). Lysozyme was the first AMP identified, and it was isolated in *Galleria mellonella* (Vilcinskas, 2010). Many of the AMPs fall within one of three categories, based on their structure (Bulet *et al.*, 1999). The first group of these is peptides with intramolecular disulfide bonds that form hairpin β-sheets and α-helical—β-sheet mixed structures. In comparison, the second group of AMPs has peptides that form amphipathic α-helices. Finally, the third group has a disproportionate amount of proline and/or glycine residues.

Various classes of Amp residue within these groups. For example, cecropins and defensins are two major classes of AMPs specific for Gram-positive or Gram-negative bacteria, respectively (Hoffmann, 1995). Proline- and glycine-rich polypeptides are another class of AMP, and these are specific for Gram-negative bacteria. Acute phase glycoproteins dTEPs are another noteworthy groups of AMP that disrupt bacterial cell walls (Govind, 2008; Strand, 2008). There are exceptions to the specificities noted above; defensins sometimes act against Gram-positive bacteria as well as fungi (Bulet *et al.*, 1999), and some proline-rich polypeptides also affect Gram-positive bacteria as well (Hoffmann, 1995). Some AMPs are very specific in where they reside and function (Hoffmann, 1995). Cecropins are membrane-active proteins residing on cell membranes, while lysozyme is ubiquitous throughout insect tissues. Other AMPs are restricted to the genital tracts of insects. Finally, PGRPs that are involved in identifying PAMPs also sometimes degrade the peptidoglycan present in bacterial cell walls, and so act as antimicrobial agents as well as signaling agents (Strand, 2008). In short, new antimicrobial peptides are frequently revealed, and our understanding of this rapidly expanding source of innate immunity would benefit from some synthesis in the future.

Regulation of Humoral Defenses

Antimicrobial peptides are produced in the fat body, and to a lesser extent in other somatic cells and by circulating granulocytes and plasmatocytes (Hoffmann, 1995; Lavine and Strand, 2002; Govind, 2008; Strand, 2008). Upon encountering a foreign target, usually one of two signal transduction pathways (Toll and Imd, although there are others) that resist different pathogen classes is triggered (Govind, 2008). The Toll pathway is mobilized in response to Gram-positive and fungal infections, while Imd is mobilized by Gram-negative bacteria and infections of multiple classes associated with barrier membranes. The specificity of the pathways is truly remarkable given the genetic divergence observed in pathogen groups that trigger the same response pathway. For instance, fungi and Gram-positive bacteria share only 30% genetic similarity, but somehow both activate the Toll pathway (Hoffmann, 2003). Both the Toll and Imd pathways are triggered upon exposure to certain viruses, although the mechanisms involved are poorly understood (Sabin *et al.*, 2010). Both of these pathways belong to the intracellular nuclear factor-κB (NF-κB) related signal transduction pathway, which ultimately produces the AMPs. The Toll pathway is not mobilized by direct interactions with microbes themselves, but rather by interacting with the cytokine intermediate Spätzle, which is activated through cleavage by a serine protease (Hoffmann, 2003; Govind, 2008). Serine proteases are responsible for cleaving the proSpätzle into its active form (Royet, 2004a). During fungal infections, the gene *persephone* is involved in the production of these serine proteases (Brennan and Anderson, 2004). The Imd pathway

is regulated by *Relish* (Royet, 2004a). It appears that hemocytes are particularly important in the Imd-mediated response of *Drosophila* (Brennan and Anderson, 2004). When hemocytes function as elicitors of the Imd pathway, PGRP-LC is a transmembrane protein on the hemocytes. When the hemocytes bind to a foreign target, nitric oxide (NO) produced by the hemocytes triggers the fat body to produce the necessary AMPs (e.g., diptericin). Other signaling pathways, such as the Jak-STAT and JNK pathways, also initiate the humoral defense system in immune-challenged insects (Brennan and Anderson, 2004). The immunity provided by AMPs generally lasts for two to three days (Kanost and Nardi, 2010). Although it is known that they block the growth of microbes in the hemolymph (Hoffmann, 2003), the mechanisms by which the AMPs kill foreign targets remain poorly understood.

13.3.3. Cellular Response System

Inherently tied to humoral-based immunity is the cellular defense system of insects which, as the name suggests, involves various types of hemocytes. These hemocytes kill and remove pathogenic targets through phagocytosis of small or unicellular targets, or through nodulation/encapsulation of larger elements, groups of pathogens, or multicellular organisms. Aspects of the cellular defense system are also coupled with the humoral defense network in melanization of pathogens.

Hemocytes

Several hemocytes have been characterized by their morphology, the presence of specific antigenic markers, and their functional responses to entomopathogens (Strand, 2008). All types of hemocyte are derived from prohemocytes, which are a type of stem cell (Lavine and Strand, 2002; Kanost and Nardi, 2010). Hemocytes are produced during two life stages of the insect, during embryogenesis and during late larval development (Strand, 2008). The latter prohemocytes produced by larvae are of mesodermal origin and are derived from hematopoietic organs (Carton *et al.*, 2008). Hematopoietic organs are lymph glands (Carton *et al.*, 2008; Strand, 2008); in *Drosophila*, pairs of these glands form along the anterior area of the dorsal vessel during embryogenesis. There are three groups of these organs in *Drosophila*: the posterior signaling center, the medullary zone, and the cortical zone. The first two of these groups contain only quiescent prohemocytes, whereas the cortical zone has active production of both plasmatocytes and crystal cells (Strand, 2008). Some lepidopteran larvae have four hematopoietic organs located in the thorax near the wing imaginal disks (Strand, 2008). Some insects also have secondary hematopoietic organs and free-living prohemocytes that are involved in the synthesis of plasmatocytes (Strand, 2008). Knowledge on the generation of hemocytes comes entirely from studies on *Drosophila*. In this insect, some genes have been identified that regulate the initial production and proliferation of prohemocytes (*Srp* and *Pvf2*). Once the cells are produced, they differentiate from prohemocytes into their specific forms, and several genes that regulate this process have also been identified. In *Drosophila*, *gcm* and *gcm2* regulate the differentiation of plasmatocytes, and *Notch* and *Lz* genes regulate the differentiation of crystal cells (Carton *et al.*, 2008; Strand, 2008). Several signaling pathways have also been implicated in hemocyte proliferation and differentiation, including Toll, Jak/Stat, Jun kinase, and Ras-mitogen pathways (Carton *et al.*, 2008; Strand, 2008). Once the hemocytes have been created and have differentiated, they are able to proliferate in response to pathogens, although the mechanisms deserve further study.

Various hemocyte types comprise the hemolymph. The most abundant hemocytes in many insects are granulocytes, followed in abundance by plasmatocytes, oenocytoids, and spherulocytes (Lavine and Strand, 2002). Sometimes, granulocytes differentiate after foreign invasion. In *Manduca sexta* larvae, one group of enlarged granulocytes, called phagocytes, behaves differently from other granulocytes in response to invasion, spreading out asymmetrically during encapsulation and being capable of forming capsules around foreign targets (Strand, 2008). To confound the literature on cellular responses, the hemocytes of *Drosophila* follow a different nomenclature altogether. In this insect, plasmatocytes (90% of hemocytes) dominate in number, and lamellocytes and crystal cells are less abundant (Lavine and Strand, 2002). There are clear analogies to the hemocytes of the other insects (granulocytes and plasmatocytes, oenocytoids and crystal cells, plasmatocytes and lamellocytes), although the two nomenclature systems are not perfectly interchangeable (Strand, 2008). Not all hemocytes are found in all insects, and even congeners can vary in the composition of hemocyte types in their hemolymph (Carton *et al.*, 2008). Also, foreign invasion can induce differentiation of prohemocytes; healthy *Drosophila* do not have lamellocytes and are produced only following infection (Strand, 2008).

Each of these hemocytes has a fairly distinct function within the immune system of insects. Plasmatocytes and granulocytes are the only hemocytes that adhere to foreign molecules and pathogens, and thus are the primary agents involved in phagocytosis and encapsulation of pathogens (Lavine and Strand, 2002). Oenocytoids produce phenoloxidase components that are involved in melanization, which will be discussed below (Lavine and Strand, 2002). The function of spherulocytes is less well understood, but they appear to transport cuticular materials and have an unknown function in cellular immunity (Lavine and Strand, 2002). In *Drosophila*, lamellocytes are adhesive and play a role in encapsulation. Crystal cells are not adhesive, but they produce phenoloxidase components (Brennan and Anderson, 2004). Similar to other insects, plasmatocytes of *Drosophila* adhere to foreign targets and play a role in phagocytosis (Brennan and Anderson, 2004). In sum, these hemocytes work together in mounting the various cellular responses to invasion by pathogenic agents, which are fairly easily categorized as involving either phagocytosis or nodulation/encapsulation, depending on the size and type of pathogen involved.

Phagocytosis

The word "phagocytosis" has a Greek origin and when translated to English it means "cell eating", which clearly describes this process. In phagocytosis, granulocytes adhere to a foreign molecule or cell and envelop the invader, thereby killing it. During phagocytosis, the pathogen is encased intracellularly in a phagosome. The hemocyte releases reactive oxygen intermediates (ROIs) and reactive nitrogen intermediates (RNIs), which are toxic to the pathogen (Lavine and Strand, 2002). These ROIs and RNIs may also trigger the humoral signaling pathway that leads to the production of antimicrobial peptides (Lavine and Strand, 2002). For example, NO activates the NF-κB pathway discussed above. Regardless of the mechanism, the result is death for the pathogen. This process is constrained by the size of the hemocyte and pathogen involved, and the cellular immune system must overcome larger invaders or groups of invaders using other means.

Nodulation and Encapsulation

Insects rely on nodulation and encapsulation to ensnare larger invaders that cannot be killed with a single granulocyte (Strand, 2008; Vilcinskas, 2010); these two processes differ in their degree rather than in approach. When a pathogen is located in the hemocoel, granulocytes (and sometimes plasmatocytes), or lamellocytes in *Drosophila*, adhere to a PAMP of the invading organism. It is common for the hemocytes to spread out and apoptose when they encounter a foreign target (Brennan and Anderson, 2004; Strand, 2008; Kanost and Nardi, 2010). This process is initiated by one or several activator molecules; one such molecule is plasmatocyte spreading peptide (PSP), which belongs to the ENF-peptide family (Kanost and Nardi, 2010). ProPSP is cleaved to generate the active form which elicits spreading of the hemocyte. Another protein class known to elicit spreading is Rac GTPases (Carton *et al.*, 2008), which function in cytoskeletal organization, regulation of cellular adhesion, and related transcriptional activation (Carton *et al.*, 2008).

Spreading hemocytes overlap each other to encapsulate the target (Lavine and Strand, 2002). Different insects produce different patterns in how the hemocytes aggregate to infections. Sometimes, granulocytes form the initial sheath around the target, and plasmatocytes and granulocytes build upon this. In other species, plasmatocytes and granulocytes arrange randomly, and granulocytes are not ubiquitously necessary for encapsulation to occur. Encapsulation is almost always reinforced by melanization, a process discussed further below. Encapsulation almost always kills the target, either through asphyxiation or through the production of toxic chemicals during the melanization process.

13.3.4. Melanization

The humoral and cellular components of innate immunity often both contribute to melanization of a foreign target. Foremost, encapsulation is often accompanied by melanization, but not in all insect species (Lavine and Strand, 2002; Christensen *et al.*, 2005; Carton *et al.*, 2008; Strand, 2008; Kanost and Nardi, 2010). Also, when hemocytes are scarce, melanization can still occur in some insects (Christensen *et al.*, 2005). Melanization is a series of chemical reactions that results in the production of darkened pigment around a wound or pathogen, and there are three general steps in this chemical process (Carton *et al.*, 2008).

The first step is the activation of the phenoloxidase cascade, which occurs within minutes of infection (Royet, 2004b). Phenoloxidases are a key chemical in the cascade of reactions that ultimately leads to melanization. The few insects in which this process has been studied to date have been found to contain at least six active phenoloxidases (Christensen *et al.*, 2005). The process begins when serine proteases, which are produced with the involvement of immunolectin-2 in *M. sexta* (Eleftherianos *et al.*, 2006), trigger the hydroxylation of tyrosine (derived from phenylalanine) into dihydroxyphenylalanine (DOPA), and ends with the production of eumelanins and indolequinones (Christensen *et al.*, 2005; Carton *et al.*, 2008; Jiang, 2008). To protect themselves from random melanization events, phenoloxidases circulate in the hemolymph as prophenoloxidase (Kanost and Nardi, 2010). As mentioned above, this prophenoloxidase is produced by oenocytoids (or crystal cells in *Drosophila*), a type of hemocyte (Strand, 2008; Kanost and Nardi, 2010). The cleavage of prophenoloxidase into phenoloxidase is blocked by the protein Spn27A, which inhibits a prophenoloxidase activating enzyme (PPAE) (Brennan and Anderson, 2004). When cleavage is required, the insect produces an inhibitor to Spn27A, or it overproduces PPAE (Brennan and Anderson, 2004). Genes that regulate the creation of prophenoloxidase have been identified and different genes are expressed in response to various pathogens. For example, in *Heliothis virescens* larvae, only one of two genes (the one producing HvPPO-1) is upregulated when the host is challenged by

a bacterial invader (*Micrococcus lysodeikticus*); the other gene product is produced constitutively. The prophenoloxidase is activated by one of several prophenoloxidase-activating proteases.

The next step in melanization is the proliferation and physical alterations of hemocytes at the site of the wound or invader. Much of this process is discussed above in the section on encapsulation. In brief, hemocytes adhere to receptors on the foreign target, spread, and aggregate to form a multicellular sheath that asphyxiates the foreign body.

The final step in melanization is the production of toxic chemicals that kill the offending agent. The melanization process produces several chemicals that are known to be toxic to microbes. Specifically, ROIs (O_2^-, H_2O_2), RNIs (NO), quinones, hydroquinones, and most importantly, L-DOPA are produced (Lavine and Strand, 2002; Christensen *et al.*, 2005; Carton *et al.*, 2008). Antimicrobial peptides such as defensins also aggregate to sites of melanization, adding another lethal agent to the milieu (Hillyer and Christensen, 2005).

13.3.5. Intracellular Defenses

Defending against intracellular pathogens requires additional tools on the part of the host than those previously discussed. Indeed, innate immunity has received much more attention from the perspective of extracellular pathogens, and the mechanisms that cells use to defend themselves against intracellular bacteria and viruses remain to be fully resolved. Nevertheless, a growing body of research is beginning to shed light on how xenophagy and RNA interference (RNAi) combine with antimicrobial molecules to defend against this major class of pathogens.

Involvement of the Extracellular Immune Systems Against Intracellular Pathogens

Outside the host cell, intracellular bacteria often do not elicit the innate immune responses of the hosts, although the humoral and cellular responses are often effective against these pathogens. Phenoloxidase is antiviral, although injection with baculovirus does not necessarily elicit the production of phenoloxidase (Shelby and Popham, 2008). Likewise, antimicrobial genes are not upregulated when *Drosophila* is infected with the intracellular bacterium *Spiroplasma poulsonii* (Hurst *et al.*, 2003). This pathogen resembles the Gram-positive bacteria, but they lack a cell wall and presumably the PAMP necessary to instigate the humoral and cellular responses. However, *Drosophila* mutants that produce seven different antimicrobial peptides constitutively are able to repel *S. poulsonii* infections, as are *Drosophila* that have their immune systems "primed" by injection with dead bacteria

(Hurst *et al.*, 2003). Similarly, *Wolbachia*, one of the most well known of the intracellular bacteria (see Chapter 9), does not trigger the Gram-negative specific immune response in *Anopheles* or *Aedes* mosquitoes, even though this microbe resides within this guild of pathogens (Bourtzis *et al.*, 2000; Brennan *et al.*, 2008). In *Drosophila*, *Wolbachia pipientis* infection also does not instigate the production of two antimicrobial peptides (diptericin and cecropin, which are both specific to the Gram-negative bacteria) (Siozos *et al.*, 2008). An exception is a virulent strain of *W. pipientis* named "popcorn", which ruptures the host cells and presumably leaves the pathogen exposed to the extracellular innate immune response (Siozos *et al.*, 2008). In any case, it seems clear that prior exposure to other pathogens influences the success of these intracellular disease agents.

RNA Interference

Insect cells respond to viral infection by initiating an antiviral program involving an RNA silencing pathway; this RNAi is one of the most important components of the antiviral defenses of insects (Sabin *et al.*, 2010). Some of the key genes involved in the biogenesis of RNAi that functions against viruses are *Dcr-2*, *r2d2*, and *AGO2*, and it is likely that additional contributing genes that govern this process will be discovered (Sabin *et al.*, 2010). The protein Dcr-2 attaches to viral dsRNA, and it functions as a PAMP recognizable by other antiviral molecules (Sabin *et al.*, 2010). Vago is one such PRR that recognizes Dcr-2, and Vago also kills the virus (Sabin *et al.*, 2010). So far, the Jak-STAT signaling pathway appears to be the most important for antiviral defenses of cells. Research on antiviral defenses involving RNAi of insects is still in its infancy, and it is also unknown whether RNAi pathways function in defending against other pathogenic agents apart from viruses.

Xenophagy

Under specific adverse conditions, cells can undergo autophagy, a catabolic process by which cells digest old or damaged organelles (Amano *et al.*, 2006). The process whereby the cell directs its autophagy machinery toward killing pathogens is termed xenophagy (Levine, 2005). Xenophagy is thought to play a role in antiviral defenses (Sabin *et al.*, 2010), although there are few data on these interactions. While the specific molecules controlling this process are poorly understood, they are expected to share similarities with general steps described for autophagy processes. During autophagy, old organelles are engulfed by autophagosomes, which are essentially multimembrane structures involving cisternae and isolation membranes (Amano *et al.*, 2006). Lysosomes then fuse with the autophagosome and degrade the contents (Amano *et al.*, 2006).

The role of autophagy in maintaining healthy cells has been known for some time, although its role in destroying pathogens was discovered only recently.

Resistance Against Intracellular Host Defenses

To avoid or circumvent detection, intracellular pathogens either hide from the defensive system or co-opt it entirely at a genetic level. To hide from the immune response, intracellular bacteria will sometimes encase themselves in membranes of the host cell, called endosomes or phagosomes (Amano *et al.*, 2006). *Wolbachia* is a good example of this (McGraw and O'Neill, 2004), but the lysosomes can recognize these endosomes and degrade them through acidification (Amano *et al.*, 2006). In turn, the pathogen resists this process. Obligate endosymbionts usually have reduced genomes and rely on their host's genes for key functions, including defense. In the case of *Wolbachia*, the bacterium actually co-opts the production of antioxidants to defend against the host's reactive oxygen species, a major aspect of the humoral defense system of insects (Brennan *et al.*, 2008). Indeed, *Wolbachia* infection of *An. gambiae* upregulates more than 700 host genes, and many (but not all) of the immune, stress, and detoxification-related transcripts are downregulated upon *Wolbachia* infections (Hughes *et al.*, 2011). Viruses that are targeted by RNAi can evolve the ability to suppress these nucleic acids, and the resulting arms race between viruses and their host cells has made RNAi genes some of the fastest evolving within the *Drosophila* genome (Sabin *et al.*, 2010).

13.4. MANAGING RESISTANCE TO ENTOMOPATHOGENS

Understanding and predicting how insect hosts become resistant to pathogens or their toxins is important as land managers (especially farmers) have come to rely on entomopathogens or their toxins more heavily for controlling pests. Declaring an insect resistant can be controversial, since commercial products that rely on entomopathogens as their active ingredient could then lose applicability and market share. Thus, having a clear definition of resistance becomes imperative. As mentioned throughout this chapter, most insect populations have a range of defensive capabilities against particular pathogens. For the purposes of this discussion, the general definition of Fuxa (2004) is adopted. Fuxa (2004) describes resistance as "the development of an ability in a strain of insects to tolerate doses of a [pathogen] that would cause disease or prove lethal in the majority of individuals in a normal population of the same species." Heritability is also a key aspect to resistance; the trait must be passed on to the offspring in order for a strain to be considered resistant (Gassmann *et al.*, 2009a). Thus, we recognize that there is a continuum in resistance, and use the background or average resistance in a population as a benchmark against which to compare comparing selected strains.

Given the diversity of defensive mechanisms that pathogens must overcome for successful infection, there are remarkably few well-studied examples of host resistance to pathogens when using the definition provided above. The few instances where the evolution of resistance has been well studied are in ecosystems where tremendous selective pressure has been placed upon particular target pests. For example, the velvetbean caterpillar, *Anticarsia gemmatalis*, a key pest of soybeans in Brazil, is managed using the nucleopolyhedrosis virus AgNPV. Although field resistance in the pest has not been discovered, laboratory strains of resistant pests have been selected for in the laboratory and resistance mechanisms and stability are being revealed (Fuxa and Richter, 1998; Piubelli *et al.*, 2006; Levy *et al.*, 2007). Another system where resistance has been well studied is the case of *Bacillus sphaericus*, which is applied to control mosquitoes of medical importance. High levels of field resistance have been documented in *Culex* spp. in Asia and Europe (Charles *et al.*, 1996; Park *et al.*, 2010), which is probably the result of only a single insecticidal toxin being produced by this species of *Bacillus* (rather than the multiple toxins produced by *B. thuringiensis*). The mechanisms underlying these cases of resistance to *B. sphaericus* have been fairly well studied, and usually involve the inability of the Bin toxin (a potent protein toxin with two components, BinA and BinB) to adhere to the binding sites on the midgut epithelium of resistant individuals (Darboux *et al.*, 2002, 2007). This is certainly not always the case, and behavioral avoidance of the toxin (Rodcharoen and Mulla, 1995) and other, unidentified, mechanisms (Nielsen-Leroux *et al.*, 1997, 2002) may also be at play.

With regard to host resistance and its implications for pathogen–insect dynamics, no species has received more attention than *B. thuringiensis*. Various strains of *B. thuringiensis* have been widely applied to entire landscapes for controlling dipteran and lepidopteran pests, and now various genetically modified crops express the Cry and/or Vip toxins. Cry toxins are the crystal proteins produced by *B. thuringiensis* that destroy the midgut epithelial cells; Vip toxins are vegetative insecticidal proteins produced by *B. thuringiensis* that also target the midgut epithelium, but adhere to different receptors on the midgut wall than the Cry toxins. The wide use of these toxins increases the selective pressure towards evolving resistance in the target pests. For these reasons, this organism will be used as a model for discussing more in depth how resistance evolves within host insects, and how information on this process is crucial to preserve *B. thuringiensis* as a management option.

13.4.1. Resistance Mechanisms to *Bacillus thuringiensis*

Insecticidal products containing *B. thuringiensis* as the active ingredient have been used extensively in pest control, mostly against dipteran pests of medical concern and lepidopteran pests of agriculture. For years it had been assumed that because *B. thuringiensis* was a biological agent, capable of evolving alongside its host, host resistance was of minimal concern (Shelton *et al.*, 2007). As will be discussed below, this was an incorrect assumption. In 1996, the first transgenic crops genetically engineered to constitutively produce δ-endotoxins of *B. thuringiensis* (Bt crops) were commercialized and have quickly come to dominate cotton and corn acreages throughout North America and other countries. In the USA, this level of selection pressure — 4.2 million ha of cotton and 24.3 million ha of corn expressed one or more Bt toxins in 2011 (NASS, 2011) — has resulted in tremendous selection pressure against not only the Cry toxins expressed in the plants, but possibly the pathogen itself. Efforts to mitigate this resistance have advanced our understanding of the ecology and dynamics of how hosts evolve defenses against pathogens.

Resistance to *B. thuringiensis* was first reported in grain storage bins (McGaughey, 1985) and was subsequently found under greenhouse (Janmaat *et al.*, 2004; Meihls *et al.*, 2008) and field conditions (Tabashnik *et al.*, 1990; Tang *et al.*, 1997); these instances include resistance to transgenic Bt crops (van Rensburg, 2007; Storer *et al.*, 2010). However, resistance to *B. thuringiensis* toxins may be seen as a rare event when considering the level of adoption of these technologies (Tabashnik *et al.*, 2009). A diverse group of insect species has been selected to be resistant to *B. thuringiensis* toxins in the laboratory (Schnepf *et al.*, 1998; Ferré and Van Rie, 2002), and these strains have facilitated the characterization of resistance mechanisms that may potentially evolve in field populations. Alterations of host defenses in resistant strains are limited to Cry toxins; there are no data yet available on resistance against Vip toxins.

The most well-studied mechanism of resistance is the alteration of receptors on the midgut brush border membranes and consequent disruption of toxin binding (Ferré and Van Rie, 2002). This type of resistance mechanism was described for strains of the diamondback moth, *Plutella xylostella* (Ferré *et al.*, 1991; Wright *et al.*, 1997), and the Indian meal moth, *Plodia interpunctella* (van Rie *et al.*, 1990), which had developed resistance to commercial *B. thuringiensis* formulations, although the specific altered receptor was not reported. Since diverse Cry toxins may recognize the same midgut receptors, cross-resistance to toxins that were not present in the environment of selection is possible. For instance, selection of lepidopteran larvae with Cry1Ac toxin usually results in cross-resistance to all Cry1A toxins. One specific resistance phenotype commonly described in diverse lepidopteran species and designated as "Mode 1 resistance" (Tabashnik *et al.*, 1998) is characterized by reduced binding by the toxin and more than 500-fold resistance to at least one Cry1A toxin, recessive inheritance, and little or no cross-resistance to Cry1C. The first gene linked to this Mode 1 resistance was a cadherin from *H. virescens* (Gahan *et al.*, 2001). This cadherin gene appeared disrupted by a retrotransposon insertion, resulting in lack of a full-length cadherin protein on the surface of midgut cells in Cry1Ac-resistant larvae (Jurat-Fuentes *et al.*, 2004). Alterations in cadherin genes linked to resistance against Cry toxins have also been confirmed in the pink bollworm, *Pectinophora gossypiella* (Morin *et al.*, 2003), and the cotton bollworm, *Helicoverpa armigera* (Xu *et al.*, 2005; Yang *et al.*, 2007). Cadherin resistance alleles are not detected in field populations of *P. gossypiella* (Tabashnik *et al.*, 2006) and *H. virescens* (Gahan *et al.*, 2007), probably owing to the low frequency of resistance alleles. This observation suggests that other mechanisms may be involved in resistance, a conclusion supported by the finding that field resistance in *P. xylostella* is not linked to cadherin (Baxter *et al.*, 2005). Recently, a mutation in an ABC transporter gene has been reported to be genetically linked to resistance and lack of Cry1Ac binding in strains of *H. virescens* (Gahan *et al.*, 2010). However, further work would be necessary to characterize the role of this transporter protein in the Cry1A intoxication process.

Although alteration of the midgut receptors sites has received the most attention from researchers, this is not the only mechanism by which insects may evolve resistance to *B. thuringiensis*. Because of the importance of toxin activation during the intoxication process, alterations in the midgut protease composition of the host can result in resistance. For example, resistance to *B. thuringiensis* subsp. *entomocidus* or subsp. *aizawai* in two strains of *P. interpunctella* was associated with the loss of a major trypsin-like protease (Oppert *et al.*, 1997). Although alterations in protease gene expression in other insect-resistant strains have been reported (Karumbaiah *et al.*, 2007; Khajuria *et al.*, 2009; Rajagopal *et al.*, 2009), their genetic linkage to resistance has not been established. Since activation is a common step in the mode of action of diverse Cry toxins, cross-resistance to other Cry toxins would be expected from alteration of this process. However, not all cases of cross-resistance to diverse toxins correlate with alterations in proteases. For example, a strain of *S. exigua* selected for resistance with Cry1Ab displayed cross-resistance to toxins not expected to share receptors with Cry1A toxins, such as Cry1D and Cry1Ca, but no protease alterations were detected compared to susceptible larvae (Hernández-Martínez *et al.*, 2009).

Another potential mechanism that would result in resistance to diverse Cry toxins is a midgut regenerative response that would prevent disruption of the midgut epithelium barrier and colonization of the hemocoel by vegetative *B. thuringiensis* or other bacterial cells. Early reports suggested that larvae of *H. virescens* infected with *B. thuringiensis* could recover from intoxication when presented with a toxin-free diet after a short exposure (Dulmage and Martinez, 1973; Dulmage *et al.*, 1978). A midgut regenerative response to intoxication with *B. thuringiensis* was described in larvae of the rice moth *Corcyra cephalonica* (Chiang *et al.*, 1986) and *M. sexta* (deLello *et al.*, 1984). This defensive midgut regenerative mechanism has also been demonstrated *in vitro* (Loeb *et al.*, 2001) and was proposed to be involved in resistance to Cry toxins in strains of *H. virescens* (Forcada *et al.*, 1999; Martínez-Ramírez *et al.*, 1999). In these insects, faster regeneration of damaged midgut epithelium correlated with survival on a diet containing Cry1Ac toxin. However, the specific genes involved in this regenerative process and their linkage with resistance have not been reported.

While feeding avoidance has been reported and could also potentially result in resistance to diverse toxins, there are no documented cases of resistance correlated with selective feeding. Cross-resistance to diverse toxins can also be a result of the existence of multiple mechanisms. For instance, strains of *H. virescens* resistant to Cry1Ac and Cry2A toxins were found to present altered toxin binding and protease patterns (Jurat-Fuentes *et al.*, 2003; Karumbaiah *et al.*, 2007). Resistance to Cry1Ab in strains of *P. interpunctella* (Herrero *et al.*, 2001), and Cry3Aa in the Colorado potato beetle, *Leptinotarsa decemlineata* (Loseva *et al.*, 2002), has also been reported to correlate with alterations in both protease activity and toxin binding. Remarkably, cross-resistance to Cry1Ac toxin and the chemical pesticide deltamethrin has been described in strains of *P. xylostella* (Sayyed *et al.*, 2008). Tests looking for complementary mechanisms in these *P. xylostella* strains suggest that a common genetic locus or loci controlled resistance to both insecticides, although the specific mechanism of cross-resistance has not been identified. If common, this type of resistance would preclude the use of some chemical pesticides to delay or control episodes of resistance to *B. thuringiensis* toxins.

13.4.2. Managing Resistance to *Bacillus thuringiensis*

Once resistance to a pathogen evolves in an insect, effective resistance control practices become crucial to preserve the future utility of the pathogen for insect control. The importance of *B. thuringiensis* as a biopesticide, especially for organic farmers who have few equivalent options for managing pests, and the high adoption rate of Bt crops, made proactively mitigating the spread of Bt-resistant pest populations a major priority during the early regulation of Bt crops (Ferré and Van Rie, 2002; Tabashnik *et al.*, 2003; Carrière *et al.*, 2010). Transferring methods from the toxicology literature, an insect resistance management (IRM) strategy was developed to delay resistance evolution and spread (Bourguet *et al.*, 2005; Sivasupramaniam *et al.*, 2007). However, the evolution of resistance is difficult to predict, and several factors (including fitness costs associated with resistance and how stress factors influence these dynamics) complicate these interactions between the host and its pathogens.

Delaying the Evolution of Resistance and Slowing its Spread

As has been discussed in Sections 13.2 and 13.3, the hierarchy of host defenses that must be overcome by a pathogen is substantial and dynamic. Therefore, it is impossible to predict which host defense phenotypes will evolve in response to selection pressure from a pathogen. To overcome this hurdle, IRM strategies have partially avoided trying to predict what resistance will look like by focusing on the genetics that underlie the resistant phenotypes. After all, every phenotype is simply the product of one or several gene alleles, and approaching IRM from the perspective of nameless resistant and susceptible alleles is more manageable than trying to visualize what actual resistance will look like. The assumption frequently made (at least in the Bt literature) is that field resistance will be produced by a single autosomal recessive allele, producing a recessive trait (but see Gassmann *et al.*, 2009b). Based on this assumption, homozygotes are expected to be the only resistant members of the population. Thus, current IRM strategies for Bt crops are based on a high-dose/refuge strategy to target heterozygous individuals for resistance alleles in the population (Gould, 1988; Caprio, 1994). The high-dose approach involves making plants express high doses of Cry toxins so that any insects that are heterozygous for resistance are killed; this dose amounts to 25 times the LD_{99} of the pathogen against the targeted pest. This high-dose strategy ensures that 100% of the pest population is killed, but it also puts substantial pressure on the pest population to evolve resistance to the pathogen (Gould, 1998; Andow and Ives, 2002; Mendelsohn *et al.*, 2003; Bourguet *et al.*, 2005; Tabashnik *et al.*, 2009). It is notable that this high-dose strategy is sometimes not achieved owing to low susceptibility of some of the targeted insect pests and levels of toxin production being dependent on plant development and physiology (Abel and Adamczyk, 2004; Olsen *et al.*, 2005). In these cases, the use of plants producing combinations of Bt toxins with diverse mode of action (toxin pyramiding) can increase the product's

effectiveness and delay the onset of resistance (Stewart *et al.*, 2001; Zhao *et al.*, 2003; Onstad and Meinke, 2010).

An essential component of the high-dose strategy for delaying resistance to pathogens is to incorporate an untreated refuge of non-Bt plants so that Bt-susceptible individuals can mate with the resistant individuals emerging from Bt plants and dilute the recessive resistance alleles in the population (Ives and Andow, 2002). This refuge component of IRM for Bt crops is difficult to implement and monitor. Depending on the crop and the country, these refuge mandates vary in size, type, and distance from the Bt field. With the introduction of pyramided Bt crops, refuge regulations have been relaxed or altered substantially. Given that host resistance to Bt has been documented at several stages of the infection process, the presumption that resistance is exclusively tied to simple alterations to key midgut receptors seems short sighted, and alterations to the refuge strategy have drawn concern from some stakeholders (Alyokhin, 2011; Onstad *et al.*, 2011).

Fitness Costs Associated with Resistance

Evolving a high level of resistance to a pathogen often comes at a cost to the insect host. Often, these costs are manifested in slower development, smaller size, lower fecundity, reduced overwintering success, and reduced survival (Fuxa and Richter, 1998; Carrière *et al.*, 2001; Gassmann *et al.*, 2009a). For example, field resistance to *B. sphaericus* was associated with slower development and reduced fecundity in the mosquito *Culex quinquefasciatus* (de Oliveira *et al.*, 2003). Thirty-four percent of published studies reveal reductions in fitness (survival, development rates, and body mass) in individuals resistant to *B. thuringiensis* relative to susceptible strains (Gassmann *et al.*, 2009a). In one study involving *B. thuringiensis* subsp. *kurstaki* sprays against *P. xylostella*, the 1,500-fold resistance diminished to only 300-fold resistance after exposure was ceased. This reduced level of resistance persisted without obvious fitness costs, and resistance quickly increased to 1,000-fold by the fourth generation of exposure (Tang *et al.*, 1997). This case illustrates that resistance level in a host is balanced against fitness costs of this resistance, and maintaining low levels of resistance to a pathogen may provide some flexibility for the host when the pathogen re-emerges. Because of fitness costs, a population that is resistant to a pathogen often reverts to susceptibility once the selection pressure is removed or relaxed. Gassmann *et al.* (2009a) showed that insects that were resistant to *B. thuringiensis* lost this resistance once selection pressure was reduced in 62% of published studies. In these studies, the average degree of resistance loss was 10-fold over seven generations of the pest. These examples aside, host resistance does not always lower the fitness of the host (Amorim *et al.*, 2010). Under these circumstances, the resistance can become fixed in the population for many generations. For instance, larvae of the cabbage looper (*Trichoplusia ni*) that show one form of resistance to its NPV (TnSNPV) remain resistant to the virus for up to 22 generations after the selection pressure is removed in the laboratory (Milks and Myers, 2000; Milks *et al.*, 2002). A question that remains to be answered in depth is why or how these forms of resistance cause harm to the host. In any case, fitness costs associated with resistance must be considered as models are developed to predict how host resistance to pathogens spreads through a population.

Healthy insects are better at defending themselves from pathogens than are stressed insects, and there are myriad potential stressors that affect the dynamics of host resistance to pathogens. These external factors influence the performance of the pathogen or the host, and thereby disrupt the normal dynamics of the two organisms. Climatic conditions are one of the best documented forms of stressor. The optimal environmental conditions for pathogen performance have largely been discussed in the chapters on the different pathogen groups. Suffice it to say here that temperature, humidity, winter conditions, and sunlight are a few of the conditions known to influence the performances of both the pathogen and the host insect (Ignoffo, 1992; Fuxa *et al.*, 1999; Jaronski, 2010). In fact, insects seek out or produce higher temperatures in order to create suboptimal environments for a pathogen and fend off infections, a term called behavioral fever (Watson *et al.*, 1993; Evans and Spivak, 2010). Age of the insect also affects its ability to resist pathogens (Hochberg, 1991b; Armitage and Boomsma, 2010). For example, larval insects are often killed by NPVs, whereas these infections are sometimes sublethal in the adult stage (Fuxa, 2004). Older larvae can be more resistant to pathogens such as NPVs (Fuxa, 2004) or *B. thuringiensis* toxins (Beegle *et al.*, 1981; Kouassi *et al.*, 2001), although this is not always the case.

13.5. FUTURE RESEARCH DIRECTIONS

For a pathogen to mount a successful infection on an insect, it must circumvent or overcome three strategic components of the insect antipathogen defenses: avoid recognition, overcome barriers, and withstand the innate immune response. The antipathogen defenses of insects are inherently tied to the insect's ability to recognize the pathogen on several levels. First, pathogens may be entirely avoided if the insect can use its senses to proactively detect and avoid potential disease agents. Once the pathogen invades the insect, the ability to distinguish pathogens from "self" is critical to alerting the innate immune system of the host. Next, there are several physical barriers that deter entomopathogens from invading the host. The cuticle, peritrophic matrix, and basal lamina are particularly effective at filtering the microbiota that infiltrates the hemocoel.

Cellular membranes are another barrier that limit intracellular pathogens from establishing infections. Entomopathogens must have adaptations that allow them to overcome these barriers if they are to survive in the host. Finally, the innate immune system possesses several weapons, including AMPs and the chemicals associated with the phenoloxidase cascade, hemocytes capable of phagocytosis and encapsulation, and xenophagy and RNAi-based defenses designed to kill intracellular pathogens. The pathogen must be able to hide from or disable the defenses, or offer counter-defenses in order to persist and reproduce in the host.

The in-depth examination of *B. thuringiensis* with regard to host-resistance evolution and management illustrates several key points and challenges. First, it is clear from this system that resistance can arise from alterations of the host at multiple points along the infection process, and predicting what form resistance will take is virtually impossible. Nevertheless, designing IRM plans that reduce the evolution of insect resistance and the spread of this resistance through the population is necessary in modern agricultural systems. Current IRM strategies are predicated on using a high dose of the pathogen alongside a non-Bt refuge that increases the likelihood that highly resistant individuals will mate with susceptible conspecifics. But these strategies are changing rapidly and the implications of these alterations and reductions on resistance spread are unknown. Clearly, numerous knowledge gaps remain in predicting resistance to entomopathogens even in the well-studied system of *B. thuringiensis*.

There is no shortage of questions that can be confronted in future research efforts. Immunochallenged insects upregulate hundreds of genes, and in most cases we know almost nothing regarding their function (Hoffmann, 2003; Baton *et al.*, 2008). Thus, although our understanding of host defenses as a branch of insect pathology is growing rapidly, our concept of immune responses may still be somewhat limited. One topic that has received a fair amount of research, but little synthesis, is that insects and pathogens live in communities, and we know very little about how community interactions affect host resistance and defensive capabilities against pathogens.

A future frontier is understanding the interaction between the biological communities and their effect on insect resistance. Insects seldom live in isolation, and their interactions within their community affect their ability to defend themselves against pathogens. On the smallest scale, microbial symbionts of insects form complex communities in the gut and on the cuticle of insects that resist the establishment of pathogenic microbes (discussed briefly in Section 13.2). In part, this is through the production of antimicrobial compounds that kill invading pathogens; but competitive interactions are also likely to play a role. Pathogens can be an important part of this community, and only attain pathogen status when the normal milieu is disrupted. For example, *Enterococcous faecalis* is a common gut bacterium that is believed to be pathogenic under some circumstances, but may facilitate herbivory in insects under other conditions (Lundgren and Lehman, 2010). The key players in these microbial communities, how they restrict or promote pathogenicity, and what the implications are for host resistance remain compelling questions.

The host plant, as a form of either diet or habitat, has a great effect on the success of infection (Cory and Hoover, 2006). Indeed, some plants form symbioses with entomo-pathogens to resist herbivorous insects (Clay and Schardl, 2002; Arnold and Lewis, 2005; Lundgren, 2009). The morphology and chemistry of leaves affect the suitability of the phylloplane as a habitat for potential microorganisms (Duetting *et al.*, 2003; Jaronski, 2010). The nutritional suitability of the plant affects the rate at which herbivorous insects ingest phylloplane resident microbes (Verkerk and Wright, 1996). In addition, ingesting phytochemicals can enhance or diminish the infection, depending on the circumstances (Ekesi *et al.*, 2000; Cory and Hoover, 2006). For example, the aphid *Sitobion avenae* is more susceptible to the fungus *Erynia neoaphidis* when the infection occurs on aphid-resistant wheat plants than on susceptible plants (Fuentes-Contreras *et al.*, 1998).

Finally, insects are often the target of multiple infections simultaneously, and the intraguild interactions among the pathogens and parasites become very important in the dynamics of hosts and pathogens (Furlong and Pell, 2005). In general, studies on intraguild interactions are interested in identifying the victor. More often than not, the rule of precedence comes into play. When two infectious agents invade the same host, the winner is the one that got there first (Hochberg, 1991a; Milks *et al.*, 2001; Fuxa, 2004). In some cases, prior infection "primes" the immune system and makes the host insect more resistant to subsequent pathogens (Vilcinskas, 2010). Fytrou *et al.* (2006) found that *Drosophila simulans* encapsulated larvae of the hymenopteran parasitoid *Leptopilina heterotoma* more efficiently when the fly was previously infected with *Wolbachia*, but not with *B. bassiana*. This beneficial effect of *Wolbachia* on host resistance has been documented in other insects as well (Hughes *et al.*, 2011). Finally, pathogens sometimes specialize on certain tissues of the host, and at least one example suggests that pathogens that are able to live in more than one tissue may be more competitive in direct competition with tissue specialists. *Wolbachia pipientis* and *Spiroplasma* both live intracellularly in *D. melanogaster*, but *Spiroplasma* can also live extracellularly. When both are present in the same host, *Spiroplasma* dominates the host more frequently than *W. pipientis* (Goto *et al.*, 2006). There is a growing understanding of the physiology and genetics that control insect—pathogen interactions, and this understanding will provide the basis for asking more complex questions about how other organisms with which insects live affect the dynamics of insect pathology.

REFERENCES

Abel, C. A., & Adamczyk, J. J., Jr. (2004). Relative concentration of Cry1A in maize leaves and cotton bolls with diverse chlorophyll content and corresponding larval development of fall armyworm (Lepidoptera: Noctuidae) and southwestern corn borer (Lepidoptera: Crambidae) on maize whorl leaf profiles. *J. Econ. Entomol., 97,* 1737—1744.

Adamo, S. A. (1998). The specificity of behavioral fever in the cricket *Acheta domesticus. J. Parasitol., 84,* 529—533.

Alma, C. R., Gillespie, D. R., Roitberg, B. D., & Goettel, M. S. (2010). Threat of infection and threat-avoidance behavior in the predator *Dicyphus hesperus* feeding on whitefly nymphs infected with an entomopathogen. *J. Insect Behav., 23,* 90—99.

Alyokhin, A. (2011). Scant evidence supports EPA's pyramided *Bt* corn refuge size of 5%. *Nat. Biotechnol., 29,* 577—578.

Amano, A., Nakagawa, I., & Yoshimori, T. (2006). Autophagy in innate immunity against intracellular bacteria. *J. Biochem., 140,* 161—166.

Amorim, L. B., de Barros, R. A., de Melo Chalegre, K. D., de Oliveira, C. M. F., Regis, L. N., & Lobo Silva-Filha, M. H. N. (2010). Stability of *Culex quinquefasciatus* resistance to *Bacillus sphaericus* evaluated by molecular tools. *Insect Biochem. Mol. Biol., 40,* 311—316.

Andow, D. A., & Ives, A. R. (2002). Monitoring and adaptive resistance management. *Ecol. Appl., 12,* 1378—1390.

Armitage, S. A. O., & Boomsma, J. J. (2010). The effects of age and social interactions on innate immunity in a leaf-cutting ant. *J. Insect Physiol., 56,* 780—787.

Arnold, A., & Lewis, L. C. (2005). Ecology and evolution of fungal endophytes and their roles against insects. In F. E. Vega & M. Blackwell (Eds.), *Insect—Fungal Associations: Ecology and Evolution* (pp. 74—96). Oxford: Oxford University Press.

Baton, L. A., Garver, L., Xi, Z., & Dimopoulos, G. (2008). Functional genomics studies on the innate immunity of disease vectors. *Insect Sci., 15,* 15—27.

Baxter, S. W., Zhao, J. Z., Gahan, L. J., Shelton, A. M., Tabashnik, B. E., & Heckel, D. G. (2005). Novel genetic basis of field-evolved resistance to Bt toxins in *Plutella xylostella. Insect Mol. Biol., 14,* 327—334.

Beckage, N. E. (2007). *Insect Immunology.* San Diego: Academic Press.

Beegle, C. C., Lewis, L. C., Lynch, R. E., & Martinez, A. J. (1981). Interaction of larval age and antibiotic on the susceptibility of 3 insect species to *Bacillus thuringiensis. J. Invertebr. Pathol., 37,* 143—153.

Berdegué, M., Trumble, J. T., & Moar, W. J. (1996). Effect of CryIC toxin from *Bacillus thuringiensis* on larval feeding behavior of *Spodoptera exigua. Entomol. Exp. Appl., 80,* 389—401.

Boguś, M. I., Czygier, M., Gołebiowski, M., Kedra, E., Kucińska, J., Mazgajska, J., Samborski, J., Wieloch, W., & Włóka, E. (2010). Effects of insect cuticular fatty acids on *in vitro* growth and pathogenicity of the entomopathogenic fungus *Conidiobolus coronatus. Exp. Parasitol., 125,* 400—408.

Boots, M. (1998). Cannibalism and the stage-dependent transmission of a viral pathogen of the Indian meal moth, *Plodia interpunctella. Ecol. Entomol., 23,* 118—122.

Bourguet, D., Desquilbet, M., & Lemarié, S. (2005). Regulating insect resistance management: the case of non-Bt corn refuges in the US. *J. Environ. Manag., 76*, 210–220.

Bourtzis, K., Pettigrew, M. M., & O'Neill, S. L. (2000). *Wolbachia* neither induces nor suppresses transcripts encoding antimicrobial peptides. *Insect Mol. Biol., 9*, 635–639.

Brennan, C. A., & Anderson, K. V. (2004). *Drosophila*: the genetics of innate immune recognition and response. *Annu. Rev. Immunol., 22*, 457–483.

Brennan, L. J., Keddie, B. A., Braig, H. R., & Harris, H. L. (2008). The endosymbiont *Wolbachia pipientis* induces the expression of host antioxidant proteins in an *Aedes albopictus* cell line. *PLoS ONE, 3*. e2083.

Bulet, P., Hetru, C., Dimarcq, J.-L., & Hoffmann, D. (1999). Antimicrobial peptides in insects; structure and function. *Dev. Comp. Immunol., 23*, 329–344.

Caprio, M. A. (1994). *Bacillus thuringiensis* gene deployment and resistance management in single- and multi-tactic environments. *Biocontrol Sci. Technol., 4*, 487–497.

Carrière, Y., Ellers, K. C., Patin, A. L., Sims, M. A., Meyer, S., Liu, Y. B., Dennehy, T. J., & Tabashnik, B. E. (2001). Overwintering cost associated with resistance to transgenic cotton in the pink bollworm (Lepidoptera: Gelechiidae). *J. Econ. Entomol., 94*, 935–941.

Carrière, Y., Crowder, D. W., & Tabashnik, B. E. (2010). Evolutionary ecology of insect adaptation to Bt crops. *Evol. Appl., 3*, 561–573.

Carton, Y., Poirié, M., & Nappi, A. J. (2008). Insect immune resistance to parasitoids. *Insect Sci., 15*, 67–87.

Centofanti, M. (1995). Playing by the rules. How and why organisms turn nasty. *Sci. News, 148*, 382–383.

Chapman, R. F. (1998). *The Insects: Structure and Function* (4th ed.). Cambridge: Cambridge University Press.

Charles, J.-F., Nielsen-LeRoux, C., & Delécluse, A. (1996). *Bacillus sphaericus* toxins: molecular biology and mode of action. *Annu. Rev. Entomol., 41*, 451–472.

Chen, K., Weng, Z.-H., & Zheng, L. (2008). Innate immunity against malaria parasites in *Anopheles gambiae*. *Insect Sci., 15*, 45–52.

Chiang, A. S., Yen, D. F., & Peng, W. K. (1986). Defense reaction of midgut epithelial cells in the rice moth larva (*Corcyra cephalonica*) infected with *Bacillus thuringiensis*. *J. Invertebr. Pathol., 47*, 333–339.

Chouvenc, T., Su, N.-Y., & Robert, A. (2010). Inhibition of the fungal pathogen *Metarhizium anisopliae* in the alimentary tracts of five termite (Isoptera) species. *Fla. Entomol., 93*, 467–469.

Christensen, B. M., Li, J., Chen, C.-C., & Nappi, A. J. (2005). Melanization immune responses in mosquito vectors. *Trends Parasitol., 21*, 192–199.

Clay, K., & Schardl, C. (2002). Evolutionary origins and ecological consequences of endophyte symbioses with grasses. *Am. Nat., 160*, S99–S127.

Cory, J. S., & Hoover, K. (2006). Plant-mediated effects in insect–pathogen interactions. *Trends Ecol. Evol., 21*, 278–286.

Cremer, S., Armitage, S. A., & Schmid-Hempel, P. (2007). Social immunity. *Curr. Biol., 17*, R693–R702.

Darboux, I., Pauchet, Y., Castella, C., Silva-Filha, M. H., Nielsen-LeRoux, C., Charles, J.-F., & Pauron, D. (2002). Loss of the membrane anchor of the target receptor is a mechanism of bioinsecticide resistance. *Proc. Natl. Acad. Sci. USA, 99*, 5830–5835.

Darboux, I., Charles, J.-F., Pauchet, Y., Warot, S., & Pauron, D. (2007). Transposon-mediated resistance to *Bacillus sphaericus* in a field-evolved population of *Culex pipiens* (Diptera: Culicidae). *Cell. Microbiol., 9*, 2022–2029.

Dillon, R. J., & Dillon, V. M. (2004). The gut bacteria of insects: nonpathogenic interactions. *Annu. Rev. Entomol., 49*, 71–92.

Duetting, P. S., Ding, H., Neufeld, J., & Eigenbrode, S. D. (2003). Plant waxy bloom on peas affects infection of pea aphids by *Pandora neoaphidis*. *J. Invertebr. Pathol., 84*, 149–158.

Dulmage, H. T., & Martinez, E. (1973). The effects of continuous exposure to low concentrations of the δ endotoxin of *Bacillus thuringiensis* on the development of the tobacco budworm, *Heliothis virescens*. *J. Invertebr. Pathol., 22*, 14–22.

Dulmage, H. T., Graham, H. M., & Martinez, E. (1978). Interactions between the tobacco budworm, *Heliothis virescens*, and the δ-endotoxin produced by the HD-1 isolate of *Bacillus thuringiensis* var. *kurstaki*: relationship between length of exposure to the toxin and survival. *J. Invertebr. Pathol., 32*, 40–50.

Ekesi, S., Maniania, N. K., & Lwande, W. (2000). Susceptibility of the legume flower thrips to *Metarhizium anisopliae* on different varieties of cowpea. *BioControl, 45*, 79–95.

Eleftherianos, I., Millichap, P. J., ffrench-Constant, R. H., & Reynolds, S. E. (2006). RNAi suppression of recognition protein mediated immune responses in the tobacco hornworm *Manduca sexta* causes increased susceptibility to the insect pathogen *Photorhabdus*. *Dev. Comp. Immunol., 30*, 1099–1107.

Evans, J. D., & Spivak, M. (2010). Socialized medicine: individual and communal disease barriers in honey bees. *J. Invertebr. Pathol., 103*, S62–S72.

Evans, J. D., Aronstein, K., Chen, Y. P., Hetru, C., Imler, J.-L., Jiang, H., Kanost, M., Thompson, G. J., Zou, Z., & Hultmark, D. (2006). Immune pathways and defence mechanisms in honey bees *Apis mellifera*. *Insect Mol. Biol., 15*, 645–656.

Ewald, P. W. (1995). The evolution of virulence: a unifying link between parasitology and ecology. *J. Parasitol., 81*, 659–669.

Fernández-Marín, H., Zimmerman, J. K., Rehner, S. A., & Wcislo, W. T. (2006). Active use of the metapleural glands by ants in controlling fungal infection. *Proc. R. Soc. B., 273*, 1689–1695.

Ferré, J., & Van Rie, J. (2002). Biochemistry and genetics of insect resistance to *Bacillus thuringiensis*. *Annu. Rev. Entomol., 47*, 501–533.

Ferré, J., Real, M. D., Van Rie, J., Jansens, S., & Peferoen, M. (1991). Resistance to the *Bacillus thuringiensis* bioinsecticide in a field population of *Plutella xylostella* is due to a change in a midgut membrane receptor. *Proc. Natl. Acad. Sci. USA, 88*, 5119–5123.

Forcada, C., Alcácer, E., Garcerá, M. D., Tato, A., & Martínez, R. (1999). Resistance to *Bacillus thuringiensis* Cry1Ac toxin in three strains of *Heliothis virescens*: proteolytic and SEM study of the larval midgut. *Arch. Insect Biochem. Physiol., 42*, 51–63.

Fuentes-Contreras, E., Pell, J. K., & Niemeyer, H. M. (1998). Influence of plant resistance at the third trophic level: interactions between parasitoids and entomopathogenic fungi of cereal aphids. *Oecologia, 117*, 426–432.

Furlong, M. J., & Pell, J. K. (2005). Interactions between entomopathogenic fungi and arthropod natural enemies. In F. E. Vega & M. Blackwell (Eds.), *Insect–Fungal Associations: Ecology and Evolution* (pp. 51–73). Oxford: Oxford University Press.

Fuxa, J. R. (2004). Ecology of insect nucleopolyhedroviruses. *Agric. Ecosyst. Environ., 103*, 27–43.

Fuxa, J. R., & Richter, A. R. (1998). Repeated reversion of resistance to nucleopolyhedrovirus by *Anticarsia gemmatalis. J. Invertebr. Pathol., 71*, 159–164.

Fuxa, J. R., Sun, J.-Z., Weidner, E. H., & LaMotte, L. R. (1999). Stressors and rearing diseases of *Trichoplusia ni*: evidence of vertical transmission of NPV and CPV. *J. Invertebr. Pathol., 74*, 149–155.

Fytrou, A., Schofield, P. G., Kraaijeveld, A. R., & Hubbard, S. F. (2006). *Wolbachia* infection suppresses both host defence and parasitoid counter-defence. *Proc. R. Soc. B., 273*, 791–796.

Gahan, L. J., Gould, F., & Heckel, D. G. (2001). Identification of a gene associated with Bt resistance in *Heliothis virescens. Science, 293*, 857–860.

Gahan, L. J., Gould, F., López, J. D., Jr., Micinski, S., & Heckel, D. G. (2007). A polymerase chain reaction screen of field populations of *Heliothis virescens* for a retrotransposon insertion conferring resistance to *Bacillus thuringiensis* toxin. *J. Econ. Entomol., 100*, 187–194.

Gahan, L. J., Pauchet, Y., Vogel, H., & Heckel, D. G. (2010). An ABC transporter mutation is correlated with insect resistance to *Bacillus thuringiensis* Cry1Ac toxin. *PLoS Genet, 6*. e1001248.

Gassmann, A. J., Carrière, Y., & Tabashnik, B. E. (2009a). Fitness costs of insect resistance to *Bacillus thuringiensis. Annu. Rev. Entomol., 54*, 147–163.

Gassmann, A. J., Onstad, D. W., & Pittendrigh, B. R. (2009b). Evolutionary analysis of herbivorous insects in natural and agricultural environments. *Pest Manag. Sci., 65*, 1174–1181.

Goto, S., Anbutsu, H., & Fukatsu, T. (2006). Asymmetrical interactions between *Wolbachia* and *Spiroplasma* endosymbionts coexisting in the same insect host. *Appl. Environ. Microbiol., 72*, 4805–4810.

Gould, F. (1988). Evolutionary biology and genetically engineered crops: consideration of evolutionary theory can aid in crop design. *Bioscience, 38*, 26–33.

Gould, F. (1998). Sustainability of transgenic insecticidal cultivars: integrating pest genetics and ecology. *Annu. Rev. Entomol., 43*, 701–726.

Govind, S. (2008). Innate immunity in *Drosophila*: pathogens and pathways. *Insect Sci., 15*, 29–43.

Haider, M. Z., Knowles, B. H., & Ellar, D. J. (1986). Specificity of *Bacillus thuringiensis* var. *colmeri* insecticidal δ-endotoxin is determined by differential proteolytic processing of the protoxin by larval gut proteases. *Eur. J. Biochem., 156*, 531–540.

Hernández-Martínez, P., Ferré, J., & Escriche, B. (2009). Broad-spectrum cross-resistance in *Spodoptera exigua* from selection with a marginally toxic Cry protein. *Pest Manag. Sci., 65*, 645–650.

Herre, E. A. (1993). Population structure and the evolution of virulence in nematode parasites of fig wasps. *Science, 259*, 1442–1445.

Herrero, S., Oppert, B., & Ferré, J. (2001). Different mechanisms of resistance to *Bacillus thuringiensis* toxins in the indianmeal moth. *Appl. Environ. Microbiol., 67*, 1085–1089.

Hillyer, J. F., & Christensen, B. M. (2005). Mosquito phenoloxidase and defensin colocalize in melanization innate immune responses. *J. Histochem. Cytochem., 53*, 689–698.

Hochberg, M. E. (1991a). Intra-host interactions between a braconid endoparasitoid, *Apantales glomeratus*, and *a baculovirus for larvae of Pieris brassicae. J. Anim. Ecol., 60*, 51–63.

Hochberg, M. E. (1991b). Extra-host interactions between a braconid endoparasitoid, *Apanteles glomeratus*, and *a baculovirus for larvae of Pieris brassicae. J. Anim. Ecol., 60*, 65–77.

Hoffmann, J. A. (1995). Innate immunity of insects. *Curr. Opin. Immunol., 7*, 4–10.

Hoffmann, J. A. (2003). The immune response of *Drosophila. Nature, 426*, 33–38.

Hoover, K., Humphries, M. A., Gendron, A. R., & Slavicek, J. M. (2010). Impact of viral *enhancin* genes on potency of *Lymantria dispar* multiple nucleopolyhedrovirus in *L. dispar* following disruption of the peritrophic matrix. *J. Invertebr. Pathol., 104*, 150–152.

Hughes, G. L., Ren, R., Ramirez, J. L., Sakamoto, J. M., Bailey, J. A., Jedlicka, A. E., & Rasgon, J. L. (2011). *Wolbachia* infections in *Anopheles gambiae* cells: transcriptomic characterization of a novel host–symbionet interaction. *PLoS ONE, 7*. e1001296.

Hurst, G. D. D., Anbutsu, H., Kutsukake, M., & Fukatsu, T. (2003). Hidden from the host: *Spiroplasma* bacteria infecting *Drosophila* do not cause an immune response, but are suppressed by ectopic immune activation. *Insect Mol. Biol., 12*, 93–97.

Ignoffo, C. M. (1992). Environmental factors affecting persistence of entomopathogens. *Fla. Entomol., 75*, 516–525.

Ives, A. R., & Andow, D. A. (2002). Evolution of resistance to Bt crops: directional selection in structured environments. *Ecol. Lett., 5*, 792–801.

Janmaat, A. F., Wang, P., Kain, W., Zhao, J.-Z., & Myers, J. (2004). Inheritance of resistance to *Bacillus thuringiensis* subsp. *kurstaki* in *Trichoplusia ni. Appl. Environ. Microbiol., 70*, 5859–5867.

Jaronski, S. (2010). Ecological factors in the inundative use of fungal entomopathogens. *BioControl, 55*, 159–185.

Jiang, H. (2008). The biochemical basis of antimicrobial responses in *Manduca sexta. Insect Sci., 15*, 53–66.

Jurat-Fuentes, J. L., Gould, F. L., & Adang, M. J. (2003). Dual resistance to *Bacillus thuringiensis* Cry1Ac and Cry2Aa toxins in *Heliothis virescens* suggests multiple mechanisms of resistance. *Appl. Environ. Microbiol., 69*, 5898–5906.

Jurat-Fuentes, J. L., Gahan, L. J., Gould, F. L., Heckel, D. G., & Adang, M. J. (2004). The HevCaLP protein mediates binding specificity of the Cry1A class of *Bacillus thuringiensis* toxins in *Heliothis virescens. Biochemistry, 43*, 14299–14305.

Kanost, M. R., & Nardi, J. B. (2010). Innate immune responses of *Manduca sexta*. In M. R. Goldsmith & F. Marec (Eds.), *Molecular Biology and Genetics of the Lepidoptera* (pp. 271–291). Boca Raton: CRC Press.

Karumbaiah, L., Oppert, B., Jurat-Fuentes, J. L., & Adang, M. J. (2007). Analysis of midgut proteinases from *Bacillus thuringiensis*-susceptible and -resistant *Heliothis virescens* (Lepidoptera: Noctuidae). *Comp. Biochem. Physiol. B Biochem. Mol. Biol., 146*, 139–146.

Khajuria, C., Zhu, Y. C., Chen, M. S., Buschman, L. L., Higgins, R. A., Yao, J. X., Crespo, A. L. B., Siegfried, B. D., Muthukrishnan, S., & Zhu, K. Y. (2009). Expressed sequence tags from larval gut of the European corn borer (*Ostrinia nubilalis*): exploring candidate genes potentially involved in *Bacillus thuringiensis* toxicity and resistance. *BMC Genomics, 10*, 286.

Klinger, E., Groden, E., & Drummond, F. A. (2006). Beauveria bassiana horizontal infection between cadavers and adults of the Colorado potato beetle, *Leptinotarsa decemlineata* (Say). *Environ. Entomol., 35*, 992–1000.

Kouassi, K. C., Lorenzetti, F., Guertin, C., Cabana, J., & Mauffette, Y. (2001). Variation in the susceptibility of the forest tent caterpillar (Lepidoptera: Lasiocampidae) to *Bacillus thuringiensis* variety kurstaki HD 1: effect of the host plant. *J. Econ. Entomol., 94*, 1135–1141.

Lavine, M. D., & Strand, M. R. (2001). Surface characteristics of foreign targets that elicit an encapsulation response by the moth *Pseudoplusia includens*. *J. Insect Physiol., 47*, 965–974.

Lavine, M. D., & Strand, M. R. (2002). Insect hemocytes and their role in immunity. *Insect Biochem. Mol. Biol., 32*, 1295–1309.

Le Conte, Y., Alaux, C., Martin, J.-F., Harbo, J. R., Harris, J. W., Dantec, C., Séverac, D., Cros-Arteil, S., & Navajas, M. (2011). Social immunity in honeybees (*Apis mellifera*): transcriptome analysis of varroa-hygienic behaviour. *Insect Mol. Biol., 20*, 399–408.

Lehane, M. J. (1997). Peritrophic matrix structure and function. *Annu. Rev. Entomol., 42*, 525–550.

deLello, E., Hanton, W. K., Bishoff, S. T., & Mish, D. W. (1984). Histopathological effects of *Bacillus thuringiensis* on the midgut of tobacco hornworm larvae (*Manduca sexta*): low doses compared with fasting. *J. Invertebr. Pathol., 43*, 169–181.

Levine, B. (2005). Eating oneself and uninvited guests: autophagy-related pathways in cellular defense. *Cell, 120*, 159–162.

Levy, S. M., Falleiros, A. M. F., Moscardi, F., & Gregório, E. A. (2007). Susceptibility/resistance of *Anticarsia gemmatalis* larvae to its nucleopolyhedrovirus (AgMNPV): structural study of the peritrophic membrane. *J. Invertebr. Pathol., 96*, 183–186.

Loeb, M. J., Martin, P. A., Hakim, R. S., Goto, S., & Takeda, M. (2001). Regeneration of cultured midgut cells after exposure to sublethal doses of toxin from two strains of *Bacillus thuringiensis*. *J. Insect Physiol., 47*, 599–606.

Lord, J. C. (2001). Response of the wasp *Cephalonomia tarsalis* (Hymenoptera: Bethylidae) to *Beauveria bassiana* (Hyphomycetes: Moniliales) as free conidia or infection in its host, the sawtoothed grain beetle, *Oryzaephilus surinamensis* (Coleoptera: Silvanidae). *Biol. Control, 21*, 300–304.

Loseva, O., Ibrahim, M., Candas, M., Koller, C. N., Bauer, L. S., & Bulla, L. A., Jr. (2002). Changes in protease activity and Cry3Aa toxin binding in the Colorado potato beetle: implications for insect resistance to *Bacillus thuringiensis* toxins. *Insect Biochem. Mol. Biol., 32*, 567–577.

Lundgren, J. G. (2009). *Relationships of Natural Enemies and Non-prey Foods*. Dordrecht: Springer International.

Lundgren, J. G., & Lehman, R. M. (2010). Bacterial gut symbionts contribute to seed digestion in an omnivorous beetle. *PLoS ONE, 5*. e10831.

Martínez-Ramírez, A. C., Gould, F., & Ferré, J. (1999). Histopathological effects and growth reduction in a susceptible and a resistant strain of *Heliothis virescens* (Lepidoptera: Noctuidae) caused by sublethal doses of pure Cry1A crystal proteins from *Bacillus thuringiensis*. *Biocontrol Sci. Technol., 9*, 239–246.

Mburu, D. M., Ochola, L., Maniania, N. K., Njagi, P. G. N., Gitonga, L. M., Ndung'u, M. W., Wanjoya, A. K., & Hassanali, A. (2009). Relationship between virulence and repellency of entomopathogenic isolates of *Metarhizium anisopliae* and *Beauveria bassiana* to the termite *Macrotermes michaelseni*. *J. Insect Physiol., 55*, 774–780.

McGaughey, W. H. (1985). Insect resistance to the biological insecticide *Bacillus thuringiensis*. *Science, 229*, 193–195.

McGraw, E. A., & O'Neill, S. L. (2004). *Wolbachia pipientis*: intracellular infection and pathogenesis in *Drosophila*. *Curr. Opin. Microbiol., 7*, 67–70.

Meihls, L. N., Higdon, M. L., Siegfried, B. D., Miller, N. J., Sappington, T. W., Ellersieck, M. R., Spencer, T. A., & Hibbard, B. E. (2008). Increased survival of western corn rootworm on transgenic corn within three generations of onplant greenhouse selection. *Proc. Natl. Acad. Sci. USA, 105*, 19177–19182.

Mendelsohn, M., Kough, J., Vaituzis, Z., & Matthews, K. (2003). Are Bt crops safe? *Nat. Biotechnol., 21*, 1003–1009.

Meyling, N. V., & Pell, J. K. (2006). Detection and avoidance of an entomopathogenic fungus by a generalist insect predator. *Ecol. Entomol., 31*, 162–171.

Milks, M. L., & Myers, J. H. (2000). The development of larval resistance to a nucleopolyhedrovirus is not accompanied by an increased virulence in the virus. *Evol. Ecol., 14*, 645–664.

Milks, M. L., Lepitch, M. K., & Theilmann, D. A. (2001). Recombinant and wild-type nucleopolyhedroviruses are equally fit in mixed infections. *Environ. Entomol., 30*, 972–981.

Milks, M. L., Myers, J. H., & Leptich, M. K. (2002). Costs and stability of cabbage looper resistance to a nucleopolyhedrovirus. *Evol. Ecol., 16*, 369–385.

Mitsuhashi, W., Kawakita, H., Murakami, R., Takemoto, Y., Saiki, T., Miyamoto, K., & Wada, S. (2007). Spindles of an entomopoxvirus facilitate its infection of the host insect by disrupting the peritrophic membrane. *J. Virol., 81*, 4235–4243.

Morin, S., Biggs, R. W., Sisterson, M. S., Shriver, L., Ellers-Kirk, C., Higginson, D., Holley, D., Gahan, L. J., Heckel, D. G., Carrière, Y., Dennehy, T. J., Brown, J. K., & Tabashnik, B. E. (2003). Three cadherin alleles associated with resistance to *Bacillus thuringiensis* in pink bollworm. *Proc. Natl. Acad. Sci. USA, 100*, 5004–5009.

Moussain, B. (2010). Recent advances in understanding mechanisms of insect cuticle differentiation. *Insect Biochem. Mol. Biol., 40*, 363–375.

NASS. (2011). *National Agriculture Statistics Service*. USDA. http://www.nass.usda.gov/

Nielsen-Leroux, C., Pasquier, F., Charles, J. F., Sinegre, G., Gaven, B., & Pasteur, N. (1997). Resistance to *Bacillus sphaericus* involves different mechanisms in *Culex pipiens* (Diptera: Culicidae) larvae. *J. Med. Entomol., 34*, 321–327.

Nielsen-Leroux, C., Pasteur, N., Pretre, J., Charles, J. F., Sheikh, H. B., & Chevillon, C. (2002). High resistance to *Bacillus sphaericus* binary toxin in *Culex pipiens* (Diptera: Culicidae): the complex situation of west Mediterranean countries. *J. Med. Entomol., 39*, 729–735.

de Oliveira, C. M. F., Costa Filho, F., Beltràn, J. F. N., Silva-Filha, M. H., & Regis, L. (2003). Biological fitness of a *Culex quinquefasciatus* population and its resistance to *Bacillus sphaericus*. *J. Am. Mosq. Control Assoc., 19*, 125–129.

Olsen, K. M., Daly, J. C., Holt, H. E., & Finnegan, E. J. (2005). Season-long variation in expression of Cry1Ac gene and efficacy of *Bacillus thuringiensis* toxin in transgenic cotton against *Helicoverpa armigera* (Lepidoptera: Noctuidae). *J. Econ. Entomol., 98*, 1007–1017.

Onstad, D. W., & Meinke, L. J. (2010). Modeling evolution of *Diabrotica virgifera virgifera* (Coleoptera: Chrysomelidae) to transgenic corn with two insecticidal traits. *J. Econ. Entomol., 103*, 849–860.

Onstad, D. W., Mitchell, P. D., Hurley, T. M., Lundgren, J. G., Porter, R. P., Krupke, C. H., Spencer, J. L., DiFonzo, C. D., Baute, T. S., Hellmich, R. L., Buschmann, L. L., Hutchison, W. D., & Tooker, J. F. (2011). Seeds of change: corn seed mixtures for resistance management and IPM. *J. Econ. Entomol., 104*, 343–352.

Oppert, B., Kramer, K. J., Beeman, R. W., Johnson, D., & McGaughey, W. H. (1997). Proteinase-mediated insect resistance to *Bacillus thuringiensis* toxins. *J. Biol. Chem., 272,* 23473–23476.

Ormond, E. L., Thomas, A. P. M., Pell, J. K., Freeman, S. N., & Roy, H. E. (2011). Avoidance of a generalist entomopathogenic fungus by the ladybird, *Coccinella septempunctata. FEMS Microbiol. Ecol., 77,* 229–237.

Oxley, P. R., Spivak, M., & Oldroyd, B. P. (2010). Six quantitative trait loci influence task thresholds for hygienic behaviour in honeybees (*Apis mellifera*). *Mol. Ecol., 19,* 1452–1461.

Park, H. W., Bideshi, D. K., & Federici, B. A. (2010). Properties and applied use of the mosquitocidal bacterium, *Bacillus sphaericus. J. Asia Pac. Entomol., 13,* 159–168.

Parker, B. J., Elderd, B. D., & Dwyer, G. (2010). Host behaviour and exposure risk in an insect–pathogen interaction. *J. Anim. Ecol., 79,* 863–870.

Passarelli, A. L. (2011). Barriers to success: how baculoviruses establish efficient systemic infections. *Virology, 411,* 383–392.

Piubelli, G. C., Hoffmann-Campo, C. B., Moscardi, F., Miyakubo, S. H., & de Oliveira, M. C. N. (2006). Baculovirus-resistant *Anticarsia gemmatalis* responds differently to dietary rutin. *Entomol. Exp. Appl., 119,* 53–60.

Plymale, R., Grove, M. J., Cox-Foster, D., Ostiguy, N., & Hoover, K. (2008). Plant-mediated alteration of the peritrophic matrix and baculovirus infection in lepidopteran larvae. *J. Insect Physiol., 54,* 737–749.

Poulsen, M., Bot, A. N. M., & Boomsma, J. J. (2003). The effect of metapleural gland secretion on the growth of a mutualistic bacterium on the cuticle of leaf-cutting ants. *Naturwissenschaften, 90,* 406–409.

Rajagopal, R., Arora, N., Sivakumar, S., Rao, N. G., Nimbalkar, S. A., & Bhatnagar, R. K. (2009). Resistance of *Helicoverpa armigera* to Cry1Ac toxin from *Bacillus thuringiensis* is due to improper processing of the protoxin. *Biochem. J., 419,* 309–316.

van Rensburg, J. B. J. (2007). First report of field resistance by stem borer, *Busseola fusca* (Fuller) to Bt-transgenic maize. *S. Afr. J. Plant Soil, 24,* 147–151.

van Rie, J., McGaughey, W. H., Johnson, D. E., Barnett, B. D., & Van Mellaert, H. (1990). Mechanism of insect resistance to the microbial insecticide *Bacillus thuringiensis. Science, 247,* 72–74.

Rodcharoen, J., & Mulla, M. S. (1995). Comparative ingestion rates of *Culex quinquefasciatus* (Diptera, Culicidae) susceptible and resistant to *Bacillus sphaericus. J. Invertebr. Pathol., 66,* 242–248.

Rolff, J. & Reynolds, S. E. (Eds.) (2009). *Insect Infection and Immunity: Evolution, Ecology, and Mechanisms.* Oxford: Oxford University Press.

Royet, J. (2004a). Infectious non-self recognition in invertebrates: lessons from *Drosophila* and other insect models. *Mol. Immunol., 41,* 1063–1075.

Royet, J. (2004b). *Drosophila melanogaster* innate immunity: an emerging role for peptidoglycan recognition proteins in bacteria detection. *Cell. Mol. Life Sci., 61,* 537–546.

Ruepell, O., Hayworth, M. K., & Ross, N. P. (2010). Altruistic self-removal of health-compromised honey bee workers from their hive. *J. Evol. Biol., 23,* 1538–1546.

Sabin, L. R., Hanna, S. L., & Cherry, S. (2010). Innate antiviral immunity in *Drosophila. Curr. Opin. Immunol., 22,* 4–9.

Sayyed, A. H., Moores, G., Crickmore, N., & Wright, D. J. (2008). Cross-resistance between a *Bacillus thuringiensis* Cry toxin and non-Bt insecticides in the diamondback moth. *Pest Manag. Sci., 64,* 813–819.

Schnepf, E., Crickmore, N., Van Rie, J., Lereclus, D., Baum, J., Feitelson, J., Zeigler, D. R., & Dean, D. H. (1998). *Bacillus thuringiensis* and its pesticidal crystal proteins. *Microbiol. Mol. Biol. Rev., 62,* 775–806.

Shelby, K. S., & Popham, H. J. R. (2008). Cloning and characterization of the secreted hemocytic prophenoloxidases of *Heliothis virescens. Arch. Insect Biochem. Physiol., 69,* 127–142.

Shelton, A. M., Wang, P., Zhao, J.-Z., & Roush, R. T. (2007). Resistance to insect pathogens and strategies to manage resistance: an update. In L. A. Lacey & H. K. Kaya (Eds.), *Field Manual of Techniques in Invertebrate Pathology* (pp. 793–811). Dordrecht: Springer.

Siozos, S., Sapountzis, P., Loannidis, P., & Bourtzis, K. (2008). *Wolbachia* symbiosis and insect immune response. *Insect Sci., 15,* 89–100.

Sivasupramaniam, S., Head, G. P., English, L., Li, Y. J., & Vaughn, T. T. (2007). A global approach to resistance monitoring. *J. Invertebr. Pathol., 95,* 224–226.

Stewart, S. D., Adamczyk, J. J., Jr., Knighten, K. S., & Davis, F. M. (2001). Impact of Bt cottons expressing one or two insecticidal proteins of *Bacillus thuringiensis* Berliner on growth and survival of noctuid (Lepidoptera) larvae. *J. Econ. Entomol., 94,* 752–760.

Storer, N. P., Babcock, J. M., Schlenz, M., Meade, T., Thompson, G. D., Bing, J. W., & Huckaba, R. M. (2010). Discovery and characterization of field resistance to Bt maize: *Spodoptera frugiperda* (Lepidoptera: Noctuidae) in Puerto Rico. *J. Econ. Entomol., 103,* 1031–1038.

Stow, A., & Beattie, A. J. (2008). Chemical and genetic defenses against disease in insect societies. *Brain Behav. Immun., 22,* 1009–1013.

Strand, M. R. (2008). The insect cellular immune response. *Insect Sci., 15,* 1–14.

Swanson, J. A. I., Torto, B., Kells, S. A., Mesce, K. A., Tumlinson, J. H., & Spivak, M. (2009). Odorants that induce hygienic behavior in honeybees: identification of volatile compounds in chalkbrood-infected honeybee larvae. *J. Chem. Ecol., 35,* 1108–1116.

Tabashnik, B. E., Cushing, N. L., Finson, N., & Johnson, M. W. (1990). Field development of resistance to *Bacillus thuringiensis* in diamondback moth (Lepidoptera: Plutellidae). *J. Econ. Entomol., 83,* 1671–1676.

Tabashnik, B. E., Liu, Y.-B., Malvar, T., Heckel, D. G., Masson, L., & Ferré, J. (1998). Insect resistance to *Bacillus thuringiensis*: uniform or diverse? *Philos. Trans. R. Soc. Lond. B., 353,* 1751–1756.

Tabashnik, B. E., Carrière, Y., Dennehy, T. J., Morin, S., Sisterson, M. S., Roush, R. T., Shelton, A. M., & Zhao, J.-Z. (2003). Insect resistance to transgenic Bt crops: lessons from the laboratory and field. *J. Econ. Entomol., 96,* 1031–1038.

Tabashnik, B. E., Fabrick, J. A., Henderson, S., Biggs, R. W., Yafuso, C. M., Nyboer, M. E., Manhardt, N. M., Coughlin, L. A., Sollome, J., Carrière, Y., Dennehy, T. J., & Morin, S. (2006). DNA screening reveals pink bollworm resistance to Bt cotton remains rare after a decade of exposure. *J. Econ. Entomol., 99,* 1525–1530.

Tabashnik, B. E., Van Rensburg, J. B. J., & Carrière, Y. (2009). Field-evolved insect resistance to Bt crops: definition, theory, and data. *J. Econ. Entomol., 102,* 2011–2025.

Takehana, A., Yano, T., Mita, S., Kotani, A., Oshima, Y., & Kurata, S. (2004). Peptidoglycan recognition protein (PGRP)-LE and PGRP-LC act synergistically in *Drosophila* immunity. *EMBO J., 23,* 4690–4700.

Tang, J. D., Gilboa, S., Roush, R. T., & Shelton, A. M. (1997). Inheritance, stability, and lack-of-fitness costs of field-selected resistance to *Bacillus thuringiensis* in diamondback moth (Lepidoptera: Plutellidae) from Florida. *J. Econ. Entomol., 90*, 732–741.

Thompson, S. R., Brandenburg, R. L., & Roberson, G. T. (2007). Entomopathogenic fungi detection and avoidance by mole crickets (Orthoptera: Gryllotalpidae). *Environ. Entomol., 36*, 165–172.

Toledo, A. V., Alippi, A. M., & de Remes Lenicov, A. M. M. (2011). Growth inhibition of *Beauveria bassiana* by bacteria isolated from the cuticular surface of the corn leafhopper, *Dalbulus maidis* and the planthopper, *Delphacodes kuscheli*, two important vectors of maize pathogens. *J. Insect Sci., 11*, 29.

Ugelvig, L. V., & Cremer, S. (2007). Social prophylaxis: group interaction promotes collective immunity in ant colonies. *Curr. Biol., 17*, 1967–1971.

Verkerk, R. H. J., & Wright, D. J. (1996). Effects of interactions between host plants and selective insecticides on larvae of *Plutella xylostella* L. (Lepidoptera: Yponomeutidae) in the laboratory. *Pestic. Sci., 46*, 171–181.

Vilcinskas, A. (2010). Lepidopterans as model mini-hosts for human pathogens and as a resource for peptide antibiotics. In M. R. Goldsmith & F. Marec (Eds.), *Molecular Biology and Genetics of the Lepidoptera* (pp. 293–306). Boca Raton: CRC Press.

Watson, D. W., Mullens, B. A., & Petersen, J. J. (1993). Behavioral fever response of *Musca domestica* (Diptera: Muscidae) to infection by *Entomophthora muscae* (Zygomycetes: Entomophthrorales). *J. Invertebr. Pathol., 61*, 10–16.

Wright, D. J., Iqbal, M., Granero, F., & Ferré, J. (1997). A change in a single midgut receptor in the diamondback moth (*Plutella xylostella*) is only in part responsible for field resistance to *Bacillus thuringiensis* subsp. *kurstaki* and *B. thuringiensis* subsp. *aizawai*. *Appl. Environ. Microbiol., 63*, 1814–1819.

Xu, X., Yu, L., & Wu, Y. (2005). Disruption of a cadherin gene associated with resistance to Cry1Ac δ-endotoxin of *Bacillus thuringiensis* in *Helicoverpa armigera*. *Appl. Environ. Microbiol., 71*, 948–954.

Yang, Y., Chen, H., Wu, Y., Yang, Y., & Wu, S. (2007). Mutated cadherin alleles from a field population of *Helicoverpa armigera* confer resistance to *Bacillus thuringiensis* toxin Cry1Ac. *Appl. Environ. Microbiol., 73*, 6939–6944.

Zhao, J. Z., Cao, J., Li, Y., Collins, H. L., Roush, R. T., Earle, E. D., & Shelton, A. M. (2003). Transgenic plants expressing two *Bacillus thuringiensis* toxins delay insect resistance evolution. *Nat. Biotechnol., 21*, 1493–1497.

Index